This text is dedicated to my loving wife, Sheila.

THE INTEL MICROPROCESSORS

8086/8088, 80186/80188, 80286, 80386, 80486, Pentium, Pentium Pro Processor, Pentium II, Pentium III, and Pentium 4

Architecture, Programming, and Interfacing

Sixth Edition

BARRY B. BREY
DeVry Institute of Technology

Prentice
Hall

Upper Saddle River, New Jersey
Columbus, Ohio

Library of Congress Cataloging-in-Publication Data

Brey, Barry B.
   The Intel microprocessors : 8086/8088, 80186/80188, 80286, 80386, 80486, Pentium,
Pentium Pro processor, Pentium II, Pentium III, and Pentium 4 : architecture,
programming, and interfacing / Barry B. Brey. -- 6th ed.
      p.   cm.
   ISBN 0-13-060714-2
   1. Intel 80xxx series microprocessors.   2. Pentium (Microprocessor)   I. Title.
QA76.8.I2674 B76 2003
004.165--dc21   2002019820

Editor in Chief: Stephen Helba
Assistant Vice President and Publisher: Charles E. Stewart, Jr.
Production Editor: Alexandrina Benedicto Wolf
Production Coordination: Custom Editorial Productions, Inc.
Design Coordinator: Diane Ernsberger
Cover Designer: Rod Harris
Production Manager: Matt Ottenweller

This book was set in Times Roman by Custom Editorial Productions, Inc. It was printed and bound by Courier Kendallville, Inc. The cover was printed by The Lehigh Press, Inc.

Pearson Education Ltd.
Pearson Education Australia Pty. Limited
Pearson Education Singapore Pte. Ltd.
Pearson Education North Asia Ltd.
Pearson Education Canada, Ltd.
Pearson Educación de Mexico, S.A. de C.V.
Pearson Education—Japan
Pearson Education Malaysia Pte. Ltd.
Pearson Education, *Upper Saddle River, New Jersey*

10 9 8 7 6 5 4 3 2 1
ISBN 0-13-060714-2

PREFACE

This practical reference text is written for students in a course that requires a thorough knowledge of programming and interfacing of the Intel family of microprocessors. Today, anyone functioning or striving to function in a field of study that uses computers must understand assembly language programming and interfacing. Intel microprocessors have gained wide applications in many areas of electronics, communications, and control systems, particularly in desktop computer systems. A major addition to this sixth edition explains how to interface C/C++ with assembly language for both the older DOS and the Windows environments. Updated sections that detail new events in the fields of microprocessors and microprocessor interfacing have been added.

ORGANIZATION AND COVERAGE

To cultivate a comprehensive approach to learning, each chapter begins with a set of objectives that briefly define its content. Each chapter includes many programming applications that illustrate the main topics; ends with a numerical summary, which doubles as a study guide, and reviews the information just presented. Finally, questions and problems are provided for reinforcement and practice.

This text contains many example programs using the Microsoft Macro Assembler program and the Visual C++ environment, which provide a learning opportunity to program the Intel family of microprocessors. Operation of the programming environment includes the linker, library, macros, DOS function, BIOS functions, and Visual C/C++ program development. The in-line assembler (C/C++) is illustrated for both the 16- and 32-bit programming environments of various versions of Visual C++.

This text also provides a thorough description of family members, memory systems, and various I/O systems that include disk memory, ADC and DAC, 16550 UART, PIAs, timers, keyboard/display controllers, arithmetic coprocessors, and video display systems. Also discussed are the personal computer system buses (AGP, ISA, PCI, VESA, and USB). Through these systems, a practical approach to microprocessor interfacing can be learned.

APPROACH

Because the Intel family of microprocessors is quite diverse, this text initially concentrates on real-mode programming, which is compatible with all versions of the Intel family of microprocessors. Instructions for each family member, which includes the 80386, 80486, Pentium, Pentium Pro, Pentium II, Pentium III, and Pentium 4 processors, are compared and contrasted with the 8086/8088 microprocessors. This entire series of microprocessors is very similar, which allows more advanced versions to be learned once the basic 8086/8088 is understood. Please note that the 8086/8088 are still used in controllers along with their updated counterparts: the 80186/80188 and 80386EX embedded controllers.

This text also explains the programming and operation of the numeric coprocessor, which functions in a system to provide access to floating-point calculations that are important in control systems, video graphics, and computer aided design (CAD) applications. The numeric coprocessor allows a program to access complex arithmetic operations that are otherwise difficult to achieve with normal microprocessor programming.

This text also describes the pin-outs and function of the 8086–80486 and all versions of the Pentium microprocessor. First interfacing is explained using the 8086/8088 with some of the more common peripheral components. After explaining the basics, a more advanced emphasis is placed on the 80186/80188, 80386, 80486, and Pentium through Pentium 4 microprocessors. Coverage of the 80286, because of its similarity to the 8086 and 80386, is minimized so the 80386, 80486, and Pentium versions can be covered in complete detail.

Through this approach, the operation of the microprocessor and programming with the advanced family members, along with interfacing all family members, provides a working and practical background of the Intel family of microprocessors. Upon completing a course using this text, you will be able to:

1. Develop control software to control an application interface microprocessor. Generally, the software developed will also function on all versions of the microprocessor. This software also includes DOS-based and Windows-based applications.
2. Program using DOS function calls to control the keyboard, video display system, and disk memory in assembly language.
3. Use the BIOS functions to control the keyboard, display, and various other components in the computer system.
4. Develop software that uses macro sequences, procedures, conditional assembly, and flow control assembler directives.
5. Develop software that uses interrupt hooks and hot keys to gain access to terminate and stay resident software.
6. Program the numeric coprocessor to solve complex equations.
7. Explain the differences between the family members and highlight the features of each member.
8. Describe and use real and protected mode operation of the microprocessor.
9. Interface memory and I/O systems to the microprocessor.
10. Provide a detailed and comprehensive comparison of all family members and their software and hardware interfaces.
11. Explain the function of the real time operating system in an embedded application.
12. Explain the operation of disk and video systems.
13. Interface small systems to the ISA, VESA local, PCI, parallel port, and USB bus in a personal computer system.

CONTENT OVERVIEW

Chapter 1 introduces the Intel family of microprocessors with an emphasis on the microprocessor-based computer system: its history, operation, and the methods used to store data in a microprocessor-based system. In this edition number systems are also included. Chapter 2 explores the programming model of the microprocessor and system architecture. Both real and protected mode operations are explained.

Once an understanding of the basic machine is grasped, Chapters 3–6 explain how each instruction functions with the Intel family of microprocessors. As instructions are explained, simple applications are presented to illustrate the operation of the instructions and develop basic programming concepts.

After the basis for programming is developed, Chapter 7 provides applications using the assembler program. These applications include programming using DOS and BIOS function calls and the mouse function calls. Disk files are explained, as well as keyboard and video operation on a personal computer system. This chapter provides the tools required to develop virtually any program on a personal computer system. It also introduces the concept of interrupt hooks and hot keys.

Chapter 8 introduces the use of C/C++ with the in-line assembler and separate assembly language programming modules.

Chapter 9 introduces the 8086/8088 family as a basis for learning basic memory and I/O interfacing, which follow in later chapters. This chapter shows the buffered system as well as the system timing.

Chapter 10 explains memory interface using both integrated decoders and programmable logic devices. Parity and dynamic memory systems are illustrated. The 8-, 16-, 32-, and 64-bit memory systems are provided so the 8086–80486 and the Pentium through Pentium 4 microprocessors can be interfaced to memory.

Chapter 11 provides a detailed look at basic I/O interfacing, including PIAs, timers, keyboard/display interfaces, 16550 UART, and ADC/DAC. It also describes the interface of both DC and stepper motors.

Once these basic I/O components and their interface to the microprocessor is understood, Chapters 12 and 13 provide detail on advanced I/O techniques that include interrupts and direct memory access (DMA). Applications include a printer interface, real-time clock, disk memory, and video systems.

Chapter 14 details the operation and programming for the 8087–Pentium 4 family of arithmetic coprocessors, as well as MMX instruction. Today few applications function efficiently without the power of the arithmetic coprocessor. Remember that all Intel microprocessors since the 80486 contain a coprocessor.

Chapter 15 shows how to interface small systems to the personal computer through the use of the parallel port and the ISA, VESA, and PCI bus interfaces. This new chapter discusses the many cards being designed for use in the personal computer embedded in control systems.

Chapters 16 and 17 cover the advanced 80186/80188–80486 microprocessors and explores their differences with the 8086/8088, as well as their enhancements and features. Cache memory, interleaved memory, and burst memory are described with the 80386 and 80486 microprocessors. Chapter 17 also describes memory management and memory paging.

Chapter 18 details the Pentium and Pentium Pro microprocessors. These new microprocessors are based upon the original 8086/8088.

Chapter 19 introduces the Pentium II, Pentium III, and Pentium 4 microprocessors. It covers some of the new features, package styles, and the SIMD instructions that are added to the original instruction set.

Appendices are included to enhance the text.

Appendix A provides a complete listing of the DOS INT 21H function calls. It also details the use of the assembler program and many of the BIOS function calls, including BIOS function call INT 10H as well as mouse function calls and DPMI calls.

A complete listing of all 8086–Pentium 4 instructions, including many example instructions and machine coding in hexadecimal as well as clock timing information, is found in Appendix B. The SIMD instruction and examples of its usage are also provided.

Appendix C provides a compact list of all the instructions that change the flag bits.

Answers for the even-numbered questions and problems are provided in Appendix D.

ACKNOWLEDGMENTS

I appreciate the valuable comments and suggestions from the following reviewers: Robert L. Douglas, University of Memphis, TN; Isaac Ghansah, California State University; Raynaud F. Henton, Southern University and A & M College, LA; and Paul A. Wheeler, Utah State University.

STAY IN TOUCH

You can stay in touch through the Internet. My Internet site contains information about all of my textbooks and many important links that are specific to the personal computer, microprocessors, hardware, and software. Also available is a weekly lesson that details many of the aspects of the personal computer. Of particular interest is the "Technical Section," which presents many notes on topics that are not covered in this text.

http://members.ee.net/brey

CONTENTS

CHAPTER 1

Introduction to the Microprocessor and Computer

INTRODUCTION

This chapter provides an overview of the Intel family of microprocessors. Included is a discussion of the history of computers and the function of the microprocessor in the microprocessor-based computer system. Also introduced are terms and jargon used in the computer field, so **computerese** is understood and applied when discussing microprocessors and computers.

The block diagram and a description of the function of each block detail the operation of a computer system. The chapter also shows how the memory and input/output (I/O) system of the personal computer function. Finally, the way that data are stored in the memory is provided, so that each data type can be used as software is developed. Numeric data are stored as integers, floating-point, and binary-coded decimal (BCD); alphanumeric data are stored by using the ASCII (American Standard Code for Information Interchange) code.

CHAPTER OBJECTIVES

Upon completion of this chapter, you will be able to:

1. Converse by using appropriate computer terminology such as bit, byte, data, real memory system, expanded memory system (EMS), extended memory system (XMS), DOS, BIOS, I/O, and so forth.
2. Briefly detail the history of the computer and list applications performed by computer systems.
3. Provide an overview of the various 80X86 and Pentium–Pentium 4 family members.
4. Draw the block diagram of a computer system and explain the purpose of each block.
5. Describe the function of the microprocessor and detail its basic operation.
6. Define the contents of the memory system in the personal computer.
7. Convert between binary, decimal, and hexadecimal numbers.
8. Differentiate and represent numeric and alphabetic information as integers, floating-point, BCD, and ASCII data.

1–1   A HISTORICAL BACKGROUND

This first section outlines the historical events leading to the development of the microprocessor and, specifically, the extremely powerful and current 80X86,[1] Pentium, Pentium Pro, Pentium III, and Pentium 4[2] microprocessors. Although a study of history is not essential to understand the microprocessor, it furnishes interesting reading and provides a historical perspective of the fast-paced evolution of the computer.

The Mechanical Age

The idea of a computing system is not new—it has been around long before modern electrical and electronic devices were developed. The idea of calculating with a machine dates to 500 B.C. when the Babylonians invented the **abacus,** the first mechanical calculator. The abacus, which used strings of beads to perform calculations, was used by the ancient Babylonian priests to keep track of their vast storehouses of grain. The abacus, which was used extensively and is still in use today, was not improved until 1642, when mathematician Blaise Pascal invented a calculator that was constructed of gears and wheels. Each gear contained 10 teeth that, when moved one complete revolution, advanced a second gear one place. This is the same principal that is used in the automobile's odometer mechanism and is the basis of all mechanical calculators. Incidentally, the PASCAL programming language is named in honor of Blaise Pascal for his pioneering work in mathematics and with the mechanical calculator.

The arrival of the first practical geared, mechanical machines used to automatically compute information dates to the early 1800s. This is before humans invented the light bulb or before much was known about electricity. In this dawn of the computer age, humans dreamed of mechanical machines that could compute numerical facts with a program—not merely calculating facts, as with a calculator.

In 1937 it was discovered through plans and journals that one early pioneer of mechanical computing machinery was Charles Babbage, aided by Augusta Ada Byron, the Countess of Lovelace. Babbage was commissioned in 1823 by the Royal Astronomical Society of Great Britain to produce a programmable calculating machine. This machine was to generate navigational tables for the Royal Navy. He accepted the challenge and began to create what he called his **Analytical Engine.** This engine was a mechanical computer that stored 1000 20-digit decimal numbers and a variable program that could modify the function of the machine to perform various calculating tasks. Input to his engine was through punched cards, much as computers in the 1950s and 1960s used punched cards. It is assumed that he obtained the idea of using punched cards from Joseph Jacquard, a Frenchman who used punched cards as input to a weaving machine he invented in 1801, which is today called Jacquard's loom. Jacquard's loom used punched cards to select intricate weaving patterns in the cloth that it produced. The punched cards programmed the loom.

After many years of work, Babbage's dream began to fade when he realized that the machinists of his day were unable to create the mechanical parts needed to complete his work. The Analytical Engine required more than 50,000 machined parts, which could not be made with enough precision to allow his engine to function reliably.

The Electrical Age

The 1800s saw the advent of the electric motor (conceived by Michael Faraday); with it came a multitude of motor-driven adding machines, all based on the mechanical calculator developed by

---

[1]80X86 is shorthand notation that includes the 8086, 8088, 80188, 80286, 80386, and 80486 microprocessors.

[2]Pentium, Pentium Pro, Pentium II, Pentium III, and Pentium 4 are registered trademarks of Intel Corporation.

Blaise Pascal. These electrically driven mechanical calculators were common pieces of office equipment until well into the early 1970s, when the small hand-held electronic calculator, first introduced by Bomar, appeared. Monroe was also a leading pioneer of electronic calculators, but its machines were desktop, four-function models the size of cash registers.

In 1889, Herman Hollerith developed the punched card for storing data. Like Babbage, he too apparently borrowed the idea of a punched card from Jacquard. He also developed a mechanical machine—driven by one of the new electric motors—that counted, sorted, and collated information stored on punched cards. The idea of calculating by machinery intrigued the United States government so much that Hollerith was commissioned to use his punched-card system to store and tabulate information for the 1890 census.

In 1896, Hollerith formed a company called the Tabulating Machine Company, which developed a line of machines that used punched cards for tabulation. After a number of mergers, the Tabulating Machine Company was formed into the International Business Machines Corporation, now referred to more commonly as IBM, Inc. The punched cards used in computer systems are often called **Hollerith cards,** in honor of Herman Hollerith. The 12-bit code used on a punched card is called the **Hollerith code.**

Mechanical machines driven by electric motors continued to dominate the information processing world until the construction of the first electronic calculating machine in 1941 by a German inventor named Konrad Zuse. His Z3 calculating computer, as pictured in Figure 1–1, was used in aircraft and missile design during World War II for the German war effort. Had Zuse been given adequate funding by the German government, he most likely would have developed a much more powerful computer system. Zuse is today finally receiving some belated honor for his pioneering work in the area of digital electronics which began in the 1930s and for his Z3 computer system.

FIGURE 1–1   The Z3 computer developed by Konrad Zuse used a 5.33 Hertz clocking frequency. (Photo courtesy of Horst Zuse, the son of Konrad.)

It has recently been discovered (through the declassification of British military documents) that the first electronic computer was placed into operation in 1943 to break secret German military codes. This first electronic computing system, which used vacuum tubes, was invented by Alan Turing. Turing called his machine **Colossus,** probably because of its size. A problem with Colossus was that although its design allowed it to break secret German military codes generated by the mechanical **Enigma machine,** it could not solve other problems. Colossus was not programmable—it was a fixed-program computer system, which today is often called a **special-purpose computer.**

The first general-purpose, programmable electronic computer system was developed in 1946 at the University of Pennsylvania. This first modern computer was called the ENIAC (**Electronics Numerical Integrator and Calculator).** The ENIAC was a huge machine, containing over 17,000 vacuum tubes and over 500 miles of wires. This massive machine weighed over 30 tons, yet performed only about 100,000 operations per second. The ENIAC thrust the world into the age of electronic computers. The ENIAC was programmed by rewiring its circuits—a process that took many workers several days to accomplish. The workers changed the electrical connections on plug-boards that looked like early telephone switchboards. Another problem with the ENIAC was the life of the vacuum tube components, which required frequent maintenance.

Breakthroughs that followed were the development of the transistor in 1948 at Bell Labs, followed by the 1958 invention of the integrated circuit by Jack Kilby of Texas Instruments. The integrated circuit led to the development of digital integrated circuits (RTL, or resistor-to-transistor logic) in the 1960s and the first microprocessor at Intel Corporation in 1971. At that time, Intel and one of its engineers, Marcian E. Hoff, developed the 4004 microprocessor—the device that started the microprocessor revolution that continues today at an ever-accelerating pace.

Programming Advancements

Now that programmable machines were developed, programs and programming languages began to appear. As mentioned earlier, the first programmable electronic computer system was programmed by rewiring its circuits. Because this proved too cumbersome for practical application, early in the evolution of computer systems, computer languages began to appear in order to control the computer. The first such language, **machine language,** was constructed of ones and zeros using binary codes that were stored in the computer memory system as groups of instructions called programs. This was more efficient than rewiring a machine to program it, but it was still extremely time-consuming to develop a program because of the sheer number of codes that were required. Mathematician John von Neumann was the first person to develop a system that accepted instructions and stored them in memory. Computers are often called **von Neumann machines** in honor of John von Neumann. (Remember that Babbage also had developed the concept long before von Neumann.)

Once computer systems such as the UNIVAC became available in the early 1950s, **assembly language** was used to simplify the chore of entering binary code into a computer as its instructions. The assembler allowed the programmer to use mnemonic codes, such as ADD for addition, in place of a binary number such as 01000111. Although assembly language was an aid to programming, it wasn't until 1957, when Grace Hopper developed the first high-level programming language called **FLOW-MATIC,** that computers became easier to program. In the same year, IBM developed FORTRAN (**FORmula TRANslator)** for its computer systems. The FORTRAN language allowed programmers to develop programs that used formulas to solve mathematical problems. Note that FORTRAN is still used by some scientists for computer programming. Another similar language, introduced about a year after FORTRAN, was ALGOL (**ALGOrithmic Language).**

The first truly successful and widespread programming language for business applications was COBOL (**COmputer Business Oriented Language).** Although COBOL usage has diminished somewhat in recent years, it is still a major player in many large business systems. Another once-popular business language is RPG (**Report Program Generator**), which allows programming by specifying the form of the input, output, and calculations.

Since these early days of programming, additional languages have appeared. Some of the more common are BASIC, C/C++, PASCAL, and ADA. The BASIC and PASCAL languages were both designed as teaching languages, but have escaped the classroom and are used in many computer systems. The BASIC language is probably the easiest of all to learn. Some estimates indicate that the BASIC language is used in the personal computer for 80 percent of the programs written by users. Recently, a new version of BASIC, VISUAL BASIC, has made programming in the WINDOWS environment easier. The VISUAL BASIC language may eventually supplant C/C++ and PASCAL.

In the scientific community, C/C++ and (occasionally) PASCAL appear as control programs. Both languages, especially C/C++, allow the programmer almost complete control over the programming environment and computer system. In many cases, C/C++ is replacing some of the low-level, machine control software normally reserved for assembly language. Even so, assembly language still plays an important role in programming. Most video games written for the personal computer are written almost exclusively in assembly language. Assembly language is also interspersed with C/C++ and PASCAL to perform machine control functions efficiently.

The ADA language is used heavily by the Department of Defense. The ADA language was named in honor of Augusta Ada Byron, Countess of Lovelace. The Countess worked with Charles Babbage in the early 1800s in the development of his Analytical Engine.

The Microprocessor Age

The world's first microprocessor, the Intel 4004, was a 4-bit microprocessor—a programmable controller on a chip. It addressed a mere 4096 4-bit wide memory locations. (A **bit** is a binary digit with a value of one or zero. A 4-bit wide memory location is often called a **nibble.**) The 4004 instruction set contained only 45 instructions. It was fabricated with the then-current state-of-the-art P-channel MOSFET technology that only allowed it to execute instructions at the slow rate of 50 KIPs (**kilo-instructions per second**). This was slow when compared to the 100,000 instructions executed per second by the 30-ton ENIAC computer in 1946. The main difference was that the 4004 weighed much less than an ounce.

At first, applications abounded for this device. The 4-bit microprocessor debuted in early video game systems and small microprocessor-based control systems. One such early video game, a shuffleboard game, was produced by Balley. The main problems with this early microprocessor were its speed, word width, and memory size. The evolution of the 4-bit microprocessor ended when Intel released the 4040, an updated version of the earlier 4004. The 4040 operated at a higher speed, although it lacked improvements in word width and memory size. Other companies, particularly Texas Instruments (TMS-1000), also produced 4-bit microprocessors. The 4-bit microprocessor still survives in low-end applications such as microwave ovens and small control systems, and is still available from some microprocessor manufacturers. Most calculators are still based on 4-bit microprocessors that process 4-bit BCD (**binary-coded decimal**) codes.

Later in 1971, realizing that the microprocessor was a commercially viable product, Intel Corporation released the 8008—an extended 8-bit version of the 4004 microprocessor. The 8008 addressed an expanded memory size (16K bytes) and contained additional instructions (a total of 48) that provided an opportunity for its application in more advanced systems. (A **byte** is generally an 8-bit wide binary number and a **K** is 1024. Often, memory size is specified in K bytes.)

As engineers developed more demanding uses for the 8008 microprocessor, they discovered that its somewhat small memory size, slow speed, and instruction set limited its usefulness. Intel recognized these limitations and introduced the 8080 microprocessor in 1973—the first of the modern 8-bit microprocessors. About six months after Intel released the 8080 microprocessor,

TABLE 1–1   Early 8-bit
microprocessors.

Manufacturer	Part Number
Fairchild	F-8
Intel	8080
MOS Technology	6502
Motorola	MC6800
National Semiconductor	IMP-8
Rockwell International	PPS-8
Zilog	Z-8

Motorola Corporation introduced its MC6800 microprocessor. The floodgates opened and the 8080—and, to a lesser degree, the MC6800—ushered in the age of the microprocessor. Soon, other companies began to introduce their own versions of the 8-bit microprocessor. Table 1–1 lists several of these early microprocessors and their manufacturers. Of these early micro- processor producers, only Intel and Motorola continue successfully to create newer and im- proved versions of the microprocessor. Zilog still manufactures microprocessors, but has remained in the background, concentrating on microcontrollers and embedded controllers in- stead of general-purpose microprocessors. Rockwell has all but abandoned microprocessor de- velopment in favor of modem circuitry. Motorola has declined from having nearly 50 percent share of the microprocessor market to a much smaller share.

What Was Special about the 8080?

Not only could the 8080 address more memory and execute additional instructions, but it executed them 10 times faster than the 8008. An addition that took 20 μs (50,000 instructions per second) on an 8008-based system required only 2.0 μs (500,000 in- structions per second) on an 8080-based system. Also, the 8080 was compatible with TTL (tran- sistor-transistor logic), whereas the 8008 was not directly compatible. This made interfacing much easier and less expensive. The 8080 also addressed four times more memory (64K bytes) than the 8008 (16K bytes). These improvements are responsible for ushering in the era of the 8080 and the continuing saga of the microprocessor. Incidentally, the first personal computer, the MITS Altair 8800, was released in 1974. (Note that the number 8800 was probably chosen to avoid copyright violations with Intel.) The BASIC language interpreter, written for the Altair 8800 computer, was developed by Bill Gates, the founder of Microsoft Corporation. The assem- bler program for the Altair 8800 was written by Digital Research Corporation, which once pro- duced DR-DOS for the personal computer.

The 8085 Microprocessor.

In 1977, Intel Corporation introduced an updated version of the 8080—the 8085. This was to be the last 8-bit, general-purpose microprocessor developed by Intel. Although only slightly more advanced than an 8080 microprocessor, the 8085 executed software at an even higher speed. An addition that took 2.0 μs (500,000 instructions per second) on the 8080 required only 1.3 μs (769,230 instructions per second) on the 8085. The main ad- vantages of the 8085 were its internal clock generator, internal system controller, and higher clock frequency. This higher level of component integration reduced the 8085's cost and in- creased its usefulness. Intel has managed to sell well over 100 million copies of the 8085 micro- processor, its most successful 8-bit, general-purpose microprocessor. Because the 8085 is also manufactured (second-sourced) by many other companies, there are over 200 million of these microprocessors in existence. Applications that contain the 8085 will likely continue to be pop- ular well into the future. Another company that sold 500 million 8-bit microprocessors is Zilog Corporation, which produced the Z-80 microprocessor. The Z-80 is machine language code- compatible with the 8085, which means that there are over 700 million microprocessors that ex- ecute 8085/Z-80 compatible code!

The Modern Microprocessor

In 1978, Intel released the 8086 microprocessor; a year or so later, it released the 8088. Both devices are 16-bit microprocessors, which executed instructions in as little as 400 ns (2.5 MIPs, or 2.5 millions of instructions per second). This represented a major improvement over the execution speed of the 8085. In addition, the 8086 and 8088 addressed 1M bytes of memory, which was 16 times more memory than the 8085. (A **1M byte memory** contains 1024K byte-sized memory locations, or 1,048,576 bytes.) This higher execution speed and larger memory size allowed the 8086 and 8088 to replace smaller minicomputers in many applications. One other feature found in the 8086/8088 was a small 4- or 6-byte instruction cache or queue that prefetched a few instructions before they were executed. The queue sped the operation of many sequences of instructions and proved to be the basis for the much larger instruction caches found in modern microprocessors.

The increased memory size and additional instructions in the 8086 and 8088 have led to many sophisticated applications for microprocessors. Improvements to the instruction set included a multiply-and-divide instruction, which was missing on earlier microprocessors. In addition, the number of instructions increased from 45 on the 4004, to 246 on the 8085, to well over 20,000 variations on the 8086 and 8088 microprocessors. Note that these microprocessors were called CISC (**complex instruction set computers**) because of the number and complexity of instructions. The additional instructions eased the task of developing efficient and sophisticated applications, even though the number of instructions was at first overwhelming and time-consuming to learn. The 16-bit microprocessor also provided more internal register storage space than the 8-bit microprocessor. The additional registers allowed software to be written more efficiently.

The 16-bit microprocessor evolved mainly because of the need for larger memory systems. The popularity of the Intel family was ensured in 1981, when IBM Corporation decided to use the 8088 microprocessor in its personal computer. Applications such as spreadsheets, word processors, spelling checkers, and computer-based thesauruses were memory-intensive and required more than the 64K bytes of memory found in 8-bit microprocessors to execute efficiently. The 16-bit 8086 and 8088 provided 1M bytes of memory for these applications. Soon, even the 1M byte memory system proved limiting for large databases and other applications. This led Intel to introduce the 80286 microprocessor, an updated 8086, in 1983.

The 80286 Microprocessor. The 80286 microprocessor (also a 16-bit architecture microprocessor) was almost identical to the 8086 and 8088, except it addressed a 16M byte memory system instead of a 1M byte system. The instruction set of the 80286 was almost identical to the 8086 and 8088, except for a few additional instructions that managed the extra 15M bytes of memory. The clock speed of the 80286 was increased, so it executed some instructions in as little as 250 ns (4.0 MIPs) with the original release 8.0 MHz version. Some changes also occurred in the internal execution of the instructions, which led to an eight-fold increase in speed for many instructions when compared to 8086/8088 instructions.

The 32-bit Microprocessor. Applications began to demand faster microprocessor speeds, more memory, and wider data paths. This led to the arrival of the 80386 in 1986, by Intel Corporation. The 80386 represented a major overhaul of the 16-bit 8086–80286 architecture. The 80386 was Intel's first practical 32-bit microprocessor that contained a 32-bit data bus and a 32-bit memory address. (Note that Intel produced an earlier, although unsuccessful, 32-bit microprocessor called the iapx-432.) Through these 32-bit buses, the 80386 addressed up to 4G bytes of memory. (**1G of memory** contains 1024M, or 1,073,741,824 locations.) A 4G byte memory can store an astounding 1,000,000 typewritten, double-spaced pages of ASCII text data. The 80386 was available in a few modified versions such as the 80386SX, which addressed 16M bytes of memory through a 16-bit data and 24-bit address bus, and the 80386SL/80386SLC, which addressed 32M bytes of memory through a 16-bit data and 25-bit address bus. An 80386SLC version contained an internal cache memory that allowed it to process data at even higher rates. In 1995, Intel released the 80386EX microprocessor. The 80386EX microprocessor is called an **embedded PC** because it contains all the components of the AT class personal computer on a single integrated circuit. The 80386EX also contains 24 lines for input/output data, a 26-bit address bus, a 16-bit data bus, a DRAM refresh controller, and programmable chip selection logic.

Applications that require higher microprocessor speeds and large memory systems include software systems that use a GUI, or **graphical user interface.** Modern graphical displays often contain 256,000 or more picture elements (**pixels,** or **pels**). The least sophisticated VGA (**variable graphics array**) video display has a resolution of 640 pixels per scanning line with 480 scanning lines. To display one screen of information, each picture element must be changed, which requires a high-speed microprocessor. Many new software packages use this type of video interface. These GUI-based packages require high microprocessor speeds and accelerated video adapters for quick and efficient manipulation of video text and graphical data. The most striking system, which requires high-speed computing for its graphical display interface, is Microsoft Corporation's Windows.[3] We often call a GUI a WYSIWYG (**what you see is what you get**) display.

The 32-bit microprocessor is needed because of the size of its data bus, which transfers real (single-precision floating-point) numbers that require 32-bit wide memory. In order to efficiently process 32-bit real numbers, the microprocessor must efficiently pass them between itself and memory. If the numbers pass through an 8-bit data bus, it takes four read or write cycles; when passed through a 32-bit data bus, however, only one read or write cycle is required. This significantly increases the speed of any program that manipulates real numbers. Most high-level languages, spreadsheets, and database management systems use real numbers for data storage. Real numbers are also used in graphical design packages that use vectors to plot images on the video screen. These include such CAD (**computer aided drafting/design**) systems as AUTOCAD, ORCAD, and so forth.

Besides providing higher clocking speeds, the 80386 included a memory management unit that allowed memory resources to be allocated and managed by the operating system. Earlier microprocessors left memory management completely to the software. The 80386 included hardware circuitry for memory management and memory assignment, which improved its efficiency and reduced software overhead.

The instruction set of the 80386 microprocessor was upward-compatible with the earlier 8086, 8088, and 80286 microprocessors. Additional instructions referenced the 32-bit registers and managed the memory system. Note that memory management instructions and techniques used by the 80286 were also compatible with the 80386 microprocessor. These features allowed older, 16-bit software to operate on the 80386 microprocessor.

The 80486 Microprocessor.   In 1989, Intel released the 80486 microprocessor, which incorporated an 80386-like microprocessor, an 80387-like numeric coprocessor, and an 8K byte cache memory system into one integrated package. Although the 80486 microprocessor was not radically different from the 80386, it did include one substantial change. The internal structure of the 80486 was modified from the 80386 so that about half of its instructions executed in one clock instead of two clocks. Because the 80486 was available in a 50 MHz version, about half of the instructions executed in 25ns (50 MIPs). The average speed improvement for a typical mix of instructions was about 50 percent over the 80386 that operated at the same clock speed. Later versions of the 80486 executed instructions at even higher speeds with a 66 MHz double-clocked version (80486DX2). The double-clocked 66 MHz version executed instructions at the rate of 66 MHz, with memory transfers executing at the rate of 33 MHz. (This is why it was called a double-clocked

---

[3]Windows is a registered trademark of Microsoft Corporation and is currently available as Windows 95, Windows 98, Windows 2000, Windows ME, and Windows XP.

microprocessor.) A triple-clocked version from Intel, the 80486DX4, improved the internal execution speed to 100 MHz with memory transfers at 33 MHz. Note that the 80486DX4 microprocessor executed instructions at about the same speed as the 60 MHz Pentium. It also contained an expanded 16K byte cache in place of the standard 8K byte cache found on earlier 80486 microprocessors. Advanced Micro Devices (AMD) has produced a triple-clocked version that runs with a bus speed of 40 MHz and a clock speed of 120 MHz. The future promises to bring microprocessors that internally execute instructions at rates of up to 1 GHz or higher.

Other versions of the 80486 were called Overdrive[4] processors. The Overdrive processor was actually a double-clocked version of the 80486DX that replaced an 80486SX or slower-speed 80486DX. When the Overdrive processor was plugged into its socket, it disabled or replaced the 80486SX or 80486DX, and functioned as a doubled-clocked version of the microprocessor. For example, if an 80486SX, operating at 25 MHz, was replaced with an Overdrive microprocessor, it functioned as an 80486DX2 50 MHz microprocessor using a memory transfer rate of 25 MHz.

Table 1–2 lists many microprocessors produced by Intel and Motorola with information about their word and memory sizes. Other companies produce microprocessors, but none have attained the success of Intel and, to a lesser degree, Motorola.

The Pentium Microprocessor. The Pentium, introduced in 1993, was similar to the 80386 and 80486 microprocessors. This microprocessor was originally labeled the P5 or 80586, but Intel decided not to use a number because it appeared to be impossible to copyright a number. The two introductory versions of the Pentium operated with a clocking frequency of 60 MHz and 66 MHz, and a speed of 110 MIPs, with a higher-frequency 100 MHz one and one-half clocked version that operated at 150 MIPs. The double-clocked Pentium, operating at 120 MHz and 133 MHz, was also available, as were higher-speed versions. (The fastest version produced by Intel is the 233 MHz Pentium, which is a three and one-half clocked version.) Another difference was that the cache size was increased to 16K bytes from the 8K cache found in the basic version of the 80486. The Pentium contained an 8K byte instruction cache and an 8K byte data cache, which allowed a program that transfers a large amount of memory data to still benefit from a cache. The memory system contained up to 4G bytes, with the data bus width increased from the 32 bits found in the 80386 and 80486 to a full 64 bits. The data bus transfer speed was either 60 MHz or 66 MHz, depending on the version of the Pentium. (Recall that the bus speed of the 80486 was 33 MHz.) This wider data bus width accommodated double-precision floating-point numbers used for modern high-speed, vector-generated graphical displays. These higher bus speeds should allow virtual reality software to operate at more realistic rates on current and future Pentium-based platforms. The widened data bus and higher execution speed of the Pentium allow full-frame video displays to operate at scan rates of 30 Hz or higher—comparable to commercial television. Recent versions of the Pentium also included additional instructions, called multimedia extensions, or MMX instructions. Although Intel hoped that the MMX instructions would be widely used, it appears that few software companies have used them.

Recently, Intel released the long-awaited Pentium OverDrive (P24T) for older 80486 systems that operate at either 63 MHz or 83 MHz clock. The 63 MHz version upgrades older 80486DX2 50 MHz systems; the 83 MHz version upgrades the 80486DX2 66 MHz systems. The upgraded 83 MHz system performs at a rate somewhere between a 66 MHz Pentium and a 75 MHz Pentium. If older VESA local bus video and disk-caching controllers seem too expensive to toss out, the Pentium OverDrive represents an ideal upgrade path from the 80486 to the Pentium.

Probably the most ingenious feature of the Pentium is its dual integer processors. The Pentium executes two instructions, which are not dependent on each other, simultaneously because it contains two independent internal integer processors called superscaler technology. This allows

---

[4]Overdrive is a registered trademark of Intel Corporation.

TABLE 1–2   Many modern Intel and Motorola microprocessors.

Manufacturer	Part	Data Bus Width	Memory Size
Intel	8048	8	2K internal
	8051	8	8K internal
	8085A	8	64K
	8086	16	1M
	8088	8	1M
	8096	16	8K internal
	80186	16	1M
	80188	8	1M
	80251	8	16K internal
	80286	16	16M
	80386EX	16	64M
	80386DX	32	4G
	80386SL	16	32M
	80386SLC	16	32M + 1K cache
	80386SX	16	16M
	80486DX/DX2	32	4G + 8K cache
	80486SX	32	4G + 8K cache
	80486DX4	32	4G + 16K cache
	Pentium	64	4G + 16K cache
	Pentium Overdrive (P24T) (replaces 80486)	32	4G + 16K cache
	Pentium Pro processor	64	64G + 16K L1 cache + 256K L2 cache
	Pentium II	64	64G + 32K L1 cache + 512K L2 cache
	Pentium II Xeon	64	64G + 32K L1 cache + 512K or 1M L2 cache
	Pentium III, Pentium 4	64	64G + 32K L1 cache + 256K L2 cache
Motorola	6800	8	64K
	6805	8	2K
	6809	8	64K
	68000	16	16M
	68008Q	8	1M
	68008D	8	4M
	68010	16	16M
	68020	32	4G
	68030	32	4G + 256 cache
	68040	32	4G + 8K cache
	68050	32	Proposed, but never released
	68060	64	4G + 16K cache
	PowerPC	64	4G + 32K cache

the Pentium to often execute two instructions per clocking period. Another feature that enhances performance is a jump prediction technology that speeds the execution of programs that include loops. As with the 80486, the Pentium also employs an internal floating-point coprocessor to handle floating-point data, albeit at five times the speed improvement. These features portend continued success for the Intel family of microprocessors. They also may allow the Pentium to

FIGURE 1–2 The Intel
iCOMP index.

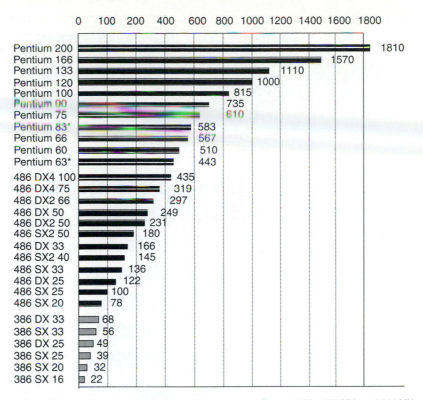

Note: * = Pentium OverDrive, the first part of the scale is not linear, and the 166 MHz and 200 MHz are MMX technology

replace some of the RISC (**reduced instruction set computer**) machines that currently execute one instruction per clock. Note that some newer RISC processors execute more than one instruction per clock through the introduction of superscaler technology. Motorola, Apple, and IBM have recently produced the PowerPC, a RISC microprocessor that has two integer units and one floating-point unit. The PowerPC certainly boosts the performance of the Apple Macintosh, but at present is slow to emulate the Intel family of microprocessors. Tests indicate that the current emulation software executes DOS and Windows applications at speed slower than the 80486SX 25 MHz microprocessor. Because of this, the Intel family should survive for many years in personal computer systems. Note that there are currently 6 million Apple Macintosh[5] systems and well over 260 million personal computers based on Intel microprocessors. In 1998, reports stated that 96 percent of all PCs were shipped with the Windows operating system.

In order to compare the speeds of various microprocessors, Intel devised the iCOMP-rating index. This index is a composite of SPEC92, ZD Bench, and Power Meter. The iCOMP1 rating was used to rate the speed of all Intel microprocessors through the Pentium. Figure 1–2 shows the relative speeds of the 80386DX 25 MHz version at the low end to the Pentium 233 MHz version at the high end of the spectrum.

Since the release of the Pentium Pro and Pentium II, Intel has switched to the iCOMP2 index, which is scaled by a factor of 10 from the iCOMP1 index. A microprocessor with an index of 1000 using iCOMP1 is rated as 100 using iCOMP2. Another difference is the benchmarks used for the scores. Figure 1–3 shows the iCOMP2 index listing the Pentium II at speed up to 450 MHz.

---

[5]Macintosh is a registered trademark of Apple Computer Corporation.

FIGURE 1–3   The Intel
iCOMP2 index.

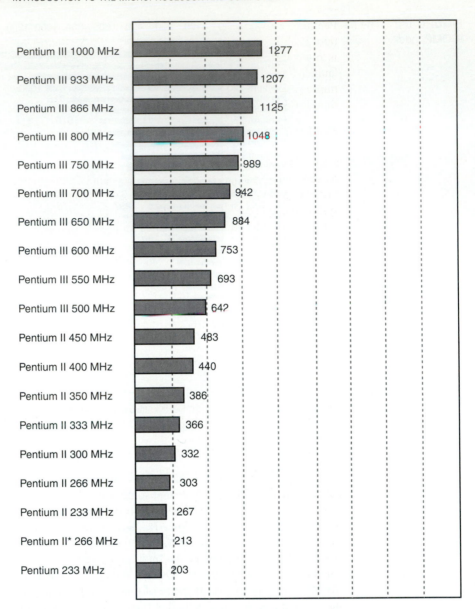

Note: *Pentium II Celeron, no cache.
iCOMP2 numbers are shown above, to
convert to iCOMP3 multiply by 2.568

Pentium Pro Processor.   A recent entry from Intel is the Pentium Pro processor, formerly code-named the P6 microprocessor. The Pentium Pro processor contains 21 million transistors, 3 integer units, as well as a floating-point unit to increase the performance of most software. The basic clock frequency was 150 MHz and 166 MHz in the initial offering made available in late 1995. In addition to the internal 16K level-one (L1) cache (8K for data and 8K for instructions), the Pentium Pro processor also contains a 256K level-two (L2) cache. One other significant change is that the Pentium Pro processor uses three execution engines, so it can execute up to three instructions at a time, which can conflict and still execute in parallel. This represents a change from the

Pentium, which executes two instructions simultaneously as long as they do not conflict. The Pentium Pro microprocessor has been optimized to efficiently execute 32-bit code; for this reason, it is often bundled with Windows NT rather than with normal versions of Windows 95. Intel launched the Pentium Pro processor for the server market. Still another change is that the Pentium Pro can address either a 4G byte memory system or a 64G byte memory system. The Pentium Pro has a 36-bit address bus if configured for a 64G memory system.

Pentium II and Pentium II Xeon Microprocessors. The Pentium II microprocessor (released in 1997) represents a new direction for Intel. Instead of being an integrated circuit as with prior versions of the microprocessor, Intel has placed the Pentium II on a small circuit board. The main reason for the change is that the L2 cache found on the main circuit board of the Pentium was not fast enough to justify a new microprocessor. On the Pentium system, the L2 cache operates at the system bus speed of 60 MHz or 66 MHz. The L2 cache and microprocessor are on a circuit board called the **Pentium II module.** This on-board, L2 cache operates at a speed of 133 MHz and stores 512K bytes of information. The microprocessor on the Pentium II module is actually a Pentium Pro with MMX extensions, which has no internal L2 cache.

In 1998, Intel changed the bus speed of the Pentium II. Because the 266 MHz through the 333 MHz Pentium II microprocessors used an external bus speed of 66 MHz, there was a bottleneck, so the newer Pentium II microprocessors use a 100 MHz bus speed. The Pentium II microprocessors rated at 350 MHz, 400 MHz, and 450 MHz all use this higher 100 MHz memory bus speed. The higher speed memory bus requires the use of 8 ns SDRAM in place of the 10 ns SDRAM found in use with the 66 MHz bus speed.

In mid-1998 Intel announced a new version of the Pentium II called the Xeon,[6] which was specifically designed for high-end workstation and server applications. The main difference between the Pentium II and the Pentium II Xeon is that the Xeon is available with a L1 cache size of 32K bytes and a L2 cache size of either 512K, 1M, or 2M bytes. The Xeon functions with the 440GX chip set. The Xeon is also designed to function with four Xeons in the same system, which is similar to the Pentium Pro. This newer product represents a change in Intel's strategy: Intel now produces a professional version and a home/business version of the Pentium II microprocessor.

Pentium III Microprocessor. The Pentium III microprocessor uses a faster core than the Pentium II, but it is still a P6 or Pentium Pro processor. It is also available in the slot 1 version mounted on a plastic cartridge and a socket 370 version called a flip-chip, which looks like the older Pentium package. Intel claims the flip-chip version costs less. Another diference is that the Pentium III is available to clock frequencies of 1 GHz. The slot 1 version contains a 512K cache and the flip-chip version contains a 256K cache. The speeds are comparable because the cache in the slot 1 version runs at one-half the clock speed, while the cache in the flip-chip version runs at the clock speed. Both versions use a memory bus speed of 100 MHz, while the Celeron[7] uses a memory bus clock speed of 66 MHz.

The speed of the front side bus, the connection from the microprocessor to the memory controller, PCI controller, and AGP controller, is now either 100 MHz or 133 MHz. Although the memory still runs at 100 MHz, this change has improved performance.

Pentium 4 Microprocessor. The Pentium 4 microprocessor, the latest Intel release, was first available in late 2000. The Pentium 4, like the Pentium Pro through the Pentium III (PIII), uses the Intel P-6 architecture. The main difference is that the Pentium 4 is available in a 1.3, 1.4, and 1.5 GHz-speed version, and the chipset that supports the Pentium 4 uses the RAMBUS memory technology

---

[6]The Pentium II Xeon is copyrighted by Intel Corporation.

[7]Celeron is a trademark of Intel Corporation.

in place of SDRAM technology. These higher microprocessor speeds are made available by an improvement in the size of the internal integration. It is also interesting to note that Intel has changed the level 1 cache size from 32K to 8K bytes. Research must have shown that this size is large enough for the initial release version of the microprocessor, with future versions possibly containing a 32K L1 cache. The level 2 cache remains at 256K bytes as in the PIII coppermine version.

Another change we are likely to see is a shift from aluminum to copper interconnections. Because copper is a better conductor, it should increase the clock frequencies for the microprocessor. This is especially true now that a method for using copper has surfaced. We may also see the front side bus speed increase from the current maximum of 133 MHz to 200 MHz or higher.

The Future of Microprocessors.   No one can really make accurate predictions, but the success of the Intel family should continue for quite a few years. What may occur is a change to RISC technology, but more likely a change to a new technology being developed jointly by Intel and Hewlett-Packard will occur. Even this new technology will embody the CISC instruction set of the 80X86 family of microprocessors, so that software for the system will survive. The basic premise behind this technology is that many microprocessors will communicate directly with each other, allowing parallel processing without any change to the instruction set or program. Currently, the superscaler technology uses many microprocessors, but they all share the same register set. This new untried technology, to be used in the next version of the Intel microprocessor, will contain many microprocessors, each containing its own register set that is linked with the other microprocessors' registers. This technology should offer true parallel processing without writing any special program.

Beginning in late 2002, Intel plans to release a new microprocessor architecture that is 64 bits in width and has a 128 bit data bus. This new architecture, code-named Merced,[8] is a joint venture called EPIC (Explicitly Parallel Instruction Computing) of Intel and Hewlett-Packard. The Merced architecture allows greater parallelism than traditional architectures, such as the Pentium Pro or Pentium II. These changes include 128 general-purpose integer registers, 128 floating-point registers, 64 predicate registers, and many execution units to ensure enough hardware resources for software.

Figure 1–4 is a conceptual view, comparing the 80486 through Pentium 4 microprocessors. Each view shows the internal structure of these microprocessors: the CPU, coprocessor, and cache memory. This illustration shows the complexity and level of integration in each version of the microprocessor.

FIGURE 1–4 Conceptual views of the 80486, Pentium Pro, Pentium II, Pentium III, and Pentium 4 microprocessors.

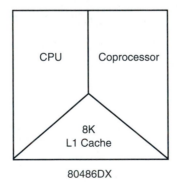

80486DX

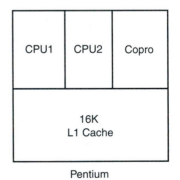

Pentium

---

[8]Merced is a trademark of Intel Corporation.

FIGURE 1–4
(continued)

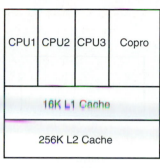

Pentium Pro

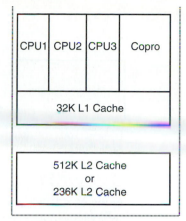

Pentium II, Pentium III,
or Pentium 4 Module

1–2   THE MICROPROCESSOR-BASED PERSONAL COMPUTER SYSTEM

Computer systems have undergone many changes recently. Machines that once filled large areas have been reduced to small desktop computer systems because of the microprocessor. Although these desktop computers are compact, they possess computing power that was only dreamed of a few years ago. Million-dollar mainframe computer systems, developed in the early 1980s, are not as powerful as the 80486- through Pentium 4-based computers of today. In fact, many smaller companies are replacing their mainframe computers with microprocessor-based systems. Companies such as DEC (Digital Equipment Corporation) have stopped producing mainframe computer systems in order to concentrate their resources on microprocessor-based computer systems.

This section shows the structure of the microprocessor-based personal computer system. This structure includes information about the memory and operating system used in many microprocessor-based computer systems.

See Figure 1–5 for the block diagram of the personal computer. This diagram also applies to any computer system, from the early mainframe computers to the latest microprocessor-based systems. The block diagram is composed of three blocks that are interconnected by buses. (A **bus** is set of common connections that carry the same type of information. For example, the address bus, which contains 20 or more connections, conveys the memory address to the memory.) These blocks and their function in a personal computer are outlined in this section of the text.

The Memory and I/O System

The memory structure of all Intel 80X86–Pentium 4 personal computer systems are similar. This includes the first personal computers based upon the 8088 introduced in 1981 by IBM to the most powerful high-speed versions of today based on the Pentium 4. Figure 1–6 illustrates the memory map of a personal computer system. This map applies to any IBM personal computer or to any of the many IBM-compatible clones that are in existence.

The memory system is divided into three main parts: TPA (**transient program area**), system area, and XMS (**extended memory system).** The type of microprocessor in your computer determines whether an extended memory system exists. If the computer is based upon

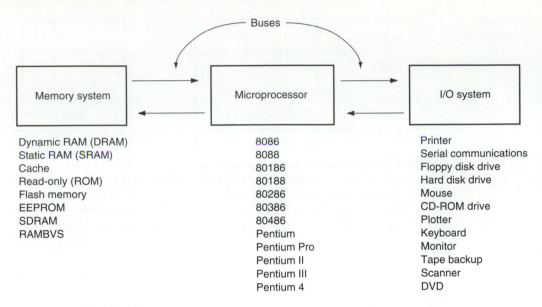

FIGURE 1–5   The block diagram of a microprocessor-based computer system.

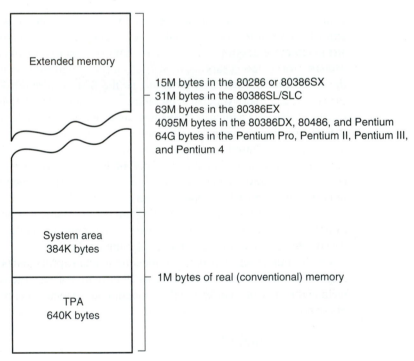

FIGURE 1–6   The memory map of the personal computer.

an older 8086 or 8088 (a PC[9] or XT[10]), the TPA and system areas exist, but there is no extended memory area. The PC and XT contain 640K bytes of TPA and 384K bytes of system memory, for a total memory size of 1M bytes. We often call the first 1M byte of memory the

---

real or conventional memory system because each Intel microprocessor is designed to function in this area by using its real mode of operation.

Computer systems based on the 80286 through the Pentium 4 not only contain the TPA (640K bytes) and **system area** (384K bytes), they also contain extended memory. These machines are often called AT[11] class machines. The PS/1 and PS/2, produced by IBM, are other versions of the same basic memory design. Sometimes, these machines are also referred to as ISA **(industry standard architecture)** or EISA **(extended ISA)** machines. The PS/2 is referred to as a micro-channel[12] architecture system, or ISA system, depending on the model number.

Recently, a new bus called the PCI **(peripheral conponent interconnect)** bus is being used in almost all Pentium–Pentium 4 systems. Extended memory contains up to 15M bytes in the 80286- and 80386SX-based computer, and up to 4095M bytes in the 80386DX, 80486, and Pentium microprocessors, in addition to the first 1M byte of real or conventional memory. The Pentium Pro through Pentium 4 computer systems have up to 1M less than 4G or 64G of extended memory. The ISA machine contains an 8-bit peripheral bus that is used to interface 8-bit devices to the computer in the 8086/8088-based PC or XT computer system. The AT class machine, also called an ISA machine, uses a 16-bit peripheral bus for interface and may contain the 80286 or above microprocessor. The EISA bus is a 32-bit peripheral interface bus found in a few older 80386DX- and 80486-based systems. Note that each of these buses is compatible with the earlier versions. That is, the 8-bit interface card functions in the 8-bit ISA, 16-bit ISA, or 32-bit EISA bus system. Likewise, a 16-bit interface card functions in the 16-bit ISA or 32-bit EISA system.

Another bus type found in many 80486-based personal computers is called the **VESA**[13] local bus, or VL bus. The **local bus** interfaces disk and video to the microprocessor at the local bus level, which allows 32-bit interfaces to function at the same clocking speed as the microprocessor. A recent modification to the VESA local bus supports the 64-bit data bus of the Pentium microprocessor and competes directly with the PCI bus, although it has generated little, if any, interest. The ISA and EISA standards function at only 8 MHz, which reduces the performance of the disk and video interfaces using these standards. The PCI bus is either a 32- or 64-bit bus that is specifically designed to function with the Pentium through Pentium 4 microprocessors at a bus speed of 33 MHz.

Two newer buses have appeared in the newest systems. The first to appear was the USB **(universal serial bus).** The universal serial bus is intended to connect peripheral devices such as keyboards, a mouse, modems, and sound cards to the microprocessor through a serial data path and a twisted pair of wires. The main idea is to reduce system cost by reducing the number of wires. Another advantage is that the sound system can have a separate power supply from the PC, which means much less noise. The data transfer rates through the USB are 10 Mbps at present; in the near future, they will increase to 100 Mbps in USB-2.

The second new bus is the AGP **(advanced graphics port)** for video cards. The advanced graphics port transfers data between the video card and the microprocessor at higher speeds (66 MHz, with a 64-bit data path, or 533M bytes per second) than were possible through any other bus or connection. This video subsystem change has been made to accommodate the new DVD players for the PC.

The TPA. The transient program area (TPA) holds the DOS operating system and other programs that control the computer system. The TPA also stores any currently active or inactive DOS application programs. The length of the TPA is 640K bytes. As mentioned, this area of memory holds the DOS operating system, which requires a portion of the TPA to function. In

---

[11]AT is a trademark of IBM Corporation used to designate an advanced class computer system.

[12]Micro-channel is a registered trademark of IBM Corporation.

[13]VESA is the Video Electronic Standards Association.

practice, the amount of memory remaining for application software is about 628K bytes if MSDOS[14] version 7.X is used as an operating system. Earlier versions of DOS required more of the TPA area and often left only 530K bytes or less for application programs. Another operating system found in some personal computers is PCDOS.[15] Both PCDOS and MSDOS are compatible, so either functions in the same manner with application programs. Windows and OS/2[16] are other operating systems that are compatible with DOS and allow DOS programs to execute. The DOS **(disk operating system)** controls the way that the disk memory is organized and controlled, as well as the function and control of the some of the I/O devices connected to the system. Windows also performs these functions for Windows-based applications. Figure 1–7 shows the organization of the TPA in a computer system.

FIGURE 1–7 The memory map of the TPA in a personal computer. (Note that this map will vary between systems.)

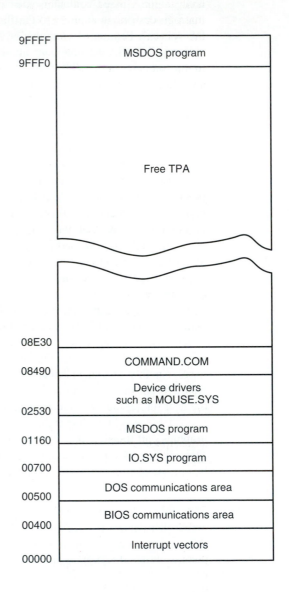

---

[14]MSDOS (Microsoft Disk Operating System) is a registered trademark of Microsoft Corporation.

[15]PCDOS (Personal Computer Disk Operating System) is a registered trademark of IBM Corporation.

[16]OS/2 (Operating System version 2) is a registered trademark of IBM Corporation.

The memory map shows how the many areas of the TPA are used for system programs, data, and drivers. It also shows a large area of memory available for application programs. To the left of each area is a hexadecimal number that represents the memory addresses that begin and end each data area. Hexadecimal **memory addresses** or **memory locations** are used to number each byte of the memory system. (A **hexadecimal number** is a number represented in radix 16 or base 16, with each digit representing a value from 0–9 and A–F. We often end a hexadecimal number with an H to indicate that it is a hexadecimal value. For example, 1234H is 1234 hexadecimal. We also represent hexadecimal data as 0x1234 for a 1234 hexadecimal.)

The Interrupt vectors access various features of the DOS, BIOS (**basic I/O system**), and applications. The system BIOS is a collection of programs stored in either a read-only (ROM) or flash memory that operates many of the I/O devices connected to your computer system. Note that a flash memory is an EEPROM (**electrically erasable read-only memory**) that is erased in the system electrically, while the ROM is a device that must be programmed in a special machine called an EPROM programmer for an EPROM (**erasable/programmable read-only memory**) or at the factory when a ROM is fabricated. These programs are stored in the system area that is defined later in this section of the chapter.

The system BIOS and DOS communications areas contain transient data used by programs to access I/O devices and the internal features of the computer system. (Refer to Appendix A for a complete listing of the BIOS and DOS communications areas.) These are stored in the TPA so they can be changed as the system operates. Note that the TPA contains read/write memory (called RAM, or **random access memory**), so it can change as a program executes.

The IO.SYS is a program that loads into the TPA from the disk whenever an MSDOS or PCDOS system is started. The IO.SYS contains programs that allow DOS to use the keyboard, video display, printer, and other I/O devices often found in the computer system. The IO.SYS program links DOS to the programs stored on the system BIOS ROM.

DOS occupies two areas of memory: One area is 16 bytes in length and is located at the top of the TPA, and the other is much larger and is located near the bottom of the TPA. The DOS program controls the operation of the computer system. The size of the DOS area depends on the version of DOS installed in the computer and how it is installed. If DOS is installed in high memory with the HIMEM.SYS driver, most of the TPA is free to hold application programs. (Note that high memory is described later in this text and only applies to 80286 or newer microprocessors.)

The size of the driver area and number of drivers change from one computer to another. Drivers are programs that control installable I/O devices such as a mouse, disk cache, hand scanner, CD-ROM memory (**Compact Disk Read-Only Memory**), DVD (**Digital Versatile Disk**), or installable devices, as well as programs. Installable drivers are programs that control or drive devices or programs that are added to the computer system. Drivers are normally files that have an extension of .SYS, such as MOUSE.SYS; in DOS version 3.2 and later, the files have an extension of .EXE, such as EMM386.EXE. Because few computer systems are identical, the driver area varies in size and contains different numbers and types of drivers. Note that even though these files are not used by Windows, they are still used to execute DOS applications, even with Windows 98. Windows uses a file called SYSTEM.INI to load drivers used by Windows. In newer versions of Windows, a registry is added to contain information about the system and the drivers used by the system. You can view the registry with the REGEDIT program.

The COMMAND.COM program (**command processor**) controls the operation of the computer from the keyboard when operated in the DOS mode. The COMMAND.COM program processes the DOS commands as they are typed from the keyboard. For example, if DIR is typed, the COMMAND.COM program displays a directory of the disk files in the current disk directory. If the COMMAND.COM program is erased, the computer cannot be used from the keyboard in DOS mode. Never erase COMMAND.COM, IO.SYS, or MSDOS.SYS to make room for other software, or your computer will not function. Note that these programs can be reloaded to the disk if erased by using the SYS.COM program located in the DOS directory (or in the WINDOWS/ COMMAND directory in Windows 95 and newer versions such as Windows XP).

The free TPA area holds DOS application programs as they are executed. These application programs include word processors, spreadsheet programs, CAD programs, and so forth. The TPA also holds TSR (**terminate and stay resident**) programs that remain in memory in an inactive state until activated by a hot-key sequence or another event such as an interrupt. A calculator program is an example of a TSR program that activates whenever an ALT-C key (hot key) is typed. A hot key is a combination of keys on the keyboard that activate a TSR program. A TSR program is often also called a pop-up program, because, when activated, it appears to pop up inside another program. If Windows is installed and in use, it uses a portion of the TPA to store information that allows it to access extended memory.

The System Area.   The system area, although smaller than the TPA, is just as important. The **system area** contains programs on either a read-only memory (ROM) or flash memory, and areas of read/write (RAM) memory for data storage. Figure 1–8 shows the system area of a typical computer system. As with the map of the TPA, this map also includes the hexadecimal memory addresses of the various areas.

The first area of the system space contains video display RAM and video control programs on ROM or flash memory. This area starts at location A0000H and extends to location C7FFFH. The size and amount of memory used depends on the type of video display adapter attached to

FIGURE 1–8   The system area of a typical personal computer.

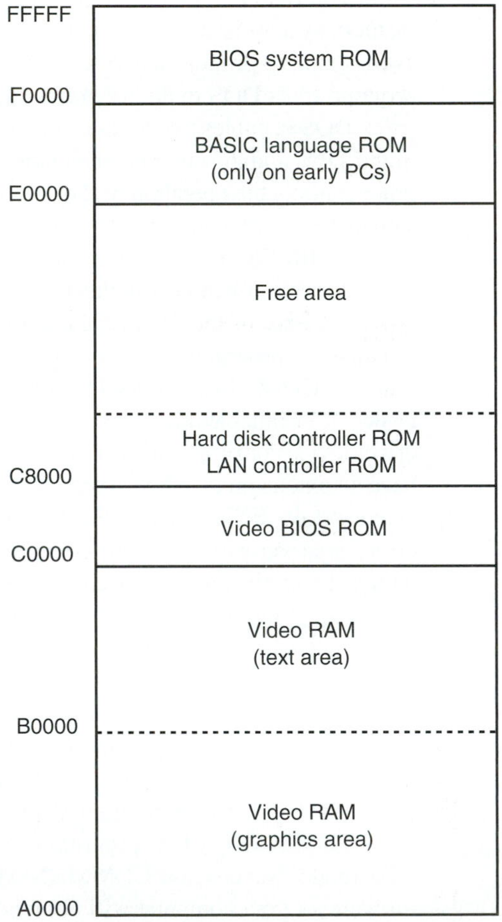

FFFFF	BIOS system ROM
F0000	BASIC language ROM (only on early PCs)
E0000	Free area
C8000	Hard disk controller ROM LAN controller ROM
C0000	Video BIOS ROM
B0000	Video RAM (text area)
A0000	Video RAM (graphics area)

the system. Display adapters that are often attached to a computer include the earlier CGA (**color graphics adapter**) and EGA (**enhanced graphics adapter**) or one of the many newer forms of VGA (**variable graphics array**). Generally, the video RAM located at A0000H–AFFFFH stores graphical or bit-mapped data, and the memory at B0000H–BFFFFH stores text data. Note that some newer video cards relocate the memory to wider, higher areas in the memory system for use under the Windows operating system. The video card in some computers uses memory locations E1800000H–E2FFFFFF in Windows for its video memory aperture. The video BIOS, located on a ROM or flash memory, is at locations C0000H–C7FFFH and contains programs that control the DOS video display.

If a hard disk memory is attached to the computer, the interface card might contain a ROM and a disk BIOS. This ROM, often found with older MFM or RLL hard disk drives, holds low-level format software at location C8005H. The size, location, and presence of the ROM depend on the type of hard disk adapter attached to the computer. Note that most IDE drives do not have this ROM, but many SCSI disk interfaces do contain a ROM.

The area at locations C8000H–DFFFFH is often open or free. This area is used for the expanded memory system (EMS) in a PC or XT system, or for the upper memory system in an AT system. Its use depends on the system and its configuration. The **expanded memory system** allows a 64K-byte page frame of memory to be used by application programs. This 64K-byte **page frame** (usually location D0000H through DFFFFH) is used to expand the memory system by switching in pages of memory from the EMS into this range of memory addresses. Note that the information is addressed in the page frame as 16K byte-sized pages of data that are swapped with pages from the EMS. Figure 1–9 shows the expanded memory system. Most application programs that state they are LIM 4.0 driver-compatible can use expanded memory. The LIM 4.0 memory management driver is the result of Lotus, Intel, and Microsoft standardizing access to expanded memory systems. Note that expanded memory is slow because the change to a new 16K-byte memory page requires action by the driver. Also note that expanded memory was designed to expand the memory system of the early 8086/8088-based computer systems. In most cases, except for some DOS-based games that use the sound card, expanded memory should be avoided in 80386- through Pentium 4-based systems.

FIGURE 1–9 The expanded memory system showing a page frame.

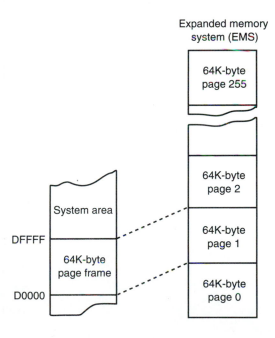

Memory locations E0000H–EFFFFH contain the cassette BASIC language on ROM found in early IBM personal computer systems. This area is often open or free in newer computer systems. In newer systems, we often back-fill this area with extra RAM for use by DOS programs, called **upper memory** or **upper memory blocks.** Each upper memory block is 4K bytes in length.

Finally, the system BIOS ROM is located in the top 64K bytes of the system area (F0000H–FFFFFH). This ROM controls the operation of the basic I/O devices connected to the computer system. It does not control the operation of the video system, which has its own BIOS ROM at location C0000H. The first part of the system BIOS (F0000H–F7FFFH) often contains programs that set up the computer; the second part contains procedures that control the basic I/O system. Once the system is set up, upper memory blocks at locations F0000H–F7FFFH are available if EMM386.EXE is installed. Also available for use as upper memory blocks are locations B0000H–B7FFFH, provided that black and white video is not needed in the CGA mode.

I/O Space.   The I/O (**input/output**) space in a computer system extends from I/O port 0000H to port FFFFH. (An **I/O port address** is similar to a memory address, except that instead of addressing memory, it addresses an I/O device.) The **I/O devices** allow the microprocessor to communicate between itself and the outside world. The I/O space allows the computer to access up to 64K different 8-bit I/O devices. A great number of these locations are available for expansion in most newer computer systems. Figure 1–10 shows the I/O map found in many personal computer systems.

The I/O area contains two major sections. The area below I/O location 0400H is considered reserved for system devices; many are depicted in Figure 1–10. The remaining area is available I/O space for expansion on newer systems that extends from I/O port 0400H through FFFFH. Some of the main boards in newer computer systems can also use other addresses above 0400H. Generally,

FIGURE 1–10   The I/O map of a personal computer showing some of the many areas of I/O devices.

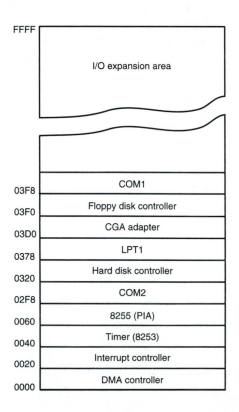

I/O addresses between 0000H and 00FFH address components on the main board of the computer, while addresses between 0100H and 03FFH address devices located on plug-in cards. Note that the limitation of I/O addresses between 0000 and 03FFH comes from the original PC standard, as specified by IBM. When using the ISA bus, you must only use addresses between 0000H and 03FFH.

Various I/O devices that control the operation of the system are usually not directly addressed. Instead, the system BIOS ROM addresses these basic devices, which can vary slightly in location and function from one computer to the next. Access to most I/O devices should always be made through DOS or BIOS function calls to maintain compatibility from one computer system to another. The map shown in Figure 1–10 is provided as a guide to illustrate the I/O space in the system.

The DOS Operating System

The operating system is the program that operates the computer. This text assumes that the operating system is DOS, which is by far the most common operating system found in most personal computer systems (85 percent, according to a recent *PC Magazine*[17] article). The Windows operating system, which contains DOS, is available on a great majority of personal computers, according to the same article. Windows XP will undoubtedly replace Windows 98 as an important operating system. The operating system is stored on a disk that is either placed in one of the floppy disk drives or found on a hard disk drive, either resident to the computer or to a local area network (LAN). Some dedicated systems store the DOS on a ROM. Newer systems that use Windows CE as an operating system also store it on a ROM. An example is the Tandy Corporation personal computer. Each time that the computer is powered up or reset, the operating system is read from the disk or LAN. We call this operation booting the system. Once DOS is installed in the memory by the boot, it controls the operation of the computer system, its I/O devices, and the application programs. In addition to the DOS operating system, other operating systems are sometimes used to control or operate the computer. The first task of the DOS operating system, after loading into memory, is to use a file called the **CONFIG.SYS** file, which should not be erased. This file specifies various drivers that load into the memory, setting up or configuring the machine for operation under DOS. Note that this file does not define the operation of Windows, which is defined in its registry file. Example 1–1 lists an example CONFIG.SYS file for DOS. Note that the statements in this file vary from machine-to-machine, and the one illustrated is just an example. If operating under Windows, this file will be shorter and will probably load only the CD-ROM driver software.

EXAMPLE 1–1

```
REM DOS CONFIG.SYS FILE
DEVICE=C:\DOS\HIMEM.SYS
DEVICE=C:\DOS\EMM386.EXE NOEMS i=c800-efff
DOS=UMB
FILES=30
BUFFERS=10
SHELL=C:\DOS\COMMAND.COM C:\DOS\ /E:2048 /P
DOS=HIGH
DEVICEHIGH C:\LASERLIB\SONY_CDU.SYS
DEVICEHIGH C:\DOS\SETVER.EXE
DEVICEHIGH C:\MOUSE1\MOUSE.SYS
LASTDRIVE=F
```

---

[17]*PC Magazine* is a Ziff-Davis publication.

The first statement (REM) is a remark that merely identifies this file and is optional. The second line informs the system to load the HIMEM.SYS driver that allows upper memory blocks to be used and provides high memory to the system. (A **driver** is a program that controls an I/O device or some other function and is loaded before other programs by the CONFIG.SYS file.) **High memory** is a section of memory that begins at location 100000H and ends at 10FEFFH, just above the first 1M byte of memory. It is to be used for programs in the 80286- through the Pentium II-based system. The HIMEM.SYS driver allows a DOS-based system access to 1M plus 64K–16 bytes of memory. This extra section of high memory holds most of DOS, freeing additional space in the TPA for DOS applications.

The first four statements in this CONFIG.SYS file set up the number of files, buffers, stacks, and file control blocks required to execute various programs. These settings should be adequate for almost any program loaded into memory using DOS. In general, if a program requires more buffers, and so forth, the documentation indicates that the CONFIG.SYS file must be changed to reflect the increased need. Many modern programs automatically adjust the CONFIG.SYS file when it is installed by changing these parameters or by adding additional statements. If the computer uses Windows as its main operating system, Windows provides most of these settings for DOS applications.

In order to enable the upper memory blocks, available only in an 80386-, 80486-, or Pentium-based system, the EMM386.EXE **(extended memory manager)** program is loaded into memory. The extended memory manager is a driver that emulates expanded memory in extended memory and also in the extended memory system. This program back-fills free areas of memory within the system area so that programs can be loaded into this area and accessed directly by DOS applications. The I=c800-efff switch tells EMM386.EXE to use memory area C8000H–EFFFFH for upper memory or upper memory blocks (UMBs). Drivers and programs are loaded into upper memory, freeing even more area in the TPA for application programs. Before using the I=c800-efff switch, make sure that your computer does not contain any system ROM/RAM in this area of the memory. The NOEMS command tells EMM386.EXE to exclude expanded memory. Expanded memory can also be installed by replacing NOEMS with a number that indicates how much extended memory to allocate to LIM 4.0 expanded memory system (EMS). Today, most systems should not use expanded memory. If expanded memory is required, the NOEMS is replaced with RAM 1024 to enable 1024 bytes of expanded memory. The FRAME=D000 statement places the page frame for expanded memory at location D0000H–DFFFFH if expanded memory is enabled. If DOS version 6.0 or higher is in use, the NOEMS statement still provides access to EMS, on an as-needed basis, through a driver called VCPI **(virtual control program interface)**. The VCPI program dynamically allocates EMS if a program requires it, and then releases it after the program completes execution. If operating under Windows, the DPMI **(DOS protected mode interface)** resource provided by Windows performs the same function as VCPI.

The next two commands tell the system to provide upper memory blocks (DOS=UMB) and dictate how many files can be opened at the same time (FILES=30). The BUFFERS statement informs DOS to allocate 10 buffers to open for file transfer areas.

The SHELL command specifies the command processor used with DOS. In this example, the COMMAND.COM file is the command processor (also selected by default) using the E:2048 /P switches. The E:2048 switch sets the environment size to 2048 bytes. The **environment space** stores shared keywords and other information used by all applications in the system. An example is the statement SET TEMP=C:\TEMP, which informs applications that the temporary directory is on the C disk drive in a directory called TEMP. The SET command can be used to assign keywords at the DOS command prompt or in the AUTOEXEC.BAT file. The /P switch tells the command processor to make COMMAND.COM permanent. If COMMAND.COM is not permanent, it must be loaded into memory from the disk each time that DOS returns to the command prompt. Although this may free a small amount of memory, the constant access to the disk for the COMMAND.COM program increases wear and tear on the disk drive, and also lengthens the time required to return to the DOS prompt. This is followed by the command DOS=HIGH, which informs the system to load DOS into the high memory area created by the HIMEM.SYS driver.

The next three lines use the DEVICEHIGH command, which loads drivers and programs into the upper memory blocks allocated by the EMM386.EXE driver. In the illustrated CONFIG.SYS file, three drivers are loaded into upper memory blocks, beginning at location C8000H. The first is a program that operates a SONY CD-ROM drive, the second loads a program called SETVER, and the third loads the MOUSE driver.

The last statement in the illustrated CONFIG.SYS file shows the LASTDRIVE statement, which tells DOS which is the last disk drive connected to your computer system. By using the LASTDRIVE statement, more memory can be freed for use in the TPA. Each drive requires a buffer area; if you use the actual last drive with this statement, extra memory is made available. Other drivers may also be loaded into memory by using the CONFIG.SYS file, such as a PRINT.SYS driver, an ANSI.SYS driver, or any other program that functions as a driver. Driver programs normally contain the DOS extension .SYS used to indicate a system file. Be very careful when changing the CONFIG.SYS file because an error locks up the computer system (except for DOS 6.0 and above, which can exit this type of system lockup by typing a Control-C). Once the computer is locked up by a CONFIG.SYS error, the only way to recover is to boot off a DOS floppy disk that contains the operating system with a functioning CONFIG.SYS file.

Once the operating system completes its configuration, as dictated by CONFIG.SYS, the AUTOEXEC.BAT **(automatic execution batch)** file is executed by the computer. If none exists, the computer asks for the time and date. Example 1–2 shows a typical AUTOEXEC.BAT file. (This is only an example; variations often occur from system to system.) The AUTOEXEC.BAT file contains commands that execute when power is first applied to the computer. These are the same commands that could be typed from the keyboard, but the AUTOEXEC.BAT file saves us from doing so each time the computer is powered. Even a Windows system often contains an AUTOEXEC.BAT file to set various labels and paths for DOS applications that often still exist.

EXAMPLE 1–2

```
PATH C:\DOS;C:\;C:\MASM\BIN;C:\MASM\BINB\;C:\UTILITY
SET BLASTER=A220 I7 D1 T3
SET INCLUDE=C:\MASM\INCLUDE\
SET HELPFILES=C:\MASM\HELP\*.HLP
SET INIT=C:\MASM\INIT\
SET ASMEX=C:\MASM\SAMPLES\
SET TMP=C:\MASM\TMP
SET SOUND=C:\SB
LOADHIGH C:\LASERLIB\MSCDEX.EXE /D:SONY_001 /L:F /M:10
LOADHIGH C:\LASERLIB\LLTSR.EXE ALT-Q
LOADHIGH C:\DOS\FASTOPEN C:=256
LOADHIGH C:\DOS\DOSKEY /BUFSIZE=1024
LOADHIGH C:\LASERLIB\PRINTF.COM
DOSKEY GO=C:\WINDOWS\WIN.EXE
GO
```

The PATH statement specifies the search paths whenever a program name is typed at the command line. The order of the path search is the same as the order of the paths in the PATH statement. For example, if PROG is typed at the command line, the machine first searches C:\DOS, then the root directory C:\, then C:\MASM\BIN and so forth until the program named PROG is found. If it isn't found, the command interpreter (COMMAND.COM) informs the user that the program is not found.

The SET statement, as introduced earlier, sets a variable name to a path. This allows names to be associated with paths for batch programs. It's also used to set command strings (environments) for various programs. The first SET command sets the environment for the sound blaster card. The second SET command sets INCLUDE to the path C:\MASM\INCLUDE\. Note that the SET statements are stored in the DOS environment space that was reserved in the CONFIG.SYS file using the SHELL statement. If the environment becomes too large, you must change the SHELL statement to allow more space.

LOADHIGH, or LH, places programs into upper memory blocks defined by EMM386.EXE program. LOADHIGH is used at any DOS command prompt to load a program into the high (upper) memory area, as long as the computer is an 80386 or above. The second-to-last command in this AUTOEXEC.BAT file uses the DOSKEY program to define a macro for the word GO. Here, the word GO is assigned the character string C:\WINDOWS\WIN.EXE. The COMMAND.COM program will then interpret the word GO so that the Windows program executes anytime that the word GO is typed from the keyboard at the DOS prompt. This last command then runs Windows.

The Microprocessor

At the heart of the microprocessor-based computer system is the microprocessor integrated circuit. The microprocessor, sometimes referred to as the CPU (central processing unit), is the controlling element in a computer system. The microprocessor controls memory and I/O through a series of connections called buses. The buses select an I/O or memory device, transfer data between an I/O device or memory and the microprocessor, and control the I/O and memory system. Memory and I/O are controlled through instructions that are stored in the memory and executed by the microprocessor.

The microprocessor performs three main tasks for the computer system: (1) data transfer between itself and the memory or I/O systems, (2) simple arithmetic and logic operations, and (3) program flow via simple decisions. Albeit these are simple tasks, but through them, the microprocessor performs virtually any series of operations or tasks.

The power of the microprocessor is in its capability to execute hundreds of millions of instructions per second from a program or software (**group of instructions**) stored in the memory system. This stored program concept has made the microprocessor and computer system very powerful devices. (Recall that Babbage also wanted to use the stored program concept in his Analytical Engine.)

Table 1–3 shows the arithmetic and logic operations executed by the Intel family of microprocessors. These operations are very basic, but through them, very complex problems are solved. Data are operated upon from the memory system or internal registers. Data widths are variable and include a **byte** (8 bits), **word** (16 bits), and **doubleword** (32 bits). Note that only the 80386 through the Pentium II directly manipulate 8-, 16-, and 32-bit numbers. The earlier 8086–80286 directly manipulated 8- and 16-bit numbers, but not 32-bit numbers. Beginning with the 80486, the microprocessor contained a numeric coprocessor that allowed it to perform complex arithmetic using floating-point arithmetic. The numeric coprocessor, which is similar to a calculator chip, was an additional component in the 8086- through the 80386-based personal computer.

Another feature that makes the microprocessor powerful is its capability to make simple decisions based upon numerical facts. For example, a microprocessor can decide if a number is zero, if it is positive, and so forth. These simple decisions allow the microprocessor to modify the program flow, so that programs appear to think through these simple decisions. Table 1–4 lists the decision-making capabilities of the Intel family of microprocessors.

Buses. A **bus** is a common group of wires that interconnect components in a computer system. The buses that interconnect the sections of a computer system transfer address, data, and control

TABLE 1–3 Simple arithmetic and logic operations.

Operation	Comment
Addition	
Subtraction	
Multiplication	
Division	
AND	Logical multiplication
OR	Logical addition
NOT	Logical inversion
NEG	Arithmetic inversion
Shift	
Rotate	

TABLE 1–4 Decisions found in 8086–80486 and Pentium/Pentium Pro microprocessors.

Decision	Comment
Zero	Test a number for zero or not-zero
Sign	Test a number for positive or negative
Carry	Test for a carry after addition or a borrow after subtraction
Parity	Test a number for an even or an odd number of ones
Overflow	Test for an overflow that indicates an invalid signed result after addition or subtraction

information between the microprocessor and its memory and I/O systems. In the microprocessor-based computer system, three buses exist for this transfer of information: address, data, and control. Figure 1–11 shows how these buses interconnect various system components, such as the microprocessor, read/write memory (RAM), read-only memory (ROM), and a few I/O devices.

The address bus requests a memory location from the memory or an I/O location from the I/O devices. If I/O is addressed, the address bus contains a 16-bit I/O address from 0000H through FFFFH. The 16-bit I/O address, or port number, selects one of 64K different I/O devices. If memory is addressed, the address bus contains a memory address, which varies in width with the different versions of the microprocessor. The 8086 and 8088 address 1M byte of memory, using a 20-bit address that selects locations 00000H–FFFFFH. The 80286 and 80386SX address 16M bytes of memory using a 24-bit address that selects locations 000000H–FFFFFFH. The 80386SL, 80386SLC, and 80386EX address 32M bytes of memory, using a 25-bit address that selects locations 0000000H–1FFFFFFH. The 80386DX, 80486SX, and 80486DX address 4G bytes of memory, using a 32-bit address that selects locations 00000000H–FFFFFFFFH. The Pentium also addresses 4G bytes of memory, but it uses a 64-bit data bus to access up to 8 bytes of memory at a time. The Pentium Pro through Pentium 4 microprocessors have a 64-bit data bus and a 32-bit address bus that address 4G of memory from location 00000000H–FFFFFFFFH, or a 36-bit address bus that addresses 64G of memory at locations 000000000H–FFFFFFFFFH, depending on their configuration. Refer to Table 1–5 for a complete listing of bus and memory sizes of the Intel family of microprocessors.

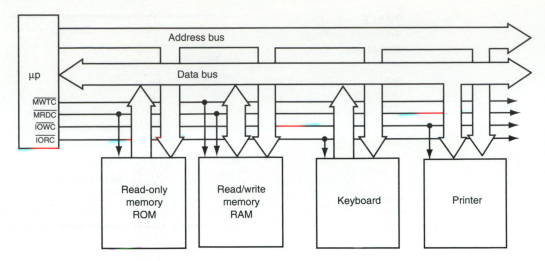

FIGURE 1–11   The block diagram of a computer system showing the address, data, and control bus structure.

TABLE 1–5   The Intel family of microprocessor bus and memory sizes

Microprocessor	Data Bus Width	Address Bus Width	Memory Size
8086	16	20	1M
8088	8	20	1M
80186	16	20	1M
80188	8	20	1M
80286	16	24	16M
80386SX	16	24	16M
80386DX	32	32	4G
80386EX	16	26	64M
80486	32	32	4G
Pentium	64	32	4G
Pentium OverDrive	32	32	4G
Pentium Pro	64	32	4G
Pentium Pro	64	36	64G
Pentium II	64	32	4G
Pentium II, Pentium III, Pentium 4	64	36	64G

The data bus transfers information between the microprocessor and its memory and I/O address space. Data transfers vary in size, from 8 bits wide to 64 bits wide in various members of the Intel microprocessor family. For example, the 8088 has an 8-bit data bus that transfers 8 bits of data at a time. The 8086, 80286, 80386SL, 80386SX, and 80386EX transfer 16 bits of data through their data buses; the 80386DX, 80486SX, and 80486DX transfer 32 bits of data; and the Pentium through Pentium 4 microprocessors transfer 64 bits of data. The advantage of a wider data bus is speed in applications that use wide data. For example, if a 32-bit number is stored in memory, it takes the 8088 microprocessor four transfer operations to complete because its data bus is only 8 bits wide. The 80486 accomplishes the same task with one transfer because its data bus is 32 bits wide. Figure 1–12 shows the memory widths and sizes of the 8086–80486 and Pentium–Pentium 4 microprocessors. Notice how the memory sizes and organizations differ between various members of the Intel microprocessor family. In all family members, the memory is numbered by byte. Notice that the Pentium through Pentium 4 microprocessors all contain a 64-bit wide data bus.

The control bus contains lines that select the memory or I/O and cause them to perform a read or write operation. In most computer systems, there are four control bus connections: $\overline{\text{MRDC}}$ **(memory read control),** $\overline{\text{MWTC}}$ **(memory write control),** $\overline{\text{IORC}}$ **(I/O read control),** and $\overline{\text{IOWC}}$ **(I/O write control).** Note that the over-bar indicates that the control signal is active-low; that is, it is active when a logic zero appears on the control line. For example, if $\overline{\text{IOWC}} = 0$, the microprocessor is writing data from the data bus to an I/O device whose address appears on the address bus.

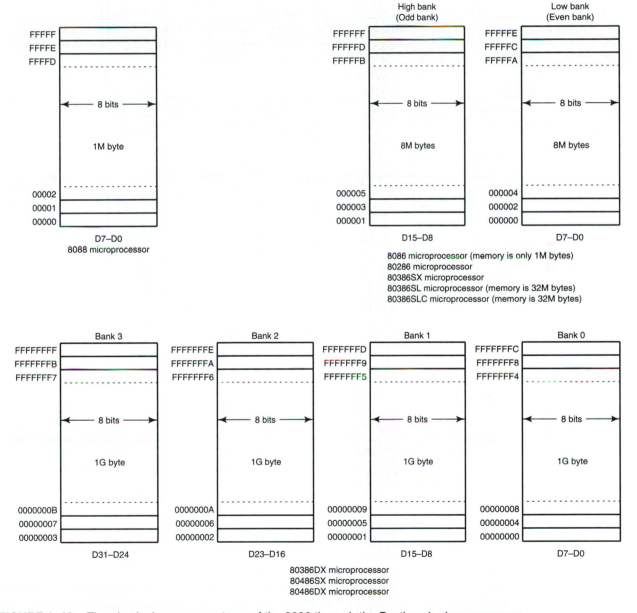

FIGURE 1–12   The physical memory systems of the 8086 through the Pentium 4 microprocessors.

(continued on next page)

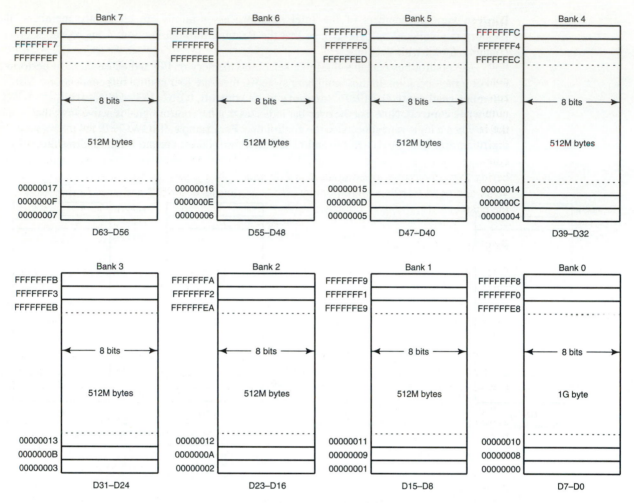

Pentium–Pentium 4 microprocessors

FIGURE 1–12   *(continued)*

The microprocessor reads the contents of a memory location by sending the memory an address through the address bus. Next, it sends the memory read control signal ($\overline{\text{MRDC}}$) to cause memory to read data. Finally, the data read from the memory are passed to the microprocessor through the data bus. Whenever a memory write, I/O write, or I/O read occurs, the same sequence ensues, except that different control signals are issued and the data flow out of the microprocessor through its data bus for a write operation.

1–3   NUMBER SYSTEMS

The use of the microprocessor requires a working knowledge of binary, decimal, and hexadecimal numbering systems. This section of the text provides a background for those who are unfamiliar with number systems. Conversions between decimal and binary, decimal and hexadecimal, and binary and hexadecimal are described.

Digits

Before numbers are converted from one number base to another, the digits of a number system must be understood. Early in our education, we learned that a decimal, or base 10, number was constructed with 10 digits: 0 through 9. The first digit in any numbering system is always a zero. For example, a base 8 (**octal**) number contains 8 digits: 0 through 7; a base 2 (**binary**) number contains 2 digits: 0 and 1. If the base of a number exceeds 10, the additional digits use the letters of the alphabet, beginning with an A. For example, a base 12 number contains 12 digits: 0 through 9, followed by A for 10 and B for 11. Note that a base 10 number does not contain a *10* digit, just as a base 8 number does not contain an *8* digit. The most common numbering systems used with computers are decimal, binary, and hexadecimal (base 16). (Many years ago octal numbers were popular.) Each system is described and used in this section of the chapter.

Positional Notation

Once the digits of a number system are understood, larger numbers are constructed by using positional notation. In grade school, we learned that the position to the left of the units position was the tens position, the position to the left of the tens position was the hundreds position, and so forth. (An example is the decimal number 132: This number has 1 hundred, 3 tens, and 2 units.) What probably was not learned was the exponential value of each position: The units position has a weight of 10^0, or 1; the tens position has weight of 10^1, or 10; and the hundreds position has a weight of 10^2, or 100. The exponential powers of the positions are critical for understanding numbers in other numbering systems. The position to the left of the radix (**number base**) point, called a **decimal point** only in the decimal system, is always the units position in any number system. For example, the position to the left of the binary point is always 2^0, or 1; the position to the left of the octal point is 8^0, or 1. In any case, any number raised to its zero power is always 1, or the units position.

The position to the left of the units position is always the number base raised to the first power; in a decimal system, this is 10^1, or 10. In a binary system, it is 2^1, or 2; and in an octal system, it is 8^1, or 8. Therefore, an 11 decimal has a different value from an 11 binary. The 11 decimal is composed of 1 ten plus 1 unit, and has a value of 11 units; while the binary number 11 is composed of 1 two plus 1 unit, for a value of 3 decimal units. The 11 octal has a value of 9 units.

In the decimal system, positions to the right of the decimal point have negative powers. The first digit to the right of the decimal point has a value of 10^{-1}, or 0.1. In the binary system, the first digit to the right of the binary point has a value of 2^{-1}, or 0.5. In general, the principles that apply to decimal numbers also apply to numbers in any other number system.

Example 1–3 shows a 110.101 in binary (often written as 110.101_2). It also shows the power and weight or value of each digit position. To convert a binary number to decimal, add the weights of each digit to form its decimal equivalent. The 110.101_2 is equivalent to a 6.625 in decimal ($4 + 2 + 0.5 + 0.125$). Notice that this is the sum of 2^2 (or 4) plus 2^1 (or 2), but 2^0 (or 1) is not added because there are no digits under this position. The fraction part is composed of 2^{-1} (0.5) plus 2^{-3} (or .125), but there is no digit under the 2^{-2} (or .25).

EXAMPLE 1–3

	2^2	2^1	2^0	2^{-1}	2^{-2}	2^{-3}		
Power	2^2	2^1	2^0	2^{-1}	2^{-2}	2^{-3}		
Weight	4	2	1	0.5	0.25	.125		
Number	1	1	0 .	1	0	1		
Numeric Value	4 +	2 +	0 +	0.5 +	0 +	.125	=	6.625

Suppose that the conversion technique is applied to a base 6 number, such as 25.2_6. Example 1–4 shows this number placed under the powers and weights of each position. In the example, there is a 2 under 6^1, which has a value of 12_{10} (2×6), and a 5 under 6^0, which has a value of 5 (5×1). The whole number portion has a decimal value of $12 + 5$, or 17. The number to the right of the hex point is a 2 under 6^{-1}, which has a value of .333 ($2 \times .167$). The number 25.2_6, therefore, has a value of 17.333.

EXAMPLE 1–4

```
Power            6¹   6⁰   6⁻¹
Weight           6    1    .167
Number           2    5  . 2
Numeric Value    12 + 5 +  .333 = 17.333
```

Conversion to Decimal

The prior examples have shown that to convert from any number base to decimal, determine the weights or values of each position of the number, and then sum the weights to form the decimal equivalent. Suppose that a 125.7_8 octal is converted to decimal. To accomplish this conversion, first write down the weights of each position of the number. This appears in Example 1–5. The value of 125.7_8 is 85.875 decimal, or 1×64 plus 2×8 plus 5×1 plus $7 \times .125$.

EXAMPLE 1–5

```
Power            8²   8¹   8⁰   8⁻¹
Weight           64   8    1    .125
Number           1    2    3  . 7
Numeric Value    64 + 16 + 5 +  .875 = 85.875
```

Notice that the weight of the position to the left of the units position is 8. This is 8 times 1. Then notice that the weight of the next position is 64, or 8 times 8. If another position existed, it would be 64 times 8, or 512. To find the weight of the next higher-order position, multiply the weight of the current position by the number base (or 8, in this example). To calculate the weights of position to the right of the radix point, divide by the number base. In the octal system, the position immediately to the right of the octal point is $^1/_8$, or .125. The next position is $^{.125}/_8$, or .015625, which can also be written as $^1/_{64}$. Also note that the number in Example 1–5 can also be written as the decimal number $85^7/_8$.

Example 1–6 shows the binary number 11011.0111 written with the weights and powers of each position. If these weights are summed, the value of the binary number converted to decimal is 27.4375.

EXAMPLE 1–6

```
Power            2⁴   2³   2²   2¹   2⁰   2⁻¹  2⁻²   2⁻³    2⁻⁴
Weight           16   8    4    2    1    0.5  0.25  .125   .0625
Number           1    1    0    1    1  . 0    1     1      1
Numeric Value    16 + 8 + 4 + 2 + 1 + 0 + .25 + .125 + .0625 = 27.4375
```

It is interesting to note that 2^{-1} is also $^1/_2$, 2^{-2} is $^1/_4$, and so forth. It is also interesting to note that 2^{-4} is $^1/_{16}$, or .0625. The fractional part of this number is $^7/_{16}$ or .4375 decimal. Notice that 0111 is a 7 in binary code for the numerator and the rightmost one is in the $^1/_{16}$ position for the denominator. Other examples: the binary fraction of .101 is $^5/_8$ and the binary fraction of .001101 is $^{13}/_{64}$.

Hexadecimal numbers are often used with computers. A 6A.CH (H for hexadecimal) is illustrated with its weights in Example 1–7. The sum of its digits is 106.75, or $106^3/_4$. The whole number part is represented with 6×16 plus 10 (A) $\times 1$. The fraction part is 12 (C) as a numerator and 16 (16^{-1}) as the denominator, or $^{12}/_{16}$, which is reduced to $^3/_4$.

EXAMPLE 1–7

```
Power           16¹   16⁰    16⁻¹
Weight          16    1      .0625
Number          6     A      C
Numeric Value   96  + 10  +  .75  =  106.75
```

Conversion From Decimal

Conversions from decimal to other number systems are more difficult to accomplish than conversion to decimal. To convert the whole number portion of a number to decimal, divide by the radix. To convert the fractional portion, multiply by the radix.

Whole Number Conversion from Decimal. To convert a decimal whole number to another number system, divide by the radix and save the remainders as significant digits of the result. An algorithm for this conversion as is follows:

1. Divide the decimal number by the radix (number base).
2. Save the remainder (first remainder is the least significant digit).
3. Repeat steps 1 and 2 until the quotient is zero.

For example, to convert a 10 decimal to binary, divide it by 2. The result is 5, with a remainder of 0. The first remainder is the units position of the result (in this example, a 0). Next, divide the 5 by 2. The result is 2, with a remainder of 1. The 1 is the value of the two's (2^1) position. Continue the division until the quotient is a zero. Example 1–8 shows this conversion process. The result is written as 1010_2, from the bottom to the top.

EXAMPLE 1–8

```
2) 10   remainder = 0
2)  5   remainder = 1
2)  2   remainder = 0
 2)1    remainder = 1          result = 1010
    0
```

To convert a 10 decimal into base 8, divide by 8, as shown in Example 1–9. A 10 decimal is a 12 octal.

EXAMPLE 1–9

```
8) 10   remainder = 2
8)  1   remainder = 1          result = 12
    0
```

Conversion from decimal to hexadecimal is accomplished by dividing by 16. The remainders will range in value from 0 through 15. Any remainder of 10 though 15 is then converted to the letters A through F for the hexadecimal number. Example 1–10 shows the decimal number 109 converted to a 6DH.

EXAMPLE 1–10

```
16) 109   remainder = 13 (D)
16)   6   remainder = 6          result = 6D
      0
```

Converting from a Decimal Fraction. Conversion from a decimal fraction to another number base is accomplished with multiplication by the radix. For example, to convert a decimal fraction into binary, multiply by 2. After the multiplication, the whole number portion of the result is saved as a significant digit of the result, and the fractional remainder is again multiplied by the radix. When the fraction remainder is zero, multiplication ends. Note that some numbers are never-ending. That is, a zero is never a remainder. An algorithm for conversion from a decimal fraction is as follows:

1. Multiply the decimal fraction by the radix (number base).
2. Save the whole number portion of the result (even if zero) as a digit. Note that the first result is written immediately to the right of the radix point.
3. Repeat steps 1 and 2, using the fractional part of step 2 until the fractional part of step 2 is zero.

Suppose that a .125 decimal is converted to binary. This is accomplished with multiplications by 2, as illustrated in Example 1–11. Notice that the multiplication continues until the fractional remainder is zero. The whole number portions are written as the binary fraction (0.001) in this example.

EXAMPLE 1–11

```
   .125
x       2
   0.25  digit is 0

   .25
x     2
   0.5   digit is 0

   .5
x   2
   1.0    digit is 1. The result is written as 0.001 binary
```

This same technique is used to convert a decimal fraction into any number base. Example 1–12 shows the same decimal fraction of .125 from Example 1–11 converted to octal by multiplying with an 8.

EXAMPLE 1–12

```
   .125
x       8
   1.0 digit is 1. The result is written as 0.1 octal
```

Conversion to a hexadecimal fraction appears in Example 1–13. Here, a decimal .046875 is converted to hexadecimal by multiplying by 16. Note that a .046875 is a 0.0CH.

EXAMPLE 1–13

```
   .046875
x       16
     0.75 digit is 0

   .75
x   16
   12.0 digit is 12 (C). The result is written as 0.0C hexadecimal
```

Binary-Coded Hexadecimal

Binary-coded hexadecimal (BCH) is used to represent hexadecimal data in binary code. A binary-coded hexadecimal number is a hexadecimal number written so that each digit is represented by a 4-bit binary number. The values for the BCH digits appear in Table 1–6.

Hexadecimal numbers are represented in BCH code by converting each digit to BCH code, with a space between each coded digit. Example 1–14 shows a 2AC converted to BCH code. Note that each BCH digit is separated by a space.

TABLE 1–6   Binary-coded hexadecimal (BCH) code.

Hexadecimal Digit	BCH Code
0	0000
1	0001
2	0010
3	0011
4	0100
5	0101
6	0110
7	0111
8	1000
9	1001
A	1010
B	1011
C	1100
D	1101
E	1110
F	1111

EXAMPLE 1–14

```
2AC = 0010 1010 1100
```

The purpose of BCH code is to allow a binary version of a hexadecimal number to be written in a form that can easily be converted between BCH and hexadecimal. Example 1–15 shows a BCH coded number converted back to hexadecimal code.

EXAMPLE 1–15

```
1000 0011 1101 . 1110 = 83D.E
```

Complements

At times, data are stored in complement form to represent negative numbers. There are two systems that are used to represent negative data: **radix** and **radix –1** complements. The earliest system was the radix –1 complement, in which each digit of the number is subtracted from the radix –1 to generate the radix –1 complement to represent a negative number.

Example 1–16 shows how the 8-bit binary number 01001100 is one's (radix –1) complemented to represent it as a negative value. Notice that each digit of the number is subtracted from one to generate the radix –1 (one's) complement. In this example, the negative of 01001100 is 10110011. The same technique can be applied to any number system, as illustrated in Example 1–17, in which the fifteen's (radix –1) complement of a 5CD hexadecimal is computed by subtracting each digit from a fifteen.

EXAMPLE 1–16

```
  1 1 1 1   1 1 1 1
- 0 1 0 0   1 1 0 0
  1 0 1 1   0 0 1 1
```

EXAMPLE 1–17

```
  15 15 15
-  5  C  D
   A  3  2
```

Today, the radix −1 complement is not used by itself; it is used as a step for finding the radix complement. The radix complement is used to represent negative numbers in modern computer systems. (The radix −1 complement was used in the early days of computer technology.) The main problem with the radix −1 complement is that a negative or a positive zero exists; in the radix complement system, only a positive zero can exist.

To form the radix complement, first find the radix −1 complement, and then add a one to the result. Example 1–18 shows how the number 0100 1000 is converted to a negative value by two's (radix) complementing it.

EXAMPLE 1–18

```
  1 1 1 1   1 1 1 1
- 0 1 0 0   1 0 0 0
  1 0 1 1   0 1 1 1   (one's complement)
+ 0 0 0 0   0 0 0 1
  1 0 1 1   1 0 0 0   (two's complement)
```

To prove that a 0100 1000 is the inverse (negative) of a 1011 0111, add the two together to form an 8-digit result. The ninth digit is dropped and the result is zero because a 0100 1000 is a positive 72, while a 1011 0111 is a negative 72. The same technique applies to any number system. Example 1–19 shows how the inverse of a 345 hexadecimal is found by first fifteen's complementing the number, and then by adding one to the result to form the sixteen's complement. As before, if the original 3-digit number 345 is added to the inverse of CBB, the result is a 3-digit 000. As before, the fourth bit (carry) is dropped. This proves that 345 is the inverse of CBB. Additional information about one's and two's complements is presented with signed numbers in the next section of the text.

EXAMPLE 1–19

```
  15 15 15
-  3  4  5
   C  B  A   (fifteen's complement)
+  0  0  1
   C  B  B   (sixteen's complement)
```

1–4   COMPUTER DATA FORMATS

Successful programming requires a precise understanding of data formats. In this section, many common computer data formats are described as they are used with the Intel family of microprocessors. Commonly, data appear as ASCII, BCD, signed and unsigned integers, and floating-point numbers (real numbers). Other forms are available, but are not presented here because they are not commonly found.

ASCII Data

ASCII (**American Standard Code for Information Interchange**) data represent alphanumeric characters in the memory of a computer system (see Table 1–7). The standard ASCII code is a 7-bit code, with the eighth and most significant bit used to hold parity in some systems. If ASCII data are used with a printer, the most significant bits are a 0 for alphanumeric printing and 1 for graphics printing. In the personal computer, an extended ASCII character set is selected by placing a logic 1 in the left-most bit. Table 1–8 shows the extended ASCII character set, using code 80H–FFH. The extended ASCII characters store some foreign letters and punctuation, Greek characters,

TABLE 1–7 ASCII code

First	*Second* X0	X1	X2	X3	X4	X5	X6	X7	X8	X9	XA	XB	XC	XD	XE	XF
0X	NUL	SOH	STX	ETX	EOT	ENQ	ACK	BEL	BS	HT	LF	VT	FF	CR	SO	SI
1X	DLE	DC1	DC2	DC3	DC4	NAK	SYN	ETB	CAN	EMS	SUB	ESC	FS	GS	RS	US
2X	SP	!	"	#	$	%	&	'	(	)	*	+	,	-	.	/
3X	0	1	2	3	4	5	6	7	8	9	:	;	<	=	>	?
4X	@	A	B	C	D	E	F	G	H	I	J	K	L	M	N	O
5X	P	Q	R	S	T	U	V	W	X	Y	Z	[	\	]	^	_
6X	`	a	b	c	d	e	f	g	h	i	j	k	l	m	n	o
7X	p	q	r	s	t	u	v	w	x	y	z	{	|	}	~	:::

TABLE 1–8 Extended ASCII code, as printed by the IBM ProPrinter.

First	*Second* X0	X1	X2	X3	X4	X5	X6	X7	X8	X9	XA	XB	XC	XD	XE	XF
0X		☺	☻	♥	♦	♣	♠	•	◘	○	◙	♂	♀	♪	♫	☼
1X	►	◄	↕	‼	¶	§	▬	↨	↑	↓	→	←	∟	↔	▲	▼
8X	Ç	ü	é	â	ä	à	å	ç	ê	ë	è	ï	î	ì	Ä	Å
9X	É	æ	Æ	ô	ö	ò	û	ù	ÿ	Ö	Ü	¢	£	¥	₧	ƒ
AX	á	í	ó	ú	ñ	Ñ	ª	º	¿	⌐	¬	½	¼	¡	«	»
BX	░	▒	▓	│	┤	╡	╢	╖	╕	╣	║	╗	╝	╜	╛	┐
CX	└	┴	┬	├	─	┼	╞	╟	╚	╔	╩	╦	╠	═	╬	╧
DX	╨	╤	╥	╙	╘	╒	╓	╫	╪	┘	┌	█	▄	▌	▐	▀
EX	α	β	Γ	π	Σ	σ	µ	γ	Φ	Θ	Ω	δ	∞	∮	∈	∩
FX	≡	±	≥	≤	⌠	⌡	÷	≈	°	•	·	√	ⁿ	²	■	

mathematical characters, box-drawing characters, and other special characters. Note that extended characters can vary from one printer to another. The list provided is designed to be use with the IBM ProPrinter[18], which also matches the special character set found with some word processors.

The ASCII control characters, also listed in Table 1–7, perform control functions in a computer system, including clear screen, backspace, line feed, and so on. To enter the control codes through the computer keyboard, hold down the Control key while typing a letter. To obtain the control code 01H, type a Control-A; a 02H is obtained by a Control-B, etc. Note that the control codes appear on the screen, from the DOS prompt, as ^A for Control-A, ^B for Control-B, and so forth. Also note that the carriage return code (CR) is the Enter key on most modern keyboards. The purpose of CR is to return the cursor or print-head to the left margin. Another code that appears in many programs is the line feed code (LF), which moves the cursor down one line.

[18]The IBM ProPrinter is a product of IBM Corporation.

To use Table 1–7 or 1–8 for converting alphanumeric or control characters into ASCII characters, first locate the alphanumeric code for conversion. Next, find the first digit of the hexadecimal ASCII code. Then find the second digit. For example, the capital letter A is ASCII code 41H, and the lowercase letter a is ASCII code 61H. Many Windows-based applications use the *Unicode* system to store alphanumeric data. This system stores each character as 16-bit data. The codes 0000H–00FFH are the same as standard ASCII code. The remaining codes 0100H–FFFFH are used to store all special characters from all world-wide character sets. This allows software written for the Windows environment to be used in any country in the world.

ASCII data are most often stored in memory by using a special directive to the assembler program called define byte(s), or DB. (The assembler is a program that is used to program a computer in its native binary machine language.) An alternative to DB is the word BYTE. The DB and BYTE directives, and several examples of their usage with ASCII-coded character strings, are listed in Example 1–20. Notice how each character string is surrounded by apostrophes (')—never use the quote ("). Also notice that the assembler lists the ASCII-coded value for each character to the left of the character string. To the far left is the hexadecimal memory location where the character string is first stored in the memory system. For example, the character string WHAT is stored beginning at memory address 001D, and the first letter is stored as 57 (W) followed by 68 (H), and so forth.

EXAMPLE 1–20

```
0000    42 61 72 72 79    NAMES   DB    'Barry B. Brey'
        20 42 2E 20 42
        72 65 79
000D    57 68 65 72 65    MESS    DB    'Where can it be?'
        20 63 61 6E 20
        69 74 20 62 65
        3F
001D    57 68 61 74 20    WHAT    DB    'What is on first.'
        69 73 20 6F 6E
        20 66 69 72 73
        74 2E
```

BCD (Binary-Coded Decimal) Data

Binary-coded decimal (BCD) information is stored in either packed or unpacked forms. **Packed BCD** data are stored as two digits per byte and **unpacked BCD** data are stored as one digit per byte. The range of a BCD digit extends from 0000_2 to 1001_2, or 0–9 decimal. Unpacked BCD data are returned from a keypad or keyboard. Packed BCD data are used for some of the instructions included for BCD addition and subtraction in the instruction set of the microprocessor.

Table 1–9 shows some decimal numbers converted to both the packed and unpacked BCD forms. Applications that require BCD data are point-of-sales terminals and almost any device that performs a minimal amount of simple arithmetic. If a system requires complex arithmetic, BCD data are seldom used because there is no simple and efficient method of performing complex BCD arithmetic.

TABLE 1–9   Packed and unpacked BCD data.

Decimal	Packed		Unpacked		
12	0001 0010		0000 0001	0000 0010	
623	0000 0110	0010 0011	0000 0110	0000 0010	0000 0011
910	0000 1001	0001 0000	0000 1001	0000 0001	0000 0000

Example 1–21 shows how to use the assembler to define both packed and unpacked BCD data. In all cases, the convention of storing the least-significant data first is followed. This means that to store an 83 into memory, the 3 is stored first, and then followed by the 8. Also note that with packed BCD data, the letter H (hexadecimal) follows the number to ensure that the assembler stores the BCD value rather than a decimal value for packed BCD data. Notice how the numbers are stored in memory as unpacked, one digit per byte; or packed, as two digits per byte.

EXAMPLE 1–21

```
                        ;Unpacked BCD data (least-significant data first)
                        ;
0000  03 04 05          NUMB1    DB      3,4,5          ;defines the number 543
0003  07 08             NUMB2    DB      7,8            ;defines the number 87
                        ;
                        ;Packed BCD data (least-significant data first)
                        ;
0005  37 34             NUMB3    DB      37H,34H        ;defines the number 3437
0007  03 45             NUMB4    DB      3,45H          ;defines the number 4503
```

Byte-Sized Data

Byte-sized data are stored as *unsigned* and *signed* integers. Figure 1–13 illustrates both the unsigned and signed forms of the byte-sized integer. The difference in these forms is the weight of the leftmost bit position. Its value is 128 for the unsigned integer and minus 128 for the signed integer. In the signed integer format, the leftmost bit represents the sign bit of the number, as well as a weight of minus 128. For example, an 80H represents a value of 128 as an unsigned number; as a signed number, it represents a value of minus 128. Unsigned integers range in value from 00H–FFH (0–255). Signed integers range in value from −128 to 0 to +127.

Although negative signed numbers are represented in this way, they are stored in the two's complement form. The method of evaluating a signed number by using the weights of each bit position is much easier than the act of two's complementing a number to find its value. This is especially true in the world of calculators designed for programmers.

Whenever a number is two's complemented, its sign changes from negative to positive or positive to negative. For example, the number 00001000 is a +8. Its negative value (−8) is found by

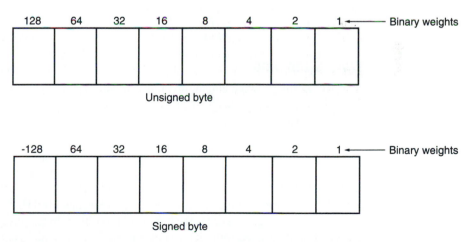

FIGURE 1–13 The unsigned and signed bytes illustrating the weights of each binary-bit position.

two's complementing the +8. To form a two's complement, first one's complement the number. To one's complement a number, invert each bit of a number from zero to one or from one to zero. Once the one's complement is formed, the two's complement is found by adding a one to the one's complement. Example 1–22 shows how numbers are two's complemented using this technique.

EXAMPLE 1–22

```
+8 = 00001000
     11110111 (one's complement)
+           1
-8 = 11111000 (two's complement)
```

Another, and probably simpler, technique for two's complementing a number starts with the rightmost digit. Start writing down the number from right to left. Write the number exactly as it appears until the first one. Write down the first one, and then invert of complement all remaining ones to its left. Example 1–23 shows this technique with the same number as in Example 1–22.

EXAMPLE 1–23

```
+8 = 00001000
         1000 (write number to first 1)
     1111     (invert the remaining bits)
-8 - 11111000
```

To store 8-bit data in memory using the assembler program, use the DB directive as in prior examples. Example 1–24 lists many forms of 8-bit numbers stored in memory using the assembler program. Notice in the example that a hexadecimal number is defined with the letter H following the number, and that a decimal number is written as is, without anything special.

EXAMPLE 1–24

```
                        ;Unsigned byte-sized data
                        ;
0000    FE              DATA1   DB      254             ;define 254 decimal
0001    87              DATA2   DB      87H             ;define 87 hexadecimal
0002    47              DATA3   DB      71              ;define 71 decimal

                        ;
                        ;Signed byte-sized data
                        ;
0003    9C              DATA4   DB      -100            ;define a -100 decimal
0004    64              DATA5   DB      +100            ;define a +100 decimal
0005    FF              DATA6   DB      -1              ;define a -1 decimal
0006    38              DATA7   DB      56              ;define a 56 decimal
```

Word-Sized Data

A word (16-bits) is formed with two bytes of data. The least significant byte is always stored in the lowest-numbered memory location, and the most significant byte is stored in the highest. This method of storing a number is called the **little endian** format. An alternate method, not used with the Intel family of microprocessors, is called the **big endian** format. In the big endian format, numbers are stored with the lowest location containing the most significant data. The big endian format is used with the Motorola family of microprocessors. Figure 1–14 (a) shows the weights of each bit position in a word of data, and Figure 1–14 (b) shows how the number 1234H appears when stored in the memory location 3000H and 3001H. The only difference between a signed and an unsigned word is the leftmost bit position. In the unsigned form, the leftmost bit is unsigned; in the signed form, its weight is a –32,768. As with byte-sized signed data, the signed word is in two's complement form when representing a negative number. Also, notice that the low-order byte is stored in the lowest-numbered memory location (3000H) and the high-order byte is stored in the highest-numbered location (3001H).

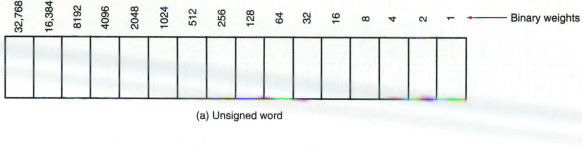

(a) Unsigned word

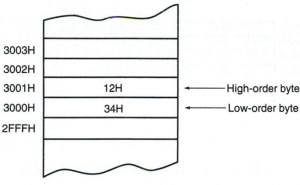

(b) The contents of memory location 3000H and 3001H are the word 1234H.

FIGURE 1–14 The storage format for a 16-bit word in (a) a register and (b) two bytes of memory.

Example 1–25 shows several signed and unsigned word-sized data stored in memory using the assembler program. Notice that the **define word(s) directive,** or DW, causes the assembler to store words in the memory instead of bytes, as in prior examples. The WORD directive can also be used to define a word. Notice that the word *data* is displayed by the assembler in the same form as entered. For example, a 1000H is displayed by the assembler as a 1000. This is for our convenience because the number is actually stored in the memory as 00 10 in two consecutive memory bytes.

EXAMPLE 1–25

```
                        ;Unsigned word-sized data
                        ;
0000    09F0            DATA1   DW      2544            ;define 2544 decimal
0002    87AC            DATA2   DW      87ACH           ;define 87AC hexadecimal
0004    02C6            DATA3   DW      710             ;define 710 decimal
                        ;
                        ;Signed word-sized data
                        ;
0006    CBA8            DATA4   DW      -13400          ;define a -13400 decimal
0008    00C6            DATA5   DW      +198            ;define a +198 decimal
000A    FFFF            DATA6   DW      -1              ;define a -1 decimal
```

Doubleword-Sized Data

Doubleword-sized data requires four bytes of memory because it is a 32-bit number. Double-word data appear as a product after a multiplication and also as a dividend before a division. In the 80386 through the Pentium 4, memory and registers are also 32 bits in width. Figure 1–15 shows the form used to store doublewords in the memory and the binary weights of each bit position.

When a doubleword is stored in memory, its least significant byte is stored in the lowest-numbered memory location, and its most significant byte is stored in the highest-numbered memory location using the little endian format. Recall that this is also true for word-sized data.

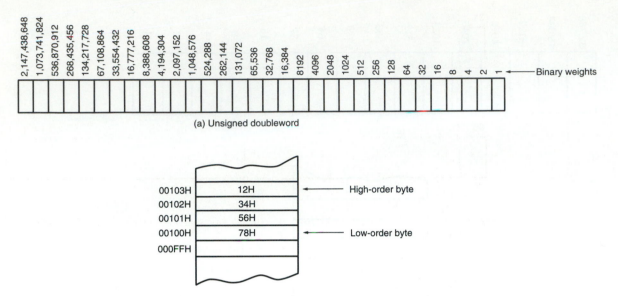

(a) Unsigned doubleword

(b) The contents of memory location 00100H–00103H are the doubleword 12345678H.

FIGURE 1–15   The storage format for a 32-bit word in (a) a register and (b) four bytes of memory.

For example, a 12345678H that is stored in memory location 00100H–00103H is stored with the 78H in memory location 00100H, the 56H in location 00101H, the 34H in location 00102H, and the 12H in location 00103H.

To define doubleword-sized data, use the assembler directive **define doubleword(s),** or DD. (You can also use the DWORD directive in place of DD.) Example 1–26 shows both signed and unsigned numbers stored in memory using the DD directive.

EXAMPLE 1–26

```
                        ;Unsigned doubleword-sized data
                        ;
0000    0003E1C0        DATA1   DD      254400      ;define 254400 decimal
0004    87AC1234        DATA2   DD      87AC1234H   ;define 87AC1234 hexadecimal
0008    00000046        DATA3   DD      70          ;define 70 decimal
                        ;
                        ;Signed doubleword-sized data
                        ;
000C    FFEB8058        DATA4   DD      -1343400    ;define a -1343400 decimal
0010    000000C6        DATA5   DD      +198        ;define a +198 decimal
0014    FFFFFFFF        DATA6   DWORD   -1          ;define a -1 decimal
```

Integers may also be stored in memory that is of any width. The forms listed here are standard forms, but that doesn't mean that a 128-byte wide integer can't be stored in the memory. The microprocessor is flexible enough to allow any size of data. When nonstandard width numbers are stored in memory, the DB directive is normally used to store them. For example, the 24-bit number 123456H is stored using a DB 56H,34H,12H directive. Note that this conforms to the little endian format.

Real Numbers

Because many high-level languages use the Intel family of microprocessors, real numbers are often encountered. A real number, or a **floating-point number,** as it is often called, contains two parts: a mantissa, significand, or fraction; and an exponent. Figure 1–16 depicts both the 4- and

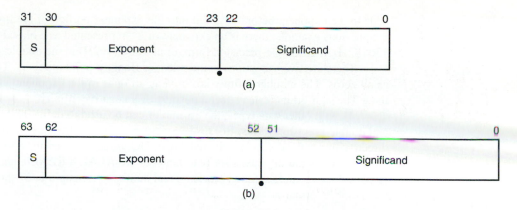

FIGURE 1–16 The floating-point numbers (a) single-precision using a bias of 7FH and (b) double-precision using a bias of 3FFH.

8-byte forms of real numbers as they are stored in any Intel system. Note that the 4-byte real number is called **single-precision** and the 8-byte form is called **double-precision.** The form presented here is the same form specified by the IEEE[19] standard, IEEE-754, version 10.0. This standard has been adopted as the standard form of real numbers with virtually all high-level programming languages and many applications packages. The standard also applies to data manipulated by the numeric coprocessor in the personal computer. Figure 1–16 (a) shows the single-precision form that contains a sign-bit, an 8-bit exponent, and a 24-bit fraction (mantissa). Note that because applications often require double-precision floating-point numbers [see Figure 1–16 (b)], the Pentium–Pentium 4 with their 64-bit data bus perform memory transfers at twice the speed of the 80386/80486 microprocessors.

Simple arithmetic indicates that it should take 33 bits to store all three pieces of data. Not true—the 24-bit mantissa contains an **implied** (hidden) one-bit that allows the mantissa to represent 24 bits while being stored in only 23 bits. The hidden bit is the first bit of the normalized real number. When normalizing a number, it is adjusted so that its value is at least 1, but less than 2. For example, if 12 is converted to binary (1100_2), it is normalized and the result is 1.1×2^3. The 1 is not stored in the 23-bit mantissa portion of the number; the 1 is the hidden one-bit. Table 1–10 shows the single-precision form of this number and others.

The exponent is stored as a **biased exponent.** With the single-precision form of the real number, the bias is 127 (7FH) and with the double-precision form, it is 1023 (3FFH). The bias

TABLE 1–10 Single-precision real numbers

Decimal	Binary	Normalized	Sign	Biased Exponent	Mantissa
+12	1100	1.1×2^3	0	10000010	1000000 00000000 00000000
−12	1100	-1.1×2^3	1	10000010	1000000 00000000 00000000
+100	1100100	1.1001×2^6	0	10000101	1001000 00000000 00000000
−1.75	1.11	-1.11×2^0	1	01111111	1100000 00000000 00000000
+0.25	.01	1.0×2^{-2}	0	01111101	0000000 00000000 00000000
+0.0	0	0	0	00000000	0000000 00000000 00000000

[19]IEEE is the Institute of Electrical and Electronic Engineers.

adds to the exponent before is stored into the exponent portion of the floating-point number. In the previous example, there is an exponent of 2^3, represented as a biased exponent of $127 + 3$ or 130 (82H) in the single-precision form, or as 1026 (402H) in the double-precision form.

There are two exceptions to the rules for floating-point numbers. The number 0.0 is stored as all zeros. The number infinity is stored as all ones in the exponent and all zeros in the mantissa. The sign-bit indicates either a positive or a negative infinity.

As with other data types, the assembler can be used to define real numbers in both single- and double-precision forms. Because single-precision numbers are 32-bit numbers, use the DD directive or use the **define quadwords(s),** or DQ, directive to define 64-bit double-precision real numbers. Optional directives for real numbers are REAL4, REAL8, and REAL10 for defining single-, double-, and extended precision real numbers. Example 1–27 shows numbers defined in real number format.

EXAMPLE 1–27

```
                ;Single-precision real numbers
                ;
0000 3F9DF3B6   NUMB1   DD      1.234       ;define 1.234
0004 C1BB3333   NUMB2   DD      -23.4       ;define -23.4
0008 43D20000   NUMB3   REAL4   4.2E2       ;define 420
000C 3F9DF3B6   NUMB4   REAL4   1.234       ;define a 4-byte real number
                ;
                ;Double-precision real numbers
                ;
0010            NUMB5   DQ      123.4       ;define 123.4
    405ED9999999999A
0018            NUMB6   REAL8   -23.4       ;define -23.4
    C1BB333333333333
0028            NUMB7   REAL8   123.4       ;define an 8-byte real number
    405ED9999999999A
                ;
                ;Extended-precision real numbers
                ;
0030            NUMB8   REAL10  123.4       ;define a 10-byte real number
    4005F6CCCCCCCCCCCCCD
```

1–5   SUMMARY

1. The mechanical computer age began with the advent of the abacus in 500 B.C. This first mechanical calculator remained unchanged until 1642, when Blaise Pascal improved it. An early mechanical computer system was the Analytical Engine developed by Charles Babbage in 1823. Unfortunately, this machine never functioned because of his inability to create the necessary machine parts.

2. The first electronic calculating machine was developed during World War II by Konrad Zuse, an early pioneer of digital electronics. His computer, the Z3, was used in aircraft and missile design for the German war effort.

3. The first electronic computer, which used vacuum tubes, was placed into operation in 1943 to break secret German military codes. This first electronic computer system, the Colossus, was invented by Alan Turing. Its only problem was that the program was fixed and could not be changed.

4. The first general-purpose, programmable electronic computer system was developed in 1946 at the University of Pennsylvania. This first modern computer was called the ENIAC (Electronics Numerical Integrator and Calculator).

5. The first high-level programming language, called FLOW-MATIC, was developed for the UNIVAC I computer by Grace Hopper in the early 1950s. This led to FORTRAN and other early programming languages such as COBOL.

6. The world's first microprocessor, the Intel 4004, was a 4-bit microprocessor—a programmable controller on a chip—that was meager by today's standards. It addressed a mere 4096 four-bit memory locations. Its instruction set contained only 45 different instructions.

7. Microprocessors that are common today include the 8086/8088, which were the first 16-bit microprocessors. Following these early 16-bit machines were the 80286, 80386, 80486, Pentium, Pentium Pro, Pentium II, Pentium III, and Pentium 4 processors. The architecture has changed from 16 bits to 32 bits and soon, with the Merced, to 64 bits. With each newer version, improvements followed that increased the processor's speed and performance. From all indications, this process of speed and performance improvement will continue.

8. Microprocessor-based personal computers contain memory systems that include three main areas: TPA (transient program area), system area, and extended memory. The TPA holds application programs, the operating system, and drivers. The system area contains memory used for video display cards, disk drives, and the BIOS ROM. The extended memory area is only available to the 80286 through the Pentium 4 microprocessor in an AT-style personal computer system.

9. The 8086/8088 address 1M byte of memory from location 00000H–FFFFFH. The 80286 and 80386SX address 16M bytes of memory from location 000000H–FFFFFFH. The 80386SL addresses 32M bytes of memory from location 0000000H–1FFFFFFH. The 80386DX and 80486 through Pentium 4 processors address 4G bytes of memory from location 00000000H–FFFFFFFFH. In addition, the Pentium Pro through the Pentium 4 can run with a 36-bit address and access up to 64G bytes of memory from location 000000000H–FFFFFFFFFH.

10. All versions of the 8086–80486 and Pentium–Pentium 4 microprocessors address 64K bytes of I/O address space. These I/O ports are numbered from 0000H–FFFFH with I/O ports 0000H–03FFH reserved for use by the personal computer system.

11. The operating system in many personal computers is either MSDOS (Microsoft disk operating system) or PCDOS (personal computer disk operating system from IBM). The operating system performs the task of operating or controlling the computer system, along with its I/O devices.

12. The microprocessor is the controlling element in a computer system. The microprocessor performs data transfers, does simple arithmetic and logic operations, and makes simple decisions. The microprocessor executes programs stored in the memory system to perform complex operations in short periods of time.

13. All computer systems contain three buses to control memory and I/O. The address bus is used to request a memory location or I/O device. The data bus transfers data between the microprocessor and its memory and I/O spaces. The control bus controls the memory and I/O, and requests reading or writing of data. Control is accomplished with $\overline{IORC}$ (I/O read control), $\overline{IOWC}$ (I/O write control), $\overline{MRDC}$ (memory read control), and $\overline{MWTC}$ (memory write control).

14. Numbers are converted from any number base to decimal by noting the weights of each position. The weight of the position to the left of the radix point is always the units position in any number system. The position to the left of the units position is always the radix times one. Succeeding positions are determined by multiplying by the radix. The weight of the position to the right of the radix point is always determined by dividing by the radix.

15. Conversion from a whole decimal number to any other base is accomplished by dividing by the radix. Conversion from a fractional decimal number is accomplished by multiplying by the radix.

16. Hexadecimal data are represented in hexadecimal form or in a code called binary-coded hexadecimal (BCH). A binary-coded hexadecimal number is one that is written with a 4-bit binary number that represents each hexadecimal digit.

17. The ASCII code is used to store alphabetic or numeric data. The ASCII code is a 7-bit code; it can have an eighth bit that is used to extend the character set from 128 codes to 256 codes. The carriage return (Enter) code returns the print head or cursor to the left margin. The line feed code moves the cursor or print head down one line.

18. Binary-coded decimal (BCD) data are sometimes used in a computer system to store decimal data. These data are stored either in packed (two digits per byte) or unpacked (one digit per byte) form.

19. Binary data are stored as a byte (8 bits), word (16 bits), or doubleword (32 bits) in a computer system. These data may be unsigned or signed. Signed negative data are always stored in the two's complement form. Data that are wider than 8 bits are always stored using the little endian format.

20. Floating-point data are used in computer systems to store whole, mixed, and fractional numbers. A floating-point number is composed of a sign, a mantissa, and an exponent.

21. The assembler directives DB or BYTE define bytes, DW or WORD define words, DD or DWORD define doublewords, and DQ or QWORD define quadwords.

22. Example 1–28 shows the assembly language formats for storing numbers as bytes, words, doublewords, and real numbers. Also shown are ASCII-coded character strings.

EXAMPLE 1–28

```
                           ;ASCII data
                           ;
0000  54 68 69 73 20 69    MES1    DB      'This is a character string in ASCII'
      73 20 61 20 63 68
      61 72 61 63 74 65
      72 20 73 74 72 69
      6E 67 20 69 6E 20
      41 53 43 49 49
0023  53 6F 20 69 73 20    MES2    DB      'So is this'
      74 68 69 73
                           ;BYTE data
                           ;
002D  17                   DATA1   DB      23         ;23 decimal
002E  DE                   DATA2   DB      -34        ;-34 decimal
002F  34                   DATA3   DB      34H        ;34 hexadecimal
                           ;
                           ;WORD data
                           ;
0030  1000                 DATA4   DW      1000H      ;1000 hexadecimal
0032  FF9C                 DATA5   DW      -100       ;-100 decimal
0034  000C                 DATA6   DW      +12        ;=12 decimal
                           ;
                           ;DOUBLEWORD data
                           ;
0036  00001000             DATA7   DD      1000H      ;1000 hexadecimal
003A  FFFFFED4             DATA8   DD      -300       ;-300 decimal
003E  00012345             DATA9   DD      12345H     ;12345 hexadecimal
                           ;
                           ;Real data
                           ;
0042  4015C28F             DATA10  REAL4   2.34       ;2.34 decimal
0046  C00CCCCD             DATA11  REAL4   -2.2       ;-2.2 decimal
004A                       DATA12  REAL8   100.3      ;100.3 decimal
      4059133333333333
```

1–6   **QUESTIONS AND PROBLEMS**

1. Who developed the Analytical Engine?
2. The 1890 census used a new device called a punched card. Who developed the punch card?
3. Who was the founder of IBM Corporation?
4. Who developed the first electronic calculator?
5. The first electronic computer system was developed for what purpose?
6. The first general-purpose, programmable computer was called the _____.
7. The world's first microprocessor was developed in 1971 by _____.
8. Who was the Countess of Lovelace?
9. Who developed the first high-level programming language called FLOW-MATIC?
10. What is a von Neumann machine?
11. Which 8-bit microprocessor ushered in the age of the microprocessor?
12. The 8085 microprocessor, introduced in 1977, has sold _____ copies.
13. Which Intel microprocessor was the first to address 1M bytes of memory?
14. The 80386SL addresses _____ bytes of memory.
15. How much memory is available to the 80486 microprocessor?
16. When did Intel introduce the Pentium microprocessor?
17. When did Intel introduce the Pentium Pro processor?
18. When did Intel introduce the Pentium 4 microprocessor?
19. Which Intel microprocessors address 64G of memory?
20. What is the acronym MIPs?
21. What is the acronym CISC?
22. A binary bit stores a(n) _____ or a(n) _____.
23. A computer K is equal to _____ bytes.
24. A computer M is equal to _____ K bytes.
25. A computer G is equal to _____ M bytes.
26. How many typewritten pages of information are stored in a 4G-byte memory system?
27. The first 1M byte of memory in a computer system contains a(n) _____ and a(n) _____ area.
28. How much memory is found in the transient program area?
29. How much memory is found in the systems area?
30. The 8086 microprocessor addresses _____ bytes of memory.
31. The Pentium 4 microprocessor addresses _____ bytes of memory.
32. Which microprocessors address 4G bytes of memory?
33. Memory above the first 1M byte is called _____ memory.
34. What is the system BIOS?
35. What is DOS?
36. What is the difference between an XT and an AT computer system?
37. What is the VESA local bus?
38. The ISA bus holds _____-bit interface cards.
39. What is the USB?
40. What is the AGP?
41. What is the XMS?
42. What is the EMS?
43. A driver is stored in the _____ area.
44. What is a TSR?
45. How is a TSR often accessed?
46. What is the purpose of the CONFIG.SYS file?

47. What is the purpose of the AUTOEXEC.BAT file?
48. The COMMAND.COM program processes what information?
49. The personal computer system addresses _____ bytes of I/O space.
50. Where is the high memory located in a personal computer?
51. The DEVICE or DEVICEHIGH statement is found in what file?
52. Where are the upper memory blocks used by MSDOS?
53. Where is the video BIOS?
54. Draw the block diagram of a computer system.
55. What is the purpose of the microprocessor in a microprocessor-based computer system?
56. List the three buses found in all computer systems.
57. Which bus transfers the memory address to the I/O device or to the memory device?
58. Which control signal causes the memory to perform a read operation?
59. What is the purpose of the $\overline{\text{IORC}}$ signal?
60. If the $\overline{\text{MRDC}}$ signal is a logic 0, which operation is performed by the microprocessor?
61. Define the purpose of the following directives:
    (a) DB
    (b) DQ
    (c) DW
    (d) DD
62. Convert the following binary numbers into decimal:
    (a) 1101.01
    (b) 111001.0011
    (c) 101011.0101
    (d) 111.0001
63. Convert the following octal numbers into decimal:
    (a) 234.5
    (b) 12.3
    (c) 7767.07
    (d) 123.45
    (e) 72.72
64. Convert the following hexadecimal numbers into decimal:
    (a) A3.3
    (b) 129.C
    (c) AC.DC
    (d) FAB.3
    (e) BB8.0D
65. Convert the following decimal integers into binary, octal, and hexadecimal:
    (a) 23
    (b) 107
    (c) 1238
    (d) 92
    (e) 173
66. Convert the following decimal numbers into binary, octal, and hexadecimal:
    (a) 0.625
    (b) .00390625
    (c) .62890625
    (d) 0.75
    (e) .9375

67. Convert the following hexadecimal numbers into binary-coded hexadecimal code (BCH):
    (a) 23
    (b) AD4
    (c) 34.AD
    (d) BD32
    (e) 234.3
68. Convert the following binary-coded hexadecimal numbers into hexadecimal:
    (a) 1100 0010
    (b) 0001 0000 1111 1101
    (c) 1011 1100
    (d) 0001 0000
    (e) 1000 1011 1010
69. Convert the following binary numbers to the one's complement form:
    (a) 1000 1000
    (b) 0101 1010
    (c) 0111 0111
    (d) 1000 0000
70. Convert the following binary numbers to the two's complement form:
    (a) 1000 0001
    (b) 1010 1100
    (c) 1010 1111
    (d) 1000 0000
71. Define byte, word, and doubleword.
72. Convert the following words into ASCII-coded character strings:
    (a) FROG
    (b) Arc
    (c) Water
    (d) Well
73. What is the ASCII code for the Enter key and what is its purpose?
74. Use an assembler directive to store the ASCII-character string 'What time is it?' in the memory.
75. Convert the following decimal numbers into 8-bit signed binary numbers:
    (a) +32
    (b) −12
    (c) +100
    (d) −92
76. Convert the following decimal numbers into signed binary words:
    (a) +1000
    (b) −120
    (c) +800
    (d) −3212
77. Use an assembler directive to store −34 into the memory.
78. Show how the following 16-bit hexadecimal numbers are stored in the memory system (use the standard Intel format):
    (a) 1234H
    (b) A122H
    (c) B100H
79. What is the difference between the big endian and little endian formats for storing numbers that are larger than eight bits in width?
80. Use an assembler directive to store a 123A hexadecimal into the memory.

81. Convert the following decimal numbers into both packed and unpacked BCD forms:
    (a) 102
    (b) 44
    (c) 301
    (d) 1000
82. Convert the following binary numbers into signed decimal numbers:
    (a) 10000000
    (b) 00110011
    (c) 10010010
    (d) 10001001
83. Convert the following BCD numbers (assume that these are packed numbers) into decimal numbers:
    (a) 10001001
    (b) 00001001
    (c) 00110010
    (d) 00000001
84. Convert the following decimal numbers into single-precision floating-point numbers:
    (a) +1.5
    (b) −10.625
    (c) +100.25
    (d) −1200
85. Convert the following single-precision floating-point numbers into decimal numbers:
    (a) 0 10000000 11000000000000000000000
    (b) 1 01111111 00000000000000000000000
    (c) 0 10000010 10010000000000000000000
86. Use the Internet to write a short report about any one of the following computer pioneers:
    (a) Charles Babbage
    (b) Konrad Zuse
    (c) Joseph Jacquard
    (d) Herman Hollerith
87. Use the Internet to write a short report about any one of the following computer languages:
    (a) COBOL
    (b) ALGOL
    (c) FORTRAN
    (d) PASCAL
88. Use the Internet to write a short report detailing the features of the Merced microprocessor.

CHAPTER 2

The Microprocessor and its Architecture

INTRODUCTION

This chapter presents the microprocessor as a programmable device by first looking at its internal programming model and then at how it addresses its memory space. The architecture of the entire family of Intel microprocessors is presented simultaneously, as are the ways that the family members address the memory system.

The addressing modes for this powerful family of microprocessors are described for both the real and protected modes of operation. Real mode memory exists at locations 00000H–FFFFFH—the first 1M byte of the memory system—and is present on all versions of the microprocessor. Protected mode memory exists at any location in the entire memory system, but is available only to the 80286–Pentium 4, not to the earlier 8086 or 8088 microprocessors. Protected mode memory for the 80286 contains 16M bytes; for the 80386–Pentium, 4G bytes; and for the Pentium Pro through the Pentium 4, either 4G or 64G bytes.

CHAPTER OBJECTIVES

Upon completion of this chapter, you will be able to:

1. Describe the function and purpose of each program-visible register in the 8086–80486 and Pentium–Pentium 4 microprocessors.
2. Detail the flag register and the purpose of each flag bit.
3. Describe how memory is accessed using real mode memory-addressing techniques.
4. Describe how memory is accessed using protected mode memory-addressing techniques.
5. Describe the program-invisible registers found within the 80286 through Pentium 4 microprocessors.
6. Detail the operation of the memory-paging mechanism.

2–1 INTERNAL MICROPROCESSOR ARCHITECTURE

Before a program is written or any instruction investigated, the internal configuration of the microprocessor must be known. This section of the chapter details the program-visible internal architecture of the 8086–80486 and the Pentium–Pentium 4 microprocessors. Also detailed are the function and purpose of each of these internal registers.

The Programming Model

The programming model of the 8086 through the Pentium 4 is considered to be **program visible** because its registers are used during application programming and are specified by the instructions. Other registers, detailed later in this chapter, are considered to be **program invisible** because they are not addressable directly during applications programming, but may be used indirectly during system programming. Only the 80286 and above contain the program-invisible registers used to control and operate the protected memory system.

Figure 2–1 illustrates the programming model of the 8086 through the Pentium 4 microprocessor. The earlier 8086, 8088, and 80286 contain **16-bit** internal architectures, a subset of the registers shown in Figure 2–1. The 80386 through the Pentium 4 microprocessors contain full **32-bit** internal architectures. The architectures of the earlier 8086 through the 80286 are fully

FIGURE 2–1   The programming model of the Intel 8086 through the Pentium 4.

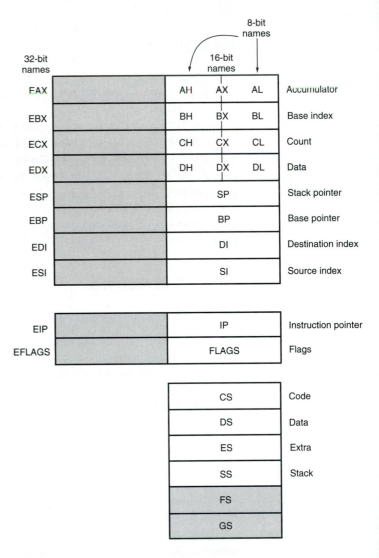

Notes:
  1. The shaded areas registers exist only on the 80386 through the Pentium 4.

  2. The FS and GS register have no special names.

upward-compatible to the 80386 through the Pentium 4. The shaded areas in this illustration represent registers that are not found in the 8086, 8088, or 80286 microprocessors and are enhancements provided on the 80386, 80486, Pentium, Pentium Pro, Pentium II, Pentium III, and Pentium 4 microprocessors.

The programming model contains 8-, 16-, and 32-bit registers. The 8-bit registers are AH, AL, BH, BL, CH, CL, DH, and DL and are referred to when an instruction is formed using these two-letter designations. For example, an ADD AL,AH instruction adds the 8-bit contents of AH to AL. (Only AL changes due to this instruction.) The 16-bit registers are AX, BX, CX, DX, SP, BP, DI, SI, IP, FLAGS, CS, DS, ES, SS, FS, and GS. These registers are also referenced with the two-letter designations. For example, an ADD DX,CX instruction adds the 16-bit contents of CX to DX. (Only DX changes due to this instruction.) The extended 32-bit registers are EAX, EBX, ECX, EDX, ESP, EBP, EDI, ESI, EIP, and EFLAGS. These 32-bit extended registers, and 16-bit registers FS and GS, are available only in the 80386 and above. These registers are referenced by the designations FS or GS for the two new 16-bit registers, and by a three-letter designation for the 32-bit registers. For example, an ADD ECX,EBX instruction adds the 32-bit contents of EBX to ECX. (Only ECX changes due to this instruction.)

Some registers are general-purpose or multipurpose registers, while some have special purposes. The **multipurpose registers** include EAX, EBX, ECX, EDX, EBP, EDI, and ESI. These registers hold various data sizes (bytes, words, or doublewords) and are used for almost any purpose, as dictated by a program.

Multipurpose Registers

EAX
(accumulator)

EAX is referenced as a 32-bit register (EAX), as a 16-bit register (AX), or as either of two 8-bit registers (AH and AL). Note that if an 8- or 16-bit register is addressed, only that portion of the 32-bit register changes without affecting the remaining bits. The accumulator is used for instructions such as multiplication, division, and some of the adjustment instructions. For these instructions, the accumulator has a special purpose, but is generally considered to be a multipurpose register. In the 80386 and above, the EAX register may also hold the offset address of a location in the memory system.

EBX
(base index)

EBX is addressable as EBX, BX, BH, or BL. The BX register sometimes holds the offset address of a location in the memory system in all versions of the microprocessor. In the 80386 and above, EBX also can address memory data.

ECX
(count)

ECX is a general-purpose register that also holds the count for various instructions. In the 80386 and above, the ECX register also can hold the offset address of memory data. Instructions that use a count are the repeated string instructions (REP/REPE/REPNE); and shift, rotate, and LOOP/LOOPD instructions. The shift and rotate instructions use CL as the count, the repeated string instructions use CX, and the LOOP/LOOPD instructions use either CX or ECX.

EDX
(data)

EDX is a general-purpose register that holds a part of the result from a multiplication or part of the dividend before a division. In the 80386 and above, this register can also address memory data.

EBP
(base pointer)

EBP points to a memory location in all versions of the microprocessor for memory data transfers. This register is addressed as either BP or EBP.

EDI
(destination index)

EDI often addresses string destination data for the string instructions. It also functions as either a 32-bit (EDI) or 16-bit (DI) general-purpose register.

ESI
(source index)

ESI is used as either ESI or SI. The source index register often addresses source string data for the string instructions. Like EDI, ESI also functions as a general-purpose register. As a 16-bit register, it is addressed as SI; as a 32-bit register, it is addressed as ESI.

Special-purpose Registers.   The special-purpose registers include EIP, ESP, EFLAGS; and the segment registers CS, DS, ES, SS, FS, and GS.

EIP
(instruction pointer)

EIP addresses the next instruction in a section of memory defined as a code segment. This register is IP (16 bits) when the microprocessor operates in the real mode and EIP (32 bits) when the 80386 and above operate in the protected mode. Note that the 8086, 8088, and 80286 do contain EIP, and only the 80286 and above operate in the protected mode. The instruction pointer, which points to the next instruction in a program, is used by the microprocessor to find the next sequential instruction in a program located within the code segment. The instruction pointer can be modified with a jump or a call instruction.

ESP
(stack pointer)

ESP addresses an area of memory called the stack. The stack memory stores data through this pointer and is explained later in the text with instructions that address stack data. This register is referred to as SP if used as a 16-bit register and ESP if referred to as a 32-bit register.

EFLAGS

EFLAGS indicate the condition of the microprocessor and control its operation. Figure 2–2 shows the flag registers of all versions of the microprocessor. Note that the flags are upward-compatible from the 8086/8088 to the Pentium 4 microprocessor. The 8086–80286 contain a FLAG register (16 bits) and the 80386 and above contain an EFLAG register (32-bit extended flag register).

The rightmost five flag bits and the overflow flag change after many arithmetic and logic instructions execute. The flags never change for any data transfer or program control operation. Some of the flags are also used to control features found in the microprocessor. Following is a list of each flag bit, with a brief description of their function. As instructions are introduced in subsequent chapters, additional detail on the flag bits is provided. The rightmost five flags and the overflow flag are changed by most arithmetic and logic operations, while data transfers do not affect them.

C (carry)

Carry holds the carry after addition or the borrow after subtraction. The carry flag also indicates error conditions, as dictated by some programs and procedures. This is especially true of the DOS function calls detailed in later chapters and Appendix A.

P (parity)

Parity is a logic 0 for odd parity and a logic 1 for even parity. Parity is a count of ones in a number expressed as even or odd. For example, if a number contains three binary one bits, it has odd parity.

FIGURE 2–2   The EFLAG and FLAG register counts for the entire 80X86 and Pentium microprocessor family.

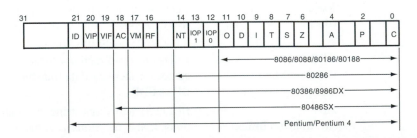

If a number contains zero one bits, it has even parity. The parity flag finds little application in modern programming and was implemented in early Intel microprocessors for checking data in data communications environments. Today parity checking is often accomplished by the data communications equipment instead of the microprocessor.

A (auxiliary carry)   The auxiliary carry holds the carry (half-carry) after addition or the borrow after subtraction between bits positions 3 and 4 of the result. This highly specialized flag bit is tested by the DAA and DAS instructions to adjust the value of AL after a BCD addition or subtraction. Otherwise, the A flag bit is not used by the microprocessor or any other instructions.

Z (zero)   The zero flag shows that the result of an arithmetic or logic operation is zero. If Z = 1, the result is zero; if Z = 0, the result is not zero.

S (sign)   The sign flag holds the arithmetic sign of the result after an arithmetic or logic instruction executes. If S = 1, the sign bit (leftmost bit of a number) is set or negative; if S = 0, the sign bit is cleared or positive.

T (trap)   The trap flag enables trapping through an on-chip debugging feature. (A program is debugged to find an error or bug.) If the T flag is enabled (1), the microprocessor interrupts the flow of the program on conditions as indicated by the debug registers and control registers. If the T flag is a logic 0, the trapping (debugging) feature is disabled. The CodeView program can use the trap feature and debug registers to debug faulty software.

I (interrupt)   The interrupt flag controls the operation of the INTR (interrupt request) input pin. If I = 1, the INTR pin is enabled; if I = 0, the INTR pin is disabled. The state of the I flag bit is controlled by the STI (**set I flag**) and CLI (**clear I flag**) instructions.

D (direction)   The direction flag selects either the increment or decrement mode for the DI and/or SI registers during string instructions. If D = 1, the registers are automatically decremented; if D = 0, the registers are automatically incremented. The D flag is set with the STD (**set direction**) and cleared with the CLD (**clear direction**) instructions.

O (overflow)   Overflows occurs when signed numbers are added or subtracted. An overflow indicates that the result has exceeded the capacity of the machine. For example, if a 7FH (+127) is added—using an 8-bit addition—to a 01H (+1), the result is 80H (−128). This result represents an overflow condition indicated by the overflow flag for signed addition. For unsigned operations, the overflow flag is ignored.

IOPL (I/O privilege level)   IOPL is used in protected mode operation to select the privilege level for I/O devices. If the current privilege level is higher or more trusted than the IOPL, I/O executes without hindrance. If the IOPL is lower than the current privilege level, an interrupt occurs, causing execution to suspend. Note that an IOPL of 00 is the highest or most trusted; if IOPL is 11, it's the lowest or least trusted.

NT (nested task)   The nested task flag indicates that the current task is nested within another task in protected mode operation. This flag is set when the task is nested by software.

RF (resume)	The resume flag is used with debugging to control the resumption of execution after the next instruction.
VM (virtual mode)	The VM flag bit selects virtual mode operation in a protected mode system. A virtual mode system allows multiple DOS memory partitions that are 1M byte in length to coexist in the memory system. Essentially, this allows the system program to execute multiple DOS programs.
AC (alignment check)	The alignment check flag bit activates if a word or doubleword is addressed on a non-word or non-doubleword boundary. Only the 80486SX microprocessor contains the alignment check bit that is primarily used by its companion numeric coprocessor, the 80487SX, for synchronization.
VIF (virtual interrupt flag)	The VIF is a copy of the interrupt flag bit available to the Pentium–Pentium 4 microprocessors.
VIP (virtual interrupt pending)	VIP provides information about a virtual mode interrupt for the Pentium–Pentium 4 microprocessors. This is used in multitasking environments to provide the operating system with virtual interrupt flags and interrupt pending information.
ID (identification)	The ID flag indicates that the Pentium–Pentium 4 microprocessors support the CPUID instruction. The CPUID instruction provides the system with information about the Pentium microprocessor, such as its version number and manufacturer.

Segment Registers.   Additional registers, called segment registers, generate memory addresses when combined with other registers in the microprocessor. There are either four or six segment registers in various versions of the microprocessor. A segment register functions differently in the real mode when compared to the protected mode operation of the microprocessor. Details on their function in real and protected mode are provided later in this chapter. Following is a list of each segment register, along with its function in the system:

CS (code)	The code segment is a section of memory that holds the code (programs and procedures) used by the microprocessor. The code segment register defines the starting address of the section of memory holding code. In real mode operation, it defines the start of a 64K-byte section of memory; in protected mode, it selects a descriptor that describes the starting address and length of a section of memory holding code. The code segment is limited to 64K bytes in the 8088–80286, and 4G bytes in the 80386 and above when these microprocessors operate in the protected mode.
DS (data)	The data segment is a section of memory that contains most data used by a program. Data are accessed in the data segment by an offset address or the contents of other registers that hold the offset address. As with the code segment and other segments, the length is limited to 64K bytes in the 8086–80286, and 4G bytes in the 80386 and above.
ES (extra)	The extra segment is an additional data segment that is used by some of the string instructions to hold destination data.
SS (stack)	The stack segment defines the area of memory used for the stack. The stack entry point is determined by the stack segment and stack pointer registers. The BP register also addresses data within the stack segment.
FS and GS	The FS and GS segments are supplemental segment registers available in the 80386, 80486, and Pentium through the Pentium 4 microprocessors to allow two additional memory segments for access by programs.

2–2 REAL MODE MEMORY ADDRESSING

The 80286 and above operate in either the real or protected mode. Only the 8086 and 8088 operate exclusively in the real mode. This section of the text details the operation of the microprocessor in the real mode. **Real mode operation** allows the microprocessor to address only the first 1M byte of memory space—even if it is the Pentium 4 microprocessor. Note that the first 1M byte of memory is called either the **real memory** or **conventional memory** system. The DOS operating system requires the microprocessor to operate in the real mode. Real mode operation allows application software written for the 8086/8088, which contain only 1M byte of memory, to function in the 80286 and above without changing the software. The upward compatibility of software is partially responsible for the continuing success of the Intel family of microprocessors. In all cases, each of these microprocessors begins operation in the real mode by default whenever power is applied or the microprocessor is reset.

Segments And Offsets

A combination of a segment address and an offset address access a memory location in the real mode. All real mode memory addresses must consist of a segment address plus an offset address. The **segment address,** located within one of the segment registers, defines the beginning address of any 64K-byte memory segment. The **offset address** selects any location within the 64K byte memory segment. Segments in the real mode always have a length of 64K bytes. Figure 2–3 shows how the segment plus offset addressing scheme selects a memory location. This illustration shows a memory segment that begins at location 10000H and ends at location 1FFFFH— 64K bytes in length. It also shows how an offset address, sometimes called a **displacement,** of F000H selects location 1F000H in the memory system. Note that the offset or displacement is the distance above the start of the segment, as shown in Figure 2–3.

The segment register in Figure 2–3 contains a 1000H, yet it addresses a starting segment at location 10000H. In the real mode, each segment register is internally appended with a **0H** on its rightmost end. This forms a 20-bit memory address, allowing it to access the start of a segment. The microprocessor must generate a 20-bit memory address to access a location within the first 1M of memory. For example, when a segment register contains a 1200H, it addresses a 64K-byte

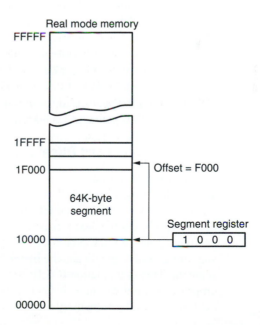

FIGURE 2–3 The real mode memory-addressing scheme, using a segment address plus an offset.

TABLE 2–1   Example segment addresses.

Segment Register	Starting Address	Ending Address
2000H	20000H	2FFFFH
2001H	20010H	3000FH
2100H	21000H	30FFFH
AB00H	AB000H	BAFFFH
1234H	12340H	2233FH

memory segment beginning at location 12000H. Likewise, if a segment register contains a 1201H, it addresses a memory segment beginning at location 12010H. Because of the internally appended 0H, real mode segments can begin only at a 16-byte boundary in the memory system. This 16-byte boundary is often called a **paragraph.**

Because a real mode segment of memory is 64K in length, once the beginning address is known, the **ending address** is found by adding FFFFH. For example, if a segment register contains 3000H, the first address of the segment is 30000H, and the last address is 30000H + FFFFH or 3FFFFH. Table 2–1 shows several examples of segment register contents, and the starting and ending addresses of the memory segments selected by each segment address.

The offset address, which is a part of the address, is added to the start of the segment to address a memory location within the memory segment. For example, if the segment address is 1000H and the offset address is 2000H, the microprocessor addresses memory location 12000H. The offset address is always added to the starting address of the segment to locate the data. The segment and offset address is sometimes written as 1000:2000 for a segment address of 1000H with an offset of 2000H.

In the 80286 (with special external circuitry), and the 80386 through the Pentium 4, an extra 64K minus 16 bytes of memory is addressable when the segment address is FFFFH and the HIMEM.SYS driver is installed in the system. This area of memory (0FFFF0H–10FFEFH) is referred to as **high memory.** When an address is generated using a segment address of FFFFH, the A20 address pin is enabled (if supported) when an offset is added. For example, if the segment address is FFFFH and the offset address is 4000H, the machine addresses memory location FFFF0H + 4000H or 103FF0H. Notice that the A20 address line is the one in address 103FF0H. If A20 is not supported, the address is generated as 03FF0H because A20 remains a logic zero.

Some addressing modes combine more than one register and an offset value to form an offset address. When this occurs, the sum of these values may exceed FFFFH. For example, the address accessed in a segment whose segment address is 4000H, and whose offset address is specified as the sum of F000H plus 3000H, will access memory location 42000H instead of location 52000H. When the F000H and 3000H are added, they form a 16-bit (**modulo 16**) sum of 2000H used as the offset address; not 12000H, the true sum. Note that the carry of 1 (F000H + 3000H = 12000H) is dropped for this addition to form the offset address of 2000H. This means that the address is generated as 4000:2000 or 42000H.

Default Segment and Offset Registers

The microprocessor has a set of rules that apply to segments whenever memory is addressed. These rules, which apply in the real and protected mode, define the segment register and offset register combination. For example, the code segment register is always used with the instruction pointer to address the next instruction in a program. This combination is **CS:IP** or **CS:EIP,** depending upon the microprocessor's mode of operation. The **code segment** register defines the start of the code segment and the **instruction pointer** locates the next instruction within the code segment. This combination (CS:IP or CS:EIP) locates the next instruction executed by the microprocessor. For example, if CS = 1400H and IP/EIP = 1200H, the microprocessor fetches its next instruction from memory location 14000H + 1200H or 15200H.

TABLE 2–2 8086–80486 and Pentium–Pentium 4 default 16-bit segment and offset address combinations.

Segment	Offset	Special Purpose
CS	IP	Instruction address
SS	SP or BP	Stack address
DS	BX, DI, SI, an 8-bit number, or a 16-bit number	Data address
ES	DI for string instructions	String destination address

Another of the default combinations is the **stack.** Stack data are referenced through the stack segment at the memory location addressed by either the stack pointer (SP/ESP) or the base pointer (BP/EBP). These combinations are referred to as SS:SP (SS:ESP) or SS:BP (SS:EBP). For example, if SS = 2000H and BP = 3000H, the microprocessor addresses memory location 23000H for the stack segment memory location. Note that in real mode, only the rightmost 16 bits of the extended register address a location within the memory segment. In the 80386–Pentium 4, never place a number larger than FFFFH into an offset register if the microprocessor is operated in the real mode. This causes the system to halt and indicate an addressing error.

Other defaults are shown in Table 2–2 for addressing memory using any Intel microprocessor with 16-bit registers. Table 2–3 shows the defaults assumed in the 80386 and above when using 32-bit registers. Note that the 80386 and above have a far greater selection of segment/ offset address combinations than do the 8086 through the 80286 microprocessors.

The 8086–80286 microprocessors allow four memory segments and the 80386 and above allow six memory segments. Figure 2–4 shows a system that contains four memory segments. Note that a memory segment can touch or even overlap if 64K bytes of memory are not required for a segment. Think of segments as windows that can be moved over any area of memory to access data or code. Also note that a program can have more than four or six segments, but can only access four or six segments at a time.

Suppose that an application program requires 1000H bytes of memory for its code, 190H bytes of memory for its data, and 200H bytes of memory for its stack. This application does not require an extra segment. When this program is placed in the memory system by DOS, it is loaded in the TPA at the first available area of memory above the drivers and other TPA programs. This area is indicated by a **free-pointer** that is maintained by DOS. Program loading is handled automatically by the **program loader** located within DOS. Figure 2–5 shows how this application is stored in the memory system. The segments show an overlap because the amount of data in them does not require 64K bytes of memory. The side view of the segments clearly shows the overlap. It also shows how segments can be moved over any area of memory by

TABLE 2–3 80386 through the Pentium 4 default 32-bit segment and offset address combinations.

Segment	Offset	Special Purpose
CS	EIP	Instruction address
SS	ESP and EBP	Stack address
DS	EAX, EBX, ECX, EDX, ESI, EDI, an 8-bit number, or a 32-bit number	Data address
ES	EDI for string instructions	String destination address
FS	No default	General address
GS	No default	General address

FIGURE 2–4 A memory system showing the placement of four memory segments.

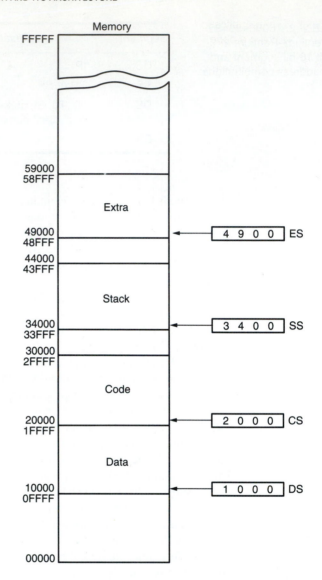

changing the segment starting address. Fortunately, the DOS program loader calculates and assigns segment starting addresses. This is explained in Chapter 7, which details the operation of the assembler, BIOS, and DOS for an assembly language program.

Segment and Offset Addressing Scheme Allows Relocation

The segment and offset addressing scheme seems unduly complicated. It is complicated, but it also affords an advantage to the system. This complicated scheme of segment plus offset addressing allows programs to be relocated in the memory system. It also allows programs written to function in the real mode to operate in a protected mode system. A relocatable program is one that can be placed into any area of memory and executed without change. Relocatable data are data that can be placed in any area of memory and used without any change to the program. The segment and offset addressing scheme allows both programs and data to be relocated without changing a thing in a program or data. This is ideal for use in a general-purpose

FIGURE 2–5 An application program containing a code, data, and stack segment loaded into a DOS system memory.

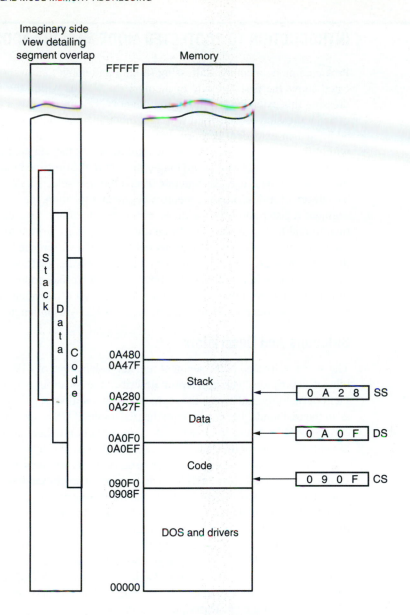

computer system in which not all machines contain the same memory areas. The personal computer memory structure is different from machine to machine, requiring relocatable software and data.

Because memory is addressed within a segment by an offset address, the memory segment can be moved to any place in the memory system without changing any of the offset addresses. This is accomplished by moving the entire program, as a block, to a new area and then changing only the contents of the segment registers. If an instruction is 4 bytes above the start of the segment, its offset address is 4. If the entire program is moved to a new area of memory, this offset address of 4 still points to 4 bytes above the start of the segment. Only the contents of the segment register must be changed to address the program in the new area of memory. Without this feature, a program would have to be extensively rewritten or altered before it is moved. This would require additional time or many versions of a program for the many different configurations of computer systems.

2–3   INTRODUCTION TO PROTECTED MODE MEMORY ADDRESSING

Protected mode memory addressing (80286 and above) allows access to data and programs located above the first 1M byte of memory, as well as within the first 1M byte of memory. Addressing this extended section of the memory system requires a change to the segment plus an offset addressing scheme used with real mode memory addressing. When data and programs are addressed in extended memory, the offset address is still used to access information located within the memory segment. One difference is that the segment address, as discussed with real mode memory addressing, is no longer present in the protected mode. In place of the segment address, the segment register contains a **selector** that selects a descriptor from a descriptor table. The **descriptor** describes the memory segment's location, length, and access rights. Because the segment register and offset address still access memory, protected mode instructions are identical to real mode instructions. In fact, most programs written to function in the real mode will function without change in the protected mode. The difference between modes is in the way that the segment register is interpreted by the microprocessor to access the memory segment. Another difference, in the 80386 and above, is that the offset address can be a 32-bit number instead of a 16-bit number in the protected mode. A 32-bit offset address allows the microprocessor to access data within a segment that can be up to 4G bytes in length.

Selectors And Descriptors

The selector, located in the segment register, selects one of 8192 descriptors from one of two tables of descriptors. The **descriptor** describes the location, length, and access rights of the segment of memory. Indirectly, the segment register still selects a memory segment, but not directly as in the real mode. For example, in the real mode, if CS = 0008H, the code segment begins at location 00080H. In the protected mode, this segment number can address any memory location in the entire system for the code segment, as explained shortly.

There are two descriptor tables used with the segment registers: one contains global descriptors and the other contains local descriptors. The **global descriptors** contain segment definitions that apply to all programs, while the **local descriptors** are usually unique to an application. You might call a global descriptor a **system descriptor** and call a local descriptor an **application descriptor.** Each descriptor table contains 8192 descriptors, so a total of 16,384 total descriptors are available to an application at any time. Because the descriptor describes a memory segment, this allows up to 16,384 memory segments to be described for each application.

Figure 2–6 shows the format of a descriptor for the 80286 through the Pentium 4. Note that each descriptor is 8 bytes in length, so the global and local descriptor tables are each a maximum of 64K bytes in length. Descriptors for the 80286 and the 80386 through the Pentium 4 differ slightly, but the 80286 descriptor is upward-compatible.

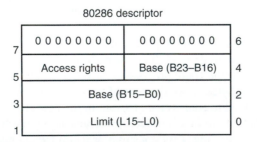
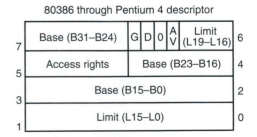

FIGURE 2–6   The descriptor formats for the 80286 and 80386 through Pentium 4 microprocessors.

The **base address** portion of the descriptor indicates the starting location of the memory segment. For the 80286 microprocessor, the base address is a 24-bit address, so segments begin at any location in its 16M bytes of memory. Note that the paragraph boundary limitation is removed in these microprocessors when operated in the protected mode. The 80386 and above use a 32-bit base address that allows segments to begin at any location in its 4G bytes of memory. Notice how the 80286 descriptor's base address is upward-compatible to the 80386 through the Pentium 4 descriptor because its most-significant 16 bits are 0000H. Refer to Chapters 18 and 19 for additional detail on the 64G memory space provided by the Pentium Pro through the Pentium 4.

The **segment limit** contains the last offset address found in a segment. For example, if a segment begins at memory location F00000H and ends at location F000FFH, the base address is F00000H and the limit is FFH. For the 80286 microprocessor, the base address is F00000H and the limit is 00FFH. For the 80386 and above, the base address is 00F00000H and the limit is 000FFH. Notice the limit the 80286 has a 16-bit limit and the 80386 through the Pentium 4 have a 20-bit limit. The 80286 accesses memory segments that are between 1 and 64K bytes in length. The 80386 and above access memory segments that are between 1 and 1M byte, or 4K and 4G bytes in length.

There is another feature found in the 80386 through the Pentium 4 descriptor that is not found in the 80286 descriptor: the G bit, or **granularity bit.** If G = 0, the limit specifies a segment limit of 00000H to FFFFFH. If G = 1, the value of the limit is multiplied by 4K bytes (appended with XXXH). The limit is then 00000XXXH to FFFFFXXXH, if G = 1. This allows a segment length of 4K to 4G bytes in steps of 4K bytes. The reason that the segment length is 64K bytes in the 80286 is that the offset address is always 16 bits because of its 16-bit internal architecture. The 80386 and above use a 32-bit architecture that allows an offset address, in the protected mode operation, of the 32 bits. This 32-bit offset address allows segment lengths of 4G bytes and the 16-bit offset address allows segment lengths of 64K bytes. Operating systems operate in a 16- or 32-bit environment. For example, DOS uses a 16-bit environment, while most Windows applications use a 32-bit environment.

Example 2–1 shows the segment start and end if the base address is 10000000H, the limit is 001FFH, and the G bit = 0.

EXAMPLE 2–1

```
Base = Start = 10000000H
G = 0
End = Base + Limit = 10000000H + 001FFH = 100001FFH
```

Example 2–2 uses the same data as Example 2–1, except that the G bit = 1. Notice that the limit is appended with XXXH to determine the ending segment address. The XXXH can be any number between 000H and FFFH. In the example, the XXXH is replaced with FFFH because that is the highest possible memory location within the segment.

EXAMPLE 2–2

```
Base = Start = 10000000H
G = 1
End = Base + Limit = 10000000H + 001FFXXXH = 101FFFFFH
```

The AV bit, in the 80386 and above descriptor, is used by some operating systems to indicate that the segment is available (AV = 1) or not available (AV = 0). The D bit indicates how the 80386 through the Pentium 4 instructions access register and memory data in the protected or real mode. If D = 0, the instructions are 16-bit instructions, compatible with the 8086–80286 microprocessors. This means that the instructions use 16-bit offset addresses and 16-bit registers by default. This mode is often called the 16-bit instruction mode. If D = 1, the instructions are 32-bit instructions. By default, the 32-bit instruction mode assumes that all offset addresses and

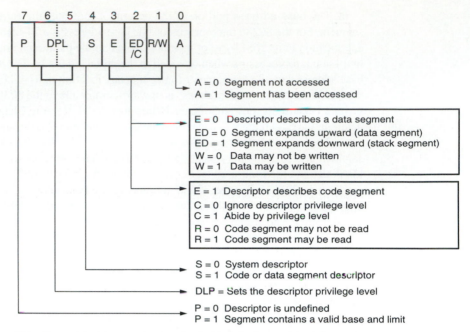

Note: Some of the letters used to describe the bits in the access rights bytes vary in Intel documentation.

FIGURE 2–7   The access rights byte for the 80286 through Pentium 4 descriptor.

all registers are 32 bits. Note that the default for register size and offset address size can be overridden in both the 16- and 32-bit instruction modes. Both the MSDOS and PCDOS operating systems require that the instructions are always used in the 16-bit instruction mode. Windows 3.1 also requires that the 16-bit instruction mode is selected. Note that the 32-bit instruction mode is accessible only in a protected-mode system such as Windows NT, Windows 95, Windows 98, or OS/2. More detail on these modes and their application to the instruction set appears in Chapters 3 and 4.

The **access rights byte** (see Figure 2–7) controls access to the protected mode memory segment. This byte describes how the segment functions in the system. The access rights byte allows complete control over the segment. If the segment is a data segment, the direction of growth is specified. If the segment grows beyond its limit, the microprocessor's program is interrupted, indicating a general protection fault. You can even specify whether a data segment can be written or is write-protected. The code segment is also controlled in a similar fashion and can have reading inhibited to protect software.

Descriptors are chosen from the descriptor table by the segment register. Figure 2–8 shows how the segment register functions in the protected mode system. The segment register contains a 13-bit selector field, a table selector bit, and a requested privilege level field. The 13-bit selector chooses one of the 8192 descriptors from the descriptor table. The **TI bit** selects either the global descriptor table (TI = 0) or the local descriptor table (TI = 1). The **requested privilege level** (RPL) requests the access privilege level of a memory segment. The highest privilege level is 00 and the lowest is 11. If the requested privilege level matches or is higher in priority than the privilege level set by the access rights byte, access is granted. For example, if the requested privilege level is 10 and the access rights byte sets the segment privilege level at 11, access is granted because 10 is higher in priority than privilege level 11. Privilege levels are used in multiuser environments. If the privilege level is violated, the system normally indicates a privilege violation.

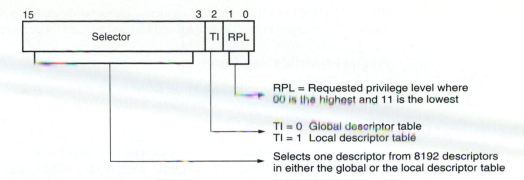

FIGURE 2–8 The contents of a segment register during protected mode operation of the 80286 through Pentium 4 microprocessors.

Figure 2–9 shows how the segment register, containing a selector, chooses a descriptor from the global descriptor table. The entry in the global descriptor table selects a segment in the memory system. In this illustration, DS contains 0008H, which accesses the descriptor number 1 from the global descriptor table by using a requested privilege level of 00. Descriptor number 1 contains a descriptor that defines the base address as 00100000H with a segment limit of 000FFH. This means that a value of 0008H loaded into DS causes the microprocessor to use memory locations

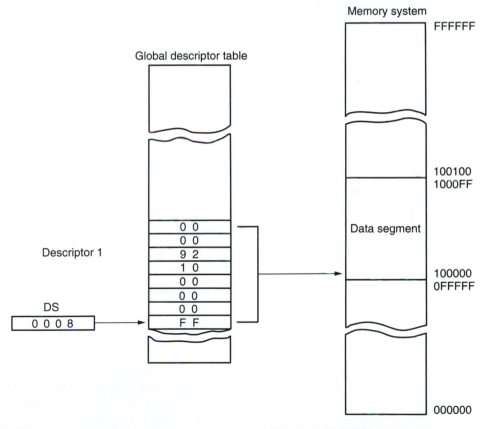

FIGURE 2–9 Using the DS register to select a descriptor from the global descriptor table. In this example, the DS register accesses memory locations 100000H–1000FFH as a data segment.

00100000H–001000FFH for the data segment with this example descriptor table. Note that descriptor zero is called the null descriptor and may not be used for accessing memory.

Program-Invisible Registers

The global and local descriptor tables are found in the memory system. In order to access and specify the address of these tables, the 80286, 80386, 80486, Pentium, Pentium Pro, and Pentium 4 contain program-invisible registers. The program-invisible registers are not directly addressed by software so they are given this name (although some of these registers are accessed by the system software). Figure 2–10 illustrates the program-invisible registers as they appear in the 80286 through the Pentium 4. These registers control the microprocessor when operated in the protected mode.

Each of the segment registers contains a program-invisible portion used in the protected mode. The program-invisible portion of these registers is often called cache memory because a cache is any memory that stores information. This cache is not to be confused with the normal level 1 or level 2 caches found with the microprocessor. The program-invisible portion of the segment register is loaded with the base address, limit, and access rights each time the number in the segment register is changed. When a new segment number is placed in a segment register, the microprocessor accesses a descriptor table and loads the descriptor into the program-invisible cache portion of the segment register. It is held there and used to access the memory segment until the segment number is again changed. This allows the microprocessor to repeatedly access a memory segment without referring to the descriptor table for each access (hence the term cache).

The GDTR (**global descriptor table register**) and IDTR (**interrupt descriptor table register**) contain the base address of the descriptor table and its limit. The limit of each descriptor table is 16 bits because the maximum table length is 64K bytes. When the protected mode operation is desired, the address of the global descriptor table and its limit are loaded into the GDTR.

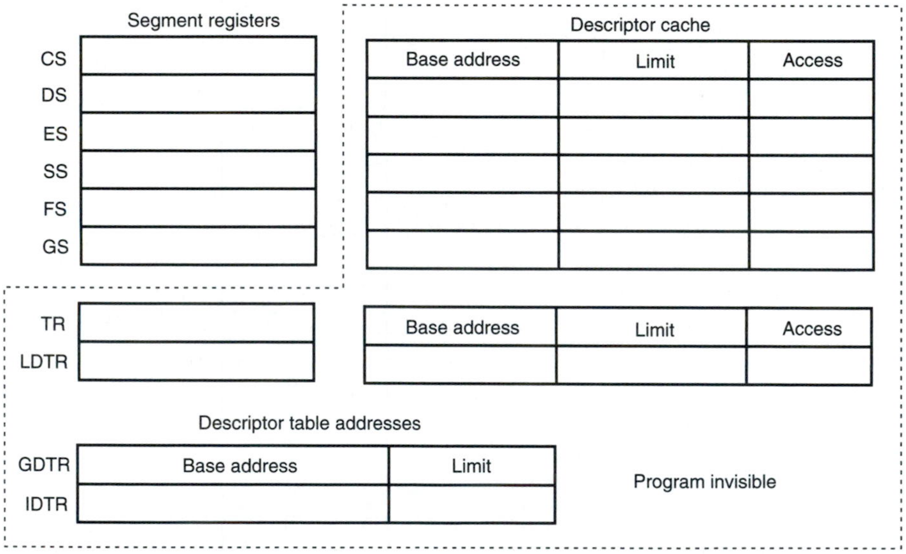

Notes:
1. The 80286 does not contain FS and GS nor the program-invisible portions of these registers.
2. The 80286 contains a base address that is 24-bits and a limit that is 16-bits.
3. The 80386/80486/Pentium/Pentium Pro contain a base address that is 32-bits and a limit that is 20-bits.
4. The access rights are 8-bits in the 80286 and 12-bits in the 80386/80486/Pentium.

FIGURE 2–10   The program-invisible register within the 80286–Pentium 4 microprocessors.

Before using the protected mode, the interrupt descriptor table and the IDTR must also be initialized. More detail is provided on protected mode operation later in the text. At this point, the programming and additional description of these registers are impossible.

The location of the local descriptor table is selected from the global descriptor table. One of the global descriptors is set up to address the local descriptor table. To access the local descriptor table, the LDTR (**local descriptor table register**) is loaded with a selector, just as a segment register is loaded with a selector. This selector accesses the global descriptor table and loads the base address, limit, and access rights of the local descriptor table into the cache portion of the LDTR.

The TR (**task register**) holds a selector, which accesses a descriptor that defines a task. A task is most often a procedure or application program. The descriptor for the procedure or application program is stored in the global descriptor table, so access can be controlled through the privilege levels. The task register allows a context or task switch in about 17 µs. Task switching allows the microprocessor to switch between tasks in a fairly short amount of time. The task switch allows multitasking systems to switch from one task to another in a simple and orderly fashion.

2–4   MEMORY PAGING

The **memory paging mechanism** located within the 80386 and above allows any physical memory location to be assigned to any linear address. The **linear address** is defined as the address generated by a program. With the memory paging unit, the linear address is invisibly translated into any **physical address,** which allows an application written to function at a specific address to be relocated through the paging mechanism. It also allows memory to be placed into areas where no memory exists. An example is the upper memory blocks provided by EMM386.EXE.

The EMM386.EXE program reassigns extended memory, in 4K blocks, to the system memory between the video BIOS and the system BIOS ROMS for upper memory blocks. Without the paging mechanism, the use of this area of memory is impossible.

Paging Registers

The paging unit is controlled by the contents of the microprocessor's control registers. See Figure 2–11 for the contents of control registers CR0 through CR3. Note that these registers are only available to the 80386 through the Pentium microprocessors. Beginning with the Pentium, an additional control register labeled CR4 controls extensions to the basic architecture provided in the Pentium and above microprocessors. One of these features is a 4M-byte page that is enabled by setting bit position 4, or CR4. Refer to Chapters 18 and 19 for additional details on 4M-byte memory paging.

The registers important to the paging unit are CR0 and CR3. The leftmost bit (PG) position of CR0 selects paging when placed at a logic 1 level. If the PG bit is cleared (0), the linear address generated by the program becomes the physical address used to access memory. If the PG bit is set (1), the linear address is converted to a physical address through the paging mechanism. The paging mechanism functions in both the real and protected modes.

CR3 contains the page directory base address, and the PCD and PWT bits. The PCD and PWT bits control the operation of the PCD and PWT pins on the microprocessor. If PCD is set (1), the PCD pin becomes a logic one during bus cycles that are not pages. This allows the external hardware to control the level 2 cache memory. (Note that the level 2 cache memory is an external high-speed memory that functions as a buffer between the microprocessor and the main DRAM memory system.) The PWT bit also appears on the PWT pin, during bus cycles that are not pages, to control the write-through cache in the system. The page directory base address locates the page directory for the page translation unit. Note that this address locates the page directory at any 4K boundary in the memory system because it is appended internally with a 000H. The page directory

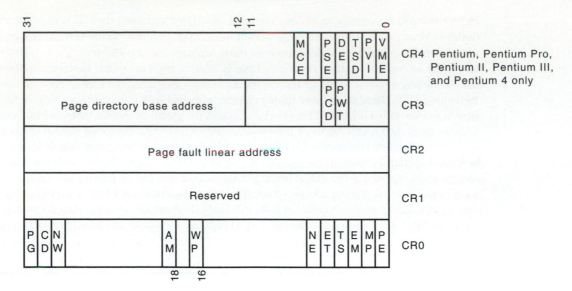

FIGURE 2–11   The control register structure of the microprocessor.

contains 1024 directory entries of 4 bytes each. Each page directory entry addresses a page table that contains 1024 entries.

The linear address, as it is generated by the software, is broken into three sections that are used to access the **page directory entry, page table entry,** and **page offset address.** Figure 2–12 shows the linear address and its makeup for paging. Notice how the leftmost 10 bits address an entry in the page directory. For linear address 00000000H–003FFFFFH, the first entry of the page directory is accessed. Each page directory entry represents or repages a 4M-byte section of the memory system. The contents of the page directory select a page table that is indexed by the

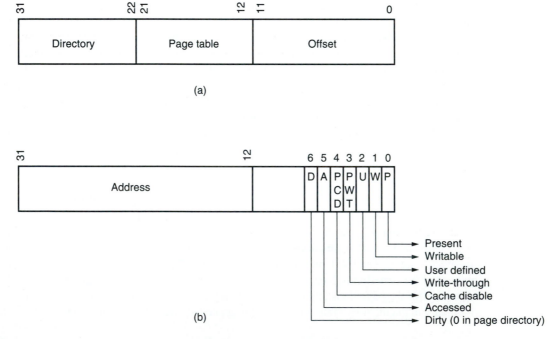

FIGURE 2–12   The format for the linear address (a) and a page directory or page table entry (b).

next 10 bits of the linear address (bit positions 12–21). This means that address 00000000H–00000FFFH selects page directory entry 0 and page table entry 0. Notice this is a 4K-byte address range. The offset part of the linear address (bit positions 0–11) next selects a byte in the 4K-byte memory page. In Figure 2–12, if the page table 0 entry contains address 00100000H, then the physical address is 00100000H–00100FFFH for linear address 00000000H–00000FFFH. This means that when the program accesses a location between 00000000H and 00000FFFH, the microprocessor physically addresses location 00100000H–00100FFFH.

Because the act of repaging a 4K-byte section of memory requires access to the page directory and a page table, which are both located in memory, Intel has incorporated a cache called the TLB **(translation look-aside buffer)**. In the 80486 microprocessor, the cache holds the 32 most recent page translation addresses. This means that the last 32 page table translations are stored in the TLB, so if the same area of memory is accessed, the address is already present in the TLB, and access to the page directory and page tables is not required. This speeds program execution. If a translation is not in the TLB, the page directory and page table must be accessed, which requires additional execution time. The Pentium, Pentium Pro, Pentium II, Pentium III, and Pentium 4 contain separate TLBs for each of their instruction and data caches.

The Page Directory and Page Table

Figure 2–13 shows the page directory, a few page tables, and some memory pages. There is only one page directory in the system. The page directory contains 1024 doubleword addresses that locate up to 1024 page tables. The page directory and each page table are 4K bytes in length. If the entire 4G byte of memory is paged, the system must allocate 4K bytes of memory for the page directory, and 4K times 1024 or 4M bytes for the 1024 page tables. This represents a considerable investment in memory resources.

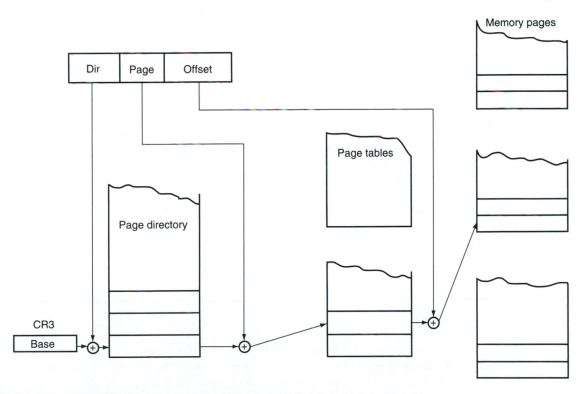

FIGURE 2–13 The paging mechanism in the 80386 through Pentium 4 microprocessors.

FIGURE 2–14   The page directory, page table 0, and two memory pages. Note how the address of page 000C8000–000C9000 has been moved to 00110000–00110FFF.

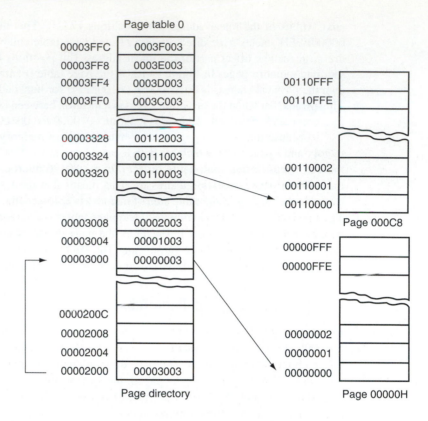

The DOS system and EMM386.EXE use page tables to redefine the area of memory between locations C8000H–EFFFFH as upper memory blocks. It does this by repaging extended memory to back-fill this part of the conventional memory system to allow DOS access to additional memory. Suppose that the EMM386.EXE program allows access to 16M bytes of extended and conventional memory through paging and locations C8000H–EFFFFH must be repaged to locations 110000–138000H, with all other areas of memory paged to their normal locations. Such a scheme is depicted in Figure 2–14.

Here, the page directory contains four entries. Recall that each entry in the page directory corresponds to 4M bytes of physical memory. The system also contains four page tables with 1024 entries each. Recall that each entry in the page table repages 4K bytes of physical memory. This scheme requires a total of 16K of memory for the four page tables and 16 bytes of memory for the page directory.

As with DOS, the Windows program also repages the memory system. At present, Windows version 3.11 supports paging for only 16M bytes of memory because of the amount of memory required to store the page tables. On the Pentium and Pentium Pro microprocessors, pages can be either 4K bytes in length or 4M bytes in length. Although no software currently supports the 4M-byte pages, as the Pentium 4 and more advanced versions pervade the personal computer, operating systems of the future will undoubtedly begin to support 4M-byte memory pages.

2–5   SUMMARY

1. The programming model of the 8086 through 80286 contains 8- and 16-bit registers. The programming model of the 80386 and above contains 8-, 16-, and 32-bit extended registers as well as two additional 16-bit segment registers: FS and GS.

2. The 8-bit registers are AH, AL, BH, BL, CH, CL, DH, and DL. The 16-bit registers are AX, BX, CX, DX, SP, BP, DI, and SI. The segment registers are CS, DS, ES, SS, FS, and GS. The 32-bit extended registers are EAX, EBX, ECX, EDX, ESP, EBP, EDI, and ESI. In addition, the microprocessor contains an instruction pointer (IP/EIP) and flag register (FLAGS or EFLAGS).

3. All real mode memory addresses are a combination of a segment address plus an offset address. The starting location of a segment is defined by the 16-bit number in the segment register that is appended with a hexadecimal zero at its rightmost end. The offset address is a 16-bit number added to the 20-bit segment address to form the real mode memory address.

4. All instructions (code) are accessed by the combination of CS (segment address) plus IP or EIP (offset address).

5. Data are normally referenced through a combination of the DS (data segment), and either an offset address or the contents of a register that contains the offset address. The 8086 through the Pentium 4 use BX, DI, and SI as default offset registers for data if 16-bit registers are selected. The 80386 and above can use the 32-bit registers EAX, EBX, ECX, EDX, EDI, and ESI as default offset registers for data.

6. Protected mode operation allows memory above the first 1M byte to be accessed by the 80286 through the Pentium 4 microprocessors. This extended memory system (XMS) is accessed via a segment address plus an offset address, just as in the real mode. The difference is that the segment address is not held in the segment register. In the protected mode, the segment starting address is stored in a descriptor that is selected by the segment register.

7. A protected mode descriptor contains a base address, limit, and access rights byte. The base address locates the starting address of the memory segment; the limit defines the last location of the segment. The access rights byte defines how the memory segment is accessed via a program. The 80286 microprocessor allows a memory segment to start at any of its 16M bytes of memory using a 24-bit base address. The 80386 and above allow a memory segment to begin at any of its 4G bytes of memory using a 32-bit base address. The limit is a 16-bit number in the 80286 and a 20-bit number in the 80386 and above. This allows an 80286 memory segment limit of 64K bytes, and an 80386 and above memory segment limit of either 1M bytes (G = 0) or 4G bytes (G = 1).

8. The segment register contains three fields of information in the protected mode. The leftmost 13 bits of the segment register address one of 8192 descriptors from a descriptor table. The TI bit accesses either the global descriptor table (TI = 0) or the local descriptor table (TI = 1). The rightmost 2 bits of the segment register select the requested priority level for the memory segment access.

9. The program-invisible registers are used by the 80286 and above to access the descriptor tables. Each segment register contains a cache portion that is used in protected mode to hold the base address, limit, and access rights acquired from a descriptor. The cache allows the microprocessor to access the memory segment without again referring to the descriptor table until the segment register's contents are changed.

10. A memory page is 4K bytes in length. The linear address, as generated by a program, can be mapped to any physical address through the paging mechanism found within the 80386 through the Pentium 4 microprocessor.

11. Memory paging is accomplished through control registers CR0 and CR3. The PG bit of CR0 enables paging and the contents of CR3 addresses the page directory. The page directory contains up to 1024 page table addresses that are used to access paging tables. The page table contains 1024 entries that locate the physical address of a 4K-byte memory page.

12. The TLB (translation look-aside buffer) caches the 32 most recent page table translations. This precludes page table translation if the translation resides in the TLB, speeding the execution of software.

2–6 QUESTIONS AND PROBLEMS

1. What are program-visible registers?
2. The 80286 addresses registers that are 8- and _____ - bits wide.
3. The extended registers are addressable by which microprocessors?
4. The extended BX register is addressed as _____
5. Which register holds a count for some instructions?
6. What is the purpose of the IP/EIP register?
7. The carry flag bit is set by which arithmetic operations?
8. Will an overflow occur if a signed FFH is added to a signed 01H?
9. A number that contains 3 one bits is said to have _____ parity.
10. Which flag bit controls the INTR pin on the microprocessor?
11. Which microprocessors contain an FS segment register?
12. What is the purpose of a segment register in the real mode operation of the microprocessor?
13. In the real mode, show the starting and ending addresses of each segment located by the following segment register values:
    (a) 1000H
    (b) 1234H
    (c) 2300H
    (d) E000H
    (e) AB00H
14. Find the memory address of the next instruction executed by the microprocessor, when operated in the real mode, for the following CS:IP combinations:
    (a) CS = 1000H and IP = 2000H
    (b) CS = 2000H and IP = 1000H
    (c) CS = 2300H and IP = 1A00H
    (d) CS = 1A00H and IP = B000H
    (e) CS = 3456H and IP = ABCDH
15. Real mode memory addresses allow access to memory below which memory address?
16. Which register or registers are used as an offset address for string instruction destinations in the microprocessor?
17. Which 32-bit register or registers are used as an offset address for data segment data in the Pentium 4 microprocessor?
18. The stack memory is addressed by a combination of the _____ segment plus _____ offset.
19. If the base pointer (BP) addresses memory, the _____ segment contains the data.
20. Determine the memory location addressed by the following real mode 80286 register combinations:
    (a) DS = 1000H and DI = 2000H
    (b) DS = 2000H and SI = 1002H
    (c) SS = 2300H and BP = 3200H
    (d) DS = A000H and BX = 1000H
    (e) SS = 2900H and SP = 3A00H
21. Determine the memory location addressed by the following real mode Pentium 4 register combinations:
    (a) DS = 2000H and EAX = 00003000H
    (b) DS = 1A00H and ECX = 00002000H
    (c) DS = C000H and ESI = 0000A000H
    (d) SS = 8000H and ESP = 00009000H
    (e) DS = 1239H and EDX = 0000A900H

22. Protected mode memory addressing allows access to which area of the memory in the 80286 microprocessor?
23. Protected mode memory addressing allows access to which area of the memory in the Pentium 4 microprocessor?
24. What is the purpose of the segment register in protected mode memory addressing?
25. How many descriptors are accessible in the global descriptor table in the protected mode?
26. For an 80286 descriptor that contains a base address of A00000H and a limit of 1000H, what starting and ending locations are addressed by this descriptor?
27. For an 80486 descriptor that contains a base address of 01000000H, a limit of 0FFFFH, and G = 0, what starting and ending locations are addressed by this descriptor?
28. For a Pentium 4 descriptor that contains a base address of 00280000H, a limit of 00010H, and G = 1, what starting and ending locations are addressed by this descriptor?
29. If the DS register contains 0020H in a protected mode system, which global descriptor table entry is accessed?
30. If DS = 0103H in a protected mode system, the requested privilege level is _____.
31. If DS = 0105H in a protected mode system, which entry, table, and requested privilege level are selected?
32. What is the maximum length of the global descriptor table in the Pentium 4 microprocessor?
33. Code a descriptor that describes a memory segment that begins at location 210000H and ends at location 21001FH. This memory segment is a code segment that can be read. The descriptor is for an 80286 microprocessor.
34. Code a descriptor that describes a memory segment that begins at location 03000000H and ends at location 05FFFFFFH. This memory segment is a data segment that grows upward in the memory system and can be written. The descriptor is for an 80386 microprocessor.
35. Which register locates the global descriptor table?
36. How is the local descriptor table addressed in the memory system?
37. Describe what happens when a new number is loaded into a segment register when the microprocessor is operated in the protected mode.
38. What are the program-invisible registers?
39. What is the purpose of the GDTR?
40. How many bytes are found in a memory page?
41. What register is used to enable the paging mechanism in the 80386, 80486, Pentium, Pentium Pro, and Pentium 4 microprocessors?
42. How many 32-bit addresses are stored in the page directory?
43. Each entry in the page directory translates how much linear memory into physical memory?
44. If the microprocessor sends linear address 00200000H to the paging mechanism, which paging directory entry is accessed, and which page table entry is accessed?
45. What value is placed in the page table to redirect linear address 20000000H–30000000H?
46. What is the purpose of the TLB located within the 80486 microprocessor?
47. Using the Internet, write a short report that details the TLB. Hint: You might want to go to the Intel Web site and search for information.

CHAPTER 3

Addressing Modes

INTRODUCTION

Efficient software development for the microprocessor requires a complete familiarity with the addressing modes employed by each instruction. In this chapter, the MOV (**move data**) instruction is used to describe the data-addressing modes. The MOV instruction transfers bytes or words of data between registers, or between registers and memory in the 8086 through the 80286 and bytes, words, or doublewords in the 80386 and above. In describing the program memory-addressing modes, the CALL and JUMP instructions show how to modify the flow of the program.

The data-addressing modes include register, immediate, direct, register indirect, base-plus-index, register relative, and base relative-plus-index in the 8086 through the 80286 microprocessor. The 80386 and above also include a scaled-index mode of addressing memory data. The program memory-addressing modes include program relative, direct, and indirect. The operation of the stack memory is explained so that the PUSH and POP instructions are understood.

CHAPTER OBJECTIVES

Upon completion of this chapter, you will be able to:

1. Explain the operation of each data-addressing mode.
2. Use the data-addressing modes to form assembly language statements.
3. Explain the operation of each program memory-addressing mode.
4. Use the program memory-addressing modes to form assembly and machine language statements.
5. Select the appropriate addressing mode to accomplish a given task.
6. Detail the difference between addressing memory data using real mode and protected mode operation.
7. Describe the sequence of events that place data onto the stack or remove data from the stack.
8. Explain how a data structure is placed in memory and used with software.

3–1 DATA-ADDRESSING MODES

Because the MOV instruction is a common and flexible instruction, it provides a basis for the explanation of the data-addressing modes. Figure 3–1 illustrates the MOV instruction and defines the

FIGURE 3–1   The MOV instruction showing the source, destination, and direction of data flow.

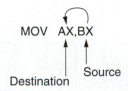

direction of data flow. The **source** is to the right and the **destination** is to the left, next to the op-code MOV. (An **opcode,** or operation code, tells the microprocessor which operation to perform.) This direction of flow, which is applied to all instructions, is awkward at first. We naturally assume that things move from left to right, whereas here they move from right to left. Notice that a comma always separates the destination from the source in an instruction. Also, note that memory-to-memory transfers are not allowed by any instruction except for the MOVS instruction.

In Figure 3–1, the MOV AX,BX instruction transfers the word contents of the source register (BX) into the destination register (AX). The source never changes, but the destination usually changes.[1] It is essential to remember that a MOV instruction always *copies* the source data and into the destination. The MOV never actually picks up the data and moves it. Also, note that the flag register remains unaffected by most data transfer instructions. The source and destination are often called **operands.**

Figure 3–2 shows all possible variations of the data-addressing modes using the MOV instruction. This illustration helps to show how each data-addressing mode is formulated with the MOV instruction and also serves as a reference. Note that these are the same data-addressing modes found with all versions of the Intel microprocessor, except for the scaled-index-addressing mode, which is found only in the 80386 through the Pentium 4. The data-addressing modes are as follows:

Register addressing	Transfers a copy of a byte or word from the source register or memory location to the destination register or memory location. (Example: the MOV CX,DX instruction copies the word-sized contents of register DX into register CX.) In the 80386 and above, a doubleword can be transferred from the source register or memory location to the destination register or memory location. (Example: the MOV ECX,EDX instruction copies the doubleword-sized contents of register EDX into register ECX.)
Immediate addressing	Transfers the source-immediate byte or word of data into the destination register or memory location. (Example: the MOV AL,22H instruction copies a byte-sized 22H into register AL.) In the 80386 and above, a doubleword of immediate data can be transferred into a register or memory location. (Example: the MOV EBX,12345678H instruction copies a doubleword-sized 12345678H into the 32-bit wide EBX register.)
Direct addressing	Moves a byte or word between a memory location and a register. The instruction set does not support a memory-to-memory transfer, except for the MOVS instruction. (Example: the MOV CX,LIST instruction copies the word-sized contents of memory location LIST into register CX.) In the 80386 and above, a doubleword-sized memory location can also be addressed. (Example: the MOV ESI,LIST instruction copies a 32-bit number, stored in four consecutive bytes of memory, from location LIST into register ESI.)

---

[1]The exceptions are the CMP and TEST instructions, which never change the destination. These instructions are described in later chapters.

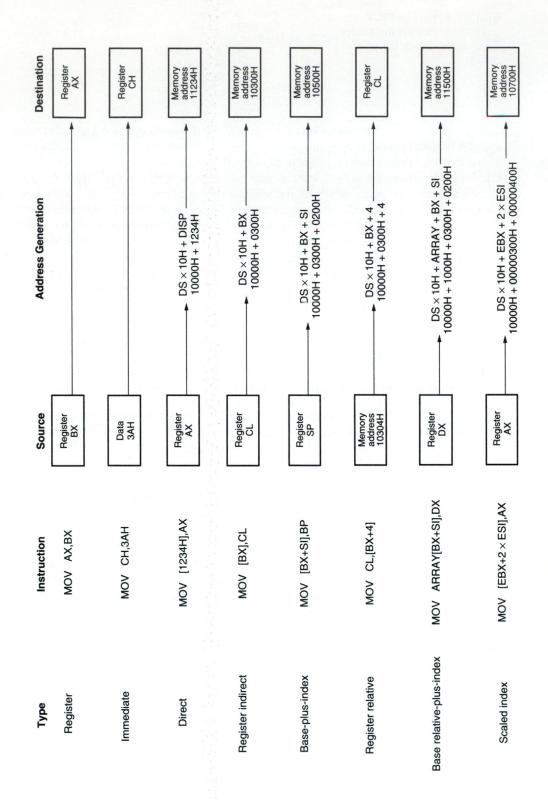

FIGURE 3–2 8086–Pentium 4 data-addressing modes.

Register indirect addressing	Transfers a byte or word between a register and a memory location addressed by an index or base register. The index and base registers are BP, BX, DI, and SI. (Example: the MOV AX,[BX] instruction copies the word-sized data from the data segment offset address indexed by BX into register AX.) In the 80386 and above, a byte, word, or double-word is transferred between a register and a memory location addressed by any register: EAX, EBX, ECX, EDX, EBP, EDI, or ESI. (Example: the MOV AL,[ECX] instruction loads AL from the data segment offset address selected by the contents of ECX.)
Base-plus-index addressing	Transfers a byte or word between a register and the memory location addressed by a base register (BP or BX) plus an index register (DI or SI). (Example: the MOV [BX+DI],CL instruction copies the byte-sized contents of register CL into the data segment memory location addressed by BX plus DI.) In the 80386 and above, any register EAX, EBX, ECX, EDX, EBP, EDI, or ESI may be combined to generate the memory address. (Example: the MOV [EAX+EBX],CL instruction copies the byte-sized contents of register CL into the data segment memory location addressed by EAX plus EBX.)
Register relative addressing	Moves a byte or word between a register and the memory location addressed by an index or base register plus a displacement. (Example: MOV AX,[BX+4] or MOV AX,ARRAY[BX]. The first instruction loads AX from the data segment address formed by BX plus 4. The second instruction loads AX from the data segment memory location in ARRAY plus the contents of BX.) The 80386 and above use any register to address memory. (Example: MOV AX,[ECX+4] or MOV AX,ARRAY[EBX]. The first instruction loads AX from the data segment address formed by ECX plus 4. The second instruction loads AX from the data segment memory location ARRAY plus the contents of EBX.)
Base relative-plus-index addressing	Transfers a byte or word between a register and the memory location addressed by a base and an index register plus a displacement. (Example: MOV AX,ARRAY[BX+DI] or MOV AX,[BX+DI+4]. These instructions load AX from a data segment memory location. The first instruction uses an address formed by adding ARRAY, BX, and DI and the second by adding BX, DI, and 4.) In the 80386 and above, MOV EAX,ARRAY[EBX+ECX] loads EAX from the data segment memory location accessed by the sum of ARRAY, EBX, and ECX.
Scaled-index addressing	Is available only in the 80386 through the Pentium 4 microprocessor. The second register of a pair of registers is modified by the scale factor of 2X, 4X, or 8X to generate the operand memory address. (Example: a MOV EDX,[EAX+4*EBX] instruction loads EDX from the data segment memory location addressed by EAX plus 4 times EBX.) Scaling allows access to word (2X), doubleword (4X), or quadword (8X) memory array data. Note that a scaling factor of 1X also exists, but it is normally implied and does not appear in the instruction. The MOV AL,[EBX+ECX] is an example in which the scaling factor is a one. Alternately, the instruction can be rewritten as MOV AL,[EBX+1*ECX]. Another example is a MOV AL,[2*EBX] instruction, which uses only one scaled register to address memory.

Register Addressing

Register addressing is the most common form of data addressing and, once the register names are learned, is the easiest to apply. The microprocessor contains the following 8-bit registers used with register addressing: AH, AL, BH, BL, CH, CL, DH, and DL. Also present are the following 16-bit registers: AX, BX, CX, DX, SP, BP, SI, and DI. In the 80386 and above, the extended 32-bit registers are EAX, EBX, ECX, EDX, ESP, EBP, EDI, and ESI. With register addressing, some MOV instructions, and the PUSH and POP instructions, also use the 16-bit segment registers (CS, ES, DS, SS, FS, and GS). It is important for instructions to use registers that are the same size. *Never* mix an 8-bit register with a 16-bit register, an 8-bit register with a 32-bit register, or a 16-bit register with 32-bit register because this is not allowed by the microprocessor and results in an error when assembled. This is even true when a MOV AX,AL or a MOV EAX,AL instruction may seem to make sense. Of course, the MOV AX,AL or MOV EAX,AL instruction is *not* allowed because these registers are of different sizes. Note that a few instructions, such as SHL DX,CL, are exceptions to this rule, as indicated in later chapters. It is also important to note that *none* of the MOV instructions affect the flag bits.

Table 3–1 shows many variations of register move instructions. It is impossible to show all combinations because there are too many. For example, just the 8-bit subset of the MOV instruction has 64 different variations. A segment-to-segment register MOV instruction is about the only type of register MOV instruction *not* allowed. Note that the code segment register is not normally changed by a MOV instruction because the address of the next instruction is found in both IP/EIP and CS. If only CS were changed, the address of the next instruction would be unpredictable. Therefore, changing the CS register with a MOV instruction is not allowed.

Figure 3–3 shows the operation of the MOV BX,CX instruction. Note that the source register's contents do not change, but the destination register's contents do change. The instruction moves *(copies)* a 1234H from register CX into register BX. This *erases* the old contents (76AFH) of register BX, but the contents of CX remain unchanged. The contents of the destination register or destination memory location change for all instructions except the CMP and TEST instructions. Note that the MOV BX,CX instruction does not affect the leftmost 16 bits of register EBX.

TABLE 3–1 Examples of the register-addressed instructions.

Assembly Language	Size	Operation
MOV AL,BL	8-bits	Copies BL into AL
MOV CH,CL	8-bits	Copies CL into CH
MOV AX,CX	16-bits	Copies CX into AX
MOV SP,BP	16-bits	Copies BP into SP
MOV DS,AX	16-bits	Copies AX into DS
MOV SI,DI	16-bits	Copies DI into SI
MOV BX,ES	16-bits	Copies ES into BX
MOV ECX,EBX	32-bits	Copies EBX into ECX
MOV ESP,EDX	32-bits	Copies EDX into ESP
MOV ES,DS	—	Not allowed (segment-to-segment)
MOV BL,DX	—	Not allowed (mixed sizes)
MOV CS,AX	—	Not allowed (the code segment register may not be the destination register)

FIGURE 3–3 The effect of executing the MOV BX, CX instruction at the point just before the BX register changes. Note that only the rightmost 16 bits of register EBX change.

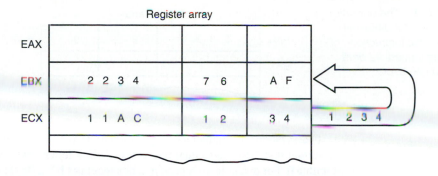

Example 3–1 shows a sequence of assembled instructions that copy various data between 8-, 16-, and 32-bit registers. As mentioned, the act of moving data from one register to another only changes the destination register, never the source. The last instruction in this example (MOV CS,AX) assembles without error, but causes problems if executed. If only the contents of CS change without changing IP, the next step in the program is unknown and therefore causes the program to go awry.

EXAMPLE 3–1

```
0000 8B C3              MOV AX,BX        ;copy contents of BX into AX
0002 8A CE              MOV CL,DH        ;copy the contents of DH into CL
0004 8A CD              MOV CL,CH        ;copy the contents of CH into CL
0006 66|8B C3           MOV EAX,EBX      ;copy the contents of EBX into EAX
0009 66|8B D8           MOV EBX,EAX      ;copy EAX into EBX, ECX, and EDX
000C 66|8B C8           MOV ECX,EAX
000F 66|8B D0           MOV EDX,EAX
0012 8C C8              MOV AX,CS        ;copy CS into DS
0014 8E D8              MOV DS,AX
0016 8E C8              MOV CS,AX        ;assembles, but will cause problems
```

Immediate Addressing

Another data-addressing mode is immediate addressing. The term *immediate* implies that the data immediately follow the hexadecimal opcode in the memory. Also note that immediate data are **constant data,** while the data transferred from a register are **variable data.** Immediate addressing operates upon a byte or word of data. In the 80386 through the Pentium 4 microprocessors, immediate addressing also operates on doubleword data. The MOV immediate instruction transfers a copy of the immediate data into a register or a memory location. Figure 3–4 shows the operation of a MOV EAX,13456H instruction. This instruction copies the 13456H from the instruction, located in the memory immediately following the hexadecimal opcode, into register EAX. As with the MOV instruction illustrated in Figure 3–3, the source data overwrites the destination data.

In symbolic assembly language, the symbol # precedes immediate data in some assemblers.[2] The MOV AX,#3456H instruction is an example. Most assemblers do not use the # symbol, but represent immediate data as in the MOV AX,3456H instruction. In this text, the # symbol is not used for immediate data. The most common assemblers—Intel ASM, Microsoft MASM,[3] and Borland TASM[4]—do not use the # symbol for immediate data, but an older assembler used with some Hewlett-Packard logic development systems do, as may others.

[2]This is true for the assembler provided by Hewlett-Packard in some development systems.

[3]MASM (MACRO assembler) is a trademark of Microsoft Corporation.

[4]TASM (Turbo assembler) is a trademark of Borland Corporation.

FIGURE 3–4 The operation of the MOV EAX,3456H instruction. This instruction copies the immediate data (13456H) into EAX.

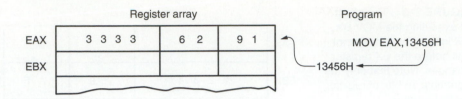

The symbolic assembler portrays immediate data in many ways. The letter H appends hexadecimal data. If hexadecimal data begin with a letter, the assembler requires that the data start with a 0. For example, to represent a hexadecimal F2, a 0F2H is used in assembly language. In some assemblers (though not in MASM, TASM, or this text), hexadecimal data are represented with an 'h, as in MOV AX,#'h1234. Decimal data are represented as is and require no special codes or adjustments. (An example is the 100 decimal in the MOV AL,100 instruction.) An ASCII-coded character or characters may be depicted in the immediate form if the ASCII data are enclosed in apostrophes. (An example is the MOV BH,'A' instruction, which moves an ASCII-coded A (41H) into register BH.) Be careful to use the apostrophe (') for ASCII data and not the single quotation mark (`). Binary data are represented if the binary number is followed by the letter B, or, in some assemblers, the letter Y. Table 3–2 shows many different variations of MOV instructions that apply immediate data.

Example 3–2 shows various immediate instructions in a short program that places a 0000H into the 16-bit registers AX, BX, and CX. This is followed by instructions that use register addressing to copy the contents of AX into registers SI, DI, and BP. This is a complete program that uses programming models for assembly and execution. The .MODEL TINY statement directs the assembler to assemble the program into a single code segment. The .CODE statement or directive indicates the start of the code segment; the .STARTUP statement indicates the starting instruction in the program; and the .EXIT statement causes the program to exit to DOS. The END statement indicates the end of the program file. This program is assembled with MASM and executed with CodeView[5] (CV) to view its execution. Note that the most recent version of TASM will also accept MASM code. To store the program into the system use either the DOS EDIT program or Programmer's WorkBench[6] (PWB). Note that a TINY program always assembles as a command (.COM) program.

TABLE 3–2 Examples of immediate addressing using the MOV instruction.

Assembly Language	Size	Operation
MOV BL,44	8-bits	Copies a 44 decimal (2CH) into BL
MOV AX,44H	16-bits	Copies a 0044H into AX
MOV SI,0	16-bits	Copies a 0000H into SI
MOV CH,100	8-bits	Copies a 100 decimal (64H) into CH
MOV AL,'A'	8-bits	Copies an ASCII A into AL
MOV AX,'AB'	16-bits	Copies an ASCII BA* into AX
MOV CL,11001110B	8-bits	Copies a 11001110 binary into CL
MOV EBX,12340000H	32-bits	Copies a 12340000H into EBX
MOV ESI,12	32-bits	Copies a 12 decimal into ESI
MOV EAX,100Y	32-bits	Copies a 100 binary into EAX

*Note: This is not an error. The ASCII characters are stored as a BA, so care should be exercised when using a word-sized pair of ASCII characters.

---

[5]CodeView is a registered trademark of Microsoft Corporation.

[6]Programmer's WorkBench is a registered trademark of Microsoft Corporation.

EXAMPLE 3–2

```
                        .MODEL TINY        ;choose single segment model
0000                    .CODE              ;indicate start of code segment

                        .STARTUP           ;indicate start of program

0100  B8 0000           MOV    AX,0        ;place 0000H into AX
0103  BB 0000           MOV    BX,0000H    ;place 0000H into BX
0106  B9 0000           MOV    CX,0        ;place 0000H into CX

0109  8B F0             MOV    SI,AX       ;copy AX into SI
010B  8B F8             MOV    DI,AX       ;copy AX into DI
010D  8B E8             MOV    BP,AX       ;copy AX into BP

                        .EXIT              ;exit to DOS
                        END                ;end of file
```

Each statement in a program consists of four parts or fields, as illustrated in Example 3–3. The leftmost field is called the *label* and it is used to store a symbolic name for the memory location that it represents. All labels must begin with a letter or one of the following special characters: @, $, _, or ?. A label may be of any length from 1 to 35 characters. The label appears in a program to identify the name of a memory location for storing data and for other purposes that are explained as they appear. The next field is called the *opcode field;* it is designed to hold the instruction, or opcode. The MOV part of the move data instruction is an example of an opcode. To the right of the opcode field is the *operand field,* which contains information used by the opcode. For example, the MOV AL,BL instruction has the opcode MOV and operands AL and BL. Note that some instructions contain between zero and three operands. The final field, the *comment field,* contains a comment about an instruction or a group of instructions. A comment always begins with a semicolon (;).

EXAMPLE 3–3

```
LABEL           OPCODE   OPERAND      COMMENT

DATA1           DB       23H          ;define DATA1 as a byte of 23H
DATA2           DW       1000H        ;define DATA2 as a word of 1000H

START:          MOV      AL,BL        ;copy BL into AL
                MOV      BH,AL        ;copy AL into BH
                MOV      CX,200       ;copy 200 decimal into CX
```

When the program is assembled and the **list** (.LST) file is viewed, it appears as the program listed in Example 3–2. The hexadecimal number at the far left is the offset address of the instruction or data. This number is generated by the assembler. The number or numbers to the right of the offset address are the machine-coded instructions or data that are also generated by the assembler. For example, if the instruction MOV AX,0 appears in a file and it is assembled, it appears in offset memory location 0100 in Example 3–2. Its hexadecimal machine language form is B8 0000. The B8 is the opcode in machine language and the 0000 is the 16-bit wide data with a value of zero. When the program was written, only the MOV AX,0 was typed into the editor; the assembler generated the machine code and addresses, and stored the program in a file ending with the extension .LST. Note that all programs shown in this text are in the form generated by the assembler.

Direct Data Addressing

Most instructions can use the direct data-addressing mode. In fact, direct data addressing is applied to many instructions in a typical program. There are two basic forms of direct data addressing: (1) **direct addressing,** which applies to a MOV between a memory location and AL, AX, or EAX, and (2) **displacement addressing,** which applies to almost any instruction in the instruction set. In either case, the address is formed by adding the displacement to the default data segment address or an alternate segment address.

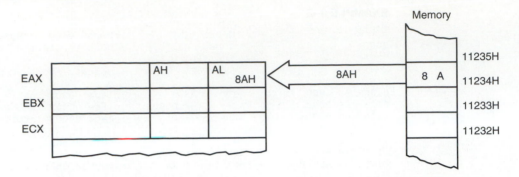

FIGURE 3–5   The operation of the MOV AL,[1234H] instruction when DS = 1000H.

Direct Addressing.   Direct addressing with a MOV instruction transfers data between a memory location, located within the data segment, and the AL (8-bit), AX (16-bit), or EAX (32-bit) register. A MOV instruction using this type of addressing is usually a 3-byte long instruction. (In the 80386 and above, a register size prefix may appear before the instruction, causing it to exceed three bytes in length.)

The MOV AL,DATA instruction, as represented by most assemblers, loads AL from data segment memory location DATA (1234H). Memory location DATA is a symbolic memory location, while the 1234H is the actual hexadecimal location. With many assemblers, this instruction is represented as a MOV AL,[1234H][7] instruction. The [1234H] is an absolute memory location that is not allowed by all assembler programs. Note that this may need to be formed as MOV AL,DS:[1234H] with some assemblers, to show that the address is in the data segment. Figure 3–5 shows how this instruction transfers a copy of the byte-sized contents of memory location 11234H into AL. The effective address is formed by adding 1234H (the offset address) to 1000H (the data segment address of 1000H) in a system operating in the real mode.

Table 3–3 lists the three direct addressed instructions. These instructions often appear in programs, so Intel decided to make them special three-byte long instructions to reduce the length

TABLE 3–3   Direct addressed instructions using EAX, AX, and AL.

Assembly Language	Size	Operation
MOV AL,NUMBER	8-bits	Copies the byte contents of data segment memory location NUMBER into AL
MOV AX,COW	16-bits	Copies the word contents of data segment memory location COW into AX
MOV EAX,WATER*	32-bits	Copies the doubleword contents of memory location WATER into EAX
MOV NEWS,AL	8-bits	Copies AL into data segment memory location NEWS
MOV THERE,AX	16-bits	Copies AX into data segment memory location THERE
MOV HOME,EAX*	32-bits	Copies EAX into data segment memory location HOME
MOV ES:[2000 H],AL	8-bits	Copies AL into extra data segment memory location 2000H

*Note: The 80386–Pentium 4 microprocessors will sometimes use more than three bytes of memory for the 32-bit move between EAX and memory.

---

[7]This form may be used with MASM, but most often appears when a program is entered or listed by DEBUG, a debugging toll provided with DOS.

of programs. All other instructions that move data from a memory location to a register, called **displacement-addressed instructions,** require four or more bytes of memory for storage in a program.

Displacement Addressing. Displacement addressing is almost identical to direct addressing, except that the instruction is four bytes wide instead of three. In the 80386 through the Pentium 4, this instruction can be up to seven bytes wide if a 32-bit register and a 32-bit displacement are specified. This type of direct data addressing is much more flexible because most instructions use it.

If the operation of the MOV CL,DS:[1234H] instruction is compared to that of the MOV AL,DS:[1234H] instruction of Figure 3–5, both basically perform the same operation except for the destination register (CL versus AL). Another difference only becomes apparent upon examining the assembled versions of these two instructions. The MOV AL,DS:[1234H] instruction is three bytes long and the MOV CL,DS:[1234H] instruction is four bytes long, as illustrated in Example 3–4. This example shows how the assembler converts these two instructions into hexadecimal machine language. You must include the segment register DS: in this example, before the [offset] part of the instruction. You may use any segment register, but, in most cases, data are stored in the data segment, so this example uses DS:[1234H].

EXAMPLE 3–4

```
0000   A0 1234 R          MOV AL,DS:[1234H]
0003   8A 0E 1234 R       MOV CL,DS:[1234H]
```

Table 3–4 lists some MOV instructions, using the displacement form of direct addressing. Not all variations are listed because there are many MOV instructions of this type. The segment registers can be stored or loaded from memory.

Example 3–5 shows a short program using models that address information in the data segment. Note that the **data segment** begins with a .DATA statement to inform the assembler where the data segment begins. The model size is adjusted from TINY, as shown in Example 3–3, to SMALL so that a data segment can be included. The **SMALL model** allows one data segment and one code segment. The SMALL model is often used whenever memory data are required for a program. A SMALL model program assembles as an execute (.EXE) program. Notice how this example allocates memory locations in the data segment by using the DB and DW directives. Here the .STARTUP statement not only indicates the start of the code, but it also loads the data segment register with the segment address of the data segment. If this program is assembled and

TABLE 3–4 Examples of direct data addressing using a displacement.

Assembly Language	Size	Operation
MOV CH,DOG	8-bits	Copies the byte contents of data segment memory location DOG into CH
MOV CH,[1000H]*	8-bits	Copies the byte contents of data segment offset address 1000H into CH
MOV ES,DATA6	16-bits	Copies the word contents of data segment memory location DATA6 into ES
MOV DATA7,BP	16-bits	Copies BP into data segment memory location DATA7
MOV NUMBER,SP	16-bits	Copies SP into data segment memory location NUMBER
MOV DATA1,EAX	32-bits	Copies EAX into data segment memory location DATA1
MOV EDI,SUM1	32-bits	Copies the doubleword contents of data segment memory location SUM1 into EDI

*Note: This form of addressing is seldom used with most assemblers because an actual numeric offset address is rarely accessed.

executed with CodeView, the instructions can be viewed as they execute and change registers and memory locations.

EXAMPLE 3–5

```
                        .MODEL  SMALL         ;select SMALL model
0000                    .DATA                 ;indicate start of DATA segment

0000 10         DATA1   DB      10H           ;place 10H in DATA1
0001 00         DATA2   DB      0             ;place 0 in DATA2
0002 0000       DATA3   DW      0             ;place 0 in DATA3
0004 AAAA       DATA4   DW      0AAAAH        ;place AAAAH in DATA4

0000                    .CODE                 ;indicate start of CODE segment
                        .STARTUP              ;indicate start of program

0017 A0 0000 R          MOV     AL,DATA1      ;copy DATA1 to AL
001A 8A 26 0001 R       MOV     AH,DATA2      ;copy DATA2 to AH
001E A3 0002 R          MOV     DATA3,AX      ;save AX at DATA3
0021 8B 1E 0004 R       MOV     BX,DATA4      ;load BX with DATA4

                        .EXIT                 ;exit to DOS
                        END                   ;end file
```

Register Indirect Addressing

Register indirect addressing allows data to be addressed at any memory location through an offset address held in any of the following registers: BP, BX, DI, and SI. For example, if register BX contains a 1000H and the MOV AX,[BX] instruction executes, the word contents of data segment offset address 1000H are copied into register AX. If the microprocessor is operated in the real mode and DS = 0100H, this instruction addresses a word stored at memory bytes 2000H and 2001H, and transfers it into register AX (see Figure 3–6). Note that the contents of 2000H are moved into AL and the contents of 2001H are moved into AH. The [ ] symbols denote indirect addressing in assembly language. In addition to using the BP, BX, DI, and SI registers to indirectly address memory, the 80386 and above allow register indirect addressing with any extended register except ESP. Some typical instructions using indirect addressing appear in Table 3–5.

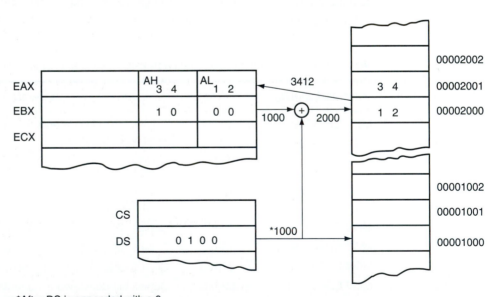

*After DS is appended with a 0.

FIGURE 3–6   The operation of the MOV AX,[BX] instruction when BX = 1000H and DS = 0100H. Note that this instruction is shown after the contents of memory are transferred to AX.

TABLE 3–5 Example of register indirect addressing.

Assembly Language	Size	Operation
MOV CX,[BX]	16-bits	Copies the word contents of the data segment memory location address by BX into CX
MOV [BP],DL*	8-bits	Copies DL into the stack segment memory location addressed by BP
MOV [DI],BH	8-bits	Copies BH into the data segment memory location addressed by DI
MOV [DI],[BX]	—	Memory-to-memory moves are not allowed except with string instructions
MOV AL,[EDX]	8-bits	Copies the byte contents of the data segment memory location addressed by EDX into AL
MOV ECX,[EBX]	32-bits	Copies the doubleword contents of the data segment memory location addressed by EBX into ECX

*Note: Data addressed by BP or EBP are by default located in the stack segment, while all other indirect addressing modes use the data segment by default.

The **data segment** is used by default with register indirect addressing or any other addressing mode that uses BX, DI, or SI to address memory. If the BP register addresses memory, the **stack segment** is used by default. These settings are considered the default for these four index and base registers. For the 80386 and above, EBP addresses memory in the stack segment by default; EAX, EBX, ECX, EDX, EDI, and ESI address memory in the data segment by default. When using a 32-bit register to address memory in the real mode, the contents of the 32-bit register must never exceed 0000FFFFH. In the protected mode, any value can be used in a 32-bit register that is used to indirectly address memory, as long as it does not access a location outside of the segment, as dictated by the access rights byte. An example 80386/80486/Pentium 4 instruction is MOV EAX,[EBX]. This instruction loads EAX with the doubleword-sized number stored at the data segment offset address indexed by EBX.

In some cases, indirect addressing requires specifying the size of the data are specified with the **special assembler directive** BYTE PTR, WORD PTR, or DWORD PTR. These directives indicate the size of the memory data addressed by the memory pointer (PTR). For example, the MOV AL,[DI] instruction is clearly a byte-sized move instruction, but the MOV [DI],10H instruction is ambiguous. Does the MOV [DI],10H instruction address a byte-, word-, or doubleword-sized memory location? The assembler can't determine the size of the 10H. The instruction MOV BYTE PTR [DI],10H clearly designates the location addressed by DI as a byte-sized memory location. Likewise, the MOV DWORD PTR [DI],10H clearly identifies the memory location as doubleword-sized. The BYTE PTR, WORD PTR, and DWORD PTR directives are used only with instructions that address a memory location through a pointer or index register with immediate data, and for a few other instructions that are described in subsequent chapters.

Indirect addressing often allows a program to refer to tabular data located in the memory system. For example, suppose that you must create a table of information that contains 50 samples taken from memory location 0000:046C. Location 0000:046C contains a counter that is maintained by the personal computer's real-time clock. Figure 3–7 shows the table and the BX register used to sequentially address each location in the table. To accomplish this task, load the starting location of the table into the BX register with a MOV immediate instruction. After initializing the starting address of the table, use register indirect addressing to store the 50 samples sequentially.

The sequence shown in Example 3–6 loads register BX with the starting address of the table and initializes the count, located in register CX, to 50. The OFFSET directive tells the assembler to load BX with the offset address of memory location TABLE, not the contents of

FIGURE 3–7 An array
(TABLE) containing 50 bytes
that are indirectly addressed
through register BX.

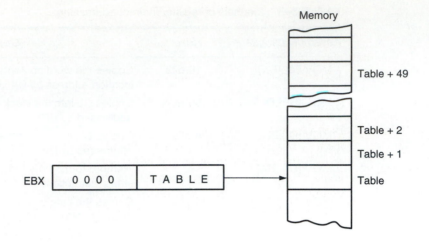

TABLE. For example, the MOV BX,DATAS instruction copies the contents of memory location
DATAS into BX, while the MOV BX,OFFSET DATAS instruction copies the offset address of
DATAS into BX. When the OFFSET directive is used with the MOV instruction, the assembler
calculates the offset address and then uses a MOV immediate instruction to load the address into
the specified 16-bit register.

EXAMPLE 3–6

```
                              .MODEL  SMALL                ;select SMALL model
0000                          .DATA                        ;start of DATA segment

0000   0032 [        DATAS   DW      50 DUP (?)            ;setup array of 50 bytes
              0000
                    ]

0000                          .CODE                        ;start of CODE segment
                              .STARTUP                     ;start of program

0017   B8 0000                MOV      AX,0
001A   8E C0                  MOV      ES,AX               ;address segment 0000 with ES

001C   BB 0000 R              MOV      BX,OFFSET DATAS     ;address DATAS array
001F   B9 0032                MOV      CX,50               ;load counter with 50
0022                AGAIN:
0022   26:A1 046C             MOV      AX,ES:[046CH]       ;get clock value
0026   89 07                  MOV      [BX],AX             ;save clock value in DATAS
0028   43                     INC      BX                  ;increment BX to next element
0029   E2 F7                  LOOP     AGAIN               ;repeat 50 times

                              .EXIT                        ;exit to DOS
                              END                          ;end file
```

Once the counter and pointer are initialized, a repeat-until CX = 0 loop executes. Here,
data are read from extra segment memory location 46CH with the MOV AX,ES:[046CH] instruction and stored in memory that is indirectly addressed by the offset address located in register BX. Next, BX is incremented (one is added to BX) to the next table location, and finally the
LOOP instruction repeats the LOOP 50 times. The LOOP instruction decrements (subtracts one
from) the counter (CX); if CX is not zero, LOOP causes a jump to memory location AGAIN. If
CX becomes zero, no jump occurs and this sequence of instructions ends. This example copies
the most recent 50 values from the clock into the memory array DATAS. This program will
often show the same data in each location because the contents of the clock are changed only
18.2 times per second. To view the program and its execution, use the CodeView program. To

use CodeView, type CV FILE.EXE or access it as DEBUG from the Programmer's WorkBench program under the RUN menu. Note that CodeView functions only with .EXE or .COM files. Some useful CodeView switches are /50 for a 50-line display and /S for use of high-resolution video displays in an application. To debug the file TEST.COM with 50 lines, type CV /50 TEST.COM at the DOS prompt.

Base-Plus-Index Addressing

Base-plus-index addressing is similar to indirect addressing because it indirectly addresses memory data. In the 8086 through the 80286, this type of addressing uses one base register (BP or BX), and one index register (DI or SI) to indirectly address memory. The base register often holds the beginning location of a memory array, while the index register holds the relative position of an element in the array. Remember that whenever BP addresses memory data, both the stack segment register and BP generate the effective address.

In the 80386 and above, this type of addressing allows the combination of any two 32-bit extended registers except ESP. For example, the MOV DL,[EAX+EBX] instruction is an example using EAX (as the base) plus EBX (as the index). If the EBP register is used, the data are located in the stack segment instead of in the data segment.

Locating Data with Base-plus-index Addressing. Figure 3–8 shows how data are addressed by the MOV DX,[BX+DI] instruction when the microprocessor operates in the real mode. In this example, BX = 1000H, DI = 0010H, and DS = 0100H, which translate into memory address 02010H. This instruction transfers a copy of the word from location 02010H into the DX register. Table 3–6 lists some instructions used for base-plus-index addressing. Note that the Intel assembler requires that this addressing mode appear as [BX][DI] instead of [BX+DI]. The MOV DX,[BX+DI] instruction is MOV DX,[BX][DI] for a program written for the Intel ASM assembler. This text uses the first form in all example programs, but the second form can be used in many assemblers, including MASM from Microsoft. Instructions like MOV DI,[BX+DI] will assemble, but will not execute correctly.

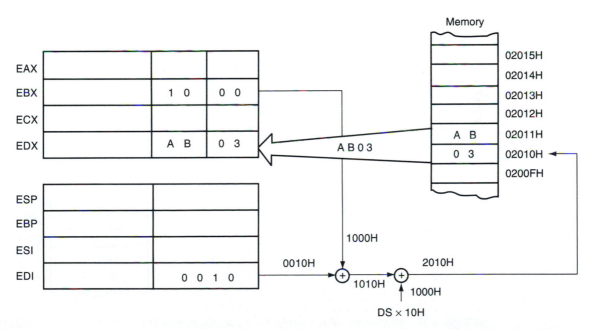

FIGURE 3–8 An example showing how the base-plus-index addressing mode functions for the MOV DX,[BX+DI] instruction. Notice that memory address 02010H is accessed because DS = 0100H, BX = 100H, and DI = 0010H.

TABLE 3–6   Examples of base-plus-index addressing.

Assembly Language	Size	Operation
MOV CX,[BX+DI]	16-bits	Copies the word contents of the data segment memory location address by BX plus DI into CX
MOV CH,[BP+SI]	8-bits	Copies the byte contents of the stack segment memory location addressed by BP plus SI into CH
MOV [BX+SI],SP	16-bits	Copies SP into the data segment memory location addresses by BX plus SI
MOV [BP+DI],AH	8-bits	Copies AH into the stack segment memory location addressed by BP plus DI
MOV CL,[EDX+EDI]	8-bits	Copies the byte contents of the data segment memory location addressed by EDX plus EDI into CL
MOV [EAX+EBX],ECX	32-bits	Copies ECX into the data segment memory location addressed by EAX plus EBX

Locating Array Data Using Base-plus-index Addressing.   A major use of the base-plus-index addressing mode is to address elements in a memory array. Suppose that the elements in an array, located in the data segment at memory location ARRAY, must be accessed. To accomplish this, load the BX register (base) with the beginning address of the array, and the DI register (index) with the element number to be accessed. Figure 3–9 shows the use of BX and DI to access an element in an array of data.

A short program, listed in Example 3–7, moves array element 10H into array element 20H. Notice that the array element number, loaded into the DI register, addresses the array element. Also notice how the contents of the ARRAY have been initialized so that element 10H contains a 29H.

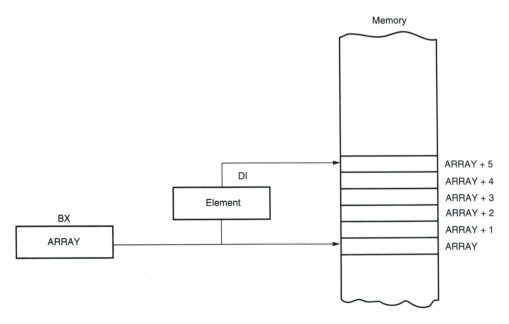

FIGURE 3–9   An example of the base-plus-index addressing mode. Here an element (DI) of an ARRAY (BX) is addressed.

EXAMPLE 3–7

```
                            .MODEL SMALL              ;select SMALL model
0000                        .DATA                     ;start of DATA segment

0000  0010 [        ARRAY   DB     16 DUP (?)         ;setup ARRAY
            00
        ]
0010  29                    DB     29H                ;sample data at element 10H
0011  001E [               DB     30 DUP (?)
            00
        ]
0000                        .CODE                     ;start of CODE segment
                            .STARTUP                  ;start of program

0017  BB 0000 R             MOV    BX,OFFSET ARRAY    ;address ARRAY
001A  BF 0010               MOV    DI,10H             ;address element 10H
001D  8A 01                 MOV    AL,[BX+DI]         ;get element 10H
001F  BF 0020               MOV    DI,20H             ;address element 20H
0022  88 01                 MOV    [BX+DI],AL         ;save in element 20H

                            .EXIT                     ;exit to DOS
                            END                       ;end of file
```

Register Relative Addressing

Register relative addressing is similar to base-plus-index addressing and displacement addressing. In register relative addressing, the data in a segment of memory are addressed by adding the displacement to the contents of a base or an index register (BP, BX, DI, or SI). Figure 3–10 shows the operation of the MOV AX,[BX+1000H] instruction. In this example, BX = 0100H and DS = 0200H, so the address generated is the sum of DS × 10H, BX, and the displacement of 1000H or 03100H. Remember that BX, DI, or SI addresses the data segment and BP addresses the stack segment. In the 80386 and above, the displacement can be a 32-bit number and the register can be any 32-bit register except the ESP register. Remember that the size of a real mode segment is 64K bytes long. Table 3–7 lists a few instructions that use register relative addressing.

The displacement can be a number added to the register within the [ ], as in the MOV AL,[DI+2] instruction, or it can be a displacement subtracted from the register, as in MOV AL,[SI–1]. A displacement also can be an offset address appended to the front of the [ ], as in MOV AL,DATA[DI]. Both forms of displacements also can appear simultaneously, as in the MOV AL,DATA[DI+3] instruction. In all cases, both forms of the displacement add to the base, or base and index register within the [ ]. In the 8086–80286 microprocessors, the value of the displacement is limited to a 16-bit signed number with a value ranging between +32,767 (7FFFH)

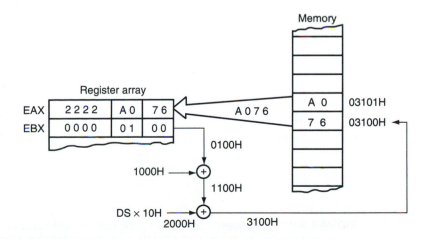

FIGURE 3–10 The operation of the MOV AX,[BX+1000H] instruction, when BX = 0100H and DS = 0200H.

TABLE 3–7   Examples of register relative addressing.

Assembly Language	Size	Operation
MOV AX,[DI+100H]	16-bits	Copies the word contents of the data segment memory location addressed by DI plus 100H into AX
MOV ARRAY[SI],BL	8-bits	Copies BL into the data segment memory location addressed by ARRAY plus SI
MOV LIST[SI+2],CL	8-bits	Copies CL into the data segment memory location addressed by sum of LIST, SI, and 2
MOV DI,SET_IT[BX]	16-bits	Copies the word contents of the data segment memory location addressed by the sum of SET_IT and BX into DI
MOV DI,[EAX+10H]	16-bits	Copies the word contents of the data segment memory location addressed by the sum of EAX and 10H into DI
MOV ARRAY[EBX],EAX	32-bits	Moves EAX into the data segment memory location addressed by the sum of ARRAY and EBX

and –32,768 (8000H); in the 80386 and above, a 32-bit displacement is allowed with a value ranging between +2,147,483,647 (7FFFFFFFH) and –2,147,483,648 (80000000H).

Addressing Array Data with Register Relative.   It is possible to address array data with register relative addressing, such as one does with base-plus-index addressing. In Figure 3–11, register relative addressing is illustrated with the same example as for base-plus-index addressing. This shows how the displacement ARRAY adds to index register DI to generate a reference to an array element.

Example 3–8 shows how this new addressing mode can transfer the contents of array element 10H into array element 20H. Notice the similarity between this example and Example 3–7. The main difference is that, in Example 3–8, register BX is not used to address memory area ARRAY; instead, ARRAY is used as a displacement to accomplish the same task.

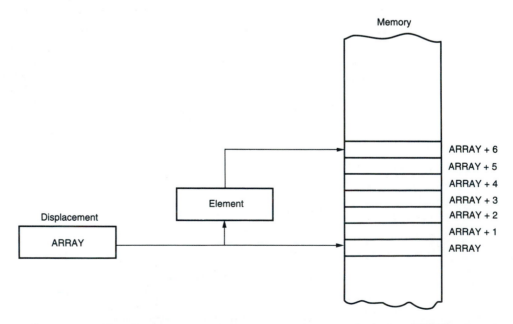

FIGURE 3–11   Register relative addressing used to address an element of ARRAY. The displacement addresses the start of ARRAY, and DI accesses an element.

EXAMPLE 3–8

```
                              .MODEL SMALL              ;select SMALL model
0000                          .DATA                     ;start of DATA segment

0000  0010 [          ARRAY   DB     16 DUP (?)          ;setup ARRAY
            00
         ]
0010  29                      DB     29H                 ;sample data at element 10H
0011  001E [                  DB     30 DUP (?)
            00
         ]
0000                          .CODE                     ;start of CODE segment
                              .STARTUP                  ;start of program

0017  BF 0010                 MOV    DI,10H              ;address element 10H
001A  8A 85 0000 R            MOV    AL,ARRAY[DI]        ;get element 10H
001E  BF 0020                 MOV    DI,20H              ;address element 20H
0021  88 85 0000 R            MOV    ARRAY[DI],AL        ;save in element 20H

                              .EXIT                     ;exit to DOS
                              END                       ;end of file
```

Base Relative-Plus-Index Addressing

The base relative-plus-index addressing mode is similar to the base-plus-index addressing mode, but it adds a displacement, besides using a base register and an index register, to form the memory address. This type of addressing mode often addresses a two-dimensional array of memory data.

Addressing Data with Base Relative-plus-index. Base relative-plus-index addressing is the least-used addressing mode. Figure 3–12 shows how data are referenced if the instruction executed by the microprocessor is a MOV AX,[BX+SI+100H]. The displacement of 100H adds to BX and SI to form the offset address within the data segment. Registers BX = 0020H, SI = 0010H, and DS = 1000H, so the effective address for this instruction is 10130H–the sum of these registers plus a displacement of 100H. This addressing mode is too complex for frequent use in a program. Some typical instructions using base relative-plus-index addressing appear in Table 3–8. Note that with the 80386 and above, the effective address is generated by the sum of two 32-bit registers plus a 32-bit displacement.

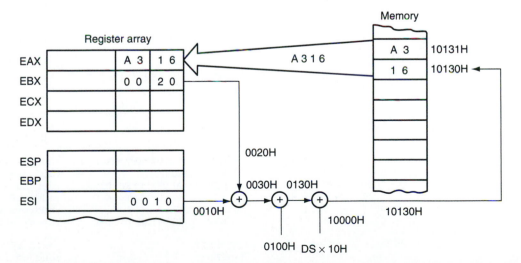

FIGURE 3–12   An example of base relative-plus-index addressing using a MOV AX,[BX+SI+100H] instruction. Note: DS = 1000H.

TABLE 3–8   Example base relative-plus-index instructions.

Assembly Language	Size	Operation
MOV DH,[BX+DI+20H]	8-bits	Copies the byte contents of the data segment memory location addressed by the sum of BX, DI, and 20H into DH
MOV AX,FILE[BX+DI]	16-bits	Copies the word contents of the data segment memory location addressed by the sum of FILE, BX, and DI into AX
MOV LIST[BP+DI],CL	8-bits	Copies CL into the stack segment memory location addressed by the sum of LIST, BP, and DI
MOV LIST[BP+SI+4],DH	8-bits	Copies DH into the stack segment memory location addressed by the sum of LIST, BP, SI, and 4
MOV EAX,FILE[EBX+ECX+2]	32-bits	Copies the doubleword contents of the data segment memory location addressed by the sum of FILE, EBX, ECX, and 2 into EAX

Addressing Arrays with Base Relative-plus-Index.   Suppose that a file of many records exists in memory and each record contains many elements. This displacement addresses the file, the base register addresses a record, and the index register addresses an element of a record. Figure 3–13 illustrates this very complex form of addressing.

Example 3–9 provides a program that copies element 0 of record A into element 2 of record C by using the base relative-plus-index mode of addressing. This example FILE contains four records and each record contains 10 elements. Notice how the THIS BYTE statement is used to define the label FILE and RECA as the same memory location.

EXAMPLE 3–9

```
                        .MODEL SMALL                    ;SMALL model
0000                    .DATA                           ;start of DATA segment

0000  = 0000      FILE  EQU      THIS BYTE              ;assign FILE to this byte

0000  000A [      RECA  DB       10 DUP (?)             ;reserve 10 bytes for RECA
         00
      ]
000A  000A [      RECB  DB       10 DUP (?)             ;reserve 10 bytes for RECB
         00
      ]
0014  000A [      RECC  DB       10 DUP (?)             ;reserve 10 bytes for RECC
         00
      ]
001E  000A [      RECD  DB       10 DUP (?)             ;reserve 10 bytes for RECD
         00
      ]

0000                    .CODE                           ;start of CODE segment
                        .STARTUP                        ;start of program

0017  BB 0000 R         MOV    BX,OFFSET RECA           ;address RECA
001A  BF 0000           MOV    DI,0                     ;address element 0
001D  8A 81 0000 R      MOV    AL,FILE[BX+DI]           ;get data
0021  BB 0014 R         MOV    BX,OFFSET RECC           ;address RECC
0024  BF 0002           MOV    DI,2                     ;address element 2
0027  88 81 0000 R      MOV    FILE[BX+DI],AL           ;save data

                        .EXIT                           ;exit to DOS
                        END                             ;end of file
```

FIGURE 3–13 Base relative-plus-index addressing used to access a FILE that contains multiple records (REC).

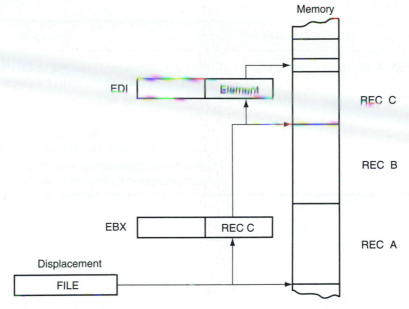

Scaled-Index Addressing

Scaled-index addressing is the last type of data-addressing mode discussed. This data-addressing mode is unique to the 80386 through the Pentium 4 microprocessors. Scaled-index addressing uses two 32-bit registers (a base register and an index register) to access the memory. The second register (index) is multiplied by a scaling factor. The scaling factor can be 1X, 2X, 4X, or 8X. A scaling factor of 1X is implied and need not be included in the assembly language instruction (MOV AL,[EBX+ECX]). A scaling factor of 2X is used to address word-sized memory arrays, a scaling factor of 4X is used with doubleword-sized memory arrays, and a scaling factor of 8X is used with quadword-sized memory arrays.

An example instruction is MOV AX,[EDI+2*ECX]. This instruction uses a scaling factor of 2X, which multiplies the contents of ECX by 2 before adding it to the EDI register to form the memory address. If ECX contains a 00000000H, word-sized memory element 0 is addressed; if ECX contains a 00000001H, word-sized memory element 1 is accessed, and so forth. This scales the index (ECX) by a factor of 2 for a word-sized memory array. Refer to Table 3–9 for some examples of scaled-index addressing. As you can imagine, there are an extremely large number of the scaled-index addressed register combinations. Scaling is also applied to instructions that use a single indirect register to access memory. The MOV EAX,[4*EDI] is a scaled-index instruction that uses one register to indirectly address memory.

Example 3–10 shows a sequence of instructions that uses scaled-index addressing to access a word-sized array of data called LIST. Note that the offset address of LIST is loaded into register EBX with the MOV EBX,OFFSET LIST instruction. Once EBX addresses array LIST, the elements (located in ECX) of 2, 4, and 7 of this word-wide array are added, using a scaling factor of 2 to access the elements. This program stores the 2 at element 2 into elements 4 and 7. Also notice the .386 directive to select the 80386 microprocessor. This directive must follow the .MODEL statement for the assembler to process 80386 instructions for DOS. If the 80486 is in use, the .486 directive appears after the .MODEL statement; if the Pentium, Pentium Pro, Pentium II, Pentium III, or Pentium 4 is in use, the .586 directive appears after .MODEL. If the microprocessor selection directive appears before the .MODEL statement, the microprocessor executes instructions in the 32-bit mode, which is not compatible with DOS.

TABLE 3–9   Examples of scaled-index addressing.

Assembly Language	Size	Operation
MOV EAX,[EBX+4*ECX]	32-bits	Copies the doubleword contents of the data segment memory location addressed by the sum of 4 times ECX plus EBX into EAX
MOV [EAX+2*EDI+100H],CX	16-bits	Copies CX into the data segment memory location addressed by the sum of EAX, 100H, and 2 times EDI
MOV AL,[EBP+2*EDI−2]	8-bits	Copies the byte contents of the stack segment memory location addressed by the sum of EBP, −2, and 2 times EDI into AL
MOV EAX,ARRAY[4*ECX]	32-bits	Copies the doubleword contents of the data segment memory location addressed by the sum of ARRAY plus 4 times ECX into EAX

EXAMPLE 3–10

```
                                        .MODEL SMALL           ;select SMALL model
                                        .386                   ;use the 80386
      0000                              .DATA                  ;start of DATA segment
      0000  0000 0001 0002   LIST   DW    0,1,2,3,4            ;define array list
            0003 0004
      000A  0005 0006 0007          DW    5,6,7,8,9
            0008 0009

      0000                              .CODE                  ;start of CODE segment
                                        .STARTUP               ;start of program
      0010  66| BB 00000000 R      MOV    EBX,OFFSET LIST       ;address array LIST

      0016  66| B9 00000002       MOV    ECX,2                 ;get element 2
      001C  67& 8B 04 4B          MOV    AX,[EBX+2*ECX]

      0020  66| B9 00000004       MOV    ECX,4                 ;store in element 4
      0026  67& 89 04 4B          MOV    [EBX+2*ECX],AX

      002A  66| B9 00000007       MOV    ECX,7                 ;store in element 7
      0030  67& 89 04 4B          MOV    [EBX+2*ECX],AX

                                        .EXIT                  ;exit to DOS
                                        END                    ;end of file
```

Data Structures

A data structure is used to specify how information is stored in a memory array and can be quite useful with applications that use arrays. It is best to think of a data structure as a template for data. The start of a structure is identified with the STRUC assembly language directive and the end with the ENDS statement. A typical data structure is defined and used three times in Example 3–11. Notice that the name of the structure appears with the STRUC and with ENDS statement.

EXAMPLE 3–11

```
                              ;Define INFO data structure
      0057                    INFO     STRUC

      0000  0020  [           NAMES    DB    32 DUP (?)          ;32 bytes for name
            00
                  ]
      0020  0020  [           STREET   DB    32 DUP (?)          ;32 bytes for street
            00
```

```
                      ]
0040  0010  [            CITY    DB     16 DUP (?)        ;16 bytes for city
              00
                      ]
0050  0002  [            STATE   DB     2 DUP (?)         ;2 bytes for state
              00
                      ]
0052  0005  [            ZIP     DB     5 DUP (?)         ;5 bytes for zip-code
              00
                      ]
                         INFO    ENDS

0000  42 6F 62 20 53 6D  NAME1   INFO   <'Bob Smith','123 Main Street','Wanda','OH','44444'>
      69 74 68
      0017 [
              00
                 ]
      31 32 33 20 4D
      61 69 6E 20 53 74
      72 65 65 74
      0011 [
              00
                 ]
      57 61 6E 64 61
      000B [
              00
                 ]
      4F 48 34 34 34
      34 34

0057  53 74 65 76 65 20  NAME2 INFO <'Steve Doe','222 Mouse Lane','Miller','PA','18100'>
      44 6F 65
      0017 [
              00
                 ]
      32 32 32 20 4D
      6F 75 73 65 20 4C
      61 6E 65
      0012 [
              00
                 ]
      4D 69 6C 6C 65
      72
      000A [
              00
                 ]
      50 41 31 38 31
      30 30

00AE  42 65 6E 20 44 6F  NAME3 INFO <'Jim Dover','303 Main Street','Orender','CA','90000'>
      76 65 72
      0017 [
              00
                 ]
      33 30 33 20 4D
      61 69 6E 20 53 74
      72 65 65 74
      0011 [
              00
                 ]
      4F 72 65 6E 64
      65 72
      0009 [
              00
                 ]
      43 41 39 30 30
      30 30
```

The data structure in Example 3–11 defines five fields of information. The first is 32 bytes long and holds a name; the second is 32 bytes long and holds a street address; the third is 16 bytes long for the city; the fourth is 2 bytes long for the state; the fifth is 5 bytes long for the ZIP Code. Once the structure is defined (INFO), it can be filled, as illustrated, with names and addresses. Three examples of uses for INFO are illustrated. Note that literals are surrounded with apostrophes and the entire field is surrounded with < > symbols when the data structure is used to define data.

When data are addressed in a structure, use the structure name and the field name to select a field from the structure. For example, to address the STREET in NAME2, use the operand NAME2.STREET, where the name of the structure is first followed by a period and then by the name of the field. Likewise, use NAME3.CITY to refer to the city in structure NAME3.

EXAMPLE 3–12

```
                          ;Clear names in array NAME1

0000   B9 0020                    MOV   CX,32
0003   B0 00                      MOV   AL,0
0005   BE 0000 R                  MOV   SI,OFFSET NAME1.NAMES
0008   F3/AA                      REP   STOSB

                          ;Clear street in array NAME2

000A   B9 0020                    MOV   CX,32
000D   B0 00                      MOV   AL,0
0010   BE 0077 R                  MOV   SI,OFFSET NAME2.STREET
0013   F3/AA                      REP   STOSB

                          ;Clear zip-code in array NAME3

0015   B9 0005                    MOV   CX,5
0018   B0 00                      MOV   AL,0
001A   BE 0100 R                  MOV   SI,OFFSET NAME3.ZIP
001D   F3/AA                      REP   STOSB
```

A short sequence of instructions appears in Example 3–12 that clears the name field in structure NAME1, the address field in structure NAME2, and the ZIP Code field in structure NAME3. The function and operation of the instructions in this program are defined in later chapters in the text. You may wish to refer to this example once these instructions are learned.

3–2   PROGRAM MEMORY-ADDRESSING MODES

Program memory-addressing modes, used with the JMP and CALL instructions, consist of three distinct forms: direct, relative, and indirect. This section introduces these three addressing forms, using the JMP instruction to illustrate their operation.

Direct Program Memory Addressing

Direct program memory addressing is what many early microprocessors used for all jumps and calls. Direct program memory addressing is also used in high-level languages, such as the BASIC language GOTO and GOSUB instructions. The microprocessor uses this form of addressing, but not as often as relative and indirect program memory addressing are used.

The instructions for direct program memory addressing store the address with the opcode. For example, if a program jumps to memory location 10000H for the next instruction, the address (10000H) is stored following the opcode in the memory. Figure 3–14 shows the direct

FIGURE 3–14 The 5-byte machine language version of a JMP [10000H] instruction.

Opcode	Offset (low)	Offset (high)	Segment (low)	Segment (high)
E A	0 0	0 0	0 0	1 0

intersegment JMP instruction and the four bytes required to store the address 10000H. This JMP instruction loads CS with 1000H and IP with 0000H to jump to memory location 10000H for the next instruction. (An **intersegment jump** is a jump to any memory location within the entire memory system.) The direct jump is often called a *far jump* because it can jump to any memory location for the next instruction. In the real mode, a far jump accesses any location within the first 1M byte of memory by changing both CS and IP. In protected mode operation, the far jump accesses a new code segment descriptor from the descriptor table, allowing it to jump to any memory location in the entire 4G-byte address range in the 80386 through Pentium 4 microprocessors.

The only other instruction that uses direct program addressing is the intersegment or far CALL instruction. Usually, the name of a memory address, called a *label,* refers to the location that is called or jumped to instead of the actual numeric address. When using a label with the CALL or JMP instruction, most assemblers select the best form of program addressing.

Relative Program Memory Addressing

Relative program memory addressing is not available in all early microprocessors, but it is available to this family of microprocessors. The term **relative** means "relative to the instruction pointer (IP)." For example, if a JMP instruction skips the next two bytes of memory, the address in relation to the instruction pointer is a 2 that adds to the instruction pointer. This develops the address of the next program instruction. An example of the relative JMP instruction is shown in Figure 3–15. Notice that the JMP instruction is a one-byte instruction, with a one-byte or a two-byte displacement that adds to the instruction pointer. A one-byte displacement is used in short jumps, and a two-byte displacement is used with near jumps and calls. Both types are considered to be intrasegment jumps. (An **intrasegment jump** is a jump anywhere within the current code segment.) In the 80386 and above, the displacement can also be a 32-bit value, allowing them to use relative addressing to any location within their 4G-byte code segments.

Relative JMP and CALL instructions contain either an 8-bit or a 16-bit signed displacement that allows a forward memory reference or a reverse memory reference. (The 80386 and above can have an 8-bit or 32-bit displacement.) All assemblers automatically calculate the distance for the displacement and select the proper one-, two- or four-byte form. If the distance is too far for a two-byte displacement in an 8086 through 80286 microprocessor, some assemblers use the direct jump. An 8-bit displacement *(short)* has a jump range of between +127 and –128 bytes from the next instruction, while a 16-bit displacement *(near)* has a range of ±32K bytes. In the 80386 and above, a 32-bit displacement allows a range of ±2G bytes. The 32-bit displacement can only be used in the protected mode.

Indirect Program Memory Addressing

The microprocessor allows several forms of program indirect memory addressing for the JMP and CALL instructions. Table 3–10 lists some acceptable program indirect jump instructions, which can use any 16-bit register (AX, BX, CX, DX, SP, BP, DI, or SI); any relative register

FIGURE 3–15 A JMP [2] instruction. This instruction skips over the two bytes of memory that follow the JMP instruction.

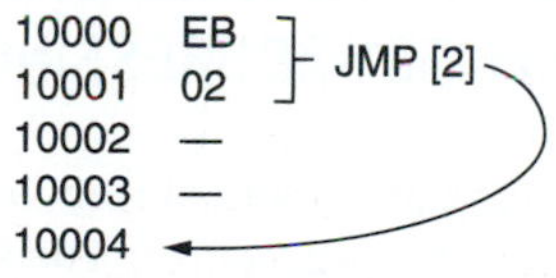

```
10000   EB  ⎤
10001   02  ⎦ JMP [2]
10002   —
10003   —
10004 ←
```

TABLE 3–10   Examples of indirect program memory addressing

Assembly Language	Operation
JMP AX	Jumps to the current code segment location addressed by the contents of AX
JMP CX	Jumps to the current code segment location addressed by the contents of CX
JMP NEAR PTR [BX]	Jumps to the current code segment location addressed by the contents of the data segment memory location addressed by BX
JMP NEAR PTR[DI+2]	Jumps to the current code segment location addressed by the contents of the data segment memory location addressed by DI plus 2
JMP TABLE[BX]	Jumps to the current code segment location addressed by the contents of the data segment memory location addressed by TABLE plus BX
JMP ECX	Jumps to the current code segment location addressed by the contents of ECX

FIGURE 3–16   A jump table that stores addresses of various programs. The exact address chosen from the TABLE is determined by an index stored with the jump instruction.

```
TABLE  DW  LOC0
       DW  LOC1
       DW  LOC2
       DW  LOC3
```

([BP], [BX], [DI], or [SI]); and any relative register with a displacement. In the 80386 and above, an extended register can also be used to hold the address or indirect address of a relative JMP or CALL. For example, the JMP EAX jumps to the location address by register EAX.

If a 16-bit register holds the address of a JMP instruction, the jump is near. For example, if the BX register contains a 1000H and a JMP BX instruction executes, the microprocessor jumps to offset address 1000H in the current code segment.

If a relative register holds the address, the jump is also considered to be an indirect jump. For example, a JMP [BX] refers to the memory location within the data segment at the offset address contained in BX. At this offset address is a 16-bit number that is used as the offset address in the intrasegment jump. This type of jump is sometimes called an *indirect-indirect* or *double-indirect jump.*

Figure 3–16 shows a jump table that is stored, beginning at memory location TABLE. This jump table is referenced by the short program of Example 3–13. In this example, the BX register is loaded with a 4 so, when it combines in the JMP TABLE[BX] instruction with TABLE, the effective address is the contents of the second entry in the jump table.

EXAMPLE 3–13

```
                          ;Using indirect addressing for a jump
                          ;
0000  BB 0004                 MOV BX,4            ;address LOC2
0003  FF A7 23A1 R            JMP TABLE[BX]       ;jump to LOC2
```

3–3   STACK MEMORY-ADDRESSING MODES

The stack plays an important role in all microprocessors. It holds data temporarily and stores return addresses for procedures. The stack memory is a LIFO (**last-in, first-out**) memory, which describes the way that data are stored and removed from the stack. Data are placed onto the stack

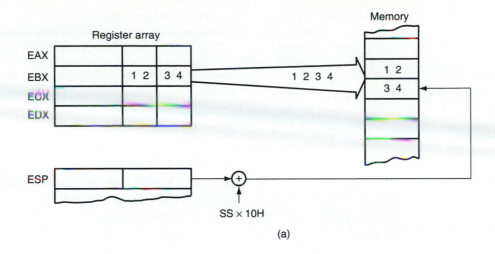

(a)

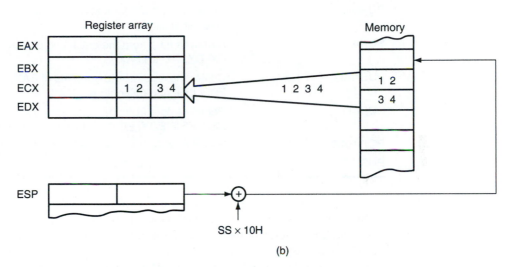

(b)

FIGURE 3–17   The PUSH and POP instructions. (a) PUSH BX places the contents of BX onto the stack, (b) POP CX removes data from the stack and places them into CX. Both instructions are shown after execution.

with a **PUSH instruction** and removed with a **POP instruction.** The CALL instruction also uses the stack to hold the return address for procedures and a RET (return) instruction to remove the return address from the stack.

The stack memory is maintained by two registers: the stack pointer (SP or ESP) and the stack segment register (SS). Whenever a word of data is pushed onto the stack [see Figure 3–17(a)], the high-order 8 bits are placed in the location addressed by SP – 1. The low-order 8 bits are placed in the location addressed by SP – 2. The SP is then decremented by 2 so that the next word of data is stored in the next available stack memory location. The SP/ESP register always points to an area of memory located within the stack segment. The SP/ESP register adds to SS × 10H to form the stack memory address in the real mode. In protected mode operation, the SS register holds a selector that accesses a descriptor for the base address of the stack segment.

Whenever data are popped from the stack [see Figure 3–17(b)], the low-order 8 bits are removed from the location addressed by SP. The high-order 8 bits are removed from the location addressed by SP + 1. The SP register is then incremented by 2. Table 3–11 lists some of the

TABLE 3–11   Example PUSH and POP instructions.

Assembly Language	Operation
POPF	Removes a word from the stack and places it into the flags
POPFD	Removes a doubleword from the stack and places it into the EFLAG register
PUSHF	Copies the flags onto the stack
PUSHFD	Copies the EFLAG register to the stack
PUSH AX	Copies AX to the stack
POP BX	Removes a word from the stack and places it into BX
PUSH DS	Copies DS to the stack
PUSH 1234H	Copies a 1234H to the stack
POP CS	Illegal instruction
PUSH WORD PTR [BX]	Copies a word from the data segment memory location addressed by BX onto the stack
PUSHA	Copies the word contents of AX, CX, DX, BX, SP, BP, DI, and SI onto the stack
POPA	Removes data from the stack and places it into SI, DI, BP, SP, BX, DX, CX, and AX
PUSHAD	Copies the doubleword contents of EAX, ECX, EDX, EBX, ESP, EBP, EDI, and ESI onto the stack
POPAD	Removes data from the stack and places it into ESI, EDI, EBP, ESP, EBX, EDX, ECX, and EAX
POP EAX	Removes data from the stack and places it into EAX
PUSH EDI	Copies EDI to the stack

PUSH and POP instructions available to the microprocessor. Note that PUSH and POP always store or retrieve words of data—never bytes—in the 8086 through the 80286 microprocessors. The 80386 and above allow words or doublewords to be transferred to and from the stack. Data may be pushed onto the stack from any 16-bit register or segment register; in the 80386 and above, from any 32-bit extended register. Data may be popped off the stack into any 16-bit register or any segment register except CS. The reason that data may not be popped from the stack into CS is that this only changes part of the address of the next instruction.

The PUSHA and POPA instructions either push or pop all of the registers, except the segment registers, on the stack. These instructions are not available on the early 8086/8088 microprocessors. The push immediate instruction is also new to the 80286 through the Pentium microprocessors. Note the examples in Table 3–11, which show the order of the registers transferred by the PUSHA and POPA instructions. The 80386 and above also allow extended registers to be pushed or popped.

Example 3–14 lists a short program that pushes the contents of AX, BX, and CX onto the stack. The first POP retrieves the value that was pushed onto the stack from CX and places it into AX. The second POP places the original value of BX into CX. The last POP places the original value of AX into BX.

EXAMPLE 3–14

```
                        .MODEL TINY        ;select TINY model
0000                    .CODE              ;start CODE segment
                        .STARTUP           ;start of program
0100  B8 1000           MOV    AX,1000H    ;load test data
```

```
0103   BB 2000        MOV      BX,2000H
0106   B9 3000        MOV      CX,3000H

0109   50             PUSH     AX              ;1000H to stack
010A   53             PUSH     BX              ;2000H to stack
010B   51             PUSH     CX              ;3000H to stack

010C   58             POP      AX              ;3000H to AX
010D   59             POP      CX              ;2000H to CX
010E   5B             POP      BX              ;1000H to BX
                      .EXIT                    ;exit to DOS
                      END                      ;end of file
```

3–4　　SUMMARY

1. The data-addressing modes include register, immediate, direct, register indirect, base-plus-index, register relative, and base relative-plus-index addressing. In the 80386 through the Pentium 4 microprocessors, an additional addressing mode, called scaled-index addressing, exists.

2. The program memory-addressing modes include direct, relative, and indirect addressing.

3. Table 3–12 lists all real mode data-addressing modes available to the 8086 through the 80286 microprocessors. Note that the 80386 and above use these modes, plus the many defined through this chapter. In the protected mode, the function of the segment register is to address a descriptor that contains the base address of the memory segment.

4. The 80386 through Pentium 4 microprocessors have additional addressing modes that allow the extended registers EAX, EBX, ECX, EDX, EBP, EDI, and ESI to address memory. Although these addressing modes are too numerous to list in tabular form, in general, any of these registers function in the same way as those listed in Table 3–12. For example, the MOV AL,TABLE[EBX+2*ECX+10H] is a valid addressing mode for the 80386–Pentium 4 microprocessors.

5. The MOV instruction copies the contents of the source operand into the destination operand. The source never changes for any instruction.

6. Register addressing specifies any 8-bit register (AH, AL, BH, BL, CH, CL, DH, or DL) or any 16-bit register (AX, BX, CX, DX, SP, BP, SI, or DI). The segment registers (CS, DS, ES, or SS) are also addressable for moving data between a segment register and a 16-bit register/memory location or for PUSH and POP. In the 80386 through the Pentium 4 microprocessors, the extended registers also are used for register addressing; they consist of EAX, EBX, ECX, EDX, ESP, EBP, EDI, and ESI. Also available to the 80386 and above are the FS and GS segment registers.

7. The MOV immediate instruction transfers the byte or word that immediately follows the opcode into a register or a memory location. Immediate addressing manipulates constant data in a program. In the 80386 and above, a doubleword immediate data may also be loaded into a 32-bit register or memory location.

8. The .MODEL statement is used with assembly language to identify the start of a file and the type of memory model used with the file. If the size is TINY, the program exists in one segment, the code segment, and is assembled as a command (.COM) program. If the SMALL model is used, the program uses a code and data segment, and assembles as an execute (.EXE) program. Other model sizes and their attributes are listed in Appendix A.

9. Direct addressing occurs in two forms in the microprocessor: (1) direct addressing and (2) displacement addressing. Both forms of addressing are identical except that direct addressing is used to transfer data between EAX, AX, or AL and memory; displacement addressing is used with any register-memory transfer. Direct addressing requires three bytes of memory, while displacement addressing requires four bytes. Note that some of these

TABLE 3–12   Example real mode data-addressing modes.

Assembly Language	Address Generation
MOV AL,BL	8-bit register addressing
MOV AX,BX	16-bit register addressing
MOV EAX,ECX	32-bit register addressing
MOV DS,CX	Segment register addressing
MOV AL,LIST	(DS x 10H) + LIST
MOV CH,DATA1	(DS x 10H) + DATA1
MOV ES,DATA2	(DS x 10H) + DATA2
MOV AL,12	Immediate data of 12H
MOV AL,[BP]	(SS x 10H) + BP
MOV AL,[BX]	{DS x 10H) + BX
MOV AL,[DI]	(DS x 10H) + DI
MOV AL,[SI]	(DS x 10H) + SI
MOV AL,[BP+2]	(SS x 10H) + BP + 2
MOV AL,[BX–4]	(DS x 10H) + BX – 4
MOV AL,[DI+1000H]	(DS x 10H) + DI + 1000H
MOV AL,[SI+300H]	(DS x 10H) + SI + 300H
MOV AL,LIST[BP]	(SS x 10H) + LIST + BP
MOV AL,LIST[BX]	(DS x 10H) + LIST + BX
MOV AL,LIST[DI]	(DS x 10H) + LIST + DI
MOV AL,LIST[SI]	(DS x 10H) + LIST + SI
MOV AL,LIST[BP+2]	(SS x 10H) + LIST + BP + 2
MOV AL,LIST[BX–6]	(DS x 10H) + LIST + BX – 6
MOV AL, LIST[DI+100H]	(DS x 10H) + LIST + DI + 100H
MOV AL,LIST[SI+200H]	(DS x 10H) + LIST + SI + 200H
MOV AL,[BP+DI]	(SS x 10H) + BP + DI
MOV AL,[BP+SI]	(SS x 10H) + BP + SI
MOV AL,[BX+DI]	(DS x 10H) + BX + DI
MOV AL,[BX+SI]	(DS x 10H) + BX + SI
MOV AL,[BP+DI+4]	(SS x 10H) + BP + DI +4
MOV AL,[BP+SI–8]	(SS x 10H) + BP + SI – 8
MOV AL,[BX+DI+10H]	(DS x 10H) + BX + DI + 10H
MOV AL,[BX+ SI–10H]	(DS x 10H) + BX + SI – 10H
MOV AL,LIST[BP+DI]	(SS x 10H) + LIST + BP + DI
MOV AL,LIST[BP+SI]	(SS x 10H) + LIST + BP + SI
MOV AL,LIST[BX+DI]	(DS x 10H) + LIST + BX + DI
MOV AL,LIST[BX+SI]	(DS x 10H) + LIST + BX + SI
MOV AL,LIST[BP+DI+2]	(SS x 10H) + LIST + BP + DI + 2
MOV AL,LIST[BP+SI–7]	(SS x 10H) + LIST + BP + SI – 7
MOV AL,LIST[BX+DI+3]	(DS x 10H) + LIST + BX + DI + 3
MOV AL,LIST[BX+SI–2]	(DS x 10H) + LIST + BX + SI – 2

instructions in the 80386 and above may require additional bytes in the form of prefixes for register and operand sizes.

10. Register indirect addressing allows data to be addressed at the memory location pointed to by either a base (BP and BX) or index register (DI and SI). In the 80386 and above, extended registers EAX, EBX, ECX, EDX, EBP, EDI, and ESI are used to address memory data.

11. Base-plus-index addressing often addresses data in an array. The memory address for this mode is formed by adding a base register, index register, and the contents of a segment register times 10H. In the 80386 and above, the base and index registers may be any 32-bit register except EIP and ESP.

12. Register relative addressing uses either a base or index register, plus a displacement to access memory data.

13. Base relative-plus-index addressing is useful for addressing a two-dimensional memory array. The address is formed by adding a base register, an index register, displacement, and the contents of a segment register times 10H.

14. Scaled-index addressing is unique to the 80386 through the Pentium 4. The second of two registers (index) is scaled by a factor of 2X, 4X, or 8X to access words, doublewords, or quadwords in memory arrays. The MOV AX,[EBX+2*ECX] and the MOV [4*ECX],EDX are examples of scaled-index instructions.

15. Data structures are templates for storing arrays of data, and are addressed by array name and field. For example, array NUMBER and field TEN of array NUMBER is addressed as NUMBER.TEN.

16. Direct program memory addressing is allowed with the JMP and CALL instructions to any location in the memory system. With this addressing mode, the offset address and segment address are stored with the instruction.

17. Relative program addressing allows a JMP or CALL instruction to branch forward or backward in the current code segment by ±32K bytes. In the 80386 and above, the 32-bit displacement allows a branch to any location in the current code segment by using a displacement value of ±2G bytes. The 32-bit displacement can be used only in protected mode.

18. Indirect program addressing allows the JMP or CALL instructions to address another portion of the program or subroutine indirectly through a register or memory location.

19. The PUSH and POP instructions transfer a word between the stack and a register or memory location. A PUSH immediate instruction is available to place immediate data on the stack. The PUSHA and POPA instructions transfer AX, CX, DX, BX, BP, SP, SI, and DI between the stack and these registers. In the 80386 and above, the extended register and extended flags can also be transferred between registers and the stack. A PUSHFD stores the EFLAGS, while a PUSHF stores the FLAGS.

20. Example 3–15 shows many of the addressing modes presented in the chapter. This example program fills the ARRAY1 from locations 0000:0000–0000:0009. It then fills ARRAY2–0 through 9. Finally, it exchanges the contents of ARRAY1 element 2 with ARRAY2 element 3.

EXAMPLE 3–15

```
                                .MODEL  SMALL                ;select SMALL model
0000                            .DATA                        ;start of DATA segment

0000  000A [        ARRAY1 DB   10 DUP (?)                   ;reserve for ARRAY1
         00
      ]
000A  000A [        ARRAY2 DB   10 DUP (?)                   ;reserve for ARRAY2
         00
      ]
0000                            .CODE                        ;start of CODE segment
                                .STARTUP                     ;start of program

0017  B8 0000                   MOV    AX,0                  ;segment ES is 0000H
001A  8E C0                     MOV    ES,AX

001C  BF 0000                   MOV    DI,0                  ;address element 0
001F  B9 000A                   MOV    CX,10                 ;count of 10
0022             LAB1:
0022  26:8A 05                  MOV    AL,ES:[DI]            ;copy data
0025  88 85 0000 R              MOV    ARRAY1[DI],AL         ;into ARRAY1
0029  47                        INC    DI
002A  E2 F6                     LOOP   LAB1

002C  BF 0000                   MOV    DI,0                  ;address element 0
002F  B9 000A                   MOV    CX,10                 ;count of 10
0032  B0 00                     MOV    AL,0                  ;initial value
0034             LAB2:
0034  88 85 000A R              MOV    ARRAY2[DI],AL         ;fill ARRAY2
```

```
0038   FE C0              INC    AL
003A   47                 INC    DI
003B   E2 F7              LOOP   LAB2

003D   BF 0003            MOV    DI,3              ;exchange array data
0040   8A 85 0000 R       MOV    AL,ARRAY1[DI]
0044   8A A5 000B R       MOV    AH,ARRAY2[DI+1]
0048   88 A5 0000 R       MOV    ARRAY1[DI],AH
004C   88 85 000B R       MOV    ARRAY2[DI+1],AL

                          .EXIT                    ;exit to DOS
                          END                      ;end of file
```

3–5   QUESTIONS AND PROBLEMS

1. What do the following MOV instructions accomplish?
   (a) MOV AX,BX
   (b) MOV BX,AX
   (c) MOV BL,CH
   (d) MOV ESP,EBP
   (e) MOV AX,CS
2. List the 8-bit registers that are used for register addressing.
3. List the 16-bit registers that are used for register addressing.
4. List the 32-bit registers that are used for register addressing in the 80386 through the Pentium 4 microprocessors.
5. List the 16-bit segment registers used with register addressing by MOV, PUSH, and POP.
6. What is wrong with the MOV BL,CX instruction?
7. What is wrong with the MOV DS,SS instruction?
8. Select an instruction for each of the following tasks:
   (a) copy EBX into EDX
   (b) copy BL into CL
   (c) copy SI into BX
   (d) copy DS into AX
   (e) copy AL into AH
9. Select an instruction for each of the following tasks:
   (a) move a 12H into AL
   (b) move a 123AH into AX
   (c) move a 0CDH into CL
   (d) move a 1000H into SI
   (e) move a 1200A2H into EBX
10. What special symbol is sometimes used to denote immediate data?
11. What is the purpose of the .MODEL TINY statement?
12. What assembly language directive indicates the start of the CODE segment?
13. What is a label?
14. The MOV instruction is placed in what field of a statement?
15. A label may begin with what characters?
16. What is the purpose of the .EXIT directive?
17. Does the .MODEL TINY statement cause a program to assemble an execute program?
18. What tasks does the .STARTUP directive accomplish in the small memory model?
19. What is a displacement? How does it determine the memory address in a MOV [2000H],AL instruction?
20. What do the symbols [ ] indicate?

21. Suppose that DS = 0200H, BX = 0300H, and DI = 400H. Determine the memory address accessed by each of the following instructions, assuming real mode operation:
    (a) MOV AL,[1234H]
    (b) MOV EAX,[BX]
    (c) MOV [DI],AL
22. What is wrong with a MOV [BX],[DI] instruction?
23. Choose an instruction that requires BYTE PTR.
24. Choose an instruction that requires WORD PTR.
25. Choose an instruction that requires DWORD PTR.
26. Explain the difference between the MOV BX,DATA instruction and the MOV BX,OFFSET DATA instruction.
27. Suppose that DS = 1000H, SS = 2000H, BP = 1000H, and DI = 0100H. Determine the memory address accessed by each of the following instructions, assuming real mode operation:
    (a) MOV AL,[BP+DI]
    (b) MOV CX,[DI]
    (c) MOV EDX,[BP]
28. What, if anything, is wrong with a MOV AL,[BX][SI] instruction?
29. Suppose that DS = 1200H, BX = 0100H, and SI = 0250H. Determine the address accessed by each of the following instructions, assuming real mode operation:
    (a) MOV [100H],DL
    (b) MOV [SI+100H],EAX
    (c) MOV DL,[BX+100H]
30. Suppose that DS = 1100H, BX = 0200H, LIST = 0250H, and SI = 0500H. Determine the address accessed by each of the following instructions, assuming real mode operation:
    (a) MOV LIST[SI],EDX
    (b) MOV CL,LIST[BX+SI]
    (c) MOV CH,[BX+SI]
31. Suppose that DS = 1300H, SS = 1400H, BP = 1500H, and SI = 0100H. Determine the address accessed by each of the following instructions, assuming real mode operation:
    (a) MOV EAX,[BP+200H]
    (b) MOV AL,[BP+SI–200H]
    (c) MOV AL,[SI–0100H]
32. Which base register addresses data in the stack segment?
33. Suppose that EAX = 00001000H, EBX = 00002000H, and DS = 0010H. Determine the addresses accessed by the following instructions, assuming real mode operation:
    (a) MOV ECX,[EAX+EBX]
    (b) MOV [EAX+2*EBX],CL
    (c) MOV DH,[EBX+4*EAX+1000H]
34. Develop a data structure that has five fields of one word each named F1, F2, F3, F4, and F5 with a structure name of FIELDS.
35. Show how field F3 of the data structure constructed in question 34 is addressed in a program.
36. What are the three program memory-addressing modes?
37. How many bytes of memory store a far direct jump instruction? What is stored in each of the bytes?
38. What is the difference between an intersegment and intrasegment jump?
39. If a near jump uses a signed 16-bit displacement, how can it jump to any memory location within the current code segment?
40. The 80386 and above use a _____-bit displacement to jump to any location within the 4G byte code segment.
41. What is a far jump?

42. If a JMP instruction is stored at memory location 100H within the current code segment, it cannot be a _____ jump if it is jumping to memory location 200H within the current code segment.

43. Show which JMP instruction assembles (short, near, or far) if the JMP THERE instruction is stored at memory address 10000H and the address of THERE is:
    (a) 10020H
    (b) 11000H
    (c) 0FFFEH
    (d) 30000H

44. Form a JMP instruction that jumps to the address pointed to by the BX register.

45. Select a JMP instruction that jumps to the location stored in memory at the location table. Assume that it is a near JMP.

46. How many bytes are stored on the stack by PUSH instructions?

47. Explain how the PUSH [DI] instruction functions.

48. What registers are placed on the stack by the PUSHA instruction? In what order?

49. What does the PUSHAD instruction accomplish?

50. Which instruction places the EFLAGS on the stack in the Pentium 4 microprocessor?

CHAPTER 4

Data Movement Instructions

INTRODUCTION

This chapter concentrates on the data movement instructions. The data movement instructions include MOV, MOVSX, MOVZX, PUSH, POP, BSWAP, XCHG, XLAT, IN, OUT, LEA, LDS, LES, LFS, LGS, LSS, LAHF, SAHF, and the string instructions MOVS, LODS, STOS, INS, and OUTS. The latest data transfer instruction implemented on the Pentium Pro through Pentium 4 is the CMOV (conditional move) instruction. The data movement instructions are presented first because they are more commonly used in programs and easy to understand.

The microprocessor requires an assembler program, which generates machine language, because machine language instructions are too complex to efficiently generate by hand. This chapter describes the assembly language syntax and some of its directives. [This text assumes that the user is developing software on an IBM personal computer or clone. It is recommended that the Microsoft MACRO assembler (MASM) be used as the development tool, but the Intel Assembler (ASM), Borland Turbo assembler (TASM), or similar software function equally as well. The most recent version of TASM completely emulates the MASM program. This text presents information that functions with the Microsoft MASM assembler, but most programs assemble without modification with other assemblers. Appendix A explains the Microsoft assembler and provides detail on the linker program and Programmer's WorkBench.]

CHAPTER OBJECTIVES

Upon completion of this chapter, you will be able to:

1. Explain the operation of each data movement instruction with applicable addressing modes.
2. Explain the purposes of the assembly language pseudo-operations and key words such as ALIGN, ASSUME, DB, DD, DW, END, ENDS, ENDP, EQU, .MODEL, OFFSET, ORG, PROC, PTR, SEGMENT, USE16, USE32, and USES.
3. Select the appropriate assembly language instruction to accomplish a specific data movement task.
4. Determine the symbolic opcode, source, destination, and addressing mode for a hexadecimal machine language instruction.
5. Use the assembler to set up a data segment, stack segment, and code segment.
6. Show how to set up a procedure using PROC and ENDP.
7. Explain the difference between memory models and full-segment definitions for the MASM assembler.

4–1   MOV REVISITED

The MOV instruction, introduced in Chapter 3, explains the diversity of 8086–Pentium 4 addressing modes. In this chapter, the MOV instruction introduces the machine language instructions available with various addressing modes and instructions. Machine code is introduced because it may occasionally be necessary to interpret machine language programs generated by an assembler. Interpretation of the machine's native language **(machine language)** allows debugging or modification at the machine language level. Occasionally, machine language patches are made by using the DEBUG program available with DOS, which requires some knowledge of machine language. Conversion between machine and assembly language instructions is illustrated in Appendix B.

Machine Language

Machine language is the **native binary code** that the microprocessor understands and uses as its instructions to control its operation. Machine language instructions for the 8086 through the Pentium 4 vary in length from one to as many as thirteen bytes. Although machine language appears complex, there is order to this microprocessor's machine language. There are well over 100,000 variations of machine language instructions, which means that there is no complete list of these variations. Because of this, some binary bits in a machine language instruction are given, and the remainder are determined for each variation of the instruction.

Instructions for the 8086 through the 80286 are 16-bit mode instructions that take the form found in Figure 4–1(a). The 16-bit mode instructions are compatible with the 80386 and above if they are programmed to operate in the 16-bit instruction mode, but they may be prefixed, as shown in Figure 4–1(b). The 80386 and above assume that all instructions are 16-bit mode instructions when the machine is operated in the *real mode.* In the *protected mode,* the upper byte of the descriptor contains the D-bit that selects either the 16- or 32-bit instruction mode. At present, only Windows XP, Windows 95, Windows 98, and OS/2 operate in the 32-bit instruction mode. The 32-bit mode instructions are in the form shown in Figure 4–1(b). These instructions occur in the 16-bit instruction mode by the use of prefixes, which are explained later in this chapter.

The first two bytes of the 32-bit instruction mode format are called **override prefixes** because they are not always present. The first modifies the size of the operand address used by the instruction and the second modifies the register size. If the 80386 through the Pentium II operate as 16-bit instruction mode machines (real or protected mode) and a 32-bit register is used, the **register-size prefix** (66H) is appended to the front of the instruction. If operated in the 32-bit instruction mode (protected mode only) and a 32-bit register is used, the register-size prefix is absent. If a 16-bit register appears in an instruction in the 32-bit instruction mode, the register-size

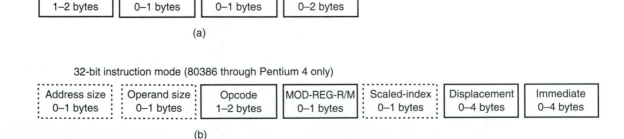

FIGURE 4–1   The formats of the 8086–Pentium 4 instructions. (a) The 16-bit form and (b) the 32-bit form.

FIGURE 4–2 Byte 1 of many machine language instructions, showing the position of the D- and W-bits.

Opcode

prefix is present to select a 16-bit register. The **address size-prefix** (67H) is used in a similar fashion, as explained later in this chapter. The prefixes toggle the size of the register and operand address from 16-bit to 32-bit or from 32-bit to 16-bit for the prefixed instruction. Note that the 16-bit instruction mode uses 8- and 16-bit registers and addressing modes, while the 32-bit instruction mode uses 8- and 32-bit registers and addressing modes by default. The prefixes override these defaults so that a 32-bit register can be used in the 16-bit mode or a 16-bit register can be used in the 32-bit mode. The **mode of operation** (16 or 32 bits) should be selected to conform with the application at hand. If 8- and 32-bit data pervade the application, the 32-bit mode should be selected; likewise, if 8- and 16-bit data pervade, the 16-bit mode should be selected. Normally, mode selection is a function of the operating system. (Remember that DOS can operate only in the 16-bit mode, however.)

The Opcode. The **opcode** selects the operation (addition, subtraction, move, and so on) that is performed by the microprocessor. The opcode is either one or two bytes long for most machine language instructions. Figure 4–2 illustrates the general form of the first opcode byte of many, but *not* all, machine language instructions. Here, the first six bits of the first byte are the binary opcode. The remaining two bits indicate the **direction** (D)—not to be confused with the instruction mode bit (16/32) or direction flag bit (used with string instructions)—of the data flow, and indicate whether the data are a byte or a word (W). In the 80386 and above, words and double-words are both specified when W = 1. The instruction mode and register-size prefix (66H) determine whether W represents a word or a doubleword.

If the direction bit (D) = 1, data flow *to* the register REG field from the R/M field located in the second byte of an instruction. If the D-bit = 0 in the opcode, data flow to the R/M field *from* the REG field. If the W-bit = 1, the data size is a *word* or *doubleword;* if the W-bit = 0, the data size is always a *byte.* The W-bit appears in most instructions, while the D-bit appears mainly with the MOV and some other instructions. Refer to Figure 4–3 for the binary bit pattern of the second opcode byte (reg-mod-r/m) of many instructions. Figure 4–3 shows the location of the MOD (mode), REG (register), and R/M (register/memory) fields.

MOD Field. The MOD field specifies the addressing mode (MOD) for the selected instruction. The MOD field selects the type of addressing and whether a displacement is present with the selected type. Table 4–1 lists the operand forms available to the MOD field for 16-bit instruction mode, unless the operand address-size override prefix (67H) appears. If the MOD field contains an 11, it selects the register addressing mode. Register addressing uses the R/M field to specify a register instead of a memory location. If the MOD field contains a 00, 01, or 10, the R/M field selects one of the data memory-addressing modes. When MOD selects a data memory-addressing mode, it indicates that the addressing mode contains no displacement (00), an 8-bit sign-extended displacement (01), or a 16-bit displacement (10). The MOV AL,[DI] instruction is

FIGURE 4–3 Byte 2 of many machine language instructions, showing the position of the MOD, REG, and R/M fields.

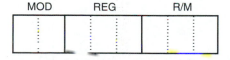

TABLE 4–1   MOD field for the 16-bit instruction mode.

MOD	Function
00	No displacement
01	8-bit sign-extended displacement
10	16-bit displacement
11	R/M is a register

TABLE 4–2   MOD field for the 32-bit instruction mode (80386–Pentium 4 only).

MOD	Function
00	No displacement
01	8-bit sign-extended displacement
10	32-bit displacement
11	R/M is a register

an example that shows no displacement, a MOV AL,[DI+2] instruction uses an 8-bit displacement (+ 2), and a MOV AL,[DI+1000H] instruction uses a 16-bit displacement (+ 1000H).

All 8-bit displacements are **sign-extended** into 16-bit displacements when the microprocessor executes the instruction. If the 8-bit displacement is 00H–7FH (positive), it is sign-extended to 0000H–007FH before adding to the offset address. If the 8-bit displacement is 80H–FFH (negative), it is sign-extended to FF80H–FFFFH. To sign-extend a number, its sign-bit is copied to the next higher-order byte, which generates either a 00H or an FFH in the next higher-order byte. Some assembler programs do not use the 8-bit displacements.

In the 80386 through the Pentium 4 microprocessors, the MOD field may be the same as shown in Table 4–1; if the instruction mode is 32-bits, the MOD field is as it appears in Table 4–2. The MOD field is interpreted as selected by the address-size override prefix or the operating mode of the microprocessor. This change in the interpretation of the MOD field and instruction supports many of the numerous additional addressing modes allowed in the 80386 through the Pentium 4. The main difference is that when the MOD field is a 10, this causes the 16-bit displacement to become a 32-bit displacement to allow any protected mode memory location (4G bytes) to be accessed. The 80386 and above only allow an 8- or 32-bit displacement when operated in the 32-bit instruction mode, unless the address-size override prefix appears. Note that if an 8-bit displacement is selected, it is sign-extended into a 32-bit displacement by the microprocessor.

Register Assignments.   Table 4–3 lists the register assignments for the REG field and the R/M field (MOD = 11). This table contains three lists of register assignments: one is used when the W-bit = 0 (bytes), and the other two are used when the W-bit = 1 (words or doublewords). Note that doubleword registers are only available to the 80386 through the Pentium 4.

TABLE 4–3   REG and R/M (when MOD = 11) assignments.

Code	W = 0 (Byte)	W = 1 (Word)	W = 1 (Doubleword)
000	AL	AX	EAX
001	CL	CX	ECX
010	DL	DX	EDX
011	BL	BX	EBX
100	AH	SP	ESP
101	CH	BP	EBP
110	DH	SI	ESI
111	BH	DI	EDI

Opcode						D	W
1	0	0	0	1	0	1	1

MOD		REG			R/M		
1	1	1	0	1	1	0	0

Opcode = MOV
D = Transfer to register (REG)
W = Word
MOD = R/M is a register
REG = BP
R/M = SP

FIGURE 4–4 The 8BEC instruction placed into Byte 1 and 2 formats from Figures 4–2 and 4–3. This instruction is a MOV BP,SP.

Suppose that a 2-byte instruction, 8BECH, appears in a machine language program. Because neither a 67H (operand address-size override prefix) nor a 66H (register-size override prefix) appears as the first byte, the first byte is the opcode. If the microprocessor is operated in the 16-bit instruction mode, this instruction is converted to binary and placed in the instruction format of bytes 1 and 2, as illustrated in Figure 4–4. The opcode is 100010. If you refer to Appendix B, which lists the machine language instructions, you will find that this is the opcode for a MOV instruction. Notice that both the D and W bits are a logic 1, which means that a word moves into the destination register specified in the REG field. The REG field contains a 101, indicating register BP, so the MOV instruction moves data into register BP. Because the MOD field contains a 11, the R/M field also indicates a register. Here, R/M = 100 (SP); therefore, this instruction moves data from SP into BP and is written in symbolic form as a MOV BP,SP instruction.

Suppose that a 668BE8H instruction appears in an 80386 or above, operated in the 16-bit instruction mode. The first byte (66H) is the register-size override prefix that selects 32-bit register operands for the 16-bit instruction mode. The remainder of the instruction indicates that the opcode is a MOV with a source operand of EAX and a destination operand of EBP. This instruction is a MOV EBP,EAX. The same instruction becomes a MOV BP,AX instruction in the 80386 and above if it is operated in the 32-bit instruction mode because the register-size override prefix selects a 16-bit register. Luckily, the assembler program keeps track of the register- and address-size prefixes and the mode of operation. Recall that if the .386 switch is placed before the .MODEL statement, the 32-bit mode is selected; if it is placed after the .MODEL statement, the 16-bit mode is selected.

R/M Memory Addressing. If the MOD field contains a 00, 01, or 10, the R/M field takes on a new meaning. Table 4–4 lists the memory-addressing modes for the R/M field when MOD is a 00, 01, or 10 for the 16-bit instruction mode.

TABLE 4–4 16-bit R/M memory-addressing modes.

R/M Code	Addressing Mode
000	DS:[BX+SI]
001	DS:[BX+DI]
010	SS:[BP+SI]
011	SS:{BP+DI]
100	DS:[SI]
101	DS:[DI]
110	SS:[BP]*
111	DS:[BX]

*Note: See text section, Special Addressing Mode.

Opcode = MOV
D = Transfer to register (REG)
W= Byte
MOD = No displacement
REG = DL
R/M = DS:[DI]

FIGURE 4–5   A MOV DL,[DI] instruction converted to its machine language form.

All of the 16-bit addressing modes presented in Chapter 3 appear in Table 4–4. The displacement, discussed in Chapter 3, is defined by the MOD field. If MOD = 00 and R/M = 101, the addressing mode is [DI]. If MOD = 01 or 10, the addressing mode is [DI+33H], or LIST [DI+22H] for the 16-bit instruction mode. This example uses LIST, 33H, and 22H as arbitrary values for the displacement.

Figure 4–5 illustrates the machine language version of the 16-bit instruction MOV DL,[DI] or instruction (8A15H). This instruction is two bytes long and has an opcode 100010, D = 1 (to REG from R/M), W = 0 (byte), MOD = 00 (no displacement), REG = 010 (DL), and R/M = 101 ([DI]). If the instruction changes to MOV DL,[DI+1], the MOD field changes to 01 for an 8-bit displacement, but the first two bytes of the instruction otherwise remain the same. The instruction now becomes 8A5501H instead of 8A15H. Notice that the 8-bit displacement appends to the first two bytes of the instruction to form a three-byte instruction instead of two bytes. If the instruction is again changed to a MOV DL,[DI+1000H], the machine language form becomes a 8A750010H. Here, the 16-bit displacement of 1000H (coded as 0010H) appends the opcode.

Special Addressing Mode.   There is a special addressing mode that does not appear in Tables 4–2, 4–3, or 4–4. It occurs whenever memory data are referenced by only the displacement mode of addressing for 16-bit instructions. Examples are the MOV [1000H],DL and MOV NUMB,DL instructions. The first instruction moves the contents of register DL into data segment memory location 1000H. The second instruction moves register DL into symbolic data segment memory location NUMB.

Whenever an instruction has only a displacement, the MOD field is always a 00 and the R/M field is always a 110. As shown in the tables, the instruction contains no displacement and uses addressing mode [BP]. You *cannot* actually use addressing mode [BP] without a displacement in machine language. The assembler takes care of this by using an 8-bit displacement (MOD = 01) of 00H whenever the [BP] addressing mode appears in an instruction. This means that the [BP] addressing mode assembles as a [BP+0], even though a [BP] is used in the instruction. The same special addressing mode is also available to the 32-bit mode.

Figure 4–6 shows the binary bit pattern required to encode the MOV [1000H],DL instruction in machine language. If the individual translating this symbolic instruction into machine language does not know about the special addressing mode, the instruction would incorrectly translate to a MOV [BP],DL instruction. Figure 4–7 shows the actual form of the MOV [BP],DL instruction. Notice that this is a three-byte instruction with a displacement of 00H.

32-bit Addressing Modes.   The 32-bit addressing modes found in the 80386 and above are obtained by either running these machines in the 32-bit instruction mode or in the 16-bit instruction mode by using the address-size prefix 67H. Table 4–5 shows the coding for R/M used to specify the 32-bit addressing modes. Notice that when R/M = 100, an additional byte appears in the instruction called a **scaled-index byte.** The scaled-index byte indicates the additional forms of scaled-index addressing that do not appear in Table 4–5. The scaled-index byte is mainly used when two registers are added to specify the memory address in an instruction. Because the

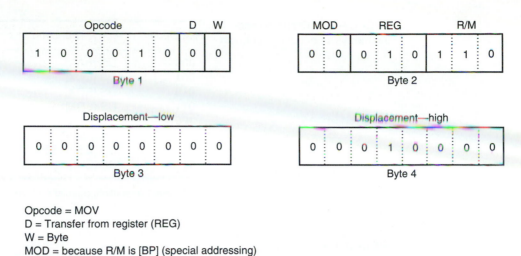

Opcode = MOV
D = Transfer from register (REG)
W = Byte
MOD = because R/M is [BP] (special addressing)
REG = DL
R/M = DS:[BP]
Displacement = 1000H

FIGURE 4–6   The MOV [1000H],DL instruction uses the special addressing mode.

scaled-index byte is added to the instruction, there are seven bits in the opcode and eight bits in the scaled-index byte to define. This means that a scaled-index instruction has 2^{15} (32K) possible combinations. There are over 32,000 different variations of the MOV instruction alone in the 80386 through the Pentium 4 microprocessors.

Figure 4–8 shows the format of the scaled-index byte, as selected by a value of 100 in the R/M field of an instruction when the 80386 and above use a 32-bit address. The leftmost two bits select a scaling factor (multiplier) of 1X, 2X, 4X, or 8X. Note that a scaling factor of 1X is implicit if none is used in an instruction that contains two 32-bit indirect address registers. The index and base fields both contain register numbers, as indicated in Table 4–3 for 32-bit registers.

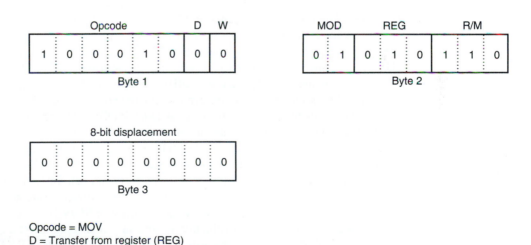

Opcode = MOV
D = Transfer from register (REG)
W = Byte
MOD = because R/M is [BP] (special addressing)
REG = DL
R/M = DS:[BP]
Displacement = 00H

FIGURE 4–7   The MOV [BP],DL instruction converted to binary machine language.

TABLE 4–5   32-bit addressing modes selected by R/M.

R/M Code	Function
000	DS:[EAX]
001	DS:[ECX[
010	DS:[EDX]
011	DS:[EBX]
100	Uses scaled-index byte
101	SS:[EBP]*
110	DS:[ESI]
111	DS:[EDI]

Note: See text section, Special Addressing Mode.

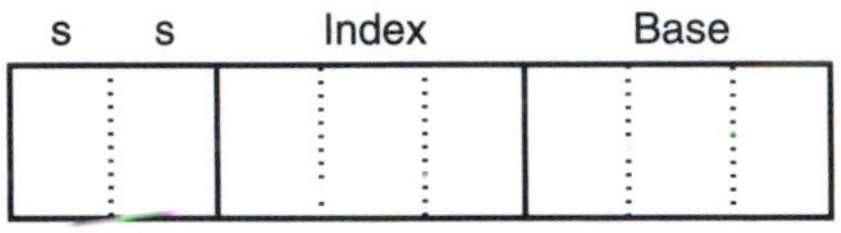

FIGURE 4–8   The scaled-
index byte.

ss
00 = × 1
01 = × 2
10 = × 4
11 = × 8

The instruction MOV EAX,[EBX+4*ECX] is encoded as 67668B048BH. Notice that both the *address size* (67H) and *register size* (66H) override prefixes appear in the instruction. This coding (67668B048BH) is used when the 80386 and above microprocessors are operated in the 16-bit instruction mode for this instruction. If the microprocessor operates in the 32-bit instruction mode, both prefixes disappear and the instruction becomes an 8B048BH instruction. The use of the prefixes depends on the mode of operation of the microprocessor. Scaled-index addressing can also use a single register multiplied by a scaling factor. An example is the MOV AL,[2*ECX] instruction. The contents of the data segment location addressed by two times ECX is copied into AL.

An Immediate Instruction.   Suppose that the MOV WORD PTR [BX+1000H],1234H instruction is chosen as an example of a 16-bit instruction using immediate addressing. This instruction moves a 1234H into the word-sized memory location addressed by the sum of 1000H, BX, and DS × 10H. This six-byte instruction uses two bytes for the opcode, W, MOD, and R/M fields. Two of the six bytes are the data of 1234H; two of the six bytes are the displacement of 1000H. Figure 4–9 shows the binary bit pattern for each byte of this instruction.

This instruction, in symbolic form, includes WORD PTR. The WORD PTR directive indicates to the assembler that the instruction uses a word-sized memory pointer. If the instruction moves a byte of immediate data, BYTE PTR replaces WORD PTR in the instruction. Likewise, if the instruction uses a doubleword of immediate data, the DWORD PTR directive replaces BYTE PTR. Most instructions that refer to memory through a pointer do not need the BYTE PTR, WORD PTR, or DWORD PTR directives. These directives are necessary only when it is not clear whether the operation is a byte or a word. The MOV [BX],AL instruction is clearly a byte move; the MOV [BX],1 instruction is not exact, and could therefore be a byte-, word-, or doubleword-sized move. Here, the instruction must be coded as MOV BYTE PTR [BX],1, MOV WORD PTR [BX],1, or MOV DWORD PTR [BX],1. If not, the assembler flags it as an error because it cannot determine the intent of this instruction.

Segment MOV Instructions.   If the contents of a segment register are moved by the MOV, PUSH, or POP instructions, a special set of register bits (REG field) selects the segment register (see Table 4–6).

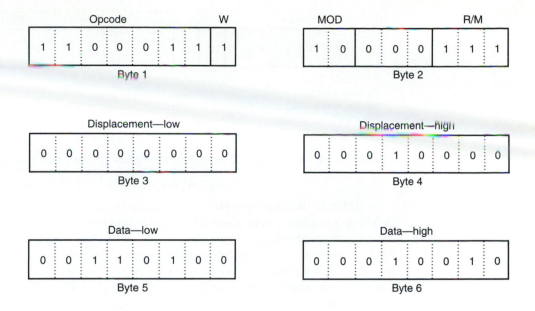

Opcode = MOV (immediate)
W = Word
MOD = 16-bit displacement
REG = 000 (not used in immediate addressing)
R/M = DS:[BX]
Displacement = 1000H
Data = 1234H

FIGURE 4-9   A MOV WORD PTR [BX+1000H],1234H instruction converted to binary machine language.

TABLE 4-6   Segment register selection.

Code	Segment Register
000	ES
001	CS*
010	SS
011	DS
100	FS
101	GS

*Note: MOV CS,R/M(16) and POP CS are not allowed by the microprocessor. The FS and GS segments are only available to the 80386–Pentium 4 microprocessors.

Figure 4–10 shows a MOV BX,CS instruction converted to binary. The opcode for this type of MOV instruction is different for the prior MOV instructions. Segment registers can be moved between any 16-bit register or 16-bit memory location. For example, the MOV [DI],DS instruction stores the contents of DS into the memory location addressed by DI in the data segment. An immediate segment register MOV is not available in the instruction set. To load a segment register with immediate data, first load another register with the data and then move it to a segment register.

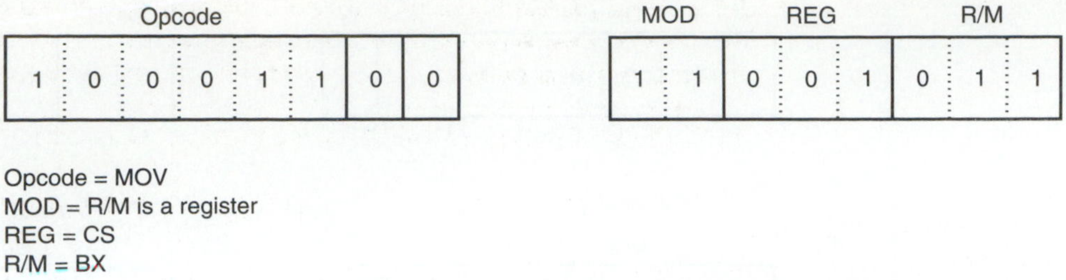

Opcode								MOD		REG			R/M		
1	0	0	0	1	1	0	0	1	1	0	0	1	0	1	1

Opcode = MOV
MOD = R/M is a register
REG = CS
R/M = BX

FIGURE 4–10   A MOV BX,CS instruction converted to binary machine language.

Although this discussion has not been a complete coverage of machine language coding, it should give you a good start in machine language programming. Remember a program written in symbolic assembly language *(assembly language)* is rarely assembled by hand into binary machine language. An assembler program converts symbolic assembly language into machine language. With the microprocessor and its over 100,000 instruction variations, let us hope that an assembler is available for the conversion because the process is very time-consuming, although not impossible.

4–2   PUSH/POP

The PUSH and POP instructions are important instructions that *store* and *retrieve* data from the LIFO (last-in, first-out) stack memory. The microprocessor has six forms of the PUSH and POP instructions: register, memory, immediate, segment register, flags, and all registers. The PUSH and POP immediate and the PUSHA and POPA (all registers) forms are not available in the earlier 8086/8088 microprocessors, but are available to the 80286 through the Pentium 4.

Register addressing allows the contents of any 16-bit register to be transferred to or from the stack. In the 80386 and above, the 32-bit extended registers and flags (EFLAGS) can also be pushed or popped from the stack. Memory addressing PUSH and POP instructions store the contents of a 16-bit memory location (or 32-bits in the 80386 and above) on the stack or stack data into a memory location. Immediate addressing allows immediate data to be pushed onto the stack, but not popped off the stack. Segment register addressing allows the contents of any segment register to be pushed onto the stack or removed from the stack (CS may be pushed, but data from the stack may never be popped into CS). The flags may be pushed or popped from that stack, and the contents of all the registers may be pushed or popped.

PUSH

The 8086–80286 PUSH instruction always transfers two bytes of data to the stack; the 80386 and above transfer two or four bytes, depending on the register or size of the memory location. The source of the data may be any internal 16- or 32-bit register, immediate data, any segment register, or any two bytes of memory data. There is also a PUSHA instruction that copies the contents of the internal register set, except the segment registers, to the stack. The PUSHA (**push all**) instruction copies the registers to the stack in the following order: AX, CX, DX, BX, SP, BP, SI, and DI. The value for SP that is pushed onto the stack is whatever it was before the PUSHA instruction executes. The PUSHF (**push flags**) instruction copies the contents of the flag register to the stack. The PUSHAD and POPAD instructions push and pop the contents of the 32-bit register set found in the 80386 through the Pentium 4.

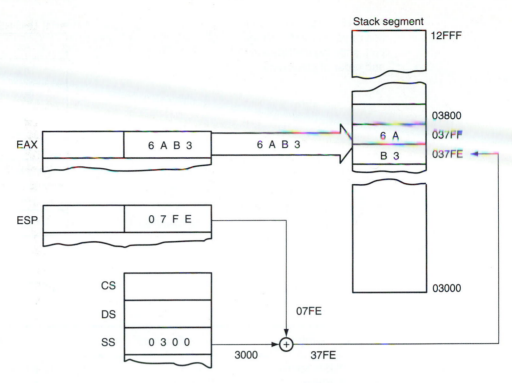

FIGURE 4–11 The effect of the PUSH AX instruction on ESP and stack memory location 37FFH and 37FEH. This instruction is shown at the point after execution.

Whenever data are pushed onto the stack, the first (most-significant) data byte moves into the stack segment memory location addressed by SP − 1. The second (least-significant) data byte moves into the stack segment memory location addressed by SP − 2. After the data are stored by a PUSH, the contents of the SP register decrement by 2. The same is true for a doubleword push, except that four bytes are moved to the stack memory (most-significant byte first), after which the stack pointer decrements by 4. Figure 4–11 shows the operation of the PUSH AX instruction. This instruction copies the contents of AX onto the stack where address SS:[SP − 1] = AH, SS:[SP − 2] = AL, and afterwards SP = SP − 2.

The PUSHA instruction pushes all the internal 16-bit registers onto the stack, as illustrated in Figure 4–12. This instruction requires 16 bytes of stack memory space to store all eight 16-bit registers. After all registers are pushed, the contents of the SP register are decremented by 16. The PUSHA instruction is very useful when the entire register set (microprocessor environment) of the 80286 and above must be saved during a task. The PUSHAD instruction places the 32-bit register set on the stack in the 80386 through the Pentium 4. PUSHAD requires 32 bytes of stack storage space.

The PUSH immediate data instruction has two different opcodes, but in both cases, a 16-bit immediate number moves onto the stack; if PUSHD is used, a 32-bit immediate datum is pushed. If the value of the immediate data are 00H–FFH, the opcode is a 6AH; if the data are 0100H–FFFFH, the opcode is 68H. The PUSH 8 instruction, which pushes a 0008H onto the stack, assembles as a 6A08H. The PUSH 1000H instruction assembles as 680010H. Another example of PUSH immediate is the PUSH 'A' instruction, which pushes a 0041H onto the stack. Here, the 41H is the ASCII code for the letter A.

Table 4–7 lists the forms of the PUSH instruction that include PUSHA and PUSHF. Notice how the instruction set is used to specify different data sizes with the assembler.

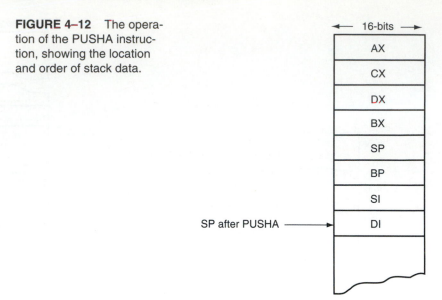

FIGURE 4–12 The operation of the PUSHA instruction, showing the location and order of stack data.

TABLE 4–7 The PUSH instructions.

Symbolic	Example	Note
PUSH reg16	PUSH BX	16-bit register
PUSH reg32	PUSH EDX	32-bit register
PUSH mem16	PUSH WORD PTR [BX]	16-bit pointer
PUSH mem32	PUSH DWORD PTR [EBX]	32-bit pointer
PUSH seg	PUSH DS	Segment register
PUSH imm8	PUSH ','	8-bit immediate
PUSHW imm16	PUSHW 1000H	16-bit immediate
PUSHD imm32	PUSHD 20	32-bit immediate
PUSHA	PUSHA	Save all 16-bit registers
PUSHAD	PUSHAD	Save all 32-bit registers
PUSHF	PUSHF	Save flags
PUSHFD	PUSHFD	Save EFLAGs

POP

The POP instruction performs the inverse operation of a PUSH instruction. The POP instruction removes data from the stack and places it into the target 16-bit register, segment register, or a 16-bit memory location. In the 80386 and above, a POP can also remove 32-bit data from the stack and use a 32-bit address. The POP instruction is not available as an immediate POP. The POPF **(pop flags)** instruction removes a 16-bit number from the stack and places it into the flag register; the POPFD removes a 32-bit number from the stack and places it into the extended flag register. The POPA **(pop all)** instruction removes 16 bytes of data from the stack and places it into the following registers, in the order shown: DI, SI, BP, SP, BX, DX, CX, and AX. This is the reverse order from the way they were placed on the stack by the PUSHA instruction, causing the same data to return to the same registers. In the 80386 and above, a POPAD instruction reloads the 32-bit registers from the stack.

Suppose that a POP BX instruction executes. The first byte of data removed from the stack (the memory location addressed by SP in the stack segment) moves into register BL. The second byte is removed from stack segment memory location SP + 1 and is placed into register BH.

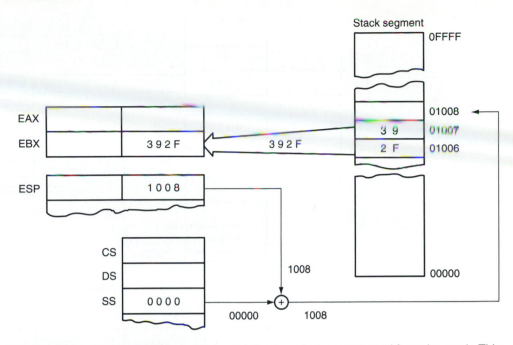

FIGURE 4–13 The POP BX instruction, showing how data are removed from the stack. This instruction is shown after execution.

TABLE 4–8 The POP instructions.

Symbolic	Example	Note
POP reg16	POP CX	16-bit register
POP reg32	POP EBP	32-bit register
POP mem16	POP WORD PTR[BX+1]	16-bit pointer
POP mem32	POP DATA3	32-bit memory address
POP seg	POP FS	Segment register
POPA	POPA	Pop all 16-bit registers
POPAD	POPAD	Pop all 32-bit registers
POPF	POPF	Pop flags
POPFD	POPFD	Pop EFLAGs

After both bytes are removed from the stack, the SP register increments by 2. Figure 4–13 shows how the POP BX instruction removes data from the stack and places them into register BX.

The opcodes used for the POP instruction and all of its variations appear in Table 4–8. Note that a POP CS instruction is not a valid instruction in the instruction set. If a POP CS instruction executes, only a portion of the address (CS) of the next instruction changes. This makes the POP CS instruction unpredictable and therefore not allowed.

Initializing the Stack

When the stack area is initialized, load both the stack segment (SS) register and the stack pointer (SP) register. It is normal to designate an area of memory as the stack segment by loading SS with the bottom location of the stack segment.

For example, if the stack segment is to reside in memory locations 10000H–1FFFFH, load SS with a 1000H. (Recall that the rightmost end of the stack segment register is appended

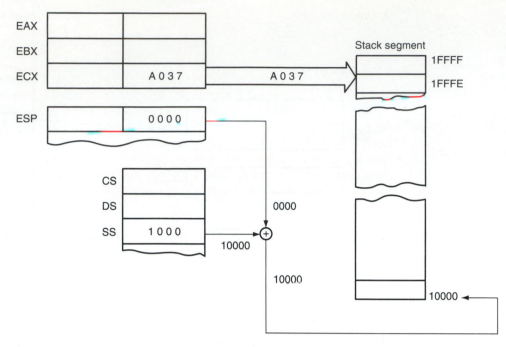

FIGURE 4–14   The PUSH CX instruction, showing the cyclical nature of the stack segment. This instruction is shown just before execution, to illustrate that the stack bottom is contiguous to the top.

with a 0H for real mode addressing.) To start the stack at the top of this 64K-byte stack segment, the stack pointer (SP) is loaded with a 0000H. Likewise, to address the top of the stack at location 10FFFH, use a value of 1000H in SP. Figure 4–14 shows how this value causes data to be pushed onto the top of the stack segment with a PUSH CX instruction. Remember that all segments are cyclic in nature—that is, the top location of a segment is contiguous with the bottom location of the segment.

In assembly language, a stack segment is set up as illustrated in Example 4–1. The first statement identifies the start of the stack segment and the last statement identifies the end of the stack segment. The assembler and linker programs place the correct stack segment address in SS and the length of the segment (top of the stack) into SP. There is no need to load these registers in your program unless you wish to change the initial values for some reason.

EXAMPLE 4–1

```
0000                    STACK_SEG     SEGMENT STACK

0000 0100[                      DW    100H DUP (?)
          ????
         ]
0200                    STACK_SEG     ENDS
```

An alternative method for defining the stack segment is used with one of the memory models for the MASM assembler only (refer to Appendix A). Other assemblers do not use models; if they do, the models are not exactly the same as with MASM. Here, the .STACK statement, followed by the number of bytes allocated to the stack, defines the stack area (see Example 4–2). The function is identical to Example 4–1. The .STACK statement also initializes both SS and SP. Note that this text uses memory models that are designed for the Microsoft Macro Assembler program MASM.

EXAMPLE 4–2

```
.MODEL SMALL
.STACK 200H ;set stack size
```

If the stack is not specified by using either method, a warning will appear when the program is linked. The warning may be ignored if the stack size is 128 bytes or fewer. The system automatically assigns (through DOS) at least 128 bytes of memory to the stack. This memory section is located in the program segment prefix (PSP), which is appended to the beginning of each program file. If you use more memory for the stack, you will erase information in the PSP that is critical to the operation of your program and the computer. This error often causes the computer program to crash. If the TINY memory model is used, the stack is automatically located at the very end of the segment, which allows for a larger stack area.

4–3   LOAD-EFFECTIVE ADDRESS

There are several load-effective address instructions in the microprocessor instruction set. The LEA instruction loads any 16-bit register with the address, as determined by the addressing mode selected for the instruction. The LDS and LES variations load any 16-bit register with the offset address retrieved from a memory location, and then load either DS or ES with a segment address retrieved from memory. In the 80386 and above, LFS, LGS, and LSS are added to the instruction set, and a 32-bit register can be selected to receive a 32-bit offset from memory. Table 4–9 lists the load-effective address instructions.

LEA

The LEA instruction loads a 16- or 32-bit register with the offset address of the data specified by the operand. As the first example in Table 4–9 shows, the operand address NUMB is loaded into register AX, not the contents of address NUMB.

By comparing LEA with MOV, it is observed that LEA BX,[DI] loads the offset address specified by [DI] (contents of DI) into the BX register; MOV BX,[DI] loads the data stored at the memory location addressed by [DI] into register BX.

Earlier in the text, several examples are presented by using the OFFSET directive. The OFFSET directive performs the same function as an LEA instruction if the operand is a displacement. For example, the MOV BX,OFFSET LIST performs the same function as LEA BX,LIST. Both instructions load the offset address of memory location LIST into the BX register. See Example 4–3 for a short program that loads SI with the address of DATA1 and DI with the

TABLE 4–9   Load-effective address instructions.

Assembly Language	Operation
LEA AX,NUMB	Loads AX with the address of NUMB
LEA EAX,NUMB	Loads EAX with the address of NUMB
LDS DI,LIST	Loads DS and DI with the 32-bit contents of data segment memory location LIST
LDS EDI,LIST	Loads DS and EDI with the 48-bit contents of data segment memory location LIST
LES BX,CAT	Loads ES and BX with the 32-bit contents of data segment memory location CAT
LFS DI,DATA1	Loads FS and DI with the 32-bit contents of data segment memory location DATA1
LGS SI,DATA5	Loads GS and SI with the 32-bit contents of data segment memory location DATA5
LSS SP,MEM	Loads SS and SP with the 32-bit contents of memory location MEM

address of DATA2. It then exchanges the contents of these memory locations. Note that the LEA and MOV with OFFSET instructions are both the same length (three bytes).

EXAMPLE 4–3

```
                                   .MODEL SMALL                 ;select SMALL model
0000                               .DATA                        ;start of DATA segment
0000    2000       DATA1   DW      2000H                        ;define DATA1
0002    3000       DATA2   DW      3000H                        ;define DATA2
0000                               .CODE                        ;start of CODE segment
                                   .STARTUP                     ;start of program
0017    BE 0000 R          LEA     SI,DATA1                     ;address DATA1 with SI
001A    BF 0002 R          MOV     DI,OFFSET DATA2              ;address DATA2 with DI

001D    8B 1C             MOV     BX,[SI]                       ;exchange DATA1 with DATA2
001F    8B 0D             MOV     CX,[DI]
0021    89 0C             MOV     [SI],CX
0023    89 1D             MOV     [DI],BX
                                   .EXIT                        ;exit to DOS
                                   END                          ;end of file
```

But why is the LEA instruction available if the OFFSET directive accomplishes the same task? First, OFFSET only functions with simple operands such as LIST. It may not be used for an operand such as [DI], LIST [SI], and so on. The OFFSET directive is more efficient than the LEA instruction for simple operands. It takes the microprocessor longer to execute the LEA BX,LIST instruction than the MOV BX,OFFSET LIST. The 80486 microprocessor, for example, requires two clocks to execute the LEA BX,LIST instruction and only one clock to execute MOV BX,OFFSET LIST. The reason that the MOV BX,OFFSET LIST instruction executes faster is because the assembler calculates the offset address of LIST, while the microprocessor calculates the LEA instruction. The MOV BX,OFFSET LIST instruction is actually assembled as a move immediate instruction and is more efficient.

Suppose that the microprocessor executes an LEA BX,[DI] instruction and DI contains a 1000H. Because DI contains the offset address, the microprocessor transfers a copy of DI into BX. A MOV BX,DI instruction performs this task in less time and is often preferred to the LEA BX,[DI] instruction.

Another example is LEA SI,[BX + DI]. This instruction adds BX to DI and stores the sum in the SI register. The sum generated by this instruction is a modulo-64K sum. If BX = 1000H and DI = 2000H, the offset address moved into SI is 3000H. If BX = 1000H and DI = FF00H, the offset address is 0F00H instead of 10F00H. Notice that the second result is a modulo-64K sum of 0F00H. (A **modulo-64K sum** drops any carry out of the 16-bit result.)

LDS, LES, LFS, LGS, and LSS

The LDS, LES, LFS, LGS, and LSS instructions load any 16-bit or 32-bit register with an offset address, and the DS, ES, FS, GS, or SS segment register with a segment address. These instructions use any of the memory-addressing modes to access a 32-bit or 48-bit section of memory that contains both the segment and offset address. The 32-bit section of memory contains a 16-bit offset and segment address, while the 48-bit section contains a 32-bit offset and a segment address. These instructions may not use the register addressing mode (MOD = 11). Note that the LFS, LGS, and LSS instructions are only available on 80386 and above, as are the 32-bit registers.

Figure 4–15 illustrates an example LDS BX,[DI] instruction. This instruction transfers the 32-bit number, addressed by DI in the data segment, into the BX and DS registers. The LDS, LES, LFS, LGS, and LSS instructions obtain a new far address from memory. The offset address appears first, followed by the segment address. This format is used for storing all 32-bit memory addresses.

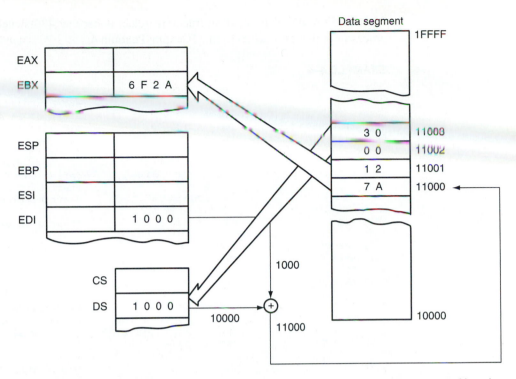

FIGURE 4–15 The LDS BX,[DI] instruction loads register BX from addresses 11000H and 11001H and register DS from locations 11002H and 11003H. This instruction is shown at the point just before DS changes to 3000H and BX changes to 127AH.

A far address can be stored in memory by the assembler. For example, the ADDR DD FAR PTR FROG instruction stores the offset and segment address (far address) of FROG in 32-bits of memory at location ADDR. The DD directive tells the assembler to store a doubleword (32-bit number) in memory address ADDR.

In the 80386 and above, an LDS EBX,[DI] instruction loads EBX from the 4-byte section of memory addressed by DI in the data segment. Following this 4-byte offset is a word that is loaded to the DS register. Notice that instead of addressing a 32-bit section of memory, the 80386 and above address a 48-bit section of the memory whenever a 32-bit offset address is loaded to a 32-bit register. The first four bytes contain the offset value loaded to the 32-bit register and the last two bytes contain the segment address.

The most useful of the load instructions is the LSS instruction. Example 4–4 shows a short program that creates a new stack area after saving the address of the old stack area. After executing some dummy instructions, the old stack area is reactivated by loading both SS and SP with the LSS instruction. Note that the CLI (disable interrupts) and STI (enable interrupts) instructions must be included to disable interrupts. (This topic is discussed near the end of this chapter.) Because the LSS instruction functions in the 80386 or above, the .386 statement appears after the .MODEL statement to select the 80386 microprocessor. Notice how the WORD PTR directive is used to override the doubleword (DD) definition for the old stack memory location. If an 80386 or newer microprocessor is in use, it is suggested that the .386 switch is used to develop software for the 80386 microprocessor. This is true even if the microprocessor is a Pentium, Pentium Pro, Pentium II, Pentium III, or Pentium 4. The reason is that the 80486–Pentium 4 microprocessors add only a few additional instructions to the 80386 instruction set, which are seldom used in software development. If the need arises to use any of the CMPXCHG, CMPXCHG8 (new to the

Pentium), XADD, or BSWAP instructions, select either the .486 switch for the 80486 micro-processor or the .586 switch for the Pentium–Pentium 4.

EXAMPLE 4–4

```
                            .MODEL  SMALL               ;select SMALL model
                            .386                        ;select 80386
0000                        .DATA                       ;start of DATA segment
0000  00000000   SADDR   DD     ?                       ;old stack address
0004  1000 [     SAREA   DW     1000H DUP (?)            ;new stack area
             0000
             ]
2004 = 2004      STOP    EQU    THIS WORD               ;define to of new stack
0000                        .CODE                       ;start of CODE segment
                            .STARTUP                    ;start of program

0010  FA                CLI                             ;disable interrupt

0011  8B C4             MOV    AX,SP                    ;save old SP
0013  A3 0000 R         MOV    WORD PTR SADDR,AX
0016  8C D0             MOV    AX,SS                    ;save old SS
0018  A3 0002 R         MOV    WORD PTR SADDR+2,AX

001B  8C D8             MOV    AX,DS                    ;load new SS
001D  8E D0             MOV    SS,AX
001F  B8 2004 R         MOV    AX,OFFSET STOP           ;load new SP
0022  8B E0             MOV    SP,AX

0024  FB                STI                             ;enable interrupt

0025  8B C0             MOV    AX,AX                    ;do dummy instructions
0027  8B C0             MOV    AX,AX
0029  0F B2 26 0000 R   LSS    SP,SADDR                 ;load old SS and SP

                            .EXIT                       ;exit to DOS
                            END                         ;end of file
```

4–4   ## STRING DATA TRANSFERS

There are five string data transfer instructions: LODS, STOS, MOVS, INS, and OUTS. Each string instruction allows data transfers that are either a single byte, word, or doubleword (or if repeated, a block of bytes, words, or doublewords). Before the string instructions are presented, the operation of the D flag-bit (direction), DI, and SI must be understood as they apply to the string instructions.

The Direction Flag

The direction flag (D) (located in the flag register) selects the auto-increment (D = 0) or the auto-decrement (D = 1) operation for the DI and SI registers during string operations. The direction flag is used only with the string instructions. The CLD instruction clears the D flag (D = 0) and the STD instruction sets it (D = 1). Therefore, the CLD instruction selects the auto-increment mode (D = 0) and STD selects the auto-decrement mode (D = 1).

Whenever a string instruction transfers a byte, the contents of DI and/or SI increment or decrement by 1. If a word is transferred, the contents of DI and/or SI increment or decrement by 2. Doubleword transfers cause DI and/or SI to increment or decrement by 4. Only the actual reg-isters used by the string instruction increment or decrement. For example, the STOSB instruction uses the DI register to address a memory location. When STOSB executes, only DI increments or decrements without affecting SI. The same is true of the LODSB instruction, which uses the SI register to address memory data. LODSB only increments/decrements SI without affecting DI.

DI and SI

During the execution of a string instruction, memory accesses occur through either or both of the DI and SI registers. The DI offset address accesses data in the extra segment for all string instructions that use it. The SI offset address accesses data, by default, in the data segment. The segment assignment of SI may be changed with a segment override prefix, as described later in this chapter. The DI segment assignment is always in the extra segment when a string instruction executes. This assignment cannot be changed. The reason that one pointer addresses data in the extra segment and the other in the data segment is so the MOVS instruction can move 64K bytes of data from one segment of memory to another.

LODS

The LODS instruction loads AL, AX, or EAX with data stored at the data segment offset address indexed by the SI register. (Note that only the 80386 and above use EAX.) After loading AL with a byte, AX with a word, or EAX with a doubleword, the contents of SI increment, if D = 0 or decrement, if D = 1. A 1 is added to or subtracted from SI for a byte-sized LODS, a 2 is added or subtracted for a word-sized LODS, and a 4 is added or subtracted for a doubleword-sized LODS.

Table 4–10 lists the permissible forms of the LODS instruction. The LODSB (**loads a byte**) instruction causes a byte to be loaded into AL, the LODSW (**loads a word**) instruction causes a word to be loaded into AX, and the LODSD (**loads a doubleword**) instruction causes a doubleword to be loaded into EAX. Although rare, as an alternative to LODSB, LODSW, and LODSD, the LODS instruction may be followed by a byte-, word- or doubleword-sized operand to select a byte, word, or doubleword transfer. Operands are often defined as bytes with DB, as words with DW, and as doublewords with DD. The DB pseudo-operation defines byte(s), the DW pseudo-operation defines word(s), and the DD pseudo-operations define doubleword(s).

Figure 4–16 shows the effect of executing the LODSW instruction if the D flag = 0, SI = 1000H, and DS = 1000H. Here, a 16-bit number stored at memory locations 11000H and 11001H moves into AX. Because D = 0 and this is a word transfer, the contents of SI increment by 2 after AX loads with memory data.

STOS

The STOS instruction stores AL, AX, or EAX at the extra segment memory location addressed by the DI register. (Note that only the 80386–Pentium 4 use EAX and doublewords.) Table 4–11 lists all forms of the STOS instruction. As with LODS, a STOS instruction may be appended with a B, W, or D for byte, word, or doubleword transfers. The STOSB (**stores a byte**) instruction stores the byte in AL at the extra segment memory location addressed by DI. The STOSW (**stores a word**) instruction stores AX in the extra segment memory location addressed by DI. A doubleword is stored in the extra segment location addressed by DI with the STOSD (**stores a**

TABLE 4–10 Forms of the LODS instruction.

Assembly Language	Operation
LODSB	AL = DS:[SI]; SI = SI ± 1
LODSW	AX = DS:[SI]; SI = SI ± 2
LODSD	EAX = DS:[SI]; SI = SI ± 4
LODS LIST	AL = DS:[SI]; SI = SI ± 1 (if LIST is a byte)
LODS DATA1	AX = DS:[SI], SI = SI ± 2 (if DATA1 is a word)
LODS FROG	EAX = DS:[SI]; SI = SI ± 4 (if FROG is a doubleword)

Note: The segment can be overridden with a segment override prefix as in LODS ES:DATA4.

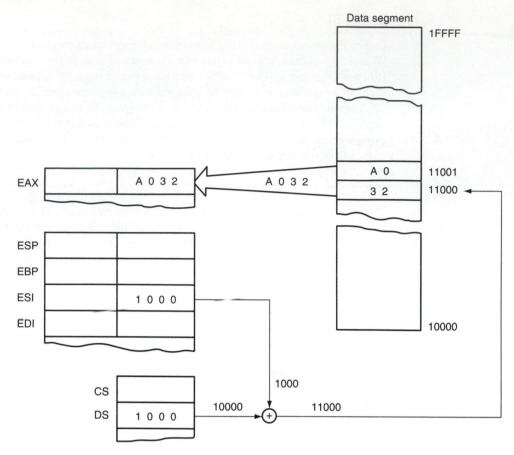

FIGURE 4–16   The operation of the LODSW instruction if DS = 1000H, D = 0, 11000H = 32, and 11001H = A0. This instruction is shown after AX is loaded from memory, but before SI increments by 2.

TABLE 4–11 Forms of the STOS instruction.

Assembly Language	Operation
STOSB	ES:[DI] = AL; DI = DI ± 1
STOSW	ES:[DI] = AX; DI = DI ± 2
STOSD	ES:[DI] = EAX; DI = DI ± 4
STOS LIST	ES:[DI] = AL; DI = DI ± 1 (if list is a byte)
STOS DATA3	ES:[DI] = AX; DI = DI ± 2 (if DATA3 is a word)
STOS DATA4	ES:[DI] = EAX; DI = DI ± 4 (if DATA4 is a doubleword)

doubleword) instruction. After the byte (AL), word (AX), or doubleword (EAX) is stored, the contents of DI increments or decrements.

STOS with a REP.   The **repeat prefix** (REP) is added to any string data transfer instruction, except the LODS instruction. It doesn't make any sense to perform a repeated LODS operation. The REP prefix causes CX to decrement by 1 each time the string instruction executes. After CX decrements, the string instruction repeats. If CX reaches a value of 0, the instruction terminates and the program continues with the next sequential instruction. Thus, if CX is loaded with a 100,

TABLE 4–12   Common operand operators.

Operator	Example	Comment
+	MOV AL,6+3	Copies 9 into AL
−	MOV AL,8–2	Copies 6 into AL
*	MOV AL,4*3	Copies 12 into AL
/	MOV AX,12/5	Copies 2 into AX (remainder is lost)
MOD	MOV AX, 12 MOD 7	Copies 5 into AX (quotient is lost)
AND	MOV AX,12 AND 4	Copies 4 into AX (1100 AND 0100 = 0100)
OR	MOV AX,12 OR 1	Copies 13 into AX (1100 OR 0001 = 1101)
NOT	MOV AL,NOT 1	Copies 254 into AL (0000 0001 NOT equals 1111 1110 or 254)

and a REP STOSB instruction executes, the microprocessor automatically repeats the STOSB instruction 100 times. Because the DI register is automatically incremented or decremented after each datum is stored, this instruction stores the contents of AL in a block of memory instead of a single byte of memory.

Suppose that the STOSW instruction is used to clear the video text display (see Example 4–5). This is accomplished by addressing video text memory that begins at memory location B800:0000. Each character position on the 25-line-by-80-character per line display comprises two bytes. The first byte contains the ASCII-coded character, and the second contains the color and attributes of the character. In this example, AL is the ASCII-coded space (20H) and AH is the color code for white text on a black background (07H). Notice how this program uses a count of 25 * 80 and the REP STOSW instruction to clear the screen with ASCII spaces.

The operands in a program can be modified by using arithmetic or logic operators such as multiplication (*). Other operators appear in Table 4–12.

EXAMPLE 4–5

```
                        .MODEL TINY              ;select TINY model
0000                    .CODE                    ;start of CODE segment
                        .STARTUP                 ;start of program
0100   FC                       CLD              ;select increment mode
0101   B8 B800                  MOV     AX,0B800H ;address segment B800
0104   8E C0                    MOV     ES,AX

0106   BF 0000                  MOV     DI,0      ;address offset 0000
0109   B9 07D0                  MOV     CX,25*80  ;load count
010C   B8 0720                  MOV     AX,0720H  ;load data

010F   F3/AB                    REP     STOSW     ;clear the screen
                        .EXIT                     ;exit to DOS
                        END                       ;end of file
```

The REP prefix precedes the STOSW instruction in both assembly language and hexadecimal machine language. In machine language, the F3H is the REP prefix and ABH is the STOSW opcode.

If the value loaded to AX is changed to 0731H, the video display fills with white ones on a black background. If AX is changed to 0132H, the video display fills with blue twos on a black background. By changing the value loaded to AX, the display can be filled with any character and any color combination. More information on accessing the video display appears in a later chapter.

MOVS

One of the more useful string data transfer instructions is MOVS because it transfers data from one memory location to another. This is the only memory-to-memory transfer allowed in the 8086–Pentium 4 microprocessors. The MOVS instruction transfers a byte, word, or doubleword

TABLE 4–13   Forms of the MOVS instruction.

Assembly Language	Operation
MOVSB	ES:[DI] = DS:[SI]; DI = DI ± 1; SI = SI ± 1 (byte transferred)
MOVSW	ES:[DI] = DS:[SI]; DI = DI ± 2; SI = SI ± 2 (word transferred)
MOVSD	ES:[DI] = DS:[SI]; DI = DI ± 4; SI = SI ± 4 (doubleword transferred)
MOVS BYTE1,BYTE2	ES:[DI] = DS:[SI]; DI = DI ± 1; SI = SI ± 1 (if BYTE1 and BYTE2 are bytes)
MOVS WORD1,WORD2	ES:[DI] = DS:[SI]; DI = DI ± 2, SI = SI ± 2 (if WORD1 and WORD2 are words)
MOVS DWORD1, DWORD2	ES:[DI] = DS:[SI]; DI = DI ± 4; SI = SI ± 4 (if DWORD1 and DWORD2 are doublewords)

from the data segment location addressed by SI to the extra segment location addressed by DI. As with the other string instructions, the pointers then increment or decrement, as dictated by the direction flag. Table 4–13 lists all the permissible forms of the MOVS instruction. Note that only the source operand (SI), located in the data segment, may be overridden so that another segment may be used. The destination operand (DI) must always be located in the extra segment.

Suppose that the video display needs to be scrolled up one line. Because we now know the location of the video display, a repeated MOVSW instruction can be used to scroll the video display up a line. Example 4–6 lists a short program that addresses the video text display, beginning at location B800:0000 with the DS:SI register combination and location B800:00A0 with the ES:DI register combination. Next, the REP MOVSW instruction is executed 24 * 80 times to scroll the display up a line. This is followed by a sequence that addresses the last line of the display so it can be cleared. The last line is cleared in this example by storing spaces on a black background. The last line could be cleared by changing only the ASCII code to a space, without modifying the attribute, by reading the code and attribute into a register. Once in a register, the code is modified, and both the code and attribute are stored in memory.

EXAMPLE 4–6

```
                        .MODEL TINY             ;select TINY model
0000                    .CODE                   ;indicate start of CODE segment
                        .STARTUP                ;indicate start of program
0100  FC                CLD                      ;select increment
0101  B8 B800           MOV    AX,0B800H         ;load ES and DS with B800
0104  8E C0             MOV    ES,AX
0106  8E D8             MOV    DS,AX

0108  BE 00A0           MOV    SI,160            ;address line 1
010B  BF 0000           MOV    DI,0              ;address line 0
010E  B9 0780           MOV    CX,24*80          ;load count
0111  F3/A5             REP    MOVSW             ;scroll screen

0113  BF 0F00           MOV    DI,24*80*2        ;clear bottom line
0116  B9 0050           MOV    CX,80
0119  B8 0720           MOV    AX,0720H
011C  F3/AB             REP    STOSW
                  .EXIT                          ;exit to DOS
                  END                            ;end of file
```

INS

The INS (**input string**) instruction (not available on the 8086/8088 microprocessors) transfers a byte, word, or doubleword of data from an I/O device into the extra segment memory location addressed by the DI register. The I/O address is contained in the DX register. This instruction is

TABLE 4–14 Forms of the INS instruction.

Assembly Language	Operation
INSB	ES:[DI] = [DX]; DI = DI ± 1 (byte transferred)
INSW	ES:[DI] = [DX]; DI = DI ± 2 (word transferred)
INSD	ES:[DI] = [DX]; DI = DI ± 4 (doubleword transferred)
INS LIST	ES:[DI] = [DX]; DI = DI ± 1 (if LIST is a byte)
INS DATA4	ES:[DI] = [DX]; DI = DI ± 2 (if DATA4 is a word)
INS DATA5	ES:[DI] = [DX]; DI = DI ± 4 (if DATA5 is a doubleword)

Note: [DX] indicates that DX contains the I/O device address. These instructions are not available on the 8086/8088 microprocessors.

useful for inputting a block of data from an external I/O device directly into the memory. One application transfers data from a disk drive to memory. Disk drives are often considered and interfaced as I/O devices in a computer system.

As with the prior string instructions, there are two basic forms of the INS. The INSB instruction inputs data from an 8-bit I/O device and stores it in the byte-sized memory location indexed by SI. The INSW instruction inputs 16-bit I/O data and stores it in a word-sized memory location. The INSD instruction inputs a doubleword. These instructions can be repeated using the REP prefix. This allows an entire block of input data to be stored in the memory from an I/O device. Table 4–14 lists the various forms of the INS instruction.

Example 4–7 shows a sequence of instructions that input 50 bytes of data from an I/O device whose address is 03ACH and stores the data in extra segment memory array LISTS. This software assumes that data are available from the I/O device at all times. Otherwise, the software must check to see if the I/O device is ready to transfer data precluding the use of a REP prefix.

EXAMPLE 4–7

```
                        ;Using the REP INSB to input data to a memory array
                        ;
0000  BF 0000 R              MOV     DI,OFFSET LISTS   ;address array
0003  BA 03AC                MOV     DX,3ACH           ;address I/O
0006  FC                     CLD                       ;auto-increment
0007  B9 0032                MOV     CX,50             ;load count
000A  F3/6C                  REP INSB                  ;input data
```

OUTS

The OUTS (**output string**) instruction (not available on the 8086/8088 microprocessors) transfers a byte, word, or doubleword of data from the data segment memory location address by SI to an I/O device. The I/O device is addressed by the DX register as it was with the INS instruction. Table 4–15 shows the variations available for the OUTS instruction.

Example 4–8 shows a short sequence of instructions that transfer data from a data segment memory array (ARRAY) to an I/O device at I/O address 3ACH. This software assumes that the I/O device is always ready for data.

EXAMPLE 4–8

```
                        ;Using the REP OUTS to output data from a memory array
                        ;
0000  BE 0064 R              MOV     SI,OFFSET ARRAY   ;address array
0003  BA 03AC                MOV     DX,3ACH           ;address I/O
0006  FC                     CLD                       ;auto-increment
0007  B9 0064                MOV     CX,100            ;load count
000A  F3/6E                  REP OUTSB
```

TABLE 4–15 Forms of the
OUTS instruction.

Assembly Language	Operation
OUTSB	[DX] = DS:[SI]; SI = SI ± 1 (byte transferred)
OUTSW	[DX] = DS:[SI]; SI = SI ± 2 (word transferred)
OUTSD	[DX] = DS:[SI]; SI = SI ± 4 (doubleword transferred)
OUTS DATA7	[DX] = DS:[SI]; SI = SI ± 1 (if DATA7 is a byte)
OUTS DATA8	[DX] = DS:[SI]; SI = SI ± 2 (if DATA8 is a word)
OUTS DATA9	[DX] = DS:[SI]; SI = SI ± 4 (if DATA9 is a doubleword)

Note: [DX] indicates that DX contains the I/O device address. These instructions are not available on the 8086/8088 microprocessors.

4–5   MISCELLANEOUS DATA TRANSFER INSTRUCTIONS

Don't be fooled by the term *miscellaneous;* these instructions are used in programs. The data transfer instructions detailed in this section are XCIIG, LAHF, SAHF, XLAT, IN, OUT, BSWAP, MOVSX, MOVZX, and CMOV. Because the miscellaneous instructions are not used as often as a MOV instruction, they have been grouped together and represented in this section.

XCHG

The XCHG (**exchange**) instruction exchanges the contents of a register with the contents of any other register or memory location. The XCHG instruction cannot exchange segment registers or memory-to-memory data. Exchanges are byte-, word-, or doubleword-sized (80386 and above), and use any addressing mode discussed in Chapter 3, except immediate addressing. Table 4–16 shows some examples of the XCHG instruction.

The XCHG instruction, using the 16-bit AX register with another 16-bit register, is the most efficient exchange. This instruction occupies one byte of memory. Other XCHG instructions require two or more bytes of memory, depending on the addressing mode selected.

When using a memory addressing mode and the assembler, it doesn't matter which operand addresses memory. The XCHG AL,[DI] instruction is identical to the XCHG [DI],AL instruction, as far as the assembler is concerned.

If the 80386 through the Pentium 4 microprocessor is available, the XCHG instruction can exchange doubleword data. For example, the XCHG EAX,EBX instruction exchanges the contents of the EAX register with the EBX register.

LAHF and SAHF

The LAHF and SAHF instructions are seldom used because they were designed as bridge instructions. These instructions allowed 8085 (an early 8-bit microprocessor) software to be

TABLE 4–16   Forms of the XCHG instruction.

Assembly Language	Operation
XCHG AL,CL	Exchanges the contents of AL with CL
XCHG CX,BP	Exchanges the contents of CX with BP
XCHG EDX,ESI	Exchanges the contents of EDX with ESI
XCHG AL,DATA2	Exchanges the contents of AL with data segment memory location DATA2

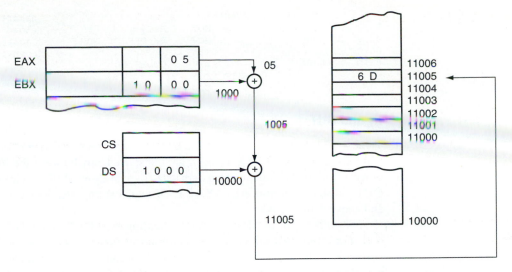

FIGURE 4–17 The operation of the XLAT instruction at the point just before 6DH is loaded into AL.

translated into 8086 software by a translation program. Because any software that required translation was probably completed many years ago, these instructions have little application today. The LAHF instruction transfers the rightmost eight bits of the flag register into the AH register. The SAHF instruction transfers the AH register into the rightmost eight bits of the flag register.

Sometimes, the SAHF instruction may find some application with the numeric coprocessor. The numeric coprocessor contains a status register that is copied into the AX register with the FSTSW AX instruction. The SAHF instruction is then used to copy from AH into the flag register. The flags are then tested for some of the conditions of the numeric coprocessor. This is detailed in Chapter 14, which explains the operation and programming of the numeric coprocessor.

XLAT

The XLAT (**translate**) instruction converts the contents of the AL register into a number stored in a memory table. This instruction performs the direct table lookup technique often used to convert one code to another. An XLAT instruction first adds the contents of AL to BX to form a memory address within the data segment. It then copies the contents of this address into AL. This is the only instruction that adds an 8-bit number to a 16-bit number.

Suppose that a 7-segment LED display lookup table is stored in memory at address TABLE. The XLAT instruction then translates the BCD number in AL to a 7-segment code in AL. Example 4–9 provides a short program that converts from a BCD code to a 7-segment code. Figure 4–17 shows the operation of this example program if TABLE = 1000H, DS = 1000H, and the initial value of AL = 05H (a 5 BCD). After the translation, AL = 6DH.

EXAMPLE 4–9

```
                        ;Using an XLAT to convert from BCD to 7-segment code
                        ;
                        .MODEL SMALL              ;select SMALL model
0000                    .DATA                     ;start of DATA segment
0000  3F 06 5B 4F   TABLE  DB    3FH,6,5BH,4FH    ;7-segment lookup table
0004  66 6D 7D 27          DB    66H,6DH,7DH,27H
0008  7F 6F               DB    7FH,6FH
000A  00            CODE7  DB    ?                ;reserve for result
0000                    .CODE                     ;start of CODE segment
```

```
                        .STARTUP                            ;start of program
0017  B0 04                     MOV    AL,4                 ;load test data
0019  BB 0000 R                 MOV    BX,OFFSET TABLE      ;address lookup table
001C  D7                        XLAT                        ;convert to 7-segment
001D  A2 000A R                 MOV    CODE7,AL             ;save 7-segment code
                        .EXIT                               ;exit to DOS
                        END                                 ;end of file
```

IN and OUT

Table 4–17 lists the forms of the IN and OUT instructions, which perform I/O operations. Notice that the contents of AL, AX, or EAX are transferred only between the I/O device and the microprocessor. An IN instruction transfers data from an external I/O device to AL, AX, or EAX; an OUT transfers data from AL, AX, or EAX to an external I/O device. (Note that only the 80386 and above contain EAX.)

Two forms of I/O device (port) addressing exist for IN and OUT: fixed-port and variable-port. *Fixed-port addressing* allows data transfer between AL, AX, or EAX using an 8-bit I/O port address. It is called fixed-port addressing because the port number follows the instruction's opcode. Often, instructions are stored in a ROM. A fixed-port instruction stored in a ROM has its port number permanently fixed because of the nature of read-only memory. A fixed-port address stored in a RAM can be modified, but such a modification does not conform to good programming practices.

The port address appears on the address bus during an I/O operation. For the 8-bit fixed-port I/O instructions, the 8-bit port address is zero-extended into a 16-bit address. For example, if the IN AL,6AH instruction executes, data from I/O address 6AH are input to AL. The address appears as a 16-bit 006AH on pins A0–A15 of the address bus. Address bus bits A16–A19 (8086/8088), A16–A23 (80286/80386SX), A16-A24 (80386SL/80386SLC/80386EX), or A16–A32 (80386–Pentium 4) are undefined for an IN or OUT instruction. Note that Intel reserves the last 16 I/O ports for use with some of its peripheral components.

Variable-port addressing allows data transfers between AL, AX, or EAX and a 16-bit port address. It is called *variable-port addressing* because the I/O port number is stored in register DX, which can be changed (varied) during the execution of a program. The 16-bit I/O port address appears on the address bus pin connections A0–A15. The IBM PC uses a 16-bit port address to access its I/O space. The I/O space for a PC is located at I/O port 0000H–03FFH. Note that some plug-in adapter cards may use I/O addresses above 03FFH.

TABLE 4–17 IN and OUT instructions.

Assembly Language	Operation
IN AL,p8	8-bits are input to AL from I/O port p8
IN AX,p8	16-bits are input to AX from I/O port p8
IN EAX,p8	32-bits are input to EAX from I/O port p8
IN AL,DX	8-bits are input to AL from I/O port DX
IN AX,DX	16-bits are input to AX from I/O port DX
IN EAX,DX	32-bits are input to EAX from I/O port DX
OUT p8,AL	8-bits are output from AL to I/O port p8
OUT p8,AX	16-bits are output from AX to I/O port p8
OUTp8,EAX	32-bits are output from EAX to I/O port p8
OUT DX,AL	8-bits are output from AL to I/O port DX
OUT DX,AX	16-bits are output from AX to I/O port DX
OUT DX,EAX	32-bits are output from EAX to I/O port DX

Note: p8 = an 8-bit I/O port number and DX = the 16-bit port address held in DX.

Microprocessor-based system

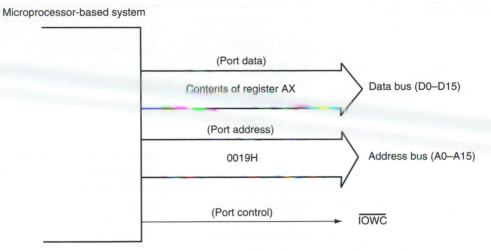

FIGURE 4–18 The signals found in the microprocessor-based system for an OUT 19H,AX instruction.

Figure 4–18 illustrates the execution of the OUT 19H,AX instruction, which transfers the contents of AX to I/O port 19H. Notice that the I/O port number appears as a 0019H on the 16-bit address bus and that the data from AX appears on the data bus of the microprocessor. The system control signal $\overline{\text{IOWC}}$ (I/O write control) is a logic zero to enable the I/O device.

A short program that clicks the speaker in the personal computer appears in Example 4–10. The speaker is controlled by accessing I/O port 61H. If the rightmost two bits of this port are set (11) and then cleared (00), a click is heard on the speaker. Note that this program uses a logical OR instruction to set these two bits and a logical AND instruction to clear them. These logic operation instructions are described in Chapter 5. The MOV CX,4000H instruction, followed by the LOOP L1 instruction, is used as a time delay. If the count is increased, the click will become longer; if shortened, the click will become shorter.

EXAMPLE 4–10

```
                        .MODEL TINY                    ;select TINY model
0000                    .CODE                          ;indicate start of code segment
                        .STARTUP                       ;indicate start of program
0100  E4 61                     IN     AL,61H          ;read port 61H
0102  0C 03                     OR     AL,3            ;set rightmost two bits
0104  E6 61                     OUT    61H,AL          ;speaker is on

0106  B9 1000                   MOV    CX,1000H        ;delay count
0109                    L1:
0109  E2 FE                     LOOP   L1              ;time delay

010B  E4 61                     IN     AL,61H          ;read port 61H
010D  24 FC                     AND    AL,0FCH         ;clear rightmost two bits
010F  E6 61                     OUT    61H,AL          ;speaker is off
                        .EXIT                          ;exit to DOS
                        END                            ;end of file
```

MOVSX and MOVZX

The MOVSX (**move and sign-extend**) and MOVZX (**move and zero-extend**) instructions are found in the 80386–Pentium 4 instruction sets. These instructions move data, and at the same time either sign- or zero-extend it. Table 4–18 illustrates these instructions with several examples of each.

TABLE 4–18   The MOVSX
and MOVZX instructions.

Assembly Language	Operation
MOVSX CX,BL	Sign-extends BL into CX
MOVSX ECX,AX	Sign-extends AX into ECX
MOVSX BX,DATA1	Sign-extends the byte at DATA1 into BX
MOVSX EAX,[EDI]	Sign-extends the word at the data segment memory location addressed by EDI into EAX
MOVZX DX,AL	Zero-extends AL into DX
MOVZX EBP,DI	Zero-extends DI into EBP
MOVZX DX,DATA2	Zero-extends the byte at data segment memory location DATA2 into DX
MOVZX EAX,DATA3	Zero-extends the word at data segment memory location DATA3 into EAX

When a number is zero-extended, the most-significant part fills with zeros. For example, if an 8-bit 34H is zero-extended into a 16-bit number, it becomes 0034H. Zero-extension is often used to convert unsigned 8- or 16-bit numbers into unsigned 16- or 32-bit numbers by using the MOVZX instruction.

A number is sign-extended when its sign-bit is copied into the most-significant part. For example, if an 8-bit 84H is sign-extended into a 16-bit number, it becomes FF84H. The sign-bit of an 84H is a one, which is copied into the most-significant part of the sign-extended result. Sign-extension is most often used to convert 8- or 16-bit signed numbers into 16- or 32-bit signed numbers by using the MOVSX instruction.

BSWAP

The BSWAP **(byte swap)** instruction is available only in the 80486 and all versions of the Pentium microprocessors. This instruction takes the contents of any 32-bit register and swaps the first byte with the fourth, and the second with the third. For example, the BSWAP EAX instruction with EAX = 00112233H swaps bytes in EAX, resulting in EAX = 33221100H. Notice that the order of all four bytes is reversed by this instruction. This instruction is used to convert data from big endian form to little endian form, or vice versa.

CMOV

The CMOV **(conditional move)** class of instruction is new to the Pentium Pro and Pentium II instruction sets. There are many variations of the CMOV instruction. Table 4–19 lists these variations of CMOV. These instructions move the data only if the condition is true. For example, the CMOVZ instruction moves data only if the result from some prior instruction was a zero. The destination is limited to only a 16- or 32-bit register, but the source can be a 16- or 32-bit register or memory location.

Because this is a new instruction, you cannot use it with the assembler until a .686 switch is provided. In the interim, the instruction can be coded in hexadecimal form by using the DB directive. The opcode for the CMOV instruction is a 0F4XH, where X is the condition code 0000–1111 (refer to Appendix B for the codes). This is followed by a mod-reg-r/m byte. Example 4–11 shows how the CMOVB instruction is coded into hexadecimal by using the DB directive.

EXAMPLE 4–11

```
0000   0F 42 C3        DB   0FH,42H,0C3H     ;same as CMOVB AX,BX
```

TABLE 4–19   The conditional move instructions.

Assembly Language	Condition Tested	Operation
CMOVB	C = 1	Move if below
CMOVAE	C = 0	Move if above or equal
CMOVBE	Z = 1 or C = 1	Move if below or equal
CMOVA	Z = 0 and C = 0	Move if above
CMOVE or CMOVZ	Z = 1	Move if equal or set if zero
CMOVNE or CMOVNZ	Z = 0	Move if not equal or set if not zero
CMOVL	S <> O	Move if less than
CMOVLE	Z = 1 or S <> O	Move if less than or equal
CMOVG	Z = 0 and S = O	Move if greater than
CMOVGE	S = O	Move if greater than or equal
CMOVS	S = 1	Move if sign (negative)
CMOVNS	S = 0	Move if no sign (positive)
CMOVC	C = 1	Move if carry
CMOVNC	C = 0	Move if no carry
CMOVO	O = 1	Move if overflow
CMOVNO	O = 0	Move if no overflow
CMOVP or CMOVPE	P = 1	Move if parity or set if parity even
CMOVNP or CMOVPO	P = 0	Move if no parity or set if parity odd

4–6   SEGMENT OVERRIDE PREFIX

The **segment override prefix,** which may be added to almost any instruction in any memory-addressing mode, allows the programmer to deviate from the default segment. The segment override prefix is an additional byte that appends the front of an instruction to select an alternate segment register. About the only instructions that cannot be prefixed are the jump and call instructions that must use the code segment register for address generation. The segment override is also used to select the FS and GS segments in the 80386 through the Pentium 4 microprocessors.

For example, the MOV AX,[DI] instruction accesses data within the data segment by default. If required by a program, this can be changed by prefixing the instruction. Suppose that the data are in the extra segment instead of in the data segment. This instruction addresses the extra segment if changed to MOV AX,ES:[DI].

Table 4–20 shows some altered instructions that address different memory segments that are different from normal. Each time an instruction is prefixed with a segment override prefix, the instruction becomes one byte longer. Although this is not a serious change to the length of the instruction, it does add to the instruction's execution time. It is usually customary to limit the use of the segment override prefix and remain in the default segments so that shorter and more efficient software can be written.

TABLE 4–20   Instructions that include segment override prefixes.

Assembly Language	Segment Accessed	Default Segment
MOV AX,DS:[BP]	Data	Stack
MOV AX,ES:[BP]	Extra	Stack
MOV AX,SS:[DI]	Stack	Data
MOV AX,CS:LIST	Code	Data
MOV AX,ES:[SI]	Extra	Data
LODS ES:DATA1	Data	Extra
MOV EAX,FS:DATA2	Data	FS
MOV BL,GS:[ECX]	Data	GS

4–7   ASSEMBLER DETAIL

The assembler[1] for the microprocessor can be used in two ways: (1) with models that are unique to a particular assembler, and (2) with full segment definitions that allow complete control over the assembly process and are universal to all assemblers. This section of the text presents both methods, and explains how to organize a program's memory space by using the assembler. It also explains the purpose and use of some of the more important directives used with this assembler. Appendix A provides additional detail about the assembler.

Directives

Before the format of an assembly language program is discussed, some details about the directives (**pseudo-operations**) that control the assembly process must be learned. Some common assembly language directives appear in Table 4–21. **Directives** indicate how an operand or section of a program is to be processed by the assembler. Some directives generate and store information in the memory, while others do not. The DB (**define byte**) directive stores bytes of data in the memory, while the BYTE PTR directive never stores data. The **BYTE PTR** directive indicates the size of the data referenced by a pointer or index register.

Note that the assembler by default accepts only 8086/8088 instructions, unless a program is preceded by the .386 or .386P directive or one of the other microprocessor selection switches. The .386 directive tells the assembler to use the 80386 instruction set in the real mode, while the .386P directive tells the assembler to use the 80386 protected mode instruction set. Most software is written assuming that the microprocessor is an 80386 or newer, so the .386 switch is often used. Windows 95 was the first major operating system to use a 32-bit architecture that conforms to the 80386.

Storing Data in a Memory Segment.   The DB (**define byte**), DW (**define word**), and DD (**define doubleword**) directives, first presented in Chapter 1, are most often used with the microprocessor to define and store memory data. If a numeric coprocessor executes software in the system, the DQ (**define quadword**) and DT (**define ten bytes**) directives are also common. These directives label a memory location with a symbolic name and indicate its size.

Example 4–12 shows a memory segment that contains various forms of data definition directives. It also shows the full segment definition with the first SEGMENT statement to indicate the start of the segment and its symbolic name. Alternately, as in past examples in this and prior chapters, the SMALL model can be used with the .DATA statement. The last statement in this example contains the ENDS directive, which indicates the end of the segment. The name of the segment (LIST_SEG) can be anything that the programmer desires to call it. This allows a program to contain as many segments as required.

EXAMPLE 4–12

```
                              ;Using the DB, DW, and DD directives
                              ;
0000                          LIST_SEG      SEGMENT

0000   01 02 03               DATA1    DB    1,2,3           ;define bytes
0003   45                              DB    45H             ;hexadecimal
0004   41                              DB    'A'             ;ASCII
0005   F0                              DB    11110000B       ;binary
0006   000C 000D              DATA2    DW    12,13           ;define words
000A   0200                            DW    LIST1           ;symbolic
000C   2345                            DW    2345H           ;hexadecimal
000E   00000300              DATA3    DD    300H            ;hexadecimal
```

---

[1]The assembler used throughout this text is the Microsoft MACRO assembler MASM, version 6.X

TABLE 4–21 Common assembler directives.

Directive	Function
.286	Selects the 80286 instruction set
.286P	Selects the 80286 protected mode instruction set
.386	Selects the 80386 instruction set
.386P	Selects the 80386 protected mode instruction set
.486	Selects the 80486 instruction set
.486P	Selects the 80486 protected mode instruction set
.586	Selects the Pentium instruction set
.586P	Selects the Pentium protected mode instruction set
.287	Selects the 80287 math coprocessor
.387	Selects the 80387 math coprocessor
.EXIT	Exits to DOS
.MODEL	Selects the programming model
.STARTUP	Indicates the start of the program when using program models
ALIGN 2	Starts data on a word boundary (4 starts data on a doubleword boundary)
ASSUME	Informs the assembler of the name of each segment for full segment definitions
BYTE	Indicates byte-sized, as in BYTE PTR
DB	Defines byte(s) (8-bits)
DD	Defines doubleword(s) (32-bits)
DQ	Defines quadword(s) (64-bits)
DT	Defines ten byte(s) (80-bits)
DUP	Generates duplicates
DW	Defines word(s) (16-bits)
DWORD	Indicates doubleword-sized, as in DWORD PTR
END	Ends a program file
ENDM	Ends a macro sequence
ENDP	Ends a procedure
ENDS	Ends a segment or data structure
EQU	Equates data to a label
FAR	Defines a far pointer
MACRO	Designates the start of a macro sequence
NEAR	Defines a near pointer
OFFSET	Specifies an offset address
ORG	Sets the origin within a segment
PROC	Starts a procedure
PTR	Designates a pointer
SEGMENT	Starts a segment
STACK	Starts a stack segment
STRUC	Defines the start of a data structure
USES	Automatically pushes and pops registers within a procedure
USE16	Directs the assembler to use 16-bit instruction mode and data sizes for the 80386–Pentium 4
USE32	Directs the assembler to use 32-bit instruction mode and data sizes for the 80386–Pentium 4
WORD	Indicates word-sized, as in WORD PTR

```
0012   4007DF3B                  DD      2.123          ;real
0016   544269E1                  DD      3.34E+12       ;real
001A   00              LISTA     DB      ?              ;reserve 1 byte
001B   000A[           LISTB     DB      10 DUP (?)     ;reserve 10 bytes
                ??
                          ]
0025   00                        ALIGN   2              ;set word boundary

0026   0100[           LISTC     DW      100H DUP (0)   ;word array
                0000
```

```
                                ]
0226  0016[           LIST_9  DD     22 DUP (?)          ;doubleword array
               ????????
                                ]
027E  0064[           SIXES   DB     100 DUP (6)         ;byte array
          06
                      ]
02E2                  LIST_SEG ENDS
```

Example 4–12 shows various forms of data storage for bytes at DATA1. More than one byte can be defined on a line in binary, hexadecimal, decimal, or ASCII code. The DATA2 label shows how to store various forms of word data. Doublewords are stored at DATA3; they include floating-point, single-precision real numbers.

Memory is **reserved** for use in the future by using a ? as an operand for a DB, DW, or DD directive. When a ? is used in place of a numeric or ASCII value, the assembler sets aside a location and does not initialize it to any specific value. (Actually, the assembler usually stores a zero into locations specified with a ?). The DUP **(duplicate)** directive creates an array, as shown in several ways in Example 4–12. A 10 DUP (?) reserves 10 locations of memory, but stores no specific value in any of the 10 locations. If a number appears within the ( ) part of the DUP statement, the assembler initializes the reserved section of memory with the data indicated. For example, the DATA1 DB 10 DUP (2) instruction reserves 10 bytes of memory for array DATA1 and initializes each location with a 02H.

The ALIGN directive, used in this example, makes sure that the memory arrays are stored on word boundaries. An ALIGN 2 places data on *word boundaries* and an ALIGN 4 places them on *doubleword boundaries.* In the Pentium–Pentium 4, quadword data for double-precision floating-point numbers should use ALIGN 8. It is important that word-sized data are placed at word boundaries and doubleword-sized data are placed at doubleword boundaries. If not, the microprocessor spends additional time accessing these data types. A word stored at an odd-numbered memory location takes twice as long to access as a word stored on an even-numbered memory location. Note that the ALIGN directive cannot be used with memory models because the size of the model determines the data alignment. If all doubleword data are defined first, followed by word and then byte-sized data, the ALIGN statement is not necessary to align data correctly.

ASSUME, EQU, and ORG.   The equate directive (EQU) equates a numeric, ASCII, or label to another label. Equates make a program clearer and simplify debugging. Example 4–13 shows several equate statements and a few instructions that show how they function in a program.

EXAMPLE 4–13

```
     ;Using equate directive
     ;
= 000A TEN EQU 10
= 0009 NINE EQU 9

0000 B0 0A MOV AL,TEN
0002 04 09 ADD AL,NINE
```

The THIS directive always appears as THIS BYTE, THIS WORD, or THIS DWORD. In certain cases, data must be referred to as both a byte and a word. The assembler can only assign either a byte or a word address to a label. To assign a byte label to a word, use the software listed in Example 4–14.

EXAMPLE 4–14

```
              ;Using the THIS and ORG directives
              ;
0000          DATA_SEG    SEGMENT

0100                ORG    100H

= 0100        DATA1   EQU    THIS BYTE
```

```
0100   0000               DATA2    DW     ?

0102                      DATA_SEG     ENDS

0000                      CODE_SEG     SEGMENT 'CODE'

                          ASSUME  CS:CODE_SEG,DS:DATA_SEG

0000   8A 1E 0100 R       MOV BL,DATA1
0004   A1 0100 R          MOV AX,DATA2
0007   8A 3E 0101 R       MOV BH,DATA1+1

000B                      CODE_SEG ENDS
```

This example also illustrates how the ORG (**origin**) statement changes the starting offset address of the data in the data segment to location 100H. At times, the origin of data or the code must be assigned to an absolute offset address with the ORG statement. The **ASSUME statement** tells the assembler what names have been chosen for the code, data, extra, and stack segments. Without the ASSUME statement, the assembler assumes nothing and automatically uses a segment override prefix on all instructions that address memory data. The ASSUME statement is only used with full-segment definitions, as described later in this section of the text.

PROC and ENDP. The PROC and ENDP directives indicate the start and end of a procedure (**subroutine**). These directives *force structure* because the procedure is clearly defined. Note that if structure is to be violated for whatever reason, use the CALLF, CALLN, RETF, and RETN instructions. Both the PROC and ENDP directives require a label to indicate the name of the procedure. The PROC directive, which indicates the start of a procedure, must also be followed with a NEAR or FAR. A NEAR procedure is one that resides in the same code segment as the program. A FAR procedure may reside at any location in the memory system. Often the call NEAR procedure is considered to be *local,* and the call FAR procedure is considered to be *global.* The term global denotes a procedure that can be used by any program, while local defines a procedure that is only used by the current program. Any labels that are defined within the procedure block are also defined as either local (NEAR) or global (FAR).

Example 4–15 shows a procedure that adds BX, CX, and DX and stores the sum in register AX. Although this procedure is short and may not be particularly useful, it does illustrate how to use the PROC and ENDP directives to delineate the procedure. Note that information about the operation of the procedure should appear as a grouping of comments that show the registers changed by the procedure and the result of the procedure.

EXAMPLE 4–15

```
                   ;A procedure that adds BX, CX, and DX with the sum
                   ;stored in AX
                   ;
0000               ADDEM    PROC    FAR            ;start procedure

0000   03 D9                ADD     BX,CX
0002   03 DA                ADD     BX,DX
0004   8B C3                MOV     AX,BX
0006   CB                   RET

0007               ADDEM ENDP                      ;end procedure
```

If version 6.X of the Microsoft MASM assembler program is available, the PROC directive specifies and automatically saves any registers used within the procedure. The USES statement indicates which registers are used by the procedure, so that the assembler can automatically save them before your procedure begins and restore them before the procedure ends with the RET instruction. For example, the ADDS PROC USES AX BX CX statement automatically pushes AX, BX, and CX on the stack before the procedure begins and pops them from the stack before the RET instruction executes at the end of the procedure. Example 4–16 illustrates a procedure written using MASM 6.X that shows the USES statement. Note that the registers in the list are not separated by commas, but by spaces, and the PUSH and POP instructions are displayed in the procedure listing because it was assembled with the .LISTALL directive. The instructions prefaced with an asterisk (*) are inserted by the assembler and were not typed in the source file. The USES statement appears elsewhere in this text, so if MASM version 5.10 is in use, the code will need to be modified.

EXAMPLE 4–16

```
                              ;A procedure that includes the USES directive to save
                              ;BX, CX, and DX on the stack and restore them before
                              ;the RET instruction.
                              ;
0000                          ADDS    PROC    NEAR USES BX CX DX

0000   53                     *       push    bx
0001   51                     *       push    cx
0002   52                     *       push    dx
0003   03 D8                          ADD     BX,AX
0005   03 CB                          ADD     CX,BX
0007   03 D1                          ADD     DX,CX
0009   8B C2                          MOV     AX,DX
                                      RET
000B   5A                     *       pop     dx
000C   59                     *       pop     cx
000D   5B                     *       pop     bx
000E   C3                     *       ret     00000h

000F                          ADDS    ENDP
```

Memory Organization

The assembler uses two basic formats for developing software: one method uses models and the other uses full-segment definitions. Memory models, as presented in this section and briefly in Chapters 2 and 3, are unique to the MASM assembler program. The TASM assembler also uses memory models, but they differ somewhat from the MASM models. The full-segment definitions are common to most assemblers, including the Intel assembler, and are often used for software development. The models are easier to use for simple tasks. The full-segment definitions offer better control over the assembly language task and are recommended for complex programs. The model was used in early chapters because it is easier to understand for the beginning programmer. Models are also used with assembly language procedures that are used by high-level languages such as C/C++. Although this text fully develops and uses the memory model definitions for its programming examples, realize that full-segment definitions offer some advantages over memory models, as discussed later in this section.

Models. There are many models available to the MASM assembler, ranging from tiny to huge. Appendix A contains a table that lists all the models available for use with the assembler. To designate a model, use the .MODEL statement followed by the size of the memory system. The **TINY model** requires that all software and data fit into one 64K-byte memory segment; it is useful for many small programs. The **SMALL model** requires that only one data segment be used with one code segment, for a total of 128K bytes of memory. Other models are available, up to the HUGE model.

Example 4–17 illustrates how the .MODEL statement defines the parameters of a short program that copies the contents of a 100-byte block of memory (LISTA) into a second 100-byte block of memory (LISTB). It also shows how to define the stack, data, and code segments. The .EXIT 0 directive returns to DOS with an error code of 0 (no error). If no parameter is added to .EXIT, it still returns to DOS, but the error code is not defined. Also note that special directives such as @DATA (see Appendix A) are used to identify various segments. If the .STARTUP directive is used (MASM version 6.X), the MOV AX,@DATA followed by MOV DS,AX statements can be eliminated. The .STARTUP directive also eliminates the need to store the starting address next to the END label. Models are important with both Microsoft C/C++ and Borland C/C++ development systems if assembly language is included with C/C++ programs. Both development systems use in-line assembly programming for adding assembly language instructions and require an understanding of programming models. Refer to the respective C/C++ language reference for each system to determine the model protocols.

EXAMPLE 4–17

```
                              .MODEL SMALL
                              .STACK 100H                     ;define stack
                              .DATA                           ;define data segment

0000   0064[                  LISTA     DB    100 DUP (?)
              ??
                        ]
0064   0064[                  LISTB     DB    100 DUP (?)
              ??
                        ]

                              .CODE                           ;define code segment

0000   B8 ---- R       HERE:  MOV       AX,@DATA              ;load ES, DS
0003   8E C0                  MOV       ES,AX
0005   8E D8                  MOV       DS,AX

0007   FC                     CLD                             ;move data
0008   BE 0000 R              MOV       SI,OFFSET LISTA
000B   BF 0064 R              MOV       DI,OFFSET LISTB
000E   B9 0064                MOV       CX,100
0011   F3/A4                  REP       MOVSB

0013                          .EXIT 0                         ;exit to DOS
                              END HERE
```

Full-segment Definitions. Example 4–18 illustrates the same program, using full segment definitions. Full-segment definitions are also used with the Borland and Microsoft C/C++ environments for procedures developed in assembly language. The program in Example 4–18 appears longer than the one pictured in Example 4–17, but it is more structured than the model method of setting up a program. The first segment defined is the STACK_SEG that is clearly delineated with the SEGMENT and ENDS directives. Within these directives, a DW 100 DUP (?) sets aside 100H words for the stack segment. Because the word STACK appears next to SEGMENT, the assembler and linker automatically load both the stack segment register (SS) and stack pointer (SP).

EXAMPLE 4–18

```
0000                          STACK_SEG     SEGMENT STACK

0000   0100[                                DW    100H DUP (?)
              ????
                        ]

0200                          STACK_SEG     ENDS

0000                          DATA_SEG      SEGMENT 'DATA'

0000   0064[                  LISTA     DB    100 DUP (?)
              ??
                        ]
0064   0064[                  LISTB     DB    100 DUP (?)
              ??
                        ]
```

```
00C8                         DATA_SEG    ENDS

0000                         CODE_SEG    SEGMENT 'CODE'

                                 ASSUME  CS:CODE_SEG,DS:DATA_SEG
                                 ASSUME  SS:STACK_SEG

0000                         MAIN    PROC    FAR

0000  B8 ---- R                      MOV     AX,DATA_SEG       ;load DS and ES
0003  8E C0                          MOV     ES,AX
0005  8E D8                          MOV     DS,AX

0007  FC                             CLD                      ;move data
0008  BE 0000 R                      MOV     SI,OFFSET LISTA
000B  BF 0064 R                      MOV     DI,OFFSET LISTB
000E  B9 0064                        MOV     CX,100
0011  F3/A4                          REP MOVSB

0013  B4 4C                          MOV     AH,4CH            ;exit to DOS
0015  CD 21                          INT     21H

0017                         MAIN    ENDP

0017                         CODE_SEG    ENDS

                                 END    MAIN
```

Next, the data are defined in the DATA_SEG. Here, two arrays of data appear as LISTA and LISTB. Each array contains 100 bytes of space for the program. The names of the segments in this program can be changed to any name. Always include the group name 'DATA', so that the Microsoft program CodeView can be effectively used to symbolically debug this software. CodeView is a part of the MASM package used to debug software. To access CodeView, type CV, followed by the file name at the DOS command line; if operating from Programmer's WorkBench, select Debug under the Run menu. If the group name is not placed in a program, CodeView can still be used to debug a program, but the program will not be debugged in symbolic form. Other group names such as 'STACK', 'CODE', and so forth are listed in Appendix A. You must at least place the word 'CODE' next to the code segment SEGMENT statement if you want to view the program symbolically in CodeView.

The CODE_SEG is organized as a far procedure because most software is procedure-oriented. Before the program begins, the code segment contains the ASSUME statement. The ASSUME statement tells the assembler and linker that the name used for the code segment (CS) is CODE_SEG; it also tells the assembler and linker that the data segment is DATA_SEG and the stack segment is STACK_SEG. Notice that the group name 'CODE' is used for the code segment for use by CodeView. Other group names appear in Appendix A with the models.

After the program loads both the extra segment register and data segment register with the location of the data segment, it transfers 100 bytes from LISTA to LISTB. Following this is a sequence of two instructions that return control back to DOS (the disk operating system). Note that the program loader does not automatically initialize DS and ES. These registers must be loaded with the desired segment addresses in the program.

The last statement in the program is END MAIN. The END statement indicates the end of the program and the location of the first instruction executed. Here, we want the machine to execute the main procedure so that a label follows the END directive.

In the 80386 through the Pentium 4 microprocessors, an additional directive is found attached to the code segment. The USE16 or USE32 directive tells the assembler to use either the 16- or 32-bit instruction modes for the microprocessor. Software developed for the DOS environment must use the USE16 directive for the 80386 through the Pentium 4 program to function correctly because MASM assumes that all segments are 32 bits and all instruction modes are 32 bits by default. In fact, any program designed to execute in the real mode must include the USE16 directive to deviate from the default 8086/8088. Example 4–19 shows how the same software listed in Example 4–18 is formed for the 80386 and above microprocessors.

EXAMPLE 4–19

```
                        .386                    ;select the 80386
0000            STACK_SEG SEGMENT STACK

0000 0100[              DW 100H DUP (?)
     ????
         ]

0200            STACK_SEG ENDS

0000            DATA_SEG  SEGMENT 'DATA'

0000 0064[      LISTA     DB    100 DUP (?)
     ??
         ]
0064 0064[      LISTB     DB    100 DUP (?)
     ??
         ]

00C8            DATA_SEG  ENDS

0000            CODE_SEG  SEGMENT USE16 'CODE'

                ASSUME CS:CODE_SEG,DS:DATA_SEG
                ASSUME SS:STACK_SEG

0000            MAIN      PROC FAR

0000 B8 ---- R            MOV   AX,DATA_SEG   ;load DS and ES
0003 8E C0                MOV   ES,AX
0005 8E D8                MOV   DS,AX

0007 FC                   CLD                 ;move data
0008 BE 0000 R            MOV   SI,OFFSET LISTA
000B BF 0064 R            MOV   DI,OFFSET LISTB
000E B9 0064              MOV   CX,100
0011 F3/A4                REP   MOVSB

0013 B4 4C                MOV   AH,4CH        ;exit to DOS
0015 CD 21                INT   21H

0017            MAIN      ENDP

0017            CODE_SEG  ENDS
                END  MAIN
```

A Sample Program

Example 4–20 provides a sample program, using full-segment definitions, that reads a character from the keyboard and displays it on the CRT screen. Although this program is trivial, it illustrates a complete workable program that functions on any personal computer using DOS, from the earliest 8088-based system to the latest Pentium 4-based system. This program also illustrates the use of a few DOS function calls. (Appendix A lists the DOS function calls with their parameters.) The BIOS function calls allow the use of the keyboard, printer, disk drives, and everything else that is available in your computer system.

This example program uses only a code segment because there is no data. A stack segment should appear, but it has been left out because DOS automatically allocates a 128-byte stack for all programs. The only time that the stack is used in this example is for the INT 21H instructions that call a procedure in DOS. Note that when this program is linked, the linker signals that no stack segment is present. This warning may be ignored in this example because the stack is fewer than 128 bytes.

Notice that the entire program is placed into a far procedure called MAIN. It is good programming practice to write all software in procedural form, which allows the program to be used as a procedure at some future time if necessary. It is also fairly important to document register use and any parameters required for the program in the program header, which is a section of comments that appear at the start of the program.

The program uses DOS functions 06H and 4CH. The function number is placed in AH before the INT 21H instruction executes. The 06H function reads the keyboard if DL = 0FFH, or displays the ASCII contents of DL if it is not 0FFH. Upon close examination, the first section of the program moves a 06H into AH and a 0FFH into DL, so that a key is read from the keyboard. The INT 21H tests the keyboard; if no key is typed, it returns equal. The JE instruction tests the equal condition and jumps to MAIN if no key is typed.

When a key is typed, the program continues to the next step, which compares the contents of AL with an @ symbol. Upon return from the INT 21H, the ASCII character of the typed key is found in AL. In this program, if an @ symbol is typed, the program ends. If the @ symbol is not typed, the program continues by displaying the character typed on the keyboard with the next INT 21H instruction.

The second INT 21H instruction moves the ASCII character into DL so it can be displayed on the CRT screen. After displaying the character, a JMP executes. This causes the program to continue at MAIN, where it repeats reading a key.

If the @ symbol is typed, the program continues at MAIN1, where it executes the DOS function code number 4CH. This causes the program to return to the DOS prompt (A>), so that the computer can be used for other tasks.

More information about the assembler and its application appears in Appendix A and in the next several chapters. Appendix A provides a complete overview of the assembler, linker, and DOS functions. It also provides a list of the BIOS (basic I/O system) functions. The information provided in the following chapters clarifies how to use the assembler for certain tasks at different levels of the text.

EXAMPLE 4–20

```
                ;An example program that reads a key and displays it.
                ;Note that an @ key ends the program.
                ;
0000            CODE_SEG    SEGMENT 'CODE'

                ASSUME CS:CODE_SEG

0000        MAIN    PROC    FAR

0000 B4 06          MOV     AH,6            ;read key
0002 B2 FF          MOV     DL,0FFH
0004 CD 21          INT     21H
0006 74 F8          JE      MAIN            ;if no key

0008 3C 40          CMP     AL,'@'          ;test for @
000A 74 08          JE      MAIN1           ;if @

000C B4 06          MOV     AH,6            ;display key
000E 8A D0          MOV     DL,AL
0010 CD 21          INT     21H
0012 EB EC          JMP     MAIN            ;repeat
0014        MAIN1:
```

```
0014 B4 4C                  MOV    AH,4CH                ;exit to DOS
0016 CD 21                  INT    21H

0018          MAIN    ENDP

0018          CODE_SEG    ENDS

                            END    MAIN
```

Example 4–21 shows the program listed in Example 4–20, except models are used instead of full-segment descriptions. Please compare the two programs to determine the differences. Notice how much shorter the models can make a program.

EXAMPLE 4–21

```
                  ;An example program that reads a key and displays it.
                  ;Note that an @ key ends the program.
                  ;
                  .MODEL TINY
0000              .CODE
                  .STARTUP

0100              MAIN:
0100 B4 06                  MOV AH,6        ;read key
0102 B2 FF                  MOV DL,0FFH
0104 CD 21                  INT 21H
0106 74 F8                  JE  MAIN        ;if no key

0108 3C 40                  CMP AL, '@'     ;text for @
010A 74 08                  JE  MAIN1       ;if no @

010C B4 06                  MOV AH,6        ;display key
010E 8A D0                  MOV DL,AL
0110 CD 21                  INT 21 H
0112 EB EC                  JMP MAIN        ;repeat

0114              MAIN1:
                  .EXIT                     ;exit to DOS
                  END
```

4–8   SUMMARY

1. Data movement instructions transfer data between registers, a register and memory, a register and the stack, memory and the stack, the accumulator and I/O, and the flags and the stack. Memory-to-memory transfers are allowed only with the MOVS instruction.
2. Data movement instructions include MOV, PUSH, POP, XCHG, XLAT, IN, OUT, LEA, LDS, LES, LSS, LGS, LFS, LAHF, SAHF; and the following string instructions: LODS, STOS, MOVS, INS, and OUTS.
3. The first byte of an instruction contains the opcode. The opcode specifies the operation performed by the microprocessor. The opcode may be preceded by one or more override prefixes in some forms of instructions.
4. The D bit, located in many instructions, selects the direction of data flow. If D = 0, the data flow from the REG field to the R/M field of the instruction. If D = 1, the data flow from the R/M field to the REG field.
5. The W bit, found in most instructions, selects the size of the data transfer. If W = 0, the data are byte-sized; if W = 1, the data are word-sized. In the 80386 and above, W = 1 specifies either a word or doubleword register.

6. MOD selects the addressing mode of operation for a machine language instruction's R/M field. If MOD = 00, there is no displacement; if MOD-01, an 8-bit sign-extended displacement appears; if MOD-10, a 16-bit displacement occurs; and if MOD-11, a register is used instead of a memory location. In the 80386 and above, the MOD bits also specify a 32-bit displacement.

7. A 3-bit binary register code specifies the REG and R/M fields when the MOD = 11. The 8-bit registers are AH, AL, BH, BL, CH, CL, DH, and DL. The 16-bit registers are AX, BX, CX, DX, SP, BP, DI, and SI. The 32-bit registers are EAX, EBX, ECX, EDX, ESP, EBP, EDI, and ESI.

8. When the R/M field depicts a memory mode, a 3-bit code selects one of the following modes: [BX+DI], [BX+SI], [BP+DI], [BP+SI], [BX], [BP], [DI], or [SI] for 16-bit instructions. In the 80386 and above, the R/M field specifies EAX, EBX, ECX, EDX, EBP, EDI, and ESI or one of the scaled-index modes of addressing memory data. If the scaled-index mode is selected (R/M = 100) an additional byte (scaled-index byte) is added to the instruction to specify the base register, index register, and the scaling factor.

9. All memory-addressing modes, by default, address data in the data segment unless BP or EBP addresses memory. The BP or EBP register addresses data in the stack segment.

10. The segment registers are addressed only by the MOV, PUSH, or POP instructions. The MOV instruction may transfer a segment register to a 16-bit register, or vice versa. MOV CS,reg or POP CS instructions are not allowed because they change only part of the address. The 80386 through the Pentium 4 include two additional segment registers, FS and GS .

11. Data are transferred between a register or a memory location and the stack by the PUSH and POP instructions. Variations of these instructions allow immediate data to be pushed onto the stack, the flags to be transferred between the stack, and all 16-bit registers can be transferred between the stack and the registers. When data are transferred to the stack, two bytes (8086–80286) always move. The most-significant byte is placed at the location addressed by SP – 1, and the least-significant byte is placed at the location addressed by SP – 2. After placing the data on the stack, SP decrements by 2. In the 80386–Pentium 4, four bytes of data from a memory location or register may also be transferred to the stack.

12. Opcodes that transfer data between the stack and the flags are PUSHF and POPF. Opcodes that transfer all the 16-bit registers between the stack and the registers are PUSHA and POPA. In the 80386 and above, PUSHFD and POPFD transfer the contents of the EFLAGS between the microprocessor and the stack.

13. LEA, LDS, and LES instructions load a register or registers with an effective address. The LEA instruction loads any 16-bit register with an effective address; LDS and LES load any 16-bit register, and either DS or ES, with the effective address. In the 80386 and above, additional instructions include LFS, LGS, and LSS, which load a 16-bit register and FS, GS, or SS.

14. String data transfer instructions use either or both DI and SI to address memory. The DI offset address is located in the extra segment, and the SI offset address is located in the data segment.

15. The direction flag (D) chooses the auto-increment or auto-decrement mode of operation for DI and SI for string instructions. To clear D to 0, use the CLD instruction to select the auto-increment mode; to set D to 1, use the STD instruction to select the auto-decrement mode. Either or both DI and SI increment/decrement by 1 for a byte operation, by 2 for a word operation, and by 4 for a doubleword operation.

16. LODS loads AL, AX, or EAX with data from the memory location addressed by SI; STOS stores AL, AX, or EAX in the memory location addressed by DI; and MOVS transfers a byte or a word from the memory location addressed by SI into the location addressed by DI.

17. INS inputs data from an I/O device addressed by DX and stores it in the memory location addressed by DI. OUTS outputs the contents of the memory location addressed by SI and sends it to the I/O device addressed by DX.

18. The REP prefix may be attached to any string instruction to repeat it. The REP prefix repeats the string instruction the number of times found in register CX.

19. Arithmetic and logic operators can be used in assembly language. An example is MOV AX,34*3, which loads AX with 102.

20. Translate (XLAT) converts the data in AL into a number stored at the memory location address by BX plus AL.

21. IN and OUT transfer data between AL, AX, or EAX and an external I/O device. The address of the I/O device is either stored with the instruction (fixed port) or in register DX (variable port).

22. The Pentium Pro and Pentium II contain a new instruction called CMOV, or conditional move. This instruction only performs the move if the condition is true.

23. The segment override prefix selects a different segment register for a memory location than the default segment. For example, the MOV AX,[BX] instruction uses the data segment, but the MOV AX,ES:[BX] instruction uses the extra segment because of the ES: prefix. The segment override prefix is the only way that the FS and GS segments are addressed in the 80386 through the Pentium 4.

24. The MOVZX (move and zero-extend) and MOVSX (move and sign-extend) instructions, found in the 80386 and above, increase the size of a byte to a word or a word to a double-word. The zero-extend version increases the size of the number by inserting leading zeros. The sign-extend version increases the size of the number by copying the sign-bit into the more significant bits of the number.

25. Assembler directives DB, (define byte), DW (define word), DD (define doubleword), and DUP (duplicate) store data in the memory system.

26. The EQU (equate) directive allows data or labels to be equated to labels.

27. The SEGMENT directive identifies the start of a memory segment and ENDS identifies the end of a segment when full-segment definitions are in use.

28. The ASSUME directive tells the assembler what segment names you have assigned to CS, DS, ES, and SS when full-segment definitions are in effect. In the 80386 and above, AS-SUME also indicates the segment name for FS and GS.

29. The PROC and ENDP directives indicate the start and end of a procedure. The USES directive (MASM version 6.X) automatically saves and restores any number of registers on the stack if they appear with the PROC directive.

30. The assembler assumes that software is being developed for the 8086/8088 microprocessors unless the .286, .386, .486, or .586 directive is used to select one of these other micropro-cessors. This directive follows the .MODEL statement to use the 16-bit instruction mode, and precedes it for the 32-bit instruction mode.

31. Memory models can be used to shorten the program slightly, but they can cause problems for larger programs. Also be aware that memory models are not compatible with all assem-bler programs.

4–9   QUESTIONS AND PROBLEMS

1. The first byte of an instruction is the _____, unless it contains one of the override prefixes.

2. Describe the purpose of the D- and W-bits found in some machine language instructions.

3. In a machine language instruction, what information is specified by the MOD field?

4. If the register field (REG) of an instruction contains a 010 and W = 0, what register is se-lected, assuming that the instruction is a 16-bit mode instruction?

5. How are the 32-bit registers selected for the 80486 microprocessor?

6. What memory-addressing mode is specified by R/M = 001 with MOD = 00 for a 16-bit instruction?

7. Identify the default segment registers assigned to the following:
   (a) SP
   (b) EBX
   (c) DI
   (d) EBP
   (e) SI
8. Convert an 8B07H from machine language to assembly language.
9. Convert an 8B1E004CH from machine language to assembly language.
10. If a MOV SI,[BX+2] instruction appears in a program, what is its machine language equivalent?
11. If a MOV ESI,[EAX] instruction appears in a program for the Pentium II microprocessor operating in the 16-bit instruction mode, what is its machine language equivalent?
12. What is wrong with a MOV CS,AX instruction?
13. Form a short sequence of instructions that load the data segment register with a 1000H.
14. The PUSH and POP instructions always transfer a(n) _____-bit number between the stack and a register or memory location in the 8086–80286 microprocessors.
15. What segment register may not be popped from the stack?
16. Which registers move onto the stack with the PUSHA instruction?
17. Which registers move onto the stack for a PUSHAD instruction?
18. Describe the operation of each of the following instructions:
   (a) PUSH AX
   (b) POP ESI
   (c) PUSH [BX]
   (d) PUSHFD
   (e) POP DS
   (f) PUSHD 4
19. Explain what happens when the PUSH BX instruction executes. Make sure to show where BH and BL are stored. (Assume that SP = 0100H and SS = 0200H.)
20. Repeat question 19 for the PUSH EAX instruction.
21. The 16-bit POP instruction (except for POPA) increments SP by _____.
22. What values appear in SP and SS if the stack is addressed at memory location 02200H?
23. Compare the operation of a MOV DI,NUMB instruction with an LEA DI,NUMB instruction.
24. What is the difference between an LEA SI,NUMB instruction and a MOV SI,OFFSET NUMB instruction?
25. Which is more efficient, a MOV with an OFFSET or an LEA instruction?
26. Describe how the LDS BX,NUMB instruction operates.
27. What is the difference between the LDS and LSS instructions?
28. Develop a sequence of instructions that move the contents of data segment memory locations NUMB and NUMB+1 into BX, DX, and SI.
29. What is the purpose of the direction flag?
30. Which instructions set and clear the direction flag?
31. The string instructions use DI and SI to address memory data in which memory segments?
32. Explain the operation of the LODSB instruction.
33. Explain the operation of the STOSW instruction.
34. Explain the operation of the OUTSB instruction.
35. What does the REP prefix accomplish and what type of instruction is it used with?
36. Develop a sequence of instructions that copy 12 bytes of data from an area of memory addressed by SOURCE into an area of memory addressed by DEST.
37. Where is the I/O address (port number) stored for an INSB instruction?
38. Select an assembly language instruction that exchanges the contents of the EBX register with the ESI register.

39. Would the LAHF and SAHF instructions normally appear in software?
40. Explain how the XLAT instruction transforms the contents of the AL register.
41. Write a short program that uses the XLAT instruction to convert the BCD numbers 0–9 into ASCII-coded numbers 30H–39H. Store the ASCII-coded data in a TABLE located within the data segment.
42. Explain what the IN AL,12H instruction accomplishes.
43. Explain how the OUT DX,AX instruction operates.
44. What is a segment override prefix?
45. Select an instruction that moves a byte of data from the memory location addressed by the BX register, in the extra segment, into the AH register.
46. Develop a sequence of instructions that exchange the contents of AX with BX, ECX with EDX, and SI with DI.
47. What is accomplished by the CMOVNE CX,DX instruction in the Pentium Pro microprocessor?
48. How is a CMOVNS ECX,EBX instruction encoded and stored in a program if the assembler does not recognize this new Pentium 4 instruction?
49. What is an assembly language directive?
50. Describe the purpose of the following assembly language directives: DB, DW, and DD.
51. Select an assembly language directive that reserves 30 bytes of memory for array LIST1.
52. Describe the purpose of the EQU directive.
53. What is the purpose of the .386 directive?
54. What is the purpose of the .MODEL directive?
55. If the start of a segment is identified with .DATA, what type of memory organization is in effect?
56. If the SEGMENT directive identifies the start of a segment, what type of memory organization is in effect?
57. What does the INT 21H accomplish if AH contains a 4CH?
58. What directives indicate the start and end of a procedure?
59. Explain the purpose of the USES statement as it applies to a procedure with version 6.X of MASM.
60. How is the Pentium II microprocessor instructed to use the 16-bit instruction mode?
61. Develop a near procedure that stores AL in four consecutive memory locations, within the data segment, as addressed by the DI register.
62. Develop a far procedure that copies contents of the word-sized memory location CS:DATA1 into AX, BX, CX, DX, and SI.

CHAPTER 5

Arithmetic and Logic Instructions

INTRODUCTION

In this chapter, the arithmetic and logic instructions are examined. The arithmetic instructions include addition, subtraction, multiplication, division, comparison, negation, increment, and decrement. The logic instructions include AND, OR, Exclusive-OR, NOT, shifts, rotates, and the logical compare (TEST). Also presented are the 80386 through the Pentium 4 instructions XADD, SHRD, SHLD, bit tests, and bit scans. The chapter concludes with a discussion of string comparison instructions, which are used for scanning tabular data and for comparing sections of memory data. Both tasks perform efficiently with the string scan (SCAS) and string compare (CMPS) instructions.

If you are familiar with an 8-bit microprocessor, you will recognize that the 8086 through the Pentium 4 instruction set is superior to most 8-bit microprocessors because most of the instructions have two operands instead of one. Even if this is your first microprocessor, you will quickly learn that this microprocessor possesses a powerful and easy-to-use set of arithmetic and logic instructions.

CHAPTER OBJECTIVES

Upon completion of this chapter, you will be able to:

1. Use arithmetic and logic instructions to accomplish simple binary, BCD, and ASCII arithmetic.
2. Use AND, OR, and Exclusive-OR to accomplish binary bit manipulation.
3. Use the shift and rotate instructions.
4. Explain the operation of the 80386 through the Pentium 4 exchange and add, compare and exchange, double precision shift, bit test, and bit scan instructions.
5. Check the contents of a table for a match with the string instructions.

5–1   ADDITION, SUBTRACTION, AND COMPARISON

The bulk of the arithmetic instructions found in any microprocessor include addition, subtraction, and comparison. In this section, addition, subtraction, and comparison instructions are illustrated. Also shown are their uses in manipulating register and memory data.

Addition

Addition (ADD) appears in many forms in the microprocessor. This section details the use of the ADD instruction for 8-, 16-, and 32-bit binary addition. A second form of addition, called **add-with-carry,** is introduced with the ADC instruction. Finally, the increment instruction (INC) is presented. Increment is a special type of addition that adds a one to a number. In Section 5–3, other forms of addition are examined, such as BCD and ASCII. Also described is the XADD instruction, found in the 80486 through the Pentium 4.

Table 5–1 illustrates the addressing modes available to the ADD instruction. (These addressing modes include almost all those mentioned in Chapter 3.) However, because there are over 32,000 variations of the ADD instruction in the instruction set, it is impossible to list them all in this table. The only types of addition not allowed are memory-to-memory and segment register. The segment registers can only be moved, pushed, or popped. Note that, as with all other instructions, the 32-bit registers are available only with the 80386 through the Pentium 4.

Register Addition.   Example 5–1 shows a simple procedure that uses register addition to add the contents of several registers. In this example, the contents of AX, BX, CX, and DX are added to form a 16-bit result stored in the AX register. Here, a procedure is used because assembly language is procedure-oriented, as are most languages.

TABLE 5–1   Addition instructions.

Assembly Language	Operation
ADD AL,BL	AL = AL + BL
ADD CX,DI	CX = CX + DI
ADD EBP,EAX	EBP = EBP + EAX
ADD CL,44H	CL = CL + 44H
ADD BX,245FH	BX = BX + 245FH
ADD EDX,12345H	EDX = EDX + 00012345H
ADD [BX],AL	AL adds to the contents of the data segment memory location addressed by BX with the sum stored in the same memory location
ADD CL,[BP]	The byte contents of the stack segment memory location addressed by BP add to CL with the sum stored in CL
ADD AL,[EBX]	The byte contents of the data segment memory location addressed by EBX add to AL with the sum stored in AL
ADD BX,[SI + 2]	The word contents of the data segment memory location addressed by the sum of SI plus 2 add to BX with the sum stored in BX
ADD CL,TEMP	The byte contents of the data segment memory location TEMP add to CL with the sum stored in CL
ADD BX,TEMP[DI]	The word contents of the data segment memory location addressed by TEMP plus DI add to BX with the sum stored in BX
ADD [BX + DI],DL	DL adds to the contents of the data segment memory location addressed by BX plus DI with the sum stored in the same memory location
ADD BYTE PTR [DI],3	A 3 adds to the byte contents of the data segment memory location addressed by DI
ADD BX,[EAX + 2*ECX]	The word contents of the data segment memory location addressed by the sum of 2 times ECX plus EAX add to BX with the sum stored in BX

EXAMPLE 5–1

```
                         ;A procedure that sums AX, BX, CD, and DX;
                         ;the result is returned in AX.
                         ;
0000                     ADDS    PROC    NEAR

0000  03 C3                      ADD     AX,BX
0002  03 C1                      ADD     AX,CX
0004  03 C2                      ADD     AX,DX
0006  C3                         RET

0007                     ADDS    ENDP
```

Whenever arithmetic and logic instructions execute, the contents of the flag register **change.** Note that the contents of the interrupt, trap, and other flags do not change due to arithmetic and logic instructions. Only the flags located in the rightmost eight bits of the flag register and the overflow flag change. These rightmost flags denote the result of the arithmetic or a logic operation. Any ADD instruction modifies the contents of the sign, zero, carry, auxiliary carry, parity, and overflow flags. The flag bits never change for most of the data transfer instructions presented in Chapter 4.

Immediate Addition.   Immediate addition is employed whenever constant or known data are added. An 8-bit immediate addition appears in Example 5–2. In this example, load DL is first loaded with a 12H by using an immediate move instruction. Next, a 33H is added to the 12H in DL by using an immediate addition instruction. After the addition, the sum (45H) moves into register DL and the flags change, as follows:

$$Z = 0 \text{ (result not zero)}$$

$$C = 0 \text{ (no carry)}$$

$$A = 0 \text{ (no half-carry)}$$

$$S = 0 \text{ (result positive)}$$

$$P = 0 \text{ (odd parity)}$$

$$O = 0 \text{ (no overflow)}$$

EXAMPLE 5–2

```
0006  B2 12        MOV   DL,12H
0008  80 C2 33     ADD   DL,33H
```

Memory-to-Register Addition.   Suppose that an application requires that memory data are added to the AL register. Example 5–3 shows an example that adds two consecutive bytes of data, stored at the data segment offset locations NUMB and NUMB+1, to the AL register.

EXAMPLE 5–3

```
                         ;A procedure that sums data in locations NUMB and NUMB+1;
                         ;the result is returned in AX.
                         ;
0000                     SUMS    PROC    NEAR

0000  BF 0000 R          MOV     DI,OFFSET NUMB    ;address NUMB
0003  B0 00              MOV     AL,0              ;clear sum
0005  02 05              ADD     AL,[DI]           ;add NUMB
0007  02 45 01           ADD     AL,[DI+1]         ;add NUMB+1
000A  C3                 RET

000B                     SUMS    ENDP
```

Procedure SUMS first loads the destination index register (DI) with offset address NUMB. The DI register, used in this example, addresses data in the data segment beginning at memory location NUMB. In most cases, loading the address inside of a procedure is poor programming practice. It is usually better to load the address outside of the procedure, and then CALL the procedure with the address in place. Next, the ADD AL,[DI] instruction adds the contents of memory location NUMB to AL. Note that AL is initialized to zero, which occurs because DI addresses memory location NUMB, and then the ADD instruction adds its contents to AL. Finally, the ADD AL,[DI+1] instruction adds the contents of memory location NUMB plus one byte to the AL register. After both ADD instructions execute, the result appears in the AL register as the sum of the contents of NUMB plus the contents of NUMB+1.

Array Addition. Memory arrays are sequential lists of data. Suppose that an array of data (ARRAY) contains 10 bytes, numbered from element 0 through element 9. Example 5–4 shows a procedure that adds the contents of array elements 3, 5, and 7. (The procedure and the array elements it adds are chosen to demonstrate the use of some of the addressing modes for the microprocessor.)

This example first clears AL to 0, so it can be used to accumulate the sum. Next, register SI is loaded with a 3 to initially address array element 3. The ADD AL,ARRAY[SI] instruction adds the contents of array element 3 to the sum in AL. The instructions that follow add array elements 5 and 7 to the sum in AL, using a 3 in SI plus a displacement of 2 to address element 5, and a displacement of 4 to address element 7.

EXAMPLE 5–4

```
                        ;A procedure that sums ARRAY elements 3, 5, and 7;
                        ;the result is returned in AL.
                        ;
                        ;Note this procedure destroys the contents of SI.
                        ;
0000                    SUM     PROC    NEAR

0000  B0 00                     MOV     AL,0              ;clear sum
0002  BE 0003                   MOV     SI,3              ;address element 3
0005  02 84 0002 R              ADD     AL,ARRAY[SI]      ;add element 3
0009  02 84 0004 R              ADD     AL,ARRAY[SI+2]    ;add element 5
000D  02 84 0006 R              ADD     AL,ARRAY[SI+4]    ;add element 7
0011  C3                        RET

0012                    SUM     ENDP
```

Suppose that an array of data contains 16-bit numbers used to form a 16-bit sum in register AX. Example 5–5 shows a procedure written for the 80386 and above, showing the scaled-index form of addressing to add elements 3, 5, and 7 of an area of memory called ARRAY. In this example, EBX is loaded with the address ARRAY, and ECX holds the array element number. Note how the scaling factor is used to multiply the contents of the ECX register by 2 to address words of data. (Recall that words are two bytes long.)

EXAMPLE 5–5

```
                        ;A procedure that sums ARRAY elements 3, 5 and 7;
                        ;the result is returned in AX.
                        ;
                        ;Note that the contents of registers EBX and ECX are
                        ;destroyed.
0000                    SUM     PROC    NEAR

0000  66| BB 00000000 R MOV     EBX,OFFSET ARRAY   ;address ARRAY
0006  66| B9 00000003   MOV     ECX,3              ;address element 3
000C  67& 8B 04 4B      MOV     AX,[EBX+2*ECX]     ;get element 3
0010  66| B9 00000005   MOV     ECX,5              ;address element 5
```

```
0016   67& 03 04 4B        ADD    AX,[EBX+2*ECX]      ;add element 5
001A   66| B9 00000007     MOV    ECX,7               ;address element 7
0020   67& 03 04 4B        ADD    AX,[EBX+2*ECX]      ;add element 7
0024   C3                  RET

0025                SUM    ENDP
```

Increment Addition. Increment addition (INC) adds 1 to a register or a memory location. The INC instruction can add 1 to any register or memory location, except a segment register. Table 5–2 illustrates some of the possible forms of the increment instruction available to the 8086–Pentium 4 processors. As with other instructions presented thus far, it is impossible to show all variations of the INC instruction because of the large number available.

With indirect memory increments, the size of the data must be described by using the BYTE PTR, WORD PTR, or DWORD PTR directives. The reason is that the assembler program cannot determine if, for example, the INC [DI] instruction is a byte-, word-, or doubleword-sized increment. The INC BYTE PTR [DI] instruction clearly indicates byte-sized memory data; the INC WORD PTR [DI] instruction unquestionably indicates a word-sized memory data; and the INC DWORD PTR [DI] instruction indicates doubleword-sized data.

Example 5–6 shows how to modify the procedure of Example 5–3 to use the increment instruction for addressing NUMB and NUMB+1. Here, an INC DI instruction changes the contents of register DI from offset address NUMB to offset address NUMB+1. Both procedures shown in Examples 5–3 and 5–6 add the contents of NUMB and NUMB+1. The difference between these programs is the way that this data's address is formed through the contents of the DI register using the increment instruction.

EXAMPLE 5–6

```
                 ;A procedure that sums NUMB and NUMB+1;
                 ;the result is returned in AL.
                 ;
                 ;Note that the contents of DI are destroyed.
                 ;
0000             SUMS    PROC    NEAR

0000   BF 0000 R        MOV    DI,OFFSET NUMB    ;address NUMB
0003   B0 00            MOV    AL,0              ;clear sum
0005   02 05            ADD    AL,[DI]           ;add NUMB
0007   47               INC    DI                ;address NUMB+1
0008   02 05            ADD    AL,[DI]           ;add NUMB+1
000A   C3               RET

000B             SUMS    ENDP
```

TABLE 5–2 Increment instructions.

Assembly Language	Operation
INC BL	BL = BL + 1
INC SP	SP = SP + 1
INC EAX	EAX = EAX + 1
INC BYTE PTR [BX]	Adds 1 to the byte contents of the data segment memory location addressed by BX
INC WORD PTR [SI]	Adds 1 to the word contents of the data segment memory location addressed by SI
INC DWORD PTR [ECX]	Adds 1 to the doubleword contents of the data segment memory location addressed by ECX
INC DATA1	Increments the contents of data segment memory location DATA1

Increment instructions affect the flag bits, as do most other arithmetic and logic operations. The difference is that increment instructions do not affect the carry flag bit. Carry doesn't change because we often use increments in programs that depend upon the contents of the carry flag. Note that increment is used to point to the next memory element in a byte-sized array of data only. If word-sized data are addressed, it is better to use an ADD DI,2 instruction to modify the DI pointer in place of two INC DI instructions. For doubleword arrays, use the ADD DI,4 instruction to modify the DI pointer. In some cases, the carry flag must be preserved, which may mean that two or four INC instructions might appear in a program to modify a pointer.

Addition-with-Carry. An addition-with-carry instruction (ADC) adds the bit in the carry flag (C) to the operand data. This instruction mainly appears in software that adds numbers that are wider than 16 bits in the 8086–80286 or wider than 32-bits in the 80386–Pentium 4.

Table 5–3 lists several add-with-carry instructions, with comments that explain their operation. Like the ADD instruction, ADC affects the flags after the addition.

Suppose that a program is written for the 8086–80286 to add the 32-bit number in BX and AX to the 32-bit number in DX and CX. Figure 5–1 illustrates this addition so that the placement and function of carry flag can be understood. This addition cannot be easily performed without adding the carry flag bit because the 8086–80286 only adds 8- or 16-bit numbers. Example 5–7 shows how the addition occurs with a procedure. Here, the contents of registers AX and CX add to form the least-significant 16 bits of the sum. This addition may or may not generate a carry. A carry appears in the carry flag if the sum is greater than FFFFH. Because it is impossible to predict a

TABLE 5–3 Add-with-carry instructions.

Assembly Language	Operation
ADC AL,AH	AL = AL + AH + carry
ADC CX,BX	CX = CX + BX + carry
ADC EBX,EDX	EBX = EBX + EDX + carry
ADC DH,[BX]	The byte contents of the data segment memory location addressed by BX add to DH with carry with the sum stored in DH
ADC BX,[BP + 2]	The word contents of the stack segment memory location addressed by BP plus 2 add to BX with carry with the sum stored in BX
ADC ECX,[EBX]	The doubleword contents of the data segment memory location addressed by EBX add to ECX with carry with the sum stored in ECX

FIGURE 5–1 Addition-with-carry showing how the carry flag (C) links the two 16-bit additions into one 32-bit addition.

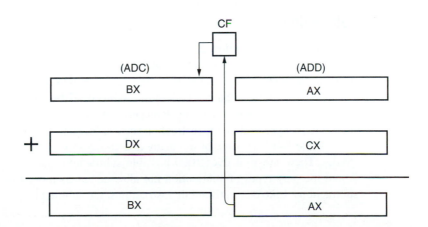

carry, the most-significant 16 bits of this addition are added with the carry flag, using the ADC instruction. The ADC instruction adds the 1 or the 0 in the carry flag to the most-significant 16 bits of the result. This program adds BX–AX to DX–CX, with the sum appearing in BX–AX.

EXAMPLE 5–7

```
                    ;A procedure that sums BX-AX and DX-CX;
                    ;the result is returned in BX-AX.
                    ;
0000                SUM32   PROC    NEAR

0000  03 C1                 ADD     AX,CX
0002  13 DA                 ADC     BX,DX
0004  C3                    RET

0005                SUM32   ENDP
```

Suppose the same procedure is rewritten for the 80386 through the Pentium 4, but modified to add two 64-bit numbers. The changes required for this operation are the use of the extended registers to hold the data, and modifications of the instructions for the 80386 and above. These changes are shown in Example 5–8, which adds two 64-bit numbers.

EXAMPLE 5–8

```
                    ;A procedure that sums EBX-EAX and EDX-ECX;
                    ;the result is returned in EBX-EAX.
                    ;
0000                SUM64   PROC    NEAR

0000 66| 03 C1              ADD     EAX,ECX
0003 66| 13 DA              ADC     EBX,EDX
0006 C3                     RET

0007                SUM64   ENDP
```

Exchange and Add for the 80486–Pentium 4 Processors.   A new type of addition called **exchange and add** (XADD) appears in the 80486 instruction set and continues through the Pentium 4. The XADD instruction adds the source to the destination and stores the sum in the destination, as with any addition. The difference is that after the addition takes place, the original value of the destination is copied into the source operand. This is one of the few instructions that change the source.

For example, if BL = 12H and DL = 02H, and the XADD BL,DL instruction executes, the BL register contains the sum of 14H and DL becomes 12H. The sum of 14H is generated and the original destination of 12H replaces the source. This instruction functions with any register size and any memory operand, just as with the ADD instruction.

Subtraction

Many forms of **subtraction** (SUB) appear in the instruction set. These forms use any addressing mode with 8-, 16-, or 32-bit data. A special form of subtraction (decrement, or DEC) subtracts a 1 from any register or memory location. Section 5–3 shows how BCD and ASCII data subtract. As with addition, numbers that are wider than 16 bits or 32 bits must occasionally be subtracted. The subtract-with-borrow instruction (SBB) performs this type of subtraction. In the 80486 through the Pentium 4 processors, the instruction set also includes a compare and exchange instruction.

Table 5–4 lists some of the many addressing modes allowed with the subtract instruction (SUB). There are well over 1000 possible subtraction instructions, far too many to list here. About the only types of subtraction not allowed are memory-to-memory and segment register subtractions. Like other arithmetic instructions, the subtract instruction affects the flag bits.

TABLE 5-4 Subtraction instructions.

Assembly Language	Operation
SUB CL,BL	CL = CL – BL
SUB AX,SP	AX = AX – SP
SUB ECX,EBP	ECX = ECX – EBP
SUB DH,6FH	DH = DH – 6FH
SUB AX,0CCCCH	AX = AX – CCCCH
SUB ESI,2000300H	ESI = ESI – 2000300H
SUB [DI],CH	Subtracts the contents of CH from the contents of the data segment memory location addressed by DI
SUB CH,[BP]	Subtracts the byte contents of the stack segment memory location addressed by BP from CH
SUB AH,TEMP	Subtracts the byte contents of the data segment memory location TEMP from AH
SUB DI,TEMP[ESI]	Subtracts the word contents of the data segment memory location addressed by TEMP plus ESI from DI
SUB ECX,DATA1	Subtracts the doubleword contents of the data segment memory location addressed by DATA1 from ECX

Register Subtraction. Example 5–9 shows a sequence of instructions that perform register subtraction. This example subtracts the 16-bit contents of registers CX and DX from the contents of register BX. After each subtraction, the microprocessor modifies the contents of the flag register. The flags change for most arithmetic and logic operations.

EXAMPLE 5–9

```
0000  2B D9          SUB   BX,CX
0002  2B DA          SUB   BX,DX
```

Immediate Subtraction. As with addition, the microprocessor also allows immediate operands for the subtraction of constant data. Example 5–10 presents a short sequence of instructions that subtract a 44H from a 22H. Here, we first load the 22H into CH using an immediate move instruction. Next, the SUB instruction, using immediate data 44H, subtracts a 44H from the 22H. After the subtraction, the difference (DEH) moves into the CH register. The flags change as follows for this subtraction:

$$Z = 0 \text{ (result not zero)}$$
$$C = 1 \text{ (borrow)}$$
$$A = 1 \text{ (half-borrow)}$$
$$S = 1 \text{ (result negative)}$$
$$P = 1 \text{ (even parity)}$$
$$O = 0 \text{ (no overflow)}$$

EXAMPLE 5–10

```
0000  B5 22          MOV   CH,22H
0002  80 ED 44       SUB   CH,44H
```

Both carry flags (C and A) hold borrows after a subtraction instead of carries, as after an addition. Notice in this example that there is no overflow. This example subtracted a 44H (+68) from a 22H (+34), resulting in a DEH (–34). Because the correct 8-bit signed result is a –34, there is no overflow in this example. An 8-bit overflow occurs only if the signed result is greater than +127 or less than –128.

Decrement Subtraction.   Decrement subtraction (DEC) subtracts a 1 from a register or the contents of a memory location. Table 5–5 lists some decrement instructions that illustrate register and memory decrements.

The decrement indirect memory data instructions require BYTE PTR, WORD PTR, or DWORD PTR because the assembler cannot distinguish a byte from a word or doubleword when an index register addresses memory. For example, DEC [SI] is vague because the assembler cannot determine whether the location addressed by SI is a byte, word, or doubleword. Using DEC BYTE PTR [SI], DEC WORD PTR [DI], or DEC DWORD PTR [SI] reveals the size of the data to the assembler.

Subtraction-with-borrow.   A subtraction-with-borrow (SBB) instruction functions as a regular subtraction, except that the **carry flag** (C), which holds the borrow, also subtracts from the difference. The most common use for this instruction is for subtractions that are wider than 16 bits in the 8086–80286 microprocessors or wider than 32 bits in the 80386–Pentium 4. Wide subtractions require that borrows propagate through the subtraction, just as wide additions propagate the carry.

Table 5–6 lists several SBB instructions, with comments that define their operations. Like the SUB instruction, SBB affects the flags. Notice that the immediate subtract from memory instruction in this table requires a BYTE PTR, WORD PTR, or DWORD PTR directive.

When the 32-bit number held in BX and AX is subtracted from the 32-bit number held in SI and DI, the carry flag propagates the borrow between the two 16-bit subtractions. The carry flag holds the borrow for subtraction. Figure 5–2 shows how the borrow propagates through the carry flag (C) for this task. Example 5–11 shows how this subtraction is performed by a program. With wide subtraction, the least-significant 16- or 32-bit data are subtracted with the SUB instruction. All subsequent and more significant data are subtracted by using the SBB instruction. The example uses the SUB instruction to subtract DI from AX; then uses SBB to subtract-with-borrow SI from BX.

EXAMPLE 5–11

```
0004   2B C7            SUB  AX,DI
0006   1B DE            SBB  BX,SI
```

TABLE 5–5   Decrement instructions.

Assembly Language	Operation
DEC BH	BH = BH – 1
DEC CX	CX = CX – 1
DEC EDX	EDX = EDX – 1
DEC BYTE PTR [DI]	Subtracts 1 from the byte contents of the data segment memory location addressed by DI
DEC WORD PTR[BP]	Subtracts 1 from the word contents of the stack segment memory location addressed by BP
DEC DWORD PTR[EBX]	Subtracts 1 from the doubleword contents of the data segment memory location addressed by EBX
DEC NUMB	Subtracts 1 from the contents of the data segment memory location NUMB

TABLE 5–6   Subtraction-with-borrow instructions.

Assembly Language	Operation
SBB AH,AL	AH = AH – AL – carry
SBB AX,BX	AX = AX – BX – carry
SBB EAX,ECX	EAX = EAX – ECX – carry
SBB CL,2	CL = CL – 2 – carry
SBB BYTE PTR[DI],3	Both a 3 and carry subtract from the contents of the data segment memory location addressed by DI
SBB [DI],AL	Both AL and carry subtract from the data segment memory location addressed by DI
SBB DI,[BP + 2]	Both carry and the word contents of the stack segment memory location addressed by the sum of BP and 2 subtract from DI
SBB AL,[EBX + ECX]	Both carry and the byte contents of the data segment memory location addressed by the sum of EBX and ECX subtract from AL

FIGURE 5–2   Subtraction-with-borrow showing how the carry flag (C) propagates the borrow.

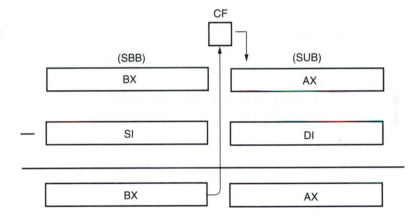

Comparison

The comparison instruction (CMP) is a subtraction that changes only the flag bits; the destination operand never changes. A comparison is useful for checking the entire contents of a register or a memory location against another value. A CMP is normally followed by a conditional jump instruction, which tests the condition of the flag bits.

Table 5–7 lists a variety of comparison instructions that use the same addressing modes as the addition and subtraction instructions already presented. Similarly, the only disallowed forms of compare are memory-to-memory and segment register compares.

Example 5–12 shows a comparison followed by a conditional jump instruction. In this example, the contents of AL are compared with a 10H. Conditional jump instructions that often follow the comparison are JA (**jump above**) or JB (**jump below**). If the JA follows the comparison, the jump occurs if the value in AL is above 10H. If the JB follows the comparison, the jump occurs if the value in AL is below 10H. In this example, the JAE instruction follows the comparison. This instruction causes the program to continue at memory location SUBER if the value in AL is 10H or above. There is also a JBE (**jump below or equal**) instruction that could follow the comparison to jump if the outcome is below or equal to 10H. Chapter 6 provides more detail on the comparison and conditional jump instructions.

TABLE 5–7   Comparison instructions.

Assembly Language	Operation
CMP CL,BL	CL – BL
CMP AX,SP	AX – SP
CMP EBP,ESI	EBP – ESI
CMP AX,2000H	AX – 2000H
CMP [DI],CH	CH subtracts from the contents of the data segment memory location addressed by DI
CMP CL,[BP]	The byte contents of the stack segment memory location addressed by BP subtract from CL
CMP AH,TEMP	The byte contents of the data segment memory location TEMP subtract from AH
CMP DI,TEMP[BX]	The word contents of the data segment memory location addressed by the sum of TEMP plus BX subtract from DI
CMP AL,[EDI + ESI]	The byte contents of the data segment memory location addressed by the sum of EDI plus ESI subtract from AL

EXAMPLE 5–12

```
0000   3C 10          CMP   AL,10H          ;compare with 10H
0002   73 1C          JAE   SUBER           ;if 10H or above
```

Compare and Exchange (80486–Pentium 4 Processors Only).   The compare and exchange instruction (CMPXCHG), found only in the 80486 through the Pentium 4 instruction sets, compares the destination operand with the accumulator. If they are equal, the source operand is copied into the destination; if they are not equal, the destination operand is copied into the accumulator. This instruction functions with 8-, 16-, or 32-bit data.

The CMPXCHG CX,DX instruction is an example of the compare and exchange instruction. This instruction first compares the contents of CX with AX. If CX equals AX, DX is copied into AX; if CX is not equal to AX, CX is copied into AX. This instruction also compares AL with 8-bit data and EAX with 32-bit data if the operands are either 8- or 32-bit.

In the Pentium–Pentium 4 processors, a CMPXCHG8B instruction is available that compares two quadwords. This is the only new data manipulation instruction provided in the Pentium–Pentium 4 when they are compared with prior versions of the microprocessor. The compare-and-exchange-8-bytes instruction compares the 64-bit value located in EDX:EAX with a 64-bit number located in memory. An example is CMPXCHG8B TEMP. If TEMP equals EDX:EAX, TEMP is replaced with the value found in ECX:EBX; if TEMP does not equal EDX:EAX, the number found in TEMP is loaded into EDX:EAX. The 0 flag bit indicates that the values are equal after the comparison.

5–2   MULTIPLICATION AND DIVISION

Only modern microprocessors contain multiplication and division instructions. Earlier 8-bit microprocessors could not multiply or divide without the use of a program that multiplied or divided by using a series of shifts and additions or subtractions. Because microprocessor manufacturers were aware of this inadequacy, they incorporated multiplication and division instructions into the instruction sets of the newer microprocessors. The Pentium–Pentium 4 processors contain special circuitry that performs a multiplication in as little as one clocking period, while it took over 40 clocking periods to perform the same multiplication in earlier Intel microprocessors.

Multiplication

Multiplication is performed on bytes, words, or doublewords, and can be signed integer (IMUL) or unsigned integer (MUL). Note that only the 80386 through the Pentium 4 processors multiply 32-bit doublewords. The product after a multiplication is always a double-width product. If two 8-bit numbers are multiplied, they generate a 16-bit product; if two 16-bit numbers are multiplied, they generate a 32-bit product; and if two 32-bit numbers are multiplied, a 64-bit product is generated.

Some flag bits (O and C) change when the multiply instruction executes, and produce predictable outcomes. The other flags also change, but their results are unpredictable and therefore are unused. In an 8-bit multiplication, if the most-significant 8 bits of the result are 0, both C and O flag bits equal 0. These flag bits show that the result is 8-bits wide (C = 0) or 16-bits wide (C = 1). In a 16-bit multiplication, if the most-significant 16-bits part of the product is 0, both C and O clear to 0. In a 32-bit multiplication, both C and O indicate that the most-significant 32 bits of the product are zero.

8-bit Multiplication. With 8-bit multiplication, the multiplicand is always in the AL register, whether signed or unsigned. The multiplier can be any 8-bit register or any memory location. Immediate multiplication is not allowed unless the special signed immediate multiplication instruction, discussed later in this section, appears in a program. The multiplication instruction contains one operand because it always multiplies the operand times the contents of register AL. An example is the MUL BL instruction, which multiplies the unsigned contents of AL by the unsigned contents of BL. After the multiplication, the unsigned product is placed in AX—a double-width product. Table 5–8 illustrates some 8-bit multiplication instructions.

Suppose that BL and CL each contain two 8-bit unsigned numbers, and these numbers must be multiplied to form a 16-bit product stored in DX. This procedure cannot be accomplished by a single instruction because we can only multiply a number times the AL register for an 8-bit multiplication. Example 5–13 shows a short program that generates DX = BL × CL. This example loads register BL and CL with example data 5 and 10. The product, a 50, moves into DX from AX after the multiplication by using the MOV DX,AX instruction.

EXAMPLE 5–13

```
0000  B3 05        MOV    BL,5        ;load data
0002  B1 0A        MOV    CL,10
0004  8A C1        MOV    AL,CL       ;position data
0006  F6 E3        MUL    BL          ;multiply
0008  8B D0        MOV    DX,AX       ;position product
```

TABLE 5–8   8-bit multiplication instructions.

Assembly Language	Operation
MUL CL	AL is multiplied by CL; the unsigned product is in AX
IMUL DH	AL is multiplied by DH; the signed product is in AX
IMUL BYTE PTR[BX]	AL is multiplied by the byte contents of the data segment memory location addressed by BX; the signed product is in AX
MUL TEMP	AL is multiplied by the byte contents of the data segment memory location addressed by TEMP; the unsigned product is in AX

TABLE 5–9   16-bit multiplication instructions.

Assembly Language	Operation
MUL CX	AX is multiplied by CX; the unsigned product is in DX–AX
IMUL DI	AX is multiplied by DI; the signed product is in DX–AX
MUL WORD PTR[SI]	AX is multiplied by the word contents of the data segment memory location addressed by SI; the unsigned product is in DX–AX

For signed multiplication, the product is in true binary form, if positive, and in two's complement form, if negative. These are the same forms used to store all positive and negative signed numbers used by the microprocessor. If the program of Example 5–13 multiplies two signed numbers, only the MUL instruction is changed to IMUL.

16-bit Multiplication.   Word multiplication is very similar to byte multiplication. The difference is that AX contains the multiplicand instead of AL, and the product appears in DX–AX instead of AX. The DX register always contains the most-significant 16 bits of the product, and AX contains the least-significant 16 bits. As with 8-bit multiplication, the choice of the multiplier is up to the programmer. Table 5–9 shows several different 16-bit multiplication instructions.

A Special Immediate 16-bit Multiplication.   The 8086/8088 microprocessors could not perform immediate multiplication; the 80186 through the Pentium 4 processors can do so by using a special version of the multiply instruction. Immediate multiplication must be signed multiplication, and the instruction format is different because it contains three operands. The first operand is the 16-bit destination register; the second operand is a register or memory location that contains the 16-bit multiplicand; and the third operand is either an 8-bit or a 16-bit immediate data used as the multiplier.

The IMUL CX,DX,12H instruction multiplies 12H times DX and leaves a 16-bit signed product in CX. If the immediate data are 8-bits, they sign-extend into a 16-bit number before the multiplication occurs. Another example is IMUL BX,NUMBER,1000H, which multiplies NUMBER times 1000H and leaves the product in BX. Both the destination and multiplicand must be 16-bit numbers. Although this is immediate multiplication, the restrictions placed upon it limit its utility, especially the fact that it is a signed multiplication and the product is 16 bits wide.

32-bit Multiplication.   In the 80386 and above, 32-bit multiplication is allowed because these microprocessors contain 32-bit registers. As with 8- and 16-bit multiplication, 32-bit multiplication can be signed or unsigned by using the IMUL and MUL instructions. With 32-bit multiplication, the contents of EAX are multiplied by the operand specified with the instruction. The product (64 bits wide) is found in EDX–EAX, where EAX contains the least-significant 32 bits of the product. Table 5–10 lists some of the 32-bit multiplication instructions found in the 80386 and above instruction set.

TABLE 5–10   32-bit multiplication instructions.

Assembly Language	Operation
MUL ECX	EAX is multiplied by ECX; the unsigned product is in EDX–EAX
IMUL EDI	EAX is multiplied by EDI; the signed product is in EDX–EAX
MUL DWORD PTR[ECX]	EAX is multiplied by the doubleword contents of the data segment memory location addressed by ECX; the unsigned product is in EDX–EAX

Division

As with multiplication, **division** occurs on 8- or 16-bit numbers in the 8086–80286 processors, and on 32-bit numbers in the 80386–Pentium 4. These numbers are signed (IDIV) or unsigned (DIV) integers. The dividend is always a double-width dividend that is divided by the operand. This means that an 8-bit division divides a 16-bit number by an 8-bit number; a 16-bit division divides a 32-bit number by a 16-bit number; and a 32-bit division divides a 64-bit number by a 32-bit number. There is no immediate division instruction available to any microprocessor.

None of the flag bits change predictably for a division. A division can result in two different types of errors; one is an attempt to divide by zero and the other is a divide overflow. A divide overflow occurs when a small number divides into a large number. For example, suppose that AX = 3000 and that it is divided by 2. Because the quotient for an 8-bit division appears in AL, the result of 1500 causes a divide overflow because the 1500 does not fit into AL. In either case, the microprocessor generates an interrupt if a divide error occurs. In most systems, a divided error interrupt displays an error message on the video screen. The divide-error-interrupt and all other interrupts for the microprocessor are explained in Chapter 6.

8-bit Division. An 8-bit division uses the AX register to store the dividend that is divided by the contents of any 8-bit register or memory location. The quotient moves into AL after the division with AH containing a whole number remainder. For a signed division, the quotient is positive or negative; the remainder always assumes the sign of the dividend and is always an integer. For example, if AX = 0010H (+16) and BL = FDH (–3) and the IDIV BL instruction executes, AX = 01FBH. This represents a quotient of –5 (AL) with a remainder of 1 (AH). If, on the other hand, a –16 is divided by a +3, the result will be a quotient of –5 (AL) with a remainder of –1 (AH). Table 5–11 lists some of the 8-bit division instructions.

With 8-bit division, the numbers are usually 8 bits wide. This means that one of them, the **dividend,** must be converted to a 16-bit wide number in AX. This is accomplished differently for signed and unsigned numbers. For the unsigned number, the most-significant 8 bits must be cleared to zero **(zero-extended).** The MOVZX instruction described in Chapter 4 can be used to zero-extend a number in the 80386 through the Pentium 4 processors. For signed numbers, the least-significant 8 bits are sign-extended into the most-significant 8 bits. In the microprocessor, a special instruction sign-extends AL into AH, or converts an 8-bit signed number in AL into a 16-bit signed number in AX. The CBW **(convert byte to word)** instruction performs this conversion. In the 80386 through the Pentium 4, a MOVSX instruction (see Chapter 4) sign-extends a number.

Example 5–14 illustrates a short program that divides the unsigned byte contents of memory location NUMB by the unsigned contents of memory location NUMB1. Here, the quotient is stored in location ANSQ and the remainder is stored in location ANSR. Notice how the contents of location NUMB are retrieved from memory and then zero-extended to form a 16-bit unsigned number for the dividend.

TABLE 5–11 8-bit division instructions.

Assembly Language	Operation
DIV CL	AX is divided by CL; the unsigned quotient is in AL and the remainder is in AH
IDIV BL	AX is divided by BL; the signed quotient is in AL and the remainder is in AH
DIV BYTE PTR[BP]	AX is divided by the byte contents of the stack segment memory location addressed by BP; the unsigned quotient is in AL and the remainder is in AH

EXAMPLE 5–14

```
0000  A0 0000 R           MOV   AL,NUMB      ;get NUMB
0003  B4 00               MOV   AH,0         ;zero-extend
0005  F6 36 0002 R        DIV   NUMB1        ;divide by NUMB1
0009  A2 0003 R           MOV   ANSQ,AL      ;save quotient
000C  88 26 0004 R        MOV   ANSR,AH      ;save remainder
```

Example 5–15 shows the same basic program except that the numbers are signed numbers. This means that instead of zero-extending AL into AH, it is sign-extended with the CBW instruction.

EXAMPLE 5–15

```
0000  A0 0000 R           MOV   AL,NUMB      ;get NUMB
0003  98                  CBW                ;sign-extend
0004  F6 3E 0002 R        IDIV  NUMB1        ;divide by NUMB1
0008  A2 0003 R           MOV   ANSQ,AL      ;save quotient
000B  88 26 0004 R        MOV   ANSR,AH      ;save remainder
```

16-bit Division.　　16-bit division is similar to 8-bit division, except that instead of dividing into AX, the 16-bit number is divided into DX–AX, a 32-bit dividend. The quotient appears in AX and the remainder appears in DX after a 16-bit division. Table 5–12 lists some of the 16-bit division instructions.

As with 8-bit division, numbers must often be converted to the proper form for the dividend. If a 16-bit unsigned number is placed in AX, DX must be cleared to 0. In the 80386 and above, the number is zero-extended by using the MOVZX instruction. If AX is a 16-bit signed number, the CWD (**convert word to doubleword**) instruction sign-extends it into a signed 32-bit number. If the 80386 and above is available, the MOVSX instruction can also be used to sign-extend a number.

Example 5–16 shows the division of two 16-bit signed numbers. Here, a –100 in AX is divided by a +9 in CX. The CWD instruction converts the –100 in AX to a –100 in DX (AX before the division). After the division, the results appear in DX–AX as a quotient of –11 in AX and a remainder of –1 in DX.

EXAMPLE 5–16

```
0000  B8 FF9C             MOV   AX,-100      ;load -100
0003  B9 0009             MOV   CX,9         ;load +9
0006  99                  CWD                ;sign-extend
0007  F7 F9               IDIV  CX
```

32-bit Division.　　The 80386 through the Pentium 4 processors perform 32-bit division on signed or unsigned numbers. The 64-bit contents of EDX–EAX are divided by the operand specified by the instruction, leaving a 32-bit quotient in EAX and a 32-bit remainder in EDX. Other

TABLE 5–12　　16-bit division instructions.

Assembly Language	Operation
DIV CX	DX–AX is divided by CX; the unsigned quotient is in AX and the remainder is in DX
IDIV SI	DX–AX is divided by SI; the signed quotient is in AX and the remainder is in DX
DIV NUMB	AX is divided by the contents of the data segment memory location NUMB; the unsigned quotient is in AX and the remainder is in DX

TABLE 5–13   32-bit division instructions.

Assembly Language	Operation
DIV ECX	EDX–EAX is divided by ECX; the unsigned quotient is in EAX and the remainder is in EDX
DIV DATA2	EDX–EAX is divided by the doubleword contents of data segment memory location DATA2; the unsigned quotient is in EAX and the remainder is in EDX
IDIV DWORD PTR[EDI]	EDX–EAX is divided by the doubleword contents of the data segment memory location addressed by EDI; the signed quotient is in EAX and the remainder is in EAX

than the size of the registers, this instruction functions in the same manner as the 8- and 16-bit divisions. Table 5–13 shows some 32-bit division instructions. The CDQ (**convert doubleword to quadword**) instruction is used before a signed division to convert the 32-bit contents of EAX into a 64-bit signed number in EDX–EAX.

The Remainder.   What is done with the remainder after a division? There are a few possible choices. The remainder could be used to **round** the result or just dropped to **truncate** the result. If the division is unsigned, rounding requires that the remainder be compared with half the divisor to decide whether to round up the quotient. The remainder could also be converted to a fractional remainder.

Example 5–17 shows a sequence of instructions that divide AX by BL, and round the result. This program doubles the remainder before comparing it with BL to decide whether to round the quotient. Here, an INC instruction rounds the contents of AL after the comparison.

EXAMPLE 5–17

```
0000   F6 F3              DIV    BL            ;divide
0002   02 E4              ADD    AH,AH         ;double remainder
0004   3A E3              CMP    AH,BL         ;test for rounding
0006   72 02              JB     NEXT
0008   FE C0              INC    AL            ;round
000A            NEXT:
```

Suppose that a fractional remainder is required instead of an integer remainder. A fractional remainder is obtained by saving the quotient. Next, the AL register is cleared to zero. The number remaining in AX is now divided by the original operand to generate a fractional remainder.

Example 5–18 shows how a 13 is divided by a 2. The 8-bit quotient is saved in memory location ANSQ, and then AL is cleared. Next, the contents of AX are again divided by 2 to generate a fractional remainder. After the division, the AL register equals an 80H. This is a 10000000_2. If the binary point (radix) is placed before the leftmost bit of AL, the fractional remainder in AL is 0.10000000_2, or 0.5 decimal. The remainder is saved in memory location ANSR in this example.

EXAMPLE 5–18

```
0000   B8 000D            MOV    AX,13         ;load 13
0003   B3 02              MOV    BL,2          ;load 2
0005   F6 F3              DIV    BL            ;13/2
0007   A2 0003 R          MOV    ANSQ,AL       ;save quotient
000A   B0 00              MOV    AL,0          ;clear AL
000C   F6 F3              DIV    BL            ;generate remainder
000E   A2 0004 R          MOV    ANSR,AL       ;save remainder
```

5-3   BCD AND ASCII ARITHMETIC

The microprocessor allows arithmetic manipulation of both BCD (**binary-coded decimal**) and ASCII (**American Standard Code for Information Interchange**) data. This is accomplished by instructions that adjust the numbers for BCD and ASCII arithmetic.

The BCD operations occur in systems such as point-of-sales terminals (e.g., cash registers) and others that seldom require arithmetic. The ASCII operations are performed on ASCII data used by many programs. In many cases, BCD or ASCII arithmetic is rarely used today.

BCD Arithmetic

Two arithmetic techniques operate with BCD data: addition and subtraction. The instruction set provides two instructions that correct the result of a BCD addition and a BCD subtraction. The DAA (**decimal adjust after addition**) instruction follows BCD addition, and the DAS (**decimal adjust after subtraction**) follows BCD subtraction. Both instructions correct the result of the addition or subtraction so that it is a BCD number.

For BCD data, the numbers always appear in the packed BCD form and are stored as two BCD digits per byte. The adjustment instructions function only with the AL register after BCD addition and subtraction.

DAA Instruction.   The DAA instruction follows the ADD or ADC instruction to adjust the result into a BCD result. Suppose that DX and BX each contain 4-digit packed BCD numbers. Example 5–19 provides a short sample program that adds the BCD numbers in DX and BX, and stores the result in CX.

EXAMPLE 5–19

```
0000   BA 1234        MOV    DX,1234H        ;load 1,234
0003   BB 3099        MOV    BX,3099H        ;load 3,099
0006   8A C3          MOV    AL,BL           ;sum BL with DL
0008   02 C2          ADD    AL,DL
000A   27             DAA                    ;adjust
000B   8A C8          MOV    CL,AL           ;answer to CL
000D   8A C7          MOV    AL,BH           ;sum BH, DH, and carry
000F   12 C6          ADC    AL,DH
0011   27             DAA                    ;adjust
0012   8A E8          MOV    CH,AL           ;answer to CH
```

Because the DAA instruction functions only with the AL register, this addition must occur eight bits at a time. After adding the BL and DL registers, the result is adjusted with a DAA instruction before being stored in CL. Next, add BH and DH registers with carry; the result is then adjusted with DAA before being stored in CH. In this example, a 1234 adds to a 3099 to generate a sum of 4333 that moves into CX after the addition. Note that 1234 BCD is the same as 1234H.

DAS Instruction.   The DAS instruction functions as does the DAA instruction, except that it follows a subtraction instead of an addition. Example 5–20 is the same as Example 5–19, except that it subtracts instead of adds DX and BX. The main difference in these programs is that the DAA instructions change to DAS, and the ADD and ADC instructions change to SUB and SBB instructions.

EXAMPLE 5–20

```
0000   BA 1234        MOV    DX,1234H        ;load 1,234
0003   BB 3099        MOV    BX,3099H        ;load 3,099
0006   8A C3          MOV    AL,BL           ;subtract DL from BL
0008   2A C2          SUB    AL,DL
```

```
000A   2F              DAS                     ;adjust
000B   8A C8           MOV    CL,AL            ;answer to CL
000D   8A C7           MOV    AL,BH            ;subtract DH
000F   1A C6           SBB    AL,DH
0011   2F              DAS                     ;adjust
0012   8A E8           MOV    CH,AL            ;answer to CH
```

ASCII Arithmetic

The ASCII arithmetic instructions function with ASCII-coded numbers. These numbers range in value from 30H to 39H for the numbers 0–9. There are four instructions used with ASCII arithmetic operations: AAA (**ASCII adjust after addition**), AAD (**ASCII adjust before division**), AAM (**ASCII adjust after multiplication**), and AAS (**ASCII adjust after subtraction**). These instructions use register AX as the source and as the destination.

AAA Instruction. The addition of two one-digit ASCII-coded numbers will not result in any useful data. For example, if 31H and 39H are added, the result is 6AH. This ASCII addition (1 + 9) should produce a two-digit ASCII result equivalent to a 10 decimal, which is a 31H and a 30H in ASCII code. If the AAA instruction is executed after this addition, the AX register will contain a 0100H. Although this is not ASCII code, it can be converted to ASCII code by adding 3030H, which generates 3130H. The AAA instruction clears AH if the result is less than 10, and adds a 1 to AH if the result is greater than 10.

Example 5–21 shows the way ASCII addition functions in the microprocessor. Please note that AH is cleared before the addition by using the MOV AX,31H instruction. The operand of 0031H places a 00H in AH and a 31H into AL.

EXAMPLE 5–21

```
0000   B8 0031         MOV    AX,31H           ;load ASCII 1
0003   04 39           ADD    AL,39H           ;add ASCII 9
0005   37              AAA                     ;adjust
0006   05 3030         ADD    AX,3030H         ;answer to ASCII
```

AAD Instruction. Unlike all other adjustment instructions, the AAD instruction appears before a division. The AAD instruction requires that the AX register contain a two-digit unpacked BCD number (not ASCII) before executing. After adjusting the AX register with AAD, it is divided by an unpacked BCD number to generate a single-digit result in AL with any remainder in AH.

Example 5–22 illustrates how a 72 in unpacked BCD is divided by 9 to produce a quotient of 8. The 0702H loaded into the AX register is adjusted by the AAD instruction to 0048H. Notice that this converts a two-digit unpacked BCD number into a binary number so it can be divided with the binary division instruction (DIV). The AAD instruction converts the unpacked BCD numbers between 00 and 99 into binary.

EXAMPLE 5–22

```
0000   B3 09           MOV    BL,9             ;load divisor
0002   B8 0702         MOV    AX,0702H         ;load dividend
0005   D5 0A           AAD                     ;adjust
0007   F6 F3           DIV    BL
```

AAM Instruction. The AAM instruction follows the multiplication instruction after multiplying two one-digit unpacked BCD numbers. Example 5–23 shows a short program that multiplies 5 times 5. The result after the multiplication is 0019H in the AX register. After adjusting the result with the AAM instruction, AX contains a 0205H. This is an unpacked BCD result of 25. If 3030H is added to 0205H, it has an ASCII result of 3235H.

EXAMPLE 5–23

```
0000  B0 05          MOV    AL,5          ;load multiplicand
0002  B1 05          MOV    CL,5          ;load multiplier
0004  F6 E1          MUL    CL
0006  D4 0A          AAM                  ;adjust
```

The AAM instruction accomplishes this conversion by dividing AX by 10. The remainder is found in AL, and the quotient is in AH. It has been noted that the second byte of the instruction contains a 0AH. If the 0AH is changed to another value, AAM divides by the new value. For example, if the second byte is changed to a 0BH, the AAM instruction divides by an 11.

One side benefit of the AAM instruction is that AAM converts from binary to unpacked BCD. If a binary number between 0000H and 0063H appears in the AX register, the AAM instruction converts it to BCD. For example, if AX contains a 0060H before AAM, it will contain a 0906H after AAM executes. This is the unpacked BCD equivalent of 96 decimal. If 3030H is added to 0906H, the result changes to ASCII code.

Example 5–24 shows how the 16-bit binary content of AX is converted to a four-digit ASCII character string by using division and the AAM instruction. Note that this works for numbers between 0 and 9999. First DX is cleared and then DX–AX is divided by 100. For example, if AX = 245_{10}, AX = 2 and DX = 45 after the division. These separate halves are converted to BCD using AAM, and then a 3030H is added to convert to ASCII code.

EXAMPLE 5–24

```
0000  33 D2          XOR    DX,DX         ;clear DX register
0002  B9 0064        MOV    CX,100        ;divide DX–AX by 100
0005  F7 F1          DIV    CX
0007  D4 0A          AAM                  ;convert quotient to BCD
0009  05 3030        ADD    AX,3030H      ;convert to ASCII
000C  92             XCHG   AX,DX         ;repeat for remainder
000D  D4 0A          AAM
000F  05 3030        ADD    AX,3030H
```

Example 5–25 uses the DOS 21H function AH = 02H to display a sample number in decimal on the video display using the AAM instruction. Notice how AAM is used to convert AL into BCD. Next, ADD AX,3030H converts the BCD code in AX into ASCII, for display with DOS INT 21H. Once the data are converted to ASCII code, they are displayed by loading DL with the most-significant digit from AH. Next, the least-significant digit is displayed from AL. Note that the DOS INT 21H function calls change AL.

EXAMPLE 5–25

```
                     ;A program that displays the number loaded into AL,
                     ;with the first instruction (48H), as a decimal number.
                     ;
                            .MODEL TINY           ;select TINY model
0000                        .CODE                 ;start of CODE segment
                            .STARTUP              ;indicate start of program
0100  B0 48                 MOV    AL,48H         ;load AL with test data
0102  B4 00                 MOV    AH,0           ;clear AH
0104  D4 0A                 AAM                   ;convert to BCD
0106  05 3030               ADD    AX,3030H;      ;convert to ASCII
0109  8A D4                 MOV    DL,AH          ;display most-significant digit
010B  B4 02                 MOV    AH,2
010D  50                    PUSH   AX             ;save least-significant digit
010E  CD 21                 INT    21H
0110  58                    POP    AX             ;restore AL
0111  8A D0                 MOV    DL,AL          ;display least-significant digit
0113  CD 21                 INT    21H
                            .EXIT                 ;exit to DOS
                     END                          ;end of file
```

AAS Instruction. Like other ASCII adjust instructions, AAS adjusts the AX register after an ASCII subtraction. For example, suppose that a 35H subtracts from a 39H. The result will be a 04H, which requires no correction. Here, AAS will modify neither AH nor AL. On the other hand, if 38H is subtracted from 37H, then AL will equal 09H, and the number in AH will decrement by 1. This decrement allows multiple-digit ASCII numbers to be subtracted from each other.

5–4 BASIC LOGIC INSTRUCTIONS

The basic logic instructions include AND, OR, Exclusive-OR, and NOT. Another logic instruction is TEST, which is explained in this section of the text because the operation of the TEST instruction is a special form of the AND instruction. Also explained is the NEG instruction, which is similar to the NOT instruction.

Logic operations provide binary bit control in low-level software. The logic instructions allow bits to be set, cleared, or complemented. Low-level software appears in machine language or assembly language form, and often controls the I/O devices in a system. All logic instructions affect the flag bits. Logic operations always clear the carry and overflow flags, while the other flags change to reflect the condition of the result.

When binary data are manipulated in a register or a memory location, the rightmost bit position is always numbered bit 0. Bit position numbers increase from bit 0 toward the left, to bit 7 for a byte, and to bit 15 for a word. A doubleword (32 bits) uses bit position 31 as its leftmost bit.

AND

The AND operation performs logical multiplication, as illustrated by the truth table in Figure 5–3. Here, two bits, A and B, are ANDed to produce the result X. As indicated by the truth table, X is a logic 1 only when both A and B are logic 1s. For all other input combinations of A and B, X is a logic 0. It is important to remember that 0 AND anything is always 0, and 1 AND 1 is always 1.

The AND instruction can replace discrete AND gates if the speed required is not too great, although this is normally reserved for embedded control applications. (Note that Intel has released the 80386EX embedded controller, which embodies the basic structure of the personal computer system.) With the 8086 microprocessor, the AND instruction often executes in about a microsecond. With newer versions, the execution speed is greatly increased. If the circuit that the AND instruction replaces operates at a much slower speed than the microprocessor, the AND instruction is a logical replacement. This replacement can save a considerable amount of money. A single AND gate integrated circuit (7408) costs approximately 40¢, while it costs less than 1/100¢ to store the AND instruction in read-only memory. Note that a logic circuit replacement such as this only appears in control systems based on microprocessors, and does not generally find application in the personal computer.

FIGURE 5–3 (a) The truth table for the AND operation and (b) the logic symbol of an AND gate.

A	B	T
0	0	0
0	1	0
1	0	0
1	1	1

(a)

(b)

FIGURE 5–4   The operation
of the AND function showing
how bits of a number are
cleared to zero.

```
  x x x x  x x x x   Unknown number
• 0 0 0 0  1 1 1 1   Mask
  ─────────────────
  0 0 0 0  x x x x   Result
```

The AND operation clears bits of a binary number. The task of clearing a bit in a binary number is called **masking.** Figure 5–4 illustrates the process of masking. Notice that the leftmost four bits clear to 0 because 0 AND anything is 0. The bit-positions that AND with 1s do not change. This occurs because if a 1 ANDs with a 1, a 1 results; if a 1 ANDs with a 0, a 0 results.

The AND instruction uses any addressing mode except memory-to-memory and segment register addressing. Table 5–14 lists some AND instructions and comments about their operations.

An ASCII-coded number can be converted to BCD by using the AND instruction to mask off the leftmost four binary bit positions. This converts the ASCII 30H to 39H to 0–9. Example 5–26 shows a short program that converts the ASCII contents of BX into BCD. The AND instruction in this example converts two digits from ASCII to BCD simultaneously.

EXAMPLE 5–26

```
0000 BB 3135            MOV  BX,3135H         ;load ASCII
0003 81 E3 0F0F         AND  BX,0F0FH         ;mask DX
```

OR

The **OR operation** performs logical addition and is often called the *Inclusive-OR* function. The OR function generates a logic 1 output if any inputs are 1. A 0 appears at the output only when all inputs are 0. The truth table for the OR function appears in Figure 5–5. Here, the inputs A and B OR together to produce the X output. It is important to remember that 1 ORed with anything yields a 1.

In embedded controller applications, the OR instruction can also replace discrete OR gates. This results in considerable savings because a quad, 2-input OR gate (7432) costs about 40¢, while the OR instruction costs less than 1/100¢ to store in a read-only memory.

Figure 5–6 shows how the OR gate sets (1) any bit of a binary number. Here, an unknown number (XXXX XXXX) ORs with a 0000 1111 to produce a result of XXXX 1111. The rightmost

TABLE 5–14   AND instructions.

Assembly Language	Operation
AND AL,BL	AL = AL AND BL
AND CX,DX	CX = CX AND DX
AND ECX,EDI	ECX = ECX AND EDI
AND CL,33H	CL = CL AND 33H
AND DI,4FFFH	DI = DI AND 4FFFH
AND ESI,34H	ESI = ESI AND 00000034H
AND AX,[DI]	AX is ANDed with the word contents of the data segment memory location addressed by DI
AND ARRAY[SI],AL	The byte contents of the data segment memory location addressed by the sum of ARRAY plus SI is ANDed with AL; the result moves to memory
AND [EAX],CL	CL is ANDed with the byte contents of the data segment memory location addressed by EAX; the result moves to memory

FIGURE 5–5 (a) The truth table for the OR operation and (b) the logic symbol of an OR gate.

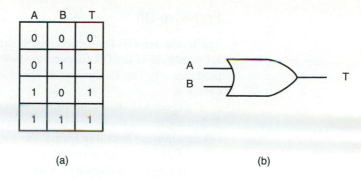

A	B	T
0	0	0
0	1	1
1	0	1
1	1	1

(a)   (b)

FIGURE 5–6 The operation of the OR function showing how bits of a number are set to one.

```
  x x x x  x x x x   Unknown number
+ 0 0 0 0  1 1 1 1   Mask
  ─────────────────
  x x x x  1 1 1 1   Result
```

four bits set, while the leftmost four bits remain unchanged. The OR operation sets any bit; the AND operation clears any bit.

The OR instruction uses any of the addressing modes allowed to any other instruction except segment register addressing. Table 5–15 illustrates several example OR instructions with comments about their operation.

Suppose that two BCD numbers are multiplied and adjusted with the AAM instruction. The result appears in AX as a two-digit unpacked BCD number. Example 5–27 illustrates this multiplication and shows how to change the result into a two-digit ASCII-coded number using the OR instruction. Here, OR AX,3030H converts the 0305H found in AX to 3335H. The OR operation can be replaced with an ADD AX,3030H to obtain the same results.

EXAMPLE 5–27

```
0000  B0 05        MOV   AL,5          ;load data
0002  B3 07        MOV   BL,7
0004  F6 E3        MUL   BL
0006  D4 0A        AAM                 ;adjust
0008  0D 3030      OR    AX,3030H      ;to ASCII
```

TABLE 5–15 OR instructions.

Assembly Language	Operation
OR AH,BL	AH = AH OR BL
OR SI,DX	SI = SI OR DX
OR EAX,EBX	EAX = EAX OR EBX
OR DH,0A3H	DH = DH OR A3H
OR SP,990DH	SP = SP OR 990DH
OR EBP,10	EBP = EBP OR 0000000AH
OR DX,[BX]	DX is ORed with the word contents of the data segment memory location addressed by BX
OR DATES[DI + 2],AL	The byte contents of the data segment memory location addressed by the sum of DATES, DI, and 2 are ORed with AL

Exclusive-OR

The **Exclusive-OR** instruction (XOR) differs from Inclusive-OR (OR). The difference is that a 1,1 condition of the OR function produces a 1; the 1,1 condition of the Exclusive-OR operation produces a 0. The Exclusive-OR operation excludes this condition, while the Inclusive-OR includes it.

Figure 5–7 shows the truth table of the Exclusive-OR function. (Compare this with Figure 5–5 to appreciate the difference between these two OR functions.) If the inputs of the Exclusive-OR function are both 0 or both 1, the output is 0. If the inputs are different, the output is 1. Because of this, the Exclusive-OR is sometimes called a comparator.

The XOR instruction uses any addressing mode except segment register addressing. Table 5–16 lists several Exclusive-OR instructions and their operations.

As with the AND and OR functions, Exclusive-OR can replace discrete logic circuitry in embedded applications. The 7486 quad, 2-input Exclusive-OR gate is replaced by one XOR instruction. The 7486 costs about 40¢, while the instruction costs less than 1/100¢ to store in the memory. Replacing just one 7486 saves a considerable amount of money, especially if many systems are built.

The Exclusive-OR instruction is useful if some bits of a register or memory location must be inverted. This instruction allows part of a number to be inverted or complemented. Figure 5–8 shows how just part of an unknown quantity can be inverted by XOR. Notice that when a 1 Exclusive-ORs with X, the result is X. If a 0 Exclusive-ORs with X, the result is X.

FIGURE 5–7   (a) The truth table for the Exclusive-OR operation and (b) the logic symbol of an Exclusive-OR gate.

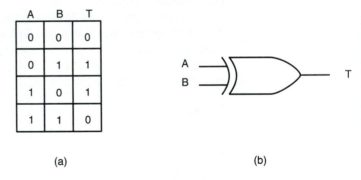

A	B	T
0	0	0
0	1	1
1	0	1
1	1	0

(a)   (b)

TABLE 5–16   Exclusive-OR instructions.

Assembly Language	Operation
XOR CH,DL	CH = CH XOR DL
XOR SI,BX	SI = SI XOR BX
XOR EBX,EDI	EBX = EBX XOR EDI
XOR AH,0EEH	AH = AH XOR EEH
XOR DI,0DDH	DI = DI XOR 00DDH
XOR ESI,100	ESI = ESI XOR 00000064H
XOR DX,[SI]	DX is Exclusive-ORed with the word contents of the data segment memory location addressed by SI
XOR DATES[DI + 2],AL	AL is Exclusive-ORed with the byte contents of the data segment memory location addressed by the sum of DATES, DI, and 2

FIGURE 5–8 The operation of the Exclusive-OR function showing how bits of a number are inverted.

```
     x x x x  x x x x    Unknown number
⊕    0 0 0 0  1 1 1 1    Mask
     ─────────────────
     x x x x  x̄ x̄ x̄ x̄    Result
```

Suppose that the leftmost 10 bits of the BX register must be inverted without changing the rightmost six bits. The XOR BX,0FFC0H instruction accomplishes this task. The AND instruction clears (0) bits, the OR instruction sets (1) bits, and now the Exclusive-OR instruction inverts bits. These three instructions allow a program to gain complete control over any bit, stored in any register or memory location. This is ideal for control system applications in which equipment must be turned on (1), turned off (0), and toggled from on to off or off to on.

A common use for the Exclusive-OR instruction is to clear a register to zero. For example, the XOR CH,CH instruction clears register CH to 00H and requires two bytes of memory to store the instruction. Likewise, the MOV CH,00H instruction also clears CH to 00H, but requires three bytes of memory. Because of this saving, the XOR instruction is used to clear a register in place of a move immediate.

Example 5–28 shows a short sequence of instructions that clears bits 0 and 1 of CX, sets bits 9 and 10 of CX, and inverts bit 12 of CX. The OR instruction is used to set bits, the AND instruction is used to clear bits, and the XOR instruction inverts bits.

EXAMPLE 5–28

```
0000  81 C9 0600      OR    CX,0600H      ;set bits 9 and 10
0004  83 E1 FC        AND   CX,0FFFCH     ;clear bits 0 and 1
0007  81 F1 1000      XOR   CX,1000H      ;invert bit 12
```

Test and Bit Test Instructions

The **TEST instruction** performs the AND operation. The difference is that the AND instruction changes the destination operand, while the TEST instruction does not. A TEST only affects the condition of the flag register, which indicates the result of the test. The TEST instruction uses the same addressing modes as the AND instruction. Table 5–17 lists some TEST instructions and their operations.

The TEST instruction functions in the same manner as a CMP instruction. The difference is that the TEST instruction normally tests a single bit (or occasionally multiple bits), while the CMP instruction tests the entire byte or word. The zero flag (Z) is a logic 1 (indicating a zero result) if the bit under test is a zero, and Z = 0 (indicating a non-zero result) if the bit under test is not zero.

Usually the TEST instruction is followed by either the JZ (**jump if zero**) or JNZ (**jump if not zero**) instruction. The destination operand is normally tested against immediate data. The value of immediate data is 1 to test the rightmost bit position, 2 to test the next bit, 4 for the next, and so on.

TABLE 5–17 TEST instructions.

Assembly Language	Operation
TEST DL,DH	DL is ANDed with DH
TEST CX,BX	CX is ANDed with BX
TEST EDX,ECX	EDX is ANDed with ECX
TEST AH,4	AH is ANDed with 4
TEST EAX,256	EAX is ANDed with 256

Example 5–29 lists a short program that tests the rightmost and leftmost bit positions of the AL register. Here, 1 selects the rightmost bit and 128 selects the leftmost bit. (Note: A 128 is an 80H.) The JNZ instruction follows each test to jump to different memory locations, depending on the outcome of the tests. The JNZ instruction jumps to the operand address (RIGHT or LEFT in the example) if the bit under test is not zero.

EXAMPLE 5–29

```
0000   A8 01          TEST   AL,1        ;test right bit
0002   75 1C          JNZ    RIGHT       ;if set
0004   A8 80          TEST   AL,128      ;test left bit
0006   75 38          JNZ    LEFT        ;if set
```

The 80386 through the Pentium 4 processors contain additional test instructions that test single bit positions. Table 5–18 lists the four different bit test instructions available to these microprocessors.

All four forms of the bit test instruction test the bit position in the destination operand selected by the source operand. For example, the BT AX,4 instruction tests bit position 4 in AX. The result of the test is located in the carry flag bit. If bit position 4 is a 1, carry is set; if bit position 4 is a 0, carry is cleared.

The remaining three-bit test instructions also place the bit under test into the carry flag, and change the bit under test afterward. The BTC AX,4 instruction complements bit position 4 after testing it, the BTR AX,4 instruction clears it (0) after the test, and the BTS AX,4 instruction sets it (1) after the test.

Example 5–30 repeats the sequence of instructions listed in Example 5–28. Here, the BTR instruction clears bits in CX, BTS sets bits in CX, and BTC inverts bits in CX.

EXAMPLE 5–30

```
0000   0F BA E9 09    BTS    CX,9        ;set bit 9
0004   0F BA E9 0A    BTS    CX,10       ;set bit 10
0008   0F BA F1 00    BTR    CX,0        ;clear bit 0
000C   0F BA F1 01    BTR    CX,1        ;clear bit 1
0010   0F BA F9 0C    BTC    CX,12       ;invert bit 12
```

NOT and NEG

Logical inversion, or the **one's complement** (NOT); and arithmetic sign inversion, or the **two's complement** (NEG) are the last two logic functions presented (except for shift and rotate in the next section of the text). These are two of a few instructions that contain only one operand. Table 5–19 lists some variations of the NOT and NEG instructions. As with most other instructions, NOT and NEG can use any addressing mode except segment register addressing.

TABLE 5–18   Bit test instructions.

Assembly Language	Operation
BT	Tests a bit in the destination operand specified by the source operand
BTC	Tests and complements a bit in the destination operand specified by the source operand
BTR	Tests and resets a bit in the destination operand specified by the source operand
BTS	Tests and sets a bit in the destination operand specified by the source operand

TABLE 5–19 NOT and NEG instructions.

Assembly Language	Operation
NOT CH	CH is one's complemented
NEG CH	CH is two's complemented
NEG AX	AX is two's complemented
NOT EBX	EBX is one's complemented
NEG ECX	ECX is two's complemented
NOT TEMP	The contents of the data segment memory location TEMP is one's complemented
NOT BYTE PTR[BX]	The byte contents of the data segment memory location addressed by BX is one's complemented

The NOT instruction inverts all bits of a byte, word, or doubleword. The NEG instruction two's complements a number, which means that the arithmetic sign of a signed number changes from positive to negative or from negative to positive. The NOT function is considered logical, and the NEG function is considered an arithmetic operation.

5–5 SHIFT AND ROTATE

Shift and rotate instructions manipulate binary numbers at the binary bit level, as did the AND, OR, Exclusive-OR, and NOT instructions. Shifts and rotates find their most common applications in low-level software used to control I/O devices. The microprocessor contains a complete set of shift and rotate instructions that are used to shift or rotate any memory data or register.

Shift

Shift instructions position or move numbers to the left or right within a register or memory location. They also perform simple arithmetic such as multiplication by powers of 2^{+n} (**left shift**) and division by powers of 2^{-n} (**right shift**). The microprocessor's instruction set contains four different shift instructions: two are logical shifts and two are arithmetic shifts. All four shift operations appear in Figure 5–9.

Notice in Figure 5–9 that there are two right shifts and two left shifts. The logical shifts move a 0 into the rightmost bit position for a logical left shift and a 0 into the leftmost bit position for a logical right shift. There are also two arithmetic shifts. The arithmetic and logical left shifts are identical. The arithmetic and logical right shifts are different because the arithmetic right shift copies the sign-bit through the number, while the logical right shift copies a 0 through the number.

Logical shift operations function with unsigned numbers, and arithmetic shifts function with signed numbers. Logical shifts multiply or divide unsigned data, and arithmetic shifts multiply or divide signed data. A shift left always multiplies by 2 for each bit position shifted, and a shift right always divides by 2 for each bit position shifted. Shifting a number 2 places multiplies or divides by 4.

Table 5–20 illustrates some addressing modes allowed for the various shift instructions. There are two different forms of shifts that allow any register (except the segment register) or

FIGURE 5–9 The shift instructions showing the operation and direction of the shift.

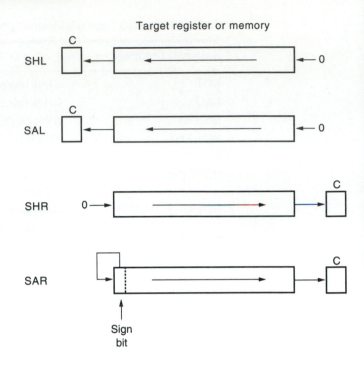

memory location to be shifted. One mode uses an immediate shift count, and the other uses register CL to hold the shift count. Note that CL must hold the shift count. When CL is the shift count, it does not change when the shift instruction executes. Note that the shift count is a modulo-32 count, which means that a shift count of 33 will shift the data one place (33/32 = remainder of 1).

Example 5–31 shows how to shift the DX register left 14 places in two different ways. The first method uses an immediate shift count of 14. The second method loads a 14 into CL and then uses CL as the shift count. Both instructions shift the contents of the DX register logically to the left 14 binary bit positions or places.

EXAMPLE 5–31

```
0000  C1 E2 0E            SHL   DX,14

                or

0003  B1 0E               MOV   CL,14
0005  D3 E2               SHL   DX,CL
```

TABLE 5–20 Shift instructions.

Assembly Language	Operation
SHL AX,1	AX is logically shifted left 1 place
SHR BX,12	BX is logically shifted right 12 places
SHR ECX,10	ECX is logically shifted right 10 places
SAL DATA1,CL	The contents of the data segment memory location DATA1 is arithmetically shifted left the number of places specified by CL
SAR SI,2	SI is arithmetically shifted right 2 places
SAR EDX,14	EDX is arithmetically shifted right 14 places

Suppose that the contents of AX must be multiplied by 10, as shown in Example 5–32. This can be done in two ways: by the MUL instruction or by shifts and additions. A number is doubled when it shifts left one place. When a number is doubled, and then added to the number times 8, the result is 10 times the number. The number 10 decimal is 1010 in binary. A logic 1 appears in both the 2's and 8's positions. If 2 times the number is added to 8 times the number, the result is 10 times the number. Using this technique, a program can be written to multiply by any constant. This technique often executes faster than the multiply instruction found in earlier versions of the Intel microprocessor.

EXAMPLE 5–32

```
                          ;Multiply AX by 10 (1010)
                          ;
0000  D1 E0                       SHL    AX,1              ;AX times 2
0002  8B D8                       MOV    BX,AX
0004  C1 E0 02                    SHL    AX,2              ;AX times 8
0007  03 C3                       ADD    AX,BX             ;10 times AX
                          ;
                          ;Multiply AX by 18 (10010)
                          ;
0009  D1 E0                       SHL    AX,1              ;AX times 2
000B  8B D8                       MOV    BX,AX
000D  C1 E0 03                    SHL    AX,3              ;AX times 16
0010  03 C3                       ADD    AX,BX             ;18 times AX
                          ;
                          ;Multiply AX by 5 (101)
                          ;
0012  8B D8                       MOV    BX,AX
0014  D1 E0                       SHL    AX,1              ;AX times 2
0016  D1 E0                       SHL    AX,1              ;AX times 4
0018  03 C3                       ADD    AX,BX             ;5 times AX
```

Double-precision Shifts (80386–Pentium 4 Only).   The 80386 and above contain two double-precision shifts: SHLD (**shift left**) and SHRD (**shift right**). Each instruction contains three operands, instead of the two found with the other shift instructions. Both instructions function with two 16- or 32-bit registers, or with one 16- or 32-bit memory location and a register.

The SHRD AX,BX,12 instruction is an example of the double-precision shift right instruction. This instruction logically shifts AX right by 12 bit positions. The rightmost 12 bits of BX shift into the leftmost 12 bits of AX. The contents of BX remain unchanged by this instruction. The shift count can be an immediate count, as in this example, or it can be found in register CL, as with other shift instructions.

The SHLD EBX,ECX,16 instruction shifts EBX left. The leftmost 16 bits of ECX fill the rightmost 16 bits of EBX after the shift. As before, the contents of ECX, the second operand, remain unchanged. This instruction, as well as SHRD, affect the flag bits.

Rotate

Rotate instructions position binary data by rotating the information in a register or memory location, either from one end to another or through the carry flag. They are often used to shift or position numbers that are wider than 16-bits in the 8086–80286 microprocessors or wider than 32-bit in the 80386 through the Pentium 4. The four available rotate instructions appear in Figure 5–10.

Numbers rotate through register or memory location, through the C flag (carry), or through a register or memory location only. With either type of rotate instruction, the programmer can select either a left or a right rotate. Addressing modes used with rotate are the same as those used with shifts. A rotate count can be immediate or located in register CL. Table 5–21 lists some of the possible rotate instructions. If CL is used for a rotate count, it does not change. As with shifts, the count in CL is a modulo-32 count.

FIGURE 5–10   The rotate instructions showing the direction and operation of each rotate.

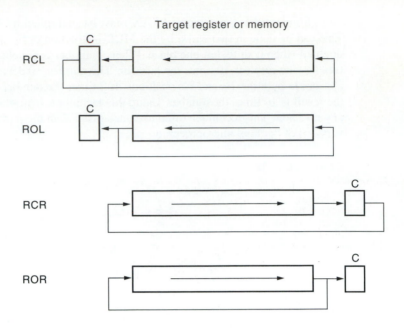

Rotate instructions are often used to shift wide numbers to the left or right. The program listed in Example 5–33 shifts the 48-bit number in registers DX, BX, and AX left one binary place. Notice that the least-significant 16 bits (AX) shift left first. This moves the leftmost bit of AX into the carry flag bit. Next, the rotate BX instruction rotates carry into BX, and its leftmost bit moves into carry. The last instruction rotates carry into DX, and the shift is complete.

EXAMPLE 5–33

```
0000   D1 E0          SHL   AX,1
0002   D1 D3          RCL   BX,1
0004   D1 D2          RCL   DX,1
```

Bit Scan Instructions

Although the bit scan instructions don't shift or rotate numbers, they do scan through a number searching for a 1 bit. Because this is accomplished within the microprocessor by shifting the number, bit scan instructions are included in this section of the text.

The bit scan instructions BSF (**bit scan forward**) and BSR (**bit scan reverse**) are available only in the 80386–Pentium 4 processors. Both forms scan through the source number, searching for the first 1-bit. The BSF instruction scans the number from the leftmost bit toward

TABLE 5–21   Rotate instructions.

Assembly Language	Operation
ROL SI,14	SI rotates left 14 places
RCL BL,6	BL rotates left through carry 6 places
ROL ECX,18	ECX rotates left 18 places
RCR AH,CL	AH rotates right through carry the number of places specified by CL
ROR WORD PTR[BP],2	The word contents of the stack segment memory location addressed by BP rotate right 2 places

the right, and BSR scans the number from the rightmost bit toward the left. If a 1-bit is encountered, the zero flag is set and the bit position number of the 1-bit is placed into the destination operand. If no 1-bit is encountered (i.e., the number contains all zeros), the zero flag is cleared. Thus, the result is not-zero if no 1-bit is encountered.

For example, if EAX = 60000000H and the BSF EBX,EAX instruction executes, the number is scanned from the leftmost bit toward the right. The first 1-bit encountered is at bit position 30, which is placed into EBX and the zero flag bit is set. If the same value for EAX is used for the BSR instruction, the EBX register is loaded with a 29 and the zero flag bit is set.

5–6 STRING COMPARISONS

As illustrated in Chapter 4, the string instructions are very powerful because they allow the programmer to manipulate large blocks of data with relative ease. Block data manipulation occurs with the string instructions MOVS, LODS, STOS, INS, and OUTS.

In this section, additional string instructions that allow a section of memory to be tested against a constant or against another section of memory are discussed. To accomplish these tasks, use the SCAS (**string scan**) or CMPS (**string compare**) instructions.

SCAS

The SCAS (string scan instruction) compares the AL register with a byte block of memory, the AX register with a word block of memory, or the EAX register (80386–Pentium 4) with a doubleword block of memory. The SCAS instruction subtracts memory from AL, AX, or EAX without affecting either the register or the memory location. The opcode used for byte comparison is SCASB, the opcode used for the word comparison is SCASW, and the opcode used for a doubleword comparison is SCASD. In all cases, the contents of the extra segment memory location addressed by DI is compared with AL, AX, or EAX. Recall that this default segment (ES) cannot be changed with a segment override prefix.

Like the other string instructions, SCAS instructions use the direction flag (D) to select either auto-increment or auto-decrement operation for DI. They also repeat if prefixed by a conditional repeat prefix.

Suppose that a section of memory is 100 bytes long and begins at location BLOCK. This section of memory must be tested to see whether any location contains a 00H. The program in Example 5–34 shows how to search this part of memory for a 00H using the SCASB instruction. In this example, the SCASB instruction has an REPNE (**repeat while not equal**) prefix. The REPNE prefix causes the SCASB instruction to repeat until either the CX register reaches 0, or until an equal condition exists as the outcome of the SCASB instruction's comparison. Another conditional repeat prefix is REPE (**repeat while equal**). With either repeat prefix, the contents of CX decrements without affecting the flag bits. The SCASB instruction and the comparison it makes change the flags.

EXAMPLE 5–34

```
0000  BF 0011 R        MOV    DI,OFFSET BLOCK    ;address data
0003  FC               CLD                       ;auto-increment
0004  B9 0064          MOV    CX,100             ;load counter
0007  32 C0            XOR    AL,AL              ;clear AL
0009  F2/AE            REPNE  SCASB              ;search
```

Suppose that you must develop a program that skips ASCII-coded spaces in a memory array. (This task appears in the procedure listed in Example 5–35.) This procedure assumes that the DI register already addresses the ASCII-coded character string, and that the length of the string is 256 bytes or fewer. Because this program is to skip spaces (20H), the REPE (**repeat while equal**) prefix is used with a SCASB instruction. The SCASB instruction repeats the comparison, searching for a 20H, as long as an equal condition exists.

EXAMPLE 5–35

```
0000                     SKIP    PROC FAR

0000 FC                          CLD             ;auto-increment
0001 B9 0100                     MOV   CX,256    ;counter
0004 B0 20                       MOV   AL,20H    ;get space
0006 F3/AE                       REPE SCASB      ;search
0008 CB                          RET

0009                     SKIP    ENDP
```

CMPS

The CMPS (compare strings instruction) always compares two sections of memory data as bytes (CMPSB), words (CMPSW), or doublewords (CMPSD). Note that only the 80386 through Pentium 4 can use doublewords. The contents of the data segment memory location addressed by SI is compared with the contents of the extra segment memory location addressed by DI. The CMPS instruction increments or decrements both SI and DI. The CMPS instruction is normally used with either the REPE or REPNE prefix. Alternates to these prefixes are REPZ (**repeat while zero**) and REPNZ (**repeat while not zero**), but usually the REPE or REPNE prefixes are used in programming.

Example 5–36 illustrates a short procedure that compares two sections of memory searching for a match. The CMPSB instruction is prefixed with a REPE. This causes the search to continue as long as an equal condition exists. When the CX register becomes 0 or an unequal condition exists, the CMPSB instruction stops execution. After the CMPSB instruction ends, the CX register is 0 or the flags indicate an equal condition when the two strings match. If CX is not 0 or the flags indicate a not-equal condition, the strings do not match.

EXAMPLE 5–36

```
0000                     MATCH   PROC    FAR

0000 BE 0075 R                   MOV     SI,OFFSET LINE    ;address LINE
0003 BF 007F R                   MOV     DI,OFFSET TABLE   ;address TABLE
0006 FC                          CLD                       ;auto-increment
0007 B9 000A                     MOV     CX,10             ;counter
000A F3/A6                       REPE    CMPSB             ;search
000C CB                          RET

000D                     MATCH   ENDP
```

5–7   SUMMARY

1. Addition (ADD) can be 8-, 16-, or 32-bit. The ADD instruction allows any addressing mode except segment register addressing. Most flags (C, A, S, Z, P, and O) change when the ADD instruction executes. A different type of addition, add-with-carry (ADC), adds two operands and the contents of the carry flag (C). The 80486 through the Pentium 4 processors have an additional instruction (XADD) that combines an addition with an exchange.

2. The increment instruction (INC) adds 1 to the byte, word, or doubleword contents of a register or memory location. The INC instruction affects the same flag bits as ADD, except the

carry flag. The BYTE PTR, WORD PTR, and DWORD PTR directives appear with the INC instruction when the contents of a memory location are addressed by a pointer.

3. Subtraction (SUB) is a byte, word, or doubleword and is performed on a register or a memory location. The only form of addressing not allowed by the SUB instruction is segment register addressing. The subtract instruction affects the same flags as ADD, and subtracts carry if the SBB form is used.

4. The decrement (DEC) instruction subtracts 1 from the contents of a register or a memory location. The only addressing modes not allowed with DEC are immediate or segment register addressing. The DEC instruction does not affect the carry flag and is often used with BYTE PTR, WORD PTR, or DWORD PTR.

5. The comparison (CMP) instruction is a special form of subtraction that does not store the difference; instead, the flags change to reflect the difference. Comparison is used to compare an entire byte or word located in any register (except segment) or memory location. An additional comparison instruction (CMPXCHG), which is a combination of comparison and exchange instructions, is found in the 80486–Pentium 4 processors. In the Pentium–Pentium 4 processors, the CMPXCHG8B instruction compares and exchanges quadword data.

6. Multiplication is byte, word, or doubleword, and it can be signed (IMUL) or unsigned (MUL). The 8-bit multiplication always multiplies register AL by an operand with the product found in AX. The 16-bit multiplication always multiplies register AX by an operand with the product found in DX–AX. The 32-bit multiply always multiplies register EAX by an operand with the product found in EDX–EAX. A special IMUL immediate instruction exists on the 80186–Pentium 4 processors that contains three operands. For example, the IMUL BX,CX,3 instruction multiplies CX by 3 and leaves the product in BX.

7. Division is byte, word, or doubleword, and it can be signed (IDIV) or unsigned (DIV). For an 8-bit division, the AX register divides by the operand, after which the quotient appears in AL and the remainder appears in AH. In the 16-bit division, the DX–AX register divides by the operand, after which the AX register contains the quotient and DX contains the remainder. In the 32-bit division, the EDX–EAX register is divided by the operand, after which the EAX register contains the quotient and the EDX register contains the remainder. Note that the remainder after a signed division always assumes the sign of the dividend.

8. BCD data add or subtract in packed form by adjusting the result of the addition with DAA or the subtraction with DAS. ASCII data are added, subtracted, multiplied, or divided when the operations are adjusted with AAA, AAS, AAM, and AAD.

9. The AAM instruction has an interesting added feature that allows it to convert a binary number into unpacked BCD. This instruction converts a binary number between 00H–63H into unpacked BCD in AX. The AAM instruction divides AX by 10, and leaves the remainder in AL and quotient in AH.

10. The AND, OR, and Exclusive-OR instructions perform logic functions on a byte, word, or doubleword stored in a register or memory location. All flags change with these instructions, with carry (C) and overflow (O) cleared.

11. The TEST instruction performs the AND operation, but the logical product is lost. This instruction changes the flag bits to indicate the outcome of the test.

12. The NOT and NEG instructions perform logical inversion and arithmetic inversion. The NOT instruction one's complements an operand, and the NEG instruction two's complements an operand.

13. There are eight different shift and rotate instructions. Each of these instructions shifts or rotates a byte, word, or doubleword register or memory data. These instructions have two operands: the first is the location of the data shifted or rotated, and the second is an immediate shift or rotate count or CL. If the second operand is CL, the CL register holds the shift or rotate count. In the 80386 through the Pentium 4 processors, two additional double-precision shifts (SHRD and SHLD) exist.

14. The scan string (SCAS) instruction compares AL, AX, or EAX with the contents of the extra segment memory location addressed by DI.

15. The string compare (CMPS) instruction compares the byte, word, or doubleword contents of two sections of memory. One section is addressed by DI in the extra segment, and the other is addressed by SI in the data segment.

16. The SCAS and CMPS instructions repeat with the REPE or REPNE prefixes. The REPE prefix repeats the string instruction while an equal condition exists, and the REPNE repeats the string instruction while a not-equal condition exists.

17. Example 5–37 illustrates a program that uses some of the instructions in this chapter to search the video display (beginning at address B800:000) to find whether it contains the word BUG. If the word BUG is found, the program displays a Y. If BUG is not found it displays an N. Notice how the CMPSB instruction is used to search for BUG.

EXAMPLE 5–37

```
                    ;program that tests the video display for the word BUG
                    ;if BUG appears anywhere on the display, a Y is
                    ;displayed
                    ;if BUG does not appear, the program displays N
                    ;
                    .MODEL SMALL                     ;select SMALL model
0000                .DATA                            ;start of DATA segment
0000 42 55 47       DATA1   DB      'BUG'            ;define BUG
0000                .CODE                            ;start of CODE segment
                    .STARTUP                         ;start of program
0017 B8 B800                MOV     AX,0B800H        ;address segment B800 with ES
001A 8E C0                  MOV     ES,AX
001C B9 07D0                MOV     CX,25*80         ;set count
001F FC                     CLD                      ;select increment
0020 BF 0000                MOV     DI,0             ;address display
0023                L1:
0023 BE 0000 R              MOV     SI,OFFSET DATA1  ;address BUG
0026 57                     PUSH    DI               ;save display address
0027 A6                     CMPSB                    ;test for B
0028 75 0A                  JNE     L2               ;if display is not B
002A 47                     INC     DI               ;address next position
002B A6                     CMPSB                    ;test for U
002C 75 06                  JNE     L2               ;if display is not U
002E 47                     INC     DI               ;address next position
002F A6                     CMPSB                    ;test for G
0030 B2 59                  MOV     DL,'Y'           ;load Y for possible BUG
0032 74 09                  JE      L3               ;if BUG is found
0034                L2:
0034 5F                     POP     DI               ;restore display address
0035 83 C7 02               ADD     DI,2             ;point to next position
0038 E2 E9                  LOOP    L1               ;repeat for whole screen
003A 57                     PUSH    DI               ;save display address
003B B2 4E                  MOV     DL,'N'           ;indicate N if no BUG
003D                L3:
003D 5F                     POP     DI               ;clear stack
003E B4 02                  MOV     AH,2             ;display DL function
0040 CD 21                  INT     21H              ;display ASCII from DL
                    .EXIT                            ;exit to DOS
                    END                              ;end of file
```

5–8   QUESTIONS AND PROBLEMS

1. Select an ADD instruction that will:
   (a) add BX to AX
   (b) add 12H to AL

    (c) add EDI and EBP

    (d) add 22H to CX

    (e) add the data addressed by SI to AL

    (f) add CX to the data stored at memory location FROG

2. What is wrong with the ADD ECX,AX instruction?

3. Is it possible to add CX to DS with the ADD instruction?

4. If AX = 1001H and DX = 20FFH, list the sum and the contents of each flag register bit (C, A, S, Z, and O) after the ADD AX,DX instruction executes.

5. Develop a short sequence of instructions that adds AL, BL, CL, DL, and AH. Save the sum in the DH register.

6. Develop a short sequence of instructions that adds AX, BX, CX, DX, and SP. Save the sum in the DI register.

7. Develop a short sequence of instructions that adds ECX, EDX, and ESI. Save the sum in the EDI register.

8. Select an instruction that adds BX to DX, and also adds the contents of the carry flag (C) to the result.

9. Choose an instruction that adds a 1 to the contents of the SP register.

10. What is wrong with the INC [BX] instruction?

11. Select a SUB instruction that will:

    (a) subtract BX from CX

    (b) subtract 0EEH from DH

    (c) subtract DI from SI

    (d) subtract 3322H from EBP

    (e) subtract the data address by SI from CH

    (f) subtract the data stored 10 words after the location addressed by SI from DX

    (g) subtract AL from memory location FROG

12. If DL = 0F3H and BH = 72H, list the difference after BH subtract from DL, and show the contents of the flag register bits.

13. Write a short sequence of instructions that subtracts the numbers in DI, SI, and BP from the AX register. Store the difference in register BX.

14. Choose an instruction that subtracts 1 from register EBX.

15. Explain what the SBB [DI–4],DX instruction accomplishes.

16. Explain the difference between the SUB and CMP instruction.

17. When two 8-bit numbers are multiplied, where is the product found?

18. When two 16-bit numbers are multiplied, what two registers hold the product? Show the registers that contain the most- and least-significant portions of the product.

19. When two numbers multiply, what happens to the O and C flag bits?

20. Where is the product stored for the MUL EDI instruction?

21. What is the difference between the IMUL and MUL instructions?

22. Write a sequence of instructions that cube the 8-bit number found in DL. Load DL with a 5 initially, and make sure that your result is a 16-bit number.

23. Describe the operation of the IMUL BX,DX,100H instruction.

24. When 8-bit numbers are divided, in which register is the dividend found?

25. When 16-bit numbers are divided, in which register is the quotient found?

26. What errors are detected during a division?

27. Explain the difference between the IDIV and DIV instructions.

28. Where is the remainder found after an 8-bit division?

29. Write a short sequence of instructions that divides the number in BL by the number in CL, and then multiplies the result by 2. The final answer must be a 16-bit number stored in the DX register.

30. Which instructions are used with BCD arithmetic operations?

31. Which instructions are used with ASCII arithmetic operations?

32. Explain how the AAM instruction converts from binary to BCD.
33. Develop a sequence of instructions that converts the unsigned number in AX (values of 0–65535) into a 5-digit BCD number stored in memory, beginning at the location addressed by the BX register in the data segment. Note that the most-significant character is stored first and no attempt is made to blank leading zeros.
34. Develop a sequence of instructions that adds the 8-digit BCD number in AX and BX to the 8-digit BCD number in CX and DX. (AX and CX are the most-significant registers. The result must be found in CX and DX after the addition.)
35. Select an AND instruction that will:
    (a) AND BX with DX and save the result in BX
    (b) AND 0EAH with DH
    (c) AND DI with BP and save the result in DI
    (d) AND 1122H with EAX
    (e) AND the data addressed by BP with CX and save the result in memory
    (f) AND the data stored in four words before the location addressed by SI with DX and save the result in DX
    (g) AND AL with memory location WHAT and save the result at location WHAT
36. Develop a short sequence of instructions that clears (0) the three leftmost bits of DH without changing the remainder DH and stores the result in BH.
37. Select an OR instruction that will:
    (a) OR BL with AH and save the result in AH
    (b) OR 88H with ECX
    (c) OR DX with SI and save the result in SI
    (d) OR 1122H with BP
    (e) OR the data addressed by BX with CX and save the result in memory
    (f) OR the data stored 40 bytes after the location addressed by BP with AL and save the result in AL
    (g) OR AH with memory location WHEN and save the result in WHEN
38. Develop a short sequence of instructions that sets (1) the rightmost five bits of DI without changing the remaining bits of DI. Save the results in SI.
39. Select the XOR instruction that will:
    (a) XOR BH with AH and save the result in AH
    (b) XOR 99H with CL
    (c) XOR DX with DI and save the result in DX
    (d) XOR 1A23H with ESP
    (e) XOR the data addressed by EBX with DX and save the result in memory
    (f) XOR the data stored 30 words after the location addressed by BP with DI and save the result in DI
    (g) XOR DI with memory location WELL and save the result in DI
40. Develop a sequence of instructions that sets (1) the rightmost four bits of AX; clears (0) the leftmost three bits of AX; and inverts bits 7, 8, and 9 of AX.
41. Describe the difference between the AND and TEST instructions.
42. Select an instruction that tests bit position 2 of register CH.
43. What is the difference between the NOT and the NEG instruction?
44. Select the correct instruction to perform each of the following tasks:
    (a) shift DI right three places, with zeros moved into the leftmost bit
    (b) move all bits in AL left one place, making sure that a 0 moves into the rightmost bit position
    (c) rotate all the bits of AL left three places
    (d) rotate carry right one place through EDX
    (e) move the DH register right one place, making sure that the sign of the result is the same as the sign of the original number

45. What does the SCASB instruction accomplish?
46. For string instructions, DI always addresses data in the _____ segment.
47. What is the purpose of the D flag bit?
48. Explain what the REPE prefix does when coupled with the SCASB instruction.
49. What condition or conditions will terminate the repeated string instruction REPNE SCASB?
50. Describe what the CMPSB instruction accomplishes.
51. Develop a sequence of instructions that scans through a 300H-byte section of memory called LIST, located in the data segment searching for a 66H.
52. What happens if AH = 02H and DL = 43H when the INT 21H instruction is executed?

CHAPTER 6

Program Control Instructions

INTRODUCTION

The program control instructions direct the flow of a program and allow the flow to change. A change in flow often occurs after a decision, made with the CMP or TEST instruction, is followed by a conditional jump instruction. This chapter explains the program control instructions, including the jumps, calls, returns, interrupts, and machine control instructions.

Also presented in this chapter are the relational assembly language statements (.IF, .ELSE, .ELSEIF, .ENDIF, .WHILE, .ENDW, .REPEAT, and .UNTIL) that are available in version 6.X and above of MASM or TASM, with version 5.X set for MASM compatibility. These relational assembly language commands allow the programmer to develop control flow portions of the program with C/C++ language efficiency.

CHAPTER OBJECTIVES

Upon completion of this chapter, you will be able to:

1. Use both conditional and unconditional jump instructions to control the flow of a program.
2. Use the relational assembly language statements .IF, .REPEAT, .WHILE, and so forth in programs.
3. Use the call and return instructions to include procedures in the program structure.
4. Explain the operation of the interrupts and interrupt control instructions.
5. Use machine control instructions to modify the flag bits.
6. Use ENTER and LEAVE to enter and leave programming structures.

6–1 THE JUMP GROUP

The main program control instruction, **jump** (JMP), allows the programmer to skip sections of a program and branch to any part of the memory for the next instruction. A conditional jump instruction allows the programmer to make decisions based upon numerical tests. The results of numerical tests are held in the flag bits, which are then tested by conditional jump instructions. Another instruction similar to the conditional jump, the conditional set, is explained with the conditional jump instructions in this section.

In this section of the text, all jump instructions are illustrated with their uses in sample programs. Also revisited are the LOOP and conditional LOOP instructions, first presented in Chapter 3, because they are also forms of the jump instruction.

Unconditional Jump (JMP)

Three types of unconditional jump instructions (see Figure 6–1) are available to the microprocessor: short jump, near jump, and far jump. The **short jump** is a two-byte instruction that allows jumps or branches to memory locations within +127 and –128 bytes from the address following the jump. The three-byte **near jump** allows a branch or jump within ±32K bytes (or anywhere in the current code segment) from the instruction in the current code segment. Remember that segments are cyclic in nature, which means that one location above offset address FFFFH is offset address 0000H. For this reason, if you jump two bytes ahead in memory and the instruction pointer addresses offset address FFFFH, the flow continues at offset address 0001H. Thus, a displacement of ±32K bytes allows a jump to any location within the current code segment. Finally, the five-byte **far jump** allows a jump to any memory location within the real memory system. The short and near jumps are often called **intrasegment** jumps, and the far jumps are often called **intersegment** jumps.

In the 80386 through the Pentium 4 processors, the near jump is within ±2G if the machine is operated in the protected mode, with a code segment that is 4G bytes long. If operated in the real mode, the near jump is within ±32K bytes. In the protected mode, the 80386 and above use a 32-bit displacement that is not shown in Figure 6–1.

Short Jump.   Short jumps are called **relative jumps** because they can be moved, along with their related software, to any location in the current code segment without a change. This is because the jump address is not stored with the opcode. Instead of a jump address, a **distance,** or displacement, follows the opcode. The short jump displacement is a distance represented by a one-byte signed number whose value ranges between +127 and –128. The short jump instruction appears in Figure 6–2. When the microprocessor executes a short jump, the displacement is sign-extended and added to the instruction pointer (IP/EIP) to generate the jump address within the current code segment. The short jump instruction branches to this new address for the next instruction in the program.

Example 6–1 shows how short jump instructions pass control from one part of the program to another. It also illustrates the use of a **label** (a symbolic name for a memory address) with the jump instruction. Notice how one jump (JMP SHORT NEXT) uses the SHORT directive to force a short jump, while the other does not. Most assembler programs choose the best form of

FIGURE 6–1   The three main forms of the JMP instruction. Note that Disp is either an 8- or 16-bit signed displacement or distance.

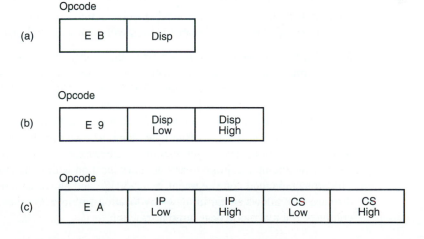

FIGURE 6–2   A short jump
to four memory locations
beyond the address of the
next instruction.

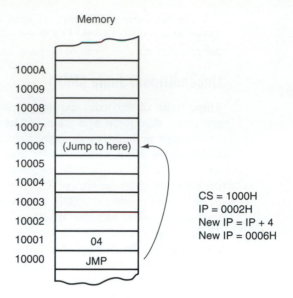

the jump instruction so the second jump instruction (JMP START) also assembles as a short jump. If the address of the next instruction (0009H) is added to the sign-extended displacement (0017H) of the first jump, the address of NEXT is at location 0017H + 0009H or 0020H.

EXAMPLE 6–1

```
0000   33 DB                         XOR    BX,BX

0002   B8 0001          START:       MOV    AX,1
0005   03 C3                         ADD    AX,BX
0007   EB 17                         JMP    SHORT NEXT

0020   8B D8            NEXT:        MOV    BX,AX
0022   EB DE                         JMP    START
```

Whenever a jump instruction references an address, a label normally identifies the address. The JMP NEXT instruction is an example; it jumps to label NEXT for the next instruction. It is very rare to ever use an actual hexadecimal address with any jump instruction, but the assembler supports addressing in relation to the instruction pointer by using the $ + a displacement. For example, a JMP $+2 jumps over the next two memory locations following the JMP instruction. The label NEXT must be followed by a colon (NEXT:) to allow an instruction to reference it for a jump. If a colon does not follow a label, you cannot jump to it. Note that the only time a colon is used after a label is when the label is used with a jump or call instruction.

Near Jump.   The near jump is similar to the short jump, except that the distance is farther. A **near jump** passes control to an instruction in the current code segment located within ±32K bytes from the near jump instruction. The distance is ±2G in the 80386 and above when operated in protected mode. The near jump is a three-byte instruction that contains an opcode followed by a signed 16-bit displacement. In the 80386 through the Pentium 4 processors, the displacement is 32 bits and the near jump is five bytes long. The signed displacement adds to the instruction pointer (IP) to generate the jump address. Because the signed displacement is in the range of ±32K, a near jump can jump to any memory location within the current real mode code segment. The protected mode code segment in the 80386 and above can be 4G bytes long, so the 32-bit displacement allows a near jump to any location within ±2G bytes. Figure 6–3 illustrates the operation of the real mode near jump instruction.

FIGURE 6–3 A near jump that adds the displacement (0002H) to the contents of IP.

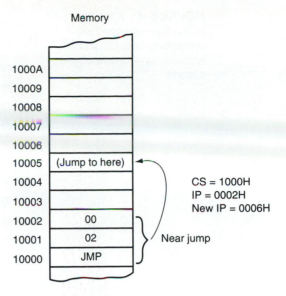

Memory

1000A	
10009	
10008	
10007	
10006	
10005	(Jump to here)
10004	
10003	
10002	00
10001	02
10000	JMP

CS = 1000H
IP = 0002H
New IP = 0006H

Near jump

The near jump is also relocatable (as was the short jump) because it is also a relative jump. If the code segment moves to a new location in the memory, the distance between the jump instruction and the operand address remains the same. This allows a code segment to be relocated by simply moving it. This feature, along with the relocatable data segments, makes the Intel family of microprocessors ideal for use in a general-purpose computer system. Software can be written and loaded anywhere in the memory and function without modification because of the relative jumps and relocatable data segments.

Example 6–2 shows the same basic program that appeared in Example 6–1, except that the jump distance is greater. The first jump (JMP NEXT) passes control to the instruction at offset memory location 0200H within the code segment. Notice that the instruction assembles as an E9 0200 R. The letter R denotes a **relocatable jump** address of 0200H. The relocatable address of 0200H is for the assembler program's internal use only. The **actual machine language instruction** assembles as an E9 F6 01, which does not appear in the assembler listing. The actual displacement is a 01F6H for this jump instruction. The assembler lists the jump address as 0200 R, so the address is easier to interpret as software is developed. If the linked execution file (.EXE) or command file (.COM) is displayed in hexadecimal code, the jump instruction appears as an E9 F6 01.

EXAMPLE 6–2

```
0000   33 DB                    XOR    BX,BX

0002   B8 0001       START:     MOV    AX,1
0005   03 C3                    ADD    AX,BX
0007   E9 0200 R                JMP    NEXT

0200   8B D8         NEXT:      MOV    BX,AX
0202   E9 0002 R                JMP    START
```

Far Jump. A far jump instruction (see Figure 6–4) obtains a new segment and offset address to accomplish the jump. Bytes 2 and 3 of this five-byte instruction contain the new offset address; bytes 4 and 5 contain the new segment address. If the microprocessor (80286 through the Pentium 4) is operated in the protected mode, the segment address accesses a descriptor that contains the base address of the far jump segment. The offset address, which is either 16- or 32-bits, contains the offset location within the new code segment.

FIGURE 6–4   A far jump instruction replaces the contents of both CS and IP with four bytes following the opcode.

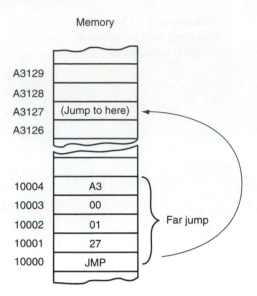

Example 6–3 lists a short program that uses a far jump instruction. The far jump instruction sometimes appears with the FAR PTR directive, as illustrated. Another way to obtain a far jump is to define a label as a **far label.** A label is far only if it is external to the current code segment or procedure. The JMP UP instruction in the example references a far label. The label UP is defined as a far label by the EXTRN UP:FAR directive. External labels appear in programs that contain more than one program file. Another way of defining a label as global is to use a double colon (LABEL::), following the label in place of the single colon. This is required inside procedure blocks that are defined as near if the label is accessed from outside the procedure block.

EXAMPLE 6–3

```
                                   EXTRN    UP:FAR

0000    33 DB                      XOR      BX,BX

0002    B8 0001        START:      MOV      AX,1
0005    03 C3                      ADD      AX,BX
0007    E9 0200 R                  JMP      NEXT

0200    8B D8          NEXT:       MOV      BX,AX
0202    EA 0002 ---- R             JMP      FAR PTR START

0207    EA 0000 ---- E             JMP      UP
```

When the program files are joined, the linker inserts the address for the UP label into the JMP UP instruction. It also inserts the segment address in the JMP START instruction. The segment address in JMP FAR PTR START is listed as – – – – R for relocatable; the segment address in JMP UP is listed as – – – – E for external. In both cases, the – – – – is filled in by the linker when it links or joins the program files.

Jumps with Register Operands.   The jump instruction can also use a 16- or 32-bit register as an operand. This automatically sets up the instruction as an **indirect jump.** The address of the jump is in the register specified by the jump instruction. Unlike the displacement associated with the near jump, the contents of the register are transferred directly into the instruction pointer. An indirect jump does not add to the instruction pointer, as with short and near jumps. The JMP AX instruction, for example, copies the contents of the AX register into the IP when the jump occurs.

This allows a jump to any location within the current code segment. In the 80386 and above, a JMP EAX instruction also jumps to any location within the current code segment; the difference is that in protected mode the code segment can be 4G bytes long, so a 32-bit offset address is needed.

Example 6–4 shows how the JMP AX instruction accesses a jump table in the code segment. This program reads a key from the keyboard and then modifies the ASCII code to a 00H in AL for a '1', a 01H for a '2', and a 02H for a '3'. If a '1', '2', or '3' is typed, AH is cleared to 00H. Because the jump table contains 16-bit offset addresses, the contents of AX are doubled to 0, 2, or 4, so a 16-bit entry in the table can be accessed. Next, the offset address of the start of the jump table is loaded to SI, and AX is added to form the reference to the jump address. The MOV AX,[SI] instruction then fetches an address from the jump table, so the JMP AX instruction jumps to the addresses (ONE, TWO, or THREE) stored in the jump table.

EXAMPLE 6–4

```
                        ;A program that reads 1, 2, or 3 from the keyboard
                        ;if a 1, 2, or 3 is typed, a 1, 2, or 3 is displayed.
                        ;
                        .MODEL SMALL                    ;select SMALL model
0000                    .DATA                           ;start of DATA segment
0000    0030 R   TABLE  DW      ONE                     ;define lookup table
0002    0034 R          DW      TWO
0004    0038 R          DW      THREE
0000                    .CODE                           ;start of CODE segment
                        .STARTUP                        ;start of program
0017            TOP:
0017    B4 01           MOV     AH,1                    ;read key into AL
0019    CD 21           INT     21H

001B    2C 31           SUB     AL,31H                  ;convert to binary
001D    72 F8           JB      TOP                     ;if below '1' typed
001F    3C 02           CMP     AL,2
0021    77 F4           JA      TOP                     ;if above '3' typed

0023    B4 00           MOV     AH,0                    ;double to 0, 2, or 4
0025    03 C0           ADD     AX,AX
0027    BE 0000 R       MOV     SI,OFFSET TABLE         ;address lookup table
002A    03 F0           ADD     SI,AX                   ;form lookup address
002C    8B 04           MOV     AX,[SI]                 ;get ONE, TWO, or THREE
002E    FF E0           JMP     AX                      ;jump address
0030            ONE:
0030    B2 31           MOV     DL,'1'                  ;load '1' for display
0032    EB 06           JMP     BOT                     ;go display '1'
0034            TWO:
0034    B2 32           MOV     DL,'2'                  ;load '2' for display
0036    EB 02           JMP     BOT                     ;go display '2'
0038            THREE:
0038    B2 33           MOV     DL,'3'                  ;load '3' for display
003A            BOT:
003A    B4 02           MOV     AH,2                    ;display number
003C    CD 21           INT     21H
                        .EXIT                           ;exit to DOS
                        END                             ;end of file
```

Indirect Jumps Using an Index. The jump instruction may also use the [ ] form of addressing to directly access the jump table. The jump table can contain offset addresses for near indirect jumps, or segment and offset addresses for far indirect jumps. (This type of jump is also known as a **double-indirect** jump if the register jump is called an **indirect jump.**) The assembler assumes that the jump is near unless the FAR PTR directive indicates a far jump instruction. Here Example 6–5 repeats Example 6–4 by using the JMP TABLE [SI] instead of JMP AX. This reduces the length of the program.

EXAMPLE 6–5

```
                                .MODEL  SMALL                   ;select SMALL model
0000                            .DATA                           ;start of DATA segment
0000   002D R        TABLE      DW      ONE                     ;lookup table
0002   0031 R                   DW      TWO
0004   0035 R                   DW      THREE
0000                            .CODE                           ;start of CODE segment
                                .STARTUP                        ;start of program
0017                   TOP:
0017   B4 01                    MOV     AH,1                    ;read key to AL
0019   CD 21                    INT     21H

001B   2C 31                    SUB     AL,31H                  ;test for below '1'
001D   72 F8                    JB      TOP                     ;if below '1'
001F   3C 02                    CMP     AL,2
0021   77 F4                    JA      TOP                     ;if above '3'
0023   B4 00                    MOV     AH,0                    ;calculate table address
0025   03 C0                    ADD     AX,AX
0027   03 F0                    ADD     SI,AX
0029   FF A4 0000 R             JMP     TABLE [SI]              ;jump to ONE, TWO, or THREE
002D                   ONE:
002D   B2 31                    MOV     DL,'1'                  ;load DL with '1'
002F   EB 06                    JMP     BOT
0031                   TWO:
0031   B2 32                    MOV     DL,'2'                  ;load DL with '2'
0033   EB 02                    JMP     BOT
0035                   THREE:
0035   B2 33                    MOV     DL,'3'                  ;load DL with '3'
0037                   BOT:
0037   B4 02                    MOV     AH,2                    ;display ONE, TWO, or THREE
0039   CD 21                    INT     21H
                                .EXIT                           ;exit to DOS
                                END                             ;end of file
```

The mechanism used to access the jump table is identical with a normal memory reference. The JMP TABLE [SI] instruction points to a jump address stored at the code segment offset location addressed by SI. It jumps to the address stored in the memory at this location. Both the register and indirect indexed jump instructions usually address a 16-bit offset. This means that both types of jumps are near jumps. If a JMP FAR PTR [SI] or JMP TABLE [SI], with TABLE data defined with the DD directive, appears in a program, the microprocessor assumes that the jump table contains doubleword, 32-bit addresses (IP and CS).

Conditional Jumps and Conditional Sets

Conditional jump instructions are always short jumps in the 8086 through the 80286 microprocessors. This limits the range of the jump to within +127 bytes and −128 bytes from the location following the conditional jump. In the 80386 and above, conditional jumps are either short or near jumps. This allows these microprocessors to use a conditional jump to any location within the current code segment. Table 6–1 lists all the conditional jump instructions with their test conditions. Note that the Microsoft MASM version 6.X assembler automatically adjusts conditional jumps if the distance is too great.

The conditional jump instructions test the following flag bits: sign (S), zero (Z), carry (C), parity (P), and overflow (O). If the condition under test is true, a branch to the label associated with the jump instruction occurs. If the condition is false, the next sequential step in the program executes. For example, a JC will jump if the carry bit is set.

The operation of most conditional jump instructions is straightforward because they often test just one flag bit, although some test more than one. Relative magnitude comparisons require more complicated conditional jump instructions that test more than one flag bit.

TABLE 6–1 Conditional jump instructions.

Assembly Language	Condition Tested	Operation
JA	Z = 0 and C = 0	Jump if above
JAE	C = 0	Jump if above or equal
JB	C = 1	Jump if below
JBE	Z = 1 or C = 1	Jump if below or equal
JC	C = 1	Jump if carry set
JE or JZ	Z = 1	Jump if equal or jump if zero
JG	Z = 0 and S = O	Jump if greater than
JGE	S = O	Jump if greater than or equal
JL	S <> O	Jump if less than
JLE	Z = 1 or S <> O	Jump if less than or equal
JNC	C = 0	Jump if no carry
JNE or JNZ	Z = 0	Jump if not equal or jump if not zero
JNO	O = 0	Jump if no overflow
JNS	S = 0	Jump if no sign
JNP or JPO	P = 0	Jump if no parity or jump if parity odd
JO	O = 1	Jump if overflow set
JP or JPE	P = 1	Jump if parity set or jump if parity even
JS	S = 1	Jump if sign is set
JCXZ	CX = 0	Jump if CX is zero
JECXZ	ECX = 0	Jump if ECX is zero

Because both signed and unsigned numbers are used in programming, and because the order of these numbers is different, there are two sets of conditional jump instructions for magnitude comparisons. Figure 6–5 shows the order of both signed and unsigned 8-bit numbers. The 16- and 32-bit numbers follow the same order as the 8-bit numbers, except that they are larger. Notice that an FFH (255) is above the 00H in the set of unsigned numbers, but an FFH (−1) is less than 00H for signed numbers. Therefore, an unsigned FFH is above 00H, but a signed FFH is less than 00H.

When signed numbers are compared, use the JG, JL, JGE, JLE, JE, and JNE instructions. The terms *greater than* and *less than* refer to signed numbers. When unsigned numbers are compared, use the JA, JB, JAE, JBE, JE, and JNE instructions. The terms *above* and *below* refer to unsigned numbers.

FIGURE 6–5 Signed and unsigned numbers follow different orders.

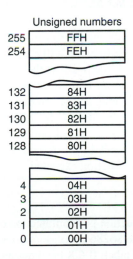
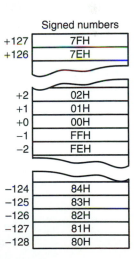

The remaining conditional jumps test individual flag bits, such as overflow and parity. Notice that JE has an alternative opcode JZ. All instructions have alternates, but many aren't used in programming because they don't usually fit the condition under test. (The alternates appear in Appendix B with the instruction set listing.) For example, the JA instruction (jump if above) has the alternative JNBE (jump if not below or equal). A JA functions exactly as a JNBE, but a JNBE is awkward in many cases when compared to a JA.

The conditional jump instructions all test flag bits except for JCXZ (jump if CX = 0) and JECXZ (jump if ECX = 0). Instead of testing flag bits, JCXZ directly tests the contents of the CX register without affecting the flag bits, and JECXZ tests the contents of the ECX register. For the JCXZ instruction, if CX = 0, a jump occurs, and if CX <> 0, no jump occurs. Likewise for the JECXZ instruction, if ECX = 0, a jump occurs; if CX <> 0, no jump occurs.

A program that uses JCXZ appears in Example 6–6. Here, the SCASB instruction searches a table for a 0AH. Following the search, a JCXZ instruction tests CX to see if the count has reached zero. If the count is zero, the 0AH is not found in the table. The carry flag is used in this example to pass the not found condition back to the calling program. Another method used to test to see if the data are found is the JNE instruction. If JNE replaces JCXZ, it performs the same function. After the SCASB instruction executes, the flags indicate a not-equal condition if the data were not found in the table.

EXAMPLE 6–6

```
                        ;A procedure that searches a table of 100 bytes for 0AH.
                        ;The address, TABLE, is transferred to the procedure
                        ;through the SI register.
                        ;
0017                    SCAN    PROC    NEAR

0017  B9 0064                   MOV     CX,100          ;load count of 100
001A  B0 0A                     MOV     AL,0AH          ;load AL with 0AH
001C  FC                        CLD                     ;select increment
001D  F2/AE                     REPNE   SCASB           ;test 100 bytes for 0AH
001F  F9                        STC                     ;set carry for not found
0020  E3 01                     JCXZ    NOT_FOUND       ;if not found
0022  F8                        CLC                     ;clear carry if found

0023                    NOT_FOUND:

0023  C3                        RET                     ;return from procedure

0024                    SCAN    ENDP
```

The Conditional Set Instructions.   In addition to the conditional jump instructions, the 80386 through the Pentium 4 processors also contain conditional set instructions. The conditions tested by conditional jumps are put to work with the conditional set instructions. The conditional set instructions set a byte to either a 01H or clear a byte to 00H, depending on the outcome of the condition under test. Table 6–2 lists the available forms of the conditional set instructions.

These instructions are useful where a condition must be tested at a point much later in the program. For example, a byte can be set to indicate that the carry is cleared at some point in the program by using the SETNC MEM instruction. This instruction places a 01H into memory location MEM if carry is cleared, and a 00H into MEM if carry is set. The contents of MEM can be tested at a later point in the program to determine if carry is cleared at the point where the SETNC MEM instruction executed.

LOOP

The LOOP instruction is a combination of a decrement CX and the JNZ conditional jump. In the 8086 through the 80286 processors, LOOP decrements CX; if CX <> 0, it jumps to the address indicated by the label. If CX becomes a 0, the next sequential instruction executes. In the 80386

TABLE 6–2 The conditional set instructions.

Assembly Language	Condition Tested	Operation
SETB	C = 1	Set if below
SETAE	C = 0	Set if above or equal
SETBE	Z = 1 or C = 1	Set if below or equal
SETA	Z = 0 and C = 0	Set if above
SETE or SETZ	Z = 1	Set if equal or set if zero
SETNE or SETNZ	Z = 0	Set if not equal or set if not zero
SETL	S <> O	Set if less than
SETLE	Z = 1 or S <> O	Set if less than or equal
SETG	Z = 0 and S = O	Set if greater than
SETGE	S = O	Set if greater than or equal
SETS	S = 1	Set if sign (negative)
SETNS	S = 0	Set if no sign (positive)
SETC	C = 1	Set if carry
SETNC	C = 0	Set if no carry
SETO	O = 1	Set if overflow
SETNO	O = 0	Set if no overflow
SETP or SETPE	P = 1	Set if parity or set if parity even
SETNP or SETPO	P = 0	Set if no parity or set if parity odd

and above, LOOP decrements either CX or ECX, depending upon the instruction mode. If the 80386 and above operate in the 16-bit instruction mode, LOOP uses CX; if operated in the 32-bit instruction mode, LOOP uses ECX. This default is changed by the LOOPW (using CX) and LOOPD (using ECX) instructions in the 80386 through the Pentium 4.

Example 6–7 shows how data in one block of memory (BLOCK1) adds to data in a second block of memory (BLOCK2), using LOOP to control how many numbers add. The LODSW and STOSW instructions access the data in BLOCK1 and BLOCK2. The ADD AX,ES:[DI] instruction accesses the data in BLOCK2 located in the extra segment. The only reason that BLOCK2 is in the extra segment is that DI addresses extra segment data for the STOSW instruction. The .STARTUP directive only loads DS with the address of the data segment. In this example, the extra segment also addresses data in the data segment, so the contents of DS are copied to ES through the accumulator. Unfortunately, there is no direct move from segment register-to-segment register instruction.

EXAMPLE 6–7

```
                        ;A program that sums the contents of BLOCK1 and BLOCK2
                        ;and stores the results over top of the data in BLOCK2.
                        ;
                        .MODEL SMALL            ;select SMALL model
0000                    .DATA                   ;start of DATA segment
0000    0064 [          BLOCK1 DW     100 DUP (?)   ;100 bytes for BLOCK1
          0000
             ]
00C8    0064 [          BLOCK2 DW     100 DUP (?)   ;100 bytes for BLOCK2
          0000
             ]
0000                    .CODE                   ;start of CODE segment
                        .STARTUP                ;start of program
0017    8C D8                  MOV    AX,DS        ;overlap DS and ES
0019    8E C0                  MOV    ES,AX

001B    FC                     CLD                  ;select increment
001C    B9 0064                MOV    CX,100        ;load count of 100
001F    BE 0000 R              MOV    SI,OFFSET BLOCK1  ;address BLOCK1
```

```
0022   BF 00C8 R              MOV    DI,OFFSET BLOCK2   ;address BLOCK2

0025                   L1:
0025   AD                     LODSW                     ;load AX with BLOCK1
0026   26:03 05               ADD    AX,ES:[DI]         ;add BLOCK2 data to AX
0029   AB                     STOSW                     ;store sum in BLOCK2
002A   E2 F9                  LOOP   L1                 ;repeat 100 times
                       .EXIT                            ;exit to DOS
                       END                              ;end of file
```

Conditional LOOPs.   As with REP, the LOOP instruction also has conditional forms: LOOPE and LOOPNE. The LOOPE **(loop while equal)** instruction jumps if CX <> 0 while an equal condition exists. It will exit the loop if the condition is not equal or if the CX register decrements to 0. The LOOPNE **(loop while not equal)** instruction jumps if CX <> 0 while a not-equal condition exists. It will exit the loop if the condition is equal or if the CX register decrements to 0. In the 80386 through the Pentium 4 processors, the conditional LOOP instruction can use either CX or ECX as the counter. The LOOPEW/LOOPED or LOOPNEW/LOOPNED instructions override the instruction mode if needed.

As with the conditional repeat instructions, alternates exist for LOOPE and LOOPNE. The LOOPE instruction is the same as LOOPZ, and the LOOPNE instruction is the same as LOOPNZ. In most programs, only the LOOPE and LOOPNE apply.

6–2   ## CONTROLLING THE FLOW OF AN ASSEMBLY LANGUAGE PROGRAM

It is much easier to use the assembly language statements .IF, .ELSE, .ELSEIF, and .ENDIF to control the flow of the program than it is to use the correct conditional jump statement. These statements always indicate a special assembly language command to MASM. Note that the control flow assembly language statements beginning with a period are only available to MASM version 6.X, and not to earlier versions of the assembler such as 5.10. Other statements developed in this chapter include the .REPEAT-.UNTIL and .WHILE-.ENDW statements.

Example 6–8 shows how these statements are used to control the flow of a program by testing the system for the version of DOS. Notice that in this example DOS INT 21H, function number 30H is used to read the DOS version. The version is tested to determine whether it is above or below version 3.3. If it is below version 3.3, the program terminates, using DOS INT 21H function number 4CH.

EXAMPLE 6–8(a)

```
;source program sequence
;
        MOV    AH,30H
        INT    21H                ;get DOS version

.IF    AL<3 && AH<30

        MOV    AH,4CH     ;terminate program
        INT    21H

.ENDIF
```

EXAMPLE 6–8(b)

```
;assembled listing file of Example 6-8 (a)
;
0000   B4 30          MOV    AH,30H
0002   CD 21          INT    21H                ;get DOS version
```

```
                                 .IF    AL<3 && AH<30
0004   3C 03         *           cmp    al,003h
0006   73 09         *           jae    @C0001
0008   80 FC 1E      *           cmp    ah,01Eh
000B   73 04         *           jae    @C0001

000D   B4 4C                     MOV    AH,4CH          ;terminate program
000F   CD 21                     INT    21H

                                 .ENDIF
0011             *   @C0001:
```

Example 6–8(a) shows the source program sequence as it was typed; Example 6–8(b) shows the fully expanded assembled output generated by the assembler program. Notice that assembler-generated and -inserted statements begin with an asterisk (*) in the listing. The .IF AL<3 && AH<30 statement tests for DOS version 3.30. If the major version number (AL) is less than 3 AND the minor version number (AH) is less than 30, the MOV AH,0 and INT 21H instructions execute. Notice how the && symbol represents the word AND in the .IF statement. See Table 6–3 for a complete list of relation operators used with the .IF statement. Note that many of these conditions (such as &&) are also used by many high-level languages such as C/C++.

Example 6–9 shows another example of the conditional .IF directive that converts all ASCII-coded letters to uppercase. First, the keyboard is read without echo using DOS INT 21H function 06H, and then the .IF statement converts the character into uppercase, if needed. In this example, the logical AND function (&&) is used to determine if the character is in lowercase. This program reads a key from the keyboard and converts it to uppercase before displaying it. Notice also how the program terminates when the control C key (ASCII = 03H) is typed. The .LISTALL directive causes all assembler-generated statements to be listed, including the label @Startup generated by the .STARTUP directive. The .EXIT directive also is expanded by .LISTALL to show the use of the DOS INT 21H function 4CH, which returns control to DOS.

EXAMPLE 6–9

```
                         ;A program that reads the keyboard and converts all
                         ;lowercase data to uppercase before displaying it.
                         ;
                         ;This program uses a control C for termination.
                         ;
                                 .MODEL TINY            ;select TINY model
                                 .LISTALL               ;list all statements
0000                             .CODE                  ;start CODE segment
                                 .STARTUP               ;start program
0100             *       @Startup:
0100                     MAIN1:
0100   B4 06                     MOV    AH,6            ;read key without echo
```

TABLE 6–3 Relational operators used with the .IF statement.

Operator	Function
==	Equal or the same as
!=	Not equal
>	Greater than
>=	Greater than or equal
<	Less than
<=	Less than or equal
&	Bit test
!	Logical inversion
&&	Logical AND
||	Logical OR

```
0102  B2 FF                    MOV     DL,0FFH
0104  CD 21                    INT     21H
0106  74 F8                    JE      MAIN1           ;if no key typed
0108  3C 03                    CMP     AL,3            ;test for control C key
010A  74 10                    JE      MAIN2           ;if control C key

                               .IF     AL>='a' && AL<='z'

010C  3C 61          *         cmp     al, 'a'
010E  72 06          *         jb      @C0001
0110  3C 7A          *         cmp     al, 'z'
0112  77 02          *         ja      @C0001

0114  2C 20                    SUB     AL,20H

                               .ENDIF
0116           * @C0001:

0116  8A D0                    MOV     DL,AL           ;echo character to display
0118  CD 21                    INT     21H
011A  EB E4                    JMP     MAIN1           ;repeat
011C              MAIN2:
                               .EXIT                   ;exit to DOS on control C

011C  B4 4C          *         mov     ah, 04Ch
011E  CD 21          *         int     021h

                               END                     ;end of file
```

In this program, a lowercase letter is converted to uppercase by the use of the .IF AL >= 'a' && AL <= 'z' statement. If AL contains a value that is greater than or equal to a lowercase a, and less than or equal to a lowercase z (a value of a–z), the statement between the .IF and .ENDIF executes. This statement (SUB AL,20H) subtracts 20H from the lowercase letter to change it to an uppercase letter. Notice how the assembler program implements the .IF statement (see lines that begin with *). The label @C0001 is an assembler-generated label used by the conditional jump statements placed in the program by the .IF statement.

Another example that uses the conditional .IF statement appears in Example 6–10. This program reads a key from the keyboard, and then converts it to hexadecimal code. This program is not listed in expanded form.

In this example, the .IF AL >='a' && AL<='f' statement causes the next instruction (SUB AL,57H) to execute if AL contains letters a through f, converting them to hexadecimal. If it is not between letters a and f, the next .ELSEIF statement tests it for the letters A through F. If it is the letters A through F, a 37H is subtracted from AL. If neither of these are true, a 30H is subtracted from AL before AL is stored at data segment memory location TEMP.

EXAMPLE 6–10

```
                   ;A program that reads a key and stores its hexadecimal
                   ;value in memory location TEMP.
                   ;
                               .MODEL SMALL            ;select SMALL model
0000                           .DATA                   ;start DATA segment
0000  00           TEMP        DB      ?               ;define TEMP
0000                           .CODE                   ;start CODE segment
                               .STARTUP                ;start program
0017  B4 01                    MOV     AH,1            ;read key
0019  CD 21                    INT     21H

                               .IF     AL>='a' && AL<='f'    ;if lowercase
0023  2C 57                        SUB  AL,57H

                               .ELSEIF AL>='A' && AL<='F'    ;if uppercase
002F  2C 37                        SUB  AL,37H
```

```
                                 .ELSE                                  ;otherwise
0033   2C 30                             SUB    AL,30H

                                 .ENDIF

0035   A2 0000 R                 MOV    TEMP,AL
                                 .EXIT                                  ;exit to DOS
                                 END                                    ;end of file
```

DO-WHILE Loops

As with most high-level languages, the assembler also provides the DO-WHILE loop construct,
available to MASM version 6.X. The .WHILE statement is used with a condition to begin the
loop, and the .ENDW statement ends the loop.

Example 6–11 shows how the .WHILE statement is used to read data from the keyboard
and store it into an array called BUF until the enter key (0DH) is typed. This program assumes
that BUF is stored in the extra segment because the STOSB instruction is used to store the keyboard data in memory. Note that the .WHILE loop portion of the program is shown in expanded
form so that the statements inserted by the assembler (beginning with a *) can be studied. After
the enter key (0DH) is typed, the string is appended with a $ so it can be displayed with DOS
INT 21H function number 9.

EXAMPLE 6–11

```
                    ;A program that reads a character string from the
                    ;keyboard and, after enter is typed, displays it again.
                    ;
                                 .MODEL SMALL                ;select small model
0000                             .DATA                       ;indicate DATA segment
0000   0D 0A         MES         DB     13,10                ;return & line feed
0002   0100 [        BUF         DB     256 DUP (?)          ;character string buffer
          00
          ]
0000                             .CODE                       ;start of CODE segment
                                 .STARTUP                    ;start of program
0017   8C D8                     MOV    AX,DS                ;make ES overlap DS
0019   8E C0                     MOV    ES,AX

001B   FC                        CLD                         ;select increment
001C   BF 0002 R                 MOV    DI,OFFSET BUF         ;address buffer

                                 .WHILE AL != 0DH            ;loop while AL not enter

001F   EB 05        *            jmp    @C0001
0021                *  @C0002:

0021   B4 01                     MOV    AH,1                 ;read key with echo
0023   CD 21                     INT    21H
0025   AA                        STOSB                       ;store key code

                                 .ENDW                       ;end while loop
0026                *  @C0001:

0026   3C 0D        *            cmp    al,00Dh
0028   75 F7        *            jne    @C0002

002A   C6 45 FF 24               MOV    BYTE PTR [DI-1],'$'   ;make it $ string
002E   BA 0000 R                 MOV    DX,OFFSET MES         ;address MES
0031   B4 09                     MOV    AH,9                 ;display MES
0033   CD 21                     INT    21H
                                 .EXIT                        ;exit to DOS
                                 END
```

The program in Example 6–11 functions perfectly, as long as we arrive at the .WHILE
statement with AL containing some other value except 0DH. This can be corrected by adding a

MOV AL,0DH instruction before the .WHILE statement in Example 6–11. A better way of handling this problem is illustrated in Example 6–12. In this example, the .BREAK statement is used to break out of the .WHILE loop. A .WHILE 1 creates an infinite loop and the .BREAK statement tests for a value of 0DH (enter) in AL. If AL = 0DH, the program breaks out of the infinite loop, correcting the problem exhibited in Example 6–11. Note that the .BREAK statement causes the break to occur at the point where it appears in the program. This is important because it allows the point of the break to be selected by the programmer.

Not illustrated in this example is the .CONTINUE statement, which can be used to allow the DO-WHILE loop to continue if a certain condition is met. For example, a .CONTINUE .IF AL == 15 allows the loop to continue if AL equals 15. Note that the .BREAK and .CONTINUE commands function in the same manner in a C-language program.

EXAMPLE 6–12

```
                              .MODEL SMALL
0000                          .DATA
0000 0D 0A        MES         DB   13,10              ;define string
0002 0100 [       BUF         DB   256 DUP (?)        ;memory for string
          00
             ]
0000                          .CODE
                              .STARTUP
0017 8C D8                    MOV  AX,DS              ;make ES overlap DS
0019 8E C0                    MOV  ES,AX

001B FC                       CLD                    ;select increment
001C BF 0002 R                MOV  DI,OFFSET BUF      ;address BUF

                              .WHILE 1               ;create an infinite loop
001F          * @C0001:

001F B4 01                        MOV  AH,1          ;read key
0021 CD 21                        INT  21H
0023 AA                           STOSB              ;store key code in BUF

                                  .BREAK .IF AL == 0DH    ;breaks loop for a 0DH
0024 3C 0D      *                 cmp  al,00Dh
0026 74 02      *                 je   @C0002

                                  .ENDW
0028 EB F5      *                 jmp  @C0001
002A          * @C0002:

002A C6 45 FF 24              MOV  BYTE PTR [DI-1],'$' ;make it a $ string
002E BA 0000 R                MOV  DX,OFFSET MES      ;display string
0031 B4 09                    MOV  AH,9
0033 CD 21                    INT  21H
                              .EXIT
                              END
```

Example 6–13 lists a practical example using the DO-WHILE construct to display the contents of EAX in decimal on the video display. Note that the EAX register is initialized with a number (123455) to test this program. Two infinite loops are used to convert EAX to decimal. The first divides EAX by 10 until the quotient is zero. After each division, the remainder is saved on the stack as a significant digit in the result. Also located within the first infinite loop is a comma counter, stored in CL. Each time that the quotient is not zero, the comma counter increments. If the comma counter reaches a 3, a comma is pushed onto the stack for later display and the comma count is reset to zero. The final infinite loop displays the result. After each POP DX instruction, the break statement checks DX to find if it contains a 10. The 10 is pushed on the stack to indicate the end of the number. If it does contain a 10, the loop breaks; if it doesn't, a decimal digit or a comma is displayed. This procedure can be added to any program where a decimal number of up to four billion must be displayed with commas at the correct places.

EXAMPLE 6–13

```
                          ;A program that displays the contents of EAX in decimal.
                          ;This program inserts commas between thousands,
                          ;millions, and billions.
                          ;
                                  .MODEL TINY
                                  .386                        ;select 80386
0000                              .CODE
                                  .STARTUP

0100   66| B8 0001E23F           MOV   EAX,123455            ;load test data
0106   E8 0004                   CALL DISPE                  ;display EAX in decimal
                                  .EXIT
                          ;
                          ;the DISPE procedure displays EAX in decimal format.
                          ;
010D              DISPE   PROC NEAR

010D   66| BB 0000000A           MOV   EBX,10                ;load 10 for decimal
0113   53                        PUSH  BX                    ;save end of number
0114   B1 00                     MOV   CL,0                  ;load comma counter

                                  .WHILE 1                    ;first infinite loop

0116   66| BA 00000000                   MOV   EDX,0          ;clear EDX
011C   66| F7 F3                          DIV   EBX           ;divide EDX:EAX by 10
011F   80 C2 30                           ADD   DL,30H        ;convert to ASCII
0122   52                                 PUSH DX             ;save remainder

                                  .BREAK .IF EAX == 0         ;break if quotient zero

0128   FE C1                              INC   CL            ;increment comma counter

                                  .IF CL == 3                 ;if comma count is 3
012F   6A 2C                              PUSH ','            ;save comma
0131   B1 00                              MOV   CL,0          ;clear comma counter
                                  .ENDIF

                                  .ENDW                       ;end first loop

                                  .WHILE 1                    ;second infinite loop

0135   5A                                 POP   DX            ;get remainder
                                  .BREAK .IF DL == 10         ;break if remainder is 10

013B   B4 02                              MOV   AH,2          ;display decimal digit
013D   CD 21                              INT   21H

                                  .ENDW

0141   C3                        RET

0142              DISPE   ENDP
                          END
```

REPEAT-UNTIL Loops

Also available to the assembler is the REPEAT-UNTIL construct. A series of instructions is repeated until some condition occurs. The .REPEAT statement defines the start of the loop; the end is defined with the .UNTIL statement, which contains a condition. Note that .REPEAT and .UNTIL are available to version 6.X of MASM.

If Example 6–11 is again reworked by using the REPEAT-UNTIL construct, this appears to be the best solution. See Example 6–14 for the program that reads keys from the keyboard and stores keyboard data into extra segment array BUF until the enter key is pressed. This program also fills the buffer with keyboard data until the enter key (0DH) is typed. Once the enter key is typed, the program displays the character string using DOS INT 21H function number 9, after appending the buffer data with the required dollar sign. Notice how the .UNTIL AL == 0DH statement generates code (statements beginning with *) to test for the enter key.

EXAMPLE 6–14

```
                                .MODEL SMALL
0000                            .DATA
0000  0D 0A         MES    DB   13,10                    ;define MES
0002  0100 [        BUF    DB   256 DUP (?)              ;reserve memory for BUF
         00
      ]
0000                            .CODE
                                .STARTUP
0017  8C D8                     MOV  AX,DS               ;overlap DS and ES
0019  8E C0                     MOV  ES,AX
001B  FC                        CLD                      ;select increment
001C  BF 0002 R                 MOV  DI,OFFSET BUF       ;address BUF

                                .REPEAT
001F                  *  @C0001:

001F  B4 01                        MOV  AH,1             ;read key with echo
0021  CD 21                        INT  21H
0023  AA                           STOSB                 ;save key code in BUF

                                .UNTIL  AL == 0DH
0024  3C 0D          *             cmp  al, 00Dh
0026  75 F7          *             jne  @C0001

0028  C6 45 FF 24                MOV  BYTE PTR [DI-1],'$' ;make $ string
002C  B4 09                      MOV  AH,9               ;display MES and BUF
002E  BA 0000 R                  MOV  DX,OFFSET MES
0031  CD 21                      INT  21H
                                .EXIT
                                END
```

There is also an .UNTILCXZ instruction available that uses the LOOP instruction to check for the until condition. The .UNTILCXZ instruction can have a condition, or it may just use the CX register as a counter to repeat a loop a fixed number of times. Example 6–15 shows a sequence of instructions that uses the .UNTILCXZ instruction used to add the contents of byte-sized array ONE to byte-sized array TWO. The sums are stored in array THREE. Note that each array contains 100 bytes of data, so the loop is repeated 100 times. This example assumes that array THREE is in the extra segment, and that arrays ONE and TWO are in the data segment.

EXAMPLE 6–15

```
012C  B9 0064                   MOV  CX,100             ;set count
012F  BF 00C8 R                 MOV  DI,OFFSET THREE    ;address arrays
0132  BE 0000 R                 MOV  SI,OFFSET ONE
0135  BB 0064 R                 MOV  BX,OFFSET TWO

                                .REPEAT

0138                  *  @C0001:

0138  AC                           LODSB
0139  02 07                        ADD  AL,[BX]
013B  AA                           STOSB
013C  43                           INC  BX

                                .UNTILCXZ

013D  E2 F9          *             loop @C0001
```

6–3   PROCEDURES

The procedure or subroutine is an important part of any computer system's architecture. A **procedure** is a group of instructions that usually performs one task. A procedure is a reusable section of the software that is stored in memory once, but used as often as necessary. This saves memory space and makes it easier to develop software. The only disadvantage of a procedure is that it takes the computer a small amount of time to link to the procedure and return from it. The CALL instruction links to the procedure, and the RET (**return**) instruction returns from the procedure.

The stack stores the return address whenever a procedure is called during the execution of a program. The CALL instruction pushes the address of the instruction following the CALL (**return address**) on the stack. The RET instruction removes an address from the stack so the program returns to the instruction following the CALL.

With the assembler, there are specific rules for storing procedures. A procedure begins with the PROC directive and ends with the ENDP directive. Each directive appears with the name of the procedure. This programming structure makes it easy to locate the procedure in a program listing. The PROC directive is followed by the type of procedure: NEAR or FAR. Example 6–16 shows how the assembler uses the definition of both a near (intrasegment) and far (intersegment) procedure. In MASM version 6.X, the NEAR or FAR type can be followed by the USES statement. The USES statement allows any number of registers to be automatically pushed to the stack and popped from the stack within the procedure. The USES statement is also illustrated in Example 6–16.

EXAMPLE 6–16

```
0000                    SUMS    PROC NEAR

0000    03 C3                   ADD    AX,BX
0002    03 C1                   ADD    AX,CX
0004    03 C2                   ADD    AX,DX
0006    C3                      RET

0007                    SUMS    ENDP

0007                    SUMS1   PROC FAR

0007    03 C3                   ADD    AX,BX
0009    03 C1                   ADD    AX,CX
000B    03 C2                   ADD    AX,DX
000D    CB                      RET

000E                    SUMS1   ENDP

000E                    SUMS2   PROC NEAR    USES BX CX DX

0011    03 C3                   ADD    AX,BX
0013    03 C1                   ADD    AX,CX
0015    03 C2                   MOV    AX,DX
                                RET

001B                    SUMS2   ENDP
```

When these two procedures are compared, the only difference is the opcode of the return instruction. The near return instruction uses opcode C3H and the far return uses opcode CBH. A near return removes a 16-bit number from the stack and places it into the instruction pointer to return from the procedure in the current code segment. A far return removes a 32-bit number from the stack and places it into both IP and CS to return from the procedure to any memory location.

Procedures that are to be used by all software (**global**) should be written as far procedures. Procedures that are used by a given task (**local**) are normally defined as near procedures.

CALL

The CALL instruction transfers the flow of the program to the procedure. The CALL instruction differs from the jump instruction because a CALL saves a return address on the stack. The return address returns control to the instruction that immediately follows the CALL in a program when a RET instruction executes.

Near CALL.   The near CALL instruction is three bytes long; the first byte contains the opcode, and the second and third bytes contain the displacement, or distance of ±32K in the 8086 through the 80286 processors. This is identical to the form of the near jump instruction. The 80386 and above use a 32-bit displacement, when operating in the protected mode, to allow a distance of ±2G bytes. When the near CALL executes, it first pushes the offset address of the next instruction on the stack. The offset address of the next instruction appears in the instruction pointer (IP or EIP). After saving this return address, it then adds the displacement from bytes 2 and 3 to the IP to transfer control to the procedure. There is no short CALL instruction. A variation on the opcode exists as CALLN, but this should be avoided in favor of using the PROC statement to define the CALL as near.

　　　Why save the IP or EIP on the stack? The instruction pointer always points to the next instruction in the program. For the CALL instruction, the contents of IP/EIP are pushed onto the stack, so program control passes to the instruction following the CALL after a procedure ends. Figure 6–6 shows the return address (IP) stored on the stack and the call to the procedure.

Far CALL.   The far CALL instruction is like a far jump because it can call a procedure stored in any memory location in the system. The far CALL is a five-byte instruction that contains an opcode, followed by the next value for the IP and CS registers. Bytes 2 and 3 contain the new contents of the IP, and bytes 4 and 5 contain the new contents for CS.

　　　The far CALL instruction places the contents of both IP and CS on the stack before jumping to the address indicated by bytes 2–5 of the instruction. This allows the far CALL to call a procedure located anywhere in the memory and return from that procedure.

FIGURE 6–6 The effect of a near CALL on the stack and the instruction pointer.

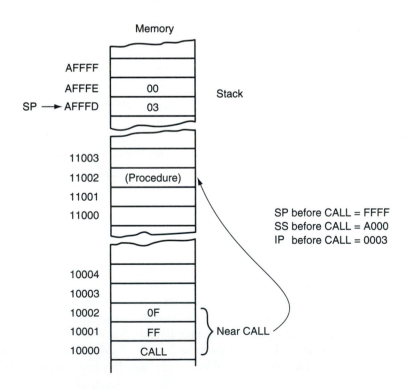

FIGURE 6–7   The effect of a far CALL instruction.

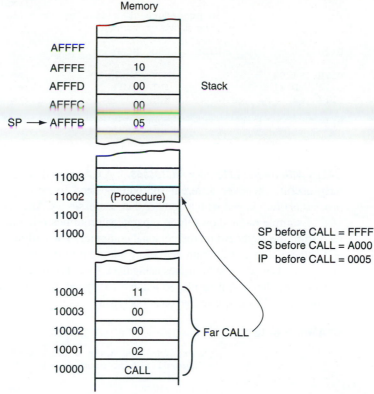

Figure 6–7 shows how the far CALL instruction calls a far procedure. Here, the contents of IP and CS are pushed onto the stack. Next, the program branches to the procedure. A variant of the far call exists as CALLF, but this should be avoided in favor of defining the type of call instruction with the PROC statement.

CALLs with Register Operands.   Like jump instructions, call instructions also may contain a register operand. An example is the CALL BX instruction, which pushes the contents of IP onto the stack. It then jumps to the offset address, located in register BX, in the current code segment. This type of CALL always uses a 16-bit offset address, stored in any 16-bit register except the segment registers.

Example 6–17 illustrates the use of the CALL register instruction to call a procedure that begins at offset address DISP. (This call could also directly call the procedure by using the CALL DISP instruction.) The OFFSET address DISP is placed into the BX register, and then the CALL BX instruction calls the procedure beginning at address DISP. This program displays an "OK" on the monitor screen.

EXAMPLE 6–17

```
                    ;A program that displays OK on the monitor screen
                    ;using procedure DISP.
                    ;
                    .MODEL TINY                 ;select TINY model
0000                .CODE                       ;start of CODE segment
                    .STARTUP                    ;start of program

0100  BB 0110 R          MOV   BX,OFFSET DISP   ;address DISP with BX
0103  B2 4F              MOV   DL,'O'           ;display 'O'
0105  FF D3              CALL  BX
0107  B2 4B              MOV   DL,'K'           ;display 'K'
0109  FF D3              CALL  BX
```

```
                            .EXIT                           ;exit to DOS
                            ;A procedure that displays the ASCII contents of DL on
                            ;the monitor screen.
                            ;
0110                        DISP    PROC NEAR

0110  B4 02                         MOV  AH,2               ;select function 02H
0112  CD 21                         INT  21H                ;execute DOS function
0114  C3                            RET                     ;return from procedure

0115                        DISP    ENDP

                            END                             ;end of file
```

CALLs with Indirect Memory Addresses.   A CALL with an indirect memory address is particularly useful whenever different subroutines need to be chosen in a program. This selection process is often keyed with a number that addresses a CALL address in a lookup table.

Example 6–18 shows three separate subroutines referenced by the number 1, 2, and 3 as read from the keyboard on the personal computer. The calling sequence adjusts the value of AL and extends it to a 16-bit number before adding it to the location of the lookup table. This references one of the three subroutines using the CALL TABLE [BX] instruction. When this program executes, the letter A is displayed when a 1 is typed, the letter B is displayed when a 2 is typed, and the letter C is displayed when a 3 is typed.

EXAMPLE 6–18

```
                            ;A program that uses a CALL lookup table to access one of
                            ;three different procedures: ONE, TWO, or THREE.
                            ;
                            .MODEL SMALL                    ;select SMALL model
0000                        .DATA                           ;start of DATA segment
0000  0000 R                TABLE   DW   ONE                ;define lookup table
0002  0007 R                        DW   TWO
0004  000E R                        DW   THREE
0000                        .CODE                           ;start of CODE segment

0000                        ONE     PROC NEAR

0000  B4 02                         MOV  AH,2               ;display a letter A
0002  B2 41                         MOV  DL,'A'
0004  CD 21                         INT  21H
0006  C3                            RET

0007                        ONE     ENDP

0007                        TWO     PROC NEAR

0007  B4 02                         MOV  AH,2               ;display letter B
0009  B2 42                         MOV  DL,'B'
000B  CD 21                         INT  21H
000D  C3                            RET

000E                        TWO     ENDP

000E                        THREE   PROC NEAR

000E  B4 02                         MOV  AH,2               ;display letter C
0010  B2 43                         MOV  DL,'C'
0012  CD 21                         INT  21H
0014  C3                            RET

0015                        THREE   ENDP

                            .STARTUP                        ;indicate start of program
002C                        TOP:
002C  B4 01                         MOV  AH,1               ;read key into AL
002E  CD 21                         INT  21H
```

```
0030   2C 31              SUB    AL,31H        ;convert to binary
0032   72 F8              JB     TOP           ;if below 0
0034   3C 02              CMP    AL,2
0036   77 F4              JA     TOP           ;if above 2

0038   B4 00              MOV    AH,0          ;form lookup address
003A   8B D8              MOV    BX,AX
003C   03 DB              ADD    BX,BX
003E   FF 97 0000 R       CALL   TABLE [BX]    ;call procedure

                          .EXIT                ;exit to DOS
                          END                  ;end of file
```

The CALL instruction also can reference far pointers if the instruction appears as a CALL FAR PTR [SI] or as a CALL TABLE [SI], if the data in the table are defined as doubleword data with the DD directive. These instructions retrieve a 32-bit address from the data segment memory location addressed by SI and use it as the address of a far procedure.

RET

The return instruction (RET) removes a 16-bit number (**near return**) from the stack and places it into IP, or removes a 32-bit number (**far return**) and places it into IP and CS. The near and far return instructions are both defined in the procedure's PROC directive, which automatically selects the proper return instruction. With the 80386 through the Pentium 4 processors operating in the protected mode, the far return removes six bytes from the stack. The first four bytes contain the new value for EIP and the last two contain the new value for CS. In the 80386 and above, a protected mode near return removes four bytes from the stack and places them into EIP.

When IP/EIP or IP/EIP and CS are changed, the address of the next instruction is at a new memory location. This new location is the address of the instruction that immediately follows the most recent CALL to a procedure. Figure 6–8 shows how the CALL instruction links to a procedure and how the RET instruction returns in the 8086–80286.

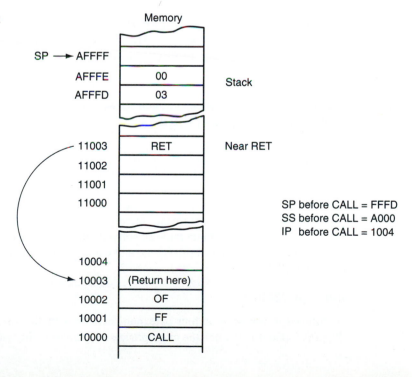

FIGURE 6–8   The effect of a near return instruction on the stack and instruction pointer.

There is one other form of the return instruction, which adds a number to the contents of the stack pointer (SP) after the return address is removed from the stack. A return that uses an immediate operand is ideal for use in a system that uses the C or Pascal calling conventions. (This is true, even though the C and PASCAL calling conventions require the caller to remove stack data for many functions.) These conventions push parameters on the stack before calling a procedure. If the parameters are to be discarded upon return, the return instruction contains a number that represents the number of bytes pushed to the stack as parameters.

Example 6–19 shows how this type of return erases the data placed on the stack by a few pushes. The RET four adds a 4 to SP after removing the return address from the stack. Because the PUSH AX and PUSH BX together place four bytes of data on the stack, this return effectively deletes AX and BX from the stack. This type of return rarely appears in assembly language programs, but it is used in high-level programs to clear stack data after a procedure. Notice how parameters are addressed on the stack by using the BP register, which by default addresses the stack segment. Parameter stacking is common in procedures written for C or PASCAL by using the C or PASCAL calling conventions.

EXAMPLE 6–19

```
0000  B8 001E              MOV   AX,30
0003  BB 0028              MOV   BX,40
0006  50                   PUSH  AX              ;stack parameter 1
0007  53                   PUSH  BX              ;stack parameter 2

0008  E8 0066              CALL  ADDM            ;add parameters from stack
                     .      .
                     .      .                    ;program continues here
0071            ADDM    PROC  NEAR

0071  55                   PUSH  BP              ;save BP
0072  8B EC                MOV   BP,SP           ;address stack with BP
0074  8B 46 04             MOV   AX,[BP+4]       ;get parameter 1
0077  03 46 06             ADD   AX,[BP+6]       ;add parameter 2
007A  5D                   POP   BP              ;restore BP
007B  C2 0004              RET   4               ;return, dump parameters

007E            ADDM    ENDP
```

As with the CALLN and CALLF instructions, there are also variants of the return instruction: RETN and RETF. As with the CALLN and CALLF instructions, these variants should also be avoided in favor of using the PROC statement to define the type of call and return.

6–4   INTRODUCTION TO INTERRUPTS

An interrupt is either a hardware-generated CALL (externally derived from a hardware signal) or a software-generated CALL (internally derived from the execution of an instruction or by some other internal event). At times, an internal interrupt is called an *exception*. Either type interrupts the program by calling an interrupt service procedure or interrupt handler.

This section explains software interrupts, which are special types of CALL instructions. This section descibes the three types of software interrupt instructions (INT, INTO, and INT 3), provides a map of the interrupt vectors, and explains the purpose of the special interrupt return instruction (IRET).

Interrupt Vectors

An **interrupt vector** is a four-byte number stored in the first 1024 bytes of the memory (000000H–0003FFH) when the microprocessor operates in the real mode. In the protected mode,

TABLE 6–4 Interrupt vectors.

Number	Address	Microprocessor	Function
0	0H–3H	All	Divide error
1	4H–7H	All	Aingle-step
2	8H–BH	All	NMI pin
3	CH–FH	All	Breakpoint
4	10H–13H	All	Interrupt on overflow
5	14H–17H	80186–Pentium 4	Bound instruction
6	18H–1BH	80186–Pentium 4	Invalid opcode
7	1CH–1FH	80186–Pentium 4	Coprocessor emulation
8	20H–23H	80386–Pentium 4	Double fault
9	24H–27H	80386	Coprocessor segment overrun
A	28H–2BH	80386–Pentium 4	Invalid task state segment
B	2CH–2FH	80386–Pentium 4	Segment not present
C	30H–33H	80386–Pentium 4	Stack fault
D	34H–37H	80386–Pentium 4	General protection fault (GPF)
E	38H–3BH	80386–Pentium 4	Page fault
F	3CH–3FH	—	Reserved
10	40H–43H	80286–Pentium 4	Floating-point error
11	44H–47H	80486SX	Alignment check interrupt
12	48H–4FH	Pentium/Pentium 4	Machine check exception
13–1F	50H–7FH	—	Reserved
20–FF	80H–3FFH	—	User interrupts

the vector table is replaced by an interrupt descriptor table that uses eight-byte descriptors to describe each of the interrupts. There are 256 different interrupt vectors, and each vector contains the address of an interrupt service procedure. Table 6–4 lists the interrupt vectors, with a brief description and the memory location of each vector for the real mode. Each vector contains a value for IP and CS that forms the address of the interrupt service procedure. The first two bytes contain the IP, and the last two bytes contain the CS.

Intel reserves the first 32 interrupt vectors for the present and future microprocessor products. The remaining interrupt vectors (32–255) are available for the user. Some of the reserved vectors are for errors that occur during the execution of software, such as the divide error interrupt. Some vectors are reserved for the coprocessor. Still others occur for normal events in the system. In a personal computer, the reserved vectors are used for system functions, as detailed later in this section. Vectors 1–6, 7, 9, 16, and 17 function in the real mode and protected mode; the remaining vectors function only in the protected mode.

Interrupt Instructions

The microprocessor has three different interrupt instructions that are available to the programmer: INT, INTO, and INT 3. In the real mode, each of these instructions fetches a vector from the vector table, and then calls the procedure stored at the location addressed by the vector. In the protected mode, each of these instructions fetches an interrupt descriptor from the interrupt descriptor table. The descriptor specifies the address of the interrupt service procedure. The interrupt call is similar to a far CALL instruction because it places the return address (IP/EIP and CS) on the stack.

INTs. There are 256 different software interrupt instructions (INTs) available to the programmer. Each INT instruction has a numeric operand whose range is 0 to 255 (00H–FFH). For example, the INT 100 uses interrupt vector 100, which appears at memory address 190H–193H.

The address of the interrupt vector is determined by multiplying the interrupt type number times 4. For example, the INT 10H instruction calls the interrupt service procedure whose address is stored beginning at memory location 40H (10H × 4) in the real mode. In the protected mode, the interrupt descriptor is located by multiplying the type number by 8 instead of 4 because each descriptor is eight bytes long.

Each INT instruction is two bytes long. The first byte contains the opcode, and the second byte contains the vector type number. The only exception to this is INT 3, a one-byte special software interrupt used for breakpoints.

Whenever a software interrupt instruction executes, it (1) pushes the flags onto the stack, (2) clears the T and I flag bits, (3) pushes CS onto the stack, (4) fetches the new value for CS from the interrupt vector, (5) pushes IP/EIP onto the stack, (6) fetches the new value for IP/EIP from the vector, and (7) jumps to the new location addressed by CS and IP/EIP.

The INT instruction performs as a far CALL except that it not only pushes CS and IP onto the stack, but it also pushes the flags onto the stack. The INT instruction performs the operation of a PUSHF, followed by a far CALL instruction.

Notice that when the INT instruction executes, it clears the interrupt flag (I), which controls the external hardware interrupt input pin INTR (interrupt request). When I = 0, the microprocessor disables the INTR pin; when I = 1, the microprocessor enables the INTR pin.

Software interrupts are most commonly used to call system procedures because the address of the system function need not be known. The system procedures are common to all system and application software. The interrupts often control printers, video displays, and disk drives. Besides relieving the program from remembering the address of the system call, the INT instruction replaces a far CALL that would otherwise be used to call a system function. The INT instruction is two bytes long whereas the far CALL is five bytes long. Each time that the INT instruction replaces a far CALL, it saves three bytes of memory in a program. This can amount to a sizable saving if the INT instruction often appears in a program, as it does for system calls.

IRET/IRETD.   The interrupt return instruction (IRET) is used only with software or hardware interrupt service procedures. Unlike a simple return instruction (RET), the IRET instruction will (1) pop stack data back into the IP, (2) pop stack data back into CS, and (3) pop stack data back into the flag register. The IRET instruction accomplishes the same tasks as the POPF, followed by a far RET instruction.

Whenever an IRET instruction executes, it restores the contents of I and T from the stack. This is important because it preserves the state of these flag bits. If interrupts were enabled before an interrupt service procedure, they are automatically re-enabled by the IRET instruction because it restores the flag register.

In the 80386 through the Pentium 4 processors, the IRETD instruction is used to return from an interrupt service procedure that is called in the protected mode. It differs from the IRET because it pops a 32-bit instruction pointer (EIP) from the stack. The IRET is used in the real mode and the IRETD is used in the protected mode.

INT 3.   An INT 3 instruction is a special software interrupt designed to function as a breakpoint. The difference between it and the other software interrupts is that INT 3 is a one-byte instruction, while the others are two-byte instructions.

It is common to insert an INT 3 instruction in software to interrupt or break the flow of the software. This function is called a **breakpoint.** A breakpoint occurs for any software interrupt, but because INT 3 is one byte long, it is easier to use for this function. Breakpoints help to debug faulty software.

INTO.   Interrupt on overflow (INTO) is a conditional software interrupt that tests the overflow flag (O). If O = 0, the INTO instruction performs no operation; if O = 1 and an INTO instruction executes, an interrupt occurs via vector type number 4.

The INTO instruction appears in software that adds or subtracts signed binary numbers. With these operations, it is possible to have an overflow. Either the JO instruction or INTO instruction detects the overflow condition.

An Interrupt Service Procedure.   Suppose that, in a particular system, a procedure is required to add the contents of DI, SI, BP, and BX and then save the sum in AX. Because this is a common task in this system, it is may occasionally be worthwhile to develop the task as a software interrupt. Realize that interrupts are usually reserved for system events and this is merely an example showing how an interrupt service procedure appears. Example 6–20 shows this software interrupt. The main difference between this procedure and a normal far procedure is that it ends with the IRET instruction instead of the RET instruction, and the contents of the flag register are saved on the stack during its execution.

EXAMPLE 6–20

```
0000                     INTS    PROC FAR

0000    03 C3                    ADD    AX,BX
0002    03 C5                    ADD    AX,BP
0004    03 C7                    ADD    AX,DI
0006    03 C6                    ADD    AX,SI
0008    CF                       IRET

0009                     INTS    ENDP
```

Interrupt Control

Although this section does not explain hardware interrupts, two instructions are introduced that control the INTR pin. The set interrupt flag instruction (STI) places a 1 into the I flag bit, which enables the INTR pin. The clear interrupt flag instruction (CLI) places a 0 into the I flag bit, which disables the INTR pin. The STI instruction enables INTR and the CLI instruction disables INTR. In a software interrupt service procedure, hardware interrupts are enabled as one of the first steps. This is accomplished by the STI instruction. The reason interrupts are enabled early in an interrupt service procedure is that just about all of the I/O devices in the personal computer are interrupt-processed. If the interrupts are disabled too long, severe system problems result.

Interrupts in the Personal Computer

The interrupts found in the personal computer differ somewhat from the ones presented in Table 6–4. The reason that they differ is that the original personal computers are 8086/8088-based systems. This meant that they only contained Intel-specified interrupts 0–4. This design is carried forward so that newer systems are compatible with the early personal computers.

Because the personal computer is operated in the real mode, the interrupt vector table is located at addresses 00000H–003FFH. The assignments used by computer system are listed in Table 6–5. Notice that these differ somewhat from the assignments in Table 6–4. Some of the interrupts shown in this table are used in example programs in later chapters. An example is the clock tick, which is extremely useful for timing events because it occurs 18.2 times per second in all personal computers.

Interrupts 00H–1FH and 70H–77H are present in the computer, no matter what operating system is installed. If DOS is installed, interrupts 20H–2FH are also present. The BIOS uses interrupts 11H through 1FH, the video BIOS uses INT 10H, and the hardware in the system uses interrupts 00H through 0FH and 70H through 77H.

TABLE 6–5 The hexadecimal interrupt assignments for the personal computer.

Number	Function
0	Divide error
1	Single-step
2	NMI pin (often parity error checks)
3	Breakpoint
4	Overflows
5	Print screen key and BOUND instruction
6	Illegal instruction
7	Coprocessor emulation
8	Clock tick (18.2 Hz)
9	Keyboard
A	IRQ2 (cascade in AT system)
B–F	IRQ3–IRQ7
10	Video BIOS
11	Equipment environment
12	Conventional memory size
13	Direct disk services
14	Serial COM port service
15	Miscellaneous
16	Keyboard service
17	Parallel port (LPT) service
18	ROM BASIC
19	Reboot
1A	Clock service
1B	Control-break handler
1C	User timer service
1D	Pointer for video parameter table
1E	Pointer for disk parameter table
1F	Pointer for graphic character pattern table
20	Terminate program (DOS 1.0)
21	DOS services
22	Program termination handler
23	Control-C handler
24	Critical error handler
25	Read disk
26	Write disk
27	Terminate and stay resident (TSR)
28	DOS idle
2F	Multiplex handler
31	DPMI (DOS protected mode interface) provided by Windows
33	Mouse driver
67	VCPI (virtual control program interface) provided by HIMEM.SYS
70–77	IRQ8–IRQ15

6–5   MACHINE CONTROL AND MISCELLANEOUS INSTRUCTIONS

The last category of real mode instructions found in the microprocessor are the machine control and miscellaneous group. These instructions provide control of the carry bit, sample the BUSY/TEST pin, and perform various other functions. Because many of these instructions are used in hardware control, they need only be explained briefly at this point.

Controlling the Carry Flag Bit

The carry flag (C) propagates the carry or borrow in multiple-word/double word addition and subtraction. It also indicates errors in procedures. There are three instructions that control the contents of the carry flag: STC (set carry), CLC (clear carry), and CMC (complement carry).

Because the carry flag is seldom used, except with multiple word addition and subtraction, it is available for other uses. The most common task for the carry flag is to indicate an error upon return from a procedure. Suppose that a procedure reads data from a disk memory file. This operation can be successful, or an error such as file-not-found can occur. Upon return from this procedure, if C = 1, an error has occurred; if C = 0, no error occurred. Most of the DOS and BIOS procedures use the carry flag to indicate error conditions.

WAIT

The WAIT instruction monitors the hardware $\overline{BUSY}$ pin on the 80286 and 80386, and the $\overline{TEST}$ pin on the 8086/8088. The name of this pin was changed in the 80286 microprocessor from $\overline{TEST}$ to $\overline{BUSY}$. If the WAIT instruction executes while the $\overline{BUSY}$ pin = 1, nothing happens and the next instruction executes. If the $\overline{BUSY}$ pin = 0 when the WAIT instruction executes, the microprocessor waits for the $\overline{BUSY}$ pin to return to a logic 1. This pin indicates a busy condition when at a logic 0 level.

The $\overline{BUSY}/\overline{TEST}$ pin of the microprocessor is usually connected to the $\overline{BUSY}$ pin of the 8087 through the 80387 numeric coprocessors. This connection allows the microprocessor to wait until the coprocessor finishes a task. Because the coprocessor is inside an 80486 through the Pentium 4, the $\overline{BUSY}$ pin is not present in these microprocessors.

HLT

The halt instruction (HLT) stops the execution of software. There are three ways to exit a halt: by an interrupt, by a hardware reset, or during a DMA operation. This instruction normally appears in a program to wait for an interrupt. It often synchronizes external hardware interrupts with the software system.

NOP

When the microprocessor encounters a no operation instruction (NOP), it takes a short time to execute. In early years, before software development tools were available, a NOP, which performs absolutely no operation, was often used to pad software with space for future machine language instructions. If you are developing machine language programs, which is extremely rare, it is recommended that you place 10 or so NOPs in your program at 50-byte intervals. This is done in case you need to add instructions at some future point. A NOP may also find application in time delays to waste time. Realize that a NOP used for timing is not very accurate because of the cache and pipelines in modern microprocessors.

LOCK Prefix

The LOCK prefix appends an instruction and causes the $\overline{LOCK}$ pin to become a logic 0. The $\overline{LOCK}$ pin often disables external bus masters or other system components. The LOCK prefix causes the $\overline{LOCK}$ pin to activate for the duration of a locked instruction. If more than one sequential instruction is locked, the $\overline{LOCK}$ pin remains a logic 0 for the duration of the sequence of locked instructions. The LOCK:MOV AL,[SI] instruction is an example of a locked instruction.

ESC

The escape (ESC) instruction passes information to the 8087–Pentium 4 numeric coprocessors. Whenever an ESC instruction executes, the microprocessor provides the memory address, if required, but otherwise performs a NOP. Six bits of the ESC instruction provide the opcode to the coprocessor and begin executing an instruction.

The ESC opcode never appears in a program as ESC and in itself is considered obsolete as an opcode. In its place are a set of coprocessor instructions (FLD, FST, FMUL, etc.) that assemble as ESC instructions for the coprocessor. More detail is provided in Chapter 13, which details the 8087–Pentium 4 numeric coprocessors.

BOUND

The BOUND instruction, first made available in the 80186 microprocessor, is a comparison instruction that may cause an interrupt (vector type number 5). This instruction compares the contents of any 16-bit or 32-bit register against the contents of two words or doublewords of memory: an upper and a lower boundary. If the value in the register compared with memory is not within the upper and lower boundary, a type 5 interrupt ensues. If it is within the boundary, the next instruction in the program executes.

For example, if the BOUND SI,DATA instruction executes, word-sized location DATA contains the lower boundary, and word-sized location DATA + 2 bytes contains the upper boundary. If the number contained in SI is less than memory location DATA or greater than memory location DATA + 2 bytes, a type 5 interrupt occurs. Note that when this interrupt occurs, the return address points to the BOUND instruction, not to the instruction following BOUND. This differs from a normal interrupt, where the return address points to the next instruction in the program.

ENTER and LEAVE

The ENTER and LEAVE instructions, first made available to the 80186 microprocessor, are used with stack frames, which are mechanisms used to pass parameters to a procedure through the stack memory. The stack frame also holds local memory variables for the procedure. Stack frames provide dynamic areas of memory for procedures in multi-user environments.

The ENTER instruction creates a stack frame by pushing BP onto the stack and then loading BP with the uppermost address of the stack frame. This allows stack frame variables to be accessed through the BP register. The ENTER instruction contains two operands: the first operand specifies the number of bytes to reserve for variables on the stack frame and the second specifies the level of the procedure.

Suppose that an ENTER 8,0 instruction executes. This instruction reserves eight bytes of memory for the stack frame and the zero specifies level 0. Figure 6–9 shows the stack frame set

FIGURE 6–9 The stack frame created by the ENTER 8.0 instruction. Notice that BP is stored beginning at the top of the stack frame. This is followed by an 8-byte area called a stack frame.

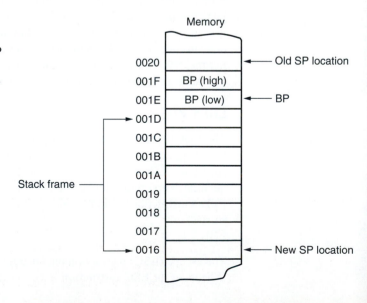

up by this instruction. Note that this instruction stores BP onto the top of the stack. It then subtracts 8 from the stack pointer, leaving eight bytes of memory space for temporary data storage. The uppermost location of this eight-byte temporary storage area is addressed by BP. The LEAVE instruction reverses this process by reloading both SP and BP with their prior values.

Example 6–21 shows how the ENTER instruction creates a stack frame so that two 16-bit parameters are passed to a system level procedure. Notice how the ENTER and LEAVE instructions appear in this program, and how the parameters pass through the stack frame to and from the procedure. This procedure uses two parameters that pass to it and returns two results through the stack frame.

EXAMPLE 6–21

```
                            ;A sequence used to call system software that
                            ;uses parameters stored in a stack frame.
                            ;
0000  C8 0004 00              ENTER 4,0               ;create 4 byte frame

0004  A1 00C8 R               MOV   AX,DATA1
0007  89 46 FC                MOV   [BP-4],AX         ;save para 1
000A  A1 00CA R               MOV   AX,DATA2
000D  89 46 FE                MOV   [BP-2],AX         ;save para 2

0010  E8 0100 R               CALL  SYS               ;call subroutine

0013  8B 46 FC                MOV   AX,[BP-4]         ;get result 1
0016  A3 00C8 R               MOV   DATA1,AX          ;save result 1
0019  8B 46 FE                MOV   AX,[BP-2]         ;get result 2
001C  A3 00CA R               MOV   DATA2,AX          ;save result 2

001F  C9                      LEAVE
                                .    .
                                .    .
                            (other software continues here)
                                .    .
                                .    .
                            ;system subroutine that uses the stack frame parameters
                            ;
0100                  SYS     PROC NEAR

0100  60                      PUSHA

0101  8B 46 FC                MOV   AX,[BP-4]         ;get para 1
0104  8B 5E FE                MOV   BX,[BP-2]         ;get para 2
                                .    .
                                .    .
                            (software that uses the parameters)
                                .    .
                                .    .
0130  89 46 FC                MOV   [BP-4],AX         ;save result 1
0133  89 5E FE                MOV   [BP-2],BX         ;save result 2

0136  61                      POPA
0137  C3                      RET

0138                  SYS     ENDP
```

6–6   SUMMARY

1. There are three types of unconditional jump instructions: short, near, and far. The short jump allows a branch to within +127 and −128 bytes. The near jump (using a displacement of ±32K) allows a jump to any location in the current code segment (intrasegment). The far jump

allows a jump to any location in the memory system (intersegment). The near jump in an 80386 through a Pentium 4 is within ± 2G bytes because these microprocessors can use a 32-bit signed displacement.

2. Whenever a label appears with a JMP instruction or conditional jump, the label, located in the label field, must be followed by a colon (LABEL:). The JMP DOGGY instruction jumps to memory location DOGGY:.

3. The displacement that follows a short or near jump is the distance from the next instruction to the jump location.

4. Indirect jumps are available in two forms: (1) jump to the location stored in a register and (2) jump to the location stored in a memory word (near indirect) or doubleword (far indirect).

5. Conditional jumps are all short jumps that test one or more of the flag bits: C, Z, O, P, or S. If the condition is true, a jump occurs; if the condition is false, the next sequential instruction executes. Note that the 80386 and above allow a 16-bit signed displacement for the conditional jump instructions.

6. A special conditional jump instruction (LOOP) decrements CX and jumps to the label when CX is not 0. Other forms of loop include LOOPE, LOOPNE, LOOPZ, and LOOPNZ. The LOOPE instruction jumps if CX is not 0 and if an equal condition exists. In the 80386 through the Pentium 4, the LOOPD, LOOPED, and LOOPNED instructions also use the ECX register as a counter.

7. The 80386 through the Pentium 4 contain conditional set instructions that either set a byte to 01H or clear it to 00H. If the condition under test is true, the operand byte is set to a 01H; if the condition under test is false, the operand byte is cleared to 00H.

8. The .IF and .ENDIF statements are useful in assembly language for making decisions. The instructions cause the assembler to generate conditional jump statements that modify the flow of the program.

9. The .WHILE and .ENDW statements allow an assembly language program to use the DO-WHILE construction, and the .REPEAT and .UNTIL statements allow an assembly language program to use the REPEAT-UNTIL construct.

10. Procedures are groups of instructions that perform one task and are used from any point in a program. The CALL instruction links to a procedure and the RET instruction returns from a procedure. In assembly language, the PROC directive defines the name and type of procedure. The ENDP directive declares the end of the procedure.

11. The CALL instruction is a combination of a PUSH and a JMP instruction. When CALL executes, it pushes the return address on the stack and then jumps to the procedure. A near CALL places the contents of IP on the stack, and a far CALL places both IP and CS on the stack.

12. The RET instruction returns from a procedure by removing the return address from the stack and placing it into IP (near return), or IP and CS (far return).

13. Interrupts are either software instructions similar to CALL or hardware signals used to call procedures. This process interrupts the current program and calls a procedure. After the procedure, a special IRET instruction returns control to the interrupted software.

14. Real mode interrupt vectors are four bytes long and contain the address (IP and CS) of the interrupt service procedure. The microprocessor contains 256 interrupt vectors in the first 1K bytes of memory. The first 32 are defined by Intel; the remaining 224 are user interrupts. In protected mode operation, the interrupt vector is eight bytes long and the interrupt vector table may be relocated to any section of the memory system.

15. Whenever an interrupt is accepted by the microprocessor, the flags IP and CS are pushed on the stack. Besides pushing the flags, the T and I flag bits are cleared to disable both the trace function and the INTR pin. The final event that occurs for the interrupt is that the interrupt vector is fetched from the vector table and a jump to the interrupt service procedure occurs.

16. Software interrupt instructions (INT) often replace system calls. Software interrupts save three bytes of memory each time they replace CALL instructions.

17. A special return instruction (IRET) must be used to return from an interrupt service procedure. The IRET instruction not only removes IP and CS from the stack, it also removes the flags from the stack.

18. Interrupt on an overflow (INTO) is a conditional interrupt that calls an interrupt service procedure if the overflow flag (O) = 1.

19. The interrupt enable flag (I) controls the INTR pin connection on the microprocessor. If the STI instruction executes, it sets I to enable the INTR pin. If the CLI instruction executes, it clears I to disable the INTR pin.

20. The carry flag bit (C) is clear, set, and complemented by the CLC, STC, and CMC instructions.

21. The WAIT instruction tests the condition of the $\overline{\text{BUSY}}$ or $\overline{\text{TEST}}$ pin on the microprocessor. If $\overline{\text{BUSY}}$ or $\overline{\text{TEST}}$ = 1, WAIT does not wait; but if $\overline{\text{BUSY}}$ or $\overline{\text{TEST}}$ = 0, WAIT continues testing the $\overline{\text{BUSY}}$ or $\overline{\text{TEST}}$ pin until it becomes a logic 1. Note that the 8086/8088 contains the $\overline{\text{TEST}}$ pin, while the 80286–80386 contain the $\overline{\text{BUSY}}$ pin. The 80486 through the Pentium 4 do not contain a $\overline{\text{BUSY}}$ or $\overline{\text{TEST}}$ pin.

22. The LOCK prefix causes the $\overline{\text{LOCK}}$ pin to become a logic 0 for the duration of the locked instruction. The ESC instruction passes instruction to the numeric coprocessor.

23. The BOUND instruction compares the contents of any 16-bit register against the contents of two words of memory: an upper and a lower boundary. If the value in the register compared with memory is not within the upper and lower boundary, a type 5 interrupt ensues.

24. The ENTER and LEAVE instructions are used with stack frames. A stack frame is a mechanism used to pass parameters to a procedure through the stack memory. The stack frame also holds local memory variables for the procedure. The ENTER instruction creates the stack frame, and the LEAVE instruction removes the stack frame from the stack. The BP register addresses stack frame data.

25. Example 6–22 lists a program that uses some of the instructions presented in this chapter, as well as those presented in prior chapters. This example contains a procedure that displays a character string on the monitor. As a test of the program, a few sample lines are displayed. Note that the character string is called a *null string* because it ends with a null (00H).

EXAMPLE 6–22

```
                        ;A program that displays a string of characters
                        ;using the procedure STRING.
                        ;
                        .MODEL SMALL               ;select SMALL model
0000                    .DATA                      ;start of DATA segment
        ;
0000  0D 0A 0A 00   MES1    DB    13,10,10,0
0004  54 68 69 73 20 MES2   DB    'This is a sample line.',0
      69 73 20 61 20
      73 61 6D 70 6C
      65 20 6C 69 6E
      65 2E 00
0000                    .CODE                      ;start of CODE segment
                        ;
                        ;A procedure that displays the character string
                        ;addressed by SI in the data segment.  The character
                        ;string must end with a null.
                        ;
                        ;This procedure changes AX, DX, and SI.
                        ;
0000                    STRING PROC NEAR

0000  AC                LODSB               ;get string character
0001  3C 00             CMP   AL,0          ;test for null
0003  74 08             JE    STRING1       ;if null
0005  8A D0             MOV   DL,AL         ;move ASCII code to DL
0007  B4 02             MOV   AH,2          ;select function 02H
0009  CD 21             INT   21H           ;access DOS
```

```
000B   EB F3                   JMP   STRING          ;repeat until null
000D                   STRING1:
000D   C3                      RET                   ;return from procedure

000E                   STRING ENDP
                       .STARTUP                      ;start of program
0025   FC                      CLD                   ;select increment
0026   BE 0000 R               MOV   SI,OFFSET MES1  ;address MES1
0029   E8 FFD4                 CALL  STRING          ;display MES1
002C   BE 0004 R               MOV   SI,OFFSET MES2  ;address MES2
002F   E8 FFCE                 CALL  STRING          ;display MES2
                               .EXIT                 ;exit to DOS
                               END                   ;end of file
```

6–7   QUESTIONS AND PROBLEMS

1. What is a short JMP?
2. Which type of JMP is used when jumping to any location within the current code segment?
3. Which JMP instruction allows the program to continue execution at any memory location in the system?
4. Which JMP instruction is five bytes long?
5. What is the range of a near jump in the 80386–Pentium 4 microprocessors?
6. Which type of JMP instruction (short, near, or far) assembles for the following:
   (a) if the distance is 0210H bytes
   (b) if the distance is 0020H bytes
   (c) if the distance is 10000H bytes
7. What can be said about a label that is followed by a colon?
8. The near jump modifies the program address by changing which register or registers?
9. The far jump modifies the program address by changing which register or registers?
10. Explain what the JMP AX instruction accomplishes. Also identify it as a near or a far jump instruction.
11. Contrast the operation of a JMP DI with a JMP [DI].
12. Contrast the operation of a JMP [DI] with a JMP FAR PTR [DI].
13. List the five flag bits tested by the conditional jump instructions.
14. Describe how the JA instruction operates.
15. When will the JO instruction jump?
16. Which conditional jump instructions follow the comparison of signed numbers?
17. Which conditional jump instructions follow the comparison of unsigned numbers?
18. Which conditional jump instructions test both the Z and C flag bits?
19. When does the JCXZ instruction jump?
20. Which SET instruction is used to set AL if the flag bits indicate a zero condition?
21. The 8086 LOOP instruction decrements register _____ and tests it for a 0 to decide if a jump occurs.
22. The Pentium 4 LOOPD instruction decrements register _____ and tests it for a 0 to decide if a jump occurs.
23. Explain how the LOOPE instruction operates.
24. Develop a short sequence of instructions that stores a 00H into 150H bytes of memory, beginning at extra segment memory location DATA. You must use the LOOP instruction to help perform this task.
25. Develop a sequence of instructions that searches through a block of 100H bytes of memory. This program must count all the unsigned numbers that are above 42H and all that are below 42H. Byte-sized data segment memory location UP must contain the count of numbers above 42H, and data segment location DOWN must contain the count of numbers below 42H.

26. Show what assembly language instructions are generated by the following sequence:

```
.IF AL==3
    ADD  AL,2
.ENDIF
```

27. What happens if the .WHILE 1 instruction is placed in a program?
28. Develop a short sequence of instructions that uses the REPEAT-UNTIL construct to copy the contents of byte-sized memory BLOCKA into byte-sized memory BLOCKB until a 00H is moved.
29. What is the purpose of the .BREAK directive?
30. Using the DO-WHILE construct, develop a sequence of instructions that add the byte-sized contents of BLOCKA to BLOCKB while the sum is not a 12H.
31. What is a procedure?
32. Explain how the near and far CALL instructions function.
33. How does the near RET instruction function?
34. The last executable instruction in a procedure must be a(n) _____.
35. Which directive identifies the start of a procedure?
36. How is a procedure identified as near or far?
37. Explain what the RET 6 instruction accomplishes.
38. Write a near procedure that cubes the contents of the CX register. This procedure may not affect any register except CX.
39. Write a procedure that multiplies DI by SI and then divides the result by 100H. Make sure that the result is left in AX upon returning from the procedure. This procedure may not change any register except AX.
40. Write a procedure that sums EAX, EBX, ECX, and EDX. If a carry occurs, place a logic 1 in EDI. If no carry occurs, place a 0 in EDI. The sum should be found in EAX after the execution of your procedure.
41. What is an interrupt?
42. Which software instructions call an interrupt service procedure?
43. How many different interrupt types are available in the microprocessor?
44. What is the purpose of interrupt vector type number 0?
45. Illustrate the contents of an interrupt vector and explain the purpose of each part.
46. How does the IRET instruction differ from the RET instruction?
47. What is the IRETD instruction?
48. The INTO instruction only interrupts the program for what condition?
49. The interrupt vector for an INT 40H instruction is stored at which memory locations?
50. What instructions control the function of the INTR pin?
51. Which personal computer interrupt services the parallel LPT port?
52. Which personal computer interrupt services the keyboard?
53. What instruction tests the $\overline{\text{BUSY}}$ pin?
54. When will the BOUND instruction interrupt a program?
55. An ENTER 16,0 instruction creates a stack frame that contains _____ bytes.
56. Which register moves to the stack when an ENTER instruction executes?
57. Which instruction passes opcodes to the numeric coprocessor?
58. What is a null string?
59. Explain how the STRING procedure operates in Example 6–22.
60. Rewrite Example 6–22 so that it displays your name.

CHAPTER 7

Programming the Microprocessor

INTRODUCTION

This chapter develops programs and programming techniques using the MASM macro assembler program, the DOS function calls, and the BIOS function calls. Many of the DOS function calls and BIOS function calls are used in this chapter, but all are explained in complete detail in Appendix A. Please scan the function calls listed in Appendix A as you read this chapter. The MASM assembler has already been explained and demonstrated in prior chapters, but there are still more features to learn at this point.

Some programming techniques explained in this chapter include macro sequences, keyboard and display manipulation, program modules, library files, using the mouse, interrupt hooks, and other important programming techniques. This chapter is meant as an introduction to programming, yet it provides valuable programming techniques that provide a wealth of background so that programs can be easily developed for the personal computer by using MSDOS as a springboard.

CHAPTER OBJECTIVES

Upon completion of this chapter, you will be able to:

1. Use the MASM assembler and linker program to create programs that contain more than one module.
2. Explain the use of EXTRN and PUBLIC as they apply to modular programming.
3. Set up a library file that contains commonly used subroutines.
4. Write and use MACRO and ENDM to develop macro sequences used with linear programming.
5. Show how both sequential and random access files are developed for use in a system.
6. Develop programs using DOS function calls.
7. Differentiate a DOS function call from a BIOS function call.
8. Show how to hook into interrupts using DOS function calls.
9. Use conditional assembly language statements in programs.
10. Use the mouse in program examples.

7–1   MODULAR PROGRAMMING

Many programs are too large to be developed by one person. This means that programs are routinely developed by teams of programmers. The linker program is provided with MSDOS so that programming modules can be linked together into a complete program. Linking is also an internal function of the Programmer's WorkBench program that is bundled with MASM version 6.X. This section of the text describes the linker, the linking task, library files, EXTRN, and PUBLIC as they apply to program modules and modular programming. It also introduces the use of Programmer's WorkBench, which is also used to manage programs generated by teams.

The Assembler and Linker

The **assembler program** converts a symbolic **source module** (file) into a hexadecimal **object file.** We have seen many examples of symbolic source files, written in assembly language, in prior chapters. Example 7–1 shows how the assembler dialog that appears as a source module named NEW.ASM is assembled. Note that this dialog is used with version 6.11 at the DOS command line. This assembler also uses the Programmer's WorkBench program for development, without resorting to the DOS command line. Whenever you create a source file, it should have an extension of ASM. Source files are created by using WorkBench, an editor that comes with the assembler, or by almost any other word processor or editor capable of generating an ASCII file.

EXAMPLE 7–1

```
C:\MASM611\FILES>ml /Flnew.lst new.asm
Microsoft (R) Macro Assembler Version 6.11
Copyright (C) Microsoft Corp 1981-1993.  All rights reserved.

 Assembling: new.asm

Microsoft (R) Segmented Executable Linker  Version 5.31.009 Jul 13 1992
Copyright (C) Microsoft Corp 1984-1992.  All rights reserved.

Object Modules [.obj]: new.obj
Run File [new.exe]: "new.exe"
List File [nul.map]: NUL
Libraries [.lib]:
Definitions File [nul.def]:

C:\MASM611\FILES>
```

The assembler program (ML) requires the source file name following ML. In the example, the /Fl switch is used to create a listing file named NEW.LST. Although this is optional, it is recommended so the output of the assembler can be viewed for troubleshooting problems. The source listing file (.LST) contains the assembled version of the source file and its hexadecimal machine language equivalent. The cross-reference file (.CRF), which is not generated in this example, lists all labels and pertinent information required for cross-referencing.

The **linker program,** which executes as the second part of ML, reads the object files that are created by the assembler program and links them together into a single execution file. An **execution file** is created with the file name extension EXE. Execution files are selected by typing the file name at the DOS prompt (A:\). An example execution file is FROG.EXE, which is executed by typing FROG at the DOS command prompt.

If a file is short enough (less than 64K bytes long) it can be converted from an execution file to a **command file** (.COM). The command file is slightly different from an execution file in that the program must be originated at location 100H before it can execute. This means that the program must be no larger than 64K–100H in length. The ML program generates a command file if the tiny model is used with a starting address of 100H. Note that Programmer's Work-Bench can also be configured to generate a command file. The main advantage of a command file is that it loads off the disk into the computer much more quickly than an execution file. It also requires less disk storage space than the equivalent execution file.

Example 7–2 shows the linker program protocol when it is used to link the files NEW, WHAT, and DONUT. The linker also links library files (LIBS) so procedures, located within LIBS, can be used with the linked execution file. To invoke the linker, type LINK at the DOS command prompt, as illustrated in Example 7–2. Note that before files are linked, they must first be assembled and they must be **error-free.** ML not only links the files, but it also assembles them prior to linking.

EXAMPLE 7–2

```
C:\MASM611\FILES>ml new.asm what.asm donut.asm
Microsoft (R) Macro Assembler Version 6.11
Copyright (C) Microsoft Corp 1981-1993.  All rights reserved.

 Assembling: new.asm
 Assembling: what.asm
 Assembling: donut.asm

Microsoft (R) Segmented Executable Linker  Version 5.31.009 Jul 13 1992
Copyright (C) Microsoft Corp 1984-1992.  All rights reserved.

Object Modules [.obj]: new.obj+
Object Modules [.obj]: "what.obj"+
Object Modules [.obj]: "donut.obj"
Run File [new.exe]: "new.exe"
List File [nul.map]: NUL
Libraries [.lib]:
Definitions File [nul.def]:

C:\MASM611\FILES>
```

In this example, after typing ML, the linker program asks for the "Object Modules," which are created by the assembler. In this example, we have three object modules: NEW, WHAT, and DONUT. If more than one object file exists, the main program file (NEW, in this example) is typed first, followed by any other supporting modules.

Library files are entered after the file name and after the switch /LINK. In this example, we did not enter a library file name. To use a library called NUMB.LIB while assembling a program called NEW.ASM, type ML NEW.ASM /LINK NUMB.LIB.

PUBLIC and EXTRN

The PUBLIC and EXTRN directives are very important to modular programming. We use PUBLIC to declare that labels of code, data, or entire segments are available to other program modules. EXTRN (external) declares that labels are external to a module. Without these statements, modules could not be linked together to create a program by using modular programming techniques. They might link, but one module would not be able to communicate to another.

The PUBLIC directive is placed in the opcode field of an assembly language statement to define a label as public, so that the label can be used by other modules. The label declared as public can be a jump address, a data address, or an entire segment. Example 7–3 shows the PUBLIC statement used to define some labels and make them public to other modules. When segments are made public, they are combined with other public segments that contain data with the same segment name.

EXAMPLE 7–3

```
                        .MODEL  SMALL
                        .DATA
                                PUBLIC    DATA1           ;declare DATA1 and DATA2 public
                                PUBLIC    DATA2

0000   0064[      DATA1         DB        100 DUP (?)
          00
       ]
0064   0064[      DATA2         DB        100 DUP (?)
          00
       ]
                        .CODE
                        .STARTUP
                                PUBLIC    READ            ;declare READ public
                  READ          PROC      FAR

0006   B4 06                    MOV       AH,6            ;read keyboard
0008   B2 FF                    MOV       DL,0FFH
000A   CD 21                    INT       21H
000C   74 F8                    JE        READ            ;if no key typed
000E   CB                       RET

                  READ          ENDP
                                END
```

The EXTRN statement appears in both data and code segments to define labels as external to the segment. If data are defined as external, their sizes must be defined as BYTE, WORD, or DWORD. If a jump or call address is external, it must be defined as NEAR or FAR. Example 7–4 shows how the external statement is used to indicate that several labels are external to the program listed. Notice in this example that any external address or data is defined with the letter E in the hexadecimal assembled listing.

EXAMPLE 7–4

```
                        .MODEL  SMALL
                        .DATA

                                EXTRN DATA1:BYTE
                                EXTRN DATA2:BYTE
                                EXTRN DATA3:WORD
                                EXTRN DATA4:DWORD
                        .CODE
                                EXTRN READ:FAR
                        .STARTUP

0005   BF 0000 E                MOV       DX,OFFSET DATA1
0008   B9 000A                  MOV       CX,10
000B                    START:
000B   9A 0000 ---- E           CALL      READ
0010   AA                       STOSB
0011   E2 F8                    LOOP      START
                        .EXIT
                        END
```

Libraries

Library files are collections of procedures that are used by many different programs. These procedures are assembled and compiled into a library file by the LIB program that accompanies the MASM assembler program. Libraries allow common procedures to be collected into one place so they can be used by many different applications. The library file (FILENAME.LIB) is invoked when a program is linked with the linker program.

Why bother with library files? A library file is a good place to store a collection of related procedures. When the library file is linked with a program, only the procedures required by the program are removed from the library file and added to the program. If any amount of assembly language programming is to be accomplished efficiently, a good set of library files is essential and saves many hours in recoding common functions.

Creating a Library File.   A library file is created with the LIB command, typed at the DOS prompt. A library file is a collection of assembled .OBJ files that each perform one procedure or task. Example 7–5 shows two separate files (READ_KEY and ECHO) that will be used to structure a library file. Please notice that the name of the procedure must be declared PUBLIC in a library file and does not necessarily need to match the file name, although it does in this example. Each procedure in this example is defined as a FAR procedure, so that the linker can place the procedures in a code segment separate from the main program. When FAR is used to define a procedure, we usually call it a global procedure.

EXAMPLE 7–5

```
                        ;The first library module is called READ_KEY.  This
                        ;procedure reads a key from the keyboard and returns with
                        ;its ASCII code in AL.
                        .MODEL TINY
                                PUBLIC READ_KEY

                        READ_KEY   PROC   FAR

0000  52                        PUSH    DX
                        READ_KEY1:
0001  B4 06                     MOV     AH,6
0003  B2 FF                     MOV     DH,0FFH
0005  CD 21                     INT     21H
0007  74 F8                     JE      READ_KEY1
0009  5A                        POP     DX
000A  CB                        RET

                        READ_KEY   ENDP
                                END
                        ;
                        ;The second library module is called ECHO.  This
                        ;procedure displays the ASCII character in AL on the
                        ;video screen.
                        .MODEL TINY
                                PUBLIC  ECHO

                        ECHO   PROC   FAR

0000  52                        PUSH    DX
0001  B4 06                     MOV     AH,6
0003  8A D0                     MOV     DL,AL
0005  CD 21                     INT     21H
0007  5A                        POP     DX
0008  CB                        RET

                        ECHO   ENDP
                                END
```

After each file is assembled (note that there are two complete example procedures in Example 7–5), the LIB program is used to combine them into a single library file. The LIB program prompts for information, as illustrated in Example 7–6, in which these files are combined to form the library IO.

EXAMPLE 7–6

```
C:\MASM611\FILES\LIB

Microsoft (R) Library Manager  Version 3.20.010
Copyright (C) Microsoft Corp. 1983-1992.  All rights reserved.

Library name: IO
Library file does not exist.  Create? Y
Operations: READ_KEY+ECHO
List file: IO
```

The LIB program begins with the copyright message from Microsoft, followed by the prompt *Library name*. The library name chosen is IO for the IO.LIB file. Because this is a new file, the library program asks if we wish to create the library file. The *Operations:* prompt is where the library module names are typed. In this case, we create a library by using two procedure files (READ_KEY and ECHO). Note that these files were created and assembled as READ_KEY.ASM and ECHO.ASM from Example 7–5. The list file shows the contents of the library and is illustrated in Example 7–7. The list file shows the size and names of the files used to create the library, and the public label (procedure name) that is used in the library file.

To add additional library modules, type the name of the library file after invoking LIB. At the *Operations*: prompt, type the new module name, preceded by a *plus sign* to add a new procedure. If you must delete a library module, use a *minus sign* before the operation file name.

EXAMPLE 7–7

```
ECHO.................ECHO        READ_KEY.................READ_KEY

READ_KEY    Offset: 00000010H Code and data size: BH
   READ_KEY

ECHO        Offset: 00000070H Code and data size: 9H
```

Once the library file is linked to your program file, only the library procedures actually used by your program are placed in the execution file. Don't forget to use the label EXTRN when specifying library calls from your program module. For example, to use the ECHO procedure in a program, type EXTRN ECHO:FAR.

Macros

A **macro** is a group of instructions that perform one task, just as a procedure performs one task. The difference is that a procedure is accessed via a CALL instruction, while a macro, and all the instructions defined in the macro, is inserted in the program at the point of usage. Creating a macro is very similar to creating a new opcode that can be used in the program. The name of the macro and any parameters associated with it are typed, and the assembler then inserts them into the program. Macro sequences execute faster than procedures because there are no CALL and RET instructions to execute. The instructions of the macro are placed in your program by the assembler at the point they are invoked.

The MACRO and ENDM directives delineate a macro sequence. The first statement of a macro is the MACRO instruction, which contains the name of the macro and any parameters associated with it. An example is MOVE MACRO A,B, which defines the macro name as MOVE. This new pseudo opcode uses two parameters: A and B. The last statement of a macro is the ENDM instruction, which is placed on a line by itself. Never place a label in front of the ENDM statement, or the macro will not assemble.

Example 7–8 shows how a macro is created and used in a program. The first six lines of code define the macro. This macro moves the word-sized contents of memory location B into word-sized memory location A. After the macro is defined in the example, it is used twice. The macro is **expanded** by the assembler in this example, so that you can see how it assembles to generate the moves. Any hexadecimal machine language statement followed by a number (1, in this example) is a macro expansion statement. The expansion statements are not typed in the source program; they are generated by the assembler to show that the assembler has inserted them into the program. Notice that the comment in the macro is preceded with ;; instead of ; as is customary. Macro sequences must always be defined before they are used in a program, so they generally appear at the top of the code segment.

EXAMPLE 7–8

```
                        MOVE    MACRO A,B

                                PUSH    AX
                                MOV     AX,B
                                MOV     A,AX
                                POP     AX

                                ENDM

                        MOVE    VAR1,VAR2    ;use the MOVE macro

0000  50            1           PUSH    AX
0001  A1 0002 R     1           MOV     AX,VAR2
0004  A3 0000 R     1           MOV     VAR1,AX
0007  58            1           POP     AX

                        MOVE    VAR3,VAR4    ;use the MOVE macro

0008  50            1           PUSH    AX
0009  A1 0006 R     1           MOV     AX,VAR4
000C  A3 0004 R     1           MOV     VAR3,AX
000F  58            1           POP     AX
```

Local Variables in a Macro.   Sometimes, macros contain local variables. A **local variable** is one that appears in the macro, but is not available outside the macro. To define a local variable, we use the LOCAL directive. Example 7–9 shows how a local variable, used as a jump address, appears in a macro definition. If this jump address is not defined as local, the assembler will flag it with errors on the second and subsequent attempts to use the macro.

EXAMPLE 7–9

```
                        READ    MACRO A             ;;reads keyboard
                                LOCAL READ1         ;;define READ1 as local

                                PUSH    DX
                        READ1:
                                MOV     AH,6
                                MOV     DL,0FFH
                                INT     21H
                                JE      READ1
                                MOV     A,AL
                                POP     DX
                                ENDM


                        READ    VAR5                ;read key into VAR5

0000  52            1           PUSH    DX
0001                1   ??0000:
0001  B4 06         1           MOV     AH,6
0003  B2 FF         1           MOV     DL,0FFH
0005  CD 21         1           INT     21H
0007  74 F8         1           JE      ??0000
0009  A2 0008 R     1           MOV     VAR5,AL
000C  5A            1           POP     DX
```

```
                               READ    VAR6         ;read key into VAR6

000D  52           1          PUSH    DX
000E               1   ??0001:
000E  B4 06        1          MOV     AH,6
0010  B2 FF        1          MOV     DL,0FFH
0012  CD 21        1          INT     21H
0014  74 F8        1          JE      ??0001
0016  A2 0009 R    1          MOV     VAR6,AL
0019  5A           1          POP     DX
```

This example reads a character from the keyboard and stores it into the byte-sized memory location indicated as a parameter with the macro. Notice how the local label READ1 is treated in the expanded macros. The assembler uses labels that start with ?? to designate them as assembler-generated labels.

The LOCAL directive must always immediately follow the MACRO directive, without any intervening spaces or comments. If a comment or space appears between MACRO and LOCAL, the assembler indicates an error and will not accept the variable as local.

Placing MACRO Definitions in Their Own Module. Macro definitions can be placed in the program file as shown, or they can be placed in their own macro module. A file can be created that contains only macros to be included with other program files. We use the INCLUDE directive to indicate that a program file will include a module that contains external macro definitions. Although this is not a library file, for all practical purposes it functions as a library of macro sequences.

When macro sequences are placed in a file (often with the extension INC or MAC), they do not contain PUBLIC statements. If a file called MACRO.MAC contains macro sequences, the INCLUDE statement is placed in the program file as INCLUDE C:\ASSM\MACRO.MAC. Notice that the macro file is on drive C, subdirectory ASSM in this example. The INCLUDE statement includes these macros, just as if you had typed them into the file. No EXTRN statement is needed to access the macro statements that have been included. Programs may contain both macro include files and library files.

Conditional Statements in Macro Sequences

Conditional assembly language statements are available to the assembler for use in the assembly process and in macro sequences. The conditional statements create instructions that control the flow of the program and are variations of the IF-THEN, IF-THEN-ELSE, DO-WHILE, and REPEAT-UNTIL constructs used in high-level language programming languages, which were presented in the last chapter. The conditional statements for macro sequence control—presented here—are also available, but they function to create instructions only at assembly time within macro sequences. The assembler distinguishes conditional statements for macro control and condition statements for program flow with a period. For example, the .IF statement is used for program flow control, while the IF statement is used for macro assembly control. Both types of conditional statements may be used in a macro, but the macro conditionals may only be used in a macro.

Conditional Assembly Statements

As mentioned, conditional assembly is implemented with the IF-THEN or IF-THEN-ELSE construct found in high-level languages. Table 7–1 shows the forms used for the IF statement in the conditional assembly process.

The IF and ENDIF statements allow portions of the program to assemble if some condition is met. Otherwise, the statements between IF and ENDIF do not assemble and generate code.

Example 7–10 shows how the IF, ELSE, and ENDIF statements are used to conditionally assemble values for the width and length of paper in a program. Note that TRUE and FALSE are defined as 1 and 0. This is important because these values are not predefined by the assembler. Next, the width and length of the paper are adjusted by using TRUE and FALSE statements. This

TABLE 7–1 Conditional assembly language IF statements.

Statement	Function
IF	If the expression is true
IFB	If argument is blank
IFE	If the expression is not true
IFDEF	If the label has been defined
IFNB	If argument is not blank
IFNDEF	If the label has not been defined
IFIDN	If argument 1 equals argument 2
IFDIF	If argument 1 does not equal argument 2

can be expanded to ask an entire series of questions about a program so that custom versions can be created. Example 7–10(a) is the original source-code, and Example 7–10(b) shows how the program assembles for TRUE answers for both the width and length. Example 7–10(c) shows the assembled output for a false width and a true length.

When Example 7–10(a) is assembled, TRUE and FALSE are equated to WIDT and LENGT to modify the way that the assembler forms the program. In Example 7–10(b), both WIDT and LENGT are defined as TRUE, which causes the assembler to modify the way the program is assembled, so that a page is 72 columns wide and the length is continuous. Example 7–10(c) is another example in which the WIDT is FALSE and LENGT is TRUE, causing the assembler to form the instructions that make the page width 80 columns and the length continuous. The only form not shown is where the page length is 66 lines.

Examples of some of the other forms listed in Table 7–1 appear later in the text. When one of these new conditional statements appears it is explained and shown with an example.

EXAMPLE 7–10(a)

```
            ;source program
            ;
TRUE    EQU   1                ;define true
FALSE   EQU   0                ;define false

WIDT    EQU   FALSE            ;set to true if 72 columns
                               ;and false if 80 columns
LENGT   EQU   TRUE             ;set to true if continuous
                               ;and false if 66 lines

        IF    WIDT             ;72 columns
WIDE    DB    72
        ELSE
WIDE    DB    80               ;80 columns
        ENDIF

        IF    LENGT            ;if continuous
LONG    DB    -1
        ELSE
LONG    DB    66               ;if 66 lines
        ENDIF
```

EXAMPLE 7–10(b)

```
            ;assembled portion with WIDT = TRUE and LENGT = TRUE
            ;
            IF    WIDT             ;72 columns
0000  48    WIDE  DB    72
            ELSE
            ENDIF

            IF    LENGT            ;if continuous
```

```
0001  FF            LONG    DB    -1
                            ELSE
                            ENDIF
```

EXAMPLE 7–10(c)

```
                    ;assembled portion with WIDT = FALSE and LENGT = TRUE
                    ;
                            IF    WIDT              ;73 columns
                            ELSE
0000  50            WIDE    DB    80                ;80 columns
                            ENDIF
                            IF    LENGT             ;if continuous
0001  FF            LONG    DB    -1
                            ELSE
                            ENDIF
```

Using Conditional Statements in Macros

Macro sequences contain their own set of conditional instructions that differ somewhat from the ones used with the assembler, as presented in Chapter 6. For example, macros can use REPEAT and WHILE, but they do so without the period in front of the keywords REPEAT and WHILE. The REPEAT has no corresponding UNTIL, and the WHILE statement has no corresponding ENDW when used in a macro. These statements are available to all versions of the assembler.

Table 7–2 lists the relational operators used with WHILE and REPEAT. These operators are also used with any of the statements listed in Table 7–1. Note that these are different from the operators specified in Table 6–3 for the .WHILE and .REPEAT statements.

REPEAT Statement in a Macro. The REPEAT statement has a parameter associated with it to repeat the macro sequence a fixed number of times. As with any macro sequence, the repeat sequence must end with the ENDM statement. The repeat sequence inserts the instructions that appear between the REPEAT statement and the ENDM statement into the program the number of times indicated with the REPEAT statement.

Example 7–11 shows a macro called TESTS and its calling program, which sends the 10 ASCII characters from 0 through 9 to the video screen. Notice how this macro is formed by using the MACRO statement to name the macro TESTS, and how the REPEAT statement appears within macro TESTS with its own ENDM statement. Notice that the macro starts by placing a 6 into AH and the ASCII code for a 0 in DL. This sets up the DOS INT 21H function call, so a 0 is displayed on the video screen. Next, the REPEAT statement appears (note that it does not contain a period, as in .REPEAT). This is a different REPEAT statement, used only in macro sequences and available to all versions of MASM.

TABLE 7–2 Relational operators used with WHILE and REPEAT in macro sequences.

Operator	Function
EQ	Equal
NE	Not equal
LE	Less than or equal
LT	Less than
GE	Greater than or equal
GT	Greater than
NOT	Logical inversion
AND	Logical AND
OR	Logical OR
XOR	Logical exclusive-OR

The repeated statements in this example are INT 21H, which display the ASCII contents of DL and INC DL, which modifies the ASCII code displayed. In this case, the REPEAT 10 causes the statements between REPEAT 10 and the first ENDM to be repeated 10 times, as illustrated. Note that the 1 and 2 to the left of the instructions are listed to show that these statements are assembler-generated and not entered as part of the source program.

EXAMPLE 7–11

```
                      TESTS    MACRO

                               MOV   AH,6
                               MOV   DL,'0'

                               REPEAT 10
                                  INT   21H
                                  INC   DL            ;;increment to next number
                               ENDM
                               ENDM

0000                  MAIN     PROC FAR

                               TESTS                  ;display 0 through 9

0000 B4 06            1        MOV   AH,6
0002 B2 30            1        MOV   DL,'0'
0004 CD 21            2        INT   21H
0006 FE C2            2        INC   DL
0008 CD 21            2        INT   21H
000A FE C2            2        INC   DL
000C CD 21            2        INT   21H
000E FE C2            2        INC   DL
0010 CD 21            2        INT   21H
0012 FE C2            2        INC   DL
0014 CD 21            2        INT   21H
0016 FE C2            2        INC   DL
0018 CD 21            2        INT   21H
001A FE C2            2        INC   DL
001C CD 21            2        INT   21H
001E FE C2            2        INC   DL
0020 CD 21            2        INT   21H
0022 FE C2            2        INC   DL
0024 CD 21            2        INT   21H
0026 FE C2            2        INC   DL
0028 CD 21            2        INT   21H
002A FE C2            2        INC   DL

                               .EXIT

0031                  MAIN     ENDP
```

WHILE Statement in a Macro.   The WHILE statement appears in macro sequences in much the same way as REPEAT appears. That is, the while loop is terminated with the ENDM statement. The expression associated with WHILE determines how many times the loop is repeated. Again, note that the WHILE statement in the macro is different from the .WHILE statement described in Chapter 6. The WHILE statement is available to all versions of MASM.

Example 7–12 shows how the WHILE statement is used to generate a table of squares from 2 squared to whatever value fits into an array of byte-sized memory called SQUARE. The first statement of the sequence defines the label SQUARE for the first byte of data generated. The WHILE RES LT 255 repeats the calculation (SEED*SEED), while the result is less than or equal to 255. Notice that the table generated contains the square of the numbers from 2 to 15, or 225 (E1H). If you look closely at Example 7–12, the value of the SEED + 1 and SEED*SEED shows the number and its square.

EXAMPLE 7–12

```
                        ;table of byte-sized squares
                        ;
0000                    SQUARE  LABEL BYTE              ;;define label
 = 0001                 SEED =   1
 = 0001                 RES  =    SEED*SEED             ;;compute square
                        WHILE  RES LT 255
                                 DB   RES
                        SEED =   SEED+1
                        RES  =   SEED*SEED
                        ENDM
0000 01             1            DB   RES
 = 0002             1 SEED =     SEED+1
 = 0004             1 RES  =     SEED*SEED
0001 04             1            DB   RES
 = 0003             1 SEED =     SEED+1
 = 0009             1 RES  =     SEED*SEED
0002 09             1            DB   RES
 = 0004             1 SEED =     SEED+1
 = 0010             1 RES  =     SEED*SEED
0003 10             1            DB   RES
 = 0005             1 SEED =     SEED+1
 = 0019             1 RES  =     SEED*SEED
0004 19             1            DB   RES
 = 0006             1 SEED =     SEED+1
 = 0024             1 RES  =     SEED*SEED
0005 24             1            DB   RES
 = 0007             1 SEED =     SEED+1
 = 0031             1 RES  =     SEED*SEED
0006 31             1            DB   RES
 = 0008             1 SEED =     SEED+1
 = 0040             1 RES  =     SEED*SEED
0007 40             1            DB   RES
 = 0009             1 SEED =     SEED+1
 = 0051             1 RES  =     SEED*SEED
0008 51             1            DB   RES
 = 000A             1 SEED =     SEED+1
 = 0064             1 RES  =     SEED*SEED
0009 64             1            DB   RES
 = 000B             1 SEED =     SEED+1
 = 0079             1 RES  =     SEED*SEED
000A 79             1            DB   RES
 = 000C             1 SEED =     SEED+1
 = 0090             1 RES  =     SEED*SEED
000B 90             1            DB   RES
 = 000D             1 SEED =     SEED+1
 = 00A9             1 RES  =     SEED*SEED
000C A9             1            DB   RES
 = 000E             1 SEED =     SEED+1
 = 00C4             1 RES  =     SEED*SEED
000D C4             1            DB   RES
 = 000F             1 SEED =     SEED+1
 = 00E1             1 RES  =     SEED*SEED
```

FOR Statement in a Macro. The FOR statement iterates a list of data. If you are familiar with BASIC, the FOR statement functions like the READ statement, and the list of data associated with it functions like the DATA statement. Example 7–13 shows how the FOR statement is used to display a series of characters on the video display. Notice that the CHR:VARARG indicates the variable name CHR that is of variable size (VARARG). The first use of DISP generates the code required to display BARRY. The second use of the DISP macro generates the code required to display BREY. The FOR statement counts the variable used after display and repeats the commands between FOR and ENDM for each variable; in this case, each ASCII character.

EXAMPLE 7–13

```
                  DISP     MACRO   CHR:VARARG
                           MOV  AH,2
                           FOR  ARG,<CHR>
                                MOV   DL,ARG
                                INT   21H
                           ENDM
                           ENDM

                           DISP  'B','A','R','R','Y',' '

0000   B4 02      1        MOV  AH,2
0002   B2 42      2        MOV  DL,'B'
0004   CD 21      2        INT  21H
0006   B2 41      2        MOV  DL,'A'
0008   CD 21      2        INT  21H
000A   B2 52      2        MOV  DL,'R'
000C   CD 21      2        INT  21H
000E   B2 52      2        MOV  DL,'R'
0010   CD 21      2        INT  21H
0012   B2 59      2        MOV  DL,'Y'
0014   CD 21      2        INT  21H
0016   B2 20      2        MOV  DL,' '
0018   CD 21      2        INT  21H

                           DISP  'B','R','E','Y'

001A   B4 02      1        MOV  AH,2
001C   B2 42      2        MOV  DL,'B'
001E   CD 21      2        INT  21H
0020   B2 52      2        MOV  DL,'R'
0022   CD 21      2        INT  21H
0024   B2 45      2        MOV  DL,'E'
0026   CD 21      2        INT  21H
0028   B2 59      2        MOV  DL,'Y'
002A   CD 21      2        INT  21H
```

IF, ELSE, and ENDIF Statements in a Macro.   The IF statement is used in a macro to make decisions, based on the parameters sent to the macro. As before, note that IF is used in a macro and .IF is used in a program. The IF statement is available to all versions of the assembler, whereas .IF is available only to version 6.X.

In Example 7–14, a macro is developed that uses a number of conditional assembly statements to read a key, display a character, or display a carriage return and line feed combination. This example illustrates the use of IF, IFB, INB, ENDIF, and ELSE. The macro is called IO. If IO is used on a line by itself, the assembler generates the code to read a key. If IO –1 appears as a statement, the assembler generates the code required to display a carriage return and line feed. If IO 'B' appears as a statement, the assembler generates the code required to display the letter B. This example is listed in expanded form, so that the code generated by the assembler can be viewed and studied. As before, the lines that contain a number between the hexadecimal code and the statement in the program are assembler-generated, and are not included in the original source program.

EXAMPLE 7–14

```
                           .MODEL TINY
0000                       .CODE
                  ;the IO macro functions in 3 ways
                  ;
                  ;(1) IO            read a key with echo
                  ;(2) IO -1         display a carriage return & line feed
                  ;(3) IO 'B'        display the letter 'B'
```

```
                         ;or  IO AL        display contents of AL
                         ;
                         IO      MACRO   CHAR
                                 IFB     <CHAR>                  ;;if CHAR is blank
                                     MOV  AH,1                   ;;read key function
                                 ENDIF

                                 IFNB    <CHAR>                  ;;if CHAR not blank
                                     MOV  AH,2                   ;;display character

                                     IF   CHAR EQ -1            ;;if CHAR equals -1
                                         MOV  DL,13             ;;display return
                                         INT  21H
                                         MOV  DL,10             ;;display line feed

                                     ELSE                       ;;if CHAR not -1

                                         MOV  DL,CHAR           ;;load CHAR to DL
                                     ENDIF

                                 ENDIF

                                 INT  21H
                                 ENDM
                                 .STARTUP
                         ;
                         ;This program does a carriage return, line feed then
                         ;displays the letters BE on the video screen.  Next it
                         ;waits for a key to be typed.  Following the key, a
                         ;carriage return/line feed is displayed.
                         ;

                                 IO      -1                      ;return & line feed

0100  B4 02      1                MOV  AH,2
0102  B2 0D      1                MOV  DL,13
0104  CD 21      1                INT  21H
0106  B2 0A      1                MOV  DL,10
0108  CD 21      1                INT  21H

                                 IO      'B'                     ;display 'B'

010A  B4 02      1                MOV  AH,2
010C  B2 42      1                MOV  DL,'B'
010E  CD 21      1                INT  21H

                                 IO      'E'                     ;display 'E'

0110  B4 02      1                MOV  AH,2
0112  B2 45      1                MOV  DL,'E'
0114  CD 21      1                INT  21H

                                 IO                              ;read key

0116  B4 01      1                MOV  AH,1
0118  CD 21      1                INT  21H

                                 IO      -1                      ;return & line feed

011A  B4 02      1                MOV  AH,2
011C  B2 0D      1                MOV  DL,13
011E  CD 21      1                INT  21H
0120  B2 0A      1                MOV  DL,10
0122  CD 21      1                INT  21H

                                 .EXIT
                                 END
```

The first part of the macro uses the IFB <CHAR> statement to test CHAR for a blank condition. If CHAR is blank, the assembler generates the MOV AH,1 instruction followed by the very last instruction in the macro, INT 21H, to read a key with echo. This is used in the program with the IO statement.

The second part of the macro contains the IFNB <CHAR> statement to test if CHAR is not blank. If CHAR is not blank, another IF-ELSE-ENDIF sequence appears to test the contents of CHAR. If CHAR is a –1, the assembler generates the code required to display a carriage return and line feed combination. If CHAR is not a –1, the ELSE statement places CHAR into DL for display. This very powerful macro can handle most keyboard and single-character display functions. It also illustrates the power of the conditional assembly statements, when used within a macro.

The Modular Programming Approach

The modular programming approach often involves a team of people with different programming tasks. This allows the team manager to assign portions of the program to different team members. Often, the team manager develops the system flowchart or shell, and then divides it into modules for team members.

A team member might be assigned the task of developing a macro definition file. This file might contain macro definitions that handle the I/O operations for the system. Another team member might be assigned the task of developing the procedures used for the system. In most cases, the procedures are organized as a library file that is linked to the program modules. Finally, several program files or modules might be used for the final system, each developed by different team members.

This approach requires considerable communications between team members and good documentation. Documentation is the key so that modules interface correctly. Communication among team members plays an essential role in this approach.

7–2

USING THE KEYBOARD AND VIDEO DISPLAY

Today, there are few programs that don't use the keyboard and video display. This section of the text explains how to use the keyboard and video display connected to the IBM PC or compatible computer running under MSDOS.

Reading the Keyboard with DOS Functions

The keyboard of the personal computer is read via a DOS function call. A complete listing of the DOS function calls appears in Appendix A. This section uses INT 21H with various DOS function calls to read the keyboard. Data read from the keyboard are either in ASCII-coded form or in extended ASCII-coded form.

The ASCII-coded data appear as outlined in Table 1–7 in Section 1–4. The extended character set of Table 1–8 applies to printed or displayed data only, and not to keyboard data. Notice that the ASCII codes in Table 1–7 correspond to most of the keys on the keyboard. Also available through the keyboard are extended ASCII-coded keyboard data. Table 7–3 lists most of the extended ASCII codes obtained with various keys and key combinations. Notice that most keys on the keyboard have alternative key codes. Each function key has four sets of codes selected by the function key alone, the shift-function key combination, the alternate-function key combination, and the control-function key combination.

TABLE 7–3   The keyboard scanning and extended ASCII codes as returned from the keyboard.

Key	Scan Code	Extended ASCII code with....			
		Nothing	Shift	Control	Alternate
Esc	01				01
1	02				78
2	03			03	79
3	04				7A
4	05				7B
5	06				7C
6	07				7D
7	08				7E
8	09				7F
9	0A				80
0	0B				81
-	0C				82
+	0D				83
Bksp	0E				0E
Tab	0F		0F	94	A5
Q	10				10
W	11				11
E	12				12
R	13				13
T	14				14
Y	15				15
U	16				16
I	17				17
O	18				18
P	19				19
[	1A				1A
]	1B				1B
Enter	1C				1C
Enter	1C				A6
Lctrl	1D				
Rctrl	1D				
A	1E				1E
S	1F				1F
D	20				20
F	21				21
G	22				22
H	23				23
J	24				24
K	25				25
L	26				26
;	27				27
'	28				28
`	29				29
Lshft	2A				
\	2B				

(continued on next page)

TABLE 7–3   *(continued)*

| Key | Scan Code | Extended ASCII code with.... | | | |
		Nothing	Shift	Control	Alternate
Z	2C				2C
X	2D				2D
C	2E				2E
V	2F				2F
B	30				30
N	31				31
M	32				32
,	33				33
.	34				34
/	35				35
Gray /	35			95	A4
Rshft	36				
PrtSc	E0 2A E0 37				
L alt	38				
R alt	38				
Space	39				
Caps	3A				
F1	3B	3B	54	5E	68
F2	3C	3C	55	5F	69
F3	3D	3D	56	60	6A
F4	3E	3E	57	61	6B
F5	3F	3F	58	62	6C
F6	40	40	59	63	6D
F7	41	41	5A	64	6E
F8	42	42	5B	65	6F
F9	43	43	5C	66	70
F10	44	44	5D	67	71
F11	57	85	87	89	8B
F12	58	86	88	8A	8C
Num	45				
Scroll	46				
Home	E0 47	47	47	77	97
Up	48	48	48	8D	98
Pgup	E0 49	49	49	84	99
Gray -	4A				
Left	4B	4B	4B	73	9B
Center	4C				
Right	4D	4D	4D	74	9D
Gray +	4E				
End	E0 4F	4F	4F	75	9F
Down	E0 50	50	50	91	A0
Pgdn	E0 51	51	51	76	A1
Ins	E0 52	52	52	92	A2
Del	E0 53	53	53	93	A3
Pause	E0 10 45				

There are three ways to read the keyboard. The first method reads a key and echoes (or displays) the key on the video screen. A second way simply tests to see if a key is pressed. If it is, it reads the key; otherwise it returns without any key. The third way allows an entire character string or line to be read from the keyboard.

Reading a Key with an Echo. Example 7–15 shows how a key is read from the keyboard and **echoed** (sent) back out to the video display by using a procedure called KEY. Although this method is the easiest way to read a key, it is also the most limited because it always echoes the character to the screen, even if it is an unwanted character. The DOS function number 01H also responds to the control-C key combination, and exits to DOS if it is typed.

EXAMPLE 7–15

```
0000                    KEY     PROC    FAR

0000  B4 01                     MOV     AH,1            ;function 01H
0002  CD 21                     INT     21H             ;read key
0004  0A C0                     OR      AL,AL           ;test for 00H, clear carry
0006  75 03                     JNZ     KEY1
0008  CD 21                     INT     21H             ;get extended
000A  F9                        STC                     ;indicate extended
000B            KEY1:
000B  CB                        RET

000C                    KEY     ENDP
```

To read and echo a character, the AH register is loaded with DOS function number 01H. This is followed by the INT 21H instruction, which calls a procedure that processes DOS function calls. Upon return from the INT 21H, the AL register contains the ASCII character typed; the video display also shows the typed character. If AL = 0 after the return, the INT 21H instruction must again be executed to obtain the extended ASCII-coded character (refer to Table 7–3). The procedure of Example 7–15 returns with carry set (1) to indicate an extended ASCII character and carry cleared (0) to indicate a normal ASCII character. When this procedure is called, the CALL instruction might be followed by a JC EXTENDED to process the extended ASCII character.

Reading a Key without an Echo. The best single character key-reading function is function number 06H. This function reads a key without an echo to the screen. It also allows extended ASCII characters and *does not* respond to the control-C key combination. This function uses AH for the function number (06H) and DL = 0FFH to indicate that the function call (INT 21H) will read the keyboard without an echo. I usually use DL = –1 instead of DL = 0FFH because it is easier to type and has the same value (because 0FFH = –1).

Example 7–16 shows a procedure that uses function number 06H to read the keyboard. This performs as shown in Example 7–15, except that no character is echoed to the video display.

EXAMPLE 7–16

```
000                     KEYS    PROC    FAR

0000  B4 06                     MOV     AH,6            ;function 06H
0002  B2 FF                     MOV     DL,0FFH
0004  CD 21                     INT     21H             ;read key
0006  74 F8                     JE      KEYS            ;if no key
0008  0A C0                     OR      AL,AL           ;test for 00H, clear carry
000A  75 03                     JNE     KEYS1
000C  CD 21                     INT     21H             ;get extended
000E  F9                        STC                     ;indicate extended
000F            KEYS1:
000F  CB                        RET

0010                    KEYS    ENDP
```

If you examine the procedure, there is one other difference. Function call number 06H returns from the INT 21H, even if no key is typed; function call 01H waits for a key to be typed. This is an important difference that should be noted. This feature allows software to perform other tasks between checking the keyboard for a character.

Read an Entire Line with an Echo.   Sometimes, it is advantageous to read an entire line of data with one function call. Function call number 0AH reads an entire line of information—up to 255 characters—from the keyboard. It continues to acquire keyboard data until either the enter key (0DH) is typed or the character count expires. This function requires that AH = 0AH, and DS:DX addresses the keyboard buffer (a memory area where the ASCII data are stored). The first byte of the buffer area must contain the maximum number of keyboard characters read by this function. If the number typed exceeds this maximum number, the function returns, just as if the enter key were typed. The second byte of the buffer contains the count of the actual number of characters typed, and the remaining locations in the buffer contain the ASCII keyboard data.

Example 7–17 shows how this function reads two lines of information into two memory buffers (BUF1 and BUF2). Before the call to the DOS function through the LINE procedure, the first byte of the buffer is loaded with a 255, so up to 255 characters can be typed. If you assemble and execute this program, the first and second lines are accepted. The only problem is that the second line appears on top of the first line. The next section of the text explains how to output characters to the video display to solve this problem.

EXAMPLE 7–17

```
                        ;A program that reads two lines of data from the keyboard
                        ;using DOS INT 21H function number 0AH.
                        ;***uses***
                        ;LINE procedure to read a line.
                        ;
                                .MODEL SMALL            ;select SMALL model
0000                            .DATA                   ;start DATA segment
0000  0101 [      BUF1    DB    257 DUP (?)             ;define BUF1
        00
            ]
0101  0101 [      BUF2    DB    257 DUP (?)             ;define BUF2
        00
            ]
0000                            .CODE                   ;start CODE segment
                                .STARTUP                ;start program
0017  C6 06 0000 R FF           MOV   BUF1,255          ;character count of 255
001C  BA 0000 R                 MOV   DX,OFFSET BUF1    ;address BUF1
001F  E8 000F                   CALL  LINE             ;read a line

0022  C6 06 0101 R FF           MOV   BUF2,255          ;character count of 255
0027  BA 0101 R                 MOV   DX,OFFSET BUF2    ;address BUF2
002A  E8 0004                   CALL  LINE             ;read a line
                                .EXIT                   ;exit to DOS
                        ;
                        ;The LINE procedure uses DOS INT 21H function 0AH to
                        ;read and echo an entire line from the keyboard.
                        ;***parameters***
                        ;DX must contain the data segment offset address of the
                        ;buffer. The first location in the buffer contains the
                        ;number of characters to be read for the line.
                        ;Upon return the second location in the buffer contains
                        ;the line length.
                        ;
0031              LINE    PROC NEAR

0031  B4 0A                     MOV   AH,0AH            ;select function 0AH
0033  CD 21                     INT   21H              ;access DOS
0035  C3                        RET                     ;return from procedure

0036              LINE    ENDP
                        END                             ;end of file
```

Writing to the Video Display with DOS Functions

With most programs, data must be displayed on the video display. Video data are displayed in a number of different ways with DOS function calls. We use function 02H or 06H for displaying one character at a time, or function 09H for displaying an entire string of characters. Because functions 02H and 06H are identical, we tend to use function 06H because it is also used to read a key and, as mentioned, does not respond to a control-C key combination.

Displaying One ASCII Character. Both DOS functions 02H and 06H are explained together because they are identical for displaying ASCII data. Example 7–18 shows how this function displays a carriage return (0DH) and a line feed (0AH). Here a macro sequence, called DISP (display), displays the carriage return and line feed. The combination of a carriage return and a line feed moves the cursor to the next line at the left margin of the video screen. This two-step process is used to correct the problem that occurred between the lines typed through the keyboard in Example 7–17.

EXAMPLE 7–18

```
                    ;A program that displays a carriage return and a line
                    ;feed using the DISP macro.
                    ;
                    .MODEL TINY               ;select TINY model
                    .CODE                     ;start CODE segment
                    DISP    MACRO A           ;;display A macro

                            MOV   AH,06H      ;;DOS function 06H
                            MOV   DL,A        ;;place parameter A in DL
                            INT   21H         ;;display parameter A

                            ENDM

                    .STARTUP                  ;start program

                            DISP 0DH          ;display carriage return

0100  B4 06    1             MOV   AH,06H
0102  B2 0D    1             MOV   DL,0DH
0104  CD 21    1             INT   21H

                            DISP 0AH          ;display line feed

0106  B4 06    1             MOV   AH,06H
0108  B2 0A    1             MOV   DL,0AH
010A  CD 21    1             INT   21H
                    .EXIT                     ;exit to DOS
                            END               ;end of file
```

Displaying a Character String. A character string is a series of ASCII-coded characters that end with a $ (24H) when used with DOS function call number 09H. Example 7–19 shows how a message is displayed at the current cursor position on the video display. Function call number 09H requires that DS:DX address the character string before executing the INT 21H instruction.

EXAMPLE 7–19

```
                            .MODEL SMALL      ;select SMALL model
0000                        .DATA             ;start DATA segment

0000  0D 0A 0A 54  MES      DB    13,10,10,'This is a test line.$'
      68 69 73 20
      69 73 20 61
      20 74 65 73
```

```
          74 20 6C 69
          6E 65 2E 24
0000                                .CODE                      ;start CODE segment
                                    .STARTUP                   ;start program

0017      B4 09                 MOV   AH,9                      ;select function 09H
0019      BA 0000 R             MOV   DX,OFFSET MES             ;address character string
001C      CD 21                 INT   21H                       ;access DOS

                                    .EXIT                      ;exit to DOS
                                    END                        ;end of file
```

This example program can be entered into the assembler, linked, and executed to produce *"This is a test line"* on the video display.

The .EXIT directive embodies the DOS function 4CH. As shown in Appendix A, DOS function 4CH terminates a program. The .EXIT directive inserts a series of two instructions in the program MOV AH,4CH, followed by an INT 21H instruction.

Using BIOS Video Function Calls

In addition to the DOS function call INT 21H, we also have video BIOS (basic I/O system) function calls at INT 10H. The DOS function calls allow a key to be read and a character to be displayed with ease, but the cursor is difficult to position at the desired screen location. The video BIOS function calls allow more control over the video display than the DOS function calls do. The video BIOS function calls also require less time to execute than the DOS function calls do. The DOS function calls do not allow cursor placement, while the video BIOS function calls do.

Cursor Position.   Before any information is placed on the video screen, the position of the cursor should be known. This allows the screen to be cleared and started at any desired location. Video BIOS function number 03H allows the cursor position to be read from the video interface. Video BIOS function number 02H allows the cursor to be placed at any screen position. Table 7–4 shows the contents of various registers for both functions 02H and 03H.

The page number in register BH should be 0 before setting the cursor position. Most software does not normally access the other pages (1–7) of the video display. The page number is often ignored after a cursor read. The 0 page is available in the CGA (color graphics adapter), EGA (enhanced graphics adapter), and VGA (variable graphics array) text modes of operation.

The cursor position assumes that the left-hand page column is column 0, progressing across a line to column 79. The row number corresponds to the character line number on the screen. Row 0 is the uppermost line, while row 24 is the last line on the screen. This assumes that the text mode selected for the video adapter is 80 characters per line by 25 lines. Other text modes are also available, such as 40×25 and 96×43.

Example 7–20 shows how the video BIOS function call INT 10H is used to clear the video screen. This is just one method of clearing the screen. Notice that the first function call positions the cursor to row 0 and column 0, which is called the **home position.** Next, we use the DOS function call to write 2000 (80 characters per line × 25 character lines) blank spaces (20H) on the video display. Finally, the cursor is again moved to the home position.

TABLE 7–4   Video BIOS function INT 10H.

AH	Description	Parameters
02H	Sets cursor position	DH = row, DL = column, and BH = page number
03H	Reads cursor position	DH = row, DL = column, and BH = page number

EXAMPLE 7–20

```
                        ;A program that clears the screen and homes the
                        ;cursor to the upper left-hand corner of the screen.
                        ;
                                .MODEL  TINY            ;select TINY model
0000                            .CODE                   ;start CODE segment
                HOME    MACRO                           ;;home cursor macro
                        MOV     AH,2                    ;;function 02H
                        MOV     BH,0                    ;;page 0
                        MOV     DX,0                    ;;row 0, line 0
                        INT     10H                     ;;home cursor
                        ENDM


                                .STARTUP                ;start program
                                HOME                    ;home cursor
0100  B4 02         1           MOV     AH,2
0102  B7 00         1           MOV     BH,0
0104  BA 0000       1           MOV     DX,0
0107  CD 10         1           INT     10H
0109  B9 07D0                   MOV     CX,25*80        ;load character count
010C  B4 06                     MOV     AH,6            ;select function 06H
010E  B2 20                     MOV     DL,' '          ;select a space
0110                MAIN1:
0110  CD 21                     INT     21H             ;display a space
0112  E2 FC                     LOOP    MAIN1           ;repeat 2000 times
                                HOME                    ;home cursor
0114  B4 02         1           MOV     AH,2
0116  B7 00         1           MOV     BH,0
0118  BA 0000       1           MOV     DX,0
011B  CD 10         1           INT     10H
                                .EXIT                   ;exit to DOS
                                END                     ;end of file
```

If this example is assembled, linked, and executed, a problem surfaces. This program is too slow to be useful in most cases. To correct this situation, another video BIOS function call is used. We can use the scroll function (06H) to clear the screen at a much higher speed.

Function 06H is used with a 00H in AL to blank the entire screen. This allows Example 7–20 to be rewritten so that the screen clears at a much higher speed. See Example 7–21 for a faster clear and home cursor program. Here, function call number 08H reads the character attributes for blanking the screen. Next, they are positioned in the correct registers and DX is loaded with the screen size, 4FH (79) and 19H (25). If this program is assembled, linked, executed, and compared with Example 7–20, there is a big difference in the speed at which the screen is cleared. (Make sure that the lines in the program that are macro expansion ending in a 1 are not typed into the program.) Please refer to Appendix A for other video BIOS INT 10H function calls that may prove useful in your applications. Also listed in Appendix A is a complete listing of all the INT functions available in most computers.

EXAMPLE 7–21

```
                        ;A program that clears the screen and homes the cursor.
                        ;
                                .MODEL  TINY            ;select TINY model
0000                            .CODE                   ;start code segment
                        HOME    MACRO                   ;;home cursor
                        MOV     AH,2
                        MOV     BH,0
                        MOV     DX,0
                        INT     10H
                        ENDM


                                .STARTUP                ;start program
```

```
0100   B7 00                     MOV   BH,0
0102   B4 08                     MOV   AH,8
0104   CD 10                     INT   10H                    ;read video attribute

0106   8A DF                     MOV   BL,BH                  ;load page number
0108   8A FC                     MOV   BH,AH
010A   B9 0000                   MOV   CX,0                   ;load attributes
010D   BA 194F                   MOV   DX,194FH               ;line 25, column 79
0110   B8 0600                   MOV   AX,600H                ;select scroll function
0113   CD 10                     INT   10H                    ;scroll screen
                                 HOME                         ;home cursor
0115   B4 02          1          MOV   AH,2
0117   B7 00          1          MOV   BH,0
0119   BA 0000        1          MOV   DX,0
011C   CD 10          1          INT   10H

                                 .EXIT                        ;exit to DOS
                                 END                          ;end program
```

Display Macro

One of the more usable macro sequences is the one illustrated in Example 7–22. Although it is simple and has been presented before, it saves much typing when creating programs that must display many individual characters. What makes this macro so useful is that a register can be specified as the argument, an ASCII character in quotes, or the numeric value for an ASCII character.

EXAMPLE 7–22

```
                                 ;A program that displays AB followed by a carriage
                                 ;return and line feed combination using the DISP macro.
                                 ;
                                         .MODEL TINY          ;select TINY model
                                         .CODE                ;start CODE segment
                                 DISP    MACRO VAR            ;;display VAR macro
                                         MOV   DL,VAR
                                         MOV   AH,6
                                         INT   21H
                                         ENDM
                                         .STARTUP             ;start program
                                         DISP  'A'            ;display 'A'
0100   B2 41          1                  MOV   DL,'A'
0102   B4 06          1                  MOV   AH,6
0104   CD 21          1                  INT   21H

0106   B0 42                             MOV   AL,'B'         ;load AL with 'B'
                                         DISP  AL             ;display 'B'
0008   8A D0          1                  MOV   DL,AL
000A   B4 06          1                  MOV   AH,6
000C   CD 21          1                  INT   21H

                                         DISP  13             ;display carriage return
000E   B2 0D          1                  MOV   DL,13
0010   B4 06          1                  MOV   AH,6
0012   CD 21          1                  INT   21H

                                         DISP  10             ;display line feed
0014   B2 0A          1                  MOV   DL,10
0016   B4 06          1                  MOV   AH,6
0018   CD 21          1                  INT   21H

                                         .EXIT                ;exit to DOS
                                         END                  ;end of file
```

The Mouse

The mouse pointing device is controlled with INT 33H. Refer to Appendix A for a list of the Microsoft-compatible mouse functions associated with INT 33H. Unlike with DOS INT 21H, the function number is selected through the AL register, and AH is usually set to 00H before the INT 33H is executed. There are a total of 50 mouse functions available, of which only the main functions are described in this section of the text.

Testing for a Mouse

To determine whether a mouse driver is installed in the system, test the contents of interrupt vector 33H. If interrupt vector 33H contains a 0000:0000, the mouse driver is not installed in the system. In some systems, a vector exists, even though no mouse driver is present. In this instance, the INT 33H vector address points to an IRET instruction (CFH). The interrupt vector address is retrieved by using the DOS INT 21H function 35H. The address is then tested for 0000:0000; if it contains another value, the contents of the address pointed to by interrupt vector 33H are tested for CFH. See Example 7–23 for a procedure that tests for the existence of the mouse driver.

Once it is determined that a mouse driver possibly exists, the mouse is reset to make certain it is connected to the system and functioning. The mouse reset is accomplished by using mouse function 00H. The return from function 00H is AX = 0000H if no mouse is present. The CHKM procedure returns if the mouse exists with carry cleared and if no mouse exists with carry set.

EXAMPLE 7–23

```
                        ;procedure that tests for a mouse driver
                        ;***Output parameters***
                        ;Carry = 1, if no mouse present
                        ;Carry = 0, if mouse is present
                        ;
0017                    CHKM    PROC    NEAR

0017  B8 3533                   MOV     AX,3533H        ;get INT 33H vector
001A  CD 33                     INT     21H             ;ES:BX
001C  8C C0                     MOV     AX,ES

                                .IF AX==0 && BX==0  ;test for 0000:0000
0026  F9                                STC           ;if no mouse driver
0027  C3                                RET
                                .ENDIF

                                .IF BYTE PTR ES:[BX]==0CFH
002E  F9                                STC           ;test for far return
002F  C3                                RET           ;if no mouse driver
                                .ENDIF

0030  B8 0000                   MOV     AX,0            ;reset mouse
0033  CD 33                     INT     33H

                                .IF AX==0             ;if no mouse
0039  F9                                STC
                                .ELSE
003C  F8                                CLC
                                .ENDIF
003D  C3                        RET

003E                    CHKM    ENDP
```

Which Mouse and Driver?

The mouse function interrupt determines both the type of mouse connected to the system and the driver version number. Example 7–24 lists a program that displays the mouse type and driver version number after a test is made to determine whether the mouse is present by using the procedure of Example 7–23. Here, mouse INT 33H, function 24H locates the mouse driver version number and mouse driver type. The return from function 24H leaves the mouse driver number in BX (BH=major and BL=minor) and the mouse type in CH. If the mouse driver version is 8.00, then BH = 08H and BL = 00H. The mouse types that are returned in register CH are currently bus = 1, serial = 2, InPort = 3, PS/2 = 4, and Hewlett-Packard = 5. As time passes, this list of mouse types may grow.

EXAMPLE 7–24

```
                              ;A program that displays the mouse driver version
                              ;number and the type of mouse installed.
                              ;
                                      .MODEL SMALL
0000                                  .DATA
0000  0D 0A 4E 6F 20 4D   MES1    DB    13,10,'No MOUSE/MOUSE DRIVER found.$'
      4F 55 53 45 2F 4D
      4F 55 53 45 20 44
      52 49 56 45 52 20
      66 6F 75 6E 64 2E
      24
001F  0D 0A 4D 6F 75 73   MES2    DB    13,10,'Mouse driver version '
      65 20 64 72 69 76
      65 72 20 76 65 72
      73 69 6F 6E 20
0036  20 20 20 20 20 20   M1      DB    '         ',13,10,'$'
      20 0D 0A 24
0040  004D R 0051 R       TYPES   DW    T1,T2,T3,T4,T5
 0058 R 005F R
      0064 R
004A  42 75 73 24         T1      DB    'Bus$'
004E  53 65 72 69 61 6C   T2      DB    'Serial$'
      24
0055  49 6E 50 6F 72 74   T3      DB    'InPort$'
      24
005C  50 53 2F 32 24      T4      DB    'PS/2$'
0061  48 50 24            T5      DB    'HP$'
0064  20 6D 6F 75 73 65   MES3    DB    ' mouse installed.',13,10,'$'
      20 69 6E 73 74 61
      6C 6C 65 64 2E 0D
      0A 24

0000                                  .CODE
                                      .STARTUP
0017  E8 0041                 CALL  CHKM                ;test for mouse
001A  73 05                   JNC   MAIN1               ;if mouse present
001C  BA 0000 R               MOV   DX,OFFSET MES1
001F  EB 32                   JMP   MAIN2               ;if no mouse
0021               MAIN1:
0021  B8 0024                 MOV   AX,24H
0024  CD 33                   INT   33H                 ;get driver version
0026  BF 0039 R               MOV   DI,OFFSET M1
0029  8A C7                   MOV   AL,BH               ;save major version
002B  E8 004D                 CALL  DISP
002E  C6 05 2E                MOV   BYTE PTR [DI],'.'  ;save period
0031  47                      INC   DI

0032  8A C3                   MOV   AL,BL               ;save minor version
0034  E8 0044                 CALL  DISP

0037  BA 0022 R               MOV   DX,OFFSET MES2      ;display version
003A  B4 09                   MOV   AH,9
003C  CD 21                   INT   21H

003E  BE 0043 R               MOV   SI,OFFSET TYPES     ;index type
0041  B4 00                   MOV   AH,0
0043  8A C5                   MOV   AL,CH
0045  48                      DEC   AX
0046  03 F0                   ADD   SI,AX
```

```
0048  03 F0                              ADD   SI,AX
004A  8B 14                              MOV   DX,[SI]                ;display type
004C  B4 09                              MOV   AH,9
004E  CD 21                              INT   21H
0050  BA 0067 R                          MOV   DX,OFFSET MES3
0053                          MAIN2:
0053  B4 09                              MOV   AH,9
0055  CD 21                              INT   21H
                                         .EXIT
                             ;A procedure that tests for the presence of a mouse.
                             ;***Output parameters***
                             ;Carry = 1, if no mouse present
                             ;Carry = 0, if mouse is present
                             ;
005B                          CHKM    PROC NEAR

005B  B8 3533                            MOV   AX,3533H               ;get INT 33H vector
005E  CD 21                              INT   21H                    ;vector in ES:BX

0060  8C C0                              MOV   AX,ES
0062  0B C3                              OR    AX,BX                  ;test for 0000:0000
0064  F9                                 STC
0065  74 13                              JZ    CHKM1                  ;if no mouse driver
0067  26: 80 3F CF                       CMP   BYTE PTR ES:[BX],0CFH
006B  F9                                 STC
006C  74 0C                              JE    CHKM1                  ;if no mouse driver
006E  B8 0000                            MOV   AX,0
0071  CD 33                              INT   33H                    ;reset mouse
0073  83 F8 00                           CMP   AX,0
0076  F9                                 STC
0077  74 01                              JZ    CHKM1                  ;if no mouse
0079  F8                                 CLC
007A                          CHKM1:
007A  C3                                 RET

007B                          CHKM    ENDP
                             ;
                             ;save the ASCII coded version number
                             ;***input parameters***
                             ;AL = version
                             ;DS:DI = address where stored
                             ;***output parameters***
                             ;ASCII version number stored at DS:DI
                             ;
007B                          DISP    PROC NEAR

007B  B4 00                              MOV   AH,0
007D  D4 0A                              AAM                          ;convert to BCD
007F  05 3030                            ADD   AX,3030H
0082  80 FC 30                           CMP   AH,30H                 ;save ASCII version
0085  74 03                              JE    DISP1                  ;suppress zero
0087  88 25                              MOV   [DI],AH
0089  47                                 INC   DI
008A                          DISP1:
008A  88 05                              MOV   [DI],AL
008C  47                                 INC   DI
008D  C3                                 RET

008E                          DISP    ENDP
                                         END
```

Using the Mouse

The mouse functions in either text mode or in graphics mode. This section illustrates how to enable the mouse for use with a text mode program. The mouse also functions in graphic mode, but instead of displaying as a block, the cursor or mouse pointer is displayed as an arrow. As with the prior examples, the first step is to check for the presence of a mouse driver. Example 7–25 uses the CHKM procedure to test for the presence of the mouse. If no mouse is present, a return from TM_ON occurs with the carry flag set. If the mouse is present, the cursor is displayed and a return the carry flag cleared is made.

EXAMPLE 7–25

```
                        ;The TM_ON procedure tests for the presence of a mouse
                        ;and enables mouse pointer.
                        ;uses the CHKM (check for mouse) procedure
                        ;
                        ;***output parameters***
                        ;Carry = 0, if mouse is present pointer enabled
                        ;Carry = 1, if no mouse present
                        ;
0000                    TM_ON    PROC NEAR

0000  E8 FFDD                    CALL CHKM           ;test for mouse
0003  72 06                      JC   TM_ON1         ;if no mouse
0005  B8 0001                    MOV  AX,1           ;show mouse pointer
0008  CD 33                      INT  33H
000A  F8                         CLC                 ;show mouse present
000B                    TM_ON1:
000B  C3                         RET

000C                    TM_ON    ENDP
```

The procedure of Example 7–25 only enables the mouse and displays the mouse cursor. To use the mouse, a program must be written that tracks the mouse and its position. Such a program appears in Example 7–26.

EXAMPLE 7–26

```
                        ;a program that displays the mouse pointer and its
                        ;X and Y position.
                        ;
                                 .MODEL SMALL
0000                             .DATA
0000  0D 58 20 50 6F 73   MES    DB   13,'X Position= '
      69 74 69 6F 6E 3D
      20
000D  20 20 20 20 20 20   MX     DB   '      '
0013  59 20 50 6F 73 69          DB   'Y Position= '
      74 69 6F 6E 3D 20
001F  20 20 20 20 20 20   MY     DB   '      $'
      24
0026  0000               X       DW   ?              ;X position
0028  0000               Y       DW   ?              ;Y position
0000                             .CODE
                                 .STARTUP
0017  E8 006D                    CALL TM_ON          ;enable mouse
001A  72 47                      JC   MAIN4          ;if no mouse
001C                    MAIN1:
001C  B8 0003                    MOV  AX,3           ;get mouse status
001F  CD 33                      INT  33H
0021  83 FB 01                   CMP  BX,1
0024  74 38                      JE   MAIN3          ;if left button

0026  3B 0E 0026 R               CMP  CX,X
002A  75 06                      JNE  MAIN2          ;if X changed
002C  3B 16 0028 R               CMP  DX,Y
0030  74 EA                      JE   MAIN1          ;if Y did not change
0032                    MAIN2:
0032  89 0E 0026 R               MOV  X,CX           ;save new position
0036  89 16 0028 R               MOV  Y,DX
```

```
003A  BF 000D R              MOV   DI,OFFSET MX
003D  8B C1                  MOV   AX,CX
003F  E8 0051                CALL  PLACE              ;store ASCII X
0042  BF 001F R              MOV   DI,OFFSET MY
0045  A1 0028 R              MOV   AX,Y
0048  E8 0048                CALL  PLACE              ;store ASCII Y

004B  B8 0002                MOV   AX,2
004E  CD 33                  INT   33H                ;hide mouse pointer

0050  B4 09                  MOV   AH,9
0052  BA 0000 R              MOV   DX,OFFSET MES
0055  CD 21                  INT   21H                ;display position

0057  B8 0001                MOV   AX,1
005A  CD 33                  INT   33H                ;show mouse pointer

005C  EB BE                  JMP   MAIN1              ;do again
005E                 MAIN3:
005E  B8 0000                MOV   AX,0               ;reset mouse
0061  CD 33                  INT   33H

0063                 MAIN4:
                             .EXIT
                     ;
                     ;A procedure that tests for the presence of a mouse.
                     ;***Output parameters***
                     ;Carry = 1, if no mouse present
                     ;Carry = 0, if mouse is present
                     ;
0067                 CHKM    PROC NEAR

0067  B8 3533                MOV   AX,3533H           ;get INT 33H vector
006A  CD 21                  INT   21H                ;vector in ES:BX

006C  8C C0                  MOV   AX,ES
006E  0B C3                  OR    AX,BX              ;test for 0000:0000
0070  F9                     STC
0071  74 13                  JZ    CHKM1              ;if no mouse driver
0073  26: 80 3F CF           CMP   BYTE PTR ES:[BX],0CFH
0077  F9                     STC
0078  74 0C                  JE    CHKM1              ;if no mouse driver
007A  B8 0000                MOV   AX,0
007D  CD 33                  INT   33H                ;reset mouse
007F  83 F8 00               CMP   AX,0
0082  F9                     STC
0083  74 01                  JZ    CHKM1              ;if no mouse
0085  F8                     CLC
0086                 CHKM1:
0086  C3                     RET

0087                 CHKM    ENDP
                     ;
                     ;The TM_ON procedure tests for the presence of a
                     ;mouse and enables mouse pointer.
                     ;uses the CHKM (check for mouse) procedure
                     ;
                     ;***output parameters***
                     ;Carry = 0, if mouse is present pointer enabled
                     ;Carry = 1, if no mouse present
                     ;
0087                 TM_ON   PROC NEAR

0087  E8 FFDD                CALL  CHKM               ;test for mouse
008A  72 06                  JC    TM_ON1
008C  B8 0001                MOV   AX,1               ;show mouse pointer
008F  CD 33                  INT   33H
0091  F8                     CLC
0092                 TM_ON1:
0092  C3                     RET
```

```
0093                            TM_ON    ENDP
                                ;
                                ;The PLACE procedure converts the contents of AX
                                ;into a decimal ASCII-coded number stored at the
                                ;memory location addressed by DS:DI.
                                ;***input parameters***
                                ;AX = number to be converted to decimal ASCII code
                                ;DS:DI = address where number is stored
                                ;
0093                            PLACE    PROC NEAR

0093  B9 0000                            MOV   CX,0             ;clear count
0096  BB 000A                            MOV   BX,10            ;set divisor
0099                            PLACE1:
0099  BA 0000                            MOV   DX,0             ;clear DX
009C  F7 F3                              DIV   BX               ;divide by 10
009E  52                                 PUSH  DX
009F  41                                 INC   CX
00A0  83 F8 00                           CMP   AX,0
00A3  75 F4                              JNE   PLACE1           ;if quotient != 0
00A5                            PLACE2:
00A5  BB 0005                            MOV   BX,5
00A8  2B D9                              SUB   BX,CX
00AA                            PLACE3:
00AA  5A                                 POP   DX
00AB  80 C2 30                           ADD   DL,30H           ;convert to ASCII
00AE  88 15                              MOV   [DI],DL          ;store digit
00B0  47                                 INC   DI
00B1  E2 F7                              LOOP  PLACE3
00B3  83 FB 00                           CMP   BX,0
00B6  74 08                              JE    PLACE5
00B8  8B CB                              MOV   CX,BX
00BA                            PLACE4:
00BA  C6 05 20                           MOV   BYTE PTR [DI],20H
00BD  47                                 INC   DI
00BE  E2 FA                              LOOP  PLACE4
00C0                            PLACE5:
00C0  C3                                 RET

00C1                            PLACE    ENDP
                                         END
```

The program in Example 7–26 displays the mouse cursor by placing a 0001H into AX, followed by the INT 33H instruction. Next, the status of the mouse is read with function AX = 0003H. The status function returns with the status of the mouse buttons in BX, the X coordinate of the mouse pointer in CX, and the Y coordinate in DX. (Refer to Appendix A for more complete information on the status for the mouse.) In this example, the program terminates if the left mouse button is pressed; otherwise, the coordinates are compared with the prior values saved in X and Y. If a change has occurred in these coordinates, the new coordinates are calculated and displayed. Notice that before the video display is accessed, the mouse pointer is hidden by using INT 33H with AX = 0002H. This is very important. If you don't hide the mouse pointer, the display will become unstable and the computer may even reboot. In most cases, a copy of the mouse pointer remains on the screen if data are displayed without turning the mouse pointer off.

7–3 DATA CONVERSIONS

In computer systems, data are seldom in the correct form. One main task of the system is to convert data from one form to another. This section of the chapter describes conversions between binary and ASCII. Binary data are removed from a register or memory and converted to ASCII for the video display. In many cases, ASCII data are converted to binary as they are typed on the keyboard. We also explain converting between ASCII and hexadecimal data.

Converting from Binary to ASCII

Conversion from binary to ASCII is accomplished in two ways: (1) by the AAM instruction if the number is less than 100, or (2) by a series of decimal divisions (divide by 10). Both techniques are presented in this section.

The AAM instruction converts the value in AX into a two-digit unpacked BCD number in AX. If the number in AX is 0062H (98 decimal) before AAM executes, AX contains a 0908H after AAM executes. This is not ASCII code, but it is converted to ASCII code by adding a 3030H to AX. Example 7–27 illustrates a program that uses the procedure DISP, which processes the binary value in AL (0–99) and displays it on the video screen as decimal. The DISP procedure blanks a leading zero, which occurs for the numbers 0–9, with an ASCII space code. This example program displays the number 74 (test data) on the video screen.

EXAMPLE 7–27

```
                        ;A program that uses the DISP procedure to display 74
                        ;decimal on the video display.
                        ;
                                .MODEL TINY             ;select TINY mode
0000                            .CODE                   ;start code segment
                                .STARTUP                ;start program
0100    B0 4A                   MOV   AL,4AH            ;load test data to AL
0102    E8 0004                 CALL DISP               ;display AL in decimal
                                .EXIT                   ;exit to DOS

                        ;
                        ;The DISP procedure displays AL (0 to 99) as a decimal
                        ;number. AX is destroyed by this procedure.
                        ;
0109                    DISP    PROC NEAR

0109    52                      PUSH DX                ;save DX
010A    B4 00                   MOV   AH,0             ;clear AH
010C    D4 0A                   AAM                    ;convert to BCD
010E    80 C4 20                ADD   AH,20H
0111    80 FC 20                CMP   AH,20H           ;test for leading zero
0114    74 03                   JE    DISP1            ;if leading zero
0116    80 C4 10                ADD   AH,10H           ;convert to ASCII
0119                    DISP1:
0119    8A D4                   MOV   DL,AH            ;display first digit
011B    B4 06                   MOV   AH,6
011D    50                      PUSH AX
011E    CD 21                   INT   21H
0120    58                      POP   AX
0121    8A D0                   MOV   DL,AL
0123    80 C2 30                ADD   DL,30H           ;convert second digit to ASCII
0126    CD 21                   INT   21H              ;display second digit
0128    5A                      POP   DX               ;restore DX
0129    C3                      RET

012A                    DISP    ENDP
                                END                     ;end of file
```

The reason that AAM converts any number between 0 and 99 to a two-digit unpacked BCD number is because it divides AX by 10. The result is left in AX so AH contains the quotient and AL the remainder. This same scheme of dividing by 10 can be expanded to convert any whole number of any number system from binary to an ASCII-coded character string that can be displayed on the video screen. For example, if AX is divided by 8 instead of 10, the number is displayed in octal.

The algorithm for converting from binary to ASCII code is:

1. Divide by the 10, then save the remainder on the stack as a significant BCD digit.
2. Repeat step 1 until the quotient is a 0.
3. Retrieve each remainder and add a 30H to convert to ASCII before displaying or printing.

Example 7–28 shows how the unsigned 16-bit content of AX is converted to ASCII and displayed on the video screen. Here, we divide AX by 10 and save the remainder on the stack after each division for later conversion to ASCII. After all the digits have been converted, the result is displayed on the video screen by removing the remainders from the stack and converting them to ASCII code. This procedure (DISPX) also blanks any leading zeros that occur.

EXAMPLE 7–28

```
                        ;A program that uses DISPX to display AX in decimal.
                        ;
                                .MODEL TINY             ;select TINY model
0000                            .CODE                   ;start CODE segment
                                .STARTUP                ;start program
0100    B8 04A3                 MOV   AX,4A3H           ;load AX with test data
0103    E8 0004                 CALL  DISPX             ;display AX in decimal
                                .EXIT                   ;exit to DOS
                        ;
                        ;The DISPX procedure displays AX in decimal.
                        ;AX is destroyed.
                        ;
010A                    DISPX   PROC NEAR

010A    52                      PUSH  DX                ;save DX, CX, and BX
010B    51                      PUSH  CX
010C    53                      PUSH  BX
010D    B9 0000                 MOV   CX,0              ;clear digit counter
0110    BB 000A                 MOV   BX,10             ;set for decimal
0113                    DISPX1:
0113    BA 0000                 MOV   DX,0              ;clear DX
0116    F7 F3                   DIV   BX                ;divide DX:AX by 10
0118    52                      PUSH  DX                ;save remainder
0119    41                      INC   CX                ;count remainder
011A    0B C0                   OR    AX,AX             ;test for quotient of zero
011C    75 F5                   JNZ   DISPX1            ;if quotient is not zero
011E                    DISPX2:
011E    5A                      POP   DX                ;get remainder
011F    B4 06                   MOV   AH,6              ;select function 06H
0121    80 C2 30                ADD   DL,30H            ;convert to ASCII
0124    CD 21                   INT   21H               ;display digit
0126    E2 F6                   LOOP  DISPX2            ;repeat for all digits

0128    5B                      POP   BX                ;restore BX, CX, and DX
0129    59                      POP   CX
012A    5A                      POP   DX
012B    C3                      RET

012C                    DISPX   ENDP
                                END                     ;end of file
```

Converting from ASCII to Binary

Conversions from ASCII to binary usually start with keyboard entry. If a single key is typed, the conversion occurs when a 30H is subtracted from the number. If more than one key is typed, conversion from ASCII to binary still requires 30H to be subtracted, but there is one additional step. After subtracting 30H, the number is added to the result after the prior result is first multiplied by 10.

The algorithm for converting from ASCII to binary is:

1. Begin with a binary result of 0.
2. Subtract 30H from the character typed on the keyboard to convert it to BCD.
3. Multiply the result by 10, and then add the new BCD digit.
4. Repeat steps 2 and 3 until the character typed is not an ASCII-coded number.

Example 7–29 illustrates a procedure (READN) used in a program that implements this algorithm. Here, the binary number returns in the AX register as a 16-bit result, which is then stored in memory location TEMP. If a larger result is required, the procedure must be reworked for a 32-bit addition. Each time this procedure is called, it reads a number from the keyboard until any key other than 0 through 9 is typed.

EXAMPLE 7–29

```
                        ;A program that reads one decimal number from the
                        ;keyboard and stores the binary value at TEMP.
                        ;
                                .MODEL SMALL            ;select TINY model
0000                            .DATA                   ;start DATA segment
0000    0000            TEMP    DW    ?                 ;define TEMP
0000                            .CODE                   ;start CODE segment
                                .STARTUP                ;start program
0017    E8 0007                 CALL READN              ;read a number
001A    A3 0000 R               MOV  TEMP,AX            ;save it in TEMP
                                .EXIT                   ;exit to DOS
                        ;
                        ;The READN procedure reads a decimal number from the
                        ;keyboard and returns its binary value in AX.
                        ;
0021                    READN   PROC NEAR

0021    53                      PUSH BX                 ;save BX and CX
0022    51                      PUSH CX
0023    B9 000A                 MOV  CX,10              ;load 10 for decimal
0026    BB 0000                 MOV  BX,0               ;clear result
0029                    READN1:
0029    B4 01                   MOV  AH,1               ;read key with echo
002B    CD 21                   INT  21H

002D    3C 30                   CMP  AL,'0'
002F    72 14                   JB   READN2             ;if below '0'
0031    3C 39                   CMP  AL,'9'
0033    77 10                   JA   READN2             ;if above '9'

0035    2C 30                   SUB  AL,'0'             ;convert to ASCII

0037    50                      PUSH AX                 ;save digit
0038    8B C3                   MOV  AX,BX              ;multiply result by 10
003A    F7 E1                   MUL  CX
003C    8B D8                   MOV  BX,AX
003E    58                      POP  AX
003F    B4 00                   MOV  AH,0
0041    03 D8                   ADD  BX,AX              ;add digit value to result
0043    EB E4                   JMP  READN1             ;repeat
0045                    READN2:
0045    8B C3                   MOV  AX,BX              ;get binary result into AX
0047    59                      POP  CX                 ;restore CX and BX
0048    5B                      POP  BX
0049    C3                      RET

004A                    READN   ENDP
                                END                     ;end of file
```

Displaying and Reading Hexadecimal

Hexadecimal data are easier to read from the keyboard and display than decimal data. These types of data are not used at the applications level, but at the system level. System-level data are often hexadecimal, and must either be displayed in hexadecimal form or read from the keyboard as hexadecimal data.

Reading Hexadecimal Data.   Hexadecimal data appear as 0 to 9 and A to F. The ASCII codes obtained from the keyboard for hexadecimal data are 30H to 39H for the numbers 0 through 9, and 41H to 46H (A–F) or 61H to 66H (a–f) for the letters. To be useful, a procedure that reads hexadecimal data must be able to accept both lowercase and uppercase letters.

Example 7–30 shows two procedures: one (CONV) converts the contents of the data in AL from ASCII code to a single hexadecimal digit, and the other (READH) reads a four-digit hexadecimal number from the keyboard and returns with it in register AX. This procedure can be modified to read any-sized hexadecimal number from the keyboard.

EXAMPLE 7–30

```
                              ;A program that reads a 4-digit hexadecimal number from
                              ;the keyboard and stores the result in word-sized
                              ;memory location TEMP.
                              ;
                                      .MODEL SMALL            ;select SMALL model
0000                                  .DATA                   ;start DATA segment
0000  0000              TEMP    DW    ?                       ;define TEMP
0000                                  .CODE                   ;start CODE segment
                                      .STARTUP                ;start program
0017  E8 0007                         CALL READH              ;read hexadecimal number
001A  A3 0000 R                       MOV  TEMP,AX            ;save it at TEMP
                                      .EXIT                   ;exit to DOS
                              ;
                              ;The READH procedure that reads a 4-digit hexadecimal
                              ;number from the keyboard and returns it in AX.
                              ;This procedure does next check for errors and uses CONV.
                              ;
0021                    READH   PROC NEAR

0021  51                              PUSH CX                 ;save BX and CX
0022  53                              PUSH BX
0023  B9 0004                         MOV  CX,4               ;load CX and SI with 4
0026  8B F1                           MOV  SI,CX
0028  BB 0000                         MOV  BX,0               ;clear result
002B                    READH1:
002B  B4 01                           MOV  AH,1               ;read a key with echo
002D  CD 21                           INT  21H
002F  E8 000A                         CALL CONV               ;convert to binary
0032  D3 E3                           SHL  BX,CL
0034  02 D8                           ADD  BL,AL              ;form result in BX
0036  4E                              DEC  SI
0037  75 F2                           JNZ  READH1             ;repeat 4 times
0039  8B C3                           MOV  AX,BX              ;move result to AX
003B  5B                              POP  BX                 ;restore BX and CX
003C  59                              POP  CX
003D  C3                              RET
003E                    READH   ENDP
                              ;
                              ;The CONV procedure converts AL into hexadecimal.
                              ;
003E                    CONV    PROC NEAR

003E  3C 39                           CMP  AL,'9'
0040  76 08                           JBE  CONV2              ;if 0 through 9
0042  3C 61                           CMP  AL,'a'
0044  72 02                           JB   CONV1              ;if uppercase A through F
0046  2C 20                           SUB  AL,20H             ;convert to uppercase
0048                    CONV1:
0048  2C 07                           SUB  AL,7
004A                    CONV2:
004A  2C 30                           SUB  AL,30H
004C  C3                              RET
004D                    CONV    ENDP
                                      END                     ;end of file
```

Displaying Hexadecimal Data. To display hexadecimal data, a number must be divided into four-bit segments that are converted into hexadecimal digits. Conversion is accomplished by adding a 30H to the numbers 0 to 9 and a 37H to the letters A to F.

A procedure (DSIPH) that displays the contents of the AX register on the video display appears in the program of Example 7–31. Here, the number is rotated left so that the leftmost digit is displayed first. Because AX contains a four-digit hexadecimal number, the procedure displays four hexadecimal digits.

EXAMPLE 7–31

```
                     ;A program that displays the hexadecimal value in AX.
                     ;This program uses DISPH to display a 4-digit value.
                     ;
                             .MODEL TINY          ;select TINY model
0000                         .CODE                ;start CODE segment
                             .STARTUP             ;start program
0100  B8 0ABC                MOV   AX,0ABCH       ;load AX with test data
0103  E8 0004                CALL  DISPH          ;display AX in hexadecimal
                             .EXIT                ;exit to DOS
                     ;
                     ;The DISPH procedure displays AX as a 4-digit hex number.
                     ;
010A                 DISPH   PROC  NEAR

010A  53                     PUSH  BX             ;save BX and CX
010B  51                     PUSH  CX
010C  B1 04                  MOV   CL,4           ;load rotate count
010E  B5 04                  MOV   CH,4           ;load digit count
0110                 DISPH1:
0110  D3 C0                  ROL   AX,CL          ;position digit
0112  50                     PUSH  AX
0113  24 0F                  AND   AL,0FH         ;convert it to ASCII
0115  04 30                  ADD   AL,30H
0117  3C 39                  CMP   AL,'9'
0119  76 02                  JBE   DISPH2
011B  04 07                  ADD   AL,7
011D                 DISPH2:
011D  B4 02                  MOV   AH,2           ;display hexadecimal digit
011F  8A D0                  MOV   DL,AL
0121  CD 21                  INT   21H
0123  58                     POP   AX
0124  FE CD                  DEC   CH
0126  75 E8                  JNZ   DISPH1         ;repeat for 4 digits
0128  59                     POP   CX             ;restore registers
0129  5B                     POP   BX
012A  C3                     RET

012B                 DISPH   ENDP
                             END                  ;end of file
```

Using Lookup Tables for Data Conversions

Lookup tables are often used to convert data from one form to another. A lookup table is formed in the memory as a list of data that is referenced by a procedure to perform conversions. In the case of many lookup tables, the XLAT instruction can often be used to look up data in a table, provided that the table contains eight-bit wide data and its length is less than or equal to 256 bytes.

Converting from BCD to 7-segment Code. One simple application that uses a lookup table is BCD to 7-segment code conversion. Example 7–32 illustrates a lookup table that contains the 7-segment codes for the numbers 0 to 9. These codes are used with the 7-segment display pictured in Figure 7–1. This 7-segment display uses active high (logic 1) inputs to light a segment.

FIGURE 7–1 The
7-segment display.

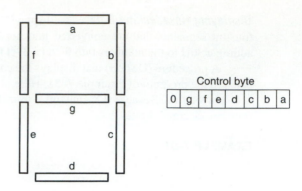

The code is arranged so that the **a** segment is in bit position 0 and the **g** segment is in bit position 6. Bit position 7 is 0 in this example, but it can be used for displaying a decimal point.

EXAMPLE 7–32

```
0000                    SEG7    PROC    FAR

0000    53                      PUSH    BX
0001    BB 0008 R               MOV     BX,OFFSET TABLE
0004    2E: D7                  XLAT    CS:TABLE        ;see text
0006    5B                      POP     BX
0007    CB                      RET

0008    3F              TABLE   DB      3FH             ;0
0009    06                      DB      6               ;1
000A    5B                      DB      5BH             ;2
000B    4F                      DB      4FH             ;3
000C    66                      DB      66H             ;4
000D    6D                      DB      6DH             ;5
000E    7D                      DB      7DH             ;6
000F    07                      DB      7               ;7
0010    7F                      DB      7FH             ;8
0011    6F                      DB      6FH             ;9

0012                    SEG7    ENDP
```

The procedure that performs the conversion contains only two instructions and assumes that AL contains the BCD digit to be converted to 7-segment code. One of the instructions addresses the lookup table by loading its address into BX, and the other performs the conversion and returns the 7-segment code in AL.

Because the lookup table is located in the code segment and the XLAT instruction accesses the data segment by default, the XLAT instruction includes a segment override. Notice that a dummy operand (TABLE) is added to the XLAT instruction so that the (CS:) code segment override prefix can be added to the instruction. Normally, XLAT does not contain an operand unless its default segment must be overridden. The LODS and MOVS instructions are also overridden in the same manner as XLAT by using a dummy operand.

Using a Lookup Table to Access ASCII Data.   Some programming techniques require that numeric codes be converted to ASCII character strings. For example, suppose that you need to display the days of the week for a calendar program. Because the number of ASCII characters in each day is different, some type of lookup table must be used to reference the ASCII-coded days of the week.

The program in Example 7–33 shows a table that references ASCII-coded character strings located in the code segment. Each character string contains an ASCII-coded day of the week. The table references each day of the week. The procedure that accesses the day of the week uses the AL register, and the numbers 0 to 6 to refer to Sunday through Saturday. If AL contains a 2 when this procedure is called, the word *"Tuesday"* is displayed on the video screen.

EXAMPLE 7–33

```
                                ;A program that displays the current day of the
                                ;week by using the system clock/calendar.
                                ;
                                   .MODEL SMALL          ;select SMALL model
0000                               .DATA                 ;start DATA segment
0000   000E R 0015 R     DTAB      DW    SUN,MON,TUE,WED,THU,FRI,SAT
       001C R 0024 R
       002E R 0037 R
       003E R
000E   53 75 6E 64 61 79 SUN       DB    'Sunday$'
       24
0015   4D 6F 6E 64 61 79 MON       DB    'Monday$'
       24
001C   54 75 65 73 64 61 TUE       DB    'Tuesday$'
       79 24
0024   57 65 64 6E 65 73 WED       DB    'Wednesday$'
       64 61 79 24
002E   54 68 75 72 73 64 THU       DB    'Thursday$'
       61 79 24
0037   46 72 69 64 61 79 FRI       DB    'Friday$'
       24
003E   53 61 74 75 72 64 SAT       DB    'Saturday$'
       61 79 24
0000                               .CODE                 ;start CODE segment
                                   .STARTUP              ;start program
0017   B4 2A                       MOV   AH,2AH          ;get day of week
0019   CD 21                       INT   21H             ;access DOS
001B   E8 0004                     CALL  DAYS            ;display day of week
                                   .EXIT                 ;exit to DOS

0022                     DAYS      PROC  NEAR

0022   52                          PUSH  DX              ;save DX and SI
0023   56                          PUSH  SI
0024   BE 0000 R                   MOV   SI,OFFSET DTAB  ;address table
0027   B4 00                       MOV   AH,0            ;find day of week
0029   03 C0                       ADD   AX,AX
002B   03 F0                       ADD   SI,AX
002D   8B 14                       MOV   DX,[SI]         ;get day of week
002F   B4 09                       MOV   AH,9            ;display string
0031   CD 21                       INT   21H
0033   5E                          POP   SI              ;restore registers
0034   5A                          POP   DX
0035   C3                          RET

0036                     DAYS      ENDP
                                   END                   ;end of file
```

This procedure first addresses the table by loading its address into the SI register. Next, the number in AL is converted into a 16-bit number and doubled because the table contains two bytes for each entry. This index is then added to SI to address the correct entry in the lookup table. The address of the ASCII character string is now loaded into DX by the MOV DX,CS:[SI] instruction.

Before the INT 21H DOS function is called, the DS register is placed on the stack and loaded with the segment address of CS. This allows DOS function number 09H (display a string) to be used to display the day of the week. This procedure converts the numbers 0 to 6 to the days of the week.

An Example Program Using Data Conversions

A program example is required to combine some of the data-conversion DOS functions. Suppose that you must display the time and date on the video screen. This example program (see Example 7–34) displays the time as 10:45 A.M. and the date as Tuesday, May 14, 2002. The program is short because it calls a procedure that displays the time and a second procedure that displays the date.

The time is available from DOS, using an INT 21H function call number 2CH. This returns with the hours in CH and minutes in CL. Also available are seconds in DH and hundredths of seconds in DL. The date is available by using INT 21H function call number 2AH. This leaves the day of the week in AL, the year in CX, the day of the month in DH, and the month in DL.

EXAMPLE 7–34

```
                                  ;A program that displays the time and date in the
                                  ;form:  10:45 A.M.,  Tuesday May 14, 2002.
                                  ;
                                          .MODEL SMALL          ;select SMALL model
                                          .NOLISTMACRO          ;don't expand macros
0000                                      .DATA                 ;start CODE segment
0000    0026 R 002F R      DTAB    DW      SUN,MON,TUE,WED,THU,FRI,SAT
        0038 R 0042 R
        004E R 0059 R
        0062 R
000E    006D R 0076 R      MTAB    DW      JAN,FEB,MAR,APR,MAY,JUN
        0080 R 0087 R
        008E R 0093 R
001A    0099 R 009F R              DW      JUL,AUG,SEP,OCT,NOV,DCE
        00A7 R 00B2 R
        00BB R 00C5 R
0026    53 75 6E 64 61 79  SUN     DB      'Sunday, $'
        2C 20 24
002F    4D 6F 6E 64 61 79  MON     DB      'Monday, $'
        2C 20 24
0038    54 75 65 73 64 61  TUE     DB      'Tuesday, $'
        79 2C 20 24
0042    57 65 64 6E 65 73  WED     DB      'Wednesday, $'
        64 61 79 2C 20 24
004E    54 68 75 72 73 64  THU     DB      'Thursday, $'
        61 79 2C 20 24
0059    46 72 69 64 61 79  FRI     DB      'Friday, $'
        2C 20 24
0062    53 61 74 75 72 64  SAT     DB      'Saturday, $'
        61 79 2C 20 24
006D    4A 61 6E 75 61 72  JAN     DB      'January $'
        79 20 24
0076    46 65 62 72 75 61  FEB     DB      'February $'
        72 79 20 24
0080    4D 61 72 63 68 20  MAR     DB      'March $'
        24
0087    41 70 72 69 6C 20  APR     DB      'April $'
        24
008E    4D 61 79 20 24     MAY     DB      'May $'
0093    4A 75 6E 65 20 24  JUN     DB      'June $'
0099    4A 75 6C 79 20 24  JUL     DB      'July $'
009F    41 75 67 75 73 74  AUG     DB      'August $'
        20 24
00A7    53 65 70 74 65 6D  SEP     DB      'September $'
        62 65 72 20 24
00B2    4F 63 74 6F 62 65  OCT     DB      'October $'
        72 20 24
00BB    4E 6F 76 65 6D 62  NOV     DB      'November $'
        65 72 20 24
```

```
00C5  44 65 63 65 6D 62    DCE       DB        'December $'
      65 72 20 24
0000                                 .CODE                      ;start CODE segment
                           DISP      MACRO  CHAR
                                     PUSH AX                    ;;save AX and DX
                                     PUSH DX
                                     MOV    DL,CHAR             ;;display character
                                     MOV    AH,2
                                     INT    21H
                                     POP    DX                  ;;restore AX and DX
                                     POP    AX
                                     ENDM
                                     .STARTUP                   ;start program
0017  E8 0007                        CALL TIMES                 ;display time
001A  E8 00A3                        CALL DATES                 ;display date
                                     .EXIT                      ;exit to DOS

0021                       TIMES     PROC NEAR

0021  B4 2C                          MOV    AH,2CH              ;get time from DOS
0023  CD 21                          INT    21H
0025  B7 41                          MOV    BH,'A'              ;set 'A' for AM
0027  80 FD 0C                       CMP    CH,12
002A  72 05                          JB     TIMES1              ;if below 12:00 noon
002C  B7 50                          MOV    BH,'P'              ;set 'P' for PM
002E  80 ED 0C                       SUB    CH,12               ;adjust to 12 hours
0031                       TIMES1:
0031  0A ED                          OR     CH,CH               ;test for 0 hour
0033  75 02                          JNE    TIMES2              ;if not 0 hour
0035  B5 0C                          MOV    CH,12               ;change 0 hour to 12
0037                       TIMES2:
0037  8A C5                          MOV    AL,CH
0039  B4 00                          MOV    AH,0
003B  D4 0A                          AAM                        ;convert hours
003D  0A E4                          OR     AH,AH
003F  74 0D                          JZ     TIMES3              ;if no tens of hours
0041  80 C4 30                       ADD    AH,'0'              ;convert tens
                                     DISP AH                    ;display tens
004E                       TIMES3:
004E  04 30                          ADD    AL,'0'              ;convert units
                                     DISP AL                    ;display units
                                     DISP ':'                   ;display colon
0064  8A C1                          MOV    AL,CL
0066  B4 00                          MOV    AH,0
0068  D4 0A                          AAM                        ;convert minutes
006A  05 3030                        ADD    AX,3030H
006D  50                             PUSH AX
                                     DISP AH                    ;display tens
0078  58                             POP    AX
                                     DISP AL                    ;display units
                                     DISP ' '                   ;display space
                                     DISP BH                    ;display 'A' or 'P'
                                     DISP '.'                   ;display .
                                     DISP 'M'                   ;display M
                                     DISP '.'                   ;display .
                                     DISP ' '                   ;display space
00BF  C3                             RET

00C0                       TIMES     ENDP

00C0                       DATES     PROC NEAR

00C0  B4 2A                          MOV    AH,2AH              ;get date from DOS
00C2  CD 21                          INT    21H
00C4  52                             PUSH DX
```

```
00C5  B4 00              MOV   AH,0              ;get day of week
00C7  03 C0              ADD   AX,AX
00C9  BE 0000 R          MOV   SI,OFFSET DTAB    ;address day table
00CC  03 F0              ADD   SI,AX
00CE  8B 14              MOV   DX,[SI]           ;address day of week
00D0  B4 09              MOV   AH,9              ;display day of week
00D2  CD 21              INT   21H
00D4  5A                 POP   DX
00D5  52                 PUSH  DX
00D6  8A C6              MOV   AL,DH             ;get month
00D8  FE C8              DEC   AL
00DA  B4 00              MOV   AH,0
00DC  03 C0              ADD   AX,AX
00DE  BE 000E R          MOV   SI,OFFSET MTAB    ;address month table
00E1  03 F0              ADD   SI,AX
00E3  8B 14              MOV   DX,[SI]           ;address month
00E5  B4 09              MOV   AH,9              ;display month
00E7  CD 21              INT   21H
00E9  5A                 POP   DX
00EA  8A C2              MOV   AL,DL             ;get day of month
00EC  B4 00              MOV   AH,0
00EE  D4 0A              AAM                     ;convert to BCD
00F0  0A E4              OR    AH,AH
00F2  74 0D              JZ    DATES1            ;if tens is 0
00F4  80 C4 30           ADD   AH,30H            ;convert tens
                         DISP  AH                ;display tens
0101             DATES1:
0101  04 30              ADD   AL,30H            ;convert units
                         DISP  AL                ;display units
                         DISP  ','               ;display comma
                         DISP  ' '               ;display space
0121  81 F9 07D0         CMP   CX,2000           ;test for year 2000
0125  72 19              JB    DATES2            ;if below year 2000
0127  83 E9 64           SUB   CX,100            ;scale to 1900 - 1999
                         DISP  '2'               ;display 2
                         DISP  '0'               ;display 0
013E  EB 14              JMP   DATES3
0140             DATES2:
                         DISP  '1'               ;display 1
                         DISP  '9'               ;display 9
0154             DATES3:
0154  81 E9 076C         SUB   CX,1900           ;scale to 00 - 99
0158  8B C1              MOV   AX,CX
015A  D4 0A              AAM                     ;convert to BCD
015C  05 3030            ADD   AX,3030H          ;convert to ASCII
                         DISP  AH                ;display tens
                         DISP  AL                ;display units
0173  C3                 RET

0174             DATES   ENDP
                         END                     ;end of file
```

This procedure uses two ASCII lookup tables that convert the day and month to ASCII character strings. It also uses the AAM instruction to convert from binary to BCD for the time and date. The displaying of data is handled in two ways: by character string (function 09H) and by single character (function 06H).

The memory model (SMALL) consists of two segments: .DATA and .CODE. The data segment contains the character strings used with the procedures that display time and date. The code segment contains TIMES and DATES procedures, and a macro (DISP) that displays an ASCII character. The main program is very short and consists of two CALL instructions. The year 2000 problem is corrected in this program, but not the year 2100 problem.

7–4   DISK FILES

Data are found stored on the disk in the form of files. The disk itself is organized in four main parts: the boot sector, the file allocation table (FAT), the root directory, and the data storage areas. The first sector on the disk is the boot sector, which is used to load the disk operating system (DOS) from the disk into the memory when power is applied to the computer.

The FAT is where the names of files/subdirectories and their locations on the disk are stored by DOS. All references to any disk file are handled through the FAT. All other subdirectories and files are referenced through the root directory. The disk files are all considered sequential access files, meaning that they are accessed, a byte at a time, from the beginning of the file toward the end.

Disk Organization

Figure 7–2 illustrates the organization of sectors and tracks on the surface of the disk. This organization applies to both floppy and hard disk memory systems. The outer track is always track 0, and the inner track is 39 (double-density) or 79 (high-density) on floppy disks. The inner track on a hard disk is determined by the disk size, and could be 10000 or higher for very large hard disks.

Figure 7–3 shows the organization of data on a disk. The length of the FAT is determined by the size of the disk. Likewise, the length of the root directory is determined by the number of files and subdirectories located within it. The boot sector is always a single 512-byte-long sector located in the outer track at sector 0, the first sector.

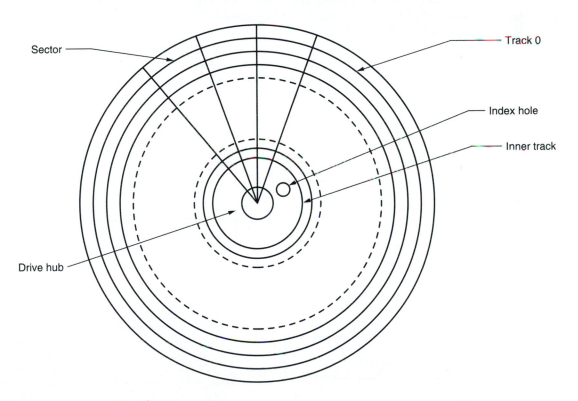

FIGURE 7–2   Structure of the 5¼" floppy disk.

FIGURE 7–3 Main data storage areas on a disk.

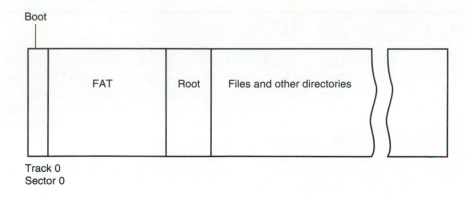

The boot sector contains a **bootstrap loader** program that is read into RAM when the system is powered. The bootstrap loader then executes and loads the IO.SYS and MSDOS.SYS programs into RAM. Next, the bootstrap loader passes control to the MSDOS control program, allowing the computer to be under the control of the DOS command processor called COMMAND.COM.

The FAT indicates which sectors are free, which are corrupted (unusable), and which contain data. The FAT table is referenced each time that DOS writes data to the disk so that it can find a free sector. Each free cluster is indicated by a 0000H in the FAT and each occupied sector is indicated by the cluster number. A **cluster** can be anything from one sector to any number of sectors in length. Many hard disk memory systems use four sectors per cluster, which means that the smallest file is 512×4, or 2048 bytes long.

Figure 7–4 shows the format of each directory entry in the root, or in any other directory or subdirectory. Each entry contains the name, extension, attribute, time, date, location, and length. The length of the file is stored as a 32-bit number. This means that a file can have a maximum length of 4G bytes. The location is the starting cluster number.

Example 7–35 shows how part of the root directory appears in a hexadecimal dump. Try to identify the date, time, location, and length of each entry. Also, identify the attribute for each entry. The listing shows hexadecimal data and ASCII data, as is customary for most computer dumps.

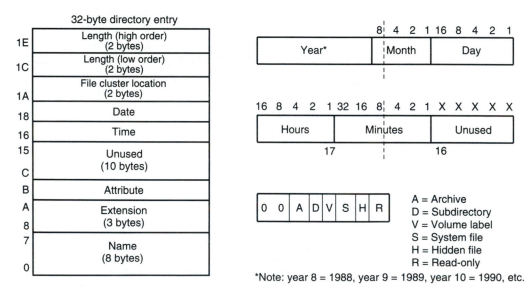

FIGURE 7–4 Format of any directory or subdirectory entry.

EXAMPLE 7–35

```
0000 49 4F 20 20 20 20 20 20 53 59 53 07 00 00 00 00    IO       SYS
0010 00 00 00 00 00 00 00 00 93 11 02 00 39 82 00 00

0020 4D 53 44 4F 53 20 20 20 53 59 53 07 00 00 00 00    MSDOS    SYS
0030 00 00 00 00 00 00 C0 44 93 12 13 00 92 00 00 00

0040 43 4F 4D 4D 41 4E 44 20 43 4F 4D 00 00 00 00 00    COMMAND  COM
0050 00 00 00 00 00 00 00 00 93 11 26 00 B5 92 00 00

0060 42 41 52 52 59 20 42 52 45 59 20 28 00 00 00 00    BARRY BREY
0070 00 00 00 00 00 00 E0 AD 6A 13 00 00 00 00 00 00

0080 50 43 54 4F 4F 4C 53 20 20 20 20 10 00 00 00 00    PCTOOLS
0090 00 00 00 00 00 00 80 AE 6A 13 5C 00 00 00 00 00

00A0 44 4F 53 20 20 20 20 20 20 20 20 10 00 00 00 00    DOS
00B0 00 00 00 00 00 00 E0 B0 6A 13 4E 00 00 00 00 00

00C0 52 55 4E 5F 46 57 20 20 42 41 54 00 00 00 00 00    FUN_FW   BAT
00D0 00 00 00 00 00 00 40 BD 6A 13 97 0F 4A 00 00 00

00E0 46 4F 4E 54 57 41 52 45 20 20 20 10 00 00 00 00    FONTWARE
00F0 00 00 00 00 00 00 60 BD 6A 13 6E 00 00 00 00 00
```

Files are usually accessed through DOS INT 21H function calls. There are two ways to access a file using INT 21H: one way uses a file control block and the other uses a file handle. Today, all software accesses files via a file handle, so this text also uses file handles for file access. File control blocks are a carryover from an earlier operating system called CP/M (control program micro), which was used with eight-bit computer systems based on the Z80 or 8080 microprocessor.

File Names

Files and programs are stored on a disk and referenced both by a file name and an extension to the file name. With the DOS operating system, the *file name* can be from one to eight characters long. The file name can contain just about any ASCII character, except for spaces or the " \ . / [ ] * , : < > | ; ? = characters. In addition to the file name, the file can have an optional one-to-three-digit *extension* to the file name. Table 7–5 shows a few file names, with and without extensions. Note that the name of a file and its extension are always separated by a period. If Windows 95 through Windows XP is in use, the file name can be of any length (up to 255 characters) and can even contain spaces. This is an improvement over the eight-character file name limitation of DOS.

Directory and Subdirectory Names. The DOS file management system arranges the data and programs on a disk into directories and subdirectories. The rules that apply to file names also apply to directory and subdirectory names: the name can be up to eight characters long and an

TABLE 7–5 DOS file names and extensions.

Name.Extension
TEST.TXT
READ-ME.DOC
ALWAY12.COM
RUN_IT~.EXE
CHAPTER.02
NOEXTEN1
1.2

extension can be up to three characters long. Most of the time, an extension is not used with a directory name, but it can appear if needed. The disk is structured so that it contains a root directory when first formatted. The root directory for a hard disk used as drive C is C:\. Any other directory is placed in the root directory. For example, C:\DATA is directory DATA in the root directory. Each directory placed in the root directory can also have subdirectories. Examples are the subdirectories C:\DATA\AREA1 and C:\DATA\AREA2, in which the directory DATA contains two subdirectories: AREA1 and AREA2. Subdirectories can also have additional subdirectories. For example, C:\DATA\AREA2\LIST depicts directory DATA, subdirectory AREA, which contains a subdirectory LIST.

Long File Names.   Example 7–36 shows how two long file names appear in the Windows 95/98 directory: the first file is CHAPTER1.DOC and the second entry is PRENTICE HALL CAREER1.DOC. Notice that the first entry starts at offset address 0020H, while the long file name is stored beginning at offset address 0000H. Notice that the file name stored in the normal directory entry is CHAPE~1.DOC. This is stored in ASCII code for compatibility with the older file system. The long file name is stored in the Unicode, which is a 16-bit code that uses ASCII as its first 256 combinations, with foreign languages using the remaining combinations. Unicode is used with most applications in place of ASCII code. The starting sector for a long file name entry contains a 0000H. This is how the long file name is detected by Windows 95 through Windows XP (or any other software that uses long file names).

EXAMPLE 7–36

```
0000   41 43 00 68 00 61 00 70-00 74 00 0F 00 42 65 00    AC.h.a.p.t...Be.
0010   72 00 20 00 31 00 2E 00-64 00 00 00 6F 00 63 00    r. .1...d...o.c.

0020   43 48 41 50 54 45 7E 31-44 4F 43 20 00 54 4D 70    CHAPTE~1DOC .TMp
0030   9B 20 9B 20 00 00 91 78-91 20 48 00 00 F6 03 00    . . ...x. H.....

0040   42 20 00 43 00 41 00 52-00 45 00 0F 00 A0 45 00    B .C.A.R.E...E.
0050   52 00 20 00 31 00 2E 00-64 00 00 00 6F 00 63 00    R. .1...d...o.c.

0060   01 50 00 52 00 45 00 4E-00 54 00 0F 00 A0 49 00    .P.R.E.N.T....I.
0070   43 00 45 00 20 00 48 00-41 00 00 00 4C 00 4C 00    C.E. .H.A...L.L.

0080   50 52 45 4E 54 49 7E 31-44 4F 43 20 00 7C CB 6D    PRENTI~1DOC .|.m
0090   9B 20 9B 20 00 00 58 7A-68 20 02 00 00 8C 00 00    . . ..Xzh ......
```

Sequential File Access

All DOS files are sequential files. A sequential file is stored and accessed from the beginning of the file toward the end, with the first byte and all bytes between it and the last accessed to read the last byte. Fortunately, files are read and written with the DOS INT 21H function calls (see Appendix A), which makes their access and manipulation easy. This section of the text describes how to create, read, write, delete, and rename a sequential access file.

File Creation.   Before a file can be used, it must exist on the disk. A file is created by the INT 21H function call number 3CH. The file name must be stored at a location addressed by DS:DX before calling the function, and CX must contain the attribute of the file (or subdirectory) created.

A file name is always stored as an ASCII-Z string, and it may contain the drive and directory path(s), if needed. Example 7–37 shows several file names that are stored in a data segment for access by the file utilities. An **ASCII-Z string** is a character string that ends with a 00H or null character.

EXAMPLE 7–37

```
0000   44 4F 47 2E 54 58   FILE1   DB      'DOG.TXT',0
       54 00
0008   43 3A 44 41 54 41   FILE2   DB      'C:DATA.DOC',0
```

```
            2E 44 4F 43 00
0013 43 3A 5C 44 52 45   FILE3   DB      'C:\DREAD\ERROR.FIL',0
     41 44 5C 45 52 52
     4F 52 2E 46 49 4C
     00
```

Suppose that you have filled a 256 memory buffer area with data that must be stored in a new file called DATA.NEW on the default disk drive. Before data can be written to this new file, the file must first be created. Example 7–38 lists a short procedure that creates this new file on the disk.

EXAMPLE 7–38

```
                            ;A program that creates file DATA.NEW.
                            ;DO NOT RUN this program because the file is not closed.
                            ;
                                    .MODEL SMALL
0000                                .DATA
0000  44 41 54 41   FILEN   DB      'DATA.NEW',0        ;file name
      2E 4E 45 57
      00
0000                                .CODE
                                    .STARTUP
0017  B4 3C                 MOV     AH,3CH              ;create file function
0019  B9 0000               MOV     CX,0                ;normal file attribute
001C  BA 0000 R             MOV     DX,OFFSET FILEN     ;address file name
001F  CD 21                 INT     21H                 ;access DOS
                                    .EXIT
                                    END
```

Whenever a file is created, the CX register must contain the attributes or characteristics of the file. Table 7–6 lists and defines the attribute bit positions. A logic 1 in a bit selects the attribute, while a logic 0 does not.

After returning from the INT 21H, the carry flag indicates whether or not an error occurred (CF = 1) during the creation of the file. Some errors that can occur (which are obtained if needed by INT 21H function call number 59H) are path not found, no file handles available, or media error. If carry is cleared, no error has occurred and the AX register contains a file handle. The **file handle** is a number that is used to refer to the file after it is created or opened. The file handle allows a file to be accessed without using the ASCII-Z string name of the file, thus speeding the operation.

Writing to a File. Now that we have created a new file called FILE.NEW, data can be written to it. Before writing to a file, the file must have been created or opened. When a file is created or opened, the file handle returns in the AX register. The file handle is used to refer to the file whenever data are written. Function number 40H is used to write data to an opened or newly created file. In addition to loading a 40H into AH, we must also load BX = the file handle, CX = the number of bytes to be written, and DS:DX = the address of the area to be written to the disk.

TABLE 7–6 File attribute definitions.

Bit Position	Value	Attribute	Function
0	01H	Read-only	A read-only file or directory
1	02H	Hidden	Prevents the file or directory from appearing in a directory listing
2	04H	System	Specifies a system file
3	08H	Volume	Specifies the name of the disk volume
4	10H	Subdirectory	Specifies a subdirectory name
5	20H	Archive	Indicates that a file has changed since the last backup

Suppose that we must write all 256 bytes of BUFFER to the file. This is accomplished, as illustrated in Example 7–39, by using function 40H. If an error occurs during a write operation, the carry flag is set. If no error occurs, the carry flag is cleared and the number of bytes written to the file is returned in the AX register. Errors that occur for writes usually indicate that the disk is full or that there is some type of media error.

EXAMPLE 7–39

```
                                      .        .
                                      .        .
                                      .        .
0010   8B D8              MOV    BX,AX           ;move handle to BX
0012   B4 40              MOV    AH,40H          ;load write function
0014   B9 0100            MOV    CX,256          ;load count
0017   BA 0009 R          MOV    DX,OFFSET BUFFER ;address BUFFER
001A   CD 21              INT    21H             ;write 256 bytes from BUFFER

001C   72 32              JC     ERROR1          ;on write error
                                      .        .
                                      .        .
                                      .        .
```

Opening, Reading, and Closing a File.   To read a file, it must be opened first. When a file is opened, DOS checks the directory to determine whether the file exists and returns the DOS file handle in register AX. The DOS file handle must be used for reading, writing, and closing a file.

Example 7–40 lists a sequence of instructions that opens a file, reads 256 bytes from the file into memory area BUFFER, and then closes the file. When a file is opened (AH = 3DH), the AL register specifies the type of operation allowed for the opened file. If AL = 00H, the file is opened for a read; if AL = 01H, the file is opened for a write; and if AL = 02H, the file is opened for a read or a write.

EXAMPLE 7–40

```
                          ;A program that opens the file TEMP.ASM and reads the
                          ;first 256 bytes into an area of memory called BUF.
                          ;
                                      .MODEL SMALL
0000                                  .DATA
0000   54 45 4D 50        FILEN   DB    'TEMP.ASM',0   ;file name
       2E 41 53 4D
       00
0009   0100 [             BUF     DB    256 DUP (?)    ;buffer area
            00
            ]
0000                                  .CODE
                                      .STARTUP
0017   B8 3D02            MOV    AX,3D02H         ;open file function
001A   BA 0000 R          MOV    DX,OFFSET FILEN  ;address file name
001D   CD 21              INT    21H              ;access DOS
001F   8B D8              MOV    BX,AX            ;file handle to BX

0021   B4 3F              MOV    AH,3FH           ;read file function
0023   B9 0100            MOV    CX,256           ;read 256 bytes
0026   BA 0009 R          MOV    DX,OFFSET BUF    ;store data at BUF
0029   CD 21              INT    21H              ;access DOS

002B   B4 3E              MOV    AH,3EH           ;close file function
002D   CD 21              INT    21H              ;access DOS
                                      .EXIT
                                      END
```

Function number 3FH causes a file to be read. As with the write function, BX contains the file handle, CX contains the number of bytes to be read, and DS:DX contains the location of a memory area where the data are stored. As with all disk functions, the carry flag indicates an error with a logic 1. If a logic 0 is indicated, the AX register indicates the number of bytes read from the file.

Closing a file is very important. If a file is left open, some serious problems can occur that can actually destroy the disk and all its data. If a file is written and not closed, the FAT can become corrupted, making it difficult or impossible to retrieve data from the disk. Always be certain to close a file after it is read or written.

The File Pointer.   When a file is opened, written, or read, the file pointer addresses the current location in the sequential file. When a file is opened, the file pointer always addresses the first byte of the file. If a file is 1024 bytes long, and a read function reads 1023 bytes, the file pointer addresses the last byte of the file, but not the end of the file.

The **file pointer** is a 32-bit number that addresses any byte in a file. Once a file is opened, the file pointer can be changed with the move file pointer function number 42H. A file pointer can be moved from the start of the file (AL = 00H), from the current location (AL = 01H), or from the end of the file (AL = 02H). In practice, all three directions of the move are used to access different parts of the file. The distance moved by the file pointer is specified by registers CX and DX. The DX register holds the least-significant part of the distance and CX holds the most-significant part of the distance. Register BX must contain the file handle before using function 42H to move the file pointer.

Suppose that a file exists on the disk and that you must append the file with 256 bytes of new information. When the file is opened, the file pointer addresses the first byte of the file. If you attempt to write without moving the file pointer to the end of the file, the new data will overwrite the first 256 bytes of the file. Example 7–41 shows a program that opens a file, moves the file pointer to the end of the file, writes 256 bytes of data, and then closes the file. This appends the file with 256 new bytes of data.

EXAMPLE 7–41

```
                              ;A program that opens FILE.NEW and appends it with 256
                              ;bytes of data from BUF.
                              ;
                                      .MODEL SMALL
0000                                  .DATA
0000  46 49 4C 45    FILEN   DB      'FILE.NEW',0        ;file name
      2E 4E 45 57
      00
0009  0100 [         BUF     DB      256 DUP (?)         ;buffer
            00
               ]
0000                                  .CODE
                                      .STARTUP
0017  B8 3D02                MOV     AX,3D02H            ;open FILE.NEW
001A  BA 0000 R             MOV     DX,OFFSET FILEN
001D  CD 21                 INT     21H
001F  8B D8                 MOV     BX,AX

0021  B8 4202                MOV     AX,4202H            ;move file pointer to end
0024  BA 0000               MOV     DX,0
0027  B9 0000               MOV     CX,0
002A  CD 21                 INT     21H

002C  B4 40                 MOV     AH,40H              ;write BUF to end of file
002E  B9 0100               MOV     CX,256
0031  BA 0009 R             MOV     DX,OFFSET BUF
0034  CD 21                 INT     21H

0036  B4 3E                 MOV     AH,3EH              ;close file
0038  CD 21                 INT     21H
                                      .EXIT
                                      END
```

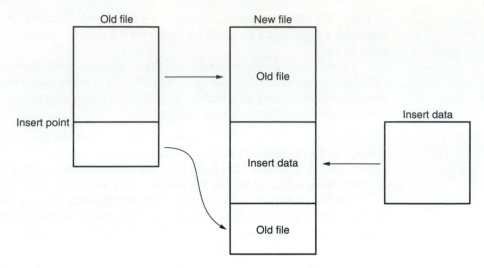

FIGURE 7–5   Inserting new data within an old file.

One of the more difficult file maneuvers is inserting new data in the middle of the file. Figure 7–5 shows how this is accomplished by creating a second file. Notice that the part of the file before the insertion point is copied into the new file. This is followed by the new information before the remainder of the file is appended after the insertion in the new file. Once the new file is complete, the old file is deleted and the new file is renamed to the old file name.

Example 7–42 shows a program that inserts new data into an old file. This program copies the DATA.NEW file into the DATA.OLD file at a point after the first 256 bytes of the DATA.OLD file.

EXAMPLE 7–42

```
                              ;A program that adds the 256 byte contents of the file
                              ;DATA.NEW to DATA.OLD at a point between the first 256
                              ;bytes of DATA.OLD and the remainder of the file.
                              ;
                                      .MODEL SMALL
0000                                  .DATA
0000  0000           HAN1     DW      ?                 ;file handle for DATA.TMP
0002  0000           HAN2     DW      ?                 ;file handle for DATA.OLD
0004  44 41 54 41    FILE1    DB      'DATA.TMP',0
      2E 54 4D 50
      00
000D  44 41 54 4     FILE2    DB      'DATA.OLD',0
      2E 4F 4C 44
      00
0016  44 41 54 41    FILE3    DB      'DATA.NEW',0
      2E 4E 45 57
      00
001F  0100 [         BUF      DB      256 DUP (?)       ;data buffer area
            00
               ]
0000                                  .CODE
                                      .STARTUP
0017  B4 3C                   MOV     AH,3CH            ;create DATA.TMP
0019  B9 0000                 MOV     CX,0
001C  BA 0004 R               MOV     DX,OFFSET FILE1
001F  CD 21                   INT     21H
0021  A3 0000 R               MOV     HAN1,AX           ;save handle at HAN1

0024  B8 3D02                 MOV     AX,3D02H          ;open DATA.OLD
0027  BA 000D R               MOV     DX,OFFSET FILE2
```

```
002A  CD 21                    INT  21H
002C  8B D8                    MOV  BX,AX
002E  A3 0002 R                MOV  HAN2,AX            ;save handle at HAN2

0031  B4 3F                    MOV  AH,3FH             ;read 256 bytes of DATA.OLD
into BUF
0033  B9 0100                  MOV  CX,256
0036  BA 001F R                MOV  DX,OFFSET BUF
0039  CD 21                    INT  21H

003B  B4 40                    MOV  AH,40H             ;write BUF to DATA.TMP
003D  8B 1E 0000 R             MOV  BX,HAN1            ;get handle
0041  B9 0100                  MOV  CX,256
0044  BA 001F R                MOV  DX,OFFSET BUF
0047  CD 21                    INT  21H

0049  B8 3D02                  MOV  AX,3D02H           ;open DATA.NEW
004C  BA 0016 R                MOV  DX,OFFSET FILE3
004F  CD 21                    INT  21H
0051  8B D8                    MOV  BX,AX

0053  B4 3F                    MOV  AH,3FH             ;read 256 bytes from DATA.NEW
to BUF
0055  B9 0100                  MOV  CX,256
0058  BA 001F R                MOV  DX,OFFSET BUF
005B  CD 21                    INT  21H

005D  B4 3E                    MOV  AH,3EH             ;close DATA.NEW
005F  CD 21                    INT  21H

0061  B4 40                    MOV  AH,40H             ;write BUF to DATA.TMP
0063  8B 1E 0000 R             MOV  BX,HAN1            ;get handle
0067  B9 0100                  MOV  CX,256
006A  BA 001F R                MOV  DX,OFFSET BUF
006D  CD 21                    INT  21H
006F             MAIN1:
006F  B4 3F                    MOV  AH,3FH             ;read 256 bytes from DATA.OLD
to BUF
0071  8B 1E 0002 R             MOV  BX,HAN2
0075  B9 0100                  MOV  CX,256
0078  BA 001F R                MOV  DX,OFFSET BUF
007B  CD 21                    INT  21H
007D  0B C0                    OR   AX,AX             ;test for zero byte read
007F  74 10                    JZ   MAIN2             ;if file empty
0081  B4 40                    MOV  AH,40H            ;write BUF to DATA.TMP
0083  8B 1E 0000 R             MOV  BX,HAN1
0087  B9 0100                  MOV  CX,256
008A  BA 001F R                MOV  DX,OFFSET BUF
008D  CD 21                    INT  21H
008F  EB DE                    JMP  MAIN1
0091             MAIN2:
0091  B4 3E                    MOV  AH,3EH             ;close DATA.OLD
0093  CD 21                    INT  21H

0095  B4 41                    MOV  AH,41H             ;delete DATA.OLD
0097  BA 000D R                MOV  DX,OFFSET FILE2
009A  CD 21                    INT  21H

009C  B4 3E                    MOV  AH,3EH             ;close DATA.TMP
009E  8B 1E 0000 R             MOV  BX,HAN1
00A2  CD 21                    INT  21H

00A4  8C D8                    MOV  AX,DS
00A6  8E C0                    MOV  ES,AX              ;overlap DS and ES

00A8  B4 56                    MOV  AH,56H             ;rename DATA.TMP to DATA.OLD
00AA  BA 0004 R                MOV  DX,OFFSET FILE1    ;old name
00AD  BF 000D R                MOV  DI,OFFSET FILE2    ;new name
```

```
00B0   CD 21                    INT    21H
                                .EXIT
                                END
```

This program uses two new INT 21H function calls. The delete and rename function calls are used to delete the old file before the temporary file is renamed to the old file name. Note that the rename function uses both the DS and ES segment registers to address the two file names.

Random Access Files

Random access files are developed through software using sequential access files. A random access file is addressed by a record number rather than by going through the file searching for data. The move pointer function call becomes very important when random access files are created. Random access files are much easier to use for large volumes of data.

Creating a Random Access File.   Planning is paramount for creating a random access file system. Suppose that a random access file is required for storing the names of customers. Each customer record requires 16 bytes for the last name, 16 bytes for the first name, and one byte for the middle initial. Each customer record contains two street address lines of 32 bytes each, a city line of 16 bytes, two bytes for the state code, and nine bytes for the Zip Code. The basic customer information alone requires 105 bytes; additional information expands the record to 256 bytes. Because the business is growing, provisions are made for 5000 customers. This means that the total random access file is 1,280,000 bytes long.

Example 7–43 illustrates a short program that creates a file called CUST.FIL and inserts 5000 blank records of 256 bytes each. A blank record contains 00H in each byte. This appears to be a large file, but it fits on a single high-density $5^1/4$" or $3^1/2$" floppy disk drive; in fact, this program assumes that the disk is in drive A.

EXAMPLE 7–43

```
                          ;A program that creates CUST.FIL and then fills 5000
                          ;records of 256 bytes each with zeros.
                          ;
                                  .MODEL SMALL
0000                              .DATA
0000   43 55 53 54   FILE1   DB    'CUST.FIL',0      ;file name
       2E 46 49 4C
       00
0009   0100 [        BUF     DB    256 DUP (0)       ;buffer
            00
       ]
0000                              .CODE
                                  .STARTUP
0017   B4 3C                 MOV    AH,3CH           ;create CUST.FIL
0019   B9 0000               MOV    CX,0
001C   BA 0000 R             MOV    DX,OFFSET FILE1
001F   CD 21                 INT    21H
0021   8B D8                 MOV    BX,AX            ;handle to BX

0023   BD 1388               MOV    BP,5000          ;record counter
0026                 MAIN1:
0026   B4 40                 MOV    AH,40H           ;write record
0028   B9 0100               MOV    CX,256
002B   BA 0009 R             MOV    DX,OFFSET BUF
002E   CD 21                 INT    21H
0030   4D                    DEC    BP               ;decrement record count
0031   75 F3                 JNZ    MAIN1            ;for 5000 records
0033   B4 3E                 MOV    AH,3EH           ;close file
0035   CD 21                 INT    21H
                             .EXIT
                             END
```

Reading and Writing a Record.   Whenever a record must be read, the record number is loaded into the BP register and the procedure listed in Example 7–44 is called. This procedure assumes that FIL contains the handle number and that the CUST.FIL remains open at all times.

Notice how the record number is multiplied by 256 to obtain a count for the move pointer function. In each case, the file pointer is moved from the start of the file to the desired record before it is read into memory area BUFFER. Although not shown, writing a record is performed in the same manner as reading a record.

EXAMPLE 7–44

```
                        ;The READ procedure reads one record from CUST.FIL.
                        ;Input parameters are:
                        ;FIL (word) = CUST.FIL handle
                        ;BP = record number
                        ;Output parameters are:
                        ;BUFFER (256 bytes) = customer record
                        ;
0000                    READ      PROC FAR

0000    8B 1E 0100 R            MOV   BX,FIL              ;get handle
0004    B8 0100                 MOV   AX,256              ;multiply by 256
0007    F7 E5                   MUL   BP
0009    8B CA                   MOV   CX,DX
000B    8B D0                   MOV   DX,AX
000D    B8 4200                 MOV   AX,4200H            ;move pointer
0010    CD 21                   INT   21H

0012    B4 3F                   MOV   AH,3FH              ;read record
0014    B9 0100                 MOV   CX,256
0017    BA 0000 R               MOV   DX,OFFSET BUFFER
001A    CD 21                   INT   21H
001C    CB                      RET

001D                    READ      ENDP
```

7–5   ## EXAMPLE PROGRAMS

Now that many of the basic programming building blocks have been discussed, we present some example application programs. Although these example programs may seem trivial, they present some additional programming techniques and illustrate programming styles for the microprocessor.

Calculator Program

This program demonstrates how data conversion plays an important part in many application programs. Example 7–45 illustrates a program that accepts two numbers and adds, subtracts, multiplies, or divides them. To limit the complexity of the program, the numbers are limited to two-digit numbers. For example, if you type a 12 + 24 followed by =, the program will calculate the result and display 36 as an answer. To further simplify the program, the numbers 0–9 must be entered as two-digit numbers 0–9.

EXAMPLE 7–45

```
                        ;calculator program
                        ;
                        .MODEL TINY
0000                    .CODE
                        DISP  MACRO PARA
                              PUSH  AX
```

```
                              MOV     AH,6
                              MOV     DL,PARA
                              INT     21H
                              POP     AX
                              ENDM

                      GET     MACRO
                              .repeat
                              MOV     AH,6
                              MOV     DL,-1
                              INT     21H
                              .UNTIL AL>='0' && AL <= '9'
                              DISP    AL
                              SUB     AL,'0'
                              ENDM

                              .STARTUP

0100  E8 001E                 CALL    READN           ;get first number
0103  8A D8                   MOV     BL,AL
                              DISP    '+'             ;display +
010D  E8 0011                 CALL    READN           ;get second number
0110  02 D8                   ADD     BL,AL           ;form sum
                              DISP    '='             ;display =
011A  E8 003F                 CALL    DISPA           ;display sum
                              .EXIT

0121                  READN PROC    NEAR USES BX

                              GET                     ;get first digit
013A  B3 0A                   MOV     BL,10
013C  F6 E3                   MUL     BL              ;multiply by 10
013E  8A D8                   MOV     BL,AL
                              GET                     ;get second digit
0158  02 C3                   ADD     AL,BL           ;form number
                              RET

015C                  READN ENDP


015C                  DISPA PROC    NEAR

015C  8A C3                   MOV     AL,BL           ;get answer
015E  BB 000A                 MOV     BX,10
0161  53                      PUSH    BX
0162  B4 00                   MOV     AH,0
                              .REPEAT
0164  BA 0000                 MOV     DX,0
0167  F7 F3                   DIV     BX
0169  52                      PUSH    DX
                              .UNTIL AX == 0
                              .WHILE 1
016E  58                         POP    AX
                                 .BREAK .IF AL == 10
0173  04 30                      ADD    AL,'0'        ;make ASCII
                                 DISP   AL
                              .ENDW
017F  C3                      RET

0180                  DISPA ENDP
                              END
```

Note that this program includes many of the techniques presented thus far in this chapter, including conditional assembly directives such as .REPEAT, .IF, .BREAK, and .WHILE. Also included are procedures and macros. This program does not support a backspace to correct an erroneous entry. To reduce the length of the program, no attempt has been made to recover from any errors.

Numeric Sort Program

At times, numbers must be sorted into numeric order. This is often accomplished with a bubble sort. Figure 7–6 shows five numbers that are sorted with a bubble sort. Notice that the set of five numbers is tested four times with four passes. For each pass, two consecutive numbers are compared and sometimes exchanged. Also notice that during the first pass, there are four comparisons, during the second three, etc.

Example 7–46 illustrates a program that accepts 10 numbers from the keyboard (0–65535). After these 16-bit numbers are accepted and stored in memory section ARRAY, they are sorted by using the bubble-sorting technique. This bubble sort uses a flag to determine whether any numbers were exchanged in a pass. If no numbers were exchanged, the numbers are in order and the sort terminates.

EXAMPLE 7–46

```
                                .MODEL SMALL
0000                            .DATA
0000  000A [                    ARRAY DW 10 DUP (?)       ;array
         0000
              ]
0014  0D 0A 45 6E 74 65         MES1  DB 13,10,'Enter 10 numbers:',13,10,10,'$'
      72 20 31 30 20 6E
      75 6D 62 65 72 73
      3A 0D 0A 0A 24
002B  0D 0A 0A 53 6F 72         MES2  DB 13,10,10,'Sorted Data:',13,10,10,'$'
      74 65 64 20 44 61
      74 61 3A 0D 0A 0A
      24
0000                            .CODE

                                DISP    MACRO PARA

                                        PUSH    AX
                                        MOV     AH,6
                                        MOV     DL,PARA
                                        INT     21H
                                        POP     AX

                                        ENDM
                                GET     MACRO

                                        .REPEAT
                                            MOV    AH,6
                                            MOV    DL,-1
```

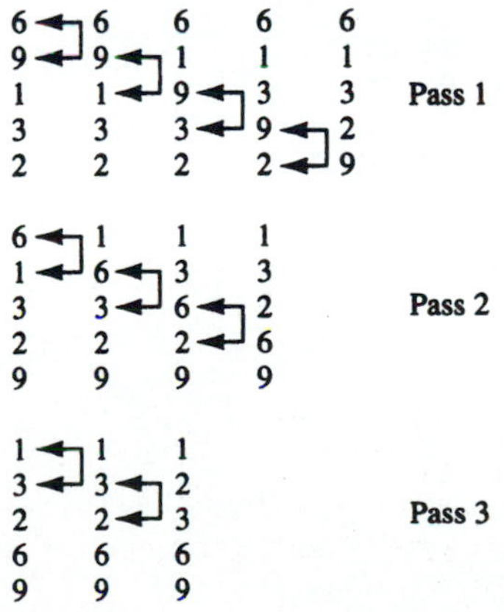

FIGURE 7–6 A bubble sort showing data as they are sorted. Note: Sorting five numbers may require four passes.

Pass 1
Pass 2
Pass 3

```
                                        INT    21H
                                        .UNTIL (AL>='0' && AL <='9') || AL==13 ||
AL==','
                                        DISP  AL
                                        .IF AL==13
                                          DISP  10
                                        .ENDIF
                                        .IF AL>='0' && AL<='9'
                                          SUB   AL,'0'
                                        .ENDIF

                                        ENDM

                              STRING    MACRO WHERE

                                        MOV   DX,OFFSET WHERE
                                        MOV   AH,9
                                        INT   21H

                                        ENDM
                              .STARTUP

                                        STRING    MES1
001E  FC                                CLD
001F  B9 000A                           MOV   CX,10
0022  BF 0000 R                         MOV   DI,OFFSET ARRAY
0025  8C D8                             MOV   AX,DS
0027  8E C0                             MOV   ES,AX
                                        .REPEAT
0029  E8 0026                             CALL  GETN      ;get 10 numbers
                                        .UNTILCXZ
                                        STRING    MES2
0035  E8 008B                           CALL  SORT        ;sort 10 numbers
0038  B9 0009                           MOV   CX,9
003B  BE 0000 R                         MOV   SI,OFFSET ARRAY
                                        .REPEAT            ;display 10 numbers
003E  E8 0061                             CALL  DISPN
                                          DISP  ','
                                        .UNTILCXZ
004B  E8 0054                           CALL  DISPN
                                        .EXIT

0052                          GETN  PROC  NEAR

0052  BD 000A                           MOV   BP,10
0055  BB 0000                           MOV   BX,0
                                        .WHILE 1
                                          GET
                                          .BREAK .IF AL==13 || AL==','
0094  93                                XCHG  .AX,BX
0095  F7 E5                             MUL   BP
0097  93                                XCHG  AX,BX
0098  B4 00                             MOV   AH,0
009A  03 D8                             ADD   BX,AX
                                        .ENDW
009E  8B C3                             MOV   AX,BX
00A0  AB                                STOSW
00A1  C3                                RET

00A2                          GETN  ENDP

00A2                          DISPN PROC  NEAR

00A2  BB 000A                           MOV   BX,10
00A5  53                                PUSH  BX
00A6  AD                                LODSW
                                        .REPEAT
```

```
00A7  BA 0000                          MOV       DX,0
00AA  F7 F3                            DIV       BX
00AC  52                               PUSH      DX
                                 .UNTIL AX==0
                                 .WHILE 1
00B1  58                               POP       AX
                                 .BREAK .IF AL==10
00B6  04 30                            ADD       AL,'0'
                                       DISP      AL
                                 .ENDW
00C2  C3                         RET

00C3                     DISPN       ENDP

00C3                     SORT        PROC      NEAR

00C3  BB 0009                         MOV       BX,9
                                 .REPEAT
00C6  8B CB                            MOV       CX,BX
00C8  BE 0000 R                        MOV       SI,OFFSET ARRAY
00CB  B2 00                            MOV       DL,0
                                   .REPEAT
00CD  AD                                 LODSW
00CE  3B 04                              CMP    AX,[SI]
                                     .IF !CARRY?
00D2  8B 2C                                MOV     BP,[SI]
00D4  89 6C FE                             MOV     [SI-2],BP
00D7  89 04                                MOV     [SI],AX
00D9  FE C2                                INC     DL
                                     .ENDIF
                                   .UNTILCXZ
00DD  4B                              DEC    BX
                                 .UNTIL BX==0 || DL==0
00E6  C3                         RET

00E7                     SORT        ENDP
                                     END
```

Once the numbers are sorted, they are displayed on the video screen in ascending numerical order. No provision is made for errors as each number is typed. The program terminates after sorting one set of 10 numbers and must be invoked again to sort 10 new numbers.

Hexadecimal File Dump

An example program that displays a file in hexadecimal format allows us to practice disk memory access. It also gives us the opportunity to read a parameter (the file name) from the DOS command line.

Whenever a command (program name) is typed at the DOS command line, any parameters that follow are placed in a **program segment prefix.** The program segment prefix (PSP) is listed in Appendix A, Figure A–6. The length of the command line and the command line parameters appear in the PSP, along with other information. Upon execution of a program, the DS segment register addresses the PSP, so an offset address of 80H is used to access the length (byte-sized) of the command line. After obtaining the length, the command line and its parameters can be accessed.

Example 7–47 lists a program that obtains a file name from the command line, and then displays the file in a hexadecimal listing. This program is useful for debugging faulty programs and as practice with disk file access and conversions. The parameter following the command always starts with a space (20H) at offset address 81H and always ends with a carriage return (0DH). The length of the parameter is always one greater. For example, if DUMPS FROG is typed at the command line and DUMPS is the name of the program, the parameter FROG is stored beginning with a 20H at offset 81H, and the length is 5.

EXAMPLE 7–47

```
                                    .MODEL  SMALL
0000                                .DATA
0000  0000                  SECT    DW      ?
0002  0100 [               BUFFER   DB      256 DUP (?)
         00
          ]
0102  0100 [               FILE     DB      256 DUP (?)
         00
          ]
0202  0D 0A 59 6F 75 20    MES1     DB      13,10,'You must enter a file '
      6D 75 73 74 20 65
      6E 74 65 72 20 61
      20 66 69 6C 65 20
                                    DB      'name',13,10,'$
021A  6E 61 6D 65 0D 0A
      24
0221  0D 0A 46 69 6C 65    MES2     DB      13,10,'File not found',13,10,'$'
      20 6E 6F 74 20 66
      6F 75 6E 64 0D 0A
      24
0234  0D 0A 46 69 6C 65    MES3     DB      13,10,'File corrupt',13,10,'$'
      20 63 6F 72 72 75
      70 74 0D 0A 24
0245  0D 0A 32 35 36 20    MES4     DB      13,10,'256 byte Section: $'
      62 79 74 65 20 53
      65 63 74 69 6F 6E
      3A 20 24
025A  0D 0A 24             MES5     DB      13,10,'$'
025D  0D 0A 54 79 70 65    MES6     DB      13,10,'Type a space to continue:$'
      20 61 20 73 70 61
      63 65 20 74 6F 20
      63 6F 6E 74 69 6E
      75 65 3A 20 24
0000                                .CODE
                           STRING   MACRO   WHERE
                                    MOV     AH,9
                                    MOV     DX,OFFSET WHERE
                                    INT     21H
                                    ENDM

                           OPEN     MACRO   WHERE
                                    MOV     AX,3D02H
                                    MOV     CX,0
                                    MOV     DX,OFFSET WHERE
                                    INT     21H
                                    MOV     BX,AX           ;save handle in BX
                                    MOV     SECT,-1         ;indicate first sector
                                    ENDM

                           READ     MACRO   BUF,COUNT
                                    MOV     AH,3FH
                                    MOV     CX,COUNT
                                    MOV     DX,OFFSET BUF
                                    INT     21H
                                    ENDM

                           CLOSE    MACRO
                                    MOV     AH,3EH
                                    INT     21H
                                    ENDM
                           DISP     MACRO   NUM
                                    MOV     AH,6
                                    MOV     DL,NUM
                                    INT     21H
                                    ENDM
```

```
                              ASCII     MACRO
                                        AND     AL,15
                                        ADD     AL,'0'
                                        .IF AL>'9'
                                            ADD AL,7
                                        .ENDIF
                                        DISP    AL
                                        ENDM

                              .STARTUP
0017  1E                      PUSH    DS              ;swap segments
0018  06                      PUSH    ES
0019  1F                      POP     DS
001A  07                      POP     ES
001B  BE 0082                 MOV     SI,82H          ;address command line
001E  8A 4C FE                MOV     CL,[SI-2]       ;get length
0021  B5 00                   MOV     CH,0
0023  80 F9 00                CMP     CL,0
0026  75 10                   JNE     MAIN1           ;if file name present
0028  B8 ---- R               MOV     AX,DGROUP
002B  8E D8                   MOV     DS,AX
                              STRING  MES1            ;display MES1
                .EXIT                                 ;exit to DOS
0038            MAIN1:
0038  49                      DEC     CX
0039  BF 0102 R               MOV     DI,OFFSET FILE
003C  F3/ A4                  REP     MOVSB           ;save file name
003E  88 2D                   MOV     [DI],CH         ;make it ASCII-Z
0040  8C C0                   MOV     AX,ES
0042  8E D8                   MOV     DS,AX           ;segment DGROUP
                              OPEN    FILE
0057  73 0B                   JNC     MAIN2           ;if file found
                              STRING  MES2            ;display MES2
                .EXIT                                 ;exit to DOS
0064            MAIN2:
                              .WHILE 1
0064  FF 06 0000 R            INC     SECT            ;increment sector
                              READ    BUFFER,256
0072  73 0B                   JNC     MAIN3           ;if file read
                              STRING  MES3            ;display MES3
                              .EXIT                   ;exit to DOS
007F            MAIN3:
                              .IF     AX==0           ;end of file
                                 CLOSE                ;close file
                                   .EXIT              ;exit to DOS
                              .ENDIF
008B  E8 000D                 CALL    DUMP            ;display sector
                              STRING  MES6
0095  B4 01                   MOV     AH,1            ;wait for key
0097  CD 21                   INT     21H
                              .ENDW

009B            DUMP          PROC    NEAR
009B  8B C8                   MOV     CX,AX           ;save count
                              STRING  MES4            ;display sector
00A4  8B 2E 0000 R            MOV     BP,SECT
00A8  E8 0028                 CALL    DISPA           ;display 16-bit hex
                              STRING  MES5            ;display CR & LF
00B2  BE 0002 R               MOV     SI,OFFSET BUFFER
                              .REPEAT
00B5  8B C6                   MOV     AX,SI
00B7  2D 0002 R               SUB     AX,OFFSET BUFFER
00BA  8B E8                   MOV     BP,AX
00BC  83 E0 0F                AND     AX,15
                              .IF AX==0
                                 STRING  MES5
00CA  E8 0006                    CALL    DISPA
                              .ENDIF
```

```
00CD   E8 002B                         CALL   DISPN
                                      .UNTILCXZ
00D2   C3                              RET
00D3                      DUMP         ENDP

00D3                      DISPA        PROC   NEAR      USES CX
00D4   B9 0004                         MOV    CX,4
                                      .REPEAT
00D7   D1 C5                            ROL    BP,1
00D9   D1 C5                            ROL    BP,1
00DB   D1 C5                            ROL    BP,1
00DD   D1 C5                            ROL    BP,1
00DF   8B C5                            MOV    AX,BP
                                        ASCII
                                      .UNTILCXZ
                                       DISP    ' '
                                       RET
00FB                      DISPA        ENDP

00FB                      DISPN        PROC   NEAR
00FB   AC                              LODSB
00FC   D0 C8                           ROR    AL,1
00FE   D0 C8                           ROR    AL,1
0100   D0 C8                           ROR    AL,1
0102   D0 C8                           ROR    AL,1
                                       ASCII
0114   8A 44 FF                        MOV    AL,[SI-1]
                                       ASCII
                                       DISP    ' '
012D   C3                              RET
012E                      DISPN        ENDP
                                       END
```

7–6   INTERRUPT HOOKS

Hooks are used to tap into or intercept the interrupt structure of the microprocessor. For example, we might hook into the keyboard interrupt so that we can detect a special keystroke called a *hot key*. Whenever the hot key is typed, we can access a terminate and stay resident (TSR) program that performs a special task. Some examples of hot key applications are pop-up calculators and pop-up clocks.

Intercepting an Interrupt

In order to intercept an interrupt, we must use a DOS function call that reads the current address from the interrupt vector. DOS function call number 35H is used to read the current interrupt vector and DOS function call number 25H is used to change the address of the current vector. In both DOS function calls, AL indicates the vector type number (00H–FFH) and AH indicates the DOS function call number.

When the vector is read by using function 35H, the offset address is returned in register BX and the segment address is in register ES. These two registers are saved so that they can be restored when the interrupt hook is removed from memory. When the vector is set, it is set to the address stored at the memory location addressed by DS:DX.

The process of installing an interrupt handler through a hook is illustrated in the program of Example 7–48. This program intercepts the divide error interrupt by first reading the current interrupt vector address and storing it into a double-word memory location for access by the new interrupt service procedure. Next, the address of the new interrupt service procedure, stored in DS:DX, is placed into the vector using DOS function call number 25H.

EXAMPLE 7–48

```
                         ;A sequence of instructions that show the installation
                         ;or a new interrupt for vector 0 (divide error).
                         ;Note this is not a complete program.
                         ;
                                 .MODEL TINY
0000                             .CODE
                                 .STARTUP
0100   EB 05                     JMP  MAIN              ;skip

0102   00000000  ADDR    DD   ?                         ;old interrupt vector

0106             NEW     PROC FAR                       ;new interrupt procedure

0106   CF                       IRET                    ;do nothing interrupt

0107             NEW     ENDP

0107             MAIN:

0107   8C C8                    MOV  AX,CS              ;address CS with DS
0109   8E D8                    MOV  DS,AX

                         ;get vector 0 address

010B   B8 3500                  MOV  AX,3500H
010E   CD 21                    INT  21H

                         ;save vector address at ADDR

0110   89 1E 0102 R             MOV  WORD PTR ADDRESS,BX
0114   8C 06 0104 R             MOV  WORD PTR ADDRESS+2,ES

                         ;install new interrupt vector 0 address

0118   B8 2500                  MOV  AX,2500H
011B   BA 0106 R                MOV  DX,OFFSET NEW
011E   CD 21                    INT  21H

                         ;other installation software continues here
```

Example TSR Alarm

A simple example showing an interrupt hook and TSR causes a beep on the speaker after one hour or one-half hour. We all seem to get lost in computer processing, and this program makes it easy to keep track of time because of the audible beep.

The beep is caused by using timer 2 of the timer found inside the PC in order to generate an audio tone at the speaker. (See Section 12–5 for a discussion of the timer and see Figure 7–7 for its connection in the computer.) Programming timer 2 with a particular beep frequency or tone is accomplished by programming timer 2 with 1,193,180, divided by the desired tone. For example, if we divide 1,193,180 by 800, the speaker generates an 800 Hz audio tone. See the BEEP procedure (shown in Example 7–49) for programming the timer, and turning the speaker on and off after a short wait determined by the number of clock ticks. This procedure uses six clock ticks to produce a beep lasting $1/3$ second. Note that each clock tick occurs about 18.2 times a second (the actual time is closer to 18.206). This is accomplished by using the user wait timer locations in the first segment of the memory. The user wait timer is updated 18.2 times per second by the computer so that it can be used to time events. The program that uses the BEEP procedure causes an audio tone of 1000 Hz, 1200 Hz, and 1400 Hz (each with a $1/3$-second duration) to repeat four times.

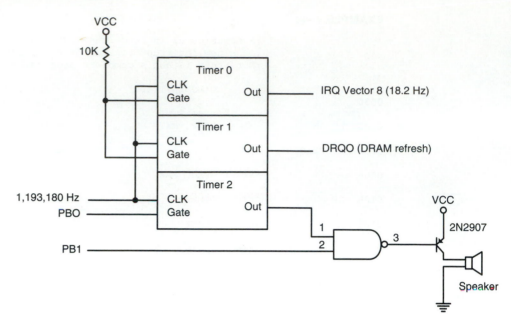

FIGURE 7–7   The speaker and timer circuit in the personal computer (I/O ports 40–43H program the timer, and I/O port 61H programs PB0 and PB1).

EXAMPLE 7–49

```
                        ;A program that beeps the speaker with some sample audio
                        ;tones that each have a duration of 1/3 second.
                        ;
                                .MODEL TINY
0000                            .CODE
                                .STARTUP
0100  B8 0000           MOV    AX,0
0103  8E D8             MOV    DS,AX             ;address segment 0000H
0105  B9 0004           MOV    CX,4              ;set count to 4
0108  E4 61             IN     AL,61H            ;enable timer and speaker
010A  0C 03             OR     AL,3              ;set PB0 and PB1
010C  E6 61             OUT    61H,AL
010E              MAIN1:
010E  BB 03E8           MOV    BX,1000           ;select 1000 Hz tone
0111  E8 0018           CALL   BEEP
0114  BB 04B0           MOV    BX,1200           ;select 1200 Hz tone
0117  E8 0012           CALL   BEEP
011A  BB 0578           MOV    BX,1400           ;select 1400 Hz tone
011D  E8 000C           CALL   BEEP
0120  E2 EC             LOOP   MAIN1             ;repeat 4 times

0122  E4 61             IN     AL,61H            ;turn speaker off
0124  34 03             XOR    AL,3              ;clear PB0 and PB1
0126  E6 61             OUT    61H,AL
                                .EXIT
                        ;
                        ;The BEEP procedure programs timer 2 to beep the speaker
                        ;for 1/3 of a second with the frequency BX.
                        ;***input parameters***
                        ;BX = desired audio tone
                        ;***uses***
                        ;WAITS procedure to wait for 1/3 second
                        ;
012C              BEEP    PROC NEAR               ;beep speaker 1/3 second
```

```
012C  B8 34DC           MOV    AX,34DCH          ;load AX with 1,193,180
012F  BA 0012           MOV    DX,12H
0132  F7 F3             DIV    BX                ;find count
0134  E6 42             OUT    42H,AL            ;program timer 2
0136  8A C4             MOV    AL,AH
0138  E6 42             OUT    42H,AL
013A  E8 0001           CALL   WAITS             ;wait 1/3 second
013D  C3                RET

013E            BEEP    ENDP
                        ;
                        ;the WAITS procedure waits 1/3 of a second
                        ;***uses***
                        ;memory doubleword location 0000:46CH to time the wait
                        ;
013E            WAITS   PROC NEAR

013E  BA 0006           MOV    DX,6              ;number of clock ticks
0141  BB 0000           MOV    BX,0
0144  03 16 046C        ADD    DX,DS:[46CH]      ;get tick count plus time
0148  13 1E 046E        ADC    BX,DS:[46EH]
014C            WAIT1:
014C  8B 2E 046C        MOV    BP,DS:[46CH]      ;test for elapsed time
0150  A1 046E           MOV    AX,DS:[46EH]
0153  2B EA             SUB    BP,DX
0155  1B C3             SBB    AX,BX
0157  72 F3             JC     WAIT1             ;keep testing

0159  C3                RET

015A            WAITS   ENDP
                        END
```

The CHIME program (see Example 7–50) hooks into interrupt vector 8 and beeps the speaker once each half-hour and twice on the hour. This program is a TSR and remains active until the computer is turned off. Note how the TSR is installed and how the interrupt vector is hooked. Also notice that the normal interrupt vector 8 procedure continues to execute, even as the beeper is activated.

EXAMPLE 7–50

```
                        ;A terminate and stay resident program that hooks into
                        ;interrupt vector 8 to beep the speaker one time per
                        ;half-hour and two times per hour.
                        ;***must be assembled as a .COM file*** for use with
                        ;version 5.10 of MASM
                                .MODEL TINY
0000                            .CODE
                                .STARTUP
0100  E9 00CE           JMP    INSTALL           ;install interrupt
   =  03E8      TONE    EQU    1000              ;set tone at 1000 Hz
0103  00        COUNT   DB     0                 ;elapsed time counter
0104  00000000  ADD8    DD     ?                 ;old vector address
0108  00        PASS    DB     0                 ;1 or 2 beeps
0109  00        BEEP    DB     0                 ;beep or silent
010A  00        FLAG    DB     0                 ;busy flag

010B            VEC8    PROC FAR                  ;interrupt procedure

010B  2E: 80 3E 010A R  CMP    CS:FLAG,0         ;test busy flag
      00
0111  74 05             JE     VEC81             ;if not busy
0113  2E: FF 2E 0104 R  JMP    CS:ADD8           ;if busy do normal INT 8
0118            VEC81:
0118  9C                PUSHF                     ;do normal INT 8
0119  2E: FF 1E 0104 R  CALL   CS:ADD8
```

```
011E    2E: C6 06 010A R        MOV   CS:FLAG,1          ;show busy
        01
0124    FB                      STI                      ;allow other interrupts
0125    2E: 80 3E 0108 R        CMP   CS:PASS,0
        00
012B    75 2C                   JNE   VEC83              ;if beep counter active
012D    50                      PUSH  AX                 ;save registers
012E    51                      PUSH  CX
012F    52                      PUSH  DX
0130    B4 02                   MOV   AH,2               ;get time from BIOS
0132    CD 1A                   INT   1AH
0134    80 FE 00                CMP   DH,0               ;is it 00 seconds
0137    75 68                   JNE   VEC86              ;not time yet, so return
0139    80 F9 00                CMP   CL,0               ;test for hour
013C    74 10                   JE    VEC82              ;if hour beep 2 times
013E    80 F9 30                CMP   CL,30H             ;test for half-hour
0141    75 5E                   JNE   VEC86              ;if not half-hour
0143    E8 0065                 CALL  BEEPS              ;start speaker beep
0146    2E: C6 06 0108 R        MOV   CS:PASS,1          ;set number of beeps to 1
        01
014C    EB 53                   JMP   VEC86              ;end it
014E                    VEC82:
014E    E8 005A                 CALL  BEEPS              ;start speaker beep
0151    2E: C6 06 0108 R        MOV   CS:PASS,2          ;set number of beeps to 2
        02
0157    EB 48                   JMP   VEC86              ;end it
0159                    VEC83:
0159    2E: 80 3E 0103 R        CMP   CS:COUNT,0         ;test for end of delay
        00
015F    74 07                   JE    VEC84              ;if time delay has elapsed
0161    2E: FE 0E 0103 R        DEC   CS:COUNT
0166    EB 3C                   JMP   VEC88              ;end it
0168                    VEC84:
0168    2E: 80 3E 0109 R        CMP   CS:BEEP,0          ;test beep on
        00
016E    75 1C                   JNE   VEC85              ;if beep is on
0170    2E: FE 0E 0108 R        DEC   CS:PASS            ;test for 2 beeps
0175    74 2D                   JZ    VEC88              ;if second beep not needed
0177    2E: C6 06 0103 R        MOV   CS:COUNT,9         ;reset count
        09
017D    2E: C6 06 0109 R        MOV   CS:BEEP,1          ;beep on for second beep
        01
0183    50                      PUSH  AX
0184    E4 61                   IN    AL,61H             ;enable speaker for beep
0186    0C 03                   OR    AL,3
0188    E6 61                   OUT   61H,AL
018A    EB 17                   JMP   VEC87              ;end it
018C                    VEC85:
018C    2E: C6 06 0103 R        MOV   CS:COUNT,9         ;reset count
        09
0192    2E: C6 06 0109 R        MOV   CS:BEEP,0          ;show beep is off
        00
0198    50                      PUSH  AX
0199    E4 61                   IN    AL,61H             ;disable speaker
019B    34 03                   XOR   AL,3
019D    E6 61                   OUT   61H,AL
019F    EB 02                   JMP   VEC87              ;end it
01A1                    VEC86:
01A1    5A                      POP   DX                 ;restore registers
01A2    59                      POP   CX
01A3                    VEC87:
01A3    58                      POP   AX
01A4                    VEC88:
01A4    2E: C6 06 010A          MOV   CS:FLAG,0          ;show not busy
        00
01AA    CF                      IRET                     ;interrupt return

01AB                    VEC8    ENDP
                    ;
```

```
                        ;The BEEPS procedure programs the speaker for the
                        ;frequency stored as TONE using an equate at assembly
                        ;time. The duration of the beep is 1/2 second.
                        ;***uses registers AX, CX, and DX***

01AB              BEEPS    PROC NEAR               ;beep speaker

01AB  2E: 8B 0E 03E8     MOV   CX,CS:TONE          ;set tone
01B0  B8 34DC            MOV   AX,34DCH            ;load AX with 1,193,180
01B3  BA 0012            MOV   DX,12H
01B6  F7 F1              DIV   CX                  ;calculate count
01B8  E6 42              OUT   42H,AL              ;program timer 2
01BA  8A C4              MOV   AL,AH
01BC  E6 42              OUT   42H,AL

01BE  E4 61              IN    AL,61H              ;speaker on
01C0  0C 03              OR    AL,3
01C2  E6 61              OUT   61H,AL
01C4  2E: C6 06 0103 R   MOV   CS:COUNT,9          ;set count for 1/2 second
      09
01CA  2E: C6 06 0109 R   MOV   CS:BEEP,1           ;indicate beep is on
      01
01D0  C3                 RET

01D1              BEEPS   ENDP

01D1              INSTALL:                         ;install interrupt VEC8
01D1  8C C8              MOV   AX,CS               ;overlap CS and DS
01D3  8E D8              MOV   DS,AX

01D5  B8 3508            MOV   AX,3508H            ;get current vector 8
01D8  CD 21              INT   21H                 ;and save it
01DA  89 1E 0104 R       MOV   WORD PTR ADD8,BX
01DE  8C 06 0106 R       MOV   WORD PTR ADD8+2,ES

01E2  B8 2508            MOV   AX,2508H
01E5  BA 010B R          MOV   DX,OFFSET VEC8      ;address interrupt VEC8
01E8  CD 21              INT   21H                 ;install vector 8

01EA  BA 01D1 R          MOV   DX,OFFSET INSTALL   ;find paragraphs
01ED  B1 04              MOV   CL,4
01EF  D3 EA              SHR   DX,CL
01F1  42                 INC   DX

01F2  B8 3100            MOV   AX,3100H            ;exit to DOS as TSR
01F5  CD 21              INT   21H
                         END
```

The CHIME program uses several memory locations as flags to signal the operation of the interrupt service procedure. The first flag tested by CHIME is the busy flag (FLAG), which indicates that a part of the interrupt service procedure is active. If FLAG = 1 (busy condition), the procedure jumps to the normal vector 8 interrupt (JMP CS:ADD8), which ends VEC8's execution. If FLAG = 0 (not busy), the interrupt service procedure continues at VEC81. The default address for all direct memory data is the data segment. In the TSR software used in this example and others, it is important to use the segment override prefix (CS:) to ensure that the program addresses data in the code segment, where it appears.

At VEC81, the normal vector 8 interrupt is executed with a forced interrupt call (PUSHF followed by a CALL CS:ADD8). Upon return from the normal vector 8 interrupt (required to keep accurate time), the busy flag is set to show a busy condition (FLAG = 1) and other interrupts are enabled with the STI instruction.

The PASS flag is now tested to see if the VEC8 procedure is currently beeping the speaker. If PASS = 0 (not beeping speaker), the time of day is retrieved from BIOS by using the INT 1AH instruction. It is important not to access DOS from within a TSR or interrupt service procedure. If DOS is accessed at this time, it may be in the process of executing an operation that affects the interrupt. This would cause the program to crash. The INT 1AH instruction returns the number of seconds (DH), minutes (CL), and hours (CH) in BCD form. After obtaining the current time, the number of seconds is tested for zero. If it is not zero seconds, the interrupt procedure ends. If it is zero seconds, then CL is tested for 00 minute (hour) and 30 minutes. If either case is true, the speaker is enabled and TONE is programmed in the timer by a call to BEEPS. If neither case is true, the interrupt ends. Notice that the BEEPS procedure programs timer 2, enables the speaker, and sets the count to 9.

The time delay counter (COUNT) is decremented each time the interrupt occurs. If the count reaches zero, the procedure tests BEEP to control the speaker. If the speaker is beeping, the procedure turns it off and resets the time delay count to 9. If the speaker is not beeping, the procedure tests PASS to determine if another beep is required on the hour. The time delay is $^1/_2$ second (COUNT = 9) in this program and cannot be less. If a delay of less than $^1/_2$ second is chosen, the speaker will beep twice, for both the hour and half hour. The reason is that the clock (INT 1AH) is checked for the zero second. If a time delay of less than $^1/_2$ second is used, the half-hour will be picked up twice.

The TSR program is loaded into memory at the DOS command line by typing the name of the program; in this case, the program is called CHIME. If DOS version 5.0 or 6.X is in use, you can load CHIME into the upper memory or high memory area by typing LOADHIGH CHIME. Once this program loads into memory, it remains in the background, beeping off the time until the power to the computer is disconnected or until it is rebooted. This is an excellent, not too annoying addition to the system to keep track of time. The next section of the text describes hot keys. If desired, a hot key could be used to enable and disable CHIME.

Example Hot Key Program

Hot keys are keystrokes that invoke terminate and stay resident programs. For example, an ALT + C key could be defined as a hot key that calls a program that displays the time. Note that the hot key is detected inside most applications, but not at the DOS command line, where it may lock up the system if used. To detect a hot key, we usually hook into interrupt vector 9, which is the keyboard interrupt that occurs if any key is typed. This allows us to test the keyboard and detect a hot key before the normal interrupt processes the keystroke.

A hot key is installed with a TSR program and an interrupt hook. To illustrate a hot key program that can be useful, a program is developed that counts keystrokes. The keystroke counter program (see Example 7–51) is useful in a business environment that uses computers for data entry or other tasks. With this type of program, productivity can be assessed. The keystroke counter program counts each keystroke and only displays the count when the ALT + K key is pressed. (It is important to note that this program spies on workers and it is the duty of any company using the program to notify the worker. It may even be the responsibility of the company to obtain permission from the worker before a program such as this is placed into service.)

This program can be modified to keep track of keystrokes by the hour or any other time unit. In this example, the keystroke count (up to four billion) accumulates keystrokes for as long as power is applied to the computer. The program also stores the installation time for security purposes. This is important because if a machine is reset, the start time for this TSR will be reset.

This program hooks into interrupt 8 and 9 to count keys. The interrupt 9 hook detects the hot key (ALT + K) and counts keystrokes. When the hot key is detected, the 18.2 Hz interrupt 8 activates the hot key program that displays the keystroke count and time of installation. This type of TSR is often called a *pop-up program* because it pops up when the hot key is typed. Notice that this program uses INT 16H to test the keyboard. Never use a DOS INT 21H function call within a TSR or interrupt hook because serious problems can arise. This program also uses direct manipulation of the video text memory that begins at location B8000H. This memory is organized with two bytes per ASCII character. The first byte contains the ASCII code, and the following byte contains the background and character color.

EXAMPLE 7–51

```
                          ;A TSR program that counts keystrokes and reports the
                          ;time of installation and number of accumulated
                          ;keystrokes when the ALT-K key combination is activated.
                          ;***requires an 80386 or newer microprocessor***
                          ;
                          .MODEL TINY
                          .386
0000                      .CODE
                          .STARTUP
0100   E9 0241            JMP  INSTALL              ;install VEC8 and VEC9

0103   00          HFLAG  DB   0                    ;Hot-key detected
0104   00000000    ADD8   DD   ?                    ;old vector 8 address
0108   00000000    ADD9   DD   ?                    ;old vector 9 address
010C   00000000    COUNT  DD   0                    ;Keystroke counter
0110   00          HOUR   DB   ?                    ;start-up time
0111   00          MIN    DB   ?
0112   0 0         SFLAG  DB   0                    ;start-up flag
0113   00          FLAG8  DB   0                    ;interrupt 8 busy
0114   25          KEY    DB   25H                  ;scan code for K
0115   08          HMASK  DB   8                    ;alternate key mask
0116   08          MKEY   DB   8                    ;alternate key
0117   00A0 [      SCRN   DB   160 DUP (?)          ;screen buffer
          00
       ]
01B7   54 69 6D 65 MES1   DB   'Time = '
       20 3D 20
01BE   20 20 20 4B MES2   DB   '   KeyStrokes = '
       65 79 53 74
       72 6F 6B 65
       73 20 3D 20

01CE               VEC9   PROC FAR                  ;keyboard intercept

01CE   FB                 STI                       ;enable interrupts
01CF   66| 50             PUSH EAX                  ;save EAX
01D1   E4 60              IN   AL,60H               ;get scan code
01D3   2E: 3A 06 0114 R   CMP  AL,CS:KEY            ;test for K
01D8   75 16              JNE  VEC91                ;no hot-key
01DA   B8 0000            MOV  AX,0                 ;address segment 0000
01DD   1E                 PUSH DS                   ;save DS
01DE   8E D8              MOV  DS,AX
01E0   A0 0417            MOV  AL,DS:[417H]         ;get shift/alternate data
01E3   1F                 POP  DS
01E4   2E: 22 06 0115 R   AND  AL,CS:HMASK          ;isolate alternate key
01E9   2E: 3A 06 0116 R   CMP  AL,CS:MKEY           ;test for alternate key
01EE   74 2A              JE   VEC93                ;if hot-key found
01F0               VEC91:
01F0   51                 PUSH CX                   ;add one to BCD COUNT
01F1   B9 0003            MOV  CX,3
01F4   66| 2E: A1 010C R  MOV  EAX,CS:COUNT
01F9   66| 83 C0 01       ADD  EAX,1
01FD   27                 DAA                       ;make result BCD
01FE               VEC92:
01FE   9C                 PUSHF
01FF   66| C1 C8 08       ROR  EAX,8
0203   9D                 POPF
0204   14 00              ADC  AL,0                 ;propagate carry
0206   27                 DAA
0207   E2 F5              LOOP VEC92
0209   66| C1 C8 08       ROR  EAX,8
020D   66| 2E: A3 010C R  MOV  CS:COUNT,EAX
0212   59                 POP  CX
0213   66| 58             POP  EAX
0215   2E: FF 2E 0108 R   JMP  CS:ADD9              ;do normal interrupt
021A               VEC93:                           ;if hot-key pressed
021A   FA                 CLI                       ;interrupts off
```

```
021B  E4 61                        IN    AL,61H          ;clear keyboard and
021D  0C 80                        OR    AL,80H          ;throw away hot key
021F  E6 61                        OUT   61H,AL
0221  24 7F                        AND   AL,7FH
0223  E6 61                        OUT   61H,AL
0225  B0 20                        MOV   AL,20H          ;reset keyboard interrupt
0227  E6 20                        OUT   20H,AL
0229  FB                           STI                   ;enable interrupts
022A  2E: C6 06 0103 R             MOV   CS:HFLAG,1      ;indicate hot-key pressed
      01
0230  66| 58                       POP   EAX
0232  CF                           IRET

0233             VEC9      ENDP

0233             VEC8      PROC FAR                       ;clock tick interrupt

0233  2E: 80 3E 0113 R             CMP   CS:FLAG8,0
      00
0239  74 05                        JZ    VEC81           ;if not busy
023B  2E: FF 2E 0104 R             JMP   CS:ADD8         ;if busy
0240             VEC81:
0240  2E: 80 3E 0103 R             CMP   CS:HFLAG,0
      00
0246  75 37                        JNZ   VEC83           ;if hot-key detected
0248  2E: 80 3E 0112 R             CMP   CS:SFLAG,0
      00
024E  74 05                        JZ    VEC82           ;if start-up
0250  2E: FF 2E 0104 R             JMP   CS:ADD8         ;if not hot-key or start
0255             VEC82:
0255  9C                           PUSHF                 ;do old interrupt 8
0256  2E: FF 1E 0104 R             CALL  CS:ADD8
025B  2E: C6 06 0113 R             MOV   CS:FLAG8,1      ;indicate busy
      01
0261  FB                           STI                   ;enable interrupts
0262  50                           PUSH  AX
0263  51                           PUSH  CX
0264  52                           PUSH  DX
0265  B4 02                        MOV   AH,2            ;get start-up time
0267  CD 1A                        INT   1AH
0269  2E: 88 2E 0110 R             MOV   CS:HOUR,CH      ;save hour
026E  2E: 88 0E 0111 R             MOV   CS:MIN,CL       ;save minute
0273  5A                           POP   DX              ;restore registers
0274  59                           POP   CX
0275  58                           POP   AX
0276  2E: C6 06 0112 R             MOV   CS:SFLAG,1      ;indicate started
      01
027C  E9 00A5                      JMP   VEC89           ;end it
027F             VEC83:                                  ;do hot-key display
027F  9C                           PUSHF                 ;do old interrupt 8
0280  2E: FF 1E 0104 R             CALL  CS:ADD8
0285  2E: C6 06 0113 R             MOV   CS:FLAG8,1      ;indicate busy
      01
028B  FB                           STI                   ;enable interrupts
028C  50                           PUSH  AX              ;save registers
028D  53                           PUSH  BX
028E  B4 0F                        MOV   AH,0FH          ;get video mode
0290  CD 10                        INT   10H
0292  3C 03                        CMP   AL,3
0294  76 05                        JBE   VEC84           ;if DOS text mode
0296  5B                           POP   BX              ;ignore if graphics mode
0297  58                           POP   AX
0298  E9 0083                      JMP   VEC88
029B             VEC84:                                  ;for text mode
029B  51                           PUSH  CX
029C  66| 52                       PUSH  EDX
029E  57                           PUSH  DI
029F  56                           PUSH  SI
```

```
02A0   1E                             PUSH  DS
02A1   06                             PUSH  ES
02A2   FC                             CLD
02A3   8C C8                          MOV   AX,CS           ;address this segment
02A5   8E C0                          MOV   ES,AX
02A7   B8 B800                        MOV   AX,0B800H       ;address text memory
02AA   8E D8                          MOV   DS,AX
02AC   B9 00A0                        MOV   CX,160          ;save top screen line
02AF   BF 0117 R                      MOV   DI,OFFSET SCRN
02B2   BE 0000                        MOV   SI,0
02B5   F3/ A4                         REP   MOVSB
02B7   1E                             PUSH  DS              ;swap segments
02B8   06                             PUSH  ES
02B9   1F                             POP   DS
02BA   07                             POP   ES
02BB   BF 0050                        MOV   DI,80           ;start display at center
02BE   BE 01B7 R                      MOV   SI,OFFSET MES1
02C1   B4 0F                          MOV   AH,0FH          ;load white on black
02C3   B9 0007                        MOV   CX,7
02C6               VEC85:
02C6   AC                             LODSB                 ;display "Time = "
02C7   AB                             STOSW
02C8   E2 FC                          LOOP  VEC85
02CA   2E: 8A 16 0111 R               MOV   DL,CS:MIN
02CF   2E: 8A 36 0110 R               MOV   DH,CS:HOUR
02D4   66| C1 E2 10                   SHL   EDX,16
02D8   B9 0002                        MOV   CX,2
02DB   B3 30                          MOV   BL,30H
02DD   E8 004B                        CALL  DISP            ;display hours
02E0   B0 3A                          MOV   AL,':'
02E2   AB                             STOSW                 ;display colon
02E3   B9 0002                        MOV   CX,2
02E6   B3 80                          MOV   BL,80H
02E8   E8 0040                        CALL  DISP            ;display minutes
02EB   BE 01BE R                      MOV   SI,OFFSET MES2  ;display KeyStrokes =
02EE   B9 0010                        MOV   CX,16
02F1               VEC86:
02F1   AC                             LODSB
02F2   AB                             STOSW
02F3   E2 FC                          LOOP  VEC86
02F5   66| 2E: 8B 16 010C R    MOV    EDX,CS:COUNT          ;get count
02FB   B9 0008                        MOV   CX,8
02FE   B3 30                          MOV   BL,30H
0300   E8 0028                        CALL  DISP            ;display count
0303               VEC87:
0303   B4 01                          MOV   AH,1            ;wait for any key (BIOS)
0305   CD 16                          INT   16H
0307   74 FA                          JZ    VEC87
0309   FC                             CLD
030A   BE 0117 R                      MOV   SI,OFFSET SCRN  ;restore text
030D   BF 0000                        MOV   DI,0
0310   B9 00A0                        MOV   CX,160
0313   F3/ A4                         REP   MOVSB
0315   07                             POP   ES
0316   1F                             POP   DS
0317   5E                             POP   SI
0318   5F                             POP   DI
0319   66| 5A                         POP   EDX
031B   59                             POP   CX
031C   5B                             POP   BX
031D   58                             POP   AX
031E               VEC88:
031E   2E: C6 06 0103 R               MOV   CS:HFLAG,0      ;kill hot-key
       00
0324               VEC89:
0324   2E: C6 06 0113 R               MOV   CS:FLAG8,0      ;indicate not busy
       00
032A   CF                             IRET
```

```
032B                     VEC8     ENDP
                         ;
                         ;The DISP procedure displays the BCD contents of EDX.
                         ;***input parameters***
                         ;CX = number of digits
                         ;BL = 30H for blank leading zeros or 80H for no blanking
                         ;ES = segment address of text mode display
                         ;DI = offset address of text mode display
                         ;
032B                     DISP     PROC NEAR                ;display

032B  66| C1 C2 04                ROL   EDX,4             ;position number
032F  8A C2                       MOV   AL,DL
0331  24 0F                       AND   AL,0FH
0333  04 30                       ADD   AL,30H            ;convert to ASCII
0335  AB                          STOSW                   ;store in text display
0336  3A C3                       CMP   AL,BL             ;test for blanking
0338  74 04                       JE    DISP1             ;if blanking needed
033A  B3 80                       MOV   BL,80H            ;turn off blanking
033C  EB 03                       JMP   DISP2             ;continue
033E               DISP1:
033E  83 EF 02                    SUB   DI,2              ;blank digit
0341               DISP2:
0341  E2 E8                       LOOP  DISP
0343  C3                          RET

0344                     DISP     ENDP

0344                     INSTALL:                         ;install VEC8 and VEC9

0344  8C C8                       MOV   AX,CS             ;load DS
0346  8E D8                       MOV   DS,AX

0348  B8 3508                     MOV   AX,3508H          ;get current vector 8
034B  CD 21                       INT   21H               ;and save it
034D  89 1E 0104 R                MOV   WORD PTR ADD8,BX
0351  8C 06 0106 R                MOV   WORD PTR ADD8+2,ES

0355  B8 3509                     MOV   AX,3509H          ;get current vector 9
0358  CD 21                       INT   21H               ;and save it
035A  89 1E 0108 R                MOV   WORD PTR ADD9,BX
035E  8C 06 010A R                MOV   WORD PTR ADD9+2,ES

0362  B8 2508                     MOV   AX,2508H
0365  BA 0233 R                   MOV   DX,OFFSET VEC8    ;address interrupt procedure
0368  CD 21                       INT   21H               ;install vector 8

036A  B8 2509                     MOV   AX,2509H
036D  BA 01CE R                   MOV   DX,OFFSET VEC9    ;address interrupt procedure
0370  CD 21                       INT   21H               ;install vector 9

0372  BA 0344 R                   MOV   DX,OFFSET INSTALL ;find paragraphs
0375  C1 EA 04                    SHR   DX,4
0378  42                          INC   DX

0379  B8 3100                     MOV   AX,3100H          ;set as a TSR
037C  CD 21                       INT   21H
                                  END
```

Note that the pop-up portion of this program only functions in the text mode and will count any unseen keystrokes that DOS generates. It also counts shift, alternate, and other keys as they are pressed and released. For example, the capital A will be counted as two or three keystrokes. This means that the count will be inflated. Even so, this program is useful for counting keystrokes by a given operator. If the operator reboots the system, the new reboot time is displayed and the count is cleared to zero.

The VEC9 interrupt service procedure intercepts all keystrokes. The IN AL,60H instruction reads the scan code from the keyboard interface within the personal computer. This is then tested for the K scan code. (Refer to Table 7–3 for the key scan codes.) If the K scan code is not found, the procedure increments the BCD count stored at location COUNT and returns to the normal keyboard interrupt handler. If the K scan code is detected, the contents of memory location 0000:0417 are tested for the alternate key. If an alternate key is detected, the program sets the HFLAG to 1, tosses away the hot key, and returns. Notice how the hot key is discarded by strobing I/O port number 61H. The keyboard is cleared by sending a logic 1 in bit position 7 of port 61H, followed by sending a logic 0 in bit position 7. The interrupt controller in the computer must also be cleared by sending a 20H out to I/O port number 20H.

The VEC8 interrupt service procedure tests the HFLAG for the hot key and the SFLAG for system startup. If the SFLAG = 0, the system has just been installed and the time is stored in HOUR and MIN. If the HFLAG = 1, a hot key was detected by VEC9. The VEC8 procedure responds to the hot key by storing the contents of the top line of the text display at memory array SCRN. Once the top line of the text display is stored, the message "Time = " is displayed, followed by the installation time. Next, the message "KeyStrokes = " is displayed, followed by the BCD number stored in COUNT. Recall that count is incremented each time VEC9 detects that a key is typed on the keyboard.

7–7 SUMMARY

1. The assembler program assembles modules that contain PUBLIC variables and segments, plus EXTRN (external) variables. The linker program links modules and library files to create a run-time program executed from the DOS command line. The run-time program usually has the extension EXE.

2. The MACRO and ENDM directives create a new opcode for use in programs. These macros are similar to procedures, except that there is no call or return. In place of them, the assembler inserts the code of the macro sequence into a program each time it is invoked. Macros can include variables that pass information and data to the macro sequence.

3. The DOS INT 21H function call provides a method of using the keyboard and video display. Function number 06H, placed into register AH, provides an interface to the keyboard and display. If DL = 0FFH, this function tests the keyboard for a keystroke. If no keystroke is detected, it returns equal. If a keystroke is detected, the standard ASCII character returns in AL. If an extended ASCII character is typed, it returns with AL = 00H, where the function must again be called to return with the extended ASCII character in AL. To display a character, DL is loaded with the character and AH with 06H before the INT 21H is used in a program.

4. Character strings are displayed by using function number 09H. The DS:DX register combination addresses the character string, which must end with a $.

5. The INT 10H instruction accesses video BIOS (basic I/O system) procedures that control the video display and keyboard. The video BIOS functions are independent of DOS and function with any operating system.

6. The mouse driver is installed at interrupt vector 33H.

7. Data conversion from binary to BCD is accomplished with the AAM instruction for numbers that are less than 100 or by repeated division by 10 for larger numbers. Once converted to BCD, a 30H is added to convert each digit to ASCII code for the video display.

8. When converting from an ASCII number to BCD, a 30H is subtracted from each digit. To obtain the binary equivalent, we multiply by 10.

9. Lookup tables are used for code conversion with the XLAT instruction if the code is an eight-bit code. If the code is wider than eight bits, a short procedure that accesses a lookup table provides the conversion. Lookup tables are also used to hold addresses so that different parts of a program or different procedures can be selected.

10. Conditional assembly language statements allow portions of a program to be assembled if a condition is met. These are useful for tailoring software to an application

11. The disk memory system contains tracks that hold information stored in sectors. Many disk systems store 512 bytes of information per sector. Data on the disk are organized in a boot sector, file allocation table, root directory, and a data storage area. The boot sector loads the DOS system from the disk into the computer memory system. The FAT indicates which sectors are present and whether they contain data. The root directory contains files names and subdirectories, through which all disk files are accessed. The data storage area contains all subdirectories and data files.

12. Files are manipulated with the DOS INT 21H function call. To read a disk file, the file must be opened, read, and then closed. To write to a disk file, it must be opened, written, and then closed. When a file is opened, the file pointer addresses the first byte of the file. To access data at other locations, the file pointer is moved before data are read or written.

13. A sequential access file is a file that is accessed sequentially from the beginning to the end. A random access file is a file that is accessed at any point. Although all disk files are sequential, they can be treated as random access files by using software procedures.

14. The program segment prefix (PSP) contains information about a program. One important part of the PSP is the command line parameters.

15. Interrupt hooks allow application software to gain access to or intercept an interrupt. We often hook into the timer click interrupt (vector 8) or the keyboard interrupt (vector 9).

16. A terminate and stay resident (TSR) program is a program that remains in the memory and is often accessed through a hooked interrupt, using either the timer click or a hot key.

17. A hot key is a key that activates a terminate and stay resident program through the keyboard interrupt hook.

7–8   QUESTIONS AND PROBLEMS

1. The assembler converts a source file to a(n) _____ file.
2. What files are generated from the source file TEST.ASM if it is processed by MASM?
3. The linker program links object files and _____ files to create an execution file.
4. What does the PUBLIC directive indicate when placed in a program module?
5. What does the EXTRN directive indicate when placed in a program module?
6. What directives appear with labels defined external?
7. Describe how a library file works when it is linked to other object files by the linker program.
8. What assembler language directives delineate a macro sequence?
9. What is a macro sequence?
10. How are parameters transferred to a macro sequence?
11. Develop a macro called ADD32 that adds the 32-bit contents of DX–CX to the 32-bit contents of BX–AX.
12. How is the LOCAL directive used within a macro sequence?
13. Develop a macro called ADDLIST PARA1,PARA2 that adds the contents of PARA1 to PARA2. Each of these parameters represents an area of memory. The number of bytes added are indicated by register CX before the macro is invoked.
14. Develop a macro that sums a list of byte-sized data invoked by the macro ADDM LIST,LENGTH. The label LIST is the starting address of the data block and length is the

number of data added. The result must be a 16-bit sum found in AX at the end of the macro sequence.

15. What is the purpose of the INCLUDE directive?

16. Develop a procedure called RANDOM. This procedure must return an eight-bit random number in register CL at the end of the subroutine. (One way to generate a random number is to increment CL each time the DOS function 06H tests the keyboard and finds *no* keystroke. In this way, a random number is generated.)

17. Develop a macro that uses the REPEAT statement to insert 10 NOP instructions in a program.

18. Develop a macro that uses the IFB/IFNB statements to test the parameter PARA in the macro DISP MACRO PARA. If PARA is blank, display a carriage return/line feed combination. If PARA is not blank, display PARA as an ASCII-coded character.

19. Develop a procedure that displays a character string that ends with a 00H. Your procedure must use the DS:DX register to address the start of the character string.

20. Develop a procedure that reads a key and displays the hexadecimal value of an extended ASCII-coded keyboard character if it is typed. If a normal character is typed, ignore it.

21. Use BIOS INT 10H to develop a procedure that positions the cursor at line 3, column 6.

22. What INT instruction is used to access the mouse?

23. Describe how to test for the existence of the mouse in a computer system.

24. How is it determined if the mouse is a serial or a bus mouse?

25. How is it determined if the right mouse button is pressed?

26. Why must the mouse be disabled when data are displayed in the video display?

27. When a number is converted from binary to BCD, the _____ instruction accomplishes the conversion, provided the number is less than 100 decimal.

28. How is a large number (over 100 decimal) converted from binary to BCD?

29. A BCD digit is converted to ASCII code by adding a(n) _____.

30. An ASCII-coded number is converted to BCD by subtracting _____.

31. Develop a procedure that reads an ASCII number from the keyboard and stores it as a BCD number into memory array DATA. The number ends when anything other than a number is typed.

32. Explain how a three-digit ASCII-coded number is converted to binary.

33. Develop a procedure that converts all lowercase ASCII-coded letters into uppercase ASCII-coded letters. Your procedure may not change any other character except the letters a–z.

34. Develop a lookup table that converts hexadecimal data 00H–0FH into the ASCII-coded characters that represent the hexadecimal digits. Make sure to show the lookup table and any software required for the conversion.

35. Develop a program sequence that jumps to memory location ONE if AL = 6, TWO if AL = 7, and THREE if AL = 8.

36. Show how to use the XLAT instruction to access a lookup table called LOOK that is located in the stack segment.

37. Develop a short sequence of instructions that place the line MOV AL,6 into a program if the contents memory location BED are true. You must use the IF statement.

38. Explain the purpose of a boot sector, FAT, and root directory.

39. The surface of a disk is divided into tracks that are further subdivided into _____.

40. What is a bootstrap loader and where is it found?

41. What is a cluster?

42. A directory entry contains an attribute byte. This byte indicates what information about the entry?

43. A directory entry contains the length of the disk file or subdirectory stored in _____ bytes of memory.

44. What is the maximum length of a file?

45. What code is used to store the name of a file when long file names are in use?

46. How many characters may appear in a long file name?

47. Develop a procedure that opens a file called TEST.LST, reads 512 bytes from the file into data segment memory area ARRAY, and closes the file.

48. Develop a procedure that renames file TEST.LST to TEST.LIS.

49. Write a program that reads any decimal number between 0 and 65,535, and displays the 16-bit binary version on the video display.

50. Write a program that displays the binary powers of 2 (in decimal) on the video screen for the powers 0 through 7. Your display shows 2^n = value for each power of 2.

51. Using the technique discussed in question number 16, develop a program that displays random numbers between 1 and 47 (or whatever) for your state's lottery.

52. Develop a program that displays the hexadecimal contents of a block of 256 bytes of memory. Your software must be able to accept the starting address as a hexadecimal number between 00000H and FFF00H.

53. Develop a program the hooks into interrupt vector 0 to display the following message on a divide error: "Oops, you have attempted to divide by 0."

CHAPTER 8

Using Assembly Language with C/C++

INTRODUCTION

Today, it is rare that a complete system is developed using only assembly language. We often use assembly language with C/C++ to develop a system. The assembly language portion usually solves tasks (difficult or inefficient to accomplish in C/C++) that often include control software for peripheral interfaces. This chapter develops the idea of mixing assembly language with C/C++. Some applications in later chapters also illustrate the use of both assembly language and C/C++ to accomplish tasks for the microprocessor.

This text uses Microsoft C/C++, but any version of C/C++ can be used, as long as it is standard ANSI format C/C++. If you want, you can use C/C++ to enter and execute all the programming applications in this text (some slight modifications might be required to programs that use macros and procedures). The 16-bit applications are written by using Microsoft Visual C/C++ version 1.52 or newer; the 32-bit applications are written using Microsoft Visual C/C++ version 4.2 or newer. If you have Visual Studio you can get version 1.52 free of cost from Micorsoft.

CHAPTER OBJECTIVES

Upon completion of this chapter, you will be able to:

1. Use assembly language in _ASM blocks within C/C+.
2. Learn the rules that apply to mixed language software development.
3. Use common C/C++ data and structures with assembly language.
4. Use both the 16-bit (DOS) interface and the 32-bit (Windows) interface with assembly language code.
5. Use assembly language objects with C/C++ programs.

8–1   USING ASSEMBLY LANGUAGE WITH C/C++ FOR 16-BIT APPLICATIONS

This section shows how to incorporate assembly language commands within a C/C++ program. This is important because the performance of a program often depends on the incorporation of assembly language sequences to speed its execution. As mentioned in the introduction to the chapter, assembly language is also used for I/O operations in embedded systems. This text assumes that you are using a version of the Microsoft C/C++ programs, but any C/C++ program should function as shown, if it supports in-line assembly commands. The only change might be setting up the C/C++ package to function with assembly language. This section of the text assumes that you are building 16-bit applications. Make sure that your software can build 16-bit applications before attempting any of the programs in this section. If you build a 32-bit application and attempt to use the DOS INT 21H function, the program will crash because DOS calls are not directly allowed. In fact, they are inefficient to use in a 32-bit application.

Basic Rules

Before assembly language code can be placed in a C/C++ program, some rules must be learned. Example 8–1 shows how to place assembly code inside an assembly language block within a short C/C++ program. Note that all the assembly code in this example is placed in the _asm block. Labels are used as illustrated by *big*: in this example.

Example 8–1 uses no C/C++ commands except for the main procedure. It is important to note that the project type must be MSDOS .EXE console application and the Windows foundation class libraries must not be activated. If you are using Microsoft C/C++, disable the Windows foundation class libraries under the project option in Programmers Workbench. The program (Example 8–1) reads one character from the console keyboard, and then filters it through assembly language so that only the numbers 0 through 9 are sent back to the video display. Although this programming example does not accomplish much, it does show how to set up and use some simple programming constructs.

EXAMPLE 8–1

```
/* Accepts and displays one character of 1 through 9,
   all others characters are ignored */

void main(void) {
    _asm {
        mov    ah,8
        int    21h
        cmp    al,'0'
        jb     big
        cmp    al,'9'
        ja     big
        mov    dx,ax
        mov    ah,2
        int    21h
    }
big:    {;}
}
```

The register AX was not saved in Example 8–1, but it was used by the program. It is very important to note that the AX, BX, CX, DX, and ES registers are never used by Microsoft C/C++. (The function of AX on a return from a procedure is explained later in this chapter.) These registers, which might be considered scratchpad registers, are available to use with assembly language. If you wish to use any of the other registers, make sure that you save them with a PUSH before they are used, and restore them with a POP afterwards. If you fail to save the registers used by a program, the program may not function correctly and can crash the computer. If you are using the 80386 processor or above (this is also selectable) as a base for your program, you need not save EAX, EBX, ECX, EDX, and ES.

Example 8–2 shows the program of Example 8–1 in its complete form (as displayed by CodeView, the debugging program provided with the Microsoft Visual C/C++ package and the assembly language package). The lines that are numbered 1:, 2:, etc., are the original program lines. The lines that are preceded by a segment and offset address (2932:0000, for example) are the lines that are generated by the compiler. Notice how efficient this program is, and note that each assembly language line corresponds to a C/C++ line. Note that the ENTER and LEAVE instructions are used to call C/C++ procedures. The main procedure ends with a LEAVE. The ENTER is not shown because that is how the operating system begins a C/C++ language program.

EXAMPLE 8–2

```
1:
2:      /* Displays one character of 1 through 9
3:         all others ignored */
4:
5:      void main(void)
6:      {
2932:0000 55                 PUSH      BP
2932:0001 8BEC               MOV       BP,SP
2932:0003 B80200             MOV       AX,0002
2932:0006 9AC2023429         CALL      __aFchkstk (2934:02C2)
2932:000B 56                 PUSH      SI
2932:000C 57                 PUSH      DI
7:              _asm {
8:                  mov     ah,8
2932:000D B408               MOV       AH,08
9:                  int     21h
2932:000F CD21               INT       21
10:                 cmp     al,'0'
2932:0011 3C30               CMP       AL,30
11:                 jb      big
2932:0013 720A               JB        big (001F)
12:                 cmp     al,'9'
2932:0015 3C39               CMP       AL,39
13:                 ja      big
2932:0017 7706               JA        big (001F)
14:                 mov     dx,ax
2932:0019 8BD0               MOV       DX,AX
15:                 mov     ah,2
2932:001B B402               MOV       AH,02
16:                 int     21h
2932:001D CD21               INT       21
17:              }
18:
19:    big:       {;}
20:    }
2932:001F 5F                 POP       DI
2932:0020 5E                 POP       SI
2932:0021 C9                 LEAVE
2932:0022 CB                 RETF
21:
```

Example 8–3 shows how to use variables from C/C++ with a short assembly language program. In this example, the char variable type (a byte in C/C++) is used to save space for a few eight-bit bytes of data. The program itself performs the operation X + Y = Z, where X and Y are two one-digit numbers, and Z is the result. As you might imagine, you could use the in-line assembly in C/C++ to learn assembly language and write all the programs in this textbook. The semicolon adds comments to the listing in the _asm block, just as with the normal assembler.

EXAMPLE 8–3

```
void main(void)
{
    char a,b;
    _asm {
        mov     ah,1
        int     21h         ;read character
        mov     a,al        ;save first number
```

```
        mov    ah,1
        int    21h            ;read plus
        cmp    al,'+'
        jne    end1
        mov    ah,1            ;read second number
        int    21h
        mov    b,al
        mov    ah,2
        mov    dl,'='          ;display =
        int    21h
        mov    al,b
        mov    ah,0
        add    al,a            ;generate sum
        aaa                    ;ASCII adjust answer
        add    ax,3030h
        cmp    ah,'0'
        je     down
        push   ax
        mov    dl,ah
        mov    ah,2
        int    21h            ;display 10s position
        pop    ax
      down:
        mov    dl,al
        mov    ah,2
        int    21h            ;display units position
      end1:
    }
}
```

What Cannot Be Used from MASM Inside an _asm Block

Although MASM contains some nice features, such as conditional commands (.IF, .WHILE. REPEAT, etc.), the in-line assembler does not include the conditional command from MASM, nor does it include the MACRO feature found in the assembler. Data allocation with the in-line assembler is handled by C/C++ instead of by using DB, DW, DD, etc. Just about all other features are supported by the in-line assembler. These omissions from the in-line assembler can cause some slight problems, as will be discussed in later sections of this chapter.

Using Character Strings

Example 8–4 illustrates a simple program that uses a character string defined with C/C++, and displays it so that each word is listed on a separate line. Notice the blend of both C/C++ statements and assembly language statements. The WHILE statement repeats the assembly language commands until the NULL (\0) is discovered at the end of the character string. If the NULL is not discovered, the assembly language instructions display a character from the string unless a space is located. For each space, the program displays a carriage return/line feed combination. This causes each word in the string to be displayed on a separate line.

EXAMPLE 8–4

```
/* Displays a character string with one word per line */

void main(void) {
   char string1[]="This is my first test program using _asm.\0";
   int sc = -1;
   while (string1[c++]!=0) {

      _asm {
         push  si                ;save pointer
         mov   si,sc             ;load pointer with index
         mov   dl,string1[si]    ;get string character
```

```
            cmp    dl,' '          ;test for a space
            jne    next            ;if not space
            mov    ah,2            ;if space
            mov    dl,10           ;do CRLF
            int    21h
            mov    dl,13
    next:   mov    ah,2
            int    21h
            pop    si              ;restore pointer
        }
    }
}
```

Suppose that you want to display more than one string in a program, but you still want to use assembly language to develop the software to display a string. Example 8–5 illustrates a program that creates a procedure displaying a character string. This procedure is called each time that a string is displayed in the program. Note that this program displays one string on each line, unlike Example 8–4.

EXAMPLE 8-5

```
/* Program ilustrating assembly language procedure
   for displaying a C/C++ Language string */

    char string1[]="This is my first test program using _asm.\0";
    char string2[]="This is the second line in this program.\0";
    char string3[]="This is the third.\0";

void main(void) {

    string (string1);
    string (string2);
    string (string3);

}

string(char *string_addr[]) {

    _asm {
            mov    bx,string_addr   ;address string
            mov    ah,2             ;set DOS function = 2
    top:
            mov    dl,[bx]          ;display string
            inc    bx
            cmp    dl,0
            je     bot
            int    21h
            jmp    top
    bot:
            mov    dl,13            ;display CRLF
            int    21h
            mov    dl,10
            int    21h
        }
    }
}
```

Using Data Structures

Data structures are an important part of most programs. This section shows how to interface a data structure, created in C/C++, with an assembly language section that manipulates the data in the structure. Example 8–6 illustrates a short program that uses a data structure to store names, ages, and salaries. The program then displays each of the entries by using a few assembly language procedures. Although the string procedure displays a character string, shown in Example 8–5, no carriage return/line feed combination is displayed—instead, a space is displayed. The crlf procedure displays a carriage return/line feed combination. The numb procedure displays an integer.

EXAMPLE 8–6

```c
/* Program illustrating assembly language procedure
   that displays the contents of a C/C++ Language data structure */

/* A simple data structure */

typedef struct records {

    char first_name[16];
    char last_name[16];
    int age;
    int salary;

} RECORD;

/* Fill some records */

RECORD record[4] =
{ {"Bill","Boyd",56,23000},
  {"Robert","Walker", 45, 23000},
  {"Bull",  "Dozer",35,14000},
  {"John","Wayne",44,12000}};

void main(void) {

    int pnt=-1;
    while (pnt++<3) {
        string(record[pnt].last_name);
        string(record[pnt].first_name);
        numb(record[pnt].age);
        numb(record[pnt].salary);
        crlf();
    }

}

string(char *string_addr[]) {

    _asm {
            mov     bx,string_addr          ;address string
            mov     ah,2                    ;set DOS function = 2
    top:
            mov     dl,[bx]                 ;display string
            inc     bx
            cmp     dl,0
            je      bot
            int     21h
            jmp     top
    bot:
            mov     dl,' '
            int     21h
    }
}

crlf() {

    _asm {

            mov     ah,2
            mov     dl,13
            int     21h
            mov     dl,10
            int     21h
    }
```

```
}

numb(int temp) {

    _asm {

            mov     ax,temp
            mov     bx,10
            pus     bx
    l1:
            mov     dx,0
            div     bx
            push    dx
            cmp     ax,0
            jnz     l1
    l2:
            pop     dx
            cmp     dl,10
            je      l3
            mov     ah,2
            add     dl,30h
            int     21h
            jmp     l2
    l3:
            mov     dl,' '
            int     21h

    }

}
```

An Example of a Mixed-Language Program

In order to show how this technique can be applied to any program in the text, Example 8–7 shows how the program in Example 7–45 can be written by using both C/C++ and assembly language. Here, the only assembly language portion of the program is the DISPN procedure that displays an integer. As with the program in Example 7–45, the program in Example 8–7 makes no attempt to detect or correct errors. Notice that this example uses assembly language to perform the I/O and data manipulation, and C/C++ to form the shell of the program.

EXAMPLE 8–7

```
/* Program that emulates Example 7-45 */

int temp, temp1;

void main(void) {

    readn();
    _asm {
        mov   ah,1
        int   21h
    }
    temp1 = temp;
    readn();
    _asm {
        mov   ah,1
        int   21h
    }
    numb(temp1+temp);
}

readn() {

    int a, b;
    _asm {
```

```
            mov     ah,1
            int     21h
            sub     al,30h
            mov     ah,0
            mov     a,al
            mov     ah,1
            int     21h
            sub     al,30h
            mov     ah,0
            mov     b,ax
    }
    temp = a * 10 + b;
}

numb(int temp) {

    _asm {

            mov     ax,temp
            mov     bx,10
            push    bx
    l1:
            mov     dx,0
            div     bx
            push    dx
            cmp     ax,0
            jnz     l1
    l2:
            pop     dx
            cmp     dl,10
            je      l3
            mov     ah,2
            add     dl,30h
            int     21h
            jmp     l2
    l3:
            mov     dl,' '
            int     21h

    }
}
```

8–2   USING ASSEMBLY LANGUAGE WITH C/C++ FOR 32-BIT APPLICATIONS

A difference exists between 16-bit and 32-bit applications. The 32-bit applications are written using Microsoft Visual C/C++ for Windows and the 16-bit applications are written using Microsoft C/C++ for DOS. The main difference is that Visual C/C++ for Windows is more common today, but does not easily call DOS functions such as INT 21H. It is suggested that embedded applications be written in 16-bit C/C++, and applications that incorporate Windows or Windows CE (available on ROM for embedded applications) use 32-bit Visual C/C++ for Windows.

A 32-bit application is written by using any of the 32-bit registers, and the memory space is essentially limited to 4G bytes or whatever Windows is willing to provide. The only thing that will appear different is that you may not use the DOS function calls; instead use the console _getch or _getche and _putch C/C++ language functions. Embedded applications use direct assembly language instructions to access I/O devices in an embedded system.

An Example that Uses Console I/O to Access the Keyboard and Display

Example 8–8 illustrates a simple example that uses the console I/O commands to read and write data from the console. Notice that instead of using the customary stdio.h library, we use the conio.h library. This example program displays any number between 0 and 1000 in all number bases between base 2 and base 16.

EXAMPLE 8–8

```c
/* Program that displays any number in all number bases
   between base 2 and base 16 */

#include <conio.h>

char *buffer = "Enter a number between 0 and 1000: ";
char *buffer1 = "Base ";
int a, b = 0;

void disps(int base, int data) {

int temp;

    _asm {
        mov     eax,data
        mov     ebx,base
        push    ebx
    top1:
        mov     edx,0
        div     ebx
        push    edx
        cmp     eax,0
        jnz     top1
    top2:
        pop     edx
        cmp     edx,ebx
        je      top4
        add     edx,30h
        cmp     edx,39h
        jbe     top3
        add     edx,7
    top3:
        mov     temp,edx
    }
        _putch(temp);
    _asm {jmp   top2}
    top4:;
}

void main(void) {

int i;

    _cputs(buffer);
    a = _getche();
    while ( a >= '0' && a <= '9' ) {
        _asm { sub a,30h}
        b = b * 10 + a;
        a = _getche();
    }
    _putch(10);
    _putch(13);
    for (i = 2; i <17; i++) {
        _cputs(buffer1);
        disps(10, i);
        _putch(' ');
        disps(i, b);
        _putch(10);
        _putch(13);
    }
}
```

This example shows a nice mixture of assembly language and C/C++ language commands. The procedure disps (base,data) does most of the work for this program. It allows any integer (unsigned) to be displayed in any number base, which can be any value between base 2 and base 36. The upper limit occurs because we run out of letters for displaying number bases after the letter Z. If you need to convert to larger number bases, a new scheme for bases over 36 has to be developed. Perhaps the letters a through z could be used for base 37 to 52. Example 8–8 displays the number that is entered in base 2 through base 16.

Directly Addressing I/O Ports

If a program is written that must access an actual port number, we can use console I/O commands such as the _inp(port) command to input byte data, and the _outp(port,byte_data) command to output byte data. When writing software for the personal computer, it is rare to directly address an I/O port, but when we write software for an embedded system, we often directly address an I/O port. An alternate to using the _inp and _out commands is assembly language, which is more efficient in most cases.

An example that directly accesses the I/O ports occurs when the contents of the system CMOS memory must be accessed and possibly displayed. The system CMOS memory is located at I/O port 70H and 71H. Port number 70H is the address port and 71H is the data port for the CMOS memory. The CMOS memory contains setup information about the computer and the system time. (Note that although this may not work on all computers, it does work on most.)

Example 8–9 lists a program that accesses the CMOS memory and displays the contents on the screen in hexadecimal format.

EXAMPLE 8–9

```
/* A program that displays the contents of the CMOS memory's
   first 64 locations */

#include <conio.h>

char *buffer = "The first 64 bytes of CMOS memory.";

void disph(int data) {

int temp1, temp2;

    _asm {
        mov     eax,data
        shr     eax,4
        and     eax,15
        add     eax,30h
        cmp     eax,39h
        jbe     fix1
        add     eax,7
fix1:
        mov     temp1,eax
        mov     eax,data
        and     eax,15
        add     eax,30h
        cmp     eax,39h
        jbe     fix2
        add     eax,7
fix2:
        mov     temp2,eax
    }
    _putch(temp1);
    _putch(temp2);
    _putch(' ');
}
```

```
void main(void) {

int i, a, b;
    _cputs(buffer);
    _putch(10);
    _putch(13);
    _asm {
        in  al,70h
        and         eax,80h
        mov         b,eax
    }
    for (i = 0; i < 64; i++) {
        _asm {
            mov         eax,i
            add         eax,b
            out         70h,al
            mov         eax,0
            in      al,71h
            mov         a,eax
        }
        if (i % 16 == 0) {
            _putch(10);
            _putch(13);
            disph(i);
        }
        disph(a);
    }
    _putch(10);
    _putch(13);
}
```

Table 8–1 shows the contents of the CMOS memory system in most computer systems. If the program is examined, the first access to port 70H reads the NMI mask bit, which must be preserved in some computers. This bit (bit 7) is then stored into variable b. In the FOR loop, when the port is addressed, the NMI bit is then added to the port address before being sent out to the CMOS memory.

TABLE 8–1   Contents of the first 64 bytes of the CMOS memory.

Address	Contents	Address	Contents
00H	Seconds	13H	Reserved
01H	Seconds (alarm)	14H	Device byte
02H	Minutes	15H	Base memory (low byte)
03H	Minutes (alarm)	16H	Base memory (high byte)
04H	Hours	17H	Extended memory (low byte)
05H	Hours (alarm)	18H	Extended memory (high byte)
06H	Day of week	19H	Extension for first HDD
07H	Day of month	1AH	Extension for second HDD
08H	Month	1BH–1FH	Reserved
09H	Year	20H–27H	HDD one parameters
0AH	Status register A	28H–2DH	Reserved
0BH	Status register B	2EH	Checksum (low byte)
0CH	Status register C	2FH	Checksum (high byte)
0DH	Status register D	30H–31H	Post extended memory
0EH	Diagnosis register	32H	Century
0FH	Shutdown status	33H–34H	Setup information
10H	Floppy types	35H–3CH	HDD 2 parameters
11H	Reserved	3DH–3FH	Reserved
12H	Hard drive types		

8–3   SEPARATE ASSEMBLY OBJECTS

As mentioned in the prior sections, the in-line assembler is limited because it cannot use MACRO sequences and the conditional program flow directives. In some cases, it is better to develop assembly language modules that are then linked with C/C++ for more flexibility. This is especially true if the application is being developed by a team of programmers. This section of the chapter details the use of different objects that are linked to form a program using both assembly language and C/C++. Information covered in this section applies to Microsoft Visual C/C++.

Linking Assembly Language with Visual C

Example 8–10 illustrates a flat model procedure that will be linked to a C-language program. We denote that the assembly module is a C-language module by using the letter C after the word "flat" in the model statement. The flat model allows assembly language software to be any length up to 4G bytes. Note that the .386 switch appears before the model statement, which causes the assembler to generate code that functions in the protected 32-bit mode. The Reverse procedure, shown in Example 8–10, accepts a character string from a C-language program, reverses its order, and returns to the C-language program. Notice how this program uses conditional program flow instructions, which are not available with the in-line assembler described in prior sections of this chapter. The assembly language module can be named anything and it can contain more than one procedure, as long as each procedure contains a PUBLIC statement defining the name of the procedure as public. Any parameters that are transferred between the C-language program and the assembly language program are indicated with the backslash following the name of the procedure. This names the parameter for the assembly language program (it can be a different name in C-language) and indicates the size of the parameter. In this example, the parameter is a pointer to a character string.

EXAMPLE 8–10

```
.386
.model flat, C
.stack 1024
.code

public Reverse            ;define Reverse as public

Reverse proc uses esi, \
arraychar:ptr             ;define external pointer

mov      esi,arraychar
push  0

.repeat                   ;push all characters from string to stack
   mov      al,[esi]
   push  ax
   inc      esi
.until byte ptr [esi] == 0

mov      esi,arraychar

.while 1                  ;retrieve characters in reverse order
   pop      ax
   .break .if al==0
   mov      [esi],al
   inc      esi
.endw
ret

Reverse endp
end
```

Example 8–11 illustrates a C-language program that uses the Reverse assembly language procedure. The EXTERN statement is used to indicate that an external procedure called Reverse is to be used in the C-language program. The name of the procedure is case-sensitive, so make sure that it is spelled the same in both the assembly language module and the C-language module. The EXTERN statement in Example 8–11 shows that the external assembly language procedure transfers a character string to the procedure and returns no data. If data are returned from the assembly language procedure, data are returned as a value in register EAX.

EXAMPLE 8-11

```
/* Program that reverses the order of a character string */

#include <stdio.h>

extern void Reverse(char *);

char chararray[17] = "So what is this?";

void main(void) {

    printf ("%s \n", chararray);
    Reverse (chararray);
    printf ("%s\n", chararray);

}
```

Once both the C-language program and the assembly language program are written, the Visual C/C++ development system must be set up to link the two together. For linking, we will assume that the assembly language module is called REVERSE.ASM and the C-language module is called MAIN.C. Both modules are stored in the C:\PROJECT\MINE directory. After the modules are placed in the same project workspace, the Programmers Workbench program is used to edit both assembly language and C-language modules.

To set up the Visual C/C++ developer studio to compile, assemble, and link these files, follow these steps:

1. Start the developer studio program.
2. Select NEW from the FILE menu at the top of the screen.
   a. Choose PROJECT WORKSPACE and press OK.
   b. Select CONSOLE APPLICATION for type, choose MINE for name, and then press CREATE. This creates your project called MINE.
3. To add the program modules to your project, choose FILES INTO PROJECT from the INSERT menu. Next, select the C-language module MAIN.C and click OK.
4. Repeat step 3 for the assembly language module REVERSE.ASM.
5. Choose SETTINGS from the BUILD menu and expand MINE–Win32 Debug by clicking on the + sign. Now click on the module REVERSE.ASM.
6. Select the CUSTOM BUILD tab from the expanded menu.
7. In the BUILD COMMAND(S) box enter the following command:

   C:\MASM611\BIN\ML.EXE /c /Cx /coff $(InputPath)

   This step assumes that the Microsoft Macro Assembler is in the C:\MASM611\BIN directory.
8. In the OUTPUT FILE(S) box, enter REVERSE.OBJ and press OK to save these changes.
9. Finally, you are ready to build this application. To do so, select REBUILD ALL under the BUILD menu. It should build the executable program without any errors. If an error occurs, check the program modules for errors and correct them.

At last, you can execute the program (also under the BUILD menu); you should see two lines of ASCII text data displayed. The first line is in correct forward order and the second is in reverse order. Although this is a trivial application, it does illustrate how to create and link C-language with assembly language.

Example 8–12 shows the C++ version of Example 8–11. The only difference is that the iostream.h library is used with cout instead of printf. The program name was also changed to MAIN.CPP instead of MAIN.C. Another important change is that the "C" is placed in the program following the EXTERN command. The C++ language program uses the C++ pragma. The assembler cannot assemble a module for C++. Instead, we create an assembly language module using the C pragma and link it to the C++ program, indicating that the external assembly language module is written for C-language. The linker creates the correct linkage to handle the connection because of the extern "C" part of the definition.

EXAMPLE 8–12

```
/* C++ program to reverse the order of a character string */

#include <iostream.h>

extern "C" void Reverse(char *);

char chararray[20] = "This is the trick!";

void main(void) {
    cout << chararray << '\n';
    Reverse(chararray);
    cout << chararray << '\n';
}
```

Now that we have a good understanding of interfacing assembly language with C/C++, a longer example is needed that uses a few assembly language procedures with a C-language program. Example 8–13 illustrates an assembly language package that includes a procedure (Scan) that tests a character input against a lookup table and returns a number that indicates the relative position in the table. A second procedure (Look) uses a number transferred to it and returns with a character string that represents Morse code. (The code is not important, but if you are interested, Table 8–2 lists Morse code.)

EXAMPLE 8–13

```
.386
.model flat, C
.data

table  db    2,1,4,8,4,10,3,4      ;ABCD
       db    1,0,4,2,3,6,4,0       ;EFGH
       db    2,0,4,7,3,5,4,4       ;IJKL
       db    2,3,2,2,3,7,4,6       ;MNOP
       db    4,13,3,2,3,0,1,1      ;QRST
       db    3,1,4,1,3,3,4,9       ;UVWX
       db    4,11,4,12             ;YZ

.code
```

TABLE 8–2   Morse code.

A	._	J	.___	S	...
B	_...	K	_._	T	_
C	_._.	L	._..	U	.._
D	_..	M	__	V	..._
E	.	N	_.	W	.__
F	.._.	O	___	X	_.._
G	__.	P	.__.	Y	_.__
H	....	Q	__._	Z	__..
I	..	R	._.		

```
    public  Scan
    public  Look

    Scan proc , \
    char:dword

    mov         eax,char
    .if al >= 'a' && al <= 'z'
        sub         al,32
    .endif
    sub         al,41h
    add         eax,eax
    mov         ebx,eax
    add         ebx,offset table
    mov         ax,word ptr [ebx]
    ret

    Scan endp

    Look proc , \

    numb:dword, \
    pntr:ptr

    mov         ebx,pntr
    mov         eax,numb
    mov         cl,8
    sub         cl,al
    shl         ah,cl
    mov         ecx,0
    mov         cl,al
    .repeat
        shl         ah,1
        .if carry?
            mov         byte ptr [ebx],'_'
        .else
            mov         byte ptr [ebx],'.'
        .endif
        inc         ebx
    .untilcxz
    mov         byte ptr [ebx],0
    ret

    Look endp

    end
```

   The lookup table in Example 8–13 contains two bytes for each character between A and Z. For example, the code for A is a 2 for two dashes or dots, and the 1 is the code for the letter a (. –). This lookup table is accessed by the Scan procedure to obtain the correct Morse code from the lookup table, which is returned in AX as a parameter to the C++ language call. The remaining assembly code is mundane.

   Example 8–14 lists the C++ program, which calls the two procedures listed in Example 8–13. This software is simple to understand, so we do not explain it.

EXAMPLE 8–14

```
/* Program that converts a character string into a
   Morse code sequence. */

#include <iostream.h>

extern "C" int Scan(int);
extern "C" void Look(int, char *);
```

```
char chararray[21] = "This, is the trick!\n";
char chararray1[10];

int a = 0x44, b = 0, c;
void main(void) {

    while (chararray[b] != '\n') {

        if (chararray[b] <= 'A' || chararray[b] >= 'z') {
            cout << chararray[b] << '\n';
        }
        else {

        c = chararray[b];
        c = Scan(c);
        Look(c,chararray1);
        cout << chararray [b] << " = " << chararray1 << '\n';
        }

        b++;
    }
}
```

Adding New Assembly Language Instructions to C/C++ Programs

From time to time, as new microprocessors are introduced by Intel, new assembly language instructions are also introduced. These new instructions cannot be used in C/C++ unless you develop a macro for C/C++ to include them in the program. An example is the CPUID assembly language instruction. This will not function in an _asm block within C/C++ because the in-line assembler does not recognize it. Another group of newer instructions is the MMX instructions. In order to use any of these newer instructions, we look up the machine language code from Appendix B. For example, the machine code for the CPUID instruction is OF A2. This two-byte instruction can be defined as a C/C++ macro, as illustrated in Example 8–15. To use the new macro, all we need to type is CPUID.

EXAMPLE 8–15

```
#define CPUID _asm _emit 0x0f _asm _emit 0xa2
```

8–4   SUMMARY

1. The in-line assembler is used to insert short, limited assembly language sequences into a C/C++ program. The main limitation of the in-line assembler is that it cannot use macro sequences or conditional program flow instructions.
2. Two versions of C/C++ language are available: one is designed for 16-bit DOS applications and the other for 32-bit Windows applications. The type chosen for an application depends on the environment.
3. The 16-bit assembly language applications use the DOS INT 21H commands to access devices in the system. The 32-bit assembly language applications cannot efficiently use the DOS INT 21H function calls even though many are available.
4. The most flexible and most often-used method of interfacing assembly language to a C/C++ program is through separate assembly language modules. The only difference is that these separate assembly language modules must be defined by using the C directive following the .model statement to define the module linkage as C/C++ compatible.

5. The PUBLIC statement is used in an assembly language module to indicate that the procedure name is public and available to use with another module. External parameters are defined in an assembly language module by using the name of the procedure in the PROC statement. Parameters are returned through the EAX register to the calling C/C++ procedure from the assembly language procedure.

6. Assembly language modules are declared external to the C/C++ program by using the EXTERN directive. If the EXTERN directive is followed by the letter "C," the directive is used in a C++ language program. In a C-language program, the "C" directive is not used.

8–5   QUESTIONS AND PROBLEMS

1. Does the in-line assembler support assembly language macro sequences?
2. Can a byte be defined in the in-line assembler by using the DB directive?
3. How are labels defined in the in-line assembler?
4. Which registers can be used in assembly language (either in-line or linked modules) without saving?
5. What register is used to return data to a C/C++ language caller?
6. The linkage to and from assembly language and C/C++ language is accomplished by which assembly language instructions?
7. In Example 8–2, how much additional code is added by the C-language compiler?
8. In Example 8–4, explain how the mov dl,string1[si] instruction accesses string1 data.
9. In Example 8–4, why was the SI register pushed and popped?
10. Notice in Example 8–6 that no C/C++ libraries are used. Do you think that compiled code for this program is smaller than a program to accomplish the same task in C/C++ language? Why?
11. Compile Example 8–7 and assemble Example 7–45. Then compare the sizes of the two execute files.
12. What is the main difference between the 16-bit and 32-bit versions of C/C++ when using the in-line assembler?
13. Can the INT 21H instruction, used to access DOS functions, be used in a program using the 32-bit version of the C/C++ compiler? Explain your answer.
14. What is the #include <conio.h> C/C++ library used for in a program?
15. Write a short C/C++ program that uses the _getche function to read a key and the _putch function to display the key.
16. Would an embedded application that is not written for the PC use the conio.h library?
17. In Example 8–8, what is the purpose of the sequence of instructions _punch(10); followed by _punch(13);?
18. In Example 8–8, explain how a number is displayed in any number base.
19. What is the CMOS memory?
20. If Example 8–9 is executed, can you decode the time? Explain how the time is stored in the CMOS memory.
21. Which is more flexible in its application, the in-line assembler or assembly language modules that are linked to C/C++?
22. What is the purpose of a PUBLIC statement in an assembly code module?
23. How is an assembly code module prepared as a module that is used with C-language?
24. In a C-language program, the **extern void GetIt(int);** statement indicates what about function GetIt?
25. In Example 8–10, what type of parameter is arraychar?

26. Can the edit screen of C/C++ be used to enter and edit an assembly language programming module?

27. How are external procedures that are written in assembly language indicated in C and C++?

28. In Example 8–13, explain what data type is used by Scan.

29. Write a short assembly language module that clears the display for a 16-bit DOS application. Call your procedure CLS. (You may need to refer to Chapter 7 for the required technique.)

30. Write a short assembly language module that receives a parameter (byte-sized) and returns a byte-sized result to a caller. Your procedure must take this byte and convert it into an uppercase letter. If a lowercase letter appears, the byte should not be modified.

CHAPTER 9

8086/8088 Hardware Specifications

INTRODUCTION

In this chapter, we describe the pin functions of both the 8086 and 8088 microprocessors and provide details on the following hardware topics: clock generation, bus buffering, bus latching, timing, wait states, and minimum mode operation versus maximum mode operation. The simple microprocessors are explained first, because of their simple structures, as an introduction to the Intel microprocessor family.

Before it is possible to connect or interface anything to the microprocessor, it is necessary to understand the pin functions and timing. Thus, the information in this chapter is essential to a complete understanding of memory and I/O interfacing, which we cover in the later chapters of the text.

CHAPTER OBJECTIVES

Upon completion of this chapter, you will be able to:

1. Describe the function of each 8086 and 8088 pin.
2. Understand the microprocessor's DC characteristics and indicate its fan-out to common logic families.
3. Use the clock generator chip (8284A) to provide the clock for the microprocessor.
4. Connect buffers and latches to the buses.
5. Interpret the timing diagrams.
6. Describe wait states and connect the circuitry required to cause various numbers waits.
7. Explain the difference between minimum and maximum mode operation.

9–1   PIN-OUTS AND THE PIN FUNCTIONS

In this section, we explain the function and (in certain instances) the multiple functions of each of the microprocessor's pins. In addition, we discuss the DC characteristics to provide a basis for understanding the later sections on buffering and latching.

The Pin-Out

Figure 9–1 illustrates the pin-outs of the 8086 and 8088 microprocessors. As a close comparison reveals, there is virtually no difference between these two microprocessors—both are packaged in 40-pin **dual in-line packages** (DIPs).

As mentioned in Chapter 1, the 8086 is a 16-bit microprocessor with a 16-bit data bus, and the 8088 is a 16-bit microprocessor with an 8-bit data bus. (As the pin-outs show, the 8086 has pin connections AD0–AD15, and the 8088 has pin connections AD0–AD7.) Data bus width is therefore the only major difference between these microprocessors.

There is, however, a minor difference in one of the control signals. The 8086 has an M/$\overline{\text{IO}}$ pin, and the 8088 has an IO/$\overline{\text{M}}$ pin. The only other hardware difference appears on Pin 34 of both chips: on the 8088, it is an SSO pin, while on the 8086, it is a $\overline{\text{BHE}}$/S7 pin.

Power Supply Requirements

Both the 8086 and 8088 microprocessors require +5.0 V with a supply voltage tolerance of ±10 percent. The 8086 uses a maximum supply current of 360 mA, and the 8088 draws a maximum of 340 mA. Both microprocessors operate in ambient temperatures of between 32° F and about 180° F. This range is not wide enough to be used outdoors in the winter or even in the summer, but extended temperature-range versions of the 8086 and 8088 microprocessors are available. There is also a CMOS version, which requires a very low supply current and has an extended temperature range. The 80C88 and 80C86 are CMOS versions that require only 10 mA of power supply current and function in temperature extremes of –40° F through +225° F.

DC Characteristics

It is impossible to connect anything to the pins of the microprocessor without knowing the input current requirement for an input pin and the output current drive capability for an output pin. This knowledge allows the hardware designer to select the proper interface components for use with the microprocessor without the fear of damaging anything.

Input Characteristics.   The input characteristics of these microprocessors are compatible with all the standard logic components available today. Table 9–1 depicts the input voltage levels and the input

FIGURE 9–1   (a) The pin-out of the 8086 microprocessor; (b) the pin-out of the 8088 microprocessor.

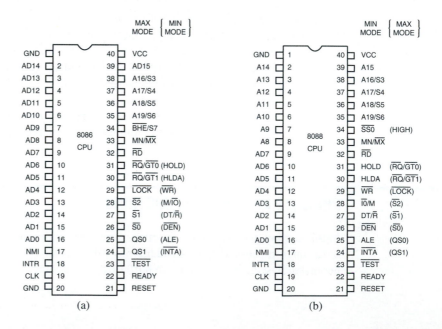

TABLE 9–1 Input characteristics of the 8086 and 8088 microprocessors.

Logic Level	Voltage	Current
0	0.8 V maximum	±10 µA maximum
1	2.0 V minimum	±10 µA maximum

current requirements for any input pin on either microprocessor. The input current levels are very small because the inputs are the gates connections of MOSFETs and represent only leakage currents.

Output Characteristics.   Table 9–2 illustrates the output characteristics of all the output pins of these microprocessors. The logic 1 voltage level of the 8086/8088 is compatible with that of most standard logic families, but the logic 0 level is not. Standard logic circuits have a maximum logic 0 voltage of 0.4 V, and the 8086/8088 has a maximum of 0.45 V. Thus, there is a difference of 0.05 V.

This difference reduces the noise immunity from a standard level of 400 mV (0.8 V – 0.45 V) to 350 mV. (The **noise immunity** is the difference between the logic 0 output voltage and the logic 0 input voltage levels.) This reduced noise immunity may result in problems with long wire connections or too many loads. It is therefore recommended that no more than 10 loads of any type or combination be connected to an output pin without buffering. If this loading factor is exceeded, noise will begin to take its toll in timing problems.

Table 9–3 lists some of the more common logic families and the recommended fan-out from the 8086/8088. The best choice of component types for the connection to an 8086/8088 output pin is a LS, 74ALS, or 74HC logic component. Note that some of the fan-out currents calculate to more than 10 unit loads. It is therefore recommended that if a fan-out of more than 10 unit loads is required, the system should be buffered.

Pin Connections

AD7–AD0   The 8088 **address/data bus** lines compose the multiplexed address data bus of the 8088 and contain the rightmost eight bits of the memory address or I/O port number whenever ALE is active (logic 1) or data whenever ALE is active (logic 0). These pins are at their high-impedance state during a hold acknowledge.

A15–A8   The 8088 **address bus** provides the upper-half memory address bits that are present throughout a bus cycle. These address connections go to their high-impedance state during a hold acknowledge.

AD15–AD8   The 8086 **address/data bus** lines compose the upper multiplexed address/data bus on the 8086. These lines contain address bits A15–A8 whenever ALE is a logic 1, and data bus connections D15–D8. These pins enter a high-impedance state whenever a hold acknowledge occurs.

A19/S6–A16/S3   The **address/status bus** bits are multiplexed to provide address signals A19–A16 and also status bits S6–S3. These pins also attain a high-impedance state during the hold acknowledge.

Status bit S6 always remains a logic 0, bit S5 indicates the condition of the IF flag bits, and S4 and S3 show which segment is accessed during the current bus cycle. See Table 9–4 for the truth table of S4 and

TABLE 9–2 Output characteristics of the 8086 and 8088 microprocessors.

Logic Level	Voltage	Current
0	0.45 V maximum	2.0 mA maximum
1	2.4 V minimum	−400 µA maximum

TABLE 9–3 Recommended fan-out from any 8086/8088 pin connection.

Family	Sink Current	Source Current	Fan-out
TTL (74)	–1.6 mA	40 µA	1
TTL (74LS)	–0.4 mA	20 µA	5
TTL (74S)	–2.0 mA	50 µA	1
TTL (74ALS)	–0.1 mA	20 µA	10
TTL (74AS)	–0.5 mA	25 µA	10
TTL (74F)	–0.5 mA	25 µA	10
CMOS (74HC)	–10 µA	10 µA	10
CMOS (CD4)	–10 µA	10 µA	10
NMOS	–10 µ	10 µA	10

S3. These two status bits can be used to address four separate 1M byte memory banks by decoding them as A21 and A20.

$\overline{RD}$
Whenever the **read signal** is a logic 0, the data bus is receptive to data from the memory or I/O devices connected to the system. This pin floats to its high-impedance state during a hold acknowledge.

READY
This input is controlled to insert wait states into the timing of the microprocessor. If the READY pin is placed at a logic 0 level, the microprocessor enters into wait states and remains idle. If the READY pin is placed at a logic 1 level, it has no effect on the operation of the microprocessor.

INTR
Interrupt request is used to request a hardware interrupt. If INTR is held high when IF = 1, the 8086/8088 enters an interrupt acknowledge cycle ($\overline{INTA}$ becomes active) after the current instruction has complete execution.

TEST
The **Test** pin is an input that is tested by the WAIT instruction. If $\overline{TEST}$ is a logic 0, the WAIT instruction functions as a NOP. If $\overline{TEST}$ is a logic 1, the WAIT instruction waits for $\overline{TEST}$ to become a logic 0. This pin is most often connected to the 8087 numeric coprocessor.

NMI
The **non-maskable interrupt** input is similar to INTR except that the NMI interrupt does not check to see whether the IF flag bit is a logic 1. If NMI is activated, this interrupt input uses interrupt vector 2.

RESET
The **reset** input causes the microprocessor to reset itself if this pin is held high for a minimum of four clocking periods. Whenever the 8086 or 8088 is reset, it begins executing instructions at memory location FFFF0H and disables future interrupts by clearing the IF flag bit.

CLK
The **clock pin** provides the basic timing signal to the microprocessor. The clock signal must have a duty cycle of 33 percent (high for one-third of the clocking period and low for two-thirds) to provide proper internal timing for the 8086/8088.

Vcc
This **power supply** input provides a +5.0 V, ±10 % signal to the microprocessor.

TABLE 9–4 Function of status bits S3 and S4.

S4	S3	Function
0	0	Extra segment
0	1	Stack segment
1	0	Code or no segment
1	1	Data segment

GND	The **ground** connection is the return for the power supply. Note that the 8086/8088 microprocessors have two pins labeled GND—both must be connected to ground for proper operation.
MN/$\overline{\text{MX}}$	The **minimum/maximum mode** pin selects either minimum mode or maximum mode operation for the microprocessor. If minimum mode is selected, the MN/$\overline{\text{MX}}$ pin must be connected directly to +5.0 V.
$\overline{\text{BHE}}$/S7	The **bus high enable** pin is used in the 8086 to enable the most-significant data bus bits (D15–D8) during a read or a write operation. The state of S7 is always a logic 1.

Minimum Mode Pins. Minimum mode operation of the 8086/8088 is obtained by connecting the MN/$\overline{\text{MX}}$ pin diretly to +5.0 V. Do not connect this pin to +5.0 V through a pull-up register, or it will not function correctly.

IO/$\overline{\text{M}}$ or M/$\overline{\text{IO}}$	The IO/$\overline{\text{M}}$ (8088) or the M/$\overline{\text{IO}}$ (8086) pin selects memory or I/O. This pin indicates that the microprocessor address bus contains either a memory address or an I/O port address. This pin is at its high-impedance state during a hold acknowledge.
$\overline{\text{WR}}$	The **write line** is a strobe that indicates that the 8086/8088 is outputting data to a memory or I/O device. During the time that the $\overline{\text{WR}}$ is a logic 0, the data bus contains valid data for memory or I/O. This pin floats to a high-impedance during a hold acknowledge.
$\overline{\text{INTA}}$	The **interrupt acknowledge** signal is a response to the INTR input pin. The $\overline{\text{INTA}}$ pin is normally used to gate the interrupt vector number onto the data bus in response to an interrupt request.
ALE	**Address latch enable** shows that the 8086/8088 address/data bus contains address information. This address can be a memory address or an I/O port number. Note that the ALE signal does not float during a hold acknowledge.
DT/$\overline{\text{R}}$	The **data transmit/receive** signal shows that the microprocessor data bus is transmitting (DT/$\overline{\text{R}}$ = 1) or receiving (DT/$\overline{\text{R}}$ = 0) data. This signal is used to enable external data bus buffers.
$\overline{\text{DEN}}$	**Data bus enable** activates external data bus buffers.
HOLD	The **hold** input requests a direct memory access (DMA). If the HOLD signal is a logic 1, the microprocessor stops executing software and places its address, data, and control bus at the high-impedance state. If the HOLD pin is a logic 0, the microprocessor executes software normally.
HLDA	**Hold acknowledge** indicates that the 8086/8088 has entered the hold state.
$\overline{\text{SS0}}$	The $\overline{\text{SS0}}$ status line is equivalent to the S0 pin in maximum mode operation of the microprocessor. This signal is combined with IO/$\overline{\text{M}}$ and DT/$\overline{\text{R}}$ to decode the function of the current bus cycle (see Table 9–5).

Maximum Mode Pins. In order to achieve maximum mode for use with external coprocessors, connect the MN/$\overline{\text{MX}}$ pin to ground.

$\overline{\text{S2}}$, $\overline{\text{S1}}$, and $\overline{\text{S0}}$	The status bits indicate the function of the current bus cycle. These signals are normally decoded by the 8288 bus controller described later in this chapter. Table 9–6 shows the function of these three status bits in the maximum mode.
$\overline{\text{RO/GT1}}$ and $\overline{\text{RO/GT0}}$	The **request/grant** pins request direct memory accesses (DMA) during maximum mode operation. These lines are bi-directional, and are used to both request and grant a DMA operation.

TABLE 9–5 Bus cycle status (8088) using $\overline{SS0}$.

$IO/\overline{M}$	$DT/\overline{R}$	$\overline{SS0}$	Function
0	0	0	Interrupt acknowledge
0	0	1	Memory read
0	1	0	Memory write
0	1	1	Halt
1	0	0	Opcode fetch
1	0	1	I/O read
1	1	0	I/O write
1	1	1	Passive

TABLE 9–6 Bus control functions generated by the bus controller (8288) using $\overline{S2}$, $\overline{S1}$, and $\overline{S0}$.

$\overline{S2}$	$\overline{S1}$	$\overline{S0}$	Function
0	0	0	Interrupt acknowledge
0	0	1	I/O read
0	1	0	I/O write
0	1	1	Halt
1	0	0	Opcode fetch
1	0	1	Memory read
1	1	0	Memory write
1	1	1	Passive

TABLE 9–7 Queue status bits.

QS1	QS0	Function
0	0	Queue is idle
0	1	First byte of opcode
1	0	Queue is empty
1	1	Subsequent byte of opcode

$\overline{\text{LOCK}}$ The **lock output** is used to lock peripherals off the system. This pin is activated by using the LOCK: prefix on any instruction.

QS1 and QS0   The **queue status** bits show the status of the internal instruction queue. These pins are provided for access by the numeric coprocessor (8087). See Table 9–7 for the operation of the queue status bits.

9–2   CLOCK GENERATOR (8284A)

This section describes the clock generator (8284A), the RESET signal, and introduces the READY signal for the 8086/8088 microprocessors. (The READY signal and its associated circuitry are treated in detail in Section 9–5.)

The 8284A Clock Generator

The 8284A is an ancillary component to the 8086/8088 microprocessors. Without the clock generator, many additional circuits are required to generate the clock (CLK) in an 8086/8088-based system. The 8284A provides the following basic functions or signals: clock generation, RESET synchronization, READY synchronization, and a TTL-level peripheral clock signal. Figure 9–2 illustrates the pin-out of the 8284A clock generator.

FIGURE 9–2 The pin-out of the 8284A clock generator.

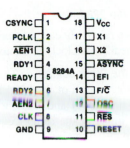

Pin Functions. The 8284A is an 18-pin integrated circuit, designed specifically for use with the 8086/8088 microprocessors. The following is a list of each pin and its function:

$\overline{\text{AEN1}}$ and $\overline{\text{AEN2}}$	The **address enable** pins are provided to qualify the bus ready signals, RDY1 and RDY2, respectively. Section 9–5 illustrates the use of these two pins, which are used to cause wait states, along with the RDY1 and RDY2 inputs. Wait states are generated by the READY pin of the 8086/8088 microprocessors, which is controlled by these two inputs.
RDY1 and RDY2	The **bus ready** inputs are provided, in conjunction with the $\overline{\text{AEN1}}$ and $\overline{\text{AEN2}}$ pins, to cause wait states in an 8086/8088-based system.
$\overline{\text{ASYNC}}$	The **ready synchronization** selection input selects either one or two stages of synchronization for the RDY1 and RDY2 inputs.
READY	**Ready** is an output pin that connects to the 8086/8088 READY input. This signal is synchronized with the RDY1 and RDY2 inputs.
X1 and X2	The **crystal oscillator** pins connect to an external crystal used as the timing source for the clock generator and all its functions.
F/$\overline{\text{C}}$	The **frequency/crystal select** input chooses the clocking source for the 8284A. If this pin is held high, an external clock is provided to the EFI input pin; if it is held low, the internal crystal oscillator provides the timing signal.
EFI	The **external frequency input** is used when the F/$\overline{\text{C}}$ pin is pulled high. EFI supplies the timing whenever the F/$\overline{\text{C}}$ pin is high.
CLK	The **clock output** pin provides the CLK input signal to the 8086/8088 microprocessors and other components in the system. The CLK pin has an output signal that is one-third of the crystal or EFI input frequency, and has a 33-percent duty cycle, which is required by the 8086/8088.
PCLK	The **peripheral clock** signal is one-sixth the crystal or EFI input frequency, and has a 50-percent duty cycle. The PCLK output provides a clock signal to the peripheral equipment in the system.
OSC	The **oscillator output** is a TTL-level signal that is at the same frequency as the crystal or EFI input. The OSC output provides an EFI input to other 8284A clock generators in some multiple-processor systems.
$\overline{\text{RES}}$	The **reset input** is an active-low input to the 8284A. The $\overline{\text{RES}}$ pin is often connected to an RC network that provides power-on resetting.
RESET	The **reset output** is connected to the 8086/8088 RESET input pin.
CSYNC	The **clock synchronization** pin is used whenever the EFI input provides synchronization in systems with multiple processors. If the internal crystal oscillator is used, this pin must be grounded.
GND	The **ground** pin connects to ground.
Vcc	This **power supply** pin connects to +5.0 V with a tolerance of ±10 percent.

Operation of the 8284A

The 8284A is a relatively easy component to understand. Figure 9–3 illustrates the internal block diagram of the 8284A clock generator.

Operation of the Clock Section.

The top half of the logic diagram represents the clock and reset synchronization section of the 8284A clock generator. As the diagram shows, the crystal oscillator has two inputs: X1 and X2. If a crystal is attached to X1 and X2, the oscillator generates a square-wave signal at the same frequency as the crystal. The square-wave signal is fed to an AND gate and also to an inverting buffer that provides the OSC output signal. The OSC signal is sometimes used as an EFI input to other 8284A circuits in a system.

An inspection of the AND gate reveals that when F/$\overline{C}$ is a logic 0, the oscillator output is steered through to the divide-by-3 counter. If F/$\overline{C}$ is a logic 1, then EFI is steered through to the counter.

The output of the divide-by-3 counter generates the timing for ready synchronization, a signal for another counter (divide-by-2), and the CLK signal to the 8086/8088 microprocessors. The CLK signal is also buffered before it leaves the clock generator. Notice that the output of the first counter feeds the second. These two cascaded counters provide the divide-by-6 output at PCLK, the peripheral clock output.

Figure 9–4 shows how an 8284A is connected to the 8086/8088. Notice that F/C and CSYNC are grounded to select the crystal oscillator; and that a 15 MHz crystal provides the normal 5 MHz clock signal to the 8086/8088, as well as a 2.5 MHz peripheral clock signal.

Operation of the Reset Section.

The reset section of the 8284A is very simple: It consists of a Schmitt trigger buffer and a single D-type flip-flop circuit. The D-type flip-flop ensures that the timing requirements of the 8086/8088 RESET input are met. This circuit applies the RESET signal to the microprocessor on the negative edge (1-to-0 transition) of each clock. The 8086/8088 microprocessors sample RESET at the positive edge (0-to-1 transition) of the clocks; therefore, this circuit meets the timing requirements of the 8086/8088.

Refer to Figure 9–4. Notice that an RC circuit provides a logic 0 to the $\overline{RES}$ input pin when power is first applied to the system. After a short time, the $\overline{RES}$ input becomes a logic 1 because the capacitor charges toward +5.0 V through the resistor. A push-button switch allows the microprocessor to be reset by the operator. Correct reset timing requires the RESET input to become a logic 1 no later than four clocks after system power is applied, and to be held high for at least 50 μs. The flip-flop makes certain that RESET goes high in four clocks, and the RC time constant ensures that it stays high for at least 50 μs.

FIGURE 9–3 The internal block diagram of the 8284A clock generator.

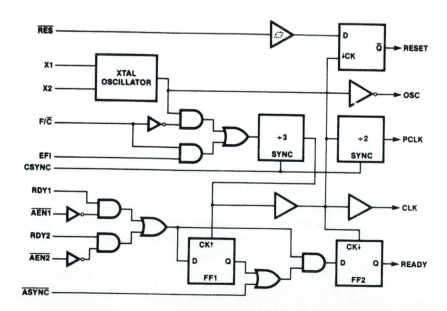

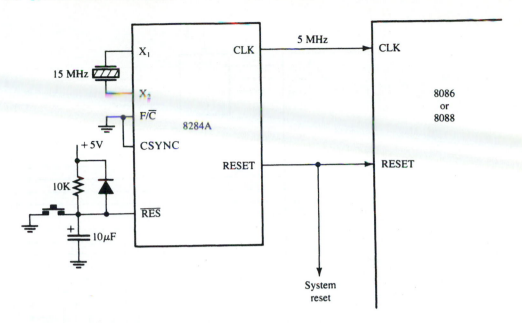

FIGURE 9–4 The clock generator (8284A) and the 8086 and 8088 microprocessor illustrating the connection for the clock and reset signals. A 15 MHz crystal provides the 5 MHz clock for the microprocessor.

9–3   BUS BUFFERING AND LATCHING

Before the 8086/8088 microprocessors can be used with memory or I/O interfaces, their multiplexed buses must be demultiplexed. This section provides the detail required to demultiplex the buses and illustrates how the buses are buffered for very large systems. (Because the maximum fan-out is 10, the system must be buffered if it contains more than 10 other components.)

Demultiplexing the Buses

The address/data bus on the 8086/8088 is multiplexed (**shared**) to reduce the number of pins required for the 8086/8088 microprocessor integrated circuit. Unfortunately, this burdens the hardware designer with the task of extracting or demultiplexing information from these multiplexed pins.

Why not leave the buses multiplexed? Memory and I/O require that the address remains valid and stable throughout a read or write cycle. If the buses are multiplexed, the address changes at the memory and I/O, which causes them to read or write data in the wrong locations.

All computer systems have three buses: (1) an address bus that provides the memory and I/O with the memory address or the I/O port number, (2) a data bus that transfers data between the microprocessor and the memory and I/O in the system, and (3) a control bus that provides control signals to the memory and I/O. These buses must be present in order to interface to memory and I/O.

Demultiplexing the 8088. Figure 9–5 illustrates the 8088 microprocessor and the components required to demultiplex its buses. In this case, two 74LS373 transparent latches are used to demultiplex the address/data bus connections AD7–AD0 and the multiplexed address/status connections A19/S6–A16/S3.

These transparent latches, which are like wires whenever the address latch enable pin (ALE) becomes a logic 1, pass the inputs to the outputs. After a short time, ALE returns to its logic 0 condition, which causes the latches to remember the inputs at the time of the change to a logic 0. In this case, A7–A0 are stored in the bottom latch and A19–A16 are stored in the top latch. This yields a separate address bus with connections A19–A0. These address connections

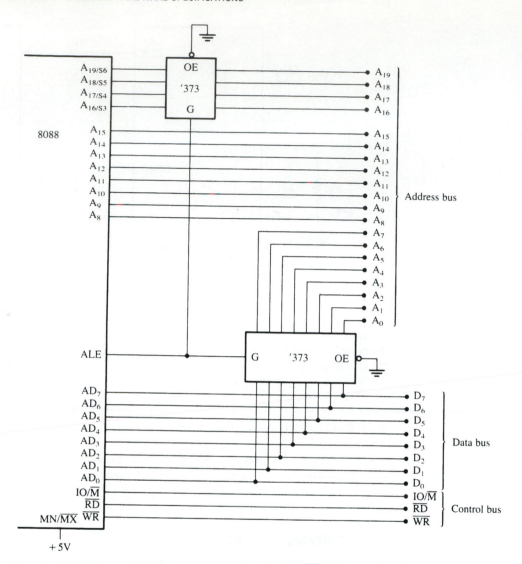

FIGURE 9–5 The 8088 microprocessor shown with a demultiplexed address bus. This is the model used to build many 8088-based systems.

allow the 8088 to address 1M bytes of memory space. The fact that the data bus is separate allows it to be connected to any eight-bit peripheral device or memory component.

Demultiplexing the 8086. Like the 8088, the 8086 system requires separate address, data, and control buses. It differs primarily in the number of multiplexed pins. In the 8088, only AD7–AD0 and A19/S6–A16/S3 are multiplexed. In the 8086, the multiplexed pins include AD15–AD0, A19/S6–A16/S3, and $\overline{BHE}$/S7. All of these signals must be demultiplexed.

Figure 9–6 illustrates a demultiplexed 8086 with all three buses: address (A19–A0 and $\overline{BHE}$), data (D15–D0), and control ($\overline{MIO}$, $\overline{RD}$, and $\overline{WR}$).

This circuit shown in Figure 9–6 is almost identical to the one pictured in Figure 9–5, except that an additional 74LS373 latch has been added to demultiplex the address/data bus pins AD15–AD8 and a $\overline{BHE}$/S7 input has been added to the top 74LS373 to select the high-order memory bank in the 16-bit memory system of the 8086. Here, the memory and I/O system see the 8086 as a device with a 20-bit address bus (A19–A0), a 16-bit data bus (D15–D0), and a three-line control bus (M/$\overline{IO}$), $\overline{RD}$, and $\overline{WR}$).

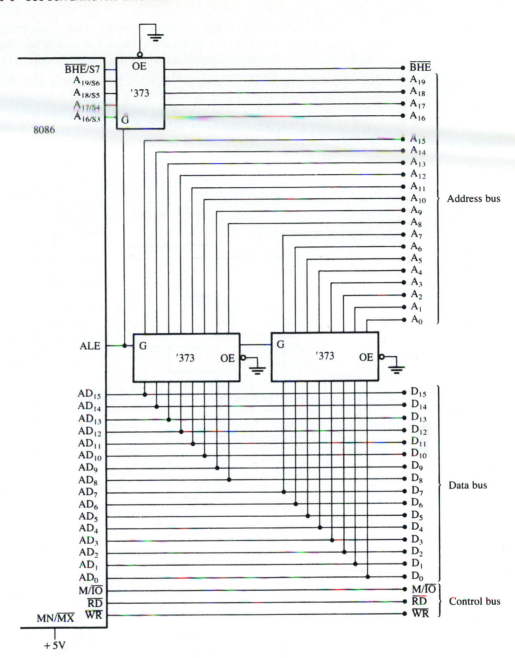

FIGURE 9–6 The 8086 microprocessor shown with a demultiplexed address bus. This is the model used to build many 8086-based systems.

The Buffered System

If more than 10 unit loads are attached to any bus pin, the entire 8086 or 8088 system must be buffered. The demultiplexed pins are already buffered by the 74LS373 latches, which have been designed to drive the high-capacitance buses encountered in microcomputer systems. The buffer's output currents have been increased so that more TTL unit loads may be driven: a logic 0 output provides up to 32 mA of sink current, and a logic 1 output provides up to 5.2 mA of source current.

A fully buffered signal will introduce a timing delay to the system. This causes no difficulty unless memory or I/O devices are used, which function at near the maximum speed of the bus. Section 9–4 discusses this problem and the time delays involved in more detail.

The Fully Buffered 8088.   Figure 9–7 depicts a fully buffered 8088 microprocessor. Notice that the remaining eight address pins, A15–A8, use a 74LS244 octal buffer; the eight data bus pins, D7–D0, use a 74LS245 octal bi-directional bus buffer; and the control bus signals, IO/$\overline{\text{M}}$, $\overline{\text{RD}}$, and $\overline{\text{WR}}$, use a 74LS244 buffer. A fully-buffered 8088 system requires two 74LS244s, one 74LS245, and two 74LS373s. The direction of the 74LS245 is controlled by the DT/$\overline{\text{R}}$ signal, and is enabled and disabled by the $\overline{\text{DEN}}$ signal.

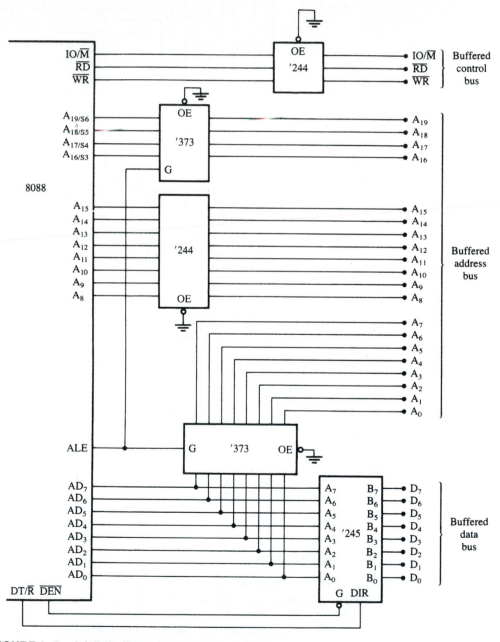

FIGURE 9–7   A fully buffered 8088 microprocessor.

The Fully Buffered 8086.   Figure 9–8 illustrates a fully buffered 8086 microprocessor. Its address pins are already buffered by the 74LS373 address latches; its data bus employs two 74LS245 octal bi-directional bus buffers; and the control bus signals, M/$\overline{\text{IO}}$, $\overline{\text{RD}}$, and $\overline{\text{WR}}$, use a 74LS244 buffer. A fully buffered 8086 system requires one 74LS244, two 74LS245s, and three 74LS373s.

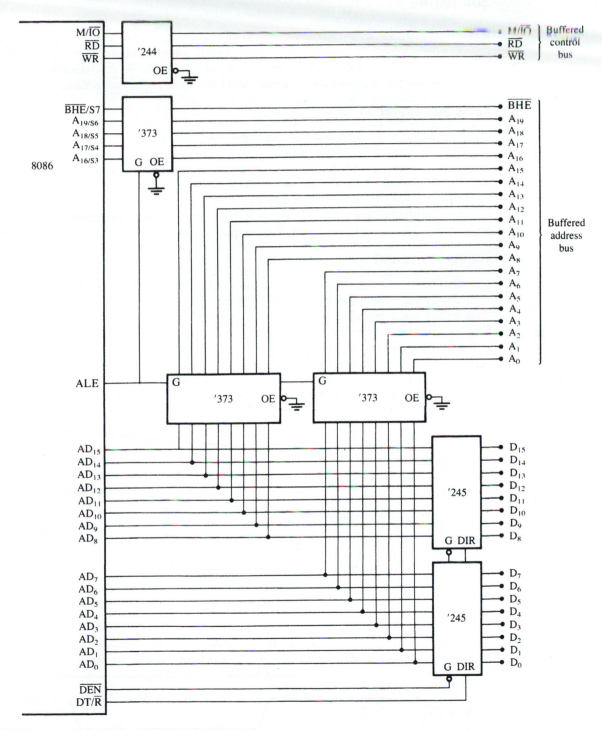

FIGURE 9–8   A fully buffered 8086 microprocessor.

The 8086 requires one more buffer than the 8088 because of the extra eight data bus connections, D15–D8. It also has a $\overline{BHE}$ signal that is buffered for memory-bank selection.

9–4   BUS TIMING

It is essential to understand system bus timing before choosing a memory or I/O device for interfacing to the 8086 or 8088 microprocessors. This section provides insight into the operation of the bus signals, and the basic read and write timing of the 8086/8088. It is important to note that we discuss only the times that affect memory and I/O interfacing in this section.

Basic Bus Operation

The three buses of the 8086 and 8088—address, data, and control—function exactly the same way as those of any other microprocessor. If data are written to the memory (see the simplified timing for write in Figure 9–9), the microprocessor outputs the memory address on the address bus, outputs the data to be written into memory on the data bus, and issues a write ($\overline{WR}$) to memory and IO/$\overline{M}$ = 0 for the 8088 and M/$\overline{IO}$ = 1 for the 8086. If data are read from the memory (see the simplified timing for read in Figure 9–10), the microprocessor outputs the memory address on the address bus, issues a read ($\overline{RD}$) memory signal, and accepts the data via the data bus.

Timing in General

The 8086/8088 microprocessors use the memory and I/O in periods called **bus cycles.** Each bus cycle equals four system-clocking periods (T states). Some new microprocessors divide the bus cycle into as few as two clocking periods. If the clock is operated at 5 MHz (the basic operating frequency for these two microprocessors), one 8086/8088 bus cycle is complete in 800 ns. This means that the microprocessor reads or writes data between itself and memory or I/O at a maximum rate of 1.25 million times a second. (Because of the internal queue, the 8086/8088 can execute 2.5 million instructions per second [MIPS] in bursts.) Other available versions of these microprocessors operate at much higher transfer rates due to higher clock frequencies.

During the first clocking period in a bus cycle, which is called T1, many things happen. The address of the memory or I/O location is sent out via the address bus and the address/data bus connections. (The address/data bus is multiplexed and sometimes contains memory-addressing

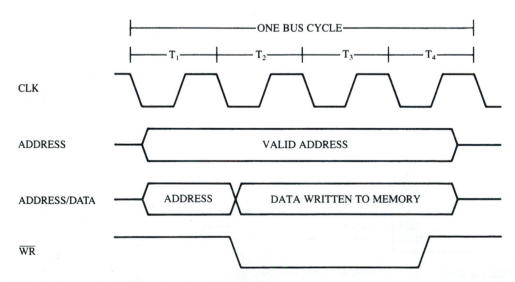

FIGURE 9–9   Simplified 8086/8088 write bus cycle.

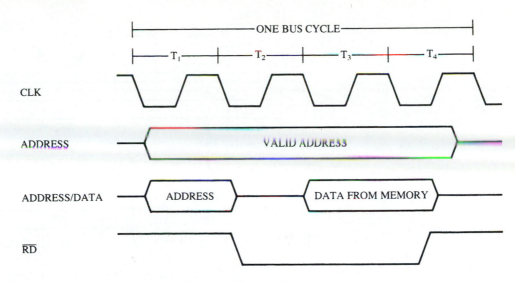

FIGURE 9–10 Simplified 8086/8088 read bus cycle.

information, sometimes data.) During T1, control signals ALE, DT/$\overline{\text{R}}$, and IO/$\overline{\text{M}}$ (8088) or M/$\overline{\text{IO}}$ (8086) are also output. The IO/$\overline{\text{M}}$ or M/$\overline{\text{IO}}$ signal indicates whether the address bus contains a memory address or an I/O device (port) number.

During T2, the 8086/8088 microprocessors issue the $\overline{\text{RD}}$ or $\overline{\text{WR}}$ signal, $\overline{\text{DEN}}$, and in the case of a write, the data to be written appear on the data bus. These events cause the memory or I/O device to begin to perform a read or a write. The $\overline{\text{DEN}}$ signal turns on the data bus buffers, if they are present in the system, so the memory or I/O can receive data to be written, or so the microprocessor can accept the data read from the memory or I/O for a read operation. If this happens to be a write bus cycle, the data are sent out to the memory or I/O through the data bus.

READY is sampled at the end of T2, as illustrated in Figure 9–11. If READY is low at this time, T3 becomes a wait state (T$_{\text{W}}$). (More detail is provided in Section 9–5.) This clocking period is provided to allow the memory time to access data. If the bus cycle happens to be a read bus cycle, the data bus is sampled at the end of T3.

In T4, all bus signals are deactivated in preparation for the next bus cycle. This is also the time when the 8086/8088 samples the data bus connections for data that are read from memory or I/O. In addition, at this point, the trailing edge of the $\overline{\text{WR}}$ signal transfers data to the memory or I/O, which activates and writes when the $\overline{\text{WR}}$ signal returns to a logic 1 level.

Read Timing

Figure 9–11 also depicts the read timing for the 8088 microprocessor. The 8086 read timing is identical except that the 8086 has 16 rather than eight data bus bits. A close look at this timing diagram should allow you to identify all the main events described for each T state.

The most important item contained in the read timing diagram is the amount of time allowed the memory or I/O to read the data. Memory is chosen by its access time, which is the fixed amount of time that the microprocessor allows it to access data for the read operation. It is therefore extremely important that the memory chosen complies with the limitations of the system.

The microprocessor timing diagram does not provide a listing for memory access time. Instead, it is necessary to combine several times to arrive at the access time. To find memory access time in this diagram, first locate the point in T3 when data are sampled. If you examine the timing diagram closely, you will notice a line that extends from the end of T3 down to the data bus. At the end of T3, the microprocessor samples the data bus.

Memory access time starts when the address appears on the memory address bus and continues until the microprocessor samples the memory data at T3. Approximately three T states elapse between these times. (See Figure 9–12 for the following times.) The address does

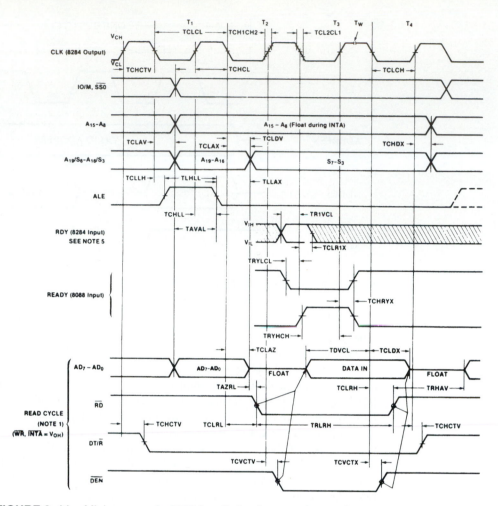

FIGURE 9–11   Minimum mode 8088 bus timing for a read operation.

not appear until T_{CLAV} time (110 ns if the clock is 5 MHz) after the start of T1. This means that T_{CLAV} time must be subtracted from the three clocking states (600 ns) that separate the appearance of the address (T1) and the sampling of the data (T3). One other time must also be subtracted: the data setup time (T_{DVCL}), which occurs before T3. Memory access time is thus three clocking states minus the sum of T_{CLAV} and T_{DVCL}. Because T_{DVCL} is 30 ns with a 5 MHz clock, the allowed memory access time is only 460 ns (access time = 600 ns – 110 ns – 30 ns).

The memory devices chosen for connection to the 8086/8088 operating at 5 MHz must be able to access data in less than 460 ns, because of the time delay introduced by the address decoders and buffers in the system. At least a 30- or 40-ns margin should exist for the operation of these circuits. Therefore, the memory speed should be no slower than about 420 ns to operate correctly with the 8086/8088 microprocessors.

The only other timing factor that may affect memory operation is the width of the $\overline{RD}$ strobe. On the timing diagram, the read strobe is given as T_{RLRH}. The time for this strobe is 325 ns (5 MHz clock rate), which is wide enough for almost all memory devices manufactured with an access time of 400 ns or less.

Write Timing

Figure 9–13 illustrates the write-timing diagram for the 8088 microprocessor. (Again, the 8086 is nearly identical, so it need not be presented here in a separate timing diagram.)

FIGURE 9–12   8088 AC characteristics.

A.C. CHARACTERISTICS (8088: T_A = 0°C to 70°C, V_{CC} = 5V ±10%)*
(8088-2: T_A = 0°C to 70°C, V_{CC} = 5V ±5%)

MINIMUM COMPLEXITY SYSTEM TIMING REQUIREMENTS

Symbol	Parameter	8088 Min.	8088 Max.	8088-2 Min.	8088-2 Max.	Units	Test Conditions
TCLCL	CLK Cycle Period	200	500	125	500	ns	
TCLCH	CLK Low Time	118		68		ns	
TCHCL	CLK High Time	69		44		ns	
TCH1CH2	CLK Rise Time		10		10	ns	From 1.0V to 3.5V
TCL2CL1	CLK Fall Time		10		10	ns	From 3.5V to 1.0V
TDVCL	Data in Setup Time	30		20		ns	
TCLDX	Data in Hold Time	10		10		ns	
TR1VCL	RDY Setup Time into 8284 (See Notes 1, 2)	35		35		ns	
TCLR1X	RDY Hold Time into 8284 (See Notes 1, 2)	0		0		ns	
TRYHCH	READY Setup Time into 8088	118		68		ns	
TCHRYX	READY Hold Time into 8088	30		20		ns	
TRYLCL	READY Inactive to CLK (See Note 3)	−8		−8		ns	
THVCH	HOLD Setup Time	35		20		ns	
TINVCH	INTR, NMI, TEST Setup Time (See Note 2)	30		15		ns	
TILIH	Input Rise Time (Except CLK)		20		20	ns	From 0.8V to 2.0V
TIHIL	Input Fall Time (Except CLK)		12		12	ns	From 2.0V to 0.8V

A.C. CHARACTERISTICS (Continued)

TIMING RESPONSES

Symbol	Parameter	8088 Min.	8088 Max.	8088-2 Min.	8088-2 Max.	Units	Test Conditions
TCLAV	Address Valid Delay	10	110	10	60	ns	
TCLAX	Address Hold Time	10		10		ns	
TCLAZ	Address Float Delay	TCLAX	80	TCLAX	50	ns	
TLHLL	ALE Width	TCLCH−20		TCLCH−10		ns	
TCLLH	ALE Active Delay		80		50	ns	
TCHLL	ALE Inactive Delay		85		55	ns	
TLLAX	Address Hold Time to ALE Inactive	TCHCL−10		TCHCL−10		ns	
TCLDV	Data Valid Delay	10	110	10	60	ns	C_L = 20–100 pF for all 8088 Outputs in addition to internal loads
TCHDX	Data Hold Time	10		10		ns	
TWHDX	Data Hold Time After WR	TCLCH−30		TCLCH−30		ns	
TCVCTV	Control Active Delay 1	10	110	10	70	ns	
TCHCTV	Control Active Delay 2	10	110	10	60	ns	
TCVCTX	Control Inactive Delay	10	110	10	70	ns	
TAZRL	Address Float to READ Active	0		0		ns	
TCLRL	RD Active Delay	10	165	10	100	ns	
TCLRH	RD Inactive Delay	10	150	10	80	ns	
TRHAV	RD Inactive to Next Address Active	TCLCL−45		TCLCL−40		ns	
TCLHAV	HLDA Valid Delay	10	160	10	100	ns	
TRLRH	RD Width	2TCLCL−75		2TCLCL−50		ns	
TWLWH	WR Width	2TCLCL−60		2TCLCL−40		ns	
TAVAL	Address Valid to ALE Low	TCLCH−60		TCLCH−40		ns	
TOLOH	Output Rise Time		20		20	ns	From 0.8V to 2.0V
TOHOL	Output Fall Time		12		12	ns	From 2.0V to 0.8V

The main differences between read and write timing are minimal. The $\overline{RD}$ strobe is replaced by the $\overline{WR}$ strobe, the data bus contains information for the memory rather than information from the memory, and DT/$\overline{R}$ remains a logic 1 instead of a logic 0 throughout the bus cycle.

When interfacing some memory devices, timing may be especially critical between the point at which $\overline{WR}$ becomes a logic 1 and the time when the data are removed from the data bus.

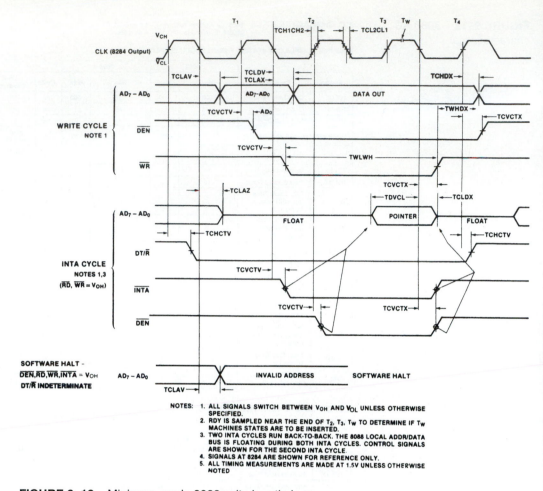

FIGURE 9–13   Minimum mode 8088 write bus timing.

This is the case because, as you will recall, memory data are written at the trailing edge of the $\overline{WR}$ strobe. According to the timing diagram, this critical period is T_{WHDX} or 88 ns when the 8088 is operated with a 5 MHz clock. Hold time is often much less than this; it is, in fact, often 0 ns for memory devices. The width of the $\overline{WR}$ strobe is T_{WLWH} or 340 ns at a 5 MHz clock rate. This rate is compatible with most memory devices that have an access time of 400 ns or less.

9–5   READY AND THE WAIT STATE

As we mentioned earlier in this chapter, the READY input causes wait states for slower memory and I/O components. A **wait state** (T_W) is an extra clocking period, inserted between T2 and T3, to lengthen the bus cycle. If one wait state is inserted, then the memory access time, normally 460 ns with a 5 MHz clock, is lengthened by one clocking period (200 ns) to 660 ns.

In this section, we discuss the READY synchronization circuitry inside the 8284A clock generator, show how to insert one or more wait states selectively into the bus cycle, and examine the READY input and the synchronization times it requires.

The READY Input

The READY input is sampled at the end of T2 and again, if applicable, in the middle of T_W. If READY is a logic 0 at the end of T2, T3 is delayed and T_W is inserted between T2 and T3. READY is next sampled at the middle of T_W to determine whether the next state is T_W or T3. It is tested for a logic 0 on the 1-to-0 transition of the clock at the end of T2, and for a 1 on the 0-to-1 transition of the clock in the middle of T_W.

The READY input to the 8086/8088 has some stringent timing requirements. The timing diagram in Figure 9–14 shows READY causing one wait state (T_W), along with the required setup and hold times from the system clock. The timing requirement for this operation is met by the internal READY synchronization circuitry of the 8284A clock generator. When the 8284A is used for READY, the RDY (ready input to the 8284A) input occurs at the end of each T state.

RDY and the 8284A

RDY is the synchronized ready input to the 8284A clock generator. The timing diagram for this input is provided in Figure 9–15. Although it differs from the timing for the READY input to the 8086/8088, the internal 8284A circuitry guarantees the accuracy of the READY synchronization provided to the 8086/8088 microprocessors.

Figure 9–16 again depicts the internal structure of the 8284A. The bottom half of this diagram is the READY synchronization circuitry. At the leftmost side, the RDY1 and $\overline{AEN1}$ inputs are ANDed, as are the RDY2 and $\overline{AEN2}$ inputs. The outputs of the AND gates are then ORed to generate the input to the one or two stages of synchronization. In order to obtain a logic 1 at the inputs to the filp-flops, RDY1 ANDed with $\overline{AEN1}$ must be active or RDY2 ANDed with $\overline{AEN2}$ must be active.

The $\overline{ASYNC}$ input selects one stage of synchronization when it is a logic 1 and two stages when it is a logic 0. If one stage is selected, then the RDY signal is kept from reaching the 8086/8088 READY pin until the next negative edge of the clock. If two stages are selected, the first

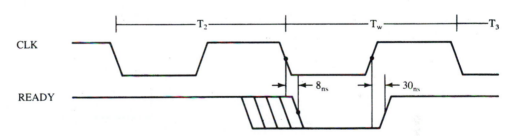

FIGURE 9–14   8086/8088 READY input timing.

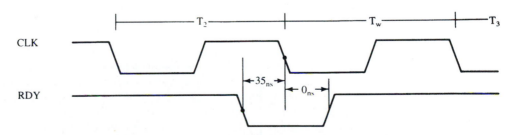

FIGURE 9–15   8284A RDY input timing.

positive edge of the clock captures RDY in the first flip-flop. The output of this flip-flop is fed to the second flip-flop, so on the next negative edge of the clock, the second flip-flop captures RDY.

Figure 9–17 illustrates a circuit used to introduce almost any number of wait states for the 8086/8088 microprocessors. Here, an eight-bit serial shift register (74LS164) shifts a logic 0 for

FIGURE 9–16 The internal block diagram of the 8284A clock generator. (Courtesy of Intel Corporation).

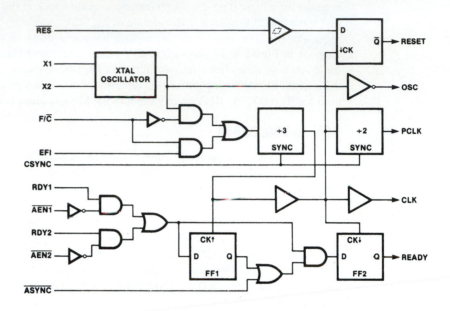

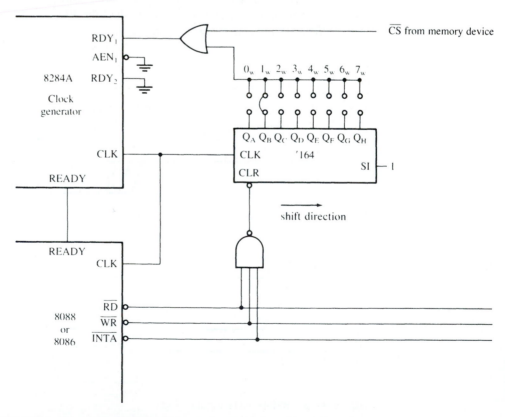

FIGURE 9–17 A circuit that will cause between 0 and 7 wait states.

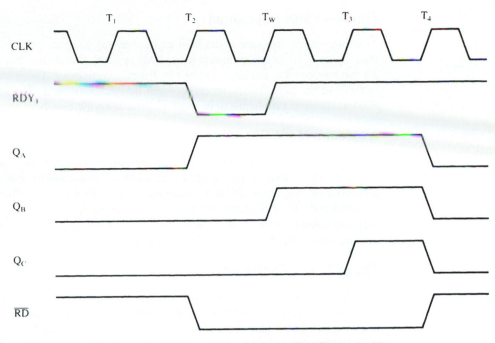

FIGURE 9–18 Wait state generation timing of the circuit of Figure 9–17.

one or more clock periods from one of its Q outputs through to the RDY1 input of the 8284A. With appropriate strapping, this circuit can provide various numbers of wait states. Notice also how the shift register is cleared back to its starting point. The output of the register is forced high when the $\overline{RD}$, $\overline{WR}$, and $\overline{INTA}$ pins are all logic 1s. These three signals are high until state T2, so the shift register shifts for the first time when the positive edge of the T2 arrives. If one wait is desired, output Q_B is connected to the OR gate. If two waits are desired, output Q_C is connected, and so forth.

Notice in Figure 9–17 that this circuit does not always generate wait states. It is enabled from the memory only for memory devices that require the insertion of waits. If the selection signal from a memory device is a logic 0, the device is selected; then this circuit will generate a wait state.

Figure 9–18 illustrates the timing diagram for this shift register wait state generator when it is wired to insert one wait state. The timing diagram also illustrates the internal contents of the shift register's flip-flops to present a more detailed view of its operation. In this example, one wait state is generated.

9–6 MINIMUM MODE VERSUS MAXIMUM MODE

There are two available modes of operation for the 8086/8088 microprocessors: minimum mode and maximum mode. Minimum mode operation is obtained by connecting the mode selection pin MN/$\overline{MX}$ to ±5.0 V, and maximum mode is selected by grounding this pin. Both modes enable different control structures for the 8086/8088 microprocessors. The mode of operation provided by minimum mode is similar to that of the 8085A, the most recent Intel eight-bit microprocessor. The maximum mode is unique and designed to be used whenever a coprocessor exists in a system. Note that the maximum mode was dropped from the Intel family, beginning with the 80286 microprocessor.

Minimum Mode Operation

Minimum mode operation is the least expensive way to operate the 8086/8088 microprocessors (see Figure 9–19 for the minimum mode 8088 system). It costs less because all the control signals for the memory and I/O are generated by the microprocessor. These control signals are identical to those of the Intel 8085A, an earlier eight-bit microprocessor. The minimum mode allows the 8085A, eight-bit peripherals to be used with the 8086/8088 without any special considerations.

Maximum Mode Operation

Maximum mode operation differs from minimum mode in that some of the control signals must be externally generated. This requires the addition of an external bus controller—the 8288 bus controller (see Figure 9–20 for the maximum mode 8088 system). There are not enough pins on the 8086/8088 for bus control during maximum mode because new pins and new features have replaced some of them. Maximum mode is used only when the system contains external coprocessors such as the 8087 arithmetic coprocessor.

The 8288 Bus Controller

An 8086/8088 system that is operated in maximum mode must have an 8288 bus controller to provide the signals eliminated from the 8086/8088 by the maximum mode operation. Figure 9–21 illustrates the block diagram and pin-out of the 8288 bus controller.

Notice that the control bus developed by the 8288 bus controller contains separate signals for I/O ($\overline{\text{IORC}}$ and $\overline{\text{IOWC}}$) and memory ($\overline{\text{MRDC}}$ and $\overline{\text{MWTC}}$). It also contains advanced memory ($\overline{\text{AMWC}}$) and I/O ($\overline{\text{AIOWC}}$) write strobes, and the $\overline{\text{INTA}}$ signal. These signals replace the minimum mode ALE, $\overline{\text{WR}}$, IO/$\overline{\text{M}}$, DT/$\overline{\text{R}}$, $\overline{\text{DEN}}$, and $\overline{\text{INTA}}$, which are lost when the 8086/8088 microprocessors are operated in the maximum mode.

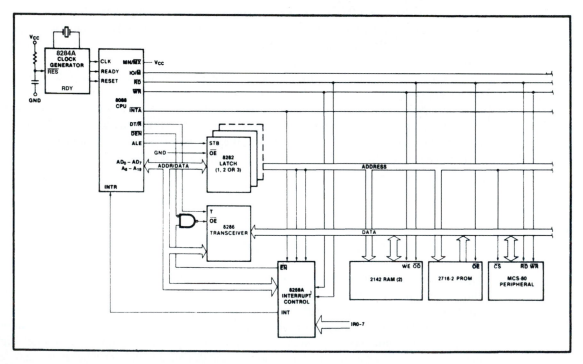

FIGURE 9–19   Minimum mode 8088 system.

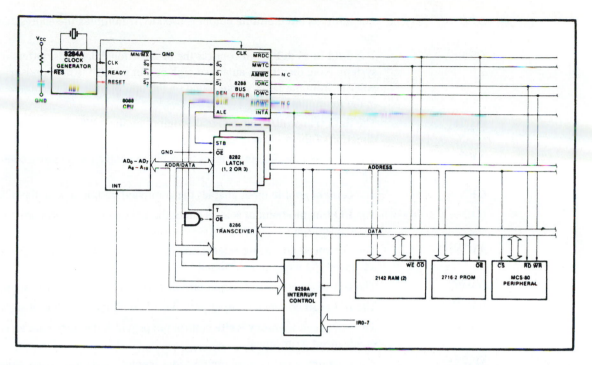

FIGURE 9–20 Maximum mode 8088 system.

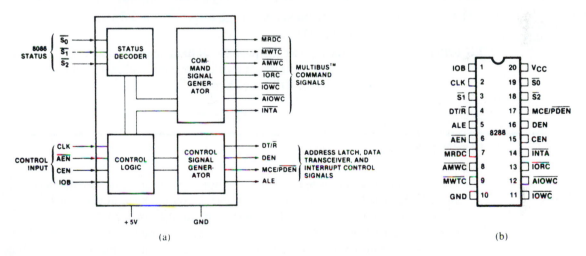

FIGURE 9–21 The 8288 bus controller: (a) block diagram and (b) pin-out.

Pin Functions. The following list provides a description of each pin of the 8288 bus controller.

S2, S1, and S0   **Status** inputs are connected to the staus output pins on the 8086/8088 microprocessor. These three signals are decoded to generate the timing signals for the system.

CLK   The **clock** input provides internal timing and must be connected to the CLK output pin of the 8284A clock generator.

ALE	The **address latch enable** output is used to demultiplex the address/data bus.
DEN	The **data bus enable** pin controls the bi-directional data bus buffers in the system. Note that this is an active high output pin that is the opposite polarity from the $\overline{\text{DEN}}$ signal found on the microprocessor when operated in the minimum mode.
DT/$\overline{\text{R}}$	The **data transmit/receive** signal is output by the 8288 to control the direction of the bi-directional data bus buffers.
$\overline{\text{AEN}}$	The **address enable** input causes the 8288 to enable the memory control signals.
CEN	The **control enable** input enables the command output pins on the 8288.
IOB	The **I/O bus mode** input selects either the I/O bus mode or system bus mode operation.
$\overline{\text{AIOWC}}$	The **advanced I/O write** is a command output used to provide I/O with an advanced I/O write control signal.
$\overline{\text{IOWC}}$	The **I/O write** command output provides I/O with its main write signal.
$\overline{\text{IORC}}$	The **I/O read** command output provides I/O with its read control signal.
$\overline{\text{AMWC}}$	The **advanced memory write** control pin provides memory with an early or advanced write signal.
$\overline{\text{MWTC}}$	The **memory write** control pin provides memory with its normal write control signal.
$\overline{\text{MRDC}}$	The **memory read** control pin provides memory with a read control signal.
$\overline{\text{INTA}}$	The **interrupt acknowledge** output acknowledges an interrupt request input applied to the INTR pin.
MCE/$\overline{\text{PDEN}}$	The **master cascade/peripheral data** output selects cascade operation for an interrupt controller if IOB is grounded, and enables the I/O bus transceivers if IOB is tied high.

9–7   SUMMARY

1. The main differences between the 8086 and 8088 are (a) an eight-bit data bus on the 8088 and a 16-bit data bus on the 8086, (b) an $\overline{\text{SS0}}$ pin on the 8088 in place of $\overline{\text{BHE}}$/S7 on the 8086, and (c) an IO/$\overline{\text{M}}$ pin on the 8088 instead of an M/$\overline{\text{IO}}$ on the 8086.

2. Both the 8086 and 8088 require a single +5.0 V power supply with a tolerance of ±10 %.

3. The 8086/8088 microprocessors are TTL-compatible if the noise immunity figure is de-rated to 350 mV from the customary 400 mV.

4. The 8086/8088 microprocessors can drive one 74XX, five 74LSXX, one 74SXX, ten 74ALSXX, and ten 74HCXX unit loads.

5. The 8284A clock generator provides the system clock (CLK), READY synchronization, and RESET synchronization.

6. The standard 5 MHz 8086/8088 operating frequency is obtained by attaching a 15 MHz crystal to the 8284A clock generator. The PCLK output contains a TTL-compatible signal at one-half the CLK frequency.

7. Whenever the 8086/8088 microprocessors are reset, they begin executing software at memory location FFFF0H (FFFF:0000) with the interrupt request pin disabled.

8. Because the 8086/8088 buses are multiplexed and most memory and I/O devices aren't, the system must be demultiplexed before interfacing with memory or I/O. Demultiplexing is accomplished by an eight-bit latch whose clock pulse is obtained from the ALE signal.

9. In a large system, the buses must be buffered because the 8086/8088 microprocessors are capable of driving only ten unit loads, and large systems often have many more.

10. Bus timing is very important to the remaining chapters in the text. A bus cycle that consists of four clocking periods acts as the basic system timing. Each bus cycle is able to read or write data between the microprocessor and the memory or I/O system.

11. A bus cycle is broken into four states, or T periods: T1 is used by the microprocessor to send the address to the memory or I/O and the ALE signal to the demultiplexers; T2 is used to send data to memory for a write and to test the READY pin and activate control signals $\overline{RD}$ or $\overline{WR}$; T3 allows the memory time to access data and allows data to be transferred between the microprocessor and the memory or I/O; and T4 is where data are written.

12. The 8086/8088 microprocessors allow the memory and I/O 460 ns to access data when they are operated with a 5 MHz clock.

13. Wait states (T_W) stretch the bus cycle by one or more clocking periods to allow the memory and I/O additional access time. Wait states are inserted by controlling the READY input to the 8086/8088. READY is sampled at the end of T2 and during T_W.

14. Minimum mode operation is similar to that of the Intel 8085A microprocessor, while maximum mode operation is new and specifically designed for the operation of the 8087 arithmetic coprocessor.

15. The 8288 bus controller must be used in the maximum mode to provide the control bus signals to the memory and I/O. This is because the maximum mode operation of the 8086/8088 removes some of the system's control signal lines in favor of control signals for the coprocessors. The 8288 reconstructs these removed control signals.

9–8   QUESTIONS AND PROBLEMS

1. List the differences between the 8086 and the 8088 microprocessors.
2. Is the 8086/8088 TTL-compatible? Explain your answer.
3. What is the fan-out from the 8086/8088 to the following devices:
   (a) 74XXX TTL
   (b) 74ALSXXX TTL
   (c) 74HCXXX CMOS
4. What information appears on the address/data bus of the 8088 while ALE is active?
5. What are the purposes of status bits S3 and S4?
6. What condition does a logic 0 on the 8086/8088 $\overline{RD}$ pin indicate?
7. Explain the operation of the $\overline{TEST}$ pin and the WAIT instruction.
8. Describe the signal that is applied to the CLK input pin of the 8086/8088 microprocessors.
9. What mode of operation is selected when MN/$\overline{MX}$ is grounded?
10. What does the $\overline{WR}$ strobe signal from the 8086/8088 indicate about the operation of the 8086/8088?
11. When does ALE float to its high-impedance state?
12. When DT/$\overline{R}$ is a logic 1, what condition does it indicate about the operation of the 8086/8088?
13. What happens when the HOLD input to the 8086/8088 is placed at its logic 1 level?
14. What three minimum mode 8086/8088 pins are decoded to discover whether the processor is halted?

15. Explain the operation of the $\overline{\text{LOCK}}$ pin.
16. What conditions do the QS1 and QS0 pins indicate about the 8086/8088?
17. What three housekeeping chores are provided by the 8284A clock generator?
18. By what factor does the 8284A clock generator divide the crystal oscillator's output frequency?
19. If the F/$\overline{\text{C}}$ pin is placed at a logic 1 level, the crystal oscillator is disabled. Where is the timing input signal attached to the 8284A under this condition?
20. The PCLK output of the 8284A is _____ MHz if the crystal oscillator is operating at 14 MHz.
21. The $\overline{\text{RES}}$ input to the 8284A is placed at a logic _____ level in order to reset the 8086/8088.
22. Which bus connections on the 8086 microprocessor are typically demultiplexed?
23. Which bus connections on the 8088 microprocessor are typically demultiplexed?
24. Which TTL-integrated circuit is often used to demultiplex the buses on the 8086/8088?
25. What is the purpose of the demultiplexed $\overline{\text{BHE}}$ signal on the 8086 microprocessor?
26. Why are buffers often required in an 8086/8088-based system?
27. What 8086/8088 signal is used to select the direction of the data flows through the 74LS245 bi-directional bus buffer?
28. A bus cycle is equal to clocking ___ periods.
29. If the CLK input to the 8086/8088 is 4 MHz, how long is one bus cycle?
30. What two 8086/8088 operations occur during a bus cycle?
31. How many MIPS is the 8086/8088 capable of obtaining when operated with a 10 MHz clock?
32. Briefly describe the purpose of each T state listed:
    (a) T1
    (b) T2
    (c) T3
    (d) T4
33. How much time is allowed for memory access when the 8086/8088 is operated with a 5 MHz clock?
34. How wide is $\overline{\text{DEN}}$ if the 8088 is operated with a 5 MHz clock?
35. If the READY pin is grounded, it will introduce _____ states into the bus cycle of the 8086/8088.
36. What does the $\overline{\text{ASYNC}}$ input to the 8284A accomplish?
37. What logic levels must be applied to $\overline{\text{AEN1}}$ and RDY1 to obtain a logic 1 at the READY pin? (Assume that $\overline{\text{AEN2}}$ is at a logic 1 level.)
38. Contrast minimum and maximum mode 8086/8088 operation.
39. What main function is provided by the 8288 bus controller when used with 8086/8088 maximum mode operation?

CHAPTER 10

Memory Interface

INTRODUCTION

Whether simple or complex, every microprocessor-based system has a memory system. The Intel family of microprocessors is no different from any other in this respect.

Almost all systems contain two main types of memory: read-only memory (ROM) and random access memory (RAM) or read/write memory. Read-only memory contains system software and permanent system data, while RAM contains temporary data and application software. This chapter explains how to interface both memory types to the Intel family of microprocessors. We demonstrate memory interface to an 8-, 16-, and 32-bit data bus by using various memory address sizes. This allows virtually any microprocessor to be interfaced to any memory system.

CHAPTER OBJECTIVES

Upon completion of this chapter, you will be able to:

1. Decode the memory address and use the outputs of the decoder to select various memory components.
2. Use programmable logic devices (PLDs) to decode memory addresses.
3. Explain how to interface both RAM and ROM to a microprocessor.
4. Explain how parity can detect memory errors.
5. Interface memory to an 8-, 16-, 32-, and 64-bit data bus.
6. Explain the operation of a dynamic RAM controller.
7. Interface dynamic RAM to the microprocessor.

10–1 MEMORY DEVICES

Before attempting to interface memory to the microprocessor, it is essential to completely understand the operation of memory components. In this section, we explain the functions of the four common types of memory: **read-only memory** (ROM), **flash memory** (EEPROM), **static random access memory** (SRAM), and **dynamic random access memory** (DRAM).

Memory Pin Connections

Pin connections common to all memory devices are the address inputs, data outputs or input/outputs, some type of selection input, and at least one control input used to select a read or write operation. See Figure 10–1 for ROM and RAM generic-memory devices.

Address Connections.   All memory devices have address inputs that select a memory location within the memory device. Address inputs are almost always labeled from A0, the least-significant address input, to A_n, where subscript n can be any value but is always labeled as one less than the total number of address pins. For example, a memory device with 10 address pins has its address pins labeled from A0 to A9. The number of address pins found on a memory device is determined by the number of memory locations found within it.

Today, the more common memory devices have between 1K (1024) to 64M (67,108,864) memory locations, with 256M memory location devices on the horizon. A 1K memory device has 10 address pins (A0–A9); therefore, 10 address inputs are required to select any of its 1024 memory locations. It takes a 10-bit binary number (1024 different combinations) to select any single location on a 1024-location device. If a memory device has 11 address connections (A0–A10), it has 2048 (2K) internal memory locations. The number of memory locations can thus be extrapolated from the number of address pins. For example, a 4K memory device has 12 address connections, an 8K device has 13, and so forth. A device that contains 1M locations requires a 20-bit address (A0–A19).

A 400H represents a 1K-byte section of the memory system. If a memory device is decoded to begin at memory address 10000H and it is a 1K device, its last location is at address 103FFH—one location less than 400H. Another important hexadecimal number to remember is a 1000H, because 1000H is 4K. A memory device that contains a starting address of 14000H that is 4K bytes long, ends at location 14FFFH—one location less than 1000H. A third number is 64K, or 10000H. A memory that starts at location 30000H and ends at location 3FFFFH is a 64K byte memory. Finally, because 1M of memory is common, a 1M memory contains 100000H memory locations.

Data Connections.   All memory devices have a set of data outputs or input/outputs. The device illustrated in Figure 10–1 has a common set of input/output (I/O) connections. Today, many memory devices have bi-directional common I/O pins.

The data connections are the points at which data are entered for storage or extracted for reading. Data pins on memory devices are labeled D0 through D7 for an 8-bit-wide memory

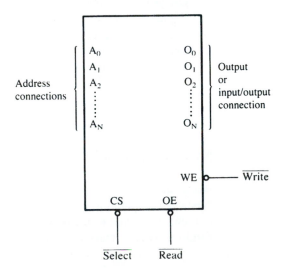

FIGURE 10–1   A pseudo-memory component illustrating the address, data, and control connections.

device. In this sample memory device, there are eight I/O connections, which means that the memory device stores eight bits of data in each of its memory locations. An 8-bit-wide memory device is often called a byte-wide memory. Although most devices are currently eight bits wide, some devices are 16 bits, four bits, or just one bit wide.

Catalog listings of memory devices often refer to memory locations times bits per location. For example, a memory device with 1K memory locations and eight bits in each location is often listed as a 1K × 8 by the manufacturer. A 16K × 1 is a memory device containing 16K 1-bit memory locations. Memory devices are often classified according to total bit capacity. For example, a 1K × 8-bit memory device is sometimes listed as an 8K memory device, or a 64K × 4 memory is listed as a 256K device. These variations occur from one manufacturer to another.

Selection Connections. Each memory device has an input—sometimes more than one—that selects or enables the memory device. This type of input is most often called a **chip select** ($\overline{\text{CS}}$), **chip enable** ($\overline{\text{CE}}$), or simply **select** ($\overline{\text{S}}$) input. RAM memory generally has at least one $\overline{\text{CS}}$ or $\overline{\text{S}}$ input, and ROM has at least one $\overline{\text{CE}}$. If the $\overline{\text{CE}}$, $\overline{\text{CS}}$, or $\overline{\text{S}}$ input is active (a logic 0, in this case, because of the over-bar), the memory device performs a read or write operation; if it is inactive (a logic 1, in this case), the memory device cannot do a read or a write because it is turned off or disabled. If more than one $\overline{\text{CS}}$ connection is present, all must be activated to read or write data.

Control Connections. All memory devices have some form of control input or inputs. A ROM usually has only one control input, while a RAM often has one or two control inputs.

The control input most often found on a ROM is the **output enable** ($\overline{\text{OE}}$) or **gate** ($\overline{\text{G}}$) connection, which allows data to flow out of the output data pins of the ROM. If $\overline{\text{OE}}$ and the selection input ($\overline{\text{CS}}$) are both active, the output is enabled; if $\overline{\text{OE}}$ is inactive, the output is disabled at its high-impedance state. The $\overline{\text{OE}}$ connection enables and disables a set of three-state buffers located within the memory device and must be active to read data.

A RAM memory device has either one or two control inputs. If there is only one control input, it is often called R/$\overline{\text{W}}$. This pin selects a read operation or a write operation only if the device is selected by the selection input ($\overline{\text{CS}}$). If the RAM has two control inputs, they are usually labeled $\overline{\text{WE}}$ (or $\overline{\text{W}}$), and $\overline{\text{OE}}$ (or $\overline{\text{G}}$). Here, $\overline{\text{WE}}$ (**write enable**) must be active to perform a memory write, and $\overline{\text{OE}}$ must be active to perform a memory read operation. When these two controls ($\overline{\text{WE}}$ and $\overline{\text{OE}}$) are present, they must never both be active at the same time. If both control inputs are inactive (logic 1s), data are neither written nor read, and the data connections are at their high-impedance state.

ROM Memory

The **read-only memory** (ROM) permanently stores programs and data that are resident to the system and must not change when power supply is disconnected. The ROM is permanently programmed so that data are always present, even when power is disconnected. This type of memory is often called **nonvolatile memory.**

Today, the ROM is available in many forms. A device we call a ROM is purchased in mass quantities from a manufacturer and programmed during its fabrication at the factory. The EPROM (**erasable programmable read-only memory**), a type of ROM, is more commonly used when software must be changed often or when too few are in demand to make the ROM economical. For a ROM to be practical, we usually must purchase at least 10,000 devices to recoup the factory programming charge. An EPROM is programmed in the field on a device called an *EPROM programmer.* The EPROM is also erasable if exposed to high-intensity ultraviolet light for about 20 minutes or less, depending on the type of EPROM.

PROM memory devices are also available, although they are not as common today. The PROM (**programmable read-only memory**) is also programmed in the field by burning open tiny NI-chrome or silicon oxide fuses; but once it is programmed, it cannot be erased.

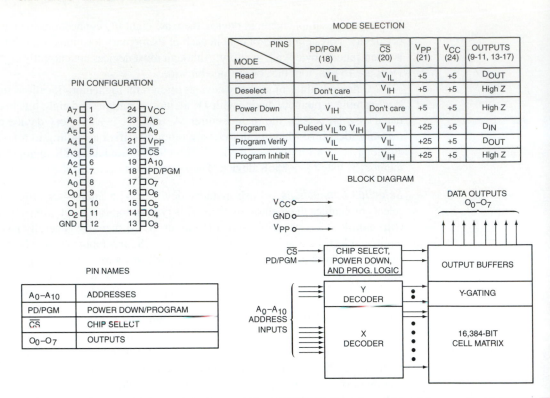

PIN CONFIGURATION

A_7	1	24	V_{CC}
A_6	2	23	A_8
A_5	3	22	A_9
A_4	4	21	V_{PP}
A_3	5	20	$\overline{CS}$
A_2	6	19	A_{10}
A_1	7	18	PD/PGM
A_0	8	17	O_7
O_0	9	16	O_6
O_1	10	15	O_5
O_2	11	14	O_4
GND	12	13	O_3

MODE SELECTION

PINS / MODE	PD/PGM (18)	$\overline{CS}$ (20)	V_{PP} (21)	V_{CC} (24)	OUTPUTS (9-11, 13-17)
Read	V_{IL}	V_{IL}	+5	+5	D_{OUT}
Deselect	Don't care	V_{IH}	+5	+5	High Z
Power Down	V_{IH}	Don't care	+5	+5	High Z
Program	Pulsed V_{IL} to V_{IH}	V_{IH}	+25	+5	D_{IN}
Program Verify	V_{IL}	V_{IL}	+25	+5	D_{OUT}
Program Inhibit	V_{IL}	V_{IH}	+25	+5	High Z

PIN NAMES

A_0-A_{10}	ADDRESSES
PD/PGM	POWER DOWN/PROGRAM
$\overline{CS}$	CHIP SELECT
O_0-O_7	OUTPUTS

FIGURE 10–2   The pin-out of the 2716, 2K × 8 EPROM. (Courtesy of Intel Corporation.)

Still another, newer type of **read-mostly memory** (RMM) is called the **flash memory.** The flash memory[1] is also often called an EEPROM **(electrically erasable programmable ROM),** EAROM **(electrically alterable ROM),** or a NOVRAM **(nonvolatile RAM).** These memory devices are electrically erasable in the system, but they require more time to erase than a normal RAM. The flash memory device is used to store setup information for systems such as the video card in the computer. It may also soon replace the EPROM in many computer systems for the BIOS memory. Some systems contain a password stored in the flash memory device.

Figure 10–2 illustrates the 2716 EPROM, which is representative of most common EPROMs. This device contains 11 address inputs and eight data outputs. The 2716 is a 2K × 8 read-only memory device. The 27XXX series of the EPROMs includes the following part numbers: 2704 (512 × 8), 2708 (1K × 8), 2716 (2K × 8), 2732 (4K × 8), 2764 (8K × 8), 27128 (16K × 8), 27256 (32K × 8), 27512 (64K × 8), and 271024 (128K × 8). Each of these parts contains address pins, eight data connections, one or more chip selection inputs ($\overline{CE}$), and an output enable pin ($\overline{OE}$).

Figure 10–3 illustrates the timing diagram for the 2716 EPROM. Data appear on the output connections only after a logic 0 is placed on both $\overline{CE}$ and $\overline{OE}$ pin connections. If $\overline{CE}$ and $\overline{OE}$ are not both logic 0s, the data output connections remain at their high-impedance or off states. Note that the VPP pin must be placed at a logic 1 level for data to be read from the EPROM. In some cases, the VPP pin is in the same position as the $\overline{WE}$ pin on the SRAM. This can allow a single socket to hold either an EPROM or an SRAM. An example is the 27256 EPROM and the 62256 SRAM, both 32K × 8 devices that have the same pin-out, except for VPP on the EPROM and $\overline{WE}$ on the SRAM.

---

[1]Flash memory is a registered trademark of Intel Corporation.

A.C. Characteristics

$T_A = 0°C$ to $70°C$, $V_{CC}^{[1]} = +5V \pm 5\%$, $V_{PP}^{[2]} = V_{CC} \pm 0.6V^{[3]}$

Symbol	Parameter	Limits			Unit	Test Conditions
		Min.	Typ.[4]	Max.		
t_{ACC1}	Address to Output Delay		250	450	ns	PD/PGM = $\overline{CS}$ = V_{IL}
t_{ACC2}	PD/PGM to Output Delay		280	450	ns	$\overline{CS}$ = V_{IL}
t_{CO}	Chip Select to Output Delay			120	ns	PD/PGM = V_{IL}
t_{PF}	PD/PGM to Output Float	0		100	ns	$\overline{CS}$ = V_{IL}
t_{DF}	Chip Deselect to Output Float	0		100	ns	PD/PGM = V_{IL}
t_{OH}	Address to Output Hold	0			ns	PD/PGM = $\overline{CS}$ = V_{IL}

Capacitance[5] $T_A = 25°C$, $f = 1$ MHz

Symbol	Parameter	Typ.	Max.	Unit	Conditions
C_{IN}	Input Capacitance	4	6	pF	V_{IN} = 0V
C_{OUT}	Output Capacitance	8	12	pF	V_{OUT} = 0V

A.C. Test Conditions:

Output Load: 1 TTL gate and C_L = 100 pF
Input Rise and Fall Times: ⩽20 ns
Input Pulse Levels: 0.8V to 2.2V
Timing Measurement Reference Level:
  Inputs   1V and 2V
  Outputs 0.8V and 2V

WAVEFORMS

A. Read Mode
PD/PGM = V_{IL}

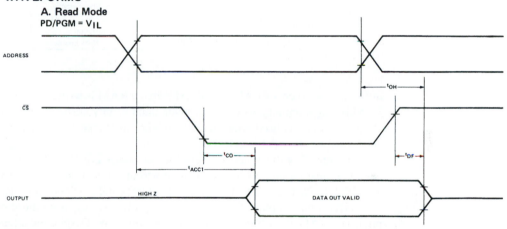

FIGURE 10–3 The timing diagram of AC characteristics of the 2716 EPROM. (Courtesy of Intel Corporation.)

One important piece of information provided by the timing diagram and data sheet is the memory access time—the time that it takes the memory to read information. As Figure 10–3 illustrates, memory access time (T_{ACC}) is measured from the appearance of the address at the address inputs until the appearance of the data at the output connections. This is based on the assumption that the $\overline{CE}$ input goes low at the same time that the address inputs become stable. Also, $\overline{OE}$ must be a logic 0 for the output connections to become active. The basic speed of this EPROM is 450 ns. (Recall that the 8086/8088 operated with a 5 MHz clock allowed memory 460 ns to access data.) This type of memory component requires wait states to operate properly with the 8086/8088 microprocessors because of its rather long access time. If wait states are not desired, higher-speed versions of the EPROM are available at an additional cost. Today, EPROM memory is available with access times of as little as 100 ns.

Static RAM (SRAM) Devices

Static RAM memory devices retain data for as long as DC power is applied. Because no special action (except power) is required to retain stored data, these devices are called **static memory.** They are also called **volatile memory** because they will not retain data without power. The main

FIGURE 10–4 The pin-out of the TMS4016, 2K × 8 static RAM (SRAM). (Courtesy of Texas Instruments Incorporated.)

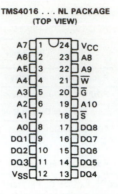

TMS4016 . . . NL PACKAGE
(TOP VIEW)

A7	1		24	Vcc
A6	2		23	A8
A5	3		22	A9
A4	4		21	W̄
A3	5		20	Ḡ
A2	6		19	A10
A1	7		18	S̄
A0	8		17	DQ8
DQ1	9		16	DQ7
DQ2	10		15	DQ6
DQ3	11		14	DQ5
Vss	12		13	DQ4

PIN NOMENCLATURE	
A0 – A10	Addresses
DQ1 – DQ8	Data In/Data Out
Ḡ	Output Enable
S̄	Chip Select
Vcc	+ 5-V Supply
Vss	Ground
W̄	Write Enable

difference between a ROM and a RAM is that a RAM is written under normal operation, while a ROM is programmed outside the computer and is only normally read. The SRAM, which stores temporary data, is used when the size of the read/write memory is relatively small. Today, a small memory is one that is less than 1M byte.

Figure 10–4 illustrates the 4016 SRAM, which is a 2K × 8 read/write memory. This device has 11 address inputs and eight data input/output connections. This device is representative of all SRAM devices, except for the number of address and data connections.

The control inputs of this RAM are slightly different from those presented earlier. The $\overline{OE}$ pin is labeled $\overline{G}$, the $\overline{CS}$ pin is $\overline{S}$, and the $\overline{WE}$ pin is $\overline{W}$. Despite the altered designations, the control pins function exactly the same as those outlined previously. Other manufacturers make this popular SRAM under the part numbers 2016 and 6116.

Figure 10–5 depicts the timing diagram for the 4016 SRAM. As the read cycle timing reveals, the access time is t_a (A). On the slowest version of the 4016, this time is 250 ns, which is fast enough to connect directly to an 8088 or an 8086 operated at 5 MHz without wait states. Again, it is important to remember that the access time must be checked to determine the compatibility of memory components with the microprocessor.

Figure 10–6 illustrates the pin-out of the 62256, 32K × 8 static RAM. This device is packaged in a 28-pin integrated circuit, and is available with access times of 120 ns or 150 ns. Other common SRAM devices are available in 8K × 8, 128K × 8, and 256K × 8 sizes, with access times of as little as 10 ns for SRAM used in computer cache memory systems.

Dynamic RAM (DRAM) Memory

About the largest static RAM available today is a 128K × 8. Dynamic RAM, on the other hand, is available in much larger sizes: up to 64M × 1. In all other respects, DRAM is essentially the same as SRAM, except that it retains data for only 2 or 4 ms on an integrated capacitor. After 2 or 4 ms, the contents of the DRAM must be completely rewritten (*refreshed*) because the capacitors, which store a logic 1 or logic 0, lose their charges.

Instead of requiring the almost impossible task of reading the contents of each memory location with a program and then rewriting them, the manufacturer has internally constructed the DRAM differently from the SRAM. In the DRAM, the entire contents of the memory is refreshed with 256 reads in a 2- or 4-ms interval. Refreshing also occurs during a write, a read, or during a special refresh cycle. Much more information about DRAM refreshing is provided in Section 10–6.

Another disadvantage of DRAM memory is that it requires so many address pins that the manufacturers have decided to multiplex the address inputs. Figure 10–7 illustrates a 64K × 4 DRAM, the TMS4464, which stores 256K bits of data. Notice that it contains only eight address inputs where it should contain 16—the number required to address 64K memory locations. The

electrical characteristics over recommended operating free-air temperature range (unless otherwise noted)

	PARAMETER	TEST CONDITIONS		MIN	TYP†	MAX	UNIT
V_{OH}	High level voltage	$I_{OH} = -1$ mA,	$V_{CC} = 4.5$ V	2.4			V
V_{OL}	Low level voltage	$I_{OL} = 2.1$ mA,	$V_{CC} = 4.5$ V			0.4	V
I_I	Input current	$V_I = 0$ V to 5.5 V				10	µA
I_{OZ}	Off-state output current	$\overline{S}$ or $\overline{G}$ at 2 V or $\overline{W}$ at 0.8 V, $V_O = 0$ V to 5.5 V				10	µA
I_{CC}	Supply current from V_{CC}	$I_O = 0$ mA, $T_A = 0°C$ (worst case)	$V_{CC} = 5.5$ V,		40	70	mA
C_i	Input capacitance	$V_I = 0$ V,	f = 1 MHz			8	pF
C_O	Output capacitance	$V_O = 0$ V,	f = 1 MHz			12	pF

†All typical values are at $V_{CC} = 5$ V, $T_A = 25°C$.

timing requirements over recommended supply voltage range and operating free-air temperature range

	PARAMETER	TMS4016-12 MIN MAX	TMS4016-15 MIN MAX	TMS4016-20 MIN MAX	TMS4016-25 MIN MAX	UNIT
$t_{c(rd)}$	Read cycle time	120	150	200	250	ns
$t_{c(wr)}$	Write cycle time	120	150	200	250	ns
$t_{w(W)}$	Write pulse width	60	80	100	120	ns
$t_{su(A)}$	Address setup time	20	20	20	20	ns
$t_{su(S)}$	Chip select setup time	60	80	100	120	ns
$t_{su(D)}$	Data setup time	50	60	80	100	ns
$t_{h(A)}$	Address hold time	0	0	0	0	ns
$t_{h(D)}$	Data hold time	5	10	10	10	ns

switching characteristics over recommended voltage range, $T_A = 0°C$ to $70°C$

	PARAMETER	TMS4016-12 MIN MAX	TMS4016-15 MIN MAX	TMS4016-20 MIN MAX	TMS4016-25 MIN MAX	UNIT
$t_{a(A)}$	Access time from address	120	150	200	250	ns
$t_{a(S)}$	Access time from chip select low	60	75	100	120	ns
$t_{a(G)}$	Access time from output enable low	50	60	80	100	ns
$t_{v(A)}$	Output data valid after address change	10	15	15	15	ns
$t_{dis(S)}$	Output disable time after chip select high	40	50	60	80	ns
$t_{dis(G)}$	Output disable time after output enable high	40	50	60	80	ns
$t_{dis(W)}$	Output disable time after write enable low	50	60	60	80	ns
$t_{en(S)}$	Output enable time after chip select low	5	5	10	10	ns
$t_{en(G)}$	Output enable time after output enable low	5	5	10	10	ns
$t_{en(W)}$	Output enable time after write enable high	5	5	10	10	ns

NOTES: 3. $C_L = 100$pF for all measurements except $t_{dis(W)}$ and $t_{en(W)}$.
$C_L = 5$ pF for $t_{dis(W)}$ and $t_{en(W)}$.
4. t_{dis} and t_{en} parameters are sampled and not 100% tested.

FIGURE 10–5   (a) The AC characteristics of the TMS4016 SRAM. (b) The timing diagrams of the TMS4016 SRAM. (Courtesy of Texas Instruments Incorporated.)

(continued on next page)

timing waveform of read cycle (see note 5)

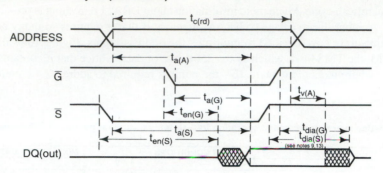

timing waveform of write cycle no. 1 (see note 6)

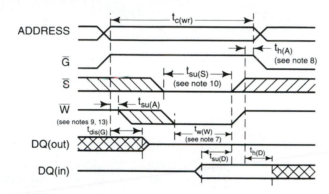

timing waveform of write cycle no. 2 (see notes 6 and 11)

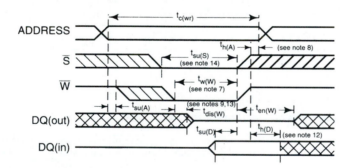

NOTES: 5. $\overline{W}$ is high Read Cycle.
6. $\overline{W}$ must be high during all address transitions.
7. A write occurs during the overlap of a low $\overline{S}$ and a low $\overline{W}$.
8. $t_{h(A)}$ is measured from the earlier of $\overline{S}$ or $\overline{W}$ going high to the end of the write cycle.
9. During this period, I/O pins are in the output state so that the input signals of opposite phase to the outputs must not be applied.
10. If the Slow transition occurs simultaneously with the $\overline{W}$ low transitions or after the $\overline{W}$ transition, output remains in a high impedance state.
11. G is continuously low ($G = V_{IL}$).
12. If $\overline{S}$ is low during this period, I/O pins are in the output state. Data input signals of opposite phase to the outputs must not be applied.
13. Transition is measured ± 200 mV from steady-state voltage.
14. If the $\overline{S}$ low transition occurs before the W low transition, then the data input signals of opposite phase to the outputs must not be applied for the duration of $t_{dis(W)}$ after the $\overline{W}$ low transition.

FIGURE 10–5   *(continued)*

FIGURE 10–6 Pin diagram of the 62256, 32K × 8 static RAM.

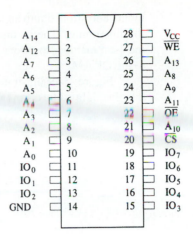

A_{14}	1		28	V_{CC}
A_{12}	2		27	$\overline{WE}$
A_7	3		26	A_{13}
A_6	4		25	A_8
A_5	5		24	A_9
A_4	6		23	A_{11}
A_3	7		22	$\overline{OE}$
A_2	8		21	A_{10}
A_1	9		20	$\overline{CS}$
A_0	10		19	IO_7
IO_0	11		18	IO_6
IO_1	12		17	IO_5
IO_2	13		16	IO_4
GND	14		15	IO_3

PIN FUNCTION

$A_0 - A_{14}$	Addresses
$IO_0 - IO_7$	Data connections
$\overline{CS}$	Chip select
$\overline{OE}$	Output enable
$\overline{WE}$	Write enable
V_{CC}	+5V Supply
GND	Ground

FIGURE 10–7 The pin-out of the TMS4464, 64K × 4 dynamic RAM (DRAM). (Courtesy of Texas Instruments Incorportated.)

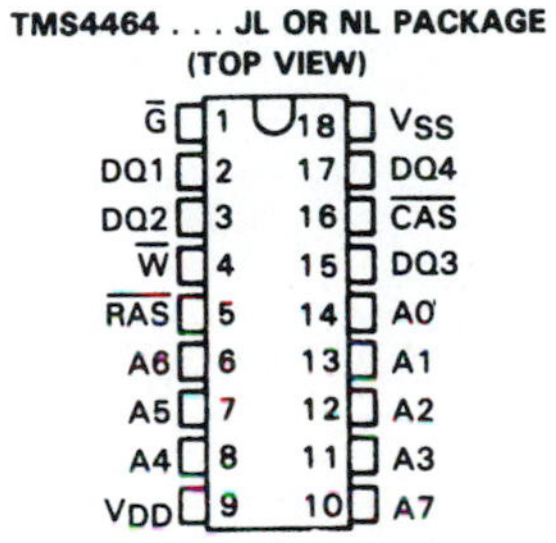

TMS4464 . . . JL OR NL PACKAGE
(TOP VIEW)

$\overline{G}$	1		18	V_{SS}
DQ1	2		17	DQ4
DQ2	3		16	$\overline{CAS}$
$\overline{W}$	4		15	DQ3
$\overline{RAS}$	5		14	A0
A6	6		13	A1
A5	7		12	A2
A4	8		11	A3
V_{DD}	9		10	A7

(a)

PIN NOMENCLATURE	
A0–A7	Address Inputs
$\overline{CAS}$	Column Address Strobe
DQ1–DQ4	Data-In/Data-Out
$\overline{G}$	Output Enable
$\overline{RAS}$	Row Address Strobe
V_{DD}	+5-V Supply
V_{SS}	Ground
$\overline{W}$	Write Enable

(b)

only way that 16 address bits can be forced into eight address pins is in two 8-bit increments. This operation requires two special pins: the **column address strobe** ($\overline{CAS}$) and **row address strobe** ($\overline{RAS}$). First, A0–A7 are placed on the address pins and strobed into an internal row latch by $\overline{RAS}$ as the row address. Next, the address bits A8–A15 are placed on the same eight address inputs and strobed into an internal column latch by $\overline{CAS}$ as the column address (see Figure 10–8 for this timing). The 16-bit address held in these internal latches addresses the contents of one of the 4-bit memory locations. Note that $\overline{CAS}$ also performs the function of the chip selection input to the DRAM.

Figure 10–9 illustrates a set of multiplexers used to strobe the column and row addresses into the eight address pins on a pair of TMS4464 DRAMs. Here, the $\overline{RAS}$ signal not only strobes

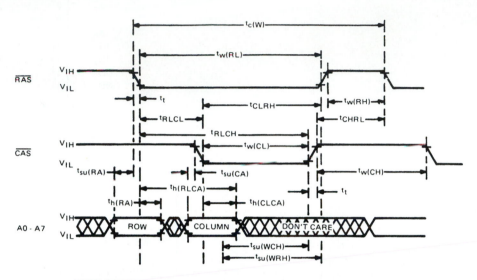

FIGURE 10–8 $\overline{RAS}$, $\overline{CAS}$, and address input timing for the TMS4464 DRAM. (Courtesy of Texas Instruments Incorporated.)

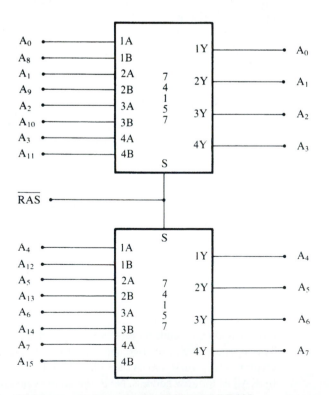

FIGURE 10–9 Address multiplexer for the TMS4464 DRAM.

FIGURE 10–10   The 41256 dynamic RAM organized as a 256K × 1 memory device.

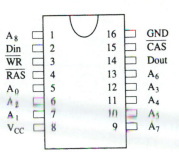

Pin			Pin
A_8	1	16	GND
Din	2	15	$\overline{CAS}$
$\overline{WR}$	3	14	Dout
$\overline{RAS}$	4	13	A_6
A_0	5	12	A_3
A_2	6	11	A_4
A_1	7	10	A_5
V_{CC}	8	9	A_7

PIN FUNCTIONS

A_0 - A_8	Addresses
Din	Data in
Dout	Data out
$\overline{CAS}$	Column Address Strobe
$\overline{RAS}$	Row Address Strobe
$\overline{WR}$	Write enable
V_{CC}	+5V Supply
GND	Ground

the row address into the DRAMs, but it also selects which part of the address is applied to the address inputs. This is possible due to the long propagation-delay time of the multiplexers. When $\overline{RAS}$ is a logic 1, the B inputs are connected to the Y outputs of the multiplexers; when the $\overline{RAS}$ input goes to a logic 0, the A inputs connect to the Y outputs. Because the internal row address latch is edge-triggered, it captures the row address before the address at the inputs changes to the column address. More detail on DRAM and DRAM interfacing is provided in Section 10–6.

As with the SRAM, the R/$\overline{W}$ pin writes data to the DRAM when a logic 0, but there is no pin labeled $\overline{G}$ or enable. There also is no $\overline{S}$ (select) input to the DRAM. As mentioned, the $\overline{CAS}$ input selects the DRAM. If selected, the DRAM is written if R/$\overline{W}$ = 0 and read if R/$\overline{W}$ = 1.

Figure 10–10 shows the pin-out of the 41256 dynamic RAM. This device is organized as a 256K × 1 memory, requiring as little as 70 ns to access data.

More recently, larger DRAMs have become available that are organized as a 1M × 1, 4M × 1, 16M × 1, and 64M × 1 memory. On the horizon is the 1G × 1 memory, which is in the planning stages. Because DRAM memory is often placed on small circuit boards called *SIMMs (Single in-line Memory Modules)*, Figure 10–11 shows the pin-outs of the two most common SIMMs. The 30-pin SIMM is organized most often as 1M × 8 or 1M × 9, and 4M × 8 or 4M × 9. (Illustrated in Figure 10–11 is a 4M × 9.) The ninth bit is the parity bit. Also shown is a newer 72-pin SIMM. The 72-pin SIMMs are often organized as 1M × 32 or 1M × 36 (with parity). Other sizes are 2M × 32, 4M × 32, 8M × 32, and 16M × 32. These are also available with parity. Illustrated in Figure 10–11 is a 4M × 36 SIMM, which has 16M bytes of memory.

Lately, many systems are using the Pentium–Pentium 4 microprocessors. These microprocessors have a 64-bit wide data bus, which precludes the use of the 8-bit wide SIMMs described here. Even the 72-pin SIMMs are cumbersome to use because they must be used in pairs to obtain a 64-bit wide data connection. Today, the 64-bit wide DIMMs (Dual In-line Memory Modules) are becoming the standard in most systems. The memory on these modules is organized as 64-bits wide. The common sizes available are 16M bytes (2M × 64), 32M bytes (4M × 64), 64M bytes (8M × 64), and 128M bytes (16M × 64). The pin-out of the DIMM is illustrated in Figure 10–12. The DIMM module is available in DRAM, EDO, and SDRAM forms, with or without an EPROM. The EPROM provides information to the system on the size and the speed of the memory device for plug-n-play applications.

(a)

(TOP VIEW)

(b)

(TOP VIEW)

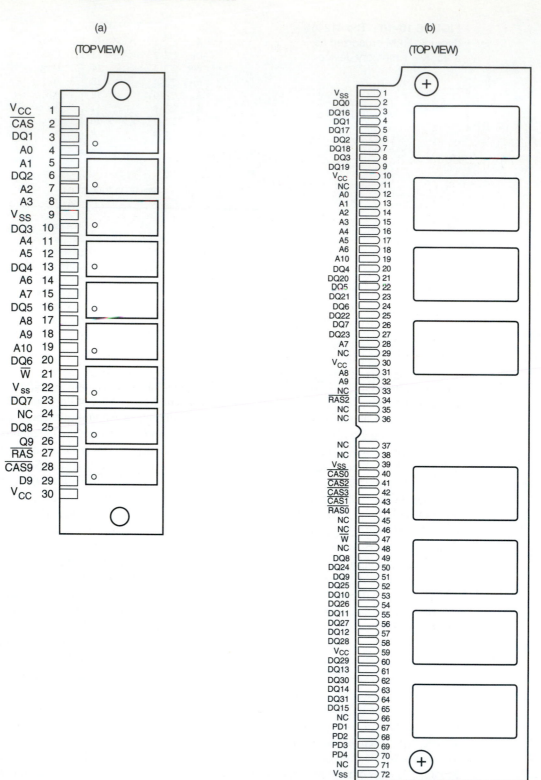

FIGURE 10–11 The pin-outs of the 30-pin and 72-pin SIMM. (a) A 30-pin SIMM organized as 4M × 9 and (b) a 72-pin SIMM organized as 4M × 36.

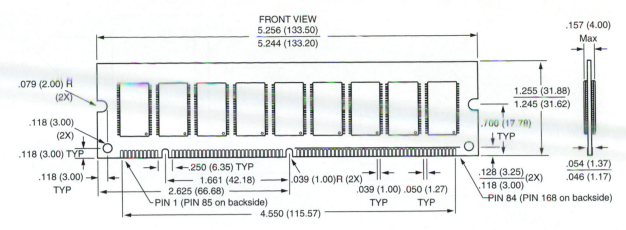

FIGURE 10–12 The pin-out of a 168-pin DIMM.

A new addition to the memory market is the RIMM memory module from RAMBUS Corporation. Like the SDRAM, the RIMM contains 168 pins, but each pin is a two-level pin, bringing the total number of connections to 336. The fastest SDRAM currently available is the PC-2400, which operates at a rate of 2.4G bytes per second. By comparison, the 800-MHz RIMM operates at a rate of 3.2G bytes per second. The RIMM module is organized as a 32-bit-wide device. This means that to populate a Pentium 4 memory, RIMM memory is used in pairs. Intel claims that the Pentium 4 system using RIMM modules is 300 percent faster than a Pentium III using PC-100 memory. According to RAMBUS, the current 800-MHz RIMM will increase to a speed of 1200 MHz in the future.

10–2 ADDRESS DECODING

In order to attach a memory device to the microprocessor, it is necessary to decode the address sent from the microprocessor. Decoding makes the memory function at a unique section or partition of the memory map. Without an address decoder, only one memory device can be connected to a microprocessor, which would make it virtually useless. In this section, we describe a few of the more common address-decoding techniques, as well as the decoders that are found in many systems.

Why Decode Memory?

When the 8088 microprocessor is compared to the 2716 EPROM, a difference in the number of address connections is apparent—the EPROM has 11 address connections and the microprocessor has 20. This means that the microprocessor sends out a 20-bit memory address whenever it reads or writes data. Because the EPROM has only 11 address pins, there is a mismatch that must be corrected. If only 11 of the 8088's address pins are connected to the memory, the 8088 will see only 2K bytes of memory instead of the 1M bytes that it "expects" the memory to contain. The decoder corrects the mismatch by decoding the address pins that do not connect to the memory component.

Simple NAND Gate Decoder

When the 2K × 8 EPROM is used, address connections A10–A0 of the 8088 are connected to address inputs A10–A0 of the EPROM. The remaining nine address pins (A19–A11) are connected to

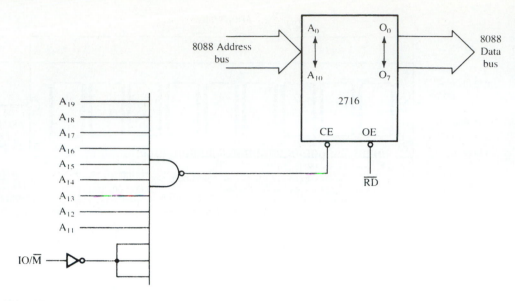

FIGURE 10–13   A simple NAND gate decoder used to select a 2716 EPROM memory component for memory locations FF800H–FFFFFH.

the inputs of a NAND gate decoder (see Figure 10–13). The decoder selects the EPROM from one of the many 2K-byte sections of the entire 1M-byte address range of the 8088 microprocessor.

In this circuit, a single NAND gate decodes the memory address. The output of the NAND gate is a logic 0 whenever the 8088 address pins attached to its inputs (A19–A11) are all logic 1s. The active low, logic 0 output of the NAND gate decoder is connected to the $\overline{CE}$ input pin that selects (**enables**) the EPROM. Recall that whenever $\overline{CE}$ is a logic 0, data will be read from the EPROM only if $\overline{OE}$ is also a logic 0. The $\overline{OE}$ pin is activated by the 8088 $\overline{RD}$ signal or the $\overline{MRDC}$ (**memory read control**) signal of other family members.

If the 20-bit binary address, decoded by the NAND gate, is written so that the leftmost nine bits are 1s and the rightmost 11 bits are don't cares (X), the actual address range of the EPROM can be determined. (A *don't care* is a logic 1 or a logic 0, whichever is appropriate.)

Example 10–1 illustrates how the address range for this EPROM is determined by writing down the externally decoded address bits (A19–A11) and the address bits decoded by the EPROM (A10–A0) as don't cares. As the example illustrates, the don't cares are first written as 0s to locate the lowest address and then as 1s to find the highest address. Example 10–1 also shows these binary boundaries as hexadecimal addresses. Here, the 2K EPROM is decoded at memory address locations FF800H–FFFFFH. Notice that this is a 2K-byte section of the memory and is also located at the reset location for the 8086/8088 (FFFF0H), the most likely place for an EPROM.

EXAMPLE 10–1

```
1111 1111 1XXX XXXX XXXX

          or

1111 1111 1000 0000 0000 = FF800H
          to
1111 1111 1111 1111 1111 = FFFFFH
```

Although this example serves to illustrate decoding, NAND gates are rarely used to decode memory because each memory device requires its own NAND gate decoder. Because of the excessive cost of the NAND gate decoder and inverters that are often required, this option requires that an alternate be found.

The 3-to-8 Line Decoder (74LS138)

One of the more common, although not only, integrated circuit decoders found in many microprocessor-based systems is the 74LS138 3-to-8 line decoder. Figure 10–14 illustrates this decoder and its truth table.

The truth table shows that only one of the eight outputs ever goes low at any time. For any of the decoder's outputs to go low, the three enable inputs ($\overline{G2A}$, $\overline{G2B}$, and G1) must all be active. To be active, the $\overline{G2A}$ and $\overline{G2B}$ inputs must both be low (logic 0), and G1 must be high (logic 1). Once the 74LS138 is enabled, the address inputs (C, B, and A) select which output pin goes low. Imagine eight EPROM $\overline{CE}$ inputs connected to the eight outputs of the decoder! This is a very powerful device because it selects eight different memory devices at the same time.

Sample Decoder Circuit. Notice that the outputs of the decoder, illustrated in Figure 10–15, are connected to eight different 2764 EPROM memory devices. Here, the decoder selects eight 8K-byte blocks of memory for a total memory capacity of 64K bytes. This figure also illustrates the address range of each memory device and the common connections to the memory devices. Notice that all of the address connections from the 8088 are connected to this circuit. Also, notice that the decoder's outputs are connected to the $\overline{CE}$ inputs of the EPROMs, and the $\overline{RD}$ signal from the 8088 is connected to the $\overline{OE}$ inputs of the EPROMs. This allows only the selected EPROM to be enabled and to send its data to the microprocessor through the data bus whenever $\overline{RD}$ becomes a logic 0.

In this circuit, a 3-input NAND gate is connected to address bits A19–A17. When all three address inputs are high, the output of this NAND gate goes low and enables input $\overline{G2B}$ of the 74LS138. Input G1 is connected directly to A16. In other words, in order to enable this decoder, the first four address connections (A19–A16) must all be high.

FIGURE 10–14 The 74LS138, 3-to-8 line decoder and function table.

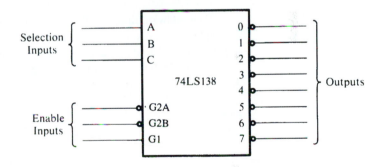

Inputs						Outputs							
Enable			Select										
$\overline{G2A}$ $\overline{G2B}$		G1	C	B	A	$\overline{0}$	$\overline{1}$	$\overline{2}$	$\overline{3}$	$\overline{4}$	$\overline{5}$	$\overline{6}$	$\overline{7}$
1	X	X	X	X	X	1	1	1	1	1	1	1	1
X	1	X	X	X	X	1	1	1	1	1	1	1	1
X	X	0	X	X	X	1	1	1	1	1	1	1	1
0	0	1	0	0	0	0	1	1	1	1	1	1	1
0	0	1	0	0	1	1	0	1	1	1	1	1	1
0	0	1	0	1	0	1	1	0	1	1	1	1	1
0	0	1	0	1	1	1	1	1	0	1	1	1	1
0	0	1	1	0	0	1	1	1	1	0	1	1	1
0	0	1	1	0	1	1	1	1	1	1	0	1	1
0	0	1	1	1	0	1	1	1	1	1	1	0	1
0	0	1	1	1	1	1	1	1	1	1	1	1	0

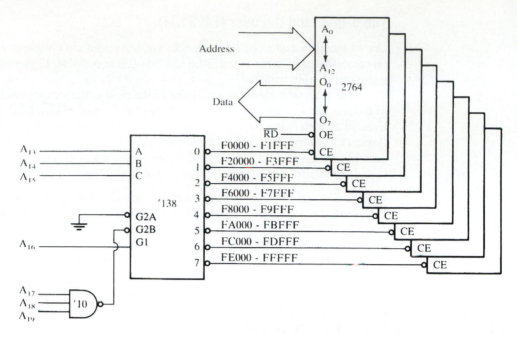

FIGURE 10–15   A circuit that uses eight 2764 EPROMs for a 64K × 8 section of memory in an 8088 microprocessor-based system. The addresses selected in this circuit are F0000H–FFFFFH.

The address inputs C, B, and A connect to microprocessor address pins A15–A13. These three address inputs determine which output pin goes low and which EPROM is selected whenever the 8088 outputs a memory address within this range to the memory system.

Example 10–2 shows how the address range of the entire decoder is determined. Notice that the range is location F0000H–FFFFFH. This is a 64K-byte span of the memory.

EXAMPLE 10–2

```
1111 XXXX XXXX XXXX XXXX

        or

1111 0000 0000 0000 0000 = F0000H
          to
1111 1111 1111 1111 1111 = FFFFFH
```

How is it possible to determine the address range of each memory device attached to the decoder's outputs? Again, the binary bit pattern is written down; this time the C, B, and A address inputs are not don't cares. Example 10–3 shows how output 0 of the decoder is made to go low to select the EPROM attached to that pin. Here, C, B, and A are shown as logic 0s.

EXAMPLE 10–3

```
    CBA
1111 000X XXXX XXXX XXXX

        or

1111 0000 0000 0000 0000 = F0000H
          to
1111 0001 1111 1111 1111 = F1FFFH
```

If the address range of the EPROM connected to output 1 of the decoder is required, it is determined in exactly same way as that of output 0. The only difference is that now the C, B, and A inputs contain a 001 instead of a 000 (see Example 10–4). The remaining output address ranges are determined in the same manner by substituting the binary address of the output pin into C, B, and A.

EXAMPLE 10–4

```
      CBA
1111 001X XXXX XXXX XXXX

           or

1111 0010 0000 0000 0000 = F2000H
            to
1111 0011 1111 1111 1111 = F3FFFH
```

The Dual 2-to-4 Line Decoder (74LS139)

Another decoder that finds some application is the 74LS139 dual 2-to-4 line decoder. Figure 10–16 illustrates both the pin-out and the truth table for this decoder. The 74LS139 contains two separate 2-to-4 line decoders—each with its own address, enable, and output connections.

PROM Address Decoder

Another, once common, address decoder is the bipolar PROM, used because of its larger number of input connections, which reduces the number of other circuits required in a system memory address decoder. The 74LS138 decoder has six inputs used for address connections. The PROM decoder may have many more inputs for address decoding.

FIGURE 10–16   The pin-out and truth table of the 74LS139, dual 2-to-4 line decoder.

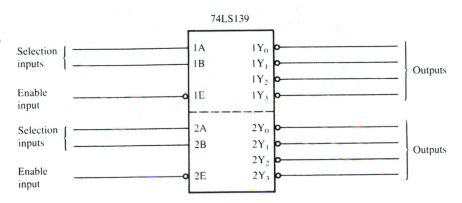

Inputs			Outputs			
$\overline{E}$	A	B	$\overline{Y_0}$	$\overline{Y_1}$	$\overline{Y_2}$	$\overline{Y_3}$
0	0	0	0	1	1	1
0	0	1	1	0	1	1
0	1	0	1	1	0	1
0	1	1	1	1	1	0
1	X	X	1	1	1	1

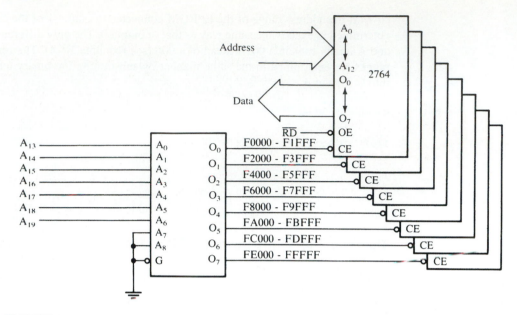

FIGURE 10–17   A memory system using the TPB28L42, 512 × 8 PROM as an address.

For example, the 82S147 (512 × 8) PROM used as an address decoder has 10 input connections and eight output connections. It can replace the circuit in Figure 10–15 without the extra 3-input NAND gate. This saves space on the printed circuit board and reduces the cost of a system.

Figure 10–17 illustrates this address decoder with the PROM in place. The PROM is a memory device that must be programmed with the correct binary bit pattern to select the eight EPROM memory devices. The PROM itself has nine address inputs that select one of the 512 internal 8-bit memory locations. The remaining input ($\overline{CE}$) must be grounded because if this PROM's outputs float to their high-impedance state, one or more of the EPROMs might be selected by noise impulses in the system.

Table 10–1 illustrates the binary bit pattern programmed into each PROM location in order to select the eight different EPROMs. The main advantage of using a PROM is that the address map is easily changed in the field. Because the PROM comes with all the locations programmed as logic 1s, only eight of the 512 locations must be programmed. This saves valuable time for the manufacturer.

TABLE 10–1   The 82S147 PROM programming pattern for the circuit of Figure 10–17.

				Inputs										Outputs			
$\overline{OE}$	A8	A7	A6	A5	A4	A3	A2	A1	A0	O0	O1	O2	O3	O4	O5	O6	O7
0	0	0	1	1	1	1	0	0	0	0	1	1	1	1	1	1	1
0	0	0	1	1	1	1	0	0	1	1	0	1	1	1	1	1	1
0	0	0	1	1	1	1	0	1	0	1	1	0	1	1	1	1	1
0	0	0	1	1	1	1	0	1	1	1	1	1	0	1	1	1	1
0	0	0	1	1	1	1	1	0	0	1	1	1	1	0	1	1	1
0	0	0	1	1	1	1	1	0	1	1	1	1	1	1	0	1	1
0	0	0	1	1	1	1	1	1	0	1	1	1	1	1	1	0	1
0	0	0	1	1	1	1	1	1	1	1	1	1	1	1	1	1	0
all	other	combinations								1	1	1	1	1	1	1	1

PLD Programmable Decoders

This section of the text explains the use of the programmable logic device, or PLD, as a decoder. The PAL has replaced PROM address decoders in the latest memory interfaces. There are three PLD devices that function in the same manner but have different names: PLA (**programmable logic array**), PAL (**programmable array logic**), and GAL (**gated array logic**). Although these devices have been in existence since the mid-1970s, they have only appeared in memory system and digital designs since the 1990s. The PAL and the PLA are fuse-programmed as is the PROM, and some PLD devices are erasable devices (as are EPROMs). In essence, all three devices are arrays of logic elements that are programmable.

Combinatorial Programmable Logic Arrays. One of the two basic types of PALs is the combinatorial programmable logic array. This device is internally structured as a programmable array of combinational logic circuits. Figure 10–18 illustrates the internal structure of the PAL16L8 that is constructed with AND/OR gate logic. This device, which is very common, has 10 fixed inputs, two fixed outputs, and six pins that are programmable as inputs or outputs. Each output signal is generated from a 7-input OR gate that has an AND gate attached to each input. The outputs of the OR gates pass through a three-state inverter that defines each out as an AND/NOR function. Initially, all of the fuses connect all of the vertical/horizontal connections illustrated in Figure 10–18. Programming is accomplished by blowing fuses to connect various inputs to the OR gate array. The wired-AND function is performed at each input connection, which allows a product term of up to 16 inputs. A logic expression using the PAL16L8 can have up to seven product terms with up to 16 inputs NORed together to generate the output expression. This device is ideal as a memory address decoder because of its structure. It is also ideal because the outputs are active low.

Fortunately, we don't have to choose the fuses by number for programming, as was customary when this device was first introduced. Today, we program the PAL by using a software package such as PALASM, the PAL assembler program. The PALASM program and its syntax are an industry standard for programming PAL devices. Example 10–5 shows a program that decodes the same areas of memory as decoded in Figure 10–17. Note that this program was developed by using a text editor such as EDIT, available with Microsoft DOS version 7.1 with XP or Notepad in Windows 98 or Windows XP. The program can also be developed by using an editor than comes with the PALASM package or any other PAL assembler program. Various editors attempt to ease the task of defining the pins, but we believe it is easier to use EDIT and the listing as shown.

EXAMPLE 10–5

```
TITLE        Address Decoder
PATTERN      Test 1
REVISION     A
AUTHOR       Barry B. Brey
COMPANY      BreyCo
DATE         6/6/99
CHIP         DECODER1 PAL16L8

;pins 1    2    3    4    5    6    7    8   9   10
      A19  A18  A17  A16  A15  A14  A13  NC  NC  GND

;pins 11 12 13 14 15 16 17 18 19   20
      NC  O8 O7 O6 O5 O4 O3 O2 O1   VCC

EQUATIONS

/O1 = A19 * A18 * A17 * A16 * /A15 * /A14 * /A13
/O2 = A19 * A18 * A17 * A16 * /A15 * /A14 * A13
/O3 = A19 * A18 * A17 * A16 * /A15 * A14 * /A13
/O4 = A19 * A18 * A17 * A16 * /A15 * A14 * A13
/O5 = A19 * A18 * A17 * A16 * A15 * /A14 * /A13
/O6 = A19 * A18 * A17 * A16 * A15 * /A14 * A13
/O7 = A19 * A18 * A17 * A16 * A15 * A14 * /A13
/O8 = A19 * A18 * A17 * A16 * A15 * A14 * A13
```

Logic Diagram   16L8

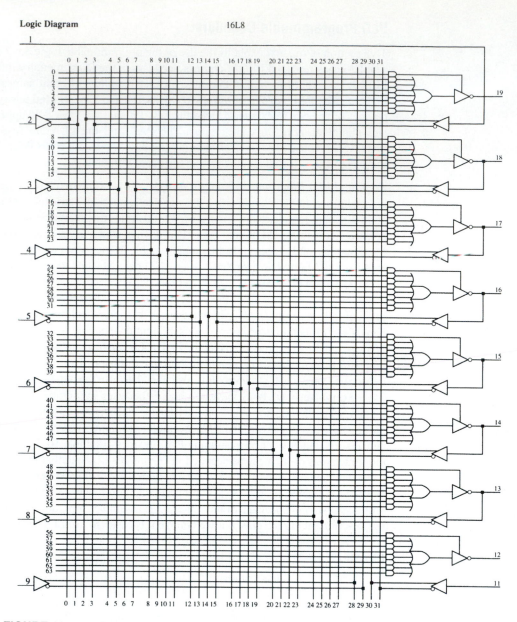

FIGURE 10–18   The PAL 16L8. (Copyright Advanced Micro Devices, Inc., 1988. Reprinted with permission of copyright owner. All rights reserved.)

The first eight lines of the program illustrated in Example 10–5 identify the program title, pattern, revision, author, company, date, and chip type with the program name. Although it is normal to find entries in each heading, the only entry that is necessary is the CHIP statement. In this example, the chip type is a PAL16L8 and the program is called DECODER1. After the program is identified, a comment statement (;pins) lists the pin numbers. Below the comment the pins appear, as defined for this application. Once all the pins are defined, we use the EQUATIONS statement to indicate that the equations for this application follow. In this example, the equations define the eight chip enable outputs for the eight EPROM memory devices. See Figure 10–19 for the complete schematic diagram of this PAL decoder.

Note that each equation specifies one of the active low output pins, as defined by the / in front of the pin name—e.g., /O1 is used in place of $\overline{O1}$. Although we normally place an over-bar on top

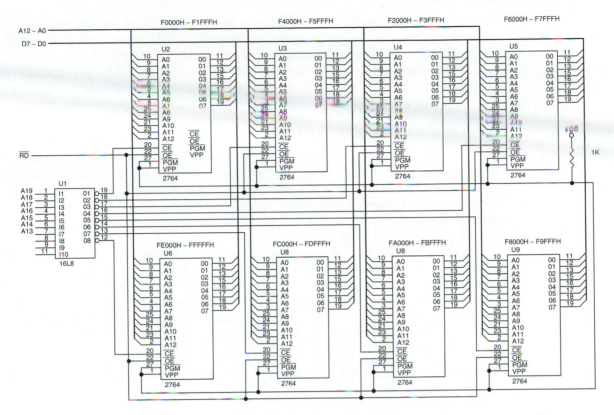

FIGURE 10–19   A PAL16L8 that decodes eight 2764 (8K × 8) memory devices.

of an active low output, that is not possible when typing, so the slash in front of a pin name is used to indicate active low outputs. In this example, all outputs are active low because of the PAL 16L8, which has only active low outputs. Other PAL devices are available with active high outputs, if needed. Normally all pins are defined before the equation statement as active high.

Logic symbols used in PAL equations include the * for the AND operation and the + for the OR operation. This example illustrates only AND operations. If a single input is inverted, the / is placed in front of the pin name. If a group must be inverted, a slash is placed in front of the group that is surrounded by parentheses—e.g., /(A + B) is the same as $\overline{A+B}$, the NOR function. A NAND function would then be /(A * B), which is the same as $\overline{A*B}$.

Let us examine the very first equation in Example 10–5. The output pin /O1 is active low, which means that it becomes a logic 0, enabling the EPROM in Figure 10–19 when the equation, on the other side of the equal sign, is true. The equation for this output contains a number of inputs that are ANDed together. In this example, when A19, A18, A17, and A16 are all ones while A15, A14, and A13 are all low, the output pin /O1 becomes a logic 0. This binary number corresponds to memory locations F0000H–F1FFFH.

10–3   8088 AND 80188 (8-BIT) MEMORY INTERFACE

This text contains separate sections on memory interfacing for the 8088 and 80188 with their 8-bit data buses; the 8086, 80186, 80286, and 80386SX with their 16-bit data buses; the 80386DX and 80486 with their 32-bit data buses; the Pentium–Pentium 4 with their 64-bit data buses. Separate sections are provided because the methods used to address the memory are slightly different in microprocessors that contain different data bus widths. Hardware engineers or technicians who wish to broaden their expertise in interfacing 16-bit, 32-bit, and 64-bit memory interface should cover all sections. This section is much more complete than the section on the 16- and 32-bit memory interface, which covers only material not covered in the 8088/80188 section.

In this section, we examine the memory interface to both RAM and ROM and explain parity checking, which is still commonplace in many microprocessor-based computer systems. We also briefly mention the error-correction schemes currently available to memory system designers.

Basic 8088/80188 Memory Interface

The 8088 and 80188 microprocessors have an 8-bit data bus, which makes them ideal to connect to the common 8-bit memory devices available today. The 8-bit memory size makes the 8088, and especially the 80188, ideal as a simple controller. For the 8088/80188 to function correctly with the memory, however, the memory system must decode the address to select a memory component. It must also use the $\overline{RD}$, $\overline{WR}$, and $IO/\overline{M}$ control signals provided by the 8088/80188 to control the memory system.

The minimum mode configuration is used in this section and is essentially the same as the maximum mode system for memory interface. The main difference is that, in maximum mode, the $IO/\overline{M}$ signal is combined with $\overline{RD}$ to generate the $\overline{MRDC}$ signal, and $IO/\overline{M}$ is combined with $\overline{WR}$ to generate the $\overline{MWTC}$ signal. The maximum mode control signals are developed inside the 8288 bus controller. In minimum mode, the memory sees the 8088 or the 80188 as a device with 20 address connections (A19–A0), eight data bus connections (AD7–AD0), and the control signals $IO/\overline{M}$, $\overline{RD}$, and $\overline{WR}$.

Interfacing EPROM to the 8088.   You will find this section very similar to Section 10–2 on decoders. The only difference is that, in this section, we discuss wait states and the use of the $IO/\overline{M}$ signal to enable the decoder.

Figure 10–20 illustrates an 8088/80188 microprocessor connected to eight 2732 EPROMs, 4K × 8 memory devices. The 2732 has one more address input (A11) than the 2716, and twice

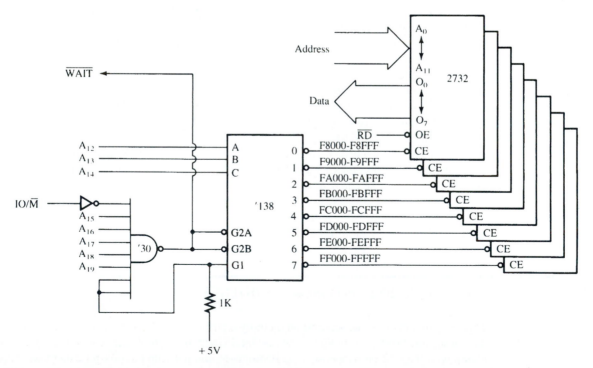

FIGURE 10–20   Eight 2732 EPROMs interfaced to the 8088 microprocessor. Note that the output of the NAND gate is used to cause a wait state whenever this section of the memory is selected.

the memory. The device in this illustration decodes eight 4K × 8 blocks of memory, for a total of 32K × 8 bits of the physical address space for the 8088/80188.

The decoder (74LS138) is connected a little differently than might be expected because the slower version of this type of EPROM has a memory access time of 450 ns. Recall from Chapter 9 that when the 8088 is operated with a 5 MHz clock, it allows 460 ns for the memory to access data. Because of the decoder's added time delay (12 ns), it is impossible for this memory to function within 460 ns. In order to correct this problem, we must add a NAND gate to generate a signal to enable the decoder and a signal for the wait state generator, covered in Chapter 9. (Note that the 80188 can internally insert from 0–15 wait states without any additional external hardware, so it does not require this NAND gate.) With a wait state inserted every time this section of the memory is accessed, the 8088 will allow 660 ns for the EPROM to access data. Recall that an extra wait state adds 200 ns (1 clock) to the access time. The 660 ns is ample time for a 450 ns memory component to access data, even with the delays introduced by the decoder and any buffers added to the data bus.

Notice that the decoder is selected for a memory address range that begins at location F8000H and continues through location FFFFFH—the upper 32K bytes of memory. This section of memory is an EPROM because FFFF0H is where the 8088 starts to execute instructions after a hardware reset. We often call location FFFF0H the **cold-start** location. The software stored in this section of memory would contain a JMP instruction at location FFFF0H that jumps to location F8000H so the remainder of the program can execute.

Interfacing RAM to the 8088.   RAM is a little easier to interface than EPROM because most RAM memory components do not require wait states. An ideal section of the memory for the RAM is the very bottom, which contains vectors for interrupts. Interrupt vectors (discussed in more detail in Chapter 12) are often modified by software packages, so it is rather important to encode this section of the memory with RAM.

In Figure 10–21, 16 62256 32K × 8 static RAMs are interfaced to the 8088, beginning at memory location 00000H. This circuit board uses two decoders to select the 16 different RAM memory components and a third to select the other decoders for the appropriate memory sections. Sixteen 32K RAMs fill memory from location 00000H through location 7FFFFH, for 512K bytes of memory.

The first decoder (U4) in this circuit selects the other two decoders. An address beginning with 00 selects decoder U3 and an address that begins with 01 selects decoder U9. Notice that extra pins remain at the output of decoder U4 for future expansion. These pins allow more 256K × 8 blocks of RAM, for a total of 1M × 8, simply by adding the RAM and the additional secondary decoders.

Also notice from the circuit in Figure 10–21 that all the address inputs to this section of memory are buffered, as are the data bus connections and control signals $\overline{RD}$ and $\overline{WR}$. Buffering is important when many devices appear on a single board or in a single system. Suppose that three other boards like this are plugged into a system. Without the buffers on each board, the load on the system address, data, and control buses would be enough to prevent proper operation. (Excessive loading causes the logic 0 output to rise above the 0.8 V maximum allowed in a system.) Buffers are normally used if the memory will contain additions at some future date. If the memory will never grow, then buffers may not be needed.

Interfacing Flash Memory

Flash memory (EEPROM) is becoming commonplace for storing setup information on video cards, as well as for storing the system BIOS in the personal computer. Flash memory is also found in many other applications to store information that is only changed occasionally.

The only difference between a flash memory device and SRAM is that the flash memory device requires a 12 V programming voltage to erase and write new data. The 12 V can be available either at the power supply or a 5 V-to-12 V converter designed for use with flash memory can be obtained.

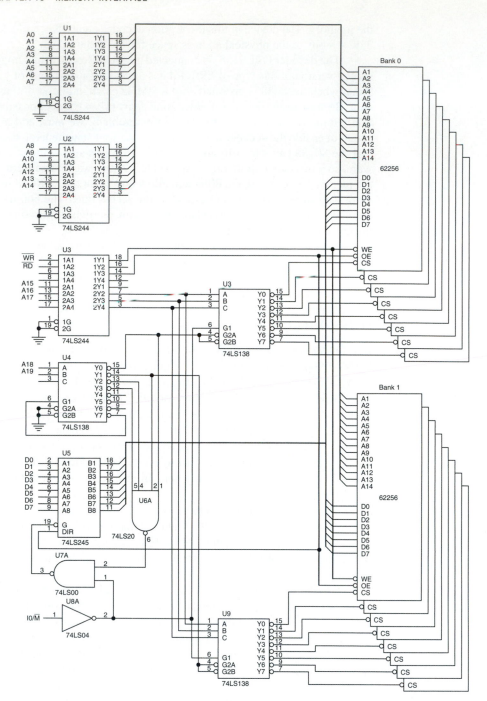

FIGURE 10–21   A 512K byte static memory system using 16 62255 SRAMs.

Figure 10–22 illustrates the 28F400 Intel flash memory device interfaced to the 8088 microprocessor. The 28F400 can be used as either a $512K \times 8$ memory device or as a $256K \times 16$ memory device. Because it is interfaced to the 8088, its configuration is $512K \times 8$. Notice that the control connections on this device are identical to that of an SRAM: $\overline{CE}$, $\overline{OE}$, and $\overline{WE}$. The only new pins are VPP, which is connected to 12 V for erase and programming; $\overline{PWD}$, which selects the power-down mode when a logic 0 and is also used for programming; and $\overline{BYTE}$, which selects

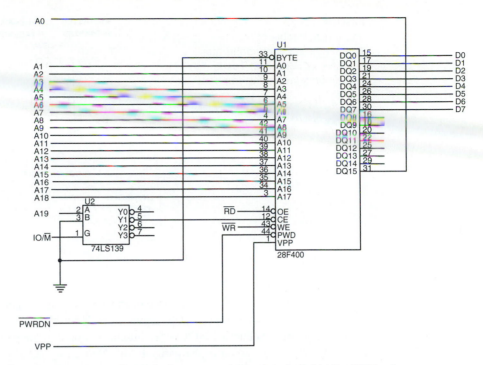

FIGURE 10–22   The 28F400 flash memory device interfaced to the 8088 microprocessor.

byte (0) or word (1) operation. Note that the pin DQ15 functions as the least-significant address input when operated in the byte mode. Another difference is the amount of time required to accomplish a write operation. The SRAM can perform a write operation in as little as 10 ns, but the flash memory requires approximately 0.4 seconds to erase a byte. The topic of programming the flash memory device is covered in Chapter 11, along with I/O devices. The flash memory device has internal registers that are programmed by using I/O techniques not yet explained. This chapter concentrates on its interface to the microprocessor.

Notice in Figure 10–22 that the decoder chosen is the 74LS139 because only a simple decoder is needed for a flash memory device this large. The decoder uses address connection A19 and IO/$\overline{M}$ as inputs. The A15 signal selects the flash memory for locations 80000H through FFFFFH, and IO/$\overline{M}$ enables the decoder.

Parity For Memory Error Detection

Because such large memories are available in today's systems, and because circuit costs are minimal, many memory board manufacturers have added parity checking to their RAM memory boards, although recently there seems to be a trend away from parity. Parity checking counts the number of 1s in data and indicates whether there is an even or odd number. If all data are stored with even parity (with an even number of 1 bits), a 1-bit error can be detected. Memory that contains parity is nine bits wide or 36 bits wide for the newer 72-pin SIMM (**single-in-line memory module**) components found in computer systems.

Figure 10–23 illustrates the 74AS280 parity generator/detector integrated circuit. This circuit has nine inputs, and generates even or odd parity for the 9-bit number placed on its inputs. It also checks the parity of a 9-bit number connected to its inputs.

Figure 10–24 illustrates a 64K × 8 static RAM system using two 62556 32K × 8 SRAM devices for data storage that has parity generation and detection. Notice that a 74AS280 generates a parity bit stored in a 6287 64K × 1 SRAM. This circuit decodes the memory at locations

FIGURE 10–23   The pin-out and function table of the 74AS280 9-bit parity generator/detector. (Courtesy of Texas Instruments Incorporated.)

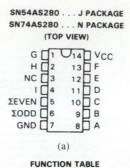

SN54AS280 . . . J PACKAGE
SN74AS280 . . . N PACKAGE
(TOP VIEW)

```
          G  ┌─┐1  ∪  14┌─┐  VCC
          H  └─┘2      13└─┘  F
         NC  ┌─┐3      12┌─┐  E
          I  └─┘4      11└─┘  D
      ΣEVEN  ┌─┐5      10┌─┐  C
       ΣODD  └─┘6       9└─┘  B
        GND  ┌─┐7       8└─┘  A
```

(a)

FUNCTION TABLE

NUMBER OF INPUTS A	OUTPUTS	
THRU I THAT ARE HIGH	Σ EVEN	Σ ODD
0,2,4,6,8	H	L
1,3,5,7,9	L	H

(b)

80000H–8FFFFH by using a 74LS138 decoder. Here, the eight data bus connections are attached to the parity generator's (U6) inputs A–H. Input I is grounded so if an even number of 1s appear on the data bus, a 1 (at the even output) is stored in the parity RAM; if an odd number of 1s appears, a 0 is stored in the parity RAM. Odd parity is stored for each byte of data, including the parity bit written to the memory.

When data are read from the memory, each datum is connected to another 74AS280 (U7) to check its parity. In this case, all the inputs to the checker are connected. Inputs A–H are connected to the data RAM's outputs, and input I is connected to the parity RAM. Note that the parity RAM control pins are different. This SRAM reads data from its output pin when selected and writes data if selected with $\overline{\text{WE}} = 0$. It does not have an $\overline{\text{OE}}$ connection for $\overline{\text{RD}}$. If parity is odd, as it is if everything is correct, the even parity output of the 74AS280 (U7) is a logic 0. If a bit of the information read from the memory changes for any reason, then the even output pin of the 74AS280 will become a logic 1. The parity output pin is connected to a special input of the 8088 called the *non-maskable interrupt (NMI) input*. The NMI input can never be turned off. If it is placed at its logic 1 level, the program being executed is interrupted and a special subroutine indicates that a parity error has been detected by the memory system. (More detail on interrupts is provided in Chapter 12.)

The application of the parity error is timed so that the data read from the memory are settled to their final state before an NMI input occurs. The operation is timed by a D-type flip-flop that latches the output of the parity checker at the end of an $\overline{\text{RD}}$ cycle from this section of the memory. In this way, the memory has enough time to read the information and pass it through the generator before the output of the generator is sampled by the NMI input.

Error Correction

Error-correction schemes have been around for a long time, but integrated circuit manufacturers have only recently started to produce error-correcting circuits. One such circuit is the 74LS636, an 8-bit error correction and detection circuit that corrects any single-bit memory read error and flags any 2-bit error. This device is found in very high-end computer systems because of the cost of implementing a system that uses error correction.

The newest computer systems are now starting to use SDRAM with ECC (error correction code). The scheme to correct the errors that might occur in these memory devices is identical to the scheme discussed in this text.

The 74LS636 corrects errors by storing five parity bits with each byte of memory data. This does increase the amount of memory required, but it also provides automatic error correction for single-bit errors. If more than two bits are in error, this circuit may not detect it. Fortunately, this

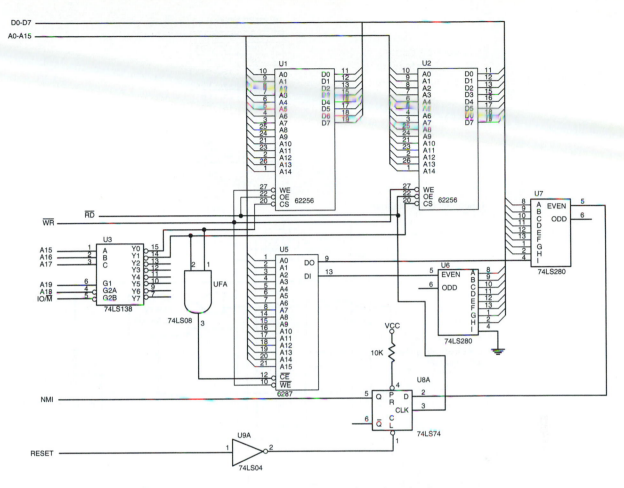

FIGURE 10–24   A 64K memory system that contains a parity error detection circuit.

is rare, and the extra effort required to correct more than a single-bit error is very expensive and not worth the effort. Whenever a memory component fails completely, its bits are all high or all low. In this case, the circuit flags the processor with a multiple-bit error indication.

Figure 10–25 depicts the pin-out of the 74LS636. Notice that it has eight data I/O pins, five check bit I/O pins, two control inputs (S0 and S1), and two error outputs: single error flag (SEF) and double error flag (DEF). The control inputs select the type of operation to be performed and are listed in the truth table of Table 10–2.

When a single error is detected, the 74LS636 goes through an error-correction cycle: it places a 01 on S0 and S1 by causing a wait and then a read following error correction.

Figure 10–26 illustrates a circuit used to correct single-bit errors with the 74LS636 and to interrupt the processor through the NMI pin for double-bit errors. To simplify the illustration, we depict only one 2K × 8 RAM and a second 2K × 8 RAM to store the 5-bit check code.

The connection of this memory component is different from that of the previous example. Notice that the $\overline{S}$ or $\overline{CS}$ pin is grounded, and data bus buffers control the flow to the system bus. This is necessary if the data are to be accessed from the memory before the $\overline{RD}$ strobe goes low.

On the next negative edge of the clock after an $\overline{RD}$, the 74LS636 checks the single-error flag (SEF) to determine whether an error has occurred. If it has, a correction cycle causes the single-error defect to be corrected. If a double error occurs, an interrupt request is generated by the double-error flag (DEF) output, which is connected to the NMI pin of the microprocessor.

pin assignments

J, N PACKAGES			
1	DEF	11	CB4
2	DB0	12	nc
3	DB1	13	CB3
4	DB2	14	CB2
5	DB3	15	CB1
6	DB4	16	CB0
7	DB5	17	S0
8	DB6	18	S1
9	DB7	19	SEF
10	GND	20	V_{CC}

(a)

functional block diagram

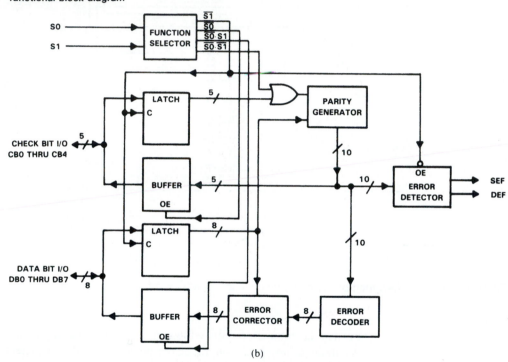

(b)

FIGURE 10–25   (a) The pin connections of the 74LS636. (b) The block diagram of the 74LS636. (Courtesy of Texas Instruments Incorportated.)

TABLE 10–2   Control bits S0 and S1.

S0	S1	Function	SEF	DEF
0	0	Write check word	0	0
0	1	Correct data word	*	*
1	0	Read data	0	0
1	1	Latch data	*	*

*Note: These levels are determined by the type of error.

FIGURE 10–26   An error
detection and correction circuit
using the 74LS636.

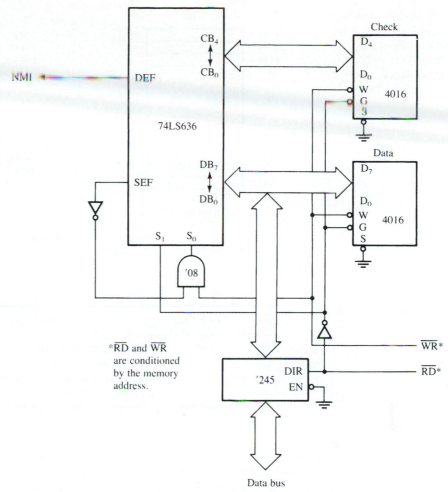

Data bus

10–4   8086, 80186, 80286, AND 80386SX (16-BIT) MEMORY INTERFACE

The 8086, 80186, 80286, and 80386SX microprocessors differ from the 8088/80188 in three ways: (1) the data bus is 16 bits wide instead of 8 bits wide as on the 8088, (2) the IO/$\overline{\text{M}}$ pin of the 8088 is replaced with an M/$\overline{\text{IO}}$ pin, and (3) there is a new control signal called bus high enable ($\overline{\text{BHE}}$). The address bit A0 or $\overline{\text{BLE}}$ is also used differently. (Because this section is based on information provided in Section 10–3, it is extremely important that you read the previous section first.) A few other differences exist between the 8086/80186 and the 80286/80386SX. The 80286/80386SX contains a 24-bit address bus (A23–A0) instead of the 20-bit address bus (A19–A0) of the 8086/80186. The 8086/80186 contain an M/$\overline{\text{IO}}$ signal while the 80286 system and 80386SX microprocessor contain control signals $\overline{\text{MRDC}}$ and $\overline{\text{MWTC}}$ instead of $\overline{\text{RD}}$ and $\overline{\text{WR}}$.

16-Bit Bus Control

The data bus of the 8086, 80186, 80286, and 80386SX is twice as wide as the bus for the 8088/80188. This wider data bus presents us with a unique set of problems that have not been

FIGURE 10–27 The high (odd) and low (even) 8-bit memory banks of the 8086/80286/80386SX microprocessors.

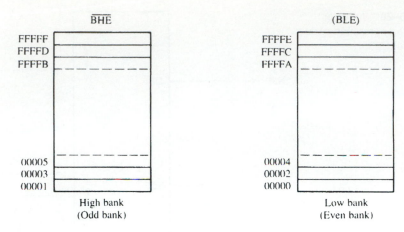

Note: A_0 is labeled $\overline{BLE}$ (Bus low enable) on the 80386SX.

TABLE 10–3 Memory bank selection using $\overline{BHE}$ and $\overline{BLE}$ (A0).

$\overline{BHE}$	$\overline{BLE}$ (A0)	Function
0	0	Both banks enabled for a 16-bit transfer
0	1	High bank enabled for an 8-bit transfer
1	0	Low bank enabled for an 8-bit transfer
1	1	No banks enabled

encountered before. The 8086, 80186, 80286, and 80386SX must be able to write data to any 16-bit location—or any 8-bit location. This means that the 16-bit data bus must be divided into two separate sections (**banks**) that are eight bits wide so that the microprocessor can write to either half (8-bit) or both halves (16-bit). Figure 10–27 illustrates the two banks of the memory. One bank (**low bank**) holds all the even-numbered memory locations, and the other bank (**high bank**) holds all the odd-numbered memory locations.

The 8086, 80186, 80286, and 80386SX use the $\overline{BHE}$ signal (high bank) and the A0 address bit or $\overline{BLE}$ (Bus low enable) to select one or both banks of memory used for the data transfer. Table 10–3 depicts the logic levels on these two pins and the bank or banks selected.

Bank selection is accomplished in two ways: (1) a separate write signal is developed to select a write to each bank of the memory, or (2) separate decoders are used for each bank. As a careful comparison reveals, the first technique is by far the least costly approach to memory interface for the 8086, 80186, 80286, and 80386SX microprocessors.

Separate Bank Decoders. The use of separate bank decoders is often the least effective way to decode memory addresses for the 8086, 80186, 80286, and 80386SX microprocessors. This method is sometimes used, but it is difficult to understand why in most cases. One reason may be to conserve energy, because only the bank or banks selected are enabled. This is not always the case with the separate bank read and write signals that are discussed later.

Figure 10–28 illustrates two 74LS138 decoders used to select 64K RAM memory components for the 80386SX microprocessor (24-bit address). Here, decoder U2 has the $\overline{BLE}$ pin (A0) attached to $\overline{G2A}$, and decoder U3 has the $\overline{BHE}$ signal attached to its $\overline{G2A}$ input. Because the decoder will not activate until all of its enable inputs are active, decoder U2 activates only for a 16-bit operation or an 8-bit operation from the low bank. Decoder U3 activates for a 16-bit

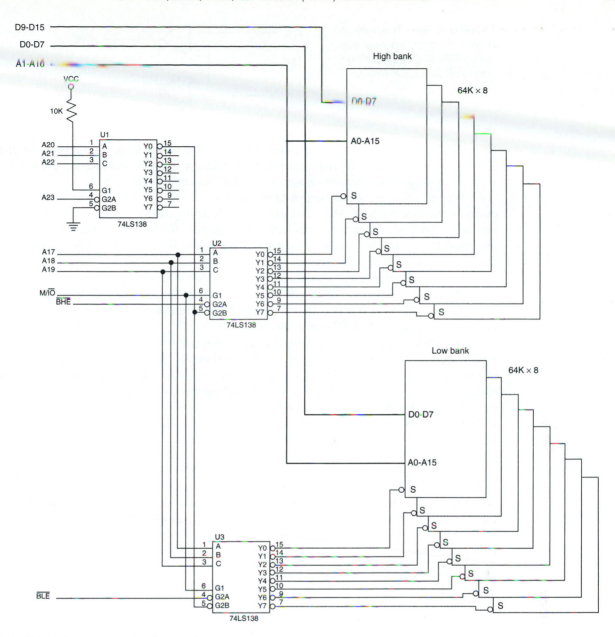

FIGURE 10–28 Separate bank decoders.

operation or an 8-bit operation to the high bank. These two decoders and the 16 64K-byte RAMs they control represent a 1M range of the 80386SX memory system. Decoder U1 enables U2 and U3 for memory address range 000000H–0FFFFFH.

Notice in Figure 10–28 that the A0 address pin does not connect to the memory because it does not exist on the 80386SX microprocessor. Also notice that address bus bit position A1 is connected to memory address input A0, A2 is connected to A1, and so forth. The reason is that A0 from the 8086/80186 (or $\overline{BLE}$ from the 80286/80386SX) is already connected to decoder U2 and does not need to be connected again to the memory. If A0 or $\overline{BLE}$ is attached to the A0 address pin of memory, every other memory location in each bank of memory would be used. This means that half of the memory is wasted if A0 or $\overline{BLE}$ is connected to A0.

FIGURE 10–29 The memory bank write selection input signals: $\overline{HWR}$ (high bank write) and $\overline{LWR}$ (low bank write).

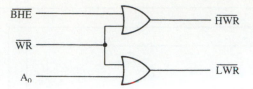

Separate Bank Write Strobes. The most effective way to handle bank selection is to develop a separate write strobe for each memory bank. This technique requires only one decoder to select a 16-bit wide memory, which often saves money and reduces the number of components in a system.

Why not also generate separate read strobes for each memory bank? This is usually unnecessary because the 8086, 80186, 80286, and 80386SX microprocessors read only the byte of data that they need at any given time from half of the data bus. If 16-bit sections of data are always presented to the data bus during a read, the microprocessor ignores the 8-bit section that it doesn't need, without any conflicts or special problems.

Figure 10–29 depicts the generation of separate 8086 write strobes for the memory. Here, a 74LS32 OR gate combines A0 with $\overline{WR}$ for the low bank selection signal ($\overline{LWR}$), and $\overline{BHE}$ combines with $\overline{WR}$ for the high bank selection signal ($\overline{HWR}$). Write strobes for the 80286/80386SX are generated by using the $\overline{MWTC}$ signal instead of $\overline{WR}$.

A memory system that uses separate write strobes is constructed differently from either the 8-bit system (8088) or the system using separate memory banks. Memory in a system that uses separate write strobes is decoded as 16-bit wide memory. For example, suppose that a memory system will contain 64K bytes of SRAM memory. This memory requires two 32K byte memory devices (62256) so that a 16-bit wide memory can be constructed. Because the memory is 16 bits wide and another circuit generates the bank write signals, address bit A0 becomes a don't care. In fact, A0 is not even a pin on the 80386SX microprocessor.

Example 10–6 shows how a 16-bit wide memory stored at locations 060000H–06FFFFH is decoded for the 80286 or 80386 microprocessor. Memory in this example is decoded, so bit A0 is a don't care for the decoder. Bit positions A1–A15 are connected to memory component address pins A0–A14. The decoder (PAL16L8) enables both memory devices by using address connection A23–A15 to select memory whenever address 06XXXXH appears on the address bus.

EXAMPLE 10–6

```
0000 0110 0000 0000 0000 0000 = 060000H
                to
0000 0110 1111 1111 1111 1111 = 06FFFFH

0000 0110 XXXX XXXX XXXX XXXX = 06XXXXH
```

Figure 10–30 illustrates this simple circuit by using a PAL16L8 to both decode memory and generate the separate write strobe. The program for the PAL16L8 decoder is illustrated in Example 10–7. Notice that not only is the memory selected, but both the lower and upper write strobes are also generated by the PAL.

EXAMPLE 10–7

```
TITLE       Address Decoder
PATTERN     Test 2
REVISION    A
AUTHOR      Barry B. Brey
COMPANY     BreyCo
DATE        6/7/99
CHIP        DECODER2 PAL16L8

;pins 1    2    3    4    5    6    7    8    9    10
      A23  A22  A21  A20  A19  A18  A17  A16  A0   GND
```

FIGURE 10–30   A 16-bit memory decoder that places memory at locations 060000H–06FFFFH.

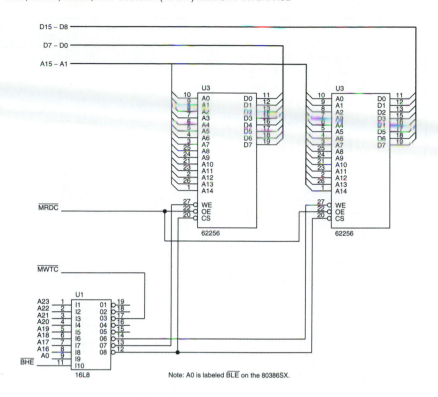

```
;pins 11   12   13   14   15  16  17    18  19  20
      BHE  SEL  LWR  HWR  NC  NC  MWTC   NC  NC  VCC

EQUATIONS

/SEL = /A23 * /A22 * /A21 * /A20 * /A19 * A18 * A17 * /A16
/LWR = /MWTC * /AO
/HWR = /MWTC * /BHE
```

Figure 10–31 depicts a small memory system for the 8086 microprocessor that contains an EPROM section and a RAM section. Here, there are four 27128 EPROMs (16K × 8) that compose a 32K × 16-bit memory at location F0000–FFFFFH and four 62256 (32K × 8) RAMs that compose an 64K × 16-bit memory at location 00000H–1FFFFH. (Remember that even though the memory is 16 bits wide, it is still numbered in bytes.)

This circuit uses a 74LS139 dual 2-to-4 line decoder that selects EPROM with one half and RAM with the other half. It decodes memory that is 16 bits wide; not eight bits, as before. Notice that the $\overline{RD}$ strobe is connected to all the EPROM $\overline{OE}$ inputs and all RAM $\overline{G}$ input pins. This is done because even if the 8086 is reading only eight bits of data, the application of the remaining eight bits to the data bus has no effect on the operation of the 8086.

The $\overline{LWR}$ and $\overline{HWR}$ strobes are connected to different banks of the RAM memory. Here, it does matter whether the microprocessor is doing a 16-bit or an 8-bit write. If the 8086 writes a 16-bit number to memory, both $\overline{LWR}$ and $\overline{HWR}$ go low and enable the $\overline{W}$ pins in both memory banks. But if the 8086 does an 8-bit write, only one of the write strobes goes low, writing to only one memory bank. Again, the only time that the banks make a difference is for a memory write operation.

Notice that an EPROM decoder signal is sent to the 8086 wait state generator because EPROM memory usually requires a wait state. The signal comes from the NAND gate used to select the EPROM decoder section, so that if EPROM is selected, a wait state is requested.

Figure 10–32 illustrates a memory system connected to the 80386SX microprocessor by using a PAL16L8 as a decoder. This interface contains 256K bytes of EPROM in the form

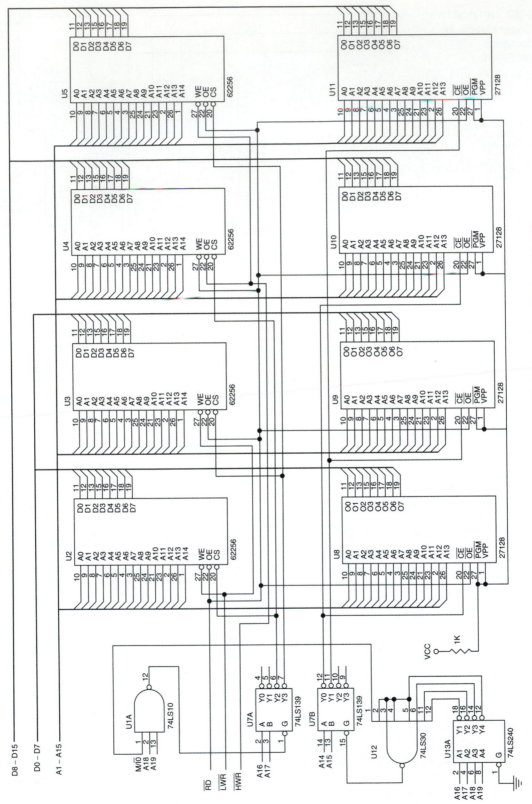

FIGURE 10–31 A memory system for the 8086 that contains a 64K-byte EPROM and a 128K-byte SRAM.

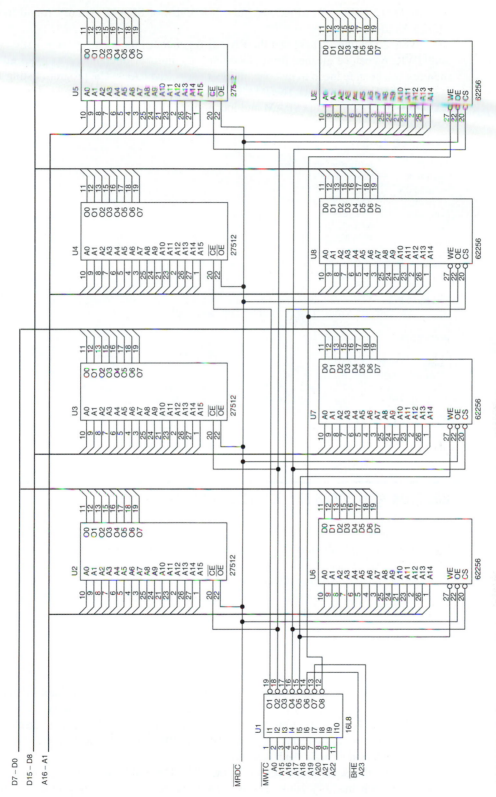

FIGURE 10–32 An 80386SX memory system containing 256K of EPROM and 128K of SRAM.

of four 27512 (64K × 8) EPROMs and 128K bytes of SRAM memory found in four 62256 (32K × 8) SRAMs.

Notice in Figure 10–32 that the PAL also generates the memory bank write signals $\overline{\text{LWR}}$ and $\overline{\text{HWR}}$. As can be gleaned from this circuit, the number of components required to interface memory has been reduced to just one, in most cases (the PAL). The program listing for the PAL is located in Example 10–8. The PAL decodes the 16-bit wide memory addresses at locations 000000H–01FFFFH for the SRAM and locations FC0000H–FFFFFFH for the EPROM.

EXAMPLE 10–8

```
TITLE       Address Decoder
PATTERN     Test 3
REVISION    A
AUTHOR      Barry B. Brey
COMPANY     BreyCo
DATE        6/8/99
CHIP        DECODER3 PAL16L8

;pins  1   2   3    4    5    6    7    8    9    10
       MWTC A0  A16  A17  A18  A19  A20  A21  A22  GND

;pins 11  12  13   14   15   16   17   18   19   20
      NC  HWR A23  BHE  LWR  RB0  RB1  EB0  EB1  VCC

EQUATIONS

/LWR = /MWTC * /A0
/HWR = /MWTC * /BHE
/RB0 = /A23 * /A22 * /A21 * /A20 * /A19 * /A18 * /A17 * /A16
/RB1 = /A23 * /A22 * /A21 * /A20 * /A19 * /A18 * /A17 * A16
/EB0 = A23 * A22 * A21 * A20 * A19 * A18 * /A17
/EB1 = A23 * A22 * A21 * A20 * A19 * A18 * A17
```

10–5   ## 80386DX AND 80486 (32-BIT) MEMORY INTERFACE

As with 8- and 16-bit memory systems, the microprocessor interfaces to memory through its data bus and control signals that select separate memory banks. The only difference with a 32-bit memory system is that the microprocessor has a 32-bit data bus and four banks of memory, instead of one or two. Another difference is that both the 80386DX and 80486 (both SX and DX) contain a 32-bit address bus that usually requires PLD decoders instead of integrated decoders because of the sizable number of address bits.

Memory Banks

The memory banks for both the 80386DX and 80486 microprocessors are illustrated in Figure 10–33. Notice that these large memory systems contain four 8-bit wide banks that each contain up to 1G bytes of memory. Bank selection is accomplished by the bank selection signals $\overline{\text{BE3}}$, $\overline{\text{BE2}}$, $\overline{\text{BE1}}$, and $\overline{\text{BE0}}$. If a 32-bit number is transferred, all four banks are selected; if a 16-bit number is transferred, two banks (usually $\overline{\text{BE3}}$ and $\overline{\text{BE2}}$, or $\overline{\text{BE1}}$ and $\overline{\text{BE0}}$) are selected; and if 8-bits are transferred a single bank is selected.

As with the 8086/80286/80386SX, the 80386DX and 80486 require separate write strobe signals for each memory bank. These separate write strobes are developed, as illustrated in Figure 10–34, by using a simple OR gate or other logic component.

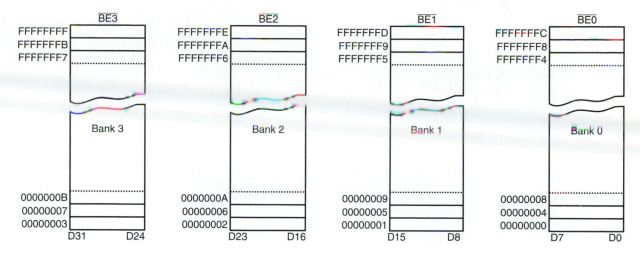

FIGURE 10–33   The memory organization for the 80386DX and 80486 microprocessors.

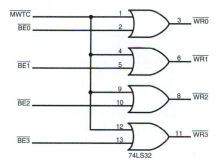

FIGURE 10–34   Bank write signals for the 80386DX and 80486 microprocessors.

32-Bit Memory Interface

As can be gathered from the prior discussion, a memory interface for the 80386DX or 80486 requires that we generate four bank write strobes and decode a 32-bit address. There are no integrated decoders, such as the 74LS138, which can easily accommodate a memory interface for the 80386DX or 80486 microprocessors. Note that address bits A0 and A1 are don't cares when 32-bit wide memory is decoded. These address bits are used within the microprocessor to generate the bank enable signals. Notice that address bus connected A2 connects to memory address pin A0.

Figure 10–35 shows a 256K × 8 memory system for the 80486 microprocessor. This interface uses eight 32K × 8 SRAM memory devices and two PAL16L8 devices. Two devices are required because of the number of address connections found on the microprocessor. This system places the SRAM memory at locations 02000000H–0203FFFFH. The programs for the PAL devices are found in Example 10–9.

EXAMPLE 10–9

```
TITLE      Address Decoder
PATTERN    Test 4 (PAL U1)
REVISION   A
AUTHOR     Barry B. Brey
COMPANY    BreyCo
```

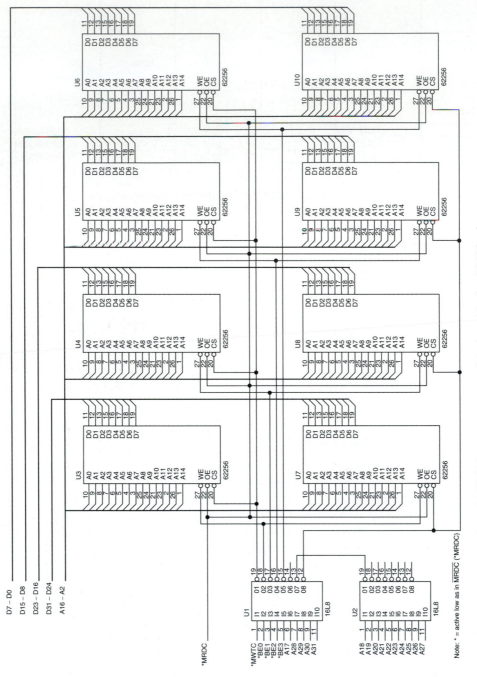

FIGURE 10–35 A small 256K SRAM memory system interfaced to the 80486 microprocessor.

Note: * = active low as in MRDC (*MRDC)

```
DATE            6/9/99
CHIP            DECODER4 PAL16L8

;pins 1    2    3    4    5    6    7    8    9   10
      MWTC BE0  BE1  BE2  BE3  A17  A28  A29  A30 GND

;pins 11   12   13  14  15   16   17   18   19   20
      A31  RB1  U2  NC  WR0  WR1  WR2  WR3  RB0  VCC

EQUATIONS

/WR0 = /MWTC * /BE0
/WR1 = /MWTC * /BE1
/WR2 = /MWTC * /BE2
/WR3 = /MWTC * /BE3
/RB0 = /A31 * /A30 * /A29 * /A28 * /A17 * /U2
/RB1 = /A31 * /A30 * /A29 * /A28 *  A17 * /U2

        TITLE       Address Decoder
        PATTERN     Test 5 (PAL U2)
        REVISION    A
        AUTHOR      Barry B. Brey
        COMPANY     BreyCo
        DATE        6/10/99
        CHIP        DECODER5 PAL16L8

;pins 1    2    3    4    5    6    7    8    9   10
      A18  A19  A20  A21  A22  A23  A24  A25  A26 GND

;pins 11   12  13  14  15  16  17  18  19  20
      A27  U2  NC  NC  NC  NC  NC  NC  NC  VCC

EQUATIONS

/U2 = /A27 * /A26 * A25 * /A24 * /A23 * /A22 * /A21 * /A20 * /A19 * /A18
```

Although not mentioned in this section of the text, the 80386DX and 80486 microprocessors operate with very high clock rates that usually require wait states for memory access. Access time calculations for these microprocessors are discussed in Chapters 17 and 18. The interface provides a signal used with the wait state generator that is not illustrated in this section of the text. Other devices with these higher speed microprocessors are **cache memory** and **interleaved memory** systems. These also are presented in Chapter 17 with the 80386DX and 80486 microprocessors.

10–6   PENTIUM THROUGH PENTIUM 4 (64-BIT) MEMORY INTERFACE

The Pentium through Pentium 4 microprocessors (except for the P24T version of the Pentium) contain a 64-bit data bus, which requires either eight decoders (one per bank) or eight separate write signals. In most systems, separate write signals are used with this microprocessor when interfacing memory. Figure 10–36 illustrates the Pentium's memory organization and its eight memory banks. Notice that this is almost identical to the 80486, except that it contains eight banks instead of four.

As with earlier versions of the Intel microprocessor, this organization is required for upward memory compatibility. The separate write strobe signals are obtained by combining the bank enable signals with the $\overline{\text{MWTC}}$ signal, which is generated by combining the M/$\overline{\text{IO}}$ with W/$\overline{\text{R}}$. The circuit employed for bank write signals appears in Figure 10–37. As can be imagined, we often find a PAL used for bank write signal generation.

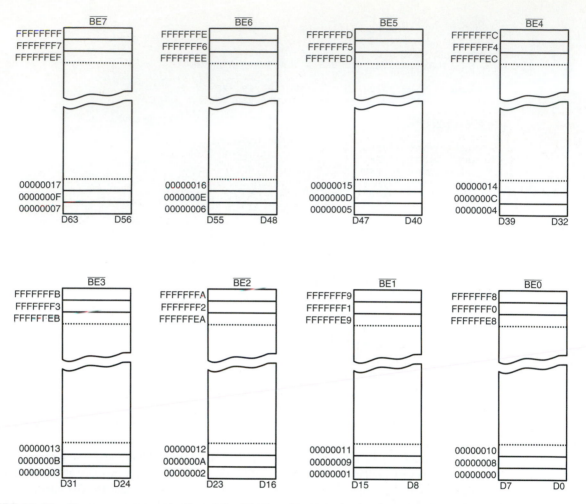

FIGURE 10–36   The memory organization of the Pentium–Pentium 4 microprocessors.

64-Bit Memory Interface

Figure 10–38 illustrates a small Pentium–Pentium 4 memory system. This system uses a PAL16L8 to decode the memory address and a second PAL16L8 to generate the separate bank write signals. This system contains sixteen 27512 EPROM memory devices (64K × 8), interfaced to the Pentium–Pentium 4 at locations FFF80000H through FFFFFFFFH. This is a total memory size of 512K bytes organized so that each bank contains two memory components. Note that the Pentium Pro through the Pentium 4 can be configured with 36 address connections, allowing up to 64G of memory.

Memory decoding, as illustrated in Example 10–10, is similar to the earlier examples, except that with the Pentium–Pentium 4 the rightmost three address bits (A2–A0) are ignored. In this case, the decoder selects two sections of memory that are 64-bits wide and contain 512K bytes of EPROM memory.

The A0 address input of each memory device connects to the A3 address output of the Pentium and above. This A1 address input of each memory device connects to the A4 address output of the Pentium and above. This skewed address connection continues until the A15 address input to the memory is connected to the A18 address output of the Pentium. Address positions A18–A28 are decoded by PAL16L8 U2, and A29–A31 are decoded by PAL16L8 U1. The program for both PAL devices is listed in Example 10–10 for memory locations FFF80000H–FFFFFFFFH.

FIGURE 10–37   The generation of the write strobes for the Pentium–Pentium 4 microprocessors.

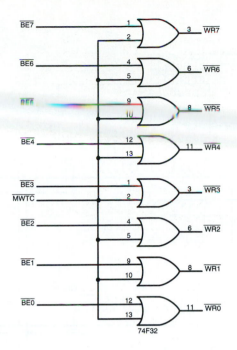

EXAMPLE 10–10

```
TITLE          Address Decoder
PATTERN        Test 6 (PAL U1)
REVISION       A
AUTHOR         Barry B. Brey
COMPANY        BreyCo
DATE           6/11/99
CHIP           DECODER6 PAL16L8

;pins 1    2    3    4   5   6   7   8   9   10
      A29  A30  A31  NC  NC  NC  NC  NC  NC  GND

;pins 11 12 13 14 15 16 17 18 19 20
      U2 CE NC NC NC NC NC NC NC VCC

EQUATIONS

/CE = /U2 * A29 * A30 * A31

TITLE          Address Decoder
PATTERN        Test 7 (PAL U2)
REVISION       A
AUTHOR         Barry B. Brey
COMPANY        BreyCo
DATE           6/12/99
CHIP           DECODER7 PAL16L8

;pins 1    2    3    4    5    6    7    8    9    10
      A19  A20  A21  A22  A23  A24  A25  A26  A27  GND

;pins 11   12 13 14 15 16 17 18 19 20
      A28  U2 NC NC NC NC NC NC U2 VCC

EQUATIONS

/U2 = A19 * A20 * A20 * A21 * A22 * A23 * A24 * A25 * A26 * A27 * A28
```

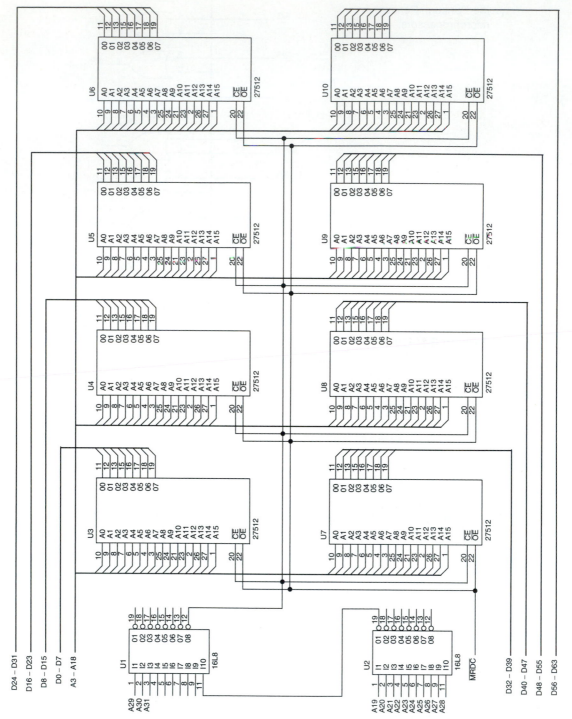

FIGURE 10-38 A small 512K-byte EPROM memory interfaced to the Pentium–Pentium 4 microprocessors.

10–7   DYNAMIC RAM

Because RAM memory is often very large, it requires many SRAM devices at a great cost or just a few DRAMs (dynamic RAMs) at a much reduced cost. The DRAM memory, as briefly discussed in Section 10–1, is fairly complex because it requires address multiplexing and refreshing. Luckily, the integrated circuit manufacturers have provided a dynamic RAM controller that includes the address multiplexers and all the timing circuitry necessary for refreshing.

This section of the text covers the DRAM memory device in much more detail than in Section 10–1 and provides information on the use of a dynamic controller in a memory system.

DRAM Revisited

As mentioned in Section 10–1, a DRAM retains data for only 2–4 ms and requires the multiplexing of address inputs. Although we have already covered address multiplexers in Section 10–1, we will examine the operation of the DRAM during refresh in detail here.

As previously mentioned, a DRAM must be refreshed periodically because it stores data internally on capacitors that lose their charge in a short period of time. In order to refresh a DRAM, the contents of a section of the memory must periodically be read or written. Any read or write automatically refreshes an entire section of the DRAM. The number of bits that are refreshed depends on the size of the memory component and its internal organization.

Refresh cycles are accomplished by doing a read, a write, or a special refresh cycle that doesn't read or write data. The refresh cycle is internal to the DRAM and is accomplished while other memory components in the system operate. This type of memory refresh is called either *hidden refresh, transparent refresh,* or sometimes *cycle stealing.*

In order to accomplish a hidden refresh while other memory components are functioning, an $\overline{RAS}$-only cycle strobes a row address into the DRAM to select a row of bits to be refreshed. The $\overline{RAS}$ input also causes the selected row to be read out internally and rewritten into the selected bits. This recharges the internal capacitors that store the data. This type of refresh is hidden from the system because it occurs while the microprocessor is reading or writing to other sections of the memory.

The DRAM's internal organization contains a series of rows and columns. A 256K × 1 DRAM has 256 columns, each containing 256 bits; or rows organized into four sections of 64K bits each. Whenever a memory location is addressed, the column address selects a column (or internal memory word) of 1024 bits (one per section of the DRAM). Refer to Figure 10–39 for the internal structure of a 256K × 1 DRAM. Note that larger memory devices are structured similarly to the 256K × 1 device. The difference usually lies in either the size of each section or the number of sections in parallel.

Figure 10–40 illustrates the timing for an $\overline{RAS}$-only refresh cycle. The difference between the $\overline{RAS}$ and a read or write is that it applies only a refresh address, which is usually obtained from a 7- or 8-bit binary counter. The size of the counter is determined by the type of DRAM being refreshed. The refresh counter is incremented at the end of each refresh cycle so all the rows are refreshed in 2 or 4 ms, depending on the type of DRAM.

If there are 256 rows to be refreshed within 4 ms, as in a 256K × 1 DRAM, then the refresh cycle must be activated at least once every 15.6 µs in order to meet the refresh specification. For example, it takes the 8086/8088, running at a 5 MHz clock rate, 800 ns to do a read or a write. Because the DRAM must have a refresh cycle every 15.6 µs, for every 19 memory reads or writes, the memory system must run a refresh cycle or else memory data will be lost. This represents a loss of five percent of the computer's time, a small price to pay for the savings represented by using the dynamic RAM.

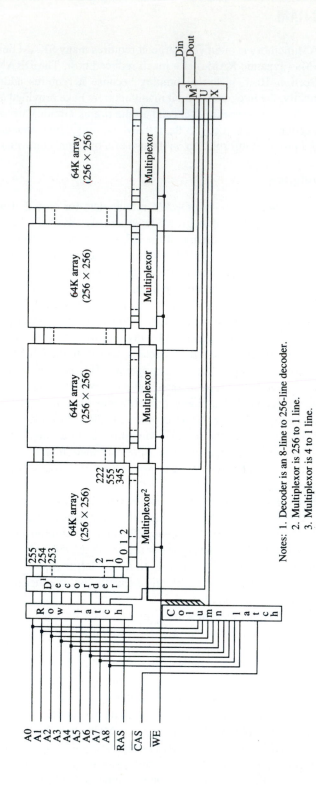

FIGURE 10–39 The internal structure of a 256K × 1 DRAM. Note that each of the internal 256 words are 1025-bits wide.

Notes: 1. Decoder is an 8-line to 256-line decoder.
2. Multiplexor is 256 to 1 line.
3. Multiplexor is 4 to 1 line.

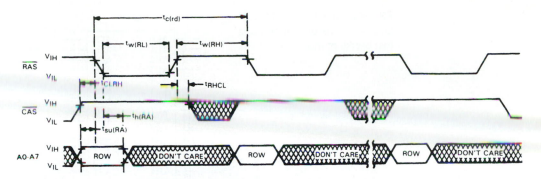

FIGURE 10–40 The timing diagram of the $\overline{RAS}$ refresh cycle for the TMS4464 DRAM. (Courtesy of Texas Instruments Corporation.)

EDO Memory

A slight modification to the structure of the DRAM changes the device into an EDO (**extended data output**) DRAM device. In the EDO memory, any memory access, including a refresh, stores the 256 bits selected by $\overline{RAS}$ into latches. These latches hold the next 256 bits of information, so in most programs, which are sequentially executed, the data are available without any wait states. This slight modification to the internal structure of the DRAM increases system performance by about 15 to 25 percent.

SDRAM

Synchronous dynamic RAM (**SDRAM**) is used with most newer systems because of its speed. Versions are available with access times of 10 ns for use with a 66 MHz system bus and 8 ns for use with a 100 MHz system bus. At first, the access time may lead one to think that these devices operate without wait states, but that is not true. After all, DRAM access time is 60 ns and SDRAM access time is 10 ns. The 10 ns access time is misleading because it only applies to the second, third, and fourth 64-bit reads from the device. The first read requires the same number of waits as a standard DRAM.

When a burst transfer occurs to the SDRAM from the microprocessor, it takes three or four bus clocks before the first 64-bit number is read. Each subsequent number is read without wait states and in one bus cycle each. Because SDRAM bursts read four 64-bit numbers, and the second through the fourth require no waits and can be read in one bus cycle each, SDRAM outperforms standard DRAM or even EDO memory. This means that if it takes three bus cycles for the first number and three more for the next three, it takes a total of seven bus clocks to read four 64-bit numbers. If this is compared to DRAM, which takes three clocks per number, or 12 clocks, you can see the increase in speed. Most estimates place SDRAM at about a 10 percent performance increase over EDO memory.

DRAM Controllers

In most systems, a DRAM controller-integrated circuit performs the task of address multiplexing and the generation of the DRAM control signals. Some newer embedded microprocessors, such as the 80186/80188, include the refresh circuitry as a part of the microprocessor. Because of the extreme complexity of some of the newer DRAM controllers, this text details the Intel 82C08, which controls up to two banks of 256K × 16 DRAM memory. With the 8086 or the 80286/80386SX microprocessors, this can be up to 1 M bytes of memory.

FIGURE 10–41   The 82C08 DRAM controller that controls two banks of memory.

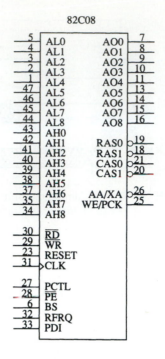

The 82C08 contains an address multiplexer that multiplexes an 18-bit address onto 9 address connections for 256K memory devices. Figure 10–41 shows the pin-out of the DRAM controller. The address inputs are labeled AL0–AL8 and AH0–AH8. The address outputs to the DRAM are labeled AO0–AO8. The 82C08 contains circuitry that generates the $\overline{CAS}$ and $\overline{RAS}$ signals for the DRAM. These signals are developed internally by the CLK, $\overline{S1}$, and $\overline{S2}$ signals. Note that $\overline{S1}$ and $\overline{S0}$, provided by the system, are used as $\overline{RD}$ and $\overline{WR}$ inputs to the 82C08 DRAM controller. The $\overline{AACK}/\overline{XACK}$ signal is an acknowledge output that is used to indicate the ready condition for the microprocessor. This pin normally connects to the READY input of the microprocessor or the clock generator SRDY input.

Figure 10–42 illustrates this device connected to a series of four 256K × 8 bit SIMM memory modules (41256A8) that comprise a 1 M-byte memory system for the 80286 microprocessor. Memory circuits U3 and U5 form the high memory bank, and U4 and U6 form the low bank. The PAL 16L8 combines the $\overline{WE}$ signal from the 82C08 with A0 to generate the bank write signal for U4 and U6 and it combines $\overline{WE}$ with $\overline{BHE}$ to generate a bank write signal for U3 and U5. The PAL also develops the controller selection signal ($\overline{PE}$) by combining M/$\overline{IO}$ with address lines A20–A23 to decode the memory at locations 000000H–0FFFFFH. The A19 signal selects the upper bank (U3 and U4) or the lower bank (U5 and U6) through the BS (**bank select**) input to the 82C08 DRAM controller. Example 10–11 lists the program for the PAL 16L8 used for bank and memory selection.

EXAMPLE 10–11

```
TITLE       Address Decoder
PATTERN     Test 7
REVISION    A
AUTHOR      Barry B. Brey
COMPANY     BreyCo
DATE        6/14/99
```

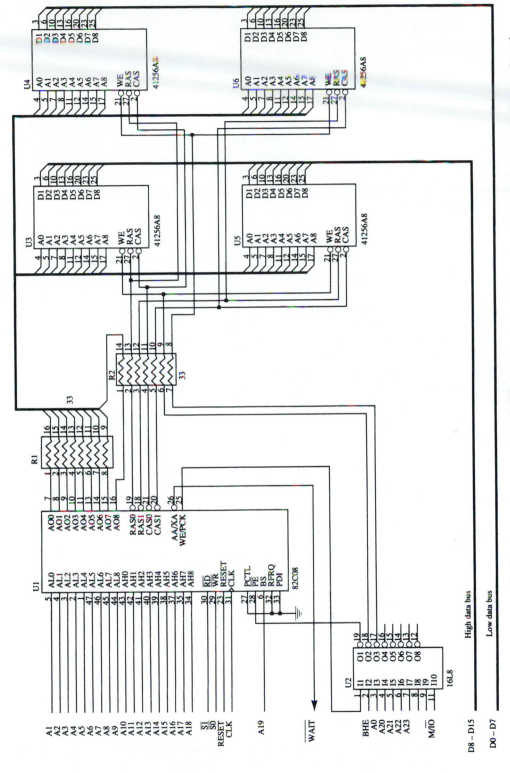

FIGURE 10–42 A 1M-byte memory system using four 256K SIMM memory devices and the 82C08 DRAM controller. This section of memory is decoded at locations 000000H–0FFFFFH by the PAL 16L8.

```
CHIP         DECODER7 PAL16L8

;pins 1    2  3   4   5   6   7   8  9 10
      WE  BHE A0 A20 A21 A22 A23 NC NC GND

;pins 11  12 13 14 15 16  17  18 19  20
      MIO NC NC NC NC NC HWR LWR PE VCC

EQUATIONS

/HWR = /BHE * /WE
/LWR = /A0  * /WE
/PE  = /A20 * /A21 * /A22 * /A23 * MIO
```

10–8   SUMMARY

1. All memory devices have address inputs; data inputs and outputs, or just outputs; a pin for selection; and one or more pins that control the operation of the memory.

2. Address connections on a memory component are used to select one of the memory locations within the device. Ten address pins have 1024 combinations and therefore are able to address 1024 different memory locations.

3. Data connections on a memory are used to enter information to be stored in a memory location and also to retrieve information read from a memory location. Manufacturers list their memory as, for example, 4K × 4, which means that the device has 4K memory locations (4096) and that four bits are stored in each location.

4. Memory selection is accomplished via a chip selection pin ($\overline{CS}$) on many RAMs or a chip enable pin ($\overline{CE}$) on many EPROM or ROM memories.

5. Memory function is selected by an output enable pin ($\overline{OE}$) for reading data, which normally connects to the system read signal ($\overline{RD}$ or $\overline{MRDC}$). The write enable pin ($\overline{WE}$), for writing data, normally connects to the system write signal ($\overline{WR}$ or $\overline{MWTC}$).

6. An EPROM memory is programmed by an EPROM programmer and can be erased if exposed to ultraviolet light. Today, EPROMs are available in sizes from 1K × 8 all the way up to 128K × 8 and larger.

7. The flash memory (EEPROM) is programmed in the system by using a 12 V programming pulse.

8. Static RAM (SRAM) retains data for as long as the system power supply is attached. These memory types are available in sizes up to 128K × 8.

9. Dynamic RAM (DRAM) retains data for only a short period, usually 2–4 ms. This creates problems for the memory system designer because the DRAM must be refreshed periodically. DRAMs also have multiplexed address inputs that require an external multiplexer to provide each half of the address at the appropriate time.

10. Memory address decoders select an EPROM or RAM at a particular area of the memory. Commonly found address decoders include the 74LS138 3-to-8 line decoder, the 74LS139 2-to-4 line decoder, and programmed selection logic in the form of a PROM or PLD.

11. The PROM and PLD address decoders for microprocessors like the 8088 through the Pentium 4 reduce the number of integrated circuits required to complete a functioning memory system.

12. The 8088 minimum mode memory interface contains 20 address lines, eight data lines, and three control lines: $\overline{RD}$, $\overline{WR}$, and IO/$\overline{M}$. The 8088 memory functions correctly only when all these lines are used for memory interface.

13. The access speed of the EPROM must be compatible with the microprocessor to which it is interfaced. Many EPROMs available today have an access time of 450 ns, which is too slow for the 5 MHz 8088. In order to circumvent this problem, a wait state is inserted to increase memory access time to 660 ns.

14. Parity checkers are becoming commonplace today in many microprocessor-based microcomputer systems. An extra bit is stored with each byte of memory, making the memory nine bits wide instead of eight.

15. Error-correction features are also available for memory systems, but these require the storage of many more bits. If an 8-bit number is stored with an error-correction circuit, it actually takes 13 bits of memory: five for an error checking code and eight for the data. Most error-correction integrated circuits are able to correct only a single-bit error.

16. The 8086/80286/80386SX memory interface has a 16-bit data bus and contains an $M/\overline{IO}$ control pin, whereas the 8088 has an 8-bit data bus and contains an $IO/\overline{M}$ pin. In addition to these changes, there is an extra control signal, bus high enable ($\overline{BHE}$).

17. The 8086/80386/80386SX memory is organized in two 8-bit banks: high bank and low bank. The high bank of memory is enabled by the $\overline{BHE}$ control signal and the low bank is enabled by the A0 address signal or by the $\overline{BLE}$ control signal.

18. Two common schemes for selecting the banks in an 8086/80286/80386SX-based system include (1) a separate decoder for each bank and (2) separate $\overline{WR}$ control signals for each bank with a common decoder.

19. Memory interfaced to the 80386DX and 80486 is 32 bits wide, as selected by a 32-bit address bus. Because of the width of this memory, it is organized in four memory banks that are each eight bits wide. Bank selection signals are provided by the microprocessor as $\overline{BE3}$, $\overline{BE2}$, $\overline{BE1}$, and $\overline{BE0}$.

20. Memory interfaced to the Pentium–Pentium 4 is 64 bits wide, as selected by a 32-bit address bus. Because of the width of the memory, it is organized in eight banks that are each eight bits wide. Bank-selection signals are provided by the microprocessor as $\overline{BE7}$–$\overline{BE0}$.

21. Dynamic RAM controllers are designed to control DRAM memory components. Many DRAM controllers today contain address multiplexers, refresh counters, and the circuitry required to do a periodic DRAM memory refresh.

10–9 QUESTIONS AND PROBLEMS

1. What types of connections are common to all memory devices?
2. List the number of words found in each memory device for the following numbers of address connections:
   (a) 8
   (b) 11
   (c) 12
   (d) 13
3. List the number of data items stored in each of the following memory devices and the number of bits in each datum:
   (a) $2K \times 4$
   (b) $1K \times 1$
   (c) $4K \times 8$
   (d) $16K \times 1$
   (e) $64K \times 4$
4. What is the purpose of the $\overline{CS}$ or $\overline{CE}$ pin on a memory component?

5. What is the purpose of the $\overline{OE}$ pin on a memory device?

6. What is the purpose of the $\overline{WE}$ pin on a RAM?

7. How many bytes of storage do the following EPROM memory devices contain?
   (a) 2708
   (b) 2716
   (c) 2732
   (d) 2764
   (e) 27128

8. Why won't a 450 ns EPROM work directly with a 5 MHz 8088?

9. What can be stated about the amount of time to erase and write a location in a flash memory device?

10. SRAM is an acronym for what type of device?

11. The 4016 memory has a $\overline{G}$ pin, an $\overline{S}$ pin, and a $\overline{W}$ pin. What are these pins used for in this RAM?

12. How much memory access time is required by the slowest 4016?

13. DRAM is an acronym for what type of device?

14. The TMS4464 has eight address inputs, yet it is a 64K DRAM. Explain how a 16-bit memory address is forced into eight address inputs.

15. What are the purposes of the $\overline{CAS}$ and $\overline{RAS}$ inputs of a DRAM?

16. How much time is required to refresh the typical DRAM?

17. Why are memory address decoders important?

18. Modify the NAND gate decoder of Figure 10–13 to select the memory for address range DF800H–DFFFFH.

19. Modify the NAND gate decoder in Figure 10–13 to select the memory for address range 40000H–407FFH.

20. When the G1 input is high, and $\overline{G2A}$ and $\overline{G2B}$ are both low, what happens to the outputs of the 74LS138 3-to-8 line decoder?

21. Modify the circuit of Figure 10–15 to address memory range 70000H–7FFFFH.

22. Modify the circuit of Figure 10–15 to address memory range 40000H–4FFFFH.

23. Describe the 74LS139 decoder.

24. Why is a PROM address decoder often found in a memory system?

25. Reprogram the PROM in Table 10–1 to decode memory address range 80000H–8FFFFH.

26. Reprogram the PROM in Table 10–1 to decode memory address range 30000H–3FFFFH.

27. Modify the circuit of Figure 10–19 by rewriting the PAL program to address memory at locations 40000H–4FFFFH.

28. Modify the circuit of Figure 10–19 by rewriting the PAL program to address memory at locations B0000H–BFFFFH.

29. The $\overline{RD}$ and $\overline{WR}$ minimum mode control signals are replaced by what two control signals in the 8086 maximum mode?

30. Modify the circuit of Figure 10–20 to select memory at location 68000–6FFFFH.

31. Modify the circuit of Figure 10–20 to select eight 2764 8K × 8 EPROMs at memory location 10000H–1FFFFH.

32. Add another decoder to the circuit of Figure 10–21 so that an additional eight 62256 SRAMs are added at location C0000H–FFFFFH.

33. Explain how odd parity is stored in a memory system and how it is checked.

34. The 74LS636 error correction and detection circuit stores a check code with each byte of data. How many bits are stored for the check code?

35. What is the purpose of the SEF pin on the 74LS636?

36. The 74LS636 will correct _____ bits that are in error.

37. Outline the major difference between the buses of the 8086 and 8088 microprocessors.

38. What is the purpose of the $\overline{\text{BHE}}$ and A0 pins on the 8086 microprocessor?
39. What is the $\overline{\text{BLE}}$ pin and what other pin has it replaced?
40. What two methods are used to select the memory in the 8086 microprocessor?
41. If $\overline{\text{BHE}}$ is a logic 0, then the _____ memory bank is selected.
42. If A0 is a logic 0, then the _____ memory bank is selected.
43. Why don't separate bank read ($\overline{\text{RD}}$) strobes need to be developed when interfacing memory to the 8086?
44. Modify the circuit of Figure 10–31 so that the EPROM is located at memory range C0000H–CFFFFH and the RAM is located at memory range 30000H–4FFFFH.
45. Develop a 16-bit wide memory interface that contains SRAM memory at locations 200000H–21FFFFH for the 80386SX microprocessor.
46. Develop a 32-bit wide memory interface that contains EPROM memory at locations FFFF0000H–FFFFFFFFH.
47. Develop a 64-bit wide memory for the Pentium–Pentium 4 that contains EPROM at locations FFF00000H–FFFFFFFFH and SRAM at locations 00000000H–003FFFFFH.
48. What is an $\overline{\text{RAS}}$-only cycle?
49. When DRAM is refreshed, can it be done while other sections of the memory operate?
50. If a 1M × 1 DRAM requires 4 ms for a refresh and has 256 rows to be refreshed, no more than _____ of time must pass before another row is refreshed.
51. Where is the memory address applied to the 82C08 DRAM controller?
52. What is the purpose of the BS pin on the 82C08?
53. What is normally connected to the $\overline{\text{WR}}$ pin of the 82C08?

CHAPTER 11

Basic I/O Interface

INTRODUCTION

A microprocessor is great at solving problems, but if it can't communicate with the outside world, it is of little worth. This chapter outlines some of the basic methods of communications, both serial and parallel, between humans or machines and the microprocessor.

In this chapter, we first introduce the basic I/O interface and discuss decoding for I/O devices. Then, we provide detail on parallel and serial interfacing, both of which have a variety of applications. As applications, we connect analog-to-digital and digital-to-analog converters, as well as both DC and stepper motors to the microprocessor.

CHAPTER OBJECTIVES

Upon completion of this chapter, you will be able to:

1. Explain the operation of the basic input and output interfaces.
2. Decode an 8-, 16-, and 32-bit I/O device so that they can be used at any I/O port address.
3. Define handshaking and explain how to use it with I/O devices.
4. Interface and program the 82C55 programmable parallel interface.
5. Interface LCD displays, LED displays, keyboards, ADC, DAC, and various other devices to the 82C55.
6. Interface and program the 8279 programmable keyboard/display controller.
7. Interface and program the 16550 serial communications interface adapter.
8. Interface and program the 8254 programmable interval timer.
9. Interface an analog-to-digital converter and a digital-to-analog converter to the microprocessor.
10. Interface both DC and stepper motors to the microprocessor.

11–1 INTRODUCTION TO I/O INTERFACE

In this section of the text, we explain the operation of the I/O instructions (IN, INS, OUT, and OUTS). We also explain the concept of isolated (sometimes called direct or I/O mapped I/O) and memory-mapped I/O, the basic input and output interfaces, and handshaking. A working knowledge of these topics will make it easier to understand the connection and operation of the programmable interface components and I/O techniques presented in the remainder of this chapter and text.

I/O Instructions

The instruction set contains one type of instruction that transfers information to an I/O device (OUT) and another to read information from an I/O device (IN). Instructions (INS and OUTS, found on all versions except the 8086/8088) are also provided to transfer strings of data between the memory and an I/O device. Table 11–1 lists all versions of each instruction found in the microprocessor's instruction set.

Both the IN and OUT instructions transfer data between an I/O device and the microprocessor's accumulator (AL, AX, or EAX). The I/O address is stored in register DX as a 16-bit I/O address or in the byte (p8) immediately following the opcode as an 8-bit I/O address. Intel calls the 8-bit form (p8) a **fixed address** because it is stored with the instruction, usually in a ROM. The 16-bit I/O address in DX is called a **variable address** because it is stored in a DX, and then used to address the I/O device. Other instructions that use DX to address I/O are the INS and OUTS instructions.

Whenever data are transferred by using the IN or OUT instruction, the I/O address, often called a **port number** (or simply port), appears on the address bus. The external I/O interface

TABLE 11–1 Input/output instructions.

Instruction	Data Width	Function
IN AL, p8	8	A byte is input from port p8 into AL
IN AX, p8	16	A word is input from port p8 into AX
IN EAX, p8	32	A doubleword is input from port p8 into EAX
IN AL,DX	8	A byte is input from the port addressed by DX into AL
IN AX,DX	16	A word is input from the port addressed by DX into AX
IN EAX,DX	32	A word is input from the port addressed by DX into EAX
INSB	8	A byte is input from the port addressed by DX into the extra segment memory location addressed by DI, then DI = DI ± 1
INSW	16	A word is input from the port addressed by DX into the extra segment memory location addressed by DI, then DI = DI ± 2
INSD	32	A doubleword is input from the port addressed by DX into the extra segment memory location addressed by DI, then DI ± 4
OUT p8,AL	8	A byte is output from AL to port p8
OUT p8,AX	16	A word is output from AX to port p8
OUT p8,EAX	32	A doubleword is output from EAX to port p8
OUT DX,AL	8	A byte is output from AL to the port addressed by DX
OUT DX,AX	16	A word is output from AX to the port addressed by DX
OUT DX,EAX	32	A doubleword is output from EAX to the port addressed by DX
OUTSB	8	A byte is output from the data segment memory location addressed by SI to the port addressed by DX, then SI = SI ± 1
OUTSW	16	A word is output from the data segment memory locations addressed by SI to the port addressed by DX, then SI = SI ± 2
OUTSD	32	A doubleword is output from the data segment memory locations addressed by SI to the port addressed by DX, then SI = SI ± 4

decodes the port number in the same manner that it decodes a memory address. The 8-bit fixed port number (p8) appears on address bus connections A7–A0 with bits A15–A8 equal to 00000000_2. The address connections above A15 are undefined for an I/O instruction. The 16-bit variable port number (DX) appears on address connection A15–A0. This means that the first 256 I/O port addresses (00H–FFH) are accessed by both the fixed and variable I/O instructions, but any I/O address from 0100H–FFFFH is only accessed by the variable I/O address. In many dedicated task systems, only the rightmost eight bits of the address are decoded, thus reducing the amount of circuitry required for decoding. In a PC computer, all 16 address bus bits are decoded with locations 0000H–03XXH, are the I/O addresses used for I/O inside the PC for the ISA (**industry standard architecture**) bus.

The INS and OUTS instructions address an I/O device by using the DX register, but do not transfer data between the accumulator and the I/O device as IN and OUT. Instead, these instructions transfer data between memory and the I/O device. The memory address is located by ES:DI for the INS instruction and by DS:SI for the OUTS instruction. As with other string instructions, the contents of the pointers are incremented or decremented, as dictated by the state of the direction flag (DF). Both INS and OUTS can be prefixed with the REP prefix, allowing more than one byte, word, or doubleword to be transferred between I/O and memory.

Isolated and Memory-Mapped I/O

There are two different methods of interfacing I/O to the microprocessor: **isolated I/O** and **memory-mapped I/O.** In the isolated I/O scheme, the IN, INS, OUT, and OUTS instructions transfer data between the microprocessor's accumulator or memory and the I/O device. In the memory-mapped I/O scheme, any instruction that references memory can accomplish the transfer. Both isolated and memory-mapped I/O are in use, so both are discussed in this text.

Isolated I/O.   The most common I/O transfer technique used in the Intel microprocessor-based system is isolated I/O. The term *isolated* describes how the I/O locations are isolated from the memory system in a separate I/O address space. (Figure 11–1 illustrates both the isolated and memory-mapped address spaces for any Intel 80X86 or Pentium–Pentium 4 microprocessor.) The addresses for isolated I/O devices, called **ports,** are separate from the memory. Because the ports are separate, the user can expand the memory to its full size without using any of memory space for I/O devices. A disadvantage of isolated I/O is that the data transferred between I/O and the microprocessor must be accessed by the IN, INS, OUT, and OUTS instructions. Separate control signals for the I/O space are developed (using M/$\overline{IO}$ and W/$\overline{R}$), which indicate an I/O read ($\overline{IORC}$) or an I/O write ($\overline{IOWC}$) operation. These signals indicate that an I/O port address, which appears on the address bus, is used to select the I/O device. In the personal computer, isolated I/O ports are used for controlling peripheral devices. An 8-bit port address is used to access devices located on the system board, such as the timer and keyboard interface, while a 16-bit port is used to access serial and parallel ports as well as video and disk drive systems.

Memory-Mapped I/O.   Unlike isolated I/O, memory-mapped I/O does not use the IN, INS, OUT, or OUTS instructions. Instead, it uses any instruction that transfers data between the microprocessor and memory. A memory-mapped I/O device is treated as a memory location in the memory map. The main advantage of memory-mapped I/O is that any memory transfer instruction can be used to access the I/O device. The main disadvantage is that a portion of the memory system is used as the I/O map. This reduces the amount of memory available to applications. Another advantage is that the $\overline{IORC}$ and $\overline{IOWC}$ signals have no function in a memory-mapped I/O system and may reduce the amount of circuitry required for decoding.

Personal Computer I/O Map

The personal computer uses part of the I/O map for dedicated functions. Figure 11–2 shows the I/O map for the PC. Note that I/O space between ports 0000H and 03FFH are normally reserved

FIGURE 11–1 The memory and I/O maps for the 8086/8088 microprocessors. (a) Isolated I/O (b) Memory-mapped I/O.

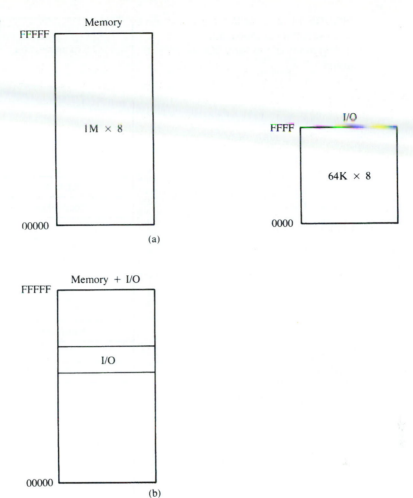

for the computer system and the ISA bus. The I/O ports located at 0400H–FFFFH are generally available for user applications, main-board functions, and the PCI bus. Note that the 80287 arithmetic coprocessor uses I/O address 00F8H–00FFH for communications. For this reason, Intel reserves I/O ports 00F0H–00FFH. The 80386–Pentium 4 use I/O ports 800000F8–800000FFH for communications to their coprocessors. The I/O ports located between 0000H and 00FFH are accessed via the fixed port I/O instructions; the ports located above 00FFH are accessed via the variable I/O port instructions.

Basic Input and Output Interfaces

The basic input device is a set of three-state buffers. The basic output device is a set of data latches. The term IN refers to moving data from the I/O device into the microprocessor and the term OUT refers to moving data out of the microprocessor to the I/O device.

The Basic Input Interface. Three-state buffers are used to construct the 8-bit input port depicted in Figure 11–3. The external TTL data (simple toggle switches in this example) are connected to the inputs of the buffers. The outputs of the buffers connect to the data bus. The exact data bus connections depend on the version of the microprocessor. For example, the 8088 has data bus connections D7–D0, the 80486 has D31–D0, and the Pentium–Pentium 4 have D63–D0. The circuit of Figure 11–3 allows the microprocessor to read the contents of the 8 switches that

FIGURE 11–2 The I/O map of a personal computer illustrating many of the fixed I/O areas.

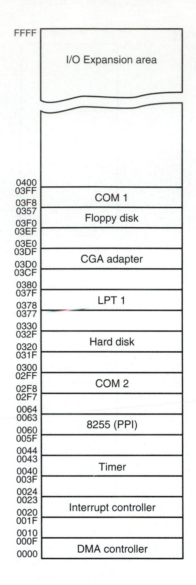

connect to any 8-bit section of the data bus when the select signal $\overline{\text{SEL}}$ becomes a logic 0. Thus, whenever the IN instruction executes, the contents of the switches are copied into the AL register.

When the microprocessor executes an IN instruction, the I/O port address is decoded to generate the logic 0 on $\overline{\text{SEL}}$. A 0 placed on the output control inputs ($\overline{\text{1G}}$ and $\overline{\text{2G}}$) of the 74ALS244 buffer causes the data input connections (A) to be connected to the data output (Y) connections. If a logic 1 is placed on the output control inputs of the 74ALS244 buffer, the device enters the three-state high-impedance mode that effectively disconnects the switches from the data bus.

This basic input circuit is not optional and must appear any time that input data are interfaced to the microprocessor. Sometimes it appears as a discrete part of the circuit, as shown in Figure 11–3; sometimes it is built into a programmable I/O device.

16- or 32-bit data can also be interfaced to various versions of the microprocessor, but this is not nearly as common as using 8-bit data. To interface 16 bits of data, the circuit in Figure 11–3 is doubled to include two 74ALS244 buffers that connect 16 bits of input data to the 16-bit data bus. To interface 32 bits of data, the circuit is expanded by a factor of 4.

FIGURE 11–3   The basic input interface illustrating the connection of eight switches. Note that the 74ALS244 is a three-state that controls the application of the switch data to the data bus.

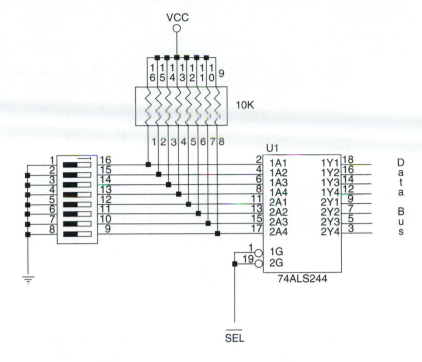

The Basic Output Interface.

The basic output interface receives data from the microprocessor and must usually hold it for some external device. Its latches or flip-flops, like the buffers found in the input device, are often built into the I/O device.

Figure 11–4 shows how eight simple light-emitting diodes (LEDs) connect to the microprocessor through a set of eight data latches. The latch stores the number output by the microprocessor from the data bus so that the LEDs can be lit with any 8-bit binary number. Latches are needed to hold the data because when the microprocessor executes an OUT instruction, the data are only present on the data bus for less than 1.0 μs. Without a latch, the viewer would never see the LEDs illuminate.

When the OUT instruction executes, the data from AL, AX, or EAX are transferred to the latch via the data bus. Here, the D inputs of a 74ALS374 octal latch are connected to the data bus to capture the output data, and the Q outputs of the latch are attached to the LEDs. When a Q output becomes a logic 0, the LED lights. Each time that the OUT instruction executes, the $\overline{\text{SEL}}$ signal to the latch activates, capturing the data output to the latch from any 8-bit section of the data bus. The data are held until the next OUT instruction executes. Thus, whenever the output instruction is executed in this circuit, the data from the AL register appear on the LEDs.

Handshaking

Many I/O devices accept or release information at a much slower rate than the microprocessor. Another method of I/O control, called **handshaking** or **polling,** synchronizes the I/O device with the microprocessor. An example device that requires handshaking is a parallel printer that prints 100 characters per second **(CPS).** It is obvious that the microprocessor can send more than 100 CPS to the printer, so a way to slow the microprocessor down to match speeds with the printer must be developed.

Figure 11–5 illustrates the typical input and output connections found on a printer. Here, data are transferred through a series of data connections (D7–D0), BUSY indicates that the printer is busy and $\overline{\text{STB}}$ is a clock pulse used to send data into the printer for printing.

FIGURE 11–4 The basic output interface connected to a set of LED displays.

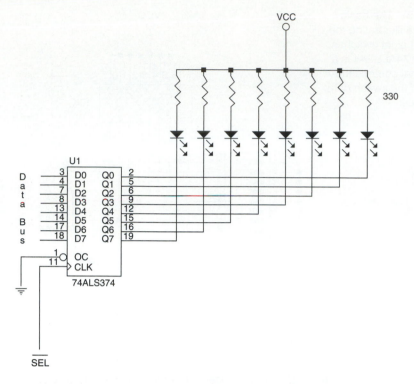

The ASCII data to be printed by the printer are placed on D7–D0, and a pulse is then applied to the $\overline{STB}$ connection. The strobe signal sends or clocks the data into the printer so that it can be printed. As soon as the printer receives the data, it places a logic 1 on the BUSY pin, indicating that the printer is busy printing data. The microprocessor software polls or tests the BUSY pin to decide whether the printer is busy. If the printer is busy, the microprocessor waits; if it is not busy, the microprocessor sends the next ASCII character to the printer. This process of interrogating the printer is called **handshaking** or **polling.** Example 11–1 illustrates a simple procedure that tests the printer BUSY flag and then sends data to the printer if it is not busy. The PRINT procedure prints the ASCII-coded contents of BL only if the BUSY flag is a logic 0, indicating that the printer is not busy. This procedure is called each time a character is to be printed.

EXAMPLE 11–1

```
                        ;A procedure that prints the ASCII contents of BL.
                        ;
0000                    PRINT  PROC   NEAR

0000   E4 4B                   IN     AL,BUSY          ;get BUSY flag
0002   A8 04                   TEST   AL,BUSY_BIT      ;test BUSY bit
0004   75 FA                   JNE    PRINT            ;if printer busy
0006   8A C3                   MOV    AL,BL            ;get data from BL
0008   E6 4A                   OUT    PRINTER,AL       ;send data to printer
000A   CB                      RET                     ;return from procedure

000B                    PRINT  ENDP
```

Notes About Interfacing Circuitry

A certain part of interfacing requires some knowledge about electronics. This portion of the introduction to interfacing examines some facts about electronic interfacing. Before a circuit or

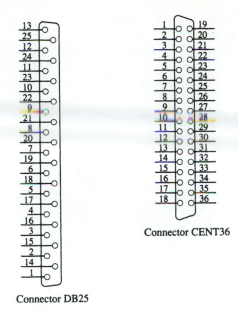

Connector CENT36

Connector DB25

DB25 Pin number	CENT36 Pin number	Function	DB25 Pin number	CENT36 Pin number	Function
1	1	$\overline{\text{Data Strobe}}$	12	12	Paper empty
2	2	Data 0 (D0)	13	13	Select
3	3	Data 1 (D1)	14	14	Afd
4	4	Data 2 (D2)	15	32	$\overline{\text{Error}}$
5	5	Data 3 (D3)	16	—	$\overline{\text{RESET}}$
6	6	Data 4 (D4)	17	31	Select in
7	7	Data 5 (D5)	18—25	19—30	Ground
8	8	Data 6 (D6)	—	17	Frame ground
9	9	Data 7 (D7)	—	16	Ground
10	10	$\overline{\text{Ack}}$	—	33	Ground
11	11	Busy			

FIGURE 11–5 The DB25 connector found on computers and the Centronics 36-pin connector found on printers for the Centronics parallel printer interface.

device can be interfaced to the microprocessor, the terminal characteristics of the microprocessor and its associated interfacing components must be known. (This was introduced at the start of Chapter 9.)

Input Devices. Input devices are already TTL and compatible, and therefore can be connected to the microprocessor and its interfacing components or they are switch-based. Most switch-based devices are either open or connected. These are not TTL levels—TTL levels are a logic 0 (0.0 V–0.8 V) or a logic 1 (2.0 V–5.0 V).

For a switch-based device to be used as a TTL-compatible input device, some conditioning must be applied. Figure 11–6 shows a simple toggle switch that is properly connected to function as an input device. Notice that a pull-up resistor is used to ensure that when the switch is open, the output signal is a logic 1; when the switch is closed, it connects to ground, producing a valid logic 0 level. The value of the pull-up resistor is not critical—it merely assures that the signal is at a logic 1 level. A standard range of values for pull-up resistors is usually anywhere between 1K Ω to 10K Ω.

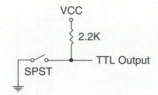

FIGURE 11–6 A single-pole, single-throw switch interfaced as a TTL device.

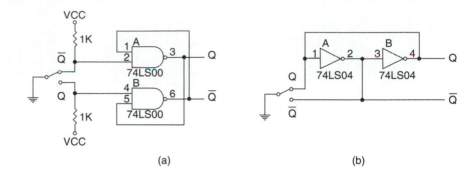

FIGURE 11–7 Debouncing switch contacts: (a) conventional debouncing and (b) practical debouncing.

(a)   (b)

Mechanical switch contacts physically bounce when they are closed, which can create a problem if a switch is used as a clocking signal for a digital circuit. To prevent problems with bounces, one of the two circuits depicted in Figure 11–7 can be constructed. The first circuit (a) is a classical textbook bounce eliminator; the second (b) is a more practical version of the same circuit. Because the first version costs more money to construct, in practice, the second would be used because it requires no pull-up resistors and only two inverters instead of two NAND gates.

You may notice that both circuits in Figure 11–7 are asynchronous flip-flops. The circuit of (b) functions in the following manner: Suppose that the switch is currently at position $\overline{Q}$. If it is moved toward Q but does not yet touch Q, the Q output of the circuit is a logic 0. The logic 0 state is remembered by the inverters. The output of inverter B connects to the input of inverter A. Because the output of inverter B is a logic 0, the output of inverter A is a logic 1. The logic 1 output of inverter A maintains the logic 0 output of inverter B. The flip-flop remains in this state until the moving switch-contact first touches the Q connection. As soon as the Q input from the switch becomes a logic 0, it changes the state of the flip-flop. If the contact bounces back away from the Q input, the flip-flop remembers and no change occurs, thus eliminating any bounce.

Output Devices. Output devices are far more diverse than input devices, but many are interfaced in a uniform manner. Before any output device can be interfaced, we must understand what the voltages and currents are from the microprocessor or a TTL interface component. The voltages are TTL-compatible from the microprocessor of the interfacing element. (Logic 0 = 0.0 V to 0.4 V; logic 1 = 2.4 V to 5.0 V.) The currents for a microprocessor and many microprocessor-interfacing components are less than for standard TTL components. (Logic 0 = 0.0 to 2.0 mA; logic 1 = 0.0 to 400 μA.)

Once the output currents are known, we can now interface a device to one of the outputs. Figure 11–8 shows how to interface a simple LED to a microprocessor peripheral pin. Notice that a transistor driver is used in Figure 11–8(a) and a TTL inverter is used in Figure 11–8(b). The TTL inverter (standard version) provides up to 16 mA of current at a logic 0 level, which is more than enough to drive a standard LED. A standard LED requires 10 mA of forward bias current to light. In both circuits, we assume that the voltage drop across the LED is about 2.0 V. The data sheet for an LED states that the nominal drop is 1.65 V, but we know from experience that the drop is anywhere between 1.5 V and 2.0 V. This means that the value of the current-limiting resistor is $^{3.0\,V}/_{10\,mA}$ or 300 Ω. Since 300 Ω is not a standard resistor value, we choose to use a 330 Ω resistor.

FIGURE 11–8 Interfacing an LED: (a) using a transistor and (b) using an inverter.

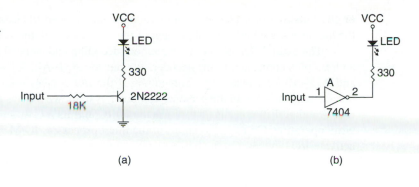

(a)

(b)

FIGURE 11–9 A DC motor interfaced to a system by using a Darlington-pair.

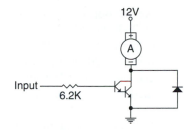

In the circuit of Figure 11–8(a), we elected to use a switching transistor in place of the TTL buffer. The 2N2222 is a good general-purpose switching transistor that has a minimum gain of 100. In this circuit, the collector current is 10 mA, so the base current will be 1/100 of the collector current of 0.1 mA. To determine the value of the base current-limiting resistor, we use the 0.1 mA current and a voltage drop of 1.7 V. The TTL input signal has a minimum value of 2.4 V and the drop across the emitter-base junction is 0.7 V. The difference is 1.7 V, which is the voltage drop across the resistor. The value of the resistor is $^{1.7\,V}/_{0.1\,mA}$ or 17K Ω. Because 17K Ω is not a standard value, we used an 18K Ω resistor.

Suppose that we need to interface a 12 V DC motor to the microprocessor and the motor current is 1A. Obviously, we cannot use a TTL inverter for two reasons: the 12 V signal would burn out the inverter and the amount of current far exceeds the 16 mA maximum current from the inverter. We cannot use a 2N2222 transistor either, because the maximum amount of current is 250 mA to 500 mA, depending on the package style. The solution is to use a Darlington-pair.

Figure 11–9 illustrates the motor connected to the Darlington-pair. The Darlington-pair has a minimum current gain of 7000 and a maximum current of 4A. The value of the bias resistor is calculated like the one used in the LED driver. The current through the resistor is 1A/7000, or about 0.143 mA. The voltage drop across the resistor is 0.9 V because we have two diode drops instead of one. The value of the bias resistor is $^{0.9\,V}/_{0.143\,mA}$ or 6.29KΩ. The standard value of 6.2 KΩ is used in the circuit. The Darlington-pair must be heat-sinked because of the amount of current going through it and the diode must also be present to prevent the Darlington-pair from being destroyed by the inductive kick-back from the motor. This circuit is also used to interface mechanical relays or just about any device that requires a large amount of current or a change in voltage.

11–2

I/O PORT ADDRESS DECODING

I/O port address decoding is very similar to memory address decoding, especially for memory-mapped I/O devices. In fact, we do not discuss memory-mapped I/O decoding because it is treated the same as memory (except that the $\overline{IORC}$ and $\overline{IOWC}$ are not used because there is no IN or OUT instruction). The decision to use memory-mapped I/O is often determined by the size of the memory system and the placement of the I/O devices in the system.

The main difference between memory decoding and isolated I/O decoding is the number of address pins connected to the decoder. We decode A31–A0, A23–A0, or A19–A0 for memory; and A15–A0 for isolated I/O. Sometimes, if the I/O devices use only fixed I/O addressing, we decode only A7–A0. In the personal computer system, we always decode all 16 I/O port address bits. Another difference is that we use the $\overline{\text{IORC}}$ and $\overline{\text{IOWC}}$ to activate I/O devices for a read or write operation. On earlier versions of the microprocessor, IO/$\overline{\text{M}}$ = 1 and $\overline{\text{RD}}$ or $\overline{\text{WR}}$ are used to activate I/O devices. On the newest versions of the microprocessor, the M/$\overline{\text{IO}}$ = 0 and W/$\overline{\text{R}}$ are used to activate I/O devices.

Decoding 8-Bit I/O Addresses

As mentioned, the fixed I/O instruction uses an 8-bit I/O port address that appears on A15–A0 as 0000H–00FFH. If a system will never contain more than 256 I/O devices, we often decode only address connections A7–A0 for an 8-bit I/O port address. Thus, we ignore address connection A15–A8. Embedded systems often use 8-bit port addresses. Please note that the DX register can also address I/O ports 00H–FFH. If the address is decoded as an 8-bit address, we can never include I/O devices that use a 16-bit I/O address.

Figure 11–10 illustrates a 74ALS138 decoder that decodes 8-bit I/O ports F0H through F7H. (We assume that this system will only use I/O ports 00H–FFH for this decoder example.) This decoder is identical to a memory address decoder except we only connect address bits A7–A0 to the inputs of the decoder. Figure 11–11 shows the PLD version, using a PAL for this decoder. The PAL is a better decoder circuit because the number of integrated circuits has been reduced to one device. The program for the PAL appears in Example 11–2.

EXAMPLE 11–2

```
AUTHOR    Barry B. Brey
COMPANY   BreyCo
DATE      7/1/99
CHIP      DECODER8 PAL16L8

;pins  1  2  3  4  5  6  7  8  9  10
       A0 A1 A2 A3 A4 A5 A6 A7 NC GND

;pins 11 12 13 14 15 16 17 18 19  20
      NC F7 F6 F5 F4 F3 F2 F1 F0 VCC

EQUATIONS

/F0 = A7 * A6 * A5 * A4 * A3 * /A2 * /A1 * /A0
/F1 = A7 * A6 * A5 * A4 * A3 * /A2 * /A1 * A0
/F2 = A7 * A6 * A5 * A4 * A3 * /A2 * A1 * /A0
/F3 = A7 * A6 * A5 * A4 * A3 * /A2 * A1 * A0
/F4 = A7 * A6 * A5 * A4 * A3 * A2 * /A1 * /A0
/F5 = A7 * A6 * A5 * A4 * A3 * A2 * /A1 * A0
/F6 = A7 * A6 * A5 * A4 * A3 * A2 * A1 * /A0
/F7 = A7 * A6 * A5 * A4 * A3 * A2 * A1 * A0
```

Decoding 16-Bit I/O Addresses

We also decode 16-bit I/O addresses, especially in a personal computer system. The main difference between decoding an 8-bit I/O address and a 16-bit I/O address is that eight additional address lines (A15–A8) must be decoded. Figure 11–12 illustrates a circuit that contains a PAL16L8 and an 8-input NAND gate used to decode I/O ports EFF8H–EFFFH.

FIGURE 11–10 A port decoder that decodes 8-bit I/O ports. This decoder generates active low outputs for ports F0H–F7H.

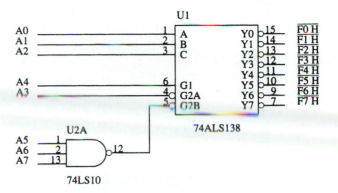

FIGURE 11–11 A PAL16L8 decoder that generates I/O port signals for port F0H–F7H.

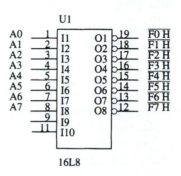

FIGURE 11–12 A PAL16L8 decoder that decodes 16-bit address EFF8H–EFFFH.

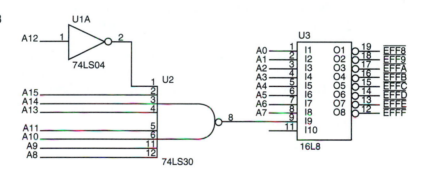

The NAND gate decodes the first eight bits of the I/O port address (A15–A8). The output of the NAND gate generates a signal to enable the PAL16L8 for any I/O address between EF00H and EFFFH. The PAL16L8 further decodes the I/O address to produce eight active low output strobes $\overline{\text{EFF8H}}$–$\overline{\text{EFFFH}}$ The program for the PAL16L8 decoder appears in Example 11–3.

EXAMPLE 11–3

```
AUTHOR    Barry B. Brey
COMPANY   BreyCo
DATE      7/2/99
CHIP      DECODER9 PAL16L8

;pins  1  2  3  4  5  6  7  8  9    10
```

```
            A0 A1 A2 A3 A4 A5 A6 A7 NAND GND

;pins 11    12    13    14    15    16    17    18    19  20
      NC  EFFFH EFFEH EFFDH EFFCH EFFBH EFFAH EFF9H EFF8H VCC


EQUATIONS

/EFF8H = A7 * A6 * A5 * A4 * A3 * /A2 * /A1 * /A0 * /NAND
/EFF9H = A7 * A6 * A5 * A4 * A3 * /A2 * /A1 *  A0 * /NAND
/EFFAH = A7 * A6 * A5 * A4 * A3 * /A2 *  A1 * /A0 * /NAND
/EFFBH = A7 * A6 * A5 * A4 * A3 * /A2 *  A1 *  A0 * /NAND
/EFFCH = A7 * A6 * A5 * A4 * A3 *  A2 * /A1 * /A0 * /NAND
/EFFDH = A7 * A6 * A5 * A4 * A3 *  A2 * /A1 *  A0 * /NAND
/EFFEH = A7 * A6 * A5 * A4 * A3 *  A2 *  A1 * /A0 * /NAND
/EFFF7 = A7 * A6 * A5 * A4 * A3 *  A2 *  A1 *  A0 * /NAND
```

8- And 16-Bit I/O Ports

Now that we understand that decoding the I/O port address is probably simpler than decoding a memory address (because of the number of bits), we explain how data are transferred between the microprocessor and 8- or 16-bit I/O devices. Data transferred to an 8-bit I/O device exist in one of the I/O banks in a 16-bit microprocessor such as the 80386SX. The I/O system contains two 8-bit memory banks, just as memory does. This is illustrated in Figure 11–13, which shows the separate I/O banks for a 16-bit system such as the 80386SX.

Because two I/O banks exist, any 8-bit I/O write requires a separate write strobe to function correctly. I/O reads do not require separate read strobes. As with memory, the microprocessor reads only the byte it expects and ignores the other byte. The only time that a read can cause problems is when the I/O device responds incorrectly to a read operation. In the case of an I/O device that responds to a read from the wrong bank, we may need to include separate read signals. This is discussed if the case arises later in this chapter.

Figure 11–14 illustrates a system that contains two different 8-bit output devices, located at 8-bit I/O address 40H and 41H. Because these are 8-bit devices and because they appear in different I/O banks, we generate separate I/O write signals. Note that all I/O ports use 8-bit addresses. Thus, ports 40H and 41H can each be addressed as separate 8-bit ports, or together as one 16-bit port. The program for the PAL16L8 decoder used in Figure 11–14 is illustrated in Example 11–4.

EXAMPLE 11–4

```
AUTHOR   Barry B. Brey
COMPANY  BreyCo
DATE     7/3/99
CHIP     DECODERA PAL16L8

;pins  1     2  3  4  5  6  7  8  9  10
      BHE IOWC A0 A1 A2 A3 A4 A5 A6 GND

;pins 11 12 13 14 15 16 17 18 19 20
      A7 NC NC NC NC NC NC 40 41 VCC


EQUATIONS

/40 = /BLE * /IOWC * /A7 * A6 * /A5 * /A4 * /A3 * /A2 * /A1
/41 = /BHE * /IOWC * /A7 * A6 * /A5 * /A4 * /A3 * /A2 * /A1
```

When selecting 16-bit wide I/O devices, the $\overline{\text{BLE}}$ (A0) and $\overline{\text{BHE}}$ pins have no function because both I/O banks are selected together. Although 16-bit I/O devices are relatively rare, a few do exist for analog-to-digital and digit-to-analog converters, as well as for some video and disk memory interfaces.

FIGURE 11–13 The I/O banks found in the 8086, 80186, 80286, and 80386SX.

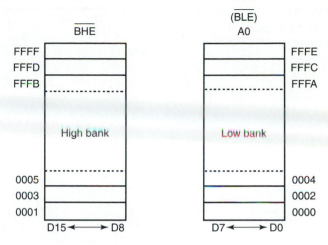

FIGURE 11–14 An I/O port decoder that selects ports 40H and 41H for output data.

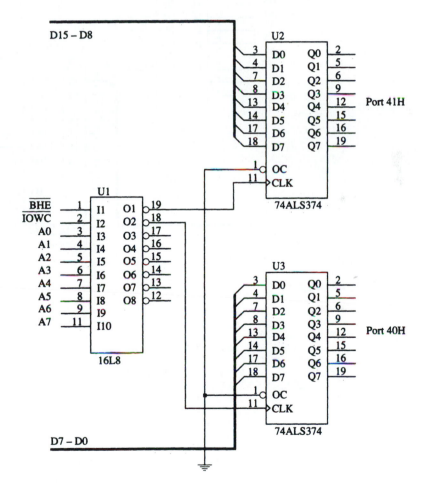

Figure 11–15 illustrates a 16-bit input device connected to function at 8-bit I/O addresses 64H and 65H. Notice that the PAL16L8 decoder does not have a connection for address bit $\overline{BLE}$ (A0) and $\overline{BHE}$ because these signals do not apply to 16-bit wide I/O devices. The program for the PAL16L8, illustrated in Example 11–5, shows how the enable signals are generated for the three-state buffers (74ALS244) used as input devices.

FIGURE 11–15 A 16-bit
I/O port decoded at I/O
addresses 64H and 65H.

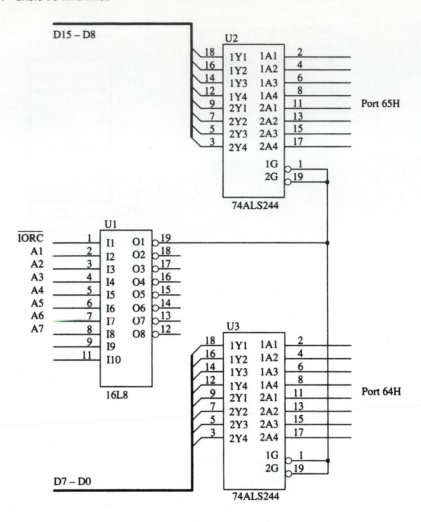

EXAMPLE 11–5

```
AUTHOR    Barry B. Brey
COMPANY   BreyCo
DATE      7/5/99
CHIP      DECODERB PAL16L8

;pins   1  2  3  4  5  6  7  8  9  10
        IORC A1 A2 A3 A4 A5 A6 A7 NC GND
;pins  11 12 13 14 15 16 17 18 19  20
        NC NC NC NC NC NC NC NC O6X VCC

EQUATIONS

/O64 = /IORC * /A7 * A6 * A5 * /A4 * /A3 * A2 * /A1
```

32-Bit Wide I/O Ports

Although 32-bit wide I/O ports are not common, they may eventually become commonplace because of newer buses found in computer systems. The once-promising EISA system bus supports 32-bit I/O as well as the VESA local and current PCI bus, but these are found only in some IO systems.

FIGURE 11–16 A 32-bit input port decoded at bytes 70H–73H for the 80386DX through the Pentium 4.

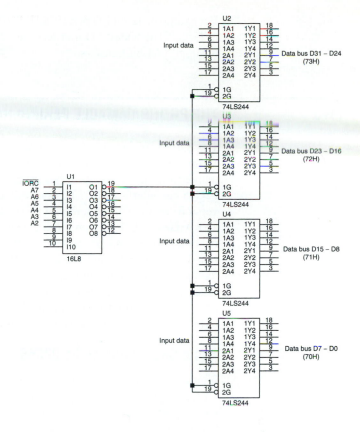

The circuit of Figure 11–16 illustrates a 32-bit input port for the 80386DX or 80486 microprocessor. As with prior interfaces, this circuit uses a single PAL to decode the I/O ports and four 74LS244 buffers to connect the I/O data to the data bus. The I/O ports decoded by this interface are the 8-bit ports 70H–73H, as illustrated by the PAL program in Example 11–6. Again, we only decode an 8-bit I/O port address.

EXAMPLE 11–6

```
AUTHOR    Barry B. Brey
COMPANY   BreyCo
DATE      7/6/99
CHIP      DECODERC PAL16L8

;pins    1   2   3   4   5   6   7   8   9   10
         IORC A7  A6  A5  A4  A3  A2  NC  NC  GND

;pins   11  12  13  14  15  16  17  18  19   20
        NC  NC  NC  NC  NC  NC  NC  NC  SEL  VCC

EQUATIONS

/SEL = /IORC * /A7 * A6 * A5 * A4 * /A3 * /A2
```

With the Pentium–Pentium 4 microprocessors and their 64-bit data buses, I/O ports appear in various banks, as determined by the I/O port address. For example, 8-bit I/O port 0034H appears in Pentium I/O bank 5, while the 16-bit I/O port 0034H–0035H appears in Pentium banks 5 and 6. A 32-bit I/O access in the Pentium system can appear in any four consecutive I/O banks.

For example, 32-bit I/O port 0100H–0103H appears in banks 0–3. How is a 64-bit I/O device interfaced? The widest I/O transfers are 32 bits, and currently there are no 64-bit I/O instructions to support 64-bit transfers.

11–3   THE PROGRAMMABLE PERIPHERAL INTERFACE

The 82C55 **programmable peripheral interface** (PPI) is a very popular, low-cost interfacing component found in many applications. The PPI, which has 24 pins for I/O that are programmable in groups of 12 pins, has groups that operate in three distinct modes of operation. The 82C55 can interface any TTL-compatible I/O device to the microprocessor. The 82C55 (CMOS version) requires the insertion wait states if operated with a microprocessor using higher than an 8 MHz clock. It also provides at least 2.5 mA of sink (logic 0) current at each output, with a maximum of 4.0 mA. Because I/O devices are inherently slow, wait states used during I/O transfers do not impact significantly upon the speed of the system. The 82C55 still finds application (compatible for programming, although it may not appear in the system as a discrete 82C55), even in the latest Pentium 4-based computer system. The 82C55 is used for interface to the keyboard and the parallel printer port in many personal computers, but it is found as a function within a interfacing chipset. The chipset also controls the timer and reads data from the keyboard interface.

Basic Description of the 82C55

Figure 11–17 illustrates the pin-out diagram of the 82C55. Its three I/O ports (labeled A, B, and C) are programmed as groups. Group A connections consist of port A (PA7–PA0) and the upper half of port C (PC7–PC4), and group B consists of port B (PB7–PB0) and the lower half of port C (PC3–PC0). The 82C55 is selected by its $\overline{CS}$ pin for programming and for reading or writing to a port. Register selection is accomplished through the A1 and A0 input pins that select an internal register for programming or operation. Table 11–2 shows the I/O port assignments used

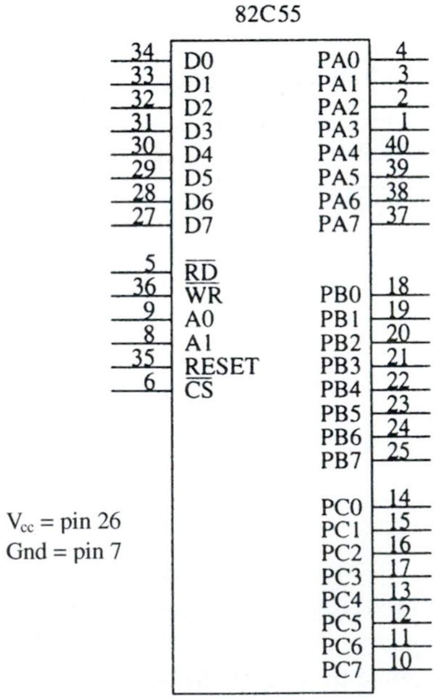

FIGURE 11–17   The pin-out of the 82C55 peripheral interface adapter (PPI).

TABLE 11–2 I/O port assignments for the 82C55.

A_1	A_0	Function
0	0	Port A
0	1	Port B
1	0	Port C
1	1	Command Register

for programming and access to the I/O ports. In the personal computer, an 82C55 or its equivalent is decoded at I/O ports 60H–63H.

The 82C55 is a fairly simple device to interface to the microprocessor and program. For the 82C55 to be read or written, the $\overline{CS}$ input must be a logic 0 and the correct I/O address must be applied to the A1 and A0 pins. The remaining port address pins are don't cares as far as the 82C55 is concerned, and are externally decoded to select the 82C55.

Figure 11–18 shows an 82C55 connected to the 80386SX so that it functions at 8-bit I/O port addresses C0H (port A), C2H (port B), C4H (port C), and C6H (command register). This interface uses the low bank of the 80386SX I/O map. Notice from this interface that all the 82C55 pins are direct connections to the 80386SX, except for the $\overline{CS}$ pin. The $\overline{CS}$ pin is decoded and selected by a 74ALS138 decoder.

The RESET input to the 82C55 initializes the device whenever the microprocessor is reset. A RESET input to the 82C55 causes all ports to be set up as simple input ports using mode 0 operation. Because the port pins are internally programmed as input pins after a RESET, damage is

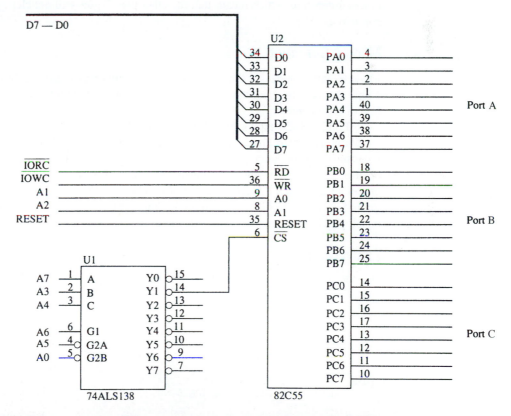

FIGURE 11–18 The 82C55 interfaced to the low bank of the 80386SX microprocessor.

prevented when the power is first applied to the system. After a RESET, no other commands are needed to program the 82C55, as long as it is used as an input device for all three ports. Note that an 82C55 is interfaced to the personal computer at port addresses 60H–63H for keyboard control; and also for controlling the speaker, timer, and other internal devices such as memory expansion. This is true for any AT or earlier style personal computer system.

Programming the 82C55

The 82C55 is programmed through the two internal command registers that are illustrated in Figure 11–19. Notice that bit position 7 selects either command byte A or command byte B. Command byte A programs the function of group A and B, while command byte B sets (1) or resets (0) bits of port C only if the 82C55 is programmed in mode 1 or 2.

Group B pins (port B and the lower part of port C) are programmed as either input or output pins. Group B operates in either mode 0 or mode 1. Mode 0 is the basic input/output mode that allows the pins of group B to be programmed as simple input and latched output connections. Mode 1 operation is the strobed operation for group B connections, where data are transferred through port B and handshaking signals are provided by port C.

Group A pins (port A and the upper part of port C) are programmed as either input or output pins. The difference is that group A can operate in modes 0, 1, and 2. Mode 2 operation is a bi-directional mode of operation for port A.

If a 0 is placed in bit position 7 of the command byte, command byte B is selected. This command allows any bit of port C to be set (1) or reset (0), if the 82C55 is operated in either mode 1 or 2. Otherwise, this command byte is not used for programming. We often use the bit set/reset function in a control system to set or clear a control bit at port C. The bit set/reset function is glitch-free, which means that the other port C pins will not change during the bit set/reset command.

Mode 0 Operation

Mode 0 operation causes the 82C55 to function either as a buffered input device or as a latched output device. These are the same as the basic input and output circuits discussed in the first section of this chapter.

Figure 11–20 shows the 82C55 connected to a set of eight 7-segment LED displays. In this circuit, both ports A and B are programmed as (mode 0) simple latched output ports. Port A provides the segment data inputs to the display and port B provides a means of selecting one display position at a time for multiplexing the displays. The 82C55 is interfaced to an 8088 microprocessor through a PAL16L8 so that it functions at I/O port numbers 0700H–0703H. The program for the PAL16L8 is listed in Example 11–7. The PAL decodes the I/O address and develops the write strobe for the $\overline{WR}$ pin of the 82C55.

EXAMPLE 11–7

```
AUTHOR    Barry B. Brey
COMPANY   BreyCo
DATE      7/6/99
CHIP      DECODERD PAL16L8

;pins   1   2   3   4   5   6   7   8   9    10
        A2  A3  A4  A5  A6  A7  A8  A9  A10  GND

;pins  11   12   13   14   15   16   17  18  19   20
       A11  CS   IOM  A12  A13  A14  A15 NC  NC   VCC

EQUATIONS

/CS = /A15 * /A14 * /A13 * /A12 * /A11 * A10 * A9 * A8 * /A6 * /A5 * /A4 * /A3 *
/A2 * /IOM
```

FIGURE 11–19 The command byte of the command register in the 82C55. (a) Programs ports A, B, and C (b) Sets or resets the bit indicated in the select a bit field.

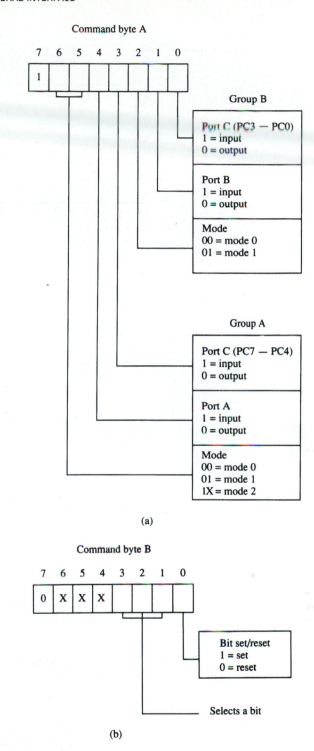

Command byte A

Group B

Port C (PC3 — PC0)
1 = input
0 = output

Port B
1 = input
0 = output

Mode
00 = mode 0
01 = mode 1

Group A

Port C (PC7 — PC4)
1 = input
0 = output

Port A
1 = input
0 = output

Mode
00 = mode 0
01 = mode 1
1X = mode 2

(a)

Command byte B

Bit set/reset
1 = set
0 = reset

Selects a bit

(b)

The resistor values are chosen in Figure 11–20 so that the segment current is 80 mA. This current is required to produce an average current of 10 mA per segment as the displays are multiplexed. A 6-digit display uses a segment current of 60 mA, for an average of 10 mA per segment. In this type of display system, only one of the eight display positions is on at any given instant. The peak anode current in an 8-digit display is 560 mA (seven segments × 80 mA), but

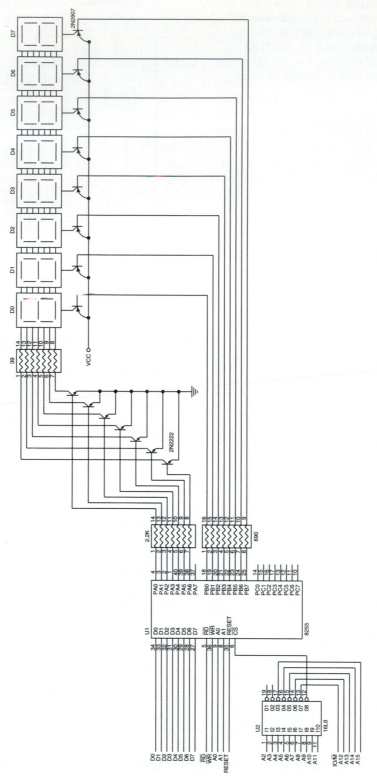

FIGURE 11–20 An 8-digit LED display interfaced to the 8088 microprocessor through an 82C55 PIA.

the average anode current is 80 mA. In a 6-digit display, the peak current would be 420 mA (seven segments × 60 mA). Whenever displays are multiplexed, we increase the segment current from 10 mA (for a display that uses 10 mA per segment as the nominal current) to a value equal to the number of display positions times 10 mA. This means that a 4-digit display uses 40 mA per segment, a 5-digit display uses 50 mA, and so on.

In this display, the segment load resistor passes 80 mA of current and has a voltage of approximately 3.0 V across it. The LED (1.65 V nominally) and a few tenths are dropped across the anode switch and the segment switch, hence a voltage of 3.0 V appears across the segment load resistor. The value of the resistor is $^{3.0\ V}/_{80\ mA} = 37.5\ \Omega$. The closest standard resistor value of 39 Ω is used in Figure 11–20 for the segment load.

The resistor in series with the base of the segment switch assumes that the minimum gain of the transistor is 100. The base current is therefore $^{80\ mA}/_{100} = 0.8\ mA$. The voltage across the base resistor is approximately 3.0 V (the minimum logic 1 voltage level of the 82C55), minus the drop across the emitter-base junction (0.7 V), or 2.3 V. The value of the base resistor is then 2.3 V/0.8 mA = 2.875 KΩ. The closest standard resistor value is 2.7 KΩ, but a 2.2 KΩ is chosen for this circuit.

The anode switch has a single resistor on its base. The current through the resistor is 560 mA/100 = 5.6 mA because the minimum gain of the transistor is 100. This exceeds the maximum current of 4.0 mA from the 82C55, but this is small enough so that it will work without problems. The maximum current assumes that you are using the port pin as a TTL input to another circuit. If the amount of current were over 8.0–10.0 mA, then appropriate circuitry (in the form of either a Darlington-pair or another transistor switch) would be required. Here, the voltage across the base resistor is 5.0 V, minus the drop across the emitter-base junction (0.7 V), minus the voltage at the port pin (0.4 V), for a logic 0 level. The value of the resistor is 3.9 V/5.6 mA = 696 W. The closest standard resistor value is 690 W, which is chosen for this example.

Before software to operate the display is examined, we must first program the 82C55. This is accomplished with the short sequence of instructions listed in Example 11–8. Here, port A and B are both programmed as outputs.

EXAMPLE 11–8

```
                        ;programming the 82C55 PIA
                        ;
0000  B0 80             MOV    AL,10000000B
0002  BA 0703           MOV    DX,703H           ;address command
0005  EE                OUT    DX,AL             ;program 82C55
```

The procedure to multiplex the displays is listed in Example 11–9. For the display system to function correctly, we must call this procedure often. Notice that the procedure calls another procedure (DELAY) that causes a 1 ms time delay. The time delay is not illustrated in this example, but it is used to allow time for each display position to turn on. It is recommended by the manufacturers of LED displays that the display flash be between 100 Hz and 1500 Hz. Using a 1 ms time delay, we light each digit for 1 ms for a total display flash rate of 1000 Hz/8 display, or a flash rate of 125 Hz.

EXAMPLE 11–9

```
                        ;Procedure that multiplexes the 8-digit LED display.
                        ;This procedure must be called from a program at
                        ;whenever possible to display 7-segment
                        ;coded data from memory.
                        ;
0006                    DISP  PROC   NEAR USES AX BX DX SI

000A  9C                      PUSHF                    ;save flag register
```

```
                                ;setup registers for display

0000B BB 0008               MOV    BX,8                 ;load count
000E B4 7F                  MOV    AH,7FH               ;load selection pattern
0010 BE 00FF R              MOV    SI,OFFSET MEM-1       ;address data
0013 BA 0701                MOV    DX,701H              ;address Port B


                       ;display 8 digits

0016                    DISP1:
0016 8A C4                  MOV    AL,AH                ;select a digit
0018 EE                     OUT    DX,AL
0019 4A                     DEC    DX                   ;address Port A
001A 8A 00                  MOV    AL,[BX+SI]           ;get 7-segment data
001C EE                     OUT    DX,AL
001D E8 029A R              CALL   DELAY                ;wait one millisecond
0020 D0 CC                  ROR    AH,1                 ;address next digit
0022 42                     INC    DX                   ;address Port B
0023 4B                     DEC    BX                   ;adjust count
0024 75 F0                  JNZ    DISP1                ;repeat 8 times


0026 5E                     POPF                        ;restore registers
                            RET


002C                    DISP   ENDP
```

The display procedure (DISP) addresses an area of memory where the data, in 7-segment code, is stored for the eight display digits called MEM. The AH register is loaded with a code (7FH) that initially addresses the most-significant display position. Once this position is selected, the contents of memory location MEM +7 is addressed and sent to the most-significant digit. The selection code is then adjusted to select the next display digit, as is the address. This process repeats eight times to display the contents of location MEM through MEM +7 on the eight display digits.

An LCD Display Interfaced to the 82C55.   LCDs (liquid crystal displays) are quickly replacing LED displays in many applications. The only disadvantage of the LCD display is that it is difficult to see in low-light situations in which the LED is still in limited use.

Figure 11–21 illustrates the connection of the Optrex DMC–20481 LCD display to an 82C55. The DMC–20481 is a 4-line by 20-characters-per-line display that accepts ASCII code as

FIGURE 11–21 The DMC-20481 LCD display interfaced to the 82C55.

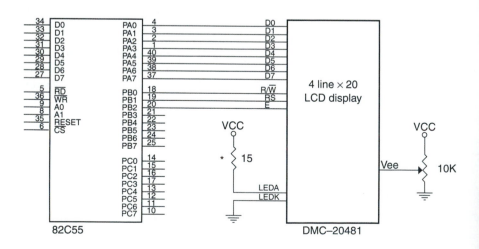

*Note: Current max is 480 mA, nominal 260 mA

input data. It also accepts commands that initialize it and control its application. As you can see in Figure 11–21, the LCD display has few connections. The data connections, which are attached to the 82C55 Port A, are used to input display data and to read information from the display.

There are four control pins on the display. The V$_{EE}$ connection is used to adjust the contrast of the LED display and is normally connected to a 10 KΩ potentiometer, as illustrated. The RS (register select) input selects data (RS = 1) or instructions (RS = 0). The E (enable) input must be a logic 1 for the DMC–20481 to read or write information. Finally, the R/$\overline{W}$ pin selects a read or a write operation. Normally, the RS pin is placed at a 1 or 0, the R/$\overline{W}$ pin is set or cleared, data are placed on the data input pins, and then the E pin is pulsed to access the DMC–20481. This display also has two inputs for back-lighting LED diodes, which are not shown in the illustration.

In order to program the DMC–20481 we must first initialize it. This applies to any display that uses the HD44780 (Hitachi) display driver integrated circuit. The entire line of small display panels from Optrex is programmed in the same manner. Initialization is accomplished via the following steps:

1. Wait at least 15 ms after V$_{CC}$ rises to 5.0 V.
2. Output the function set command (30H), and wait at least 4.1 ms.
3. Output the function set command (30H) a second time, and wait at least 100 µs.
4. Output the function set command (30H) a third time, and wait at least 40 µs.
5. Output the function set command (38H) a fourth time, and wait at least 40 µs.
6. Output a 08H to disable the display, and wait at least 40 µs.
7. Output a 01H to home the cursor and clear the display, and wait at least 1.64 ms.
8. Output the enable display cursor off (0CH), and wait at least 40 µs.
9. Output a 06H to select auto-increment, shift the cursor, and wait at least 40 µs.

The software to accomplish the initialization of the LCD display is listed in Example 11–10. It is long, but the display controller requires the long initialization dialog. Note that the software for the three time delays is not included in the listing. If you are interfacing to a PC, you can use the clock tick discussed in Chapter 7 for the time delay. One clock tick can be used for all timing in this software, even though the LCD display will function faster than your eye at 1/18 seconds. If you are developing the interface for another application, then you must write three separate time delays, which must provide the delay times indicated in the initialization dialog.

EXAMPLE 11–10

```
                    ;procedure to initialize the LCD display
                    ;
0010                INIT  PROC  NEAR

0010 BA 0303              MOV   DX,CMD8255  ;address 8255 command register
0013 B0 80                MOV   AL,80H      ;all ports are output ports
0015 EE                   OUT   DX,AL
0016 B0 00                MOV   AL,0        ;clear Port B
0018 BA 0301              MOV   DX,PORTB
001B EE                   OUT   DX,AL
001C E8 004A              CALL  DELAY15     ;wait 15 ms
001F B0 30                MOV   AL,30H      ;first function set command
0021 E8 002A              CALL  OUTCMD      ;send it
0024 E8 0056              CALL  DELAY41     ;wait 4.1 ms
0027 B0 30                MOV   AL,30H
0029 E8 0022              CALL  OUTCMD      ;second function set command
002C B0 30                MOV   AL,30H      ;third function set command
002E E8 001D              CALL  OUTCMD
0031 B0 38                MOV   AL,38H      ;fourth function set command
0033 E8 0018              CALL  OUTCMD
```

```
0036 E8 0058          CALL    DELAY100
0039 B0 08            MOV     AL,08H          ;display off
003B E8 0010          CALL    OUTCMD
003E B0 01            MOV     AL,01H          ;clear display
0040 E8 000B          CALL    OUTCMD
0043 B0 0C            MOV     AL,0CH          ;display on, cursor off
0045 E8 0006          CALL    OUTCMD
0048 B0 06            MOV     AL,06H          ;auto-increment, shift cursor
004A E8 0001          CALL    OUTCMD
004D C3               RET
004E            INIT   ENDP
                 ;
                 ;procedure to output a command
                 ;
004E            OUTCMD   PROC   NEAR
004E 50               PUSH    AX              ;save command
004F BA 0303          MOV     DX,CMD8255      ;select 8255 command resgiter
0052 B0 80            MOV     AL,80H          ;all ports are outputs
0054 EE               OUT     DX,AL
0055 BA 0300          MOV     DX,PORTA
0058 58               POP     AX
0059 EE               OUT     DX,AL           ;command to port A
005A 42               INC     DX              ;address Port B
005B B0 04            MOV     AL,4            ;put 1 on E, 0 on R/W# and 0 on S
005D EE               OUT     DX,AL
005E 90               NOP                     ;extra time so E = 1 longer
005F 90               NOP
0060 B0 00            MOV     AL,0
0062 EE               OUT     DX,AL           ;clear E
0063 E8 0029          CALL    DELAY100        ;wait 100 us
0066 C3               RET
0067            OUTCMD   ENDP
```

The NOP instructions are added in the OUTCMD procedure to ensure that the E bit remains a logic 1 long enough to activate the LCD display. This process should work in most systems at most clock frequencies, but additional NOP instructions may be needed to lengthen this time in some cases.

Before programming data to the display, the commands used in the initialization dialog must be explained. See Table 11–3 for a complete listing of the commands or instructions for the LCD display. Compare the commands sent to the LCD display in the initialization program to Table 11–3.

Once the LCD display is initialized, a few procedures are needed to display information and control the display. After initialization, time delays are no longer needed when sending data or many commands to the display. The clear display command still needs a time delay because the busy flag is not used with that command. Instead of a time delay, the busy flag is tested to see whether the display has completed an operation. A procedure to test the busy flag appears in Example 11–11. The BUSY procedure tests the LCD display and only returns when the display has completed a prior instruction.

EXAMPLE 11–11

```
                 ;procedure to test busy and return if not busy
;0010            BUSY   PROC   NEAR
                      .REPEAT
0010 BA 0303          MOV     DX,CMD8255      ;select 8255 command register
☐
0013 B0 90            MOV     AL,90H          ;port A input
0015 EE               OUT     DX,AL
0016 BA 0301          MOV     DX,PORTB        ;select port B
0019 B0 01            MOV     AL,1            ;R/W# = 1
001B EE               OUT     DX,AL
```

TABLE 11–3 Instructions for most Optrex LCD displays.

Instruction	Code	Description	Execution Time
Clear display	0000 0001	Clears display and homes the cursor	1.64 ms
Cursor home	0000 0010	Homes the cursor	1.64 ms
Entry mode set	0000 0AS	Sets cursor movement direction (A=1 increment) and shift (S=1 shift)	40 μs
Display on/off	0000 1DCB	Sets display on/off (D=1 on) (C=1 cursor on) (B=1 cursor blink)	40 μs
Cursor/display shift	0001 SR00	Sets cursor movement and display shift (S=1 shift display) (S=0 move cursor) (R=1 right)	40 μs
Function set	001L NF00	Programs chip (L=1 8-bit, L=0 4-bits) (N=1 2 lines) (F=1 5x10, F=0 5x7)	40 μs
Set CGRAM address	01XX XXXX	Sets character generator RAM address	40 μs
Set DRAM address	10XX XXXX	Sets display RAM address	40 μs
Read busy flag	B000 0000	Reads busy flag (B=1 busy)	0
Write data	Data	Writes data to display or character generator RAM	40 μs
Read data	Data	Reads data from display or character generator RAM	40 μs

```
001C B0 05        MOV   AL,5         ;R/W# = 1, E = 1, RS = 0
001E EE           OUT   DX,AL
001F 90           NOP                ;delay to allow access
0020 90           NOP
0021 BA 0300      MOV   DX,PORTA     ;select port A
0024 EC           IN    AL,DX        ;get status of busy flag
0025 50           PUSH  AX
0026 BA 0301      MOV   DX,PORTB
0029 B0 00        MOV   AL,0
002B EE           OUT   DX,AL
002C BA 0303      MOV   DX,CMD8255
002F B0 80        MOV   AL,80H
0031 EE           OUT   DX,AL
0032 58           POP   AX
0033 D0 E0        SHL   AL,1
                  .UNTIL !CARRY?     ;until not busy
0037 C3           RET
0038         BUSY  ENDP
```

Once the BUSY procedure is available, data can be sent to the display by writing another procedure called WRITE. The WRITE procedure uses BUSY to test before trying to write new data to the display. Example 11–12 shows the WRITE procedure, which transfers the ASCII character from the BL register to the current cursor position of the display. Note that the initialization dialog has sent the cursor for auto-increment, so if WRITE is called more than once, the characters written to the display will appear one next to the other, as they would on a video display.

EXAMPLE 11–12

```
                        ;procedure that writes the ASCII contents of the BL
                        ;register to the display
                        ;
003A                    WRITE  PROC   NEAR
003A BA 0303                   MOV    DX,CMD8255
003D B0 80                     MOV    AL,80H
003F EE                        OUT    DX,AL
0040 BA 0300                   MOV    DX,PORTA    ;data to port A
0043 8A C3                     MOV    AL,BL
0045 EE                        OUT    DX,AL
0046 BA 0301                   MOV    DX,PORTB
0049 B0 02                     MOV    AL,2        ;RS = 1, R/W# = 0, E = 0
004B EE                        OUT    DX,AL
004C 90                        NOP
004D 90                        NOP
004E B0 06                     MOV    AL,6        ;RS = 1, R/W# = 0, E = 1
0050 EE                        OUT    DX,AL
0051 90                        NOP
0052 90                        NOP
0053 B0 00                     MOV    AL,0        ;RS = 0, R/W# = 0, E = 0
0055 EE                        OUT    DX,AL
0056 E8 FFB7                   CALL   BUSY        ;wait for LCD
0059 C3                        RET
005A                    WRITE  ENDP
```

The only other procedure that is needed for a basic display is the clear and home cursor procedure called CLS, shown in Example 11–13. With CLS and the procedures presented thus far, you can display any message on the display, clear it, display another message, and basically operate the display. As mentioned earlier, the clear command requires a time delay (at least 1.64 ms) instead of a call to BUSY for proper operation. In this procedure, we used the 4.1 ms time delay.

EXAMPLE 11–13

```
                        ;procedure to clear the display and home the cursor
                        ;
005A                    CLS    PROC   NEAR
005A BA 0303                   MOV    DX,CMD8255
005D B0 80                     MOV    AL,80H
005F EE                        OUT    DX,AL
0060 BA 0300                   MOV    DX,PORTA
0063 B0 01                     MOV    AL,1        ;clear instruction
0065 EE                        OUT    DX,AL
0066 BA 0301                   MOV    DX,PORTB
0069 B0 04                     MOV    AL,4        ;RS = 0, R/W# = 0, E = 1
006B EE                        OUT    DX,AL
006C 90                        NOP
006D 90                        NOP
006E B0 00                     MOV    AL,0        ;RS = 0, R/W# = 0, E = 0
0070 EE                        OUT    DX,AL
0071 E8 0044                   CALL   DELAY41
0074 C3                        RET
0075                    CLS    ENDP
```

Additional procedures that could be developed might select a display RAM position. The display RAM address starts at 0 and progresses across the display until the last character address on the first line is location 19, location 20 is the first display position of the second line, and so forth. Once you can move the display address, you can change individual characters on the display and even read data from the display. These procedures are for you to develop if they are needed.

A word about the display RAM inside of the LCD display. The LCD contains 128 bytes of memory, addresssed from 00H to 7FH. Not all of this memory is used. For example, the one-line × 20-character display uses only the first 20 bytes of memory (00H–13H.) The first line of any of these displays always starts at address 00H. The second line of any display powered by the

HD44780 always begins at address 40H. For example, a two-line × 40-character display uses addresses 00H–27H to store ASCII-coded data from the first line. The second line is stored at addresses 40H–67H for this display. In the four-line displays, the first line is at 00H, the second is at 40H, the third is at 14H, and the last line is at 54H. The largest display device that uses the HD44780 is a two-line × 40 character display. The four-line by 40-character display uses an M50530 or a pair of HD44780s. Because information on these devices can be readily found on the Internet, they are not covered in the text.

A Stepper Motor Interfaced to the 82C55.   Another device often interfaced to a computer system is the *stepper motor*. A stepper motor is a digital motor because it is moved in discrete steps as it traverses through 360°. A common stepper motor is geared to move perhaps 15° per step in an inexpensive stepper motor, to 1° per step in a more costly high-precision stepper motor. In all cases, these steps are gained through many magnetic poles and/or gearing. Notice that two coils are energized in Figure 11–22. If less power is required, one coil may be energized at a time, causing the motor to step at 45°, 135°, 225°, and 315°.

Figure 11–22 shows a four-coil stepper motor that uses an armature with a single pole. Notice that the stepper motor is shown four times with the armature (permanent magnetic) rotated to four discrete places. This is accomplished by energizing the coils, as shown. This is an illustration of full stepping. The stepper motor is driven by using NPN Darlington amplifier pairs to provide a large current to each coil.

A circuit that can drive this stepper motor is illustrated in Figure 11–23, with the four coils shown in place. This circuit uses the 82C55 to provide it with the drive signals that are used to rotate the armature of the motor in either the right-hand or left-hand direction.

A simple procedure that drives the motor (assuming that port A is programmed in mode 0 as an output device) is listed in Example 11–14. This subroutine is called, with CX holding the number of steps and direction of the rotation. If CX > 8000H, the motor spins in the right-hand

FIGURE 11–22   The stepper motor showing full-step operation. (a) 45° (b) 135° (c) 225° (d) 315°.

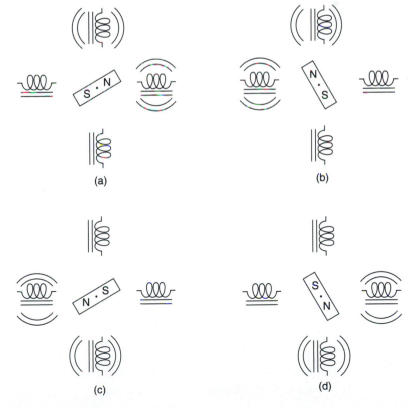

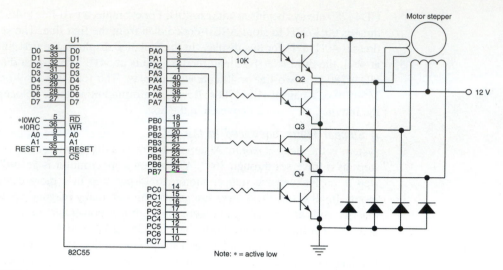

FIGURE 11–23   A stepper motor interfaced to the 82C55. This illustration does not show the decoder.

direction; if CX < 8000H, it spins in the left-hand direction. The leftmost bit of CX is removed and the remaining 15 bits contain the number of steps. Notice that the procedure uses a time delay (not illustrated) that causes a 1 ms time delay. This time delay is required to allow the stepper-motor armature time to move to its next position.

EXAMPLE 11–14

```
= 0040                      PORT  EQU   40H           ;assign Port A
                            ;
                            ;A procedure to control stepper motor.
                            ;
0000                        STEP  PROC  NEAR

0000  A0 0000 R                   MOV   AL,POS         ;get position
0003  81 F9 8000                  CMP   CX,8000H
0007  77 10                       JA    RH             ;if right-hand direction
0009  83 F9 00                    CMP   CX,0
000C  74 14                       JE    STEP_OUT       ;if no steps
000E                        STEP1:
000E  D0 C0                       ROL   AL,1           ;step left
0010  E6 40                       OUT   PORT,AL
0012  E8 0011                     CALL  DELAY          ;wait one millisecond
0015  E2 F7                       LOOP  STEP1          ;repeat until CX = 0
0017  EB 09                       JMP   STEP_OUT
0019                        RH:
0019  81 E1 7FFF                  AND   CX,7FFFH       ;clear bit 15
001D                        RH1:
001D  D0 C8                       ROR   AL,1           ;step right
001F  E6 40                       OUT   PORT,AL
0021  E8 0006                     CALL  DELAY          ;wait one millisecond
0024  E2 F7                       LOOP  RH1            ;repeat until CX = 0
0026                        STEP_OUT:
0026  A2 0000                     MOV   POS,AL         ;save position
0029  C3                          RET

0029                        STEP  ENDP
```

The current position is stored in memory location POS, which must be initialized with 33H, 66H, 0CCH, or 99H. This allows a simple ROR (step right) or ROL (step left) instruction to rotate the binary bit pattern for the next step.

Stepper motors can also be operated in the half-step mode, which allows eight steps per sequence. This is accomplished by using the full-step sequence described with a half step obtained by energizing one coil interspersed between the full steps. Half-stepping allows the armature to be positioned at 0°, 90°, 180°, and 270°. The half-step position codes are 11H, 22H, 44H, and 88H. A complete sequence of eight steps would follow as: 11H, 33H, 22H, 66H, 44H, 0CCH, 88H, and 99H. This sequence could be either output from a lookup table or generated with software.

Key Matrix Interface.   Keyboards come in a vast variety of sizes, from the standard 101-key QWERTY keyboards interfaced to the microprocessor to small specialized keyboards that may contain only four to 16 keys. This section of the text concentrates on the smaller keyboards that may be purchased preassembled or may be constructed from individual key switches.

Figure 11–24 illustrates a small key-matrix that contains 16 switches interfaced to ports A and B of an 82C55. In this example, the switches are formed into a 4 × 4 matrix, but any matrix could be used such as a 2 × 8. Notice how the keys are organized into four rows (ROW0–ROW3) and four columns (COL0–COL3). Each row is connected to 5.0 V through a 10 KΩ pull-up resistor to ensure that the row is pulled high when no push-button switch is closed.

The 82C55 is decoded (PAL program is not shown) at I/O ports 50H–53H for an 8088 microprocessor. Port A is programmed as an input port to read the rows and port B is programmed as an output port to select a column. For example, if 1110 is output to port B pins PB3–PB0, column 0 has a logic 1, so the four keys in column 0 are selected. Notice that with a logic 0 on PB0, the only switches that can place a logic 0 onto port A are switches 0–3. If switches 4–F are closed, the corresponding port A pins remain a logic 1. Likewise, if a 1101 is output to port B, switches 4–7 are selected, and so forth.

A flowchart of the software required to read a key from the keyboard matrix and de-bounce the key is illustrated in Figure 11–25. Keys must be de-bounced, which is normally accomplished with a short time delay of from 10–20 ms. The flowchart contains three main sections. The first waits for the release of a key. This seems awkward, but software executes very quickly in a microprocessor and there is a possibility that the program will return to the top of this program before the key is released, so we must wait for a release first. Next, the flowchart shows that we wait for a keystroke. Once the keystroke is detected, the position of the key is calculated in the final part of the flowchart.

The software uses a procedure called SCAN to scan the keys and another called DELAY to waste 10 ms of time for de-bouncing. The main keyboard procedure is called KEY and it appears with the others in Example 11–15. Note that the SCAN procedure is generic, so it can handle any keyboard configuration from a 2 × 2 matrix to an 8 × 8 matrix. Changing the two equates at the start of the program (ROW and COL) will change the configuration of the software for any size keyboard. Also note that the steps required to initialize the 82C55 so that port A is an input port and port B is an output port are not shown.

EXAMPLE 11–15

```
                         ;A keyboard procedure that scans the keyboard and
                         ;returns with the numeric code of the key in AL.
                         ;
= 0004                   ROWS  EQU   4          ;number of rows
= 0004                   COLS  EQU   4          ;number of columns
= 0050                   PORTA EQU   50H        ;port A address
= 0051                   PORTB EQU   51H        ;port B address

0000              KEY   PROC   NEAR USES CX

0001  E8 002F           CALL  SCAN             ;test all keys
0004  75 FA             JNZ   KEY              ;if key closed
0006  E8 0048           CALL  DELAY            ;wait for about 10 ms
0009  E8 0027           CALL  SCAN             ;test all keys
000C  75 F2             JNZ   KEY              ;if key closed
000E              KEY1:
```

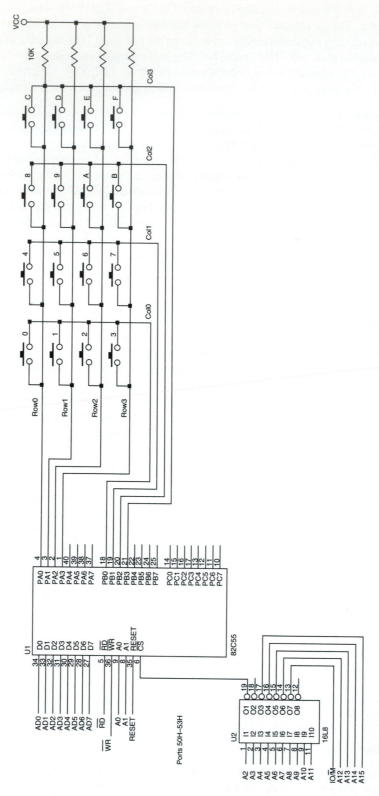

FIGURE 11–24 A 4 × 4 keyboard matrix connected to an 8088 microprocessor through the 82C55 PIA.

```
000E   E8 0022          CALL   SCAN        ;test all keys
0011   74 FB            JZ     KEY1        ;if no key closed
0013   E8 003B          CALL   DELAY       ;wait for about 10 ms
0016   E8 001A          CALL   SCAN        ;test all keys
0019   74 F3            JZ     KEY1        ;if no key closed
001B   50               PUSH   AX          ;save row codes
001C   B0 04            MOV    AL,COLS     ;calculate starting row key
001E   2A C1            SUB    AL,CL
0020   B5 04            MOV    CH,ROWS
0022   F6 E5            MUL    CH
0024   8A C8            MOV    CL,AL
0026   FE C9            DEC    CL
0028   58               POP    AX
0029            KEY2:
0029   D0 C8            ROR    AL,1        ;find row position
002B   FE C1            INC    CL
002D   72 FA            JC     KEY2
002F   8A C1            MOV    AL,CL       ;mode code to AL
                        RET

0033            KEY    ENDP
0033            SCAN   PROC   NEAR USES BX
0034   B1 04            MOV    CL,ROWS     ;form row mask
0036   B7 FF            MOV    BH,0FFH
0038   D2 E7            SHL    BH,CL
003A   B9 0004          MOV    CX,COLS     ;load column count
003D   B3 FE            MOV    BL,0FEH     ;get selection code
003F            SCAN1:
003F   8A C3            MOV    AL,BL       ;select column
0041   E6 51            OUT    PORTB,AL
0043   D0 C3            ROL    BL,1
0045   E4 50            IN     AL,PORTA    ;read rows
0047   0A C7            OR     AL,BH
0049   3C FF            CMP    AL,0FFH     ;test for a key
004B   75 02            JNZ    SCAN2
004D   E2 F0            LOOP   SCAN1
004F            SCAN2:
                        RET

0051            SCAN   ENDP

0051            DELAY PROC   NEAR USES CX

0052 B9 1388            MOV    CX,5000     ;10ms (8MHz clock)
0055            DELAY1:
0055 E2 FE              LOOP   DELAY1
                        RET

0059            DELAY ENDP
```

A note about the SCAN procedure. The time between where the keyboard column is selected and where the rows are read is very short. In a very high-speed system, a small time delay must be placed between these two points for the data at port A to settle to its final state. In most cases, this is not needed—the SCAN procedure should not scan the display at a rate higher than 30 KHz. If it does, the Federal Communications Commission (FCC) will not approve its application in any accepted system.

Mode 1 Strobed Input

Mode 1 operation causes port A and/or port B to function as latching input devices. This allows external data to be stored into the port until the microprocessor is ready to retrieve it. Port C is also used in mode 1 operation—not for data, but for control or handshaking signals that help operate either or both port A and port B as strobed input ports. Figure 11–26 shows how both ports are structured for mode 1 strobed input operation and the timing diagram.

FIGURE 11–25 The flowchart of a keyboard-scanning procedure.

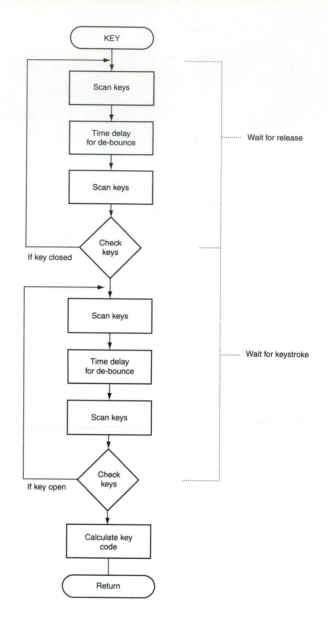

The strobed input port captures data from the port pins when the strobe ($\overline{STB}$) is activated. Note that the strobe captures the port data on the 0-to-1 transition. The $\overline{STB}$ signal causes data to be captured in the port and it activates the IBF **(input buffer full)** and INTR **(interrupt request)** signals. Once the microprocessor, through software (IBF) or hardware (INTR), notices that data are strobed into the port, it executes an IN instruction to read the port ($\overline{RD}$). The act of reading the port restores both IBF and INTR to their inactive states until the next datum is strobed into the port.

Signal Definitions for Mode 1 Strobed Input

$\overline{STB}$	The **strobe** input loads data into the port latch, which holds the information until it is input to the microprocessor via the IN instruction.
IBF	**Input buffer full** is an output indicating that the input latch contains information.
INTR	**Interrupt request** is an output that requests an interrupt. The INTR pin becomes a logic 1 when the $\overline{STB}$ input returns to a logic 1, and is cleared when the data are input from the port by the microprocessor.

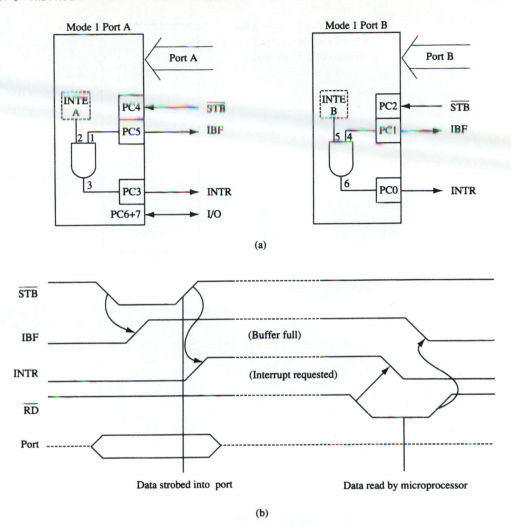

FIGURE 11–26 Strobed input operation (mode 1) of the 82C55. (a) Internal structure and (b) timing diagram.

INTE	The **interrupt enable** signal is neither an input nor an output; it is an internal bit programmed via the port PC4 (port A) or PC2 (port B) bit position.
PC7, PC6	The port C pins 7 and 6 are general-purpose I/O pins that are available for any purpose.

Strobed Input Example. An excellent example of a strobed input device is a keyboard. The keyboard encoder de-bounces the key-switches, and provides a strobe signal whenever a key is depressed and the data output contain the ASCII-coded key code. Figure 11–27 illustrates a keyboard connected to strobed input port A. Here $\overline{DAV}$ (**data available**) is activated for 1.0 μs each time that a key is typed on the keyboard. This causes data to be strobed into port A because $\overline{DAV}$ is connected to the $\overline{STB}$ input of port A. Each time a key is typed, therefore, it is stored into port A of the 82C55. The $\overline{STB}$ input also activates the IBF signal, indicating that data are in port A.

Example 11–16 shows a procedure that reads data from the keyboard each time a key is typed. This procedure reads the key from port A and returns with the ASCII code in AL. To detect a key, port C is read and the IBF bit (bit position PC5) is tested to see whether the buffer is full. If the buffer is empty (IBF = 0), then the procedure keeps testing this bit, waiting for a character to be typed on the keyboard.

FIGURE 11–27 Using the
82C55 for strobed input
operation of a keyboard.

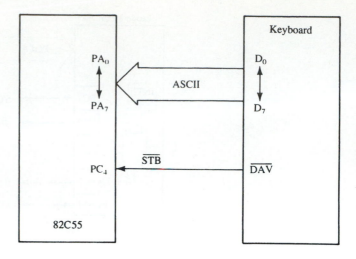

EXAMPLE 11–16

```
                              ;A procedure that reads the keyboard encoder
                              ;and returns the ASCII character in AL.
                              ;
= 0020                        BIT5  EQU   20H
= 0022                        PORTC EQU   22H
= 0020                        PORTA EQU   20H

0000                          READ  PROC  NEAR

0000   E4 22                        IN    AL,PORTC     ;read Port C
0002   A8 20                        TEST  AL,BIT5      ;test IBF
0004   74 FA                        JZ    READ         ;if IBF = 0
0006   E4 20                        IN    AL,PORTA     ;read data
0008   C3                           RET

0009                          READ  ENDP
```

Mode 1 Strobed Output

Figure 11–28 illustrates the internal configuration and timing diagram of the 82C55 when it is
operated as a strobed output device under mode 1. Strobed output operation is similar to mode 0
output operation, except that control signals are included to provide handshaking.

   Whenever data are written to a port programmed as a strobed output port, the $\overline{OBF}$ (**output
buffer full**) signal becomes a logic 0 to indicate that data are present in the port latch. This signal
indicates that data are available to an external I/O device that removes the data by strobing the
$\overline{ACK}$ (**acknowledge**) input to the port. The $\overline{ACK}$ signal returns the $\overline{OBF}$ signal to a logic 1, indicating that the buffer is not full.

Signal Definitions for Mode 1 Strobed Output

$\overline{OBF}$ **Output buffer full** is an output that goes low whenever data are output (OUT)
              to the port A or port B latch. This signal is set to a logic 1 whenever the $\overline{ACK}$
              pulse returns from the external device.

$\overline{ACK}$ The **acknowledge** signal causes the $\overline{OBF}$ pin to return to a logic 1 level. The
              $\overline{ACK}$ is a response from an external device, indicating that it has received the
              data from the 82C55 port.

INTR   **Interrupt request** is a signal that often interrupts the microprocessor when the
              external device receives the data via the $\overline{ACK}$ signal. This pin is qualified by
              the internal INTE (**interrupt enable**) bit.

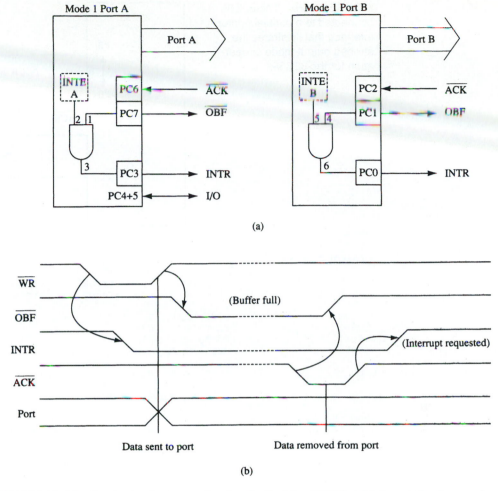

FIGURE 11-28   Strobed output operation (mode 1) of the 82C55. (a) Internal structure and (b) timing diagram.

INTE   **Interrupt enable** is neither an input nor an output; it is an internal bit programmed to enable or disable the INTR pin. The INTE A bit is programmed as PC6 and INTE B is programmed as PC2.

PC5, PC4   Port C pins 5 and 4 are general-purpose I/O pins. The bit set and reset command may be used to set or reset these two pins.

Strobed Output Example.   The printer interface discussed in Section 11–1 is used here to demonstrate how to achieve strobed output synchronization between the printer and the 82C55. Figure 11–29 illustrates port B connected to a parallel printer, with eight data inputs for receiving ASCII-coded data, a $\overline{DS}$ (**data strobe**) input to strobe data into the printer, and an $\overline{ACK}$ output to acknowledge the receipt of the ASCII character.

In this circuit, there is no signal to generate the $\overline{DS}$ signal to the printer, so PC4 is used with software that generates the $\overline{DS}$ signal. The $\overline{ACK}$ signal that is returned from the printer acknowledges the receipt of the data and is connected to the $\overline{ACK}$ input of the 82C55.

Example 11–17 lists the software that sends the ASCII-coded character in AH to the printer. The procedure first tests $\overline{OBF}$ to decide whether the printer has removed the data from port B. If not, the procedure waits for the $\overline{ACK}$ signal to return from the printer. If $\overline{OBF} = 1$, then the procedure sends the contents of AH to the printer through port B and also sends the $\overline{DS}$ signal.

FIGURE 11–29   The 82C55 connected to a parallel printer interface that illustrates the strobed output mode of operation for the 82C55.

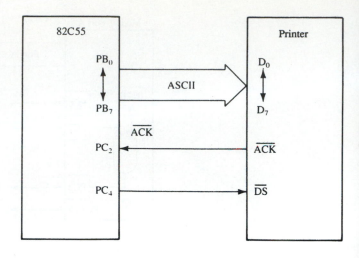

EXAMPLE 11–17

```
                          ;A procedure that transfers the ASCII character
                          ;from AH to the printer via port B.
                          ;
= 0002                    BIT1  EQU   2
= 0062                    PORTC EQU   62H
= 0061                    PORTB EQU   61H
= 0063                    CMD   EQU   63H

0000                      PRINT PROC  NEAR

                          ;check printer ready

0000  E4 62                       IN    AL,PORTC    ;get OBF
0002  A8 02                       TEST  AL,BIT1     ;test OBF
0004  74 FA                       JZ    PRINT       ;if OBF = 0

                          ;send character to printer

0006  8A C4                       MOV   AL,AH       ;get data
0008  E6 61                       OUT   PORTB,AL    ;print data

                          ;send data strobe to printer

000A  B0 08                       MOV   AL,8        ;clear DS
000C  E6 63                       OUT   CMD,AL
000E  B0 09                       MOV   AL,9        ;set DS
0010  E6 63                       OUT   CMD,AL
0012  C3                          RET

0013                      PRINT ENDP
```

Mode 2 Bi-directional Operation

In mode 2, which is allowed with group A only, port A becomes bi-directional, allowing data to be transmitted and received over the same eight wires. Bi-directional bused data are useful when interfacing two computers. It is also used for the IEEE–488 parallel high-speed GPIB (**general purpose instrumentation bus**) interface standard. Figure 11–30 shows the internal structure and timing diagram for mode 2 bi-directional operation.

Signal Definitions for Bi-directional Mode 2

INTR   **Interrupt request** is an output used to interrupt the microprocessor for both input and output conditions.

$\overline{\text{OBF}}$ **Output buffer full** is an output indicating that the output buffer contains data for the bi-directional bus.

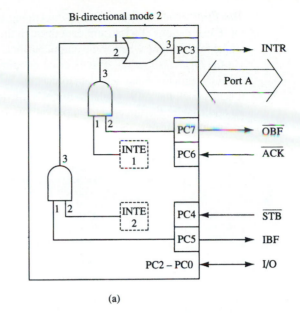

(a)

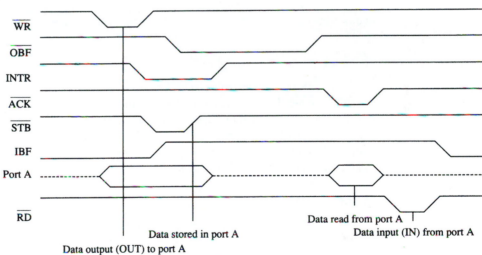

FIGURE 11–30   Mode 2 operation of the 82C55. (a) Internal structure and (b) timing diagram.

$\overline{ACK}$	**Acknowledge** is an input that enables the three-state buffers so that data can appear on port A. If $\overline{ACK}$ is a logic 1, the output buffers of port A are at their high-impedance state.
$\overline{STB}$	The **strobe** input loads the port A input latch with external data from the bi-directional port A bus.
IBF	**Input buffer full** is an output used to signal that the input buffer contains data for the external bi-directional bus.
INTE	**Interrupt enable** are internal bits (INTE1 and INTE2) that enable the INTR pin. The state of the INTR pin is controlled through port C bits PC6 (INTE1) and PC4 (INTE2).
PC2, PC1, and PC0	These pins are general-purpose I/O pins in mode 2 controlled by the bit set and reset command.

The Bi-directional Bus.   The bi-directional bus is used by referencing port A with the IN and OUT instructions. To transmit data through the bi-directional bus, the program first tests the $\overline{OBF}$ signal to determine whether the output buffer is empty. If it is, then data are sent to the output buffer via the OUT instruction. The external circuitry also monitors the $\overline{OBF}$ signal to decide whether the microprocessor has sent data to the bus. As soon as the output circuitry sees a logic 0 on $\overline{OBF}$, it sends back the $\overline{ACK}$ signal to remove it from the output buffer. The $\overline{ACK}$ signal sets the $\overline{OBF}$ bit and enables the three-state output buffers so that data may be read. Example 11–18 lists a procedure that transmits the contents of the AH register through bi-directional port A.

EXAMPLE 11–18

```
                        ;A procedure that transmits AH through the bi-
                        ;directional bus of port A.
                        ;
= 0080                  BIT7  EQU   80H
= 0062                  PORTC EQU   62H
= 0060                  PORTA EQU   60H

0000                    TRANS PROC  NEAR

0000   E4 62                  IN    AL,PORTC    ;get OBF
0002   A8 80                  TEST  AL,BIT7     ;test OBF
0004   74 FA                  JZ    TRANS       ;if OBF = 1

0006   8A C4                  MOV   AL,AH       ;get data
0008   E6 60                  OUT   PORTA,AL    ;send data
000A   C3                     RET

000B                    TRANS ENDP
```

To receive data through the bi-directional port A bus, the IBF bit is tested with software to decide whether data have been strobed into the port. If IBF = 1, then data are input using the IN instruction. The external interface sends data into the port by using the $\overline{STB}$ signal. When $\overline{STB}$ is activated, the IBF signal becomes a logic 1 and the data at port A are held inside the port in a latch. When the IN instruction executes, the IBF bit is cleared and the data in the port are moved into AL. Example 11–19 lists a procedure that reads data from the port.

EXAMPLE 11–19

```
                        ;A procedure that reads data from the bi-
                        ;directional port A and returns it in AL.
                        ;
= 0020                  BIT5  EQU   20H
= 0062                  PORTC EQU   62H
= 0060                  PORTA EQU   60H

0000                    READ  PROC  NEAR

0000   E4 62                  IN    AL,PORTC    ;get IBF
0002   A8 20                  TEST  AL,BIT5     ;test IBF
0004   74 FA                  JZ    READ        ;if IBF = 0
0006   E4 60                  IN    AL,PORTA    ;get data
0008   C3                     RET

0009                    READ  ENDP
```

The INTR **(interrupt request)** pin can be activated from both directions of data flow through the bus. If INTR is enabled by both INTE bits, then the output and input buffers both cause interrupt requests. This occurs when data are strobed into the buffer using $\overline{STB}$ or when data are written using OUT.

FIGURE 11–31 A summary of the port connections for the 82C55 PIA.

		Mode 0		Mode 1		Mode 2
Port A		IN	OUT	IN	OUT	I/O
Port B		IN	OUT	IN	OUT	Not used
	0			$INTR_B$	$INTR_B$	I/O
	1			IBF_B	$\overline{OBF}_B$	I/O
	2			$\overline{STB}_B$	$\overline{ACK}_B$	I/O
Port C	3	IN	OUT	$INTR_A$	$INTR_A$	INTR
	4			$\overline{STB}_A$	I/O	$\overline{STB}$
	5			IBF_A	I/O	IBF
	6			I/O	$\overline{ACK}_A$	$\overline{ACK}$
	7			I/O	$\overline{OBF}_A$	$\overline{OBF}$

82C55 Mode Summary

Figure 11–31 shows a graphical summary of the three modes of operation for the 82C55. Mode 0 provides simple I/O, mode 1 provides strobed I/O, and mode 2 provides bi-directional I/O. As mentioned, these modes are selected through the command register of the 82C55.

11–4   THE 8279 PROGRAMMABLE KEYBOARD/DISPLAY INTERFACE

The 8279 is a programmable keyboard and display interfacing component that scans and encodes up to a 64-key keyboard and controls up to a 16-digit numerical display. The keyboard interface has a built-in first-in, first-out (FIFO) buffer that allows it to store up to eight keystrokes before the microprocessor must retrieve a character. The display section controls up to 16 numeric displays from an internal 16×8 RAM that stores the coded display information.

Basic Description of the 8279

As we shall see, the 8279 is designed to easily interface with any microprocessor. Figure 11–32 illustrates the pin-out of this device. The definition of each pin connection follows.

Pin Definitions for the 8279

A0	The A0 address input selects data or control for reads and writes between the microprocessor and the 8279. A logic 0 selects data and a logic 1 selects control or status register.
$\overline{BD}$	**Blank** is an output used to blank the displays.
CLK	**Clock** is an input that generates the internal timing for the 8279. The maximum allowable frequency on the CLK pin is 3.125 MHz for the 8279-5 and 2.0 MHz for the 8279. Other timings require wait states in microprocessors executing at above 5 MHz.
CN/ST	**Control/strobe** is an input normally connected to the Control key on a keyboard.
$\overline{CS}$	**Chip select** is an input that enables the 8279 for programming, reading the keyboard and status information, and writing control and display data.

PIN CONFIGURATION

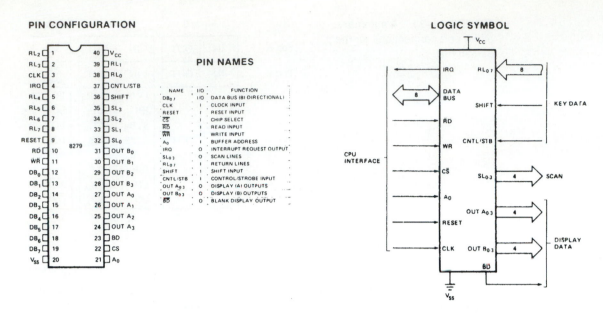

LOGIC SYMBOL

PIN NAMES

FIGURE 11–32   The pin-out and logic symbol of the 8279 programmable keyboard/display interface. (Courtesy of Intel Corporation.)

DB7–DB0	The **data bus** consists of bi-directional pins that connect to the data bus on the microprocessor.
IRQ	**Interrupt request** is an output that becomes a logic 1 whenever a key is pressed on the keyboard. This signal indicates that keyboard data are available for the microprocessor.
OUTA3–OUTA0	Outputs that send data to the displays (most-significant).
OUTB3–OUTB0	Outputs that send data to the displays (least-significant).
$\overline{RD}$	The **read** input is directly connected to the $\overline{IORC}$ or $\overline{RD}$ signal from the system. The $\overline{RD}$ input causes, when $\overline{CS}$ is a logic 0, a read from the data registers or status register.
RESET	The **reset** input connects to the system RESET signal.
RL7–RL0	**Return lines** are inputs used to sense any key depression in the keyboard matrix.
SHIFT	The **shift** input normally connects to the Shift key on a keyboard.
SL3–SL0	The **scan line** outputs scan both the keyboard and the displays.
$\overline{WR}$	**Write** is an input that connects to either the write strobe signal that is developed with external logic. The $\overline{WR}$ input causes data to be written to either the data registers or control registers within the 8279.
Vcc	A power supply pin connected to the system +5.0 V bus.
Vss	A ground pin connected to the system ground.

Interfacing the 8279 to the Microprocessor

In Figure 11–33, the 8279 is connected to the 8088 microprocessor. The 8279 is decoded to function at 8-bit I/O address 10H and 11H, where port 10H is the data port and 11H is the control

FIGURE 11–33 The 8279 interfaced to the 8088 microprocessor to function at 8-bit I/O ports 10H and 11H.

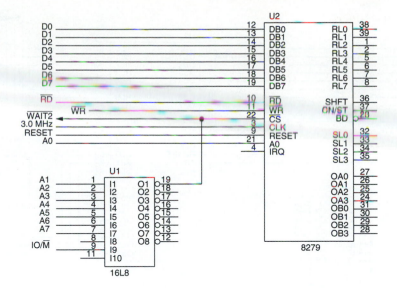

port. This circuit uses a PAL16L8 (see Example 11–20) to decode the I/O address for the 8279. Address bus bit A0 selects either the data or control port. Notice that the $\overline{CS}$ signal selects the 8279 and also provides a signal called $\overline{WAIT2}$ that is used to cause two wait states so that this device can function with an 8 MHz 8088.

The only signal not connected to the microprocessor is the IRQ output. This is an interrupt request pin and is beyond the scope of this section of the text. Chapter 12 explains interrupts and where they operate and function in a system.

EXAMPLE 11–20

```
TITLE        Address Decoder
PATTERN      Test 14
REVISION     A
AUTHOR       Barry B. Brey
COMPANY      BreyCo
DATE         7/10/99
CHIP         DECODERE PAL16L8

;pins  1   2   3   4   5   6   7   8    9    10
       A1  A2  A3  A4  A5  A6  A7  NC   IOM  GND

;pins  11  12  13  14  15  16  17  18  19  20
       NC  NC  NC  NC  NC  NC  NC  NC  CS  VCC

EQUATIONS

/CS = /A7 * /A6 * /A5 * A4 * /A3 * /A2 * /A1 * IOM
```

Keyboard Interface

Suppose that a 64-key keyboard (with no numeric displays) is connected through the 8279 to the 8088 microprocessor. Figure 11–34 shows this connection, as well as the keyboard. With the 8279, the keyboard matrix is any size from a 2 × 2 matrix (four keys) to an 8 × 8 matrix (64 keys). (Note that each crossover point in the matrix contains a normally open push-button switch that connects one vertical column with one horizontal row when a key is pressed.)

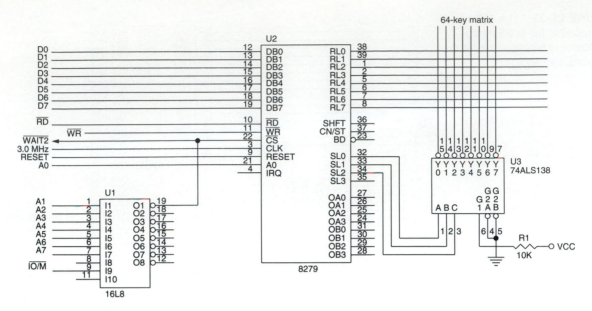

FIGURE 11–34   A 64-key keyboard interfaced to the 8088 microprocessor through the 8279.

The I/O port number decoded is the same as that decoded for Figure 11–33. The I/O port number is 10H for the data port and 11H for the control port in this circuit.

The 74ALS138 decoder generates eight active low column strobe signals for the keyboard. The selection pins SL2–SL0 sequentially scan each column of the keyboard, and the internal circuitry of the 8279 scans the RL pins, searching for a key switch closure. Pull-up resisters, normally found on input lines of a keyboard, are not required because the 8279 contains its own internal pull-ups on the RL inputs.

Programming the Keyboard Interface.   Before any keystroke is detected, the 8279 must be programmed—a more involved procedure than with the 82C55. The 8279 has eight control words to consider before it is programmed. The first three bits of the number sent to the control port (11H, in this example) select one of the eight different control words. Table 11–4 lists all eight control words and briefly describes them.

Control Word Descriptions.   Following is a list of the control words that program the 8279. Note that the first three bits are the control register number from Table 11–4, which are followed by other binary bits of information as they apply to each control.

000DDMMM   **Mode set** is a command with an opcode of 000 and two fields programmed to select the mode of operation for the 8279. The DD field selects the mode of operation for the displays (see Table 11–5), and the MMM field selects the mode of operation for the keyboard (see Table 11–6).

The DD field selects either an 8- or 16-digit display, and determines whether new data are entered to the rightmost or leftmost display position. The MMM field is quite a bit more complex: it provides encoded, decoded, or strobed keyboard operation.

In encoded mode, the SL outputs are active-high, and follow the binary bit pattern 0 through 7 or 0 through 15, depending whether 8- or 16-digit displays are selected. In decoded mode, the SL outputs are active-low, and only one of the four outputs is low at any given instant. The decoded outputs repeat the pattern: 1110, 1101, 1011, and 0111. In strobed mode, an

TABLE 11–4 The 8279 control word summary.

D_7	D_6	D_5	Function	Purpose
0	0	0	Mode set	Selects the number of display positions, left or right entry, and type of keyboard scan
0	0	1	Clock	Programs the internal clock and sets the scan and de-bounce times
0	1	0	Read FIFO	Selects the type of FIFO read and the address of the read
0	1	1	Read display	Selects the type of display read and the address of the read
1	0	0	Write display	Selects the type of write and the address of the write
1	0	1	Display write inhibit	Allows half-bytes to be blanked
1	1	0	Clear	Clears the display or FIFO
1	1	1	End interrupt	Clears the IRQ signal to the microprocessor

TABLE 11–5 Binary bit assignment for DD of the mode set control word.

DD	Function
00	8-digit display with left entry
01	16-digit display with left entry
10	8-digit display with right entry
11	16-digit display with right entry

TABLE 11–6 Binary bit assignment for MMM of the mode set control word.

MMM	Function
000	Encoded keyboard with 2-key lockout
001	Decoded keyboard with 2-key lockout
010	Encoded keyboard with N-key rollover
011	Decoded keyboard with N-key rollover
100	Encoded sensor matrix
101	Decoded sensor matrix
110	Strobed keyboard, encoded display scan
111	Strobed keyboard, decoded display scan

active-high pulse on the CN/ST input pin strobes data from the RL pins into an internal FIFO, where they are held for the microprocessor.

It is also possible to select either 2-key lockout or N-key rollover. 2-key lockout prevents two keys from being recognized, if pressed simultaneously. N-key rollover will accept all keys pressed simultaneously, from first to last.

001PPPPP The **clock command** word programs the internal clock divider. The code PPPPP is a prescaler that divides the clock input pin (CLK) to achieve the desired operating frequency of approximately 100 KHz. An input clock of 1 MHz thus requires a prescaler of 01010_2 for PPPPP.

010Z0AAA   The **read FIFO** control word selects the address of a keystroke from the internal FIFO buffer. Bit positions AAA select the desired FIFO location from 000 to 111, and Z selects auto-increment for the address. Under normal operation, this control word is used only with the sensor matrix operation of the 8279.

011ZAAAA   The **display read** control word selects the read address of one of the display RAM positions for reading through the data port. AAAA is the address of the position to be read and Z selects auto-increment mode. This command is used if the information stored in the display RAM must be read.

100ZAAAA   The **write display** control word selects the write address of one of the displays. AAAA addresses the position to be written to through the data port, and Z selects auto-increment so subsequent writes through the data port are to subsequent display positions.

1010WWBB   The **display write inhibit** control word inhibits writing to either half of each display RAM location. The leftmost W inhibits writing to the leftmost four bits of the display RAM location, and the rightmost W inhibits the rightmost four bits. The BB field functions in a like manner, except that they blank (turn off) either half of the output pins.

1100CCFA   The **clear** control word clears the display, the FIFO, or both the display and FIFO. Bit F clears the FIFO and the display RAM status, and sets the address pointer to 000. If the CC bits are 00 or 01, all of the display RAM locations become 0000000; if CC = 10, all locations become 00100000; and if CC = 11, all locations become 11111111.

111E000   The **end of interrupt** control word is issued to clear the IRQ pin to zero in the sensor matrix mode. If E is a 1, the special error mode is used. In the special error mode, the status register indicates if multiple key closures have occurred.

The large number of control words make programming the keyboard interface appear complex. Before anything is programmed, the clock divider rate must be determined. In the circuit illustrated in Figure 11–34, we use a 3.0 MHz clock input signal. To program the prescaler to generate a 100 KHz internal rate, we program PPPPP of the clock control word with a 30 or 11110_2.

The next step involves programming the keyboard type. The example keyboard in Figure 11–34 is an encoded keyboard. Notice that the circuit includes an external decoder that converts the encoded data from the SL pins into decoded column-selection signals. We are free in this example to choose either 2-key lockout or N-key rollover, but most applications use 2-key lockout.

Finally, we program the operation of the FIFO. Once the FIFO is programmed, it never needs to be reprogrammed unless we need to read prior keyboard codes. Each time a key is typed, the data are stored in the FIFO; if they are read from the FIFO before the FIFO is full (eight characters), the data from the FIFO follows the same order as the typed data. Example 11–21 provides the software required to initialize the 8279 to control the keyboard illustrated in Figure 11–34.

EXAMPLE 11–21

```
                            ;Initialization dialog for the keyboard interface
                            ;of Figure 11-34.
                            ;
        0000  B0 3E                 MOV     AL,00111110B    ;program clock
        0002  E6 11                 OUT     11H,AL

        0004  B0 00                 MOV     AL,0
        0006  E6 11                 OUT     11H,AL          ;program mode

        0008  B0 50                 MOV     AL,01010000B
        000A  E6 11                 OUT     11H,AL          ;program FIFO
```

FIGURE 11–35 The 8279–5 FIFO status register.

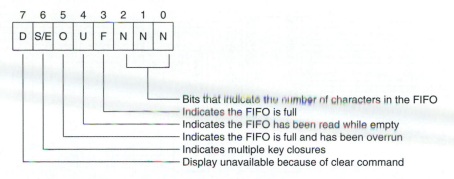

Once the 8279 is initialized, a procedure is required to read data from the keyboard. We determine whether a character is typed in the keyboard by looking at the FIFO status register. Whenever the control port is addressed by the IN instruction, the contents of the FIFO status word is copied into the AL register. Figure 11–35 shows the contents of the FIFO status register and defines the purpose of each status bit.

The procedure listed in Example 11–22 first tests the FIFO status register to see whether it contains any data. If NNN = 000, the FIFO is empty. Upon determining that the FIFO is not empty, the procedure inputs data to AL and returns with the keyboard code in AL.

EXAMPLE 11–22

```
                         ;A procedure that reads data from the FIFO and
                         ;returns it in AL.
                         ;
= 0007                   MASKS EQU   7

0000                     READ  PROC  NEAR

0000   E4 11                   IN    AL,11H    ;read status
0002   A8 07                   TEST  AL,MASKS  ;test NNN
0004   74 FA                   JZ    READ      ;if NNN = 0
0006   E4 10                   IN    AL,10H    ;read FIFO data
0008   C3                      RET

0009                     READ  ENDP
```

The data found in AL upon returning from the subroutine contains raw data from the keyboard. Figure 11–36 shows the format of this data for both the scanned and strobed modes of operation. The scanned code is returned from our keyboard interface and is converted to ASCII code by using the XLAT instruction with an ASCII code lookup table. The scanned code is returned with the row and column number occupying the rightmost six bits.

The SH bit shows the state of the shift pin and the CT bit shows the state of the control pin. In the strobed mode, the contents of the eight RL inputs appear as they are sampled by placing a logic 1 on the strobe input pin to the 8279.

Six-Digit Display Interface

Figure 11–37 depicts the 8279 connected to the 8088 microprocessor and a 6-digit numeric display. This interface uses a PAL16L8 (program not shown) to decode the 8279 at I/O ports 20H

FIGURE 11–36 The (a) scanned keyboard code and (b) strobed keyboard code for the 8279–5 FIFO.

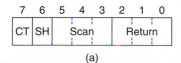

(a)

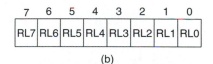

(b)

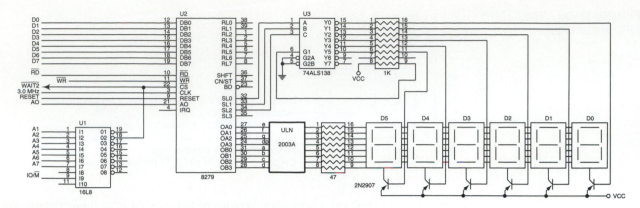

FIGURE 11–37   A 6-digit numeric display interfaced to the 8279.

(data) and 21H (control/status). The segment data are supplied to the displays through the OUTA and OUTB pins of the 8279. These bits are buffered by a segment driver (ULN2003A) to drive the segment inputs to the display.

A 74ALS138 3-to-8 line decoder enables the anode switches of each display position. The SL2–SL0 pins supply the decoder with the encoded display position from the 8279. Notice that the left-hand display is at position 0101 and the right-hand display is at position 0000. These are the addresses of the display positions, as indicated in control words for the 8279.

It is necessary to choose resistor values that allow 60 mA of current flow per segment. In this circuit, we use 47 Ω resistors. If we allow 60 mA of segment current, then the average segment current is 10 mA, or one-sixth of 60 mA because current only flows for one-sixth of the time through a segment. The anode switches must supply the current for all seven segments plus the decimal point. Here, the total anode current is 8×60 mA, or 480 mA.

Example 11–23 lists the initialization dialog for programming the 8279 to function with this 6-digit display. This software programs the display and clears the display RAM.

EXAMPLE 11–23

```
                        ;Initialization dialog for the 6-digit display of
                        ;Figure 11-37.
                        ;
0000  B0 3E             MOV   AL,00111110B   ;program clock
0002  E6 21             OUT   21H,AL

0004  B0 00             MOV   AL,0           ;program mode set
0006  E6 21             OUT   21H,AL

0008  B0 C1             MOV   AL,11000001B   ;clear display
000A  E6 21             OUT   21H,AL
```

Example 11–24 lists a procedure for displaying information on the displays. Data are transferred to the procedure through the AX register. AH contains the 7-segment display code and AL contains the address of the displayed digit.

EXAMPLE 11–24

```
                        ;A procedure that displays AH on the display
                        ;position addressed by AL.
                        ;
= 0080                  MASKS EQU  80H

0000                    DISP  PROC NEAR
```

```
0000  50                PUSH  AX          ;save data
0001  0C 80             OR    AL,MASKS    ;select digit
0003  E6 21             OUT   21H,AL
0005  8A C4             MOV   AL,AH       ;display data
0007  E6 20             OUT   20H,AL
0009  58                POP   AX          ;restore data
000A  C3                RET

000B            DISP    ENDP
```

11–5   8254 PROGRAMMABLE INTERVAL TIMER

The 8254 programmable interval timer consists of three independent 16-bit programmable counters **(timers).** Each counter is capable of counting in binary or binary-coded decimal (BCD). The maximum allowable input frequency to any counter is 10 MHz. This device is useful wherever the microprocessor must control real-time events. Some examples of usage include real-time clock, events counter, and motor speed and direction control.

This timer also appears in the personal computer decoded at ports 40H–43H to do the following:

1. Generate a basic timer interrupt that occurs at approximately 18.2 Hz.
2. Cause the DRAM memory system to be refreshed.
3. Provide a timing source to the internal speaker and other devices. The timer in the personal computer is an 8253 instead of an 8254.

8254 Functional Description

Figure 11–38 shows the pin-out of the 8254, which is a higher-speed version of the 8253, and a diagram of one of the three counters. Each timer contains a CLK input, a gate input, and an output (OUT) connection. The CLK input provides the basic operating frequency to the timer, the gate pin controls the timer in some modes, and the OUT pin is where we obtain the output of the timer.

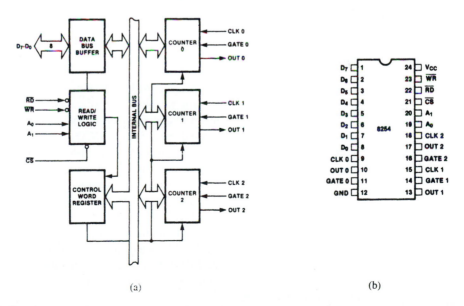

(a)   (b)

FIGURE 11–38   The 8254 programmable interval timer. (a) Internal structure and (b) pin-out. (Courtesy of Intel Corporation.)

TABLE 11–7 Address selection inputs to the 8254.

A_1	A_0	Function
0	0	Counter 0
0	1	Counter 1
1	0	Counter 2
1	1	Control word

The signals that connect to the microprocessor are the data bus pins (D7–D0), $\overline{RD}$, $\overline{WR}$, $\overline{CS}$, and address inputs A1 and A0. The address inputs are present to select any of the four internal registers used for programming, reading, or writing to a counter. The personal computer contains an 8253 timer or its equivalent, decoded at I/O ports 40H–43H. Timer zero is programmed to generate an 18.2 Hz signal that interrupts the microprocessor at interrupt vector 8 for a clock tick. The tick is often used to time programs and events. Timer 1 is programmed for 15 μs, which is used on the PC/XT personal computer to request a DMA action used to refresh the dynamic RAM. Timer 2 is programmed to generate tone on the personal computer speaker.

Pin Definitions

A1, A0   The **address inputs** select one of four internal registers within the 8254. See Table 11–7 for the function of the A1 and A0 address bits.

CLK   The **clock** input is the timing source for each of the internal counters. This input is often connected to the PCLK signal from the microprocessor system bus controller.

$\overline{CS}$ **Chip select** enables the 8254 for programming, and reading or writing a counter.

G   The **gate** input controls the operation of the counter in some modes of operation.

GND   **Ground** connects to the system ground bus.

OUT   A **counter output** is where the wave-form generated by the timer is available.

$\overline{RD}$ **Read** causes data to be read from the 8254 and often connects to the $\overline{IORC}$ signal.

Vcc   **Power** connects to the +5.0 V power supply.

$\overline{WR}$ **Write** causes data to be written to the 8254 and often connects to the write strobe ($\overline{IOWC}$).

Programming the 8254

Each counter is individually programmed by writing a control word, followed by the initial count. Figure 11–39 lists the program control word structure of the 8254. The **control word** allows the programmer to select the counter, mode of operation, and type of operation (read/write). The control word also selects either a binary or BCD count. Each counter may be programmed with a count of 1 to FFFFH. A count of 0 is equal to FFFFH+1 (65,536) or 10,000 in BCD. The minimum count of 1 applies to all modes of operation except modes 2 and 3, which have a minimum count of 2. Timer 0 is used in the personal computer with a divide by count of 64K (FFFFH) to generate the 18.2 Hz (18.196 Hz) interrupt clock tick. Timer 0 has a clock input frequency of 4.77 MHz ÷ 4 or 1.1925 MHz.

The control word uses the BCD bit to select a BCD count (BCD = 1) or a binary count (BCD = 0). The M2, M1, and M0 bits select one of the 6 different modes of operation (000–101) for the counter. The RW1 and RW0 bits determine how the data are read from or written to the counter. The SC1 and SC0 bits select a counter or the special read back mode of operation, discussed later in this section.

Each counter has a program control word used to select the way the counter operates. If two bytes are programmed into a counter, then the first byte (LSB) will stop the count, and the

FIGURE 11–39 The control word for the 8254–2 timer.

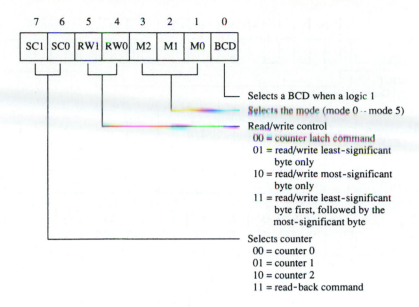

second byte (MSB) will start the counter with the new count. The order of programming is important for each counter, but programming of different counters may be interleaved for better control. For example, the control word may be sent to each counter before the counts for individual programming. Example 11–25 shows a few ways to program counter 1 and 2. The first method programs both control words, then the LSB of the count for each counter, which stops them from counting. Finally, the MSB portion of the count is programmed starting both counters with the new count. The second example shows one counter programmed before the other.

EXAMPLE 11–25

```
PROGRAM CONTROL WORD 1          ;setup counter 1
PROGRAM CONTROL WORD 2          ;setup counter 2
PROGRAM LSB 1                    ;stop counter 1 and program LSB
PROGRAM LSB 2                    ;stop counter 2 and program LSB
PROGRAM MSB 1                    ;program MSB of counter 1 and start it
PROGRAM MSB 2                    ;program MSB of counter 2 and start it

        or

PROGRAM CONTROL WORD 1          ;setup counter 1
PROGRAM LSB 1                    ;stop counter 1 and program LSB
PROGRAM MSB 1                    ;program MSB of counter 1 and start it
PROGRAM CONTROL WORD 2          ;setup counter 2
PROGRAM LSB 2                    ;stop counter 2 and program LSB
PROGRAM MSB 2                    ;program MSB of counter 2 and start it
```

Modes of Operation.   Six modes (mode 0–mode 5) of operation are available to each of the 8254 counters. Figure 11–40 shows how each of these modes functions with the CLK input, the gate (G) control signal, and OUT signal. A description of each mode follows:

Mode 0   Allows the 8254 counter to be used as an events counter. In this mode, the output becomes a logic 0 when the control word is written and remains there until N plus the number of programmed counts. For example, if a count of 5 is programmed, the output will remain a logic 0 for 6 counts beginning with N. Note that the gate (G) input must be a logic 1 to allow the counter to count. If G becomes a logic 0 in the middle of the count, the counter will stop until G again becomes a logic 1.

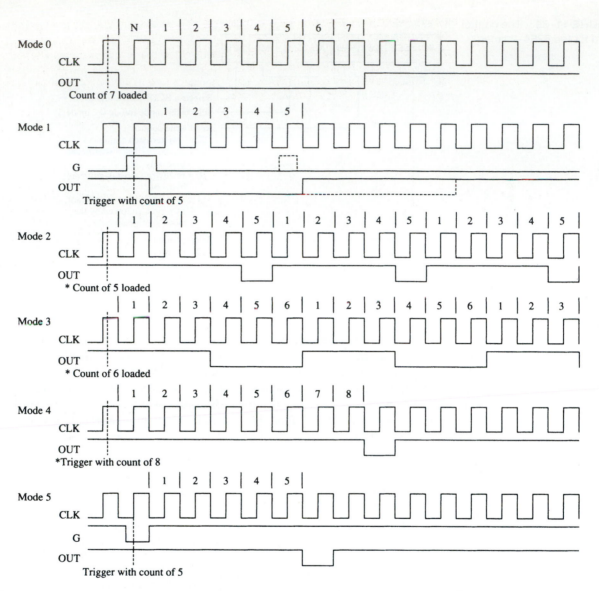

FIGURE 11–40   The six modes of operation for the 8254-2 programmable interval timer. *Note:* The G input stops the count when 0 in modes 2, 3, and 4.

Mode 1	Causes the counter to function as a retriggerable monostable multivibrator (one-shot). In this mode the G input triggers the counter so that it develops a pulse at the OUT connection that becomes a logic 0 for the duration of the count. If the count is 10, then the OUT connection goes low for 10 clocking periods when triggered. If the G input occurs within the duration of the output pulse, the counter is again reloaded with the count and the OUT connection continues for the total length of the count.
Mode 2	Allows the counter to generate a series of continuous pulses that are one clock pulse wide. The separation between pulses is determined by the count. For example, for a count of 10, the output is a logic 1 for nine clock periods and low for one clock period. This cycle is repeated until the counter is programmed with a new count or until the G pin is placed at a logic 0 level. The G input must be a logic 1 for this mode to generate a continuous series of pulses.

Mode 3   Generates a continuous square-wave at the OUT connection, provided that the G pin is a logic 1. If the count is even, the output is high for one-half of the count and low for one-half of the count. If the count is odd, the output is high for one clocking period longer than it is low. For example, if the counter is programmed for a count of 5, the output is high for three clocks and low for two clocks.

Mode 4   Allows the counter to produce a single pulse at the output. If the count is programmed as a 10, the output is high for 10 clocking periods and low for one clocking period. The cycle does not begin until the counter is loaded with its complete count. This mode operates as a software triggered one-shot. As with modes 2 and 3, this mode also uses the G input to enable the counter. The G input must be a logic 1 for the counter to operate for these three modes.

Mode 5   A hardware triggered one-shot that functions as mode 4, except that it is started by a trigger pulse on the G pin instead of by software. This mode is also similar to mode 1 because it is retriggerable.

Generating a Wave-form with the 8254.   Figure 11–41 shows an 8254 connected to function at I/O ports 0700H, 0702H, 0704H, and 0706H of an 80386SX microprocessor. The addresses are decoded by using a PAL16L8 that also generates a write strobe signal for the 8254, which is connected to the low order data bus connections. The PAL also generates a wait signal for the microprocessor that causes two wait states when the 8254 is accessed. The wait state generator connected to the microprocessor actually controls the number of wait states inserted into the timing. The program for the PAL is not illustrated here because it is the same as many of the prior examples.

Example 11–26 lists the program that generates a 100 KHz square-wave at OUT0 and a 200 KHz continuous pulse at OUT1. We use mode 3 for counter 0 and mode 2 for counter 1. The count programmed into counter 0 is 80 and the count for counter 1 is 40. These counts generate the desired output frequencies with an 8 MHz input clock.

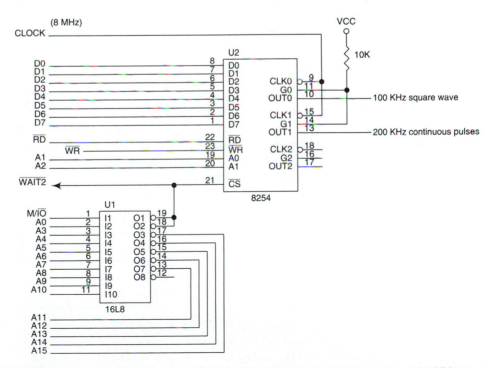

FIGURE 11–41   The 8254 interfaced to an 8 MHz 8086 so that it generates a 100 KHz square wave at OUT0 and a 200 KHz continuous pulse at OUT1.

EXAMPLE 11–26

```
                              ;A procedure that programs the 8254 timer to function
                              ;as illustrated in Figure 11-41.
                              ;
0000                  TIME    PROC    NEAR

0000  50                      PUSH    AX              ;save registers
0001  52                      PUSH    DX

0002  BA 0706                 MOV     DX,706H         ;address control word
0005  B0 36                   MOV     AL,00110110B    ;program counter 0
0007  EE                      OUT     DX,AL           ;for mode 3

0008  B0 74                   MOV     AL,01110100B    ;program counter 1
000A  EE                      OUT     DX,AL           ;for mode 2

000B  BA 0700                 MOV     DX,700H         ;address counter 0
000E  B0 50                   MOV     AL,80           ;load count of 80
0010  EE                      OUT     DX,AL
0011  32 C0                   XOR     AL,AL
0013  EE                      OUT     DX,AL

0014  BA 0702                 MOV     DX,702H         ;address counter 1
0017  B0 28                   MOV     AL,40           ;load count of 40
0019  EE                      OUT     DX,AL
001A  32 C0                   XOR     AL,AL
001C  EE                      OUT     DX,AL

001D  5A                      POP     DX              ;restore registers
001E  58                      POP     AX
001F  C3                      RET

0020                  TIME    ENDP
```

Reading a Counter.   Each counter has an internal latch that is read with the read counter port operation. These latches will normally follow the count. If the contents of the counter are needed, then the latch can remember the count by programming the counter latch control word (see Figure 11–42), which causes the contents of the counter to be held in a latch until it is read. Whenever a read from the latch or the counter is programmed, the latch tracks the contents of the counter.

When it is necessary for the contents of more than one counter to be read at the same time, we use the read-back control word, illustrated in Figure 11–43. With the read-back control word, the $\overline{\text{CNT}}$ bit is a logic 0 to cause the counters selected by CNT0, CNT1, and CNT2 to be latched. If the status register is to be latched, then the $\overline{\text{ST}}$ bit is placed at a logic 0. Figure 11–44 shows the status register, which shows the state of the output pin, whether the counter is at its null state (0), and how the counter is programmed.

FIGURE 11–42 The 8254-2 counter latch control word.

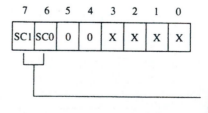

Select counter
00 = counter 0
01 = counter 1
10 = counter 2
11 = read-back command

FIGURE 11–43   The 8254-2 read-back control word.

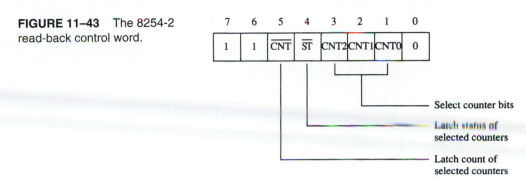

FIGURE 11–44   The 8254-2 status register.

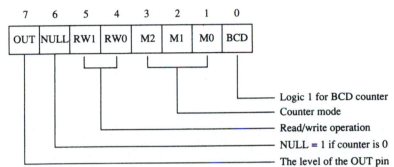

DC Motor Speed and Direction Control

One application of the 8254 timer is as a motor speed controller for a DC motor. Figure 11–45 shows the schematic diagram of the motor and its associated driver circuitry. It also illustrates the interconnection of the 8254, a flip-flop, and the motor and its driver.

The operation of the motor driver circuitry is straightforward. If the Q output of the 74ALS112 is a logic 1, the base Q2 is pulled up to +12 V through the base pull-up resistor, and the base of Q2 is open circuited. This means that Q1 is off and Q2 is on, with ground applied to the positive lead of the motor. The bases of both Q3 and Q4 are pulled low to ground through the inverters. This causes Q3 to conduction or turn on and Q4 to turn off, applying ground to the negative lead of the motor. The logic 1 at the Q output of the flip-flop therefore connects +12 V to the positive lead of the motor and ground to the negative lead. This connection causes the motor to spin in its forward direction. If the state of the Q output of the flip-flop becomes a logic 0, then the conditions of the transistors are reversed and +12 V is attached to the negative lead of the motor, with ground attached to the positive lead. This causes the motor to spin in the reverse direction.

If the output of the flip-flop is alternated between a logic 1 and 0, the motor spins in either direction at various speeds. If the duty cycle of the Q output is 50 percent, the motor will not spin at all and exhibits some holding torque because current flows through it. Figure 11–46 shows some timing diagrams and their effects on the speed and direction of the motor. Notice how each counter generates pulses at different positions to vary the duty cycle at the Q output of the flip-flop. This output is also called *pulse width modulation*.

To generate these wave forms, counters 0 and 1 are both programmed to divide the input clock (PCLK) by 30,720. We change the duty cycle of Q by changing the point at which counter 1 is started in relationship to counter 0. This changes the direction and speed of the motor. But

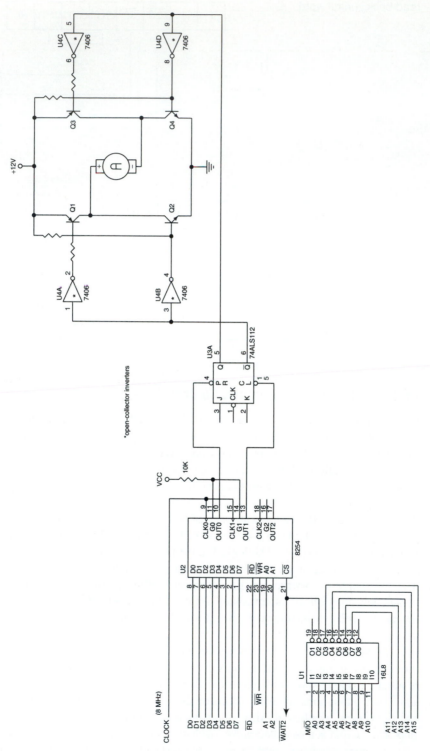

FIGURE 11–45 Motor speed and direction control using the 8254 timer.

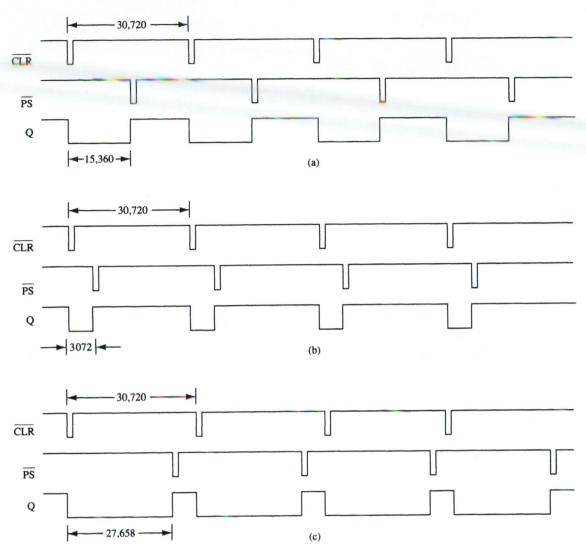

FIGURE 11–46   Timing for the motor speed and direction control circuit of Figure 11–45. (a) No rotation, (b) high-speed rotation in the reverse direction, and (c) high-speed rotation in the forward direction.

why divide the 8 MHz clock by 30,720? The divide rate of 30,720 is divisible by 256, so we can develop a short program that allows 256 different speeds. This also produces a basic operating frequency for the motor of about 260 Hz, which is low enough in frequency to power the motor. It is important to keep this operating frequency below 1000 Hz, but above 60 Hz.

Example 11–27 lists a procedure that controls the speed and direction of the motor. The speed is controlled by the value of AH when this procedure is called. Because we have an 8-bit number to represent speed, a 50 percent duty cycle, for a stopped motor, is a count of 128. By changing the value in AH when the procedure is called, we can adjust the motor speed. The speed of the motor will increase in either direction by changing the number in AH when this procedure is called. As the value in AH approaches 00H, the motor begins to increase its speed in the reverse direction. As the value of AH approaches FFH, the motor increases its speed in the forward direction.

EXAMPLE 11–27

```
                              ;A procedure that controls the speed and direction
                              ;of the motor in Figure 11-45.
                              ;
                              ;When this procedure is called, the contents of
                              ;AH determine the speed and direction of the
                              ;motor where AH is between 00H and FFH.
       ;
= 0706                        CNTR  EQU   706H
= 0700                        CNT0  EQU   700H
= 0702                        CNT0  EQU   702H
= 7800                        COUNT EQU   30720

0000                          SPEED PROC  NEAR

0000  50                            PUSH  AX            ;save registers
0001  51                            PUSH  DX
0002  53                            PUSH  BX

0003  8A CD                         MOV   BL,AL         ;calculate count
0005  B8 0078                       MOV   AX,120
0008  F6 E3                         MUL   BL
000A  8B D8                         MOV   BX,AX
000C  B8 7800                       MOV   AX,COUNT
000F  2B C3                         SUB   AX,BX
0011  8B D8                         MOV   BX,AX

0013  BA 0706                       MOV   DX,CNTR       ;program control words
0016  B0 34                         MOV   AL,00110100B
0018  EE                            OUT   DX,AL
0019  B0 74                         MOV   AL,01110100B
001B  EE                            OUT   DX,AL

001C  BA 0702                       MOV   DX,CNT1       ;program counter 1
001F  B8 7800                       MOV   AX,COUNT      ;to generate a clear
0022  EE                            OUT   DX,AL
0023  8A C4                         MOV   AL,AH
0025  EE                            OUT   DX,AL
0026                          SPE:
0026  EC                            IN    AL,DX         ;wait for counter 1
0027  86 C4                         XCHG  AL,AH         ;to reach calculated
0029  EC                            IN    AL,DX         ;count
002A  86 C4                         XCHG  AL,AH
002C  3B C3                         CMP   AX,BX
002E  72 F6                         JB    SPE

0030  BA 0700                       MOV   DX,CNT0       ;program counter 0
0033  B8 7800                       MOV   AX,COUNT      ;to generate a set
0036  EE                            OUT   DX,AL
0037  8A C4                         MOV   AL,AH
0039  EE                            OUT   DX,AL

003A  5B                            POP   BX            ;restore registers
003B  5A                            POP   DX
003C  58                            POP   AX
003D  C3                            RET

003E                          SPEED ENDP
```

The procedure adjusts the wave form at Q by first calculating the count that counter 0 is to start in relationship to counter 1. This is accomplished by multiplying AH by 120 and then subtracting it from 30,720. This is required because the counters are down-counters that count from the programmed count to 0 before restarting. Next, counter 1 is programmed with a count of

30,720 and started to generate the clear wave form for the flip-flop. After counter 1 is started, it is read and compared with the calculated count. Once it reaches this count, counter 0 is started with a count of 30,720. From this point forward, both counters continue generating the clear and set wave forms until the procedure is again called to adjust the speed and direction of the motor.

11–6 16550 PROGRAMMABLE COMMUNICATIONS INTERFACE

The National Semiconductor Corporation's PC16550D is a programmable communications interface designed to connect to virtually any type of serial interface. The 16550 is a universal asynchronous receiver/transmitter (UART) that is fully compatible with the Intel microprocessors. The 16550 is capable of operating at 0–1.5 M Baud. Baud rate is the number of bits transferred per second, including start, stop, data, and parity. The 16550 also includes a programmable Baud rate generator and separate FIFOs for input and output data to ease the load on the microprocessor. Each FIFO contains 16 bytes of storage. This is the most common communications interface found in modern microprocessor-based equipment, including the personal computer and many modems.

Asynchronous Serial Data

Asynchronous serial data are transmitted and received without a clock or timing signal. Figure 11–47 illustrates two frames of asynchronous serial data. Each frame contains a start bit, seven data bits, parity, and one stop bit. The figure shows a frame that contains one ASCII character and 10 bits. Most dial-up communications systems, such as CompuServe, Prodigy, and America Online, use 10 bits for asynchronous serial data with even parity. Most Internet and bulletin board services also use 10 bits, but they normally do not use parity. Instead, eight data bits are transferred, replacing parity with a data bit. This makes byte transfers of non-ASCII data much easier to accomplish.

16550 Functional Description

Figure 11–48 illustrates the pin-out of the 16550 UART. This device is available as a 40-pin DIP **(dual in-line package)** or as a 44-pin PLCC **(plastic lead-less chip carrier).** Two completely separate sections are responsible for data communications: the receiver and the transmitter. Because each of these sections is independent of each other, the 16550 is able to function in simplex, half-duplex, or full-duplex modes. One of the main features of the 16550 is its internal receiver and transmitter FIFO (first-in, first-out) memories. Because each is 16 bytes deep, the UART requires attention only from the microprocessor after receiving 16 bytes of data. It also holds 16 bytes before the microprocessor must wait for the transmitter. The FIFO makes this UART ideal when interfacing to high-speed systems because less time is required to service it.

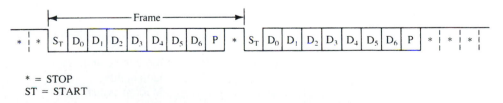

* = STOP
ST = START

FIGURE 11–47   Asynchronous serial data.

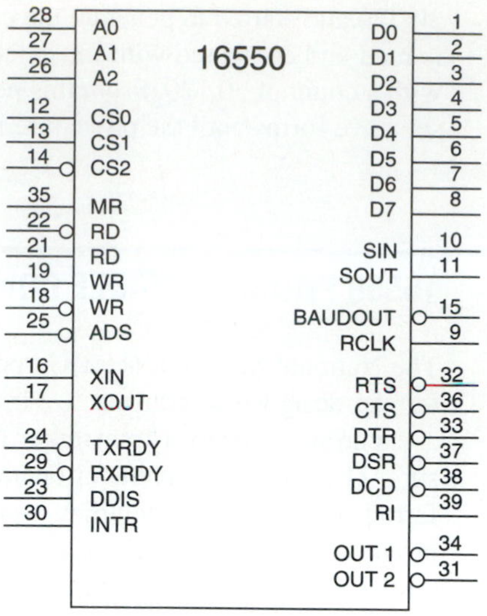

FIGURE 11–48 The pin-out of the 16550 UART.

Pin	Signal		Signal	Pin
28	A0		D0	1
27	A1	16550	D1	2
26	A2		D2	3
12	CS0		D3	4
13	CS1		D4	5
14	CS2		D5	6
			D6	7
35	MR		D7	8
22	RD			
21	RD		SIN	10
19	WR		SOUT	11
18	WR			
25	ADS		BAUDOUT	15
			RCLK	9
16	XIN			
17	XOUT		RTS	32
			CTS	36
24	TXRDY		DTR	33
29	RXRDY		DSR	37
23	DDIS		DCD	38
30	INTR		RI	39
			OUT 1	34
			OUT 2	31

An example **simplex** system is one in which the transmitter or receiver is used by itself such as in an FM (**frequency modulation**) radio station. An example **half-duplex** system is a CB (**citizens band**) radio, where we transmit and receive, but not both at the same time. The **full-duplex** system allows transmission and reception in both directions simultaneously. An example full-duplex system is the telephone.

The 16550 can control a **modem** (modulator/demodulator), which is a device that converts TTL levels of serial data into audio tones that can pass through the telephone system. Six pins on the 16650 are devoted to modem control: $\overline{\text{DSR}}$ (data set ready), $\overline{\text{DTR}}$ (data terminal ready), $\overline{\text{CTS}}$ (clear-to-send), $\overline{\text{RTS}}$ (request-to-send), $\overline{\text{RI}}$ (ring indicator), and $\overline{\text{DCD}}$ (data carrier detect). The modem is referred to as the **data set** and the 16550 is referred to as the **data terminal.**

16550 Pin Functions

A0, A1, A2 The address inputs are used to select an internal register for programming and also data transfer. See Table 11–8 for a list of each combination of the address inputs and the registers selected.

$\overline{\text{ADS}}$ The **address strobe** input is used to latch the address lines and chip select lines. If not needed (as in the Intel system), connect this pin to ground. The $\overline{\text{ADS}}$ pin is designed for use with Motorola microprocessors.

TABLE 11–8 The registers selected by A0, A1, and A2.

A2	A1	A0	Register
0	0	0	Receiver buffer (read) and transmitter holding (write)
0	0	1	Interrupt enable
0	1	0	Interrupt identification (read) and FIFO control (write)
0	1	1	Line control
1	0	0	Modem control
1	0	1	Line status
1	1	0	Modem status
1	1	1	Scratch

$\overline{\text{BAUDOUT}}$	The **Baud out** pin is where the clock signal generated by the Baud rate generator from the transmitter section is made available. It is most often connected to the RCLK input to generate a receiver clock that is equal to the transmitter clock.
CS0, CS1, $\overline{\text{CS2}}$	The **chip select** inputs must all be active to enable the 16550 UART.
$\overline{\text{CTS}}$	The **clear to send** (if low) indicates that the modem or data set is ready to exchange information. This pin is often used in a half-duplex system to turn the line around.
D7–D0	The **data bus** pins are connected to the microprocessor data bus.
$\overline{\text{DCD}}$	The **data carrier detect** input is used by the modem to signal the 16550 that a carrier is present.
DDIS	The **disable driver** output becomes a logic 0 to indicate that the microprocessor is reading data from the UART. DDIS can be used to change the direction of data flow through a buffer.
$\overline{\text{DSR}}$	**Data set ready** is an input to the 16550, indicating that the modem or data set is ready to operate.
$\overline{\text{DTR}}$	**Data terminal ready** is an output that indicates that the data terminal (16550) is ready to function.
INTR	**Interrupt request** is an output to the microprocessor used to request an interrupt (INTR = 1) whenever the 16550 has a receiver error, it has received data, and if the transmitter is empty.
MR	**Master reset** initializes the 16550 and should be connected to the system RESET signal.
$\overline{\text{OUT1}}$, $\overline{\text{OUT2}}$	User-defined output pins that can provide signals to a modem or any other device, as needed in a system.
RCLK	**Receiver clock** is the clock input to the receiver section of the UART. This input is always 16 × the desired receiver Baud rate.
RD, $\overline{\text{RD}}$	**Read** inputs (either may be used) cause data to be read from the register specified by the address inputs to the UART.
$\overline{\text{RI}}$	The **ring indicator** input is placed at the logic 0 level by the modem to indicate that the telephone is ringing.
$\overline{\text{RTS}}$	**Request-to-send** is a signal to the modem, indicating that the UART wishes to send data.
SIN, SOUT	These are the serial data pins. SIN accepts serial data and SOUT transmits serial data.
$\overline{\text{RXRDY}}$	**Receiver ready** is a signal used to transfer received data via DMA techniques (see text).
$\overline{\text{TXRDY}}$	**Transmitter ready** is a signal used to transfer transmitter data via DMA techniques (see text).
WR, $\overline{\text{WR}}$	**Write** (either may be used) connects to the microprocessor write signal to transfer commands and data to the 16550.
XIN, XOUT	These are the **main clock connections.** A crystal is connected across these pins to form a crystal oscillator, or XIN is connected to an external timing source.

Programming the 16550

Programming the 16550 is simple, although may be slightly more involved when compared to some of the other programmable interfaces described in this chapter. Programming is a two-part process that includes the initialization dialog and operational dialog.

In the personal computer, which uses the 16550 or its programming equivalent, the I/O port addresses are decoded at 3F8H through 3FFH for COM port 0 and 2F8H through 2FFH for COM port 2. Although the examples in this section of the chapter are not written specifically for the personal computer, they can be adapted by changing the port numbers to control the COM ports on the PC.

Initializing the 16550.   Initialization dialog, which occurs after a hardware or software reset, consists of two parts: programming the line control register and the Baud rate generator. The line control register selects the number of data bits, number of stop bits, and parity (whether it's even or odd, or if parity is sent as a 1 or a 0). The Baud rate generator is programmed with a divisor that determines the Baud rate of the transmitter section.

Figure 11–49 illustrates the line control register. The line control register is programmed by outputting information to I/O port 011 (A2, A1, A0). The rightmost two bits of the line control register select the number of transmitted data bits (5, 6, 7, or 8). The number of stop bits is selected by S in the line control register. If S = 0, one stop bit is used; if S = 1, 1.5 stop bits are used for five data bits, and two stop bits are used with six, seven, or eight data bits.

The next three bits are used together to send even or odd parity, to send no parity, or to send a 1 or a 0 in the parity bit position. To send even or odd parity, the ST (**stick**) bit must be placed at a logic 0 level, and parity enable must be a logic 1. The value of the parity bit then determines even or odd parity. To send no parity (common in Internet connections), ST = 0 as well as the parity enable bit. This sends and receives data without parity. Finally, if a 1 or a 0 must be sent and received in the parity bit position for all data, ST = 1 with a 1 in parity enable. To send a 1 in the parity bit position, place a 0 in the parity bit; to send a 0, place a 1 in the parity bit. (See Table 11–9 for the operation of the parity and stick bits.)

The remaining bits in the line control register are used to send a break and to select programming for the Baud rate divisor. If bit position 6 of the line control register is a logic 1, a

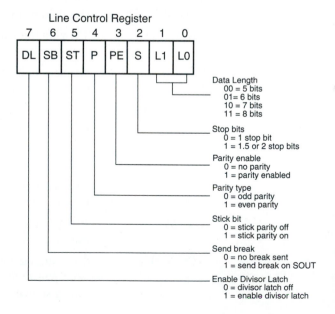

FIGURE 11–49   The contents of the 16550 line control register.

Line Control Register

7	6	5	4	3	2	1	0
DL	SB	ST	P	PE	S	L1	L0

Data Length
00 = 5 bits
01= 6 bits
10 = 7 bits
11 = 8 bits

Stop bits
0 = 1 stop bit
1 = 1.5 or 2 stop bits

Parity enable
0 = no parity
1 = parity enabled

Parity type
0 = odd parity
1 = even parity

Stick bit
0 = stick parity off
1 = stick parity on

Send break
0 = no break sent
1 = send break on SOUT

Enable Divisor Latch
0 = divisor latch off
1 = enable divisor latch

TABLE 11–9   The operation of the ST and parity bits.

ST	P	PE	Function
0	0	0	No parity
0	0	1	Odd parity
0	1	0	No parity
0	1	1	Even parity
1	0	0	Undefined
1	0	1	Send/receive 1
1	1	0	Undefined
1	1	1	Send/receive 0

TABLE 11–10   The divisor used with the Baud rate generator for an 18.432 MHz crystal illustrating common Baud rates.

Baud rate	Divisor value
110	10,473
300	3840
1200	920
2400	480
4800	240
9600	120
19,200	60
38,400	30
57,600	20
115,200	10

break is transmitted. As long as this bit is a 1, the break is sent from the SOUT pin. A break, by definition, is at least two frames of logic 0 data. The software in the system is responsible for timing the transmission of the break. To end the break, bit position 6 or the line control register is returned to a logic 0 level. The Baud rate divisor is only programmable when bit position 7 of the line control register is a logic 1.

Programming the Baud Rate.   The Baud rate generator is programmed at I/O addresses 000 and 001 (A2, A1, A0). Port 000 is used to hold the least-significant part of the 16-bit divisor, and port 001 is used to hold the most-significant part. The value used for the divisor depends on the external clock or crystal frequency. Table 11–10 illustrates common Baud rates obtainable if a 18.432 MHz crystal is used as a timing source. It also shows the divisor values programmed into the Baud rate generator to obtain these Baud rates. The actual number programmed into the Baud rate generator causes it to produce a clock that is 16 times the desired Baud rate. For example, if 240 is programmed into the Baud rate divisor, the Baud rate is $^{18.432\ MHz}/_{16 \times 240} = 4800$ Baud.

Sample Initialization.   Suppose that an asynchronous system requires seven data bits, odd parity, a Baud rate of 9600, and one stop bit. Example 11–28 lists a procedure that initializes the 16550 to function in this manner. Figure 11–50 shows the interface to the 8088 microprocessor, using a PAL16L8 to decode the 8-bit port addresses F0H and F7H. (The PAL program is not shown.) Here, port F3H accesses the line control register and F0H and F1H access the Baud rate divisor registers. The last part of Example 11–28 is described with the function of the FIFO control register in the next few paragraphs.

FIGURE 11–50 The 16550 interfaced to the 8088 micro-processor at ports 00F0H–00F7H.

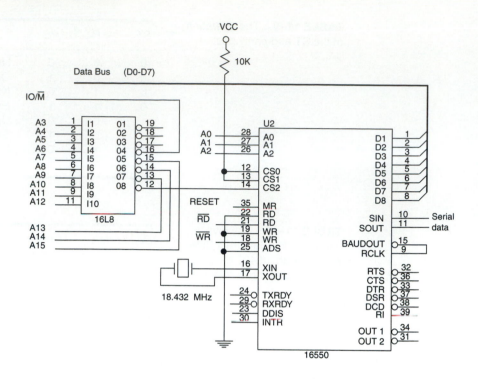

EXAMPLE 11–28

```
                            ;Initialization Dialog for Figure 11-50.
                            ;Baud rate 9600, 7 data, odd parity, one stop
                            ;
= 00F3                      LINE   EQU   0F3H
= 00F0                      LSB    EQU   0F0H
= 00F1                      MSB    EQU   0F1H
= 00F2                      FIFO   EQU   0F2H

0000                        START PROC   NEAR

0000  B0 8A                        MOV   AL,10001010B   ;enable Baud divisor
0002  E6 F3                        OUT   LINE,AL

0004  B0 78                        MOV   AL,120         ;program Baud rate
0006  E6 F0                        OUT   LSB,AL
0008  B0 00                        MOV   AL,0
000A  E6 F1                        OUT   MSB,AL

000C  B0 0A                        MOV   AL,00001010B   ;program 7-data, odd
000E  E6 F3                        OUT   LINE,AL        ;parity, one stop

0010  B0 07                        MOV   AL,00000111B   ;enable transmitter and
0012  E6 F2                        OUT   FIFO,AL        ;receiver

0014  C3                           RET

0015                        START ENDP
```

After the line control register and Baud rate divisor are programmed into the 16550, it is still not ready to function. After programming the line control register and Baud rate, we still must program the FIFO control register, which is at port F2H in the circuit of Figure 11–50.

FIGURE 11–51   The FIFO control register of the 16550 UART.

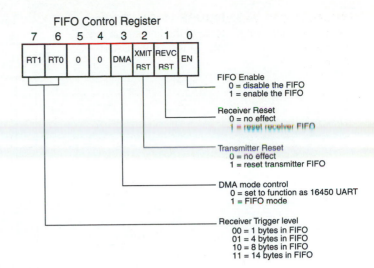

Figure 11–51 illustrates the FIFO control register for the 16550. This register enables the transmitter and receiver (bit 0 = 1), and clears the transmitter and receiver FIFOs. It also provides control for the 16550 interrupts, which are discussed in Chapter 12. Notice that the last section of Example 11–28 places a 7 into the FIFO control register. This enables the transmitter and receiver, and clears both FIFOs. The 16550 is now ready to operate, but without interrupts. Interrupts are automatically disabled when the MR (master reset) input is placed at a logic 1 by the system RESET signal.

Sending Serial Data.   Before serial data can be sent or received through the 16550, we need to know the function of the line status register (see Figure 11–52). The line status register contains information about error conditions and the state of the transmitter and receiver. This register is tested before a byte is transmitted or can be received.

Suppose that a procedure (see Example 11–29) is written to transmit the contents of AH to the 16550 and out through its serial data pin (SOUT). The TH bit is polled by software to determine whether the transmitter is ready to receive data. This procedure uses the circuit of Figure 11–50.

EXAMPLE 11–29

```
                              ;A Procedure that transmits AH via the 16550 UART
                              ;
= 00F5              LSTAT EQU  0F5H          ;line status port
= 00F0              DATA  EQU  0F0H          ;data port

0000                SEND  PROC NEAR

0000  50                  PUSH  AX           ;save AX
0001  E4 F5               IN    AL,LSTAT     ;get line status register
0003  A8 20               TEST  AL,20H       ;test TH bit
0005  74 FA               JZ    SEND         ;if transmitter not ready

0007  8A C4               MOV   AL,AH        ;get data
0009  E6 F0               OUT   DATA,AL      ;transmit data
000B  58                  POP   AX           ;restore AX
000C  C3                  RET

000D                SEND  ENDP
```

FIGURE 11–52 The contents of the line status register of the 16550 UART.

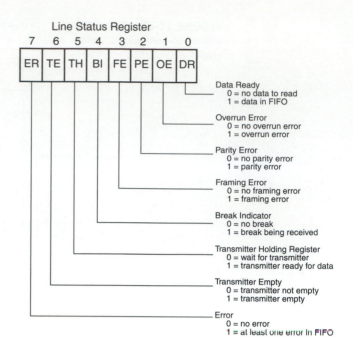

Receiving Serial Data. To read received information from the 16550, we test the DR bit of the line status register. Example 11–30 lists a procedure that tests the DR bit to decide whether the 16550 has received any data. Upon the reception of data, the procedure tests for errors. If an error is detected, the procedure returns with AL equal to an ASCII '?'. If no error has occurred, then the procedure returns with AL equal to the received character.

EXAMPLE 11–30

```
                            ;A procedure that receives data from the 16550 UART
                            ;and returns it in AL.
                            ;
= 00F5              LSTAT   EQU     0F5H            ;line status port
= 00F0              DATA    EQU     0F0H            ;data port

0000                RECV    PROC    NEAR

0000  E4 F5                 IN      AL,LSTAT        ;get line status register
0002  A8 01                 TEST    AL,1            ;test DR bit
0004  74 FA                 JZ      RECV            ;if no data in receiver

0006  A8 0E                 TEST    AL,0EH          ;test all 3 error bits
0008  75 03                 JNZ     ERR             ;for an error

000A  E4 F0                 IN      AL,DATA         ;read data from 16550
000C  C3                    RET
000D                ERR:
000D  B0 3F                 MOV     AL,'?'          ;get question mark
000F  C3                    RET

0010                REVC    ENDP
```

UART Errors. The types of errors detected by the 16550 are parity error, framing error, and overrun error. A **parity error** indicates that the received data contain the wrong parity. A **framing error** indicates that the start and stop bits are not in their proper places. An **overrun error** indicates that data have overrun the internal receiver FIFO buffer. These errors should not occur during normal operation. If a parity error occurs, it indicates that noise was encountered during reception. A framing error occurs if the receiver is receiving data at an incorrect Baud rate. An overrun error occurs only if the software fails to read the data from the UART before the receiver FIFO is full. This example does not test the BI (break indicator bit) for a break condition. Note that a break is two consecutive frames of logic 0s on the SIN pin of the UART. The remaining registers, which are used for interrupt control and modem control, are developed in Chapter 12.

11–7 ANALOG-TO-DIGITAL (ADC) AND DIGITAL-TO-ANALOG (DAC) CONVERTERS

Analog-to digital (ADC) and digital-to-analog (DAC) converters are used to interface the microprocessor to the analog world. Many events that are monitored and controlled by the microprocessor are analog events. These often include monitoring all forms of events, even speech, to controlling motors and like devices. In order to interface the microprocessor to these events, we must have an understanding of the interface and control of the ADC and DAC, which convert between analog and digital data.

The DAC0830 Digital-to-Analog Converter

A fairly common and low-cost digital-to-analog converter is the DAC0830 (a product of National Semiconductor Corporation). This device is an 8-bit converter that transforms an 8-bit binary number into an analog voltage. Other converters are available that convert from 10-, 12-, or 16-bit binary numbers into analog voltages. The number of voltage steps generated by the converter is equal to the number of binary input combinations. Therefore, an 8-bit converter generates 256 different voltage levels, a 10-bit converter generates 1024 levels, and so forth. The DAC0830 is a medium speed converter that transforms a digital input to an analog output in approximately 1.0 μs.

Figure 11–53 illustrates the pin-out of the DAC0830. This device has a set of eight data bus connections for the application of the digital input code, and a pair of analog outputs labeled Iout1 and Iout2 that are designed as inputs to an external operational amplifier. Because this is an 8-bit converter, its output step voltage is defined as $-V_{REF}$ (reference voltage), divided by 255. For example, if the reference voltage is –5.0 V, its output step voltage is +.0196 V. Note that the output voltage is the opposite polarity of the reference voltage. If an input of $1001\ 0010_2$ is applied to the device, the output voltage will be the step voltage times $1001\ 0010_2$, or, in this case +2.862 V. By changing the reference voltage to –5.1 V, the step voltage becomes +.02 V. The step voltage is also often called the **resolution** of the converter.

FIGURE 11–53   The pin-out of the DAC0830 digital-to-analog converter.

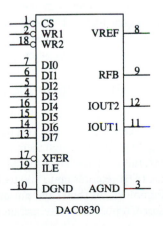

DAC0830

FIGURE 11–54   The internal structure of the DAC0830.

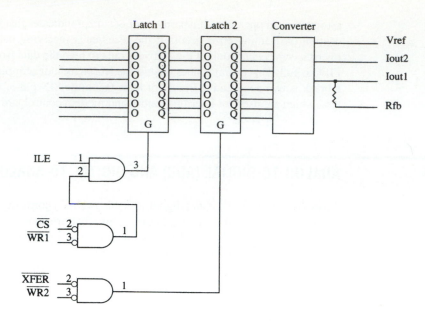

Internal Structure of the DAC0830.   Figure 11–54 illustrates the internal structure of the DAC0830. Notice that this device contains two internal registers. The first is a holding register, while the second connects to the R–2R internal ladder converter. The two latches allow one byte to be held while another is converted. In many cases, we disable the first latch and only use the second for entering data into the converter. This is accomplished by connecting a logic 1 to ILE and a logic 0 to $\overline{CS}$ (**chip select**).

Both latches within the DAC0830 are transparent latches. That is, when the G input to the latch is a logic 1, data pass through the latch, but when the G input becomes a logic 0, data are latched or held. The converter has a reference input pin (V_{REF}) that establishes the full scale output voltage. If –10 V is placed on V_{REF}, the full scale (11111111_2) output voltage is +10 V. The output of the R–2R ladder within the converter appears at Iout1 and Iout2. These outputs are designed to be applied to an operational amplifier such as a 741 or similar device.

Connecting the DAC0830 to the Microprocessor.   The DAC0830 is connected to the microprocessor, as illustrated in Figure 11–55. Here, a PAL16L8 is used to decode the DAC0830 at 8-bit I/O port address 20H. Whenever an OUT 20H,AL instruction is executed, the contents of data bus connections AD0–AD7 are passed to the converter within the DAC0830. The 741 operational amplifier, along with the –12 V zener reference voltage, causes the full scale output voltage to equal +12 V. The output of the operational amplifier feeds a driver that powers a 12 V DC motor. This driver is a Darlington amplifier for large motors. This example shows the converter driving a motor, but other devices could be used as outputs.

The ADC080X Analog-To-Digital Converter

A common, low-cost ADC is the ADC0804, which belongs to a family of converters that are all identical, except for accuracy. This device is compatible with a wide range of microprocessors such as the Intel family. Although there are faster ADCs available and some have more resolution than eight bits, this device is ideal for many applications that do not require a high degree of accuracy. The ADC0804 requires up to 100 μs to convert an analog input voltage into a digital output code.

Figure 11–56 shows the pin-out of the ADC0804 converter (a product of National Semiconductor Corporation). To operate the converter, the $\overline{WR}$ pin is pulsed with $\overline{CS}$ grounded to start the conversion process. Because this converter requires a considerable amount of time for the conversion, a pin labeled INTR signals the end of the conversion. Refer to Figure 11–57 for a timing diagram that shows the interaction of the control signals. As can be seen, we start the converter with the $\overline{WR}$ pulse, we wait for INTR to return to a logic 0 level, and then we read the

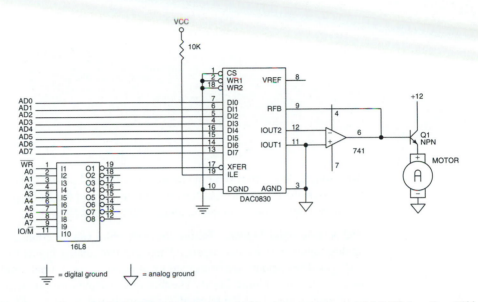

FIGURE 11–55   A DAC0830 interfaced to the 8086 microprocessor at 8-bit I/O location 20H.

FIGURE 11–56 The pin-out of the ADC0804 analog-to-digital converter.

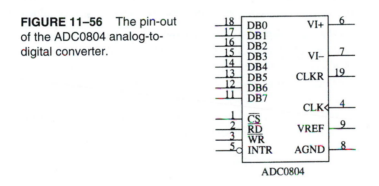

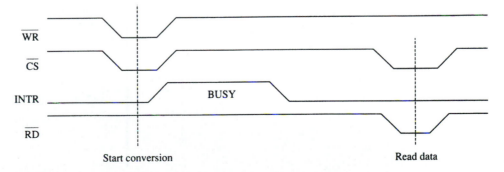

FIGURE 11–57   The timing for the ADC0804 analog-to-digital converter.

FIGURE 11–58   The analog inputs to the ADC0804 converter. (a) To sense a 0- to +5.0-V input. (b) To sense an input offset from ground.

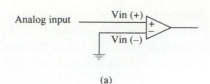

(a)

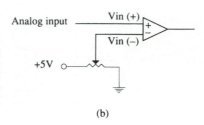

(b)

data from the converter. If a time delay is used that allows at least 100 μs of time, then we don't need to test the INTR pin. Another option is to connect the INTR pin to an interrupt input, so that when the conversion is complete, an interrupt occurs.

The Analog Input Signal.   Before the ADC0804 can be connected to the microprocessor, its analog inputs must be understood. There are two analog inputs to the ADC0804: VIN (+) and VIN (–). These inputs are connected to an internal operational amplifier and are differential inputs, as shown in Figure 11–58. The differential inputs are summed by the operational amplifier to produce a signal for the internal analog-to-digital converter. Figure 11–58 shows a few ways to use these differential inputs. The first way (see Figure 11–58a) uses a single input that can vary between 0 V and +5.0 V. The second way (see Figure 11–58b) shows a variable voltage applied to the VIN (–) pin, so the zero reference for VIN (+) can be adjusted.

Generating the Clock Signal.   The ADC0804 requires a clock source for operation. The clock can be an external clock applied to the CLK IN pin or it can be generated with an RC circuit. The permissible range of clock frequencies is between 100 KHz and 1460 KHz. It is desirable to use a frequency that is as close as possible to 1460 KHz, so conversion time is kept to a minimum.

If the clock is generated with an RC circuit, we use the CLK IN and CLK R pins connected to an RC circuit, as illustrated in Figure 11–59. When this connection is in use, the clock frequency is calculated by the following equation:

$$Fclk = \frac{1}{1.1RC}$$

FIGURE 11–59   Connecting the RC circuit to the CLK IN and CLK R pins on the ADC0804.

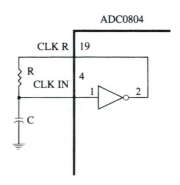

FIGURE 11–60 The ADC0804 interfaced to the microprocessor.

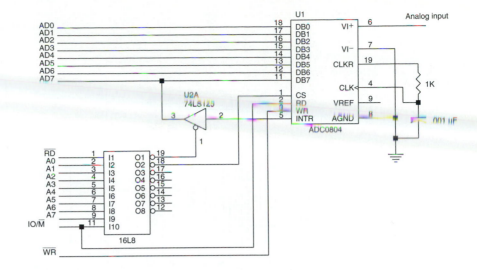

Connecting the ADC0804 to the Microprocessor.   The ADC0804 is interfaced to the 8086 microprocessor, as illustrated in Figure 11–60. Note that the V_{REF} signal is not attached to anything, which is normal. Suppose that the ADC0804 is decoded at 8-bit I/O port address 40H for the data and port address 42H for the INTR signal, and a procedure is required to start and read the data from the ADC. This procedure is listed in Example 11–31. Notice that the INTR bit is polled and if it becomes a logic 0, the procedure ends with AL, containing the converted digital code.

EXAMPLE 11–31

```
                              ;A procedure that reads data from the ADC and returns
                              ;it in AL.
                              ;
0000                          ADCX   PROC   NEAR

0000   E6 40                         OUT    40H,AL       ;start conversion
0002                          ADCX1:
0002   E4 42                         IN     AL,42H        ;read INTR
0004   A8 80                         TEST   AL,80H        ;test INTR
0006   75 FA                         JNZ    ADCX1         ;repeat until INTR = 0
0008   E4 40                         IN     AL,40H        ;get ADC data
000A   C3                            RET

000B                          ADCX   ENDP
```

Using the ADC0804 and the DAC0830

This section of the text illustrates an example that uses both the ADC0804 and the DAC0830 to capture and replay audio signals or speech. In the past, we often used a speech synthesizer to generate speech, but the quality of the speech was poor. For human quality speech, we can use the ADC0804 to capture an audio signal and store it in memory for later playback through the DAC0830.

Figure 11–61 illustrates the circuitry required to connect the ADC0804 at I/O ports 0700H and 0702H. The DAC0830 is interfaced at I/O port 704H. These I/O ports are in the low bank of a 16-bit microprocessor such as the 8086 or 80386SX. The software used to run these converters appears in Example 11–32. This software reads a 1-second burst of speech and then plays it back 10 times. This process repeats until the system is turned off. In this example, speech is sampled and stored in a section of memory called WORDS. The sample rate is chosen at 2048 samples per second, which renders acceptable-sounding speech.

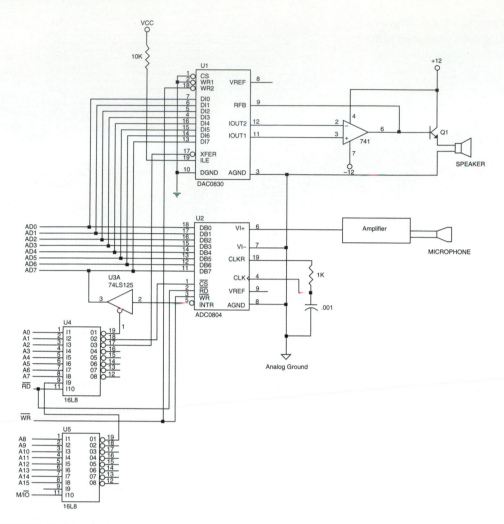

FIGURE 11–61   A circuit that stores speech and plays it back through the speaker.

EXAMPLE 11–32

```
                    ;Software that records a 1-second passage of speech
                    ;and plays it back 10 times before recording the
                    ;next 1-second passage of speech.
                    ;
                    ;Assumes a clock of 8 MHz (8086) for the time delay.
                    ;
                    .MODEL SMALL
                    .DATA
0000 0500 [         WORDS DB 2048 DUP (?)      ;space for speech
     0000
     ]
                    .CODE
                    .STARTUP
0018            AGAIN:
0018 E8 000A        CALL    READ               ;read speech
001B B9 000A        MOV     CX,10              ;set count to 10
001E            LOOP1:
001E E8 0023        CALL    WRITE              ;playback speech
0021 E2 FB          LOOP    LOOP1              ;repeat 10 times
```

```
0023    EB EA              JMP     AGAIN           ;repeat forever
0025               READ    PROC    NEAR
0025    BF 0000 R          MOV     DI,OFFSET WORDS ;address data area
0028    B9 0500            MOV     CX,2048         ;load count
002B    BA 0700            MOV     DX,0700H        ;address port
002E               READ1:
002E    EE                 OUT     DX,AL           ;start converter
002F    83 C2 C0           ADD     DX,2            ;address status port
0032               READ2:
0032    EC                 IN      AL,DX           ;get INTR
0033    A8 80              TEST    AL,80H          ;test INTR
0035    75 FB              JNZ     READ2           ;wait for INTR = 0
0037    83 EA 02           SUB     DX,2            ;address data port
003A    EC                 IN      AL,DX           ;get data from ADC
003B    88 05              MOV     [DI],AL         ;store data in array
003D    47                 INC     DI              ;address next element
003E    E8 0018            CALL    DELAY           ;wait for 1/2048 seconds
0041    E2 EB              LOOP    READ1           ;repeat 2,048 times
0043    C3                 RET

0044               READ    ENDP

0044               WRITE   PROC    NEAR

0044    51                 PUSH    CX
0045    BF 0000 R          MOV     DI,OFFSET WORDS ;address data
0048    B9 0500            MOV     CX,2048         ;load count
004B    BA 0704            MOV     DX,0704H        ;address DAC
004E               WRITE1:
004E    8A 05              MOV     AL,[DI]         ;get data from array
0050    EE                 OUT     DX,AL           ;send data to DAC
0051    47                 INC     DI              ;address next element
0052    E8 0004            CALL    DELAY           ;wait 1/2048 second
0055    E2 F7              LOOP    WRITE1          ;repeat 2,048 times
0057    59                 POP     CX
0058    C3                 RET

0059               WRITE   ENDP

0059               DELAY   PROC    NEAR

0059    51                 PUSH    CX
005A    B9 00E1            MOV     CX,225          ;approximately 1/2048 sec.
005D               DELAY1:
005D    E2 FE              LOOP    DELAY1
005F    59                 POP     CX
0060    C3                 RET

0061               DELAY   ENDP
                   .END
```

11–8   SUMMARY

1. The 8086–Pentium 4 microprocessors have two basic types of I/O instructions: IN and OUT. The IN instruction inputs data from an external I/O device into either the AL (8-bit) or AX (16-bit) register. The IN instruction is available as a fixed port instruction, a variable port instruction, or a string instruction (80286–Pentium 4) INSB or INSW. The OUT instruction outputs data from AL or AX to an external I/O device and is available as a fixed, variable, or string instruction OUTSB or OUTSW. The fixed port instruction uses an 8-bit I/O port address, while the variable and string I/O instructions use a 16-bit port number found in the DX register.

2. Isolated I/O, sometimes called direct I/O, uses a separate map for the I/O space, freeing the entire memory for use by the program. Isolated I/O uses the IN and OUT instructions to transfer data between the I/O device and the microprocessor. The control structure of the I/O map uses $\overline{\text{IORC}}$ (I/O read control) and $\overline{\text{IOWC}}$ (I/O write control), plus the bank selection signals $\overline{\text{BHE}}$ and $\overline{\text{BLE}}$ (A0 on the 8086 and 80286), to affect the I/O transfer. The early 8086/8088 uses the M/$\overline{\text{IO}}$ (IO/$\overline{\text{M}}$) signal with $\overline{\text{RD}}$ and $\overline{\text{WR}}$ to generate the I/O control signals.

3. Memory-mapped I/O uses a portion of the memory space for I/O transfers. This reduces the amount of memory available, but it negates the need to use the $\overline{\text{IORC}}$ and $\overline{\text{IOWC}}$ signals for I/O transfers. In addition, any instruction that addresses a memory location using any addressing mode can be used to transfer data between the microprocessor and the I/O device using memory-mapped I/O.

4. All input devices are buffered so that the I/O data are connected only to the data bus during the execution of the IN instruction. The buffer is either built into a programmable peripheral or located separately.

5. All output devices use a latch to capture output data during the execution of the OUT instruction. This is necessary because data appear on the data bus for less than 100 ns for an OUT instruction, and most output devices require the data for a longer time. In many cases, the latch is built into the peripheral.

6. Handshaking or polling is the act of two independent devices synchronizing with a few control lines. For example, the computer asks a printer if it is busy by inputting the BUSY signal from the printer. If it isn't busy, the computer outputs data to the printer and informs the printer that data are available with a data strobe ($\overline{\text{DS}}$) signal. This communication between the computer and the printer is a handshake or a poll.

7. Interfaces are required for most switch-based input devices and for most output devices that are not TTL-compatible.

8. The I/O port number appears on address bus connections A7–A0 for a fixed port I/O instruction and on A15–A0 for a variable port I/O instruction (note that A15–A8 contain zeros for an 8-bit port). In both cases, address bits above A15 are undefined.

9. Because the 8086/80286/80386SX microprocessors contain a 16-bit data bus and the I/O addresses reference byte-sized I/O locations, the I/O space is also organized in banks, as is the memory system. In order to interface an 8-bit I/O device to the 16-bit data bus, we often require separate write strobes (an upper and a lower) for I/O write operations. Likewise, the 80486 and Pentium–Pentium 4 also have I/O arranged in banks.

10. The I/O port decoder is much like the memory address decoder, except instead of decoding the entire address, the I/O port decoder decodes only a 16-bit address for variable port instructions and often an 8-bit port number for fixed I/O instructions.

11. The 82C55 is a programmable peripheral interface (PIA) that has 24 I/O pins that are programmable in two groups of 12 pins each (group A and group B). The 82C55 operates in three modes: simple I/O (mode 0), strobed I/O (mode 1), and bi-directional I/O (mode 2). When the 82C55 is interfaced to the 8086 operating at 8 MHz, we insert two wait states because the speed of the microprocessor is faster than the 82C55 can handle.

12. The 8279 is a programmable keyboard/display controller that can control a 64-key keyboard and a 16-digit numeric display.

13. The LCD display device requires a fair amount of software, but it displays ASCII-coded information.

14. The 8254 is a programmable interval timer that contains three 16-bit counters that count in binary or binary-coded decimal (BCD). Each counter is independent of each other, and operates in six different modes. The six modes of the counter are (1) events counter, (2) retriggerable monostable multivibrator, (3) pulse generator, (4) square-wave generator, (5) software-triggered pulse generator, and (6) hardware triggered pulse generator.

15. The 16550 is a programmable communications interface, capable of receiving and transmitting asynchronous serial data.
16. The DAC0830 is an 8-bit digital-to-analog converter that converts a digital signal to an analog voltage within 1.0 μs.
17. The ADC0804 is an 8-bit analog-to-digital converter that converts an analog signal into a digital signal within 100 μs.

11–9 QUESTIONS AND PROBLEMS

1. Explain which way the data flow for an IN and an OUT instruction.
2. Where is the I/O port number stored for a fixed I/O instruction?
3. Where is the I/O port number stored for a variable I/O instruction?
4. Where is the I/O port number stored for a string I/O instruction?
5. To which register are data input by the 16-bit IN instruction?
6. Describe the operation of the OUTSB instruction.
7. Describe the operation of the INSW instruction.
8. Contrast a memory-mapped I/O system with an isolated I/O system.
9. What is the basic input interface?
10. What is the basic output interface?
11. Explain the term *handshaking* as it applies to computer I/O systems.
12. An even-number I/O port address is found in the _____ I/O bank in the 8086 microprocessor.
13. Show the circuitry that generates the upper and lower I/O write strobes.
14. What is the purpose of a contact bounce eliminator?
15. Develop an interface to correctly drive a relay. The relay is 12 V and requires a coil current of 150 mA.
16. Develop an I/O port decoder, using a 74ALS138, which generates low-bank I/O strobes for the following 8-bit I/O port addresses: 10H, 12H, 14H, 16H, 18H, 1AH, 1CH, and 1EH.
17. Develop an I/O port decoder, using a 74ALS138, which generates high-bank I/O strobes for the following 8-bit I/O port addresses: 11H, 13H, 15H, 17H, 19H, 1BH, 1DH, and 1FH.
18. Develop an I/O port decoder, using a PAL16L8, which generates 16-bit I/O strobes for the following 16-bit I/O port addresses: 1000H–1001H, 1002H–1003H, 1004H–1005H, 1006H–1007H, 1008H–1009H, 100AH–100BH, 100CH–100DH, and 100EH–100FH.
19. Develop an I/O port decoder, using the PAL16L8, which generates the following low-bank I/O strobes: 00A8H, 00B6H, and 00EEH.
20. Develop an I/O port decoder, using the PAL16L8, which generates the following high-bank I/O strobes: 300DH, 300BH, 1005H, and 1007H.
21. Why are both $\overline{BHE}$ and $\overline{BLE}$ (A0) ignored in a 16-bit port address decoder?
22. An 8-bit I/O device, located at I/O port address 0010H, is connected to which data bus connections?
23. An 8-bit I/O device, located at I/O port address 100DH, is connected to which data bus connections?
24. The 82C55 has how many programmable I/O pin connections?
25. List the pins that belong to group A and to group B in the 82C55.
26. Which two 82C55 pins accomplish internal I/O port address selection?
27. The $\overline{RD}$ connection on the 82C55 is attached to which 8086 system control bus connection?
28. Using a PAL16L8, interface an 82C55 to the 8086 microprocessor so that it functions at I/O locations 0380H, 0382H, 0384H, and 0386H.

29. When the 82C55 is reset, its I/O ports are all initialized as _____.
30. What three modes of operation are available to the 82C55?
31. What is the purpose of the $\overline{\text{STB}}$ signal in strobed input operation of the 82C55?
32. Explain the operation of a simple four-coil stepper motor.
33. What sets the IBF pin in strobed input operation of the 82C55?
34. Write the software required to place a logic 1 on the PC7 pin of the 82C55 during strobed input operation.
35. How is the interrupt request pin (INTR) enabled in the strobed input mode of operation of the 82C55?
36. In strobed output operation of the 82C55, what is the purpose of the $\overline{\text{ACK}}$ signal?
37. What clears the $\overline{\text{OBF}}$ signal in strobed output operation of the 82C55?
38. Write the software required to decide whether PC4 is a logic 1 when the 82C55 is operated in the strobed output mode.
39. Which group of pins are used during bi-directional operation of the 82C55?
40. Which pins are general-purpose I/O pins during mode 2 operation of the 82C55?
41. Describe how the display is cleared by using the LCD display.
42. How is a display position selected in the LCD display?
43. Write a short procedure that places an ASCII-Z in display position 6 on the LCD display.
44. How is the busy flag tested in the LCD display?
45. What changes must be made to Figure 11–24 so that it functions with a keyboard matrix that contains three rows and five columns?
46. What time is usually used to de-bounce a keyboard?
47. What is normally connected to the CLK pin of the 8279?
48. How many wait states are required to interface the 8279 to the 8086 microprocessor operating with an 8 MHz clock?
49. If the 8279 CLK pin is connected to a 3.0 MHz clock, program the internal clock.
50. What is an overrun error in the 8279?
51. What is the difference between encoded and decoded, as defined for the 8279?
52. Interface the 8279 so that it functions at 8-bit I/O ports 40H–7FH. Use the 74ALS138 as a decoder, and use either the upper or lower data bus.
53. Interface a 16-key keyboard and an 8-digit numeric display to the 8279.
54. The 8254 interval timer functions from DC to _____ Hz.
55. Each counter in the 8254 functions in how many different modes?
56. Interface an 8254 to function at I/O port addresses XX10H, XX12H, XX14H, and XX16H. Write the software that programs counter 2 to generate an 80 KHz square-wave if the CLK input to counter 2 is 8 MHz.
57. What number is programmed in an 8254 counter to count 300 events?
58. If a 16-bit count is programmed into the 8254, which byte of the count is programmed first?
59. Explain how the read-back control word functions in the 8254.
60. Program counter 1 of the 8254 so that it generates a continuous series of pulses that have a high time of 100 μs and a low time of 1 μs. Make sure to indicate the CLK frequency required for this task.
61. Why does a 50 percent duty cycle cause the motor to stand still in the motor speed and direction control circuit presented in this chapter?
62. What is asynchronous serial data?
63. What is Baud rate?
64. Program the 16550 for operation using six data bits, even parity, one stop bit, and a Baud rate of 19,200 using a 18.432 MHz clock. (Assume that the I/O ports are numbered 20H and 22H.)
65. If the 16550 is to generate a serial signal at a Baud rate of 2400 Baud and the Baud rate divisor is programmed for 16, what is the frequency of the signal?

66. Describe the following terms: *simplex, half-duplex,* and *full-duplex.*
67. How is the 16550 reset?
68. Write a procedure for the 16550 that transmits 16 bytes from a small buffer in the data segment address (DS is loaded externally) by SI (SI is loaded externally).
69. The DAC0830 converts an 8-bit digital input to an analog output in approximately

    _____.

70. What is the step voltage at the output of the DAC0830 if the reference voltage is –2.55 V?
71. Interface a DAC0830 to the 8086 so that it operates at I/O port 400H.
72. Develop a program for the interface of Question number 71 so the DAC0830 generates a triangular voltage wave-form. The frequency of this wave-form must be approximately 100 Hz.
73. The ADC080X requires approximately _____ to convert an analog voltage into a digital code.
74. What is the purpose of the INTR pin on the ADC080X?
75. The $\overline{\text{WR}}$ pin on the ADC080X is used for what purpose?
76. Interface an ADC080X at I/O port 0260H for data and 0270H to test the INTR pin.
77. Develop a program for the ADC080X in Question 76 so that it reads an input voltage once per 100 ms and stores the results in a memory array that is 100H bytes long.

CHAPTER 12

Interrupts

INTRODUCTION

In this chapter, we expand our coverage of basic I/O and programmable peripheral interfaces by examining a technique called interrupt-processed I/O. An **interrupt** is a hardware-initiated procedure that interrupts whatever program is currently executing.

This chapter provides examples and a detailed explanation of the interrupt structure of the entire Intel family of microprocessors.

CHAPTER OBJECTIVES

Upon completion of this chapter, you will be able to:

1. Explain the interrupt structure of the Intel family of microprocessors.
2. Explain the operation of software interrupt instructions INT, INTO, INT 3, and BOUND.
3. Explain how the interrupt enable flag bit (IF) modifies the interrupt structure.
4. Describe the function of the trap interrupt flag-bit (TF) and the operation of trap-generated tracing.
5. Develop interrupt-service procedures that control lower-speed, external peripheral devices.
6. Expand the interrupt structure of the microprocessor by using the 8259A programmable interrupt controller and other techniques.
7. Explain the purpose and operation of a real-time clock.

12–1   BASIC INTERRUPT PROCESSING

In this section, we discuss the function of an interrupt in a microprocessor-based system, and the structure and features of interrupts available to the Intel family of microprocessors.

The Purpose of Interrupts

Interrupts are particularly useful when interfacing I/O devices that provide or require data at relatively low data-transfer rates. In Chapter 11, for instance, we showed a keyboard example

FIGURE 12–1 A time line that indicates interrupt usage in a typical system.

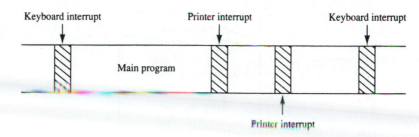

using strobed input operation of the 82C55. In that example, software polled the 82C55 and its IBF bit to decide whether data were available from the keyboard. If the person using the keyboard typed one character per second, the software for the 82C55 waited an entire second between each keystroke for the person to type another key. This process was such a tremendous waste of time that designers developed another process, *interrupt processing,* to handle this situation.

Unlike the polling technique, interrupt processing allows the microprocessor to execute other software while the keyboard operator is thinking about what key to type next. As soon as a key is pressed, the keyboard encoder de-bounces the switch and puts out one pulse that interrupts the microprocessor. In this way, the microprocessor executes other software until the key is actually pressed when it reads a key and returns to the program that was interrupted. As a result, the microprocessor can print reports or complete any other task while the operator is typing a document and thinking about what to type next.

Figure 12–1 shows a time line that indicates a typist typing data on a keyboard, a printer removing data from the memory, and a program executing. The program is the main program that is interrupted for each keystroke and each character that is to print on the printer. Note that the keyboard interrupt service procedure, called by the keyboard interrupt, and the printer interrupt service procedure each take little time to execute.

Interrupts

The interrupts of the entire Intel family of microprocessors include two hardware pins that request interrupts (INTR and NMI), and one hardware pin ($\overline{\text{INTA}}$) that acknowledges the interrupt requested through INTR. In addition to the pins, the microprocessor also has software interrupts INT, INTO, INT 3, and BOUND. Two flag bits, IF (interrupt flag) and TF (trap flag), are also used with the interrupt structure and a special return instruction IRET (or IRETD in the 80386, 80486, or Pentium–Pentium 4).

Interrupt Vectors. The interrupt vectors and vector table are crucial to an understanding of hardware and software interrupts. The **interrupt vector table** is located in the first 1024 bytes of memory at addresses 000000H–0003FFH. It contains 256 different 4-byte interrupt vectors. An **interrupt vector** contains the address (segment and offset) of the interrupt service procedure.

Figure 12–2 illustrates the interrupt vector table for the microprocessor. The first five interrupt vectors are identical in all Intel microprocessor family members, from the 8086 to the Pentium. Other interrupt vectors exist for the 80286 that are upward-compatible to the 80386, 80486, and Pentium–Pentium 4, but not downward-compatible to the 8086 or 8088. Intel reserves the first 32 interrupt vectors for their use in various microprocessor family members. The last 224 vectors are available as user interrupt vectors. Each vector is four bytes long and contains the **starting address** of the interrupt service procedure. The first two bytes of the vector contain the offset address, and the last two bytes contain the segment address.

FIGURE 12–2   (a) The interrupt vector table for the microprocessor and (b) the contents of an interrupt vector.

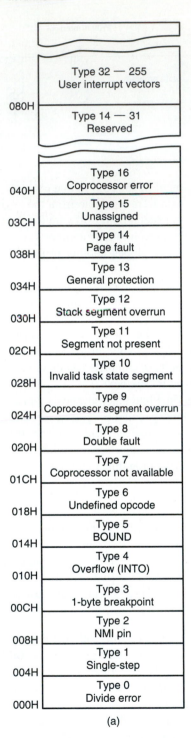

(a)

Any interrupt vector

3	Segment (high)
2	Segment (low)
1	Offset (high)
0	Offset (low)

(b)

The following list describes the function of each dedicated interrupt in the microprocessor:

Type 0   Divide Error—Occurs whenever the result of a division overflows or whenever an attempt is made to divide by zero.

Type 1   Single-step or Trap—Occurs after the execution of each instruction if the trap (TF) flag bit is set. Upon accepting this interrupt, the TF bit is cleared so that the interrupt service procedure executes at full speed. (More detail is provided about this interrupt later in this section of the chapter.)

Type 2   Non-maskable Hardware Interrupt—A result of placing a logic 1 on the NMI input pin to the microprocessor. This input is non-maskable, which means that it cannot be disabled.

Type 3   One-Byte Interrupt—A special one-byte instruction (INT 3) that uses this vector to access its interrupt-service procedure. The INT 3 instruction is often used to store a breakpoint in a program for debugging.

Type 4   Overflow—A special vector used with the INTO instruction. The INTO instruction interrupts the program if an overflow condition exists, as reflected by the overflow flag (OF).

Type 5   BOUND—An instruction that compares a register with boundaries stored in the memory. If the contents of the register are greater than or equal to the first word in memory and less than or equal to the second word, no interrupt occurs because the contents of the register is within bounds. If the contents of the register are out-of-bounds, a type 5 interrupt ensues.

Type 6   Invalid Opcode—Occurs whenever an undefined opcode is encountered in a program.

Type 7   Coprocessor Not Available—Occurs when a coprocessor is not found in the system, as dictated by the machine status word (MSW) coprocessor control bits. If an ESC or WAIT instruction executes and the coprocessor is not found, a type 7 exception or interrupt occurs.

Type 8   Double Fault—Activated whenever two separate interrupts occur during the same instruction.

Type 9   Coprocessor Segment Overrun—Occurs if the ESC instruction (coprocessor opcode) memory operand extends beyond offset address FFFFH.

Type 10   Invalid Task State Segment—Occurs if the TSS is invalid because the segment limit field is not 002BH or higher. In most cases, this is caused because the TSS is not initialized.

Type 11   Segment not Present—Occurs when the P bit (P = 0) in a descriptor indicates that the segment is not present or not valid.

Type 12   Stack Segment Overrun—Occurs if the stack segment is not present (P = 0) or if the limit of the stack segment is exceeded.

Type 13   General Protection—Occurs for most protection violations in the 80286–Pentium 4 protected mode system. (These errors occur in Windows as **general protection faults.**) A list of these protection violations follows:
   a. Descriptor table limit exceeded
   b. Privilege rules violated
   c. Invalid descriptor segment type loaded
   d. Write to code segment that is protected
   e. Read from execute-only code segment
   f. Write to read-only data segment
   g. Segment limit exceeded
   h. CPL = IOPL when executing CTS, HLT, LGDT, LIDT, LLDT, LMSW, or LTR
   i. CPL > IOPL when executing CLI, IN, INS, LOCK, OUT, OUTS, and STI

Type 14   Page Fault—Occurs for any page fault memory or code access in the 80386, 80486, and Pentium–Pentium 4 microprocessors.

Type 16   Coprocessor Error—Takes effect whenever a coprocessor error ($\overline{\text{ERROR}} = 0$) occurs for the ESCape or WAIT instructions for the 80386, 80486, and Pentium–Pentium 4 microprocessors only.

Type 17   Alignment Check—Indicates that word and doubleword data are addressed at an odd memory location (or an incorrect location, in the case of a doubleword). This interrupt is active in the 80486 and Pentium–Pentium 4 microprocessors.

Type 18   Machine Check—Activates a system memory management mode interrupt in the Pentium–Pentium 4 microprocessors.

Interrupt Instructions: BOUND, INTO, INT, INT 3, and IRET

Of the five software interrupt instructions available to the microprocessor, INT and INT 3 are very similar, BOUND and INTO are conditional, and IRET is a special interrupt return instruction.

The BOUND instruction, which has two operands, compares a register with two words of memory data. For example, if the instruction BOUND AX,DATA is executed, AX is compared with the contents of DATA and DATA+1 and also with DATA+2 and DATA+3. If AX is less than the contents of DATA and DATA+1, a type 5 interrupt occurs. If AX is greater than DATA+2 and DATA+3, a type 5 interrupt occurs. If AX is within the bounds of these two memory words, no interrupt occurs.

The INTO instruction checks the overflow flag (OF). If OF = 1, the INTO instruction calls the procedure whose address is stored in interrupt vector type number 4. If OF = 0, then the INTO instruction performs no operation and the next sequential instruction in the program executes.

The INT n instruction calls the interrupt service procedure that begins at the address represented in vector number n. For example, an INT 80H or INT 128 calls the interrupt service procedure whose address is stored in vector type number 80H (000200H–00203H). To determine the vector address, just multiply the vector type number (n) by 4, which gives the beginning address of the 4-byte long interrupt vector. For example, an INT $5 = 4 \times 5$ or 20 (14H). The vector for INT 5 begins at address 000014H and continues to 000017H. Each INT instruction is stored in two bytes of memory: the first byte contains the opcode, and the second byte contains the interrupt type number. The only exception to this is the INT 3 instruction, a 1-byte instruction. The INT 3 instruction is often used as a breakpoint-interrupt because it is easy to insert a 1-byte instruction into a program. Breakpoints are often used to debug faulty software.

The IRET instruction is a special return instruction used to return for both software and hardware interrupts. The IRET instruction is much like a far RET, because it retrieves the return address from the stack. It is unlike the near return because it also retrieves a copy of the flag register from the stack. An IRET instruction removes six bytes from the stack: two for the IP, two for the CS, and two for the flags.

In the 80386–Pentium 4, there is also an IRETD instruction because these microprocessors can push the EFLAG register (32 bits) on the stack, as well as the 32-bit EIP in the protected mode. If operated in the real mode, we use the IRET instruction with the 80386–Pentium 4 microprocessors.

The Operation of a Real Mode Interrupt

When the microprocessor completes executing the current instruction, it determines whether an interrupt is active by checking (1) instruction executions, (2) single-step, (3) NMI, (4) coprocessor segment overrun, (5) INTR, and (6) INT instruction in the order presented. If one or more of these interrupt conditions are present, the following sequence of events occurs:

1. The contents of the flag register are pushed onto the stack.
2. Both the interrupt (IF) and trap (TF) flags are cleared. This disables the INTR pin and the trap or single-step feature.

3. The contents of the code segment register (CS) are pushed onto the stack.
4. The contents of the instruction pointer (IP) are pushed onto the stack.
5. The interrupt vector contents are fetched, and then placed into both IP and CS so that the next instruction executes at the interrupt service procedure addressed by the vector.

Whenever an interrupt is accepted, the microprocessor stacks the contents of the flag register, CS and IP; clears both IF and TF; and jumps to the procedure addressed by the interrupt vector. After the flags are pushed onto the stack, IF and TF are cleared. These flags are returned to the state prior to the interrupt when the IRET instruction is encountered at the end of the interrupt service procedure. Therefore, if interrupts were enabled prior to the interrupt service procedure, they are automatically re-enabled by the IRET instruction at the end of the procedure.

The return address (in CS and IP) is pushed onto the stack during the interrupt. Sometimes, the return address points to the next instruction in the program; sometimes it points to the instruction or point in the program where the interrupt occurred. Interrupt type numbers 0, 5, 6, 7, 8, 10, 11, 12, and 13 push a return address that points to the offending instruction, instead of to the next instruction in the program. This allows the interrupt service procedure to possibly retry the instruction in certain error cases.

Some of the protected mode interrupts (types 8, 10, 11, 12, and 13) place an error code on the stack following the return address. The error code identifies the selector that caused the interrupt. In cases where no selector is involved, the error code is a 0.

Operation of a Protected Mode Interrupt

In the protected mode, interrupts have exactly the same assignments as in the real mode, but the interrupt vector table is different. In place of interrupt vectors, protected mode uses a set of 256 interrupt descriptors that are stored in an interrupt descriptor table (IDT). The interrupt descriptor table is 256×8 (2K) bytes long, with each descriptor containing eight bytes. The interrupt descriptor table is located at any memory location in the system by the interrupt descriptor table address register (IDTR).

Each entry in the IDT contains the address of the interrupt service procedure in the form of a segment selector and a 32-bit offset address. It also contains the P bit (present) and DPL bits to describe the privilege level of the interrupt. Figure 12–3 shows the contents of the interrupt descriptor.

Real mode interrupt vectors can be converted into protected mode interrupts by copying the interrupt procedure addresses from the interrupt vector table and converting them to 32-bit offset addresses that are stored in the interrupt descriptors. A single selector and segment descriptor can be placed in the global descriptor table that identifies the first 1M byte of memory as the interrupt segment.

Other than the IDT and interrupt descriptors, the protected mode interrupt functions like the real mode interrupt. We return from both interrupts by using the IRET or IRETD instruction. The only difference is that in protected mode the microprocessor accesses the IDT instead of the interrupt vector table.

FIGURE 12–3 The protected mode interrupt descriptor.

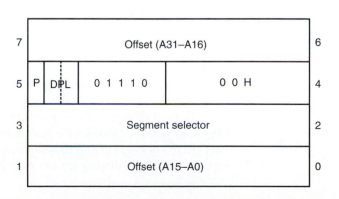

FIGURE 12–4   The flag register. (Courtesy of Intel Corporation.)

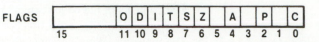

Interrupt Flag Bits

The interrupt flag (IF) and the trap flag (TF) are both cleared after the contents of the flag register are stacked during an interrupt. Figure 12–4 illustrates the contents of the flag register and the location of IF and TF. When the IF bit is set, it allows the INTR pin to cause an interrupt; when the IF bit is cleared, it prevents the INTR pin from causing an interrupt. When TF = 1, it causes a trap interrupt (type number 1) to occur after each instruction executes. This is why we often call trap a *single-step*. When TF = 0, normal program execution occurs. This flag bit allows debugging, as explained in Chapters 17–19, which detail the 80386–Pentium 4.

The interrupt flag is set and cleared by the STI and CLI instructions, respectively. There are no special instructions that set or clear the trap flag. Example 12–1 shows an interrupt service procedure that turns tracing on by setting the trap flag bit on the stack from inside the procedure. Example 12–2 shows an interrupt service procedure that turns tracing off by clearing the trap flag on the stack from within the procedure.

EXAMPLE 12–1

```
                        ;A procedure that sets TF to enable trap.
                        ;
0000                    TRON  PROC  NEAR

0000  50                      PUSH  AX          ;save registers
0001  55                      PUSH  BP
0002  8B EC                   MOV   BP,SP       ;get SP
0004  8B 46 08                MOV   AX,[BP+8]   ;get flags from stack
0007  80 CC 01                OR    AH,1        ;set TF
000A  89 46 08                MOV   [BP+8],AX   ;save flags
000D  5D                      POP   BP          ;restore registers
000E  58                      POP   AX
000F  CF                      IRET

0010                    TRON  ENDP
```

EXAMPLE 12–2

```
                        ;A procedure that clears TF to disable trap.
                        ;
0000                    TROFF PROC  NEAR

0000  50                      PUSH  AX          ;save registers
0001  55                      PUSH  BP
0002  8B EC                   MOV   BP,SP       ;get SP
0004  8B 46 08                MOV   AX,[BP+8]   ;get TF
0007  80 E4 FE                AND   AH,0FEH     ;clear TF
000A  89 46 08                MOV   [BP+8],AX   ;save flags
000D  5D                      POP   BP          ;restore registers
000E  58                      POP   AX
000F  CF                      IRET

0010                    TROFF ENDP
```

In both examples, the flag register is retrieved from the stack by using the BP register, which, by default, addresses the stack segment. After the flags are retrieved, the TF bit is either set (TRON) or clears (TROFF) before returning from the interrupt service procedure. The IRET instruction restores the flag register with the new state of the trap flag.

Trace Procedure. Assuming that TRON is accessed by an INT 40H instruction and TROFF is accessed by an INT 41H instruction, Example 12–3 traces through a program immediately following the INT 40H instruction. The interrupt service procedure illustrated in Example 12–3 responds to interrupt type number 1 or a trap interrupt. Each time that a trap occurs—after each instruction executes following INT 40H—the TRACE procedure displays the contents of all the 16-bit microprocessor registers on the CRT screen. This provides a register trace of all the instructions between the INT 40H (TRON) and INT 41H (TROFF).

EXAMPLE 12–3

```
                                        .MODEL TINY
0000                                    .CODE
0000   41 58 20 3D 20 42        RNAME   DB   'AX = ','BX = ','CX = ','DX = '
       58 20 3D 20 43 58
       20 3D 20 44 58 20
       3D 20
0014   53 50 20 3D 20 42                DB   'SP = ','BP = ','SI = ','DI = '
       50 20 3D 20 53 49
       0 3D 20 44 49 20
       3D 20
0028   49 50 20 3D 20 46                DB   'IP = ','FL = ','CS = ','DS = '
       4C 20 3D 20 43 53
       20 3D 20 44 53 20
       3D 20
003C   45 53 20 3D 20 53                DB   'ES = ','SS = '
       53 20 3D 20

                                DISP    MACRO PAR1
                                        PUSH  AX
                                        PUSH  DX
                                        MOV   DL,PAR1
                                        MOV   AH,6
                                        INT   21H
                                        POP   DX
                                        POP   AX
                                        ENDM

                                CRLF    MACRO
                                        DISP  13
                                        DISP  10
                                        ENDM

0046                            TRACE   PROC  FAR USES AX BP BX

0049   BB 0000 R                        MOV   BX,OFFSET RNAME    ;address names
                                        CRLF
0060   E8 004D                          CALL  DREG               ;display AX
0063   58                               POP   AX                 ;get BX
0064   50                               PUSH  AX
0065   E8 0048                          CALL  DREG               ;display BX
0068   8B C1                            MOV   AX,CX
006A   E8 0043                          CALL  DREG               ;display CX
006D   8B C2                            MOV   AX,DX
006F   E8 003E                          CALL  DREG               ;display DX
0072   8B C4                            MOV   AX,SP
0074   83 C0 0C                         ADD   AX,12
0077   E8 0036                          CALL  DREG               ;display SP
007A   8B C5                            MOV   AX,BP
007C   E8 0031                          CALL  DREG               ;display BP
007F   8B C6                            MOV   AX,SI
0081   E8 002C                          CALL  DREG               ;display SI
0084   8B C7                            MOV   AX,DI
0086   E8 0027                          CALL  DREG               ;display DI
0089   8B EC                            MOV   BP,SP
008B   8B 46 06                         MOV   AX,[BP+6]
```

```
008E   E8 001F                    CALL   DREG            ;display IP
0091   8B 46 0A                   MOV    AX,[BP+10]
0094   E8 0019                    CALL   DREG            ;display Flags
0097   8B 46 08                   MOV    AX,[BP+8]
009A   E8 0013                    CALL   DREG            ;display CX
009D   8C D8                      MOV    AX,DS
009F   E8 000E                    CALL   DREG            ;display DS
00A2   8C C0                      MOV    AX,ES
00A4   E8 0009                    CALL   DREG            ;display ES
00A7   8C D0                      MOV    AX,SS
00A9   E8 0004                    CALL   DREG            ;display SS
                                  IRET

00B0                       TRACE  ENDP

00B0                       DREG   PROC   NEAR USES CX
00B1   B9 0005                    MOV    CX,5            ;load count
00B4             DREG1:
                                  DISP   CS:[BX]         ;display character
00BF   43                         INC    BX              ;address next
00C0   E2 F2                      LOOP   DREG1           ;repeat 5 times
00C2   B9 0004                    MOV    CX,4            ;load count
00C5             DREG2:
00C5   D3 C8                      ROL    AX,1            ;position digit
00C7   D3 C8                      ROL    AX,1
00C9   D3 C8                      ROL    AX,1
00CB   D3 C8                      ROL    AX,1
00CD   50                         PUSH   AX
00CE   24 0F                      AND    AL,0FH          ;convert to ASCII
                                  .IF AL > 9
00D4   04 07                         ADD AL,7
                                  .ENDIF
00D6   04 30                      ADD    AL,30H
                                  DISP   AL
00E2   58                         POP    AX
00E3   E2 E0                      LOOP   DREG2           ;repeat 4 times
                                  DISP   ' '
                                  RET

00F1                       DREG   ENDP
                                  END
```

Storing an Interrupt Vector in the Vector Table

In order to install an interrupt vector—sometimes called a **hook**—the assembler must address absolute memory. Example 12–4 shows how a new vector is added to the interrupt vector table by using the assembler and a DOS function call. Here, INT 21H function call number 25H initializes the interrupt vector. Notice that the first thing done in this procedure is to save the old interrupt vector number by using DOS INT 21H function call number 35H to read the current vector. See Appendix A for more detail on DOS INT 21H function calls.

EXAMPLE 12–4

```
                           .MODEL TINY
                           .CODE
                           ;A program that installs NEW40 at INT 40H.
                           ;
                           .STARTUP
0100   EB 05                      JMP    START
0102   00000000          OLD      DD     ?
                           ;
                           ;new interrupt procedure
                           ;
0106                      NEW40 PROC  FAR
```

```
0106  CF                          IRET

0107            NEW40 ENDP

0107          START:
0107  8C C8                       MOV   AX,CS      ;get data segment
0109  8E D8                       MOV   DS,AX
010B  B4 35                       MOV   AH,35H     ;get old interrupt vector
010D  B0 40                       MOV   AL,40H
010F  CD 21                       INT   21H
0111  89 1E 0102 R                MOV   WORD PTR OLD,BX
0115  8C 06 0104 R                MOV   WORD PTR OLD+2,ES
                            ;
                            ;install new interrupt vector 40H
                            ;
0119  BA 0106 R                   MOV   DX,OFFSET NEW40
011C  B4 25                       MOV   AH,25H
011E  B0 40                       MOV   AL,40H
0120  CD 21                       INT   21H
                            ;
                            ;leave NEW40 in memory
                            ;
0122  BA 0107 R                   MOV   DX,OFFSET START
0125  D1 EA                       SHR   DX,1
0127  D1 EA                       SHR   DX,1
0129  D1 EA                       SHR   DX,1
012B  D1 EA                       SHR   DX,1
012D  42                          INC   DX
012E  B8 3100                     MOV   AX,3100H
0131  CD 21                       INT   21H
                                  END
```

12–2   HARDWARE INTERRUPTS

The microprocessor has two hardware interrupt inputs: non-maskable interrupt (NMI) and interrupt request (INTR). Whenever the NMI input is activated, a type 2 interrupt occurs because NMI is internally decoded. The INTR input must be externally decoded to select a vector. Any interrupt vector can be chosen for the INTR pin, but we usually use an interrupt type number between 20H and FFH. Intel has reserved interrupts 00H through 1FH for internal and future expansion. The $\overline{INTA}$ signal is also an interrupt pin on the microprocessor, but it is an output that is used in response to the INTR input to apply a vector type number to the data bus connections D7–D0. Figure 12–5 shows the three user interrupt connections on the microprocessor.

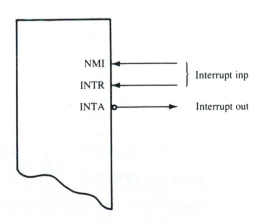

FIGURE 12–5   The interrupt pins on all versions of the Intel microprocessor.

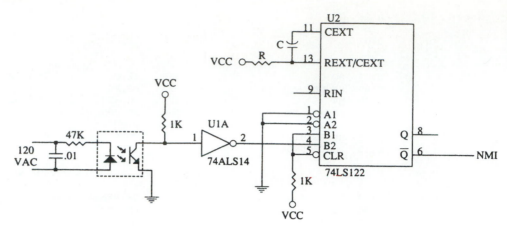

FIGURE 12–6   A power failure detection circuit.

The **non-maskable interrupt** (NMI) is an edge-triggered input that requests an interrupt on the positive edge (0-to-1 transition). After a positive edge, the NMI pin must remain a logic 1 until it is recognized by the microprocessor. Note that before the positive edge is recognized, the NMI pin must be a logic 0 for at least two clocking periods.

The NMI input is often used for parity errors and other major system faults, such as power failures. Power failures are easily detected by monitoring the AC power line and causing an NMI interrupt whenever AC power drops out. In response to this type of interrupt, the microprocessor stores all of the internal register in a battery backed-up-memory or an EEPROM. Figure 12–6 shows a power failure detection circuit that provides a logic 1 to the NMI input whenever AC power is interrupted.

In this circuit, an optical isolator provides isolation from the AC power line. The output of the isolator is shaped by a Schmitt-trigger inverter that provides a 60 Hz pulse to the trigger input of the 74LS122 retriggerable monostable multivibrator. The values of R and C are chosen so that the 74LS122 has an active pulse width of 33 ms or 2 AC input periods. Because the 74LS122 is retriggerable, as long as AC power is applied, the Q output remains triggered at a logic 1 and $\overline{Q}$ remains a logic 0.

If the AC power fails, the 74LS122 no longer receives trigger pulses from the 74ALS14, which means that Q returns to a logic 0 and $\overline{Q}$ returns to a logic 1, interrupting the microprocessor through the NMI pin. The interrupt service procedure, not shown here, stores the contents of all internal registers and other data into a battery-backed-up memory. This system assumes that the system power supply has a large enough filter capacitor to provide energy for at least 75 ms after the AC power ceases.

Figure 12–7 shows a circuit that supplies power to a memory after the DC power fails. Here, diodes are used to switch supply voltages from the DC power supply to the battery. The diodes used are standard silicon diodes because the power supply to this memory circuit is elevated above +5.0 V to +5.7 V. The resistor is used to trickle-charge the battery, which is either NiCAD, Lithium, or a gel cell.

When DC power fails, the battery provides a reduced voltage to the Vcc connection on the memory device. Most memory devices will retain data with Vcc voltages as low as 1.5 V, so the battery voltage does not need to be +5.0 V. The $\overline{WR}$ pin is pulled to Vcc during a power outage, so no data will be written to the memory.

INTR and $\overline{INTA}$

The interrupt request input (INTR) is level-sensitive, which means that it must be held at a logic 1 level until it is recognized. The INTR pin is set by an external event and cleared inside the

FIGURE 12–7 A battery-backed-up memory system using a NiCad, lithium, or gel cell.

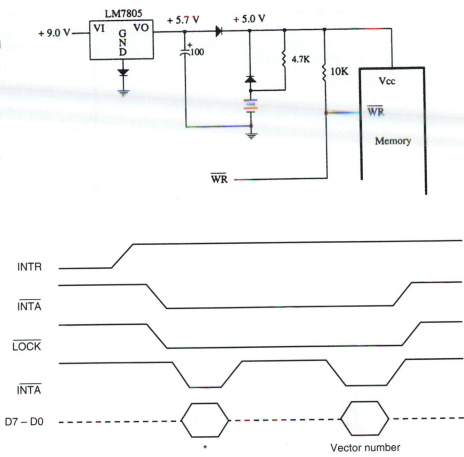

FIGURE 12–8 The timing of the INTR input and $\overline{\text{INTA}}$ output. *Note: This portion of the data bus is ignored and usually contains the vector number.

interrupt service procedure. This input is automatically disabled once it is accepted by the microprocessor and re-enabled by the IRET instruction at the end of the interrupt service procedure. The 80386–Pentium 4 use the IRETD instruction in the protected mode of operation.

The microprocessor responds to the INTR input by pulsing the $\overline{\text{INTA}}$ output in anticipation of receiving an interrupt vector type number on data bus connection D7–D0. Figure 12–8 shows the timing diagram for the INTR and $\overline{\text{INTA}}$ pins of the microprocessor. There are two $\overline{\text{INTA}}$ pulses generated by the system that are used to insert the vector type number on the data bus.

Figure 12–9 illustrates a simple circuit that applies interrupt vector type number FFH to the data bus in response to an INTR. Notice that the $\overline{\text{INTA}}$ pin is not connected in this circuit. Because resistors are used to pull the data bus connections (D0–D7) high, the microprocessor automatically sees vector type number FFH in response to the INTR input. This is possibly the least expensive way to implement the INTR pin on the microprocessor.

Using a Three-state Buffer for INTA. Figure 12–10 shows how interrupt vector type number 80H is applied to the data bus (D0–D7) in response to an INTR. In response to the INTR, the microprocessor outputs the $\overline{\text{INTA}}$ that is used to enable a 74ALS244 three-state octal buffer. The octal buffer applies the interrupt vector type number to the data bus in response to the $\overline{\text{INTA}}$ pulse. The vector type number is easily changed with the DIP switches that are shown in this illustration.

FIGURE 12–9   A simple method for generating interrupt vector type number FFH in response to INTR.

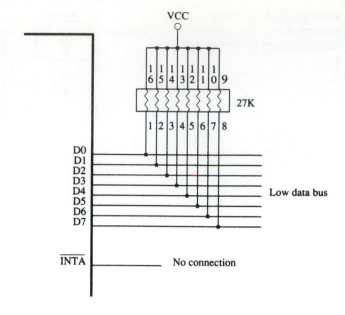

FIGURE 12–10   A circuit that applies any interrupt vector type number in response to INTA. Here the circuit is applying type number 80H.

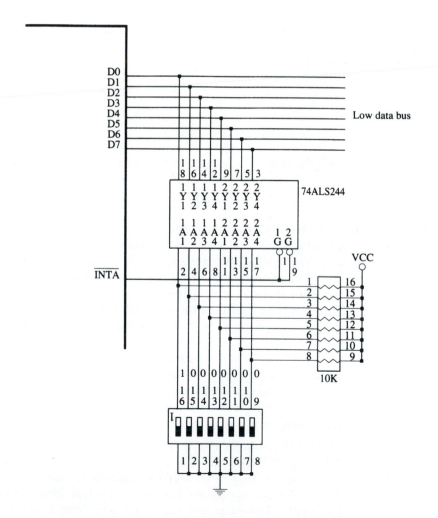

FIGURE 12–11 Converting INTR into an edge-triggered interrupt request input.

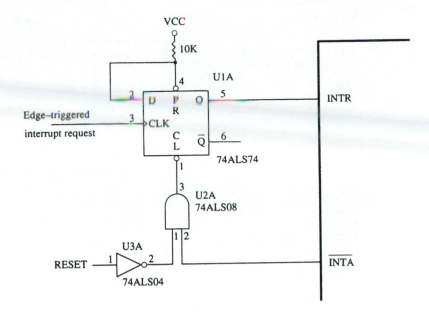

Making the INTR Input Edge-triggered. Often, we need an edge-triggered input instead of a level-sensitive input. The INTR input can be converted to an edge-triggered input by using a D-type flip-flop, as illustrated in Figure 12–11. Here, the clock input becomes an edge-triggered interrupt request input, and the clear input is used to clear the request when the $\overline{\text{INTA}}$ signal is output by the microprocessor. The RESET signal initially clears the flip-flop so that no interrupt is requested when the system is first powered.

The 82C55 Keyboard Interrupt

The keyboard example presented in Chapter 11 provides a simple example of the operation of the INTR input and an interrupt. Figure 12–12 illustrates the interconnection of the 82C55 with the microprocessor and the keyboard. It also shows how a 74ALS244 octal buffer is used to provide the microprocessor with interrupt vector type number 40H in response to the keyboard interrupt during the $\overline{\text{INTA}}$ pulse.

The 82C55 is decoded at 80386SX I/O port address 0500H, 0502H, 0504H, and 0506H by a PAL16L8 (the program is not illustrated). The 82C55 is operated in mode 1 (strobed input mode), so whenever a key is typed, the INTR output (PC3) becomes a logic 1 and requests an interrupt through the INTR pin on the microprocessor. The INTR pin remains high until the ASCII data are read from port A. In other words, every time a key is typed, the 82C55 requests a type 40H interrupt through the INTR pin. The $\overline{\text{DAV}}$ signal from the keyboard causes data to be latched into port A and causes INTR to become a logic 1.

Example 12–5 illustrates the interrupt service procedure for the keyboard. It is very important that all registers affected by an interrupt are saved before they are used. In the software required to initialize the 82C55 (not shown here), the FIFO is initialized so that both pointers are equal, the INTR request pin is enabled through the INTE bit inside the 82C55, and the mode of operation is programmed.

EXAMPLE 12–5

```
                        ;An interrupt service procedure that reads a key
                        ;from the keyboard in Figure 12–12.
                        ;
= 0500                  PORTA   EQU     500H
= 0506                  CNTR    EQU     506H
```

```
0000   0100 [                   FIFO   DB  256 DUP (?)       ;queue
            00
              ]
0100   0000                     INP    DW  ?                 ;input pointer
0102   0000                     OUTP   DW  ?                 ;output pointer

0104                            KEY    PROC  FAR USES AX BX DI DX

0108   2E: 8B 1E 0100 R         MOV    BX,CS:INP             ;load input pointer
010D   2E: 8B 3E 0102 R         MOV    DI,CS:OUTP            ;load output pointer

0112   FE C3                    INC    BL                    ;test for queue = full
0114   3B DF                    CMP    BX,DI
0116   74 11                    JE     FULL                  ;if queue is full

0118   FE CB                    DEC    BL
011A   BA 0500                  MOV    DX,PORTA
011D   EC                       IN     AL,DX                 ;get data from 82C55
011E   2E: 88 07                MOV    CS:[BX],AL            ;save data in queue
0121   2E: FE 06 0100 R         INC    BYTE PTR INP
0126   EB 07 90                 JMP    DONE
0129                     FULL:
0129   B0 08                    MOV    AL,8                  ;disable 82C55 interrupt
012B   BA 0506                  MOV    DX,CNTR
012E   EE                       OUT    DX,AL
012F                    DONE:
012F                            IRET

0134                            KEY    ENDP
```

The procedure is short because the 80386SX already knows that keyboard data are available when the procedure is called. Data are input from the keyboard and then stored in the FIFO (first-in, first-out) buffer. Most keyboard interfaces contain a FIFO that is at least 16 bytes in depth. The FIFO in this example is 256 bytes, which is more than adequate for a keyboard interface. Take note at how the INC BYTE PTR INP is used to add one to the input pointer and also make sure that it always addressed data in the queue.

This procedure first checks to see whether the FIFO is full. A full condition is indicated when the input pointer (INP) is one byte below the output pointer (OUTP). If the FIFO is full, the interrupt is disabled with a bit set/reset command to the 82C55, and a return from the interrupt occurs. If the FIFO is not full, the data are input from port A, and the input pointer is incremented before a return occurs.

Example 12–6 shows the procedure that removes data from the FIFO. This procedure first determines whether the FIFO is empty by comparing the two pointers. If the pointers are equal, the FIFO is empty, and the software waits at the EMPTY loop where it continuously tests the pointers. The EMPTY loop is interrupted by the keyboard interrupt, which stores data into the FIFO so that it is no longer empty. This procedure returns with the character in register AH.

EXAMPLE 12–6

```
                            ;A procedure that reads data from the queue of
                            ;Example 12-5 and returns with it in AH.
                            ;
0134                        READ   PROC  FAR USES BX DI DX

0137                EMPTY:
0137   2E: 8B 1E 0100 R     MOV    BX,CS:INP             ;load input pointer
013D   2E: 8B 3E 0102 R     MOV    DI,CS:OUTP            ;load output pointer
0142   3B DF                CMP    BX,DI
0144   74 F2                JE     EMPTY                 ;if queue is empty

0146   2E: 8A 25            MOV    AH,CS:[DI]            ;get data
0149   B0 09                MOV    AL,9                  ;enable 82C55 interrupt
```

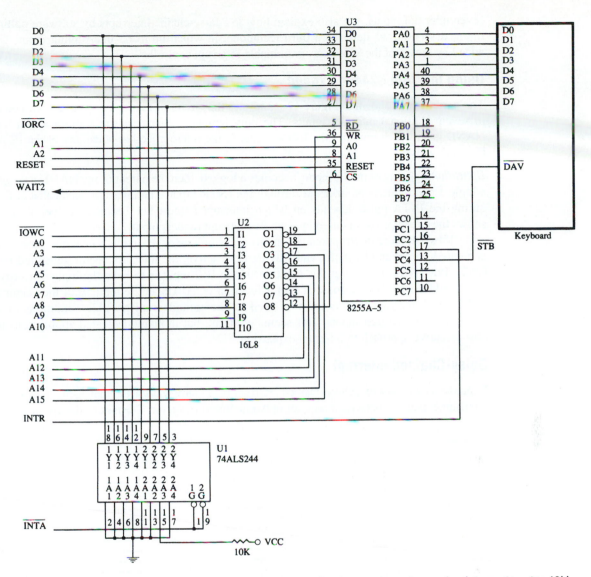

FIGURE 12–12 An 82C55 interfaced to a keyboard from the microprocessor system using interrupt vector 40H.

```
014B    BA 0506              MOV     DX,CNTR
014E    EE                   OUT     DX,AL
014F    2E: FE 06 0102 R     INC     BYTE PTR CS:OUTP
                             RET

0157                 READ    ENDP
```

12–3   EXPANDING THE INTERRUPT STRUCTURE

This text covers three of the more common methods of expanding the interrupt structure of the microprocessor. In this section, we explain how, with software and some hardware modification of the circuit shown in Figure 12–10, it is possible to expand the INTR input so that it accepts seven interrupt inputs. We also explain how to "daisy-chain" interrupts by software polling. In the next section, we describe a third technique, in which up to 63 interrupting inputs can be added by means of the 8259A programmable interrupt controller.

Using the 74ALS244 to Expand

The modification shown in Figure 12–13 allows the circuit of Figure 12–10 to accommodate up to seven additional interrupt inputs. The only hardware change is the addition of an 8-input NAND gate, which provides the INTR signal to the microprocessor when any of the $\overline{IR}$ inputs becomes active.

Operation.   If any of the $\overline{IR}$ inputs becomes a logic 0, then the output of the NAND gate goes to a logic 1 and requests an interrupt through the INTR input. The interrupt vector that is fetched during the $\overline{INTA}$ pulse depends on which interrupt request line becomes active. Table 12–1 shows the interrupt vectors used by a single interrupt request input.

If two or more interrupt request inputs are simultaneously active, a new interrupt vector is generated. For example, if $\overline{IR1}$ and $\overline{IR0}$ are both active, the interrupt vector generated is FCH (252). Priority is resolved at this location. If the $\overline{IR0}$ input is to have the higher priority, the vector address for $\overline{IR0}$ is stored at vector location FCH. The entire top half of the vector table and its 128 interrupt vectors must be used to accommodate all possible conditions of these seven interrupt request inputs. This seems wasteful, but in many dedicated applications it is a cost-effective approach to interrupt expansion.

Daisy-Chained Interrupt

Expansion by means of a daisy-chained interrupt is in many ways better than using the 74ALS244 interrupt expansion because it requires only one interrupt vector. The task of determining priority

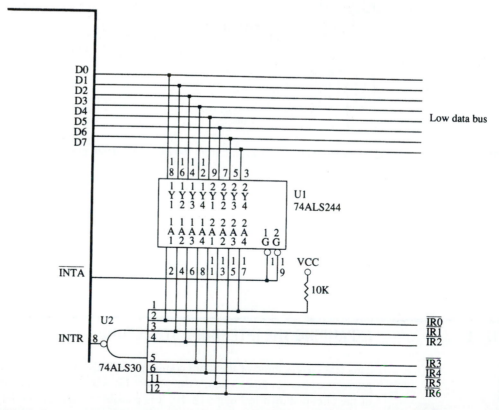

FIGURE 12–13   Expanding the INTR input from one to seven interrupt request lines.

TABLE 12–1 Single interrupt request for Figure 12–13.

$\overline{IR6}$	$\overline{IR5}$	$\overline{IR4}$	$\overline{IR3}$	$\overline{IR2}$	$\overline{IR1}$	$\overline{IR0}$	Vector
1	1	1	1	1	1	0	FEH
1	1	1	1	1	0	1	FDH
1	1	1	1	0	1	1	FBH
1	1	1	0	1	1	1	F7H
1	1	0	1	1	1	1	EFH
1	0	1	1	1	1	1	DFH
0	1	1	1	1	1	1	BFH

Note: Although not illustrated, the IR inputs are all active low.

is left to the interrupt service procedure. Setting priority for a daisy-chain does require additional software execution time, but in general this is a much better approach to expanding the interrupt structure of the microprocessor.

Figure 12–14 illustrates a set of two 82C55 peripheral interfaces with their four INTR outputs daisy-chained and connected to the single INTR input of the microprocessor. If any interrupt output becomes a logic 1, so does the INTR input to the microprocessor causing an interrupt.

When a daisy-chain is used to request an interrupt, it is better to pull the data bus connections (D0–D7) high by using pull-up resistors so interrupt vector FFH is used for the chain. Any interrupt vector can be used to respond to a daisy-chain. In the circuit, any of the four INTR outputs from the two 82C55s will cause the INTR pin on the microprocessor to go high, requesting an interrupt.

When the INTR pin does go high with a daisy-chain, the hardware gives no direct indication as to which 82C55 or which INTR output caused the interrupt. The task of locating which INTR output became active is up to the interrupt service procedure, which must poll the 82C55s to determine which output caused the interrupt.

Example 12–7 illustrates the interrupt service procedure that responds to the daisy-chain interrupt request. The procedure polls each 82C55 and each INTR output to decide which interrupt service procedure to utilize.

EXAMPLE 12–7

```
                              ;A procedure that services the daisy-chain interrupt
                              ;of Figure 12-14.
                              ;
= 0504                        C1     EQU    504H      ;first 82C55
= 0604                        C2     EQU    604H      ;second 82C55
= 0001                        MASK1  EQU    1         ;INTRB
= 0008                        MASK2  EQU    8         ;INTRA

0000                          POLL   PROC   FAR USES AX DX

0002  BA 0504                        MOV    DX,C1      ;address first 82C55
0005  EC                             IN     AL,DX      ;get port C
0006  A8 01                          TEST   AL,MASK1
0008  75 0F                          JNZ    LEVEL_0    ;if INTRB is set
000A  A8 08                          TEST   AL,MASK2
000C  75 13                          JNZ    LEVEL_1    ;if INTRA is set

000E  BA 0604                        MOV    DX,C2      ;address second 82C55
0011  EC                             IN     AL,DX      ;get port C
0012  A8 01                          TEST   AL,MASK1
0014  75 1B                          JNZ    LEVEL_2    ;if INTRB is set
0016  EB 29 00                       JMP    LEVEL_3    ;for INTRA

0019                          POLL   ENDP
```

FIGURE 12–14 Two 82C55 PIAs connected to the INTR outputs are daisy-chained to produce an INTR signal.

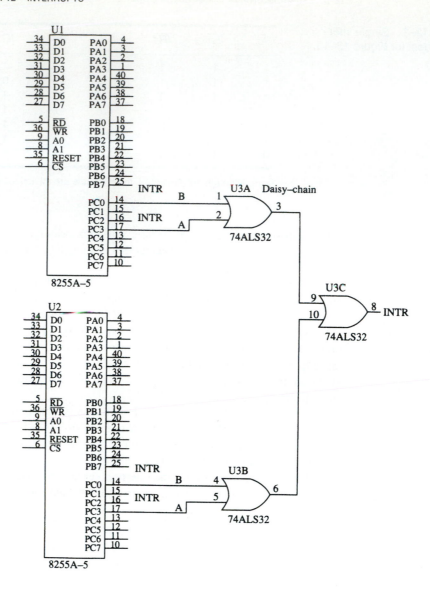

12–4   8259A PROGRAMMABLE INTERRUPT CONTROLLER

The 8259A programmable interrupt controller (PIC) adds eight vectored priority encoded interrupts to the microprocessor. This controller can be expanded, without additional hardware, to accept up to 64 interrupt requests. This expansion requires a master 8259A and eight 8259A slaves.

General Description of the 8259A

Figure 12–15 shows the pin-out of the 8259A. The 8259A is easy to connect to the microprocessor because all of its pins are direct connections except the $\overline{CS}$ pin, which must be decoded, and the $\overline{WR}$ pin, which must have an I/O bank write pulse. Following is a description of each pin on the 8259A:

D7–D0   The **bi-directional data connections** are normally connected to either the upper or lower data bus on the 80386SX microprocessor or the data bus on

FIGURE 12–15   The pin-out
of the 8259A programmable
interrupt controller (PIC).

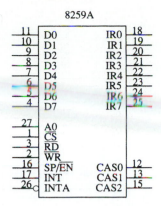

	the 8088. If an 80486 or Pentium–Pentium 4 is used, then they connect to any 8-bit bank.	
IR7–IR0	**Interrupt request** inputs are used to request an interrupt and to connect to a slave in a system with multiple 8259As.	
$\overline{WR}$	The **write** input connects to either the lower or upper write strobe signal in a 16-bit system, or to any other bus write strobe in any size system.	
$\overline{RD}$	The **read** input connects to the $\overline{IORC}$ signal.	
INT	The **interrupt** output connects to the INTR pin on the microprocessor from the master, and is connected to a master IR pin on a slave.	
$\overline{INTA}$	**Interrupt acknowledge** is an input that connects to the $\overline{INTA}$ signal on the system. In a system with a master and slaves, only the master $\overline{INTA}$ signal is connected.	
A0	The A0 address input selects different command words within the 8259A.	
$\overline{CS}$	**Chip select** enables the 8259A for programming and control.	
SP/$\overline{EN}$	**Slave program/enable buffer** is a dual-function pin. When the 8259A is in buffered mode, this is an output that controls the data bus transceivers in a large microprocessor-based system. When the 8259A is not in the buffered mode, this pin programs the device as a master (1) or a slave (0).	
CAS2–CAS0	The **cascade** lines are used as outputs from the master to the slaves for cascading multiple 8259As in a system.	

Connecting a Single 8259A

Figure 12–16 shows a single 8259A connected to the 8086 microprocessor. Here the SP/$\overline{EN}$ pin is pulled high to indicate that it is a master. The 8259A is decoded at I/O ports 0400H and 0402H by the PAL16L8 (no program shown). Like other peripherals discussed in Chapter 11, the 8259A requires four wait states for it to function properly with a 16 MHz 80386SX and more for some other versions of the Intel microprocessor family.

Cascading Multiple 8259As

Figure 12–17 shows two 8259As connected to the 80386SX microprocessor in a way that is often found in the AT-style computer, which has two 8259As for interrupts. The XT- or PC-style computer uses an 8259A controller at interrupt vectors 08H–0FH. The AT-style computer uses interrupt vector 0AH as a cascade input from a second 8259A located at vectors 70H through 77H. Appendix A contains a table that lists the functions of all the interrupt vectors used in the PC-, XT-, and AT-style computers.

FIGURE 12–16   An 8259A interfaced to the 8086 microprocessor.

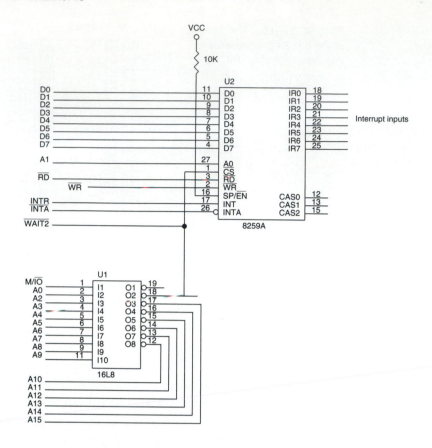

This circuit uses vectors 08H–0FH and I/O ports 0300H and 0302H for U1, the master; and vectors 70H–77H and I/O ports 0304H and 0306H for U2, the slave. Notice that we also include data bus buffers to illustrate the use of the SP/$\overline{\text{EN}}$ pin on the 8259A. These buffers are used only in very large systems that have many devices connected to their data bus connections. In practice, we seldom find these buffers.

Programming the 8259A

The 8259A is programmed by initialization and operation command words. **Initialization command words** (ICWs) are programmed before the 8259A is able to function in the system and dictate the basic operation of the 8259A. **Operation command words** (OCWs) are programmed during the normal course of operation. The OCWs control the operation of the 8259A.

Initialization Command Words.   There are four initialization command words (ICWs) for the 8259A that are selected when the A0 pin is a logic one. When the 8259A is first powered up, it must be sent ICW1, ICW2, and ICW4. If the 8259A is programmed in cascade mode by ICW1, then we also must program ICW3. So if a single 8259A is used in a system, ICW1, ICW2, and ICW4 must be programmed. If cascade mode is used in a system, then all four ICWs must be programmed. Refer to Figure 12–18 for the format of all four ICWs. The following is a description of each ICW:

ICW1   Programs the basic operation of the 8259A. To program this ICW for 8086–Pentium 4 operation, we place a logic 1 in bit IC4. Bits ADI, A7, A6, and A5 are don't cares for microprocessor operation and only apply to the 8259A when used with an 8-bit 8085 microprocessor (not covered in this textbook). This ICW selects single or cascade operation by programming the SNGL bit. If cascade operation is selected, we must also program ICW3. The LTIM bit determines whether the interrupt request inputs are positive edge-triggered or level-triggered.

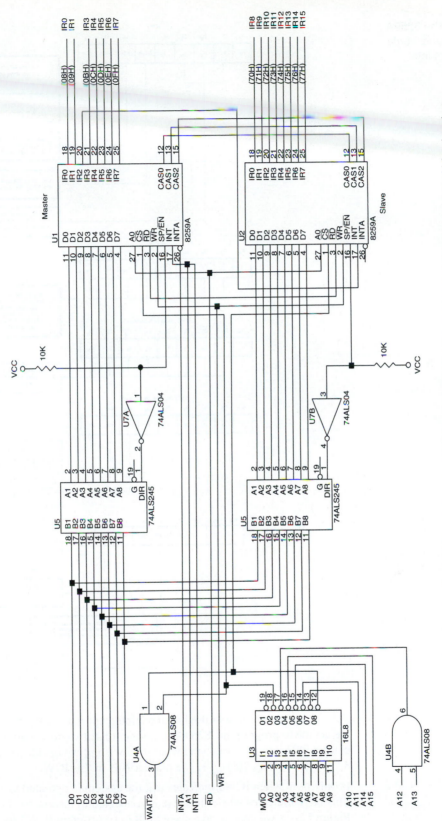

FIGURE 12–17 Two 8259As interfaced to the 8259A at I/O ports 0300H and 0302H for the master and 0304H and 0306H for the slave.

FIGURE 12–18   The 8259A initialization command words (ICWs) (Courtesy of Intel Corporation.)

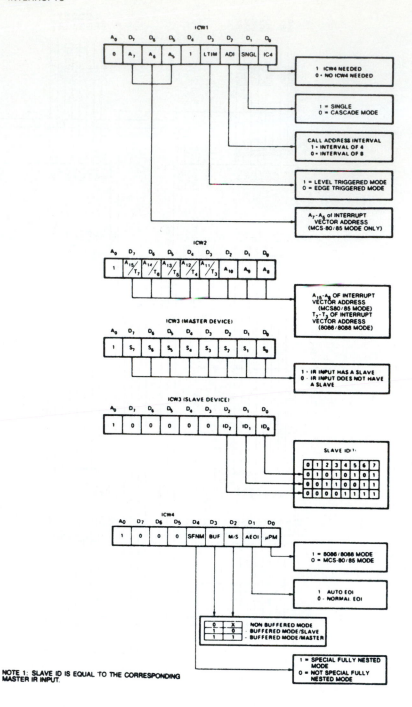

ICW2 Selects the vector number used with the interrupt request inputs. For example, if we decide to program the 8259A so it functions at vector locations 08H–0FH, we place a 08H into this command word. Likewise, if we decide to program the 8259A for vectors 70H–77H, we place a 70H in this ICW.

ICW3 Is only used when ICW1 indicates that the system is operated in cascade mode. This ICW indicates where the slave is connected to the master. For example, in Figure 12–18 we connected a slave to IR2. To program ICW3 for this connection, in both master and slave, we place a 04H in ICW3. Suppose we have two slaves connected to a master using IR0 and IR1. The master is programmed with an ICW3 of 03H; one slave is programmed with an ICW3 of 01H and the other with an ICW3 of 02H.

ICW4   Is programmed for use with the 8086–Pentium 4 microprocessors, but is not programmed in a system that functions with the 8085 microprocessor. The rightmost bit must be a logic 1 to select operation with the 8086–Pentium 4 microprocessors, and the remaining bits are programmed as follows:

SFNM—Selects the special fully-nested mode of operation for the 8259A if a logic 1 is placed in this bit. This allows the highest-priority interrupt request from a slave to be recognized by the master while it is processing another interrupt from a slave. Normally, only one interrupt request is processed at a time and others are ignored until the process is complete.

BUF and M/S—Buffer and master slave are used together to select buffered operation or nonbuffered operation for the 8259A as a master or a slave.

AEOI—Selects automatic or normal end of interrupt (discussed more fully under operation command words). The EOI commands of OCW2 are used only if the AEOI mode is not selected by ICW4. If AEOI is selected, the interrupt automatically resets the interrupt request bit and does not modify priority. This is the preferred mode of operation for the 8259A and reduces the length of the interrupt service procedure.

Operation Command Words.   The operation command words (OCWs) are used to direct the operation of the 8259A once it is programmed with the ICW. The OCWs are selected when the A0 pin is at a logic 0 level, except for OCW1, which is selected when A0 is a logic 1. Figure 12–19 lists the binary bit patterns for all three operation command words of the 8259A. Following is a list describing the function of each OCW:

OCW1   Is used to set and read the interrupt mask register. When a mask bit is set, it will turn off (mask) the corresponding interrupt input. The mask register is read when OCW1 is read. Because the state of the mask bits are unknown when the 8259A is first initialized, OCW1 must be programmed after programming the ICW upon initialization.

OCW2   Is programmed only when the AEOI mode is not selected for the 8259A. In this case, this OCW selects the way that the 8259A responds to an interrupt. The modes are listed as follows:

Nonspecific End-of-Interrupt—A command sent by the interrupt service procedure to signal the end of the interrupt. The 8259A automatically determines which interrupt level was active and resets the correct bit of the interrupt status register. Resetting the status bit allows the interrupt to take action again or a lower priority interrupt to take effect.

Specific End-of-Interrupt—A command that allows a specific interrupt request to be reset. The exact position is determined with bits L2–L0 of OCW2.

Rotate-on-Nonspecific EOI—A command that functions exactly like the Nonspecific End-of-Interrupt command, except that it rotates interrupt priorities after resetting the interrupt status register bit. The level reset by this command becomes the lowest-priority interrupt. For example, if IR4 was just serviced by this command, it becomes the lowest-priority interrupt input and IR5 becomes the highest priority.

FIGURE 12–19 The 8259A operation command words (OCWs). (Courtesy of Intel Corporation.)

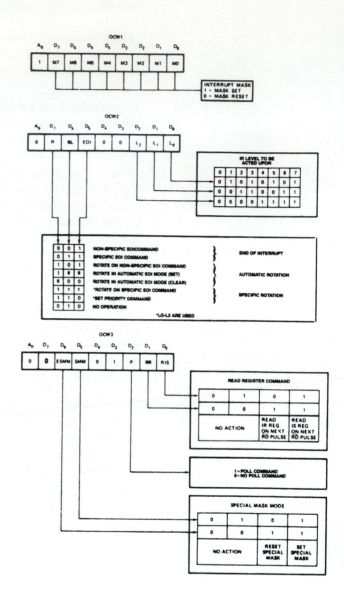

Rotate-on-Automatic EOI—A command that selects automatic EOI with rotating priority. This command must only be sent to the 8259A once if this mode is desired. If this mode must be turned off, use the clear command.

Rotate-on-Specific EOI—Functions as the specific EOI, except that it selects rotating priority.

Set priority—Allows the programmer to set the lowest priority interrupt input using the L2–L0 bits.

OCW3   Selects the register to be read, the operation of the special mask register, and the poll command. If polling is selected, the P bit must be set and then output to the 8259A. The next read operation will read the poll word. The rightmost three bits of the poll word indicate the active interrupt request with the highest priority. The leftmost bit indicates whether there is an interrupt and must be checked to determine whether the rightmost three bits contain valid information.

FIGURE 12–20 The 8259A in-service register (ISR). (a) Before IR4 is accepted and (b) after IR4 is accepted. (Courtesy of Intel Corporation.)

Status Register. Three status registers are readable in the 8259A: interrupt request register (IRR), in-service register (ISR), and interrupt mask register (IMR). (See Figure 12–20 for all three status registers; they all have the same bit configuration.) The IRR is an 8-bit register that indicates which interrupt request inputs are active. The ISR is an 8-bit register that contains the level of the interrupt being serviced. The IMR is an 8-bit register that holds the interrupt mask bits and indicates which interrupts are masked off.

Both the IRR and ISR are read by programming OCW3 and IMR is read through OCW1. To read the IMR, A0 = 1; to read either IRR or ISR, A0 = 0. Bit positions D0 and D1 of OCW3 select which register (IRR or ISR) is read when A0 = 0.

8259A Programming Example

Figure 12–21 illustrates the 8259A programmable interrupt controller connected to a 16550 programmable communications controller. In this circuit, the INTR pin from the 16550 is connected to the programmable interrupt controller's interrupt request input IR0. An IR0 occurs whenever (1) the transmitter is ready to send another character, (2) the receiver has received a character, (3) an error is detected while receiving data, and (4) a modem interrupt occurs. Notice that the 16550 is decoded at I/O ports 40H and 47H, and the 8259A is decoded at 8-bit I/O ports 48H and 49H. Both devices are interfaced to data bus of an 8088 microprocessor.

Initialization Software. The first portion of the software for this system must program both the 16550 and the 8259A, and then enable the INTR pin on the 8088 so that interrupts can take effect. Example 12–8 lists the software required to program both devices and enable INTR. This software uses two memory FIFOs that hold data for the transmitter and for the receiver. Each memory FIFO is 16K bytes long and is addressed by a pair of pointers (input and output).

EXAMPLE 12–8

```
                          ;Initialization software for the 16550 and 8259A
                          ;of the circuit in Figure 12-21.
                          ;
= 0048                    PIC1    EQU    48H         ;8259A control A0 = 0
= 0049                    PIC1    EQU    49H         ;8259A control A0 = 1
= 001B                    ICW1    EQU    1bH         ;8259A ICW1
= 0080                    ICW2    EQU    80H         ;8259A ICW2
= 0003                    ICW4    EQU    3           ;8259A ICW4
= 00FE                    OCW1    EQU    0FEH        ;8259A OCW1
= 0043                    LINE    EQU    43H         ;16550 line register
= 0040                    LSB     EQU    40H         ;16550 Baud divisor LSB
= 0041                    MSB     EQU    41H         ;16550 Baud divisor MSB
= 0042                    FIFO    EQU    42H         ;16550 FIFO register
= 0041                    ITR     EQU    41H         ;16550 interrupt register

0000                      START   PROC   NEAR
                          ;
                          ;Program 16550, but do not enable interrupts yet
                          ;
0000   B0 8A                      MOV    AL,10001010B   ;enable Baud divisor
```

```
0002   E6 43                    OUT    LINE,AL

0004   B0 78                    MOV    AL,120            ;program Baud rate
0006   E6 40                    OUT    LSB,AL            ;9600 Baud rate
0008   B0 00                    MOV    AL,0
000A   E6 41                    OUT    MSB,AL

000C   B0 0A                    MOV    AL,00001010B      ;program 7-data, odd
000E   E6 43                    OUT    LINE,AL           ;parity, one stop
0010   B0 07                    MOV    AL,00000111B      ;enable transmitter and
0012   E6 42                    OUT    FIFO,AL           ;and receiver
                             ;
                             ;Program 8259A
                             ;
0014   B0 1B                    MOV    AL,ICW1           ;program ICW1
0016   E6 48                    OUT    PIC1,AL

0018   B0 80                    MOV    AL,ICW2           ;program ICW2
001A   E6 49                    OUT    PIC2,AL

001C   B0 03                    MOV    AL,ICW4           ;program ICW4
001E   E6 49                    OUT    PIC2,AL

0020   B0 FE                    MOV    AL,OCW1           ;program OCW1
0022   E6 49                    OUT    PIC2,AL
0024   FB                       STI                      ;enable system INTR pin
                             ;
                             ;enable 16550 interrupts
                             ;
0025   B0 07                    MOV    AL,5              ;enable receiver and
0027   E6 41                    OUT    ITR,AL            ;error interrupts
0029   C3                       RET

002A                     START ENDP
```

The first portion of the procedure (START) programs the 16550 UART for operation with seven data bits, odd parity, one stop bit, and a Baud rate clock of 9600. The FIFO control register also enables both the transmitter and receiver.

The second part of the procedure programs the 8259A, with its three ICWs and its one OCW. The 8259A is set up so that it functions at interrupt vectors 80H–87H and operates with automatic EOI. The ICW enables the interrupt for the 16550 UART. The INTR pin of the microprocessor is also enabled by using the STI instruction.

The final part of the software enables the receiver and error interrupts of the 16550 UART through the interrupt control register. The transmitter interrupt is not enabled until data are available for transmission. See Figure 12–22 for the contents of the interrupt control register of the 16550 UART. Notice that the control register can enable or disable the receiver, transmitter, line status (error), and modem interrupts.

Handling the 16550 UART Interrupt Request. Because the 16550 generates only one interrupt request for various interrupts, the interrupt handler must poll the 16550 to determine what type of interrupt has occurred. This is accomplished by examining the interrupt identification register (see Figure 12–23). Note that the interrupt identification register (read-only) shares the same I/O port as the FIFO control register (write-only).

The interrupt identification register indicates whether an interrupt is pending, the type of interrupt, and whether the transmitter and receiver FIFO memories are enabled. See Table 12–2 for the contents of the interrupt control bits.

The interrupt service procedure must examine the contents of the interrupt identification register to determine what event caused the interrupt and pass control to the appropriate procedure for the event. Example 12–9 shows the first part of an interrupt handler that passes control to RECV for a receiver data interrupt, TRANS for a transmitter data interrupt, and ERR for a line status error interrupt. Note that the modem status is not tested in this example.

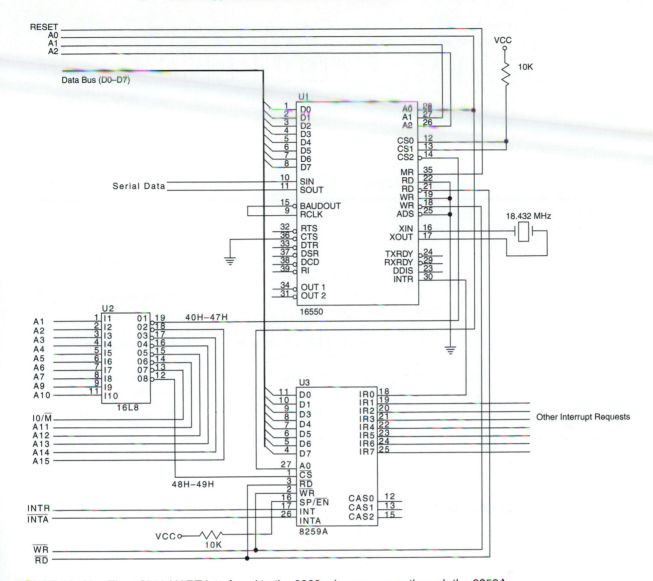

FIGURE 12–21   The 16550 UART interfaced to the 8088 microprocessor through the 8259A.

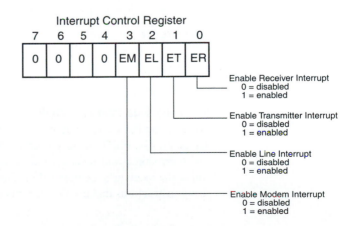

FIGURE 12–22   The 16550 interrupt control register.

FIGURE 12–23   The 16550
interrupt identification register.

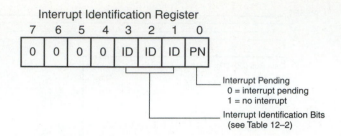

Interrupt Identification Register

TABLE 12–2   The interrupt control bits of the 16550.

Bit3	Bit2	Bit1	Bit0	Priority	Type	Reset Control
0	0	0	1	—	No interrupt	—
0	1	1	0	1	Receiver error (parity, framing, overrun, or break)	Reset by reading the line register
0	1	0	0	2	Receiver data available	Reset by reading the data
1	1	0	0	2	Character time-out, nothing has been removed from the receiver FIFO for at least four character times	Reset by reading the data
0	0	1	0	3	Transmitter empty	Reset by writing to the transmitter
0	0	0	0	4	Modem status	Reset by reading the modem status

Note: 1 is the highest priority and 4 the lowest.

EXAMPLE 12–9

```
                        ;Interrupt handler for the 16550 UART of
                        ;Figure 12-21.
                        ;
0000                    INT80 PROC   FAR

0000  50                      PUSH  AX
0001  E4 42                   IN    AL,42H      ;input interrupt ID reg
0003  3C 06                   CMP   AL,6        ;test for error
0005  74 20                   JE    ERR         ;for receiver error

0007  3C 02                   CMP   AL,2        ;test for transmitter
0009  74 55                   JE    TRANS       ;for transmitter ready

000B  3C 04                   CMP   AL,4        ;test for receiver
000D  74 11                   JE    RECV        ;for receiver ready
```

Receiving Data from the 16550.   The data received by the 16550 are stored, not only in the FIFO within the UART, but also in a FIFO memory until the software in the main program can use them. The FIFO memory used for received data is 16K bytes long, so many characters can easily be stored and received before any intervention from the microprocessor is required to empty the receiver's memory FIFO. The receiver memory FIFO is stored in the extra segment so string instructions that use the DI register can be used to access it.

Receiving data from the 16550 requires two procedures. One procedure reads the data register of the 16550 each time that the INTR pin requests an interrupt, and stores it into the memory FIFO. The other procedure reads data from the memory FIFO from the main program.

Example 12–10 lists the procedure used to read data from the memory FIFO from the main program. This procedure assumes that the pointers (IIN and IOUT) are initialized in the initialization dialog for the system (not shown). The READ procedure returns with AL containing a character read from the memory FIFO. If the memory FIFO is empty, the procedure returns with the carry flag bit set to a logic one. If AL contains a valid character, the carry flag bit is cleared upon return from READ.

Notice how the FIFO is reused by changing the address from the top of the FIFO to the bottom whenever it exceeds the start of the FIFO plus 16K. This is located at the CMP instruction at offset address 0015. Also notice that interrupts are enabled at the end of this procedure, in case they are disabled by a full memory FIFO condition by the RECV interrupt procedure.

EXAMPLE 12–10

```
                         ;A procedure that reads one character from the memory
                         ;FIFO and returns with it in AL.
                         ;If the FIFO is empty the return occurs with Carry = 1.
                         ;
0000                     READ  PROC  NEAR USES BX DI

0002  26: 8B 3E 4002 R         MOV   DI,IOUT           ;get output pointer
0007  26: 8B 1E 4000 R         MOV   BX,IIN            ;get input pointer

000C  3B DF                    CMP   BX,DI             ;compare pointers
000E  F9                       STC                     ;set carry flag
000F  74 16                    JE    DONE1             ;if empty

0011  26: BA 06                MOV   AL,ES:[DI]        ;get data from FIFO
0014  47                       INC   DI                ;address next byte
0015  81 FF 4000 R             CMP   DI,OFFSET FIFO+16*1024
0019  26: 89 3E 4002 R         MOV   IOUT,DI           ;save pointer
001E  76 07                    JBE   DONE              ;if within bounds
0020  26: C7 06 4002 R         MOV   IOUT,OFFSET FIFO
      0000   R
0027                     DONE:
0027  F8                       CLC                     ;clear carry flag
0028                     DONE1:
0028  9C                       PUSHF                   ;save carry flag
0029  E4 41                    IN    AL,41H            ;read interrupt control
002B  06 05                    OR    AL,5              ;enable receiver interrupts
002D  E6 41                    OUT   41H,AL
002F  9D                       POPF
                               RET

0033                     READ  ENDP
```

Example 12–11 lists the RECV interrupt service procedure that is called each time the 16550 receives a character for the microprocessor. In this example, the interrupt uses vector type number 80H, which must address the interrupt handler of Example 12–9. Each time that this interrupt occurs, the REVC procedure is accessed by the interrupt handler reading a character from the 16550. The RECV procedure stores the character into the memory FIFO. If the memory FIFO is full, the receiver interrupt is disabled by the interrupt control register within the 16550. This may result in lost data, but at least it will not cause the interrupt to overrun valid data already stored in the memory FIFO. Any error conditions detected by the 8251A store a ? (3FH) in the memory FIFO. Note that errors are detected by the ERR portion of the interrupt handler (not shown).

EXAMPLE 12–11

```
                        ;RECV portion of the interrupt handler in Example
                        ;12-9.
                        ;
0020                    RECV:                   ;continues from Example 12-9
0020 53                     PUSH  BX            ;save registers
0021 57                     PUSH  DI
0022 56                     PUSH  SI
0023 26: 8B 1E 4002 R       MOV   BX,IOUT       ;load output pointer
0028 26: 8B 36 4000 R       MOV   SI,IIN        ;load input pointer
002D 8B FE                  MOV   DI,SI
002F 46                     INC   SI
0030 81 FE 4000 R           CMP   SI,OFFSET FIFO+16*1024
0034 76 03                  JBE   NEXT
0036 BE 0000 R              MOV   SI,OFFSET FIFO
0039              NEXT:
0039 3B DE                  CMP   BX,SI         ;is FIFO full?
003B 74 0B                  JE    FULL          ;if it is full
003D E4 40                  IN    AL,40H        ;read 16550 receiver
003F AA                     STOSB               ;save it in FIFO
0040 26: 89 36 4000 R       MOV   IIN,SI        ;save input pointer
0045 EB 06 90               JMP   DONE          ;end up
0048              FULL
0048 E4 41                  IN    AL,41H        ;read interrupt control
004A 24 FA                  AND   AL,0FAH       ;disable receiver
004C E6 41                  OUT   41H,AL
004E              DONE:
004E B0 20                  MOV   AL,20H        ;signal 8259A EOI
0050 E6 49                  OUT   49H,AL
0052 5E                     POP   SI            ;restore registers
0053 5F                     POP   DI
0054 5B                     POP   BX
0055 58                     POP   AX
0056 CF                     IRET
```

Transmitting Data to the 16550.   Data are transmitted to the 16550 in much the same manner as they are received, except that the interrupt service procedure removes transmit data from a second 16K-byte memory FIFO.

Example 12–12 lists the procedure that fills the output FIFO. It is similar to the procedure listed in Example 12–10, except it determines whether the FIFO is full instead of empty.

EXAMPLE 12–12

```
                        ;A procedure that places data into the memory FIFO for
                        ;transmission by the transmitter interrupt.
                        ;AL = character to be transmitted.
                        ;
0000                    SAVE  PROC  NEAR USES BX DI SI

0003 26: 8B 36 8004 R       MOV   SI,OIN    ;get input pointer
0008 26: 8B 1E 8006 R       MOV   BX,OOUT   ;get output pointer
000D 8B FE                  MOV   DI,SI
000F 46                     INC   SI
0010 81 FE 8004 R           CMP   SI,OFFSET OFIFO+16*1024
0014 76 03                  JBE   NEXT
0016 BE 4004 R              MOV   SI,OFFSET OFIFO
0019              NEXT:
0019 3B DE                  CMP   BX,SI
001B 74 06                  JE    DONE      ;if full
001D AA                     STOSB           ;save data in OFIFO
001E 26: 89 36 8004 R       MOV   OIN,SI
0023              DONE:
0023 E4 41                  IN    AL,41H    ;read interrupt control
0025 06 01                  OR    AL,1      ;enable transmitter
0027 E6 41                  OUT   41H,AL
                            RET
002D                    SAVE  ENDP
```

Example 12–13 lists the interrupt service subroutine for the 16550 UART transmitter. This procedure is a continuation of the interrupt handler presented in Example 12–9 and is similar to the RECV procedure of Example 12–11, except that it determines whether the FIFO is empty rather than full. Note that we do not include an interrupt service procedure for the break interrupt or any errors.

EXAMPLE 12–13

```
                                ;Interrupt service procedure for the 16550
                                ;transmitter.
                                ;
0060                            TRANS:
0060   53                               PUSH    BX              ;save registers
0061   57                               PUSH    DI
0062   26: 8B 1E 8004 R                 MOV     BX,OIN          ;load input pointer
0068   26: 8B 3E 8006 R                 MOV     DI,OOUT         ;load output pointer
006D   3B DF                            CMP     BX,DI
006F   74 17                            JE      EMPTY           ;if empty
0071   26: 8A 05                        MOV     AL,ES:[DI]      ;get character
0074   E6 40                            OUT     40H,AL          ;send it to UART
0076   47                               INC     DI
0077   81 FF 8004 R                     CMP     DI,OFFSET OFIFO+16*1024
007B   76 03                            JBE     NEXT1
007D   BF 4004 R                        MOV     DI,OFFSET OFIFO
0080                            NEXT1:
0080   26: 89 3E 8006 R                 MOV     OOUT,DI
0085   EB 07 90                         JMP     DONES
0088                            EMPTY:
0088   E4 41                            IN      AL,41H          ;read interrupt control
008A   24 FD                            AND     AL,0FDH         ;disable transmitter
008C   E6 41                            OUT     41H,AL
008E                            DONES:
008E   B0 20                            MOV     AL,20H          ;signal 8259A EOI
0090   E6 49                            OUT     49H,AL
0092   5F                               POP     DI
0093   5B                               POP     BX
0094   58                               POP     AX
0095   CF                               IRET
```

The 16550 also contains a scratch register, which is a general-purpose register that can be used in any way deemed necessary by the programmer. Also contained within the 16550 are a modem control register and a modem status register. These registers allow the modem to cause interrupt and control the operation of the 16550 with a modem. See Figure 12–24 for the contents of both the modem status register and the modem control register.

The modem control register uses bit positions 0–3 to control various pins on the 16550. Bit position 4 enables the internal loop-back test for testing purposes. The modem status register allows the status of the modem pins to be tested; it also allows the modem pins to be checked for a change or, in the case of $\overline{RI}$, a trailing edge.

Figure 12–25 illustrates the 16550 UART, connected to an RS-232C interface that is often used to control a modem. Included in this interface are line driver and receiver circuits used to convert between TTL levels on the 16550 to RS-232C levels found on the interface. Note that RS-232C levels are usually +12 V for a logic 0 and –12 V for a logic 1 level.

In order to transmit or receive data through the modem, the $\overline{DTR}$ pin is activated (logic 0) and the UART then waits for the $\overline{DSR}$ pin to become a logic 0 from the modem, indicating that the modem is ready. Once this handshake is complete, the UART sends the modem a logic 0 on the $\overline{RTS}$ pin. When the modem is ready, it returns the $\overline{CTS}$ signal (logic 0) to the UART. Communications can now commence. The $\overline{DCD}$ signal from the modem is an indication that the modem has detected a carrier. This signal must also be tested before communications can begin.

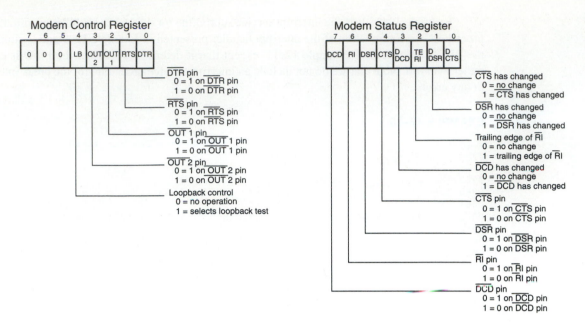

FIGURE 12–24   The 16550 modem control and modem status registers.

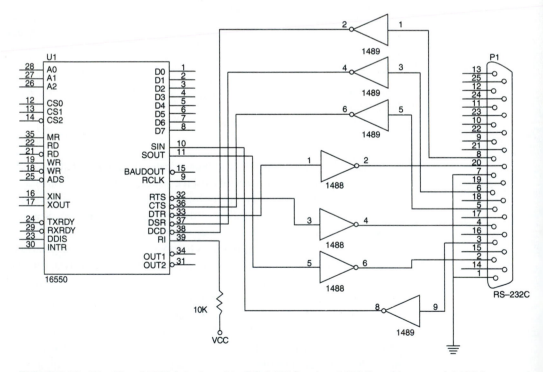

FIGURE 12–25   The 16550 interfaced to RS-2332C using 1488 line drivers and 1489 line receivers.

12-5   INTERRUPT EXAMPLES

This section of the text presents a real-time clock and an interrupt processed keyboard as examples of interrupt applications. A **real-time clock** keeps time in real time—that is, in hours and minutes. The example illustrated here keeps time in hours, minutes, seconds, and 1/60 seconds, using four memory locations to hold the BCD time of day. The interrupt processed keyboard uses a periodic interrupt to scan through the keys of the keyboard.

Real-Time Clock

Figure 12-26 illustrates a simple circuit that uses the 60 Hz AC power line to generate a periodic interrupt request signal for the NMI interrupt input pin. Although we are using a signal from the AC power line, which varies slightly in frequency from time to time, it is accurate over a period of time.

The circuit uses a signal from the 120 VAC power line that is conditioned by a Schmitt trigger inverter before it is applied to the NMI interrupt input. Note that you must make certain that the power line ground is connected to the system ground in this schematic. The power line ground (neutral) connection is the large flat pin on the power line. The narrow flat pin is the hot side or 120 VAC side of the line.

The software for the real-time clock contains an interrupt service procedure that is called 60 times per second and a procedure that updates the count located in four memory locations. Example 12-14 lists both procedures, along with the four bytes of memory used to hold the BCD time of day.

EXAMPLE 12-14

```
                    .MODEL TINY
0000                .CODE
                    .STARTUP
0100 EB 04          JMP     TIMES
0102 00     TIME    DB      ?               ;1/60 second counter
0103 00             DB      ?               ;seconds counter
0104 00             DB      ?               ;minutes counter
0105 00             DB      ?               ;hours counter
            ;
            ;Interrupt handler for NMI
            ;
0106        TIMES PROC    FAR

0106 50             PUSH    AX              ;save registers
0107 56             PUSH    SI

0108 B4 60          MOV     AH,60H          ;load modulus 60
010A BE 0102 R      MOV     SI,OFFSET TIME  ;address time
010D E8 0014        CALL    UP              ;increment 1/60 counter
0110 75 0F          JNZ     DONE
0112 E8 000F        CALL    UP              ;increment seconds
0115 75 0A          JNZ     DONE
0117 E8 000A        CALL    UP              ;increment minutes
011A 75 05          JNZ     DONE
011C B4 24          MOV     AH,24H          ;load modulus 24
011E E8 0003        CALL    UP              ;increment minutes
```

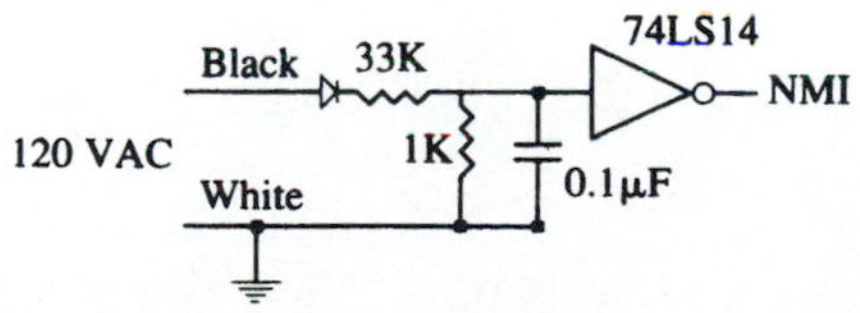

FIGURE 12-26 Converting the AC power line to a 60 Hz TTL signal for the NMI input.

```
0121              DONE:
0121   5E                  POP    SI                 ;reload registers
0122   58                  POP    AX
0123   CF                  IRET


0124              TIMES    ENDP


0124              UP       PROC   NEAR


0124   2E: 8A 04           MOV    AL,CS:[SI]         ;get count
0127   46                  INC    SI                 ;address next counter
0128   04 01               ADD    AL,1               ;increment count
012A   27                  DAA                       ;make it BCD
012B   2E: 88 44 FF        MOV    CS:[SI-1],AL       ;save count
012F   2A C4               SUB    AL,AH              ;test modulus
0131   75 04               JNZ    UP1
0133   2E: 88 44 FF        MOV    CS:[SI-1],AL       ;clear count
0137              UP1:
0137   C3                  RET


0138              UP       ENDP


                           END
```

Interrupt-Processed Keyboard

The interrupt-processed keyboard scans through the keys on a keyboard through a periodic interrupt. Each time the interrupt occurs, the interrupt-service procedure tests for a key or debounces the key. Once a valid key is detected, the interrupt-service procedure stores the key-code into a keyboard queue for later reading by the system. The basis for this system is a periodic interrupt that can be caused by a timer or other device in the system. Note that most systems already have a periodic interrupt for the real-time clock. In this example, we assume that the interrupt calls the interrupt-service procedure every 10 ms.

Figure 12–27 shows the keyboard interfaced to an 8255. It does not show the timer or other circuitry required to call the interrupt, once in every 10 ms. (Not shown in the software is

FIGURE 12–27 A telephone style keypad interfaced to the 82C55.

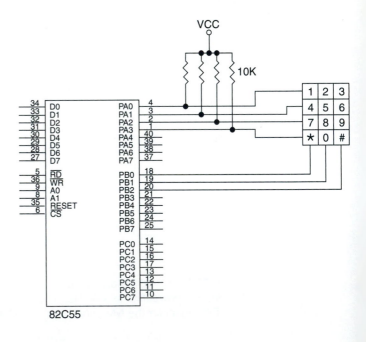

programming of the 82C55.) The 82C55 must be programmed so that port A is an input port, port B is an output port, and the initialization software must store a 00H at port B. This interfaces uses memory that is stored in the code segment for a queue and a few bytes that keep track of the keyboard scanning. Example 12–15 lists the interrupt service procedure for the keyboard.

EXAMPLE 12–15

```
                        ;interrupt procedure for the keyboard in
                        ;figure 12-27
                        ;
                        .MODEL TINY
                        .386
0000                    .CODE

= 1000                  PORTA EQU   1000H            ;define port A address
= 1001                  PORTB EQU   1001H            ;define Port B address


                        .STARTUP
0100                    INTKEY  PROC  FAR USES AX DX ;keyboard interrupt
0102 BA 1000                    MOV   DX,PORTA
0105 EC                         IN AL,DX             ;test for any key
0106 0C F0                      OR AL,0F0H
                                .IF AL != 0FFH       ;if key found
010C FE 06 0192 R                   INC DBCNT        ;increment bounce count
                                    .IF DBCNT==3     ;if > 20 ms
0117 C6 06 0192 R 02                    MOV DBCNT,2
                                        .IF DBF==0
0123 53                                     PUSH BX
0124 C6 06 0193 R 01                        MOV   DBF,1
0129 BB 00FB                                MOV   BX,0FBH
                                            .WHILE 1      ;find key
012C BA 1001                                    MOV DX,PORTB
012F 8A C3                                      MOV AL,BL
0131 EE                                         OUT DX,AL
0132 D0 CB                                      ROR BL,1
0134 BA 1000                                    MOV DX,PORTA
0137 EC                                         IN  AL,DX
0138 0C F0                                      OR  AL,0F0H
                                                .BREAK .IF AL != 0FFH
013E 80 C7 04                                   ADD BH,4
                                            .ENDW
0143 8A D8                                  MOV BL,AL
0145 BA 1001                                MOV DX,PORTB
0148 B0 00                                  MOV AL,0
014A EF                                      OUT DX,AX    ;clear port B pins
014B FE CF                                  DEC BH
                                            .REPEAT       ;find key code
014D D0 EB                                      SHR BL,1
014F FE C7                                      INC BH
                                            .UNTIL !CARRY?
0153 8A C7                                  MOV AL,BH
0155 8B 1E 0194 R                           MOV BX,PNTR
0159 2E: 88 07                              MOV CS:[BX],AL ;key code to queue
015C FF 06 0194 R                           INC PNTR
                                            .IF BX == OFFSET DBCNT-1
0166 C7 06 0194 R 0182 R                        MOV PNTR, OFFSET QUEUE
                                            .ENDIF
016C 5B                                     POP BX
                                        .ENDIF
                                    .ENDIF

                                .ELSE                ;no key found
016F FE 0E 0192 R                   DEC DBCNT        ;decrement bounce count
```

```
                                             .IF SIGN?
0175 C6 06 0192 R 00                             MOV   DBCNT,0   ;clear count and flag
017A C6 06 0193 R 00                             MOV   DBF,0
                                             .ENDIF
                                         .ENDIF
                                         IRET
0182                          INTKEY ENDP
0182 0010 [                   QUEUE DB 16 DUP(?)              ;keyboard queue
       00
         ]
0192 00                       DBCNT DB 0                      ;debounce count
0193 00                       DBF   DB 0                      ;debounce flag
0194 0182 R                   PNTR  DW QUEUE                  ;pointer to queue
                              END
```

The keyboard-interrupt finds the key and stores the key-code in the queue. The code stored in the queue is a raw code that does not indicate the key number. For example, the key code for the 1-key is a 00H, the key code for a 4-key is a 01H, etc. There is no provision for a queue over-flow in this software. It could be added, but in almost all cases it is difficult to out-type a 16 bytes queue.

Example 12–16 illustrates a procedure that removes data from the keyboard queue. This procedure is not interrupt-driven and is called only when information from the keyboard is needed in a program. Example 12–17 shows the caller software for the key procedure.

EXAMPLE 12–16

```
0198 01 04 07 0A      LOOK  DB 1,4,7,10              ;look up table
019C 02 05 08 00            DB 2,5,8,0
01A0 03 06 09 0B            DB 3,6,9,11

01A4                  KEY   PROC   NEAR   USES BX
01A5 8B 1E 0196 R           MOV    BX,OPNTR
                            .IF BX == PNTR           ;queue empty
01AF F9                         STC
                            .ELSE
01B2 2E: 8A 07                  MOV AL,CS:[BX]       ;get code from queue
01B5 43                         INC BX
                                .IF BX == OFFSET DBCNT-1
01BC BB 0182 R                     MOV BX,OFFSET QUEUE
                                .ENDIF
01BF 89 1E 0196 R               MOV OPNTR,BX
01C3 BB 0198 R                  MOV BX,OFFSET LOOK
01C6 2E: D7                     XLAT CS:LOOK
01C8 F8                         CLC
                            .ENDIF
                            RET
01CB                  KEY   ENDP
```

EXAMPLE 12–17

```
                      .REPEAT
01CB E8 FFD6            CALL    KEY
                      .UNTIL !CARRY?
```

12–6   SUMMARY

1. An interrupt is a hardware- or software-initiated call that interrupts the currently executing program at any point and calls a procedure. The procedure is called by the interrupt handler or an interrupt service procedure.

2. Interrupts are useful when an I/O device needs to be serviced only occasionally at low data-transfer rates.

3. The microprocessor has five instructions that apply to interrupts: BOUND, INT, INT 3, INTO, and IRET. The INT and INT 3 instructions call procedures with addresses stored in the interrupt vector whose type is indicated by the instruction. The BOUND instruction is a conditional interrupt that uses interrupt vector type number 5. The INTO instruction is a conditional interrupt that interrupts a program only if the overflow flag is set. Finally, the IRET instruction is used to return from interrupt service procedures.

4. The microprocessor has three pins that apply to its hardware interrupt structure: INTR, NMI, and $\overline{INTA}$. The interrupt inputs are INTR and NMI, which are used to request interrupts and $\overline{INTA}$ is an output used to acknowledge the INTR interrupt request.

5. Real mode interrupts are referenced through a vector table that occupies memory locations 00000H–003FFH. Each interrupt vector is four bytes long and contains the offset and segment addresses of the interrupt service procedure. In protected mode, the interrupts reference the interrupt descriptor table (IDT) that contains 256 interrupt descriptors. Each interrupt descriptor contains a segment selector and a 32-bit offset address.

6. Two flag bits are used with the interrupt structure of the microprocessor: trap (TF) and interrupt enable (IF). The IF flag bit enables the INTR interrupt input, and the TF flag bit causes interrupts to occur after the execution of each instruction, as long as TF is active.

7. The first 32 interrupt vector locations are reserved for Intel use, with many predefined in the microprocessor. The last 224 interrupt vectors are for user use and can perform any function desired.

8. Whenever an interrupt is detected, the following events occur: (1) the flags are pushed onto the stack, (2) the IF and TF flag bits are both cleared, (3) the IP and CS registers are both pushed onto the stack, and (4) the interrupt vector is fetched from the interrupt vector table and the interrupt service subroutine is accessed through the vector address.

9. Tracing or single-stepping is accomplished by setting the TF flag bit. This causes an interrupt to occur after the execution of each instruction for debugging.

10. The non-maskable interrupt input (NMI) calls the procedure whose address is stored at interrupt vector type number 2. This input is positive-edge triggered.

11. The INTR pin is not internally decoded, as is the NMI pin. Instead, $\overline{INTA}$ is used to apply the interrupt vector type number to data bus connections D0–D7 during the $\overline{INTA}$ pulse.

12. Methods of applying the interrupt vector type number to the data bus during $\overline{INTA}$ vary widely. One method uses resisters to apply interrupt type number FFH to the data bus, while another uses a three-state buffer to apply any vector type number.

13. The 8259A programmable interrupt controller (PIC) adds at least eight interrupt inputs to the microprocessor. If more interrupts are needed, this device can be cascaded to provide up to 64 interrupt inputs.

14. Programming the 8259A is a two-step process. First, a series of initialization command words (ICWs) are sent to the 8259A, then a series of operation command words (OCWs) are sent.

15. The 8259A contains three status registers: IMR (interrupt mask register), ISR (in-service register), and IRR (interrupt request register).

16. A real-time clock is used to keep time in real-time. In most cases, time is stored in either binary or BCD form in several memory locations.

12–7   QUESTIONS AND PROBLEMS

1. What is interrupted by an interrupt?
2. Define the term *interrupt*.

3. What is called by an interrupt?
4. Why do interrupts free up time for the microprocessor?
5. List the interrupt pins found on the microprocessor.
6. List the five interrupt instructions for the microprocessor.
7. What is an interrupt vector?
8. Where are the interrupt vectors located in the microprocessor's memory?
9. How many different interrupt vectors are found in the interrupt vector table?
10. Which interrupt vectors are reserved by Intel?
11. Explain how a type 0 interrupt occurs.
12. Where is the interrupt descriptor table located for protected mode operation?
13. Each protected mode interrupt descriptor contains what information?
14. Describe the differences between a protected and real mode interrupt.
15. Describe the operation of the BOUND instruction.
16. Describe the operation of the INTO instruction.
17. What memory locations contain the vector for an INT 44H instruction?
18. Explain the operation of the IRET instruction.
19. What is the purpose of interrupt vector type number 7?
20. List the events that occur when an interrupt becomes active.
21. Explain the purpose of the interrupt flag (IF).
22. Explain the purpose of the trap flag (TF).
23. How is IF cleared and set?
24. How is TF cleared and set?
25. The NMI interrupt input automatically vectors through which vector type number?
26. Does the $\overline{\text{INTA}}$ signal activate for the NMI pin?
27. The INTR input is _____ -sensitive.
28. The NMI input is _____ -sensitive.
29. When the $\overline{\text{INTA}}$ signal becomes a logic 0, it indicates that the microprocessor is waiting for an interrupt _____ number to be placed on the data bus (D0–D7).
30. What is a FIFO?
31. Develop a circuit that places interrupt type number 86H on the data bus in response to the INTR input.
32. Develop a circuit that places interrupt type number CCH on the data bus in response to the INTR input.
33. Explain why pull-up resistors on D0–D7 cause the microprocessor to respond with interrupt vector type number FFH for the $\overline{\text{INTA}}$ pulse.
34. What is a daisy-chain?
35. Why must interrupting devices be polled in a daisy-chained interrupt system?
36. What is the 8259A?
37. How many 8259As are required to have 64 interrupt inputs?
38. What is the purpose of the IR0–IR7 pins on the 8259A?
39. When are the CAS2–CAS0 pins used on the 8259A?
40. Where is a slave INT pin connected on the master 8259A in a cascaded system?
41. What is an ICW?
42. What is an OCW?
43. How many ICWs are needed to program the 8259A when operated as a single master in a system?
44. Where is the vector type number stored in the 8259A?
45. Where is the sensitivity of the IR pins programmed in the 8259A?

46. What is the purpose of ICW1?
47. What is a non-specific EOI?
48. Explain priority rotation in the 8259A.
49. What is the purpose of IRR in the 8259A?
50. At which I/O ports is the master 8259A PIC found in the personal computer?
51. At which I/O ports is the slave 8259A found in the personal computer?

CHAPTER 13

Direct Memory Access and DMA-Controlled I/O

INTRODUCTION

In previous chapters, we discussed basic and interrupt-processed I/O. Now we turn to the final form of I/O called **direct memory access** (DMA). The DMA I/O technique provides direct access to the memory while the microprocessor is temporarily disabled. This allows data to be transferred between memory and the I/O device at a rate that is limited only by the speed of the memory components in the system or the DMA controller. The DMA transfer speed can approach 32–40 M-byte transfer rates with today's high-speed RAM memory components.

DMA transfers are used for many purposes, but more common are DRAM refresh, video displays for refreshing the screen, and disk memory system reads and writes. The DMA transfer is also used to do high-speed memory-to-memory transfers.

This chapter also explains the operation of disk memory systems and video systems that are often DMA-processed. Disk memory includes floppy, fixed, and optical disk storage. Video systems include digital and analog monitors.

CHAPTER OBJECTIVES

Upon completion of this chapter, you will be able to:

1. Describe a DMA transfer.
2. Explain the operation of the HOLD and HLDA direct memory access control signals.
3. Explain the function of the 8237 DMA controller when used for DMA transfers.
4. Program the 8237 to accomplish DMA transfers.
5. Describe the disk standards found in personal computer systems.
6. Describe the various video interface standards that are found in the personal computer.

13–1   BASIC DMA OPERATION

Two control signals are used to request and acknowledge a direct memory access (DMA) transfer in the microprocessor-based system. The HOLD pin is an input that is used to request a DMA action and the HLDA pin is an output that acknowledges the DMA action. Figure 13–1 shows the timing that is typically found on these two DMA control pins.

FIGURE 13–1 HOLD and HLDA timing for the microprocessor.

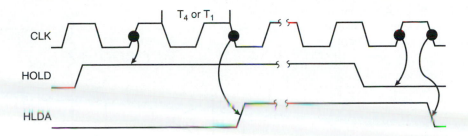

Whenever the HOLD input is placed at a logic 1 level, a DMA action (hold) is requested. The microprocessor responds, within a few clocks, by suspending the execution of the program and by placing its address, data, and control bus at their high-impedance states. The high-impedance state causes the microprocessor to appear as if it has been removed from its socket. This state allows external I/O devices or other microprocessors to gain access to the system buses so that memory can be accessed directly.

As the timing diagram indicates, HOLD is sampled in the middle of any clocking cycle. Thus, the hold can take effect any time during the operation of any instruction in the microprocessor's instruction set. As soon as the microprocessor recognizes the hold, it stops executing software and enters hold cycles. Note that the HOLD input has a higher priority than the INTR or NMI interrupt inputs. Interrupts take effect at the end of an instruction, while a HOLD takes effect in the middle of an instruction. The only microprocessor pin that has a higher priority than a HOLD is the RESET pin. Note that the HOLD input may not be active during a RESET or the reset is not guaranteed.

The HLDA signal becomes active to indicate that the microprocessor has indeed placed its buses at their high-impedance state, as can be seen in the timing diagram. Note that there are a few clock cycles between the time that HOLD changes and until HLDA changes. The HLDA output is a signal to the external requesting device that the microprocessor has relinquished control of its memory and I/O space. You could call the HOLD input a DMA request input and the HLDA output a DMA grant signal.

Basic DMA Definitions

Direct memory accesses normally occur between an I/O device and memory without the use of the microprocessor. A **DMA read** transfers data from the memory to the I/O device. A **DMA write** transfers data from an I/O device to memory. In both operations, the memory and I/O are controlled simultaneously, which is why the system contains separate memory and I/O control signals. This special control bus structure of the microprocessor allows DMA transfers. A DMA read causes both the $\overline{\text{MRDC}}$ and $\overline{\text{IOWC}}$ signals to simultaneously activate, transferring data from the memory to the I/O device. A DMA write causes the $\overline{\text{MWTC}}$ and $\overline{\text{IORC}}$ signals to both activate. These control bus signals are available to all microprocessors in the Intel family except the 8086/8088 system. The 8086/8088 require their generation with either a system controller or a circuit such as the one illustrated in Figure 13–2. The DMA controller provides the memory with its address and a signal from the controller ($\overline{\text{DACK}}$) selects the I/O device during the DMA transfer.

The data transfer speed is determined by the speed of the memory device or a DMA controller that often controls DMA transfers. If the memory speed is 100 ns, DMA transfers occur at rates of up to 1/100 ns or 10 M-bytes per second. If the DMA controller in a system functions at a maximum rate of 5 MHz and we still use 100 ns memory, the maximum transfer rate is 5 MHz because the DMA controller is slower than the memory. In many cases, the DMA controller slows the speed of the system when DMA transfers occur.

FIGURE 13–2   A circuit that generates system control signals in a DMA environment.

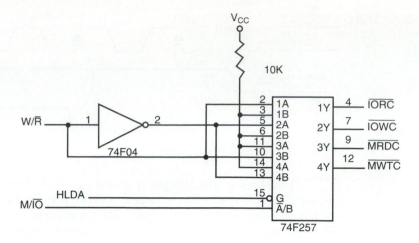

13–2   THE 8237 DMA CONTROLLER

The 8237 DMA controller supplies the memory and I/O with control signals and memory address information during the DMA transfer. The 8237 is actually a special-purpose microprocessor whose job is high-speed data transfer between memory and the I/O. Figure 13–3 shows the pin-out and block diagram of the 8237 programmable DMA controller. Although this device may not appear as a discrete component in modern microprocessor-based systems, it does appear within system controller chip-sets found in most systems. Although not described because of its complexity, the chip set (82357 ISP or integrated system peripheral controller) and its integral set of two DMA controllers are programmed exactly as the 8237. The ISP also provides a pair of 8259A programmable interrupt controllers for the system.

The 8237 is a four-channel device that is compatible with the 8086/8088 microprocessors. The 8237 can be expanded to include any number of DMA channel inputs, although four channels seem to be adequate for many small systems. The 8237 is capable of DMA transfers at rates of up to 1.6M bytes per second. Each channel is capable of addressing a full 64K-byte section of memory and can transfer up to 64K bytes with a single programming.

Pin Definitions

CLK	The **clock** input is connected to the system clock signal as long as that signal is 5 MHz or less. In the 8086/8088 system, the clock must be inverted for the proper operation of the 8237.
$\overline{\text{CS}}$	**Chip select** enables the 8237 for programming. The $\overline{\text{CS}}$ pin is normally connected to the output of a decoder. The decoder does not use the 8086/8088 control signal IO/$\overline{\text{M}}$ (M/$\overline{\text{IO}}$) because it contains the new memory and I/O control signals ($\overline{\text{MEMR}}$, $\overline{\text{MEMW}}$, $\overline{\text{IOR}}$, and $\overline{\text{IOW}}$).
RESET	The **reset** pin clears the command, status, request, and temporary registers. It also clears the first/last flip-flop and sets the mask register. This input primes the 8237 so it is disabled until programmed otherwise.
READY	A logic 0 on the **ready** input causes the 8237 to enter wait states for slower memory and/or I/O components.

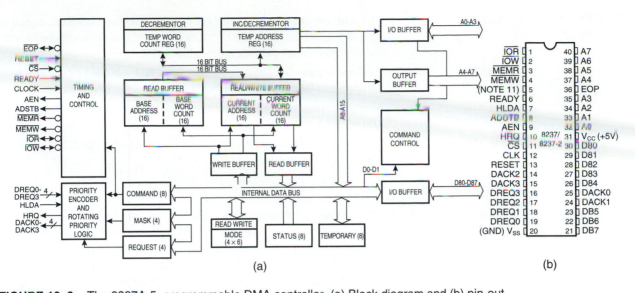

FIGURE 13–3   The 8237A-5 programmable DMA controller. (a) Block diagram and (b) pin-out.
(Courtesy of Intel Corporation.)

HLDA	A **hold acknowledge** signals the 8237 that the microprocessor has relinquished control of the address, data, and control buses.
DREQ3–DREQ0	The **DMA request** inputs are used to request a DMA transfer for each of the four DMA channels. Because the polarity of these inputs is programmable, they are either active-high or active-low inputs.
DB7–DB0	The **data bus** pins are connected to the microprocessor data bus connections and are used during the programming of the DMA controller.
IOR	**I/O read** is a bi-directional pin used during programming and during a DMA write cycle.
IOW	**I/O write** is a bi-directional pin used during programming and during a DMA read cycle.
EOP	**End-of-process** is a bi-directional signal that is used as an input to terminate a DMA process or as an output to signal the end of the DMA transfer. This input is often used to interrupt a DMA transfer at the end of a DMA cycle.
A3–A0	These **address pins** select an internal register during programming and also provide part of the DMA transfer address during a DMA action.
A7–A4	These **address pins** are outputs that provide part of the DMA transfer address during a DMA action.
HRQ	**Hold request** is an output that connects to the HOLD input of the microprocessor in order to request a DMA transfer.
DACK3–DACK0	**DMA channel acknowledge** outputs acknowledge a channel DMA request. These outputs are programmable as either active-high or active-low signals. The DACK outputs are often used to select the DMA controlled I/O device during the DMA transfer.

AEN   The **address enable** signal enables the DMA address latch connected to the DB7–DB0 pins on the 8237. It is also used to disable any buffers in the system connected to the microprocessor.

ADSTB   **Address strobe** functions as ALE, except that it is used by the DMA controller to latch address bits A15–A8 during the DMA transfer.

MEMR   **Memory read** is an output that causes memory to read data during a DMA read cycle.

MEMW   **Memory write** is an output that causes memory to write data during a DMA write cycle.

Internal Registers

CAR   The **current address register** is used to hold the 16-bit memory address used for the DMA transfer. Each channel has its own current address register for this purpose. When a byte of data is transferred during a DMA operation, the CAR is either incremented or decremented, depending on how it is programmed.

CWCR   The **current word count register** programs a channel for the number of bytes (up to 64K) transferred during a DMA action. The number loaded into this register is one less than the number of bytes transferred. For example, if a 10 is loaded into the CWCR, then 11 bytes are transferred during the DMA action.

BA and BWC   The **base address** (BA) and **base word count** (BWC) registers are used when auto-initialization is selected for a channel. In the auto-initialization mode, these registers are used to reload both the CAR and CWCR after the DMA action is completed. This allows the same count and address to be used to transfer data from the same memory area.

CR   The **command register** programs the operation of the 8237 DMA controller. Figure 13–4 depicts the function of the command register.
   The command register uses bit position 0 to select the memory-to-memory DMA transfer mode. Memory-to-memory DMA transfers use

FIGURE 13–4   8237A-5 command register. (Courtesy of Intel Corporation.)

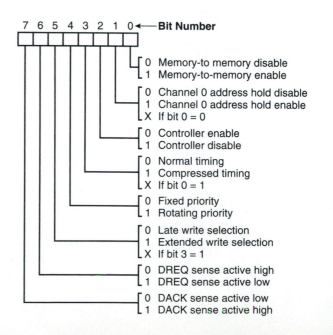

DMA channel 0 to hold the source address and DMA channel 1 to hold the destination address. (This is similar to the operation of a MOVSB instruction.) A byte is read from the address accessed by channel 0 and saved within the 8237 in a temporary holding register. Next, the 8237 initiates a memory write cycle, in which the contents of the temporary holding register are written into the address selected by DMA channel 1. The number of bytes transferred is determined by the channel 1 count register.

The channel 0 address hold enable bit (bit position 1) programs channel 0 for memory-to-memory transfers. For example, if you must fill an area of memory with data, channel 0 can be held at the same address while channel 1 changes for memory-to-memory transfer. This copies the contents of the address accessed by channel 0 into a block of memory accessed by channel 1.

The controller enable/disable bit (bit position 2) turns the entire controller on and off. The normal and compressed bit (bit position 3) determine whether a DMA cycle contains 2 (compressed) or 4 (normal) clocking periods. Bit position 5 is used in normal timing to extend the write pulse so it appears one clock earlier in the timing for I/O devices that require a wider write pulse.

Bit position 4 selects priority for the four DMA channel DREQ inputs. In the fixed priority scheme, channel 0 has the highest priority and channel 3 has the lowest. In the rotating priority scheme, the most recently serviced channel assumes the lowest priority. For example, if channel 2 just had access to a DMA transfer, it assumes the lowest priority and channel 3 assumes the highest priority position. Rotating priority is an attempt to give all channels equal priority.

The remaining two bits (bit positions 6 and 7) program the polarities of the DREQ inputs and the DACK outputs.

MR   The **mode register** programs the mode of operation for a channel. Note that each channel has its own mode register (see Figure 13–5), as selected by bit positions 1 and 0. The remaining bits of the mode register select the operation, auto-initialization, increment/decrement, and mode for the channel. Verification operations generate the DMA addresses without generating the DMA memory and I/O control signals.

FIGURE 13–5   8237A-5 mode register (Courtesy of Intel Corporation).

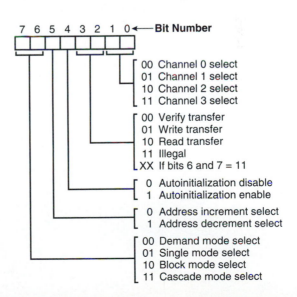

7 6 5 4 3 2 1 0 ←— Bit Number

00 Channel 0 select
01 Channel 1 select
10 Channel 2 select
11 Channel 3 select

00 Verify transfer
01 Write transfer
10 Read transfer
11 Illegal
XX If bits 6 and 7 = 11

0 Autoinitialization disable
1 Autoinitialization enable

0 Address increment select
1 Address decrement select

00 Demand mode select
01 Single mode select
10 Block mode select
11 Cascade mode select

The modes of operation include demand mode, single mode, block mode, and cascade mode. Demand mode transfers data until an external EOP is input or until the DREQ input becomes inactive. Single mode releases the HOLD after each byte of data is transferred. If the DREQ pin is held active, the 8237 again requests a DMA transfer through the DRQ line to the microprocessor's HOLD input. Block mode automatically transfers the number of bytes indicated by the count register for the channel. DREQ need not be held active through the block mode transfer. Cascade mode is used when more than one 8237 is present in a system.

RR   The **request register** is used to request a DMA transfer via software (see Figure 13–6). This is very useful in memory-to-memory transfers, where an external signal is not available to begin the DMA transfer.

MRSR   The **mask register set/reset** sets or clears the channel mask, as illustrated in Figure 13–7. If the mask is set, the channel is disabled. Recall that the RESET signal sets all channel masks to disable them.

MSR   The **mask register** (see Figure 13–8) clears or sets all of the masks with one command instead of individual channels, as with the MRSR.

FIGURE 13–6   8237A-5 request register. (Courtesy of Intel Corporation.)

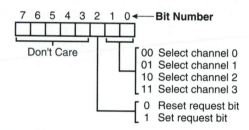

FIGURE 13–7   8237A-5 mask register set/reset mode. (Courtesy of Intel Corporation.)

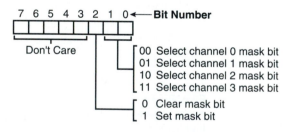

FIGURE 13–8   8237A-5 mask register. (Courtesy of Intel Corporation.)

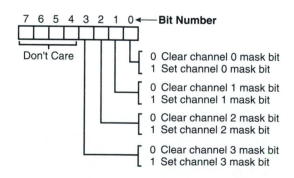

FIGURE 13–9 8237A-5 status register. (Courtesy of Intel Corporation.)

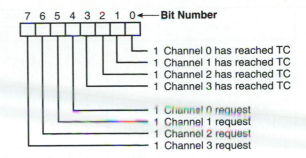

SR   The **status register** shows the status of each DMA channel (see Figure 13–9). The TC bits indicate whether the channel has reached its terminal count (transferred all its bytes). Whenever the terminal count is reached, the DMA transfer is terminated for most modes of operation. The request bits indicate whether the DREQ input for a given channel is active.

Software Command

Three software commands are used to control the operation of the 8237. These commands do not have a binary bit pattern, as do the various control registers within the 8237. A simple output to the correct port number enables the software command. Figure 13–10 shows the I/O port assignments that access all registers and the software commands.

The function of the software commands are explained in the following list:

1. **Clear the first/last flip-flop**—Clears the first/last (F/L) flip-flop within the 8237. The F/L flip-flop selects which byte (low or high order) is read/written in the current address and current count registers. If F/L = 0, the low order byte is selected; if F/L = 1, the high order byte is selected. Any read or write to the address or count register automatically toggles the F/L flip-flop.
2. **Master clear**—Acts exactly the same as the RESET signal to the 8237. As with the RESET signal, this command disables all channels.
3. **Clear mask register**—Enables all four DMA channels.

FIGURE 13–10 8237A-5 command and control port assignments. (Courtesy of Intel Corporation.)

Signals						Operation
A3	A2	A1	A0	IOR	IOW	
1	0	0	0	0	1	Read Status Register
1	0	0	0	1	0	Write Command Register
1	0	0	1	0	1	Illegal
1	0	0	1	1	0	Write Request Register
1	0	1	0	0	1	Illegal
1	0	1	0	1	0	Write Single Mask Register Bit
1	0	1	1	0	1	Illegal
1	0	1	1	1	0	Write Mode Register
1	1	0	0	0	1	Illegal
1	1	0	0	1	0	Clear Byte Pointer Flip/Flop
1	1	0	1	0	1	Read Temporary Register
1	1	0	1	1	0	Master Clear
1	1	1	0	0	1	Illegal
1	1	1	0	1	0	Clear Mask Register
1	1	1	1	0	1	Illegal
1	1	1	1	1	0	Write All Mask Register Bits

Channel	Register	Operation	CS	IOR	IOW	A3	A2	A1	A0	Internal Flip-Flop	Data Bus DB0-DB7
0	Base and Current Address	Write	0	1	0	0	0	0	0	0	A0-A7
			0	1	0	0	0	0	0	1	A8-A15
	Current Address	Read	0	0	1	0	0	0	0	0	A0-A7
			0	0	1	0	0	0	0	1	A8-A15
	Base and Current Word Count	Write	0	1	0	0	0	0	1	0	W0-W7
			0	1	0	0	0	0	1	1	W8-W15
	Current Word Count	Read	0	0	1	0	0	0	1	0	W0-W7
			0	0	1	0	0	0	1	1	W8-W15
1	Base and Current Address	Write	0	1	0	0	0	1	0	0	A0-A7
			0	1	0	0	0	1	0	1	A8-A15
	Current Address	Read	0	0	1	0	0	1	0	0	A0-A7
			0	0	1	0	0	1	0	1	A8-A15
	Base and Current Word Count	Write	0	1	0	0	0	1	1	0	W0-W7
			0	1	0	0	0	1	1	1	W8-W15
	Current Word Count	Read	0	0	1	0	0	1	1	0	W0-W7
			0	0	1	0	0	1	1	1	W8-W15
2	Base and Current Address	Write	0	1	0	0	1	0	0	0	A0-A7
			0	1	0	0	1	0	0	1	A8-A15
	Current Address	Read	0	0	1	0	1	0	0	0	A0-A7
			0	0	1	0	1	0	0	1	A8-A15
	Base and Current Word Count	Write	0	1	0	0	1	0	1	0	W0-W7
			0	1	0	0	1	0	1	1	W8-W15
	Current Word Count	Read	0	0	1	0	1	0	1	0	W0-W7
			0	0	1	0	1	0	1	1	W8-W15
3	Base and Current Address	Write	0	1	0	0	1	1	0	0	A0-A7
			0	1	0	0	1	1	0	1	A8-A15
	Current Address	Read	0	0	1	0	1	1	0	0	A0-A7
			0	0	1	0	1	1	0	1	A8-A15
	Base and Current Word Count	Write	0	1	0	0	1	1	1	0	W0-W7
			0	1	0	0	1	1	1	1	W8-W15
	Current Word Count	Read	0	0	1	0	1	1	1	0	W0-W7
			0	0	1	0	1	1	1	1	W8-W15

FIGURE 13–11   8237A-5 DMA channel I/O port addresses. (Courtesy of Intel Corporation.)

Programming the Address and Count Registers

Figure 13–11 illustrates the I/O port locations for programming the count and address registers for each channel. Notice that the state of the F/L flip-flop determines whether the LSB or MSB is programmed. If the state of the F/L flip-flop is unknown, the count and address could be programmed incorrectly. It is also important that the DMA channel be disabled before its address and count are programmed.

There are four steps required to program the 8237: (1) the F/L flip-flop is cleared using a clear F/L command, (2) the channel is disabled, (3) the LSB and then MSB of the address are programmed, and (4) the LSB and MSB of the count are programmed. Once these four operations are performed, the channel is programmed and ready to use. Additional programming is required to select the mode of operation before the channel is enabled and started.

The 8237 Connected to the 80X86 Microprocessor

Figure 13–12 shows an 80X86-based system that contains the 8237 DMA controller.

The address enable (AEN) output of the 8237 controls the output pins of the latches and the outputs of the 74LS257 (E). During normal 80X86 operation (AEN = 0), latches A and C and the multiplexer (E) provide address bus bits A19–A16 and A7–A0. The multiplexer provides the system control signals as long as the 80X86 is in control of the system. During a DMA action (AEN = 1), latches A and C are disabled along with the multiplexer (E). Latches D and B now provide address bits A19–A16 and A15–A8. Address bus bits A7–A0 are provided directly by the 8237 and contain a part of the DMA transfer address. The control signals $\overline{MEMR}$, $\overline{MEMW}$, $\overline{IOR}$, and $\overline{IOW}$ are provided by the DMA controller.

The address strobe output (ADSTB) of the 8237 clocks the address (A15–A8) into latch D during the DMA action so that the entire DMA transfer address becomes available on the address bus. Address bus bits A19–A16 are provided by latch B, which must be programmed with these four address bits before the controller is enabled for the DMA transfer. The DMA operation of the 8237 is limited to a transfer of not more than 64K bytes within the same 64K-byte section of the memory.

The decoder (F) selects the 8237 for programming and the 4-bit latch (B) for the upper-most four address bits. The latch in a PC is called the DMA page register (8-bits) that holds address bits A16–A23 for a DMA transfer. A high page register also exists, but its address is chip-dependent. The port numbers for the DMA page registers are listed in Table 13–1 (these are for the Intel ISP). The decoder in this system enables the 8237 for I/O port addresses XX60H–XX7FH, and the I/O latch (B) for ports XX00H–XX1FH. Notice that the decoder output is combined with the $\overline{IOW}$ signal to generate an active-high clock for the latch (B).

During normal 80X86 operation, the DMA controller and integrated circuits B and D are disabled. During a DMA action, integrated circuits A, C, and E are disabled so that the 8237 can take control of the system through the address, data, and control buses.

In the personal computer, the two DMA controllers are programmed at I/O ports 0000H–000FH for DMA channels 0–3, and at ports 00C0H–00DFH for DMA channels 4–7. Note that the second controller is programmed at even addresses only, so the channel 4 base and current address is programmed at I/O port 00C0H and the channel 4 base and current count is programmed at port 00C2H. The page register, which holds address bits A23–A16 of the DMA address, are located at I/O ports 0087H (CH–0), 0083H (CH–1), 0081H (CH–2), 0082H (CH–3), (no channel 4), 008BH (CH–5), 0089H (CH–6) and 008AH (CH–7). The page register functions as the address latch described with the examples in this text.

Memory-to-Memory Transfer with the 8237

The memory-to-memory transfer is much more powerful than even the automatically repeated MOVSB instruction. While the repeated MOVSB instruction tables the 8088 4.2 μs per byte, the 8237 requires only 2.0 μs per byte, which is over twice as fast as a software data transfer. This is not true if an 80386, 80846, or Pentium through Pentium 4 is in use in the system.

Sample Memory-to-Memory DMA Transfer. Suppose that the contents of memory locations 10000H–13FFFH are to be transferred into memory locations 14000H–17FFFH. This is accomplished with a repeated string move instruction, or, at a much faster rate, with the DMA controller.

Example 13–1 illustrates the software required to initialize the 8237 and program latch B in Figure 13–12 for this DMA transfer. This software is written for an embedded application. For it to function in the PC, you must use the port addresses listed in Table 13–1 for the page registers.

EXAMPLE 13–1

```
;A procedure that transfers a block of data using the
;8237A DMA controller in Figure 13-12. This is a
;memory-to-memory block transfer.
;
;Calling parameters:
;    SI = source address
;    DI = destination address
;    CX = count
```

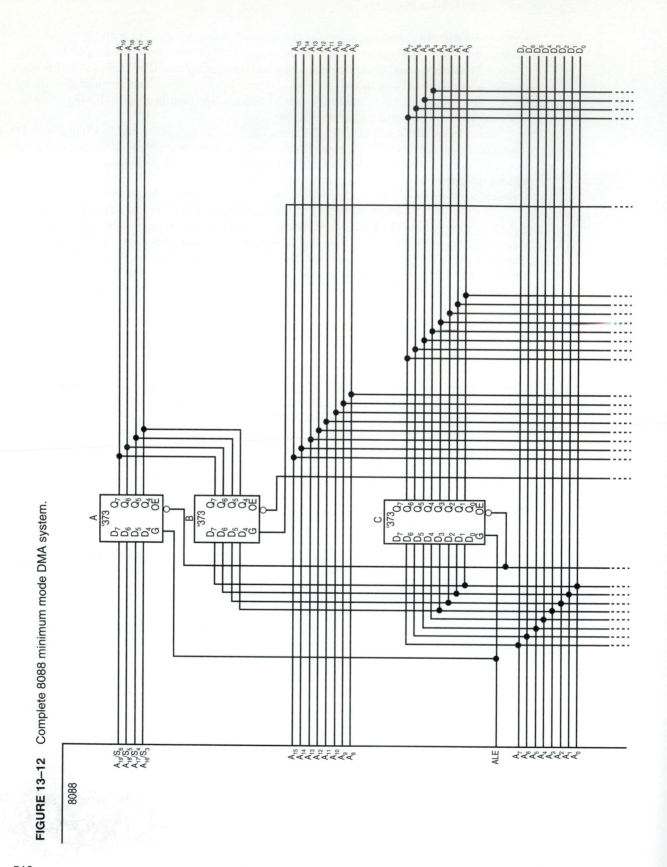

FIGURE 13–12 Complete 8088 minimum mode DMA system.

512

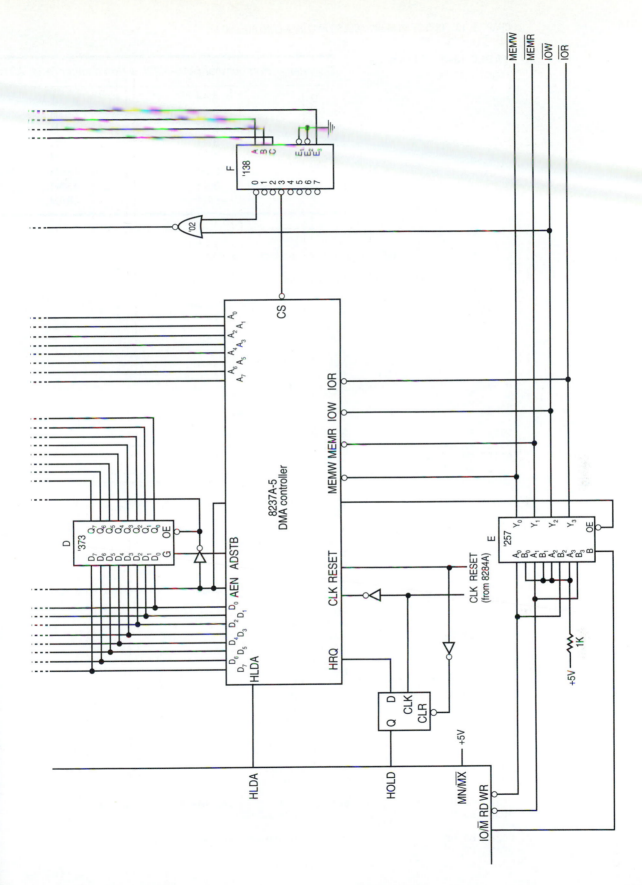

TABLE 13–1   DMA page register ports.

Channel	Port number (A16–A23)	Port number (A24–A31)
0	87H	487H
1	83H	483H
2	81H	481H
3	82H	482H
4	8FH	48FH
5	8BH	48BH
6	89H	489H
7	8AH	48AH

```
                   ;      ES = segment of source and destination
       ;
= 0010             LATCHB  EQU    10H         ;latch B
= 007C             CLEAR_F EQU    7CH         ;F/L flip flop
= 0070             CH0_A   EQU    70H         ;channel 0 address
= 0072             CH1_A   EQU    72H         ;channel 1 address
= 0073             CH1_C   EQU    73H         ;channel 1 count
= 007B             MODE    EQU    7BH         ;mode
= 0078             CMMD    EQU    78H         ;command
= 007F             MASKS   EQU    7FH         ;masks
= 0079             REQ     EQU    79H         ;request register
= 0078             STATUS  EQU    78H         ;status register

0000               TRANS   PROC   FAR USES AX

0001  8C C0                MOV    AX,ES       ;program latch B
0003  8A C4                MOV    AL,AH
0005  C0 E8 04             SHR    AL,4
0008  E6 10                OUT    LATCHB,AL
000A  E6 7C                OUT    CLEAR_F,AL  ;clear F/L flip-flop

000C  8C C0                MOV    AX,ES       ;program source address
000E  C1 E0 04             SHL    AX,4
0011  03 C6                ADD    AX,SI       ;form source offset
0013  E6 70                OUT    CH0_A,AL
0015  8A C4                MOV    AL,AH
0017  E6 70                OUT    CH0_A,AL

0019  8C C0                MOV    AX,ES       ;program destination address
001B  C1 E0 04             SHL    AX,4
001E  03 C7                ADD    AX,DI       ;form destination offset
0020  E6 72                OUT    CH1_A,AL
0022  8A C4                MOV    AL,AH
0024  E6 72                OUT    CH1_A,AL

0026  8B C1                MOV    AX,CX       ;program count
0028  48                   DEC    AX          ;adjust count
0029  E6 73                OUT    CH1_C,AL
002B  8A C4                MOV    AL,AH
002D  E6 73                OUT    CH1_C,AL

002F  B0 88                MOV    AL,88H      ;program mode
0031  E6 7B                OUT    MODE,AL
0033  B0 85                MOV    AL,85H
0035  E6 7B                OUT    MODE,AL

0037  B0 01                MOV    AL,1        ;enable block transfer
0039  E6 78                OUT    CMMD,AL

003B  B0 0E                MOV    AL,0EH      ;unmask channel 0
003D  E6 7F                OUT    MASKS,AL
```

```
003F  B0 04              MOV    AL,4          ;start DMA transfer
0041  E6 79              OUT    REQ,AL

                         .REPEAT              ;wait until DMA complete
0043  E4 78              IN     AL,STATUS
                         .UNTIL AL &1

                         RET

         TRANS           ENDP
```

Programming the DMA controller requires a few steps, as illustrated in Example 13–1. The leftmost digit of the 5-digit address is sent to latch B. Next, the channels are programmed after the F/L flip-flop is cleared. Note that we use channel 0 as the source and channel 1 as the destination for a memory-to-memory transfer. The count is next programmed with a value that is one less than the number of bytes to be transferred. Next, the mode register of each channel is programmed, the command register selects a block move, channel 0 is enabled, and a software DMA request is initiated. Before return is made from the procedure, the status register is tested for a terminal count. Recall that the terminal count flag indicates that the DMA transfer is completed. The TC also disables the channel, preventing additional transfers.

Sample Memory Fill Using the 8237.   In order to fill an area of memory with the same data, the channel 0 source register is programmed to point to the same address throughout the transfer. This is accomplished with the channel 0 hold mode. The controller copies the contents of this single memory location to an entire block of memory addressed by channel 1. This has many useful applications.

For example, suppose that a video display must be cleared. This operation can be performed using the DMA controller with the channel 0 hold mode and a memory-to-memory transfer. If the video display contains 80 columns and 25 lines, it has 2000 display positions that must be set to 20H (an ASCII space) to clear the screen.

Example 13–2 shows a procedure that clears an area of memory addressed by ES:DI. The CX register transfers the number of bytes to be cleared to the CLEAR procedure. Notice that this procedure is nearly identical to Example 13–1, except that the command register is programmed so the channel 0 address is held. The source address is programmed as the same address as ES:DI, and then the destination is programmed as one location beyond ES:DI. Also note that this program is designed to function with the hardware in Figure 13–12 and will not function in the personal computer unless you have the same hardware.

EXAMPLE 13–2

```
                         ;A procedure that clears an area of memory using the
                         ;8237A DMA controller in Figure 13-12. This is a
                         ;memory-to-memory block transfer with a channel 0 hold.
                         ;
                         ;Calling parameters:
                         ;    DI = offset address of area cleared
                         ;    ES = segment address of area cleared
                         ;    CX = number of bytes cleared
                         ;
= 0010                   LATCHB   EQU   10H        ;latch B
= 007C                   CLEAR_F  EQU   7CH        ;F/L flip flop
= 0070                   CH0_A    EQU   70H        ;channel 0 address
= 0072                   CH1_A    EQU   72H        ;channel 1 address
= 0073                   CH1_C    EQU   73H        ;channel 1 count
= 007B                   MODE     EQU   7BH        ;mode
= 0078                   CMMD     EQU   78H        ;command
= 007F                   MASKS    EQU   7FH        ;masks
= 0079                   REQ      EQU   79H        ;request register
```

```
= 0078                    STATUS  EQU   78H          ;status register
= 0000                    ZERO    EQU   0H           ;zero

0000                      CLEAR   PROC  FAR USES AX

0001  8C C0                       MOV   AX,ES        ;program latch B
0003  8A C4                       MOV   AL,AH
0005  C0 E8 04                    SHR   AL,4
0008  E6 10                       OUT   LATCHB,AL

000A  E6 7C                       OUT   CLEAR_F,AL   ;clear F/L flip-flop

000C  2E: A0 0000                 MOV   AL,CS:ZERO
0010  26: 88 05                   MOV   ES:[DI],AL   ;save zero in first byte

0013  8C C0                       MOV   AX,ES        ;program source address
0015  C1 E0 04                    SHL   AX,4
0018  03 C7                       ADD   AX,DI        ;form source offset
001A  E6 70                       OUT   CH0_A,AL
001C  8A C4                       MOV   AL,AH
001E  E6 70                       OUT   CH0_A,AL

0020  8C C0                       MOV   AX,ES        ;program destination address
0022  C1 E0 04                    SHL   AX,4
0025  03 C7                       ADD   AX,DI        ;form destination offset
0027  48                          INC   AX
0028  E6 72                       OUT   CH1_A,AL
002A  8A C4                       MOV   AL,AH
002C  E6 72                       OUT   CH1_A,AL

002E  8B C1                       MOV   AX,CX        ;program count
0030  48                          DEC   AX           ;adjust count
0031  48                          DEC   AX
0032  E6 73                       OUT   CH1_C,AL
0034  8A C4                       MOV   AL,AH
0036  E6 73                       OUT   CH1_C,AL

0038  B0 88                       MOV   AL,88H       ;program mode
003A  E6 7B                       OUT   MODE,AL
003C  B0 85                       MOV   AL,85H
003E  E6 7B                       OUT   MODE,AL

0040  B0 03                       MOV   AL,3         ;enable block hold transfer
0042  E6 78                       OUT   CMMD,AL

0044  B0 0E                       MOV   AL,0EH       ;unmask channel 0
0046  E6 7F                       OUT   MASKS,AL

0048  B0 04                       MOV   AL,4         ;start DMA transfer
004A  E6 79                       OUT   REQ,AL

                                  .REPEAT            ;wait until DMA complete
004C  E4 78                       IN    AL,STATUS
                                  .UNTIL AL &1
                                  RET

0054                      CLEAR   ENDP
```

DMA-Processed Printer Interface

Figure 13–13 illustrates the hardware added to Figure 13–12 for a DMA-controlled printer interface. Little additional circuitry is added for this interface to a Centronics-type parallel printer. The
latch is used to capture the data as it is sent to the printer during the DMA transfer. The write pulse
passed through to the latch during the DMA action also generates the data strobe ($\overline{DS}$) signal to the
printer through the single-shot. The $\overline{ACK}$ signal returns from the printer each time it is ready for
additional data. In this circuit, $\overline{ACK}$ is used to request a DMA action through a flip-flop.

FIGURE 13–13 DMA-processed printer interface.

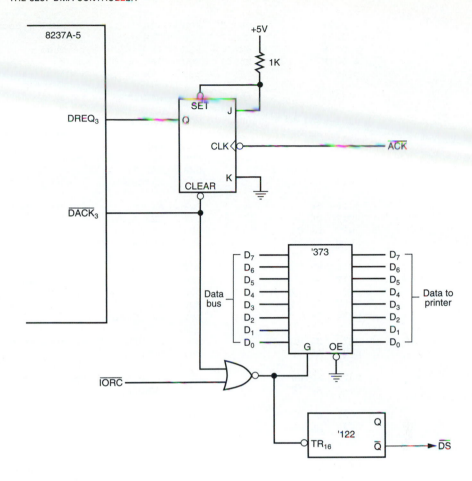

Notice that the I/O device is not selected by decoding the address on the address bus. During the DMA transfer, the address bus contains the memory address and cannot contain the I/O port address. In place of the I/O port address, the $\overline{DACK3}$ output from the 8237 selects the latch by gating the write pulse through an OR gate.

Software that controls this interface is simple because only the address of the data and the number of characters to be printed are programmed. Once programmed, the channel is enabled, and the DMA action transfers a byte at a time to the printer interface each time that the interface receives the $\overline{ACK}$ single from the printer.

The procedure that prints data from the current data segment is illustrated in Example 13–3. This procedure programs the 8237, but doesn't actually print anything. Printing is accomplished by the DMA controller and the printer interface.

EXAMPLE 13–3

```
                     ;A procedure that prints data via the printer interface
                     ;in Figure 13-13.
                     ;
                     ;Calling parameters:
                     ;    BX = offset address of printer data
                     ;    DS = segment address of printer data
                     ;    CX = number of bytes to print
                     ;
     = 0010          LATCHB   EQU   10H        ;latch B
     = 007C          CLEAR_F  EQU   7CH        ;F/L flip flop
     = 0076          CH3_A    EQU   76H        ;channel 0 address
```

```
= 0077              CH3_C   EQU    77H          ;channel 1 count
= 007B              MODE    EQU    7BH          ;mode
= 0078              CMMD    EQU    78H          ;command
= 007F              MASKS   EQU    7FH          ;masks
= 0079              REQ     EQU    79H          ;request register

0000                PRINT   PROC   FAR USES AX CX BX

0003  66| B8 00000000      MOV    EAX,0
0009  8C D8                MOV    AX,DS        ;program latch B
000B  66| C1 E8 04         SHR    EAX,4
000F  50                   PUSH   AX
0010  66| C1 E8 10         SHR    EAX,16
0014  E6 10                OUT    LATCHB,AL

0016  58                   POP    AX           ;program address
0017  E6 76                OUT    CH3_A,AL
0019  8A C4                MOV    AL,AH
001B  E6 76                OUT    CH3_A,AL

001D  8B C1                MOV    AX,CX        ;program count
001F  48                   DEC    AX           ;adjust count
0020  E6 77                OUT    CH3_C,AL
0022  8A C4                MOV    AL,AH
0024  E6 77                OUT    CH3_C,AL

0026  B0 0B                MOV    AL,0BH       ;program mode
0028  E6 7B                OUT    MODE,AL

002A  B0 00                MOV    AL,0         ;enable block hold transfer
002C  E6 78                OUT    CMMD,AL

002E  B0 07                MOV    AL,7         ;unmask channel 3
0030  E6 7F                OUT    MASKS,AL
                           RET

0036                PRINT   ENDP
```

A secondary procedure is needed to determine whether the DMA action has been completed. Example 13–4 lists the secondary procedure that tests the DMA controller to see whether the DMA transfer is complete. The TEST_P procedure is called before programming the DMA controller to see whether the prior transfer is complete.

EXAMPLE 13–4

```
              ;A procedure that tests for a complete DMA action

= 0078              STATUS EQU    78H          ;status register

0000                TEST_P PROC   NEAR USES AX

                           .REPEAT
0001  E4 78                  IN    AL,STATUS
                           .UNTIL AL &8

                           RET

0009                TEST_P   ENDP
```

Printed data can be double-buffered by first loading buffer 1 with data to be printed. Next, the PRINT procedure is called to begin printing buffer 1. Because it takes very little time to program the DMA controller, a second buffer (buffer 2) can be filled with new printer data while the first buffer (buffer 1) is printed by the printer interface and DMA controller. This process is repeated until all data are printed.

13–3 SHARED-BUS OPERATION

Complex present-day computer systems have so many tasks to perform that some systems are using more than one microprocessor to accomplish the work. This is called a **multiprocessing** system. We also sometimes call this a **distributed** system. A system that performs more than one task is called a **multitasking** system. In systems that contain more than one microprocessor, some method of control must be developed and employed. In a distributed, multiprocessing, multitasking environment, each microprocessor accesses two buses: (1) the **local bus** and (2) the **remote** or **shared** bus.

This section of the text describes shared bus operation for the 8086 and 8088 microprocessors using the 8289 bus arbiter. The 80286 uses the 82289 bus arbiter and the 80386/80486 uses the 82389 bus arbiter. The Pentium–Pentium 4 directly support a multiuser environment, as described in Chapters 16, 17, and 18. These systems are much more complex and difficult to illustrate at this point in the text, but their terminology and operation is essentially the same as for the 8086/8088.

The local bus is connected to memory and I/O devices that are directly accessed by a single microprocessor without any special protocol or access rules. The remote (shared) bus contains memory and I/O that are accessed by any microprocessor in the system. Figure 13-14 illustrates this idea with a few microprocessors. Note that the personal computer is also configured in the same manner as the system in Figure 13–14. The bus master is the main microprocessor in the personal computer. What we call the local bus in the personal computer is the shared bus in this illustration. The ISA bus is operated as a slave to the personal computer's microprocessor as well as any other devices attached to the shared bus.

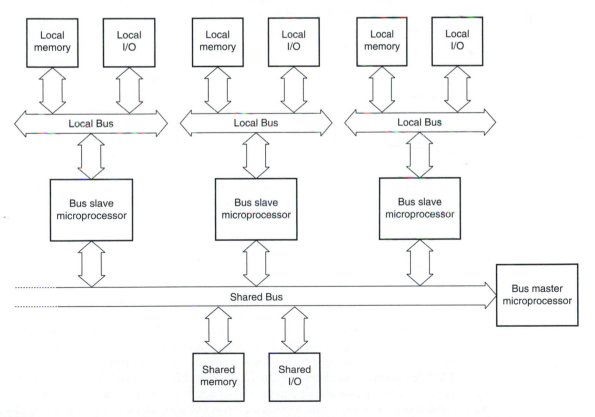

FIGURE 13–14 A block diagram illustrating the shared and local buses.

Types of Buses Defined

The **local bus** is the bus that is resident to the microprocessor. The local bus contains the resident or local memory and I/O. All microprocessors studied thus far in this text are considered to be local bus systems. The local memory and local I/O are accessed by the microprocessor that is directly connected to them.

A **shared bus** is one that is connected to all microprocessors in the system. The shared bus is used to exchange data between microprocessors in the system. A shared bus may contain memory and I/O devices that are accessed by all microprocessors in the system. Access to the shared bus is controlled by some form or arbiter that allows only a single microprocessor to access the system's shared bus space. As mentioned, the shared bus in the personal computer is what we often call the local bus in the personal computer because it is local to the microprocessor in the personal computer.

Figure 13–15 shows an 8088 microprocessor that is connected as a remote bus master. The term **bus master** applies to any device (microprocessor or otherwise) that can control a bus containing memory and I/O. The 8237 DMA controller presented earlier in the chapter is an example of a remote bus master. The DMA controller gained access to the system memory and I/O space to cause a data transfer. Likewise, a remote bus master gains access to the shared bus for the same purpose. The difference is that the remote bus master microprocessor can execute variable software, whereas the DMA controller can only transfer data.

Access to the shared bus is accomplished by using the HOLD pin on the microprocessor for the DMA controller. Access to the shared bus for the remote bus master is accomplished via a **bus arbiter,** which functions to resolve priority between bus masters and allows only one device at a time to access the shared bus.

Notice in Figure 13–15 that the 8088 microprocessor has an interface to both a local, resident bus and the shared bus. This configuration allows the 8088 to access local memory and I/O or, through the bus arbiter and buffers, the shared bus. The task assigned to the microprocessor might be data communications. It may, after collecting a block of data from the communications interface, pass those data on to the shared bus and shared memory so that other microprocessors attached to the system can access the data. This allows many microprocessors to share common data. In the same manner, multiple microprocessors can be assigned various tasks in the system, drastically improving throughput.

The Bus Arbiter

Before Figure 13–15 can be fully understood, the operation of the bus arbiter must be grasped. The 8289 bus arbiter controls the interface of a bus master to a shared bus. Although the 8289 is not the only bus arbiter, it is designed to function with the 8086/8088 microprocessors, so it is presented here. Each bus master or microprocessor requires an arbiter for the interface to the shared bus, which Intel calls the Multibus and IBM calls the Micro Channel.

The shared bus is used only to pass information from one microprocessor to another; otherwise, the bus masters function in their own local bus modes by using their own local programs, memory, and I/O space. Microprocessors connected in this kind of system are often called **parallel** or **distributed** processors because they can execute software and perform tasks in parallel.

8289 Architecture.   Figure 13–16 illustrates the pin-out and block diagram of the 8289 bus arbiter. The left side of the block diagram depicts the connections to the microprocessor. The right side denotes the 8289 connection to the shared (remote) bus or Multibus.

The 8289 controls the shared bus by causing the READY input to the microprocessor to become a logic 0 (not ready) if access to the shared bus is denied. The **blocking** occurs whenever another microprocessor is accessing the shared bus. As a result, the microprocessor requesting access is blocked by the logic 0 applied to its READY input. When the READY pin is a logic 0, the microprocessor and its software wait until access to the shared bus is granted by the arbiter. In this manner, one microprocessor at a time gains access to the shared bus. No special instructions are required for bus arbitration with the 8289 bus arbiter because arbitration is accomplished strictly by the hardware.

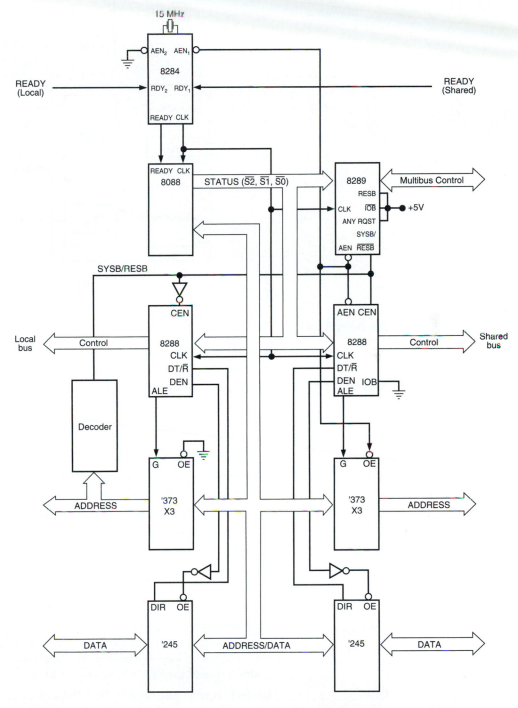

FIGURE 13–15 The 8088 operated in the remote mode, illustrating the local and shared bus connections.

FIGURE 13–16 The 8289 pin-out and block diagram. (Courtesy of Intel Corporation.)

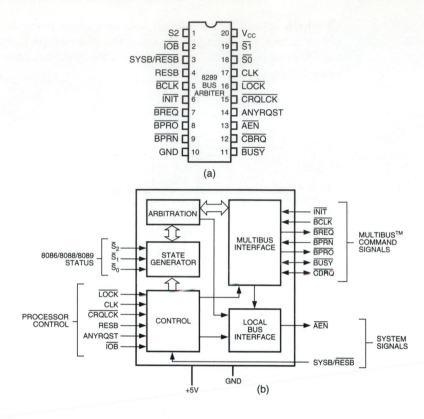

(a)

(b)

Pin Definitions

$\overline{\text{AEN}}$	The **address enable** output causes the bus drivers in a system to switch to their three-state, high-impedance state.
ANYRQST	The **any request** input is a strapping option that prevents a lower-priority microprocessor from gaining to the shared bus. If tied to a logic 0, normal arbitration occurs and a lower priority microprocessor can gain access to the shared bus if $\overline{\text{CBRQ}}$ is also a logic 0.
$\overline{\text{BCLK}}$	The **bus clock** input synchronizes all shared-bus masters.
$\overline{\text{BPRN}}$	The **bus priority input** allows the 8289 to acquire the shared bus on the next falling edge of the $\overline{\text{BLCK}}$ signal.
BPRO	The **bus priority output** is a signal that is used to resolve priority in a system that contains multiple bus masters.
$\overline{\text{BREQ}}$	The **bus request** output is used to request access to the shared bus.
$\overline{\text{BUSY}}$	The **busy input/output** indicates, as an output, that an 8289 has acquired the shared bus. As an input, $\overline{\text{BUSY}}$ is used to detect that another 8289 has acquired the shared bus.
$\overline{\text{CBRQ}}$	The **common bus request** input/output is used when a lower-priority microprocessor is asking for the use of the shared bus. As an output, $\overline{\text{CBRQ}}$ becomes a logic 0 whenever the 8289 requests the shared bus and remains low until the 8289 obtains access to the shared bus.
CLK	The **clock** input is generated by the 8284A clock generator and provides the internal timing source to the 8289.

$\overline{\text{CRQLCK}}$	The **common request lock** input prevents the 8289 from surrendering the shared bus to any of the 8289s in the system. This signal functions in conjunction with the $\overline{\text{CBRQ}}$ pin.
$\overline{\text{INIT}}$	The **initialization** input resets the 8289 and is normally connected to the system RESET signal.
$\overline{\text{IOB}}$	The **I/O bus** input selects whether the 8289 operates in a shared-bus system (if selected by RESB) with I/O ($\overline{\text{IOB}} = 0$) or with memory and I/O ($\overline{\text{IOB}} = 1$).
$\overline{\text{LOCK}}$	The **lock** input prevents the 8289 from allowing any other microprocessor from gaining access to the shared bus. An 8086/8088 instruction that contains a LOCK prefix will prevent other microprocessors from accessing the shared bus.
RESB	The **resident-bus** input is a strapping connection that allows the 8289 to operate in systems that have either a shared-bus or resident-bus system. If RESB is a logic 1, the 8289 is configured as a shared-bus master. If RESB is a logic 0, the 8289 is configured as a local-bus master. When configured as a shared-bus master, access is requested through the SYSB/$\overline{\text{RESB}}$ input pin.
S2, S1, and S0	The **status** inputs initiate shared-bus requests and surrenders. These pins connect to the 8288 system bus controller status pins.
SYSB/$\overline{\text{RESB}}$	The **system bus/resident bus** input selects the shared-bus system when placed at a logic 1 or the resident local bus when placed at a logic 0.

General 8289 Operation. As the pin descriptions demonstrate, the 8289 can be operated in three basic modes: (1) I/O peripheral-bus mode, (2) resident-bus mode, and (3) single-bus mode. See Table 13–2 for the connections required to operate the 8289 in these modes. In the **I/O peripheral-bus mode**, all devices on the local bus are treated as I/O, including memory, and are accessed by I/I instructions. All memory references access the shared bus and all I/O access the resident-local bus. The **resident-bus mode** allows memory and I/O accesses on both the local and shared buses. Finally, the **single-bus mode** interfaces a microprocessor to a shared bus, but the microprocessor has no local memory or local I/O. In many systems, one microprocessor is set up as the shared-bus master (single-bus mode) to control the shared bus and become the shared-bus master. The **shared-bus master** controls the system through shared memory and I/O. Additional microprocessors are connected to the shared bus as resident- or I/O peripheral-bus masters. These additional bus masters usually perform independent tasks that are reported to the shared-bus master through the shared bus.

System Illustrating Single-bus and Resident-bus Connections. Single-bus operation interfaces a microprocessor to a shared bus that contains both I/O and memory resources that are shared by other microprocessors. Figure 13–17 illustrates three 8088 microprocessors, each connected to a shared bus. Two of the three microprocessors operate in the resident-bus mode, while the third operates in the single-bus mode. Microprocessor A, in Figure 13–17, operates in the single-bus mode and has no local bus. This microprocessor accesses only the shared memory and I/O space.

TABLE 13–2 8289 modes of operation.

Mode	Pin Connections
Single bus	$\overline{\text{IOB}} = 1$ and RESB = 0
Resident bus	$\overline{\text{IOB}} = 1$ and RESB = 1
I/O bus	$\overline{\text{IOB}} = 0$ and RESB = 0
I/O bus and resident bus	$\overline{\text{IOB}} = 0$ and RESB = 1

Microprocessor A is often referred to as the **system-bus master** because it is responsible for coordinating the main memory and I/O tasks. The remaining two microprocessors (B and C) are connected in the resident-bus mode, which allows them access to both the shared bus and their own local buses. These resident-bus microprocessors are used to perform tasks that are independent from the system-bus master. In fact, the only time that the system-bus master is interrupted from performing its tasks are when one of the two resident-bus microprocessors needs to transfer data between itself and the shared bus. This connection allows all three microprocessors to perform tasks simultaneously, yet data can be shared between microprocessors when needed.

In Figure 13–17, the bus master (A) allows the user to operate with a video terminal that allows the execution of programs and generally controls the system. Microprocessor B handles all telephone communications and passes this information to the shared memory in blocks. This means that microprocessor B waits for each character to be transmitted or received and controls the protocol used for the transfers. For example, suppose that a 1K byte-block of data

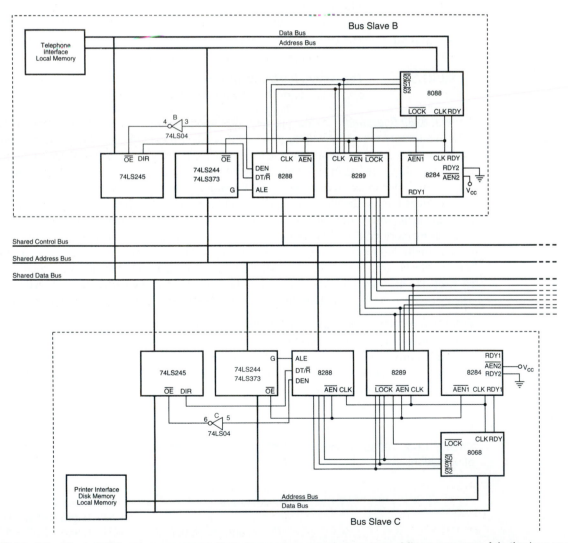

FIGURE 13–17   Three 8088 microprocessors that share a common bus system. Microprocessor A is the bus master in control of the shared memory and CRT terminal. Microprocessor B is a bus slave controlling its local telephone interface and memory. Microprocessor C is also a slave that controls a printer, disk memory system, and local memory.

are transmitted across the telephone interface and this occurs at the rate of 100 characters per second. This means that the transfer requires 10 seconds. Rather than tie up the bus master for 10 seconds, microprocessor B patiently performs the data transfer from its own local memory and the local communications interface. This frees the bus master for other tasks. The only time the microprocessor B interrupts the bus master is to transfer data between the shared memory and its local memory system. This data transfer, betweem microporcessor B and the bus master, requires only a few hundred microseconds.

Microprocessor C is used as a print spooler. Its only task is to print data on the printer. Whenever the bus master requires printed output, it transfers the task to microprocessor C. Microprocessor C then accesses the shared memory and captures the data to be printed and stores it in its own local memory. Data are then printed from the local memory, freeing the bus master to perform other tasks. This allows the system to execute a program with the bus master, transfer data through the communications interface with microprocessor B, and print information on the printer with microprocessor C. These tasks all execute simultaneously. There is no limit to the

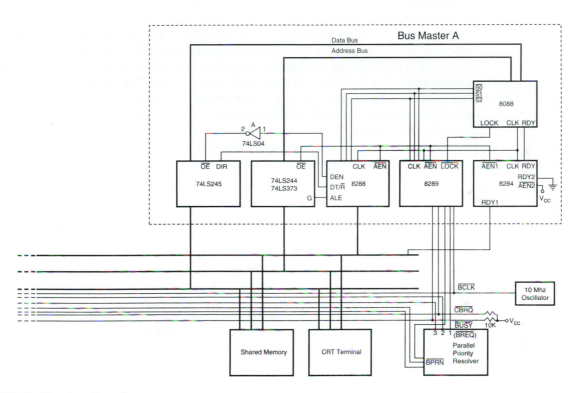

FIGURE 13–17 *(continued)*

number of microprocessors connected to a system or the number of tasks performed simultaneously using this technique. The only limit is that introduced by the system design and the designer's ingenuity.

Priority Logic Using the 8289

In applications that use the 8289, there is always more than one microprocessor connected to a shared bus. Because only one can access the shared bus at a time, some method of resolving priority must be employed. Priority prevents more than one microprocessor from accessing the bus at a time. There are two methods for resolving priority with the 8289 bus arbiter: the daisy-chain (serial) scheme and the parallel-priority scheme.

FIGURE 13–18 Daisy-chain 8289 priority resolver.

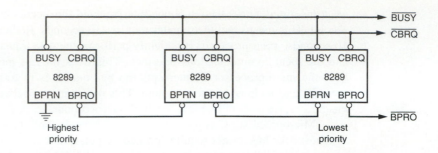

Daisy-chain Priority.   The daisy-chain priority scheme connects the $\overline{\text{BPRO}}$ output to the $\overline{\text{BPRN}}$ input of the next-lower priority 8289 and is the least expensive to implement. Figure 13–18 illustrates the daisy-chain scheme for connecting several 8289s in a system. Because the $\overline{\text{BPRN}}$ input of the higher priority 8289 is grounded, it gets an immediate acknowledgment whenever its microprocessor requests access to the shared bus. The $\overline{\text{BPRO}}$ output is a logic 0 if the 8289 and its microprocessor are inactive, and a logic 1 if the 8289 and its microprocessor are actively using the shared bus. If no requests are active, all $\overline{\text{BPRN}}$ inputs will see a logic 0. As soon as the highest-priority 8289s receives a bus acknowledgment, its $\overline{\text{BPRO}}$ output goes high, blocking all lower-priority 8289s from accessing the shared bus. If more than one 8289 receives an acknowledgment, more than one 8289 will function at the same time. This scheme is seldom used because two microprocessors can access the shared bus at a time. For this reason, Intel recommends that this scheme be limited to no more than three 8289s in a system that uses a bus clock of 10 MHz or less. With this frequency bus clock, no conflict occurs and only one microprocessor accesses the shared bus at a time. If more arbiters are connected, Intel suggests that the priority be resolved using the parallel scheme.

Parallel Priority

Figure 13–19 illustrates a parallel-priority scheme in which four 8289 bus arbiters are connected with a parallel circuit. Here, a 74LS148, eight-input priority encoder is used to resolve priority conflicts in parallel. In this example, only four of the inputs are used to resolve priority for the four 8289s. The four unused encoder inputs are pulled up to a logic 1 to disable these unneeded inputs to the encoder. Note that this circuit can be expanded to provide priority for up to eight 8289s with their microprocessors.

The circuit in Figure 13–19 functions as follows. If all 8289 arbiters are idle (no requests for the shared bus through the SYSB/$\overline{\text{RESB}}$ input), all $\overline{\text{BREQ}}$ outputs are high and the outputs of the 74LS148 are logic 1s ($\overline{\text{A}}$ and $\overline{\text{B}}$ are both 1). This means that the highest-priority 8289 will gain access to the shared bus if it is requested by its microprocessor. On the other hand, if a lower priority request is made, the $\overline{\text{BREQ}}$ output becomes a logic 0. This causes the priority encoder to place a logic 0 on the corresponding $\overline{\text{BPRN}}$ input pin of the 8289, allowing access to the shared bus. For example, if the rightmost 8289 places a logic 0 on its $\overline{\text{BREQ}}$ output pin, the priority encoder will have a zero on input $\overline{3}$. This causes the 74LS148 to generate a 00 on its output pins. The 00 causes the 74LS138 to activate the $\overline{\text{BPRN}}$ input of the rightmost 8289, giving it access to the shared bus. This also locks out any other requests because the $\overline{\text{BUSY}}$ signal becomes a logic 0. If simultaneous requests occur, they are automatically prioritized by the 74LS148, preventing conflicts no matter how many 8289s are connected in a system. For this reason, this priority scheme is desirable.

Print Spooler and Interface

Figure 13–20 shows the block diagram of a printer interface and spooler (print queue), which is controlled by an 8088 microprocessor. Here, two microprocessors are placed in a system with one (the system bus master) operated in single-bus mode and a second operated in the resident-bus mode. Because two microprocessors exist, one can print data while the other is used to process new information in the interim.

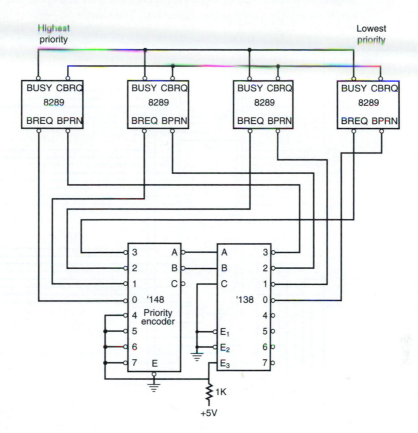

FIGURE 13–19 Parallel-priority resolver for the 8289.

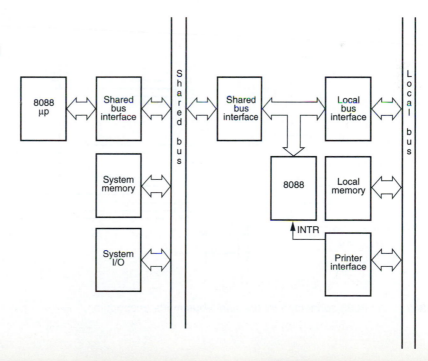

FIGURE 13–20 Block diagram of two 8088s used to control a printer interface and print spooler.

In this interface, the slave microprocessor transfers data to the printer from its local memory without intervention from the system-bus master. Data are transferred to the local memory of the slave microprocessor whenever it accesses the shared memory for additional data.

Single-mode Bus-master Interface. Figure 13–21 illustrates the 8088 bus master, operated in single mode, interfaced to the shared bus. The shared-bus master has access to every memory lo-

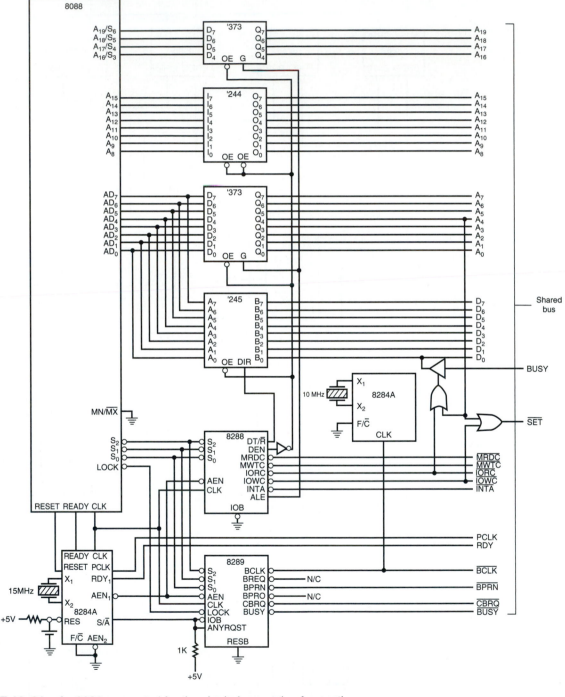

FIGURE 13–21 An 8088 connected for the single-bus mode of operation.

cation and I/O device on the shared bus. The $\overline{\text{BCLK}}$ signal is generated by an 8284A clock generator used as a 10-MHz oscillator.

The 8289 arbiter is operated with no I/O bus, no resident bus, and ANYRQST pulled high so that the shared bus can be accessed for all memory and I/O transfers. This system uses the daisy-chain priority scheme, which means that the $\overline{\text{BREQ}}$ signal is not connected. This interface allows the slave microprocessor to access shared memory whenever necessary and thus prevents the slave microprocessor from locking the bus for too long.

In addition to the normal bus signals, this circuit also provides the shared bus with a 2.5 MHz PCLK signal for any I/O device. The RDY input is shown, in case system memory and I/O require wait states.

Resident-bus Operation of the Slave 8088. Figure 13–22 illustrates the slave 8088 microprocessor that functions as a print spooler. This microprocessor is connected as a resident-bus master. This illustration depicts the resident-local-bus interface as well as the shared-bus interface.

Whenever the slave 8088 microprocessor accesses memory above location 7FFFFH, it places a logic 1 on the SYS/$\overline{\text{RESB}}$ input to the 8289 and the CEN pin of the shared-bus master's 8288. This requests access to the shared bus through the 8289 bus arbiter. If the address is below 80000H, the SYS/$\overline{\text{RESB}}$ pin on the 8289 is grounded and a logic 1 is placed on the CEN pin of the local bus 8288. This requests access to the resident-local bus for control of the printer and resident-local bus memory. Notice that no attempt is made to access the shared-bus I/O because the purpose of the slave 8088 is to access the printer interface on its local bus as local I/O.

Figure 13–23 shows the memory maps of both the slave 8088 local memory and the shared memory and the local memory of the 8088 bus master. The bus master has access to all of the shared memory, while the slave can only access locations 80000H–FFFFFH. Transfers to the print spooler are made through the upper half of the shared memory.

The resident-local bus in this system contains EPROM, DRAM, and a printer interface. The EPROM contains the program that controls the slave 8088, the DRAM contains the data for the printer, and the printer interface controls the printer. Figure 13–24 illustrates these three devices on the resident-local bus of the slave 8088 microprocessor.

The decoder in this illustration selects EPROM for locations 00000H–0FFFFH, the printer for locations 10000H–1FFFFH, and the DRAM for locations 20000H–7FFFFH. This allows the spooler to hold up to 384K bytes of data (see Figure 13–25). If the printer can produce 1,000 lines per minute (high-speed printer), then this represents about three minutes of printing and about 50–75 printed pages of data. If the system is expanded with a fixed memory system, the capacity of the spooler can become almost boundless because of the huge capacity of a modern fixed-disk memory system. For simplicity's sake, this is not added to this example.

The printer is interfaced to the slave microprocessor through a 74LS244 that is operated as a strobed-output interrupting device. Whenever the printer completes printing a character, as signaled by the $\overline{\text{ACK}}$ signal, an interrupt is generated, causing the slave microprocessor to retrieve data from the print queue for the printer. This process continues until the queue is empty, when the slave microprocessor disables interrupt requests until new information is placed in the queue for printing.

Print Spooler Software. The software for the print spooler is fairly straightforward. The entire software listing is provided in sections so that a study can be made of initialization, data transfer, and interrupt-controlled printer service procedures. The only software not illustrated is that which are required to program and initialize the system. When the system is initialized, the input and output pointers of the queue are both set up with address 20000H. This condition (equal pointers) indicates that the queue is empty. The segment portion of the pointers contains 2000H and the offset portion contains 0000H.

Example 13–5 lists the software that transfers data into the print spooler from the shared bus. In this software, the bus master loads a block of shared memory with printer data, called a

FIGURE 13-22 The 8088 shown with both a shared and a local bus.

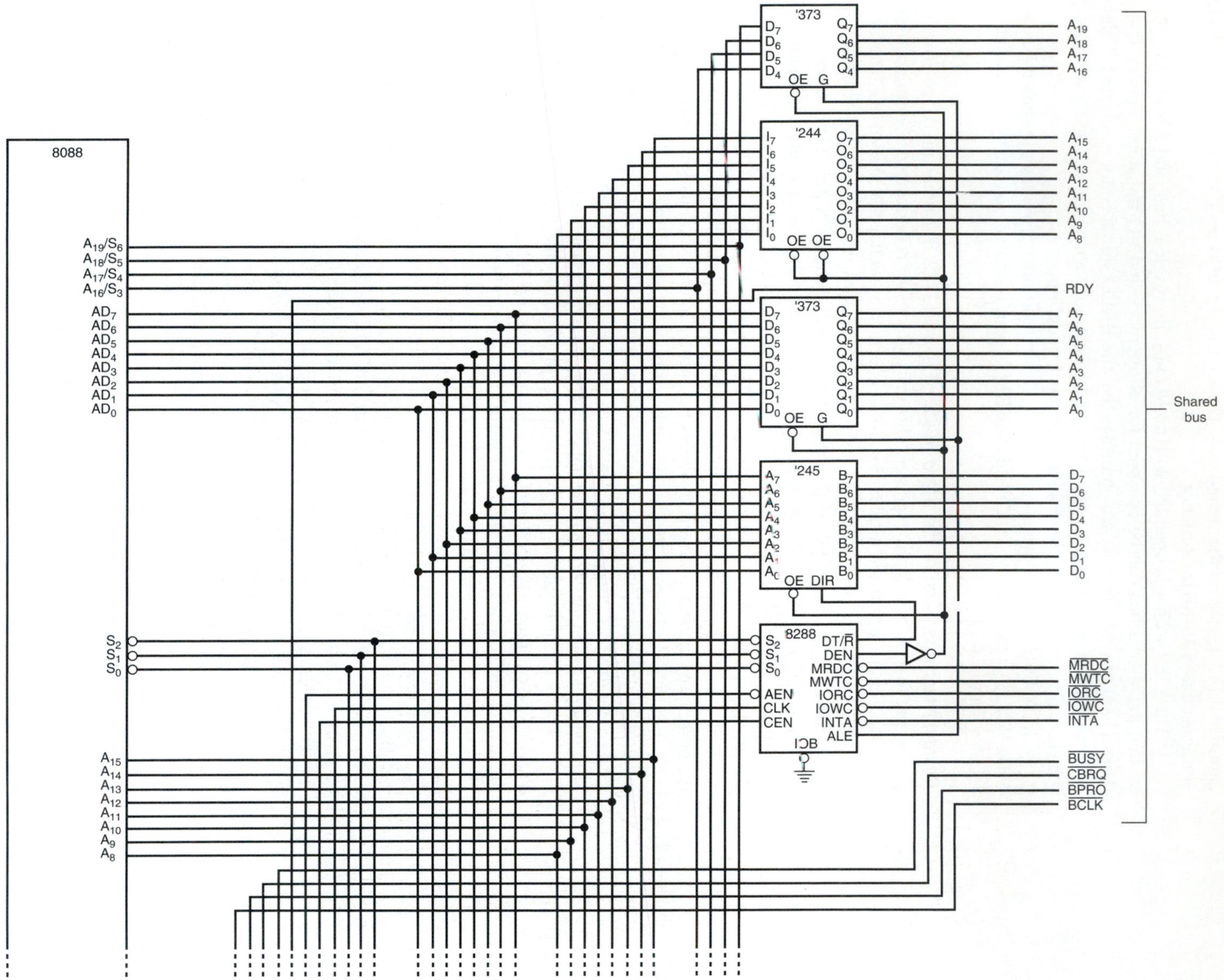

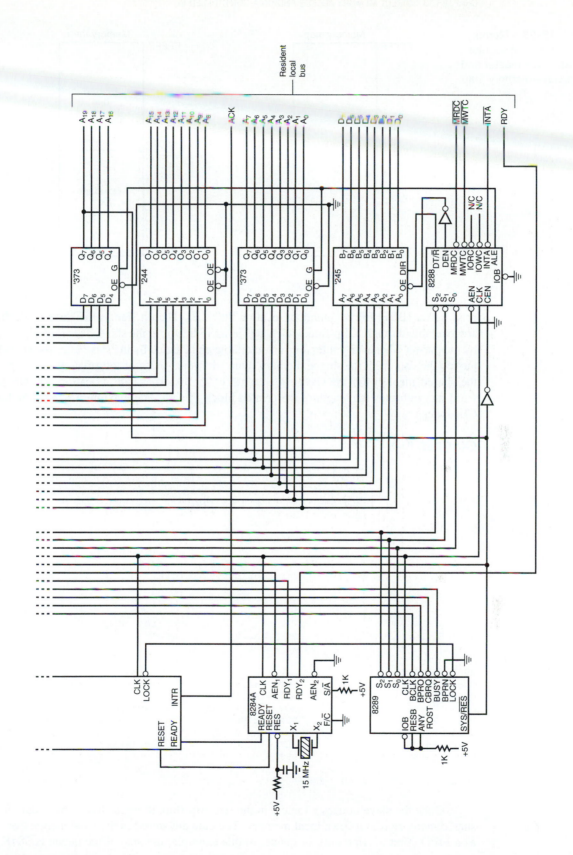

FIGURE 13–23 Memory maps for the print spooler. (a) Shared-bus master and (b) bus-slave memory map.

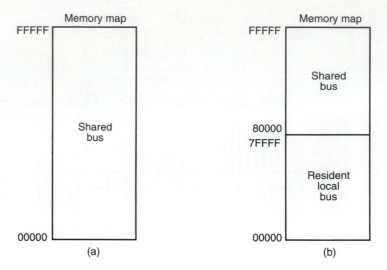

print buffer (BUFFER beginning at location 80002H). Then it signals the slave 8088 that data are available through a flip-flop, which acts as an indicator to the slave. This flag flip-flop is set by the master whenever printer data are available and cleared by the slave after the slave has removed the data from the shared memory. One additional piece of information is also placed in the shared memory for the slave—the length of the block of data. Location 80000H holds a word that indicates the length of the printed block of data. The maximum size of the buffer is 64K bytes.

EXAMPLE 13–5

```
                    ;A procedure found in the master system that tests the
                    ;FLAG to determine if the slave is busy. If the slave
                    ;is not busy, data are transferred to the print buffer
                    ;and the FLAG is set to indicate that data are available
                    ;for printing.
                    ;
                    ;Calling parameters:
                    ;
                    ;    DS:SI = address of printer data
                    ;    CX = number of bytes to print
                    ;
0000                LOAD    PROC    FAR USES ES

                        .REPEAT                 ;test FLAG
0001 E4 00                  IN    AL,0
                        .UNTIL AL==0

0007 BF 0002                MOV   DI,2          ;offset of buffer
000A B8 8000                MOV   AX,8000H      ;segment of buffer & count
000D 8E C0                  MOV   ES,AX
000F 26: 89 0E 0000         MOV   ES:[0],CX     ;save count
0014 F3/ A4                 REP   MOVSB         ;save data to buffer
0016 B0 FF                  MOV   AL,0FFH
0018 E6 00                  OUT   0,AL          ;set FLAG
                            RET

001C                LOAD    ENDP
```

Once the slave notices a logic 1 in the flag flip-flop, it begins transferring data from the shared memory into its own local memory. The data are stored in the local memory organized as a FIFO (first in, first out), or queue. In this example, the size of the queue is 384K bytes.

FIGURE 13–24 Resident-local bus for the print spooler.

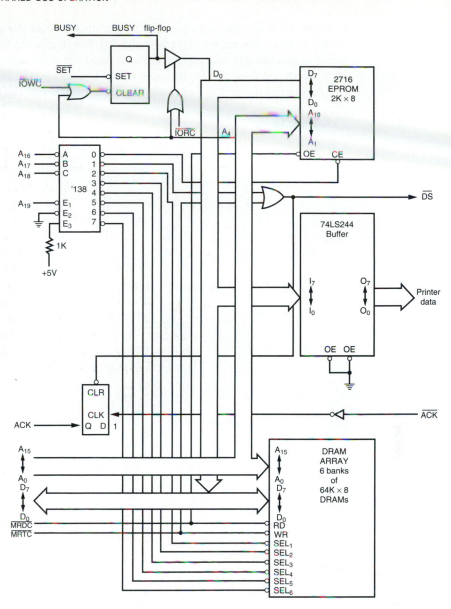

FIGURE 13–25 8088 resident-local bus memory map.

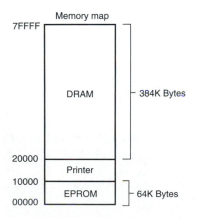

Example 13–6 lists the software used by the slave to load the queue from the shared memory buffer. In this software, the slave tests the flag flip-flop to see whether the master has filled the buffer in the shared memory. If the buffer is filled, the slave microprocessor transfers its contents form the shared memory into its queue. Once the transfer is complete, the slave clears the flag flip-flop so that the master may begin filling the buffer with additional printer data.

EXAMPLE 13–6

```
                        ;A procedure found in the slave system that tests the
                        ;FLAG to determine if data are available for printing.
                        ;If data are available, data are transferred into the
                        ;queue for the printer.
                        ;
0000                    TRANS  PROC  FAR

                          .REPEAT                     ;test FLAG
0000 E4 00                  IN     AL,0
0002 A8 01                  TEST   AL,1
                          .UNTIL AL == 0              ;if not finished

0008 B8 8000              MOV    AX,8000H             ;segment of buffer & count
000B 8E D8                MOV    DS,AX
000D 8B 0E 0000           MOV    CX,DS:[0]            ;get count
0011 BE 0002              MOV    SI,2                 ;address buffer

0014 E8 000D              CALL   FILL                 ;fill queue

0017 FB                   STI
0018 E6 00                OUT    0,AL                 ;clear FLAG
001A EB E4                JMP    TRANS

001C            TRANS  ENDP
                ;
                ;Data used with FILL
                ;
001C 0000       IN_PS  DW   ?                         ;input pointer
001E 0000       IN_PO  DW   ?
0020 0000       OUT_PS DW   ?
0022 0000       OUT_PO DW   ?                         ;output pointer
                ;
                ;Procedure that fills the printer queue.
                ;
0024            FILL   PROC   NEAR

0024 FC                   CLD                         ;clear direction
0025 2E: A1 001C R        MOV    AX,CS:IN_PS
0029 8E C0                MOV    ES,AX
002B 2E: 8B 3E 001E R     MOV    DI,CS:IN_PO          ;load input address
                          .REPEAT
0030 E8 0014                CALL  TESTF               ;test for full queue
                          .UNTIL !ZERO?               ;while full
0035 A4                     MOVSB                     ;store byte
0036 E8 0024                CALL  INCP                ;increment pointer
0039 8C C0                  MOV    AX,ES
003B 2E: A3 001C R          MOV    CS:IN_PS,AX
003F 2E: 89 3E 001E R       MOV    CS:IN_PO,DI        ;save input pointer
                          .UNTILCXZ
0046 C3                   RET
0047            FILL   ENDP

0047             TESTF  PROC   NEAR USES DI DS

0047 E8 0011              CALL   INCP                 ;increment pointer
004C 8C D8                MOV    AX,DS
004E 2E: 3B 06 0020 R     CMP    AX,CS:OUT_PS
```

```
0053   75 05                    JNE     TESTF1
0055   2E: 3B 3E 0022 R         CMP     DI,CS:OUT_PO
005A              TESTF1:
                               RET

005D              TESTF   ENDP

005D              INCP    PROC    NEAR

005D   47                       INC     DI
                               .IF DI == 0
0062   8C C0                    MOV     AX,ES
0064   05 1000                  ADD     AX,1000H
                                 .IF AX == 8000H
006C   B8 2000                  MOV     AX,2000H
                                 .ENDIF
006F   8E C0                    MOV     ES,AX
                               .ENDIF
0071   C3                       RET

0072              INCP    ENDP
```

Example 13–7 lists the software that prints data from the queue. This software is interrupt driven, and therefore runs as a background program virtually hidden from the software listed in Example 13–6, except for the interrupt enable instruction. Whenever the printer interface, through a flip-flop, indicates that it is ready to accept additional data, an interrupt occurs calling this interrupt service procedure. The procedure extracts data from the queue and sends them to the printer.

EXAMPLE 13–7

```
0072              PRINT   PROC    FAR USES AX BP DS DI ES

0077   E8 0023            CALL    TESTE
                         .IF ZERO?                 ;if empty
007C   8B EC                    MOV     BP,SP         ;interrupt off
007E   8B 46 0C                 MOV     AX,[BP+12]
0081   80 E4 FD                 AND     AH,0FDH
0084   89 46 0C                 MOV     [BP+12],AX
                         .ELSE
0089   B8 1000                  MOV     AX,1000H
008C   8E C0                    MOV     ES,AX
008E   8A 05                    MOV     AL,[DI]
0090   26: A2 0000              MOV     ES:[0],AL     ;print data
0094   E8 001E                  CALL    INCO
                         .ENDIF
                         IRET

009D              PRINT   ENDP



009D              TESTE   PROC    NEAR

009D   2E: A1 0020 R            MOV     AX,CS:OUT_PS
00A1   8E D8                    MOV     DS,AX
00A3   2E: 8B 3E 0022 R         MOV     DI,CS:OUT_PO
00A8   2E: 3B 06 001C R         CMP     AX,CS:IN_PS
00AD   75 05                    JNE     TESTE1
00AF   2E: 3B 3E 001E R         CMP     DI,CS:IN_PO
00B4              TESTE1:
00B4   C3                       RET

00B5              TESTE   ENDP
```

```
00B5                    INCO    PROC    NEAR

00B5    47                      INC     DI
                                .IF DI == 0
00BA    8C D8                       MOV     AX,DS
00BC    05 1000                     ADD     AX,1000H
                                    .IF AX == 8000H
00C4    B8 2000                         MOV   AX,2000H
                                    .ENDIF
00C7    8E D8                       MOV     DS,AX
00C9    2E: A3 0020 R               MOV     CS:OUT_PS,AX
                                .ENDIF
00CD    2E: 89 3E 0022 R        MOV     CS:OUT_PO,DI
00D2    C3                      RET

00D3                    INCO    ENDP
```

13–4   DISK MEMORY SYSTEMS

Disk memory is used to store long-term data. Many types of disk storage systems are available today and they use magnetic media, except the optical disk memory that stores data on a plastic disk. Optical disk memory is either a **CD-ROM** (compact disk/read only memory) that is read, but never written, or a **WORM** (write once/read mostly) that is read most of the time, but can be written once by a laser beam. Also becoming available is optical disk memory that can be read and written many times, but there is still a limitation on the number of write operations allowed. The latest optical disk technology is called DVD **(digital versatile disk).** This section of the chapter provides an introduction to disk memory systems so that they may be used with computer systems. It also provides details of their operation.

Floppy Disk Memory

The most common and the most basic form of disk memory is the floppy, or flexible disk. This magnetic recording media is available in three sizes: the 8" **standard,** $5^1/4$" **mini-floppy,** and the $3^1/2$" **micro-floppy.** Today, the 8" standard version has all but disappeared, giving way to the mini- and micro-floppy disks. The 8" disk is too large and difficult to handle and stockpile. To solve this problem, industry developed the $5^1/4$" mini-floppy disk. Today, the micro-floppy disk has just about replaced the mini-floppy in newer systems because of its reduced size, ease of storage, and durability. Even so, a few systems are still marketed with both the mini- and micro-floppy disk drives.

All disks have several things in common. They are all organized so that data are stored in tracks. A **track** is a concentric ring of data that is stored on a surface of a disk. Figure 13–26 illustrates the surface of a $5^1/4$" mini-floppy disk, showing a track that is divided into sectors. A **sector** is a common subdivision of a track that is designed to hold a reasonable amount of data. In many systems, a sector holds either 512 or 1024 bytes of data. The size of a sector can vary from 128 bytes to the length of one entire track.

Notice in Figure 13–26 that there is a hole through the disk that is labeled an index hole. The **index hole** is designed so that the electronic system that reads the disk can find the beginning of a track and its first sector (00). Tracks are numbered from track 00, the outermost track, in increasing value toward the center or innermost track. Sectors are often numbered from sector 00 on the outermost track, to whatever value is required to reach the innermost track and its last sector.

The $5^1/4$" Mini-floppy Disk.   Today, the $5^1/4$" floppy is probably the most popular disk size used with older microcomputer systems. Figure 13–27 illustrates this mini-floppy disk. The floppy disk is rotated at 300 RPM inside its semi-rigid plastic jacket. The head mechanism in a floppy disk drive makes physical contact with the surface of the disk, which eventually causes wear and damage to the disk.

Today, most mini-floppy disks are double-sided. This means that data are written on both the top and bottom surfaces of the disk. A set of tracks called a **cylinder** consists of one top and one bottom track. Cylinder 00, for example, consists of the outermost top and bottom tracks.

FIGURE 13–26 The format of a 5$^{1}/_{4}$" mini-floppy disk.

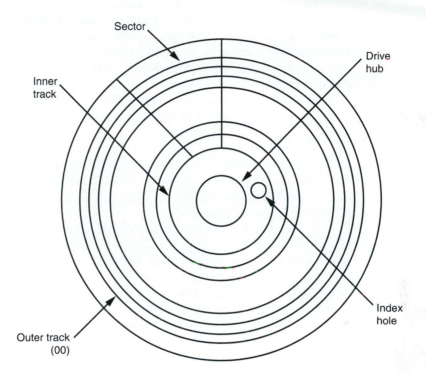

FIGURE 13–27 The 5$^{1}/_{4}$" mini-floppy disk.

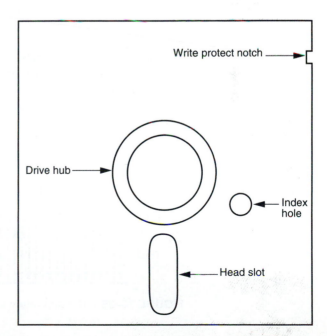

Floppy disk data are stored in the double-density format, which uses a recording technique called **MFM** (modified frequency modulation) to store the information. Double-density, double-sided (**DSDD**) disks are normally organized with 40 tracks of data on each side of the disk. A double-density disk track is typically divided into nine sectors, with each sector containing 512 bytes of information. This means that the total capacity of a double-density, double-sided disk is 40 tracks per side × 2 sides × 9 sectors per track × 512 bytes per sector, or 368,640 (360K) bytes of information.

Earlier disk memory systems used single-density and **FM** (frequency modulation) to store information in 40 tracks on one or two sides of the disk. Each of the eight or nine sectors on the single-density disk stored 256 bytes of data. This meant that a single-density disk stored 90K bytes of data per side. A single-density, double-sided disk stored 180K bytes of data.

Also common today are **high-density** (HD) mini-floppy disks. A high-density mini-floppy disk contains 80 tracks of information per side, with eight sectors per track. Each sector contains 1024 bytes of information. This gives the 5$^1/_4$" high-density, mini-floppy disk a total capacity of 80 tracks per side × 2 sides × 15 sectors per track × 512 bytes per sector, or 1,228,800 (approximately 1.2 M) bytes of information.

The magnetic recording technique used to store data on the surface of the disk is called **non-return to zero** (NRZ) recording. With NRZ recording, magnetic flux placed on the surface of the disk never returns to zero. Figure 13–28 illustrates the information stored in a portion of a track. It also shows how the magnetic field encodes the data. Note that arrows are used in this illustration to show the polarity of the magnetic field stored on the surface of the disk.

The main reason that this form of magnetic encoding was chosen is that it automatically erases old information when new information is recorded. If another technique were used, a separate erase head would be required. The mechanical alignment of a separate erase head and a separate read/write head is virtually impossible. The magnetic flux density of the NRZ signal is so intense that it completely saturates (magnetizes) the surface of the disk, erasing all prior data. It also ensures that information will not be affected by noise because the amplitude of the magnetic field contains no information. The information is stored in the placement of the changes of the magnetic field.

Data are stored in the form of MFM (modified frequency modulation) in modern floppy disk systems. The MFM recording technique stores data in the form illustrated in Figure 13–29. Notice that each bit time is 2.0 µs wide on a double-density disk. This means that data are recorded at the rate of 500,000 bits per second. Each 2.0 µs bit time is divided into two parts: one part is designated to hold a clock pulse and the other holds a data pulse. If a clock pulse is present, it is 1 µs wide, as is a data pulse. Clock and data pulses are never present at the same time in one bit period. (Note that high-density disk drives halve these times so that a bit time is 1.0 µs and a clock or data pulse is 0.5 µs wide. This also doubles the transfer rate to 1 million bits per second.)

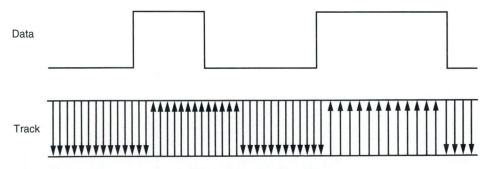

FIGURE 13–28   The non-return to zero (NRZ) recording technique.

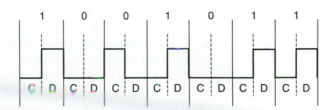

FIGURE 13–29 Modified frequency modulation (MFM) used with disk memory.

If a data pulse is present, the bit time represents a logic 1. If no data or no clock is present, the bit time represents a logic 0. If a clock pulse is present with no data pulse, the bit time also represents a logic 0. The rules followed when data are stored using MFM are as follows:

1. A data pulse is always stored for a logic 1.
2. No data and no clock is stored for the first logic 0 in a string of logic 0s.
3. The second and subsequent logic 0s in a row contain a clock pulse, but no data pulse.

The reason that a clock is inserted as the second and subsequent zero in a row is to maintain synchronization as data are read from the disk. The electronics used to recapture the data from the disk drive use a phase-locked loop to generate a clock and a data window. The phase-locked loop needs a clock or data to maintain synchronized operation.

The 3¹/₂" Micro-floppy disk. Another very popular disk size is the 3¹/₂" micro-floppy disk. Recently, this size floppy disk has begun to sell very well and is the dominant floppy disk size. The micro-floppy disk is a much improved version of the mini-floppy disk described earlier. Figure 13–30 illustrates the 3¹/₂" micro-floppy disk.

Disk designers noticed several shortcomings of the mini-floppy, which is a scaled down version of the 8" standard floppy, soon after it was released. Probably one of the biggest problems with the mini-floppy is that it is packaged in a semi-rigid plastic cover that bends easily. The micro-floppy is packaged in a rigid plastic jacket that will not bend easily. This provides a much greater degree of protection to the disk inside the jacket.

Another problem with the mini-floppy is the head slot that continually exposes the surface of the disk to contaminants. This problem is also corrected on the micro-floppy because it is con-

FIGURE 13–30 The 3¹/₂" micro-floppy disk.

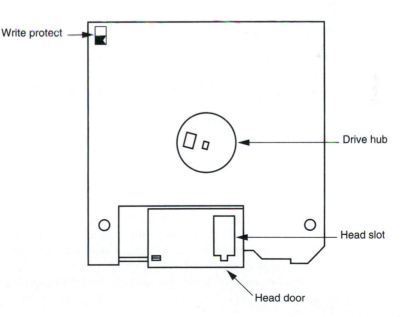

structed with a spring-loaded sliding head door. The head door remains closed until the disk is inserted into the drive. Once inside the drive, the drive mechanism slides open the door, exposing the surface of the disk to the read/write heads. This provides a great deal of protection to the surface of the micro-floppy disk.

Yet another improvement is the sliding plastic write-protection mechanism on the micro-floppy disk. On the mini-floppy disk, a piece of tape was placed over a notch on the side of the jacket to prevent writing. This plastic tape easily became dislodged inside disk drives, causing problems. On the micro-floppy, an integrated plastic slide has replaced the tape write-protection mechanism. To write-protect (prevent writing) the micro-floppy disk, the plastic slide is moved to open the hole though the disk jacket. This allows light to strike a sensor that inhibits writing.

Still another improvement is the replacement of the index hole with a different drive mechanism. The drive mechanism on the mini-floppy allows the disk drive to grab the disk at any point. This requires an index hole so that the electronics can find the beginning of a track. The index hole is another trouble spot because it collects dirt and dust. The micro-floppy has a drive mechanism that is keyed so that it only fits one way inside the disk drive. The index hole is no longer required because of this keyed drive mechanism. Because of the sliding head mechanism and the fact that no index hole exists, the micro-floppy disk has no place to catch dust or dirt.

Two types of micro-floppy disks are widely available: the double-sided, double-density (DSDD) and the high-density (HD). The double-sided, double-density micro-floppy disk has 80 tracks per side, with each track containing nine sectors. Each sector contains 512 bytes of information. This allows 80 tracks per side × 2 sides × 9 sectors × 512 bytes per sector, or 737,280 (720K) bytes of data to be stored on a double-density, double-sided floppy disk.

The high-density, double-sided micro-floppy disk stores even more information. The high-density version has 80 tracks per side, but the number of sectors is doubled to 18 per track. This format still uses 512 bytes per sector, as did the double-density format. The total number of bytes on a high-density, double-sided micro-floppy disk is 80 tracks per side × 2 sides × 18 sectors per track × 512 bytes per sector, or 1,474,560 (1.44M) bytes of information.

Recently, a new size $3^1/2''$ floppy disk has been introduced, the EHD (**extended high density**) floppy disk. This new format stores 2.88M bytes of data on a single floppy disk. At this time, this format is expensive and will take time to become common. Also available is the **floptical disk,** which stores data magnetically by using an optical tracking system. The floptical disk stores 21M bytes of data.

Hard Disk Memory

Larger disk memory is available in the form of the **hard disk drive.** The hard disk drive is often called a **fixed disk** because it is not removable like the floppy disk. A hard disk is also often called a **rigid disk.** The term **Winchester drive** is also used to describe a hard disk drive, but less commonly today. Hard disk memory has a much larger capacity than the floppy disk memory. Hard disk memory is available in sizes exceeding 1G byte of data. Common, low-cost (less than $0.30 per megabyte) sizes are presently 3.2G bytes to 12G bytes.

There are several differences between the floppy disk and the hard disk memory. The hard disk memory uses a flying head to store and read data from the surface of the disk. A flying head, which is very small and light, does not touch the surface of the disk. It flies above the surface on a film of air that is carried with the surface of the disk as it spins. The hard disk typically spins at 3000 to 10,000 R.P.M, which is more than 10 times faster than the floppy disk. This higher rotational speed allows the head to fly (just as an airplane flies) just over the top of the surface of the disk. This is an important feature because there is no wear on the hard disk's surface, as there is with the floppy disk.

Problems can arise because of flying heads. One problem is a head crash. If the power is abruptly interrupted or the hard disk drive is jarred, the head can crash onto the disk surface,

FIGURE 13–31 A hard disk drive that uses four heads per platter.

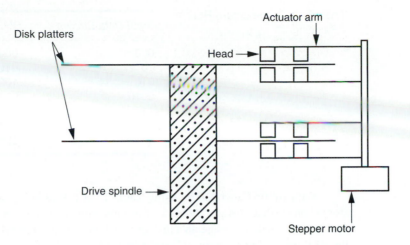

which can damage the disk surface or the head. To help prevent crashes, some drive manufacturers have included a system that automatically parks the head when power is interrupted. This type of disk drive has auto-parking heads. When the heads are parked, they are moved to a safe landing zone (unused track) when the power is disconnected. Some drives are not auto-parking; they usually require a program that parks the heads on the innermost track before power is disconnected. The innermost track is a safe landing area because it is the very last track filled by the disk drive. Parking is the responsibility of the operator in this type of disk drive.

Another difference between a floppy disk drive and a hard disk drive is the number of heads and disk surfaces. A floppy disk drive has two heads, one for the upper surface and one for the lower surface. The hard disk drive may have up to eight disk surfaces (four platters), with up to two heads per surface. Each time that a new cylinder is obtained by moving the head assembly, 16 new tracks are available under the heads. See Figure 13–31, which illustrates a hard disk system.

Heads are moved from track to track by using either a stepper motor or a voice coil. The stepper motor is slow and noisy, while the voice coil mechanism is quiet and quick. Moving the head assembly requires one step per cylinder in a system that uses a stepper motor to position the heads. In a system that uses a voice coil, the heads can be moved many cylinders with one sweeping motion. This makes the disk drive faster when seeking new cylinders.

Another advantage of the voice coil system is that a servo mechanism can monitor the amplitude of the signal as it comes from the read head and make slight adjustments in the position of the heads. This is not possible with a stepper motor, which relies strictly on mechanics to position the head. Stepper-motor-type head positioning mechanisms can often become misaligned with use, while the voice coil mechanism corrects for any misalignment.

Hard disk drives often store information in sectors that are 512 bytes long. Data are addressed in **clusters** of eight or more sectors, which contain 4096 bytes (or more) on most hard disk drives. Hard disk drives use either MFM or RLL to store information. MFM is described with floppy disk drives. **Run-length limited** (RLL) is described here.

A typical older MFM hard disk drive uses 18 sectors per track so that 18 K bytes of data are stored per track. If a hard disk drive has a capacity of 40M bytes, it contains approximately 2280 tracks. If the disk drive has two heads, this means that it contains 1140 cylinders; if it contains four heads, then it has 570 cylinders. These specifications vary from disk drive to disk drive.

RLL Storage. Run-length limited (RLL) disk drives use a different method for encoding the data than MFM. The term RLL means that the run of zeros (zeros in a row) is limited. A common RLL encoding scheme in use today is RLL 2,7. This means that the run of zeros is always between two and seven. Table 13–3 illustrates the coding used with standard RLL.

TABLE 13–3 Standard RLL 2,7 coding.

Input Data Stream	RLL Output
000	000100
10	0100
010	100100
0010	00100100
11	1000
011	001000
0011	00001000

Data are first encoded by using Table 13–3 before being sent to the drive electronics for storage on the disk surface. Because of this encoding technique, it is possible to achieve a 50 percent increase in data storage on a disk drive when compared to MFM. The main difference is that the RLL drive often contains 27 tracks instead of the 18 found on the MFM drive. (Some RLL drives also use 35 sectors per track.)

In most cases, RLL encoding requires no change to the drive electronics or surface of the disk. The only difference is a slight decrease in the pulse width using RLL, which may require slightly finer oxide particles on the surface of the disk. Disk manufacturers test the surface of the disk and grade the disk drive as either an MFM-certified or an RLL-certified drive. Other than grading, there is no difference in the construction of the disk drive or the magnetic material that coats the surface of the disks.

Figure 13–32 shows a comparison of MFM data and RLL data. Notice that the amount of time (space) required to store RLL data is reduced when compared to MFM. Here a 101001011 is coded in both MFM and RLL so that these two standards can be compared. Notice that the width of the RLL signal has been reduced so that three pulses fit in the same space as a clock and a data pulse for MFM. A 40M-byte MFM disk can hold 60M bytes of RLL-encoded data. Besides holding more information, the RLL drive can be written and read at a higher rate.

All hard disk drives use either MFM or RLL encoding. There are a number of disk drive interfaces in use today. The oldest is the ST-506 interface, which uses either MFM or RLL data.

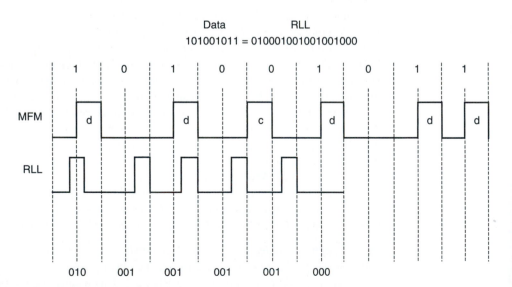

FIGURE 13–32 A comparison of MFM with RLL using data 101001011.

A disk system using this interface is also called either MFM or RLL disk system. Newer standards are also found in use today, which include ESDI, SCSI, and IDE. All of these newer standards use RLL, even though they normally do not call attention to it. The main difference is the interface between the computer and the disk drive. The IDE system is becoming the standard hard disk memory interface.

The **enhanced small disk interface** (ESDI) system, which has disappeared, is capable of transferring data between itself and the computer at rates approaching 10 M bytes per second. An ST-506 interface can approach a transfer rate of 860 K bytes per second.

The **small computer system interface** (SCSI) system is also in use because it allows up to seven different disk or other interfaces to be connected to the computer through same interface controller. SCSI is found in some PC-type computers and also in the Apple Macintosh system. An improved version, SCSI-II, has started to appear in some systems. In the future, this interface may be replaced with IDE in most applications.

The newest system is **integrated drive electronics** (IDE), which incorporates the disk controller in the disk drive and attaches the disk drive to the host system through a small interface cable. This allows many disk drives to be connected to a system without worrying about bus conflicts or controller conflicts. IDE drives are found in newer IBM PS-2 systems and many clones. Even Apple computer systems are starting to be found with IDE drives in place of the SCSI drives found in older Apple computers. The IDE interface is also capable of driving other I/O devices besides the hard disk. This interface also usually contains at least a 256K- to 2M-byte cache memory for disk data. The cache speeds disk transfers. Common access times for an IDE drive are often less than 8 ms, where the access time for a floppy-disk is about 200 ms.

Optical Disk Memory

Optical disk memory (see Figure 13–33) is commonly available in two forms: the CD-ROM (compact disk/read only memory) and the WORM (write once/read mostly). The CD-ROM is the lowest cost optical disk, but it suffers from lack of speed. Access times for a CD-ROM are typically 300 ms or longer, about the same as a floppy disk. (Note that slower CD-ROM devices are on the market and should be avoided.) Hard disk magnetic memory can have access times as little as 11 ms. A CD-ROM stores 660M bytes of data, or a combination of data and musical passages. As systems develop and become more visually active, the use of the CD-ROM drive will become even more common.

The WORM drive sees far more commercial application than the CD-ROM. The problem is that its application is very specialized due to the nature of the WORM. Because data may be written only once, the main application is in the banking industry, insurance industry, and other massive data storing organizations. The WORM is normally used to form an audit trail of transactions that are spooled onto the WORM and retrieved only during an audit. You might call the WORM an archiving device.

Many WORM and read/write optical disk memory systems are interfaced to the microprocessor by using the SCSI or ESDI interface standards used with hard disk memory. The difference is that the current optical disk drives are no faster than the most floppy drives. Some CD-ROM drives are interfaced to the microprocessor through proprietary interfaces that are not compatible with other disk drives.

The main advantage of the optical disk is its durability. Because a solid state laser beam is used to read the data from the disk, and the focus point is below a protective plastic coating, the surface of the disk may contain small scratches and dirt particles and still be read correctly. This feature allows less care of the optical disk than a comparable floppy disk. About the only way to destroy data on an optical disk is to break it or deeply scar it.

The read/write CD-ROM drive is here and its cost is dropping rapidly. In the near future, we should start seeing the read/write CD-ROM replacing floppy disk drives. The main advantage is

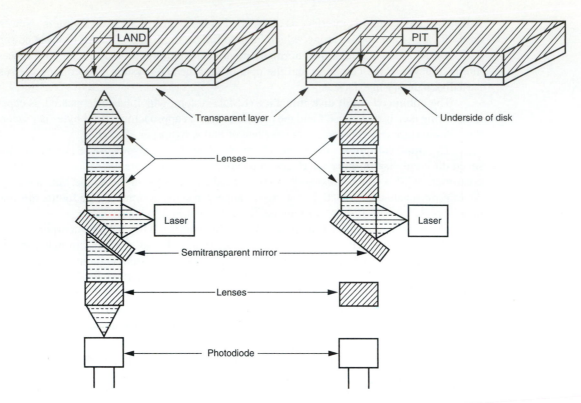

FIGURE 13–33   The optical CD-ROM memory system.

the vast storage available on the read/write CD-ROM. Soon, the format will change so that many G bytes of data will be available. The new versatile read/write CD-ROM, called a DVD, became available in late 1996 or early 1997. The DVD functions exactly as the CD-ROM except that the bit density is much higher. The CD-ROM stores 660M bytes of data, while the current-genre DVD stores 4.7G bytes or 9.4G bytes, depending on the current standard. The future standard, which requires a new drive called a DVD-RAM, stores up to 17G bytes of data.

13–5   VIDEO DISPLAYS

Modern video displays are OEM **(original equipment manufacturer)** devices that are usually purchased and incorporated into a system. Today, there are many different types of video displays available. Of the types available, either color or monochrome versions are found.

Monochrome versions usually display information using amber, green, or paper-white displays. The paper-white display is extremely popular for many applications. The most common of these applications are desktop publishing and computer-aided drafting (CAD).

The color displays are more diverse. Color display systems are available that accept information as a composite video signal, much as your home television, as TTL voltage level signals (0 or 5 V), and as analog signals (0–0.7 V). Composite video displays are disappearing because the available resolution is too low. Today, many applications require high-resolution graphics that cannot be displayed on a composite display such as a home television receiver. Early composite video displays were found with Commodore 64, Apple 2, and similar computer systems.

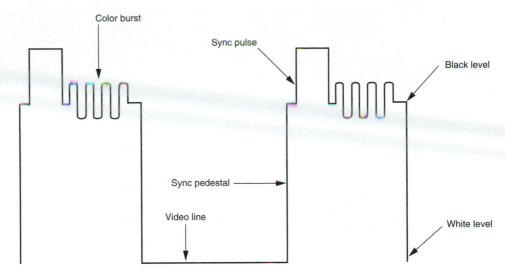

FIGURE 13–34   The composite video signal.

Video Signals

Figure 13–34 illustrates the signal sent to a composite video display. This signal is composed of several parts that are required for this type of display. The signals illustrated represent the signals sent to a color composite-video monitor. Notice that these signals include not only video, but also include sync pulses, sync pedestals, and a color burst. Notice that no audio signal is illustrated because one often does not exist. Rather than include audio with the composite video signal, audio is developed in the computer and output from a speaker inside the computer cabinet. It can also be developed by a sound system and output in stereo to external speakers. The major disadvantages of the composite video display are the resolution and color limitations. Composite video signals were designed to emulate television video signals so that a home television receiver could function as a video monitor.

Most modern video systems use direct video signals that are generated with separate sync signals. In a direct video system, video information is passed to the monitor through a cable that uses separate lines for video and also synchronization pulses. Recall that these signals were combined in a composite video signal.

A monochrome (one color) monitor uses one wire for video, one for horizontal sync, and one for vertical sync. Often, these are the only signal wires found. A color video monitor uses three video signals. One signal represents red, another green, and the third blue. These monitors are often called RGB monitors for the video primary colors of light: red (R), green (G), and blue (B).

The TTL RGB Monitor

The RGB monitor is available as either an analog or TTL monitor. The RGB monitor uses TTL level signals (0 or 5 V) as video inputs and a fourth line called intensity to allow a change in intensity. The RGB video TTL display can display a total of 16 different colors. The TTL RGB monitor is used in the CGA (**color graphics adapter**) system found in older computer systems.

Table 13–4 lists these 16 colors and also the TTL signals present to generate them. Eight of the 16 colors are generated at high intensity and the other eight at low intensity. The three video colors are red, green, and blue. These are primary colors of light. The secondary colors are cyan, magenta, and yellow. Cyan is a combination of blue and green video signals, and is blue-green in color. Magenta is a combination of blue and red video signals, and is a purple color.

TABLE 13–4   16 colors found in the CGA display.

Intensity	Red	Green	Blue	Color
0	0	0	0	Black
0	0	0	1	Blue
0	0	1	0	Green
0	0	1	1	Cyan
0	1	0	0	Red
0	1	0	1	Magenta
0	1	1	0	Brown
0	1	1	1	White
1	0	0	0	Gray
1	0	0	1	Bright Blue
1	0	1	0	Bright Green
1	0	1	1	Bright Cyan
1	1	0	0	Bright Red
1	1	0	1	Bright Magenta
1	1	1	0	Yellow
1	1	1	1	Bright White

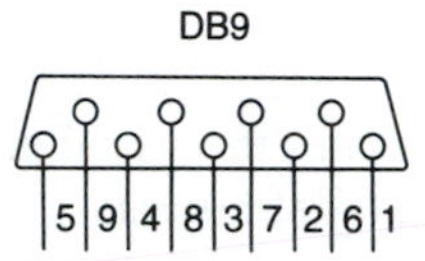

FIGURE 13–35   The 9-pin connector found on a TTL monitor.

DB9

5 | 9 | 4 | 8 | 3 | 7 | 2 | 6 | 1

Pin	Function
1	Ground
2	Ground
3	Red video
4	Green video
5	Blue video
6	Intensity
7	Normal video
8	Horizontal retrace
9	Vertical retrace

Yellow (high-intensity) and brown (low-intensity) are both a combination of red and green video signals. If additional colors are desired, TTL video is not normally used. A scheme was developed by using low and medium color TTL video signals, which provided 32 colors, but it proved of little application and never found widespread use in the field.

Figure 13–35 illustrates the connector most often found on the TTL RGB monitor or a TTL monochrome monitor. The connector illustrated is a 9–pin connector. Two of the connections are used for ground, three for video, two for synchronization or retrace signals, and one for intensity. Notice that pin 7 is labeled *normal video*. This is the pin used on a monochrome monitor for the luminance or brightness signal. Monochrome TTL monitors use the same 9-pin connector as RGB TTL monitors.

The Analog RGB Monitor

In order to display more than 16 colors, an analog video display is required. These are often called analog RGB monitors. Analog RGB monitors still have three video input signals, but don't have the intensity input. Because the video signals are analog signals instead of two-level TTL signals, they are any voltage level between 0.0 V and 0.7 V, which allows an infinite

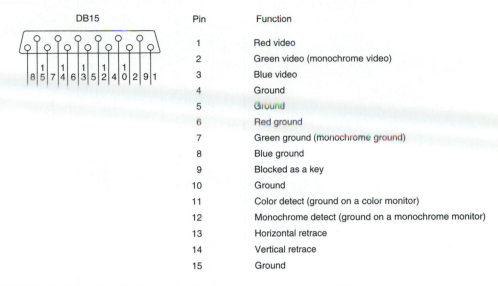

Pin	Function
1	Red video
2	Green video (monochrome video)
3	Blue video
4	Ground
5	Ground
6	Red ground
7	Green ground (monochrome ground)
8	Blue ground
9	Blocked as a key
10	Ground
11	Color detect (ground on a color monitor)
12	Monochrome detect (ground on a monochrome monitor)
13	Horizontal retrace
14	Vertical retrace
15	Ground

FIGURE 13–36 The 15-pin connector found on an analog monitor.

number of colors to be displayed. This is because an infinite number of voltage levels between the minimum and maximum could be generated. In practice, a finite number of levels are generated. This is usually either 256K, 16M, or 24M colors, depending on the standard.

Figure 13–36 illustrates the connector used for an analog RGB or analog monochrome monitor. Notice that the connector has 15 pins and supports both RGB and monochrome analog displays. The way data are displayed on an analog RGB monitor depends upon the interface standard used with the monitor. Pin 9 is a key, which means that no hole exists on the female connector for this pin.

Most analog displays use a **digital-to-analog converter** (DAC) to generate each color video voltage. A common standard uses a 6-bit DAC for each video signal to generate 64 different voltage levels between 0 V and 0.7 V. There are 64 different red video levels, 64 different green video levels, and 64 different blue video levels. This allows $64 \times 64 \times 64$, or 262,144 (256K) colors to be displayed.

Other arrangements are possible, but the speed of the DAC is critical. Most modern displays require an operating conversion time of 25 ns to 40 ns maximum. When converter technology advances, additional resolution at a reasonable price will become available. If 7-bit converters are used for generating video, $128 \times 128 \times 128$, or 2,097,152 (2M) colors are displayed. In this system, a 21-bit color code is needed so that a 7-bit code is applied to each DAC. 8-bit converters also find applications and allow $256 \times 256 \times 256$, or 16,777,216 (16M) colors.

Figure 13–37 illustrates the video generation circuit employed in many common video standards such as the short-lived EGA **(enhanced graphics adapter)** and VGA **(variable graphics array),** as used with an IBM PC. This circuit is used to generate VGA video. Notice that each color is generated with an 18-bit digital code. Six of the 18 bits are used to generate each video color voltage when applied to the inputs of a 6-bit DAC.

A high-speed palette SRAM (access time of less than 40 ns) is used to store 256 different 18-bit codes that represent 256 different hues. This 18-bit code is applied to the digital-to-analog converters. The address input to the SRAM selects one of the 256 colors stored as 18-bit binary codes. This system allows 256 colors out of a possible 256K colors to be displayed at one time. In order to select any of 256 colors, an 8-bit code that is stored in the computer's video display RAM is used to specify the color of a picture element. If more colors are used in

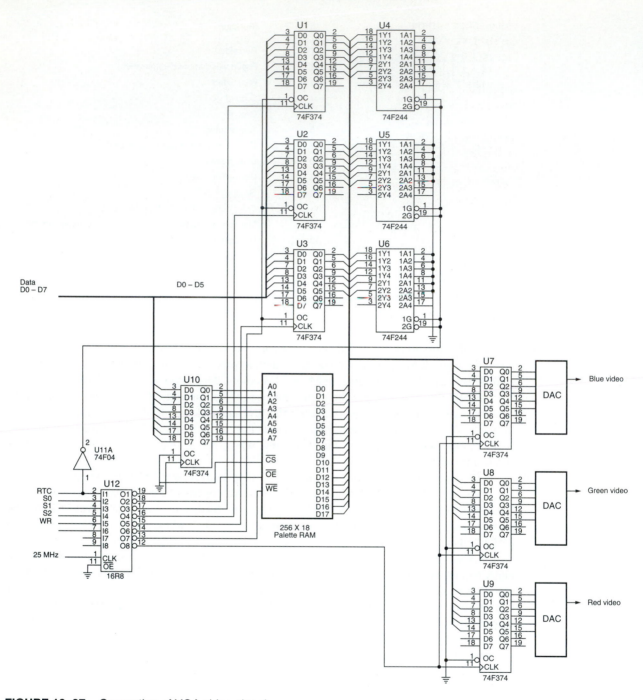

FIGURE 13–37   Generation of VGA video signals.

a system, the code must be wider. For example, a system that displays 1024 colors out of 256K colors requires a 10-bit code to address the SRAM that contains 1024 locations, each containing an 18-bit color code. Some newer systems use a larger palette SRAM to store up to 64K of different color codes.

The Apple Macintosh uses a 24-bit binary code to specify each color in its color video adapter. Each DAC is eight bits wide. This means that each converter can generate 256 different video voltage levels. There are $256 \times 256 \times 256$, or 16,777,216 different possible colors. As with the IBM VGA standard, only 256 colors are displayed at a time. The SRAM in the Apple interface is 256×24 instead of 256×18.

Whenever a color is placed on the video display, provided that RTC is a logic 0, the system sends the 8-bit code that represents a color to the D0–D7 connections. The PAL 16R8 then generates a clock pulse for U10, which latches the color code. After 40 ns (one 25 MHz clock), the PAL generates a clock pulse for the DAC latches (U7, U8, and U9). This amount of time is required for the palette SRAM to look up the 18-bit contents of the memory location selected by U10. Once the color code (18-bit) is latched into U7–U9, the three DACs convert it to three video voltages for the monitor. This process is repeated for each 40 ns-wide picture element (pixel) that is displayed. The pixel is 40 ns wide because a 25 MHz clock is used in this system. Higher resolution is attainable if a higher clock frequency is used with the system.

If the color codes (18-bits) stored in the SRAM must be changed, this is always accomplished during retrace when RTC is a logic 1. This prevents any video noise from disrupting the image displayed on the monitor.

In order to change a color, the systems uses the S0, S1, and S2 inputs of the PAL to select U1, U2, U3, or U10. First, the address of the color to be changed is sent to latch U10, which addresses a location in the palette SRAM. Next, each new video color is loaded into U1, U2, and U3. Finally, the PAL generates a write pulse for the $\overline{WE}$ input to the SRAM to write the new color code into the palette SRAM.

Retrace occurs 70.1 times per second in the vertical direction and 31,500 times per second in the horizontal direction for a 640×480 display. During retrace, the video signal voltage sent to the display must be 0 V, which causes black to be displayed during the retrace. Retrace itself is used to move the electron beam to the upper left-hand corner for vertical retrace and to the left margin of the screen for horizontal retrace.

The circuit illustrated causes U4–U6 buffers to be enabled so that they apply 00000 each to the DAC latch for retrace. The DAC latches capture this code and generate 0 V for each video color signal to blank the screen. By definition, 0 V is considered to be the black level for video and 0.7 V is considered to be the full intensity on a video color signal.

The resolution of the display, for example 640×480, determines the amount of memory required for the video interface card. If this resolution is used with a 256-color display (eight bits per pixel), then 640×480 bytes of memory (307,200) are required to store all of the pixels for the display. Higher-resolution displays are possible, but, as you can imagine, even more memory is required. A 640×480 display has 480 video raster lines and 640 pixels per line. A **raster line** is the horizontal line of video information that is displayed on the monitor. A pixel is the smallest subdivision of this horizontal line.

Figure 13–38 illustrates the video display, showing the video lines and retrace. The slant of each video line in this illustration is greatly exaggerated, as is the spacing between lines. This illustration shows retrace in both the vertical and horizontal directions. In the case of a VGA display, as described, the vertical retrace occurs exactly 70.1 times per second and the horizontal retrace occurs exactly 31,500 times per second. (The Apple Macintosh uses a vertical rate of 66.67 Hz and a horizontal rate of 35 KHz to generate a 640×480 color display.)

In order to generate 640 pixels across one line, it takes 40 ns $\times$ 640, or 25.6 μs. A horizontal time of 31,500 Hz allows a horizontal line time of 1/31,500, or 31.746 μs. The difference between these two times is the retrace time allowed to the monitor. (The Apple Macintosh has a horizontal line time of 28.57 μs.)

Because the vertical retrace repetition rate is 70.1 Hz, the number of lines generated is determined by dividing the vertical time into the horizontal time. In the case of a VGA display (a 640×400 display), this is 449.358 lines. Only 400 of these lines are used to display information; the rest are lost during the retrace. Because 49.358 lines are lost during the retrace, the retrace time is 49.358×31.766 μs, or 1568 μs. It is during this relatively large amount of time that the

FIGURE 13–38 A video
screen illustrating the raster
lines and retrace.

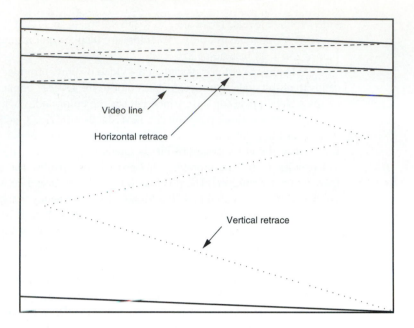

color palette SRAM is changed or the display memory system is updated for a new video display. In the Apple Macintosh computer (640×480), the number of lines generated is 525 lines. Of the total number of lines, 45 are lost during vertical retrace.

Other display resolutions are 800×600 and 1024×768. The 800×600 SVGA (super VGA) display is ideal for a 14" color monitor, while the 1024×768 EVGA or XVGA (extended VGA) is ideal for a 21" or 25" monitor used in CAD systems. These resolutions sound like just another set of numbers, but realize that an average home television receiver has a resolution approximately 400×300. The high-resolution display available on computer systems is much clearer than that available as home television. A resolution of 1024×768 approaches that found in 35 mm film. The only disadvantage of the video display on a computer screen is the number of colors displayed at a time, but as time passes, this will surely improve. Additional colors allow the image to appear more realistically because of subtle shadings that are required for a true high-quality, life-like image.

If a display system operates with a 60 Hz vertical time and a 15,600 Hz horizontal time, the number of lines generated is 15,600/60, or 260 lines. The number of usable lines in this system is most likely 240, where 20 are lost during vertical retrace. It is clear that the number of scanning lines is adjustable by changing the vertical and horizontal scanning rates. The vertical scanning rate must be greater than or equal to 50 Hz, or flickering will occur. The vertical rate must not be higher than about 75 Hz or problems with the vertical deflection coil may occur. The electron beam in a monitor is positioned by an electrical magnetic field generated by coils in a yoke that surrounds the neck of the picture tube. Because the magnetic field is generated by coils, the frequency of the signal applied to the coil is limited.

The horizontal scanning rate is also limited by the physical design of the coils in the yoke. Because of this, it is normal to find the frequency applied to the horizontal coils within a narrow range. This is usually 30,000 Hz–37,000 Hz or 15,000 Hz–17,000 Hz. Some newer monitors are called multisync monitors because the deflection coil is taped so that it can be driven with different deflection frequencies. Sometimes, both the vertical and horizontal coils are both taped for different vertical and horizontal scanning rates.

High-resolution displays use either interlaced or non-interlaced scanning. The non-interlaced scanning system is used in all standards except the highest. In the interlaced system, the video image is displayed by drawing half the image first with all of the odd scanning lines, then the other half is drawn using the even scanning lines. Obviously, this system is more complex and is only more efficient because the scanning frequencies are reduced by 50 percent in an interlaced system. For example, a video system that uses 60 Hz for the vertical scanning frequency and 15,720 Hz for the horizontal frequency generates 262 (15,720/60) lines of video at the rate of 60 full frames per second. If the horizontal frequency is changed slightly to 15,750 Hz, 262.5 (15,750/60) lines are generated, so two full sweeps are required to draw one complete picture of 525 video lines. Notice how just a slight change in horizontal frequency doubled the number of raster lines.

13–6   SUMMARY

1. The HOLD input is used to request a DMA action, and the HLDA output signals that the hold is in effect. When a logic 1 is placed on the HOLD input, the microprocessor (1) stops executing the program; (2) places its address, data, and control bus at their high-impedance state; and (3) signals that the hold is in effect by placing a logic 1 on the HLDA pin.

2. A DMA read operation transfers data from a memory location to an external I/O device. A DMA write operation transfers data from an I/O device into the memory. Also available is a memory-to-memory transfer that allows data to be transferred between two memory locations by using DMA techniques.

3. The 8237 direct memory access (DMA) controller is a four-channel device that can be expanded to include an additional channel of DMA.

4. Disk memory comes in the form of floppy disk storage that is found as either the $5^{1}/_{4}$" mini-floppy disk or $3^{1}/_{2}$" micro-floppy disk. Both disks are found as double-sided, double-density (DSDD), or as high-density (HD) storage devices. The DSDD $5^{1}/_{4}$" disk stores 360 K bytes of data and the HD $5^{1}/_{4}$" disk stores 1.2 M bytes of data. The DSDD $3^{1}/_{2}$" disk stores 720 K bytes of data and the HD $3^{1}/_{2}$" disk stores 1.44 M bytes of data.

5. Floppy disk memory data are stored using NRZ (non-return to zero) recording. This method saturates the disk with one polarity of magnetic energy for a logic 1 and the opposite polarity for a logic 0. In either case, the magnetic field never returns to 0. This technique eliminates the need for a separate erase head.

6. Data are recorded on disks by using either modified frequency modulation (MFM) or run-length limited (RLL) encoding schemes. The MFM scheme records a data pulse for a logic 1, no data or clock for the first logic 0 of a string of zeros, and a clock pulse for the second and subsequent logic 0 in a string of zeros. The RLL scheme encodes data so that 50 percent more information can be packed onto the same disk area. Most modern disk memory systems use the RLL encoding scheme.

7. Video monitors are either TTL or analog. The TTL monitor uses two discrete voltage levels of 0 V and 5.0 V. The analog monitor uses an infinite number of voltage levels between 0.0 V and 0.7 V. The analog monitor can display an infinite number of video levels, while the TTL monitor is limited to two video levels.

8. The color TTL monitor displays 16 different colors. This is accomplished through three video signals (red, green, and blue) and an intensity input. The analog color monitor can display an infinite number of colors through its three video inputs. In practice, the most common form of color analog display system (VGA) can display 256K different colors.

9. The video standards found today include VGA (640×480), SVGA (800×600), and EVGA or XVGA (1024×768). In all three cases, the video information can be 256 colors out of a total possible 256K colors.

13–7   QUESTIONS AND PROBLEMS

1. Which microprocessor pins are used to request and acknowledge a DMA transfer?
2. Explain what happens whenever a logic 1 is placed on the HOLD input pin.
3. A DMA read transfers data from _____ to _____.
4. A DMA write transfers data from _____ to _____.
5. The DMA controller selects the memory location used for a DMA transfer through what bus signals?
6. The DMA controller selects the I/O device used during a DMA transfer by which pin?
7. What is a memory-to-memory DMA transfer?
8. Describe the effect on the microprocessor and DMA controller when the HOLD and HLDA pins are at their logic 1 levels.
9. Describe the effect on the microprocessor and DMA controller when the HOLD and HLDA pins are at their logic 0 levels.
10. The 8237 DMA controller is a _____ channel DMA controller.
11. If the 8237 DMA controller is decoded at I/O ports 2000H–200FH, what ports are used to program channel 1?
12. Which 8237 DMA controller register is programmed to initialize the controller?
13. How many bytes can be transferred by the 8237 DMA controller?
14. Write a sequence of instructions that transfer data from memory location 21000H–210FFH to 20000H–200FFH by using channel 2 of the 8237 DMA controller. You must initialize the 8237 and use the latch described in Section 12–1 to hold A19–A16.
15. Write a sequence of instructions that transfer data from memory to an external I/O device by using channel 3 of the 8237. The memory area to be transferred is at location 20000H–20FFFH.
16. The $5^{1}/4$" disk is known as a(n) _____- floppy disk.
17. The $3^{1}/4$" disk is known as a(n) _____- floppy disk.
18. Data are recorded in concentric rings on the surface of a disk known as a(n) _____.
19. A track is divided into sections of data called _____.
20. On a double-sided disk, the upper and lower tracks together are called a(n) _____.
21. Why is NRZ recording used on a disk memory system?
22. Draw the timing diagram generated to write a 1001010000 using MFM encoding.
23. Draw the timing diagram generated to write a 1001010000 using RLL encoding.
24. What is a flying head?
25. Why must the heads on a hard disk be parked?
26. What is the difference between a voice coil head position mechanism and a stepper motor head positioning mechanism?
27. What is a WORM?
28. What is a CD-ROM?
29. What is the difference between a TTL monitor and an analog monitor?
30. What are the three primary colors of light?
31. What are the three secondary colors of light?
32. What is a pixel?
33. A video display with a resolution of 800×600 contains _____ lines of video information, with each line divided into _____ pixels.
34. Explain how a TTL RGB monitor can display 16 different colors.
35. Explain how an analog RGB monitor can display an infinite number of colors.
36. If an analog RGB video system uses 7-bit DACs, it can generate _____ different colors.
37. Why does standard VGA allow only 256 different colors out of 256K colors to be displayed at one time?
38. If a video system uses a vertical frequency of 60 Hz and a horizontal frequency of 32,400 Hz, how many raster lines are generated?

CHAPTER 14

The Arithmetic Coprocessor and MMX Technology

INTRODUCTION

The Intel family of arithmetic coprocessors includes the 8087, 80287, 80387SX, 80387DX, and the 80487SX for use with the 80486SX microprocessor. The 80486DX–Pentium 4 microprocessors contain their own built-in arithmetic coprocessors. Be aware that some of the cloned 80486 microprocessors (from IBM and Cyrix) do not contain arithmetic coprocessors. The instruction sets and programming for all devices are almost identical; the main difference is that each coprocessor is designed to function with a different Intel microprocessor. This chapter provides detail on the entire family of arithmetic coprocessors. Because the coprocessor is a part of the 80486DX–Pentium 4, and because these microprocessors are commonplace, many programs now require or at least benefit from a coprocessor.

The family of coprocessors, which is labeled the 80X87, is able to multiply, divide, add, subtract, find the square root; and calculate the partial tangent, partial arctangent, and logarithms. Data types include 16-, 32-, and 64-bit signed-integers; 18-digit BCD data; and 32-, 64-, and 80-bit floating-point numbers. The operations performed by the 80X87 generally execute many times faster than equivalent operations written with the most efficient programs that use the microprocessor's normal instruction set. With the improved Pentium coprocessor, operations execute at about five times faster than those performed by the 80486 microprocessor with an equal clock frequency. Note that the Pentium can often execute a coprocessor instruction and two integer instructions simultaneously. The Pentium Pro through Pentium 4 coprocessors are similar in performance to the Pentium coprocessor, except that a few new instructions have been added: FMOV and FCOMI.

The multimedia extension (MMX) to the Pentium–Pentium 4 are instructions to a device similar to the arithmetic coprocessor. In fact, the MMX extension shares the arithmetic coprocessor registers. The MMX extension is a special internal processor designed to execute instructions, at high-speed, for external multimedia devices. For this reason, we decided to place the MMX instruction set and specifications in this chapter.

CHAPTER OBJECTIVES

Upon completion of this chapter, you will be able to:

1. Convert between decimal data and signed integer, BCD, and floating-point data for use by the arithmetic coprocessor.
2. Explain the operation of the 80X87 arithmetic coprocessor.

3. Explain the operation and addressing modes of each arithmetic coprocessor instruction.
4. Develop programs that solve complex arithmetic problems using the arithmetic coprocessor.
5. Describe the operation and features of the MMX extension.
6. Develop short programs that use the MMX instructions.

14–1   DATA FORMATS FOR THE ARITHMETIC COPROCESSOR

This section of the text presents the types of data used with all arithmetic coprocessor family-members. (See Table 14–1 for a listing of all Intel microprocessors and their companion co-processors.) These data types include signed-integer, BCD, and floating-point. Each has a specific use in a system, and many systems require all three data types. Note that assembly language programming with the coprocessor is often limited to modifying the coding generated by a high-level language such as C/C++. In order to accomplish any such modification, the instruction set and some basic programming concepts are required, which are presented in this chapter.

Signed Integers

The signed integers used with the coprocessor are the same as those described in Chapter 1. When used with the arithmetic coprocessor, signed integers are 16- (word), 32- (short integer), or 64-bits (long integer) wide. The long integer is new to the coprocessor and is not described in Chapter 1, but the principles are the same. Conversion between decimal and signed-integer format is handled in exactly the same manner as for the signed integers described in Chapter 1. As you will recall, positive numbers are stored in true form with a leftmost sign-bit of 0, and negative numbers are stored in two's complement form with a leftmost sign-bit of 1.

The word integers range in value from $-32,768$ to $+32,767$, the short integer range is $\pm 2 \times 10^{+9}$, and the long integer range is $\pm 9 \times 10^{+18}$. Integer data types are found in some applications that use the arithmetic coprocessor. See Figure 14–1, which shows these three forms of signed-integer data.

Data are stored in memory using the same assembler directives described and used in earlier chapters. The DW directive defines words, DD defines short integers, and DQ defines long integers. Example 14–1 shows how several different sizes of signed integers are defined for use by the assembler and arithmetic coprocessor.

TABLE 14–1   Microprocessor and Intel coprocessor compatibility.

Microprocessor	Coprocessor
8086	8087
8088	8087
80186	80187
80188	80187
80286	80287
80386SX	80387SX
80386DX	80387DX
80486SX	80487SX
80486DX	Built into microprocessor
Pentium	Built into microprocessor
Pentium Pro	Built into microprocessor
Pentium II	Built into microprocessor
Pentium III	Built into microprocessor
Pentium 4	Built into microprocessor

FIGURE 14–1 Integer formats for the 80X87 family of arithmetic coprocessors: (a) word, (b) short, and (c) long.

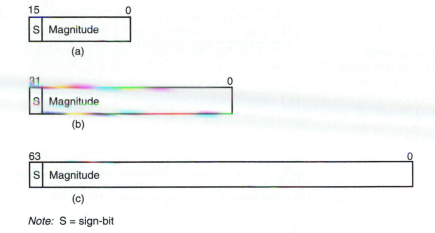

Note: S = sign-bit

EXAMPLE 14–1

```
0000    0002                DATA1    DW    +2        ;16-bit integer
0002    FFDE                DATA2    DW    -34       ;16-bit integer
0004    000004D2            DATA3    DD    +1234     ;short integer
0008    FFFFFF9C            DATA4    DD    -100      ;short integer
000C    0000000000005BA0    DATA5    DQ    +23456    ;long integer
0014    FFFFFFFFFFFFFF86    DATA6    DQ    -122      ;long integer
```

Binary-coded Decimal (BCD)

The binary-coded decimal (BCD) form requires 80 bits of memory. Each number is stored as an 18-digit packed integer in nine bytes of memory as two digits per byte. The tenth byte contains only a sign bit for the 18-digit signed BCD number. Figure 14–2 shows the format of the BCD number used with the arithmetic coprocessor. Note that both positive and negative numbers are stored in true form and never in 10's complement form. The DT directive stores BCD data in the memory as illustrated in Example 14–2.

EXAMPLE 14–2

```
0000              DATA1    DT    200      ;200 decimal stored as BCD
        00000000000000000200
000A              DATA2    DT    -10      ;-10 decimal stored as BCD
        80000000000000000010
0014              DATA3    DT    10020    ;10,020 decimal stored as BCD
        00000000000000010020
```

Floating-point

Floating-point numbers are often called *real numbers* because they hold signed integers, fractions, and mixed numbers. A floating-point number has three parts: a sign-bit, a biased exponent, and a significand. Floating-point numbers are written in scientific binary notation. The Intel family of arithmetic coprocessors supports three types of floating-point numbers: short (32 bits),

FIGURE 14–2 BCD data format for the 80X87 family of arithmetic coprocessors.

FIGURE 14–3 Floating-point (real) format for the 80X87 family of arithmetic coprocessors. (a) Short (single-precision) with a bias of 7FH, (b) long (double-precision) with a bias of 3FFH, and (c) temporary (extended-precision) with a bias of 3FFFH.

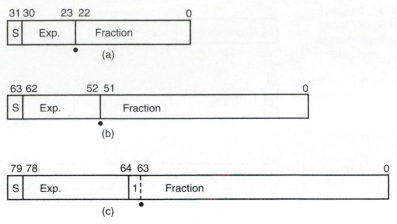

Note: S = sign-bit and Exp. = exponent

long (64 bits), and temporary (80 bits). See Figure 14–3 for the three forms of the floating-point number. Please note that the short form is also called a single-precision number and the long form is called a double-precision number. Sometimes the 80-bit temporary form is called an extended-precision number. The floating-point numbers and the operations performed by the arithmetic coprocessor conform to the IEEE-754 standard, as adopted by all major personal computer software producers. This includes Microsoft, which in 1995 stopped supporting the Microsoft floating-point format and also the ANSI floating-point standard that is popular in mainframe computer systems.

Converting to Floating-point Form. Converting from decimal to the floating-point form is a simple task that is accomplished through the following steps:

1. Convert the decimal number into binary.
2. Normalize the binary number.
3. Calculate the biased exponent.
4. Store the number in the floating-point format.

These four steps are illustrated for the decimal number 100.25_{10} in Example 14–3. Here, the decimal number is converted to a single-precision (32-bit) floating-point number.

EXAMPLE 14–3

```
Step      Result

1         100.25 = 1100100.01

2         1100100.01 = 1.10010001 x 2⁶

3         110 + 01111111 = 10000101

4         Sign = 0
          Exponent = 10000101
          Significand = 10010001000000000000000
```

In step three of Example 14–3, the biased exponent is the exponent, a 2^6 or 110, plus a bias of 01111111 (7FH) or 10000101 (85H). All single-precision numbers use a bias of 7FH, double-precision numbers use a bias of 3FFH, and extended-precision numbers use a bias of 3FFFH.

In step 4 of Example 14–3, the information found in the prior steps is combined to form the floating-point number. The leftmost bit is the sign-bit of the number. In this case, it is a 0 because the number is $+100.25_{10}$. The biased exponent follows the sign-bit. The significand is a 23-bit number with an implied one-bit. Note that the significand of a number 1.XXXX is the XXXX portion. The 1. is an **implied one-bit** that is only stored in the extended precision form of the floating-point number as an explicit one-bit.

Some special rules apply to a few numbers. The number 0, for example, is stored as all zeros except for the sign-bit, which can be a logic 1 to represent a negative zero. The plus and minus infinity is stored as logic 1s in the exponent with a significand of all zeros and the sign-bit that represents plus or minus. A NAN (not-a-number) is an invalid floating-point result that has all ones in the exponent with a significand that is not all zeros.

Converting from Floating-point Form.

Conversion to a decimal number from a floating-point number is summarized in the following steps:

1. Separate the sign-bit, biased exponent, and significand.
2. Convert the biased exponent into a true exponent by subtracting the bias.
3. Write the number as a normalized binary number.
4. Convert it to a de-normalized binary number.
5. Convert the de-normalized binary number to decimal.

These five steps convert a single-precision floating-point number to decimal, as shown in Example 14–4. Notice how the sign-bit of 1 makes the decimal result negative. Also notice that the implied one-bit is added to the normalized binary result in step 3.

EXAMPLE 14–4

```
Step      Result

1         Sign = 1
          Exponent = 10000011
          Significand = 10010010000000000000000

2         100 = 10000011 - 01111111

3         1.1001001 x 2⁴

4         11001.001

5         -25.125
```

Storing Floating-point Data in Memory.

Floating-point numbers are stored with the assembler using the DD directive for single-precision, DQ for double-precision, and DT for extended-precision. Some examples of floating-point data storage are shown in Example 14–5. The author discovered that the Microsoft version 6.0 macro assembler contains an error that does not allow a plus sign to be used with positive floating-point numbers. A +92.45 must be defined as 92.45 for the assembler to function correctly. Microsoft has assured the author that this error has been corrected in version 6.11 of MASM if the REAL4, REAL8, or REAL10 directives are used in place of DD, DQ, and DT to specify floating-point data. The assembler provides access 8087 emulator if your system does not contain a microprocessor with a coprocessor. The emulator comes with all Microsoft high-level languages or as shareware programs such as EM87. The emulator is accessed by including the OPTION EMULATOR statement immediately following the .MODEL statement in a program. Be aware that the emulator does not emulate some of the coprocessor instructions. Do not use this option if your system contains a coprocessor. In all cases, you must include the .8087, .80187, .80287, .80387, .80487, or .80587 switch to enable the generation of coprocessor instructions. Note that there is currently no switch for the Pentium 4 through Pentium Pro, but it will most likely be .80687 when Microsoft produces the next version of the assembler program.

EXAMPLE 14–5

```
0000   C377999A      DATA7      DD      -247.6      ;define single-precision
0004   40000000      DATA8      DD      2.0         ;define single-precision
0008   486F4200      DATA9      REAL4   2.45E+5     ;define single-precision
000C                 DATA10     DQ      100.25      ;define double-precision
       4059100000000000
0014                 DATA11     REAL8   0.001235    ;define double-precision
       3F543BF727136A40
001C                 DATA12     REAL10  33.9876     ;define extended-precision
       400487F34D6A161E4F76
```

14–2   THE 80X87 ARCHITECTURE

The 80X87 is designed to operate concurrently with the microprocessor. Note that the 80486DX–Pentium 4 microprocessors contain their own internal and fully-compatible versions of the 80387. With other family members, the coprocessor is an external integrated circuit that parallels most of the connections on the microprocessor. The 80X87 executes 68 different instructions. The microprocessor executes all normal instructions and the 80X87 executes arithmetic coprocessor instructions. Both the microprocessor and coprocessor can execute their respective instructions simultaneously or concurrently. The numeric or arithmetic coprocessor is a special-purpose microprocessor that is especially designed to efficiently execute arithmetic and transcendental operations.

The microprocessor intercepts and executes the normal instruction set, and the coprocessor intercepts and executes only the coprocessor instructions. Recall that the coprocessor instructions are actually escape (ESC) instructions. These instructions are used by the microprocessor to generate a memory address for the coprocessor so that the coprocessor can execute a coprocessor instruction.

Internal Structure of the 80X87

Figure 14–4 shows the internal structure of the arithmetic coprocessor. Notice that this device is divided into two major sections: the control unit and the numeric execution unit.

The **control unit** interfaces the coprocessor to the microprocessor-system data bus. Both the devices monitor the instruction stream. If the instruction is an ESCape (coprocessor) instruction, the coprocessor executes it; if not, the microprocessor executes it.

The **numeric execution unit** (NEU) is responsible for executing all coprocessor instructions. The NEU has an eight-register stack that holds operands for arithmetic instructions and the results of arithmetic instructions. Instructions either address data in specific stack data-registers or use a push-and-pop mechanism to store and retrieve data on the top of the stack. Other registers in the NEU are status, control, tag, and exception pointers. A few instructions transfer data between the coprocessor and the AX register in the microprocessor. The FSTSW AX instruction is the only instruction available to the coprocessor that allows direct communications to the microprocessor through the AX register. Note that the 8087 does not contain the FSTSW AX instruction.

The stack within the coprocessor contains eight registers that are each 80 bits wide. These stack registers always contain an 80-bit extended-precision floating-point number. The only time that data appear as any other form is when they reside in the memory system. The coprocessor converts from signed integer, BCD, single-precision, or double-precision form as the data are moved between the memory and the coprocessor register stack.

Status Register.   The status register (see Figure 14-5) reflects the overall operation of the coprocessor. The status register is accessed by executing the instruction (FSTSW), which stores the contents of the status register into a word of memory. The FSTSW AX instruction copies the status register directly into the microprocessor's AX register on the 80287 or above coprocessor. Once status is stored in memory or the AX register, the bit positions of the status register can be examined

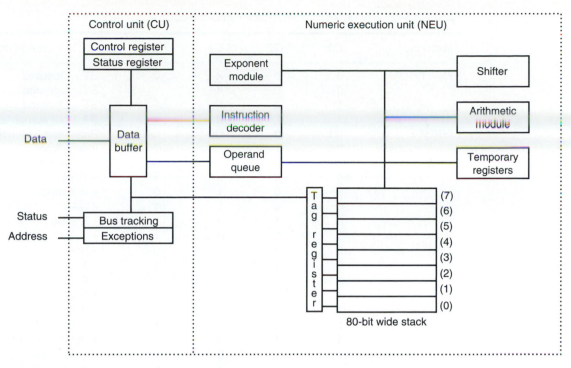

FIGURE 14–4 The internal structure of the 80X87 arithmetic coprocessor.

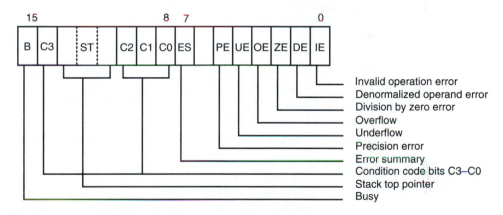

FIGURE 14–5 The 80X87 arithmetic coprocessor status register.

by normal software. The coprocessor/microprocessor communications are carried out through the I/O ports 00FAH–00FFH on the 80287, and I/O ports 800000FAH–800000FFH on the 80386 through the Pentium 4. Never use these I/O ports for interfacing I/O devices to the microprocessor.

The newer coprocessors (80287 and above) use status bit position 6 (SF) to indicate a stack overflow or underflow error. Following is a list of the status bits, except for SF, and their applications:

B   The **busy bit** indicates that the coprocessor is busy executing a task. Busy can be tested by examining the status register or by using the FWAIT instruction. Newer coprocessors automatically synchronize with the microprocessor, so the busy flag need not be tested before performing additional coprocessor tasks.

TABLE 14–2   The 80X87 status register condition code bits.

Instruction	C3	C2	C1	C0	Indication
FTST, FCOM	0	0	X	0	ST > Operand
	0	0	X	1	ST < Operand
	1	0	X	1	ST = Operand
	1	1	X	1	ST is not comparable
FPREM	Q1	0	Q0	Q2	Rightmost 3 bits of quotient
	?	1	?	?	Incomplete
FXAM	0	0	0	0	+ unnormal
	0	0	0	1	+ NAN
	0	0	1	0	− unnormal
	0	0	1	1	− NAN
	0	1	0	0	+ normal
	0	1	0	1	+ ∞
	0	1	1	0	− normal
	0	1	1	1	− ∞
	1	0	0	0	+ 0
	1	0	0	1	Empty
	1	0	1	0	− 0
	1	0	1	1	Empty
	1	1	0	0	+ denormal
	1	1	0	1	Empty
	1	1	1	0	− denormal
	1	1	1	1	Empty

Notes: Unnormal = leading bits of the significand are zero; denormal = exponent at its most negative value; normal = standard floating-point form; NAN (not-a-number) = an exponent of all ones and a significand not equal to zero, and the operand for FTST is zero.

C3–C0   The **condition code bits** indicate conditions about the coprocessor (see Table 14–2 for a complete listing of each combination of these bits and their functions). Note that these bits have different meanings for different instructions, as indicated in the table. The top of the stack is denoted as ST in this table.

TOP   The **top-of-stack (ST)** bit indicates the current register addressed as the top-of-the-stack (ST). This is normally register 0.

ES   The **error summary** bit is set if any unmasked error bit (PE, UE, OE, ZE, DE, or IE) is set. In the 8087 coprocessor, the error summary also caused a coprocessor interrupt. Since the 80287, the coprocessor interrupt has been absent from the family.

PE   The **precision error** indicates that the result or operands exceed the selected precision.

UE   An **underflow error** indicates a non-zero result that is too small to represent with the current precision selected by the control word.

OE   An **overflow error** indicates a result that is too large to be represented. If this error is masked, the coprocessor generates infinity for an overflow error.

ZE   A **zero error** indicates the divisor was zero while the dividend is a non-infinity or non-zero number.

DE   A **denormalized error** indicates that at least one of the operands is denormalized.

IE   An **invalid error** indicates a stack overflow or underflow, indeterminate form (0 ÷ 0, + ∞, − , etc.), or the use of a NAN as an operand. This flag indicates errors such as those produced by taking the square root of a negative number, etc.

There are two ways to test the bits of the status register once they are moved into the AX register with the FSTSW AX instruction. One method uses the TEST instruction to test individual bits of the status register. The other uses the SAHF instruction to transfer the leftmost eight bits of the status register into the microprocessor's flag register. Both methods are illustrated in Example 14–6. This example uses the DIV instruction to divide the top of the stack by the contents of DATA1 and the FSQRT instruction to find the square root of the top of the stack. The example also uses the FCOM instruction to compare the contents of the stack top with DATA1. Note that the conditional jump instructions are used with the SAHF instruction to test for the condition listed in Table 14–3. Although SAHF and conditional jumps cannot test all possible operating conditions of the coprocessor, they can help to reduce the complexity of certain tested conditions. Note that SAHF places C0 into the carry flag, C2 into the parity flag, and C3 into the zero flag.

EXAMPLE 14–6

```
                        ;Using TEST to isolate the divide-by-zero error bit.

0000  67& D8 35 00000000 R      FDIV  DATA1
0007  9B DF E                    FSTSW AX                 ;copy status register to AX
000A  A9 000                     TEST  AX,4               ;test bit position 2
000D  75 18                      JNZ   DIVIDE_ERROR

                        ;Using TEST to isolate the invalid operation error bit
                        ;after a FSQRT instruction.

000F  D9 FA                      FSQRT
0011  9B DF E0                    FSTSW  AX                ;copy status register to AX
0014  A9 0001                     TEST   AX,1             ;test bit position 1
0017  75 0E                       JNZ    FSQRT_ERROR

                        ;Using the SAHF instruction and conditional jumps to
                        ;test for the conditions in Table 14-3 after an FCOM.

0019  67& D8 15 00000000 R      FCOM  DATA1
0020  9B DF E0                    FSTSW AX                 ;copy status register to AX
0023  9E                          SAHF                     ;copy status bits to flags
0024  74 04                       JE    ST_EQUAL
0026  72 02                       JB    ST_BELOW
0028  77 00                       JA    ST_ABOVE
```

When the FXAM instruction and FSTSW AX are executed and followed by the SAHF instruction, the zero flag will contain C3. The FXAM instruction could be used to test a divisor before a division for a zero value by using the JZ instruction following FXAM, FSTSW AX, and SAHF.

TABLE 14–3 Coprocessor conditions tested with the conditional jump instructions and SAHF after FCOM or FTST, as illustrated in Example 14–6.

C3	C2	C0	Condition	Jump Instruction
0	0	0	ST > Operand	JA (jump if ST above)
0	0	1	ST < Operand	JB (jump if ST below)
1	0	0	ST = Operand	JE (jump if ST equal)

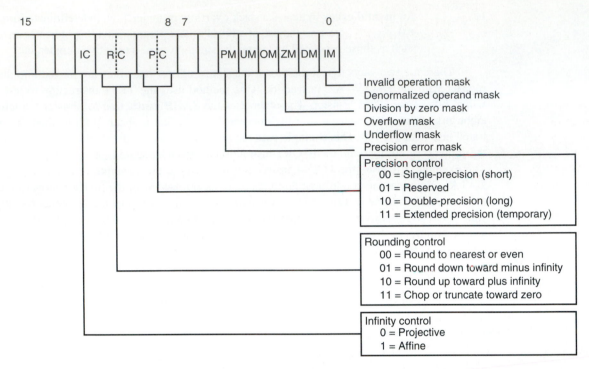

FIGURE 14–6   The 80X87 arithmetic coprocessor control register.

Control Register.   The control register is pictured in Figure 14–6. The control register selects precision, rounding control, and infinity control. It also masks and unmasks the exception bits that correspond to the rightmost six bits of the status register. The FLDCW instruction is used to load a value into the control register.

Following is a description of each bit or grouping of bits found in the control register:

IC	**Infinity control** selects either affine or projective infinity. Affine allows positive and negative infinity, while projective assumes infinity is unsigned.
RC	**Rounding control** determines the type of rounding, as defined in Figure 14–6.
PC	The **precision control** sets the precision of the result, as defined in Figure 14–6.
Exception Masks	Determine whether the error indicated by the exception affects the error bit in the status register. If a logic 1 is placed in one of the exception control bits, the corresponding status register bit is masked off.

Tag Register.   The **tag register** indicates the contents of each location in the coprocessor stack. Figure 14–7 illustrates the tag register and the status indicated by each tag. The tag indicates

FIGURE 14–7   The 80X87 arithmetic coprocessor tag register.

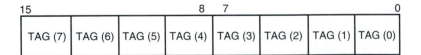

whether a register is valid; zero; invalid or infinity; or empty. The only way that a program can view the tag register is by storing the coprocessor environment using the FSTENV, FSAVE, or FRSTOR instructions. Each of these instructions stores the tag register along with other co-processor data.

14–3   INSTRUCTION SET

The arithmetic coprocessor executes over 68 different instructions. Whenever a coprocessor instruction references memory, the microprocessor automatically generates the memory address for the instruction. The coprocessor uses the data bus for data transfers during coprocessor instructions and the microprocessor uses it during normal instructions. Also note that the 80287 uses the Intel-reserved I/O ports 00F8H–00FFH for communications between the coprocessor and the microprocessor (even though the coprocessor only uses ports 00FCH–00FFH). These ports are used mainly for the FSTSW AX instruction. The 80387–Pentium 4 use I/O ports 800000F8H–800000FFH for these communications.

This section of the text describes the function of each instruction and lists its assembly language form. Because the coprocessor uses the microprocessor memory-addressing modes, not all forms of each instruction are illustrated. Each time that the assembler encounters one of the coprocessor mnemonic opcodes, it converts it into a machine language ESC instruction. The ESC instruction represents an opcode to the coprocessor.

Data Transfer Instructions

There are three basic data transfers: floating-point, signed-integer, and BCD. The only time that data ever appear in the signed-integer or BCD form is in the memory. Inside the coprocessor, data are always stored as an 80-bit extended-precision floating-point number.

Floating-point Data Transfers. There are four traditional floating-point data transfer instructions in the coprocessor instruction set: FLD (load real), FST (store real), FSTP (store real and pop), and FXCH (exchange). A new instruction is added to the Pentium Pro through Pentium 4 called a conditional floating point move instruction that uses the opcode FCMOV with a floating-point condition.

The FLD instruction loads floating-point memory data to the top of the internal stack, referred to as ST (stack top). This instruction stores the data on the top of the stack and then decrements the stack pointer by one. Data loaded to the top of the stack are from any memory location or from another coprocessor register. For example, an FLD ST(2) instruction copies the contents of register 2 to the stack top, which is ST. The top of the stack is register 0 when the coprocessor is reset or initialized. Another example is the FLD DATA7 instruction, which copies the contents of memory location DATA7 to the top of the stack. The size of the transfer is automatically determined by the assembler through the directives DD or REAL4 for single-precision, DQ or REAL 8 for double-precision, and DT or REAL10 for extended-precision.

The FST instruction stores a copy of the top of the stack into the memory location or coprocessor register indicated by the operand. At the time of storage, the internal, extended-precision floating-point number is rounded to the size of the floating point number indicated by the control register.

The FSTP (floating-point store and pop) instruction stores a copy of the top of the stack into memory or any coprocessor register, and then pops the data from the top of the stack. You might think of FST as a copy instruction and FSTP as a removal instruction.

The FXCH instruction exchanges the register indicated by the operand with the top of the stack. For example, the FXCH ST(2) instruction exchanges the top of the stack with register 2.

Integer Data Transfer Instructions.   The coprocessor supports three integer data transfer instructions: FILD (load integer), FIST (store integer), and FISTP (store integer and pop). These three instructions function as did FLD, FST, and FSTP, except that the data transferred are integer data. The coprocessor automatically converts the internal extended-precision floating-point data to integer data. The size of the data is determined by the way that the label is defined with DW, DD, or DQ in the assembly language program.

BCD Data Transfer Instructions.   Two instructions load or store BCD signed-integer data. The FBLD instruction loads the top of the stack with BCD memory data and the FBSTP stores the top of the stack and does a pop.

Example 14–7 shows how the assembler automatically adjusts the FLD, FILD, and FBLD instructions for different size operands. (Look closely at the machine-coded forms of the instructions.) In this example, it begins with the .386 and .387 directives that identify the microprocessor as an 80386 and the coprocessor as an 80387. If the 80286 microprocessor is in use with its coprocessor, the directives .286 and .287 appear. The assembler by default assumes that the software is assembled for an 8086/8088 with an 8087 coprocessor. The .486, .487, .586, and .587 switches are also available for use with the 80486 and Pentium microprocessors. Even though the program in Example 14–7 executes, CodeView or some other debugging tool must be used to view any changes to the coprocessor stack.

EXAMPLE 14–7

```
                                   .MODEL SMALL
                                   .386                    ;select 80386 microprocessor
                                   .387                    ;select 80387 coprocessor
0000                               .DATA
0000   41F00000      DATA1   DD    30.0                    ;single-precision
0004                 DATA2   DQ    100.25                  ;double-precision
       4059100000000000
000C                 DATA3   DT    33.9876                 ;extended-precision
       400487F34D6A161E4F76
0016   001E          DATA4   DW    30                      ;16-bit integer
0018   0000001E      DATA5   DD    30                      ;32-bit integer
001C                 DATA6   DQ    30                      ;64-bit integer
       000000000000001E
0024                 DATA7   DT    30H                     ;BCD 30
       00000000000000000030
0000                               .CODE
                                   .STARTUP

0010   D9 06 0000   R       FLD    DATA1
0014   DD 06 0004   R       FLD    DATA2
0018   DB 2E 000C   R       FLD    DATA3
001C   DF 06 0016   R       FILD   DATA4
0020   DB 06 0018   R       FILD   DATA5
0024   DF 2E 001C   R       FILD   DATA6

0028   DF 26 0024   R       FBLD   DATA7
                                   .EXIT
                                   END
```

The Pentium Pro through Pentium 4 FCMOV Instruction.   The Pentium Pro through the Pentium 4 contain a new instruction called FCMOV, which also contains a condition. If the condition is true, the FCMOV instruction copies the source to the destination. The conditions tested by FCMOV and the opcodes used with FCMOV appear in Table 14–4. Notice that these conditions

TABLE 14–4 The variation of the FCMOV instruction and conditions tested.

Instruction	Condition
FCMOVB	Move if below
FCMOVE	Move if equal
FCMOVBE	Move if below or equal
FCMOVU	Move if unordered
FCMOVNB	Move if not below
FCMOVNE	Move if not equal
FCMOVNBE	Move if not below or equal
FCMOVNU	Move if not ordered

check for either an ordered or unordered. The testing for NAN and de-normalized numbers are not checked with FCMOV.

Example 14–8 shows how the FCMOVB (Move if below) instruction is used to copy the contents of ST(2) to the stack top (ST) if the contents of ST(2) is below ST. Notice that the FCOM instruction must be used to perform the compare and the contents of the status register must still be copied to the flags for this instruction to function. More about the FCMOV instruction appears with the FCOMI instruction, which is also new to the Pentium Pro through the Pentium 4 microprocessors.

EXAMPLE 14–8

```
FCOM    ST(2)        ;compare ST and ST(2)
FNSTSW  AX           ;copy floating flags to AX
SAHF                 ;copy floating flags to flags
FCMOVB  ST(2)        ;copy ST(2) to ST if below
```

Arithmetic Instructions

Arithmetic instructions for the coprocessor include addition, subtraction, multiplication, division, and calculating square roots. The arithmetic-related instructions are scaling, rounding, absolute value, and changing the sign.

Table 14–5 shows the basic addressing modes allowed for the arithmetic operations. Each addressing mode is shown with an example using the FADD (real addition) instruction. All arithmetic operations are floating-point, except some cases in which memory data are referenced as an operand.

TABLE 14–5 Arithmetic addressing modes.

Mode	Form	Example
Stack	ST(1),ST	FADD
Register	ST,ST(n)	FADD ST,ST(2)
	ST(n),ST	FADD ST(2),ST
Register pop	ST(n),ST	FADDP ST(3),ST
Memory	Operand	FADD DATA2

Note: Stack addressing is fixed as ST(1),ST and includes a pop, so only the result remains at the top of the stack; and n = register number 0–7. Register addressing for any instruction can use a destination of ST or ST(n), as illustrated.

The classic stack form of addressing operand data (stack addressing) uses the top of the stack as the source operand and the next to the top of the stack as the destination operand. Afterward, a pop removes the source datum from the stack and only the result in the destination register remains at the top of the stack. To use this addressing mode, the instruction is placed in the program without any operands such as FADD or FSUB. The FADD instruction adds ST to ST(1) and stores the answer at the top of the stack; it also removes the original two data from the stack by popping. Note carefully that FSUB subtracts ST from ST(1) and leaves the difference at ST. Therefore, a reverse subtraction (FSUBR) subtracts ST(1) from ST and leaves the difference at ST. (Note that an error exists in Intel documentation, including the Pentium data book, which describes the operation of some reverse instructions.) Another use for reverse operations is for finding a reciprocal (1/X). This is accomplished, if X is at the top of the stack, by loading a 1.0 to ST (FLD1), followed by the FDIVR instruction. The FDIVR instruction divides ST(1) into ST or X into 1, and leaves the reciprocal (1/X) at ST.

The register-addressing mode uses ST for the top of the stack and ST(n) for another location, where n is the register number. With this form, one operand must be ST and the other is ST(n). Note that to double the top of the stack, the FADD ST,ST(0) instruction is used where ST(0) also addresses the top of the stack. One of the two operands in the register-addressing mode must be ST, while the other must be in the form ST(n), where n is a stack register 0–7. For many instructions, either ST or ST(n) can be the destination. It is fairly important that the top of the stack be ST(0). This is accomplished by resetting or initializing the coprocessor before using it in a program. Another example of register-addressing is FADD ST(1),ST where the contents of ST are added to ST(1) and the result is placed into ST(1).

The top of the stack is always used as the destination for the memory-addressing mode because the coprocessor is a stack-oriented machine. For example, the FADD DATA instruction adds the real number contents of memory location DATA to the top of the stack.

Arithmetic Operations. The letter P in an opcode specifies a register pop after the operation (FADDP compared to FADD). The letter R in an opcode (subtraction and division only) indicates reverse mode. The reverse mode is useful for memory data because memory data normally subtract from the top of the stack. A reversed subtract instruction subtracts the top of the stack from memory and stores the result in the top of the stack. For example, if the top of the stack contains a 10 and memory location DATA1 contains a 1, the FSUB DATA1 instruction results in a +9 on the stack top, and the FSUBR instruction results in a –9. Another example is FSUBR ST,ST(1), which will subtract ST from ST(1) and store the result on ST. A variant is FSUBR ST(1),ST, which will subtract ST(1) from ST and store the result on ST(1).

The letter I as a second letter in an opcode indicates that the memory operand is an integer. For example, the FADD DATA instruction is a floating-point addition, while the FIADD DATA is an integer addition that adds the integer at memory location DATA to the floating-point number at the top of the stack. The same rules apply to FADD, FSUB, FMUL, and FDIV instructions.

Arithmetic-related Operations. Other operations that are arithmetic in nature include FSQRT (square root), FSCALE (scale a number), FPREM/FPREM1 (find partial remainder), FRNDINT (round to integer), FXTRACT (extract exponent and significand), FABS (find absolute value), and FCHG (change sign). These instructions and the functions that they perform follow:

FSQRT	Finds the square root of the top of the stack and leaves the resultant square root at the top of the stack. An invalid error occurs for the square root of a negative number. For this reason, the IE bit of the status register should be tested whenever an invalid result can occur. The IE bit can be tested by loading the status register to AX with the FSTSW AX instruction, followed by TEST AX,1 to test the IE status bit.

FSCALE	Adds the contents of ST(1) (interpreted as an integer) to the exponent at the top of the stack. FSCALE multiplies or divides rapidly by powers of two. The value in ST(1) must be between 2^{-15} and 2^{+15}.
FPREM/FPREM1	Performs modulo division of ST by ST(1). The resultant remainder is found in the top of the stack and has the same sign as the original dividend. Note that a modulo division results in a remainder without a quotient. Note also that FPREM is supported for the 8086 and 80287, and FPREM1 should be used in newer coprocessors.
FRNDINT	Rounds the top of the stack to an integer.
FXTRACT	Decomposes the number at the top of the stack into two separate parts that represent the value of the unbiased exponent and the value of the significand. The extracted significand is found at the top of the stack and the unbiased exponent at ST(1). This instruction is often used to convert a floating-point number into a form that can be printed as a mixed number.
FABS	Changes the sign of the top of the stack to positive.
FCHS	Changes the sign from positive to negative or negative to positive.

Comparison Instructions

The comparison instructions all examine data at the top of the stack in relation to another element and return the result of the comparison in the status register condition code bits C3–C0. Comparisons that are allowed by the coprocessor are FCOM (floating-point compare), FCOMP (floating-point compare with a pop), FCOMPP (floating-point compare with two pops), FICOM (integer compare), FICOMP (integer compare and pop), FSTS (test), and FXAM (examine). New with the introduction of the Pentium Pro is the floating compare and move results to flags or FCOMI instruction. Following is a list of these instructions with a description of their functions:

FCOM	Compares the floating-point data at the top of the stack with an operand, which may be any register or any memory operand. If the operand is not coded with the instruction, the next stack element ST(1) is compared with the stack top ST.
FCOMP and FCOMPP	Both instructions perform as FCOM, but they also pop one or two data from the stack.
FICOM and FICOMP	The top of the stack is compared with the integer stored at a memory operand. In addition to the compare, FICOMP also pops the top of the stack.
FTST	Tests the contents of the top of the stack against a zero. The result of the comparison is coded in the status register condition code bits, as illustrated in Table 14–2 with the status register. Also, refer to Table 14–3 for a way of using SAHF and the conditional jump instruction with FTST.
FXAM	Examines the stack top and modifies the condition code bits to indicate whether the contents are positive, negative, normalized, etc. Refer to the status register in Table 14–2.
FCOMI/ FUCOMI	New to the Pentium Pro through the Pentium 4, this instruction compares in exactly the same manner as the FCOM instruction, with one additional feature: it moves the floating-point flags into the flag register, just as the FNSTSW AX and SAHF instructions do in Example 14–8. Intel has

combined the FCOM, FNSTSW AX, and SAHF instructions to form FCOMI. Also available is the unordered compare or FUCOMI. Each is also available with a pop by appending the opcode with a P.

Transcendental Operations

The transcendental instructions include FPTAN (partial tangent), FPATAN (partial arctangent), FSIN (sine), FCOS (cosine), FSINCOS (sine and cosine), F2XM1 ($2^x - 1$), FYL2X (Y $\log_2$ X), and FYL2XP1 (Y $\log_2$ (X + 1)). A list of these operations follows with a description of each transcendental operation:

FPTAN   Finds the partial tangent of Y/X = tan θ. The value of θ is at the top of the stack. It must be between 0 and n/4 radians for the 8087 and 80287; and must be less than 2^{63} for the 80387, 80486/7, and Pentium–Pentium 4 microprocessors. The result is a ratio found as ST = X and ST(1) = Y. If the value is outside of the allowable range, an invalid error occurs, as indicated by the status register IE bit. Also note that ST(7) must be empty for this instruction to function properly.

FPATAN   Finds the partial arctangent as θ = ARCTAN X/Y. The value of X is at the top of the stack and Y is at ST(1). The values of X and Y must be as follows: $0 \le Y < X < \infty$. The instruction pops the stack and leaves θ at the top of the stack.

F2XM1   Finds the function $2^x - 1$. The value of X is taken from the top of the stack and the result is returned to the top of the stack. To obtain 2^x, add one to the result at the top of the stack. The value of X must be in the range of –1 and +1. The F2XM1 instruction is used to derive the functions listed in Table 14–6. Note that the constants $\log_2$ 10 and $\log_2$ e are built-in as standard values for the coprocessor.

FSIN/FCOS   Finds the sine or cosine of the argument located in ST expressed in radians (360° = 2π radians), with the result found in ST. The values of ST must be less than 2^{63}.

FSINCOS   Finds the sine and cosine of ST, expressed in radians, and leaves the results as ST = sine and ST(1) = cosine. As with FSIN or FCOS, the initial value of ST must be less than 2^{63}.

FYL2X   Finds Y $\log_2$ X. The value X is taken from the stack top, and Y is taken from ST(1). The result is found at the top of the stack after a pop. The value of X must range between 0 and ∞, and the value of Y must be between –∞ and +∞. A logarithm with any positive base (b) is found by the equation $LOG_b X = (LOG_2 b)^{-1} \times LOG_2 X$

FYL2XP1   Finds Y $\log_2$ (X + 1). The value of X is taken from the stack top and Y is taken from ST(1). The result is found at the top of the stack after a pop. The value of X must range between 0 and $1 - \sqrt{2}/2$ and the value of Y must be between –∞ and +∞.

TABLE 14–6 Exponential functions.

Function	Equation
10^Y	$2^Y \times \log_2 10$
ε^Y	$2^Y \times \log_2 \varepsilon$
X^Y	$2^Y \times \log_2 X$

TABLE 14–7 Constant operations.

Instruction	Constant pushed to ST
FLDZ	+0.0
FLD1	+1.0
FLDPI	π
FLDL2T	$\log_2 10$
FLDL2E	$\log_2 \varepsilon$
FLDLG2	$\log_{10} 2$
FLDLN2	$\log_\varepsilon 2$

Constant Operations

The coprocessor instruction set includes opcodes that return constants to the top of the stack. A list of these instructions appears in Table 14–7.

Coprocessor Control Instructions

The coprocessor has control instructions for initialization, exception handling, and task switching. The control instructions have two forms. For example, FINIT initializes the coprocessor, as does FNINIT. The difference is that FNINIT does not cause any wait states, while FINIT does cause waits. The microprocessor waits for the FINIT instruction by testing the BUSY pin on the coprocessor. All control instructions have these two forms. Following is a list of each control instruction with its function:

FINIT/FNINIT   Performs a reset operation on the arithmetic coprocessor (see Table 14–8 for the reset conditions). The coprocessor operates with a closure of projective (unsigned infinity), rounds to the nearest or even, and uses extended-precision when reset or initialized. It also sets register 0 as the top of the stack.

FSETPM   Changes the addressing mode of the coprocessor to the protected-addressing mode. This mode is used when the microprocessor is also operated in the protected mode. As with the microprocessor, protected mode can only be exited by a hardware reset or, in the case of the 80386 through the Pentium 4, with a change to the control register.

TABLE 14–8 Coprocessor state after a reset or initialization.

Field	Value	Condition
Infinity	0	Projective
Rounding	00	Round to nearest
Precision	11	Extended-precision
Error masks	11111	Error bits disabled
Busy	0	Not busy
C3–C0	????	Unknown
TOP	000	Register 000 or ST(0)
ES	0	No errors
Error bits	00000	No errors
All tags	11	Empty
Registers	ST(0)–ST(7)	Not changed

FIGURE 14–8 Memory format when the 80X87 registers are saved with the FSAVE instruction.

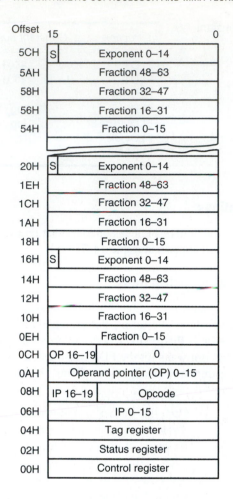

	Instruction	Description
FLDCW		Loads the control register with the word addressed by the operand.
FSTCW/FNSTCW		Stores the control register into the word-sized memory operand.
FSTSW AX/ FNSTSW AX		Copies the contents of the control register to the AX register. This instruction is not available to the 8087 coprocessor.
FCLEX/FNCLEX		Clears the error flags in the status register and also the busy flag.
FSAVE/FNSAVE		Writes the entire state of the machine to memory. Figure 14–8 shows the memory layout for this instruction.
FRSTOR		Restores the state of the machine from memory. This instruction is used to restore the information saved by FSAVE/FNSAVE.
FSTENV/FNSTENV		Stores the environment of the coprocessor, as shown in Figure 14–9.
FLDENV		Reloads the environment saved by FSTENV/FNSTENV.
FINCST		Increments the stack pointer.
FDECSTP		Decrements the stack pointer.
FFREE		Frees a register by changing the destination register's tag to empty. It does not affect the contents of the register.
FNOP		Floating-point coprocessor NOP.

FIGURE 14–9 Memory format for the FSTENV instruction: (a) real mode and (b) protected mode.

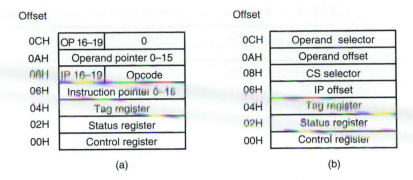

Offset	
0CH	OP 16–19 / 0
0AH	Operand pointer 0–15
08H	IP 16–19 / Opcode
06H	Instruction pointer 0–15
04H	Tag register
02H	Status register
00H	Control register

(a)

Offset	
0CH	Operand selector
0AH	Operand offset
08H	CS selector
06H	IP offset
04H	Tag register
02H	Status register
00H	Control register

(b)

FWAIT   Causes the microprocessor to wait for the coprocessor to finish an operation. FWAIT should be used before the microprocessor accesses memory data that are affected by the coprocessor.

Coprocessor Instructions

Although the microprocessor circuitry has not been discussed, the instruction sets of these co-processors and their differences from the other versions of the coprocessor can be discussed. These newer coprocessors contain the same basic instructions provided by the earlier versions, with a few additional instructions.

The 80387, 80486, 80487SX, and Pentium through the Pentium 4 contain the following additional instructions: FCOS (cosine), FPREM1 (partial remainder), FSIN (sine), FSINCOS (sine and cosine), and FUCOM/FUCOMP/FUCOMPP (unordered compare). The sine and cosine instructions are the most-significant addition to the instruction set. In the earlier versions of the coprocessor, the sine and cosine is calculated from the tangent. The Pentium Pro through the Pentium 4 contain two new floating-point instructions: FCMOV (a conditional move) and FCOMI (a compare and move to flags).

Table 14–9 lists the instruction sets for all versions of the coprocessor. It also lists the number of clocking periods required to execute each instruction. Execution times are listed for the 8087, 80287, 80387, 80486, 80487, and Pentium–Pentium 4. (The timings for the Pentium through the Pentium 4 are the same because the coprocessor is identical in each of these microprocessors.) To determine the execution time of an instruction, the clock time is multiplied times the listed execution time. The FADD instruction requires 70–143 clocks for the 80287. Suppose that an 8 MHz clock is used with the 80287. The clocking period is 1/8 MHz, or 125 ns. The FADD instruction requires between 8.75 μs and 17.875 μs to execute. Using a 33 MHz (33 ns) 80486DX2, this instruction requires between 0.264 μs and 0.66 μs to execute. On the Pentium the FADD instruction requires from 1–7 clocks, so if operated at 133 MHz (7.52 ns), the FADD requires between 0.00752 μs and 0.05264 μs. The Pentium Pro through the Pentium 4 are even faster than the Pentium.

Table 14–9 uses some shorthand notations to represent the displacement that may or may not be required for an instruction that uses a memory-addressing mode. It also uses the abbreviation *mmm,* to represent a register/memory addressing mode, and uses *rrr* to represent one of the floating-point coprocessor registers ST(0)–ST(7). The d (destination) bit that appears in some instruction opcodes defines the direction of the data flow, as in FADD ST,ST(2) or FADD ST(2),ST. The d bit is a logic 0 for flow toward ST, as in FADD ST,ST(2), where ST holds the sum after the addition; and a logic 1 for FADD ST(2),ST, where ST(2) holds the sum.

Also note that some instructions allow a choice of whether a wait is inserted. For example, the FSTSW AX instruction copies the status register into AX. The FNSTSW AX instruction also copies the status register to AX, but without a wait.

TABLE 14–9   The instruction set of the arithmetic coprocessor (pp. 572–587).

F2XM1 $2^{ST} - 1$		
11011001 11110000		
Example		Clocks
F2XM1	8087	310–630
	80287	310–630
	80387	211–476
	80486/7	140–279
	Pentium–Pentium 4	13–57

FABS Absolute value of ST		
11011001 11100001		
Example		Clocks
FABS	8087	10–17
	80287	10–17
	80387	22
	80486/7	3
	Pentium–Pentium 4	1

FADD/FADDP/FIADD Addition			
11011000 oo000mmm disp	32-bit memory (FADD)		
11011100 oo000mmm disp	64-bit memory (FADD)		
11011d00 11000rrr	FADD ST,ST(rrr)		
11011110 11000rrr	FADDP ST,ST(rrr)		
11011110 oo000mmm disp	16-bit memory (FIADD)		
11011010 oo000mmm disp	32-bit memory (FIADD)		
Format	Examples		Clocks
FADD	FADD DATA	8087	70–143
FADDP	FADD ST,ST(1)	80287	70–143
FIADD	FADDP	80387	23–72
	FIADD NUMBER		
	FADD ST,ST(3)	80486/7	8–20
	FADDP ST,ST(2)		
	FADD ST(2),ST	Pentium–Pentium 4	1–7

FCLEX/FNCLEX   Clear errors

11011011 11100010

Example		Clocks
FCLEX	8087	2–8
FNCLEX	80287	2–8
	80387	11
	80486/7	7
	Pentium–Pentium 4	9

FCOM/FCOMP/FCOMPP/FICOM/FICOMP   Compare

11011000 oo010mmm disp	32-bit memory (FCOM)	
11011100 oo010mmm disp	64-bit memory (FCOM)	
11011000 11010rrr	FCOM ST(rrr)	
11011000 oo011mmm disp	32-bit memory (FCOMP)	
11011100 oo011mmm disp	64-bit memory (FCOMP)	
11011000 11011rrr	FCOMP ST(rrr)	
11011110 11011001	FCOMPP	
11011110 oo010mmm disp	16-bit memory (FICOM)	
11011010 oo010mmm disp	32-bit memory (FICOM)	
11011110 oo011mmm disp	16-bit memory (FICOMP)	
11011010 oo011mmm disp	32-bit memory (FICOMP)	

Format	Examples		Clocks
FCOM	FCOM ST(2)	8087	40–93
FCOMP	FCOMP DATA	80287	40–93
FCOMPP	FCOMPP	80387	24–63
FICOM	FICOM NUMBER	80486/7	15–20
FICOMP	FICOMP DATA3	Pentium–Pentium 4	1–8

FCOMI/FUCOMI/COMIP/FUCOMIP   Compare and Load Flags

11011011 11110rrr	FCOMI ST(rrr)	
11011011 11101rrr	FUCOMI ST(rrr)	
11011111 11110rrr	FCOMIP ST(rrr)	
11011111 11101rrr	FUCOMIP ST(rrr)	

Format	Examples		Clocks
FCOM	FCOMI ST(2)	8087	—
FUCOMI	FUCOMI ST(4)	80287	—
FCOMIP	FCOMIP ST(0)	80387	—
FUCOMIP	FUCOMIP ST(1)	80486/7	—
		Pentium–Pentium 4	—

FCMOVcc   Conditional Move

11011010 11000rrr	FCMOVB ST(rrr)
11011010 11001rrr	FCMOVE ST(rrr)
11011010 11010rrr	FCMOVBE ST(rrr)
11011010 11011rrr	FCMOVU ST(rrr)
11011011 11000rrr	FCMOVNB ST(rrr)
11011011 11001rrr	FCMOVNE ST(rrr)
11011011 11010rrr	FCMOVENBE ST(rrr)
11011011 11011rrr	FCMOVNU ST(rrr)

Format	Examples		Clocks
FCMOVB FCMOVB ST(2)		8087	—
FCMOVE FCMOVE ST(3)		80287	—
		80387	—
		80486/7	—
		Pentium–Pentium 4	—

FCOS   Cosine of ST

11011001 11111111

Example		Clocks
FCOS	8087	—
	80287	—
	80387	123–772
	80486/7	193–279
	Pentium–Pentium 4	18–124

FDECSTP   Decrement stack pointer

11011001 11110110

Example		Clocks
FDECSTP	8087	6–12
	80287	6–12
	80387	22
	80486/7	3
	Pentium–Pentium 4	1

FDISI/FNDISI   Disable interrupts

11011011 11100001

(Ignored on the 80287, 80387, 80486/7, Pentium–Pentium 4)

Example		Clocks
FDISI	8087	2–8
FNDISI	80287	—
	80387	—
	80486/7	—
	Pentium–Pentium 4	—

FDIV/FDIVP/FIDIV   Division

11011000 oo110mmm disp	32-bit memory (FDIV)	
11011100 oo100mmm disp	64-bit memory (FDIV)	
11011d00 11111rrr	FDIV ST,ST(rrr)	
11011110 11111rrr	FDIVP ST,ST(rrr)	
11011110 oo110mmm disp	16-bit memory (FIDIV)	
11011010 oo110mmm disp	32-bit memory (FIDIV)	

Format	Examples		Clocks
FDIV	FDIV DATA	8087	191–243
FDIVP	FDIV ST,ST(3)	80287	191–243
FIDIV	FDIVP	80387	88–140
	FIDIV NUMBER	80486/7	8–89
	FDIV ST,ST(5)		
	FDIVP ST,ST(2)	Pentium–Pentium 4	39–42
	FDIV ST(2),ST		

FDIVR/FDIVRP/FIDIVR   Division reversed

11011000 oo111mmm disp	32-bit memory (FDIVR)	
11011100 oo111mmm disp	64-bit memory (FDIVR)	
11011d00 11110rrr	FDIVR ST,ST(rrr)	
11011110 11110rrr	FDIVRP ST,ST(rrr)	
11011110 oo111mmm disp	16-bit memory (FIDIVR)	
11011010 oo111mmm disp	32-bit memory (FIDIVR)	

Format	Examples		Clocks
FDIVR	FDIVR DATA	8087	191–243
FDIVRP	FDIVR ST,ST(3)	80287	191–243
FIDIVR	FDIVRP	80387	88–140
	FIDIVR NUMBER	80486/7	8–89
	FDIVR ST,ST(5)		
	FDIVRP ST,ST(2)	Pentium–Pentium 4	39–42
	FDIVR ST(2),ST		

FENI/FNENI   Disable interrupts

11011011 11100000

(Ignored on the 80287, 80387, 80486/7, Pentium–Pentium 4)

Example		Clocks
FENI FNENI	8087	2–8
	80287	—
	80387	—
	80486/7	—
	Pentium–Pentium 4	—

FFREE   Free register

11011101 11000rrr

Format	Examples		Clocks
FFREE	FFREE FFREE ST(1) FFREE ST(2)	8087	9–16
		80287	9–16
		80387	18
		80486/7	3
		Pentium–Pentium 4	1

FINCSTP   Increment stack pointer

11011001 11110111

Example		Clocks
FINCSTP	8087	6–12
	80287	6–12
	80387	21
	80486/7	3
	Pentium–Pentium 4	1

FINIT/FNINIT   Initialize coprocessor

11011001 11110110

Example		Clocks
FINIT	8087	2–8
FNINIT	80287	2–8
	80387	33
	80486/7	17
	Pentium–Pentium 4	12–16

FLD/FILD/FBLD   Load data to ST(0)

11011001 oo000mmm disp	32-bit memory (FLD)	
11011101 oo000mmm disp	64-bit memory (FLD)	
11011011 oo101mmm disp	80-bit memory (FLD)	
11011111 oo000mmm disp	16-bit memory (FILD)	
11011011 oo000mmm disp	32-bit memory (FILD)	
11011111 oo101mmm disp	64-bit memory (FILD)	
11011111 oo100mmm disp	80-bit memory (FBLD)	

Format	Examples		Clocks
FLD	FLD DATA	8087	17–310
FILD	FILD DATA1	80287	17–310
FBLD	FBLD DEC_DATA	80387	14–275
		80486/7	3–103
		Pentium–Pentium 4	1–3

FLD1   Load +1.0 to ST(0)

11011001 11101000

Example		Clocks
FLD1	8087	15–21
	80287	15–21
	80387	24
	80486/7	4
	Pentium–Pentium 4	2

FLDZ	Load +0.0 to ST(0)		
11011001 11101110			
Example			Clocks
FLDZ		8087	11–17
		80287	11–17
		80387	20
		80486/7	4
		Pentium–Pentium 4	2

FLDPI	Load π to ST(0)		
11011001 11101011			
Example			Clocks
FLDPI		8087	16–22
		80287	16–22
		80387	40
		80486/7	8
		Pentium–Pentium 4	3–5

FLDL2E	Load $\log_2 e$ to ST(0)		
11011001 11101010			
Example			Clocks
FLDL2E		8087	15–21
		80287	15–21
		80387	40
		80486/7	8
		Pentium–Pentium 4	3–5

FLDL2T	Load $\log_2 10$ to ST(0)		
11011001 11101001			
Example			Clocks
FLDL2T		8087	16–22
		80287	16–22
		80387	40
		80486/7	8
		Pentium–Pentium 4	3–5

FLDLG2   Load log$_{10}$2 to ST(0)

11011001 11101000

Example		Clocks
FLDLG2	8087	18–24
	80287	18–24
	80387	41
	80486/7	8
	Pentium–Pentium 4	3–5

FLDLN2   Load log$_e$2 to ST(0)

11011001 11101101

Example		Clocks
FLDLN2	8087	17–23
	80287	17–23
	80387	41
	80486/7	8
	Pentium–Pentium 4	3–5

FLDCW   Load control register

11011001 oo101mmm disp

Format	Examples		Clocks
FLDCW	FLDCW DATA FLDCW STATUS	8087	7–14
		80287	7–14
		80387	19
		80486/7	4
		Pentium–Pentium 4	7

FLDENV   Load environment

11011001 oo100mmm disp

Format	Examples		Clocks
FLDENV	FLDENV ENVIRON FLDENV DATA	8087	35–45
		80287	25–45
		80387	71
		80486/7	34–44
		Pentium–Pentium 4	32–37

FMUL/FMULP/FIMUL   Multiplication

```
11011000  oo001mmm  disp        32-bit memory (FMUL)
11011100  oo001mmm  disp        64-bit memory (FMUL)
11011d00  11001rrr              FMUL ST,ST(rrr)
11011110  11001rrr              FMULP ST,ST(rrr)
11011110  oo001mmm  disp        16-bit memory (FIMUL)
11011010  oo001mmm  disp        32-bit memory (FIMUL)
```

Format	Examples		Clocks
FMUL	FMUL DATA	8087	110–168
FMULP	FMUL ST,ST(2)		
FIMUL	FMUL ST(2),ST	80287	110–168
	FMULP	80387	29–82
	FIMUL DATA3	80486/7	11–27
		Pentium–Pentium 4	1–7

FNOP   No operation

```
11011001  11010000
```

Example		Clocks
FNOP	8087	10–16
	80287	10–16
	80387	12
	80486/7	3
	Pentium–Pentium 4	1

FPATAN   Partial arctangent of ST(0)

```
11011001  11110011
```

Example		Clocks
FPATAN	8087	250–800
	80287	250–800
	80387	314–487
	80486/7	218–303
	Pentium–Pentium 4	17–173

FPREM   Partial remainder

11011001 11111000

Example		Clocks
FPREM	8087	15–190
	80287	15–190
	80387	74–155
	80486/7	70–138
	Pentium–Pentium 4	16–64

FPREM1   Partial remainder (IEEE)

11011001 11110101

Example		Clocks
FPREM1	8087	—
	80287	—
	80387	95–185
	80486/7	72–167
	Pentium–Pentium 4	20–70

FPTAN   Partial tangent of ST(0)

11011001 11110010

Example		Clocks
FPTAN	8087	30–450
	80287	30–450
	80387	191–497
	80486/7	200–273
	Pentium–Pentium 4	17–173

FRNDINT   Round ST(0) to an integer

11011001 11111100

Example		Clocks
FRNDINT	8087	16–50
	80287	16–50
	80387	66–80
	80486/7	21–30
	Pentium–Pentium 4	9–20

FRSTOR   Restore state

11011101 oo110mmm disp

Format	Examples		Clocks
FRSTOR	FRSTOR DATA FRSTOR STATE FRSTOR MACHINE	8087	197–207
		80287	197–207
		80387	308
		80486/7	120–131
		Pentium–Pentium 4	70–95

FSAVE/FNSAVE   Save machine state

11011101 oo110mmm disp

Format	Examples		Clocks
FSAVE FNSAVE	FSAVE STATE FNSAVE STATUS FSAVE MACHINE	8087	197–207
		80287	197–207
		80387	375
		80486/7	143–154
		Pentium–Pentium 4	124–151

FSCALE   Scale ST(0) by ST(1)

11011001 11111101

Example		Clocks
FSCALE	8087	32–38
	80287	32–38
	80387	67–86
	80486/7	30–32
	Pentium–Pentium 4	20–31

FSETPM   Set protected mode

11011011 11100100

Example		Clocks
FSETPM	8087	—
	80287	2–18
	80387	12
	80486/7	—
	Pentium–Pentium 4	—

FSIN	Sine of ST(0)		
11011001 11111110			
Example			Clocks
FSIN		8087	—
		80287	—
		80387	122–771
		80486/7	193–279
		Pentium–Pentium 4	16–126

FSINCOS	Find sine and cosine of ST(0)		
11011001 11111011			
Example			Clocks
FSINCOS		8087	—
		80287	—
		80387	194–809
		80486/7	243–329
		Pentium–Pentium 4	17–137

FSQRT	Square root of ST(0)		
11011001 11111010			
Example			Clocks
FSQRT		8087	180–186
		80287	180–186
		80387	122–129
		80486/7	83–87
		Pentium–Pentium 4	70

FST/FSTP/FIST/FISTP/FBSTP   Store

11011001 oo010mmm disp	32-bit memory (FST)
11011101 oo010mmm disp	64-bit memory (FST)
11011101 11010rrr	FST ST(rrr)
11011011 oo011mmm disp	32-bit memory (FSTP)
11011101 oo011mmm disp	64-bit memory (FSTP)
11011011 oo111mmm disp	80-bit memory (FSTP)
11011101 11001rrr	FSTP ST(rrr)
11011111 oo010mmm disp	16-bit memory (FIST)
11011011 oo010mmm disp	32-bit memory (FIST)
11011111 oo011mmm disp	16-bit memory (FISTP)
11011011 oo011mmm disp	32-bit memory (FISTP)
11011111 oo111mmm disp	64-bit memory (FISTP)
11011111 oo110mmm disp	80-bit memory (FBSTP

Format	Examples		Clocks
FST	FST DATA	8087	15–540
FSTP	FST ST(3)	80287	15–540
FIST	FST	80387	11–534
FISTP	FSTP	80486/7	3–176
FBSTP	FIST DATA2	Pentium–Pentium 4	1–3
	FBSTP DATA6		
	FISTP DATA9		

FSTCW/FNSTCW   Store control register

11011001 oo111mmm disp

Format	Examples		Clocks
FSTCW	FSTCW CONTROL	8087	12–18
FNSTCW	FNSTCW STATUS	80287	12–18
	FSTCW MACHINE	80387	15
		80486/7	3
		Pentium–Pentium 4	2

FSTENV/FNSTENV   Store environment

11011001 oo110mmm disp

Format	Examples		Clocks
FSTENV	FSTENV CONTROL	8087	40–50
FNSTENV	FNSTENV STATUS	80287	40–50
	FSTENV MACHINE	80387	103–104
		80486/7	58–67
		Pentium–Pentium 4	48–50

FSTSW/FNSTSW   Store status register

11011101 oo111mmm disp

Format	Examples		Clocks
FSTSW	FSTSW CONTROL	8087	12–18
FNSTSW	FNSTSW STATUS	80287	12–18
	FSTSW MACHINE	80387	15
	FSTSW AX	80486/7	3
		Pentium–Pentium 4	2–5

FSUB/FSUBP/FISUB   Subtraction

11011000	oo100mmm disp	32-bit memory (FSUB)
11011100	oo100mmm disp	64-bit memory (FSUB)
11011d00	11101rrr	FSUB ST,ST(rrr)
11011110	11101rrr	FSUBP ST,ST(rrr)
11011110	oo100mmm disp	16-bit memory (FISUB)
11011010	oo100mmm disp	32-bit memory (FISUB)

Format	Examples		Clocks
FSUB	FSUB DATA	8087	70–143
FSUBP	FSUB ST,ST(2)	80287	70–143
FISUB	FSUB ST(2),ST	80387	29–82
	FSUBP	80486/7	8–35
	FISUB DATA3	Pentium–Pentium 4	1–7

FSUBR/FSUBRP/FISUBR   Reverse subtraction

11011000	oo101mmm disp	32-bit memory (FSUBR)
11011100	oo101mmm disp	64-bit memory (FSUBR)
11011d00	11100rrr	FSUBR ST,ST(rrr)
11011110	11100rrr	FSUBRP ST,ST(rrr)
11011110	oo101mmm disp	16-bit memory (FISUBR)
11011010	oo101mmm disp	32-bit memory (FISUBR)

Format	Examples		Clocks
FSUBR	FSUBR DATA	8087	70–143
FSUBRP	FSUBR ST,ST(2)	80287	70–143
FISUBR	FSUBR ST(2),ST	80387	29–82
	FSUBRP	80486/7	8–35
	FISUBR DATA3	Pentium–Pentium 4	1–7

FTST   Compare ST(0) with + 0.0

11011001 11100100

Example		Clocks
FTST	8087	38–48
	80287	38–48
	80387	28
	80486/7	4
	Pentium–Pentium 4	1–4

FUCOM/FUCOMP/FUCOMPP   Unordered compare

11011101 11100rrr   FUCOM ST,ST(rrr)
11011101 11101rrr   FUCOMP ST,ST(rrr)
11011101 11101001   FUCOMPP

Format	Examples		Clocks
FUCOM	FUCOM ST,ST(2)	8087	—
FUCOMP	FUCOM	80287	—
FUCOMPP	FUCOMP ST,ST(3)	80387	24–26
	FUCOMP	80486/7	4–5
	FUCOMPP	Pentium–Pentium 4	1–4

FWAIT   Wait

10011011

Example		Clocks
FWAIT	8087	4
	80287	3
	80387	6
	80486/7	1–3
	Pentium–Pentium 4	1–3

FXAM   Examine ST(0)

11011001 11100101

Example		Clocks
FXAM	8087	12–23
	80287	12–23
	80387	30–38
	80486/7	8
	Pentium–Pentium 4	21

FXCH	Exchange ST(0) with another register		

11011001 11001rrr   FXCH ST,ST(rrr)

Format	Examples		Clocks
FXCH	FXCH ST,ST(1)	8087	10–15
	FXCH	80287	10–15
	FXCH ST,ST(4)	80387	18
		80486/7	4
		Pentium–Pentium 4	1

FXTRACT	Extract components of ST(0)		

11011001 11110100

Example		Clocks
FXTRACT	8087	27–55
	80287	27–55
	80387	70–76
	80486/7	16–20
	Pentium–Pentium 4	13

FYL2X	$ST(1) \times \log_2 ST(0)$		

11011001 11110001

Example		Clocks
FYL2X	8087	900–1100
	80287	900–1100
	80387	120–538
	80486/7	196–329
	Pentium–Pentium 4	22–111

FXL2XP1	$ST(1) \times \log_2 [ST(0) + 1.0]$		

11011001 11111001

Example		Clocks
FXL2XP1	8087	700–1000
	80287	700–1000
	80387	257–547
	80486/7	171–326
	Pentium–Pentium 4	22–103

Notes: d = direction, where d = 0 for ST as the destination, and d = 1 for ST as the source; rrr = floating-point register number; oo = mode; mmm = r/m field; and disp = displacement

14–4   PROGRAMMING WITH THE ARITHMETIC COPROCESSOR

This section of the chapter provides programming examples for the arithmetic coprocessor. Each example is chosen to illustrate a programming technique for the coprocessor.

Calculating the Area of a Circle

This first programming example illustrates a simple method of addressing the coprocessor stack. First, recall that the equation for calculating the area of a circle is $A = \pi R^2$. A program that performs this calculation is listed in Example 14–9. Note that this program takes test data from array RAD that contains five sample radii. The five areas are stored in a second array called AREA. No attempt is made in this program to use the data from the AREA array.

EXAMPLE 14–9

```
                        ;A short program that finds the area of five circles whose
                        ;radii are stored in array RAD.
                        ;
                                .MODEL SMALL
                                .386                    ;select 80386
                                .387                    ;select 80387
0000                            .DATA
0000  4015C28F      RAD     DD   2.34,5.66,9.33,234.5,23.4
      40B51EB8
      411547AE
      436A8000
      41BB3333
0014  0005 [        AREA    DD   5 DUP (?)
      00000000
                    ]
0000                            .CODE
                                .STARTUP
0010  BE 0000               MOV  SI,0                   ;source element 0
0013  BF 0000               MOV  DI,0                   ;destination element 0
0016  B9 0005               MOV  CX,5                   ;count of 5
0019             MAIN1:
0019  D9 84 0000 R          FLD  RAD [SI]               ;radius to ST
001D  D8 C8                 FMUL ST,ST(0)               ;square radius
001F  D9 EB                 FLDPI                       ;π to ST
0021  DE C9                 FMUL                        ;multiply ST = ST x ST(1)
0023  D9 9D 0014 R          FSTP AREA [DI]              ;save area
0027  46                    INC  SI
0028  47                    INC  DI
0029  E2 EE                 LOOP MAIN1
                                .EXIT
                                END
```

Although this is a simple program, it does illustrate the operation of the stack. To provide a better understanding of the operation of the stack, Figure 14–10 shows that the contents of the stack after each instruction of Example 14–9 executes. Note only one pass through the loop is illustrated because the program calculates five areas and each pass is identical.

The first instruction loads the contents of memory location RAD [SI], one of the elements of the array, to the top of the stack. Next, the FMUL ST,ST(0) instruction squares the radius on the top of the stack. The FLDPI instruction loads n to the stack top. The FMUL instructions use the classic stack addressing mode to multiply ST by ST(1). After the multiplication, the prior values of ST and ST(1) are removed from the stack and the product replaces them at the top of the stack. Finally, the FSTP [DI] instruction copies the top of the stack, the area, to an array memory location AREA and clears the stack.

FIGURE 14–10 Operation of the stack for Example 14–9. Note that the stack is shown after the execution of the indicated instruction.

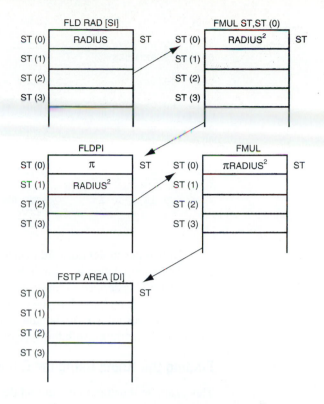

Notice how care is taken to always remove all stack data. This is important because if data remain on the stack at the end of the procedure, the stack top will no longer be register 0. This could cause problems because software assumes that the top of the stack is register 0. Another way of ensuring that the coprocessor is initialized is to place the FINIT (initialization) instruction at the start of the program.

Finding the Resonant Frequency

An equation commonly used in electronics is the formula for determining the resonant frequency of an LC circuit. The equation solved by the program illustrated in Example 14–10 is

$$Fr = \frac{1}{2\pi\sqrt{LC}}$$

This example uses L1 for the inductance L, C1 for the capacitor C, and RESO for the resultant resonant frequency.

EXAMPLE 14–10

```
                        ;A sample program that finds the resonant frequency of an LC
                        ;tank circuit.
                        ;
                                .MODEL SMALL
                                .386
                                .387
0000                            .DATA
0000   00000000         RESO    DD      ?               ;resonant frequency
0004   358637BD         L1      DD      0.000001        ;inductance
0008   358637BD         C1      DD      0.000001        ;capacitance
000C   40000000         TWO     DD      2.0             ;constant
```

```
0000                              .CODE
                                  .STARTUP
0010   D9 06 0004 R               FLD  L1                ;get L
0014   D8 0E 0008 R               FMUL C1                ;find LC

0018   D9 FA                      FSQRT                  ;find √LC

001A   D8 0E 000C R               FMUL TWO               ;find 2√LC

001E   D9 EB                      FLDPI                  ;get π
0020   DE C9                      FMUL                   ;get 2π√LC

0022   D9 E8                      FLD1                   ;get 1
0024   DE F1                      FDIVR                  ;form 1/(2π√LC)

0026   D9 1E 0000 R               FSTP RESO              ;save frequency
                                  .EXIT
                                  END
```

Notice the straightforward manner in which the program solves this equation. Very little extra data manipulation is required because of the stack inside the coprocessor. Notice how the constant TWO is defined for the program and how the DIVRP, using classic stack addressing, is used to form the reciprocal. If you own a reverse-polish entry calculator, such as those produced by Hewlett-Packard, you are familiar with stack addressing. If not, using the coprocessor will increase your experience with this type of entry.

Finding the Roots Using the Quadratic Equation

This example illustrates how to find the roots of a polynomial expression $(ax^2 + bx + c = 0)$ by using the quadratic equation. The quadratic equation is

$$b\pm \frac{\sqrt{b^2 - 4ac}}{2a}$$

Example 14–11 illustrates a program that finds the roots (R1 and R2) for the quadratic equation. The constants are stored in memory locations A1, B1, and C1. Note that no attempt is made to determine the roots if they are imaginary. This example tests for imaginary roots and exits to DOS with a zero in the roots (R1 and R2), if it finds them. In practice, imaginary roots could be solved for and stored in a separate set of result memory locations.

EXAMPLE 14–11

```
                       ;A program that finds the roots of a polynomial equation using
                       ;the quadratic equation. Note imaginary roots are indicated if
                       ;both root 1 (R1) and root 2 (R2) are zero.
                       ;
                               .MODEL SMALL
                               .386
                               .387
0000                           .DATA
0000   40000000        TWO     DD    2.0
0004   40800000        FOUR    DD    4.0
0008   3F800000        A1      DD    1.0
000C   00000000        B1      DD    0.0
0010   C1100000        C1      DD    -9.0
0014   00000000        R1      DD    ?
0018   00000000        R2      DD    ?
0000                           .CODE
                               .STARTUP
0010   D9 EE                   FLDZ
0012   D9 16 0014 R            FST   R1               ;clear roots
0016   D9 1E 0018 R            FSTP  R2
001A   D9 06 0000 R            FLD   TWO
001E   D8 0E 0008 R            FMUL  A1               ;form 2a
```

```
0022  D9 06 0004 R           FLD   FOUR
0026  D8 0E 0008 R           FMUL  A1
002A  D8 0E 0010 R           FMUL  C1              ;form 4ac
002E  D9 06 000C R           FLD   B1
0032  D8 0E 000C R           FMUL  B1              ;form b²
0036  DE E1                  FSUBR                 ;form b²-4ac
0038  D9 E4                  FTST                  ;test b²-4ac for zero
003A  9B DF E0               FSTSW AX              ;copy status register to AX
003D  9E                     SAHF                  ;move to flags
003E  74 0E                  JZ    ROOTS1          ;if b²-4ac is zero
0040  D9 FA                  FSQRT                 ;find square root of b²-4ac
0042  9B DF E0               FSTSW AX
0045  A9 0001                TEST  AX,1            ;test for invalid error (negative)
0048  74 04                  JZ ROOTS1
004A  DE D9                  FCOMPP                ;clear stack
004C  EB 18                  JMP   ROOTS2 ;end
004E              ROOTS1:
004E  D9 06 000C R           FLD   B1
0052  D8 E1                  FSUB  ST,ST(1)
0054  D8 F2                  FDIV  ST,ST(2)
0056  D9 1E 0014 R           FSTP  R1              ;save root 1
005A  D9 06 000C R           FLD   B1
005E  DE C1                  FADD
0060  DE F1                  FDIVR
0062  D9 1E 0018 R           FSTP  R2              ;save root 2
0066              ROOTS2:
                             .EXIT
                             END
```

Using a Memory Array to Store Results

The next programming example illustrates the use of a memory array and the scaled-indexed addressing mode to access the array. Example 14–12 shows a program that calculates 100 values of inductive reactance. The equation for inductive reactance is $XL = 2\pi FL$. In this example, the frequency range is from 10 Hz to 1000 Hz for F and an inductance of 4H. Notice how the instruction FSTP DWORD PTR CS:[EDI+4*ECX] is used to store the reactance for each frequency, beginning with the last at 1000 Hz and ending with the first at 10 Hz. Also notice how the FCOMP instruction is used to clear the stack just before the RET instruction.

EXAMPLE 14–12

```
                       ;A program that calculates the inductive reactance of L
                       ;at a frequency range of 10Hz to 1000Hz and stores them
                       ;in array XL. Note that the increment is 10Hz.
                             .MODEL SMALL
                             .386
                             .387
0000                         .DATA
0000  40800000      L    DD   4.0                 ;4.0H test value
0004  0064 [        XL   DD   100 DUP (?)
         00000000
              ]
0194  447A0000      F    DD   1000.0              ;start at 1000Hz
0198  41200000      TEN  DD   10.0                ;increment of 10Hz
0000                         .CODE
                             .STARTUP
0010  66| B9 00000064        MOV   ECX,100        ;load count
0016  66| BF 00000000 R      MOV   EDI,OFFSET XL-4 ;address result
001C  D9 EB                  FLDPI                ;get π
001E  D8 C0                  FADD  ST,ST(0)       ;form 2π
0020  D8 0E 0000 R           FMUL  L              ;form 2πL
0024              L1:
0024  D9 06 0194 R           FLD   F              ;get F
0028  D8 C9                  FMUL  ST,ST(1)
002A  67& D9 1C 8F           FSTP  DWORD PTR [EDI+4*ECX]
002E  D9 06 0194 R           FLD   F
```

```
0032    D8 26 0198 R      FSUB  TEN                 ;change frequency
0036    D9 1E 0194 R      FSTP  F
003A    E2 E8             LOOP  L1
003C    D8 D9             FCOMP
                          .EXIT
                          END
```

Displaying a Single-precision Floating-point Number

This section of the text shows how to take the floating-point contents of a 32-bit single-precision floating-point number and display it on the video display. The procedure displays the floating-point number as a mixed number with an integer part and a fractional part, separated by a decimal point. In order to simplify the procedure, a limit is placed on the display size of the mixed number so the integer portion is a 32-bit binary number and the fraction is a 24-bit binary number. The procedure will not function properly for larger or smaller numbers.

Example 14–13 lists a program that calls a procedure for displaying the contents of memory location NUMB on the video display at the current cursor position. The procedure first tests the sign of the number and displays a minus sign for a negative number. After displaying the minus sign, if needed, the number is made positive by the FABS instruction. Next, it is divided into an integer and fractional part and stored at WHOLE and FRACT. Notice how the FRNDINT instruction is used to round (using the chop mode) the top of the stack to form the whole number part of NUMB. The whole number part is then subtracted from the original number to generate the fractional part. This is accomplished with the FSUB instruction that subtracts the contents of ST(1) from ST.

EXAMPLE 14–13

```
                   ;A program that displays the floating-point contents of NUMB
                   ;as a mixed decimal number.
                         .MODEL SMALL
                         .386
                         .387
0000                     .DATA
0000  C50B0200    NUMB  DD   -2224.125            ;test data
0004  0000        TEMP  DW   ?
0006  00000000    WHOLE DD   ?
000A  00000000    FRACT DD   ?
0000                     .CODE
                          .STARTUP
0010  E8 000B            CALL DISP                ;display NUMB
                         .EXIT
              ;
              ;A procedure that displays the ASCII code from AL.
              ;
0017          DISPS PROC NEAR

0017  B4 06            MOV   AH,6                ;display AL
0019  8A D0            MOV   DL,AL
001B  CD 21            INT   21H
001D  C3               RET

001E          DISPS ENDP
              ;
              ;A procedure that displays the floating-point contents of NUMB
              ;in decimal form.
              ;
001E          DISP  PROC NEAR

001E  9B D9 3E 0004 R   FSTCW TEMP               ;save current control word
0023  81 0E 0004 R 0C00 OR    TEMP,0C00H         ;set rounding to chop
0029  D9 2E 0004 R      FLDCW TEMP
002D  D9 06 0000 R      FLD   NUMB ;get NUMB
```

```
0031   D9 E4              FTST                      ;test NUMB
0033   9B DF E0           FSTSW AX                  ;status to AX
0036   25 4500            AND   AX,4500H            ;get C3, C2, and C0
                          .IF AX == 0100H
003E   B0 2D                MOV AL,'-'
0040   E8 FFD4              CALL DISPS
0043   D9 E1                FABS
                          .ENDIF
0045   D9 C0              FLD ST
0047   D9 FC              FRNDINT                   ;get integer part
0049   DB 16 0006 R       FIST WHOLE
004D   DE E1              FSUBR
004F   D9 E1              FABS
0051   D9 1E 000A R       FSTP FRACT                ;save fraction
0055   66| A1 0006 R      MOV   EAX,WHOLE
0059   66| BB 0000000A    MOV   EBX,10
005F   B9 0000            MOV   CX,0
0062   53                 PUSH BX
                          .WHILE 1                  ;divide until quotient = 0
0063   66| BA 00000000      MOV   EDX,0
0069   66| F7 F3            DIV   EBX
006C   80 C2 30            ADD   DL,30H
006F   52                  PUSH  DX
                          .BREAK    .IF EAX == 0
0075   41                  INC   CX
                          .IF   CX == 3
007B   6A 2C                  PUSH   ','
007D   B9 0000                MOV    CX,0
                            .ENDIF
                          .ENDW
                          .WHILE  1                 ;display whole number part
0082   5A                  POP   DX
                          .BREAK   .IF DX == BX
0087   8A C2                MOV   AL,DL
0089   E8 FF8B              CALL  DISPS
                          .ENDW
008E   B0 2E              MOV AL,'.'                ;display decimal point
0090   E8 FF84            CALL DISPS
0093   66| A1 000A R      MOV EAX,FRACT
0097   9B D9 3E 0004 R    FSTCW TEMP                ;save current control word
009C   81 36 0004 R 0C00  XOR TEMP,0C00H            ;set rounding to nearest
00A2   D9 2E 0004 R       FLDCW TEMP
00A6   D9 06 000A R       FLD FRACT
00AA   D9 F4              FXTRACT
00AC   D9 1E 000A R       FSTP FRACT
00B0   D9 E1              FABS
00B2   DB 1E 0006 R       FISTP WHOLE
00B6   66| 8B 0E 0006 R   MOV   ECX,WHOLE
00BB   66| A1 000A R      MOV   EAX,FRACT
00BF   66| C1 E0 09       SHL   EAX,9
00C3   66| D3 D8          RCR   EAX,CL
                          .REPEAT
00C6   66| F7 E3            MUL   EBX
00C9   66| 50              PUSH EAX
00CB   66| 92              XCHG EAX,EDX
00CD   04 30               ADD  AL,30H
00CF   E8 FF45             CALL DISPS
00D2   66| 58              POP EAX
                          .UNTIL EAX == 0
00D9   C3                 RET

00DA               DISP ENDP
                   END
```

The last part of the procedure displays the whole number part, followed by the fractional part. The techniques are the same as introduced earlier in the text—dividing a number by ten and displaying the remainders in reverse order converts and displays an integer. A multiplication by

10 converts a fraction to decimal for displaying. Note that the fractional part may contain a rounding error for certain values. This occurs because the number has not been adjusted to remove the rounding error that is inherent in floating-point fractional numbers.

Reading a Mixed Number from the Keyboard

If floating-point arithmetic is used in a program, a method of reading the number from the keyboard and converting it to a floating-point number must be developed. The procedure listed in Example 14–14 reads a signed mixed number from the keyboard and converts it to a floating-point number located at memory location NUMB.

EXAMPLE 14–14

```
                          ;A program that reads a mixed number from the keyboard.
                          ;The result is stored at memory location NUMB as a
                          ;double-precision floating-point number.
                          ;
                                  .MODEL SMALL
                                  .386
                                  .387
0000                              .DATA
0000   00          SIGN    DB    ?              ;sign indicator
0001   0000        TEMP1   DW    ?              ;temporary storage
0003   41200000    TEN     DD    10.0 ;10.0
0007   00000000    NUMB    DD    ?              ;result
0000                              .CODE
                    GET     MACRO                ;;read key macro
                            MOV AH,1
                            INT 21H
                            ENDM
                                  .STARTUP
0010   D9 EE               FLDZ                  ;clear ST
                           GET                   ;read a character
                           .IF  AL == '+'        ;test for +
001A   C6 06 0000 R 00         MOV   SIGN,0      ;clear sign indicator
                           GET
                           .ENDIF
                           .IF  AL == '-'        ;test for -
0027   C6 06 0000 R 01         MOV   SIGN,1      ;set sign indicator
                               GET
                           .ENDIF
                           .REPEAT
0030   D8 0E 0003 R            FMUL   TEN        ;multiply result by 10
0034   B4 00                   MOV    AH,0
0036   2C 30                   SUB    AL,30H     ;convert from ASCII
0038   A3 0001 R               MOV    TEMP1,AX
003B   DE 06 0001 R            FIADD TEMP1
                               GET                ;get next character
                           .UNTIL  AL < '0' || AL > '9'
                           .IF AL == '.'          ;do if -
004F   D9 E8                   FLD1               ;get one
                               .WHILE 1
0051   D8 36 0003 R                FDIV   TEN
                                   GET
                               .BREAK .IF AL < '0' || AL > '9'
0061   B4 00                       MOV    AH,0
0063   2C 30                       SUB    AL,30H   ;convert from ASCII
0065   A3 0001 R                   MOV    TEMP1,AX
0068   DF 06 0001 R                FILD   TEMP1
006C   D8 C9                       FMUL ST,ST(1)
006E   DC C2                       FADD ST(2),ST
0070   D8 D9                       FCOMP
                               .ENDW
0074   D8 D9                   FCOMP              ;clear stack
                           .ENDIF
                           .IF SIGN == 1
007D   D9 E0                   FCHS               ;make negative
```

```
                                      .ENDIF
007F  D9 1E 0007 R                   FSTP NUMB                    ;save result
                                      .EXIT
                                      END
```

Unlike other examples in this chapter, Example 14–14 uses some of the high-level language constructs presented in earlier chapters to reduce its size. Here, the sign is first read from the keyboard, if present, and saved for later use as a 0 for positive and a 1 for negative, in adjusting the sign of the resultant floating-point number. Next, the integer portion of the number is read. The .REPEAT–.UNTIL loop is used to read the number until something other than a number (0–9) is typed. This portion terminates with a period, space, or carriage return. If a period is typed, then the procedure continues and reads a fractional part by using an .IF–.ENDIF construct. If a space or carriage return is entered, the number is converted to floating-point form and stored at NUMB. The .WHILE–.ENDW loop converts the fractional part of the number. The whole number portion is converted with a multiply by 10, and the fractional portion is converted with a divide by 10.

14–5   INTRODUCTION TO MMX TECHNOLOGY

MMX[1] **(multimedia extensions)** technology adds 57 new instructions to the instruction set of the Pentium–Pentium 4 microprocessors. The MMX technology also introduces new general-purpose instructions. The new MMX instructions are designed for applications such as motion video, combined graphics with video, image processing, audio synthesis, speech synthesis and compression, telephony, video conferencing, 2D graphics, and 3D graphics. These new instructions operate in parallel with other operations as the instruction for the arithmetic coprocessor.

Data Types

The MMX architecture introduces new packed data types. The data types are eight packed, consecutive 8-bit bytes; four packed, consecutive 16-bit words; and two packed, consecutive 32-bit doublewords. Bytes in this multibyte format have consecutive memory addresses and use the little endian form, as with other Intel data. See Figure 14–11 for the format for these new data types.

The MMX technology registers have the same format as a 64-bit quantity in memory and have two data access modes: 64-bit access mode and 32-bit access mode. The 64-bit access mode is used for 64-bit memory and registers transfers for most instructions. The 32-bit access mode is used for 32-bit memory and registers transfers for most instructions. The 32-bit transfers occur between microprocessor registers, and the 64-bit transfers occur between floating-point coprocessor registers.

Figure 14–12 illustrates the internal register set of the MMX technology extension and how it uses the floating-point coprocessor register set. This technique is called **aliasing** because the floating-point registers are shared as the MMX registers. That is, the MMX registers (MM0–MM7) are the same as the floating-point registers. Note that the MMX register set is 64 bits wide and uses the rightmost 64-bits of the floating-point register set.

Instruction Set

The instruction for MMX technology includes arithmetic, comparison, conversion, logical, shift, and data transfer instructions. Although the instruction types are similar to the microprocessor's instruction set, the main difference is that the MMX instructions use the data types shown in Figure 14–11 instead of the normal data types used with the microprocessor.

---

[1]MMX™ is a registered trademark of the Intel Corporation.

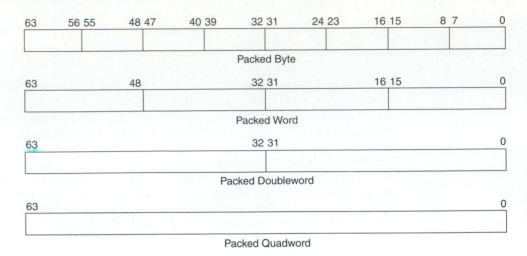

FIGURE 14–11   The structure of data stored in the MMX registers.

FIGURE 14–12   The structure of the MMX register set.
Note that MM0 and FF0
through MM7 and FP7 interchange with each other.

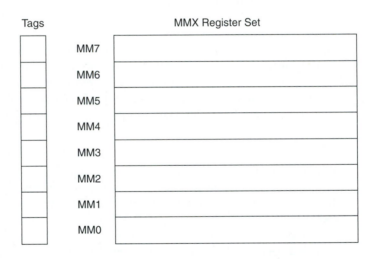

Arithmetic Instructions.   The set of arithmetic instructions includes addition, subtraction, multiplication, and a special multiplication with an addition. Three additions exist. The PADD and PSUB instructions add or subtract packed signed or unsigned packed bytes, packed words, or packed doubleword data. The add instructions are appended with a B, W, or D to select the size, as in PADDB for a byte, PADDW for a word, and PADDD for a doubleword. The same is true for the PSUB instruction. The PMULHW and the PMULLW instructions perform multiplication on four pairs of 16-bit operands, producing 32-bit results. The PMULHW instruction multiplies the high-order 16-bits, and the PMULLW instruction multiplies the low-order 16-bits. The PMADDWD instruction multiplies and adds. After multiplying, the four 32-bit results are added to produce two 32-bit doubleword results.

The MMX instructions use operands just as the integer or floating-point instructions do. The difference is the register names (MM0–MM7). For example, the PADDB MM1, MM2 instruction adds the entire 64-bit contents of MM2 to MM1, byte-by-byte. The result is steered into MM1. When each 8-bit section is added, any carries generated are dropped. For example, the byte A0H added to the byte 70H produces the byte sum of 10H. The true sum is 110H, but the carry is dropped. Note that the second operand or source can be a memory location containing the 64-bit packed source or an MMX register. You might say that this instruction performs the same function as eight separate byte-sized ADD instructions! If used in an application, this certainly speeds execution of the application. Like PADD, PSUB also does not carry or borrow. The difference is that if an overflow or underflow occurs, the difference becomes a 7FH (+127) for an overflow and an 80H (−128) for an underflow. Intel calls this **saturation,** because these values represent the largest and smallest signed bytes.

Comparison Instructions. There are two comparison instructions: PCMPEQ **(equal)** and PCMPGT **(greater than).** As with PADD and PSUB, there are three versions of each compare instruction: for example, PCMPEQUB (compares bytes), PCMPEQUW (compares words), and PCMPEQUD (compares doublewords). These instructions do not change the microprocessor flag bits; instead, the result is all ones for a true condition and all zeros for a false condition. For example, if the PCMPEQB MM2, MM3 instruction is executed and the least-significant bytes of MM2 and MM3 = 10H and 11H, respectively, the result found in the least-significant byte of MM2 is 00H. This indicates that the least-significant bytes were not equal. If the least-significant byte contained an FFH, it indicates that the two bytes were equal.

Conversion Instructions. There are two basic conversion instructions: PACK and PUNPCK. PACK is available as PACKSS (signed saturation) and PACKUS (unsigned saturation). PUNPCK is available as PUNPCKH (unpack high data) and PUNPCKL (unpack low data). Similar to the prior instructions, these can be appended with B, W, or D for byte, word, and doubleword pack and unpack; but they must be used in combinations WB (word to byte) or DW (doubleword to word). For example, the PACKUSWB MM3, MM6 instruction packs the words from MM6 into bytes in MM3. If the unsigned word does not fit (too large) into a byte, the destination byte becomes an FFH. For signed saturation, we use the same values explained under addition.

Logic Instructions. The logic instructions are PAND (and), PANDN (nand), POR (or), and PXOR (exclusive-or). These instructions do not have size extensions, and perform these bit-wise operations on all 64 bits of the data. For example, the POR MM2, MM3 instruction ORs all 64 bits of MM3 with MM2. The logical sum in placed into MM2 after the OR operation.

Shift Instruction. This instruction contains logical shifts and an arithmetic shift right instruction. The logic shifts are PSLL (left) and PSRL (right). Variations are word (W), doubleword (D), and quadword (Q). For example, the PSLLQ MM3,2 instruction shifts all 64 bits in MM3 left two places. Another example is the PSLLD MM3,2 instruction that shifts the two 32-bit doublewords in MM3 left two places each.

The PSRA (arithmetic right shift) instruction functions in the same manner as the logical shifts, except that the sign-bit is preserved.

Data Transfer Instructions. There are two data transfer instructions: MOVED and MOVEQ. These instructions allow transfers between registers and between a register and memory. The MOVED instruction transfers 32 bits of data between an integer register or memory location and an MMX register. For example, the MOVED ECX, MM2 instruction copies the rightmost 32 bits of MM2 into ECX. There is no instruction to transfer the leftmost 32 bits of an MMX register. You could use a shift right before a MOVED to do the transfer.

The MOVEQ instruction copies all 64 bits of an MMX register between memory or another MMX register. The MOVEQ MM2, MM3 instruction transfers all 64 bits of MM3 into MM2.

EMMS Instruction. The EMMS (empty MMX-state) instruction sets (11) all the tags in the floating-point unit, so the floating-point registers are listed as empty. The EMMS instruction must be executed before the return instruction at the end of any MMX procedure, or a subsequent floating-point operation will cause a floating-point interrupt error, crashing Windows or any other application. If you plan to use floating-point instructions within an MMX procedure, you must use the EMMS instruction before executing the floating-point instruction. All other MMX instructions clear the tags, which indicate that all floating-point registers are in use.

Instruction Listing.   Table 14–10 lists all the MMX instructions with the machine code so these instructions can be used with the assembler. At present, MASM does not support these new instructions; to use them, you must have the machine code listing. For example, the EMMS instruction has a machine language code of 0F 77. You can use this instruction in your software with the DB 0FH, 77H coded sequence in the software.

From Table 14–10 (and because the assembler does not yet understand these new instructions), we need a tool for assembling MMX extension instructions. Tools are available, but you can also make a macro library for the new MMX instructions. Problems can occur with your own macro library: it is difficult to access the address if a memory address is specified in the macro, and the macro library would be long. Example 14–15 lists a few macros, but only the MMX registers are handled by the macros. Only the PADDB, PADDW, PAND, and EMMS instructions are implemented, but the remaining instructions can be added.

EXAMPLE 14–15

```
                    ;Illustrates a few macros and a program for
                    ;generating the MMX instructions
                    .MODEL SMALL
                    .386
0000                .DATA

0000                .CODE

                    ;macros for the MMX extensions
                    ;
                    OPS    MACRO    OP1,OP2       ;;gets MMX registers and
                    LOCAL    TEMP1,TEMP2,BYTE3
                      IFIDNI    <OP1>,<MM0>        ;;generates byte3
                          TEMP1=0
                      ELSEIFIDNI <OP1>,<MM1>
                          TEMP1=1
                      ELSEIFIDNI <OP1>,<MM2>
                          TEMP1=2
                      ELSEIFIDNI <OP1>,<MM3>
                          TEMP1=3
                      ELSEIFIDNI <OP1>,<MM4>
                          TEMP1=4
                      ELSEIFIDNI <OP1>,<MM5>
                          TEMP1=5
                      ELSEIFIDNI <OP1>,<MM6>
                          TEMP1=6
                      ELSEIFIDNI <OP1>,<MM7>
                          TEMP1=7
                      ENDIF
                      IFIDNI <OP2>,<MM0>
                          TEMP2=0
                      ELSEIFIDNI <OP2>,<MM1>
                          TEMP2=8
                      ELSEIFIDNI <OP2>,<MM2>
                          TEMP2=10H
                      ELSEIFIDNI <OP2>,<MM3>
                          TEMP2=18H
                      ELSEIFIDNI <OP2>,<MM4>
                          TEMP2=20H
                      ELSEIFIDNI <OP2>,<MM5>
                          TEMP2=28H
                      ELSEIFIDNI <OP2>,<MM6>
                          TEMP2=30H
                      ELSEIFIDNI <OP2>,<MM7>
                          TEMP2=38H
                      ENDIF
                      BYTE3=TEMP1+TEMP2+BIAS
                          DB        BYTE3
                      ENDM

                    EMMS    MACRO              ;;EMMS instruction
```

```
                                      DB        0FH,77H
                              ENDM

              PAND  MACRO OP1,OP2        ;;PAND instruction
                              DB        0FH,0DBH
                              BIAS=0C0H
                              OPS    OP1,OP2
                              ENDM

              PADDB  MACRO    OP1,OP2        ;;PADDB instruction
                              DB        0FH,0FCH
                              BIAS=0C0H
                              OPS    OP1,OP2
                              ENDM

              PADDW MACRO     OP1,OP2        ;;PADDW instruction
                              DB        0FH,0FDH
                              BIAS=0C0H
                              OPS    OP1,OP2
                              ENDM

                              .STARTUP

                              EMMS          ;test program
0010 0F 77   1                DB          0FH,77H
                              PADDB MM1,MM3
0012 0F FC   1                DB          0FH,0FCH
0014 D9      2                DB          ??0002
                              PAND   MM2,MM7
0015 0F DB   1                DB          0FH,0DBH
0017 FA      2                DB          ??0005
                              PADDW MM2,MM6
0018 0F FD   1                DB          0FH,0FDH
001A F2      2                DB          ??0008
                              .EXIT
                              END
```

Now that the instruction set is known and operation of the MMX unit is understood, an example is shown that uses some MMX instructions. The MMX instruction set can be used for any task, but many of the operations are designed for multimedia applications. The task presented here is not a multimedia application, but it does illustrate the operation of a few MMX instructions.

Because the MMX extension executes some basic instructions faster than the integer unit in the microprocessor, some of them could be used to accomplish normal tasks at very high rates. Suppose that we have two arrays of data that need to be summed. The MMX extensions are ideal for this type of application. Example 14–16 shows a short procedure that sums two byte-sized arrays of data that each contain 256 numbers.

EXAMPLE 14–16

```
;A procedure that uses MMX instructions to sum the contents of
;two byte-sized arrays
;EBX addresses one array and EDX addresses the second
;sums are placed in the array addressed by EDX
;
SUMS   PROC   NEAR
        MOV   CX,32
         .REPEAT
             MOVEQ   MM0,[EBX+8*ECX-8]
             PADDB   MMO,[EDX+8*ECX-8]
             MOVEQ   [EDX+8*ECX-8]
         .UNTILCXZ
         EMMS
         RET
SUMS   ENDP
```

TABLE 14–10   The MMX instruction set extension (pp. 600–607).

EMMS Empty MMX state	
0000 1111 0111 1111	
Example	
EMMS	

MOVED Move doubleword	
0000 1111 0110 1110 11 xxx rrr	reg → xreg
Examples	
MOVED MM3, EDX MOVED MM4, EAX	
0000 1111 0111 1110 11 xxx rrr	xreg → reg
Examples	
MOVED EAX, MM3 MOVED EBP, MM7	
0000 1111 0110 1110 oo xxx mmm	mem → xreg
Examples	
MOVED MM3, DATA1 MOVED MM5, BIG_ONE	
0000 1111 0111 1110 oo xxx mmm	xreg → mem
Examples	
MOVED DATA2, MM3 MOVED SMALL_POTS, MM7	

MOVEQ Move quadword	
0000 1111 0110 1111 11 xxx1 xxx2	xreg2 → xreg1
Examples	
MOVEQ MM3, MM2 ;copies MM2 to MM3 MOVEQ MM7, MM3	
0000 1111 0111 1111 11 xxx1 xxx2	xreg1 → reg2
Examples	
MOVEQ MM3, MM2 ;copies MM3 to MM2 MOVEQ MM7, MM3	
0000 1111 0110 1111 oo xxx mmm	mem → xreg
Examples	
MOVEQ MM3, DATA1 MOVEQ MM5, DATA3	
0000 1111 0111 1111 oo xxx mmm	xreg → mem
Examples	
MOVEQ DATA2, MM0 MOVEQ SMALL_POTS, MM3	

PACKSSDW	Pack signed doubleword to word	
0000 1111 0110 1011 11 xxx1 xxx2		xreg2 → xreg1
Examples		
PACKSSDW MM1, MM2 PACKSSDW MM7, MM3		
0000 1111 0111 1011 oo xxx mmm		mem → xreg
Examples		
PACKSSDW MM3, BUTTON PACKSSDW MM7, SOUND		

PACKSSWB	Pack signed word to byte	
0000 1111 0110 0011 11 xxx1 xxx2		xreg2 → xreg1
Examples		
PACKSSWB MM1, MM2 PACKSSWB MM7, MM3		
0000 1111 0111 0011 oo xxx mmm		mem → xreg
Examples		
PACKSSWB MM3, BUTTON PACKSSWB MM7, SOUND		

PACKUSWB	Pack unsigned word to byte	
0000 1111 0110 0111 11 xxx1 xxx2		xreg2 → xreg1
Examples		
PACKUSWB MM1, MM2 PACKUSWB MM7, MM3		
0000 1111 0111 0111 oo xxx mmm		mem → xreg
Examples		
PACKUSWB MM3, BUTTON PACKUSWB MM7, SOUND		

PADD	Add with truncation	Byte, word, and doubleword
0000 1111 1111 11gg 11 xxx1 xxx2		xreg2 → xreg1
Examples		
PADDB MM1, MM2 PADDW MM7, MM3 PADDD MM3, MM4		
0000 1111 1111 11gg oo xxx mmm		mem → xreg
Examples		
PADDB MM3, BUTTON PADDW MM7, SOUND PADDD MM3, BUTTER		

PADDS Add with signed saturation	Byte and word
0000 1111 1110 11gg 11 xxx1 xxx2 Examples	xreg2 → xreg1
PADDSB MM1, MM2 PADDSW MM7, MM3	
0000 1111 1110 11gg oo xxx mmm Examples	mem → xreg
PADDSB MM3, BUTTON PADDSW MM7, SOUND	

PADDUS Add with unsigned saturation	Byte and word
0000 1111 1101 11gg 11 xxx1 xxx2 Examples	xreg2 → xreg1
PADDUSB MM1, MM2 PADDUSW MM7, MM3	
0000 1111 1101 11gg oo xxx mmm Examples	mem → xreg
PADDUSB MM3, BUTTON PADDUSW MM7, SOUND	

PAND And	
0000 1111 1101 1011 11 xxx1 xxx2 Examples	xreg2 → xreg1
PAND MM1, MM2 PAND MM7, MM3	
0000 1111 1101 1011 oo xxx mmm Examples	mem → xreg
PAND MM3, BUTTON PAND MM7, SOUND	

PANDN Nand	
0000 1111 1101 1111 11 xxx1 xxx2 Examples	xreg2 → xreg1
PANDN MM1, MM2 PANDN MM7, MM3	
0000 1111 1101 1111 oo xxx mmm Examples	mem → xreg
PANDN MM3, BUTTON PANDN MM7, SOUND	

PCMPEQU	Compare for equality	Byte, word, and doubleword
0000 1111 0111 01gg 11 xxx1 xxx2		xreg2 → xreg1
Examples		
PCMPEQUB MM1, MM2 PCMPEQUW MM7, MM3 PCMPEQUD MM0, MM5		
0000 1111 0111 01gg oo xxx mmm		mem → xreg
Examples		
PCMPEQUB MM3, BUTTON PCMPEQUW MM7, SOUND PCMPEQUD MM0, FROG		

PCMPGT	Compare for greater than	Byte, word, and doubleword
0000 1111 0110 01gg 11 xxx1 xxx2		xreg2 → xreg1
Examples		
PCMPGTB MM1, MM2 PCMPGTW MM7, MM3 PCMPGTD MM0, MM5		
0000 1111 0110 01gg oo xxx mmm		mem → xreg
Examples		
PCMPGTB MM3, BUTTON PCMPGTW MM7, SOUND PCMPGTD MM0, FROG		

PMADD	Multiply and add	
0000 1111 1111 0101 11 xxx1 xxx2		xreg2 → xreg1
Examples		
PMADD MM1, MM2 PMADD MM7, MM3		
0000 1111 1111 0101 oo xxx mmm		mem → xreg
Examples		
PMADD MM3, BUTTON PMADD MM7, SOUND		

PMULH	Multiplication–high	
0000 1111 1110 0101 11 xxx1 xxx2		xreg2 → xreg1
Examples		
PMULH MM1, MM2 PMULH MM7, MM3		
0000 1111 1110 0101 oo xxx mmm		mem → xreg
Examples		
PMULH MM3, BUTTON PMULH MM7, SOUND		

PMULL Multiplication–low	
0000 1111 1101 0101 11 xxx1 xxx2	xreg2 → xreg1
Examples	
PMULL MM1, MM2 PMULL MM7, MM3	
0000 1111 1101 0101 oo xxx mmm	mem → xreg
Examples	
PMULL MM3, BUTTON PMULL MM7, SOUND	

POR Or	
0000 1111 1110 1011 11 xxx1 xxx2	xreg2 → xreg1
Examples	
POR MM1, MM2 POR MM7, MM3	
0000 1111 1110 1011 oo xxx mmm	mem → xreg
Examples	
POR MM3, BUTTON POR MM7, SOUND	

PSLL Shift left	Word, doubleword, and quadword
0000 1111 1111 00gg 11 xxx1 xxx2	xreg2 → xreg1
Examples	
PSLLW MM1, MM2 PSLLD MM7, MM3 PSLLQ MM6, MM5	
0000 1111 1111 00gg oo xxx mmm	mem → xreg shift count in memory
Examples	
PSLLW MM3, BUTTON PSLLD MM7, SOUND PSLLQ MM2, COUNT1	
0000 1111 0111 00gg 11 110 mmm data8	xreg by count shift count is data8
Examples	
PSLLW MM3, 2 PSLLD MM0, 6 PSLLQ MM7, 1	

PSRA	Shift arithmetic right	Word, doubleword, and quadword
0000 1111 1110 00gg 11 xxx1 xxx2		xreg2 → xreg1
Examples		
PSRAW MM1, MM2 PSRAD MM7, MM3 PSRAQ MM6, MM5		
0000 1111 1110 00gg oo xxx mmm		mem → xreg shift count in memory
Examples		
PSRAW MM3, BUTTON MM7, SOUND PSRAQ MM2, COUNT1		
0000 1111 0111 00gg 11 100 mmm data8		xreg by count shift count is data8
Examples		
PSRAW MM3, 2 PSRAD MM0, 6 PSRAQ MM7, 1		

PSRL	Shift right	Word, doubleword, and quadword
0000 1111 1101 00gg 11 xxx1 xxx2		xreg2 → xreg1
Examples		
PSRLW MM1, MM2 PSRLD MM7, MM3 PSRLQ MM6, MM5		
0000 1111 1101 00gg oo xxx mmm		mem → xreg shift count in memory
Examples		
PSRLW MM3, BUTTON PSRLD MM7, SOUND PSRLQ MM2, COUNT1		
0000 1111 0111 00gg 11 010 mmm data8		xreg by count shift count is data8
Examples		
PSRLW MM3, 2 PSRLD MM0, 6 PSRLQ MM7, 1		

PSUB	Subtract with truncation	Byte, word, and doubleword

0000 1111 1111 10gg 11 xxx1 xxx2	xreg2 → xreg1
Examples	

PSUBB MM1, MM2
PSUBW MM7, MM3
PSUBD MM3, MM4

0000 1111 1111 10gg oo xxx mmm	mem → xreg
Examples	

PSUBB MM3, BUTTON
PSUBW MM7, SOUND
PSUBD MM3, BUTTER

PSUBS	Subtract with signed saturation	Byte, word, and doubleword

0000 1111 1110 10gg 11 xxx1 xxx2	xreg2 → xreg1
Examples	

PSUBSB MM1, MM2
PSUBSW MM7, MM3
PSUBSD MM3, MM4

0000 1111 1110 10gg oo xxx mmm	mem → xreg
Examples	

PSUBSB MM3, BUTTON
PSUBSW MM7, SOUND
PSUBSD MM3, BUTTER

PSUBUS	Subtract with unsigned saturation	Byte, word, and doubleword

0000 1111 1101 10gg 11 xxx1 xxx2	xreg2 → xreg1
Examples	

PSUBUSB MM1, MM2
PSUBUSW MM7, MM3
PSUBUSD MM3, MM4

0000 1111 1101 10gg oo xxx mmm	mem → xreg
Examples	

PSUBUSB MM3, BUTTON
PSUBUSW MM7, SOUND
PSUBUSD MM3, BUTTER

PUNPCKH	Unpack–high	Byte, word, and doubleword
0000 1111 0110 10gg 11 xxx1 xxx2		xreg2 → xreg1
Examples		
PUNPCKHB MM1, MM2 PUNPCKHW MM7, MM3 PUNPCKHD MM3, MM4		
0000 1111 0110 10gg oo xxx mmm		mem → xreg
Examples		
PUNPCKHB MM3, BUTTON PUNPCKHW MM7, SOUND PUNPCKHD MM3, BUTTER		

PUNPCHL	Unpack–low	Byte, word, and doubleword
0000 1111 0110 00gg 11 xxx1 xxx2		xreg2 → xreg1
Examples		
PUNPCKLB MM1, MM2 PUNPCKLW MM7, MM3 PUNPCKLD MM3, MM4		
0000 1111 0110 00gg oo xxx mmm		mem → xreg
Examples		
PUNPCKLB MM3, BUTTON PUNPCKLW MM7, SOUND PUNPCKLD MM3, BUTTER		

PXOR	Exclusive–Or	
0000 1111 1110 1111 11 xxx1 xxx2		xreg2 → xreg1
Examples		
PXOR MM1, MM2 PXOR MM7, MM3		
0000 1111 1110 1111 oo xxx mmm		mem → xreg
Examples		
PXOR MM3, BUTTON PXOR MM7, SOUND		

Notes: gg = 00 (byte), = 01 (word), = 10 (doubleword), = 11 (quadword) and; xxx = MMX register number or for MOVED 32-bit register number, oo = mode, and mmm = r/m field.

14–6   SUMMARY

1. The arithmetic coprocessor functions in parallel with the microprocessor. This means that the microprocessor and coprocessor can execute their respective instructions simultaneously.
2. The data types manipulated by the coprocessor include signed-integer, floating-point, and binary-coded decimal (BCD).
3. There are three forms of integers used with the coprocessor: word (16 bits), short (32 bits), and long (64 bits). Each integer contains a signed number in true magnitude for positive numbers and two's complement form for negative numbers.
4. A BCD number is stored as an 18-digit number in 10 bytes of memory. The most-significant byte contains the sign-bit, and the remaining 9 bytes contain an 18-digit packed BCD number.
5. The coprocessor supports three types of floating-point numbers: single-precision (32 bits), double-precision (64 bits), and extended-precision (80 bits). A floating-point number has three parts: the sign, biased exponent, and significand. In the coprocessor, the exponent is biased with a constant and the integer bit of the normalized number is not stored in the significand, except in the extended-precision form.
6. Decimal numbers are converted to floating-point numbers by (a) converting the number to binary, (b) normalizing the binary number, (c) adding the bias to the exponent, and (d) storing the number in floating-point form.
7. Floating-point numbers are converted to decimal by (a) subtracting the bias from the exponent, (b) unnormalizing the number, and (c) converting it to decimal.
8. The 80287 uses I/O space for the execution of some of its instructions. This space is invisible to the program and is used internally by the 80286/80287 system. These 16-bit I/O addresses (00F8H–00FFH) must not be used for I/O data transfers in a system that contains an 80287. The 80387, 80486/7, and Pentium through Pentium 4 use I/O addresses 800000F8H–800000FFH.
9. The coprocessor contains a status register that indicates busy, what conditions follow a compare or test, the location of the top of the stack, and the state of the error bits. The FSTSW AX instruction, followed by SAHF, is often used with conditional jump instructions to test for some coprocessor conditions.
10. The control register of the coprocessor contains control bits that select infinity, rounding, precision, and error masks.
11. The following directives are often used with the coprocessor for storing data: DW (define word), DD (define doubleword), DQ (define quadword), and DT (define 10 bytes).
12. The coprocessor uses a stack to transfer data between itself and the memory system. Generally, data are loaded to the top of the stack or removed from the top of the stack for storage.
13. All internal coprocessor data are always in the 80-bit extended-precision form. The only time that data are in any other form is when they are stored or loaded from the memory.
14. The coprocessor addressing modes include the classic stack mode, register, register with a pop, and memory. Stack addressing is implied and the data at ST become the source, ST(1) the destination, and the result is found in ST after a pop.
15. The coprocessor's arithmetic operations include addition, subtraction, multiplication, division, and square root calculation.
16. There are transcendental functions in the coprocessor's instruction set. These functions find the partial tangent or arctangent, $2^x - 1$, $Y \log_2 X$, and $Y \log_2 (X + 1)$. The 80387, 80486/7, and Pentium–Pentium 4 also include sine and cosine functions.
17. Constants are stored inside the coprocessor that provide +0.0, +1.0, π, $\log_2 10$, $\log_2 \varepsilon$, $\log_{10} 2$, and $\log_\varepsilon 2$.
18. The 80387 functions with the 80386 microprocessor and the 80487SX functions with the 80486SX microprocessor, but the 80486DX and Pentium–Pentium 4 contain their own

internal arithmetic coprocessor. The instructions performed by the earlier versions are available on these coprocessors. In addition to these instructions, the 80387, 80486/7, and Pentium–Pentium 4 also can find the sine and cosine.

19. The Pentium Pro through Pentium 4 contain two new floating-point instructions: FCMOV and FCOMI. The FCMOV instruction is a conditional move and the FCOMI performs the same task as FCOM, but it also places the floating-point flags into the system flag register.

20. The MMX extension uses the arithmetic coprocessor registers for MM0–MM7. Therefore, it is important that coprocessor software and MMX software do not try to use them at the same time.

21. The instructions for the MMX extensions perform arithmetic and logic operations on bytes (eight at a time), words (four at a time), doublewords (two at a time), and quadwords (two at a time), and quadwords. The operations performed are addition, subtraction, multiplication, AND, OR, Exclusive-OR, and NAND.

14–7   QUESTIONS AND PROBLEMS

1. List the three types of data that are loaded or stored in memory by the coprocessor.
2. List the three integer data types, the range of the integers stored in them, and the number of bits allotted to each.
3. Explain how a BCD number is stored in memory by the coprocessor.
4. List the three types of floating-point numbers used with the coprocessor and the number of binary bits assigned to each.
5. Convert the following decimal numbers into single-precision floating-point numbers:
   (a) 28.75
   (b) 624
   (c) –0.615
   (d) +0.0
   (e) –1000.5
6. Convert the following single-precision floating-point numbers into decimal:
   (a) 11000000   11110000   00000000   00000000
   (b) 00111111   00010000   00000000   00000000
   (c) 01000011   10011001   00000000   00000000
   (d) 01000000   00000000   00000000   00000000
   (e) 01000001   00100000   00000000   00000000
   (f) 00000000   00000000   00000000   00000000
7. Explain what the coprocessor does when a normal microprocessor instruction executes.
8. Explain what the microprocessor does when a coprocessor instruction executes.
9. What is the purpose of the C3–C0 bits in the status register?
10. What operation is accomplished with the FSTSW AX instruction?
11. What is the purpose of the IE bit in the status register?
12. How can SAHF and a conditional jump instruction be used to determine whether the top of the stack (ST) is equal to register ST(2)?
13. How is the rounding mode selected in the 80X87?
14. What coprocessor instruction uses the microprocessor's AX register?
15. What I/O ports are reserved for coprocessor use with the 80287?
16. How are data stored inside the coprocessor?
17. What is a NAN?
18. Whenever the coprocessor is reset, the top of the stack register is register number _____.
19. What does the term *chop* mean in the rounding control bits of the control register?
20. What is the difference between affine and projective infinity control?
21. What microprocessor instruction forms the opcodes for the coprocessor?

22. The FINIT instruction selects _____-precision for all coprocessor operations.
23. Using assembler pseudo-opcodes, form statements that accomplish the following:
    (a) Store a 23.44 into a double-precision floating-point memory location named FROG.
    (b) Store a −123 into a 32-bit signed integer location named DATA3.
    (c) Store a −23.8 into a single-precision floating-point memory location named DATA1.
    (d) Reserve a double-precision memory location named DATA2.
24. Describe how the FST DATA instruction functions. Assume that DATA is defined as a 64-bit memory location.
25. What does the FILD DATA instruction accomplish?
26. Form an instruction that adds the contents of register 3 to the top of the stack.
27. Describe the operation of the FADD instruction.
28. Choose an instruction that subtracts the contents of register 2 from the top of the stack and stores the result in register 2.
29. What is the function of the FBSTP DATA instruction?
30. What is the difference between a forward and a reverse division?
31. What is the purpose of the Pentium Pro FCOMI instruction?
32. What does a Pentium Pro FCMOVB instruction accomplish?
33. What must occur before executing any FCMOV instruction?
34. Develop a procedure that finds the reciprocal of the single-precision floating-point number. The number is passed to the procedure in EAX and must be returned as a reciprocal in EAX.
35. What is the difference between the FTST instruction and FXAM?
36. Explain what the F2XM1 instruction calculates.
37. Which coprocessor status register bit should be tested after the FSQRT instruction executes?
38. Which coprocessor instruction pushes n onto the top of the stack?
39. Which coprocessor instruction places a 1.0 at the top of the stack?
40. What will FFREE ST(2) accomplish when executed?
41. Which instruction stores the environment?
42. What does the FSAVE instruction save?
43. Develop a procedure that finds the area of a rectangle ($A = L \times W$). Memory locations for this procedure are single-precision floating-point locations A, L, and W.
44. Write a procedure that finds the capacitive reactance ($XC = {}^1\!/2\pi FC1$). Memory locations for this procedure are single-precision floating-point locations XC, F, and $C1$.
45. Develop a procedure that generates a table of square roots for the integers 2 through 10. The results must be stored as single-precision floating-point numbers in an array called ROOTS.
46. When is the FWAIT instruction used in a program?
47. What is the difference between the FSTSW and FNSTSW instructions?
48. Given the series/parallel circuit and equation illustrated in Figure 14–13, develop a program using single-precision values for R1, R2, R3, and R4 that finds the total resistance and stores the result at single-precision location RT.

FIGURE 14–13   The series/parallel circuit (Question 48).

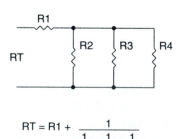

$$RT = R1 + \cfrac{1}{\cfrac{1}{R2} + \cfrac{1}{R3} + \cfrac{1}{R4}}$$

49. Develop a procedure that finds the cosine of a single-precision floating-point number. The angle, in degrees, is passed to the procedure in EAX and the cosine is returned in EAX. Recall that FCOS finds the cosine of an angle expressed in radians.

50. Given two arrays of double-precision floating-point data (ARRAY1 and ARRAY2) that each contain 100 elements, develop a procedure that finds the product of ARRAY1 times ARRAY2, and then stores the double-precision floating-point result in a third array (ARRAY3).

51. Develop a procedure that takes the single-precision contents of register EBX times n and stores the result in register EBX as a single-precision floating-point number. You must use memory to accomplish this task.

52. Write a procedure that raises a single-precision floating-point number X to the power Y. Parameters are passed to the procedure with $EAX = X$ and $EBX = Y$. The result is passed back to the calling sequence in ECX.

53. Given that the $LOG_{10} X = (LOG_2 10)^{-1} \times LOG_2 X$, write a procedure called LOG_{10} that finds the LOG_{10} of the value at the stack top. Return the LOG_{10} at the stack top at the end of the procedure.

54. Use the procedure developed in Question 53 to solve the equation Gain in decibels = 20 LOG_{10} (VOUT/VIN). The program should take arrays of single-precision values for VOUT and VIN and store the decibel gains in a third array called DBG. These are 100 values for VOUT and VIN.

55. What is the MMX extension to the Pentium–Pentium 4 microprocessors?

56. What is the purpose of the EMMS instruction?

57. Where are the MM0–MM7 registers found in the microprocessor?

58. What is signed saturation?

59. What is unsigned saturation?

60. How could all of the MMX registers be stored in the memory with one instruction?

61. Write a short program that uses MMX instruction to multiply the word-size numbers in two arrays and stores the 32-bit results in a third array. The source arrays are 256 words long.

CHAPTER 15

Bus Interface

INTRODUCTION

Many applications require knowledge of the bus systems located within the personal computer. At times, main boards from personal computers are used as core systems in industrial applications. These systems often require custom interfaces that are attached to one of the buses on the main board. This chapter presents the ISA (industry standard architecture) bus, the VESA local bus, the PCI (peripheral component interconnect) bus, the USB (universal serial bus), and the AGP (advanced graphics port). Also provided are some simple interfaces to many of these bus systems as design guides.

CHAPTER OBJECTIVES

Upon completion of this chapter, you will be able to:

1. Detail the pin connections and signal bus connections on the parallel port and on the ISA, VESA local, and PCI buses.
2. Develop simple interfaces that connect to the parallel port and on the ISA, VESA local, and PCI buses.
3. Program interface places on boards that connect to the ISA, VESA local, and PCI buses.
4. Describe the operation of the USB and develop some short programs that transfer data.
5. Explain how the AGP increases the efficiency of the graphics subsystem.

15–1 THE ISA BUS

The ISA, or **industry standard architecture,** bus has been around since the very start of the IBM-compatible personal computer system (circa 1982). In fact, any card from the very first personal computer will plug into and function in any of the most modern Pentium 4-based computers. This is all made possible by the ISA bus interface found in all these machines, which is still compatible with the early personal computers.

Evolution of the ISA Bus

The ISA bus has changed from its early days. Over the years, the ISA bus has evolved from its original 8-bit standard to the 16-bit standard found in most systems today. Along the way, there was even a 32-bit version called the EISA bus (**extended ISA**), but that seems to have all but disappeared. What remains today in most personal computers is an ISA slot (**connection**) on the main board that can accept either an 8-bit ISA card or a 16-bit ISA printed circuit card. The 32-bit printed circuit cards are more often PCI or, in some older 80486-based machines, the VESA cards. The ISA bus has all but vanished recently in home computers, but it is available as a special order in most main boards. The ISA bus is still found in many industrial applications, but its days now seem limited.

The 8-Bit ISA Bus Ouput Interface

Figure 15–1 illustrates the 8-bit ISA connector found on the main board of all personal computer systems (again, this may be combined with a 16-bit connector). The ISA bus connector contains the entire de-multiplexed address bus (A19–A0) for the 1M byte 8088 system, the 8-bit data bus (D7–D0), and the four control signals $\overline{MEMR}$, $\overline{MEMW}$, $\overline{IOR}$, and $\overline{IOW}$ for controlling I/O and any memory that might be placed on the printed circuit card. Memory is seldom added to any ISA bus card today because the ISA card only operates at an 8 MHz rate. There might be an EPROM or Flash memory used for setup information on some ISA cards, but never any RAM.

Other signals, which are useful for I/O interface, are the interrupt request line IRQ2–IRQ7. Note that IRQ2 is redirected to IRQ9 on modern systems and is so labeled on the connector in Figure 15–1. The DMA channels 0–3 control signals are also present on the connector. The **DMA request inputs** are labeled DRQ1–DRQ3 and the **DMA acknowledge** outputs are labeled $\overline{DACK0}$–$\overline{DACK3}$. Notice that the DRQ0 input pin is missing because the early personal computers used the $\overline{DACK0}$ output as a refresh signal to refresh any DRAM that might be located on the ISA card. Today, this output pin contains a 15.2 μs clock signal. The remaining pins are for power and RESET.

FIGURE 15–1   The 8-bit ISA bus.

Back of Computer

Pin #

#	Solder Side	Component Side
1	GND	IO CHK
2	RESET	D7
3	+5V	D6
4	IRQ9	D5
5	−5V	D4
6	DRQ2	D3
7	−12V	D2
8	OWS	D1
9	+12V	D0
10	GND	IO RDY
11	$\overline{MEMW}$	AEN
12	$\overline{MEMR}$	A19
13	$\overline{IOW}$	A18
14	$\overline{IOR}$	A17
15	$\overline{DACK3}$	A16
16	DRQ3	A15
17	$\overline{DACK1}$	A14
18	DRQ1	A13
19	$\overline{DACK0}$	A12
20	CLOCK	A11
21	IRQ7	A10
22	IRQ6	A9
23	IRQ5	A8
24	IRQ4	A7
25	IRQ3	A6
26	$\overline{DACK2}$	A5
27	T/C	A4
28	ALE	A3
29	+5V	A2
30	OSC	A1
31	GND	A0

Solder Side

Component Side

Suppose that a series of four 8-bit latches must be interfaced to the personal computer for 32 bits of parallel data. This is accomplished by purchasing an ISA interface card (part number 4713-1) from a company like Vector Electronics or other companies. In addition to the edge connector for the ISA bus, the card also contains room at the back for interface connectors. A 37-pin sub-miniature D-type connector can be placed on the back of the card to transfer the 32 bits of data to the external source.

Figure 15–2 shows a simple interface for the ISA bus, which provides 32 bits of parallel TTL data. This example system illustrates some important points about any system interface. First, it is extremely important that the loading to the ISA bus is kept to one low power (LS) TTL load. In this circuit, a 74LS244 buffer is used to reduce the loading on the data bus. If the 74LS244 were not present, this system would present the data bus with four unit loads. If all bus cards were to provide heavy loads, the system would not operate properly (or perhaps not at all).

Output from the ISA card is provided in this circuit by a 37-pin connector labeled P1. The output pins from the circuit connect to P1, and a ground wire is attached. You must provide ground to the outside world, or else the TTL data on the parallel ports are useless. If needed, the output control pins ($\overline{OC}$) on each of the 74LS374 latch chips can also be removed from ground and connected to the four remaining pins on P1. This allows an external circuit to control the outputs from the latches.

A small DIP switch is placed on two of the outputs of U7, so the address can be changed if an address conflict occurs with another card. This is unlikely, unless you plan to use two of these cards in the same system. Address connection A2 is not decoded in this system so it becomes a

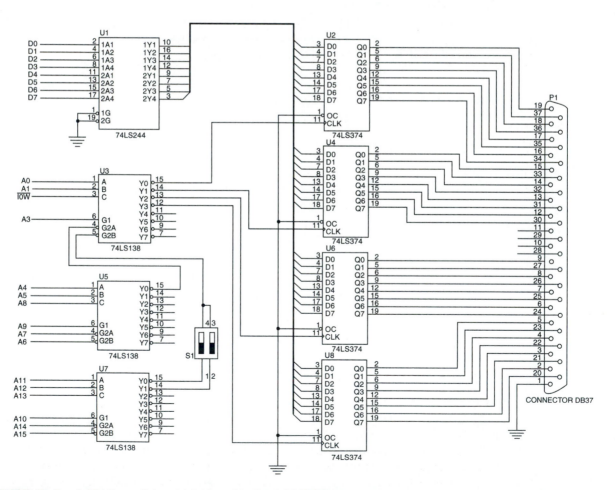

FIGURE 15–2 A 32-bit parallel port interfaced to the 8-bit ISA bus.

TABLE 15–1   The I/O port assignments for Figure 15–2

DIP Switch	Latch U2	Latch U4	Latch U6	Latch U8
1–4 On	0608H or 060CH	0609H or 060DH	060AH or 060EH	060BH or 060FH
2–3 On	0E08H or 0E0CH	0E09H or 0E0DH	0E0AH or 0E0EH	0E0BH or 0E0FH

don't care. See Table 15–1 for the addresses of each latch and each position of the S1. Note that only one of the two switches may be on at a time and that each port has two possible addresses for each switch setting because A2 is not connected.

In the personal computer, the ISA bus is designed to operate at I/O address 0000H through 03FFH. Depending on the version and manufacturer of the main-board, ISA cards may or may not function above these locations. Newer systems often allow ISA ports at locations above 03FFH, but older systems may not. The ports in this example may need to be changed for some systems. Some older cards only decode I/O addresses 0000H–03FFH and may have address conflicts if the port addresses above 03FFH conflict. The ports are decoded in this example by three 74LS138 decoders. It would be more efficient and cost-effective to decode the ports with a programmable logic device.

Figure 15–3 shows the circuit of Figure 15–2 reworked using a PAL16L8 to decode the addresses for the system. Notice that address bits A15–A4 are decoded by the PAL and the switch is connected to two of the PAL inputs. This change allows four different I/O port addresses for each latch, making the circuit more flexible. Table 15–2 shows the port number selected by switch 1–4 and switch 2–3. Example 15–1 shows the program for the PAL16L8 that causes the port assignments of Table 15–2.

EXAMPLE 15–1

```
AUTHOR    Barry B. Brey
COMPANY   BreyCo
DATE      7/20/99
CHIP      ISA1 PAL16L8

;pins  1  2  3  4  5  6  7   8   9   10
       A4 A5 A6 A7 A8 A9 A10 A11 A12 GND

;pins  11   12 13  14  15 16 17 18 19  20
       A13  NC A14 A15 S1 S2 NC NC DC VCC

EQUATIONS

/DC = /A15*/A14*/A13*/A12*/A11*A10*A9*/A8*/A7*/S1*/S2*/A6*/A5*/A4
    + /A15*/A14*/A13*/A12*/A11*A10*A9*/A8*/A7*/S1*S2*/A6*A5*/A4
    + /A15*/A14*/A13*/A12*/A11*A10*A9*/A8*/A7*S1*/S2*A6*/A5*/A4
    + /A15*/A14*/A13*/A12*/A11*A10*A9*/A8*/A7*S1*S2*A6*A5*/A4
```

Notice in Example 15–1 how the first product term generates a logic 0 on the output to the decoder only when both switches are in their off positions for I/O ports 0600H–060FH. The 74LS138 further refines the ports to 604H, 605H, 606H, or 607H for the latches. The second product term is active when switch 1–4 is off and switch 2–3 is on. The other two combinations on the switch select the last two port address assignments in Table 15–2.

Example 15–2 shows a small program that sends data to the ports in a pattern that could be used for testing. The pattern selected places a logic 1 on bit 0 of U2, and all zeros on the remaining latches. This pattern is then rotated through all 32 bits until one minute has elapsed. The timing is handled by the counter located at the 32-bit memory location starting at 0000:046C, which increments 18.2 times per second. This test program assumes that the latches appear at I/O ports 0604H–0607H, as selected by the switches.

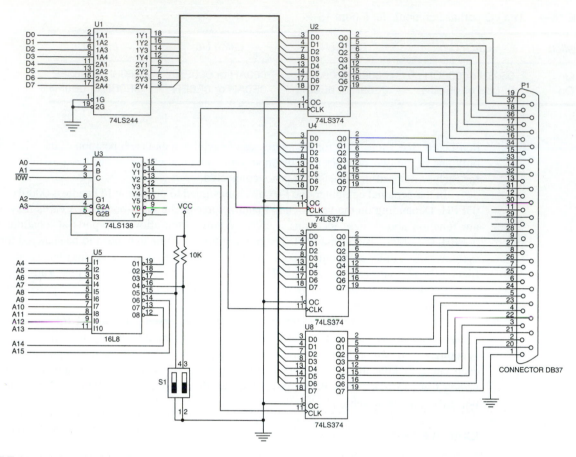

FIGURE 15–3   A 32-bit parallel port interfaced to the 8-bit ISA bus using a PAL16L8 for a decoder.

TABLE 15–2   Port assignments in Figure 15–3.

Switch 1–4	Switch 2–3	Latch U2	Latch U4	Latch U6	Latch U8
Off	Off	0604H	0605H	0606H	0607H
Off	On	0624H	0625H	0626H	0627H
On	Off	0644H	0645H	0646H	0647H
On	On	0664H	0665H	0666H	0667H

Note: On = a closed switch and Off = an open switch.

EXAMPLE 15–2

```
                    ;A program that sends a test pattern to the I/O ports of
                    ;Figure 15-3.
                    ;
                    .MODEL TINY
0000                .CODE
                    .STARTUP
0100    B8 0000         MOV     AX,0
0103    8E D8           MOV     DS,AX           ;address segment 0000H
0105    BB 0001         MOV     BX,1            ;setup starting bit pattern
0108    BA 0000         MOV     DX,0            ;in registers DX-BX
```

```
010B    B9 0444             MOV    CX,1092         ;set count for 1 minute
                            .REPEAT
010E    BA 0604             MOV    DX,0604H        ;address latch U2
0111    8A C3               MOV    AL,BL           ;send BL to U2
0113    EE                  OUT    DX,AL
0114    42                  INC    DX              ;address latch U4
0115    8A C7               MOV    AL,BH           ;send BH to U4
0117    EE                  OUT    DX,AL
0118    42                  INC    DX              ;address latch U6
0119    8A C2               MOV    AL,DL           ;send DL to U6
011B    EE                  OUT    DX,AL
011C    42                  INC    DX              ;address latch U8
011D    8A C6               MOV    AL,DH           ;send DH to U8
011F    D1 E2               SHL    DX,1            ;rotate number in DX-BX
0121    D1 D3               RCL    BX,1
                            .IF CARRY?
0125    83 C2 01            ADD    DX,1
                            .ENDIF
0128    B8 046C             MOV    AX,[46CH]       ;get counter
012B    BD 046E             MOV    BP,[46EH]
012E    40                  INC    AX
012F    83 D5 00            ADC    BP,0
                            .REPEAT
                            .UNTIL    AX==[46CH] && BP==[46EH]
                    .UNTILCXZ
            .EXIT
                    END
```

The 8-Bit ISA Bus Input Interface

To illustrate the input interface to the ISA bus, a pair of ADC804 analog-to-digital converters are interfaced to the ISA bus in Figure 15–4. The connections to the converters are made through a 9-pin DB9 connector. The task of decoding the I/O port addresses is more complex, because each converter needs a write pulse to start a conversion, a read pulse to read the digital data once it has been converted from the analog input data, and a pulse to enable the selection of the $\overline{INTR}$ output. Notice that the $\overline{INTR}$ output is connected to data bus bit position D0. When $\overline{INTR}$ is input to the microprocessor, the rightmost bit of AL is tested to see whether the converter is busy.

As before, great care is taken so that the connections to the ISA bus present one unit load to the system. Table 15–3 illustrates the I/O port assignment decoded by the PAL16L8 in Example 15–3.

EXAMPLE 15–3

```
AUTHOR     Barry B. Brey
COMPANY    BreyCo
DATE       7/21/99
CHIP       ISA2 PAL16L8

;pins  1   2   3   4   5   6   7   8   9   10
       A3  A4  A5  A6  A7  A8  A9  A10 A11 GND

;pins  11  12  13  14  15  16  17  18  19  20
       A12 NC  A13 A14 A15 IOR G   NC  DC  VCC

EQUATIONS

/DC = /A15*/A14*/A13*A12*/A11*/A10*A9*/A8*/A7*/A6*/A5*/A4*/A3
/G  = /A15*/A14*/A13*A12*/A11*/A10*A9*/A8*/A7*/A6*/A5*/A4*/A3*/IOR
```

Example 15–4 lists a macro that can be used to read either ADC U3 or U5. The address is generated by passing either a 0 for U3 or a 1 for U5 to the macro as a parameter. The macro starts the converter by writing to it, and then waits until the $\overline{INTR}$ pin returns to a logic 0, indicating that the conversion is complete before the data are read and returned in the AL register.

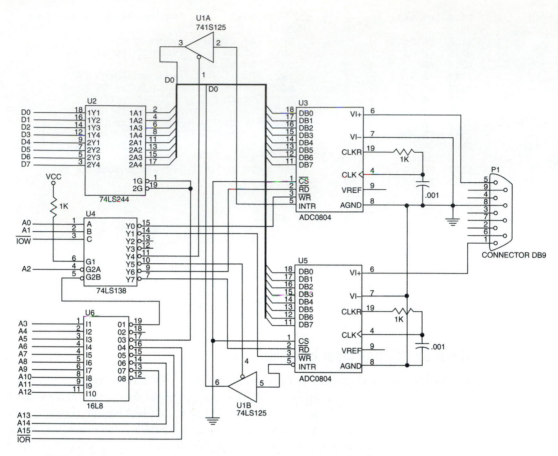

FIGURE 15–4   A pair of analog-to-digital converters interfaced to the ISA bus.

TABLE 15–3   I/O port
assignments for Figure 15–4.

Device	Port Number
Start ADC (U3)	1200H
Read INTR (U3)	1200H
Read ADC (U3)	1202H
Start ADC (U5)	1201H
Read INTR (U5)	1201H
Read ADC (U5)	1203H

EXAMPLE 15–4

```
;A macro that operates either converter 0 or converter 1 and
;returns the digital data in AL.  Note that converter 0 is U3
;and converter 1 is U5.
;
ADC    MACRO   WHICH
       MOV     DX,1200H
       ADD     DX,WHICH            ;address converter
       OUT     DX,AL              ;start converter
       .REPEAT
          IN      AL,DX           ;get busy signal
          TEST    AL,1
       .UNTIL  ZERO?
       ADD     DX,2               ;address data port
       IN      AL,DX
       ENDM
```

FIGURE 15–5 The 16-bit ISA bus. (a) Both 8- and 16-bit connectors and (b) the pin-out of the 16-bit connector.

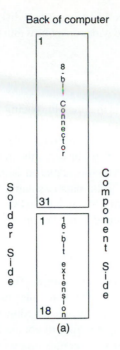

Pin#		
1	MCS16	BHE
2	IOCS16	A23
3	IRQ10	A22
4	IRQ11	A21
5	IRQ12	A20
6	IRQ15	A19
7	IRQ14	A18
8	DACK0	A17
9	DRQ0	MEMR
10	DACK5	MEMW
11	DRQ5	D8
12	DACK6	D9
13	DRQ6	D10
14	DACK7	D11
15	DRQ7	D12
16	+5V	D13
17	MASTER	D14
18	GND	D15

(b)

The 16-Bit ISA Bus

The only difference between the 8- and 16-bit ISA bus is that an additional connector is attached behind the 8-bit connector. A 16-bit ISA card contains two edge connectors: one plugs into the original 8-bit connector and the other plugs into the new 16-bit connector. Figure 15–5 shows the pin-out of the additional connector and its placement in the computer in relation to the 8-bit connector. Unless additional memory is added on the ISA card, the extra address connections A23–A20 do not serve any function for I/O operations. The added features that are most often used are the additional interrupt request inputs and the DMA request signals. In some systems, 16-bit I/O uses the additional eight data bus connections (D8–D15), but more often today, the EISA bus, VESA local bus, or PCI bus is used for peripherals that are wider than eight bits. The ISA bus is beginning to be used less than it once was and will probably be replaced in the near future by the USB **(universal serial bus).** About the only recent interfaces found for the ISA bus are modems and sound cards. Modems are serial devices that lend themselves well to the USB; sound cards are affected by noise. The computer power supply is a switching supply that generates noise that cannot be filtered. If the sound function is placed external to the computer with its own separate power supply, this annoying noise can be removed from the audio signal. The sound card will benefit greatly when placed on the USB.

15–2   THE EXTENDED ISA (EISA) AND VESA LOCAL BUS ARCHITECTURES

The **extended ISA architecture** (EISA) is a 32-bit modification to the ISA bus. As computers became larger and had wider data buses (80386–Pentium 4), a new bus was needed that would transfer 32-bit data. Although the EISA bus seems to be fading, it is a steppingstone in the evolution of the computer system bus. The main problem with the EISA bus is that even though the data bus width is increased to 32 bits, the clocking speed remains at 8 MHz, which is why this interface standard has all but vanished. Note that the newer VESA local bus and PCI bus both operate at a higher speeds. The VESA local and PCI buses both presently operate at 33 MHz.

The most common application for the EISA bus is as a disk controller or as a video graphics adapter. These applications benefit from the wider data bus width because the data transfer rates for these devices are high.

EISA Bus Pin-Out

One interesting change from ISA to EISA is that the pin spacing is 0.05" instead of 0.1", as on the ISA bus edge connector. The new pins for the EISA bus are interspersed with the older pins in the standard 16-bit ISA connector pair. This makes insertion a little difficult, but it preserves compatibility with the older ISA standard. Figure 15–6 illustrates the pin-out of the EISA bus connectors and details the way that this new set of connectors is placed with the old ISA connector. Most of the new EISA connections are used for a 32-bit data bus and a latched 32-bit address bus. Also present are the bank enable signals used in the 80386 and 80486 microprocessors. The EISA standard does not support the 64-bit data bus found in the Pentium–Pentium 4 microprocessors.

EISA Bus Interface Example

Figure 15–7 shows the interface to a 32-bit events counter. This system uses four 74LS590 octal (8-bit) binary counters with built-in latches. Each counter also has a direct clear input that is used to asynchronously clear the counter before any events are counted. The counter itself is wired as a 32-bit synchronous counter that uses ripple output to enable the next octal counter when the counter rolls over to 00000000, so the next stage counts up by one.

Back of Computer

Pin#

1	GND	CMD
2	+5V	START
3	+5V	EXRDY
4	NS	EX32
5	NS	GND
6	KEY	KEY
7	NS	EX16
8	NS	SLBRST
9	+12V	MSBRST
10	M/IO	W/R
11	LOCK	GND
12	NS	NS
13	GND	NS
14	NS	NS
15	BE3	GND
16	KEY	KEY
17	BE2	BE1
18	BE0	LA31
19	GND	GND
20	+5V	LA30
21	LA29	LA28
22	GND	LA27
23	LA26	LA25
24	LA24	GND
25	KEY	KEY
26	LA16	LA15
27	LA14	LA13
28	+5V	LA12
29	+5V	LA11
30	GND	GND
31	LA10	LA9

Solder Side / Component Side

This connector is combined with the 8-bit connector in Figure 15–1.

Bottom of 8-bit connector

Pin#

1	LA8	LA7
2	LA6	GND
3	LA5	LA4
4	+5V	LA3
5	LA2	GND
6	KEY	KEY
7	D16	D17
8	D18	D19
9	GND	D20
10	D21	D22
11	D23	GND
12	D24	D25
13	GND	D26
14	D27	D28
15	KEY	KEY
16	D29	GND
17	–5V	D30
18	+5V	D31
19	MAKX	MREQX

Solder Side / Component Side

This connector is combined with the 16-bit connector in Figure 15–5.

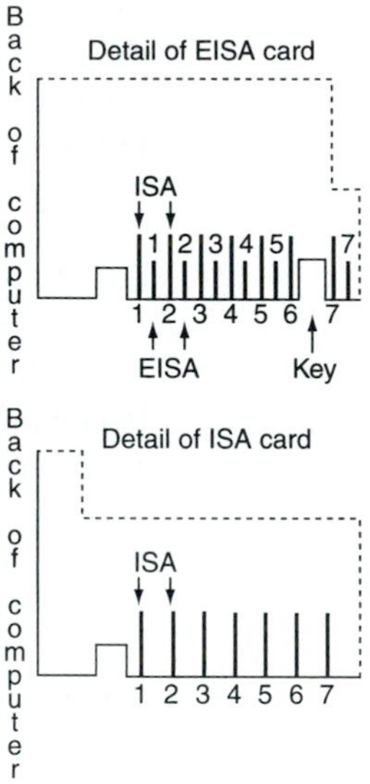

Detail of EISA card

Detail of ISA card

FIGURE 15–6   The EISA connectors and detail comparing the EISA card with the ISA card.

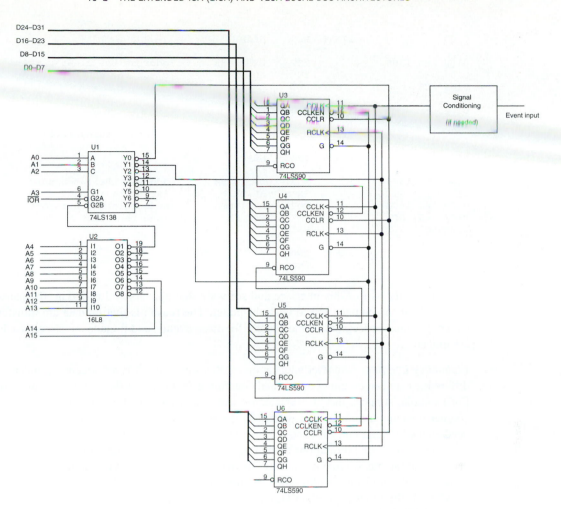

FIGURE 15–7   A 32-bit events and general purpose counter interfaced to the EISA bus.

A PAL16L8 and 74LS138 decoder are used to decode the I/O addresses at ports 0308H, 0309H, and 030CH. Port 0308H is used to clear the counter, port 0309H latches the count, and port 030CH enables the three-state output buffers to apply the contents of the 32-bit latch to the microprocessor data bus. Port 030CH is used because 32-bit data should fall at a 32-bit address boundary for proper and efficient I/O operation. Note that the PAL16L8 program is not illustrated for this example because it is essentially the same as other examples in this chapter.

Events Counter.   Example 15–5 illustrates software required to use this system as an events counter. Because the counter is 32-bits wide, it can accumulate up to 4 billion or so events before the count becomes incorrect. This is large enough for most applications that require event counts.

EXAMPLE 15–5

```
            .MODEL TINY
            .386
0000        .CODE
            .STARTUP
            ;
            ;Procedure that starts the events counter.
```

```
                              ;
0100                          START  PROC    NEAR

0100   BA 1308                       MOV     DX,1308H       ;address the clear port
0103   EC                            IN      AL,DX          ;clear the counter
0104   C3                            RET

0105                          START  ENDP
                              ;
                              ;Procedure that reads the count and returns it in EAX.
                              ;
0105                          READC  PROC    NEAR

0105   BA 1309                       MOV     DX,1309H       ;address the latch port
0108   EC                            IN      AL,DX          ;latch the count
0109   BA 130C                       MOV     DX,130CH       ;address the count port
010C   66| ED                        IN      EAX,DX         ;read the count to EAX
010E   C3                            RET

010F                          READC  ENDP
                                     END
```

The only part of this interface and software that appears odd is that an IN instruction is used to clear the counter and to latch the count. The reason is that no data are transferred between the microprocessor and the counter for these events, and it made the circuitry a little less complicated.

Frequency Counter.   Another application for this same circuit is a frequency counter. The only difference between counting events and counting a frequency is the way the circuit is operated. For example, suppose that the counter is cleared and then read exactly one second later. The counter contains a number that represents the frequency at the events input. If a frequency is needed in kilo-Hertz range instead of Hertz, the counter is reset and then read exactly one ms later.

There are some limitations to the frequency, which can be measured by this circuitry. The accuracy of the measurement can be off by one count. The highest frequency is due to the speed of the counter and the conditioning circuitry, if needed. The 74LS590 can count to at least 32 MHz. If the input is purely TTL (no conditioning), then this circuit can measure frequency up to 32 MHz. The conditioning circuit chosen may also limit the upper frequency limit. The accuracy is determined by the time between the reset and the latching of the count. If the clock in the computer system is very accurate, the frequency is very accurate. It is assumed that the clock in the personal computer is about 0.1 percent accurate, which would be the accuracy of any frequency measured by this system. This is good enough for many applications that require frequency measurements.

Example 15–6 shows how the system can measure a frequency if you are willing to wait for one second. Notice how the SYNC macro is used to synchronize to the clock tick and how the result is scaled by multiplying the count by 18.2. To obtain a more accurate reading, the scaling factor could be changed by measuring the frequency with a laboratory frequency counter and then modifying the contents of NUM. The sample time (point where the clear and latch operations occur) can be reduced to make more samples per second.

EXAMPLE 15–6

```
                              .MODEL SMALL
                              .386
                              .387
0000                          .DATA
0000   4191999A               NUM    DD   18.2
0004   00000000               TEMP   DD   ?
0000                          .CODE
```

```
              .STARTUP
              ;
              ;A simple frequency counter.
              ;
              ;This procedure returns the frequency in EAX.
              ;The accuracy should be fairly good.
              ;
              SYNC    MACRO              ;sync to clock tick
                      MOV    AH,0
                      INT    16H         ;get clock tick information into CX:DX
                      MOV    BX,CX
                      MOV    BP,DX
                      .REPEAT            ;sync to clock tick
                         MOV    AH,0
                         INT    16H
                      .UNTIL CX!=BX || DX!=BP
                      ENDM

0010               FREQ   PROC   NEAR

                          SYNC              ;sync to clock tick

0024  BA 1308             MOV    DX,1308H   ;address clear counter port
0027  EC                  IN     AL,DX      ;clear counter

                          SYNC              ;wait for next tick

003C  BA 1309             MOV    DX,1309H   ;address latch port
003F  EC                  IN     AL,DX      ;latch count
0040  BA 130C             MOV    DX,130CH   ;address counter
0043  66| ED              IN     EAX,DX     ;get count
0045  66| A3 0004 R       MOV    TEMP,EAX
0049  DB 06 0004 R        FILD   TEMP       ;convert to Hertz
004D  D8 0E 0000 R        FMUL   NUM
0051  DB 1E 0004 R        FISTP  TEMP       ;save as integer
0055  66| A1 0004 R       MOV    EAX,TEMP
0059  C3                  RET
005A               FREQ   ENDP
                          END
```

The VESA Local Bus

A much better approach to 32-bit interfacing is the VESA[1] local bus. The EISA bus only operates at 8 MHz, while the VESA local bus operates at 33 MHz. This means that applications that require high-speed data transfers benefit from the VESA local bus. This section of the chapter details the VESA local bus, which is common for video and disk interfaces to the personal computer that is 80486-based.

VESA Local Bus Pin-Out

Figure 15–8 illustrates the VESA local bus (or the VL bus, as it is often called). Like the EISA bus, the VESA local bus is also an extension of the ISA bus. The difference is that the VESA local bus does not add anything to the 16-bit ISA connectors; instead, a third connector (the VESA connector) is added behind the 16-bit ISA connector.

The connections on the VESA local bus are very similar to the EISA bus card. The VESA local bus also contains a 32-bit address and data bus for interfacing memory or I/O to the microprocessor.

---

[1]VESA is the Video Electronics Standards Association.

Pin #

Pin #	Solder Side	Component Side
1	D0	D1
2	D2	D3
3	D4	GND
4	D6	D5
5	D8	D7
6	GND	D9
7	D10	D11
8	D12	D13
9	+5V	D15
10	D14	GND
11	D16	D17
12	D18	+5V
13	D20	D19
14	GND	D21
15	D22	D23
16	D24	D25
17	D26	GND
18	D28	D27
19	D30	D29
20	+5V	D31
21	A31	A30
22	GND	A28
23	A29	A26
24	A27	GND
25	A25	A24
26	A23	A22
27	A21	+5V
28	A19	A20
29	GND	A18
30	A17	A16
31	A15	A14
32	+5V	A12
33	A13	A10
34	A11	A8
35	A9	GND
36	A7	A6
37	A5	A4
38	GND	$\overline{\text{WBACK}}$
39	A3	BE0
40	A2	+5V
41	nc	$\overline{\text{BE1}}$
42	$\overline{\text{RESET}}$	$\overline{\text{BE2}}$
43	D/$\overline{\text{C}}$	GND
44	M/$\overline{\text{IO}}$	$\overline{\text{BE3}}$
45	W/$\overline{\text{R}}$	$\overline{\text{AD5}}$
46		
47		
48	$\overline{\text{RDYRTN}}$	$\overline{\text{LRDY}}$
49	GND	$\overline{\text{LDEV}}$
50	IRQ9	$\overline{\text{LREQ}}$
51	$\overline{\text{BRDY}}$	GND
52	BLAST	$\overline{\text{LGNT}}$
53	ID0	+5V
54	ID1	ID2
55	GND	ID3
56	LCLK	ID4
57	+5V	$\overline{\text{LKEN}}$
58	$\overline{\text{LBS16}}$	$\overline{\text{LEADS}}$

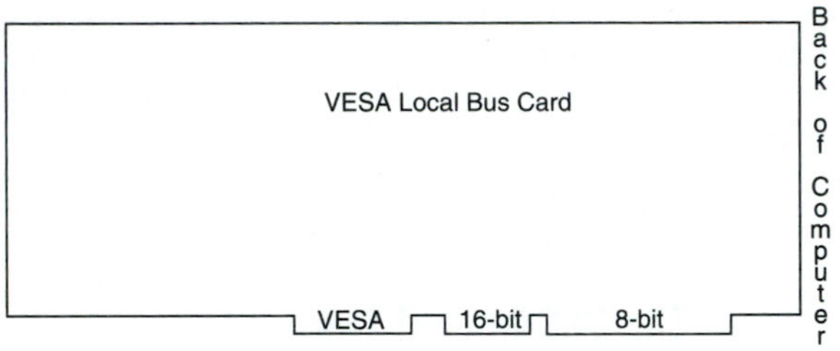

VESA Local Bus Card

Back of Computer

VESA   16-bit   8-bit

FIGURE 15–8   The VESA local bus connector and card.

15–3   THE PERIPHERAL COMPONENT INTERCONNECT (PCI) BUS

The PCI (**peripheral component interconnect**) bus is virtually the only bus found in the newest Pentium 4 systems and just about all the Pentium systems. In all of the newer systems, the ISA bus still exists, but as an interface for older 8-bit and 16-bit interface cards. Many new systems contain only two ISA bus slots or no ISA slots. In time, the ISA bus may disappear, but it is still an important interface for many applications. The PCI bus has replaced the VESA local bus. One reason is the PCI bus has plug-and-play characteristics and the ability to function with a 64-bit data bus. A PCI interface contains a series of registers, located in a small memory device on the PCI interface, that contain information about the board. This same memory can provide plug-and-play characteristics to the ISA bus or any other bus. The information in these registers allows the computer to automatically configure the PCI card. This feature, called **plug-and-play (PnP),** is probably the main reason that the PCI bus has become so popular in the newest systems.

Figure 15–9 shows the system structure for the PCI bus in a personal computer system. Notice that the microprocessor bus is separate and independent of the PCI bus. The microprocessor connects to the PCI bus through an integrated circuit called a **PCI bridge.** This means that virtually any microprocessor can be interfaced to the PCI bus, as long as a PCI controller or bridge is designed for the system. In the future, all computer systems may use the same bus. Even the Apple Macintosh system is switching to the PCI bus. Certainly, IBM will produce a Power-PC system that contains the PCI bus. Apple computers currently use the PCI bus.

The PCI Bus Pin-Out

As with the other buses described in this chapter, the PCI bus contains all of the system control signals. Unlike the other buses, the PCI bus functions with either a 32-bit or a 64-bit data bus and

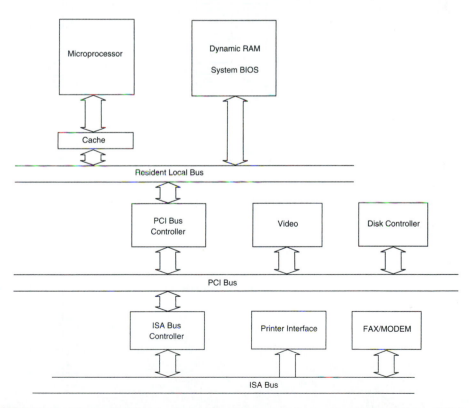

FIGURE 15–9   The system block diagram for the personal computer that contains a PCI bus.

a full 32-bit address bus. Another difference is that the address and data buses are multiplexed to reduce the size of the edge connector. These multiplexed pins are labeled AD0–AD63 on the connector. The 32-bit card has only connections 1 through 62, while the 64-bit card has all 94 connections. The 64-bit card can accommodate a 64-bit address if it is required at some point in the future. Figure 15–10 illustrates the PCI bus pin-out.

As with the other bus systems, the PCI bus is most often used for interfacing I/O components to the microprocessor. Memory could be interfaced, but it would operate only at a 33 MHz

FIGURE 15–10 The pin-out of the PCI bus.

Back of computer

Pin #	Solder Side	Component Side
1	−12V	TRST
2	TCK	+12V
3	GND	TM5
4	TD0	TD1
5	+5V	+5V
6	+5V	INTA
7	INTB	INTC
8	INTD	+5V
9	PRSNT 1	
10		+VI/O
11	PRSNT 2	
12	KEY	KEY
13	KEY	KEY
14		
15	GND	RST
16	CLK	VI/O
17	GND	VNT
18	REQ	GND
19	+V IO	
20	AD31	AD30
21	AD29	+3.3V
22	GND	AD28
23	AD27	AD26
24	AD25	GND
25	+3.3V	AD24
26	C/BE3	IDSEL
27	AD23	+3.3V
28	GND	AD22
29	AD21	AD20
30	AD19	GND
31	+3.3V	AD18
32	AD17	AD16
33	C/BE2	+3.3V
34	GND	FRAME
35	IRDY	GND
36	+3.3V	TRDY
37	DEVSEL	GND
38	GND	STOP
39	LOCK	+3.3V
40	PERR	SDONE

Pin #	Solder Side	Component Side
41	+3.3V	SBO
42	SERR	GND
43	+3.3V	PAR
44	C/BE1	AD15
45	AD14	+3.3V
46	GND	AD13
47	AD12	AD11
48	AD10	GND
49	GND	AD9
50	KEY	KEY
51	KEY	KEY
52	AD8	C/BE0
53	AD7	+3.3V
54	+3.3V	AD6
55	AD5	AD4
56	AD3	GND
57	GND	AD2
58	AD1	AD0
59	+V IO	+V IO
60	ACK64F	REQ64
61	+5V	+5V
62	+5V	+5V
63		GND
64	GND	C/BE7
65	C/BE6	C/BE5
66	C/BE4	+V IO
67	GND	PAR64
68	AD63	AD62
69	AD61	GND
70	+V IO	AD60
71	AD59	AD58
72	AD57	GND
73	GND	AD56
74	AD55	AD54
75	AD53	+V IO
76	GND	AD52
77	AD51	AD50
78	AD49	GND
79	+V IO	AD48
80	AD47	AD46
81	AD45	GND
82	GND	AD44
83	AD43	AD42
84	AD41	+VI/O
85	GND	AD40
86	AD39	AD38
87	AD37	GND
88	+VI/O	AD36
89	AD35	AD34
90	AD33	GND
91	GND	AD32
92		
93		GND
94	GND	

Notes: (1) pins 63–94 exist only on the 64-bit PCI card
(2) + VI/O is 3.3V on a 3.3V board and +5V on a 5V board
(3) blank pins are reserved

rate with the Pentium, which is half the speed of the 66 MHz resident local bus of the Pentium through Pentium 4 system. A more recent version of PCI (2.1-compliant) operates at 66 MHz and at 33 MHz for older interface cards. Pentium 4 systems use a 400 MHz system bus speed, but there is no planned modification to the PCI bus speed yet.

The PCI Address/Data Connections

The PCI address appears on AD0–AD31 and it is multiplexed with data. In some systems, there is a 64-bit data bus that uses AD32–AD63 for data transfer only. In the future, these pins can be used for extending the address to 64 bits. Figure 15–11 illustrates the timing diagram for the PCI bus, which shows the way that the address is multiplexed with data and also the control signals used for multiplexing.

During the first clocking period, the address of the memory or I/O location appears on the AD connections, and the command to a PCI peripheral appears on the C/$\overline{BE}$ pins. Table 15–4 illustrates the bus commands found on the PCI bus.

INTA Sequence	During the interrupt acknowledge sequence, an interrupt controller (the controller that caused the interrupt) is addressed and interrogated for the interrupt vector. The byte-sized interrupt vector is returned during a byte read operation.
Special Cycle	The special cycle is used to transfer data to all PCI components. During this cycle, the rightmost 16 bits of the data bus contain a 0000H, indicating a processor shutdown, a 0001H for a processor halt, or a 0002H for 80X86 specific code or data.
I/O Read Cycle	Data are read from an I/O device using the I/O address that appears on AD0–AD15. Burst reads are not supported for I/O devices.
I/O Write Cycle	As with I/O read, this cycle accesses an I/I device, but writes data.

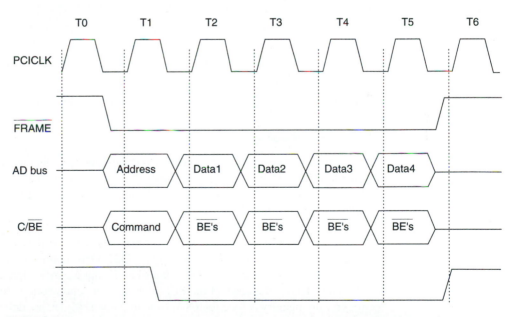

FIGURE 15–11 The basic burst mode timing for the PCI bus system. Note that this transfers either four 32-bit numbers (32-bit PCI) or four 64-bit numbers (64-bit PCI).

TABLE 15–4   PCI bus commands.

$\overline{C/BE3}$–$\overline{C/BE0}$	Command
0000	INTA sequence
0001	Special cycle
0010	I/O read cycle
0011	I/O write cycle
0100–0101	Reserved
0110	Memory read cycle
0111	Memory write cycle
1000–1001	Reserved
1010	Configuration read
1011	Configuration write
1100	Memory multiple access
1101	Dual addressing cycle
1110	Line memory access
1111	Memory write with invalidation

Memory Read Cycle	Data are read from a memory device located on the PCI bus.
Memory Write Cycle	As with memory read, data are accessed in a device located on the PCI bus. The location is written.
Configuration Read	Configuration information is read from the PCI device using the configuration read cycle.
Configuration Write	The configuration write allows data to be written to the configuration area in a PCI device. Note that the address is specified by the configuration read.
Memory Multiple Access	This is similar to the memory read access, except that it is usually used to access many data instead of one.
Dual Addressing Cycle	Used for transferring address information to a 64-bit PCI device, which only contains a 32-bit data path.
Line Memory Access	Used to read more than two 32-bit numbers from the PCI bus.
Memory Write with Invalidation	This is the same as line memory access, but it is used with a write. This write bypasses the write-back function of the cache.

Configuration Space

The PCI interface contains a 256-byte configuration memory that allows the computer to interrogate the PCI interface. This feature allows the system to automatically configure itself for the PCI plug-board. Microsoft Corporation calls this plug-and-play (PnP). Figure 15–12 illustrates the configuration memory and its contents.

The first 64 bytes of the configuration memory contain the header that holds information about the PCI interface. The first 32-bit doubleword contains the unit ID code and the vendor ID code. The unit ID code is a 16-bit number (D31–D16) that is an FFFFH if the unit is not installed, and a number between 0000H and FFFEH that identifies the unit if it is installed. The class codes identify the class of the PCI interface. The class code is found in bits D31–D16 of configuration memory at location 08H. Note that bits D15–D0 are defined by the manufacturer. The current class codes are listed in Table 15–5 and are assigned by the PCI SIG, which is the governing body for the PCI bus interface standard. The vendor ID (D15–D0) is also allocated by the PCI SIG.

FIGURE 15–12 The contents of the configuration memory on a PCI expansion board.

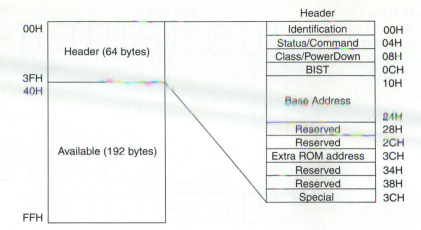

TABLE 15–5 The class codes.

Class Code	Function
0000H	Older non-VGA devices (not plug-and-play)
0001H	Older VGA devices (not plug-and-play)
0100H	SCSI controller
0101H	IDE controller
0102H	Floppy disk controller
0103H	IPI controller
0180H	Other hard/floppy controller
0200H	Ethernet controller
0201H	Token ring controller
0202H	FDDI
0280H	Other network controller
0300H	VGA controller
0301H	XGA controller
0380H	Other video controller
0400H	Video multimedia
0401H	Audio multimedia
0480H	Other multimedia controller
0500H	RAM controller
0501H	FLASH memory controller
0580H	Other memory controller
0600H	Host bridge
0601H	ISA bridge
0602H	EISA bridge
0603H	MCA bridge
0604H	PCI–PCI bridge
0605H	PCMIA bridge
0680H	Other bridge
0700H–FFFEH	Reserved
FFFFH	Unit not in any of the above classes

The status word is loaded in bits D31–D16 of configuration memory location 04H and the command is at bits D15–D0 of location 04H. Figure 15–13 illustrates the format of both the status and command registers.

The base address space consists of a base address for the memory, a second for the I/O space, and a third for the expansion ROM. The first two doublewords of the base address space

FIGURE 15–13 The contents of the status and control words in the configuration memory.

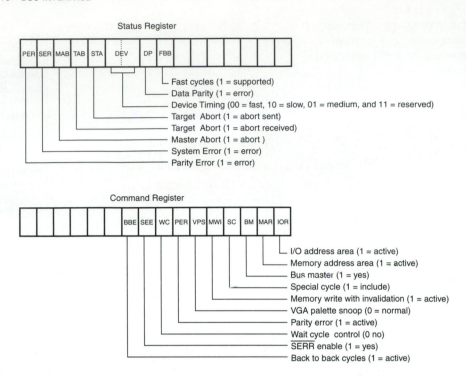

Status Register

- Fast cycles (1 = supported)
- Data Parity (1 = error)
- Device Timing (00 = fast, 10 = slow, 01 = medium, and 11 = reserved)
- Target Abort (1 = abort sent)
- Target Abort (1 = abort received)
- Master Abort (1 = abort)
- System Error (1 = error)
- Parity Error (1 = error)

Command Register

- I/O address area (1 = active)
- Memory address area (1 = active)
- Bus master (1 = yes)
- Special cycle (1 = include)
- Memory write with invalidation (1 = active)
- VGA palette snoop (0 = normal)
- Parity error (1 = active)
- Wait cycle control (0 no)
- SERR enable (1 = yes)
- Back to back cycles (1 = active)

contain either the 32- or 64-bit base address for the memory present on the PCI interface. The next doubleword contains the base address of the I/O space. Note that, even though the Intel microprocessors only use a 16-bit I/O address, there is room for expanding the I/O address to 32 bits. This allows systems that use the 680X0 family and PowerPC access to the PCI bus because they do have I/O space that is accessed via a 32-bit address. The 600X0 and PowerPC use memory mapped I/O, discussed at the beginning of Chapter 11.

Bios for PCI

Most modern Pentium–Pentium 4-based personal computers contain the PCI bus and an extension to the normal system BIOS that supports the PCI bus. These newer systems contain access to the PCI bus at interrupt vector 1AH. Table 15–6 lists the functions currently available through the INT 1AH instruction with AH = B1H for the PCI bus.

Example 15–7 show how the BIOS is used to determine whether the PCI bus extension is available.

EXAMPLE 15–7

```
                              .MODEL SMALL
0000                          .DATA
0000  50 43 49 20 42 49   MES1    DB    'PCI BIOS NOT PRESENT$'
      4F 53 20 4E 4F 54
      20 50 52 45 53 45
      4E 54 24
0015  50 43 49 20 42 49   MES2    DB    'PCI BIOS PRESENT$'
      4F 53 20 50 52 45
      53 45 4E 54 24
0000                          .CODE
                              .STARTUP
0017  B4 B1                       MOV    AH,0B1H        ;access PCI extension
```

TABLE 15–6   BIOS INT1AH functions for the PCI bus.

01H	BIOS Available?
Entry	AH = 0B1H
	AL = 01H
Exit	AH = 00H if PCI BIOS extension is available
	BX = version number
	EDX = ASCII string 'PCI '
	CARRY = 1 if no PCI extension present

02H	PCI Unit Search
Entry	AH = 0B1H
	AL = 02H
	CX = Unit
	DX = Manufacturer
	SI = index
Exit	AH = result code (see notes)
	BX = bus and unit number
	Carry = 1 for error
Notes	The result codes are:
	00H = successful search
	81H = function not supported
	83H = invalid manufacturer ID code
	86H = unit not found
	87H = invalid register number

03H	PCI Class Code Search
Entry	AH = 0B1H
	AL = 03H
	ECX = class code
	SI = index
Exit	AH = result code (see notes for function 02H)
	BX = bus and unit number
	Carry = 1 for an error

06H	Start Special Cycle
Entry	AH = 0B1H
	AL = 06H
	BX = bus and unit number
	EDX = data
Exit	AH = result code (see notes for function 02H)
	Carry = 1 for error
Notes	The value passed in EDX is sent to the PCI bus during the address phase.

08H	Configuration Byte-Sized Read
Entry	AH = 0B1H
	AL = 08H
	BX = bus and unit number
	DI = register number
Exit	AH = result code (see notes for function 02H)
	CL = data from configuration register
	Carry = 1 for error

(continued on next page)

TABLE 15–6   *(continued)*

09H	Configuration Word-Sized Read
Entry	AH = 0B1H AL = 08H BX = bus and unit number DI = register number
Exit	AH = result code (see notes for function 02H) CX = data from configuration register Carry = 1 for error

0AH	Configuration Doubleword-Sized Read
Entry	AH = 0B1H AL = 08H BX = bus and unit number DI = register number
Exit	AH = result code (see notes for function 02H) ECX = data from configuration register Carry = 1 for error

0BH	Configuration Byte-Sized Write
Entry	AH = 0B1H AL = 08H BX = bus and unit number CL = data to be written to configuration register DI = register number
Exit	AH = result code (see notes for function 02H) Carry = 1 for error

0CH	Configuration Word-Sized Write
Entry	AH = 0B1H AL = 08H BX = bus and unit number CX = data to be written to configuration register DI = register number
Exit	AH = result code (see notes for function 02H) Carry = 1 for error

0DH	Configuration Doubleword-Sized Write
Entry	AH = 0B1H AL = 08H BX = bus and unit number ECX = data to be written to configuration register DI = register number
Exit	AH = result code (see notes for function 02H) Carry = 1 for error

FIGURE 15–14   The block diagram of the PCI interface.

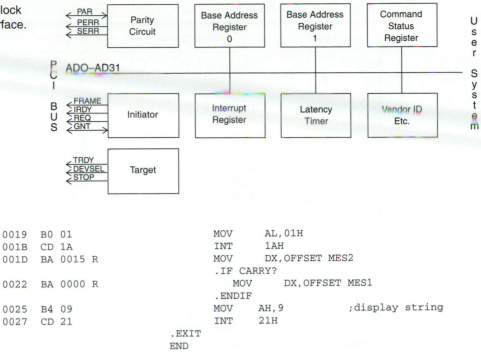

```
0019  B0 01              MOV      AL,01H
001B  CD 1A              INT      1AH
001D  BA 0015 R          MOV      DX,OFFSET MES2
                         .IF CARRY?
0022  BA 0000 R              MOV      DX,OFFSET MES1
                         .ENDIF
0025  B4 09              MOV      AH,9           ;display string
0027  CD 21              INT      21H
                         .EXIT
                         END
```

Once the presence of the BIOS is established, the contents of the configuration memory can be read using the BIOS functions. Note that the BIOS does not support data transfers between the computer and the PCI interface. Data transfers are handled by drivers that are provided with the interface. These drivers control the flow of data between the microprocessor and the component found on the PCI interface.

PCI Interface

The PCI interface is complex. It requires memory (EPROM) to store vendor information and other information, as explained earlier in this section of the chapter. The basic structure of the PCI interface is illustrated in Figure 15–14. The contents of this block diagram illustrate the required components for a functioning PCI interface; it does not illustrate the interface itself. The Registers, Parity Block, Initiator, Target, and Vendor ID EPROM are required components of any PCI interface. If a PCI interface is constructed, a PCI controller is often used because of the complexity of this interface. The PCI controller provides the structures shown in Figure 15–14.

15–4   THE PARALLEL PRINTER INTERFACE (LPT)

The parallel printer interface (LPT) is located on the rear of the personal computer, and as long as it is a part of the PC it can be used as an interface to the PC. LPT stands for Line Printer. The printer interface gives the user access to eight lines that can be programmed to receive or send data.

Port Details

The parallel port (LPT1) is normally at I/O port addresses 378H, 379H, and 37AH. The secondary (LPT2) port, if present, is located at I/O port addresses 278H, 279H, and 27AH. The following information applies to both ports, but LPT1 port addresses are used throughout.

TABLE 15–7   Parallel Port (LPT) pin and signal connections.

Signal	Description	25-pin	36-pin
#STR	Strobe to printer	1	1
D0	Data bit 0	2	2
D1	Data bit 1	3	3
D2	Data bit 2	4	4
D3	Data bit 3	5	5
D4	Data bit 4	6	6
D5	Data bit 5	7	7
D6	Data bit 6	8	8
D7	Data bit 7	9	9
#ACK	Acknowledge from printer	10	10
BUSY	Busy from printer	11	11
PAPER	Out of paper	12	12
ONLINE	Printer is online	13	13
#ALF	Low if printer issues a LF after a CR	14	14
#ERROR	Printer error	15	32
#RESET	Resets the printer	16	31
#SEL	Selects the printer	17	36
+5V	5V from printer	—	18
Protective Ground	Earth ground	—	17
Signal Ground	Signal Ground	All other pins	All other pins

Note: # indicates an active low signal.

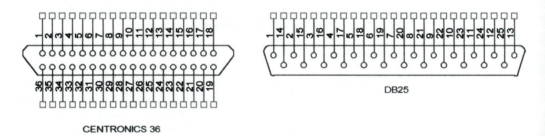

CENTRONICS 36

DB25

FIGURE 15–15   The connectors used for the parallel port.

The Centronics Interface implemented by the parallel port uses two connectors, a 25-pin D-type on the back of the PC and a 36-pin Centronics on the back of the printer. The pin-outs of these connectors are listed in Table 15–7, and the connectors are shown in Figure 15–15.

The parallel port can work as both a receiver and a transmitter at its data pins (D0–D7). This allows devices other than printers, such as CD-ROMs, to be connected to and used by the PC through the parallel port. Anything that can receive and/or send data through an 8-bit interface can and often does connect to the parallel port (LPT1) of a PC.

Figure 15–16 illustrates the contents of the data port (378H), the status register (379H), and an additional status port (37AH). Some of the status bits are true when they are a logic zero.

Using the Parallel Port Without ECP Support

In most systems since the PS/2 was released by IBM, you can basically follow the information presented in Figure 15–16 to use the parallel port without ECP. To read the port, you must first intialize it by sending a 20H to register 37AH as illustrated in Example 15–8.

FIGURE 15–16 Ports 378H, 379H, and 37AH as used by the parallel port.

Port 378H

The data port that connects to bits D0–D7 (pins 2–9)

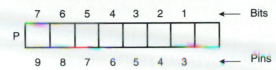

Port 379H

This is a read-only port that returns the information from the printer through signals such as BUSY, #ERROR, and so forth. (Careful! Some of the bits are inverted.)

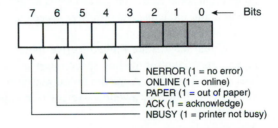

Port 37AH

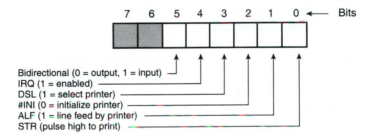

EXAMPLE 15–8

```
MOV     AL,20H
MOV     DX,37AH
OUT     DX,AL
```

Once the port is initialized, it can be read as illustrated in Example 15–9. Notice that reading the port is very easy.

EXAMPLE 15–9

```
MOV     DX,378H
IN      AL,DX
```

To write data to the port, it must be initialized as in Example 15–8, but instead of sending it a 20H we send it a 00H to set it up for writing data. Example 15–10 shows how to write data to the parallel port.

EXAMPLE 15–10

```
MOV     DX,378H
MOV     AL,WRITE_DATA
OUT     DX,AL
```

On some older systems without a bidirectional bit you may need to output an 0FFH to port 378H before you begin to input information. This action takes the place of programming the control register 37AH. Once the FFH is written, you can read the port.

THE UNIVERSAL SERIAL BUS (USB)

The universal serial bus (USB) has solved a problem with the personal computer system. The current ISA sound cards use the internal PC power supply, which generates a tremendous amount of noise. Because the USB allows the sound card to have its own power supply, the noise associated with the PC power supply can be eliminated, allowing for high-fidelity sound. Other benefits are ease of user connection and access to up to 127 different connections through a four-connection serial cable. This interface is ideal for keyboards, sound cards, simple video-retrieval devices, and modems. Data transfer speeds are 12 Mbps for full-speed operation and 1.5 Mbps for slow-speed operation.

Cable lengths are limited to five meters maximum for the full-speed interface and three meters maximum for the low-speed interface. The maximum power available through these cables is rated at 100 mA maximum current at 5.0 V. If the amount of current exceeds 100 mA, Windows will display a yellow exclamation point next to the device.

The Connector

Figure 15–17 illustrates the pin-out of the USB connector. There are two types of female connectors specified and both are in use. In either case, there are four pins on each connector, which contain the signals indicated in Table 15–8. As mentioned, the +5.0 V and ground signals can be used to power devices connected to the bus as long as the amount of current does not exceed 100 mA per device. The data signals are biphase signals. When +data are at 5.0 V, –data are at zero volts and vice versa.

USB Data

The data signals are biphase signals that are generated using a circuit such as the one illustrated in Figure 15–18. The line receiver is also illustrated in Figure 15–18. Placed on the transmission pair is a noise-suppression circuit that is available from Texas Instruments (SN75240).

FIGURE 15–17   The front view of the two common types of USB connectors.

Front View

Front View

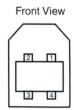

TABLE 15–8   USB pin configuration.

Pin Number	Signal
1	5.0 V
2	– Data
3	+ Data
4	Ground

FIGURE 15–18 The interface to the USB using a pair of CMOS buffers.

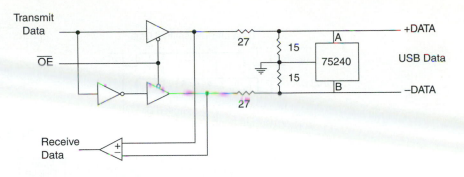

FIGURE 15–19 NRZI encoding used with the USB.

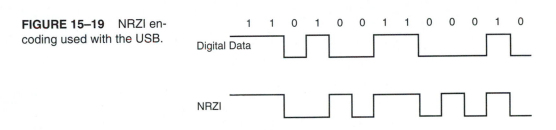

Once the transceiver is in place, interfacing to the USB is complete. The 75773 integrated circuit from Texas Instruments functions as both the differential line driver and receiver for this schematic.

The next phase is learning how the signals interact on the USB. These signals allow data to be sent and received from the host computer system. The USB uses NRZI (non-return to zero, inverted) data encoding for transmitting packets. This encoding method does not change the signal level for the transmission of a logic 1, but the signal level is inverted for each change to a logic 0. Figure 15–19 illustrates a digital data stream and the USB signal produced using this encoding method.

The actual data transmitted includes sync bits using a method called *bit stuffing*. If a logic 1 is transmitted for more than six bits in a row, the bit stuffing technique adds an extra bit (logic 0) after six continuous 1s in a row. Because this lengthens the data stream, it is called bit stuffing. Figure 15–20 shows a bit-stuffed serial data stream and the algorithm used to create it from raw digital serial data. Bit stuffing ensures that the receiver can maintain synchronization for long strings of 1s. Data are always transmitted beginning with the least-significant bit first, followed by subsequent bits.

USB Commands

Now that the USB data format is understood, the commands used to transfer data and select the receptor are discussed. To begin communications, the sync byte (80H) is transmitted first, followed by the packet identification byte (PID). The PID contains eight bits, but only the rightmost four bits contain the type of packet that follows, if any. The leftmost four bits of the PID are the ones complementing the rightmost four bits. For example, if a command of 1000 is sent, the actual byte sent for the PID is a 0111 1000. Table 15–9 shows the available 4-bit PIDs and their 8-bit codes. Notice that there are PIDs used as token indicators, as data indicators, and for handshaking.

Figure 15–21 lists the formats of the data, token, handshaking, and start-of-frame packets found on the USB. In the token packet, the ADDR (address field) contains the 7-bit address of the USB device. As mentioned earlier, there are up to 127 devices present on the USB at a time. The ENDP (endpoint) is a 4-bit number used by the USB. Endpoint 0 is used for initialization, while other endpoint numbers are unique to each USB device.

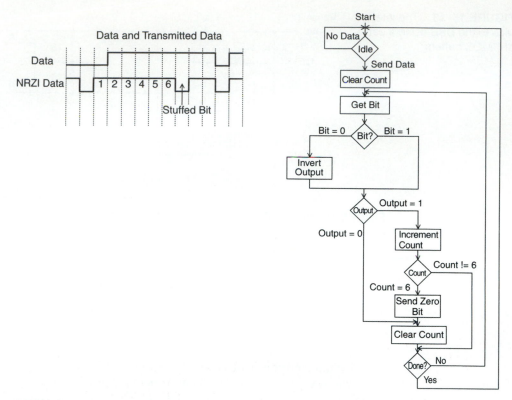

FIGURE 15–20   The data stream and the flow chart used to generate USB data.

TABLE 15–9   PID codes.

PID	Name	Type	Description
E1H	OUT	Token	Host $\rightarrow$ function transaction
D2H	ACK	Handshake	Receiver accepts packet
C3H	Data0	Data	Data packet PID even
A5H	SOF	Token	Start of frame
69H	IN	Token	Function $\rightarrow$ host transaction
5AH	NAK	Handshake	Receiver does not accept data
4BH	Data1	Data	Data packet PID odd
3CH	PRE	Special	Host preamble
2DH	Setup	Token	Setup command
1EH	Stall	Token	Stalled

There are two types of CRC (cyclic redundancy checks) used on the USB: one is a 5-bit CRC and the other (used for data packets) is a 16-bit CRC. The 5-bit CRC is generated with the $X^5 + X^2 + 1$ polynomial; the 16-bit CRC is generated with the $X^{16} + X^{15} + X^2 + 1$ polynomial. When constructing circuitry to generate or detect the CRC, the plus signs represent exclusive-OR circuits. Note that a CRC is a serial checking mechanism. When using the 5-bit CRC, a residual of 01100 is received for no error in all five bits of the CRC and the data bits. With the 16-bit CRC, the residual is 1000000000001101 for no error.

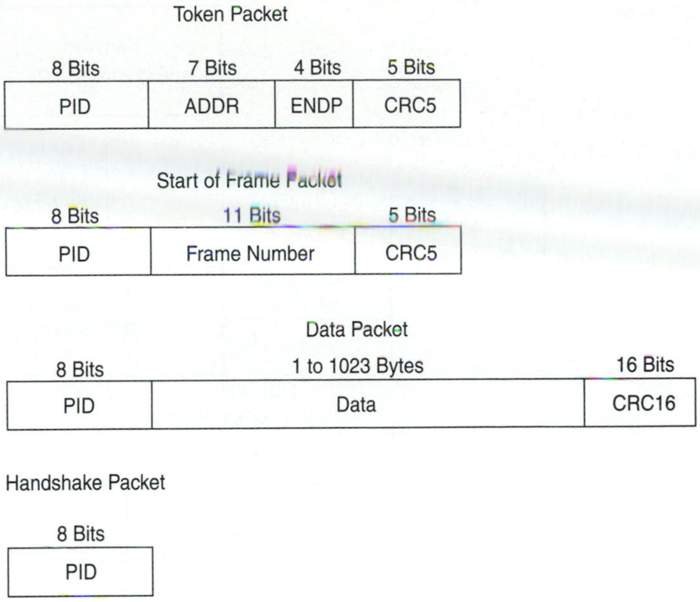

FIGURE 15–21 The types of packets and contents found on the USB.

Token Packet

8 Bits	7 Bits	4 Bits	5 Bits
PID	ADDR	ENDP	CRC5

Start of Frame Packet

8 Bits	11 Bits	5 Bits
PID	Frame Number	CRC5

Data Packet

8 Bits	1 to 1023 Bytes	16 Bits
PID	Data	CRC16

Handshake Packet

8 Bits
PID

The USB uses the ACK and NAK tokens to coordinate the transfer of data packets between the host system and the USB device. Once a data packet is transferred from the host to the USB device, the USB device either transmits an ACK (acknowledge) or a NAK (not acknowledge) token back to the host. If the data and CRC are received correctly, the ACK is sent; if not, the NAK is sent. If the host receives a NAK token, it retransmits the data packet until the receiver finally receives it correctly. This method of data transfer is often called **stop and wait flow control.** The host must wait for the client to send an ACK or NAK before transferring additional data packets.

15–6 ACCELERATED GRAPHICS PORT (AGP)

The latest addition to many computer systems is the inclusion of the **accelerated graphics port** (AGP). The AGP operates at the bus clock frequency of the microprocessor. It is designed so that a transfer between the video card and the system memory can progress at a maximum speed. The AGP can transfer data at a maximum rate of 528M bytes per second. This port probably will never be used for any devices other than the video card, so not much space is devoted to its coverage.

Figure 15–22 illustrates the interface of the AGP to a Pentium 4 system and the placement of other buses in the system. The main advantage of the AGP bus over the PCI bus is that the AGP can sustain transfers (using the 2X compliant system) at speeds up to 528M bytes per second. (The 1X system is now obsolete.) The 4X system, which will be available soon, transfers data at rates of over 1G bytes per second. The PCI bus has a maximum transfer speed of about 100M bytes per second. The AGP is designed specifically to allow high-speed transfers between the video card frame buffer and the system memory through the chip set.

The future will bring the demise of the ISA bus and the incorporation of the USB into the chip set. At present, the system requires the 440LX or 440BX chip set and the ISA-USB bridge chip (PII4X).

FIGURE 15–22 Structure of
a modern computer, illus-
trating all the buses.

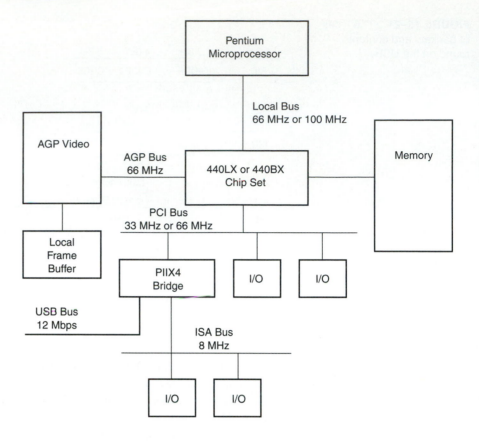

15–7   SUMMARY

1. The bus systems (ISA, EISA, VESA, PCI, and USB) allow I/O and memory systems to be interfaced to the personal computer.
2. The ISA bus is either 8- or 16-bits, and supports either memory or I/O transfers at rates of 8 MHz.
3. The EISA bus is an extended version of the ISA bus that supports 8-, 16-, and 32-bit transfers between the personal computer and memory or I/O at rates of 8 MHz.
4. The VESA (Video Electronics Standards Association) local bus supports 32-bit transfers between the personal computer and I/O or memory at rates of 33 MHz.
5. The PCI (peripheral component interconnect) supports 32- or 64-bit transfers between the personal computer and memory or I/O at rates of 33 MHz. This bus also allows virtually any microprocessor to be interfaced to the PCI bus via the use of a bridge interface.
6. A plug-and-play (PnP)interface is one that contains a memory that holds configuration information for the system.
7. The universal serial bus (USB) promises to replace the ISA bus in the most advanced system. The USB has two data transfer rates: 1.5 Mbps and 12 Mbps.
8. The USB uses the NRZI system to encode data, and uses bit stuffing for logic 1 transmission more than six bits long.
9. The accelerated graphics port (AGP) is a high-speed connection between the memory system and the video graphics card.

15–8 QUESTIONS AND PROBLEMS

1. The letters ISA are an acronym for what phrase?
2. The ISA bus system supports what size data transfers?
3. Is the ISA bus interface often used for memory expansion?
4. Develop an ISA bus interface that is decoded at addresses 800H–803H. This interface must contain an 8255 accessed via these port addresses. (Don't forget to buffer all inputs to the ISA bus card.)
5. Develop an ISA bus interface that decodes ports 0820H–0823H to control a single 8254 timer.
6. Develop a 16-bit ISA bus interface that adds a 27C256 EPROM at memory addresses FF0000H–FF7FFFH.
7. Given a 74LS244 buffer and a 74LS374 latch, develop an ISA bus interface that contains an 8-bit input port at I/O address 808H and an 8-bit output port at I/O address 80AH.
8. Create an ISA bus interface that allows four channels of analog output signals of from 0–5.0 V each. These four channels must be decoded at I/O addresses 800H, 810H, 820H, and 830H. Also develop software that supports the four channels.
9. Redo Question 8, but instead of four output channels, use four ADCs to create four analog input channels at the same addresses.
10. Using an 8254 timer or timers, develop a darkroom timer on an ISA bus card. Your timer must generate a logic 0 for 1/100-second intervals from 1/100 second to five minutes. Use the system clock of 8 MHz as a timing source. The software you develop must allow the user to select the time from the keyboard. The output signal from the timer must be a logic 0 for the duration of the selected time and must be passed through an inverter to enable a solid-state relay that controls the photographic enlarger.
11. Interface a 16550 UART to the personal computer on an ISA bus interface. Develop software that transmits and receives data at Baud rates of 300, 1200, 9600, and 19,200. The UART must respond to I/O ports 0E3XH.
12. The EISA bus can transfer data that are _____ wide at the rate of 8 MHz.
13. Describe how the ISA connector is modified to accommodate the EISA interface.
14. The VESA local bus operates at what rate?
15. Does the VESA local bus use the contacts on the ISA bus?
16. What is the difference between the VESA local bus and the PCI bus?
17. Describe how the address can be captured from the PCI bus.
18. What is the purpose of the configuration memory found on the PCI bus interface?
19. Define the term "plug-and-play."
20. What is the purpose of the C/$\overline{BE}$ connection on the PCI bus system?
21. How is the BIOS tested for the PCI BIOS extension?
22. Develop a short program that interrogates the PCI bus, using the extension to the BIOS, that reads the 32-bit contents of configuration register 08H. For this problem, consider that the bus and unit numbers are 0000H.
23. What advantage does the PCI bus exhibit over the ISA bus?
24. What data rates are available for use on the USB?
25. How are data encoded on the USB?
26. What is the maximum cable length for use with the USB?
27. Will the USB ever replace the ISA bus?
28. How many device addresses are available on the USB?
29. What is NRZI encoding?

30. What is a stuffed bit?
31. If the following raw data are sent on the USB, draw the waveform of the signal found on the USB: (1100110000110011011010)
32. How long can a data packet be on the USB?
33. What is the purpose of the NAK and ACK tokens on the USB?
34. Describe the difference in data transfer rates on the PCI bus when compared with the AGP.
35. What is the transfer rate in a system using a 4X AGP video card?

CHAPTER 16

The 80186, 80188, and 80286 Microprocessors

INTRODUCTION

The Intel 80186/80188 and the 80286 are enhanced versions of the earlier 8086/8088 microprocessors. The 80186/80188 and 80286 are all 16-bit microprocessors that are upward-compatible to the 8086/8088. Even the hardware of these microprocessors is similar to the earlier versions. This chapter presents an overview of each microprocessor, and points out the differences or enhancements that are present in each version. The first part of the chapter describes the 80186/80188 microprocessors, and the last part shows the 80286 microprocessor.

New to this edition is an expanded coverage of the 80186/80188 family. Intel has added four new versions of each of these embedded controllers to its lineup of microprocessors. Each is a CMOS version and designated with a two-letter suffix: XL, EA, EB, and EC. The 80C186XL and 80C188XL models are most similar to the earlier 80186/80188 models.

CHAPTER OBJECTIVES

Upon completion of this chapter, you will be able to:

1. Describe the hardware and software enhancements of the 80186/80188 and the 80286 microprocessors as compared to the 8086/8088.
2. Detail the differences between the various versions of the 80186 and 80188 embedded controllers.
3. Interface the 80186/80188 and the 80286 to memory and I/O.
4. Develop software using the enhancements provided in these microprocessors.
5. Describe the operation of the memory management unit (MMU) within the 80286 microprocessor.
6. Define and detail the operation of a Real Time Operating System (RTOS).

16–1 80186/80188 ARCHITECTURE

The 80186 and 80188, like the 8086 and 8088, are nearly identical. The only difference between the 80186 and 80188 is the width of their data buses. The 80186 (like the 8086) contains a 16-bit data bus, while the 80188 (like the 8088) contains an 8-bit data bus. The internal register structure of the 80186/80188 is virtually identical to the 8086/8088. About the only difference is that the 80186/80188 contain additional reserved interrupt vectors and some very powerful built-in I/O features. The 80186 and 80188 are often called **embedded controllers** because of their application as a controller, not as a microprocessor-based computer.

Versions of the 80186/80188

As mentioned, the 80186 and 80188 are available in four different versions, which are all CMOS microprocessors. Table 16–1 lists each version and the major features provided. The 80C186XL and 80C188XL are the most basic versions of the 80186/80188, while the 80C186EC and 80C188EC are the most advanced. This text details the 80C186XL/80C188XL, and then describes the additional features and enhancements provided in the other versions.

80186 Basic Block Diagram

Figure 16–1 provides the block diagram of the 80188 microprocessor that generically represents all versions except for the enhancements and additional features outlined in Table 16-1. Notice that this microprocessor has a great deal more internal circuitry than the 8088. The block diagrams of the 80186 and 80188 are identical except for the pre-fetch queue, which is four bytes in the 80188 and six bytes in the 80186. Like the 8088, the 80188 contains a bus interface unit (BIU) and an execution unit (EU).

In addition to the BIU and EU, the 80186/80188 family contains a clock generator, a programmable interrupt controller, programmable timers, a programmable DMA controller, and a programmable chip selection unit. These enhancements greatly increase the utility of the 80186/80188 and reduce the number of peripheral components required to implement a system. Many popular subsystems for the personal computer use the 80186/80188 microprocessors as caching

TABLE 16–1   The four versions of the 80186/80188 embedded controller.

Feature	80C186XL 80C188XL	80C186EA 80C188EA	80C186EB 80C188EB	80C186EC 80C188EC
80286-like instruction set	✔	✔	✔	✔
Power-save (green mode)	✔	✔		✔
Power down mode		✔	✔	✔
80C187 interface	✔	✔	✔	✔
ONCE mode	✔	✔	✔	✔
Interrupt controller	✔	✔	✔	✔ 8259-like
Timer unit	✔	✔	✔	✔
Chip selection unit	✔	✔	✔ enhanced	✔ enhanced
DMA controller	✔ 2-channel	✔ 2-channel		✔ 4-channel
Serial communications unit			✔	✔
Refresh controller	✔	✔	✔ enhanced	✔ enhanced
Watchdog timer				✔
I/O ports			✔ 16-bits	✔ 22-bits

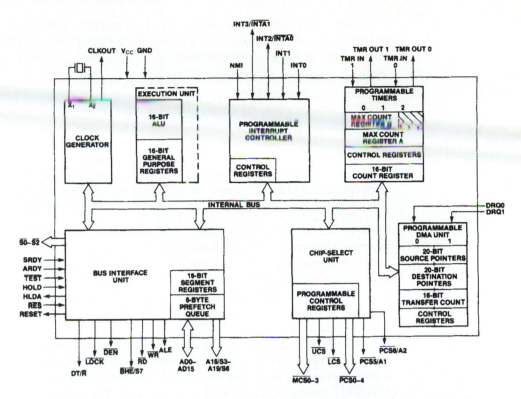

FIGURE 16–1   The block diagram of the 80186 microprocessor. Note that the block diagram of the 80188 is identical, except that $\overline{BHE}$/S7 is missing and AD15–AD8 are relabeled A15–A8. (Courtesy of Intel Corporation.)

disk controllers, local area network (LAN) controllers, etc. The 80186/80188 also finds application in the cellular telephone network as a switcher.

Software for the 80186/80188 is identical to the 80286 microprocessor, without the memory management instructions. This means that the 80286-like instructions immediate multiplication, immediate shift counts, string I/O, PUSHA, POPA, BOUND, ENTER, and LEAVE all function on the 80186/80188 microprocessors.

80186/80188 Basic Features

In this segment of the text, we introduce the enhancements of the 80186/80188 microprocessors or embedded controllers that apply to all versions except where noted, but we do not provide an exclusive coverage. More details on the operation of each enhancement and details of each advanced version are provided later in the chapter.

Clock Generator.   The internal clock generator replaces the external 8284A clock generator used with the 8086/8088 microprocessors. This reduces the component count in a system.

The internal clock generator has three pin connections: X1, X2, and CLKOUT (or on some versions: CLKIN, OSCOUT, and CLKOUT). The X1 (CLKIN) and X2 (OSCOUT) pins are connected to a crystal that resonates at twice the operating frequency of the microprocessor. In the 8 MHz version of the 80186/80188, a 16 MHz crystal is attached to X1 (CLKIN) and X2 (OSCOUT). The 80186/80188 is available in 6 MHz, 8 MHz, 12 MHz, 16 MHz, or 25 MHz versions.

The CLKOUT pin provides a system clock signal that is one-half the crystal frequency, with a 50 percent duty cycle. The CLKOUT pin drives other devices in a system and provides a timing source to additional microprocessors in the system.

In addition to these external pins, the clock generator provides the internal timing for synchronizing the READY input pin, whereas in the 8086/8088 system, READY synchronization is provided by the 8284A clock generator.

Programmable Interrupt Controller.   The programmable interrupt controller (PIC) arbitrates the internal and external interrupts, and controls up to two external 8259A PICs. When an external 8259 is attached, the 80186/80188 microprocessors function as the master and the 8259 functions as the slave. The 80C186EC and 80C188EC models contain an 8259A-compatible interrupt controller in place of the one described here for the other versions (XL, EA, and EB).

If the PIC is operated without the external 8259, it has five interrupt inputs: INT0–INT3 and NMI. Note that the number of available interrupts depends on the version: the EB version has six interrupt inputs and the EC version has 16. This is an expansion from the two interrupt inputs available on the 8086/8088 microprocessors. In many systems, the five interrupt inputs are adequate.

Timers.   The timer section contains three fully programmable 16-bit timers. Timers 0 and 1 generate waveforms for external use, and are driven by either the master clock of the 80186/80188 or by an external clock. They are also used to count external events. The third timer, timer 2, is internal and clocked by the master clock. The output of timer 2 generates an interrupt after a specified number of clocks and can provide a clock to the other timers. Timer 2 can also be used as a watchdog timer because it can be programmed to interrupt the microprocessor after a certain length of time.

The 80C186EC and 80C188EC models have an additional timer called a *watchdog*. The watchdog timer is a 32-bit counter that is clocked internally by the CLKOUT signal (one-half the crystal frequency). Each time the counter hits zero, it reloads and generates a pulse on the $\overline{\text{WDTOUT}}$ pin that is four CLKOUT periods wide. This output can be used for any purpose: it can be wired to the reset input to cause a reset or to the NMI input to cause an interrupt. Note that if it is connected to the reset or NMI inputs, it is periodically reprogrammed so that it never counts down to zero. The purpose of a watchdog timer is to reset or interrupt the system if the software goes awry.

Programmable DMA Unit.   The programmable DMA unit contains two DMA channels or four DMA channels in the 80C186EC/80C188EC models. Each channel can transfer data between memory locations, between memory and I/O, or between I/O devices. This DMA controller is similar to the 8237 DMA controller discussed in Chapter 13. The main difference is that the 8237 DMA controller has four DMA channels, as does the EC model.

Programmable Chip Selection Unit.   The chip selection is a built-in programmable memory and I/O decoder. It has six output lines to select memory, seven lines to select I/O on the XL and EA models, and 10 lines that select either memory or I/O on the EB and EC models.

On the XL and EA models, the memory selection lines are divided into three groups that select memory for the major sections of the 80186/80188 memory-map. The lower memory select signal enables memory for the interrupt vectors, the upper memory select signal enables memory for reset, and the middle memory select signals enable up to four middle memory devices. The boundary of the lower memory begins at location 00000H and the boundary of the upper memory ends at location FFFFFH. The sizes of the memory areas are programmable, and wait states (0–3 waits) can be automatically inserted with the selection of an area of memory.

On the XL and EA models, each programmable I/O selection signal addresses a 128-byte block of I/O space. The programmable I/O area starts at a base I/O address programmed by the user, and all seven 128-byte blocks are contiguous.

On the EB and EC models, there is an upper and lower memory chip selection pin, and eight general-purpose memory or I/O chip selection pins. Another difference is that from 0–15 wait states can be programmed in these two versions of the 80186/80188 embedded controllers.

Power Save/Power Down Feature.   The power save feature allows the system clock to be divided by 4, 8, or 16 to reduce power consumption. The power-saving feature is started by software and exited by a hardware event such as an interrupt. The power down feature stops the clock completely, but it is not available on the XL version. The power down mode is entered by executing a HLT instruction and is exited by any interrupt.

Refresh Control Unit.   The refresh control unit generates the refresh row address at the interval programmed. The refresh control unit does not multiplex the address for the DRAM—this is still the responsibility of the system designer. The refresh address is provided to the memory system at the end of the programmed refresh interval, along with the $\overline{\text{RFSH}}$ control signal. The memory system must run a refresh cycle during the active time of the $\overline{\text{RFSH}}$ control signal. More on memory and refreshing is provided in the section that explains the chip selection unit.

Pin-out

Figure 16–2 illustrates the pin-out of the 80C186XL microprocessor. Note that the 80C186XL is packaged in either a 68-pin lead-less chip carrier (LCC) or in a pin grid array (PGA). The LCC package and PGA packages are illustrated in Figure 16–3.

Pin Definitions.   The following list defines each 80C186XL pin, and notes any differences between the 80C186XL and 80C188XL microprocessors. The enhanced versions are described later in this chapter.

Vcc	This is the system **power** supply connection for ±10%, +5.0 V.
Vss	This is the system **ground** connection.
X1 and X2	These pins are generally connected to a fundamental-mode parallel resonant crystal that operates an internal crystal oscillator. An external clock signal may be connected to the X1 pin. The internal master clock operates at one-half the external crystal or clock input signal. Note that these pins are labeled CLKIN (X1) and OSCOUT (X2) on some versions of the 80186/80188.

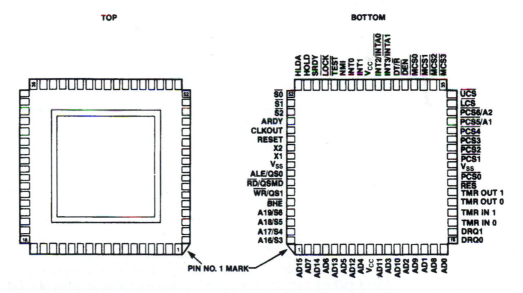

FIGURE 16–2   Pin-out of the 80186 microprocessor. (Courtesy of Intel Corporation.)

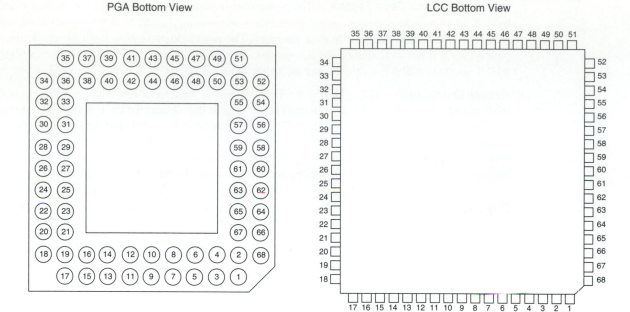

FIGURE 16–3   The bottom views of the PGA and LCC style versions of the 80C188XL microprocessor.

CLKOUT	This pin provides a **timing signal** to system peripherals at one-half the clock frequency with a 50 percent duty cycle.
$\overline{\text{RES}}$	This pin **resets** the 80186/80188. For a proper reset, the $\overline{\text{RES}}$ must be held low for at least 50 ms after power is applied. This pin is often connected to an RC circuit that generates a reset signal after power is applied. The reset location is identical to that of the 8086/8088 microprocessor—FFFF0H.
RESET	The companion **reset output** pin (goes high for a reset) connects to system peripherals to initialize them whenever the $\overline{\text{RES}}$ input goes low.
$\overline{\text{TEST}}$	This **test** pin connects to the BUSY output of the 80187 numeric coprocessor. The $\overline{\text{TEST}}$ pin is interrogated with the WAIT instruction.
TMRIN0, TMRIN1	These pins are used as **external clocking sources** to timers 0 and 1.
TMROUT0 and TMROUT1	These pins provide the **output signals** from timers 0 and 1, which can be programmed to provide square waves or pulses.
DRQ0 and DRQ1	These pins are active-high level triggered **DMA request lines** for DMA channels 0 and 1.
NMI	This is a **non-maskable interrupt** input. It is positive edge-triggered and always active. When NMI is activated, it uses interrupt vector 2.
INT0, INT1, INT2/$\overline{\text{INTA0}}$, and INT3/$\overline{\text{INTA1}}$	These are **maskable interrupt** inputs. They are active-high, and are programmed as either level or edge-triggered. These pins are configured as four interrupt inputs if no external 8259 is present, or as two interrupt inputs if 8259s are present.

A19/$\overline{\text{ONCE}}$, A18, A17, A16	These are multiplexed **address status connections** that provide the address (A19–A16) and status (S6–S3). Status bits found on address pins A18–A16 have no system function and are used during manufacturing for testing. The A19 pin is an input for the $\overline{\text{ONCE}}$ function on a reset. If $\overline{\text{ONCE}}$ is held low on a reset, the microprocessor enters a testing mode.
AD15–AD0	These are multiplexed **address/data bus connections.** During T1, the 80186 places A15–A0 on these pins; during T2, T3, and T4, the 80186 uses these pins as the data bus for signals D15–D0. Note that the 80188 has pins AD7–AD0 and A15–A8.
$\overline{\text{BHE}}$	This pin indicates (when a logic 0) that valid data are transferred through data bus connections D15–D8.
ALE	This is a multiplexed output pin that contains ALE one-half clock cycle earlier than in the 8086. It is used to de-multiplex the address/data and address/status buses. (Even though the status bits on A19–A16 are not used in the system, they must still be de-multiplexed.)
$\overline{\text{WR}}$	This **write** pin causes data to be written to memory or I/O.
$\overline{\text{RD}}$	This **read** pin causes data to be read from memory or I/O.
ARDY	The **asynchronous READY** input informs the 80186/80188 that the memory or I/O is ready for the 80186/80188 to read or write data. If this pin is tied to +5.0 V, the microprocessor functions normally; if it is grounded, the microprocessor enters wait states.
SRDY	The **synchronous READY** input is synchronized with the system clock to provide a relaxed timing for the ready input. Like ARDY, SRDY is tied to +5.0 V for no wait states.
$\overline{\text{LOCK}}$	This **lock** pin is an output controlled by the LOCK prefix. If an instruction is prefixed with LOCK, the $\overline{\text{LOCK}}$ pin becomes a logic 0 for the duration of the locked instruction.
S2, S1, and S0	These are **status bits** that provide the system with the type of bus transfer in effect. See Table 16-2 for the states of the status bits.
$\overline{\text{UCS}}$	The **upper-memory chip select** pin selects memory on the upper portion of the memory map. This output is programmable to enable memory sizes of 1K–256K bytes ending at location FFFFFH. Note that this pin is programmed differently on the EB and EC versions and enables memory between 1K and 1M long.

TABLE 16–2 The S2, S1, and S0 status bits.

S2	S1	S0	Function
0	0	0	Interrupt acknowledge
0	0	1	I/O read
0	1	0	I/O write
0	1	1	Halt
1	0	0	Opcode fetch
1	0	1	Memory read
1	1	0	Memory write
1	1	1	Passive

$\overline{\text{LCS}}$	The **lower-memory chip select** pin enables memory beginning at location 00000H. This pin is programmed to select memory sizes from 1K–256K bytes. Note that this pin functions differently for the EB and EC versions and enables memory between 1K and 1M bytes long.
$\overline{\text{MCS0}}$–$\overline{\text{MCS3}}$	The **middle-memory chip select** pins enable four middle memory devices. These pins are programmable to select an 8K–512K byte block of memory, containing four devices. Note that these pins are not present on the EB and EC versions.
$\overline{\text{PCS0}}$–$\overline{\text{PCS4}}$	These are five different **peripheral selection lines.** Note that the lines are not present on the EB and EC versions.
$\overline{\text{PCS5}}$/A1 and $\overline{\text{PCS6}}$/A2	These are programmed as **peripheral selection lines** or as internally latched address bits A2 and A1. These lines are not present on the EB and EC versions.
DT/$\overline{\text{R}}$	This pin controls the direction of data bus buffers if attached to the system.
$\overline{\text{DEN}}$	This pin **enables** the external data bus buffers.

DC Operating Characteristics

It is necessary to know the DC operating characteristics before attempting to interface or operate the microprocessor. The 80C186/801C88 microprocessors require between 42 mA and 63 mA of power-supply current. Each output pin provides 3.0 mA of logic 0 current and –2 mA of logic 1 current.

80186/80188 Timing

The timing diagram for the 80186 is provided in Figure 16–4. Timing for the 80188 is identical except for the multiplexed address connection, which are AD7–AD0 instead of AD15–AD0, and the $\overline{\text{BHE}}$, which does not exist on the 80188.

The basic timing for the 80186/80188 is composed of four clocking periods just as in the 8086/8088. A bus cycle for the 8 MHz version requires 500 ns, while the 16 MHz version requires 250 ns.

There are very few differences between the timing for the 80186/80188 and the 8086/8088. The most noticeable difference is that ALE appears one-half clock cycle earlier in the 80186/80188.

Memory Access Time.   One of the more important points in any microprocessor's timing diagram is the memory access time. Access time calculations for the 80186/80188 are identical to that of the 8086/8088. Recall that the access time is the time allotted to the memory and I/O to provide data to the microprocessor after the microprocessor sends the memory or I/O its address.

A close examination of the timing diagram reveals that the address appears on the address bus T_{CLAV} time after the start of T1. T_{CLAV} is listed as 44 ns for the 8 MHz version. (See Figure 16–5.) Data are sampled from the data bus at the end of T3, but a setup time is required before the clock defined as T_{DVCL}. The times listed for T_{DVCL} are 20 ns for both versions of the microprocessor. Access time is therefore equal to three clocking periods minus both T_{CLAV} and T_{DVCL}. Access time for the 8 MHz microprocessor is 375 ns – 44 ns – 20 ns, or 311 ns. The access time for the 16 MHz version is calculated in the same manner, except that T_{CLAV} is 25 ns and T_{DVCL} is 15 ns.

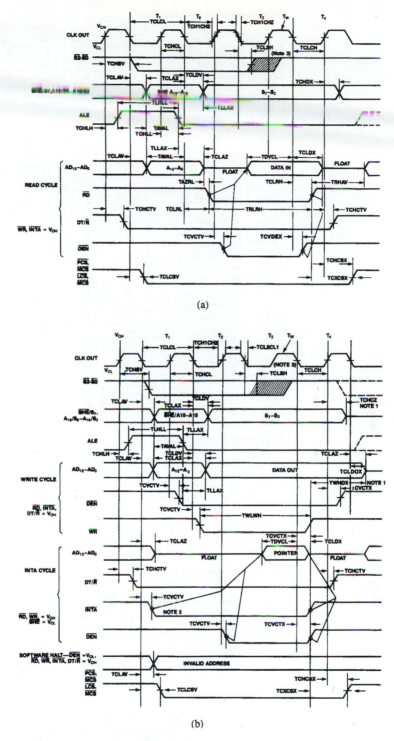

(a)

(b)

FIGURE 16–4   80186/80188 timing. (a) Read cycle timing and (b) write cycle timing. (Courtesy of Intel Corporation.)

80186 Master Interface Timing Responses

Symbol	Parameters	80188 (8 MHz) Min.	Max.	80188-6 (6 MHz) Min.	Max.	Units	Test Conditions
T_{CLAV}	Address Valid Delay	5	44	5	63	ns	C_L = 20-200 pF all outputs
T_{CLAX}	Address Hold	10		10		ns	
T_{CLAZ}	Address Float Delay	T_{CLAX}	35	T_{CLAX}	44	ns	
T_{CHCZ}	Command Lines Float Delay		45		56	ns	
T_{CHCV}	Command Lines Valid Delay (after float)		55		76	ns	
T_{LHLL}	ALE Width	$T_{CLCL-35}$		$T_{CLCL-35}$		ns	
T_{CHLH}	ALE Active Delay		35		44	ns	
T_{CHLL}	ALE Inactive Delay		35		44	ns	
T_{LLAX}	Address Hold to ALE Inactive	$T_{CHCL-25}$		$T_{CHCL-30}$		ns	
T_{CLDV}	Data Valid Delay	10	44	10	55	ns	
T_{CLDOX}	Data Hold Time	10		10		ns	
T_{WHDX}	Data Hold after WR	$T_{CLCL-40}$		$T_{CLCL-50}$		ns	
T_{CVCTV}	Control Active Delay 1	5	70	5	87	ns	
T_{CHCTV}	Control Active Delay 2	10	55	10	76	ns	
T_{CVCTX}	Control Inactive Delay	5	55	5	76	ns	
T_{CVDEX}	DEN Inactive Delay (Non-Write Cycle)		70		87	ns	
T_{AZRL}	Address Float to RD Active	0		0		ns	
T_{CLRL}	RD Active Delay	10	70	10	87	ns	
T_{CLRH}	RD Inactive Delay	10	55	10	76	ns	
T_{RHAV}	RD Inactive to Address Active	$T_{CLCL-40}$		$T_{CLCL-50}$		ns	
T_{CLHAV}	HLDA Valid Delay	10	50	10	67	ns	
T_{RLRH}	RD Width	$2T_{CLCL-50}$		$2T_{CLCL-50}$		ns	
T_{WLWH}	WR Width	$2T_{CLCL-40}$		$2T_{CLCL-40}$		ns	
T_{AVAL}	Address Valid to ALE Low	$T_{CLCH-25}$		$T_{CLCH-45}$		ns	
T_{CHSV}	Status Active Delay	10	55	10	76	ns	
T_{CLSH}	Status Inactive Delay	10	55	10	76	ns	
T_{CLTMV}	Timer Output Delay		60		75	ns	100 pF max
T_{CLRO}	Reset Delay		60		75	ns	
T_{CHQSV}	Queue Status Delay		35		44	ns	

80186 Chip-Select Timing Responses

Symbol	Parameter	Min.	Max.	Min.	Max.	Units	Test Conditions
T_{CLCSV}	Chip-Select Active Delay		66		80	ns	
T_{CXCSX}	Chip-Selct Hold from Command Inactive	35		35		ns	
T_{CHCSX}	Chip-Select Inactive Delay	5	35	5	47	ns	

Symbol	Parameter	Min.	Max.	Units	Test Conditions
TDVCL	Data in Setup (A/D)	20		ns	
TCLDX	Data in Hold (A/D)	10		ns	
TARYHCH	Asynchronous Ready (AREADY) active setup time*	20		ns	
TARYLCL	AREADY inactive setup time	35		ns	
TCHARYX	AREADY hold time	15		ns	
TSRYCL	Synchronous Ready (SREADY) transition setup time	35		ns	
TCLSRY	SREADY transition hold time	15		ns	
THVCL	HOLD Setup*	25		ns	
TINVCH	INTR, NMI, TEST, TIMERIN, Setup*	25		ns	
TINVCL	DRQ0, DRQ1, Setup*	25		ns	

*To guarantee recognition at next clock.

FIGURE 16–5 80186 AC characteristics. (Courtesy of Intel Corporation.)

16-2   PROGRAMMING THE 80186/80188 ENHANCEMENTS

This section provides detail on the programming and operation of the 80186/80188 enhancements of all versions (XL, EA, EB, and EC). The next section details the use of the 80C188EB in a system that uses many of the enhancements discussed here. The only new feature not discussed here is the clock generator, which is described in the previous section on architecture.

Peripheral Control Block

All internal peripherals are controlled by a set of 16-bit wide registers located in the **peripheral control block** (PCB). The PCB (see Figure 16–6) is a set of 256 registers located in the I/O or memory space. Note that this set applies to the XL and EA versions. Later in this section, the EB and EC versions of the PCB are defined and described.

Whenever the 80186/80188 is reset, the peripheral control block is automatically located at the top of the I/O map (I/O addresses FF00H–FFFFH). In most cases, it stays in this area of I/O space, but the PCB may be relocated at any time to any other area of memory or I/O. Relocation is accomplished by changing the contents of the relocation register (see Figure 16–7) located at offset addresses FEH and FFH.

The relocation register is set to a 20FFH when the 80186/80188 is reset. This locates the PCB at I/O addresses FF00H–FFFFH afterwards. To relocate the PCB, the user need only send a

FIGURE 16–6 Peripheral control block (PCB) of the 80186/80188. (Courtesy of Intel Corporation.)

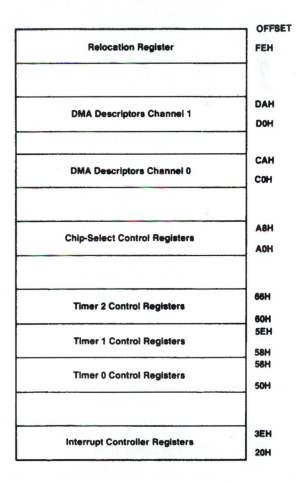

	OFFSET
Relocation Register	FEH
DMA Descriptors Channel 1	DAH
	D0H
DMA Descriptors Channel 0	CAH
	C0H
Chip-Select Control Registers	A8H
	A0H
Timer 2 Control Registers	66H
	60H
Timer 1 Control Registers	5EH
	58H
Timer 0 Control Registers	56H
	50H
Interrupt Controller Registers	3EH
	20H

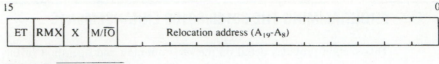

ET = ESC/NO ESC TRAP
RMX = iRM × 86 mode/master mode
M/$\overline{\text{IO}}$ = Memory/IO space
X = Unused

FIGURE 16–7 Peripheral control register.

word OUT to I/O address FFFEH with a new bit pattern. For example, to relocate the PCB to memory locations 20000H–200FFH, a 1200H is sent to I/O address FFFEH. Notice that M/$\overline{\text{IO}}$ is a logic 1 to select memory, and that a 200H selects memory address 20000H as the base address of the PCB. Note that all accesses to the PCB must be word accesses because it is organized as 16-bit wide registers. Example 16–1 shows the software required to relocate the PCB to memory location 20000H–200FFH. Note that either an 8- or 16-bit output can be used to program the 80186; in the 80188, never use the OUT DX,AX instruction because it takes additional clocking periods to execute.

EXAMPLE 16–1

```
0100  BA FFFE        MOV    DX,0FFFEH     ;address relocation register
0103  B8 1200        MOV    AX,1200H      ;code for new PCB location
0106  EE             OUT    DX,AL         ;this can also be OUT DX,AX
```

The EB and EC versions use a different address for programming the PCB location. Both versions have the PCB relocation register stored at offset XXA8H, instead of at offset XXFEH for the XL and EA versions. The bit patterns of these versions is the same as for the XL and EA versions, except that the RMX bit is missing.

Interrupts in the 80186/80188

The interrupts in the 80186/80188 are identical to the 8086/8088, except that there are additional interrupt vectors defined for some of the internal devices. A complete listing of the reserved interrupt vectors appears in Table 16–3. The first five are identical to the 8086/8088.

The array BOUND instruction interrupt is requested if the boundary of an index register is outside the values set up in the memory. The unused opcode interrupt occurs whenever the 80186/80188 executes any undefined opcode. This is important if a program begins to run awry. Note that the unused opcode interrupt can be accessed by an instruction, but the assembler does not include it in the instruction set. On the Pentium Pro/Pentium 4 and some earlier Intel microprocessors, the 0F0BH or 0FB9H instruction will cause the program to call the procedure whose address is stored at the unused opcode interrupt vector.

The ESC opcode interrupt occurs if ESC opcodes D8H–DFH are executed. This occurs only if the ET (escape trap) bit of the relocation register is set. If an ESC interrupt occurs, the address stored on the stack by the interrupt points to the ESC instruction or to its segment override prefix, if one is used.

The internal hardware interrupts must be enabled by the I flag bit and must be unmasked to function. The I flag bit is set (enabled) with STI and cleared (disabled) with CLI. The remaining internally decoded interrupts are discussed with the timers and DMA controller, later in this section.

Interrupt Controller

The interrupt controller inside the 80186/80188 is a sophisticated device. It has many interrupt inputs that arrive from the five external interrupt inputs, the DMA controller, and the three

TABLE 16–3 80186/80188 interrupt vectors.

Name	Type	Address	Priority
Divide error	0	00000H–00003H	1
Single-step	1	00004H–00007H	1A
NMI pin	2	00008H–0000BH	1
Breakpoint	3	0000CH–0000FH	1
Overflow	4	00010H–00013H	1
BOUND instruction	5	00014H–00017H	1
Unused opcode	6	00018H–0001BH	1
ESCape opocde	7	0001CH–0001FH	1
Timer 0	8	00020H–00023H	2A
Reserved	9	00024H–00027H	
DMA 0	10	00028H–0002BH	4
DMA 1	11	0002CH–0002FH	5
INT0	12	00030H–00033H	6
INT1	13	00034H–00037H	7
INT2	14	00038H–0003BH	8
INT3	15	0003CH–0003FH	9
80187	16	00040H–00043H	1
Reserved	17	00044H–00047H	
Timer 1	18	00048H–0004BH	2B
Timer 2	19	0004CH–0004FH	2C
Serial 0 receiver (EB only)	20	00050H–00053H	3A
Serial 0 transmitter (EB only)	21	00054H–00057H	3B

Note: Priority level 1 has the highest priority and level 9 the lowest. Some interrupts have the same priority. Only the EB and EC versions contain the serial unit.

FIGURE 16–8 80186/80188 programmable interrupt controller. (Courtesy of Intel Corporation.)

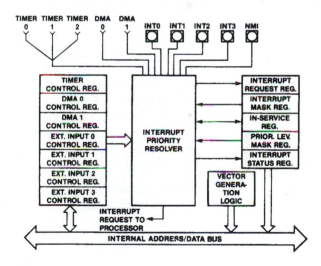

timers. Figure 16–8 provides a block diagram of the interrupt structure of the 80186/80188 interrupt controller. This controller appears in the XL, EA, and EB versions, but the EC version contains the exact equivalent to a pair of 8259As, as found in Chapter 12. In the EB version, the DMA inputs are replaced with inputs from the serial unit for receive and transmit.

The interrupt controller operates in two modes: master and slave mode. The mode is selected by a bit in the interrupt control register (EB and EC versions) called the *CAS bit.* If the CAS bit is a logic 1, the interrupt controller connects to external 8259A programmable interrupt

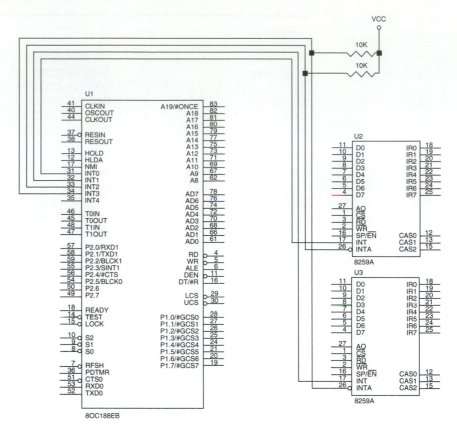

FIGURE 16–9   The interconnection between the 80C188EB and two 8259A programmable interrupt controllers. *Note:* Only the connections vital for this interface are shown.

controllers (see Figure 16–9); if CAS is a logic 0, the internal interrupt controller is selected. In many cases, there are enough interrupts within the 80186/80188, so the slave mode is not normally used. Note that in the XL and EA versions, the master and slave modes are selected in the peripheral control register at offset address FEH.

This portion of the text does not detail the programming of the interrupt controller. Instead, it is limited to a discussion of the internal structure of the interrupt controller. The programming and application of the interrupt controller is discussed in the sections that describe the timer and DMA controller.

Interrupt Controller Registers.   Figure 16–10 illustrates the interrupt controller's registers. These registers are located in the peripheral control block beginning at offset address 22H. For the EC version, which is compatible with the 8259A, the interrupt controller ports are at offset addresses 00H and 02H for the master and ports 04H and 06H for the slave. In the EB version, the interrupt controller is programmed at offset address 02H. Note that the EB version has an additional interrupt input (INT4).

Slave Mode.   When the interrupt controller operates in the slave mode, it uses up to two external 8259A programmable interrupt controllers for interrupt input expansion. Figure 16–9 shows how the external interrupt controllers connect to the 80186/80188 interrupt input pins for slave operation. Here, the INT0 and INT1 inputs are used as external connections to the interrupt request outputs of the 8259s, and $\overline{INTA0}$ (INT2) and $\overline{INTA1}$ (INT3) are used as interrupt acknowledge signals to the external controllers.

XL and EA Versions   EB Version

3EH	INT3 Control Register		1EH	INT3 Control Register
3CH	INT2 Control Register		1CH	INT2 Control Register
3AH	INT1 Control Register		1AH	INT1 Control Register
38H	INT0 Control Register		18H	INT0 Control Register
36H	DMA1 Control Register		16H	INT4 Control Register
34H	DMA0 Control Register		14H	Serial Control Register
32H	Timer Control Register		12H	Timer Control Register
30H	Interrupt Status		10H	Interrupt Status
2EH	Request		0EH	Request
2CH	In Service		0CH	In Service
2AH	PRIMSK		0AH	PRIMSK
28H	Interrupt Masks		08H	Interrupt Masks
26H	POLL Status		06H	POLL Status
24H	POLL		04H	POLL
22H	EOI		02H	EOI

FIGURE 16–10   The I/O offset port assignment for the interrupt control unit.

Interrupt Control Registers.   There are interrupt control registers in both modes of operation, which each control a single interrupt source. Figure 16–11 depicts the binary bit pattern of each of these interrupt control registers. The mask bit enables (0) or disables (1) the interrupt input represented by the control word, and the priority bits set the priority level of the interrupt source. The highest priority level is 000, and the lowest is 111. The CAS bit is used to enable slave or cascade mode (0 enables slave mode), and the SFNM bit selects the special fully nested mode. The SFNM allows the priority structure of the 8259A to be maintained.

Interrupt Request Register.   The interrupt request register contains an image of the interrupt sources in each mode of operation. Whenever an interrupt is requested, the corresponding interrupt request bit becomes a logic 1, even if the interrupt is masked. The request is cleared whenever the 80186/80188 acknowledges the interrupt. Figure 16–12 illustrates the binary bit pattern of the interrupt request register for both the master and slave modes.

Mask and Priority Mask Registers.   The interrupt mask register has the same format as the interrupt register illustrated in Figure 16–12. If a source is masked (disabled), the corresponding bit of the interrupt mask register contains a logic 1; if enabled, it contains a logic 0. The interrupt mask register is read to determine which interrupt sources are masked and which are enabled. A source is masked by setting the source's mask bit in its interrupt control register.

The priority mask register, illustrated in Figure 16–13, shows the priority of the interrupt currently being serviced by the 80186/80188. The level of the interrupt is indicated by priority bits P2–P0. Internally, these bits prevent an interrupt by a lower priority source. These bits are automatically set to the next lower level at the end of an interrupt, as issued by the 80186/80188. If no other interrupts are pending, these bits are set (111) to enable all priority levels.

Timer and Serial Control Registers
15
0
MASK
P2
P1
P0
INT2, INT3, and INT4 Control Registers
15
0
LVL
MASK
P2
P1
P0
INT0 and INT1 Control Registers
15
0
SFNM
CAS
LVL
MASK
P2
P1
P0
P2–P0 = Priority Level
Mask = 0 enables interrupt
LVL = 0 = edge and 1 = level triggering
CAS = 1 selects slave mode
SFNM = 1 selects special fully nested mode

FIGURE 16–11 The interrupt control registers.

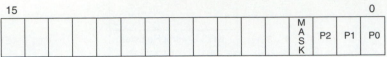

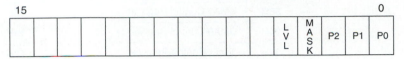

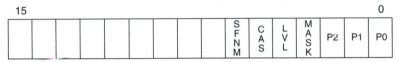

FIGURE 16–12 The interrupt request register.

Interrupt Request Register (EB version)
15
0
INT3
INT2
INT1
INT0
INT4
SER
TIM
Interrupt Request Register (XL and EA versions)
15
0
INT3
INT2
INT1
INT0
DMA1
DMA0
TIM

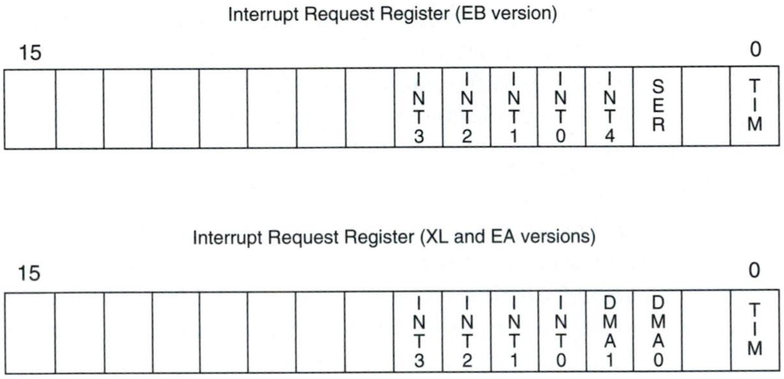

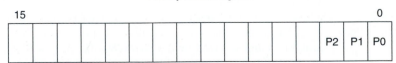

FIGURE 16–13 The priority mask register.

Priority Mask Register
15
0
P2
P1
P0
P2–P0 = Priority Level

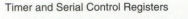

FIGURE 16–14 The poll and poll status registers.

Poll and Poll Status Registers

15											VT4	VT3	VT2	VT1	VT0 0
IREQ											V T 4	V T 3	V T 2	V T 1	V T 0

IREQ = 1 = Interrupt pending
VT4–VT0 = Interrupt type number of highest priority pending interrupt

In-Service Register. The in-service register has the same binary bit pattern as the request register of Figure 16–12. The bit that corresponds to the interrupt source is set if the 80186/80188 is currently acknowledging the interrupt. The bit is reset at the end of an interrupt.

The Poll and Poll Status Registers. Both the interrupt poll and interrupt poll status registers share the same binary bit patterns as those illustrated in Figure 16–14. These registers have a bit (INT REQ) that indicates an interrupt is pending. This bit is set if an interrupt is received with sufficient priority, and cleared when an interrupt is acknowledged. The S bits indicate the interrupt vector type number of the highest priority pending interrupt.

The poll and poll status registers may appear to be identical because they contain the same information. However, they differ in function. When the interrupt poll register is read, the interrupt is acknowledged. When the interrupt poll status register is read, no acknowledge is sent. These registers are used only in the master mode, not in the slave mode.

End-of-interrupt Register. The end-of-interrupt (EOI) register causes the termination of an interrupt when written by a program. Figure 16–15 shows the contents of the EOI register for both the master and slave mode.

In the master mode, writing to the EOI register ends either a specific interrupt level or whatever level is currently active (nonspecific). In the nonspecific mode, the NSPEC bit must be set before the EOI register is written to end a nonspecific interrupt. The nonspecific EOI clears the highest level interrupt bit in the in-service register. The specific EOI clears the selected bit in the in-service register. The nonspecific mode is used unless there is a special circumstance that requires a different order for interrupt acknowledges.

In the slave mode, the level of the interrupt to be terminated is written to the EOI register. The slave mode does not allow a nonspecific EOI.

Interrupt Status Register. The format of interrupt status register is depicted in Figure 16–16. In the master mode, T2–T0 indicates which timer (timer 0, timer 1, or timer 2) is causing an interrupt. This is necessary because all three timers have the same interrupt priority level. These bits are set when the timer requests an interrupt and are cleared when the interrupt is acknowledged. The DHLT (DMA halt) bit is only used in the master mode; when set, it stops a DMA action. Note that the interrupt status register is different for the EB version.

Interrupt Vector Register. The interrupt vector register is present only in the slave mode, and only in the XL and EA versions at offset address 20H. It is used to specify the most significant five bits of the interrupt vector type number. Figure 16–17 illustrates the format of this register.

FIGURE 16–15 The end-of-interrupt (EOI) register.

End-of-Interrupt Register

15											VT4	VT3	VT2	VT1	VT0 0
NSPEC											V T 4	V T 3	V T 2	V T 1	V T 0

FIGURE 16–16 The interrupt status register.

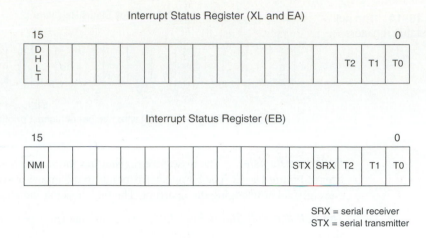

Interrupt Status Register (XL and EA)

15													T2	T1	T0
DHLT															0

Interrupt Status Register (EB)

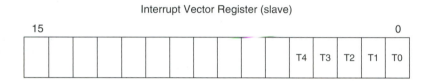

15										STX	SRX	T2	T1	T0
NMI														0

SRX = serial receiver
STX = serial transmitter

FIGURE 16–17 The interrupt vector register.

Interrupt Vector Register (slave)

15											T4	T3	T2	T1	T0
															0

Timers

The 80186/80188 microprocessors contain three fully programmable 16-bit timers and each is totally independent of the others. Two of the timers (timer 0 and timer 1) have input and output pins that allow them to count external events or generate wave-forms. The third timer (timer 2) connects to the 80186/80188 clock. Timer 2 is used as a DMA request source, as a prescaler for other timers, or as a watchdog timer.

Figure 16–18 shows the internal structure of the timer unit. Notice that the timer unit contains one counting element that is responsible for updating all three counters. Each timer is actually a register that is rewritten from the counting element (a circuit that reads a value from

FIGURE 16–18 Internal structure of the 80186/80188 timers. (Courtesy of Intel Corporation.)

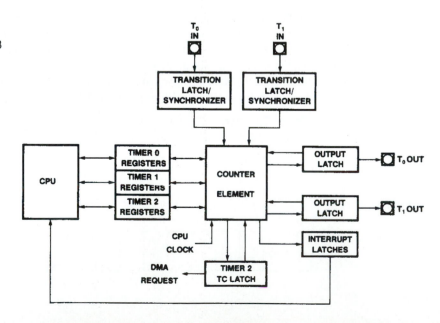

a timer register and increments it before returning it). The counter element is also responsible for generating the outputs on the pins T0OUT and T1OUT, reading the T0IN and T1IN pins, and causing a DMA request from the terminal count (TC) of timer 2 if timer 2 is programmed to request a DMA action.

Timer Register Operation.   The timers are controlled by a block of registers in the peripheral control block (see Figure 16–19). Each timer has a count register, maximum-count register or registers, and a control register. These registers may all be read or written at any time because the 80186/80188 microprocessors ensure that the contents never change during a read or write.

The timer count register contains a 16-bit number that is incremented whenever an input to the timer occurs. Timers 0 and 1 are incremented at the positive edge on an external input pin, every fourth 80186/80188 clock, or by the output of timer 2. Timer 2 is clocked on every fourth 80186/80188 clock pulse and has no other timing source. This means that in the 8 MHz version of the 80186/80188, timer 2 operates at 2 MHz, and the maximum counting frequency of timers 0 and 1 is 2 MHz. Figure 16–20 depicts these four clocking periods, which are not related to the bus timing.

FIGURE 16–19 The offset locations and contents of the registers used to control the timers.

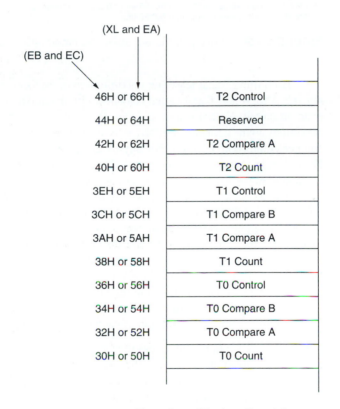

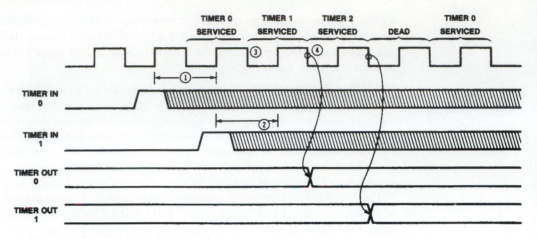

1. Timer in 0 resolution time
2. Timer in 1 resolution time
3. Modified count value written into 80186 timer 0 count register
4. Modified count value written into 80186 time

FIGURE 16–20   Timing for the 80186/80188 timers. (Courtesy of Intel Corporation.)

Each timer has at least one maximum-count register, called a **compare** register (compare register A for timers 0 and 1) that is loaded with the maximum count of the count register to generate an output. Note that a timer is an up counter. Whenever the count register is equal to the maximum-count compare register, it is cleared to 0. With a maximum count of 0000H, the counter counts 65,536 times. For any other value, the timer counts the true value of the count. For example, if the maximum count is 0002H, then the counter will count from 0 to 1 and then be cleared to 0—a modulus 2 counter has 2 states.

Timers 0 and 1 each have a second maximum-count compare register (compare register B) that is selected by the control register for the timer. Either maximum-count compare register A or both maximum-count compare registers A and B are used with these timers, as programmed by the ALT bit in the control register for the timer. When both maximum-count compare registers are used, the timer counts up to the value in maximum-count compare register A, clears to 0, and then counts up to the count in maximum-count compare register B. This process is then repeated. Using both maximum-count registers allows the timer to count up to 131,072.

The control register (refer to Figure 16–19) of each timer is 16 bits wide and specifies the operation of the timer. A definition of each control bit follows:

EN   The **enable** bit allows the timer to start counting. If EN is cleared, the timer does not count; if it is set, the timer counts.

INH   The **inhibit** bit allows a write to the timer control register to affect the enable bit (EN). If INH is set, then the EN bit can be set or cleared to control the counting. If INH is cleared, EN is not affected by a write to the timer control register. This allows other features of the timer to be modified without enabling or disabling the timer.

INT   The **interrupt** bit allows an interrupt to be generated by the timer. If INT is set, an interrupt will occur each time that the maximum count is reached in either maximum-count compare register. If this bit is cleared, no interrupt is generated. When the interrupt request is generated, it remains in force, even if the EN bit is cleared after the interrupt request.

RIU   The **register in use** bit indicates which maximum-count compare register is currently in use by the timer. If RIU is a logic 0, then maximum-count compare register A is in use. This bit is a read-only bit, and writes do not affect it.

MC   The **maximum count** bit indicates that the timer has reached its maximum count. This bit becomes a logic 1 when the timer reaches its maximum count and remains a logic 1 until the MC bit is cleared by writing a logic 0. This allows the maximum count to be detected by software.

RTG   The **re-trigger** bit is active only for external clocking (EXT = 0). The RTG bit is used only with timers 0 and 1 to select the operation of the timer input pins (T0IN and T1IN). If RTG is a logic 0, the external input will cause the timer to count if it is a logic 1; the timer will hold its count (stop counting) if it is a logic 0. If RTG is a logic 1, the external input pin clears the timer count to 0000H each time a positive-edge occurs.

P   The **prescaler** bit selects the clocking source for timers 0 and 1. If $EXT = 0$ and $P = 0$, the source is one-fourth the system clock frequency. If $P = 1$, the source is timer 2.

EXT   The **external** bit selects internal timing ($EXT = 0$) or external timing ($EXT = 1$). If $EXT = 1$, the timing source is applied to the T0IN or T1IN pins. In this mode, the timer increments after each positive-edge on the timer input pin. If $EXT = 0$, the clocking source is from one of the internal sources.

ALT   The **alternate** bit selects single maximum-count mode (maximum-count compare register A) if a logic 0, or alternate maximum-count mode (maximum-count compare registers A and B) if a logic 1.

CONT   The **continuous** bit selects continuous operation if a logic 1. In continuous operation, the counter automatically continues counting after it reaches its maximum count. If CONT is a logic 0, the timer will automatically stop counting and clear the EN bit. Note that whenever the 80186/80188 are reset, the timers are automatically disabled.

Timer Output Pin.   Timers 0 and 1 have an output pin used to generate either square waves or pulses. To produce pulses, the timer is operated in single maximum-count mode ($ALT = 0$). In this mode, the output pin goes low for one clock period when the counter reaches its maximum count. By controlling the CONT bit in the control register, either a single pulse or continuous pulses can be generated.

To produce square waves or varying duty cycles, the alternate mode ($ALT = 1$) is selected. In this mode, the output pin is a logic 1 while maximum-count compare register A controls the timer; it is a logic 0 while maximum-count compare register B controls the timer. As with the single maximum-count mode, the timer can generate either a single square wave or continuous square waves. See Table 16–4 for the function of the ALT and CONT control bits.

Almost any duty cycle can be generated in the alternate mode. For example, suppose that a 10 percent duty cycle is required at a timer output pin. Maximum-count register A is loaded with a 10 and maximum-count register B is loaded with a 90 to produce an output that is a logic 1 for 10 clocks and a logic 0 for 90 clocks. This also divides the frequency of the timing source by a factor of 100.

Real-Time Clock Example.   Many systems require the time of day. This is often called a **real-time clock.** A timer within the 80186/80188 can provide the timing source for software that maintains the time of day.

TABLE 16–4 Function of ALT and CONT in the timer control register.

ALT	CONT	Mode
0	0	Single pulse
0	1	Continuous pulses
1	0	Single square wave
1	1	Continuous square waves

The hardware required for this application is not illustrated. All that is required is that the T1IN pin be connected to +5.0 V through a pull-up resistor to enable timer 1. In the example, timers 1 and 2 are used to generate a 1-second interrupt that provides the software with a timing source.

The software required to implement a real-time clock is listed in Examples 16–2 and 16–3. Example 16–2 illustrates the software required to initialize the timers. Example 16–3 shows an interrupt service procedure, which keeps time. There is a another procedure in Example 16–3 that increments a BCD modulus counter. None of the software required install the interrupt vector, and time of day is illustrated here.

EXAMPLE 16–2

```
;software for a real-time clock using the 80C188EA microprocessor

= FF62              T2_CA   EQU 0FF62H          ;address of timer 2 compare A
= FF66              T2_CON  EQU 0FF66H          ;address of timer 2 control
= FF60              T2_CNT  EQU 0FF60H          ;address of timer 2 count
= FF5A              T1_CA   EQU 0FF5AH          ;address of timer 1 compare A
= FF58              T1_CON  EQU 0FF58H          ;address of timer 1 control
= FF5E              T1_CNT  EQU 0FF5EH          ;address of timer 1 count

0010                CLOCK_UP    PROC    FAR

0010  B8 4E20               MOV     AX,20000    ;count for timer 2
0013  BA FF62               MOV     DX,T2_CA    ;address timer 2 compare A
0016  EE                    OUT     DX,AL       ;program for 10 ms

0017  B8 0064               MOV     AX,100      ;count for timer 1
001A  BA FF5A               MOV     DX,T1_CA    ;address timer 1 compare A
001D  EE                    OUT     DX,AL       ;program for 1 second

001E  B8 0000               MOV     AX,0
0021  BA FF60               MOV     DX,T2_CNT   ;address timer 2 count
0024  EE                    OUT     DX,AL       ;clear count

0025  BA FF5E               MOV     DX,T1_CNT   ;address timer 1 count
0028  EE                    OUT     DX,AL       ;clear count
0029  B8 C001               MOV     AX,0C001H   ;enable timer 2
002C  BA FF66               MOV     DX,T2_CON   ;address timer 2 control

002F  EE                    OUT     DX,AL
0030  B8 E009               MOV     AX,0E009H   ;enable timer 1
0033  BA FF58               MOV     DX,T1_CON   ;address timer 1 control
0036  EE                    OUT     DX,AL

0037  CB                    RET

0038                CLOCK_UP    ENDP

                    END
```

Timer 2 is programmed to divide by a factor of 20,000. This causes the clock (2 MHz on the 8 MHz version of the 80186/80188) to be divided down to one pulse every 10 ms. The clock for timer 1 is derived internally from the timer 2 output. Timer 1 is programmed to divide by 100 and generates a pulse once per second. The control register of timer 1 is programmed so that the one-second pulse internally generates an interrupt.

The interrupt service procedure is called once per second to keep time. The interrupt service procedure adds a one to the content of memory location SECONDS. Once every 60 seconds, the content of the next memory location (SECONDS + 1) is incremented. Finally, once per hour, the content of memory location SECONDS + 2 is incremented. The time is stored in these three consecutive memory locations in BCD, so the system software can easily access the time.

EXAMPLE 16–3

```
0000   00            SECONDS   DB    ?                      ;time
0001   00            MINUTES   DB    ?
0002   00            HOURS     DB    ?

0100                 INTRS     PROC  FAR USES AX SI

0102   BE 0000 R               MOV   SI,OFFSET SECONDS ;address time
0105   B4 60                   MOV   AH,60H
0107   E8 000F                 CALL  UP_COUNT               ;increment seconds
010A   75 0A                   JNZ   ENDI
010C   E8 000A                 CALL  UP_COUNT               ;increment minutes
010F   75 05                   JNZ   ENDI
0111   B4 24                   MOV   AH,24H
0113   E8 0003                 CALL  UP_COUNT               ;increment hours
0116                 ENDI:
                               IRET

0119                 INTRS     ENDP
0119                 UP_COUNT  PROC  NEAR

0119   2E: 8A 04               MOV   AL,CS:[SI]             ;get count
011C   46                      INC   SI
011D   04 01                   ADD   AL,1                   ;increment count
011F   27                      DAA                          ;make it BCD
0120   2E: 88 44 FF            MOV   CS:[SI-1],AL           ;save new count
0124   2A C4                   SUB   AL,AH                  ;test modulus
0126   75 04                   JNE   ENDU                   ;if no roll-over needed
0128   2E: 88 44 FF            MOV   CS:[SI-1],AL           ;clear count
012C                 ENDU:
012C   C3                      RET

012D                 UP_COUNT  ENDP

                               END
```

DMA Controller

The DMA controller within the 80186/80188 has two fully independent DMA channels. Each has its own set of 20-bit address registers, so any memory or I/O location is accessible for a DMA transfer. In addition, each channel is programmable for auto-increment or auto-decrement to either source or destination registers. This controller is not available in the EB or EC versions. The EC version contains a modified four-channel DMA controller, while the EB version contains no DMA controller. This text does not describe the DMA controller within the EC version.

Figure 16–21 illustrates the internal register structure of the DMA controller. These registers are located in the peripheral control block at offset addresses C0H–DFH.

Notice that both DMA channel register sets are identical; each channel contains a control word, a source and destination pointer, and a transfer count. The transfer count is 16 bits wide and allows unattended DMA transfers of bytes (80188/80186) and words (80186 only). Each time that a byte or word is transferred, the count is decremented by one until it reaches 0000H—the terminal count.

The source and destination pointers are each 20 bits wide, so DMA transfers can occur to any memory location or I/O address without concern for segment and offset addresses. If the source or destination address is an I/O port, bits A19–A16 must be 0000 or a malfunction may occur.

Channel Control Register. Each DMA channel contains its own channel control register (refer to Figure 16–21), which defines its operation. The leftmost six bits specify the operation of the source and destination registers. The M/$\overline{\text{IO}}$ bit indicates a memory or I/O location, DEC causes the pointer to be decremented, and INC causes the pointer to be incremented. If both the INC and DEC bits are 1, then the pointer is unchanged after each DMA transfer. Notice that memory-to-memory transfers are possible with this DMA controller.

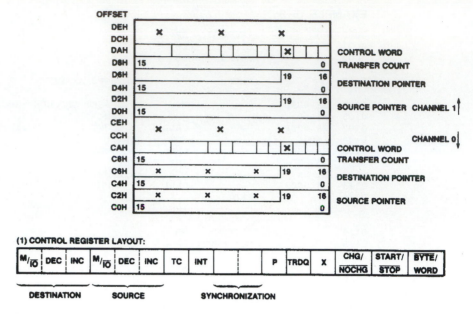

FIGURE 16–21   Register structure of the 80186/80188 DMA controller. (Courtesy of Intel Corporation.)

The TC (terminal count) bit causes the DMA channel to stop transfers when the channel count register is decremented to 0000H. If this bit is a logic 1, the DMA controller continues to transfer data, even after the terminal count is reached.

The INT bit enables interrupts to the interrupt controller. If set, this bit causes an interrupt to be issued when the terminal count of the channel is reached.

The SYN bit selects the type of synchronization for the channel: 00 = no synchronization, 01 = source synchronization, and 10 = destination synchronization. When either unsynchronized or source synchronization is selected, data are transferred at the rate of 2M bytes per second. These two types of synchronization allow transfers to occur without interruption. If destination synchronization is selected, the transfer rate is slower (1.3M bytes per second), and the controller relinquishes control to the 80186/80188 after each DMA transfer.

The P bit selects the channel priority. If $P = 1$, the channel has the highest priority. If both channels have the same priority, the controller alternates transfers between channels.

The TRDQ bit enables DMA transfers from timer 2. If this bit is a logic 1, the DMA request originates from timer 2. This can prevent the DMA transfers from using all of the microprocessor's time for the transfer.

The CHG/$\overline{\text{NOCHG}}$ bit determines whether START/$\overline{\text{STOP}}$ changes for a write to the control register. The START/$\overline{\text{STOP}}$ bit starts or stops the DMA transfer. To start a DMA transfer, both CHG/$\overline{\text{NOCHG}}$ and START/$\overline{\text{STOP}}$ are placed at a logic 1 level.

The $\overline{\text{BYTE}}$/WORD selects whether the transfer is byte- or word-sized.

Sample Memory-to-Memory Transfer.   The built-in DMA controller is capable of performing memory-to-memory transfers. The procedure used to program the controller and start the transfer is listed in Example 16–4.

EXAMPLE 16–4

```
            .MODEL SMALL
            .186
0000        .CODE
```

```
                      ;Memory-to-memory DMA transfer procedure
                      ;
                      ;Calling parameters:
                      ;
                      ;   DS:SI = source address
                      ;   ES:DI = destination address
                      ;   CX = count
                      ;

              GETA    MACRO   SEGA,OFFA,DMAA
                      MOV     AX,SEGA         ;;get segment
                      SHL     AX,4            ;shift segment left 4 places
                      ADD     AX,OFFA         ;;add in offset
                      MOV     DX,DMAA         ;;address DMA controller
                      OUT     DX,AL           ;;program rightmost 16-bits
                      PUSHF                   ;;save possible carry
                      MOV     AX,SEGA         ;;get segment
                      SHR     AX,12           ;;for leftmost 4-bits
                      POPF
                      ADD     AX,0            ;;add in possible carry
                      ADD     DX,2
                      OUT     DX,AL
                      ENDM

0000          MOVES   PROC    FAR

              GETA    DS,SI,0FFC0H            ;program source address
              GETA    ES,DI,0FFC4H            ;program destination address

0032 BA FFC8          MOV     DX,0FFC8H       ;program count
0035 8B C1            MOV     AX,CX
0037 EE              OUT     DX,AL

0038 BA FFCA          MOV     DX,0FFCAH       ;program control
003B B8 B606          MOV     AX,0B606H
003E EE              OUT     DX,AL            ;start transfer

003F CB              RET

0040          MOVES   ENDP
                      END
```

The procedure in Example 16–4 transfers data from the data segment location addressed by SI into the extra segment location addressed by DI. The number of bytes transferred is held in register CX. This operation is identical to the REP MOVSB instruction, but execution occurs at a much higher speed.

Chip Selection Unit

The chip selection unit simplifies the interface of memory and I/O to the 80186/80188. This unit contains programmable chip selection logic. In small- and medium-sized systems, no external decoder is required to select memory and I/O. Large systems, however, may still require external decoders. There are two forms of the chip selection unit; one form found in the XL and EA versions differs from the unit found in the EB and EC versions.

Memory Chip Selects. Six pins (XL and EA versions) or 10 pins (EB and EC versions) are used to select different external memory components in a small- or medium-sized 80186/80188-based system. The $\overline{UCS}$ (upper chip select) pin enables the memory device located in the upper portion of the memory map that is most often populated with ROM. This programmable pin allows the size of the ROM to be specified and the number of wait states required. Note that the ending address of the ROM is FFFFFH. The $\overline{LCS}$ (lower chip select) pin selects the memory device (usually a RAM) that begins at memory location 00000H. As with the $\overline{UCS}$ pin, the memory size and number of wait states are programmable. The remaining four or eight pins select middle memory devices. The four pins in the XL and EA version ($\overline{MCS3}$–$\overline{MCS0}$) are programmed for both the starting (base) address and memory size. Note that all devices must be of the same size. The 8 pins ($\overline{GCS7}$–$\overline{GCS0}$) in the EB and EC versions are programmed by size and also by starting address, and can represent a memory device or an I/O device.

Peripheral Chip Selects.   The 80186/80188 addresses up to seven external peripheral devices with pins $\overline{PCS6}$–$\overline{PCS0}$ (in the XL and EA versions). The $\overline{GCS}$ pins are used in the EB and EC versions to select up to eight memory or I/O devices. The base I/O address is programmed at any 1K-byte interval with port address block sizes of 128 bytes.

Programming the Chip Selection Unit for XL and EA Versions.   The number of wait states in each section of the memory and the I/O are programmable. The 80186/80188 microprocessors have a built-in wait state generator that can introduce between 0–3 wait states. Table 16–5 lists the logic levels required on bits R2–R0 in each programmable register to select various numbers of wait states. These three lines also select if an external READY signal is required to generate wait states. If READY is selected, the external READY signal is in parallel with the internal wait state generator. For example, if READY is a logic 0 for three clocking periods but the internal wait state generator is programmed to insert two wait states, three wait states are inserted.

Suppose that a 64K-byte EPROM is located at the top of the memory system and requires two wait states for proper operation. To select this device for this section of memory, the $\overline{UCS}$ pin is programmed for a memory range of F0000H–FFFFFH with two wait states. Figure 16–22 lists the control registers for all memory and I/O selections in the peripheral control block at offset addresses A0–A9H. Notice that the rightmost three bits of these control registers are from Table 16–5. The control register for the upper memory area is at location PCB offset address A0H. This 16-bit register is programmed with the starting address of the memory area (F0000H, in this case) and the number of wait states. Please note that the upper two bits of the address must be programmed as 00, and that only address bits A17–A10 are programmed into the control register. See Table 16–6 for examples illustrating the codes for various memory sizes. Because our example requires two wait states, the basic address is the same as in the table for a 64K device, except that the rightmost three bits are 110 instead of 100. The datum sent to the upper memory control register is 3006H.

Suppose that a 32K-byte SRAM that requires no waits and no READY input is located at the bottom of the memory system. To program the $\overline{LCS}$ pin to select this device, register A2 is loaded in exactly the same manner as register A0H. In this example, a 07FCH is sent to register A2H. Table 16–7 lists the programming values for the lower chip-selection output.

The central part of the memory is programmed via two registers: A6H and A8H. Register A6H programs the beginning or base address of the middle memory select lines ($\overline{MCS3}$—$\overline{MCS0}$) and number of waits. Register A8H defines the size of the block of memory and the individual memory device size (see Table 16–8). In addition to block size, the number of peripheral wait states are programmed as with other areas of memory. The EX (bit 7) and MS (bit 6) specify the peripheral selection lines, and will be discussed shortly.

TABLE 16–5 Wait state control bits R2, R1, and R0 (XL and EA versions).

R2	R1	R0	Number of Waits	READY required
0	X	X	–	Yes
1	0	0	0	No
1	0	1	1	No
1	1	0	2	No
1	1	1	3	No

For example, suppose that four 32K-byte SRAMs are added to the middle memory area, beginning at location 80000H and ending at location 9FFFFH with no wait states. To program the middle memory selection lines for this area of memory, we place the leftmost seven address bits in register A6H, with bits 8–3 containing logic 0s, and the rightmost three bits containing the ready control bits. For this example, register A6H is loaded with 8004H. Register A8H is programmed with a 1F44H, assuming that $EX = 0$ and $MS = 1$ and no wait states and no READY are required for the peripherals.

FIGURE 16–22 The chip selection registers for the XL and EA versions of the 80186/80188.

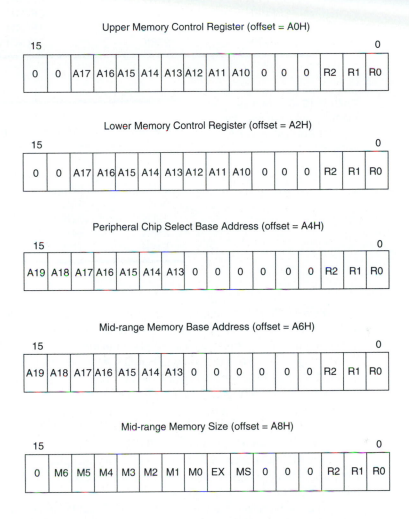

Upper Memory Control Register (offset = A0H)

15															0
0	0	A17	A16	A15	A14	A13	A12	A11	A10	0	0	0	R2	R1	R0

Lower Memory Control Register (offset = A2H)

15															0
0	0	A17	A16	A15	A14	A13	A12	A11	A10	0	0	0	R2	R1	R0

Peripheral Chip Select Base Address (offset = A4H)

15															0
A19	A18	A17	A16	A15	A14	A13	0	0	0	0	0	0	R2	R1	R0

Mid-range Memory Base Address (offset = A6H)

15															0
A19	A18	A17	A16	A15	A14	A13	0	0	0	0	0	0	R2	R1	R0

Mid-range Memory Size (offset = A8H)

15															0
0	M6	M5	M4	M3	M2	M1	M0	EX	MS	0	0	0	R2	R1	R0

TABLE 16–6 Upper memory programming for register A0H (XL and EA versions).

Start Address	Block Size	Value for No Waits, No READY
FFC00H	1K	3FC4H
FF800H	2K	3F84H
FF000H	4K	3F04H
FE000H	8K	3E04H
FC000H	16K	3C04H
F8000H	32K	3804H
F0000H	64K	3004H
E0000H	128K	1004H
C0000H	256K	0004H

TABLE 16–7 Lower memory programming for register A2H (XL and EA versions).

Ending Address	Block Size	Value for No Waits, No READY
003FFH	1K	0004H
007FFH	2K	0044H
00FFFH	4K	00C4H
01FFFH	8K	01C4H
03FFFH	16K	03C4H
07FFFH	32K	07C4H
0FFFFH	64K	0FC4H
1FFFFH	128K	1FC4H
3FFFFH	256K	3FC4H

TABLE 16–8 Middle memory programming for register A8H (XL and EA versions).

Block Size	Chip Size	Value for No Waits, No READY, and EX=0 MS=1
8K	2K	0144H
16K	4K	0344H
32K	8K	0744H
64K	16K	0F44H
128K	32K	1F44H
256K	64K	3F44H
512K	128K	7F44H

Register A4H programs the peripheral chip selection pins ($\overline{PCS6}$–$\overline{PCS0}$) along with the EX and MS bits of register A8H. Register A4H holds the beginning or base address of the peripheral selection lines. The peripherals may be placed in memory or in the I/O map. If they are placed in the I/O map, A19–A16 of the port number must be 0000. Once the starting address is programmed on any 1K-byte I/O address boundary, the $\overline{PCS}$ pins are spaced at 128-byte intervals.

For example, if register A4H is programmed with a 0204H, with no waits and no READY synchronization, the memory address begins at 02000H or the I/O port begins at 2000H. In this case, the I/O ports are: $\overline{PCS0}$ = 2000H, $\overline{PCS1}$ = 2080H, $\overline{PCS2}$ = 2100H, $\overline{PCS3}$ = 2180H, $\overline{PCS4}$ = 2200H, $\overline{PCS5}$ = 2280H, and $\overline{PCS6}$ = 2300H.

The MS bit of register A8H selects memory-mapping or I/O mapping for the peripheral select pins. If MS is a logic 0, then the PCS lines are decoded in the memory map; if it is a logic 1, then the $\overline{PCS}$ lines are in the I/O map.

The EX bit selects the function of the $\overline{PCS5}$ and $\overline{PCS6}$ pins. If $EX = 1$, these $\overline{PCS}$ pins select I/O devices; if $EX = 0$, these pins provide the system with latched address lines A1 and A2. The A1 and A2 pins are used by some I/O devices to select internal registers and are provided for this purpose.

Programming the Chip Selection Unit for EB and EC Versions. As mentioned earlier, the EB and EC versions have a different chip selection unit. These newer versions of the 80186/80188 contain an upper and lower memory chip selection pin as do earlier versions, but they do not contain middle selection and peripheral selection pins. In place of the middle and peripheral chip selection pins, the EB and EC versions contain eight general chip selection pins ($\overline{GCS7}$–$\overline{GCS0}$) that select either a memory device or an I/O device.

Programming is also different because each of the chip selection pins contains a starting address register and an ending address register. See Figure 16–23 for the offset address of each pin and the contents of the start and end registers.

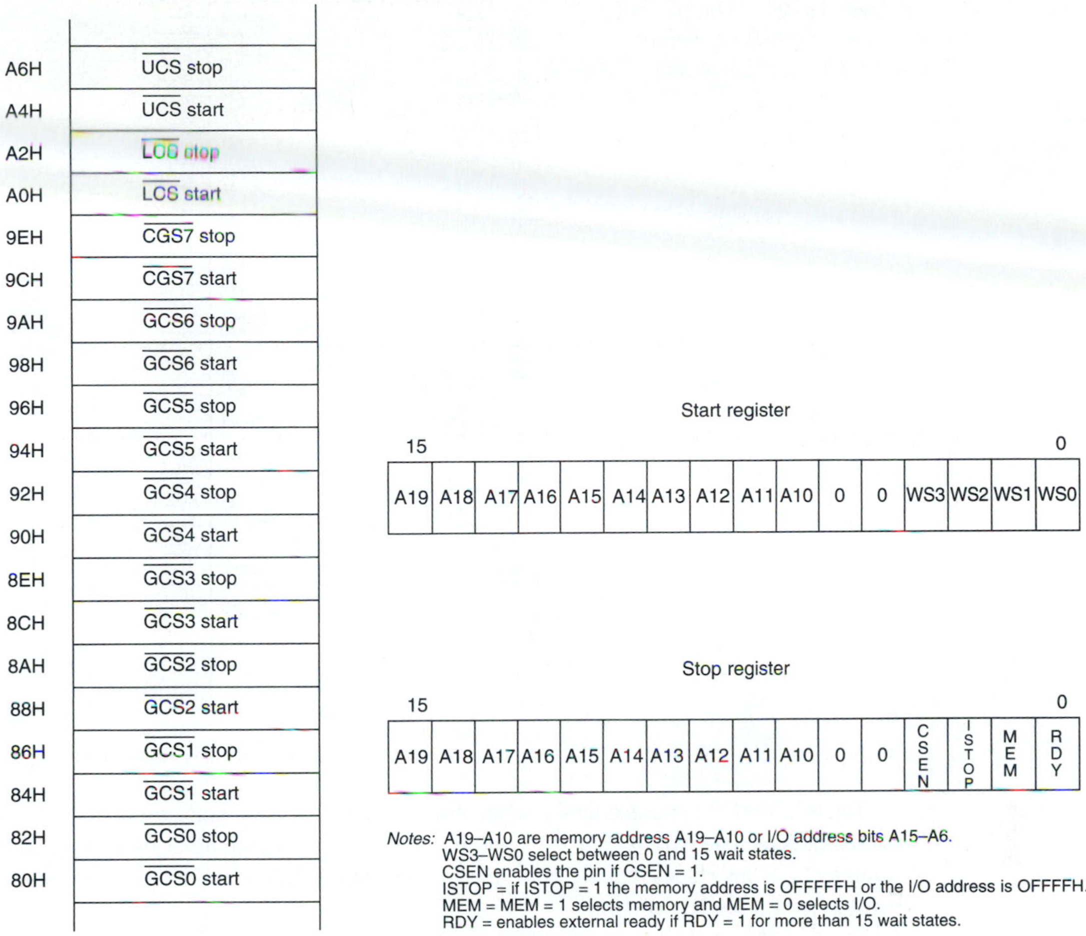

A6H	$\overline{\text{UCS}}$ stop
A4H	$\overline{\text{UCS}}$ start
A2H	$\overline{\text{LCS}}$ stop
A0H	$\overline{\text{LCS}}$ start
9EH	$\overline{\text{CGS7}}$ stop
9CH	$\overline{\text{CGS7}}$ start
9AH	$\overline{\text{GCS6}}$ stop
98H	$\overline{\text{GCS6}}$ start
96H	$\overline{\text{GCS5}}$ stop
94H	$\overline{\text{GCS5}}$ start
92H	$\overline{\text{GCS4}}$ stop
90H	$\overline{\text{GCS4}}$ start
8EH	$\overline{\text{GCS3}}$ stop
8CH	$\overline{\text{GCS3}}$ start
8AH	$\overline{\text{GCS2}}$ stop
88H	$\overline{\text{GCS2}}$ start
86H	$\overline{\text{GCS1}}$ stop
84H	$\overline{\text{GCS1}}$ start
82H	$\overline{\text{GCS0}}$ stop
80H	$\overline{\text{GCS0}}$ start

Start register

15															0
A19	A18	A17	A16	A15	A14	A13	A12	A11	A10	0	0	WS3	WS2	WS1	WS0

Stop register

15															0
A19	A18	A17	A16	A15	A14	A13	A12	A11	A10	0	0	CSEN	ISTOP	MEM	RDY

Notes: A19–A10 are memory address A19–A10 or I/O address bits A15–A6.
WS3–WS0 select between 0 and 15 wait states.
CSEN enables the pin if CSEN = 1.
ISTOP = if ISTOP = 1 the memory address is 0FFFFFH or the I/O address is 0FFFFH.
MEM = MEM = 1 selects memory and MEM = 0 selects I/O.
RDY = enables external ready if RDY = 1 for more than 15 wait states.

FIGURE 16–23   The chip selection unit in the EB and EC versions of the 80186/80188.

Notice that programming for the EB and EC versions of the 80186/80188 are much easier than for the earlier XL and XA versions. For example, to program the $\overline{\text{UCS}}$ pin for an address that begins at location F0000H and ends at location FFFFFH (64K bytes), the starting address register (*offset* = A4H) is programmed with F002H for a starting address of F0000H with two wait states. The ending address register (*offset* = A6H) is programmed with 000EH for an ending address of FFFFFH for memory with no external ready synchronization. The other chip selection pins are programmed in a similar fashion.

16–3   80C188EB EXAMPLE INTERFACE

Because the 80186/80188 microprocessors are designed as embedded controllers, this section of the text provides an example of such an application. The example illustrates simple memory and I/O attached to the 80C188EB microprocessor. It also lists the software required to program the 80C188EB and its internal registers after a system reset. The software to control the system itself is not provided. Figure 16–24 illustrates the pin-out of the 80C188EB version of the 80188 microprocessor. Note the differences between this version and the XL version presented earlier in the text.

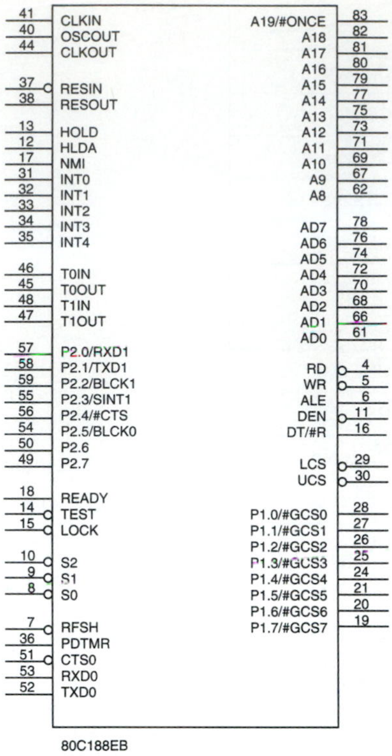

FIGURE 16–24 The pin-out of the 80C188EB version of the 80188 microprocessor.

```
 41   CLKIN            A19/#ONCE   83
 40   OSCOUT           A18         82
 44   CLKOUT           A17         81
                       A16         80
 37   RESIN            A15         79
 38   RESOUT           A14         77
                       A13         75
 13   HOLD             A12         73
 12   HLDA             A11         71
 17   NMI              A10         69
 31   INT0             A9          67
 32   INT1             A8          62
 33   INT2
 34   INT3             AD7         78
 35   INT4             AD6         76
                       AD5         74
 46   T0IN             AD4         72
 45   T0OUT            AD3         70
 48   T1IN             AD2         68
 47   T1OUT            AD1         66
                       AD0         61
 57   P2.0/RXD1
 58   P2.1/TXD1        RD           4
 59   P2.2/BLCK1       WR           5
 55   P2.3/SINT1       ALE          6
 56   P2.4/#CTS        DEN         11
 54   P2.5/BLCK0       DT/#R       16
 50   P2.6
 49   P2.7             LCS         29
                       UCS         30
 18   READY
 14   TEST             P1.0/#GCS0  28
 15   LOCK             P1.1/#GCS1  27
                       P1.2/#GCS2  26
 10   S2               P1.3/#GCS3  25
  9   S1               P1.4/#GCS4  24
  8   S0               P1.5/#GCS5  21
                       P1.6/#GCS6  20
  7   RFSH             P1.7/#GCS7  19
 36   PDTMR
 51   CTS0
 53   RXD0
 52   TXD0

          80C188EB
```

The 80C188EB version contains some new features that were not present on earlier versions. These features include two I/O ports (P1 and P2) that are shared with other functions and two serial communications interfaces that are built into the processor. This version does not contain a DMA controller, as did the XL version.

The 80188 can be interfaced with a small system designed to be used as a microprocessor trainer. The trainer illustrated in this text uses a 27256 EPROM for program storage, three 62256 SRAMs for data storage, an 8279 programmable keyboard/display interface, and one of the built-in serial ports for serial communications. Figure 16–25 illustrates a small microprocessor trainer that is based on the 80C188EB microprocessor.

Memory is selected by the $\overline{UCS}$ pin for the 27256 EPROM and the $\overline{LCS}$ pin for one of the 62256 SRAMs; the $\overline{GCS0}$ and $\overline{GCS1}$ pins select the remaining SRAM devices. The 8270 keyboard/display peripheral is selected by $\overline{GCS2}$. Note that five wait states are programmed for the EPROM, assuming a really slow 450 ns EPROM, two waits for the 250 ns SRAM, and two waits for the 8279 keyboard/display interface. Faster EPROM and SRAM reduce or eliminate the number of waits required for the memory.

The system places the EPROM at memory addresses F8000H–FFFFFH; the SRAM at 00000H–07FFFH, 80000H–87FFFH, and 88000H–8FFFFH; and the 8279 at I/O ports 1000H–107FH. In this system, as is normally the case, we do not modify the address of the peripheral control block, which resides at I/O ports FF00H–FFFFH.

Example 16–5 lists the software required to initialize the 80C188EB microprocessor. It does not list any of the software required to program the 8279, nor does it show the software required to operate the system as a microprocessor-based trainer.

EXAMPLE 16–5

```
              .MODEL SMALL
              .186
0000          .CODE
              ;A program that initializes the 80C188EB.
```

```
                          ;
                                    ORG     8000H           ;start of EPROM
                          ;
8000                      MAIN:

8000   BA FFA6            MOV     DX,0FFA6H        ;address UCS stop
8003   B8 000E            MOV     AX,000EH         ;FFFFFH

8006   EE                 OUT     DX,AL            ;set stop address for UCS

8007   BA FFA0            MOV     DX,0FFA0H        ;address LCS start
800A   B8 0002            MOV     AX,0002H         ;00000H with 2 waits
800D   EE                 OUT     DX,AL            ;set start address for SRAM U4

800E   BA FFA2            MOV     DX,0FFA2H        ;address LCS stop
8011   B8 080A            MOV     AX,080AH         ;07FFFH
8014   EE                 OUT     DX,AL            ;set stop address for SRAM U4

8015   BA FF80            MOV     DX,0FF80H        ;address GCS0 start
8018   B8 0802            MOV     AX,0802H         ;08000H with 2 waits
801B   EE                 OUT     DX,AL            ;set start address for SRAM U5

801C   BA FF82            MOV     DX,0FF82H        ;address GCS0 stop
801F   B8 100A            MOV     AX,100AH         ;0FFFFH
8022   EE                 OUT     DX,AL            ;set stop address for SRAM U5

8023   BA FF84            MOV     DX,0FF84H        ;address GCS1 start
8026   B8 1002            MOV     AX,1002H         ;10000H with 2 waits
8029   EE                 OUT     DX,AL            ;set start address for SRAM U6

802A   BA FF86            MOV     DX,0FF86H        ;address GCS1 stop
802D   B8 180A            MOV     AX,180AH         ;17FFFH
8030   EE                 OUT     DX,AL            ;set stop address for SRAM U6

8031   BA FF88            MOV     DX,0FF88H        ;address GCS2 start
8034   B8 1002            MOV     AX,1002H         ;1000H with 2 waits
8037   EE                 OUT     DX,AL            ;set start address for 8279

8038   BA FF8A            MOV     DX,0FF8AH        ;address GCS2 stop
803B   B8 1048            MOV     AX,1048H         ;103FH (I/O)
803E   EE                 OUT     DX,AL            ;set stop address for 8279

803F   BA FF60            MOV     DX,0FF60H        ;address serial baud rate
8042   B8 8067            MOV     AX,8067H         ;generate a 9600 Baud rate
8045   EE                 OUT     DX,AL            ;set Baud rate

8046   BA FF62            MOV     DX,0FF62H        ;address serial control register
8049   B8 0059            MOV     AX,59H           ;7 data, even parity, 1 stop
804C   EE                 OUT     DX,AL            ;set serial port

804D   BA FF66            MOV     DX,0FF66H        ;address serial status register
8050   ED                 IN      AX,DX            ;clear serial port

8051   BA FF62            MOV     DX,0FF62H        ;address serial control register
8054   ED                 IN      AX,DX            ;read control register
8055   83 C8 20           OR      AX,20H           ;set REN bit
8058   EE                 OUT     DX,AL            ;enable serial port
                          ;
                          ;
                          ;Remainder of system software is placed at this point.
                          ;
                          ;
                                    ORG     0FFF0H          ;reset location
                          ;
FFF0   BA FFA4            MOV     DX,0FFA4H        ;address UCS start
FFF3   B8 F805            MOV     AX,0F805H        ;F8000H with 5 waits

FFF6   EE                 OUT     DX,AL            ;set start address for UCS
FFF7   E9 8006            JMP     MAIN             ;jump to start of EPROM
                                    END
```

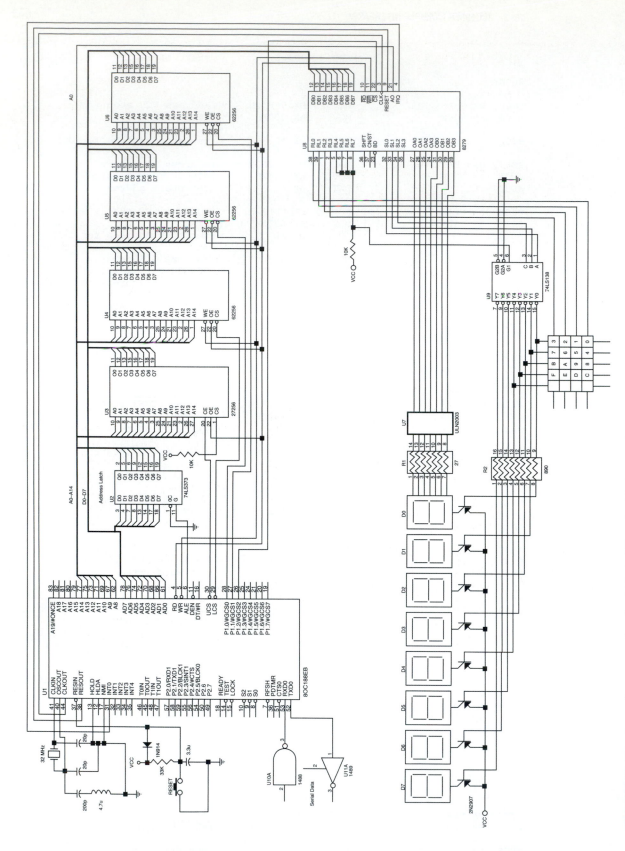

FIGURE 16–25 A small system using the 80C188EB embedded controller.

16–4   REAL TIME OPERATING SYSTEMS (RTOS)

This section of the text describes the real time operating system (RTOS). Interrupts are used to develop RTOS because they are used in embedded applications of the microprocessor. All systems, from the simplest embedded application the most sophisticated system, must have an operating system.

What Is a Real Time Operating System (RTOS)?

The RTOS is an operating system used in embedded applications that performs tasks in a predictable amount of time. Operating systems, like Windows, defer many tasks and do not guarantee their execution in a predictable time. The RTOS is much like any other operating system in that it contains the same basic sections. Figure 16–26 illustrates the basic structure of an operating system as it might be placed on an EPROM or Flash memory device.

There are three components to all operating systems: (1) initialization, (2) the kernel, (3) data and procedures. The initialization section is used to program all hardware components in the system, load drivers specific to a system, and program the contents of the microprocessor's registers. The kernel performs the basic system task, provides system calls or functions, and comprises the embedded system. The data and procedure section holds all procedures and any static data used by the operating system.

An Example System

Figure 16–27 illustrates a simple embedded system based on the 80188EB embedded microprocessor. This system contains a 2-line × 16 character-per-line LCD display that shows the time of day and the temperature. The system itself is stored on a small 32K × 8 EPROM. A 32K × 8 SRAM is included to act as a stack and store the time. A database holds the most recent temperatures and the times at which the temperatures were obtained.

The temperature sensor is located inside the LM70 digital temperature sensor manufactured by National Semiconductor Corporation for less than $1.00. The interface to the microprocessor is in serial format, and the converter has a resolution of 10 bits plus a sign bit. Figure 16–28 illustrates the pin-out of the LM70 temperature sensor.

The LM70 transfers data to and from the microprocessor through the SI/O pin, which is a bidirectional serial data pin. Information is clocked through the SI/O pin by the SC (clock) pin. The LM70 contains three 16-bit registers: the configuration register, the temperature sensor register, and the identification register. The configuration register selects either the shutdown mode (XXFFH) or continuous conversion mode (XX00). The temperature register contains the signed temperature in the leftmost 11 bits of the 16-bit data word. If the temperature is negative it is in their respective complement forms. The identification register presents an 8100 when it is read.

FIGURE 16–26   The structure of a RTOS operating system.

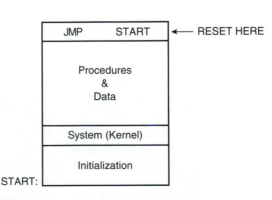

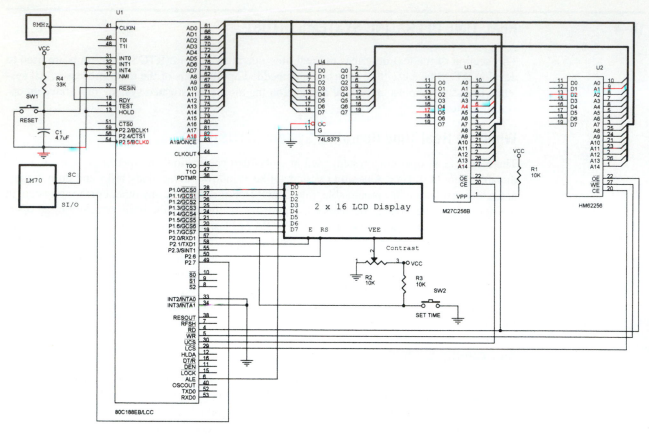

FIGURE 16–27   A simple embedded application.

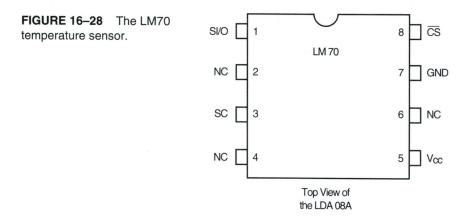

FIGURE 16–28   The LM70 temperature sensor.

Top View of
the LDA 08A

When the temperature is read from the LM70, it is read in Celsius and each step is equal to 0.25°C. For example, if the temperature register is 0000 1100 100X XXXX or a value of 100 decimal, the temperature is 25.0°C.

Example 16–6 illustrates the software for the system, with each section (initialization, kernel, and data and procedures) clearly delineated. The system samples the temperatures once per minute and stores them in a circular queue along with the day and the time in hour and minutes. The day is a number that starts at zero when the system is initialized. The size of the queue has been set to 16K bytes, so the most recent 4,096 measurements can be stored.

EXAMPLE 16–6

```
;Simple real-time operating test system for an 80188EB.
;LCD version
;
; INT 40H delays for BL milliseconds (range 1 to 99)
; INT 41H delays for BL seconds (range 1 to 59)
;      Note* delay times must be written in hexadecimal!
;      ie. 15 milliseconds is 15H.
; INT 42H displays character string on LCD
;      ES:BX addresses the NULL string
;      AL = where (80H line 1, C0H line2)
; INT 43H clears the LCD
;
.MODEL TINY
.186     ;switch to the 80186/80188 instruction set
.CODE
.STARTUP
;
;MUST USE MASM 6.11
;command line = ML /AT XXXXXXXX.ASM
;
;;;;;;;;;;;;;;;;;;;;;;;;;;;;;;;;;;;;;;;;;;;;;;;;;;;;;;
;MACROs placed here
;;;;;;;;;;;;;;;;;;;;;;;;;;;;;;;;;;;;;;;;;;;;;;;;;;;;;;
;
IO      MACRO   PORT,DATA
        MOV     DX,PORT
        MOV     AX,DATA
        OUT     DX,AX
        ENDM


CS_IO   MACRO   PORT,START,STOP
        IO  PORT,START
        IO  PORT+2,STOP
        ENDM


SEND    MACRO   VALUE,COMMAND,DELAYS
        PUSH    DX
        MOV     AL,VALUE
        MOV     DX,0FF56H
        OUT     DX,AL           ;value to Port 1
        MOV     DX,0FF5AH        ;read Port 2
        IN      AL,DX
        AND     AL,0BDH
        OR      AL,COMMAND
        OUT     DX,AL           ;send RS
        OR      AL,2
        OUT     DX,AL           ;E = 1
        AND     AL,0FDH
        OUT     DX,AL           ;E = 0
        PUSH    BX
        MOV     BL,DELAYS
        INT     40H
        POP     BX
        POP     DX
        ENDM


KEY     MACRO
        IN      AL,DX
        SHR     AL,1
        ENDM


SENDS   MACRO   WHAT
        PUSH    CX
        PUSH    BX
        MOV     BX,AX
        MOV     CX,16
        .REPEAT
            SHL     BX,1
```

```
                        IN        AL,DX
                        .IF  CARRY?
                           OR     AL,80H
                           OUT    DX,AL
                        .ELSE
                           AND    AL,7FH
                           OUT    DX,AL
                        .ENDIF
                        OR  AL,4
                        OUT DX,AL
                        AND AL,0FBH
                        OUT DX,AL
                     .UNTILCXZ
                     POP    BX
                     POP    CX
                     ENDM

          READS  MACRO
                     PUSH      BX
                     PUSH      CX
                     MOV       BX,16
                     .REPEAT
                        IN        AL,DX
                        OR        AL,80H
                        OUT       DX,AL
                        IN        AL,DX
                        TEST      AL,4
                        .IF  ZERO?
                           CLC
                        .ELSE
                           STC
                        .ENDIF
                        RLC       BX,1
                        OR        AL,7FH
                        OUT       DX,AL
                     .UNTILCXZ
                     MOV    AX,BX
                     POP    CX
                     POP    BX
                     ENDM
          ;
          ;;;;;;;;;;;;;;;;;;;;;;;;;;;;;;;;;;;;;;;;;;;;;;;;;;;;;;;;;;;
          ;Initialization placed here
          ;;;;;;;;;;;;;;;;;;;;;;;;;;;;;;;;;;;;;;;;;;;;;;;;;;;;;;;;;;;
          ;
                     IO  0FFA6H,000EH          ;UCS Stop address
                     CS_IO  0FFA0H,0,80AH      ;program LCS
                     IO  0FF54H,0              ;Port 1 control (all OUT)
                     IO  0FF5CH,0              ;Port 2 control
                     IO  0FF58H,0039H          ;port 2 direction

                     MOV    AX,0               ;address segment 0000
                     MOV    DS,AX              ;for DS, ES, and SS
                     MOV    ES,AX
                     MOV    SS,AX
                     MOV    SP,8000H           ;setup stack pointer (0000:8000)

                     MOV    BX,OFFSET INT_TABLE - 100H     ;install interrupt vectors
                     .WHILE WORD PTR CS:[BX] != 0
                        MOV    AX,CS:[BX]
                        MOV    DI,CS:[BX+2]
                        MOV    DS:[DI],AX
                        MOV    DS:[DI+2],CS
                        ADD    BX,4
                     .ENDW

                     MOV    BYTE PTR DS:[40FH],0      ;don't display time
```

```
        IO 0FF40H,0                 ;Timer 2 count
        IO 0FF42H,1000              ;Timer 2 compare
        IO 0FF46H,0E001H            ;Timer 2 control
        IO 0FF08H,00FCH             ;interrupt mask
        MOV   DX,0FF5AH
        MOV   AL,0H
        OUT   DX,AL                 ;place E at 0 for LCD
        SENDS 0                     ;turn temperature unit on
        MOV   AX,500H
        MOV   DS:[404H],AX          ;initialize queue as empty
        MOV   DX:[406H],AX
        STI                         ;enable interrupts

        CALL  INIT_LCD              ;initialize LCD


;;;;;;;;;;;;;;;;;;;;;;;;;;;;;;;;;;;;;;;;;;;;;;;;;;;;;;;;;;;;;;;;;;;;;;;;;;;
;          SYSTEM SOFTWARE HERE
;;;;;;;;;;;;;;;;;;;;;;;;;;;;;;;;;;;;;;;;;;;;;;;;;;;;;;;;;;;;;;;;;;;;;;;;;;;
;
        MOV   BYTE PTR DS:[40FH],0      ;do  not display time/temp
        MOV   WORD PTR DS:[40CH],0      ;clear clock to 00:00:00 AM
        MOV   BYTE PTR DS:[40EH],0
        MOV   AX,CS                 ;line 1 message
        MOV   ES,AX
        MOV   AL,80H
        MOV   BX,OFFSET MES1 - 100H
        INT   42h                   ;display "The Temp/Clock!"
        MOV   BL,3
        INT   41H                   ;3 second time delay
        INT   43H                   ;clear display
        MOV   AL,80H
        MOV   BX,OFFSET MES2 - 100H
        INT   42H                   ;display "Set Hours"
        MOV   AL,0C0H
        MOV   BX,OFFSET MES3 - 100H
        INT   42H                   ;display "Hours = 00"
        .REPEAT
           MOV   SI,40EH            ;address hours
           MOV   CL,24H
           MOV   CH,0C8H
           CALL  KEYS
        .UNTIL CARRY?
        MOV   AL,80H
        MOV   BL,OFFSET MES4 - 100H
        INT   42H                   ;display "Set Minutes"
        MOV   AL,0C0H
        MOV   BX,OFFSET MES5 - 100H
        INT   42H                   ;display "Minutes = 00"
        .REPEAT
           MOV   SI,40DH            ;address minutes
           MOV   CL,0CAH
           MOV   CL,60H
           CALL  KEYS
        .UNTIL CARRY?
        MOV   AX,0
        MOV   ES,0
        .WHILE 1
           INT   43H                    ;clear display
           MOV   BYTE PTR DS:[40FH],0FFH        ;display time/temp
           .REPEAT                       ;wait for key press
              .REPEAT
                KEY
              .UNTIL !CARRY?
              MOV   BL,15H
              INT   40H
              KEY
```

```
                                .UNTIL !CARRY?
                        MOV  BYTE PTR DS:[40FH],0      ;time/temp off
                        INT   42H
                        MOV  SI,DS:[406H]    ;get most ancient temp
                        .WHILE 1
                           CALL  DTEMP
                           ADD   SI,4
                           .IF SI == 5400H
                              MOV    SI,500H
                           .ENDIF
                           MOV   CH,DS:[40CH]   ;2 second timeout
                           ADD   CH,2
                           DAA
                           .IF CH >= CL
                              SUB   CH,CL
                              DAS
                           .ENDIF
                           .REPEAT
                              .REPEAT
                                 .BREAK .IF CH == DS:[40CH]
                                 KEY
                              .UNTIL CARRY?
                              MOV   BL,15H
                              INT   40H
                              KEY
                           .UNTIL CARRY?
                           .REPEAT                     ;wait for key press
                              .REPEAT
                                 .BREAK .IF CH == DS:[40CH]
                                 KEY
                              .UNTIL !CARRY?
                              MOV   BL,15H
                              INT   40H
                              KEY
                           .UNTIL !CARRY?
                        .ENDW
                     .ENDW
         ;
         ;;;;;;;;;;;;;;;;;;;;;;;;;;;;;;;;;;;;;;;;;;;;;;;;;;;;;;;;;;;;;;;;;;;;;
         ;Procedures and Static Data follow the system software
         ;;;;;;;;;;;;;;;;;;;;;;;;;;;;;;;;;;;;;;;;;;;;;;;;;;;;;;;;;;;;;;;;;;;;;
         ;
         MES1       DB   'The Temp/Clock!',0
         MES2       DB   'Set Hours',0
         MES3       DB   'Hours = 00',0
         MES4       DB   'Set Minutes',0
         MES5       DB   'Minutes = 00',0
         MES6       DB   'C at '
         ;
         ;Interrupt vector table
         ;
         INT_TABLE   DW  TIM2 - 100H
               DW  4CH
               DW  DELAYM - 100H
               DW  100H
               DW  DELAYS - 100H
               DW  104H
               DW  STRING - 100H
               DW  108H
               DW  CLEAR - 100H
               DW  10CH
               DW  0

         DTEMP PROC  NEAR

               MOV   AX,DS:[SI]
               MOV   BX,410H
```

```
        SHR     AX,7            ;convert to degrees C
        TEST    AL,80H
        .IF !ZERO?
                NOT     AL
                MOV     BYTE PTR [BX],'-'
                INC     BX
        .ENDIF
        .IF AL > 99
            SUB     AL,100
            MOV     BYTE PTR [BX],'1'
            INC     BX
        .ENDIF
        AAM
        ADD     AX,3030H
        MOV     [BX],AH
        MOV     [BX+1],AL
        ADD     BX,2
        MOV     DI,OFFSET MES6 - 100H
        MOV     CX,5
        .REPEAT
            MOV   AL,CS:[DI]
            MOV   [BX],AL
            INC   BX
            INC   DI
        .UNTILCXZ
        MOV     AL,DS:[40EH]
        CALL    STORE
        MOV     BYTE PTR [BX],':'
        INC     BX
        MOV     AL,DS:[40DH]
        CALL    STORE
        MOV     BYTE PTR [BX],0
        MOV     AL,80H
        MOV     BX,410H
        INT     42H

DTEMP ENDP

KEYS  PROC  NEAR

        MOV     DX,0FF5AH
        MOV     AX,0
        PUSH    ES
        MOV     ES,AX
        MOV     BX,400H         ;display hours/minutes
        MOV     AL,[SI]
        CALL    STORE
        MOV     BYTE PTR [BX],0
        MOV     BX,OFFSET 400H
        MOV     AL,CH
        INT     42H
        POP     ES
        MOV     CH,DS:[40CH]  ;2 second timeout
        ADD     CH,2
        DAA
        .IF CH >= CL
            SUB     CH,CL
            DAS
        .ENDIF
        .REPEAT             ;wait for release
            .REPEAT
                .IF CH == DS:[40CH]
                    STC
                    RET
                .ENDIF
                KEY
            .UNTIL CARRY?
```

```
                MOV     BL,15H
                INT     40H
                KEY
        .UNTIL CARRY?
        .REPEAT                 ;wait for key press
          .REPEAT
            .IF  CH == DS:[40CH]
                STC
                RET
            .ENDIF
            KEY
          .UNTIL !CARRY?
            MOV     BL,15H
            INT     40H
            KEY
        .UNTIL !CARRY?
        MOV     AL,[SI]
        ADD     AL,1
        DAA
        .IF     AL >= CL
          MOV     AL,0
        .ENDIF
        MOV     [SI],AL
        CLC
        RET

KEYS    ENDP
;
;Interrupt service for TIMER 2 interrupt.  (Once per millisecond)
;
TIM2    PROC    FAR USES ES DS AX BX SI DX

        MOV     AX,0
        MOV     DS,AX
        MOV     ES,AX
        MOV     BX,409H              ;address real-time clock - 1
        MOV     SI,OFFSET MODU-101H     ;address Modulus Table - 1
        .REPEAT
          INC    SI      ;point to modulus
          INC    BX      ;point to counter
          MOV    AL,[BX]  ;get counter
          ADD    AL,1     ;add 1
          DAA            ;make it BCD
          .IF AL == BYTE PTR CS:[SI]  ;check modulus
            MOV    AL,0
          .ENDIF
          MOV    [BX],AL    ;save new count
        .UNTIL !ZERO? || BX == 40EH
        IO 0FF02H,8000H        ;end of interrupt
        .IF    BYTE PTR DS:[40AH] == 0 && BYTE PTR DS:[40BH] == 0
          CALL   DISPLAY   ;start Display Thread
        .ENDIF
        IRET

TIM2    ENDP

MODU    DB 0            ;Mod 100
        DB 10H          ;Mod 10
        DB 60H          ;Mod 60
        DB 60H          ;Mod 60
        DB 24H          ;Mod 24

DISPLAY PROC NEAR        ;display time of day (once a second)

    .IF BYTE PTR DS:[40FH] != 0     ;if display time is not zero
        STI                 ;enable future interrupts
        .IF BYTE PTR DS:[40CH] == 0   ;once a minute
```

```
        CALL   TEMP          ;display and log temperature
    .ENDIF
    MOV   BX,3F0H
    MOV   SI,40EH          ;address clock
    MOV   AL,[SI]          ;get hours
    .IF AL > 12H           ; for AM / PM
        SUB   AL,12H
        DAS
    .ELSEIF AL == 0
        MOV   AL,12H
    .ENDIF

    CALL  STORE
    MOV   BYTE PTR [BX],':'
    INC   BX
    MOV   AL,[SI]          ;get minutes
    CALL  STORE
    MOV   BYTE PTR [BX],':'
    INC   BX
    MOV   AL,[SI]          ;get seconds
    CALL  STORE
    MOV   DL,'A'           ;for AM / PM
    .IF   BYTE PTR DS:[40EH] > 11H
        MOV   DL,'P'
    .ENDIF
    MOV   BYTE PTR [BX],' '
    MOV   [BX+1],DL
    MOV   BYTE PTR [BX+2],'M'
    MOV   BYTE PTR [BX+3],0   ;end of string
    MOV   BX,3F0H          ;display buffer
    MOV   AL,0C2H          ;LCD starting position
    INT   42H
    MOV   AX,DS:[SI]       ;display temp
    MOV   BX,410H
    SHR   AX,7             ;convert to degrees C
    TEST  AL,80H
    .IF   !ZERO?
        NOT   AL
        MOV   BYTE PTR [BX],'-'
        INC   BX
    .ENDIF
    .IF AL > 99
        SUB   AL,100
        MOV   BYTE PTR [BX],'1'
        INC   BX
    .ENDIF
    AAM
    ADD   AX,3030H
    MOV   [BX],AH
    MOV   [BX+1],AL
    MOV   BYTE PTR [BX+2],'C'
    MOV   BYTE PTR [BX+3],0
    MOV   AL,80H
    MOV   BX,410H
    INT   42H
 .ENDIF
  RET

DISPLAY ENDP

STORE PROC  NEAR

    PUSH  AX
    SHR   AL,4
    MOV   DL,AL
    POP   AX
    AND   AL,15
```

```
              MOV    DH,AL
              ADD    DX,3030H
              MOV    [BX],DX
              ADD    BX,2
              DEC    SI
              RET

STORE ENDP

TEMP   PROC   NEAR USES SI

              READS                    ;get current temp to AX
              MOV    SI,DS:[404H]       ;get head pointer
              MOV    DS:[SI],AX         ;save temp
              MOV    AL,DS:[40EH]       ;get hours
              MOV    DS:[SI+2],AL
              MOV    AL,DS:[40DH]       ;get minutes
              MOV    DS:[SI+3],AL
              ADD    SI,4
              .IF SI == 5400H
                 MOV   SI,500H
              .ENDIF
              MOV    DS:[404H],SI
              ADD    SI,4
              .IF SI == 5400H
                 MOV   SI,504H
              .ENDIF
              MOV    DS:[406H],SI
              RET

TEMP   ENDP

DELAYM        PROC   FAR USES DS BX AX

              STI              ;enable future interrupts
              MOV    AX,0
              MOV    DS,AX
              MOV    AL,DS:[40AH]     ;get milli counter
              ADD    AL,BL     ;BL = no. of milliseconds
              DAA
              .REPEAT
              .UNTIL AL == DS:[40AH]
              IRET

DELAYM        ENDP

DELAYS        PROC   FAR USES DS BX AX

              STI              ;enable future interrupts
              MOV    AX,0
              MOV    DS,AX
              MOV    AL,DS:[40CH]      ;get seconds

              ADD    AL,BL
              DAA
              .IF AL >= 60H
                 SUB   AL,60H
                 DAS
              .ENDIF
              .REPEAT
              .UNTIL AL == DS:[40CH]
              IRET

DELAYS        ENDP

INIT_LCD PROC  NEAR

              MOV    BL,60H       ;wait 60 milliseconds
```

```
        INT    40H
        MOV    CX,4
        .REPEAT
            SEND   38H,0,6
        .UNTILCXZ

        SEND   8,0,2
        SEND   1,0,2
        SEND   12,0,2
        SEND   6,0,2
        RET

INIT_LCD ENDP

STRING       PROC   FAR USES BX AX

        STI              ;enable future interrupts
        SEND   AL,0,1         ;send start position
        .REPEAT
            SEND   BYTE PTR ES:[BX],2,1
            INC    BX
        .UNTIL BYTE PTR ES:[BX] == 0
        IRET

STRING       ENDP

CLEAR     PROC   FAR USES AX BX

        STI              ;enable future interrupts
        SEND   1,0,2
        IRET

CLEAR     ENDP

ORG    080F0H  ;get to reset location for 32K EPROM

RESET:
        IO 0FFA4H,0F800H     ;UCS start address
        DB 0EAH              ;JMP  F800:0000     (F8000H)
        DW 0000H,0F800H
END
```

A Threaded System

At times we need to implement an operating system that can process multiple threads. Multiple threads are handled by the kernel using a real-time clock interrupt. One method for scheduling processes in a small RTOS is to use a time slice to switch between various processes. The basic time slice can be any duration and is somewhat dependent on the execution speed of the microprocessor. For example, in a 100 MHz clock many instructions execute in one or two clocks on a modern microprocessor. Assuming the machine executes one instruction every two clocks, if we were to choose a time slice of 1 ms, the machine can execute about 50,000 instructions per time slice, which should be adequate for most systems. If a lower clock frequency is employed, then a time slice of 10 ms or even 100 ms is selected.

Each time slice is activated by a timer interrupt. The interrupt service procedure must look to a queue to see whether a task is available to execute, and if it is, it must start execution of the new task. If no new task is present, it must continue executing an old task or enter an idle state and wait for a new task to be queued. The queue is circular and may contain any number of tasks for the system up to some finite limit. For example, it might be a small queue in a small system with 10 entries. The size is determined by the intended overall system needs and could be much larger or smaller.

Each scheduling queue entry must contain a pointer to the process (CS:IP) and the entire context state of the machine. Scheduling queue entries may also contain some form of a time-to-live entry in case of a deadlock, a priority entry, and an entry that can lengthen the slice activation time. In the following example we will not use the priority entry or an entry to lengthen the amount of consecutive time slices allowed a program. Our kernel will service processes strictly on a linear basis or on a round robin as they come from the queue.

To implement a scheduler for the embedded system, we need to implement procedures, or macros, to start a new application, kill an application when it completes, and to pause an application if it needs time to access I/O. Each of these macros will access a scheduling queue located in the memory system at an available address, 0500H. The scheduling queue will use the data structure in Example 16-7 to make creating the queue fairly easy, and it will have room for 10 entries. This scheduling queue allows us to start up to 10 processes at a time.

EXAMPLE 16–7

```
PRESENT   DB 0      ; 00H = not and FFH = present
RAX       DW ?      ;context
RBX       DW ?
RCX       DW ?
RDX       DW ?
RSP       DW ?
RBP       DW ?
RSI       DW ?
RDI       DW ?
RFLAG     DW ?
RIP       DW ?
RCS       DW ?
RDS       DW ?
RES       DW ?
RSS       DW ?
DUMMY1    DW ?      ;for future use and as a pad to make the
DUMMY2    DB ?      ;entry size 32 bytes (20H bytes)
```

The data structure of Example 16–7 is copied into memory 10 times to complete the queue structure during system initialization; hence, it contains no active process at initialization. We also need a queue pointer initialized to 500H. The queue pointer is stored at location 4FEH in this example. Example 16–8 provides one possible initialization.

EXAMPLE 16–8

```
PUSH    DS
MOV     AX,0
MOV     DS,AX
MOV     SI,500H
MOV     BX,OFFSET PRESENT - 100H
MOV     CX,10
.REPEAT
   MOV     BYTE PTR DS:[SI],0
   ADD     SI,32
.UNTILCXZ
MOV     DS:[4FEH],500H   ;set queue pointer
POP     DS
```

The NEW procedure (installed at INT 60H in Example 16-9) adds a process to the queue and searches through the 10 entries until it finds a zero in the first byte (PRESENT), which indicates that the entry is empty. If it finds an empty entry, it places the starting address of the process into RCS and RIP and a 0200H into the RFLAG location. A 200H in RFLAG makes sure that the interrupt is enabled when the process begins, which prevents the system from crashing. NEW waits, if 10 processes are already scheduled, until one process ends. Each process is also assigned stack space in 256-byte sections beginning at offset address 7600H, so the lowest process has stack space 7500H–75FFH, the next has stack space 7600H–76FFH, and so on. The assignment of a stack area could be allocated by a memory manager algorithm.

EXAMPLE 16–9

```
INT60H    PROC    FARUSES DS AX DX SI
          MOV     AX,0              ;BX = offset, DX = segment
          MOV     DS,AX
          .REPEAT                   ;get stuck if full
              MOV     SI,500H - 32
              MOV     CX,10
              .REPEAT               ;find free entry
                  ADD     SI,32
                  DEC     CX
              .UNTIL BYTE PTR DS:[SI] == 0 || SI == 660H
          .UNTIL BYTE PTR DS:[SI] == 0
          CALL    SAVE_STATE
          MOV     WORD PTR DS:[SI+17],200H  ;flags
          MOV     DS:[SI+19],BX       ;IP
          MOV     DS:[SI+21],DX       ;CS
          MOV     AX,SS
          MOV     DS:[SI+27],AX     ;SS
          MOV     AX,10
          SUB     CX,AX
          SHL     AX,8
          ADD     AX,7500H
          MOV     DS:[SI+9],AX     ;SP
          MOV     BYTE PTR DS:[SI],0FFH ;activate process
          IRET
INT60H    ENDP
```

The PAUSE procedure is merely a call to the time slice procedure (INT 12H) which bails out of the process and returns control to the time slice procedure, prematurely ending the time slice for the process. This early out allows other processes to continue before returning to the current process.

The final control procedure (KILL), located at interrupt vector 61H as illustrated in Example 16–10, kills an application by placing a 00H into PRESENT of the queue data structure, which removes it from the scheduling queue.

EXAMPLE 16–10

```
INT61 PROC    FARUSES DS AX DX SI

        MOV     AX,0
        MOV     DS,AX
        MOV     SI,DS:[4FEH]   ;get queue pointer
        MOV     DS:[SI],0      ;kill it
        JMP     INT12A

INT61 ENDP
```

The time slice interrupt service procedure for an 80188EB using a 10 ms time slice appears in Example 16–11. Because this is an interrupt service procedure, care has been taken to make it as efficient as possible. Example 16–11 illustrates the time slice procedure located at interrupt vector 12H for operation with Timer 1 in the 80188EB microprocessor.

EXAMPLE 16–11

```
INT12 PROC    FARUSES   DS AX DX SI   ;timer 1 service

        MOV   AX,0              ;address shared memory
        MOV   DS,AX
        MOV   SI,DS:[4FEH]      ;get current queue pointer
        CALL  SAVE_STATE        ;save the current state
INT12A:
        ADD   SI,32             ;get next process
        .IF SI == 660H          ;make queue circular
```

```
        MOV  SI,500H
        .ENDIF
        .WHILE BYTE PTR DS:[SI] != FF
           ADD  SI,32
           .IF SI == 660H
              MOV  SI,500H
           .ENDIF
        .ENDW
        MOV  DX,0FF38H        ;Timer 1 count register
        MOV  AX,0             ;set count to 0000H
        OUT  DX,AX            ;for a complete 10 ms slice
        MOV  DX,0FF02H        ;clear interrupt
        MOV  AX,8000H
        OUT  DX,AX
        JMP  LOAD_STATE       ;get next processor from queue

INT12 ENDP
```

The system startup (placed after system initialization) must be a single process inside of an infinite wait loop, as shown in Example 16–12.

EXAMPLE 16–12

```
SYSTEM_START:

        MOV  BX,OFFSET WAITLOOP - 100H
        MOV  DX,0F800H
        INT  60H             ;Call NEW to start WAITLOOP thread
        ADD  DS:[4FEH],32 ;only for the this first time
        STI                  ;enable interrupts for the first time

        ;other processes would be started here


WAITLOOP:                    ;system idle loop
        .WHILE 1
           INT 12H           ;end slice early
        .ENDW
```

Finally, we use the SAVE_STATE and LOAD_STATE procedures to load and save the machine context as we switch from one process to another. These procedures are called by the time slice interrupt and appear in Example 16–13.

EXAMPLE 16–13

```
SAVE_STATE   PROC   NEAR

        MOV  DS:[SI+3],BX     ;save BX
        MOV  DS:[SI+5],CX     ;save CX
        MOV  DS:[SI+9],SP     ;save SP
        MOV  DS:[SI+11],BP    ;save BP
        MOV  DS:[SI+15],DI    ;save DI
        MOV  DS:[SI+25],ES    ;save ES
        MOV  DS:[SI+27],SS    ;save SS
        MOV  BP,SP            ;get SP
        MOV  AX,[BP+2]        ;get SI
        MOV  DS:[SI+13]       ;save SI
        MOV  AX,[BP+4]        ;get DX
        MOV  DS:[SI+7]        ;save DX
        MOV  AX,[BP+6]        ;get AX
        MOV  DS:[SI+1]        ;save AX
        MOV  AX,[BP+8]        ;get DS
        MOV  DS:[SI+23]       ;save DS
        MOV  AX,[BP+10]       ;get flags
```

```
                    MOV     DS:[SI+17]      ;save flags
                    MOV     AX,[BP+12]      ;get CS
                    MOV     DS:[SI+21]      ;save CS
                    MOV     AX,[BP+23]      ;get IP
                    MOV     DS:[SI+19]      ;save IP
                    RET

          SAVE_STATE   ENDP


          LOAD_STATE   PROC   NEAR

                    MOV     SS,DS:[SI+27]   ;get SS
                    MOV     SP,DS:[SI+9]    ;get SP
                    MOV     AX,DS:[SI+19]   ;get IP
                    PUSH    AX              ;IP to stack
                    MOV     AX,DS:[SI+21]   ;get CS
                    PUSH    AX              ;CS to stack
                    MOV     AX,DS:[SI+21]   ;get flags
                    PUSH    AX              ;flags to stack
                    MOV     AX,DS:[SI+23]   ;get DS
                    PUSH    AX              ;PUSH DS
                    MOV     AX,DS:[SI+13]   ;get SI
                    PUSH    AX              ;PUSH SI
                    MOV     AX,DS:[SI+1]    ;get AX
                    MOV     BX,DS:[SI+3]    ;get BX
                    MOV     CX,DS:[SI+5]    ;get CX
                    MOV     DX,DS:[SI+7]    ;get DX
                    MOV     BP,DS:[SI+11]   ;get BP
                    MOV     DI,DS:[SI+15]   ;get DI
                    MOV     ES,DS:[SI+25]   ;get ES
                    POP     SI
                    POP     DX
                    POP     AX
                    POP     DS
                    IRET

          LOAD_STATE   ENDP
```

16–5   INTRODUCTION TO THE 80286

The 80286 microprocessor is an advanced version of the 8086 microprocessor that was designed for multiuser and multitasking environments. The 80286 addresses 16M bytes of physical memory and 1G bytes of virtual memory by using its memory-management system. This section of the text introduces the 80286 microprocessor, which finds use in earlier AT-style personal computers that once pervaded the computer market and still find some applications. The 80286 is basically an 8086 that is optimized to execute instructions in fewer clocking periods than the 8086. The 80286 is also an enhanced version of the 8086 because it contains a memory manager. At this time, the 80286 no longer has a place in the personal computer system, but it does find applications in control systems as an embedded controller.

Hardware Features

Figure 16–29 shows the internal block diagram of the 80286 microprocessor. Note that like the 80186/80188, the 80286 does not incorporate internal peripherals; instead, it contains a memory-management unit (MMU) that is called the **address unit** in the block diagram.

As a careful examination of the block diagram reveals, address pins A23–A0, $\overline{BUSY}$, CAP, $\overline{ERROR}$, $\overline{PEREQ}$, and $\overline{PEACK}$ are new or additional pins that do not appear on the 8086

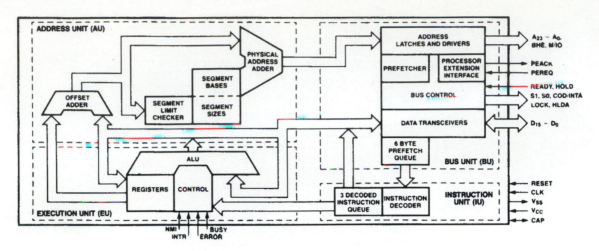

FIGURE 16–29   The block diagram of the 80286 microprocessor. (Courtesy of Intel Corporation.)

microprocessor. The $\overline{\text{BUSY}}$, $\overline{\text{ERROR}}$, $\overline{\text{PEREQ}}$, and $\overline{\text{PEACK}}$ signals are used with the microprocessor extension or coprocessor, of which the 80287 is an example. (Note that the $\overline{\text{TEST}}$ pin is now referred to as the $\overline{\text{BUSY}}$ pin.) The address bus is now 24-bits wide to accommodate the 16M bytes of physical memory. The CAP pin is connected to a 0.047 μF, ±20% capacitor that acts as a 12 V filter and connects to ground. The pin-outs of the 8086 and 80286 are illustrated in Figure 16–30 for comparative purposes. Note that the 80286 does not contain a multiplexed address/data bus.

FIGURE 16–30   The 8086 and 80286 microprocessor pin-outs. Notice that the 80286 does not have a multiplexed address/data bus.

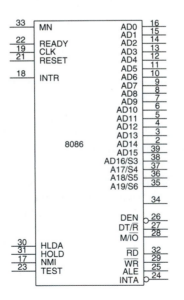

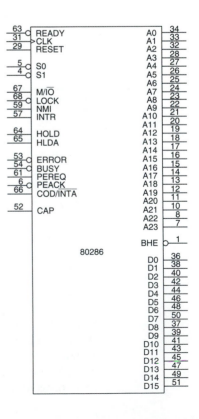

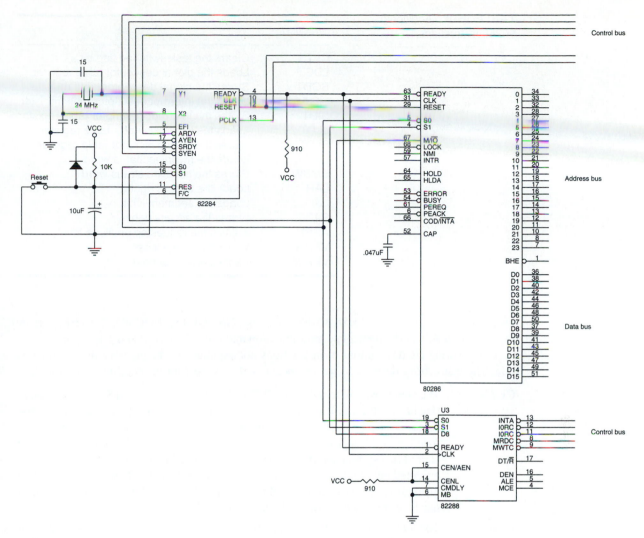

FIGURE 16–31   The interconnection of the 80286 microprocessor, 82284 clock generator, and 8288 system bus controller.

As mentioned in Chapter 1, the 80286 operates in both the real and protected modes. In the real mode, the 80286 addresses a 1M-byte memory address space and is virtually identical to the 8086. In the protected mode, the 80286 addresses a 16M byte memory space.

Figure 16–31 illustrates the basic 80286 microprocessor based-system. Notice that the clock is provided by the 82284 clock generator (similar to the 8284A) and the system control signals are provided by the 82288 system bus controller (similar to the 8288). Also, note the absence of the latch circuits used to de-multiplex the 8086 address/data bus.

Additional Instructions

The 80286 has even more instructions than its predecessors. These extra instructions control the virtual memory system through the memory manager of the 80286. Table 16–9 lists the additional 80286 instructions with a comment about the purpose of each instruction. These instructions are the only new instructions added to the 80286. Note that the 80286 contains the new

TABLE 16–9   Additional
80286 instructions.

Instruction	Purpose
CLTS	Clears the task-switched flag bit
LDGT	Loads the global descriptor table register
SGDT	Stores the global descriptor table register
LIDT	Loads the interrupt descriptor table register
SIDT	Stores the interrupt descriptor table register
LLDT	Loads the local descriptor table register
SLDT	Stores the local descriptor table register
LMSW	Loads the machine status word
SMSW	Stores the machine status word
LAR	Loads the access rights
LSL	Loads the segment limit
SAR	Stores the access rights
ARPL	Adjusts the requested privilege level
VERR	Verifies a read access
VERW	Verifies a write access

instructions added to the 80186/80188 such as INS, OUTS, BOUND, ENTER, LEAVE, PUSHA, POPA, and the immediate multiplication and immediate shift and rotate counts.

Following are descriptions of instructions not explained in the memory-management section. The instructions described here are special and only used for the conditions indicated.

CLTS   The **clear task-switched flag** (CLTS) instruction clears the TS (task-switched) flag bit to a logic 0. If the TS flag bit is a logic 1 and the 80287 numeric coprocessor is used by the task, an interrupt occurs (vector type 7). This allows the function of the coprocessor to be emulated with software. The CLTS instruction is used in a system and is considered a privileged instruction because it can be executed only in the protected mode at privilege level 0. There is no set TS flag instruction: this is accomplished by writing a logic 1 to bit position 3 (TS) of the machine status word (MSW) by using the LMSW instruction.

LAR   The **load access rights** (LAR) instruction reads the segment descriptor and places a copy of the access rights byte into a 16-bit register. An example is the LAR AX,BX instruction that loads AX with the access rights byte from the descriptor selected by the selector value found in BX. This instruction is used to get the access rights so that it can be checked before a program uses the segment of memory described by the descriptor.

LSL   The **load segment limit** (LSL) instruction loads a user-specified register with the segment limit. For example, the LSL AX,BX instruction loads AX with the limit of the segment described by the descriptor selected by the selector in BX. This instruction is used to test the limit of a segment.

ARPL   The **adjust requested privilege level** (ARPL) instruction is used to test a selector so that the privilege level of the requested selector is not violated. An example is ARPL AX,CX: AX contains the requested privilege level and CX contains the selector value to be used to access a descriptor. If the requested privilege level is of a lower priority than the descriptor under test, the zero flag is set. This may require that a program adjust the requested privilege level or indicate a privilege violation.

VERR   The **verify for read access** (VERR) instruction verifies that a segment can be read. Recall from Chapter 1 that a code segment can be read-protected. If the code segment can be read, the zero flag bit is set. The VERR AX instruction tests the descriptor selected by the AX register.

VERW   The **verify for write access** (VERW) instruction is used to verify that a segment can be written. Recall from Chapter 1 that a data segment can be write-protected. If the data segment can be written, the zero flag bit is set.

The Virtual Memory Machine

A **virtual memory machine** is a machine that maps a larger memory space (1G bytes for the 80286) into a much smaller physical memory space (16M bytes for the 80286), which allows a very large system to execute in smaller physical memory systems. This is accomplished by spooling the data and programs between the fixed disk memory system and the physical memory. Addressing a 1G-byte memory system is accomplished by the descriptors in the 80286 microprocessor. Each 80286 descriptor describes a 64K-byte memory segment and the 80286 allows 16K descriptors. This (64K × 16K) allows a maximum of 1G bytes of memory to be described for the system.

As mentioned in Chapter 1, descriptors describe the memory segment in the protected mode. The 80286 has descriptors that define codes, data, stack segments, interrupts, procedures, and tasks. Descriptor accesses are performed by loading a segment register with a selector in the protected mode. The selector accesses a descriptor that describes an area of the memory. Additional details on descriptors and their applications are defined in Chapter 1, and also Chapters 17 and 18. Please refer to these chapters for a detailed view of the protected mode memory management system.

16–6   SUMMARY

1. The 80186/80188 microprocessors contain the same basic instruction set as the 8086/8088 microprocessors, except that a few additional instructions are added. The 80186/80188 are thus enhanced versions of the 8086/8088 microprocessors. The new instructions include PUSHA, POPA, INS, OUTS, BOUND, ENTER, LEAVE, and immediate multiplication and shift/rotate counts.

2. Hardware enhancements to the 80186/80188 include a clock generator, a programmable interrupt controller, three programmable timers, a programmable DMA controller, a programmable chip selection logic unit, a watchdog timer, a dynamic RAM refresh logic circuit, and additional features on various versions.

3. The clock generator allows the 80186/80188 to operate from an external TTL-level clock source, or from a crystal attached to the X1 (CLKIN) and X2 (OSCOUT) pins. The frequency of the crystal is twice the operating frequency of the microprocessor. The 80186/80188 microprocessors are available in speeds of 6–20 MHz.

4. The programmable interrupt controller arbitrates all internal and external interrupt requests. It is also capable of operating with two external 8259A interrupt controllers.

5. There are three programmable timers located within the 80186/80188. Each timer is a fully programmable, 16-bit counter used to generate wave-forms or count events. Two of the timers, timers 0 and 1, have external inputs and outputs. The third timer, timer 2, is clocked from the system clock and is used either to provide a clock for another timer or to request a DMA action.

6. The programmable DMA controller is a fully programmable, two-channel controller. DMA transfers are made between memory and I/O, I/O and I/O, or between memory locations. DMA requests occur from software, hardware, or the output of timer 2.

7. The programmable chip selection unit is an internal decoder that provides up to 13 output pins to select memory (6 pins) and I/O (7 pins). It also inserts 0–3 wait states, with or without external READY synchronization. On the EB and EC versions, the number of wait states can be programmed from 0–15.

8. The only difference between the timing of the 80186/80188 and the 8086/8088 is that ALE appears one-half clock pulse earlier. Otherwise, the timing is identical.

9. The 6 MHz version of the 80186/80188 allows 417 ns of access time for the memory; the 8 MHz version allows 309 ns of access time.

10. The internal 80186/80188 peripherals are programmed via a peripheral control block (PCB), initialized at I/O ports FF00H–FFFFH. The PCB may be moved to any area of memory or I/O by changing the contents of the PCB relocation register at initial I/O location FFFEH and FFFFH.

11. The 80286 is an 8086 that has been enhanced to include a memory-management unit (MMU). The 80286 is capable of addressing a 16M-byte physical memory space because of the management unit.

12. The 80286 contains the same instructions as the 80186/80188, except for a handful of additional instructions that control the memory-management unit.

13. Through the memory-management unit, the 80286 microprocessor addresses a virtual memory space of 1G bytes, as specified by the 16K descriptors stored in two descriptor tables.

16–7   QUESTIONS AND PROBLEMS

1. List the differences between the 8086/8088 and the 80186/80188 microprocessors.
2. What hardware enhancements are added to the 80186/80188 that are not present in the 8086/8088?
3. The 80186/80188 is packaged in what types of integrated circuits?
4. If the 20 MHz crystal is connected to X1 and X2, what frequency signal is found at CLKOUT?
5. Describe the differences between the 80C188XL and the 80C188EB versions of the 80188 embedded controller.
6. The fan-out from any 80186/80188 pin is _____ for a logic 0.
7. How many clocking periods are found in an 80186/80188 bus cycle?
8. What is the main difference between the 8086/8088 and 80186/80188 timing?
9. What is the importance of memory access time?
10. How much memory access time is allowed by the 80186/80188 if operated with a 6 MHz clock?
11. Where is the peripheral control block located after the 80186/80188 are reset?
12. Write the software required to move the peripheral control block to memory locations 10000H–100FFH.
13. Which interrupt vector is used by the INT0 pin on the 80186/80188 microprocessors?
14. How many interrupt vectors are available to the interrupt controller located within the 80186/80188 microprocessors?
15. Which two modes of operation are available to the interrupt controller?
16. What is the purpose of the interrupt control register?
17. Whenever an interrupt source is masked, the mask bit in the interrupt mask register is a logic _____.
18. What is the difference between the interrupt poll and interrupt poll status registers?
19. What is the purpose of the end-of-interrupt (EOI) register?
20. How many 16-bit timers are found within the 80186/80188?
21. Which timers have input and output pin connections?
22. Which timer connects to the system clock?
23. If two maximum-count compare registers are used with a timer, explain the operation of the timer.
24. What is the purpose of the INH timer control register bit?
25. What is the purpose of the P timer control register bit?

26. The timer control register bit ALT selects what type of operation for timers 0 and 1?
27. Explain how the timer output pins are used.
28. Develop a program that causes timer 1 to generate a continuous signal that is a logic 1 for 123 counts and a logic 0 for 23 counts.
29. Develop a program that causes timer 0 to generate a single pulse after 105 clock pulses on its input pin have occurred.
30. How many DMA channels are controlled by the DMA controller in the 80C186XL?
31. The DMA controller's source and destination registers are each _____ bits wide.
32. How is the DMA channel started with software?
33. The chip selection unit (XL and EA) has ____ pins to select memory devices.
34. The chip selection unit (XL and EA) has ____ pins to select peripheral devices.
35. The last location of the upper memory block, as selected by the $\overline{UCS}$ pin, is location _____.
36. The middle memory chip selection pins (XL and EA) are programmed for a(n) _____ address and a block size.
37. The lower memory area, as selected by $\overline{LCS}$, begins at address _____.
38. The internal wait state generator (EB and EC versions) is capable of inserting between ____ and ____ wait states.
39. Program register A8H (XL and EA) so that the mid-range memory block size is 128K bytes.
40. What is the purpose of the EX bit in register A8H?
41. Develop the software required to program the $\overline{GCS3}$ pin so that it selects memory from location 20000H–2FFFFH and inserts two wait states.
42. Develop the software required to program the $\overline{GSC4}$ pin so that it selects an I/O device for ports 1000H–103FH and inserts one wait state.
43. The 80286 microprocessor addresses _____ bytes of physical memory.
44. When the memory manager is in use, the 80286 addresses _____ bytes of virtual memory.
45. The instruction set of the 80286 is identical to the _____, except for the memory-management instructions.
46. What is the purpose of the VERR instruction?
47. What is the purpose of the LSL instruction?
48. What is an RTOS?
49. How can multiple threads be handled with the RTOS?

CHAPTER 17

The 80386 and 80486 Microprocessors

INTRODUCTION

The 80386 microprocessor is a full 32-bit version of the earlier 8086/80286 16-bit microprocessors, and represents a major advancement in the architecture—a switch from a 16-bit architecture to a 32-bit architecture. Along with this larger word size are many improvements and additional features. The 80386 microprocessor features multitasking, memory management, virtual memory (with or without paging), software protection, and a large memory system. All software written for the early 8086/8088 and the 80286 are upward-compatible to the 80386 microprocessor. The amount of memory addressable by the 80386 is increased from the 1M bytes found in the 8086/8088 and the 16M bytes found in the 80286, to 4G bytes in the 80386. The 80386 can switch between protected mode and real mode without resetting the microprocessor. Switching from protected mode to real mode was a problem on the 80286 microprocessor because it required a hardware reset.

The 80486 microprocessor is an enhanced version of the 80386 microprocessor that executes many of its instructions in one clocking period. The 80486 microprocessor also contains an 8K-byte cache memory and an improved 80387 numeric coprocessor. (Note that the 80486DX4 contains a 16K-byte cache.) When the 80486 is operated at the same clock frequency as an 80386, it performs with about a 50 percent speed improvement. In Chapter 18, we shall see that the Pentium and Pentium Pro, which both contain a 16K cache memory, perform at better than twice the speed of the 80486 microprocessor. The Pentium and Pentium Pro also contain improved numeric coprocessors that operate five times faster than the 80486 numeric coprocessor. Chapter 19 deals with additional improvements in the Pentium II–Pentium 4 microprocessors.

CHAPTER OBJECTIVES

Upon completion of this chapter, you will be able to:

1. Contrast the 80386 and 80486 microprocessors with earlier Intel microprocessors.
2. Describe the operation of the 80386 and 80486 memory management unit and paging unit.
3. Switch between protected mode and real mode.
4. Define the operation of additional 80386/80486 instructions and addressing modes.
5. Explain the operation of a cache memory system.
6. Detail the interrupt structure and direct memory access structure of the 80386/80486.
7. Contrast the 80486 with the 80386 microprocessor.
8. Explain the operation of the 80486 cache memory.

17-1 INTRODUCTION TO THE 80386 MICROPROCESSOR

Before the 80386 or any other microprocessor can be used in a system, the function of each pin must be understood. This section of the chapter details the operation of each pin, along with the external memory system and I/O structures of the 80386.

Figure 17–1 illustrates the pin-out of the 80386DX microprocessor. The 80386DX is packaged in a 132-pin PGA (pin grid array). Two versions of the 80386 are commonly available: the 80386DX, which is illustrated and described in this chapter; the other is the 80386SX, which is a reduced bus version of the 80386. A new version of the 80386—the 80386EX—incorporates the AT bus system, dynamic RAM controller, programmable chip selection logic, 26 address pins, 16 data pins, and 24 I/O pins. Figure 17–2 illustrates the 80386EX embedded PC.

The 80386DX addresses 4G bytes of memory through its 32-bit data bus and 32-bit address. The 80386SX, more like the 80286, addresses 16M bytes of memory with its 24-bit address bus via its 16-bit data bus. The 80386SX was developed after the 80386DX for applications that didn't require the full 32-bit bus version. The 80386SX is found in many personal computers that use the same basic motherboard design as the 80286. At the time that the 80386SX was popular, most applications, including Windows, required fewer than 16M bytes of memory, so the 80386SX is a popular and a less costly version of the 80386 microprocessor. Even though the 80486 has become a less expensive upgrade path for newer systems, the 80386 still can be used for many applications. For example, the 80386EX does not appear in computer systems, but it is becoming very popular in embedded applications.

As with earlier versions of the Intel family of microprocessors, the 80386 requires a single +5.0 V power supply for operation. The power supply current averages 550 mA for the 25 MHz version of the 80386, 500 mA for the 20 MHz version, and 450 mA for the 16 MHz version. Also available is a 33 MHz version that requires 600 mA of power supply current. The power supply current for the 80386EX is 320 mA when operated at 33 MHz. Note that during some modes of

FIGURE 17–1 The pin-outs of the 80386DX and 80386SX microprocessors.

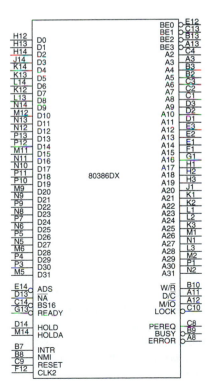

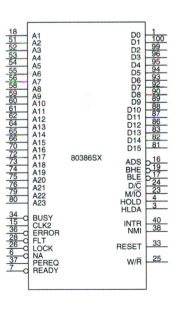

FIGURE 17–2 The 80386EX embedded PC.

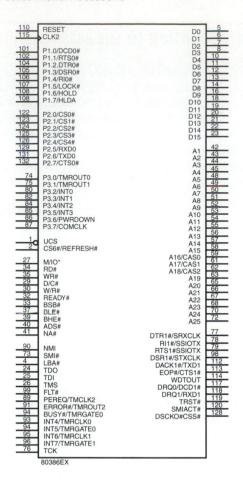

80386EX

normal operation, power supply current can surge to over 1.0 A. This means that the power supply and power distribution network must be capable of supplying these current surges. This device contains multiple Vcc and Vss connections that must all be connected to +5.0 V and grounded for proper operation. Some of the pins are labeled N/C (no connection) and must not be connected. Additional versions of the 80386SX and 80386EX are available with a +3.3 V power supply. They are often found in portable notebook or laptop computers and are usually packaged in a surface mount device.

Each 80386 output pin is capable of providing 4.0 mA (address and data connections) or 5.0 mA (other connections). This represents an increase in drive current compared to the 2.0 mA available on earlier 8086, 8088, and 80286 output pins. The output current available on most 80386EX output pins is 8.0 mA. Each input pin represents a small load, requiring only ±10 μA of current. In some systems, except the smallest, these current levels require bus buffers.

The function of each 80386DX group of pins follows:

A31–A2   **Address bus connections** address any of the $1G \times 32$ memory locations found in the 80386 memory system. Note that A0 and A1 are encoded in the bus enable ($\overline{BE3}$–$\overline{BE0}$) to select any or all of the four bytes in a 32-bit wide memory location. Also note that because the 80386SX contains a 16-bit data bus in place of the 32-bit data bus found on the 80386DX, A1 is present on the 80386SX, and the bank selection signals are replaced with $\overline{BHE}$ and $\overline{BLE}$. The $\overline{BHE}$ signal enables the upper data bus half; the $\overline{BLE}$ signal enables the lower data bus half.

D31–D0   **Data bus connections** transfer data between the microprocessor and its memory and I/O system. Note that the 80386SX contains D15–D0.

BE3–BE0   **Bank enable signals** select the access of a byte, word, or doubleword of data. These signals are generated internally by the microprocessor from address bits A1 and A0. On the 80386SX, these pins are replaced by $\overline{BHE}$, $\overline{BLE}$, and A1.

M/IO   **Memory/IO** selects a memory device when a logic 1 or an I/O device when a logic 0. During the I/O operation, the address bus contains a 16-bit I/O address on address connections A15–A2.

W/R   **Write/read** indicates that the current bus cycle is a write when a logic 1 or a read when a logic 0.

ADS   The **address data strobe** becomes active whenever the 80386 has issued a valid memory or I/O address. This signal is combined with the W/R signal to generate the separate read and write signals present in the earlier 8086–80286 microprocessor-based systems.

RESET   **Reset** initializes the 80386, causing it to begin executing software at memory location FFFFFFF0H. The 80386 is reset to the real mode, and the leftmost 12 address connections remain logic 1s (FFFH) until a far jump or far call is executed. This allows compatibility with earlier microprocessors.

CLK2   **Clock times 2** is driven by a clock signal that is twice the operating frequency of the 80386. For example, to operate the 80386 at 16 MHz, we apply a 32 MHz clock to this pin.

READY   **Ready** controls the number of wait states inserted into the timing to lengthen memory accesses.

LOCK   **Lock** becomes a logic 0 whenever an instruction is prefixed with the LOCK: prefix. This is used most often during DMA accesses.

D/C   **Data/control** indicates that the data bus contains data for or from memory or I/O when a logic 1. If D/$\overline{C}$ is a logic 0, the microprocessor is halted or executes an interrupt acknowledge.

BS16   **Bus size 16** selects either a 32-bit data bus ($\overline{BS16}$ = 1) or a 16-bit data bus ($\overline{BS16}$ = 0). In most cases, if an 80386DX is operated on a 16-bit data bus, we use the 80386SX that has a 16-bit data bus. On the 80386EX, the $\overline{BS8}$ pin selects an 8-bit data bus.

NA   **Next address** causes the 80386 to output the address of the next instruction or data in the current bus cycle. This pin is often used for pipelining the address.

HOLD   **Hold** requests a DMA action.

HLDA   **Hold acknowledge** indicates that the 80386 is currently in a hold condition.

PEREQ   The **coprocessor request** asks the 80386 to relinquish control and is a direct connection to the 80387 arithmetic coprocessor.

BUSY   **Busy** is an input used by the WAIT or FWAIT instruction that waits for the coprocessor to become not busy. This is also a direct connection to the 80387 from the 80386.

ERROR   **Error** indicates to the microprocessor that an error is detected by the coprocessor.

INTR   An **interrupt** request is used by external circuitry to request an interrupt.

NMI   A **non-maskable interrupt** requests a non-maskable interrupt as it did on the earlier versions of the microprocessor.

The Memory System

The physical memory system of the 80386DX is 4G bytes in size and is addressed as such. If virtual addressing is used, 64T bytes are mapped into the 4G bytes of physical space by the memory management unit and descriptors. (Note that virtual addressing allows a program to be larger than 4G bytes if a method of swapping with a very large hard disk drive exists.) Figure 17–3 shows the organization of the 80386DX physical memory system.

The memory is divided into four 8-bit wide memory banks, each containing up to 1G bytes of memory. This 32-bit wide memory organization allows bytes, words, or doublewords of memory data to accessed directly. The 80386DX transfers up to a 32-bit wide number in a single memory cycle, whereas the early 8088 requires four cycles to accomplish the same transfer, and the 80286 and 80386SX require two cycles. Today, the data width is important, especially with single-precision floating-point numbers that are 32 bits wide. High-level software normally uses floating-point numbers for data storage, so 32-bit memory locations speed the execution of high-level software when it is written to take advantage of this wider memory.

Each memory byte is numbered in hexadecimal as they were in prior versions of the family. The difference is that the 80386DX uses a 32-bit wide memory address, with memory bytes numbered from location 00000000H–FFFFFFFFH.

The two memory banks in the 8086, 80286, and 80386SX system are accessed via $\overline{\text{BLE}}$ (A0 on the 8086 and 80286) and $\overline{\text{BHE}}$. In the 80386DX, the memory banks are accessed via four bank enable signals $\overline{\text{BE3}}$–$\overline{\text{BE0}}$. This arrangement allows a single byte to be accessed when one bank enable signal is activated by the microprocessor. It also allows a word to be addressed when two bank enable signals are activated. In most cases, a word is addressed in bank 0 and 1, or in bank 2 and 3. Memory location 00000000H is in bank 0, location 00000001H is in bank 1, location 00000002H is in bank 2, and location 00000003H is in bank 3. The 80386DX does not contain address connections A0 and A1 because these have been encoded as the bank enable signals. Likewise, the 80386SX does not contain the A0 address pin because it is encoded in the $\overline{\text{BLE}}$ and $\overline{\text{BHE}}$ signals. The 80386EX addresses data either in two banks for a 16-bit wide memory system if $\overline{\text{BS8}}$ = 1 or as an 8-bit system if $\overline{\text{BS8}}$ = 0.

FIGURE 17–3   The memory system for the 80386 microprocessor. Notice that the memory is organized as four banks, each containing 1G byte. Memory is accessed as 8-, 16-, or 32-bit data.

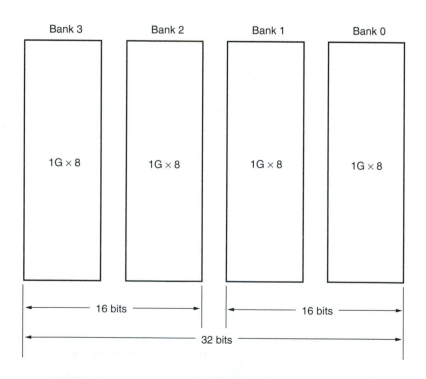

Buffered System.   Figure 17–4 shows the 80386DX connected to buffers that increase fan-out from its address, data, and control connections. This microprocessor is operated at 25 MHz using a 50 MHz clock input signal that is generated by an integrated oscillator module. Oscillator modules are usually used to provide a clock in modern microprocessor-based equipment. The HLDA signal is used to enable all buffers in a system that uses direct memory access. Otherwise, the buffer enable pins are connected to ground in a non-DMA system.

Pipelines and Caches.   The cache memory is a buffer that allows the 80386 to function more efficiently with lower DRAM speeds. A **pipeline** is a special way of handling memory accesses so the memory has additional time to access data. A 16 MHz 80386 allows memory devices with access times of 50 ns or less to operate at full speed. Obviously, there are few DRAMs currently available with these access times. In fact, the fastest DRAMs currently in use have an access time of 60 ns or longer. This means that some technique must be found to interface these memory devices, which are slower than required by the microprocessor. Three techniques are available: interleaved memory, caching, and a pipeline.

The pipeline is the preferred means of interfacing memory because the 80386 microprocessor supports pipelined memory accesses. Pipelining allows memory an extra clocking period to access data. The extra clock extends the access time from 50 ns to 81 ns on an 80386 operating with a 16 MHz clock. The **pipe,** as it is often called, is set up by the microprocessor. When an instruction is fetched from memory, the microprocessor often has extra time before the next instruction is fetched. During this extra time, the address of the next instruction is sent out from the address bus ahead of time. This extra time (one clock period) is used to allow additional access time to slower memory components.

Not all memory references can take advantage of the pipe, which means that some memory cycles are not pipelined. These non-pipelined memory cycles request one wait state if the normal pipeline cycle requires no wait states. Overall, a pipe is a cost-saving feature that reduces the access time required by the memory system in low-speed systems.

Not all systems can take advantage of the pipe. Those systems typically operate at 20, 25, or 33 MHz. In these higher-speed systems, another technique must be used to increase the memory system speed. The **cache** memory system improves overall performance of the memory systems for data that are accessed more than once. Note that the 80486 contains an internal cache called a **level one cache** and the 80386 can only contain an external cache called a **level two cache.**

A **cache** is a high-speed memory system that is placed between the microprocessor and the DRAM memory system. Cache memory devices are usually static RAM memory components with access times of less than 25 ns. In many cases, we see level 2 cache memory systems with sizes between 32K and 1M byte. The size of the cache memory is determined more by the application than by the microprocessor. If a program is small and refers to little memory data, a small cache is beneficial. If a program is large and references large blocks of memory, the largest cache size possible is recommended. In many cases, a 64K-byte cache improves speed sufficiently, but the maximum benefit is often derived from a 256K-byte cache. It has been found that increasing the cache size much beyond 256K provides little benefit to the operating speed of the system that contains an 80386 microprocessor.

Interleaved Memory Systems.   An **interleaved memory system** is another method of improving the speed of a system. Its only disadvantage is that it costs considerably more memory because of its structure. Interleaved memory systems are present in some systems, so memory access times can be lengthened without the need for wait states. In some systems, an interleaved memory may still require wait states, but may reduce their number. An interleaved memory system requires two or more complete sets of address buses and a controller that provides addresses for each bus. Systems that employ two complete buses are called a **two-way interleave;** systems that use four complete buses are called a **four-way interleave.**

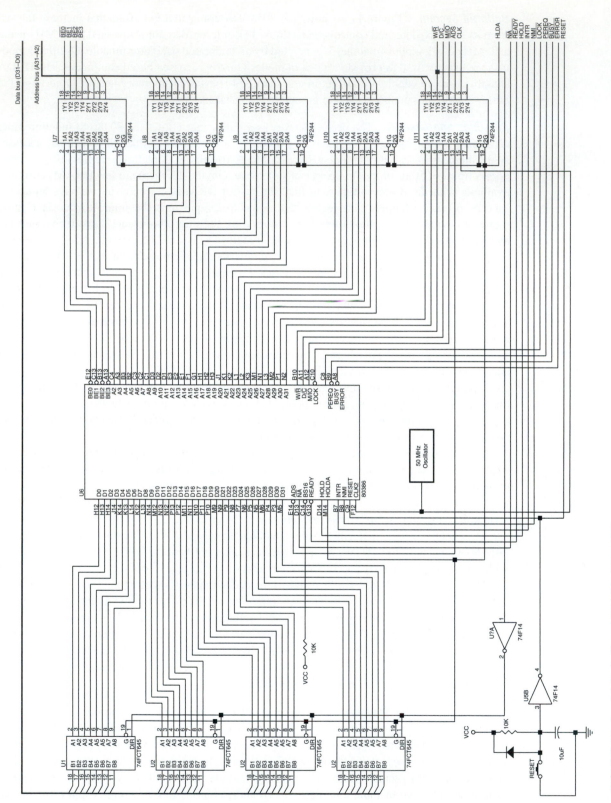

FIGURE 17–4 A fully buffered 25 MHz 80386DX.

An interleaved memory is divided into two or four parts. For example, if an interleaved memory system is developed for the 80386SX microprocessor, one part contains the 16-bit addresses 000000H–000001H, 000004H–000005H, etc.; the other part contains addresses 000002–000003, 000006H–000007H, etc. While the microprocessor accesses locations 000000H–000001H, the interleave control logic generates the address strobe signal for locations 000002H–000003H. This selects and accesses the word at location 000002H–000003H, while the microprocessor processes the word at location 000000H–000001H. This process alternates memory sections, thus increasing the performance of the memory system.

Interleaving lengthens the amount of access time provided to the memory because the address is generated to select the memory before the microprocessor accesses it. This is because the microprocessor pipelines memory addresses, sending the next address out before the data are read from the last address.

The problem with interleaving, although not major, is that the memory addresses must be accessed so that each section is alternately addressed. This does not always happen as a program executes. Under normal program execution, the microprocessor alternately addresses memory approximately 93 percent of the time. For the remaining 7 percent, the microprocessor addresses data in the same memory section, which means that in these 7 percent of the memory accesses, the memory system must cause wait states because of the reduced access time. The access time is reduced because the memory must wait until the previous data are transferred before it can obtain its address. This leaves the memory with less access time; therefore, a wait state is required for accesses in the same memory bank.

See Figure 17–5 for the timing diagram of the address as it appears at the microprocessor address pins. This timing diagram shows how the next address is output before the current data are accessed. It also shows how access time is increased by using interleaved memory addresses for each section of memory compared to a non-interleaved access, which requires a wait state.

Figure 17–6 pictures the interleave controller. Admittedly, this is a complex logic circuit, which needs some explanation. First, if the SEL input (used to select this section of the memory) is inactive (logic 0), then the $\overline{\text{WAIT}}$ signal is a logic 1. Also, both ALE0 and ALE1, used to strobe the address to the memory sections, are both logic 1s, causing the latches connected to them to become transparent.

As soon as the SEL input becomes a logic 1, this circuit begins to function. The A1 input is used to determine which latch (U2B or U5A) becomes a logic 0, selecting a section of the memory. Also the ALE pin that becomes a logic 0 is compared with the previous state of the ALE pins. If the same section of memory is accessed a second time, the $\overline{\text{WAIT}}$ signal becomes a logic 0, requesting a wait state.

Figure 17–7 illustrates an interleaved memory system that uses the circuit of Figure 17–6. Notice how the ALE0 and ALE1 signals are used to capture the address for either section of memory. The memory in each bank is 16-bits wide. If accesses to memory require 8-bit data, the system causes wait states, in most cases. As a program executes, the 80386SX fetches instruction 16-bits at a time from normally sequential memory locations. Program execution uses interleaving in most cases. If a system is going to access mostly 8-bit data, it is doubtful that memory interleaving will reduce the number of wait states.

The access time allowed by an interleaved system, such as the one shown in Figure 17–7, is increased to 112 ns from 69 ns by using a 16 MHz system clock. (If a wait state is inserted, access time with a 16 MHz clock is 136 ns, which means that an interleaved system performs at about the same rate as a system with one wait state.) If the clock is increased to 20 MHz, the interleaved memory requires 89.6 ns, where standard, non-interleaved memory interfaces allow 48 ns for memory access. At this higher clock rate, 80 ns DRAMs function properly, without wait states when the memory addresses are interleaved. If an access to the same section occurs, then a wait state is inserted.

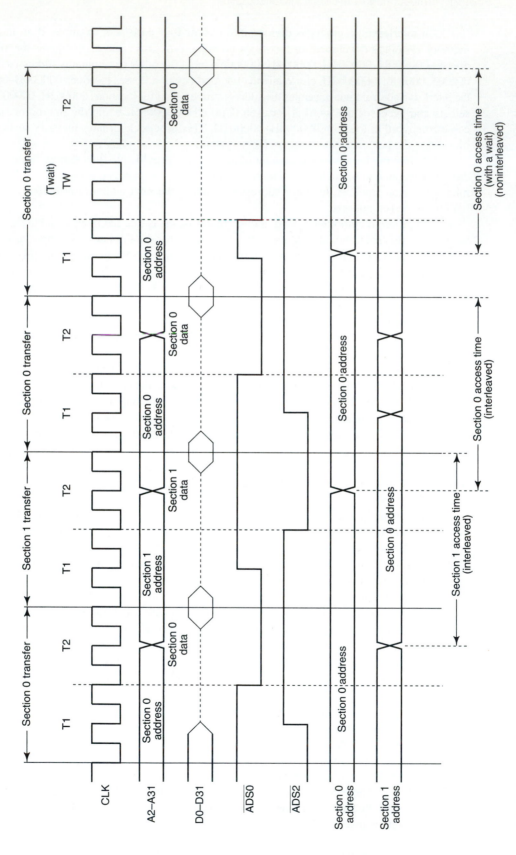

FIGURE 17–5 The timing diagram of an interleaved memory system showing the access times and address signals for both sections of memory.

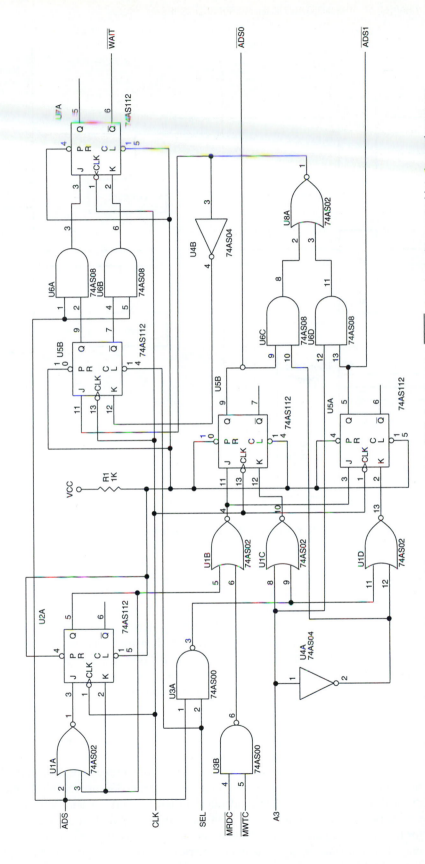

FIGURE 17–6 The interleaved control logic, which generates separate ADS signals and a $\overline{\text{WAIT}}$ signal used to control interleaved memory.

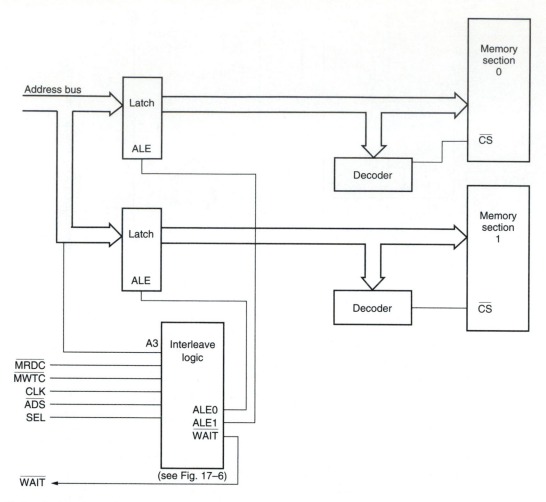

FIGURE 17–7   An interleaved memory system showing the address latches and the interleaved logic circuit.

The Input/Output System

The 80386 input/output system is the same as that found in any Intel 8086 family micro-processor-based system. There are 64K different bytes of I/O space available if isolated I/O is implemented. With isolated I/O, the IN and OUT instructions are used to transfer I/O data between the microprocessor and I/O devices. The I/O port address appears on address bus connections A15–A2, with $\overline{BE3}$–$\overline{BE0}$ used to select a byte, word, or doubleword of I/O data. If memory-mapped I/O is implemented, then the number of I/O locations can be any amount up to 4G bytes. With memory-mapped I/O, any instruction that transfers data between the microprocessor and memory system can be used for I/O transfers because the I/O device is treated as a memory device. Almost all 80386 systems use isolated I/O because of the I/O protection scheme provided by the 80386 in protected mode operation.

Figure 17–8 shows the I/O map for the 80386 microprocessor. Unlike the I/O map of earlier Intel microprocessors, which were 16-bits wide, the 80386 uses a full 32-bit wide I/O system divided into four banks. This is identical to the memory system, which is also divided into four banks. Most I/O transfers are 8-bits wide because we often use ASCII code (a 7-bit code) for transferring alphanumeric data between the microprocessor and printers and keyboards. This may change if Unicode, a 16-bit alphanumeric code, becomes common and replaces ASCII code. Recently, I/O devices that are 16- and even 32-bits wide have appeared for systems such as disk

FIGURE 17–8 The isolated I/O map for the 80386 microprocessor. Here four banks of 8-bits each are used to address 64K different I/O locations. I/O is numbered from location 0000H to FFFFH.

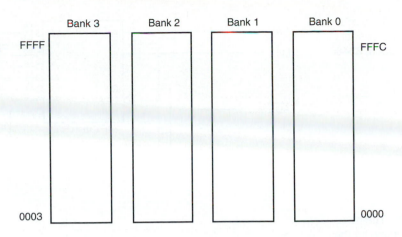

memory and video display interfaces. These wider I/O paths increase the data transfer rate between the microprocessor and the I/O device when compared to 8-bit transfers.

The I/O locations are numbered from 0000H–FFFFH. A portion of the I/O map is designated for the 80387 arithmetic coprocessor. Although the port numbers for the coprocessor are well above the normal I/O map, it is important that they be taken into account when decoding I/O space (overlaps). The coprocessor uses I/O location 800000F8H–800000FFH for communications between the 80387 and 80386. The 80287 numeric coprocessor designed to use with the 80286 uses the I/O addresses 00F8H–00FFH for coprocessor communications. Because we often decode only address connections A15–A2 to select an I/O device, be aware that the coprocessor will activate devices 00F8H–00FFH unless address line A31 is also decoded. This should present no problem because you really should not be using I/O ports 00F8H–00FFH for any purpose.

The only new feature that was added to the 80386 with respect to I/O is the I/O privilege information added to the tail end of the TSS when the 80386 is operated in protected mode. As described in the section on memory management, an I/O location can be blocked or inhibited in the protected mode. If the blocked I/O location is addressed, an interrupt (type 13, general fault) is generated. This scheme is added so that I/O access can be prohibited in a multiuser environment. Blocking is an extension of the protected mode operation, as are privilege levels.

Memory and I/O Control Signals

The memory and I/O are controlled with separate signals. The $M/\overline{IO}$ signal indicates whether the data transfer is between the microprocessor and the memory ($M/\overline{IO} = 1$) or I/O ($M/\overline{IO} = 0$). In addition to $M/\overline{IO}$, the memory and I/O systems must read or write data. The $W/\overline{R}$ signal is a logic 0 for a read operation, and a logic 1 for a write operation. The $\overline{ADS}$ signal is used to qualify the $M/\overline{IO}$ and $W/\overline{R}$ control signals. This is a slight deviation from earlier Intel microprocessors, which didn't use $\overline{ADS}$ for qualification.

See Figure 17–9 for a simple circuit that generates four control signals for the memory and I/O devices in the system. Notice that two control signals are developed for memory control ($\overline{MRDC}$ and $\overline{MWTC}$) and two for I/O control ($\overline{IORC}$ and $\overline{IOWC}$). These signals are consistent with the memory and I/O control signals generated for use in earlier versions of the Intel microprocessor.

Timing

Timing is important for understanding how to interface memory and I/O to the 80386 microprocessor. Figure 17–10 shows the timing diagram of a non-pipelined memory read cycle. Note that the timing is referenced to the CLK2 input signal and that a bus cycle consists of four clocking periods.

Each bus cycle contains two clocking states with each state (T1 and T2) containing two clocking periods. Note in Figure 17–10 that the access time is listed as time number 3. The

FIGURE 17–9 Generation of memory and I/O control signals for the 80386, 80486, and Pentium.

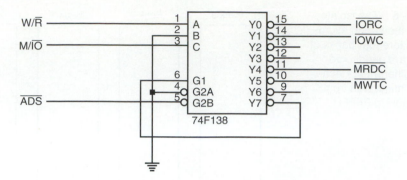

FIGURE 17–10 The non-pipelined read timing for the 80386 microprocessor.

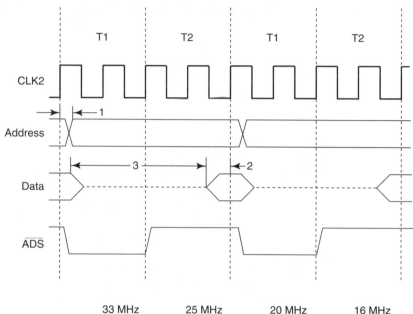

	33 MHz	25 MHz	20 MHz	16 MHz
Time 1:	4–15 ns	4–21 ns	4–30 ns	4–36 ns
Time 2:	5 ns	7 ns	11 ns	11 ns
Time 3:	46 ns	52 ns	59 ns	78 ns

16 MHz version allows memory an access time of 78 ns before wait states are inserted in this non-pipelined mode of operation. To select the non-pipelined mode, we place a logic 1 on the $\overline{\text{NA}}$ pin.

Figure 17–11 illustrates the read timing when the 80386 is operated in the pipelined mode. Notice that additional time is allowed to the memory for accessing data because the address is sent out early. Pipelined mode is selected by placing a logic 0 on the $\overline{\text{NA}}$ pin and by using address latches to capture the pipelined address. The clock pulse that is applied to the address latches comes from the $\overline{\text{ADS}}$ signal. Address latches must be used with a pipelined system, as well as with interleaved memory banks. The minimum number of interleaved banks of two and four have been successfully used in some applications.

Notice that the pipelined address appears one complete clocking state before it normally appears with non-pipelined addressing. In the 16 MHz version of the 80386, this allows an additional 62.5 ns for memory access. In a non-pipelined system, a memory access time of 78 ns is allowed to the memory system; in a pipelined system, 140.5 ns is allowed. The advantages of the pipelined system are that no wait states are required (in many, but not all bus cycles) and much lower-speed memory devices may be connected to the microprocessor. The disadvantage is that we need to interleave memory to use a pipe, which requires additional circuitry and occasional wait states.

FIGURE 17–11 The pipelined read timing for the 80386 microprocessor.

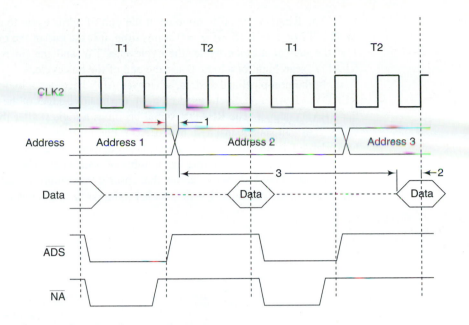

Wait States

Wait states are needed if memory access times are long compared with the time allowed by the 80386 for memory access. In a non-pipelined 33 MHz system, memory access time is only 46 ns. Currently, no DRAM memory exists that has an access time of 46 ns. This means that wait states must be introduced to access the DRAM (one wait for 60 ns DRAM) or an EPROM that has an access time of 100 ns (two waits). Note that this wait state is built into a motherboard and cannot be removed.

The $\overline{READY}$ input controls whether or not wait states are inserted into the timing. The $\overline{READY}$ input on the 80386 is a dynamic input that must be activated during each bus cycle. Figure 17–12 shows a few bus cycles with one normal (no wait) cycle and one that contains a single wait state. Notice how the $\overline{READY}$ is controlled to cause 0 or 1 wait.

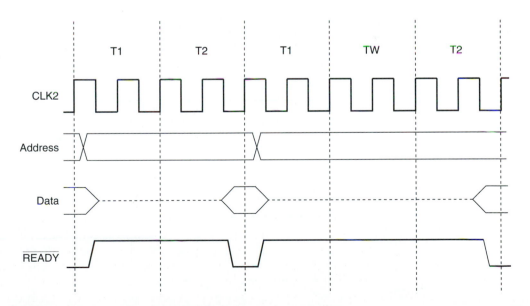

FIGURE 17–12 A non-pipelined 80386 with 0 and 1 wait states.

The $\overline{\text{READY}}$ signal is sampled at the end of a bus cycle to determine whether the clock cycle is T2 or TW. If $\overline{\text{READY}} = 0$ at this time, it is the end of the bus cycle or T2. If $\overline{\text{READY}}$ is 1 at the end of a clock cycle, the cycle is a TW and the microprocessor continues to test $\overline{\text{READY}}$, searching for a logic 0 and the end of the bus cycle.

In the non-pipelined system, whenever $\overline{\text{ADS}}$ becomes a logic 0, a wait state is inserted if $\overline{\text{READY}} = 1$. After $\overline{\text{ADS}}$ returns to a logic 1, the positive edges of the clock are counted to generate the $\overline{\text{READY}}$ signal. The $\overline{\text{READY}}$ signal becomes a logic 0 after the first clock to insert 0 wait states. If 1 wait state is inserted, the $\overline{\text{READY}}$ line must remain a logic 1 until at least two clocks have elapsed. If additional wait states are desired, then additional time must elapse before $\overline{\text{READY}}$ is cleared. This essentially allows any number of wait states to be inserted into the timing.

Figure 17–13 shows a circuit that inserts 0 through 3 wait states for various memory addresses. In the example, 1 wait state is produced for a DRAM access and 2 wait states for an

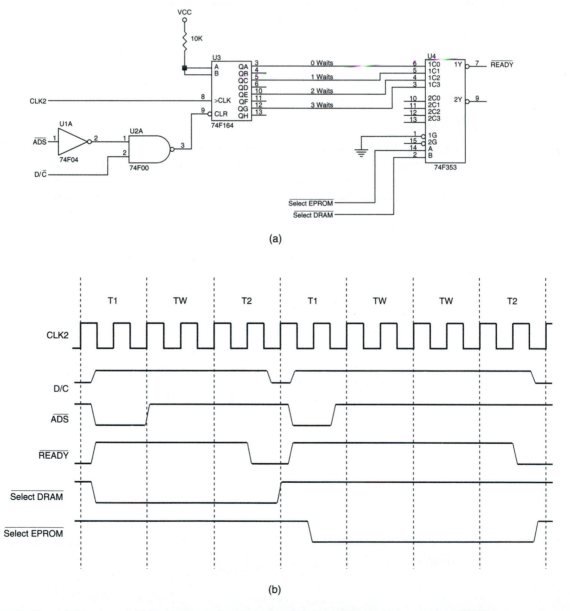

(a)

(b)

FIGURE 17–13   (a) Circuit and (b) timing that selects 1 wait state for DRAM and 2 waits for EPROM.

EPROM access. The 74F164 clears whenever $\overline{ADS}$ is low and D/$\overline{C}$ is high. It begins to shift after $\overline{ADS}$ returns to a logic 1 level. As it shifts, the 00000000 in the shift register begins to fill with logic 1s from the QA connection toward the QH connection. The four different outputs are connected to an inverting multiplexer that generates the active low $\overline{READY}$ signal.

17–2 SPECIAL 80386 REGISTERS

A new series of registers, not found in earlier Intel microprocessors, appears in the 80386 as control, debug and test registers. Control registers CR0–CR3 control various features, DR0–DR7 facilitate debugging, and registers TR6 and TR7 are used to test paging and caching.

Control Registers

In addition to the EFLAGS and EIP as described earlier, there are other control registers found in the 80386. Control register 0 (CR0) is identical to the MSW (machine status word) found in the 80286 microprocessor, except that it is 32 bits wide instead of 16 bits wide. Additional control registers are CR1, CR2, and CR3.

Figure 17–14 illustrates the control register structure of the 80386. Control register CR1 is not used in the 80386, but is reserved for future products. Control register CR2 holds the linear page address of the last page accessed before a page fault interrupt. Finally, control register CR3 holds the base address of the page directory. The rightmost 12 bits of the 32-bit page table address contain zeros and combine with the remainder of the register to locate the start of the 4K-long page table.

Register CR0 contains a number of special control bits that are defined as follows in the 80386:

PG	Selects page table translation of linear addresses into physical addresses when $PG = 1$. Page table translation allows any linear address to be assigned any physical memory location.
ET	Selects the 80287 coprocessor when $ET = 0$ or the 80387 coprocessor when $ET = 1$. This bit was installed because there was no 80387 available when the 80386 first appeared. In most systems, ET is set to indicate that an 80387 is present in the system.
TS	Indicates that the 80386 has switched tasks (in protected mode, changing the contents of TR places a 1 into TS). If $TS = 1$, a numeric coprocessor instruction causes a type 7 (coprocessor not available) interrupt.

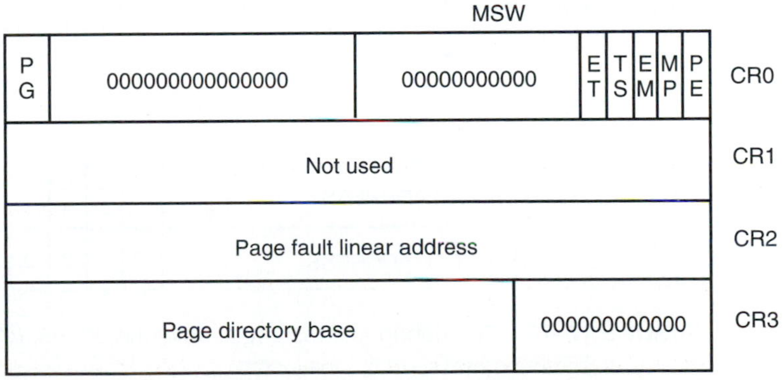

FIGURE 17–14 The control register structure of the 80386 microprocessor.

			MSW					
P G	000000000000000	00000000000		E T	T S	E M	M P	P E	CR0

Not used	CR1

Page fault linear address	CR2

Page directory base	000000000000	CR3

EM   Is set to cause a type 7 interrupt for each ESC instruction. (ESCape instructions are used to encode instructions for the 80387 coprocessor.) We often use this interrupt to emulate, with software, the function of the coprocessor. Emulation reduces the system cost, but it often requires at least 100 times longer to execute the emulated coprocessor instructions.

MP   Is set to indicate that the arithmetic coprocessor is present in the system.

PE   Is set to select the protected mode of operation for the 80386. It may also be cleared to re-enter the real mode. This bit can only be set in the 80286. The 80286 could not return to real mode without a hardware reset, which precludes its use in most systems that use protected mode.

Debug and Test Registers

Figure 17–15 shows the sets of debug and test registers. The first four debug registers contain 32-bit linear breakpoint addresses. (A **linear address** is a 32-bit address generated by a microprocessor instruction that may or may not be the same as the physical address.) The breakpoint addresses, which may locate an instruction or datum, are constantly compared with the addresses generated by the program. If a match occurs, the 80386 will cause a type 1 interrupt (TRAP or debug interrupt) to occur, if directed by debug registers DR6 and DR7. This feature is a much-expanded version of the basic trapping or tracing allowed with the earlier Intel microprocessors through the type 1 interrupt. The breakpoint addresses are very useful in debugging faulty software. The control bits in DR6 and DR7 are defined as follows:

BT   If set (1), the debug interrupt was caused by a task switch.

BS   If set, the debug interrupt was caused by the TF bit in the flag register.

BD   If set, the debug interrupt was caused by an attempt to read the debug register with the GD bit set. The GD bit protects access to the debug registers.

B3–B0   Indicate which of the four debug breakpoint addresses caused the debug interrupt.

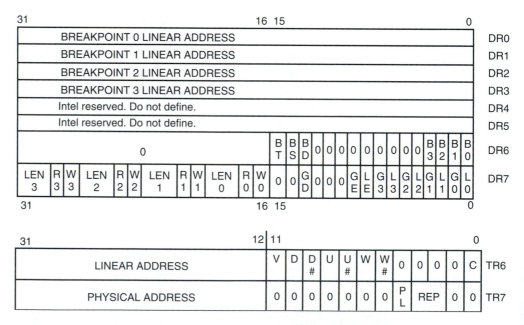

FIGURE 17–15   The debug and test registers of the 80386. (Courtesy of Intel Corporation.)

LEN   Each of the four length fields pertains to each of the four breakpoint addresses stored in DR0–DR3. These bits further define the size of access at the breakpoint address as 00 (byte), 01 (word), or 11 (doubleword).

RW   Each of the four read/write fields pertains to each of the four breakpoint addresses stored in DR0–DR3. The RW field selects the cause of action that enabled a breakpoint address as 00 (instruction access), 01 (data write), and 11 (data read and write).

GD   If set, GD prevents any read or write of a debug register by generating the debug interrupt. This bit is automatically cleared during the debug interrupt so that the debug registers can be read or changed, if needed.

GE   If set, selects a global breakpoint address for any of the four breakpoint address registers.

LE   If set, selects a local breakpoint address for any of the four breakpoint address registers.

The test registers, TR6 and TR7, are used to test the **translation look-aside buffer** (TLB). The TLB is used with the paging unit within the 80386. The TLB holds the most commonly used page table address translations. The TLB reduces the number of memory reads required for looking up page translation addresses in the page translation tables. The TLB holds the most common 32 entries from the page table, and it is tested with the TR6 and TR7 test registers.

Test register TR6 holds the tag field (linear address) of the TLB, and TR7 holds the physical address of the TLB. To write a TLB entry, perform the following steps:

1. Write TR7 for the desired physical address, PL, and REP values.
2. Write TR6 with the linear address, making sure that $C = 0$.

To read a TLB entry:

1. Write TR6 with the linear address, making sure that $C = 1$.
2. Read both TR6 and TR7. If the PL bit indicates a hit, then the desired values of TR6 and TR7 indicate the contents of the TLB.

The bits found in TR6 and TR7 indicate the following conditions:

V   Shows that the entry in the TLB is valid.

D   Indicates that the entry in the TLB is invalid or dirty.

U   A bit for the TLB.

W   Indicates that the area addressed by the TLB entry is writable.

C   Selects a write (0) or immediate lookup (1) for the TLB.

PL   Indicates a hit if a logic 1.

REP   Selects which block of the TLB is written.

Refer to the section on memory management and the paging unit for more detail on the function of the TLB.

17–3   80386 MEMORY MANAGEMENT

The **memory-management unit** (MMU) within the 80386 is similar to the MMU inside the 80286, except that the 80386 contains a paging unit not found in the 80286. The MMU performs the task of converting linear addresses, as they appear as outputs from a program, into physical addresses that access a physical memory location located anywhere within the memory system. The 80386 uses the paging mechanism to allocate any physical address to any logical address. Therefore, even though the program is accessing memory location A0000H with an instruction, the actual physical address could be memory location 100000H, or any other location if paging is enabled. This feature allows virtually any software, written to operate at any memory location, to function in an 80386 because any linear location can become any physical location. Earlier Intel microprocessors did not have this flexibility. Paging is used with DOS to relocate 80386 and 80486 memory at addresses above FFFFFH and into spaces between ROMs at locations D0000–DFFFFH and other areas as they are available. The area between ROMs is often referred to as *upper memory;* the area above FFFFFH is referred to as *extended memory.*

Descriptors and Selectors

Before the memory paging unit is discussed, we examine the descriptor and selector for the 80386 microprocessor. The 80386 uses descriptors in much the same fashion as the 80286. In both microprocessors, a **descriptor** is a series of eight bytes that describe and locate a memory segment. A **selector** (segment register) is used to index a descriptor from a table of descriptors. The main difference between the 80286 and 80386 is that the latter has two additional selectors (FS and GS) and the most-significant two bytes of the descriptor are defined for the 80386. Another difference is that 80386 descriptors use a 32-bit base address and a 20-bit limit, instead of the 24-bit base address and a 16-bit limit found on the 80286.

The 80286 addresses a 16M-byte memory space with its 24-bit base address and has a segment length limit of 64K bytes, due to the 16-bit limit. The 80386 addresses a 4G-byte memory space with its 32-bit base address and has a segment length limit of 1M bytes or 4G bytes, due to a 20-bit limit that is used in two different ways. The 20-bit limit can access a segment with a length of 1M byte if the granularity bit $(G) = 0$. If $G = 1$, the 20-bit limit allows a segment length of 4G bytes.

The granularity bit is found in the 80386 descriptor. If $G = 0$, the number stored in the limit is interpreted directly as a limit, allowing it to contain any limit between 00000H and FFFFFH for a segment size up to 1M byte. If $G = 1$, the number stored in the limit is interpreted as 00000XXXH–FFFFFXXXH, where the XXX is any value between 000H and FFFH. This allows the limit of the segment to range between 0 bytes to 4G bytes in steps of 4K bytes. A limit of 00001H indicates that the limit is 4K bytes when $G = 1$ and 1 byte when $G = 0$. An example is a segment that begins at physical address 10000000H. If the limit is 00001H and $G = 0$, this segment begins at 10000000H and ends at 10000001H. If $G = 1$ with the same limit (00001H), the segment begins at location 10000000H and ends at location 10001FFFH.

Figure 17–16 shows how the 80386 addresses a memory segment in the protected mode using a selector and a descriptor. Note that this is identical to the way that a segment is addressed by the 80286. The difference is the size of the segment accessed by the 80386. The selector uses its leftmost 13 bits to select a descriptor from a descriptor table. The TI bit indicates either the local $(TI = 1)$ or global $(TI = 0)$ descriptor table. The rightmost two bits of the selector define the requested privilege level of the access.

FIGURE 17–16   Protected mode addressing using a segment register as a selector. (Courtesy of Intel Corporation.)

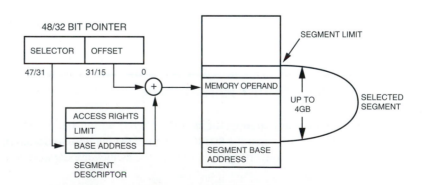

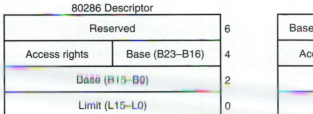
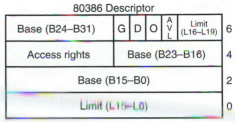

FIGURE 17–17 The descriptors for the 80286 and 80386 microprocessors.

Because the selector uses a 13-bit code to access a descriptor, there are at most 8192 descriptors in each table—local or global. Because each segment (in an 80386) can be 4G bytes in length, we can access 16,384 segments at a time with the two descriptor tables. This allows the 80386 to access a virtual memory size of 64T bytes. Of course, only 4G bytes of memory actually exist in the memory system (1T byte = 1024G bytes). If a program requires more than 4G bytes of memory at a time, it can be swapped between the memory system and a disk drive or other form of large volume storage.

The 80386 uses descriptor tables for both global (GDT) and local (LDT) descriptors. A third descriptor table appears for interrupt (IDT) descriptors or gates. The first six bytes of the descriptor are the same as in the 80286, which allows 80286 software to be upward compatible with the 80386. (An 80286 descriptor used 00H for its most significant two bytes.) See Figure 17–17 for the 80286 and 80386 descriptor. The base address is 32 bits in the 80386, the limit is 20 bits, and a G bit selects the limit multiplier (1 or 4K times). The fields in the descriptor for the 80386 are defined as follows:

Base (B31–B0) Defines the starting 32-bit address of the segment within the 4G-byte physical address space of the 80386 microprocessor.

Limit (L19–L0) Defines the limit of the segment in units of bytes if the *G bit* = 0, or in units of 4K bytes if *G* = 1. This allows a segment to be of any length from 1 byte to 1M bytes if *G* = 0, and from 4K bytes to 4G bytes if *G* = 1. Recall that the limit indicates the last byte in a segment.

Access Rights Determines privilege level and other information about the segment. This byte varies with different types of descriptors and is elaborated with each descriptor type.

G The granularity bit selects a multiplier of 1 or 4K times for the limit field. If *G* = 0, the multiplier is 1; if *G* = 1, the multiplier is 4K.

D Selects the default register size. If *D* = 0, the registers are 16-bits wide, as in the 80286; if *D* = 1, they are 32-bits wide, as in the 80386. This bit determines whether prefixes are required for 32-bit data and index registers. If *D* = 0, then a prefix is required to access 32-bit registers and to use 32-bit pointers. If *D* = 1, then a prefix is required to access 16-bit registers and 16-bit pointers. The USE16 and USE32 directives appended to the SEGMENT statement in assembly language control the setting of the D bit. In the real mode, it is always assumed that the registers are 16-bits wide, so any instruction that references a 32-bit register or pointer must be prefixed. The current version of DOS assumes *D* = 0.

AVL This bit is available to the operating system to use in any way that it sees fit. It often indicates that the segment described by the descriptor is available.

FIGURE 17–18   The format of the 80386 segment descriptor.

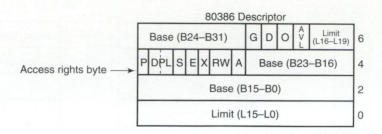

Access rights byte ⟶

Descriptors appear in two forms in the 80386 microprocessor: the segment descriptor and the system descriptor. The segment descriptor defines data, stack, and code segments; the system descriptor defines information about the system's tables, tasks, and gates.

Segment Descriptors.   Figure 17–18 shows the segment descriptor. This descriptor fits the general form, as dictated in Figure 17–17, but the access rights bits are defined to indicate how the data, stack, or code segment described by the descriptor functions. Bit position 4 of the access rights byte determines whether the descriptor is a data or code segment descriptor ($S = 1$) or a system segment descriptor ($S = 0$). Note that the labels used for these bits may vary in different versions of Intel literature, but they perform the same tasks.

Following is a description of the access rights bits and their function in the segment descriptor:

P   **Present** is a logic 1 to indicate that the segment is present. If $P = 0$ and the segment is accessed through the descriptor, a type 11 interrupt occurs. This interrupt indicates that a segment was accessed that is not present in the system.

DPL   **Descriptor privilege level** sets the privilege level of the descriptor, where 00 has the highest privilege and 11 has the lowest. This is used to protect access to segments. If a segment is accessed with a privilege level that is lower (higher in number) than the DPL, a privilege violation interrupt occurs. Privilege levels are used in multiuser systems to prevent access to an area of the system memory.

S   **Segment** indicates a data or code segment descriptor ($S = 1$), or a system segment descriptor ($S = 0$).

E   **Executable** selects a data (stack) segment ($E = 0$) or a code segment ($E = 1$). E also defines the function of the next two bits (X and RW).

X   If $E = 0$, then X indicates the direction of expansion for the data segment. If $X = 0$, the segment expands upward, as in a data segment; if $X = 1$, the segment expands downward as in a stack segment. If $E = 1$, then X indicates whether the privilege level of the code segment is ignored ($X = 0$) or observed ($X = 1$).

RW   If $E = 0$, then RW indicates that the data segment may be written ($RW = 1$) or not written ($RW = 0$). If $E = 1$, then RW indicates that the code segment may be read ($RW = 1$) or not read ($RW = 0$).

A   **Accessed** is set each time that the microprocessor accesses the segment. It is sometimes used by the operating system to keep track of which segments have been accessed.

System Descriptor.   The system descriptor is illustrated in Figure 17–19. There are 16 possible system descriptor types (see Table 17–1 for the different descriptor types), but not all are used in the 80386 microprocessor. Some of these types are defined for the 80286 so that the 80286 software is compatible with the 80386. Some of the types are new and unique to the 80386; some have yet to be defined and are reserved for future Intel products.

FIGURE 17–19 The general
format of an 80386 system
descriptor.

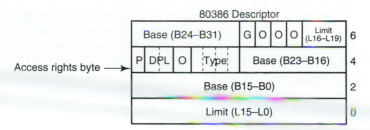

TABLE 17–1 80386 system
descriptor types.

Type	Purpose
0000	Invalid
0001	Available 80286 TSS
0010	LDT
0011	Busy 80286 TSS
0100	80286 call gate
0101	Task gate (80286 or 80386)
0110	80286 interrupt gate
0111	80286 trap gate
1000	Invalid
1001	Available 80386 TSS
1010	Reserved for future Intel products
1011	Busy 80386 TSS
1100	80386 call gate
1101	Reserved for future Intel products
1110	80386 interrupt gate
1111	80836 trap gate

Descriptor Tables

The descriptor tables define all the segments used in the 80386 when it operates in the protected mode. There are three types of descriptor tables: the global descriptor table (GDT), the local descriptor table (LDT), and the interrupt descriptor table (IDT). The registers used by the 80386 to address these three tables are called the global descriptor table register (GDTR), the local descriptor table register (LDTR), and the interrupt descriptor table register (IDTR). These registers are loaded with the LGDT, LLDT, and LIDT instructions, respectively.

The **descriptor table** is a variable-length array of data, with each entry holding an 8-byte long descriptor. The local and global descriptor tables hold up to 8192 entries each, and the interrupt descriptor table holds up to 256 entries. A descriptor is indexed from either the local or global descriptor table by the selector that appears in a segment register. Figure 17–20 shows a segment register and the selector that it holds in the protected mode. The leftmost 13 bits index a descriptor, the TI bit selects either the local ($TI = 1$) or global ($TI = 0$) descriptor table, and the RPL bits indicate the requested privilege level.

Whenever a new selector is placed into one of the segment registers, the 80386 accesses one of the descriptor tables and automatically loads the descriptor into a program-invisible cache portion

FIGURE 17–20 A segment
register showing the selector,
T1 bit, and requested privilege level (RPL) bits.

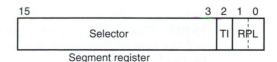

of the segment register. As long as the selector remains the same in the segment register, no additional accesses are required to the descriptor table. The operation of fetching a new descriptor from the descriptor table is program-invisible because the microprocessor automatically accomplishes this each time that the segment register contents are changed in the protected mode.

Figure 17–21 shows how a sample global descriptor table (GDT), which is stored at memory address 00010000H, is accessed through the segment register and its selector. This table contains four entries. The first is a null (0) descriptor. Descriptor 0 must always be a null descriptor. The other entries address various segments in the 80386 protected mode memory system. In this illustration, the data segment register contains a 0008H. This means that the selector is indexing descriptor location 1 in the global descriptor table ($TI = 0$), with a requested privilege level of 00. Descriptor 1 is located eight bytes above the base descriptor table address, beginning at location 00010008H. The descriptor located in this memory location accesses a base address of 00200000H and a limit of 100H. This means that this descriptor addresses memory locations 00200000H–00200100H. Because this is the DS (data segment) register, the data segment is located at these locations in the memory system. If data are accessed outside of these boundaries, an interrupt occurs.

The local descriptor table (LDT) is accessed in the same manner as the global descriptor table (GDT). The only difference in access is that the TI bit is cleared for a global access and set for a local access. Another difference exists if the local and global descriptor table registers are examined. The global descriptor table register (GDTR) contains the base address of the global

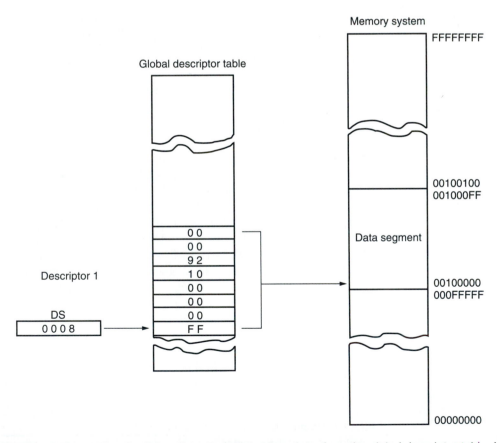

FIGURE 17–21   Using the DS register to select a descriptor from the global descriptor table. In this example, the DS register accesses memory locations 00100000H–001000FFH as a data segment.

FIGURE 17–22   The gate descriptor for the 80386 microprocessor.

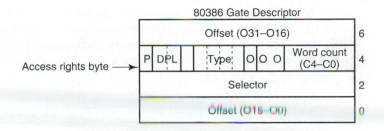

descriptor table and the limit. The local descriptor table register (LDTR) contains only a selector, and it is 16-bits wide. The contents of the LDTR addresses a type 0010 system descriptor that contains the base address and limit of the LDT. This scheme allows one global table for all tasks; but allows many local tables, one or more for each task, if necessary. Global descriptors describe memory for the system, while local descriptors describe memory for applications or tasks.

Like the GDT, the interrupt descriptor table (IDT) is addressed by storing the base address and limit in the interrupt descriptor table register (IDTR). The main difference between the GDT and IDT is that the IDT contains only interrupt gates. The GDT and LDT contain segment and system descriptors, but never contain interrupt gates.

Figure 17–22 shows the gate descriptor, a special form of the system descriptor described earlier. (Refer to Table 17–1 for the different gate descriptor types.) Notice that the gate descriptor contains a 32-bit offset address, a word count, and a selector. The 32-bit offset address points to the location of the interrupt service procedure or other procedure. The word count indicates how many words are transferred from the caller's stack to the stack of the procedure accessed by a call gate. This feature of transferring data from the caller's stack is useful for implementing high-level languages such as C/C++. Note that the word count field is not used with an interrupt gate. The selector is used to indicate the location of the task state segment (TSS) in the GDT or LDT if it is a local procedure.

When a gate is accessed, the contents of the selector are loaded into the task register (TR), causing a task switch. The acceptance of the gate depends on the privilege and priority levels. A return instruction (RET) ends a call gate procedure and a return from interrupt instruction (IRET) ends an interrupt gate procedure. Tasks are usually accessed with a CALL or an INT instruction, where the call instruction addresses a call gate in the descriptor table and the interrupt addresses an interrupt descriptor.

The difference between real mode interrupts and protected mode interrupts is that the interrupt vector table is an IDT in the protected mode. The IDT still contains up to 256 interrupt levels, but each level is accessed through an interrupt gate instead of an interrupt vector. Thus, interrupt type number 2 is located at IDT descriptor number 2 at 16 locations above the base address of the IDT. This also means that the first 1K byte of memory no longer contains interrupt vectors, as it did in the real mode. The IDT can be located at any location in the memory system.

The Task State Segment (TSS)

The **task state segment** (TSS) descriptor contains information about the location, size, and privilege level of the task state segment, just as any other descriptor. The difference is that the TSS described by the TSS descriptor does not contains data or code. It contains the state of the task and linkage so tasks can be nested (one task can call a second, which can call a third, and so forth). The TSS descriptor is addressed by the **task register** (TR). The contents of the TR are changed by the LTR instruction. Whenever the protected mode program executes a far JMP or CALL instruction, the contents of TR are also changed. The LTR instruction is used to initially access a task during system initialization. After initialization, the CALL or JUMP instructions normally switch tasks. In most cases, we use the CALL instruction to initiate a new task.

The TSS is illustrated in Figure 17–23. As can be seen, the TSS is quite a formidable section of memory, containing many different types of information. The first word of the TSS is labeled **back-link.** This is the selector that is used, on a return (RET or IRET), to link back to the prior TSS by loading the back-link selector into the TR. The following word must contain a 0.

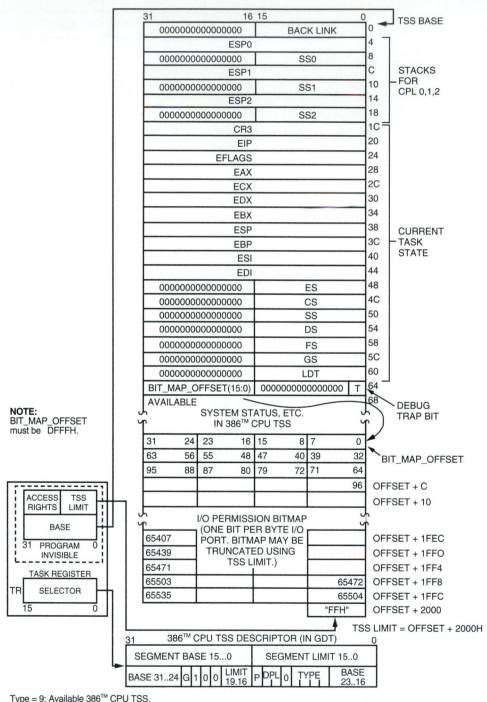

FIGURE 17–23   The task state segment (TSS) descriptor. (Courtesy of Intel Corporation.)

The second through the seventh doublewords contain the ESP and ESS values for privilege levels 0–2. These are required in case the current task is interrupted so these privilege level (PL) stacks can be addressed. The eighth word (offset 1CH) contains the contents of CR3, which stores the base address of the prior state's page directory register. This must be restored if paging is in effect. The contents of the next 17 doublewords are loaded into the registers indicated. Whenever a task is accessed, the entire state of the machine (all of the registers) is stored in these memory locations and then reloaded from the same locations in the new TSS. The last word (offset 66H) contains the I/O permission bit map base address.

The I/O permission bit map allows the TSS to block I/O operations to inhibited I/O port addresses via an I/O permission denial interrupt. The permission denial interrupt is type number 13, the general protection fault interrupt. The I/O permission bit map base address is the offset address from the start of the TSS. This allows the same permission map to be used by many TSSs.

Each I/O permission bit map is 64K bits long (8K bytes), beginning at the offset address indicated by the I/O permission bit map base address. The first byte of the I/O permission bit map contains I/O permission for I/O ports 0000H–0007H. The rightmost bit contains the permission for port number 0000H. The leftmost bit contains the permission for port number 0007H. This sequence continues for the very last port address (FFFFH) stored in the leftmost bit of the last byte of the I/O permission bit map. A logic 0 placed in an I/O permission bit map bit enables the I/O port address, while a logic 1 inhibits or blocks the I/O port address. At present, only Windows NT uses the I/O permission scheme to disable I/O ports dependent on the application or the user.

In review of the operation of a task switch, which requires only 17 μs to execute, we list the following steps:

1. The gate contains the address of the procedure or location jumped to by the task switch. It also contains the selector number of the TSS descriptor and the number of words transferred from the caller to the user stack area for parameter passing.
2. The selector is loaded into TR from the gate. (This step is accomplished by a CALL or JMP that refers to a valid TSS descriptor.)
3. The TR selects the TSS.
4. The current state is saved in the current TSS and the new TSS is accessed with the state of the new task (all the registers) loaded into the microprocessor. The current state is saved at the TSS selector currently found in the TR. Once the current state is saved, a new value (by the JMP or CALL) for the TSS selector is loaded into TR and the new state is loaded from the new TSS.

The return from a task is accomplished by the following steps:

1. The current state of the microprocessor is saved in the current TSS.
2. The back-link selector is loaded to the TR to access the prior TSS so that the prior state of the machine can be returned to and be restored to the microprocessor. The return for a called TSS is accomplished by the IRET instruction.

17–4   MOVING TO PROTECTED MODE

In order to change the operation of the 80386 from the real mode to the protected mode, several steps must be followed. Real mode operation is accessed after a hardware reset or by changing the PE bit to a logic 0 in CR0. Protected mode is accessed by placing a logic 1 into the PE bit of

FIGURE 17–24 The memory map for Example 17–1.

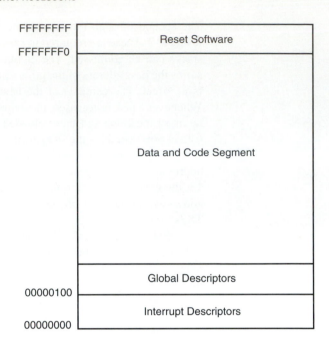

CR0; before this is done, however, some other things must be initialized. The following steps accomplish the switch from the real mode to the protected mode:

1. Initialize the interrupt descriptor table so that it contains valid interrupt gates for at least the first 32 interrupt type numbers. The IDT may (and often does) contain up to 256 8-byte interrupt gates defining all 256 interrupt types.
2. Initialize the global descriptor table (GDT) so that it contains a null descriptor at descriptor 0, and valid descriptors for at least one code, one stack, and one data segment.
3. Switch to protected mode by setting the PE bit in CR0.
4. Perform an intrasegment (near) JMP to flush the internal instruction queue and load the TR with the base TSS descriptor.
5. Load all the data selectors (segment registers) with their initial selector values.
6. The 80386 is now operating in the protected mode, using the segment descriptors that are defined in GDT and IDT.

Figure 17–24 shows the protected system memory map set up by following steps 1–5. The software for this task is listed in Example 17–1. This system contains one data segment descriptor and one code segment descriptor with each segment set to 4G bytes in length. This is the simplest protected mode system possible: loading all the segment registers, except code, with the same data segment descriptor from the GDT. The privilege level is initialized to 00, the highest level. This system is most often used where one user has access to the microprocessor and requires the entire memory space. This program is designed for use in a system that does not use DOS or shell from Windows to DOS. Later in this section, we show how to go to protected mode in a DOS environment. (Please note that the software in Example 17–1 is designed for a standalone system such as the 80386EX, and not for use in the PC.)

EXAMPLE 17–1

```
                        .MODEL SMALL
                        .386P
0000                    .DATA
0000  0040 [    IDT1    DD      64 DUP (?)      ;space for 32 interrupt vectors
```

```
                      00000000
                          ]
                      ;
                      ;Global descriptor table
                      ;
0100                  DESC0   DQ      0                   ;clear null descriptor
      0000000000000000
                      ;
                      ;code segment descriptor
                      ;
0108  FFFF            DESC1   DW      0FFFFH              ;limit = 4G
010A  0000                    DW      0                   ;base address = 00000000H
010C  0000                    DW      0
010E  9E                      DB      9EH                 ;code segment
010F  8F                      DB      8FH                 ;G = 1
0110  00                      DB      0
                      ;
                      ;data segment descriptor
                      ;
0111  FFFF            DESC2   DW      0FFFFH              ;limit = 4G
0113  0000                    DW      0                   ;base address = 00000000H
0115  0000                    DW      0
0117  92                      DB      92H                 ;data segment
0118  8F                      DB      8FH                 ;G = 1
0119  00                      DB      0
                      ;
                      ;IDT table data
                      ;
011A  00FF            IDT     DW      0FFH                ;set limit to FFH
011C  00000000        IDTA    DD      0
                      ;
                      ;GDT table data
                      ;
0120  0017            GDT     DW      17H                 ;set limit to 17H
0122  00000000        GDTA    DD      0
0000                          .CODE
                      MAK32   MACRO   SEG,OFF             ;;make a seg+off a linear address
                              MOV     EAX,0
                              MOV     EBX,0
                              MOV     AX,SEG
                              MOV     BX,OFF
                              SHL     EAX,4
                              ADD     EAX,EBX
                              ENDM
                              .STARTUP
                              MAK32   DS,OFFSET IDT1
0028  66| A3 011C R          MOV     IDTA,EAX            ;save IDT address
                              MAK32   DS,OFFSET DESC0
0044  66| A3 0122 R          MOV     GDTA,EAX            ;save GDT address
0048  B9 0020                 MOV     CX,32
004B  BF 0000 R               MOV     DI,OFFSET IDT1
004E  BE 0000                 MOV     SI,0
0051  B8 0000                 MOV     AX,0
0054  8E C0                   MOV     ES,AX
                              .REPEAT                     ;setup first 32 interrupts
                              MAK32   ES:[SI+2],ES:[SI]
0070  89 05                   MOV     [DI],AX
0072  66| C1 E8 10            SHR     EAX,16
0076  89 45 06                MOV     [DI+6],AX
0079  C7 45 02 0008           MOV     WORD PTR [DI+2],8
007E  C7 45 04 8F00           MOV     WORD PTR [DI+4],8F00H
0083  83 C7 08                ADD     DI,8
0086  83 C6 04                ADD     SI,4
                              .UNTILCXZ
008B  0F 01 1E 011A R  LIDT    FWORD PTR IDT   ;load IDT
0090  0F 01 16 0120 R  LGDT    FWORD PTR GDT   ;load GDT

0095  0F 20 C0                MOV     EAX,CR0             ;set PE
```

```
0098    66| 83 C8 01        OR      EAX,1
009C    0F 22 C0           MOV     CR0,EAX

009F    EB 00              JMP     START           ;near jump

00A1              START:
00A1    B8 0010            MOV     AX,10H          ;set selector 2
00A4    8E D8              MOV     DS,AX
00A6    8E C0              MOV     ES,AX
00A8    8E D0              MOV     SS,AX
00AA    8E E8              MOV     GS,AX
00AC    8E E0              MOV     FS,AX
00AE    66| BC FFFFF000    MOV     ESP,0FFFFF000H

                        ;now in protected mode.

                        END
```

In more complex systems, the steps required to initialize the system in the protected mode are more involved. For complex systems that are often multiuser systems, the registers are loaded by using the task state segment (TSS). The steps required to place the 80386 into protected mode operation for a more complex system using a task switch follow:

1. Initialize the interrupt descriptor table so that it refers to valid interrupt descriptors with at least 32 descriptors in the IDT.
2. Initialize the global descriptor table so that it contains at least two task state segment (TSS) descriptors, and the initial code and data segments required for the initial task.
3. Initialize the task register (TR) so that it points to a valid TSS; when the initial task switch occurs and accesses the new TSS, the current registers are stored in the initial TSS.
4. Switch to protected mode by using an intrasegment (near) jump to flush the internal instruction queue. Load the TR with the current TSS selector.
5. Load the TR with a far jump instruction to access the new TSS and save the current state.
6. The 80386 is now operating in the protected mode under control of the first task.

Example 17–2 illustrates the software required to initialize the system and switch to protected mode by using a task switch. The initial system task operates at the highest level of protection (00) and controls the entire operating environment for the 80386. In many cases, it is used to boot (load) software that allows many users to access the system in a multiuser environment.

EXAMPLE 17–2

```
                    .MODEL  SMALL
                    .386P
                    .STACK   800H
0000                .DATA
0008                DESC    STRUC                   ;define descriptor structure

0000    0000        LIM_L   DW    0
0002    0000        BAS_L   DW    0
0004    00          BAS_M   DB    0
0005    00          ACCESS  DB    0
0006    00          LIM_M   DB    0
0007    00          BAS_H   DB    0

                    DESC    ENDS

0068                TSS     STRUC                   ;define TSS structure

0000    0000        BACK_L  DW      0
0002    0000                DW      0
0004    00000000    ESP0    DD      0
0008    0000        SS0     DW      0
000A    0000                DW      0
```

```
000C  00000000    ESP1    DD      0
0010  0000        SS1     DW      0
0014  0000                DW      0
0014  00000000    ESP2    DD      0
0018  0000        SS2     DW      0
001A  0000                DW      0
001C  00000000    CCR3    DD      0
0020  00000000    EIP     DD      0
0024  00000000    TFALGS  DD      0
0028  00000000    EEAX    DD      0
002C  00000000    EECX    DD      0
0030  00000000    EEDX    DD      0
0034  00000000    EEBX    DD      0
0038  00000000    EESP    DD      0
003C  00000000    EEBP    DD      0
0040  00000000    EESI    DD      0
0044  00000000    EEDI    DD      0
0048  0020        EES     DW      20H
004A  0000                DW      0
004C  0018        ECS     DW      18H
004E  0000                DW      0
0050  0020        ESS     DW      20H
0052  0000                DW      0
0054  0020        EDS     DW      20H
0056  0000                DW      0
0058  0020        EFS     DW      20H
005A  0000                DW      0
005C  0020        EGS     DW      20H
005E  0000                DW      0
0060  0000        ELDT    DW      0
0062  0000                DW      0
0064  0000                DW      0
0066  0000        BITM    DW      0

            TSS     ENDS

0000  0000 0000   TSS1    TSS     <>              ;task state 1
      00000000
      0000 0000 00000000
      0000 0000 00000000
      0000 0000 00000000
      00000000 00000000
      00000000 00000000
      00000000 00000000
      00000000 00000000
      00000000 00000000
      0020 0000 0018
      0000 0020 0000
      0020 0000 0020
      0000 0020 0000
      0000 0000 0000
      0000
0068  0000 0000   TSS2    TSS     <>              ;task state 2
      00000000
      0000 0000 00000000
      0000 0000 00000000
      0000 0000 00000000
      00000000 00000000
      00000000 00000000
      00000000 00000000
      00000000 00000000
      00000000 00000000
      0020 0000 0018
      0000 0020 0000
      0020 0000 0020
      0000 0020 0000
      0000 0000 0000
      0000
```

```
00D0  0040 [    IDT1    DD      64 DUP (?)               ;space for 32 interrupt vectors
      00000000
           ]
                       ;
                       ;Global descriptor table
                       ;
01D0  0000 0000 00 GDT0 DESC    <>                      ;clear null descriptor
      00 00 00
01D8  0000 0028 00 TG1  DESC    <0,28H,0,85H,0,0>       ;task gate 1
      85 00 00
01E0  0000 0030 00 TG2  DESC    <0,30H,0,85H,0,0>       ;task gate 2
      85 00 00
01E8  FFFF 0000 00 ACS  DESC    <-1,0,0,9AH,0CFH,0>     ;code segment (4G)
      9A CF 00
01F0  FFFF 0000 00 DS1  DESC    <-1,0,0,92H,0CFH,0>     ;data segment (4G)
      92 CF 00
01F8  FFFF 0000 00 TS1  DESC    <-1,0,0,89H,0CFH,0>     ;TSS1 available
      89 CF 00
0200  FFFF 0000 00 TS2  DESC    <-1,0,0,89H,0CFH,0>     ;TSS2 available
      89 CF 00
0208  2000 [    IOBP    DB      2000H DUP (0)           ;enable all I/O
         00
           ]
                       ;
                       ;IDT table data
                       ;
2208  00FF       IDT    DW      0FFH                    ;set limit to FFH
220A  00000000   IDTA   DD      0
                       ;
                       ;GDT table data
                       ;
220E  0017       GDT    DW      17H                     ;set limit to 17H
2210  00000000   GDTA   DD      0
0000              .CODE
                  MAK32  MACRO   SEG,OFF                 ;;make a seg+off a linear address
                         MOV     EAX,0
                         MOV     EBX,0
                         MOV     AX,SEG
                         MOV     BX,OFF
                         SHL     EAX,4
                         ADD     EAX,EBX
                         ENDM
                  MAKD   MACRO   TSS,DES                 ;;save base address
                         PUSH    EAX
                         MOV     EBX,0
                         MOV     BX,OFFSET TSS
                         ADD     EAX,EBX
                         MOV     DES.BAS_L,AX
                         SHR     EAX,16
                         MOV     DES.BAS_M,AL
                         MOV     DES.BAS_H,AH
                         POP     EAX
                         ENDM
                  .STARTUP
                         MAK32   DS,OFFSET IDT1
0028  66| A3 220A R     MOV     IDTA,EAX                ;save IDT address
                         MAK32   DS,OFFSET GDT0
0044  66| A3 2210 R     MOV     GDTA,EAX                ;save GDT address

0048  B9 0020          MOV     CX,32
004B  BF 00D0 R        MOV     DI,OFFSET IDT1
004E  BE 0000          MOV     SI,0
0051  B8 0000          MOV     AX,0
0054  8E C0            MOV     ES,AX
                  .REPEAT                                ;setup first 32 interrupts
                     MAK32   ES:[SI+2],ES:[SI]
0070  89 05              MOV     [DI],AX
0072  66| C1 E8 10       SHR     EAX,16
```

```
0076   89 45 06              MOV     [DI+6],AX
0079   C7 45 02 0008         MOV     WORD PTR [DI+2],8
007E   C7 45 04 8F00         MOV     WORD PTR [DI+4],8F00H
0083   83 C7 08              ADD     DI,8
0086   83 C6 04              ADD     SI,4
                             .UNTILCXZ
008B   B8 0208 R             MOV     AX,OFFSET IOBP          ;setup IO bit map
008E   A3 0066 R             MOV     TSS1.BITM,AX
0091   A3 00CE R             MOV     TSS2.BITM,AX
                             MAK32   CS,OFFSET TASK1         ;get task 1 address
00AC   66| A3 0088 R         MOV     TSS2.EIP,EAX            ;save it in task 2
                             MAKD    TSS1,TS1
                             MAKD    TSS2,TS2
00EC   0F 01 1E 2208 R       LIDT    FWORD PTR IDT           ;load IDT
00F1   0F 01 16 220E R       LGDT    FWORD PTR GDT           ;load GDT

00F6   0F 20 C0              MOV     EAX,CR0                 ;set PE
00F9   66| 83 C8 01          OR      EAX,1
00FD   0F 22 C0              MOV     CR0,EAX

0100   EB 00                 JMP     START                   ;near jump

0102   START:

0102   B8 0008               MOV     AX,8                    ;address TSS1
0105   0F 00 D8              LTR     AX
0108   B8 0010               MOV     AX,10H
010B   FF E0                 JMP     AX                      ;jump to TSS2

                             ;now in protected mode at task 1.

010D          TASK1:

                             END
```

Neither Example 17–1 nor 17–2 is written to function in the personal computer environment. The personal computer environment requires the use of either the VCPI (virtual control program interface) driver provided by the HIMEM.SYS driver in DOS or the DPMI (DOS protected mode interface) driver provided by Windows when shelling to DOS. Example 17–3 shows how to switch to protected mode using DPMI, and then display the contents of any area of memory. This includes memory in the extended memory area or anywhere else. This DOS application allows the contents of any memory location to be displayed in hexadecimal format on the monitor. This includes locations above the first 1M byte of the memory system.

EXAMPLE 17–3

```
                        ;A program that displays the contents of any area of memory
                        ;including extended memory.
                        ;***command line syntax***
                        ;EDUMP XXXX,YYYY  where XXXX is the start address and YYYY is
                        ;the end address.
                        ;Note:  this program must be executed from WINDOWS.
                        ;
                                .MODEL SMALL
                                .386
                                .STACK 1024             ;stack area of 1,024 bytes
0000                            .DATA
0000   00000000         ENTRY DD     ?                  ;DPMI entry point
0004   00000000         EXIT  DD     ?                  ;DPMI exit point
0008   00000000         FIRST DD     ?                  ;first address
000C   00000000         LAST1 DD     ?                  ;last address
0010   0000             MSIZE DW     ?                  ;memory needed for DPMI
0012   0D 0A 0A 50 61   ERR1  DB      13,10,10,'Parameter error.$'
       72 61 6D 65 74
       65 72 20 65 72
```

```
          72 6F 72 2E 24
0026      0D 0A 0A 44 50   ERR2   DB    13,10,10,'DPMI not present.$'
          4D 49 20 6E 6F
          74 20 70 72 65
          73 65 6E 74 2E
          24
003B      0D 0A 0A 4E 6F   ERR3   DB    13,10,10,'Not enough real memory.$'
          74 20 65 6E 6F
          75 67 68 20 72
          65 61 6C 20 6D
          65 6D 6F 72 79
          2E 24
0056      0D 0A 0A 43 6F   ERR4   DB    13,10,10,'Could not move to protected mode.$'
          75 6C 64 20 6E
          6F 74 20 6D 6F
          76 65 20 74 6F
          20 70 72 6F 74
          65 63 74 65 64
          20 6D 6F 64 65
          2E 24
007B      0D 0A 0A 43 61   ERR5   DB    13,10,10,'Cannot allocate selector.$'
          6E 6E 6F 74 20
          61 6C 6C 6F 63
          61 74 65 20 73
          65 6C 65 63 74
          6F 72 2E 24
0098      0D 0A 0A 43 61   ERR6   DB    13,10,10,'Cannot use base address.$'
          6E 6E 6F 74 20
          75 73 65 20 62
          61 73 65 20 61
          64 64 72 65 73
          73 2E 24
00B4      0D 0A 0A 43 61   ERR7   DB    13,10,10,'Cannot allocate 64K to limit.$'
          6E 6E 6F 74 20
          61 6C 6C 6F 63
          61 74 65 20 36
          34 4B 20 74 6F
          20 6C 69 6D 69
          74 2E 24
00D5      0D 0A 24         CRLF   DB    13,10,'$'
00D8      50 72 65 73 73   MES1   DB    'Press any key...$'
          20 61 6E 79 20
          6B 65 79 2E 2E
          2E 24
                           ;
                           ;register array storage for DPMI function 0300H
                           ;
00E9      = 00E9           ARRAY EQU   THIS BYTE
00E9      00000000         REDI   DD    0                       ;EDI
00ED      00000000         RESI   DD    0                       ;ESI
00F1      00000000         REBP   DD    0                       ;EBP
00F5      00000000                DD    0                       ;reserved
00F9      00000000         REBX   DD    0                       ;EBX
00FD      00000000         REDX   DD    0                       ;EDX
0101      00000000         RECX   DD    0                       ;ECX
0105      00000000         REAX   DD    0                       ;EAX
0109      0000             RFLAG DW    0                       ;flags
010B      0000             RES    DW    0                       ;ES
010D      0000             RDS    DW    0                       ;DS
010F      0000             RFS    DW    0                       ;FS
0111      0000             RGS    DW    0                       ;GS
0113      0000             RIP    DW    0                       ;IP
0115      0000             RCS    DW    0                       ;CS
0117      0000             RSP    DW    0                       ;SP
0119      0000             RSS    DW    0                       ;SS
0000                              .CODE
                                  .STARTUP
0010      8C C0            MOV   AX,ES
0012      8C DB            MOV   BX,DS                  ;find size of program and data
0014      2B D8            SUB   BX,AX
```

```
0016  8B C4            MOV   AX,SP              ;find stack size
0018  C1 E8 04         SHR   AX,4
001B  40               INC   AX
001C  03 D8            ADD   BX,AX              ;BX = length in paragraphs
001E  B4 4A            MOV   AH,4AH
0020  CD 21            INT   21H                ;modify memory allocation
0022  E8 00D1          CALL  GETDA              ;get command line information
0025  73 0A            JNC   MAIN1              ;if parameters are good
0027  B4 09            MOV   AH,9               ;parameter error
0029  BA 0012 R        MOV   DX,OFFSET ERR1
002C  CD 21            INT   21H
002E  E9 00AA          JMP   MAINE              ;exit to DOS
0031               MAIN1:
0031  E8 00AB          CALL  ISDPMI             ;is DPMI loaded?
0034  72 0A            JC    MAIN2              ;if DPMI present
0036  B4 09            MOV   AH,9
0038  BA 0026 R        MOV   DX,OFFSET ERR2
003B  CD 21            INT   21H                ;display DPMI not present
003D  E9 009B          JMP   MAINE              ;exit to DOS
0040               MAIN2:
0040  B8 0000          MOV   AX,0               ;indicate 0 memory needed
0043  83 3E 0010 R 00  CMP   MSIZE,0
0048  74 F6            JE    MAIN2              ;if DPMI needs no memory
004A  8B 1E 0010 R     MOV   BX,MSIZE           ;get amount
004E  B4 48            MOV   AH,48H
0050  CD 21            INT   21H                ;allocate memory for DPMI
0052  73 09            JNC   MAIN3
0054  B4 09            MOV   AH,9               ;if not enough real memory
0056  BA 003B R        MOV   DX,OFFSET ERR3
0059  CD 21            INT   21H
005B  EB 7E            JMP   MAINE              ;exit to DOS
005D               MAIN3:
005D  8E C0            MOV   ES,AX
005F  B8 0000          MOV   AX,0               ;16-bit application
0062  FF 1E 0000 R     CALL  DS:ENTRY           ;switch to protected mode
0066  73 09            JNC   MAIN4
0068  B4 09            MOV   AH,9               ;if switch failed
006A  BA 0056 R        MOV   DX,OFFSET ERR4
006D  CD 21            INT   21H
006F  EB 6A            JMP   MAINE              ;exit to DOS
                   ;
                   ;PROTECTED MODE
                   ;
0071               MAIN4:
0071  B8 0000          MOV   AX,0000H           ;get local selector
0074  B9 0001          MOV   CX,1               ;only one is needed
0077  CD 31            INT   31H
0079  72 48            JC    MAIN7              ;if error
007B  8B D8            MOV   BX,AX              ;save selector
007D  8E C0            MOV   ES,AX              ;load ES with selector
007F  B8 0007          MOV   AX,0007H           ;set base address
0082  8B 0E 000A R     MOV   CX,WORD PTR FIRST+2
0086  8B 16 0008 R     MOV   DX,WORD PTR FIRST
008A  CD 31            INT   31H
008C  72 3D            JC    MAIN8              ;if error
008E  B8 0008          MOV   AX,0008H
0091  B9 0000          MOV   CX,0
0094  BA FFFF          MOV   DX,0FFFFH          ;set limit to 64K
0097  CD 31            INT   31H
0099  72 38            JC    MAIN9              ;if error
009B  B9 0018          MOV   CX,24              ;load line count
009E  BE 0000          MOV   SI,0               ;load offset
00A1               MAIN5:
00A1  E8 00F4          CALL  DADDR              ;display address, if needed
00A4  E8 00CE          CALL  DDATA              ;display data
00A7  46               INC   SI                 ;point to next data
00A8  66| A1 0008 R    MOV   EAX,FIRST          ;test for end
00AC  66| 3B 06 000C R CMP   EAX,LAST1
00B1  74 07            JE    MAIN6              ;if done
00B3  66| FF 06 0008 R INC   FIRST
```

```
00B8  EB E7                    JMP   MAIN5
00BA                  MAIN6:
00BA  B8 0001                  MOV   AX,0001H           ;release descriptor
00BD  8C C3                    MOV   BX,ES
00BF  CD 31                    INT   31H
00C1  EB 18                    JMP   MAINE              ;exit to DOS
00C3                    MAIN7:
00C3  BA 007B R                MOV   DX,OFFSET ERR5
00C6  E8 0096                  CALL  DISPS              ;display cannot allocate selector
00C9  EB 10                    JMP   MAINE              ;exit to DOS
00CB                  MAIN8:
00CB  BA 0098 R                MOV   DX,OFFSET ERR6
00CE  E8 008E                  CALL  DISPS              ;display cannot use base address
00D1  EB E7                    JMP   MAIN6              ;release descriptor
00D3                  MAIN9:
00D3  BA 00B4 R                MOV   DX,OFFSET ERR7
00D6  E8 0086                  CALL  DISPS              ;display cannot allocate 64K limit
00D9  EB DF                    JMP   MAIN6              ;release descriptor
00DB                  MAINE:
                          .EXIT
                      ;
                      ;The ISDPMI procedure tests for the presence of DPMI.
                      ;***exit parameters***
                      ;carry = 1; if DPMI is present
                      ;carry = 0; if DPMI is not present
                      ;
00DF                  ISDPMI PROC NEAR

00DF  B8 1687                  MOV   AX,1687H           ;get DPMI status
00E2  CD 2F                    INT   2FH                ;DOS multiplex
00E4  0B C0                    OR    AX,AX
00E6  75 0D                    JNZ   ISDPMI1            ;if no DPMI
00E8  89 36 0010 R             MOV   MSIZE,SI           ;save amount of memory needed
00EC  89 3E 0000 R             MOV   WORD PTR ENTRY,DI
00F0  8C 06 0002 R             MOV   WORD PTR ENTRY+2,ES
00F4  F9                       STC
00F5                  ISDPMI1:
00F5  C3                       RET

00F6                  ISDPMI ENDP
                      ;
                      ;The GETDA procedure retrieves the command line parameters
                      ;for memory display in hexadecimal.
                      ;FIRST = the first address from the command line
                      ;LAST1 = the last address from the command line
                      ;***return parameters***
                      ;carry = 1; if error
                      ;carry = 0; for no error
                      ;
00F6                  GETDA  PROC NEAR

00F6  1E                       PUSH DS
00F7  06                       PUSH ES
00F8  1F                       POP  DS
00F9  07                       POP  ES                  ;exchange ES with DS
00FA  BE 0081                  MOV  SI,81H               ;address command line
00FD                  GETDA1:
00FD  AC                       LODSB                     ;skip spaces
00FE  3C 20                    CMP  AL,' '
0100  74 FB                    JE   GETDA1               ;if space
0102  3C 0D                    CMP  AL,13
0104  74 1E                    JE   GETDA3               ;if enter = error
0106  4E                       DEC  SI                   ;adjust SI
0107                  GETDA2:
0107  E8 0020                  CALL GETNU                ;get first number
010A  3C 2C                    CMP  AL,','
010C  75 16                    JNE  GETDA3               ;if no comma = error
010E  66| 26: 89 16 0008 R MOV  ES:FIRST,EDX
```

```
0114  E8 0013              CALL GETNU              ;get second number
0117  3C 0D                CMP  AL,13
0119  75 09                JNE  GETDA3             ;if error
011B  66| 26: 89 16 000C R MOV  ES:LAST1,EDX
0121  F8                   CLC                     ;indicate no error
0122  EB 01                JMP  GETDA4             ;return no error
0124              GETDA3:
0124  F9                   STC                     ;indicate error
0125              GETDA4:
0125  1E                   PUSH DS                 ;exchange ES with DS
0126  06                   PUSH ES
0127  1F                   POP  DS
0128  07                   POP  ES
0129  C3                   RET
012A              GETDA ENDP
                  ;
                  ;The GETNU procedure extracts a number from the command line
                  ;and returns with it in EDX and last command line character in
                  ;AL as a delimiter.
                  ;
012A              GETNU  PROC NEAR

012A  66| BA 00000000      MOV  EDX,0              ;clear result
0130              GETNU1:
0130  AC                   LODSB                   ;get digit from command line
                           .IF  AL >= 'a' && AL <= 'z'
0139  2C 20                    SUB  AL,20H         ;make uppercase
                           .ENDIF
013B  2C 30                SUB  AL,'0'             ;convert from ASCII
013D  72 12                JB   GETNU2             ;if not a number
                           .IF  AL > 9             ;convert A-F from ASCII
0143  2C 07                    SUB  AL,7
                           .ENDIF
0145  3C 0F                CMP  AL,0FH
0147  77 08                JA   GETNU2             ;if not 0-F
0149  66| C1 E2 04         SHL  EDX,4
014D  02 D0                ADD  DL,AL              ;add digit to EDX
014F  EB DF                JMP  GETNU1             ;get next digit
0151              GETNU2:
0151  8A 44 FF             MOV  AL,[SI-1]          ;get delimiter
0154  C3                   RET

0155              GETNU  ENDP
                  ;
                  ;The DISPC procedure displays the ASCII character found
                  ;in register AL.
                  ;***uses***
                  ;INT21H
                  ;
0155              DISPC  PROC NEAR

0155  52                   PUSH DX
0156  8A D0                MOV  DL,AL
0158  B4 06                MOV  AH,6
015A  E8 0084              CALL INT21H             ;do real INT 21H
015D  5A                   POP  DX
015E  C3                   RET

015F              DISPC ENDP
                  ;
                  ;The DISPS procedure displays a character string from
                  ;protected mode addressed by DS:EDX.
                  ;***uses***
                  ;DISPC
                  ;
015F              DISPS  PROC NEAR

015F  66| 81 E2 0000FFFF   AND  EDX,0FFFFH
```

```
0166  67& 8A 02              MOV  AL,[EDX]           ;get character
0169  3C 24                  CMP  AL,'$'             ;test for end
016B  74 07                  JE   DISP1              ;if end
016D  66| 42                 INC  EDX                ;address next character
016F  E8 FFE3                CALL DISPC              ;display character
0172  EB EB                  JMP  DISPS              ;repeat until $
0174              DISP1:
0174  C3                     RET

0175              DISPS  ENDP
                  ;
                  ;The DDATA procedure displays a byte of data at the location
                  ;addressed by ES:SI. The byte is followed by one space.
                  ;***uses***
                  ;DIP and DISPC
                  ;
0175              DDATA PROC NEAR

0175  26: 8A 04              MOV  AL,ES:[SI]         ;get byte
0178  C0 E8 04               SHR  AL,4
017B  E8 000C                CALL DIP                ;display first digit
017E  26: 8A 04              MOV  AL,ES:[SI]         ;get byte
0181  E8 0006                CALL DIP                ;display second digit
0184  B0 20                  MOV  AL,' '             ;display space
0186  E8 FFCC                CALL DISPC
0189  C3                     RET

018A              DDATA ENDP
                  ;
                  ;The DIP procedure displays the right nibble found in AL as a
                  ;hexadecimal digit.
                  ;***uses***
                  ;DISPC
                  ;
018A              DIP   PROC NEAR

018A  24 0F                  AND  AL,0FH             ;get right nibble
018C  04 30                  ADD  AL,30H             ;convert to ASCII
                             .IF  AL > 39H           ;if A-F
0192  04 07                     ADD  AL,7
                             .ENDIF
0194  E8 FFBE                CALL DISPC              ;display digit
0197  C3                     RET

0198              DIP   ENDP
                  ;
                  ;The DADDR procedure displays the hexadecimal address found
                  ;in DS:FIRST if it is a paragraph boundary.
                  ;***uses***
                  ;DIP, DISPS, DISPC, and INT21H
                  ;
0198              DADDR PROC NEAR

0198  66| A1 0008 R          MOV  EAX,FIRST          ;get address
019C  A8 0F                  TEST AL,0FH             ;test for XXXXXXX0
019E  75 40                  JNZ  DADDR4             ;if not, don't display address
01A0  BA 00D5 R              MOV  DX,OFFSET CRLF
01A3  E8 FFB9                CALL DISPS              ;display CR and LF
01A6  49                     DEC  CX                 ;decrement line count
01A7  75 18                  JNZ  DADDR2             ;if not end of page
01A9  BA 00D8 R              MOV  DX,OFFSET MES1     ;if end of page
01AC  E8 FFB0                CALL DISPS              ;display press any key
01AF              DADDR1:
01AF  B4 06                  MOV  AH,6               ;get any key, no echo
01B1  B2 FF                  MOV  DL,0FFH
01B3  E8 002B                CALL INT21H             ;do real INT 21H
01B6  74 F7                  JZ   DADDR1             ;if nothing typed
01B8  BA 00D5 R              MOV  DX,OFFSET CRLF
```

```
01BB   E8 FFA1              CALL DISPS              ;display CRLF
01BE   B9 0018              MOV  CX,24              ;reset line count
01C1              DADDR2:
01C1   51                   PUSH CX                 ;save line count
01C2   B9 0008              MOV  CX,8               ;load digit count
01C5   66| 8B 16 0008 R     MOV  EDX,FIRST          ;get address
01CA              DADDR3:
01CA   66| C1 C2 04         ROL  EDX,4
01CE   8A C2                MOV  AL,DL
01D0   E8 FFB7              CALL DIP                ;display digit
01D3   E2 F5                LOOP DADDR3             ;repeat 8 times
01D5   59                   POP  CX                 ;retrieve line count
01D6   B0 3A                MOV  AL,':'
01D8   E8 FF7A              CALL DISPC              ;display colon
01DB   B0 20                MOV  AL,' '
01DD   E8 FF75              CALL DISPC              ;display space
01E0              DADDR4:
01E0   C3                   RET

01E1              DADDR ENDP
                  ;
                  ;The INT21H procedure gains access to the real mode DOS
                  ;INT 21H instruction with the parameters intact.
                  ;
01E1              INT21H PROC NEAR

01E1   66| A3 0105 R        MOV  REAX,EAX           ;save registers
01E5   66| 89 1E 00F9 R     MOV  REBX,EBX
01EA   66| 89 0E 0101 R     MOV  RECX,ECX
01EF   66| 89 16 00FD R     MOV  REDX,EDX
01F4   66| 89 36 00ED R     MOV  RESI,ESI
01F9   66| 89 3E 00E9 R     MOV  REDI,EDI
01FE   66| 89 2E 00F1 R     MOV  REBP,EBP
0203   9C                   PUSHF
0204   58                   POP  AX
0205   A3 0109 R            MOV  RFLAG,AX
0208   06                   PUSH ES                 ;do DOS interrupt
0209   B8 0300              MOV  AX,0300H
020C   BB 0021              MOV  BX,21H
020F   B9 0000              MOV  CX,0
0212   1E                   PUSH DS
0213   07                   POP  ES
0214   BF 00E9 R            MOV  DI,OFFSET ARRAY
0217   CD 31                INT  31H
0219   07                   POP  ES
021A   A1 0109 R            MOV  AX,RFLAG           ;restore registers
021D   50                   PUSH AX
021E   9D                   POPF
021F   66| 8B 3E 00E9 R     MOV  EDI,REDI
0224   66| 8B 36 00ED R     MOV  ESI,RESI
0229   66| 8B 2E 00F1 R     MOV  EBP,REBP
022E   66| A1 0105 R        MOV  EAX,REAX
0232   66| 8B 1E 00F9 R     MOV  EBX,REBX
0237   66| 8B 0E 0101 R     MOV  ECX,RECX
023C   66| 8B 16 00FD R     MOV  EDX,REDX
0241   C3                   RET

0242              INT21H ENDP
                  END
```

You might notice that the DOS INT 21H function call must be treated differently when operating in the protected mode. The procedure that calls a DOS INT 21H is at the end of Example 17–3. Because this is extremely long and time consuming, we have tended to move away from using the DOS interrupts from a Windows application. The best way to develop software for Windows is through the use of C/C++ with the inclusion of assembly language procedures for arduous tasks.

17–5   VIRTUAL 8086 MODE

One special mode of operation not discussed thus far is the virtual 8086 mode. This special mode is designed so that multiple 8086 real-mode software applications can execute at one time. The PC operates in this mode for DOS applications. Figure 17–25 illustrates two 8086 applications mapped into the 80386 using the virtual mode. If the operating system allows multiple applications to execute, it is usually done through a technique called **time-slicing.** The operating system allocates a set amount of time to each task. For example, if three tasks are executing, the operating system can allocate 1 ms to each task. This means that after each millisecond, a task switch occurs to the next task. In this manner, all tasks receive a portion of the microprocessor's execution time, resulting in a system that appears to execute more than one task at a time. The task times can be adjusted to give any task any percentage of the microprocessor execution time.

A system that can use this technique is a print spooler. The print spooler can function in one DOS partition and be accessed 10 percent of the time. This allows the system to print using the print spooler, but it doesn't detract for the system because it uses only 10 percent of the system time.

The main difference between 80386 protected mode operation and the virtual 8086 mode is the way the segment registers are interpreted by the microprocessor. In the virtual 8086 mode, the segment registers are used as they are in the real mode: as a segment address and an offset address capable of accessing a 1M-byte memory space from location 00000H–FFFFFH. Access to many virtual 8086 mode systems is made possible by the paging unit that is explained in the next section. Through paging, the program still accesses memory below the 1M-byte boundary, yet the microprocessor can access a physical memory space at any location in the 4G-byte range of the memory system.

Virtual 8086 mode is entered by changing the VM bit in the EFLAG register to a logic 1. This mode is entered via an IRET instruction if the privilege level is 00. This bit cannot be set in any other manner. An attempt to access a memory address above the 1M-byte boundary will cause a type-13 interrupt to occur.

FIGURE 17–25 Two tasks resident to an 80386 operated in the virtual 8086 mode.

Memory Map

FFFFFFFF	
001FFFFF	
	TASK 2
	MSDOS
00100000	
000FFFFF	
	TASK 1
	MSDOS
00000000	

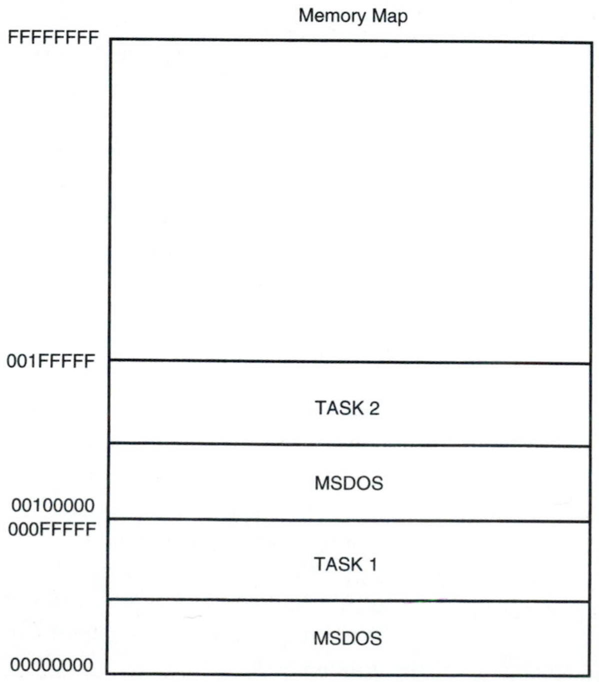

The virtual 8086 mode can be used to share one microprocessor with many users by partitioning the memory so that each user has its own DOS partition. User 1 can be allocated memory locations 00100000H–01FFFFFH, user 2 can be allocated locations 0020000H–02FFFFFH, and so forth. The system software located at memory locations 00000000H–000FFFFFH can then share the microprocessor between users by switching from one to another to execute software. In this manner, one microprocessor is shared by many users.

17–6 THE MEMORY PAGING MECHANISM

The paging mechanism allows any linear (logical) address, as it is generated by a program, to be placed into any physical memory page, as generated by the paging mechanism. A **linear memory page** is a page that is addressed with a selector and an offset in either the real or protected mode. A **physical memory page** is a page that exists at some actual physical memory location. For example, linear memory location 20000H could be mapped into physical memory location 30000H, or any other location, with the paging unit. This means that an instruction that accesses location 20000H actually accesses location 30000H.

Each 80386 memory page is 4K bytes long. Paging allows the system software to be placed at any physical address with the paging mechanism. Three components are used in page address translation: the page directory, the page table, and the actual physical memory page. Note that EEM386.EXE, the extended memory manager, uses the paging mechanism to simulate expanded memory in extended memory and to generate upper memory blocks between system ROMs.

The Page Directory

The page directory contains the location of up to 1024 page translation tables. Each page translation table translates a logic address into a physical address. The page directory is stored in the memory and accessed by the page descriptor address register (CR3) (see Figure 17–14). Control register CR3 holds the base address of the page directory, which starts at any 4K-byte boundary in the memory system. The MOV CR3,reg instruction is used to initialize CR3 for paging. In a virtual 8086 mode system, each 8086 DOS partition would have its own page directory.

The page directory contains up to 1024 entries, which are each four bytes long. The page directory itself occupies one 4K-byte memory page. Each entry in the page directory (see Figure 17–26) translates the leftmost 10 bits of the memory address. This 10-bit portion of the linear address is used to locate different page tables for different page table entries. The page table address (A32–A12), stored in a page directory entry, accesses a 4K-byte long page translation table. To completely translate any linear address into any physical address requires 1024 page tables that are each 4K bytes long, plus the page table directory, which is also 4K bytes long. This translation scheme requires up to 4M plus 4K bytes of memory for a full address translation. Only the largest operating systems support this size address translation. Many commonly found operating systems translate only the first 16M bytes of the memory system if paging is enabled. This includes programs such as Windows. This translation requires four entries in the page directory (16 bytes) and four complete page tables (16K bytes).

FIGURE 17–26 The page table directory entry.

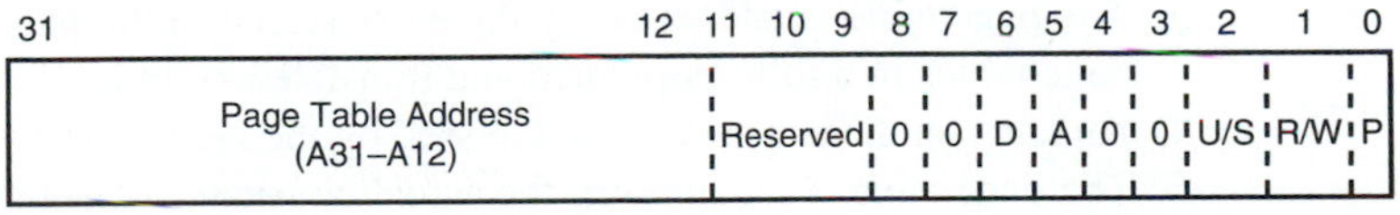

31	12	11	10	9	8	7	6	5	4	3	2	1	0
Page Table Address (A31–A12)		Reserved		0	0	D	A	0	0	U/S	R/W	P	

TABLE 17–2 Protection for level 3 using U/S and R/W.

U/S	R/W	Access Level 3
0	0	None
0	1	None
1	0	Read-only
1	1	Read/write

The page table directory entry control bits, as illustrated in Figure 17–26, each perform the following functions:

D	**Dirty** is undefined for page table directory entries by the 80386 microprocessor and is provided for use by the operating system.
A	**Accessed** is set to a logic 1 whenever the microprocessor accesses the page directory entry.
R/W and U/S	**Read/write** and **user/supervisor** are both used in the protection scheme, as listed in Table 17–2. Both bits combine to develop paging priority level protection for level 3, the lowest user level.
P	**Present,** if a logic 1, indicates that the entry can be used in address translation. If $P = 0$, the entry cannot be used for translation. A not present entry can be used for other purposes, such as indicating that the page is currently stored on the disk. If $P = 0$, the remaining bits of the entry can be used to indicate the location of the page on the disk memory system.

The Page Table

The page table contains 1024 physical page addresses, accessed to translate a linear address into a physical address. Each page table translates a 4M section of the linear memory into 4M of physical memory. The format for the page table entry is the same as for the page directory entry (refer to Figure 17–26). The main difference is that the page directory entry contains the physical address of a page table, while the page table entry contains the physical address of a 4K-byte physical page of memory. The other difference is the D (dirty bit), which has no function in the page directory entry, but indicates that a page has been written to in a page table entry.

Figure 17–27 illustrates the paging mechanism in the 80386 microprocessor. Here, the linear address 00C03FFCH, as generated by a program, is converted to physical address XXXXXFFCH, as translated by the paging mechanism. (Note: XXXXX is any 4K-byte physical page address.) The paging mechanism functions in the following manner:

1. The 4K-byte long page directory is stored as the physical address located by CR3. This address is often called the **root address.** One page directory exists in a system at a time. In the 8086 virtual mode, each task has its own page directory, allowing different areas of physical memory to be assigned to different 8086 virtual tasks.
2. The upper 10 bits of the linear address (bits 31–22), as determined by the descriptors described earlier in this chapter or by a real address, are applied to the paging mechanism to select an entry in the page directory. This maps the page directory entry to the leftmost 10 bits of the linear address.
3. The page table is addressed by the entry stored in the page directory. This allows up to 4K page tables in a fully-populated and translated system.
4. An entry in the page table is addressed by the next 10 bits of the linear address (bits 21–12).
5. The page table entry contains the actual physical address of the 4K-byte memory page.
6. The rightmost 12 bits of the linear address (bits 11–0) select a location in the memory page.

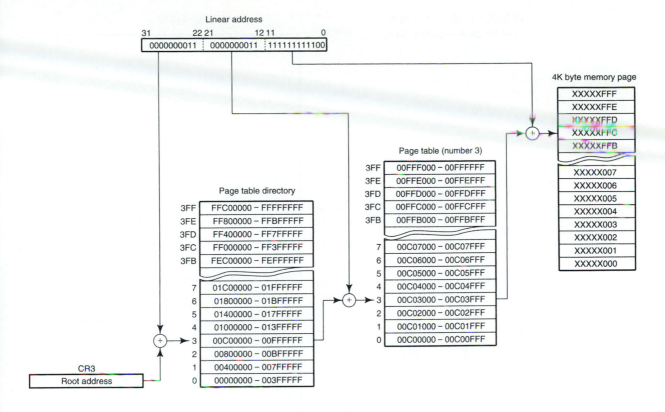

FIGURE 17–27 The translation of linear address 00C03FFCH to physical memory address XXXXXFFCH. The value of XXXXX is determined by the page table entry (not shown here).

The paging mechanism allows the physical memory to be assigned to any linear address through the paging mechanism. For example, suppose that linear address 20000000H is selected by a program, but this memory location does not exist in the physical memory system. The 4K-byte linear page is referenced as locations 20000000H–20000FFFH by the program. Because this section of physical memory does not exist, the operating system might assign an existing physical memory page such as 12000000H–12000FFFH to this linear address range.

In the address translation process, the leftmost 10 bits of the linear address select page directory entry 200H located at offset address 800H in the page directory. This page directory entry contains the address of the page table for linear addresses 20000000H–203FFFFFH. Linear address bits (21–12) select an entry in this page table that corresponds to a 4K-byte memory page. For linear addresses 2000000H–20000FFFH, the first entry (entry 0) in the page table is selected. This first entry contains the physical address of the actual memory page, or 12000000H–12000FFFH in this example.

Take, for example, a typical DOS-based computer system. The memory map for the system appears in Figure 17–28. Note from the map that there are unused areas of memory, which can be paged to a different location, giving a DOS real mode application program more memory. The normal DOS memory system begins at location 00000H and extends to location 9FFFFH, which is 640K bytes of memory. Above location 9FFFFH, we find sections devoted to video cards, disk cards, and the system BIOS ROM. In this example, an area of memory just above 9FFFFH is unused (A0000–AFFFFH). This section of the memory could be used by DOS, so

FIGURE 17–28 Memory map for an AT-style clone.

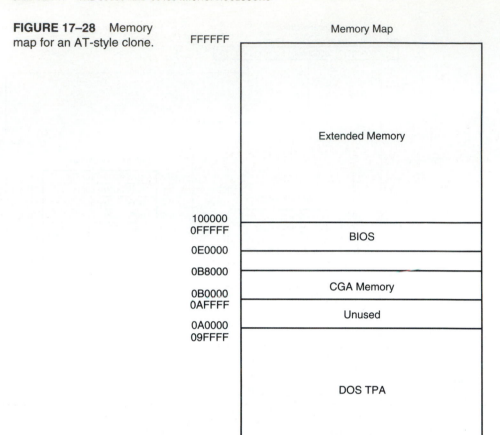

that the total application-memory area is 704K instead of 640K. Be careful when using A0000H–AFFFFH for additional RAM because the video card uses this area for bit-mapped graphics in mode 12H and 13H.

This section of memory can be used by mapping it into extended memory at locations 102000H–11FFFFH. Software to accomplish this translation and initialize the page table directory, and page tables required to set up memory are illustrated in Example 17–4. Note that this procedure initializes the page table directory, a page table, and loads CR3. It does not switch to protected mode and it does enable paging. Note that paging functions in real mode memory operation.

EXAMPLE 17–4

```
                    .MODEL SMALL
                    .386P
0000                .DATA
                    ;page directory
                    ;
0000   00000004     PDIR    DD      4
                    ;
                    ;page table 0
                    ;
0004   0400 [       TAB0    DD      1024 DUP (?)
         00000000
                ]
0000                    .CODE
                    .STARTUP
0010   66| B8 00000000   MOV     EAX,0
0016   8C C8            MOV     AX,CS
0018   66| C1 E0 04      SHL     EAX,4
```

```
001C   66|   05 00000004 R   ADD      EAX,OFFSET TAB0
0022   66|   25 FFFFF000     AND      EAX,0FFFFF000H
0028   66|   83 C0 07        ADD      EAX,7
002C   66|   A3 0000 R       MOV      PDIR,EAX              ;address page table 0
0030   B9 0100              MOV      CX,256
0033   BF 0004 R            MOV      DI,OFFSET TAB0
0036   8C D8                MOV      AX,DS
0038   8E C0                MOV      ES,AX
003A   66|   B8 00000007     MOV      EAX,7
                            .REPEAT                        ;remap 00000H-09FFFH
0040   66|   AB                     STOSD                 ;to 00000H-09FFFH
0042   66|   05 00001000     ADD      EAX,4096
                            .UNTILCXZ
004A   66|   B8 00102007     MOV      EAX,0102007H
0050   B9 0010              MOV      CX,16
                            .REPEAT                        ;remap 0A000H-0AFFFH
0053   66|   AB                     STOSD                 ;to 102000H-11FFFFH
0055   66|   05 00001000       ADD      EAX,4096
                            .UNTILCXZ
005D   66|   B8 00000000     MOV      EAX,0
0063   8C D8                MOV      AX,DS
0065   66|   C1 E0 04        SHL      EAX,4
0069   66|   05 00000000 R   ADD      EAX,OFFSET PDIR      ;load CR3 with page directory
006F   0F 22 D8             MOV      CR3,EAX

               ;additional software to remap other areas of memory

               end
```

INTRODUCTION TO THE 80486 MICROPROCESSOR

The 80486 microprocessor is a highly integrated device, containing well over 1.2 million transistors. Located within this device circuit are a memory-management unit (MMU), a complete numeric coprocessor that is compatible with the 80387, a high-speed level-one cache memory that contains 8K bytes of space, and a full 32-bit microprocessor that is upward-compatible with the 80386 microprocessor. The 80486 is currently available as a 25 MHz, 33 MHz, 50 MHz, 66 MHz, or 100 MHz device. Note that the 66 MHz version is double-clocked and the 100 MHz version is triple-clocked. In 1990, Intel demonstrated a 100 MHz version (not double-clocked) of the 80486 for *Computer Design* magazine, but it has yet to be released. Advanced Micro Devices (AMD) has produced a 40 MHz version that is also available in an 80 MHz (double-clocked) and a 120 MHz (triple-clocked) form. The 80486 is available as an 80486DX or an 80486SX. The only difference between these devices is that the 80486SX does not contain the numeric coprocessor, which reduces its price. The 80487SX numeric coprocessor is available as a separate component for the 80486SX microprocessor.

This section details the differences between the 80486 and 80386 microprocessors. These differences are few, as shall be seen. The most notable differences apply to the cache memory system and parity generator.

Pin-out of the 80486DX and 80486SX Microprocessors

Figure 17–29 illustrates the pin-out of the 80486DX microprocessor, a 168-pin PGA. The 80486SX, also packaged in a 168-pin PGA, is not illustrated because only a few differences exist. Note that pin B15 is NMI on the 80486DX and pin A15 is NMI on the 80486SX. The only other differences are that pin A15 is $\overline{\text{IGNNE}}$ on the 80486DX (not present on the 80486SX), pin C14 is $\overline{\text{FERR}}$ on the 80486DX, and pins B15 and C14 on the 80486SX are not connected.

When connecting the 80486 microprocessor, all Vcc and Vss pins must be connected to the power supply for proper operation. The power supply must be capable of supplying 5.0 V

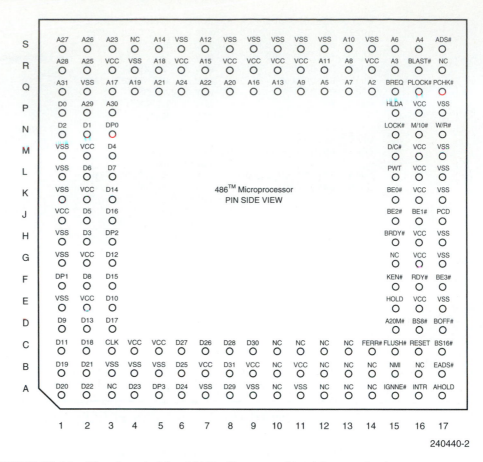

240440-2

FIGURE 17–29   The pin-out of the 80486. (Courtesy of Intel Corporation.)

±10 percent, with up to 1.2 A of surge current for the 33 MHz version. The average supply current is 650 mA for the 33 MHz version. Intel has also produced a 3.3 V version that requires an average of 500 mA at a triple-clock speed of 100 MHz. Logic 0 outputs allow up to 4.0 mA of current, and logic 1 outputs allow up to 1.0 mA. If larger currents are required, as they often are, then the 80486 must be buffered. Figure 17–30 shows a buffered 80486DX system. In the circuit shown, only the address, data, and parity signals are buffered.

Pin Definitions.

A31–A2	**Address outputs** A31–A2 provide the memory and I/O with the address during normal operation; during a cache line invalidation, A31–A4 are used to drive the microprocessor.
A20M	**Address bit 20 mask** causes the 80486 to wrap its address around from location 000FFFFFH to 00000000H, as does the 8086 microprocessor. This provides a memory system that functions like the 1M-byte real memory system in the 8086 microprocessor.
ADS	**Address data strobe** becomes a logic 0 to indicate that the address bus contains a valid memory address.
AHOLD	**Address hold input** causes the microprocessor to place its address bus connections at their high-impedance state, with the remainder of the buses staying active. It is often used by another bus master to gain access for a cache-invalidation cycle.

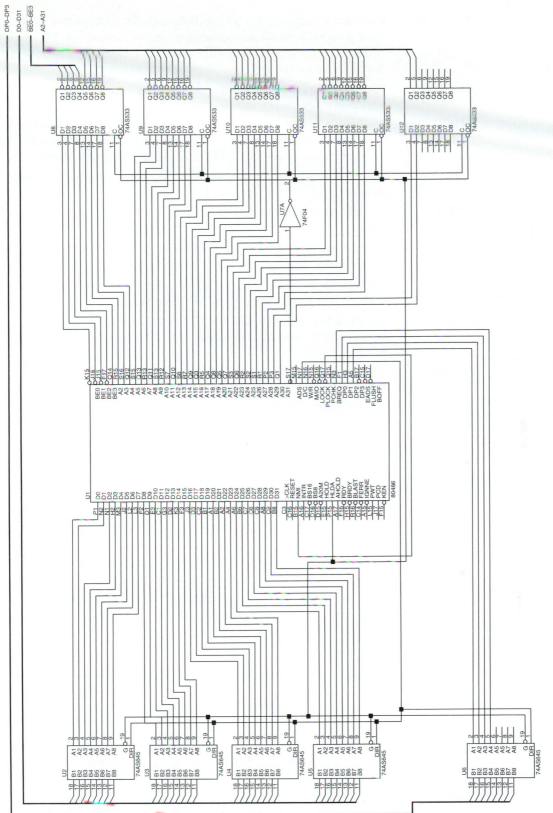

FIGURE 17–30 An 80486 microprocessor showing the buffered address, data, and parity buses.

BE3–BE0   **Byte enable** outputs select a bank of the memory system when information is transferred between the microprocessors and its memory and I/O space. The $\overline{BE3}$ signal enables D31–D24, $\overline{BE2}$ enables D23–D16, $\overline{BE1}$ enables D15–D8, and $\overline{BE0}$ enables D7–D0.

BLAST   The **burst last output** shows that the burst bus cycle is complete on the next activation of the $\overline{BRDY}$ signal.

BOFF   The **back-off** input causes the microprocessor to place its buses at their high-impedance state during the next clock cycle. The microprocessor remains in the bus hold state until the $\overline{BOFF}$ pin is placed at a logic 1 level.

BRDY   The **burst ready** input is used to signal the microprocessor that a burst cycle is complete.

BREQ   The **bus request** output indicates that the 80486 has generated an internal bus request.

BS8   The **bus size 8** input causes the 80486 to structure itself with an 8-bit data bus to access byte-wide memory and I/O components.

BS16   The **bus size 16** input causes the 80486 to structure itself with a 16-bit data bus to access word-wide memory and I/O components.

CLK   The **clock input** provides the 80486 with its basic timing signal. The clock input is a TTL-compatible input that is 25 MHz to operate the 80486 at 25 MHz.

D31–D0   The **data bus** transfers data between the microprocessor and its memory and I/O system. Data bus connections D7–D0 are also used to accept the interrupt vector type number during an interrupt acknowledge cycle.

D/$\overline{C}$ The **data/control** output indicates whether the current operation is a data transfer or control cycle. See Table 17–3 for the function of D/$\overline{C}$, M/$\overline{IO}$, and W/$\overline{R}$.

DP3–DP0   **Data parity I/O** provides even parity for a write operation and check parity for a read operation. If a parity error is detected during a read, the $\overline{PCHK}$ output becomes a logic 0 to indicate a parity error. If parity is not used in a system, these lines must be pulled high to +5.0 V or to 3.3 V in a system that uses a 3.3 V supply.

EADS   The **external address strobe** input is used with AHOLD to signal that an external address is used to perform a cache-invalidation cycle.

FERR   The **floating-point error** output indicates that the floating-point coprocessor has detected an error condition. It is used to maintain compatibility with DOS software.

TABLE 17–3 Bus cycle identification.

M/$\overline{IO}$	D/$\overline{C}$	W/$\overline{R}$	Bus Cycle Type
0	0	0	Interrupt acknowledge
0	0	1	Halt/special
0	1	0	I/O read
0	1	1	I/O write
1	0	0	Opcode fetch
1	0	1	Reserved
1	1	0	Memory read
1	1	1	Memory write

$\overline{\text{FLUSH}}$	The **cache flush** input forces the microprocessor to erase the contents of its 8K-byte internal cache.
HLDA	The **hold acknowledge** output indicates that the HOLD input is active and that the microprocessor has placed its buses at their high-impedance state.
HOLD	The **hold** input requests a DMA action. It causes the address, data, and control buses to be placed at their high-impedance state and also, once recognized, causes HLDA to become a logic 0.
$\overline{\text{IGNNE}}$	The **ignore numeric error** input causes the coprocessor to ignore floating-point errors and to continue processing data. This signal does not affect the state of the $\overline{\text{FERR}}$ pin.
INTR	The **interrupt request** input requests a maskable interrupt, as it does in all other family members.
$\overline{\text{KEN}}$	The **cache enable** input causes the current bus to be stored in the internal cache.
$\overline{\text{LOCK}}$	The **lock** output becomes a logic 0 for any instruction that is prefixed with the lock prefix.
M/$\overline{\text{IO}}$	**Memory/$\overline{\text{IO}}$** defines whether the address bus contains a memory address or an I/O port number. It is also combined with the W/$\overline{\text{R}}$ signal to generate memory, and I/O read and write control signals.
NMI	The **non-maskable interrupt** input requests a type 2 interrupt.
PCD	The **page cache disable** output reflects the state of the PCD attribute bit in the page table entry or the page directory entry.
$\overline{\text{PCHK}}$	The **parity check** output indicates that a parity error was detected during a read operation on the DP3–DP0 pins.
$\overline{\text{PLOCK}}$	The **pseudo-lock** output indicates that the current operation requires more than one bus cycle to perform. This signal becomes a logic 0 for arithmetic coprocessor operations that access 64- or 80-bit memory data.
PWT	The **page write through** output indicates the state of the PWT attribute bit in the page table entry or the page directory entry.
$\overline{\text{RDY}}$	The **ready** input indicates that a non-burst bus cycle is complete. The $\overline{\text{RDY}}$ signal must be returned, or the microprocessor places wait states into its timing until $\overline{\text{RDY}}$ is asserted.
RESET	The **reset** input initializes the 80486, as it does in other family members. Table 17–4 shows the effect of the RESET input on the 80486 microprocessor.
W/$\overline{\text{R}}$	**Write/read** signals that the current bus cycle is either a read or a write.

Basic 80486 Architecture

The architecture of the 80486DX is almost identical to the 80386. Added to the 80386 architecture inside the 80486DX is a math coprocessor and an 8K-byte level 1 cache memory. The 80486SX is almost identical to an 80386 with an 8K-byte cache, but no numeric coprocessor. Figure 17–31 illustrates the basic internal structure of the 80486 microprocessor. If this is compared to the architecture of the 80386, no differences are observed. The most prominent difference between the 80386 and the 80486 is that almost half of the 80486 instructions execute in one clocking period instead of the two clocking periods for the 80386 to execute similar instructions.

As with the 80386, the 80486 contains eight general-purpose 32-bit registers: EAX, EBX, ECX, EDX, EBP, EDI, ESI, and ESP. These registers may be used as 8-, 16-, or 32-bit data registers

TABLE 17–4 The effect of the RESET signal.

Register	Initial Value with Self Test	Initial Value without Self Test
EAX	00000000H	?
EDX	00000400H + ID*	00000400H + ID*
EFLAGS	00000002H	00000002H
EIP	0000FFF0H	0000FFF0H
ES	0000H	0000H
CS	F000H	F000H
DS	0000H	0000H
SS	0000H	0000H
FS	0000H	0000H
GS	0000H	0000H
IDTR	base = 0, limit = 3FFH	base = 0, limit = 3FFH
CR0	60000010H	60000010H
DR7	00000000H	00000000H

*Note: Revision ID number is supplied by Intel for revisions to the microprocessor.

FIGURE 17–31 The internal programming model of the 80486. (Courtesy of Intel Corporation.)

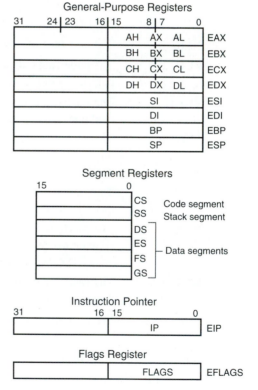

or to address a location in the memory system. The 16-bit registers are the same set as found in the 80286 and are assigned: AX, BX, CX, DX, BP, DI, SI, and SP. The 8-bit registers are AH, AL, BH, BL, CH, CL, DH, and DL.

In addition to the general-purpose registers, the 80486 also contains the same segment registers as the 80386, which are CS, DS, ES, SS, FS, and GS. Each are 16-bits wide, as in all earlier versions of the family.

The IP (instruction pointer) addresses the program located within the 1M byte of memory in combination with CS, or as EIP (extended instruction pointer) to address a program at any location

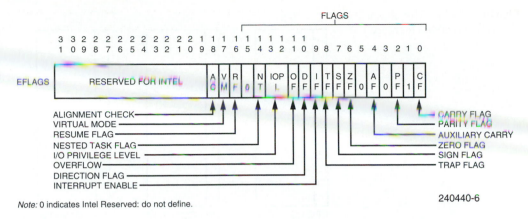

FIGURE 17–32   The EFLAG register of the 80486. (Courtesy of Intel Corporation.)

within the 4G-byte memory system. In protected mode operation, the segment registers function to hold selectors, as they did in the 80286 and 80386 microprocessors.

The 80486 also contains the global, local, and interrupt descriptor table register and memory management unit, as in the 80386. Although these registers are not illustrated in Figure 17–31, they are present as they are in the 80386. (The function of the MMU and its paging unit was described earlier in the chapter.)

The extended flag register (EFLAG) is illustrated in Figure 17–32. As with other family members, the rightmost flag bits perform the same functions for compatibility. The only new flag bit is the AC **(alignment check),** used to indicate that the microprocessor has accessed a word at an odd address or a doubleword stored at a non-doubleword boundary. Efficient software and execution require that data be stored at word or doubleword boundaries.

80486 Memory System

The memory system for the 80486 is identical to the 80386 microprocessor. The 80486 contains 4G bytes of memory, beginning at location 00000000H and ending at location FFFFFFFFH. The major change to the memory system is internal to the 80486 in the form of an 8K-byte cache memory, which speeds the execution of instructions and the acquisition of data. Another addition is the parity checker/generator built into the 80486 microprocessor.

Parity Checker/Generator.   Parity is often used to determine if data are correctly read from a memory location. To facilitate this, Intel has incorporated an internal parity generator/detector. Parity is generated by the 80486 during each write cycle. Parity is generated as even parity, and a parity bit is provided for each byte of memory. The parity check bits appear on pins DP0–DP3, which are also parity inputs as well as outputs. These are typically stored in memory during each write cycle and read from memory during each read cycle.

On a read, the microprocessor checks parity and generates a parity check error, if it occurs, on the $\overline{\text{PCHK}}$ pin. A parity error causes no change in processing unless the user applies the $\overline{\text{PCHK}}$ signal to an interrupt input. Interrupts are often used to signal a parity error in DOS-based computer systems. Figure 17–33 shows the organization of the 80486 memory system that includes parity storage. Note that this is the same as for the 80386, except for the parity bit storage. If parity is not used, Intel recommends that the DP0–DP3 pins be pulled-up to +5.0 V.

Cache Memory.   The cache memory system caches (stores) data used by a program and also the instructions of the program. The cache is organized as a 4-way set associative cache, with each location (line) containing 16 bytes or four doublewords of data. The cache operates as a write-through cache. Note that the cache changes only if a miss occurs. This means that data written to

FIGURE 17–33 The organization of the 80486 memory, showing parity.

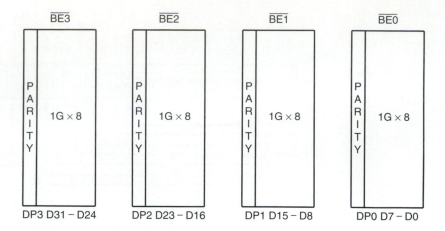

a memory location not already cached are not written to the cache. In many cases, much of the active portion of a program is found completely inside the cache memory. This causes execution to occur at the rate of one clock cycle for many of the instructions that are commonly used in a program. About the only way that these efficient instructions are slowed is when the microprocessor must fill a line in the cache. Data are also stored in the cache, but it has less of an impact on the execution speed of a program because data are not referenced repeatedly as many portions of a program are.

Control register 0 (CR0) is used to control the cache with two new control bits not present in the 80386 microprocessor. (See Figure 17–34 for CR0 in the 80486 microprocessor.) The CD (cache disable) and NW (non-cache write-through) bits are new to the 80486 and are used to control the 8K-byte cache. If the CD bit is a logic 1, all cache operations are inhibited. This setting is used only for debugging software and normally remains cleared. The NW bit is used to inhibit cache write-through operations. As with CD, cache write-through is inhibited only for testing. For normal program operation, $CD = 0$ and $NW = 0$.

Because the cache is new to the 80486 microprocessor and the cache is filled by using burst cycles not present on the 80386, some detail is required to understand bus-filling cycles. When a bus line is filled, the 80486 must acquire four 32-bit numbers from the memory system to fill a line in the cache. Filling is accomplished with a burst cycle. The burst cycle is a special memory where four 32-bit numbers are fetched from the memory system in five clocking periods. This assumes that the speed of the memory is sufficient and that no wait states are required. If the clock frequency of the 80486 is 33 MHz, we can fill a cache line in 167 ns, which is very efficient considering that a normal, non-burst 32-bit memory read operation requires two clocking periods.

Memory Read Timing.   Figure 17–35 illustrates the read timing, for the 80486, for a non-burst memory operation. Note that two clocking periods are used to transfer data. Clocking period T1 provides the memory address and control signals, and clocking period T2 is where the data are transferred between the memory and the microprocessor. Note that the $\overline{RDY}$ must become a logic 0 to cause data to be transferred and to terminate the bus cycle. Access time for a non-burst access is determined by taking two clocking periods, minus the time required for the address to appear on the address bus connection, minus a setup time for the data bus connections. For the

31							24	23								16	15							8	7						0
P G	C E	W T													A M	W P								N E		T S	E M	M P	P E

FIGURE 17–34   Control register zero (CR0) for the 80486 microprocessor.

FIGURE 17–35 The non-burst read timing for the 80486 microprocessor.

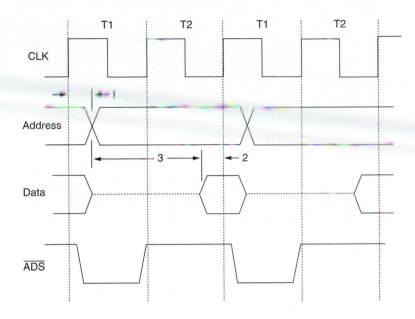

20 MHz version of the 80486, two clocking periods require 100 ns minus 28 ns for address setup time, and 6 ns for data setup time. This yields a non-burst access time of 100 ns – 34 ns, or 76 ns. Of course, if decoder time and delay times are included, the access time allowed the memory is even less for no wait-state operation. If a higher frequency version of the 80486 is used in a system, memory access time is still less.

The 80486 33 MHz, 66 MHz, and 100 MHz processors all access bus data at a 33 MHz rate. In other words, the microprocessor may operate at 100 MHz, but the system bus operates at 33 MHz. Notice that the non-burst access timing for the 33 MHz system bus allows 60 ns – 24ns = 36 ns. It is obvious that wait states are required for operation with standard DRAM memory devices.

Figure 17–36 illustrates the timing diagram for filling a cache line with four 32-bit numbers using a burst. Note that the addresses (A31–A4) appear during T1 and remain constant throughout

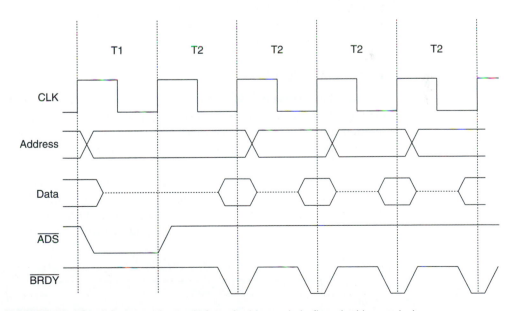

FIGURE 17–36 A burst cycle reads four doublewords in five clocking periods.

the burst cycle. Also, note that A2 and A3 change during each T2 after the first to address four consecutive 32-bit numbers in the memory system. As mentioned, cache fills using bursts require only five clocking periods (one T1 and four T2s) to fill a cache line with four doublewords of data. Access time using a 20 MHz version of the 80486 for the second and subsequent doublewords is 50 ns – 28 ns – 5 ns, or 17 ns, assuming no delays in the system. To use burst mode transfers, we need high-speed memory. Because DRAM memory access times are 40 ns at best, we are forced to use SRAM for burst cycle transfers. The 33 MHz system allows an access time of 30 ns – 19 ns – 5 ns, or 6 ns for the second and subsequent bytes. If an external counter is used in place of address bits A2 and A3, the 19 ns can be eliminated and the access time becomes 30 ns – 5 ns, or 25 ns, which is enough time for even the slowest SRAM connected to the system as a cache. This circuit is often called a *synchronous burst mode cache* if SRAM cache is used with the system. Note that the $\overline{BRDY}$ pin acknowledges a burst transfer rather than the $\overline{RDY}$ pin, which acknowledges a normal memory transfer.

80486 Memory Management

The 80486 contains the same memory-management system as the 80386. This includes a paging unit to allow any 4K-byte block of physical memory to be assigned to any 4K-byte block of linear memory. The 80486 descriptor types are the same as those for the 80386 microprocessor. The only difference between the 80386 memory-management system and the 80486 memory-management system is paging.

The 80486 paging system can be disabled for caching sections of translated memory pages, while the 80386 cannot. Figure 17–37 illustrates the page table directory entry and the page table entry. If these entries are compared with the 80386 entries, the addition of two new control bits is observed (PWT and PCD). The page write-through (PWT) and page cache disable (PCD) bits control caching.

The PWT controls how the cache functions for a write operation of the external cache memory; it does not control writing to the internal cache. The logic level of this bit is found on the PWT pin of the 80486 microprocessor. Externally, it can be used to dictate the write-though policy of the external cache.

The PCD bit controls the on-chip cache. If the *PCD* = 0, the on-chip cache is enabled for the current page of memory. Note that 80386 page table entries place a logic 0 in the PCD bit position, enabling caching. If *PCD* = 1, the on-chip cache is disabled. Caching is disabled, regardless of the condition of $\overline{KEN}$, CD, and NW.

Cache Test Registers

Although not instructions, the cache test registers are placed in this section to illustrate the use of the cache test registers and some software for the 80486 microprocessor. The 80486 cache test registers are TR3 (cache data register), TR4 (cache status test register), and TR5 (cache control test register), which are undefined for the 80386 microprocessor. These three registers are illustrated in Figure 17–38.

The cache data register (TR3) is used to access either the cache fill buffer for a write test operation or the cache read buffer for a cache read test operation. This register is a window into the 8K-byte cache memory located within the 80486 and is used for testing the cache. In order to fill or read a cache line (128 bits wide), TR3 must be written or read four times.

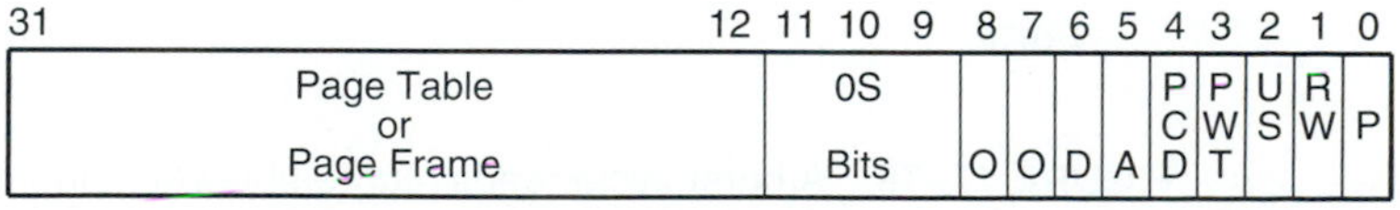

FIGURE 17–37 The page directory or page table entry for the 80486 microprocessor.

31	12	11	10	9	8	7	6	5	4	3	2	1	0
Page Table or Page Frame		0S Bits			O	O	D	A	P C D	P W T	U S	R W	P

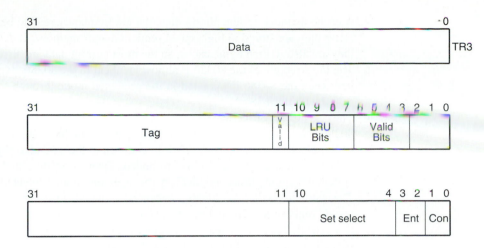

FIGURE 17–38 Cache test register of the 80486 microprocessor.

The contents of the set select field in TR5 determine which internal cache line is written or read through TR3. The 7-bit test field selects one of the 128 different 16 byte-wide cache lines. The entry select bits of TR5 select an entry in the set or the 32-bit location in the fill/read buffer. The control bits in TR5 enable the fill buffer or read buffer operation (00), perform a cache write (01), perform a cache read (10), or flush the cache (11).

The cache status register (TR4) holds the cache tag, LRU bits, and a valid bit. This register is loaded with the tag and valid bit before a cache write operation; and contains the tag, valid bit, LRU bits, and four valid bits on a cache test read.

The cache is tested each time that the microprocessor is reset if the AHOLD pin is high for two clocks prior to the RESET pin going low. This causes the 80486 to completely test itself with a built-in self-test or BIST. The BIST uses TR3, TR4, and TR5 to completely test the internal cache. Its outcome is reported in register EAX. If EAX is a zero, the microprocessor, coprocessor, and cache have passed the self-test. The value of EAX can be tested after a reset to determine if an error is detected. In most cases, we do not directly access the test registers unless we wish to perform our own tests on the cache or TLB.

17–8 SUMMARY

1. The 80386 microprocessor is an enhanced version of the 80286 microprocessor and includes a memory-management unit that is enhanced to provide memory paging. The 80386 also includes 32-bit extended registers, and a 32-bit address and data bus. A scaled-down version of the 80386DX with a 16-bit data and 24-bit address bus is available as the 80386SX microprocessor. The 80386EX is a complete AT-style personal computer on a chip.

2. The 80386 has a physical memory size of 4G bytes that can be addressed as a virtual memory with up to 64T bytes. The 80386 memory is 32 bits wide, and it is addressed as bytes, words, or doublewords.

3. When the 80386 is operated in the pipelined mode, it sends the address of the next instruction or memory data to the memory system prior to completing the execution of the current instruction. This allows the memory system to begin fetching the next instruction or data before the current is completed. This increases access time, thus reducing the speed of the memory.

4. A cache memory system allows data that are frequently read to be accessed in less time because they are stored in high-speed semiconductor memory. If data are written to memory, they are also written to the cache, so the most current data are always present in the cache.

5. The I/O structure of the 80386 is almost identical to the 80286, except that I/O can be inhibited when the 80386 is operated in the protected mode through the I/O bit protection map stored with the TSS.

6. The register set of the 80386 contains extended versions of the registers introduced on the 80286 microprocessor. These extended registers include: EAX, EBX, ECX, EDX, EBP, ESP, EDI, ESI, EIP, and EFLAGS. In addition to the extended registers, two supplemental segment registers (FS and GS) are added. Debug registers and control registers handle system debugging tasks and memory management in the protected mode.

7. The instruction set of the 80386 is enhanced to include instructions that address the 32-bit extended register set. The enhancements also include additional addressing modes that allow any extended register to address memory data. Scaling has been added so that an index register can be multiplied by 1, 2, 4, or 8. New instruction types include bit scan, string moves with sign- or zero-extension, set byte upon condition, and double-precision shifts.

8. In the 80386 microprocessor, interrupts have been expanded to include additional predefined interrupts in the interrupt vector table. These additional interrupts are used with the memory-management system.

9. The 80386 memory manager is similar to the 80286, except that the physical addresses generated by the MMU are 32 bits wide instead of 24 bits wide. The 80386 MMU is also capable of paging.

10. The 80386 is operated in the real mode (8086 mode) when it is reset. The real mode allows the microprocessor to address data in the first 1M byte of memory. In the protected mode, the 80386 addresses any location in its 4G bytes of physical address space.

11. A descriptor is a series of eight bytes that specify how a code or data segment is used by the 80386. The descriptor is selected by a selector that is stored in one of the segment registers. Descriptors are used only in the protected mode.

12. Memory management is accomplished through a series of descriptors, stored in descriptor tables. To facilitate memory management, the 80386 uses three descriptor tables: the global descriptor table (GDT), the local descriptor table (LDT), and the interrupt descriptor table (IDT). The GDT and LDT each hold up to 8192 descriptors; the IDT holds up to 256 descriptors. The GDT and LDT describe code and data segments as well as tasks. The IDT describes the 256 different interrupt levels through interrupt gate descriptors.

13. The TSS (task state segment) contains information about the current task and the previous task. Appended to the end of the TSS is an I/O bit protection map that inhibits selected I/O port addresses.

14. The memory paging mechanism allows any 4K-byte physical memory page to be mapped to any 4K-byte linear memory page. For example, memory location 00A00000H can be assigned memory location A0000000H through the paging mechanism. A page directory and page tables are used to assign any physical address to any linear address. The paging mechanism can be used in the protected mode or the virtual mode.

15. The 80486 microprocessor is an improved version of the 80386 microprocessor that contains an 8K-byte cache and an 80387 arithmetic coprocessor; it executes many instructions in one clocking period.

16. The 80486 microprocessor executes a few new instructions that control the internal cache memory and allow addition (XADD) and comparison (CMPXCHG) with an exchange and a byte swap (BSWAP) operation. Other than these few additional instructions, the 80486 is 100 percent upward-compatible with the 80386 and 80387.

17. A new feature found in the 80486 is the BIST (built-in self-test) that tests the microprocessor, coprocessor, and cache at reset time. If the 80486 passes the test, EAX contains a zero.

18. Additional test registers are added to the 80486 to allow the cache memory to be tested. These new test registers are TR3 (cache data), TR4 (cache status), and TR5 (cache control). Although we seldom use these registers, they are used by BIST each time that a BIST is performed after a reset operation.

17–9 QUESTIONS AND PROBLEMS

1. The 80386 microprocessor addresses _____ bytes of physical memory when operated in the protected mode.
2. The 80386 microprocessor addresses _____ bytes of virtual memory through its memory-management unit.
3. Describe the differences between the 80386DX and the 80386SX.
4. Draw the memory map of the 80386 when operated in the
   (a) protected mode
   (b) real mode
5. How much current is available on various 80386 output pin connections? Compare these currents with the currents available at the output pin connection of an 8086 microprocessor.
6. Describe the 80386 memory system, and explain the purpose and operation of the bank selection signals.
7. Explain the action of a hardware reset on the address bus connections of the 80386.
8. Explain how pipelining lengthens the access time for many memory references in the 80386 microprocessor-based system.
9. Briefly describe how the cache memory system functions.
10. I/O ports in the 80386 start at I/O address _____ and extend to I/O address _____.
11. What I/O ports communicate data between the 80386 and its companion 80387 coprocessor?
12. Compare and contrast the memory and I/O connections found on the 80386 with those found in earlier microprocessors.
13. If the 80386 operates at 20 MHz, what clocking frequency is applied to the CLK2 pin?
14. What is the purpose of the $\overline{BS16}$ pin on the 80386 microprocessor?
15. What two additional segment registers are found in the 80386 programming model that are not present in the 8086?
16. List the extended registers found in the 80386 microprocessor.
17. List each 80386 flag register bit and describe its purpose.
18. Define the purpose of each of the control registers (CR0, CR1, CR2, and CR3) found within the 80386.
19. Define the purpose of each 80386 debug register.
20. The debug registers cause which level of interrupt?
21. Describe the operation of the bit scan forward instruction.
22. Describe the operation of the bit scan reverse instruction.
23. Describe the operation of the SHRD instruction.
24. Form an instruction that accesses data in the FS segment at the location indirectly addressed by the DI register. The instruction should store the contents of EAX into this memory location.
25. What is scaled index addressing?
26. Is the following instruction legal? MOV AX,[EBX+ECX]
27. Explain how the following instructions calculate the memory address:
   (a) ADD [EBX+8*ECX],AL
   (b) MOV DATA[EAX+EBX],CX
   (c) SUB EAX,DATA
   (d) MOV ECX,[EBX]

28. What is the purpose of interrupt type number 7?
29. Which interrupt vector type number is activated for a protection privilege violation?
30. What is a double interrupt fault?
31. If an interrupt occurs in the protected mode, what defines the interrupt vectors?
32. What is a descriptor?
33. What is a selector?
34. How does the selector choose the local descriptor table?
35. What register is used to address the global descriptor table?
36. How many global descriptors can be stored in the GDT?
37. Explain how the 80386 can address a virtual memory space of 64T bytes when the physical memory contains only 4G bytes of memory.
38. What is the difference between a segment descriptor and a system descriptor?
39. What is the task state segment (TSS)?
40. How is the TSS addressed?
41. Describe how the 80386 switches from the real mode to the protected mode.
42. Describe how the 80386 switches from the protected mode to the real mode.
43. What is virtual 8086 mode operation of the 80386 microprocessor?
44. How is the paging directory located by the 80386?
45. How many bytes are found in a page of memory?
46. Explain how linear memory address D0000000H can be assigned to physical memory address C0000000H with the paging unit of the 80386.
47. What are the differences between an 80386 and 80486 microprocessor?
48. What is the purpose of the $\overline{\text{FLUSH}}$ input pin on the 80486 microprocessor?
49. Compare the register set of the 80386 with the 80486 microprocessor.
50. What differences exist in the flags of the 80486 when compared to the 80386 microprocessor?
51. Which pins are used for parity checking on the 80486 microprocessor?
52. The 80486 microprocessor uses _____ parity.
53. The cache inside the 80486 microprocessor is _____ K bytes.
54. A cache line is filled by reading _____ bytes from the memory system.
55. What is an 80486 burst?
56. Define the term *cache-write through*.
57. What is a BIST?
58. Can 80486 caching be disabled by software? Explain your answer.
59. Explain how the XADD EBX,EDX instruction operates.
60. The CMPXCHG CL,AL instruction compares CL with AL. What else occurs when this instruction executes?
61. Compare the INVD instruction with the WBINVD instruction.
62. What is the purpose of the PCD bit in the page table directory or page table entry?
63. Does the PWT bit in the page table directory or page table entry affect the on-chip cache?

CHAPTER 18

The Pentium and Pentium Pro Microprocessors

INTRODUCTION

The Pentium microprocessor signals an improvement to the architecture found in the 80486 microprocessor. The changes include an improved cache structure, a wider data bus width, a faster numeric coprocessor, a dual integer processor, and branch prediction logic. The cache has been reorganized to form two caches that are each 8K bytes in size, one for caching data, and the other for instructions. The data bus width has been increased from 32 bits to 64 bits. The numeric coprocessor operates at about five times faster than the 80486 numeric coprocessor. A dual-integer processor often allows two instructions per clock. Finally, the branch prediction logic allows programs that branch to execute more efficiently. Notice that these changes are internal to the Pentium, which makes software upward-compatible from earlier Intel 80X86 microprocessors. A later improvement to the Pentium was the addition of the MMX instructions.

The Pentium Pro is a still faster version of the Pentium, and it contains a modified internal architecture that can schedule up to five instructions for execution and an even faster floating-point unit. The Pentium Pro also contains a 256K-byte or 512K-byte level two cache in addition to the 16K-byte (8K for data and 8K for instruction) level one cache. The Pentium Pro includes error correction circuitry (ECC) to correct a one bit error and indicate a two bit error. Also added are four additional address lines, giving the Pentium Pro access to an astounding 64G bytes of directly addressable memory space.

CHAPTER OBJECTIVES

Upon completion of this chapter, you will be able to:

1. Contrast the Pentium and Pentium Pro with the 80386 and 80486 microprocessors.
2. Describe the organization and interface of the 64-bit wide Pentium memory system and its variations.
3. Contrast the changes in the memory-management unit and paging unit when compared to the 80386 and 80486 microprocessors.
4. Detail the new instructions found with the Pentium microprocessor.
5. Explain how the superscaler dual integers units improve performance of the Pentium microprocessor.
6. Describe the operation of the branch prediction logic.
7. Detail the improvements in the Pentium Pro when compared with the Pentium.
8. Explain how the dynamic execution architecture of the Pentium Pro functions.

18–1   INTRODUCTION TO THE PENTIUM MICROPROCESSOR

Before the Pentium or any other microprocessor can be used in a system, the function of each pin must be understood. This section of the chapter details the operation of each pin, along with the external memory system and I/O structures of the Pentium microprocessor.

Figure 18–1 illustrates the pin-out of the Pentium microprocessor, which is packaged in a huge 237-pin PGA (pin grid array). Currently, the Pentium is available in two versions: the full-blown Pentium and the P24T version called the Pentium OverDrive. The P24T version contains a 32-bit data bus, compatible for insertion into 80486 machines, which contains the P24T socket.

FIGURE 18–1   The pin-out of the Pentium microprocessor.

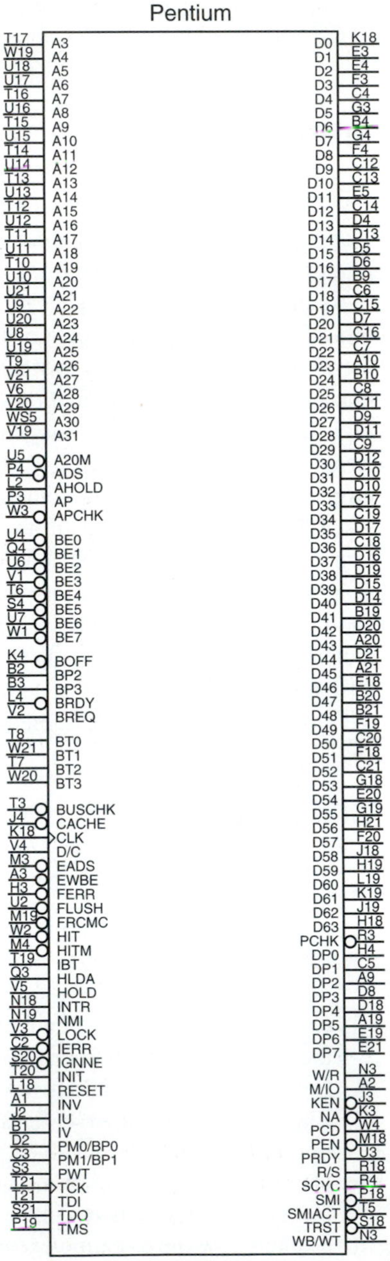

The P24T version also comes with a fan built into the unit. The most notable difference in the pin-out of the Pentium, when compared to earlier 80486 microprocessors, is that there are 64 data bus connections instead of 32, which require a larger physical footprint.

As with earlier versions of the Intel family of microprocessors, the early versions of the Pentium require a single +5.0 V power supply for operation. The power supply current averages 3.3 A for the 66 MHz version of the Pentium, and 2.91 A for the 60 MHz version. Because these currents are significant, so are the power dissipations of these microprocessors: 13 W for the 66 MHz version and 11.9 W for the 60 MHz version. The current versions of the Pentium, 90 MHz and above, use a 3.3 V power supply with reduced current consumption. At present, a good heat sink with considerable airflow is required to keep the Pentium cool. The Pentium contains multiple Vcc and Vss connections that must all be connected to +5.0 V or +3.3 V and ground for proper operation. Some of the pins are labeled N/C (no connection) and must not be connected. The latest versions of the Pentium have been improved to reduce the power dissipation. For example, the 233 MHz Pentium requires 3.4 A or current, which is only slightly more than the 3.3 A required by the early 66 MHz version.

Each Pentium output pin is capable of providing 4.0 mA of current at a logic 0 level and 2.0 mA at a logic 1 level. This represents an increase in drive current, compared to the 2.0 mA available on earlier 8086, 8088, and 80286 output pins. Each input pin represents a small load requiring only 15 μA of current. In some systems, except the smallest, these current levels require bus buffers.

The function of each Pentium group of pins follows:

$\overline{\text{A20}}$	The **address A20 mask** is an input that is asserted in the real mode to signal the Pentium to perform address wraparound, as in the 8086 microprocessor, for use of the HIMEM.SYS driver.
A31–A3	**Address bus** connections address any of the 512K × 64 memory locations found in the Pentium memory system. Note that A0, A1, and A2 are encoded in the bus enable ($\overline{\text{BE7}}$–$\overline{\text{BE0}}$), described elsewhere, to select any or all of the eight bytes in a 64-bit wide memory location.
$\overline{\text{ADS}}$	The **address data strobe** becomes active whenever the Pentium has issued a valid memory or I/O address. This signal is combined with the W/$\overline{\text{R}}$ and M/$\overline{\text{IO}}$ signals to generate the separate read and write signals present in the earlier 8086–80286 microprocessor-based systems.
AHOLD	**Address hold** is an input that causes the Pentium to hold the address and AP signals for the next clock.
AP	**Address parity** provides even parity for the memory address on all Pentium-initiated memory and I/O transfers. The AP pin must also be driven with even parity information on all inquire cycles in the same clocking period as the $\overline{\text{EADS}}$ signal.
$\overline{\text{APCHK}}$	**Address parity check** becomes a logic 0 whenever the Pentium detects an address parity error.
$\overline{\text{BE7}}$–$\overline{\text{BE0}}$	**Bank enable signals** select the access of a byte, word, doubleword, or quadword of data. These signals are generated internally by the microprocessor from address bits A0, A1, and A2.
$\overline{\text{BOFF}}$	The **back-off** input aborts all outstanding bus cycles and floats the Pentium buses until $\overline{\text{BOFF}}$ is negated. After $\overline{\text{BOFF}}$ is negated, the Pentium restarts all aborted bus cycles in their entirety.
BP[3:2] and PM/BP[1:0]	The **breakpoint pins** BP3–BP0 indicate a breakpoint match when the debug registers are programmed to monitor for matches. The **performance monitoring** pins PM1 and PM0 indicate the settings of the performance monitoring bits in the debug mode control register.

$\overline{\text{BRDY}}$	The **burst ready** input signals the Pentium that the external system has applied or extracted data from the data bus connections. This signal is used to insert wait states into the Pentium timing.
BREQ	The **bus request** output indicates that the Pentium has generated a bus request.
BT3–BT0	The **branch trace** outputs provide bits 2–0 of the branch target linear address and the default operand size on BT3. These outputs become valid during a branch trace special message cycle.
$\overline{\text{BUSCHK}}$	The **bus check** input allows the system to signal the Pentium that the bus transfer has been unsuccessful.
$\overline{\text{CACHE}}$	The **cache** output indicates that the current Pentium cycle can cache data.
CLK	The **clock** is driven by a clock signal that is at the operating frequency of the Pentium. For example, to operate the Pentium at 66 MHz, we apply a 66 MHz clock to this pin.
D63–D0	**Data bus** connections transfer byte, word, doubleword, and quadword data between the microprocessor and its memory and I/O system.
D/$\overline{\text{C}}$	**Data/control** indicates that the data bus contains data for or from memory or I/O when a logic 1. If D/$\overline{\text{C}}$ is a logic 0, the microprocessor is either halted or executing an interrupt acknowledge.
DP7–DP0	**Data parity** is generated by the Pentium and detects its eight memory banks through these connections.
$\overline{\text{EADS}}$	The **external address strobe** input signals that the address bus contains an address for an inquire cycle.
$\overline{\text{EWBE}}$	The **external write buffer empty** input indicates that a write cycle is pending in the external system.
$\overline{\text{FERR}}$	The **floating-point error** is comparable to the $\overline{\text{ERROR}}$ line in the 80386 and shows that the internal coprocessor has erred.
$\overline{\text{FLUSH}}$	The **flush cache** input causes the cache to flush all write-back lines and invalidate its internal caches. If the $\overline{\text{FLUSH}}$ input is a logic 0 during a reset operation, the Pentium enters its test mode.
$\overline{\text{FRCMC}}$	The **functional redundancy check** is sampled during a reset to configure the Pentium in the master (1) or checker mode (0).
$\overline{\text{HIT}}$	**Hit** shows that the internal cache contains valid data in the inquire mode.
$\overline{\text{HITM}}$	**Hit modified** shows that the inquire cycle found a modified cache line. This output is used to inhibit other master units from accessing data until the cache line is written to memory.
HOLD	**Hold** requests a DMA action.
HLDA	**Hold acknowledge** indicates that the 80386 is currently in a hold condition.
IBT	**Instruction branch taken** indicates that the Pentium has taken an instruction branch.
$\overline{\text{IERR}}$	The **internal error output** shows that the Pentium has detected an internal parity error or functional redundancy error.
$\overline{\text{IGNNE}}$	The **ignore numeric error** input causes the Pentium to ignore a numeric coprocessor error.
INIT	The **initialization** input performs a reset without initializing the caches, write-back buffers, and floating-point registers. This may not be used to reset the microprocessor in lieu of RESET after power-up.

INTR	The **interrupt request** is used by external circuitry to request an interrupt.
INV	The **invalidation** input determines the cache line state after an inquiry.
IU	The **U-Pipe instruction complete** output shows that the instruction in the U-pipe is complete.
IV	The **V-Pipe instruction complete** output shows that the instruction in the V-pipe is complete.
$\overline{\text{KEN}}$	The **cache enable** input enables internal caching.
$\overline{\text{LOCK}}$	**LOCK** becomes a logic 0 whenever an instruction is prefixed with the LOCK: prefix. This is most often used during DMA accesses.
M/$\overline{\text{IO}}$	**Memory/IO** selects a memory device when a logic 1 or an I/O device when a logic 0. During the I/O operation, the address bus contains a 16-bit I/O address on address connections A15–A3.
$\overline{\text{NA}}$	**Next address** indicates that the external memory system is ready to accept a new bus cycle.
NMI	The **non-maskable interrupt** requests a non-maskable interrupt, just as on the earlier versions of the microprocessor.
PCD	The **page cache disable** output shows that the internal page caching is disabled by reflecting the state of the CR3 PCD bit.
$\overline{\text{PCHK}}$	The **parity check** output signals a parity check error for data read from memory or I/O.
$\overline{\text{PEN}}$	The **parity enable** input enables the machine check interrupt or exception.
PRDY	The **probe ready** output indicates that the probe mode has been entered for debugging.
PWT	The **page write-through** output shows the state of the PWT bit in CR3.
R/$\overline{\text{S}}$	This pin is provided for use with the Intel Debugging Port and causes an interrupt.
RESET	**Reset** initializes the Pentium, causing it to begin executing software at memory location FFFFFFF0H. The Pentium is reset to the real mode and the leftmost 12 address connections remain logic 1s (FFFH) until a far jump or far call is executed. This allows compatibility with earlier microprocessors. See Table 18–1 for the state of the Pentium after a hardware reset.
SCYC	The **split cycle** output signals a misaligned LOCKed bus cycle.
$\overline{\text{SMI}}$	The **system management interrupt** input causes the Pentium to enter the system management mode of operation.
$\overline{\text{SMIACT}}$	The **system management interrupt active** output shows that the Pentium is operating in the system management mode.
TCK	The **testability clock** input selects the clocking function in accordance to the IEEE 1149.1 Boundary Scan interface.
TDI	The **test data** input is used to test data clocked into the Pentium with the TCK signal.
TDO	The **test data** output is used to gather test data and instructions shifted out of the Pentium with TCK.
TMS	The **test mode select** input controls the operation of the Pentium in test mode.
$\overline{\text{TRST}}$	The **test reset** input allows the test mode to be reset.
W/$\overline{\text{R}}$	**Write/read** indicates that the current bus cycle is a write when a logic 1 or a read when a logic 0.
WB/$\overline{\text{WT}}$	**Write-back/write-through** selects the operation for the Pentium data cache.

TABLE 18–1   State of the Pentium after a RESET.

Register	RESET Value	RESET + BIST Value
EAX	0	0 (if test passes)
EDX	0500XXXXH	0500XXXXH
EBX, ECX, ESP, EBP, ESI, and EDI	0	0
EFLAGS	2	2
EIP	0000FFF0H	0000FFF0H
CS	F000H	F000H
DS, ES, FS, GS, and SS	0	0
GDTR and TSS	0	0
CR0	60000010H	60000010H
CR2, CR3, and CR4	0	0
DR0–DR3	0	0
DR6	FFFF0FF0H	FFFF0FF0H
DR7	00000400H	00000040H

Notes: BIST = built-in self-test; XXXX = Pentium version number.

The Memory System

The memory system for the Pentium microprocessor is 4G bytes in size, just as in the 80386DX and 80486 microprocessors. The difference lies in the width of the memory data bus. The Pentium uses a 64-bit data bus to address memory organized in eight banks that each contain 512M bytes of data. See Figure 18–2 for the organization of the Pentium physical memory system.

The Pentium memory system is divided into eight banks that each store a byte of data with a parity bit. The Pentium, like the 80486, employs internal parity generation and checking logic for the memory system's data bus information. (Note that most Pentium systems do not use parity checks, but it is available.) The 64-bit wide memory is important to double-precision floating-point data. Recall that a double-precision floating-point number is 64 bits wide. Because of the change to a 64-bit wide data bus, the Pentium is able to retrieve floating-point data with one read cycle, instead of two as in the 80486. This causes the Pentium to function at a higher throughput than an 80486. As with earlier 32-bit Intel microprocessors, the memory system is numbered in bytes from byte 00000000H to byte FFFFFFFFH.

Memory selection is accomplished with the bank enable signals ($\overline{BE7}$–$\overline{BE0}$). These separate memory banks allow the Pentium to access any single byte, word, doubleword, or quadword with one memory transfer cycle. As with earlier memory selection logic, we often generate eight separate write strobes for writing to the memory system.

A new feature added to the Pentium is its capability to check and generate parity for the address bus (A31–A5) during certain operations. The AP pin provides the system with parity information

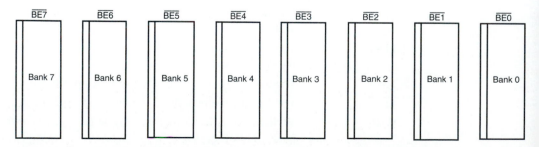

FIGURE 18–2   The 8-byte wide memory banks of the Pentium microprocessor.

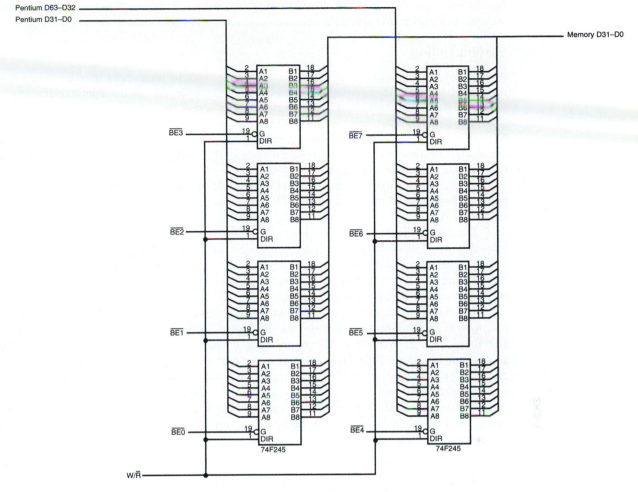

FIGURE 18–3 A circuit that generates a 32-bit memory data bus from the 64-bit Pentium data bus.

and the $\overline{\text{APCHK}}$ indicates a bad parity check for the address bus. The Pentium takes no action when an address parity error is detected. The error must be assessed by the system and the system must take appropriate action (an interrupt), if so desired.

How is a 32-bit memory system connected to the Pentium? The Pentium can function with a 32-bit wide memory system by using a multiplexer to convert the 64-bit data bus to a 32-bit data bus. Figure 18–3 shows a set of bi-directional multiplexers (bi-directional buffers are used as multiplexers) that are used to convert the Pentium's 64-bit data bus into a 32-bit data bus. Care must be taken when using this arrangement because software could access a doubleword that crosses the boundary between the lower and upper halves of the data bus. All doublewords must be stored at doubleword boundaries. Note that a doubleword boundary is an address that is divisible by 4.

Input/Output System

The input/output system of the Pentium is completely compatible with earlier Intel microprocessors. The I/O port number appears on address lines A15–A3 with the bank enable signals used to select the actual memory banks used for the I/O transfer.

Beginning with the 80386 microprocessor, I/O privilege information is added to the TSS segment when the Pentium is operated in the protected mode. Recall that this allows I/O ports to be selectively inhibited. If the blocked I/O location is accessed, the Pentium generates a type 13 interrupt to signal an I/O privilege violation.

System Timing

As with any microprocessor, the system timing signals must be understood in order to interface the microprocessor. This portion of the text details the operation of the Pentium through its timing diagrams and shows how to determine memory access times.

The basic Pentium, non-pipelined memory cycle consists of two clocking periods: T1 and T2. See Figure 18–4 for the basic non-pipelined read cycle. Notice from the timing diagram that the 66 MHz Pentium is capable of 33 million memory transfers per second. This assumes that the memory can operate at that speed.

Also notice form the timing diagram that the W/R̄ signal becomes valid if $\overline{ADS}$ is a logic 0 at the positive edge of the clock (end of T1). This clock must be used to qualify the cycle as a read or a write.

During T1, the microprocessor issues the $\overline{ADS}$, W/R̄, address, and M/IO signals. In order to qualify the W/R̄ signal and generate appropriate $\overline{MRDC}$ and $\overline{MWTC}$ signals, we use a flip-flop to generate the W/R̄ signal. Then we use a 2 line-to-1 line multiplexer to gencrate the memory and I/O control signals. See Figure 18–5 for a circuit that generates the memory and I/O control signals for the Pentium microprocessor.

During T2, the data bus is sampled in synchronization with the end of T2 at the positive transition of the clock pulse. The setup time before the clock is given as 3.8 ns, and the hold time after the clock is given as 2.0 ns. This means that the data window around the clock is 5.8 ns. The address appears on the 8.0 ns maximum after the start of T1. This means that the Pentium microprocessor operating at 66 MHz allows 30.3 ns (two clocking periods), minus the address delay time of 8.0 ns and minus the data setup time of 3.8 ns. Memory access time without any wait states is 30.3 – 8.0 – 3.8, or 18.5 ns. This is enough time to allow access to a SRAM, but not to any DRAM without inserting wait states into the timing. The SRAM is normally found in the form of an external level 2 cache.

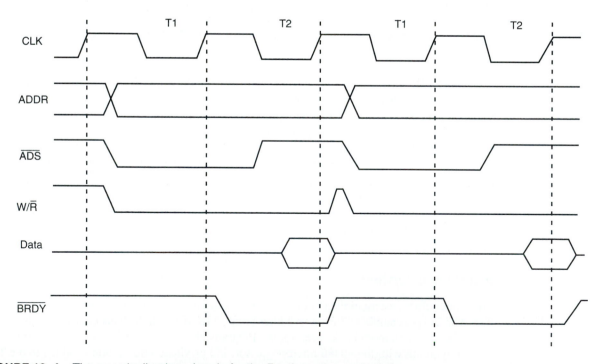

FIGURE 18–4   The non-pipelined read cycle for the Pentium microprocessor.

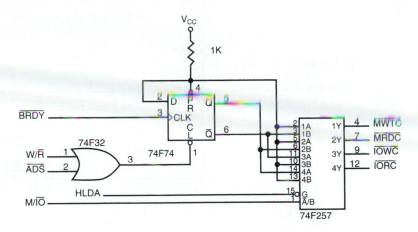

FIGURE 18–5 A circuit that generates the memory and I/O control signals.

Wait states are inserted into the timing by controlling the $\overline{\text{BRDY}}$ input to the Pentium. The $\overline{\text{BRDY}}$ signal must become a logic 0 by the end of T2 or additional T2 states are inserted into the timing. See Figure 18–6 for a read cycle timing diagram that contains wait states for slower memory. The effect of inserting wait states into the timing is to lengthen the timing, allowing additional time to the memory to access data. In the timing shown, the access time has been lengthened so that standard 60 ns DRAM can be used in a system. Note that this requires the insertion of four wait states of 15.2 ns (one clocking period) each to lengthen the access time to 79.5 ns. This is enough time for the DRAM and any decoder in the system to function.

The $\overline{\text{BRDY}}$ signal is a synchronous signal generated by using the system clock. Figure 18–7 illustrates a circuit that can be used to generate $\overline{\text{BRDY}}$ for inserting any number of wait states into the Pentium timing diagram. You may recall a similar circuit inserting wait states into the timing diagram of the 80386 microprocessor. The $\overline{\text{ADS}}$ signal is delayed between 0 and 7 clocking periods by the 74F161 shift register to generate the $\overline{\text{BRDY}}$ signal. The exact number of wait states is selected by the 74F151 8 line-to-1 line multiplexer. In this example, the multiplexer selects the 4-wait output from the shift register.

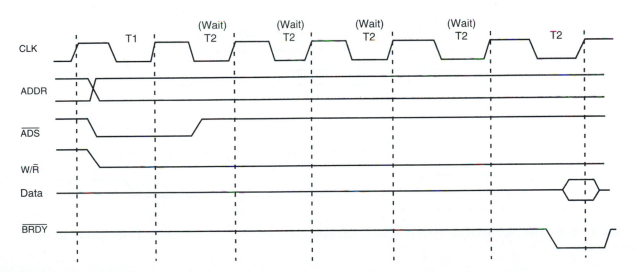

FIGURE 18–6 The Pentium timing diagram with four wait states inserted for an access time of 79.5 ns.

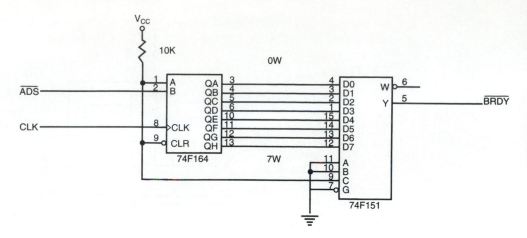

FIGURE 18–7   A circuit that generates wait states by delaying ADS. This circuit is wired to generate four wait states.

A more efficient method of reading memory data is via the burst cycle. The burst cycle in the Pentium transfers four 64-bit numbers per burst cycle in five clocking periods. A burst without wait states requires that the memory system transfers data every 15.2 ns. If a level 2 cache is in place, this speed is no problem as long as the data are read from the cache. If the cache does not contain the data, then wait states must be inserted, which will reduce the data throughput. See Figure 18–8 for the Pentium burst cycle transfer without wait states. As before, wait states can be inserted to allow more time to the memory system for accesses.

Branch Prediction Logic

The Pentium microprocessor uses a branch prediction logic to reduce the time required for a branch caused by internal delays. These delays are minimized because when a branch instruction (short or near only) is encountered, the microprocessor begins pre-fetch instruction at the branch address. The instructions are loaded into the instruction cache, so when the branch occurs, the instructions are present and allow the branch to execute in one clocking period. If for any reason

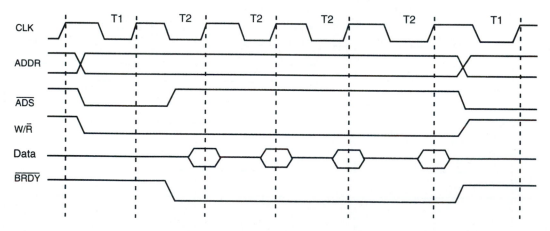

FIGURE 18–8   The Pentium burst cycle operation that transfers four 64-bit data between the microprocessor and memory.

the branch prediction logic errs, the branch requires an extra three clocking periods to execute. In most cases, the branch prediction is correct and no delay ensues.

Cache Structure

The cache in the Pentium has been changed from the one found in the 80486 microprocessor. The Pentium contains two 8K-byte cache memories instead of one as in the 80486. There is an 8K-byte data cache and an 8K-byte instruction cache. The instruction cache stores only instructions, while the data cache stores data used by instructions.

In the 80486 with its unified cache, a program that was data-intensive quickly filled the cache, allowing little room for instructions. This slowed the execution speed of the 80486 microprocessor. In the Pentium, this cannot occur because of the separate instruction cache.

Superscaler Architecture

The Pentium microprocessor is organized with three execution units. One executes floating-point instructions, and the other two (U-pipe and V-pipe) execute integer instructions. This means that it is possible to execute three instructions simultaneously. For example, the FADD ST,ST(2) instruction, MOV EAX,10H instruction, and MOV EBX,12H instruction can all execute simultaneously because none of these instructions depend on each other. The FADD ST,ST(2) instruction is executed by the coprocessor; the MOV EAX,10H is executed by the U-pipe; and the MOV EBX,12H instruction is executed by the V-pipe. Because the floating-point unit is also used for MMX instructions, if available, the Pentium can execute two integers and one MMX instruction simultaneously.

Software should be written to take advantage of this feature by looking at the instructions in a program, and then modifying them when cases are discovered in which dependent instructions can be separated by non-dependent instructions. These changes can result in up to a 40 percent execution speed improvement in some software. Make sure that any new compiler or other application package takes advantage of this new superscaler feature of the Pentium.

18–2   SPECIAL PENTIUM REGISTERS

The Pentium is essentially the same microprocessor as the 80386 and 80486, except that some additional features and changes to the control register set have occurred. This section highlights the differences between the 80386 control register structure and the flag register.

Control Registers

Figure 18–9 shows the control register structure for the Pentium microprocessor. Note that a new control register CR4 has been added to the control register array.

This section of the text only explains the new Pentium components in the control registers. See Figure 17–15 for a description and illustration of the 80386 control registers. Following is a description of the new control bits and new control register CR4:

CD   **Cache disable** controls the internal cache. If *CD* = 1, the cache will not fill with new data for cache misses, but it will continue to function for cache hits. If *CD* = 0, misses will cause the cache to fill with new data.

NW   **Not write-through** selects the mode of operation for the data cache. If *NW* = 1, the data cache is inhibited from cache write-through.

6
0
MCE
PSE
DE
TSD
PVI
VME
CR4
Page directory base
PCD
pwt
CR3
Page fault linear address
CR2
Reserved
CR1
PG
CD
NW
AM
WP
NE
ET
TS
EM
MP
PE
CR0
31
18
16
0

FIGURE 18–9 The structure of the Pentium control registers.

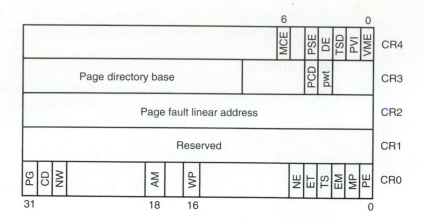

AM — **Alignment mask** enables alignment checking when set. Note that alignment checking only occurs for protected mode operation when the user is at privilege level 3.

WP — **Write protect** protects user level pages against supervisor level write operations. When *WP* = 1, the supervisor can write to user level segments.

NE — **Numeric error** enables standard numeric coprocessor error detection. If *NE* = 1, the $\overline{\text{FERR}}$ pin becomes active for a numeric coprocessor error. If *NE* = 0, any coprocessor error is ignored.

VME — **Virtual mode extension** enables support for the virtual interrupt flag in protected mode. If *VME* = 0, virtual interrupt support is disabled.

PVI — **Protected mode virtual interrupt** enables support for the virtual interrupt flag in protected mode.

TSD — **Time stamp disable** controls the RDTSC instruction.

DE — **Debugging extension** enables I/O breakpoint debugging extensions when set.

PSE — **Page size extension** enables 4M-byte memory pages when set.

MCE — **Machine check enable** enables the machine checking interrupt.

The Pentium contains new features that are controlled by CR4 and a few bits in CR0. These new features are explained in later sections of the text.

EFLAG Register

The extended flag (EFLAG) register has been changed in the Pentium microprocessor. Figure 18–10 pictures the contents of the EFLAG register. Note that four new flag bits have been added to this register to control or indicate conditions about some of the new features in the Pentium. Following is a list of the four new flags and the function of each:

ID — The **identification flag** is used to test for the CPUID instruction. If a program can set and clear the ID flag, the processor supports the CPUID instruction.

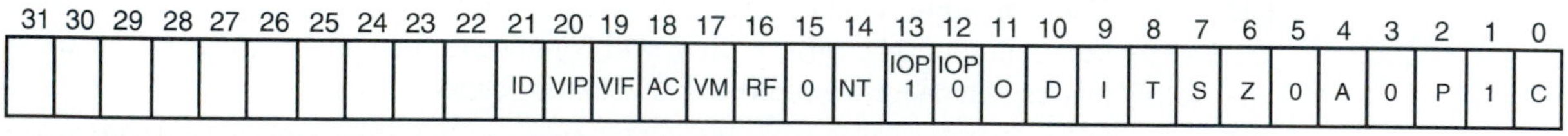

31 30 29 28 27 26 25 24 23 22 21 20 19 18 17 16 15 14 13 12 11 10 9 8 7 6 5 4 3 2 1 0
ID
VIP
VIF
AC
VM
RF
0
NT
IOP 1
IOP 0
O
D
I
T
S
Z
0
A
0
P
1
C

Note: The blank bits in the flag register are reserved for future use and must not be defined.

FIGURE 18–10 The structure of the Pentium EFLAG register.

VIP	**Virtual interrupt pending** indicates that a virtual interrupt is pending.
VIF	**Virtual interrupt** is the image of the virtual interrupt flag IF used with VIP.
AC	**Alignment check** indicates the state of the AM bit in control register 0.

Built-In Self-Test (BIST)

The built-in self-test (BIST) is accessed on power-up by placing a logic 1 on INIT while the RESET pin changes from 1 to 0. The BIST tests 70 percent of the internal structure of the Pentium in approximately 150 μs. Upon completion of the BIST, the Pentium reports the outcome in register EAX. If *EAX* = 0, the BIST passed and the Pentium is ready for operation. If EAX contains any other value, the Pentium has malfunctioned and is faulty.

18–3

PENTIUM MEMORY MANAGEMENT

The memory-management unit within the Pentium is upward-compatible with the 80386 and 80486 microprocessors. Many of the features of these earlier microprocessors are basically unchanged in the Pentium. The main change is in the paging unit and a new system memory-management mode.

Paging Unit

The paging mechanism functions with 4K-byte memory pages or with a new extension available to the Pentium with 4M byte-memory pages. As detailed in Chapters 1 and 17, the size of the paging table structure can become large in a system that contains a large memory. Recall that to fully repage 4G bytes of memory, the microprocessor requires slightly over 4M bytes of memory just for the page tables. In the Pentium, with the new 4M-byte paging feature, this is dramatically reduced to just a single page table. The new 4M-byte page sizes are selected by the PSE bit in control register 0.

The main difference between 4K paging and 4M paging is that in the 4M paging scheme there is no page table entry in the linear address. See Figure 18–11 for the 4M paging system in the Pentium microprocessor. Pay close attention to the way the linear address is used with this scheme. Notice that the leftmost 10 bits of the linear address select an entry in the page directory (just as with 4K pages). Unlike 4K pages, there are no page tables; instead, the page directory addresses a 4M-byte memory page.

Memory-Management Mode

The system memory-management mode (SMM) is on the same level as protected mode, real mode, and virtual mode, but it is provided to function as a manager. The SMM is not intended to be used as an application or a system-level feature. It is intended for high-level system functions such as power management and security, which most Pentiums use during operation.

Access to the SMM is accomplished via a new external hardware interrupt applied to the SMI pin on the Pentium. When the SMM interrupt is activated, the processor begins executing system-level software in an area of memory called the *system management RAM,* or *SMMRAM,* called the *SMM state dump record.* The SMI interrupt disables all other interrupts that are normally handled by user applications and the operating system. A return from the SMM interrupt is accomplished with a new instruction. RSM returns from the memory-management mode interrupt and returns to the interrupted program at the point of the interruption.

The SMM interrupt calls the software, initially stored at memory location 38000H, using *CS* = 3000H and *EIP* = 8000H. This initial state can be changed using a jump to any location within the first 1M byte of memory. An environment similar to real-mode memory addressing is

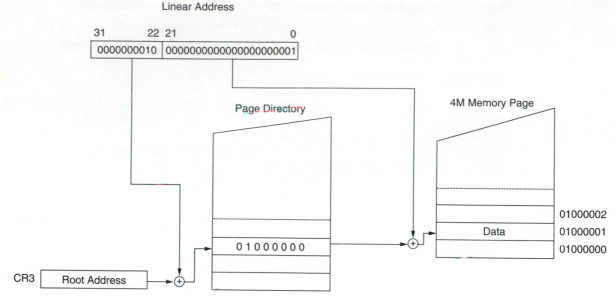

FIGURE 18–11   The linear address 00200001H repaged to memory location 01000002H in 4M-byte pages. Note that there are no page tables.

entered by the management mode interrupt, but it is different because, instead of being able to address the first 1M of memory, SMM mode allows the Pentium to treat the memory system as a flat, 4G-byte system.

In addition to executing software that begins at location 38000H, the SMM interrupt also stores the state of the Pentium in what is called a *dump record.* The dump record is stored at memory locations 3FFA8H through 3FFFFH, with an area at locations 3FE00H through 3FEF7H that is reserved by Intel. The dump record allows a Pentium-based system to enter a sleep mode and reactivate at the point of program interruption. This requires that the SMMRAM be powered during the sleep period. Many laptop computers have a separate battery to power the SMMRAM for many hours during sleep mode. Table 18–2 lists the contents of the dump record.

The Halt auto restart and I/O trap restarts are used when the SMM mode is exited by the RSM instruction. These data allow the RSM instruction to return to the halt-state or return to the interrupt I/O instruction. If neither a halt nor an I/O operation is in effect upon entering the SMM mode, the RSM instruction reloads the state of the machine from the state dump and returns to the point of interruption.

The SMM mode can be used by the system before the normal operating system is placed in the memory and executed. It can also periodically be used to manage the system, provided that normal software doesn't exist at location 38000H–3FFFFH. If the system relocates the SMRAM before booting the normal operating system, it becomes available for use in addition to the normal system.

The base address of the SMM mode SMRAM is changed by modifying the value in the state dump base address register (locations 3FEF8H through 3F3FBH) after the first memory-management mode interrupt. When the first RSM instruction is executed, returning control back to the interrupted system, the new value from these locations changes the base address of the SMM interrupt for all future uses. For example, if the state dump base address is changed to 000E8000H, all subsequent SMM interrupts use locations E8000H–EFFFFH for the Pentium state dump. These locations are compatible with DOS and Windows.

TABLE 18–2 Pentium SMM state dump record.

Offset Address	Register
FFFCH	CR0
FFF8H	CR3
FFF4H	EFLAGS
FFF0H	EIP
FFECH	EDI
FFE8H	ESI
FFE4H	EBP
FFE0H	ESP
FFDCH	EBX
FFD8H	EDX
FFD4H	ECX
FFD0H	EAX
FFCCH	DR6
FFC8H	DR7
FFC4H	TR
FFC0H	LDTR
FFBCH	GS
FFB8H	FS
FFB4H	DS
FFB0H	SS
FFACH	CS
FFA8H	ES
FF04H–FFA7H	Reserved
FF02H	Halt auto restart
FF00H	I/O trap restart
FEFCH	SMM revision identifier
FEF8H	State dump base
FE00H–FEF7H	Reserved

Note: The offset addresses are initially located at base address 00003000H.

18–4 NEW PENTIUM INSTRUCTIONS

The Pentium contains only one new instruction that functions with normal system software; the remainder of the new instructions are added to control the memory-management mode feature and serializing instructions. Table 18–3 lists the new instructions added to the Pentium instruction set.

The CMPXCHG8B instruction is an extension of the CMPXCHG instruction added to the 80486 instruction set. The CMPXCHG8B instruction compares the number 64-bit stored in EDX and EAX with the contents of a 64-bit memory location or register pair. For example, the CMPXCHG8B DATA1 instruction compared the eight bytes stored in memory location DATA1 with the 64-bit number in EDX and EAX. If DATA1 equals EDX:EAX, the 64-bit number stored in ECX:EBX is stored in memory location DATA1. If they are not equal, the contents of DATA1 are stored into EDX:EAX. Note that the zero flag bit indicates that the contents of EDX:EAX were equal or not equal to DATA1.

The CPUID instruction reads the CPU identification code and other information from the Pentium. Table 18–4 shows different information returned from the CPUID instruction for various input values for EAX. To use the CPUID instruction, first load EAX with the input value and then execute CPUID. The information is returned in the registers indicated in the table.

TABLE 18–3 New Pentium instructions.

Instruction	Function
CMPXCHG8B	Compare and exchange eight bytes
CPUID	Return the CPU identification code
RDTSC	Read time stamp counter
RDMSR	Read model specific register
WRMSR	Write model specific register
RSM	Return from system management interrupt

TABLE 18–4 CPUID instruction execution.

Input Value (EAX)	Result after CPUID Executes
0	EAX = 1 for all microprocessors
	EBX–EDX–ECX = vendor identification
1	EAX (bits 3–0) = Stepping ID
	EAX (bits 7–4) = Model
	EAX (bits 11–8) = Family
	EAX (bits 13–12) = Type
	EAX (bits 31–14) = Reserved
	EDX (bit 0) = CPU contains FPU
	EDX (bit 1) = Enhanced 8086 virtual mode supported
	EDX (bit 2) = I/O breakpoints supported
	EDX (bit 3) = Page size extensions supported
	EDX (bit 4) = Time stamp counter TSC supported
	EDX (bit 5) = Pentium-style MSR supported
	EDX (bit 6) = Reserved
	EDX (bit 7) = Machine check exception supported
	EDX (bit 8) = CMPXCHG8B supported
	EDX (bit 9) = 3.3 V microprocessor
	EDX (bits 10–31) = Reserved

If a 0 is placed in EAX before executing the CPUID instruction, the microprocessor returns the vendor identification in EBX, EDX, and EBX. For example, the Intel Pentium returns "GenuineIntel" in ASCII code with the "Genu" in the EBX, "ineI" in EDX, and "ntel" in ECX. The EDX register returns information if EAX is loaded with a 1 before executing the CPUID instruction.

Example 18–1 illustrates a short program that reads the vendor information with the CPUID instruction. It then displays it on the video screen using the DISP macro. Note that this program works with a Pentium or any of the other Pentium clones that are on the market. It also works with the Pentium Pro microprocessor and later versions of the 80486 microprocessor.

EXAMPLE 18–1

```
                .MODEL TINY
                .586                        ;select the Pentium
0000            .CODE
                DISP    MACRO               ;;display character macro
                        MOV     AH,2
                        MOV     DL,BL
                        INT     21H
                        SHR     EBX,8
                        ENDM
                .STARTUP
0100  66| B8 00000000    MOV     EAX,0
0106  0F A2             CPUID               ;get ID from Pentium
```

```
0108    66| 52          PUSH    EDX
                        DISP                    ;display first 4 letters
                        DISP
                        DISP
                        DISP
0132    66| 5B          POP     EBX             ;display next 4 letters
                        DISP
                        DISP
                        DISP
                        DISP
015C    66| 8B D9       MOV     EBX,ECX         ;display last 4 letters
                        DISP
                        DISP
                        DISP
                        DISP
                .EXIT
                        END
```

The RDTSC instruction reads the time-stamp counter into EDX:EAX. The time-stamp counter counts CPU clocks from the time the microprocessor is reset, where the time stamp counter is initialized to an unknown count. Because this is a 64-bit count, a 100 MHz microprocessor can accumulate a count of over 5800 years before the time-stamp counter rolls over. This instruction functions only in real mode or privilege level 0 in protected mode. If you are using a DOS shell from Windows or operating with a memory manager, this instruction will not function and will cause a general protection error. Windows uses privilege level 0 and operates with level 0 as a protected level.

Example 18–2 shows a macro sequence that times events to the microsecond. It makes use of the Pentium time stamp to time events and returns the elapsed time in microseconds in the EAX register. Note that there are three parameters associated with the macro. The first parameter passes the location of a quadword memory location used to store an image of the time stamp clock in the memory system. Make sure that this is defined with the DQ directive. The second parameter passes the clock frequency of the Pentium in MHz to the macro and determines whether it is to start the clock or return the elapsed time. If the final parameter is START, the clock is started; if it is READ, the elapsed time is returned in EAX in microseconds. Note that this macro can be used only in real mode or privilege level 0 of protected mode.

EXAMPLE 18–2

```
EVENT   MACRO   WHERE,SPEED,OPER
        PUSH    EDX                             ;save registers
        PUSH    ECX
        IFIDN <OPER>,<START>
                RDTSC
                MOV     DWORD PTR WHERE,EAX     ;;save current count
                MOV     DWORD PTR WHERE+4,EDX
        ENDIF
        IFIDN <OPER>,<READ>
                RDTSC
                SUB     EAX,DWORD PTR WHERE     ;;form difference
                SBB     EDX,DWORD PTR WHERE+4
                MOV     ECX,SPEED              ;;convert to microseconds
                DIV     ECX
        ENDIF
        POP     ECX
        POP     EDX
        ENDM
```

The RDMSR and WRMSR instructions allow the model-specific registers to be read or written. The model-specific registers are unique to the Pentium and are used to trace, check performance, test, and check for machine errors. Both instructions use ECX to convey the register number to the microprocessor and use EDX:EAX for the 64-bit wide read or write. Note that the

TABLE 18–5 The Pentium model-specific registers.

Address (ECX)	Size	Function
00H	64-bits	Machine check exception address
01H	5-bits	Machine check exception type
02H	14-bits	TR1 parity reversal test register
03H	—	—
04H	4-bits	TR2 instruction cache end bits
05H	32-bits	TR3 cache data
06H	32-bits	TR4 cache tag
07H	15-bits	TR4 cache control
08H	32-bits	TR6 TLB command
09H	32-bits	TR7 TLB data
0AH	—	—
0BH	32-bits	TR9 BTB tag
0CH	32-bits	TR10 BTB target
0DH	12-bits	TR11 BTB control
0EH	10-bits	TR12 new feature control
0FH	—	—
10H	64-bits	Time stamp counter (can be written)
11H	26-bits	Events counter selection and control
12H	40-bits	Events counter 0
13H	40-bits	Events counter 1

register addresses are 0H–13H. See Table 18–5 for a list of the Pentium model-specific registers and their contents. As with the RDTSC instruction, these model-specific registers operate only in the real or privilege level 0 of protected mode.

Never use an undefined value in ECX before using the RDMSR or WRMSR instructions. If $ECX = 0$ before the read or write machine-specific register instruction, the value returned EDX:EAX is the machine check exception address. If $ECX = 1$, the value is the machine check exception type; if $ECX = 0EH$, the test register 12 (TR12) is accessed. Note that these are internal registers designed for in-house testing. The contents of these registers are proprietary to Intel and should not be used during normal programming.

The RMS instruction returns form a memory-management mode interrupt. The memory-management mode interrupt is explained in Section 18–3.

18–5 INTRODUCTION TO THE PENTIUM PRO MICROPROCESSOR

Before this or any other microprocessor can be used in a system, the function of each pin must be understood. This section of the chapter details the operation of each pin, along with the external memory system and I/O structures of the Pentium Pro microprocessor.

Figure 18–12 illustrates the pin-out of the Pentium Pro microprocessor, which is packaged in an immense 387-pin PGA (pin grid array). Currently, the Pentium Pro is available in two versions: one version contains a 256K level 2 cache; the other contains a 512K level 2 cache. The notable difference in the pin-out of the Pentium Pro when compared to earlier Pentiums is that there are provisions for a 36-bit address bus, which allows access to 64G bytes of memory. This is meant for future use because no system today contains anywhere near that amount of memory.

As with most recent versions of the Pentium microprocessor, the Pentium Pro requires a single +3.3 V or +2.7 V power supply for operation. The power supply current is a maximum of 9.9 A for the 150 MHz version of the Pentium Pro, which also has a maximum power dissipation

FIGURE 18–12 The pin-out of the Pentium Pro microprocessor.

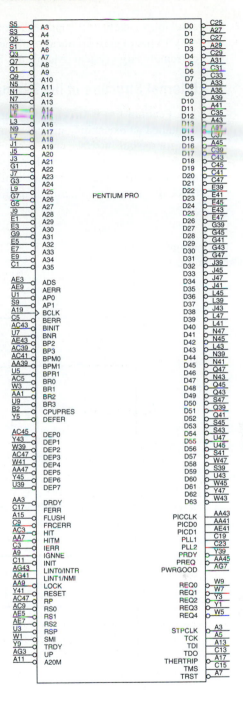

of 26.7 W. At present, a good heat sink with considerable airflow is required to keep the Pentium Pro cool. As with the Pentium, the Pentium Pro contains multiple Vcc and Vss connections that must all be connected for proper operation. The Pentium Pro contains VccP pins (primary Vcc) that connect to +3.1 V, VccS (secondary Vcc) pins that connect to +3.3 V, and Vcc5 (standard Vcc) pins that connect to +5.0 V. There are some pins are labeled N/C (no connection) and must not be connected.

Each Pentium Pro output pin is capable of providing an ample 48.0 mA of current at a logic 0 level. This represents a considerable increase in drive current, compared to the 2.0 mA available on earlier microprocessor output pins. Each input pin represents a small load, requiring only 15 μA of current. Because of the 48.0 mA of drive current available on each output, only an extremely large system requires bus buffers.

Internal Structure of the Pentium Pro

The Pentium Pro is structured differently than earlier microprocessors. Early microprocessors contained an execution unit and a bus interface unit with a small cache buffering the execution unit for the bus interface unit. This structure was modified in later microprocessors, but the modifications were just additional stages within the microprocessors. The Pentium architecture is also a modification, but more significant that earlier microprocessors. Figure 18–13 shows a block diagram of the internal structure of the Pentium Pro microprocessor.

FIGURE 18–13   The internal structure of the Pentium Pro microprocessor.

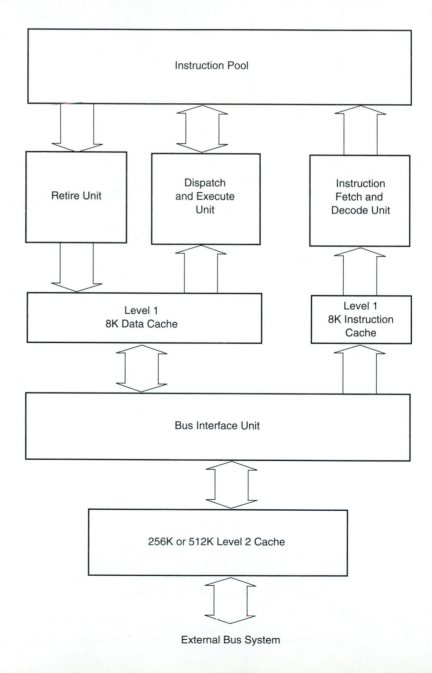

The system buses, which communicate to the memory and I/O, connect to an internal level 2 cache that is often on the main board in most other microprocessor systems. The level 2 cache in the Pentium Pro is either 256K bytes or 512K bytes. The integration of the level 2 cache speeds processing and reduces the number of components in a system.

The bus interface unit (BIU) controls the access to the system buses through the level 2 cache, as it does in most other microprocessors. Again, the difference is that the level 2 cache is integrated. The BIU generates the memory address and control signals, and passes and fetches data or instructions to either a level 1 data cache or a level 1 instruction cache. Each of these are 8K bytes in size at present and may be made larger in future versions of the microprocessor. Earlier versions of the Intel microprocessor contained a unified cache that held both instructions and data. The implementation of separate caches improves performance because data-intensive programs no longer fill the cache with data.

The instruction cache is connected to the instruction fetch and decode unit (IFDU). Although not shown, the IFDU contains three separate instruction decoders that decode three instructions simultaneously. Once decoded, the outputs of the three decoders are passed to the instruction pool, where they remain until the dispatch and execution unit or retire unit obtains them. Also included within the IFDU is a branch prediction logic section that looks ahead in code sequences that contain conditional jump instructions. If a conditional jump is located, the branch prediction logic tries to determine the next instruction in the flow of a program.

Once decoded instructions are passed to the instruction pool, they are held for processing. The instruction pool is a content-addressable memory, but Intel never states its size in the literature.

The dispatch and execute unit (DEU) retrieves decoded instructions from the instruction pool when they are complete, and then executes them. The internal structure of the DEU is illustrated in Figure 18–14. Notice that the DEU contains three instruction execution units: two for processing integer instructions and one for floating-point instructions. This means that the Pentium Pro can process two integer instructions and one floating-point instruction simultaneously. The Pentium also contains three execution units, but the architecture is different because the Pentium does not contain a jump execution unit or address generation units, as does the Pentium Pro. The reservation station (RS) can schedule up to five events for execution and process four simultaneously. Note that there are two station components connected to one of the address generation units that does not appear in the illustration of Figure 18–14.

The last internal structure of the Pentium Pro is the retire unit (RU). The RU checks the instruction pool and removes decoded instructions that have been executed. The RU can remove three decoded instructions per clock pulse.

FIGURE 18–14 The Pentium Pro dispatch and execution unit (DEU).

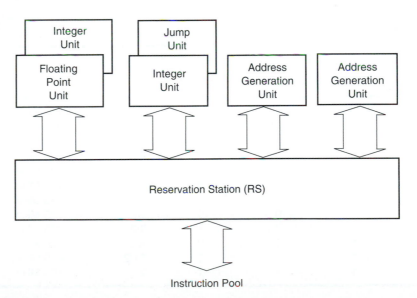

Pin Connections

The number of pins on the Pentium Pro has increased from the 237 pins on the Pentium to 387 pins on the Pentium Pro. Following is a description of each pin or grouping of pins:

$\overline{\text{A20M}}$	The **address A20 mask** is an input that is asserted in the real mode to signal the Pentium Pro to perform address wraparound, as in the 8086 microprocessor, for use of the HIMEM.SYS driver.
$\overline{\text{A35}}$–$\overline{\text{A3}}$	**Address bus** connections address any of the $1G \times 64$ memory locations found in the Pentium Pro memory system.
$\overline{\text{ADS}}$	The **address data strobe** becomes active whenever the Pentium Pro has issued a valid memory or I/O address.
$\overline{\text{AP0}}$, $\overline{\text{AP1}}$	**Address parity** provides even parity for the memory address on all Pentium Pro-initiated memory and I/O transfers. The $\overline{\text{AP0}}$ output provides parity for address connections A23–A3, and the $\overline{\text{AP1}}$ output provides parity for address connections A35–A24.
$\overline{\text{ASZ0}}$,$\overline{\text{ASZ1}}$	**Address size** inputs are driven to select the size of the memory access. Table 18–6 illustrates the size of the memory access for the binary bit patterns on these two inputs to the Pentium Pro.
BCLK	The **bus clock** input determines the operating frequency of the Pentium Pro microprocessor. For example, if BCLK is 66 MHz, various internal clocking speeds are selected by the logic levels applied to the pins in Table 18–7. A BLCK frequency of 66 MHz runs the system bus at 66 MHz.
$\overline{\text{BERR}}$	The **bus error** input/output either signals a bus error along or is asserted by an external device to cause a machine check interrupt or a non-maskable interrupt.
$\overline{\text{BINIT}}$	**Bus initialization** is active on power-up to initialize the bus system.
$\overline{\text{BNR}}$	**Block next request** is used to halt the system in a multiple microprocessor system.

TABLE 18–6 The memory size as dictated by ASZ pins.

$\overline{\text{ASZ1}}$	$\overline{\text{ASZ0}}$	Memory Size
0	0	0–4G
0	1	4G–64G
1	X	Reserved

TABLE 18–7 The BLCK signal and its effect on the Pentium Pro clock speed.

LINT1/NMI	LINT0/INT R	$\overline{\text{IGNNE}}$	$\overline{\text{A20M}}$	Ratio	Speed with BCLK = 50 MHz	Speed with BCLK = 66 MHz
0	0	0	0	2	100 MHz	133 MHz
0	0	0	1	4	200 MHz	266 MHz
0	0	1	0	3	150 MHz	200 MHz
0	0	1	1	5	250 MHz	333 MHz
0	1	0	0	5/2	125 MHz	166 MHz
0	1	0	1	9/2	225 MHz	300 MHz
0	1	1	0	7/2	175 MHz	233 MHz
0	1	1	1	11/2	275 MHz	366 MHz
1	1	1	1	2	100 MHz	133 MHz

$\overline{BP3}$, $\overline{BP2}$	The **break point** status outputs indicate the status of the Pentium Pro break points.
$\overline{BPM1}$, $\overline{BPM0}$	The **break point monitor** outputs indicate the status of the breakpoints and programmable counters.
$\overline{BPRI}$	The **priority agent bus request** is an input that causes the microprocessor to cease bus requests.
$\overline{BR3}$–$\overline{BR0}$	The **bus request** inputs allow up to four Pentium Pro microprocessors to coexist on the same bus system.
$\overline{BREQ3}$–$\overline{BREQ0}$	**Bus request** signals are used for multiple microprocessors on the same system bus.
$\overline{D63}$–$\overline{D0}$	**Data bus** connections transfer byte, word, doubleword, and quadword data between the microprocessor and its memory and I/O system.
$\overline{DBSY}$	**Data bus busy** is asserted to indicate that the data bus is busy transferring data.
$\overline{DEFER}$	The **defer** input is asserted during the snoop phase to indicate that the transaction cannot be guaranteed in-order completion.
$\overline{DEN}$	The **defer enable** signal is driven to the bus on the second phase of a request phase.
$\overline{DEP7}$–$\overline{DEP0}$	**Data bus ECC protection signals** provide error-correction codes for correcting a single-bit error and detecting a double-bit error.
$\overline{FERR}$	The **floating-point error**, comparable to the $\overline{ERROR}$ line in the 80386, shows that the internal co-processor has erred.
$\overline{FLUSH}$	The **flush cache** input causes the cache to flush all write-back lines and invalidate its internal caches. If the $\overline{FLUSH}$ input is a logic 0 during a reset operation, the Pentium enters its test mode.
$\overline{FRCERR}$	**Functional redundancy check error** is used if two Pentium Pro microprocessors are configured in a pair.
$\overline{HIT}$	**Hit** shows that the internal cache contains valid data in the inquire mode.
$\overline{HITM}$	**Hit modified** shows that the inquire cycle found a modified cache line. This output is used to inhibit other master units from accessing data until the cache line is written to memory.
$\overline{IERR}$	**Internal error** output shows that the Pentium Pro has detected an internal parity error or functional redundancy error.
$\overline{IGNNE}$	The **ignore numeric error** input causes the Pentium Pro to ignore a numeric coprocessor error.
INIT	The **initialization** input performs a reset without initializing the caches, write-back buffers, and floating-point registers. This input may not be used to reset the microprocessor in lieu of RESET after power-up.
INTR	The **interrupt request** is used by external circuitry to request an interrupt.
$\overline{LEN}$	**Length** signals (bit 0 and 1) indicate the size of the data transfer, as illustrated in Table 18–8.
$\overline{LINT}$	The **local interrupt** inputs function as NMI and INTR, and also set the clock divider frequency on reset.
$\overline{LOCK}$	$\overline{LOCK}$ becomes a logic 0 whenever an instruction is prefixed with the LOCK: prefix. This is most often used during DMA accesses.
NMI	The **non-maskable interrupt** requests a non-maskable interrupt, as it did on the earlier versions of the microprocessor.

TABLE 18–8 The $\overline{LEN}$ bits show the size of a data transfer.

LEN1	LEN0	Data Transfer Size
0	0	0–8 bytes
0	1	16 bytes
1	0	32 bytes
1	1	Reserved

PICCLK   The **clock signal** input is used for synchronous data transfers.

PICD   The **processor interface serial data** is used to transfer bi-directional serial messages between Pentium Pro microprocessors.

PWRGOOD   **Power good** is an input that is placed at a logic 1 level when the power supply and clock have stabilized.

$\overline{REQ}$ **Request** signals (bits 0–4) define the type of data-transfer operation, as illustrated in Tables 18–9 and 18–10.

$\overline{RESET}$ **Reset** initializes the Pentium Pro, causing it to begin executing software at memory location FFFFFFF0H. The Pentium Pro is reset to the real mode and the leftmost 12 address connections remain logic 1s (FFFH) until a far jump or far call is executed. This allows compatibility with earlier microprocessors.

$\overline{RP}$ **Request parity** provides a means of requesting that the Pentium Pro checks parity.

TABLE 18–9 The function of the request signals in the first clocking period.

$\overline{REQ4}$	$\overline{REQ3}$	$\overline{REQ2}$	$\overline{REQ1}$	$\overline{REQ0}$	Function
0	0	0	0	0	Deferred reply
0	0	0	0	1	Reserved
0	1	0	0	0	Case 1*
0	1	0	0	1	Case 2*
1	0	0	0	0	I/O read
1	0	0	0	1	I/O write
X	X	0	1	0	Memory read
X	X	0	1	1	Memory write
X	X	1	0	0	Memory code read
X	X	1	1	0	Memory data read
X	X	1	X	1	Memory write

*Note: See Table 18–10 for the second clock pulse for these codes.

TABLE 18–10 The second clock pulse and the request signals as they apply to case 1 and 2 from Table 18–9.

Case	$\overline{REQ4}$	$\overline{REQ3}$	$\overline{REQ2}$	$\overline{REQ1}$	$\overline{REQ0}$	Function
1	X	X	X	0	0	Interrupt acknowledge
1	X	X	X	0	1	Special transactions
1	X	X	X	1	X	Reserved
2	X	X	X	0	0	Branch trace message
2	X	X	X	0	1	Reserved
2	X	X	X	1	X	Reserved

TABLE 18–11 The operation of the Pentium Pro in response to the $\overline{RS}$ inputs.

RS2	RS1	RS0	Function	HITM	DEFER
0	0	0	Idle state	X	X
0	0	1	Retry	0	1
0	1	0	Defer	0	1
0	1	1	Reserved	0	1
1	0	0	Hard failure	X	X
1	0	1	Normal, no data	0	0
1	1	0	Implicit write-back	1	X
1	1	1	Normal with data	0	0

$\overline{RS}$	The **response status** inputs cause the Pentium Pro to perform the functions listed in Table 18–11.
$\overline{RSP}$	The **response parity** input applies a parity error signal from an external parity checker.
$\overline{SMI}$	The **system management interrupt** input causes the Pentium Pro to enter the system management mode of operation.
$\overline{SMMEM}$	The **system memory-management mode** signal becomes a logic 0 whenever the Pentium Pro is executing in the system memory-management mode interrupt and address space.
$\overline{SPCLK}$	The **split lock** signal is placed at a logic 0 level to indicate that the transfer will contain four locked transactions.
$\overline{STPCLK}$	**Stop clock** causes the Pentium Pro to enter the power-down state when placed at a logic 0 level.
TCK	The **testability clock** input selects the clocking function in accordance with the IEEE 1149.1 Boundary Scan interface.
TDI	The **test data** input is used to test data clocked into the Pentium with the TCK signal.
TDO	The **test data** output is used to gather test data and instructions shifted out of the Pentium with TCK.
TMS	The **test mode select** input controls the operation of the Pentium in test mode.
$\overline{TRDY}$	The **target ready** input is asserted when the target is ready for a data transfer operation.

The Memory System

The memory system for the Pentium Pro microprocessor is 4G bytes in size, just as in the 80386DX–Pentium microprocessors, but access to an area between 4G and 64G is made possible by additional address signals A32–A35. The Pentium Pro uses a 64-bit data bus to address memory organized in eight banks that each contain 8G bytes of data. Note that the additional memory is enabled with bit position 5 of CR4 and is accessible only when 2M paging is enabled. Note also that 2M paging is new to the Pentium Pro to allow memory above 4G to be accessed. More information is presented on Pentium Pro paging later in this chapter. Refer to Figure 18–15 for the organization of the Pentium Pro physical memory system.

The Pentium Pro memory system is divided into eight banks that each store a byte of data with a parity bit. Note that most Pentium and Pentium Pro microprocessor-based systems forgo the use of the parity bit. The Pentium Pro, like the 80486 and Pentium, employs internal parity generation and checking logic for the memory system data bus information. The 64-bit wide

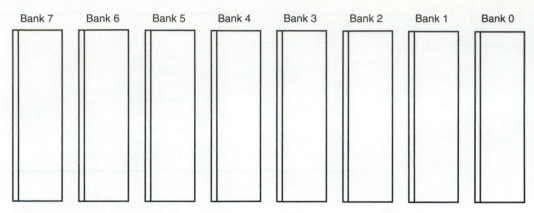

FIGURE 18–15   The eight memory banks in the Pentium Pro system. Note each bank is 8-bits wide and 8G long if 36-bit addressing is enabled.

memory is important to double-precision floating-point data. Recall that a double-precision floating-point number is 64 bits wide. As with earlier Intel microprocessors, the memory system is numbered in bytes from byte 000000000H to byte FFFFFFFFFH. This nine-digit hexadecimal address is employed in a system that addresses 64G of memory.

Memory selection is accomplished with the bank enable signals ($\overline{BE7}$–$\overline{BE0}$). In the Pentium Pro microprocessor, the bank enable signals are presented on the address bus (A15–A8) during the second clock cycle of a memory or I/O access. These must be extracted from the address bus to access memory banks. The separate memory banks allow the Pentium Pro to access any single byte, word, doubleword, or quadword with one memory transfer cycle. As with earlier memory selection logic, we often generate eight separate write strobes for writing to the memory system. Note that the memory write information is provided on the request lines from the microprocessor during the second clock phase of a memory or I/O access.

A new feature added to the Pentium and Pentium Pro is the capability to check and generate parity for the address bus during certain operations. The AP pin (Pentium) or pins (Pentium Pro) provides the system with parity information, and the $\overline{APCHK}$ (Pentium) or AP pins (Pentium Pro) indicate a bad parity check for the address bus. The Pentium Pro takes no action when an address-parity error is detected. The error must be assessed by the system, and the system must take appropriate action (an interrupt) if so desired.

New to the Pentium Pro is a built-in error correction circuit (ECC) that allows the correction of a one-bit error and the detection of a two-bit error. To accomplish the detection and correction of errors, the memory system must have room for an extra 8-bit number that is stored with each 64-bit number. The extra eight bits are used to store an error-correction code that allows the Pentium Pro to automatically correct any single-bit error. A 1M × 64 is a 64M SDRAM without ECC, and a 1M × 72 is an SDRAM with EEC support. The ECC code is much more reliable than the old parity scheme, which is rarely used in modern systems. The only drawback of the ECC scheme is the additional cost of SDRAM that is 72-bits wide.

Input/Output System

The input/output system of the Pentium Pro is completely compatible with earlier Intel microprocessors. The I/O port number appears on address lines A15–A3 with the bank-enable signals used to select the actual memory banks used for the I/O transfer.

Beginning with the 80386 microprocessor, I/O privilege information is added to the TSS segment when the Pentium is operated in the protected mode. Recall that this allows I/O ports to be selectively inhibited. If the blocked I/O location is accessed, the Pentium Pro generates a type-13 interrupt to signal an I/O privilege violation.

System Timing

As with any microprocessor, the system timing signals must be understood in order to interface the microprocessor. This portion of the text details the operation of the Pentium Pro through its timing diagrams and shows how to determine memory access times.

The basic Pentium Pro memory cycle consists of two sections: the address phase and the data phase. During the address phase, the Pentium Pro sends the address (T1) to the memory and I/O system, and also the control signals (T2). The control signals include the ATTR lines (A31–A24), the DID lines (A23–A16), the bank enable signals (A15–A8), and the EXF lines (A7–A3). See Figure 18–16 for the basic timing cycle. The type of memory cycle appears on the request pins. During the data phase, four 64-bit wide numbers are fetched or written to the memory. This operation is most common because data from the main memory are transferred between the internal 256K or 512K write-back cache and the memory system. Operations that write a byte, word, or doubleword, such as I/O transfers, use the bank selection signals and have only one clock in the data transfer phase. Notice from the timing diagram that the 66 MHz Pentium Pro is capable of 33 million memory transfers per second. (This assumes that the memory can operate at that speed.)

The setup time before the clock is given as 5.0 ns and the hold time after the clock is given as 1.5 ns. This means that the data window around the clock is 6.5 ns. The address appears on the 8.0 ns maximum after the start of T1. This means that the Pentium Pro microprocessor operating at 66 MHz allows 30 ns (two clocking periods), minus the address delay time of 8.0 ns and also minus the data setup time of 5.0 ns. Memory access time without any wait states is $30 - 8.0 - 5.0$, or 17.0 ns. This is enough time to allow access to a SRAM, but not to any DRAM without inserting wait states into the timing.

Wait states are inserted into the timing by controlling the $\overline{\text{TRDY}}$ input to the Pentium Pro. The $\overline{\text{TRDY}}$ signal must become a logic 0 by the end of T2; otherwise, additional T2 states are inserted into the timing. Note that 60 ns DRAM requires the insertion of four wait states of 15 ns (one clocking period) each to lengthen the access time to 77 ns. This is enough time for the DRAM and any decoder in the system to function. Because many EPROM memory devices require an access time of 100 ns, EPROM requires the addition of seven wait states to lengthen the access time to 122 ns.

FIGURE 18–16 The basic Pentium Pro timing.

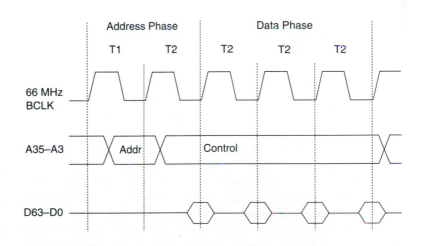

FIGURE 18–17 The new control register 4 (CR4) in the Pentium Pro microprocessor.

31			7	6	5	4	3	2	1	0
			PGE	MCE	PAE	PSE	DE	TSD	PVI	VME

18–6   SPECIAL PENTIUM PRO FEATURES

The Pentium Pro is essentially the same microprocessor as the 80386, 80486, and Pentium, except that some additional features and changes to the control register set have occurred. This section highlights the differences between the 80386 control register structure and the flag register.

Control Register 4

Figure 18–17 shows control register 4 of the Pentium Pro microprocessor. Notice that CR4 has two new control bits that are added to the control register array.

This section of the text explains only the two new Pentium Pro components in the control register 4. (Refer to Figure 18–9 for a description and illustration of the Pentium control registers.) Following is a description of the Pentium CR4 bits and the new Pentium Pro control bits in control register CR4:

VME   **Virtual mode extension** enables support for the virtual interrupt flag in protected mode. If *VME* = 0, virtual interrupt support is disabled.

PVI   **Protected mode virtual interrupt** enables support for the virtual interrupt flag in protected mode.

TSD   **Time stamp disable** controls the RDTSC instruction.

DE   **Debugging extension** enables I/O breakpoint debugging extensions when set.

PSE   **Page size extension** enables 4M-byte memory pages when set in the Pentium, or 2M-byte pages when set in the Pentium Pro whenever PSE is also set.

PAE   **Page address extension** enables address lines A35–A32 whenever a special new addressing mode, controlled by PSE, is enabled for the Pentium Pro.

MCE   **Machine check enable** enables the machine checking interrupt.

PSE   **Page size extension** controls the new, larger 64G addressing mode whenever it is set along with PAE and PSE.

18–7   SUMMARY

1. The Pentium microprocessor is almost identical to the earlier 80386 and 80486 microprocessors. The main difference is that the Pentium has been modified internally to contain a dual cache (instruction and data) and a dual integer unit. The Pentium also operates at a higher clock speed of 66 MHz.

2. The 66 MHz Pentium requires 3.3 A of current, and the 60 MHz version requires 2.91 A. The power supply must be a +5.0 V supply with a regulation of ±5 percent. Newer versions of the Pentium require a 3.3 V or 2.7 V power supply.

3. The data bus on the Pentium is 64-bits wide and contains eight byte-wide memory banks selected with bank enable signals ($\overline{BE0}$–$\overline{BE7}$).

4. Memory access time, without wait states, is only about 18 ns in the 66 MHz Pentium. In many cases, this short access time requires wait states that are introduced by controlling the $\overline{BRDY}$ input to the Pentium.

5. The superscaler structure of the Pentium contains three independent processing units: a floating-point processor and two integer processing units labeled U and V by Intel.

6. The cache structure of the Pentium is modified to include two caches. One 8K × 8 cache is designed as an instruction cache; the other 8K × 8 cache is a data cache. The data cache can be operated as either a write-through or a write-back cache.

7. A new mode of operation called the *system memory-management* (SMM) mode has been added to the Pentium. The SMM mode is accessed via the system memory-management interrupt applied to the $\overline{SMI}$ input pin. In response to $\overline{SMI}$, the Pentium begins executing software at memory location 38000H.

8. New instructions include the CMPXCHG8B, RSM, RDMSR, WRMSR, and CPUID. The CMPXCHG8B instruction is similar to the 80486 CMPXCHG instruction. The RSM instruction returns from the system memory-management interrupt. The RDMSR and WRMSR instructions read or write to the machine-specific registers. The CPUID instruction reads the CPU identification code from the Pentium.

9. The built-in self-test (BIST) allows the Pentium to be tested when power is first applied to the system. A normal power-up reset activates the RESET input to the Pentium. A BIST power-up reset activates INIT and then deactivates the RESET pin. EAX is equal to a 00000000H in the BIST passes.

10. A new proprietary Intel modification to the paging unit allows 4M-byte memory pages instead of the 4K-byte pages. This is accomplished by using the page directory to address 1024 pages that each contain 4M of memory.

11. The Pentium Pro is an enhanced version of the Pentium microprocessor that contains not only the level 1 caches found inside the Pentium, but also the level 2 cache of 256K or 512K found on most main boards.

12. The Pentium Pro operates by using the same 66 MHz bus speed as the Pentium and the 80486. It uses an internal clock generator to multiply the bus speed by various factors to obtain higher internal execution speeds.

13. The only significant software difference between the Pentium Pro and earlier microprocessors is the addition of the FCMOV and CMOV instructions.

14. The only hardware difference between the Pentium Pro and earlier microprocessors is the addition of 2M paging and four extra address lines that allow access to a memory address space of 64G bytes.

15. Error correction code has been added to the Pentium Pro, which corrects any single-bit error and detects any two-bit error.

18–8   QUESTIONS AND PROBLEMS

1. How much memory is accessible to the Pentium microprocessor?
2. How much memory is accessible to the Pentium Pro microprocessor?
3. The memory data bus width is _____ in the Pentium.
4. What is the purpose of the DP0–DP7 pins on the Pentium?
5. If the Pentium operates at 66 MHz, what frequency clock signal is applied to the CLK pin?
6. What is the purpose of the $\overline{BRDY}$ pin on the Pentium?
7. What is the purpose of the AP pin on the Pentium?
8. How much memory access time is allowed by the Pentium, without wait states, when it is operated at 66 MHz?
9. What Pentium pin is used to insert wait states into the timing?
10. A wait state is an extra ___ clocking period.
11. Explain how two integer units allow the Pentium to execute two non-dependent instructions simultaneously.

12. How many caches are found in the Pentium and what are their sizes?
13. How wide is the Pentium memory data sample window for a memory read operation?
14. Can the Pentium execute three instructions simultaneously?
15. What is the purpose of the $\overline{\text{SMI}}$ pin?
16. What is the system memory-management mode of operation for the Pentium?
17. How is the system memory-management mode exited?
18. Where does the Pentium begin to execute software for an $\overline{\text{SMI}}$ interrupt input?
19. How can the system memory-management unit dump address be modified?
20. Explain the operation of the CMPXCHG8B instruction.
21. What information is returned in register EAX after the CPUID instruction executes with an initial value of 0 in EAX?
22. What new flag bits are added to the Pentium microprocessor?
23. What new control register is added to the Pentium microprocessor?
24. Describe how the Pentium accesses 4M pages.
25. Explain how the time-stamp clock functions and how it can be used to time events.
26. Contrast the Pentium with the Pentium Pro microprocessor.
27. Where are the bank enable signals found in the Pentium Pro microprocessor?
28. How many address lines are found in the Pentium Pro system?
29. What changes have been made to CR4 in the Pentium Pro and for what purpose?
30. Compare access times in the Pentium system with the Pentium Pro system.
31. What is ECC?
32. What type of SDRAM must be purchased to use ECC?

CHAPTER 19

The Pentium II, Pentium III, and Pentium 4 Microprocessors

INTRODUCTION

The Pentium II, Pentium III, and Pentium 4 microprocessors may well signal the end to the evolution of the 32-bit architecture with the advent of the Itanium[1] microprocessor from Intel. The Itanium is a 64-bit architecture microprocessor. The Pentium II, Pentium III, and Pentium 4 architectures are extensions of the Pentium Pro architecture, with some differences. The most notable difference is that the internal cache from the Pentium Pro architecture has been moved out of the microprocessor in the Pentium II. Another major change is that the Pentium II is not available in integrated circuit form. Instead, the Pentium II is found on a small plug-in circuit board along with the level 2 cache chip. Various versions of the Pentium II are available. The Celeron[2] is a version of the Pentium II that does not contain the level 2 cache on the Pentium II circuit board. The Xeon[3] is an enhanced version of the Pentium II that contains up to a 2M-byte cache on the circuit board.

Similar to the Pentium II, early Pentium III microprocessors were packaged in a cartridge instead of an integrated circuit. More recent versions, such as the Coppermine, are again packaged in an integrated circuit (370 pins). The Pentium III Coppermine, like the Pentium Pro, contains an internal cache. The Pentium 4 is packaged in a larger integrated circuit, with 421 pins. The Pentium 4 also uses physically smaller transistors, which makes it much smaller and faster than the Pentium III. Intel to date has released versions of the Pentium 4 that operate at frequencies over 2 GHz with a limit of possibly 10 GHz at some future date.

CHAPTER OBJECTIVES

Upon completion of this chapter, you will be able to:

1. Detail the differences between the Pentium II, Pentium III, and Pentium 4 and prior Intel microprocessors.
2. Explain how the architectures of the Pentium II, Pentium III, and Pentium 4 improve system speed.

---

[1] Itanium is a registered trademark of Intel Corporation.

[2] The Celeron in a registered trademark of Intel Corporation.

[3] Xeon is a registered trademark of Intel Corporation.

3. Explain how the basic architecture of the computer system has changed by using the Pentium II, Pentium III, and Pentium 4 microprocessors.
4. Detail the changes to the CPUID instruction.
5. Describe the operation of the SYSENTER and SYSEXIT instructions.
6. Describe the operation of the FXSAVE and FXRSTOR instructions.

19–1   **INTRODUCTION TO THE PENTIUM II MICROPROCESSOR**

Before the Pentium II or any other microprocessor can be used in a system, the function of each pin must be understood. This section of the chapter details the operation of each pin, along with the external memory system and I/O structures of the Pentium II microprocessor.

Figure 19–1 illustrates the basic outline of the Pentium II microprocessor's slot 1 connector and the signals used to interface to the chipset. Figure 19–2 shows a simplified diagram of the components on the cartridge, and the placement of the Pentium II cartridge and bus components in the typical Pentium II system. There are 242 pins on the slot 1 connector for the microprocessor. (These connections are a reduction in the number of pins found on the Pentium and the Pentium II microprocessors.) The Pentium II is packaged on a printed circuit board instead of the integrated circuits of the past Intel microprocessors. The level 1 cache is 32K-bytes as it was in the Pentium Pro, but the level 2 cache is no longer inside the integrated circuit. Intel changed the architecture so that a level 2 cache could be placed very closely to the microprocessor. This change makes the microprocessor less expensive and still allows the level 2 cache to operate efficiently. The Pentium level 2 cache operates at one-half the microprocessor clock frequency, instead of the 66 MHz of the Pentium microprocessor. A 400 MHz Pentium II has a cache speed of 200 MHz. Currently, the Pentium II is available in three versions. The first is the full-blown Pentium II, which is the Pentium II for the slot 1 connector. The second is the Celeron, which is like the Pentium II, except that the slot 1 circuit board does not contain a level 2 cache; the level 2 cache in the Celeron system is located on the main board and operates at 66 MHz. The most recent version is the Xeon, which, because it uses a level 2 cache of 512K, 1M, or 2M, represents a significant speed improvement over the Pentium II. The Xeon's level 2 cache operates at the clock frequency of the microprocessor. A 400 MHz Xeon has a level 2 cache speed of 400 MHz, which is twice the speed of the regular Pentium II.

The early versions of the Pentium II require a 5.0 V, 3.3 V and variable voltage power supply for operation. The main variable power supply voltages vary from 3.5 V to as low as 1.8 V at the microprocessor. The power-supply current averages 14.2 A to 8.4 A, depending on the operating frequency and voltage of the Pentium II. Because these currents are significant, so is the power dissipation of these microprocessors. At present, a good heat sink with considerable airflow is required to keep the Pentium II cool. Luckily, the heat sink and fan are built into the Pentium II cartridge. The latest versions of the Pentium II have been improved to reduce the power dissipation.

Each Pentium II cartridge output pin is capable of providing at least 36 mA of current at a logic 0 level on the signal connections. Some of the output control signals provide only 14 mA of current. Another change to the Pentium II is that the outputs are open-drain and require an external pull-up resister for proper operation.

The function of each Pentium II group of pins follows:

$\overline{\text{A20}}$ **Address A20 mask** is an input that is asserted in the real mode to signal the Pentium II to perform address wraparound, as in the 8086 microprocessor, for use of the HIMEM.SYS driver.

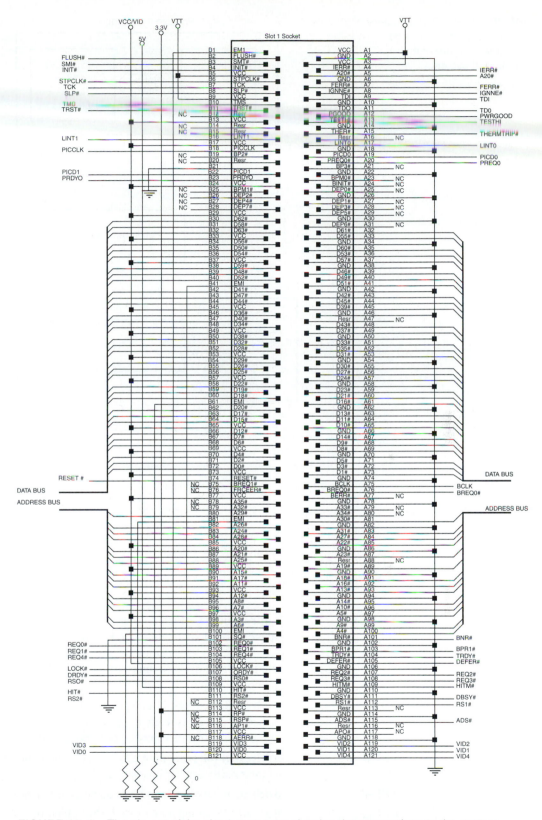

FIGURE 19–1   The pin-out of the slot 1 connector showing the connections to the system.

FIGURE 19–2 The structure of the Pentium II cartridge and the structure of the Pentium II system.

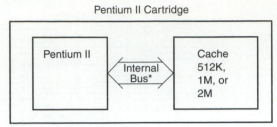

Pentium II Cartridge

* The bus speed is 1/2 Pentium speed or the same as the Pentium speed in the Xeon.

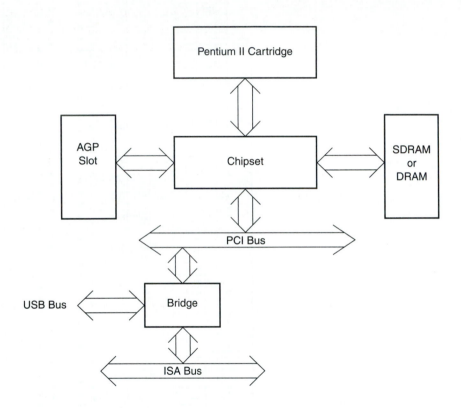

$\overline{\text{A35}}$–$\overline{\text{A3}}$	**Address buses,** which are active low connections, address any of the memory locations found in the Pentium II memory system. Note that A0, A1, and A2 are encoded in the bus enable ($\overline{\text{BE7}}$–$\overline{\text{BE0}}$), which are generated by the chipset, to select any or all of the eight bytes in a 64-bit wide memory location.
$\overline{\text{ADS}}$	**Address data strobe** is an input that is activated to indicate to the Pentium II that the system is ready to perform a memory or I/O operation. This signal causes the microprocessor to provide the address to the system.
$\overline{\text{AERR}}$	**Address error** is an input used to cause the Pentium II to check for an address parity error if it is activated.
$\overline{\text{AP1}}$–$\overline{\text{0}}$	**Address parity** inputs indicate an address parity error.
$\overline{\text{BCLK}}$	**Bus clock** is an input that sets the bus clock frequency. This is either 66 MHz or 100 MHz in the Pentium II.

$\overline{\text{BERR}}$	**Bus error** is asserted to indicate that an error has occurred on the bus system.
$\overline{\text{BINT}}$	**Bus initialization** is a logic 0 during system reset or initialization. It is an input to indicate that a bus error has occurred and the system needs to be reinitialized.
$\overline{\text{BNR}}$	**Bus not ready** is an input used to insert wait states into the timing for the Pentium II. Placing a logic 0 in this pin causes the Pentium II to enter stall states or wait states.
BP[3:2] and PM/BP[1:0]	The **breakpoint pins** BP3–BP0 indicate a breakpoint match when the debug registers are programmed to monitor for matches. The **performance monitoring pins** PM1 and PM0 indicate the settings of the performance monitoring bits in the debug mode control register.
$\overline{\text{BPRI}}$	The **bus priority request** input is used to request the system bus from the Pentium II.
$\overline{\text{BR0–1}}$	**Bus requests** indicate that the Pentium II has generated a bus request. During initialization, the BR0 pin must be activated.
BSEL	**Bus select** is currently not used by the Pentium II and must be connected to ground for proper operation.
$\overline{\text{D63–D0}}$	**Data bus** connections transfer byte, word, doubleword, and quadword data between the microprocessor and its memory and I/O system.
$\overline{\text{DEFER}}$	**Defer** is used to indicate that the external system cannot complete the bus cycle.
$\overline{\text{DEP7–DEP0}}$	**Data EEC pins** are used in the error-correction scheme of the Pentium II and normally connect to an extra 8-bit memory section.
$\overline{\text{DRDY}}$	**Data ready** is activated to indicate that the system is presenting valid data to the Pentium II.
EMI	**Electro-magnetic interference** must be grounded to prevent the Pentium II from generating or receiving noise.
$\overline{\text{FERR}}$	**Floating-point error,** comparable to the $\overline{\text{ERROR}}$ line in the 80386, shows that the internal coprocessor has erred.
$\overline{\text{FLUSH}}$	The **flush cache** input causes the cache to flush all write-back lines and invalidate its internal caches. If the $\overline{\text{FLUSH}}$ input is a logic 0 during a reset operation, the Pentium enters its test mode.
FRCERR	**Functional redundancy check** is sampled during a reset to configure the Pentium II in the master (1) or checker (0) mode.
$\overline{\text{HIT}}$	**Hit** shows that the internal cache contains valid data in the inquire mode.
$\overline{\text{HITM}}$	**Hit modified** shows that the inquire cycle found a modified cache line. This output is used to inhibit other master units from accessing data until the cache line is written to memory.
$\overline{\text{IERR}}$	The **internal error** output shows that the Pentium II has detected an internal error or functional redundancy error.
$\overline{\text{IGNNE}}$	The **ignore numeric error** input causes the Pentium II to ignore a numeric coprocessor error.
$\overline{\text{INIT}}$	The **initialization** input performs a reset without initializing the caches, write-back buffers, and floating-point registers. This input may not be used to reset the microprocessor in lieu of RESET after power-up.

INTR	**Interrupt request** is used by external circuitry to request an interrupt.
LINT1, LINT0	Local APIC interrupt signals must connect the appropriate pins of all APIC bus agents. When the APIC is disabled, the LINT0 signal becomes INTR, a maskable interrupt request signal; LINT1 becomes NMI, a non-maskable interrupt.
$\overline{\text{LOCK}}$	$\overline{\text{LOCK}}$ becomes a logic 0 whenever an instruction is prefixed with the LOCK: prefix. This is most often used during DMA accesses.
NMI	**Non-maskable interrupt** requests a non-maskable interrupt as it did on the earlier versions of the microprocessor.
PICCLK	Must be $^1/4$ the frequency of BLCK.
PICD1–PICD0	Used for serial messages between the Pentium II and APIC.
$\overline{\text{PM1}}$–$\overline{\text{PM0}}$	**Performance monitor** signals are used to test the performance of the Pentium II.
PRDY	The **probe ready** output indicates that the probe mode has been entered for debugging.
$\overline{\text{PREQ}}$	The **probe request** is used to request debugging.
PWRGOOD	An input that indicates that the system power supply is operational.
$\overline{\text{REQ4}}$–$\overline{\text{REQ0}}$	**Request signals** communicate commands between bus controllers and the Pentium II.
$\overline{\text{RESET}}$	**Reset** initializes the Pentium II, causing it to begin executing software at memory location FFFFFFF0H or 000FFFF0. The A35–A32 address bits are set as logic 0s during the reset operation. The Pentium II is reset to the real mode and the leftmost 12 address connections remain logic 1s (FFFH) until a far jump or far call is executed. This allows compatibility with earlier microprocessors. See Table 19–1 for the state of the Pentium II after a hardware reset.
$\overline{\text{RP}}$	**Request parity** is used to request parity.
$\overline{\text{RS2}}$–$\overline{\text{RS0}}$	**Request status** inputs are used to request the current status of the Pentium II.

TABLE 19–1   State of the Pentium II after a RESET.

Register	RESET Value	RESET + BIST Value
EAX	0	0 (if test passes)
EDX	0500XXXXH	0500XXXXH
EBX, ECX, ESP, EBP, ESI, and EDI	0	0
EFLAGS	2	2
EIP	0000FFF0H	0000FFF0H
CS	F000H	F000H
DS, ES, FS, GS, and SS	0	0
GDTR and TSS	0	0
CR0	60000010H	60000010H
CR2, CR3, and CR4	0	0
DR0–DR3	0	0
DR6	FFFF0FF0H	FFFF0FF0H
DR7	00000400H	00000040H

Notes: BIST = built-in self-test, XXXX = Pentium II version number.

$\overline{\text{RSP}}$	The **response parity** input is activated to request parity.
$\overline{\text{SLOTOCC}}$	The **slot occupied** output is a logic 0 if slot zero contains either a Pentium II or a dummy terminator.
$\overline{\text{SLP}}$	**Sleep** is an input that, when inserted in the stop grant state, causes the Pentium II to enter the sleep state.
$\overline{\text{SMI}}$	The **system management interrupt** input causes the Pentium II to enter the system management mode of operation.
$\overline{\text{STPCLK}}$	The **stop clock** input causes the Pentium II to enter the low-power stop grant state.
TCK	The **testability clock** input selects the clocking function in accordance with the IEEE 1149.1 Boundary Scan interface.
TDI	The **test data** input is used to test data clocked into the Pentium II with the TCK signal.
TDO	The **test data** output is used to gather test data and instruction shifted out of the Pentium II with TCK.
TESTHI	**Test high** is an input that must be connected to +2.5 V through a 1K–10K Ω resister for proper Pentium II operation.
$\overline{\text{THERMTRIP}}$	**Thermal sensor trip** is an output that that becomes a zero when the temperature of the Pentium II exceeds 130°C.
TMS	The **test mode select** input controls the operation of the Pentium in test mode.
$\overline{\text{TRDY}}$	**Target ready** is an input that is used to cause the Pentium II to perform a write-back operation.
$\overline{\text{VID4}}$–$\overline{\text{VID0}}$	**Voltage data** output pins are either open or grounded signals that indicate what supply voltage is currently required by the Pentium II. The power supply must apply the request voltage to the Pentium II, as listed in Table 19–2.

The Memory System

The memory system for the Pentium II microprocessor is 64G bytes in size, just like the Pentium Pro microprocessor. Both microprocessors address a memory system that is 64 bits wide with an address bus that is 36 bits wide. Most systems use SDRAM operating at 66 MHz or 100 MHz. The SDRAM for the 66 MHz system has an access time of 10 ns and the SDRAM for the 100 MHz system has an access time of 8 ns. The memory system, which connects to the chipset, is not illustrated in this chapter. Refer to Chapter 18 to see the organization of a 64-bit wide memory system without ECC.

The Pentium II memory system is divided into eight or nine banks that each store a byte of data. If the ninth byte is present, it stores an error checking code (ECC). The Pentium II, like the 80486–Pentium Pro, employs internal parity generation and checking logic for the memory system's data bus information. (Note that most Pentium II systems do not use parity checks, but it is available.) If parity checks are employed, each memory bank contains a ninth bit. The 64-bit wide memory is important to double-precision floating-point data. Recall that a double-precision floating-point number is 64 bits wide. As with the Pentium Pro, the memory system is numbered in bytes from byte 000000000H to byte FFFFFFFFFH. Please note that none of the current chip sets support more than 1G bytes of system memory, so the additional address connections are for future expansion. Figure 19–3 illustrates the basic memory map of the Pentium II system, using the AGP for the video card.

TABLE 19–2 Power supply voltages that must be applied to Vcc as requested by the $\overline{VID}$ pins.

$\overline{VID4}$	$\overline{VID3}$	$\overline{VID2}$	$\overline{VID1}$	$\overline{VID0}$	Vcc
0	0	0	0	0	2.05 V
0	0	0	0	1	2.00 V
0	0	0	1	0	1.95 V
0	0	0	1	1	1.90 V
0	0	1	0	0	1.85 V
0	0	1	0	1	1.80 V
0	0	1	1	0	—
0	0	1	1	1	—
0	1	0	0	0	—
0	1	0	0	1	—
0	1	0	1	0	—
0	1	0	1	1	—
0	1	1	0	0	—
0	1	1	0	1	—
0	1	1	1	0	—
0	1	1	1	1	—
1	0	0	0	0	3.5 V
1	0	0	0	1	3.4 V
1	0	0	1	0	3.3 V
1	0	0	1	1	3.2 V
1	0	1	0	0	3.1 V
1	0	1	0	1	3.0 V
1	0	1	1	0	2.9 V
1	0	1	1	1	2.8 V
1	1	0	0	0	2.7 V
1	1	0	0	1	2.6 V
1	1	0	1	0	2.5 V
1	1	0	1	1	2.4 V
1	1	1	0	0	2.3 V
1	1	1	0	1	2.2 V
1	1	1	1	0	2.1 V
1	1	1	1	1	—

The memory map for the Pentium II system is similar to the map illustrated in earlier chapters, except that an area of the memory is used for the AGP area. The AGP area allows the video card and Windows to access the video information in a linear address space. This is unlike the 128K-byte window in the DOS area for a standard VGA video card. The benefit is much faster video updates because the video card does not need to page through the 128K-byte DOS video memory.

Transfers between the Pentium II and the memory system are controlled by the 440LX or 440 BX chipset. Data transfers between the Pentium II and the chipset are eight bytes wide. The chipset communicates to the microprocessor through the five $\overline{REQ}$ signals, as listed in Table 19–3. In essence, the chipset controls the Pentium II, which is a departure from the traditional method of connecting a microprocessor to the system directly to the memory.

The Pentium II connects only directly to the cache, which is on the Pentium II cartridge. As mentioned, the Pentium II cache operates at one-half the clock frequency of the microprocessor. Therefore, a 400 MHz Pentium II cache operates at 200 MHz. The Pentium II Xeon cache operates at the same frequency as the microprocessor, which means that the Xeon, with its 512K, 1M, or 2M cache, outperforms the standard Pentium II.

FIGURE 19–3 The memory map of a Pentium II-based computer system.

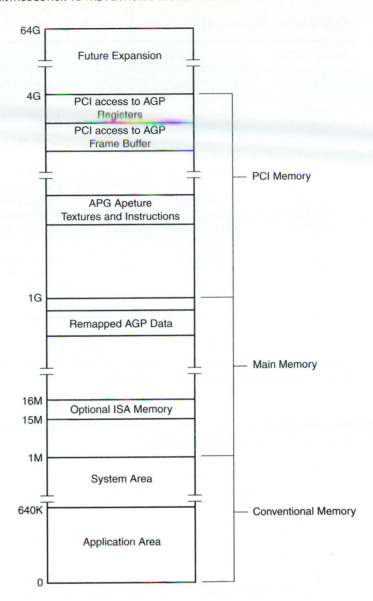

Input/Output System

The input/output system of the Pentium II is completely compatible with earlier Intel microprocessors. The I/O port number appears on address lines A15–A3 with the bank-enable signals used to select the actual memory banks used for the I/O transfer. Transfers are controlled by the chipset, which is a departure from the standard microprocessor architecture before the Pentium II.

Beginning with the 80386 microprocessor, I/O privilege information is added to the TSS segment when the Pentium II is operated in the protected mode. Recall that this allows I/O ports to be selectively inhibited. If the blocked I/O location is accessed, the Pentium II generates a type 13 interrupt to signal an I/O privilege violation.

System Timing

As with any microprocessor, the system timing signals must be understood in order to interface the microprocessor, or so it was at one time. Because the Pentium II is designed to be controlled by the

TABLE 19–3   The request ($\overline{\text{REQ}}$) signals to the Pentium II.

$\overline{\text{REQ4–REQ0}}$	Name	Comment
00000	Deferred reply	Deferred replies are issued for previously deferred transactions
00001	Reserved	Future use
00010	Memory read & invalidate	Memory read from DRAM or PCI write to DRAM from PCI
00011	Reserved	Future use
00100	Memory code read	Memory read
00101	Memory write	Memory write-back cycle
00110	Memory data read	Memory read
00111	Memory write	Normal memory write
01000	Interrupt acknowledge or special cycle	Interrupt acknowledge cycle for PCI bus
01001	Reserved	Future use
10000	I/O read	I/O read operation
10001	I/O write	I/O write operation
1100x	Reserved	Future use

chipset, the timing signals between the microprocessor and chipset have become proprietary Intel information and are not released to the public.

19–2   PENTIUM II SOFTWARE CHANGES

The Pentium II microprocessor core is a Pentium Pro. This means that the Pentium II and the Pentium Pro are essentially the same device for software. This section of the text lists the changes to the CPUID instruction; and the SYSENTER, SYSEXIT, FXSAVE, and FXRSTORE instructions (the only modifications to the software).

CPUID Instruction

Table 19–4 lists the values passed between the Pentium II and the CPUID instruction. These are changed from earlier versions of the Pentium microprocessor.

The version information returned after executing the CPUID instruction with a logic 0 in EAX is returned in EAX. The family ID is returned in bits 8 to 11; the model ID is returned in bits 4 to 7. The stepping ID is returned in bits 0 to 3. For the Pentium II, the model number is 6 and the family ID is a 3. The stepping number refers to an update number. The higher the stepping number, the newer the version.

The features are indicated in the EDX register after executing the CPUID instruction with a zero in EAX. Only two new features are returned in EDX for the Pentium II. Bit position 11 indicates whether the microprocessor supports the two new fast call instructions SYSENTER and SYSEXIT. Bit position 23 indicates whether the microprocessor supports the MMX instruction set introduced in Chapter 14. The remaining bits are identical to earlier versions of the microprocessor and are not described. Bit 16 indicates whether the microprocessor supports the page attribute table or PAT. Bit 17 indicates whether the microprocessor supports the page size extension found with the Pentium Pro and Pentium II microprocessors. The page size extension allows memory above 4G through 64G to be addressed. Finally, bit 24 indicates whether the fast floating-point save and restore instructions are implemented.

TABLE 19–4 CPUID instruction.

Input EAX	Output Register	Contents
0	EAX	Maximum value for input to EAX for CPUID instruction
0	EBX	'uneG'
0	ECX	'Inei'
0	EDX	'Iotn'
1	EAX	Version number
1	EBX	—
1	ECX	—
1	EDX	Feature information
2	EAX	Cache data
2	EBX	Cache data
2	ECX	Cache data
2	EDX	Cache data

SYSENTER and SYSEXIT Instructions

The SYSENTER and SYSEXIT instructions use the fast call facility introduced in the Pentium II microprocessor. Please note that these instructions function only in ring zero (privilege level 0) in protected mode. Windows operates in ring 0, but does not allow applications access to ring 0. These new instructions are meant for operating system software.

The SYSENTER instruction uses some of the model-specific registers to store CS, EIP, and ESP to execute a fast call to a procedure defined by the model-specific register. The fast call is different from a regular call because it does not push the return address onto the stack as a regular call. Table 19–5 illustrates the model-specific register used with SYSENTER and SYSEXIT. Note that the model-specific registers are read with the RDMSR instruction and written with the WRMSR instruction.

To use the RDMSR or WRMSR instructions, place the register number in the ECX register. If the WRMSR is used, place the new data for the register in EDS:EAX. For the SYSENTER instruction, you need use only the EAX register, but place a zero into EDX. If the RDMSR instruction is used, the data are returned in the EDX:EAX register pair. Example 19–1 illustrates a macro sequence that can be used to add the RDTSC, CPUID, RDMSR, WRMSR, SYSENTER, and SYSEXIT instructions to a program. Some of these macro definitions can also be obtained from Microsoft's Internet Web site as an update to the MASM assembler program.

EXAMPLE 19–1

```
;macro sequences to add RDTSC, CPUID, RDMSR, WRMSR, SYSENTER, and SYSEXIT
;
RDTSC     MACRO
          DB      0FH, 31H
          ENDM
;
```

TABLE 19–5 The model-specific registers used with SYSENTER and SYSEXIT.

Name	Number	Function
SYSENTER_CS	174H	SYSENTER target code segment
SYSENTER_ESP	175H	SYSENTER target stack segment
SYSENTER_EIP	176H	SYSENTER target instruction pointer

```
CPUID       MACRO
            DB      0FH, 0A2H
            ENDM
;
RDMSR       MACRO
            DB      0FH, 32H
            ENDM
;
WRMSR       MACRO
            DB      0FH, 30H
            ENDM
;
SYSENTER    MACRO
            DB      0FH, 34H
            ENDM
;
SYSEXIT     MACRO
            DB      0FH, 35H
            ENDM
```

To use the SYSENTER instruction, you must first load the model-specific registers with the address of the system entrance point into the SYSENTER_CS and SYSENTER_EIP registers. This would normally be the address of the operating system such as Windows or Windows NT. Note that this instruction is meant as a system instruction to access code or software in ring 0. The stack segment register is loaded with the value placed into SYSENTER_CS plus 8. In other words, the selector pair addressed by SYSENTER_CS selector value are loaded into CS and SS. The value of the stack offset is loaded into SYSENTER_ESP.

The SYSEXIT instruction loads CS and SS with the selector pair addressed by SYSENTER_CS plus 16 and 24. Table 19–6 illustrates the selectors from the global selector table, as addressed by SYSENTER_CS. In addition to the code and stack segment selector and the memory segments that they represent, the SYSEXIT instruction passes the value in EDX to the EIP register and the value in ECX to the ESP register. The SYSEXIT instruction returns control back to application ring 3. As mentioned, these instructions appear to have been designed for quick entrance and return from the Windows or Windows NT operating systems on the personal computer.

To use SYSENTER and SYSEXIT, the SYSENTER instruction must pass the return address to the system. This is accomplished by loading the EDX register with the return offset and by placing the segment address in the global descriptor table at location SYSENTER_CS+16. The stack segment is transferred by loading the stack segment selector into SYSENTER_CS+24 and the ESP into the ECX.

FXSAVE and FXRSTOR Instructions

The last two new instructions added to the Pentium II microprocessor are the FXSAVE and FXRSTOR instructions, which are almost identical to the FSAVE and FRSTOR instructions detailed in Chapter 14. The main difference is that the FXSAVE instruction is designed to properly store the state of the MMX machine, while the FSAVE properly stores the state of the floating-point coprocessor. The FSAVE instruction stores the entire tag field, while the FXSAVE instruction only stores the valid bits of the tag field. The valid tag field is used to reconstruct the restore tag field when the FXRSTOR instruction executes. This means that if the MMX state of

TABLE 19–6 Selectors addressed by the SYSENTER_CS select value.

SYSENTER_CS MSR	Function
SYSENTER_CS value	SYSENTER code segment selector
SYSENTER_CS value + 8	SYSENTER stack segment selector
SYSENTER_CS value + 16	SYSEXIT code segment selector
SYSENTER_CS value + 24	SYSEXIT stack segment selector

the machine is saved, use the FXSAVE instruction; if the floating-point state of the machine is saved, use the FSAVE instruction. For new applications, it is recommended that the FXSAVE and FXRSTOR instructions should be used to save the MMX state and floating-point state of the machine. Do not use the FSAVE and FRSTOR instructions in new applications. See Example 19–2 for macros that allow FXSAVE and FXRSTOR to be used in a program. This example assumes direct memory addressing, only using 32-bit offset addresses.

EXAMPLE 19–2

```
;FXSAVE and FXRSTOR macros
;
FXSAVE     MACRO     Addr
           DB        0FH, 0AEH, 6
           DD        offset Addr
           ENDM
;
FXRSTOR    MACRO     Addr
           DB        0FH, 0AEH, 0EH
           DD        offset Addr
           ENDM
```

19–3   THE PENTIUM III

The Pentium III microprocessor is an improved version of the Pentium II microprocessor. Even though it is newer than the Pentium II, it is still based on the Pentium Pro architecture.

There are two versions of the Pentium III. One version is available with a non-blocking 512K-byte cache and packaged in the slot 1 cartridge, and the other version is available with a 256K-byte advanced transfer cache and packaged in an integrated circuit. The slot 1-version cache runs at half the processor speed, and the integrated-cache version runs at the processor clock frequency. As shown in most benchmarks of cache performance, increasing the cache size from 256K bytes to 512K bytes only improves performance by a few percent.

Chip Sets

The chip set for the Pentium III is different from the Pentium II. The Pentium III uses an Intel 810, 815, or 820 chipset. The 815 is most commonly found in newer systems that use the Pentium III. A few other vendor chip sets are available, but problems with drivers for new peripherals, such as the video cards, have been reported. An 840 chip set also was developed for the Pentium III, but Intel does not make it available.

Bus

The Coppermine version of the Pentium III increases the bus speed to either 100 MHz or 133 MHz. The faster version allows transfers between the microprocessor and the memory at higher speeds.

Suppose that you have a 1-GHz microprocessor that uses a 133-MHz memory bus. You might think that the memory bus speed could be faster to improve performance, and we agree. However, the connections between the microprocessor and the memory preclude using a higher speed for the memory. If we decided to use a 200-MHz bus speed, we must recognize that a wavelength at 200 MHz is 300,000,000/200,000,000 or 3/2 meter. An antenna is 1/4 of a wavelength. At 200 MHz, an antenna is 14.8 inches. We do not want to radiate energy at 200 MHz, so we need to keep the printed circuit board connections shorter than 1/4-wavelength. In practice, we would keep the connections to no more than 1/10 of 1/4-wavelength. This means that the connections in a 200 MHz system should be no longer than 1.48 inches. This size would present the main board manufacturer with a problem when placing the sockets for a 200 MHz memory system.

Will it be possible to approach or even exceed the 200 MHz memory system? Yes, if we develop a new technology for interconnecting the microprocessor, chipset, and memory. At present the memory functions in bursts of four 64-bit numbers each time we read the main memory. This burst of 32 bytes is read into the cache. The main memory requires 3 wait states at 100 MHz to access the first 64-bit number and then zero wait states for each of the three remaining 64-bit wide numbers for a total of seven 100 MHz bus clocks. This means we are reading data at 70 ns / 32 = 2.1875 ns per byte, which is a bus speed of 457M bytes per second. This is slower than the clock on a 1GHz microprocessor, but because most programs are cyclic and the instructions are stored in an internal cache, we can and often do approach the operating frequency of the microprocessor.

Pin-out

Figure 19–4 shows the pin-out of the socket 370 version of the Pentium III microprocessor. This integrated circuit is packaged in a 370-pin, pin grid array (PGA) socket. It is designed to function with one of the chipsets available from Intel. In addition to the full version of the Pentium III, the Celeron, which uses a 66 MHz memory bus speed, is available. The Pentium III Xeon, also manufactured by Intel, allows larger cache sizes for server applications.

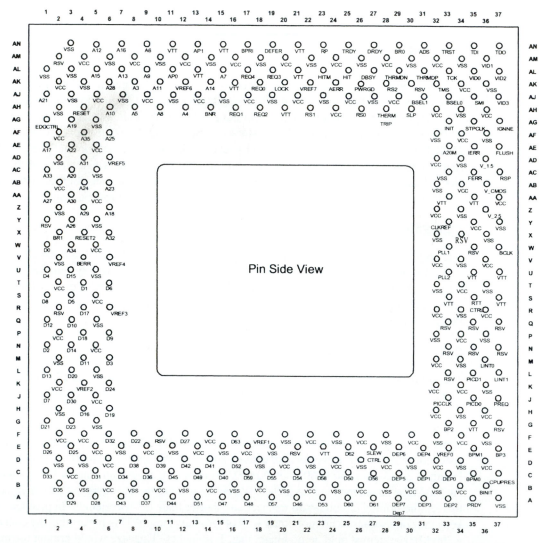

FIGURE 19–4   The pin-out of the socket 370 version of the Pentium III microprocessor. (Courtesy of Intel Corporation.)

19–4   THE PENTIUM 4

The most recent version of the Pentium Pro architecture microprocessor is the Pentium 4 microprocessor from Intel. The Pentium 4 was released initially in November 2000 with a speed of 1.3 GHz. It is currently available in speeds up to 2.0 GHz. There are two packages available for this integrated microprocessor, the 423-pin PGA and the 478-pin FC-PGA2. Both versions use the 1.8 micron technology for fabrication. As with earlier versions of the Pentium, the Pentium 4 uses a 100-MHz memory bus speed, but because it is quad pumped, the bus speed can approach 400 MHz. Figure 19–5 illustrates the pin-out of the 432-pin PGA of the Pentium 4 microprocessor.

Memory Interface

The memory interface to the Pentium 4 typically uses the Intel 850 chipset. The 850 provides a dual-pipe memory bus to the microprocessor with each pipe interfaced to a 32-bit wide section of the memory. The two pipes function together to comprise the 64-bit wide data path to the microprocessor. Because of the dual pipe arrangement, the memory must be populated with pairs of

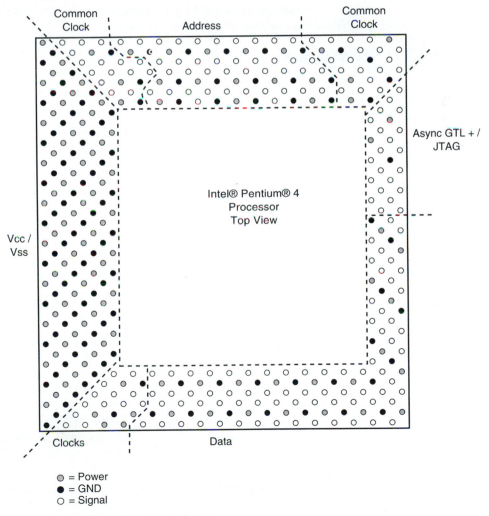

FIGURE 19–5   The pin-out of the Pentium 4, 423 PGA. (Courtesy of Intel Corporation.)

RDRAM memory devices operating at either 600 MHz or 800 MHz. According to Intel this arrangement provides a 300% increase in speed over a memory populated with PC-100 memory.

Register Set

The Pentium 4 register set is nearly identical to all other versions of the Pentium except that the MMX registers are separate entities from the floating-point registers. In addition, eight 128-bit wide XMM registers are added for use with the SIMD (single instruction multiple data) instructions and the extended 128-bit packed doubled floating-point numbers.

You might think of the XMM registers as double wide MMX registers that can hold a pair of 64-bit double-precision floating-point numbers or four single-precision floating-point numbers. Likewise they can also hold 16 byte-wide numbers as the MMX registers hold 8 byte-wide numbers. The XMM registers are double width MMX registers.

If the new patch for MASM 6.14 is downloaded from Microsoft,[4] programs can be assembled using both the MMX and XMM instructions. To assemble programs that include MMX instructions, use the .MMX switch. For programs that include the SIMD instructions, use the .XMM switch. Example 19-3 illustrates a very simple program that uses the MMX instructions to add two eight-byte-wide numbers together. Notice how the .MMX switch is used to select the MMX instruction set. The MOVQ instructions transfer numbers between memory and the MMX registers. The MMX registers are numbered from MM0 to MM7. You can also use the MMX and SIMD instructions in Microsoft Visual C using the inline assembler if you download the latest patch from Microsoft for Visual Studio version 6.0.

EXAMPLE 19–3

```
                                            .MODEL TINY
                                            .MMX
0000                                        .DATA

0000                                        DATA1   DQ   1ffh
        00000000000001FF
0008                                        DATA2   DQ   101h
        0000000000000101
0010                                        DATA3   DQ   ?
        0000000000000000
0000                                        .CODE
                                            .STARTUP

0100  9B 0F 6F 06 0000 R                        MOVQ   MM0,DATA1
0106  9B 0F 6F 0E 0008 R                        MOVQ   MM1,DATA2
010C  9B 0F FC C1                               PADDB  MM0,MM1
0110  9B 0F 7F 06 0010 R                        MOVQ   DATA3,MM0

                                            .EXIT
                                            END
```

Similarly, the XMM software can be used in a program with the .XMM switch. Most modern programs use the XXM registers and the XXM instruction set to accomplish multimedia and other high speed operations. Example 19–4 shows a short program that illustrates the use of a few XMM instructions. This program multiplies two sets of four single-precision numbers and stores the four products into the four double words at ANS. In order to enable access to octal words (128-byte wide numbers) we use the OWORD PTR directive. Also notice that the FLAT model is used with the C profile. Since the SIMD instructions only function in protected mode

---

[4] The update to any version of MASM 6.XX can be downloaded from Microsoft at www.microsoft.com.

(WIN32 model) we define the program in the FLAT model format. This means that the .686 and .XMM switches must precede the MODEL directive.

EXAMPLE 19–4

```
                                      .686
                                      .XMM
                                      .MODEL FLAT,0
00000000                              .DATA

00000000 3F800000         DATA1    DD 1.0
00000004 40000000                  DD 2.0
00000008 40400000                  DD 3.0
0000000C 40800000                  DD 4.0
00000010 40C9999A         DATA2    DD 6.3
00000014 40933333                  DD 4.6
00000018 40900000                  DD 4.5
0000001C C0133333                  DD -2.3
00000020 00000004 [       ANS      DD 4 DUP(?)
         00000000
       ]
00000000                           .CODE
00000000 0F 28 05                  MOVAPS    XMM0,OWORD PTR DATA1
         00000000 R
00000007 0F 28 0D                  MOVAPS    XMM1,OWORD PTR DATA2
         00000010 R
0000000E 0F 59 C1                  MULPS     XMM0,XMM1
00000011 0F 29 05                  MOVAPS    OWORD PTR ANS, XMM0
         00000020 R

                                   END
```

Hyper Pipelined Technology

The Pentium 4 incorporates a deeper pipelined architecture than prior versions of the Pentium microprocessor. Not only does it queue instructions for execution, but it also queues micro-instruction for execution in a special cache for the microprocessor core. This special microinstruction cache is 12K bytes deep. This technology excludes the execution unit from the main cache path to the microinstruction stream to increase performance.

CPUID

Like earlier versions of the Pentium, the CPUID instruction returns the standard vendor ID information if executed with a zero in EAX. The most significant part of the CPU serial number is returned if the EAX register is a 1 before execution of the CPUID instruction. The middle and least significant parts of serial number are returned in the EDX and ECX, middle and least respectively, after a second execution of the CPUID instruction with a 3 in EAX. The CPUID instruction is displayed as a hexadecimal value as XXXX-XXXX-XXXX-XXXX-XXXX-XXXX. Example 19–5 shows a sample of code that extracts the serial number from the microprocessor and stores it in three double words of memory. This functions in both the real mode and protected modes of operation.

EXAMPLE 19–5

```
                      .MODEL SMALL
                      .686
0000                  .DATA
0000 00000000              MOST    DD ?
0004 00000000              MID     DD ?
0008 00000000              LEAST   DD ?
0000                  .CODE
```

```
                          .STARTUP     ;read CPU serial number
0010 66| B8 00000001      MOV     EAX,1
0016 0F A2                CPUID
0018 66| A3 0000 R        MOV     MOST,EAX
001C 66| B8 00000003      MOV     EAX,3
0022 0F A2                CPUID
0024 66| 89 16 0004 R     MOV     MID,EDX
0029 66| 89 0E 0008 R     MOV     LEAST,ECX
                          .EXIT
                          END
```

Pentium 4 Mechanicals

The Pentium 4 microprocessor uses the ATX architecture, but there are some changes that should be noted. The power supply for the Pentium 4 is different from other ATX power supplies. The Pentium 4 power supply contains the standard ATX connector, a 12 V connector, and an auxiliary connector that looks similar to the AT power supply connector. All three connectors must be plugged into the Pentium 4 main board for proper operation.

Another change to the Pentium 4 system is the case. The case for the Pentium 4 main board must have four additional standoffs to support the microprocessor. Without the additional standoffs you cannot use the Pentium 4 main board because without them the heat-sink for the microprocessor cannot be attached to the main board.

The power supply for the microprocessor should also be at least 300 W to handle the additional 60 to 70 W of power required by the microprocessor. This microprocessor runs a bit hotter than prior versions of the Pentium. The system usually reports a temperature of 120°F, an increase of 25° over the normal Pentium III temperature.

19–5   SUMMARY

1. The Pentium II differs from earlier microprocessors because instead of being offered as an integrated circuit, the Pentium II is available on a plug-in cartridge or printed circuit board.
2. The level 2 cache for the Pentium II is mounted inside of the cartridge, except for the Celeron, which has no level 2 cache. The cache speed is one-half the Pentium II clock speed, except in the Xeon, where it is at the same speed as the Pentium II. All versions of the Pentium II contain an internal level 1 cache that stores 32K bytes of data.
3. The Pentium II is the first Intel microprocessor that is controlled from an external bus controller. Unlike earlier versions of the microprocessor, which issued read and write signals, the Pentium II is ordered to read or write information by an external bus controller.
4. The Pentium II operates at clock frequencies from 233 MHz to 450 MHz with bus speeds of 66 MHz or 100 MHz. The level 2 cache can be 512K-, 1M-, or 2M-bytes in size. The Pentium II contains a 64-bit data bus and a 36-bit address bus that allow up to 64G bytes of memory to be accessed.
5. The new instructions added to the Pentium II are SYSENTER, SYSEXIT, FXSAVE, and FXRSTOR.
6. The SYSENTER and SYSEXIT commands are optimized to access the operating system in privilege level 0 from a privilege level 3 access. These instructions operate at a much higher speed than a task switch or even a call and return combination.
7. The FXSAVE and FXRSTOR instructions are optimized to properly store the state of both the MMX technology unit and the floating-point coprocessor.
8. The Pentium III microprocessor is an extension of the Pentium Pro architecture with the addition of the SIMD instruction set that uses the XXM registers.

9. The Pentium 4 microprocessor is an extension of the Pentium Pro architecture, which includes enhancements that allow it to operate at higher clock frequencies than previously possible because of the 1.8 micron fabrication technology.
10. The Pentium 4 microprocessor requires a modified ATX power supply and case to function properly in a system.
11. Version 6.14 of the MASM program and Visual Studio version 6 now support the new MMX and SIMD instructions using the .686 switch with the .MMX and .XXM switches.

19–6   QUESTIONS AND PROBLEMS

1. What is the size of the level 1 cache in the Pentium II microprocessor?
2. What sizes are available for the level 2 cache in the Pentium II microprocessor? (List all versions.)
3. What is the difference between the level 2 cache on the Pentium-based system and the Pentium II-based system?
4. What is the difference between the level 2 cache in the Pentium Pro and the Pentium II?
5. The speed of the Pentium II Xeon level 2 cache is _____ times faster than the cache in the Pentium II (excluding the Celeron).
6. How much memory can be addressed by the Pentium II?
7. Is the Pentium II available in integrated circuit form?
8. How many pin connections are found on the Pentium II cartridge?
9. What is the purpose of the PID control signals?
10. What happened to the read and write pins on the Pentium II?
11. At what bus speeds does the Pentium II operate?
12. How fast is the SDRAM connected to the Pentium II system for a 100 MHz bus speed version?
13. How wide is the Pentium II memory if ECC is employed?
14. What new model-specific registers (MSR) have been added to the Pentium II microprocessor?
15. What new CPUID identification information has been added to the Pentium II microprocessor?
16. How is a model-specific register addressed and what instruction is used to read it?
17. Write software that stores a 12H into model-specific register 175H.
18. Write a short procedure that determines whether the microprocessor contains the SYSENTER and SYSEXIT instructions. Your procedure must return carry set if the instructions are present, and return carry cleared if not present.
19. How is the return address transferred to the system when using the SYSENTER instruction?
20. How is the return address retrieved when using the SYSEXIT instruction to return to the application?
21. The SYSENTER instruction transfers control to software at what privilege level?
22. The SYSEXIT instruction transfers control to software at what privilege level?
23. What is the difference between the FSAVE and the FXSAVE instructions?
24. The Pentium III is an extension of the _____architecture.
25. What new instructions appear in the Pentium III microprocessor that do not appear in the Pentium Pro microprocessor?
26. What changes to the power supply does the Pentium 4 microprocessor require?
27. Write a short program that reads and displays the serial number of the PIII microprocessor on the video screen.
28. Develop a procedure that multiplies 256 double-precision floating-point numbers and stores the products in a memory array called ANSWER.

APPENDIX A

The Assembler, Disk Operating System, Basic I/O System, Mouse, and DPMI Memory Manager

This appendix is provided so the assembler can be understood and to also show the DOS (disk operating system), BIOS (basic I/O system), mouse, and DPMI (DOS Protected Mode Interface) function calls. These function calls are used by assembly to control the personal computer. The function calls control everything from reading and writing disk data, to managing the keyboard and displays, to controlling the mouse. The assembler represented in this text is the Microsoft ML (Version 6.X) and MASM (version 5.10) macro assembler programs. It is fairly important that version 6.X be used instead of the dated version 5.10. Also presented is the DPMI memory manager, used when shelling out of Windows.

ASSEMBLER USAGE

The assembler program requires that a symbolic program first be written, using a word processor, text editor, or the workbench program provided with the assembler package. The editor provided with version 5.10 is M.EXE, and it is strictly a full-screen editor. The editor provided with version 6.X is PWB.EXE, and it is a fully integrated development system that contains extensive help. Refer to the documentation that accompanies your assembler package for details on the operation of the editor program. If at all possible, use version 6.X of the assembler because it contains a detailed help file that guides you through assembly language statements, directives, and even the DOS and BIOS interrupt function calls.

If you are using a word processor to develop your software, make sure that it is initialized to generate a pure ASCII file. The source file that you generate must use the extension .ASM, which is required for the assembler to properly identify your source program.

Once your source file is prepared, it must be assembled. If you are using the workbench provided with version 6.X, this is accomplished by selecting the compile feature with your mouse. If you are using a word processor and DOS command lines with version 6.14, see Example A–1 for the dialog for version 6.14 to assemble a file called FROG.ASM. Note that this example shows the portions typed by the user in italics.

EXAMPLE A–1

```
A>MASM

Microsoft (R) Macro Assembler Version 6.14.8444
Copyright (C) Microsoft Corp 1981, 1997. All rights reserved.

Source filename [.ASM]:FROG
Object filename [FROG.OBJ]:FROG
List filename [NUL.LST]:FROG
Cross reference [NUL.CRF]:FROG
```

Once a program is assembled, it must be linked before is can be executed. The linker converts the object file into an executable file (.EXE). Example A–2 shows the dialog required for the linker using an MASM version 5.10 object file. If the ML version 6.X assembler is in use, it automatically assembles and links a program by using the COMPILE or BUILD command from workbench. After compiling with ML, workbench allows the program to be debugged with a debugging tool called *code view*. Code view is also available with MASM, but CV must be typed at the DOS command line to access it.

EXAMPLE A–2

```
A:\>LINK

Microsoft (R) Overlay Linker Version 3.64
Copyright (C) Microsoft Corp 1983-1988. All rights reserved.

Object modules [.OBJ]:FROG
Run file [FROG.EXE]:FROG
List file [NUL.MAP]:FROG
Libraries [.LIB]:SUBR
```

If MASM version 6.X is in use, the command line syntax differs from version 5.10. Example A–3 shows the command line syntax for ML, the assembler and linker for MASM version 6.X.

EXAMPLE A–3

```
C:\>ML /FlTEST.LST TEST.ASM

Microsoft (R) Macro Assembler Version 6.14.844
Copyright (C) Microsoft Corp 1981-1997. All rights reserved.

    Assembling: TEST.ASM

Microsoft (R) Segmented-Executable Linker Version 5.13
Copyright (C) Microsoft Corp 1984-1993. All rights reserved.

Object Modules [.OBJ]: TEST.obj/t
Run File [TEST.com]: "TEST.com"
List File [NUL.MAP]: NUL
Libraries [.LIB]:
Definitions File [NUL.DEF]: ;
```

Version 6.X of the Microsoft MASM program contains the Programmer's Workbench program. Programmer's Workbench allows an assembly language program to be developed with its full screen editor and tool bar. Figure A–1 illustrates the display found with Programmer's Workbench. To access this program, type PWB at the DOS prompt. The make option allows a program to be automatically assembled and linked, making these tasks simple in comparison to version 5.10 of the assembler.

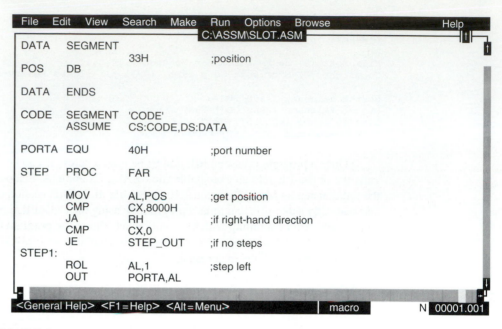

FIGURE A–1   The edit screen from Programmer's Workbench used to develop assembly language programs.

ASSEMBLER MEMORY MODELS

Memory models and the .MODEL statement are introduced in Chapter 4 and used exclusively throughout the text. Here, we completely define the memory models available for software development. Each model defines the way that a program is stored in the memory system. Table A–1 describes the different models available with both MASM and ML.

TABLE A–1   Memory models for the assembler.

Model Type	Description
Tiny	All data and code must fit into one segment. Tiny programs are written in .COM format, which means that the program must be originated at location 100H.
Small	This model contains two segments: one data segment of 64K bytes and one code segment of 64K bytes.
Medium	This model contains one data segment of 64K bytes and any number of code segments for large programs.
Compact	One code segment contains the program, and any number of data segments contain the data.
Large	The large model allows any number of code and data segments.
Huge	This model is the same as large, but the data segments may contain more than 64K bytes each.
Flat	Only available to MASM 6.X. The flat model uses one segment of 512K bytes to store all data and code. Note that this model is mainly used with Windows NT.

Note that the tiny model is used to create a .COM file instead of an execute file. The .COM file is different because all data and code fit into one code segment. A .COM file must use an origin of offset address 0100H as the start of the program. A .COM file loads from the disk and executes faster than the normal execute (.EXE) file. For most applications, we normally use the execute file (.EXE) and the small memory model.

When models are used to create a program, certain defaults apply, as illustrated in Table A–2. The directive in this table is used to start a particular type of segment for the models listed in the

TABLE A–2 Defaults for the .MODEL directive.

Model	Directives	Name	Align	Combine	Class	Group
Tiny	.CODE	_TEXT	word	PUBLIC	'CODE'	DGROUP
	.FARDATA	FAR_DATA	para	private	'FAR_DATA'	
	.FARDATA?	FAR_BSS	para	private	FAR_BSS	
	.DATA	_DATA	word	PUBLIC	DATA'	DGROUP
	.CONST	CONST	word	PUBLIC	'CONST'	DGROUP
	.DATA?	_BSS	word	PUBLIC	'BSS'	DGROUP
Small	.CODE	_TEXT	word	PUBLIC	'CODE'	
	.FARDATA	FAR_DATA	para	private	'FAR_DATA'	
	.FARDATA?	FAR_BSS	para	private	'FAR_BSS'	
	.DATA	_DATA	word	PUBLIC	'DATA'	DGROUP
	.CONST	CONST	word	PUBLIC	'CONST'	DGROUP
	.DATA?	_BSS	word	PUBLIC	'BSS'	DGROUP
	.STACK	STACK	para	STACK	'STACK'	DGROUP
Medium	CODE	name_TEXT	word	PUBLIC	'CODE'	
	.FARDATA	FAR_DATA	para	private	'FAR_DATA'	
	.FARDATA?	FAR_BSS	para	private	'FAR_BSS'	
	.DATA	_DATA	word	PUBLIC	'DATA'	DGROUP
	.CONST	CONST	word	PUBLIC	'CONST'	DGROUP
	.DATA?	_BSS	word	PUBLIC	'BSS'	DGROUP
	.STACK	STACK	para	STACK	'STACK'	DGROUP
Compact	CODE	_TEXT	word	PUBLIC	'CODE'	
	.FARDATA	FAR_DATA	para	private	'FAR_DATA'	
	.FARDATA?	FAR_BSS	para	private	'FAR_BSS'	
	.DATA	_DATA	word	PUBLIC	'DATA'	DGROUP
	.CONST	CONST	word	PUBLIC	'CONST'	DGROUP
	.DATA?	_BSS	word	PUBLIC	'BSS'	DGROUP
	.STACK	STACK	para	STACK	'STACK'	DGROUP
Large or huge	CODE	name_TEXT	word	PUBLIC	'CODE'	
	.FARDATA	FAR_DATA	para	private	'FAR_DATA'	
	.FARDATA?	FAR_BSS	para	private	'FAR_BSS'	
	.DATA	_DATA	word	PUBLIC	'DATA'	DGROUP
	.CONST	CONST	word	PUBLIC	'CONST'	DGROUP
	.DATA?	_BSS	word	PUBLIC	'BSS'	DGROUP
	.STACK	STACK	para	STACK	'STACK'	DGROUP
Flat	CODE	_TEXT	dword	PUBLIC	'CODE'	
	.FARDATA	_DATA	dword	PUBLIC	'DATA'	
	.FARDATA?	_BSS	dword	PUBLIC	'FBSS'	
	.DATA	_DATA	dword	PUBLIC	'DATA'	DGROUP
	.CONST	CONST	dword	PUBLIC	'CONST'	DGROUP
	.DATA?	_BSS	dword	PUBLIC	'BSS'	DGROUP
	.STACK	STACK	dword	STACK	'STACK'	DGROUP

table. If the .CODE directive is placed in a program, it indicates the beginning of the code segment. Likewise, .DATA indicates the start of a data segment. The name *column* indicates the name of the segment. *Align* indicates whether the segment is aligned on a word, doubleword, or a 16-byte paragraph. *Combine* indicates the type of segment created. The *class* indicates the class of the segment, such as 'CODE' or 'DATA'. The *group* indicates the group type of the segment.

The directive from Table A–2 selects the type of information in a program. For example, .CODE is placed before the code. The name column is used if full-segment descriptions are mixed with the programming models for reference. The alignment specifies how the data in the segment are aligned. A para (paragraph) alignment starts a segment at the next paragraph, i.e., the next hexadecimal address ending in a 0H. The combine column indicates how various segment are combined and labeled (PUBLIC or private). The class is the actual segment name, and the group is the grouping of segments.

Example A–4 shows a program that uses the small model. The small model is used for programs that contain one DATA and one CODE segment. This applies to many programs that are developed. Notice that not only is the program listed, but so is all the information generated by the assembler. Here, the .DATA directive and .CODE directive indicate the start of each segment. Also notice how the DS register is loaded in this program. As presented throughout the text, the .STARTUP directive can be used to load the data segment register, set up the stack, and define the starting address of a program. In this example, an alternate method (END BEGIN) is illustrated for loading the data segment register and defining the starting address of the program.

EXAMPLE A–4

```
Microsoft (R) Macro Assembler Version 6.14.8444

                              .MODEL SMALL
                              .STACK 100H
0000                          .DATA

0000 0A              FROG   DB    10
0001 0064 [          DATA1  DB    100 DUP (2)
        02
          ]

0000                          .CODE

0000 B8 —— R    BEGIN: MOV   AX,DGROUP           ;set up DS
0003 8E D8             MOV   DS,AX
                       .
                       .
                       .
                       END   BEGIN

Segments and Groups:

          N a m e                    Size    Length  Align  Combine   Class

DGROUP . . . . . . . . . . . . GROUP
_DATA  . . . . . . . . . . . . 16 Bit  0065    Word   Public    'DATA'
STACK  . . . . . . . . . . . . 16 Bit  0100    Para   Stack     'STACK'
_TEXT  . . . . . . . . . . . . 16 Bit  0005    Word   Public    'CODE'

Symbols:

          N a m e                    Type    Value   Attr

@CodeSize  . . . . . . . . . . Number  0000h
@DataSize  . . . . . . . . . . Number  0000h
@Interface . . . . . . . . . . Number  0000h
@Model . . . . . . . . . . . . Number  0002h
```

```
@code  .  .  .  .  .  .  .  .  .  .  .  .  . Text              _TEXT
@data  .  .  .  .  .  .  .  .  .  .  .  .  . Text              DGROUP
@fardata?  .  .  .  .  .  .  .  .  .  .  . Text              FAR_BSS
@fardata  .  .  .  .  .  .  .  .  .  .  .  . Text              FAR_DATA
@stack  .  .  .  .  .  .  .  .  .  .  .  . Text              DGROUP
BEGIN  .  .  .  .  .  .  .  .  .  .  .  . L Near 0000        _TEXT
DATA1  .  .  .  .  .  .  .  .  .  .  .  . Byte   0001        _DATA
FROG  .  .  .  .  .  .  .  .  .  .  .  .  . Byte   0000        _DATA

            0 Warnings
            0 Errors
```

Example A–5 lists a program that uses the large model. Notice how it differs from the small model program of Example A–4. Models can be very useful in developing software, but often we use full-segment descriptions, as depicted in most examples in the text.

EXAMPLE A–5

```
Microsoft (R) Macro Assembler Version 6.14.8444

                              .MODEL LARGE
                              .STACK 1000H
0000                          .FARDATA?

0000 00                   FROG   DB    ?
0001 0064 [               DATA1  DW    100 DUP (?)
        0000
            ]

0000                          .CONST

0000 54 68 69 73 20 69   MES1   DB    'This is a character string'
     73 20 61 20 63 68
     61 72 61 63 74 65
     72 20 73 74 72 69
     6E 67
001A 53 6F 20 69 73 20   MES2   DB    'So is this!'
     74 68 69 73 21

0000                          .DATA

0000 000C                 DATA2  DW    12
0002 00C8 [               DATA3  DB    200 DUP (1)
        01
         ]

0000                          .CODE

0000                      FUNC   PROC  FAR
                                 .
                                 .
                                 .
                                 .
0000 CB                          RET

0001                      FUNC   ENDP

                          END    FUNC

Segments and Groups:

            N a m e              Size    Length  Align Combine  Class

DGROUP  .  .  .  .  .  .  .  .  .  .  .  . GROUP
```

```
_DATA    . . . . . . . . . . . . 16 Bit   00CA    Word    Public   'DATA'
STACK    . . . . . . . . . . . . 16 Bit   1000    Para    Stack    'STACK'
CONST    . . . . . . . . . . . . 16 Bit   0025    Word    Public   'CONST'
ReadOnly
EXA_TEXT . . . . . . . . . . . . 16 Bit   0001    Word    Public   'CODE'
FAR_BSS  . . . . . . . . . . . . 16 Bit   00C9    Para    Private  'FAR_BSS'
_TEXT    . . . . . . . . . . . . 16 Bit   0000    Word    Public   'CODE'
```

Procedures,   parameters and locals:

```
                N a m e           Type    Value   Attr


FUNC . . . . . . . . . . . . . . P Far    0000    EXA_TEXT Length= 0001 Public
```

Symbols:

```
                N a m e          Type    Value   Attr


@CodeSize  . . . . . . . . . . Number 0001h
@DataSize  . . . . . . . . . . Number 0001h
@Interface . . . . . . . . . . Number 0000h
@Model . . . . . . . . . . . . Number 0005h
@code  . . . . . . . . . . . . Text            EXA_TEXT
@data  . . . . . . . . . . . . Text            DGROUP
@fardata?  . . . . . . . . . . Text            FAR_BSS
@fardata . . . . . . . . . . . Text            FAR_DATA
@stack . . . . . . . . . . . . Text            DGROUP
DATA1  . . . . . . . . . . . . Word    0001    FAR_BSS
DATA2  . . . . . . . . . . . . Word    0000    _DATA
DATA3  . . . . . . . . . . . . Byte    0002    _DATA
FROG . . . . . . . . . . . . . Byte    0000    FAR_BSS
MES1 . . . . . . . . . . . . . Byte    0000    CONST
MES2 . . . . . . . . . . . . . Byte    001A    CONST


              0 Warnings
              0 Errors
```

DOS FUNCTION CALLS

In order to use DOS function calls, always place the function number into register AH and load all other pertinent information into registers, as described in the table as entry data. Once this is accomplished, follow with an INT 21H to execute the DOS function. Example A–6 shows how to display an ASCII A on the CRT screen at the current cursor position with a DOS function call. Table A–3 is a complete listing of the DOS function calls. Note that some function calls require a segment and offset address, indicated as DS:DI, for example. This means the data segment is the segment address and DI is the offset address. All of the function calls use INT 21H, and AH contains the function call number. Note that functions marked with an @ should not be used unless DOS version 2.XX is in use. Also note that not all function numbers are implemented. As a rule, DOS function calls save all registers not used as exit data, but in certain cases some registers may change. In order to prevent problems, it is advisable to save registers where problems occur.

EXAMPLE A–6

```
0000 B4 06              MOV    AH,6        ;load function 06H
0002 B2 41              MOV    DL,'A'      ;select letter 'A'
0004 CD 21              INT    21H         ;call DOS function
```

TABLE A–3   DOS function calls (pp. 809–831).

00H	**TERMINATE A PROGRAM**
Entry	AH = 00H CS = program segment prefix address
Exit	DOS is entered

01H	**READ THE KEYBOARD**
Entry	AH = 01H
Exit	AL = ASCII character
Notes	If AL = 00H, the function call must be invoked again to read an extended ASCII character. Refer to Chapter 7, Table 7–3 for a listing of the extended ASCII keyboard codes. This function call automatically echoes whatever is typed to the video screen.

02H	**WRITE TO STANDARD OUTPUT DEVICE**
Entry	AH = 02H DL = ASCII character to be displayed
Notes	This function call normally displays data on the video display.

03H	**READ CHARACTER FROM COM1**
Entry	AH = 03H
Exit	AL = ASCII character read from the communications port
Notes	This function call reads data from the serial communications port.

04H	**WRITE TO COM1**
Entry	AH = 04H DL = character to be sent out of COM1
Notes	This function transmits data through the serial communications port. The COM port assignment can be changed to use other COM ports with functions 03H and 04H by using the DOS MODE command to reassign COM1 to another COM port.

05H	WRITE TO LPT1
Entry	AH = 05H DL = ASCII character to be printed
Notes	Prints DL on the line printer attached to LPT1. Note that the line printer port can be changed with the DOS MODE command.

06H	DIRECT CONSOLE READ/WRITE
Entry	AH = 06H DL = 0FFH or DL = ASCII character
Exit	AL = ASCII character
Notes	If DL = 0FFH on entry, then this function reads the console. If DL = ASCII character, then this function displays the ASCII character on the console (CON) video screen. If a character is read from the console keyboard, the zero flag (ZF) indicates whether a character was typed. A zero condition indicates that no key was typed, and a not-zero condition indicates that AL contains the ASCII code of the key or a 00H. If AL = 00H, the function must again be invoked to read an extended ASCII character from the keyboard. Note that the key does not echo to the video screen.

07H	DIRECT CONSOLE INPUT WITHOUT ECHO
Entry	AH = 07H
Exit	AL = ASCII character
Notes	This functions exactly as function number 06H with DL = 0FFH, but it will not return from the function until the key is typed.

08H	READ STANDARD INPUT WITHOUT ECHO
Entry	AH = 08H
Exit	AL = ASCII character
Notes	Performs as function 07H, except that it reads the standard input device. The standard input device can be assigned as either the keyboard or the COM port. This function also responds to a control-break, where function 06H and 07H do not. A control-break causes INT 23H to execute. By default, this functions as does function 07H.

09H	DISPLAY A CHARACTER STRING
Entry	AH = 09H DS:DX = address of the character string
Notes	The character string must end with an ASCII $ (24H). The character string can be of any length and may contain control characters such as carriage return (0DH) and line feed (0AH).

0AH	BUFFERED KEYBOARD INPUT
Entry	AH = 0AH DS:DX = address of keyboard input buffer
Notes	The first byte of the buffer contains the size of the buffer (up to 255). The second byte is filled with the number of characters typed upon return. The third byte through the end of the buffer contains the character string typed, followed by a carriage return (0DH). This function continues to read the keyboard (displaying data as typed) until either the specified number of characters are typed or until the enter key is typed.

0BH	TEST STATUS OF THE STANDARD INPUT DEVICE
Entry	AH = 0BH
Exit	AL = status of the input device
Notes	This function tests the standard input device to determine if data are available. If AL = 00, no data are available. If AL = 0FFH, then data are available that must be input using function number 08H.

0CH	CLEAR KEYBOARD BUFFER AND INVOKE KEYBOARD FUNCTION
Entry	AH = 0CH AL = 01H, 06H, 07H, or 0AH
Exit	See exit for functions 01H, 06H, 07H, or 0AH
Notes	The keyboard buffer holds keystrokes while programs execute other tasks. This function empties or clears the buffer and then invokes the keyboard function located in register AL.

0DH	FLUSH DISK BUFFERS
Entry	AH = 0DH
Notes	Erases all file names stored in disk buffers. This function does not close the files specified by the disk buffers, so care must be exercised in its usage.

0EH	SELECT DEFAULT DISK DRIVE
Entry	AH = 0EH DL = desired default disk drive number
Exit	AL = the total number of drives present in the system
Notes	Drive A = 00H, drive B = 01H, drive C = 02H, and so forth.

0FH	@ OPEN FILE WITH FCB
Entry	AH = 0FH DS:DX = address of the unopened file control block (FCB)
Exit	AL = 00H if file found AL = 0FFH if file not found
Notes	The file control block (FCB) is only used with early DOS software and should never be used with new programs. File control blocks do not allow path names as do the newer file access function codes presented later. Figure A–2 (p. 831) illustrates the structure of the FCB. To open a file, the file must either be present on the disk or be created with function call 16H.

10H	@ CLOSE FILE WITH FCB
Entry	AH = 10H DS:DX = address of the opened file control block (FCB)
Exit	AL = 00H if file closed AL = 0FFH if error found
Notes	Errors that occur usually indicate either that the disk is full or the media is bad.

11H	@ SEARCH FOR FIRST MATCH (FCB)
Entry	AH = 11H DS:DX = address of the file control block to be searched
Exit	AL = 00H if file found AL = 0FFH if file not found
Notes	Wild card characters (? or *) may be used to search for a file name. The ? wild card character matches any character and the * matches any name or extension.

12H	@SEARCH FOR NEXT MATCH (FCB)
Entry	AH = 12H DS:DX = address of the file control block to be searched
Exit	AL = 00H if file found AL = 0FFH if file not found
Notes	This function is used after function 11H finds the first matching file name.

13H	@DELETE FILE USING FCB
Entry	AH = 13H DS:DX = address of the file control block to be deleted
Exit	AL = 00H if file deleted AL = 0FFH if error occurred
Notes	Errors that most often occur are defective media errors.

14H	@SEQUENTIAL READ (FCB)
Entry	AH = 14H DS:DX = address of the file control block to be read
Exit	AL = 00H if read successful AL = 01H if end of file reached AL = 02H if DTA had a segment wrap AL = 03H if less than 128 bytes were read

15H	@SEQUENTIAL WRITE (FCB)
Entry	AH = 15H DS:DX = address of the file control block to be written
Exit	AL = 00H if write successful AL = 01H if disk is full AL = 02H if DTA had a segment wrap

16H	@CREATE A FILE (FCB)
Entry	AH = 16H DS:DX = address of an unopened file control block
Exit	AL = 00H if file created AL = 01H if disk is full

17H	**@RENAME A FILE (FCB)**
Entry	AH = 17H DS:DX = address of a modified file control block
Exit	AL = 00H if file renamed AL = 01H if error occurred
Notes	See Figure A–3 (p. 831) for the modified FCB used to rename a file.

19H	**RETURN CURRENT DRIVE**
Entry	AH = 19H
Exit	AL = current drive
Notes	AL = 00H for drive A, 01H for drive B, and so forth.

1AH	**SET DISK TRANSFER AREA**
Entry	AH = 1AH DS:DX = address of new DTA
Notes	The disk transfer area is normally located within the program segment prefix at offset address 80H. The DTA is used by DOS for all disk data transfers using file control blocks.

1BH	**GET DEFAULT DRIVE FILE ALLOCATION TABLE (FAT)**
Entry	AH = 1BH
Exit	AL = number of sectors per cluster DS:BX = address of the media–descriptor CX = size of a sector in bytes DX = number of clusters on drive
Notes	See Figure A–4 (p. 832) for the format of the media–descriptor byte. The DS register is changed by this function, so make sure to save it before using this function.

1CH	**GET ANY DRIVE FILE ALLOCATION TABLE (FAT)**
Entry	AH = 1CH DL = disk drive number
Exit	AL = number of sectors per cluster DS:BX = address of the media–descriptor CX = size of a sector in bytes DX = number of clusters on drive

21H	@RANDOM READ USING FCB
Entry	AH = 21H DS:DX = address of opened FCB
Exit	AL = 00H if read successful AL = 01H if end of file reached AL = 02H if the segment wrapped AL = 03H if less than 128 bytes read
22H	@RANDOM WRITE USING FCB
Entry	AH = 22H DS:DX = address of opened FCB
Exit	AL = 00H if write successful AL = 01H if disk full AL = 02H if the segment wrapped
23H	@RETURN NUMBER OF RECORDS (FCB)
Entry	AH = 23H DS:DX = address of FCB
Exit	AL = 00H number of records AL = 0FFH if file not found
24H	@SET RELATIVE RECORD SIZE (FCB)
Entry	AH = 24H DS:DX = address of FCB
Notes	Sets the record field to the value contained in the FCB.
25H	SET INTERRUPT VECTOR
Entry	AH = 25H AL = interrupt vector number DS:DX = address of new interrupt procedure
Notes	Before changing the interrupt vector, it is suggested that the current interrupt vector first be saved using DOS function 35H. This allows a back-link so the original vector can later be restored.
26H	CREATE NEW PROGRAM SEGMENT PREFIX
Entry	AH = 26H DX = segment address of new PSP
Notes	Figure A–5 (p. 832) illustrates the structure of the program segment prefix.

27H	**@RANDOM FILE BLOCK READ (FCB)**
Entry	AH = 27H CX = the number of records DS:DX = address of opened FCB
Exit	AL = 00H if read successful AL = 01H if end of file reached AL = 02H if the segment wrapped AL = 03H if less than 128 bytes read CX = the number of records read
28H	**@RANDOM FILE BLOCK WRITE (FCB)**
Entry	AH = 28H CX = the number of records DS:DX = address of opened FCB
Exit	AL = 00H if write successful AL = 01H if disk full AL = 02H if the segment wrapped CX = the number of records written
29H	**@PARSE COMMAND LINE (FCB)**
Entry	AH = 29H AL = parse mask DS:SI = address of FCB DS:DI = address of command line
Exit	AL = 00H if no file name characters found AL = 01H if file name characters found AL = 0FFH if drive specifier incorrect DS:SI = address of character after name DS:DI = address first byte of FCB
2AH	**READ SYSTEM DATE**
Entry	AH = 2AH
Exit	AL = day of the week CX = the year (1980–2099) DH = the month DL = day of the month
Notes	The day of the week is encoded as Sunday = 00H through Saturday = 06H. The year is a binary number equal to 1980 through 2099.

2BH	SET SYSTEM DATE
Entry	AH = 2BH CX = the year (1980–2099) DH = the month DL = day of the month

2CH	READ SYSTEM TIME
Entry	AH = 2CH
Exit	CH = hours (0–23) CL = minutes DH = seconds DL = hundredths of seconds
Notes	All times are returned in binary form, and hundredths of seconds may not be available.

2DH	SET SYSTEM TIME
Entry	AH = 2DH CH = hours CL = minutes DH = seconds DL = hundredths of seconds

2EH	DISK VERIFY WRITE
Entry	AH = 2EH AL = 00H to disable verify on write AL = 01H to enable verify on write
Notes	By default, disk verify is disabled.

2FH	READ DISK TRANSFER AREA ADDRESS
Entry	AH = 2FH
Exit	ES:BX = contains DTA address

30H	READ DOS VERSION NUMBER
Entry	AH = 30H
Exit	AH = fractional version number AL = whole number version number
Notes	For example, DOS version number 3.2 is returned as a 3 in AL and a 14H in AH.

31H	TERMINATE AND STAY RESIDENT (TSR)
Entry	AH = 31H AL = the DOS return code DX = number of paragraphs to reserve for program
Notes	A paragraph is 16 bytes, and the DOS return code is read at the batch file level with ERRORCODE.

33H	TEST CONTROL-BREAK
Entry	AH = 33H AL = 00H to request current control-break AL = 01H to change control-break DL = 00H to disable control-break DL = 01H to enable control-break
Exit	DL = current control-break state

34H	GET ADDRESS OF InDOS FLAG
Entry	AH = 34H
Exit	ES:BX = address of InDOS flag
Notes	The InDOS flag is available in DOS versions 3.2 or newer and indicates DOS activity. If InDOS = 00H, DOS is inactive; or 0FFH, if DOS is active and pursuing another operation.

35H	READ INTERRUPT VECTOR
Entry	AH = 35H AL = interrupt vector number
Exit	ES:BX = address stored at vector
Notes	This DOS function is used with function 25H to install/remove interrupt handlers.

36H	DETERMINE FREE DISK SPACE
Entry	AH = 36H DL = drive number
Exit	AX = FFFFH if drive invalid AX = number of sectors per cluster BX = number of free clusters CX = bytes per sector DX = number of clusters on drive
Notes	The default disk drive is DL = 00H, drive A = 01H, drive B = 02H, and so forth.

38H	RETURN COUNTRY CODE
Entry	AH = 38H AL = 00H for current country code BX = 16-bit country code DS:DX = data buffer address
Exit	AX = error code if carry set BX = counter code DS:DX = data buffer address

39H	CREATE SUBDIRECTORY
Entry	AH = 39H DS:DX = address of ASCII-Z string subdirectory name
Exit	AX = error code if carry set
Notes	The ASCII-Z string is the name of the subdirectory in ASCII code ended with a 00H instead of a carriage return/line feed.

3AH	ERASE SUBDIRECTORY
Entry	AH = 3AH DS:DX = address of ASCII-Z string subdirectory name
Exit	AX = error code if carry set

3BH	CHANGE SUBDIRECTORY
Entry	AH = 3BH DS:DX = address of new ASCII-Z string subdirectory name
Exit	AX = error code if carry set

3CH	CREATE A NEW FILE
Entry	AH = 3CH CX = attribute word DS:DX = address of ASCII-Z string file name
Exit	AX = error code if carry set AX = file handle if carry cleared
Notes	The attribute word can contain any of the following (added together): 01H read-only access, 02H = hidden file or directory, 04H = system file, 08H = volume label, 10H = subdirectory, and 20H = archive bit. In most cases, a file is created with 0000H.

3DH	OPEN A FILE
Entry	AH = 3DH AL = access code DS:DX = address of ASCII-Z string file name
Exit	AX = error code if carry set AX = file handle if carry cleared
Notes	The access code in AL = 00H for a read-only access, AL = 01H for a write-only access, and AL = 02H for a read/write access. For shared files in a network environment, bit 4 of AL = 1 will deny read/write access, bit 5 of AL = 1 will deny a write access, bits 4 and 5 of AL = 1 will deny read access, bit 6 of AL = 1 denies none, bit 7 of AL = 0 causes the file to be inherited by child; if bit 7 of AL = 1, file is restricted to current process.
3EH	CLOSE A FILE
Entry	AH = 3EH BX = file handle
Exit	AX = error code if carry set
3FH	READ A FILE
Entry	AH = 3FH BX = file handle CX = number of bytes to be read DS:DX = address of file buffer to hold data read
Exit	AX = error code if carry set AX = number of bytes read if carry cleared
40H	WRITE A FILE
Entry	AH = 40H BX = file handle CX = number of bytes to write DS:DX = address of file buffer that holds write data
Exit	AX = error code if carry set AX = number of bytes written if carry cleared
41H	DELETE A FILE
Entry	AH = 41H DS:DX = address of ASCII-Z string file name
Exit	AX = error code if carry set

42H	MOVE FILE POINTER
Entry	AH = 42H AL = move technique BX = file handle CX:DX = number of bytes pointer moved
Exit	AX = error code if carry set AX:DX = bytes pointer moved
Notes	The move technique causes the pointer to move from the start of the file if AL = 00H, from the current location if AL = 01H, and from the end of the file if AL = 02H. The count is stored so DX contains the least-significant 16-bits and either CX or AX contains the most-significant 16-bits.

43H	READ/WRITE FILE ATTRIBUTES
Entry	AH = 43H AL = 00H to read attributes AL = 01H to write attributes CX = attribute word (see function 3CH) DS:DX = address of ASCII-Z string file name
Exit	AX = error code if carry set CX = attribute word of carry cleared

44H	I/O DEVICE CONTROL (IOTCL)
Entry	AH = 44H AL = sub function code (see notes)
Exit	AX = error code (see function 59H) if carry set

Notes	The sub function codes found in AL are as follows:
	
	00H = read device status
	    Entry: BX = file handle
	    Exit: DX = status
	01H = write device status
	    Entry: BX = file handle, DH = 0, DL = device information
	    Exit: AX = error code if carry set
	02H = read control data from character device
	    Entry: BX = file handle, CX = number of bytes, DS:DX = I/O
	               buffer address
	    Exit: AX = number of bytes read
	03H = write control data to character device
	    Entry: BX = file handle, CX = number of bytes, DS:DX = I/O
	               buffer address
	    Exit: AX = number of bytes written

04H = read control data from block device
    Entry: BL = drive number (0 = default, 1 = A, 2 = B, etc),
                CX = number of bytes, DS:DX = I/O buffer address
    Exit: AX = number of bytes read
05H = write control data to block device
    Entry: BL = drive number, CX = number of bytes,
                DS:DX = I/O buffer address
    Exit: AX = number of bytes written
06H = check input status
    Entry: BX = file handle
    Exit: AL = 00H ready or FFH not ready
07H = check output status
    Entry: BX = file handle
    Exit: AL = 00H ready or FFH not ready
08H = removable media?
    Entry: BL = drive number
    Exit: AL = 00H removable, 01H fixed
09H = network block device?
    Entry: BL = drive number
    Exit: bit 12 of DX set for network block device
0AH = local or network character device?
    Entry: BX = file handle
    Exit: bit 15 of DX set for network character device
0BH = change entry count (must have SHARE.EXE loaded)
    Entry: CX = delay loop count, DX = retry count
    Exit: AX = error code if carry set
0CH = generic I/O control for character devices
    Entry: BX = file handle, CH = category, CL = function
    Categories: 00H = unknown, 01H = COM port, 02H =
                       CON, 05H = LPT ports
    Function:
         CL = 45H; set iteration count
         CL = 4AH; select code page
         CL = 4CH; start code page preparation
         CL = 4DH; end code page preparation
         CL = 5FH; set display information
         CL = 65H; get iteration count
         CL = 6AH; query selected code page
         CL = 6BH; query preparation list
         CL = 7FH; get display information
0DH = generic I/O control for block devices
    Entry: BL = drive number, CH = category, CL = function,
                DS:DX = address of parameter block
    Category: 08H = disk drive
    Function:
         CL = 40H; set device parameters
         CL = 41H; write track
         CL = 42H; format and verify track
         CL = 46H; set media ID code
         CL = 47H; set access flag
         CL = 60H; get device parameters
         CL = 61H; read track
         CL = 62H; verify track
         CL = 66H; get media ID code
         CL = 67H; get access code

	0EH = return logical device map
	    Entry: BL = drive number
	    Exit: AL = number of last device
	0FH = change logical device map
	    Entry: BL = drive number
	    Exit: AL = number of last device

45H	DUPLICATE FILE HANDLE
Entry	AH = 45H BX = current file handle
Exit	AX = error code if carry set AX = duplicate file handle
46H	FORCE DUPLICATE FILE HANDLE
Entry	AH = 46H BX = current file handle CX = new file handle
Exit	AX = error code if carry set
Notes	This function works like function 45H except that function 45H allows DOS to select the new handle, while this function allows the user to select the new handle.
47H	READ CURRENT DIRECTORY
Entry	AH = 47H DL = drive number DS:SI = address of a 64-byte buffer for directory name
Exit	DS:SI addresses current directory name if carry cleared
Notes	Drive A = 00, drive B = 01, and so forth
48H	ALLOCATE MEMORY BLOCK
Entry	AH = 48H BX = number of paragraphs to allocate CX = new file handle
Exit	BX = largest block available if carry cleared
49H	RELEASE ALLOCATED MEMORY BLOCK
Entry	AH = 49H ES = segment address of block to be released CX = new file handle
Exit	Carry indicates an error if set

4AH	MODIFY ALLOCATED MEMORY BLOCK
Entry	AH = 4AH BX = new block size in paragraphs ES = segment address of block to be modified
Exit	BX = largest block available if carry cleared

4BH	LOAD OR EXECUTE A PROGRAM
Entry	AH = 4BH AL = function code ES:BX = address of parameter block DS:DX = address ASCII-Z string command
Exit	Carry indicates an error if set
Notes	The function codes are AL = 00H to load and execute a program, AL = 01H to load a program but not execute it, AL = 03H to load a program overlay, and AL = 05H to enter the EXEC state. Figure A–6 (p. 833) shows the parameter block used with this function.

4CH	TERMINATE A PROCESS
Entry	AH = 4CH AL = error code
Exit	Returns control to DOS
Notes	This function returns control to DOS with the error code saved so it can be obtained using DOS ERROR LEVEL batch processing system. We normally use this function with an error code of 00H to return to DOS.

4DH	READ RETURN CODE
Entry	AH = 4DH
Exit	AX = return error code
Notes	This function is used to obtain the return status code created by executing a program with DOS function 4BH. The return codes are AX = 0000H for a normal-no error-termination, AX = 0001H for a control-break termination, AX = 0002H for a critical device error, and AX = 0003H for a termination by an INT 31H.

4EH	FIND FIRST MATCHING FILE
Entry	AH = 4EH CX = file attributes DS:DX = address ASCII-Z string file name
Exit	Carry is set for file not found
Notes	This function searches the current or named directory for the first matching file. Upon exit, the DTA contains the file information. See Figure A–7 (p. 833) for the disk transfer area (DTA).
4FH	FIND NEXT MATCHING FILE
Entry	AH = 4FH
Exit	Carry is set for file not found
Notes	This function is used after the first file is found with function 4EH.
50H	SET PROGRAM SEGMENT PREFIX (PSP) ADDRESS
Entry	AH = 50H BX = offset address of the new PSP
Notes	Extreme care must be used with this function because no error recovery is possible.
51H	GET PSP ADDRESS
Entry	AH = 51H
Exit	BX = current PSP segment address
54H	READ DISK VERIFY STATUS
Entry	AH = 54H
Exit	AL = 00H if verify off AL = 01H if verify on
56H	RENAME FILE
Entry	AH = 56H ES:DI = address of ASCII-Z string containing new file name DS:DX = address of ASCII-Z string containing file to be renamed
Exit	Carry is set for error condition

57H	READ FILE'S DATE AND TIME STAMP
Entry	AH = 57H AL = function code BX = file handle CX = new time DX = new date
Exit	Carry is set for error condition CX = time if carry cleared DX = date if carry cleared
Notes	AL = 00H to read date and time or 01H to write date and time.

59H	GET EXTENDED ERROR INFORMATION
Entry	AH = 59H BX = 0000H for DOS version 3.X
Exit	AX = extended error code BH = error class BL = recommended action CH = locus
Notes	Following are the extended error codes found in AX: 0001H = invalid function number 0002H = file not found 0003H = path not found 0004H = no file handles available 0005H = access denied 0006H = file handle invalid 0007H = memory control block failure 0008H = insufficient memory 0009H = memory block address invalid 000AH = environment failure 000BH = format invalid 000CH = access code invalid 000DH = data invalid 000EH = unknown unit 000FH = disk drive invalid 0010H = attempted to remove current directory 0011H = not same device 0012H = no more files 0013H = disk write-protected 0014H = unknown unit 0015H = drive not ready 0016H = unknown command 0017H = data error (CRC check error) 0018H = bad request structure length 0019H = seek error 001AH = unknown media type 001BH = sector not found

001CH = printer out of paper
001DH = write fault
001EH = read fault
001FH = general failure
0020H = sharing violation
0021H = lock violation
0022H = disk change invalid
0023H = FCB unavailable
0024H = sharing buffer exceeded
0025H = code page mismatch
0026H = handle end of file operation not completed
0027H = disk full
0028H–0031H reserved
0032H = unsupported network request
0033H = remote machine not listed
0034H = duplicate name on network
0035H = network name not found
0036H = network busy
0037H = device no longer exists on network
0038H = netBIOS command limit exceeded
0039H = error in network adapter hardware
003AH = incorrect response from network
003BH = unexpected network error
003CH = remote adapter is incompatible
003DH = print queue is full
003EH = not enough room for print file
003FH = print file was deleted
0040H = network name deleted
0041H = network access denied
0042H = incorrect network device type
0043H = network name not found
0044H = network name exceeded limit
0045H = netBIOS session limit exceeded
0046H = temporary pause
0047H = network request not accepted
0048H = print or disk redirection pause
0049H–004FH reserved
0050H = file already exists
0051H = duplicate FCB
0052H = cannot make directory
0053H = failure in INT 24H (critical error)
0054H = too many re-directions
0055H = duplicate redirection
0056H = invalid password
0057H = invalid parameter
0058H = network write failure
0059H = function not supported by network
005AH = required system component not installed
0065H = device not selected

Following are the error class codes as found in BH:

01H = no resources available
02H = temporary error
03H = authorization error

	04H = internal software error 05H = hardware error 06H = system failure 07H = application software error 08H = item not found 09H = invalid format 0AH = item blocked 0BH = media error 0CH = item already exists 0DH = unknown error Following is the recommended action as found in BL: 01H = retry operation 02H = delay and retry operation 03H = user retry 04H = abort processing 05H = immediate exit 06H = ignore error 07H = retry with user intervention Following is a list of locus in CH: 01H = unknown source 02H = block device error 03H = network area 04H = serial device error 05H = memory error
5AH	CREATE UNIQUE FILE NAME
Entry	AH = 5AH CX = attribute code DS:DX = address of the ASCII-Z string directory path
Exit	Carry is set for error condition AX = file handle if carry cleared DS:DX = address of the appended directory name
Notes	The ASCII-Z file directory path must end with a backslash (\). On exit, the directory name is appended with a unique file name.
5BH	CREATE A DOS FILE
Entry	AH = 5BH CX = attribute code DS:DX = address of the ASCII-Z string contain the file name
Exit	Carry is set for error condition AX = file handle if carry cleared
Notes	The function works only in DOS version 3.X or higher. It is almost identical to function 3CH, except that function 3CH erases the file, if it already exists, while function 5BH reports that the file exists without erasing it.

5CH	LOCK/UNLOCK FILE CONTENTS
Entry	AH = 5CH BX = file handle CX:DX = offset address of locked/unlocked area SI:DI = number of bytes to lock or unlock beginning at offset
Exit	Carry is set for error condition

5DH	SET EXTENDED ERROR INFORMATION
Entry	AH = 5DH AL = 0AH DS:DX = address of the extended error data structure
Notes	This function is used by DOS version 3.1 or higher to store extended error information.

5EH	NETWORK/PRINTER
Entry	AH = 5EH AL = 00H (get network name) DS:DX = address of the ASCII-Z string containing network name
Exit	Carry is set for error condition CL = netBIOS number if carry cleared
Entry	AH = 5EH AL = 02H (define network printer) BX = redirection list CX = length of setup string DS:DX = address of printer setup buffer
Exit	Carry is set for error condition
Entry	AH = 5EH AL = 03H (read network printer setup string) BX = redirection list DS:DX = address of printer setup buffer
Exit	Carry is set for error condition CX = length of setup string if carry cleared ES:DI = address of printer setup buffer

62H	GET PSP ADDRESS
Entry	AH = 62H
Exit	BX = segment address of the current program
Notes	The function works only in DOS version 3.0 or higher.

65H	GET EXTENDED COUNTRY INFORMATION
Entry	AH = 65H AL = function code ES:DI = address of buffer to receive information
Exit	Carry is set for error condition CX = length of country information
Notes	The function works only in DOS version 3.3 or higher.

66H	GET/SET CODE PAGE
Entry	AH = 66H AL = function code BX = code page number
Exit	Carry is set for error condition BX = active code page number DX = default code page number
Notes	A function code in AL of 01H gets the code page number, and a code of 02H sets the code page number.

67H	SET HANDLE COUNT
Entry	AH = 67H BX = number of handles desired
Exit	Carry is set for error condition
Notes	This function is available for DOS version 3.3 or higher.

68H	COMMIT FILE
Entry	AH = 68H BX = handle number
Exit	Carry is set for error condition; otherwise, the date and time stamp is written to directory.
Notes	This function is available for DOS version 3.3 or higher.

6CH	EXTENDED OPEN FILE
Entry	AH = 6CH AL = 00H BX = open mode CX = attributes DX = open flag DS:SI = address of ASCII-Z string file name
Exit	AX = error code if carry is set AX = handle if carry is cleared CX = 0001H file existed and was opened CX = 0002H file did not exist and was created
Notes	This function is available for DOS version 4.0 or higher.

FIGURE A–2 Contents of the file control block (FCB).

Offset	Contents
00H	Drive
01H	8-character filename
09H	3-character file extension
0CH	Current block number
0EH	Record size
10H	File size
14H	Creation date
16H	Reserved space
20H	Current record number
21H	Relative record number

FIGURE A–3 Contents of the modified file control block (FCB).

Offset	Contents
00H	Drive
01H	8-character filename
09H	3-character extension
0CH	Current block number
0EH	Record size
10H	File size
14H	Creation date
16H	Second file name

FIGURE A–4   Contents of
the media-descriptor byte.

7	6	5	4	3	2	1	0
?	?	?	?	?	?	?	?

Bit 0 = 0 if not two-sided
      = 1 if two-sided

Bit 1 = 0 if not eight sectors per track
      = 1 if eight sectors per track

Bit 2 = 0 if nonremovable
      = 1 if removable

FIGURE A–5   Contents of
the program segment prefix
(PSP).

Offset	Contents
00H	INT 20H
02H	Top of memory
04H	Reserved
05H	Opcode
06H	Number of bytes in segment
0AH	Terminate address (offset)
0CH	Terminate address (segment)
0EH	Control-break address (offset)
10H	Control-break address (segment)
12H	Critical error address (offset)
14H	Critical error address (segment)
16H	Reserved
2CH	Environment address (segment)
2EH	Reserved
50H	DOS call
52H	Reserved
5CH	File control block 1
6CH	File control block 2
80H	Command line length
81H	Command line

FIGURE A–6 The parameter blocks used with function 4BH (EXEC). (a) For function code 00H. (b) For function code 03H.

(a)

Offset	Contents
00H	Environment address (segment)
02H	Command line address (offset)
04H	Command line address (segment)
06H	File control block 1 address (offset)
08H	File control block 1 address (segment)
0AH	File control block 2 address (segment)
0CH	File control block 2 address (offset)

(b)

Offset	Contents
00H	Overlay destination segment address
02H	Relocation factor

FIGURE A–7 Data transfer area (DTA) used to find a file.

Offset	Contents
15H	Attributes
16H	Creation time
18H	Creation date
1AH	Low word file size
1CH	High word file size
1EH	Search file name

BIOS FUNCTION CALLS

In addition to DOS function call INT 21H, some other BIOS function calls are useful in controlling the I/O environment of the computer. Unlike INT 21H, which exists in the DOS program, the BIOS function calls are found stored in the system and video BIOS ROMs. These BIOS ROM functions directly control the I/O devices, with or without DOS loaded into a system.

INT 10H

The INT 10H BIOS interrupt is often called the *video services interrupt* because it directly controls the video display in a system. The INT 10H instruction uses register AH to select the video service provided by this interrupt. The video BIOS ROM is located on the video board and varies from one video card to another.

Video Mode Selection.   The mode of operation for the video display is accomplished by placing a 00H into AH, followed by one of many mode numbers in AL. Table A-4 lists the modes of operation found in video display systems using standard video modes. The VGA can use any mode listed, whereas the other displays are more restrictive in use. Additional higher resolution modes are explained later in this section.

TABLE A–4   Video display modes.

Mode	Type	Columns	Rows	Resolution	Standard	Colors
00H	Text	40	25	320 × 200	CGA	2
00H	Text	40	25	320 × 250	EGA	2
00H	Text	40	25	360 × 400	VGA	2
01H	Text	40	25	320 × 200	CGA	16
01H	Text	40	25	320 × 350	EGA	16
01H	Text	40	25	360 × 640	VGA	16
02H	Text	80	25	640 × 200	CGA	2
02H	Text	80	25	640 × 350	EGA	2
02H	Text	80	25	720 × 400	VGA	2
03H	Text	80	25	640 × 200	CGA	16
03H	Text	80	25	640 × 350	EGA	16
03H	Text	80	25	720 × 400	VGA	16
04H	Graphics	80	25	320 × 200	CGA	4
05H	Graphics	80	25	320 × 350	CGA	2
06H	Graphics	80	25	640 × 200	CGA	2
07H	Text	80	25	720 × 350	EGA	4
07H	Text	80	25	720 × 400	VGA	4
0DH	Graphics	80	25	320 × 200	CGA	16
0EH	Graphics	80	25	640 × 200	CGA	16
0FH	Graphics	80	25	640 × 350	EGA	4
10H	Graphics	80	25	640 × 350	EGA	16
11H	Graphics	80	30	640 × 480	VGA	2
12H	Graphics	80	30	640 × 480	VGA	16
13H	Graphics	40	25	320 × 200	VGA	256

Example A–7 lists a short sequence of instructions that place the video display into mode 03H operation. This mode is available on CGA, EGA, and VGA displays. This mode allows the display to draw test data with 16 colors at various resolutions, dependent upon the display adapter.

EXAMPLE A–7

```
0000 B4 00              MOV     AH,0            ;select mode
0002 B0 03              MOV     AL,3            ;mode is 03H
0004 CD 10              INT     10H
```

Cursor Control and Other Standard Features.   Table A–5 shows the function codes (placed in AH) used to control the cursor on the video display. These cursor control functions will work on any video display, from the CGA display to the latest super VGA display. It also lists the functions used to display data and change to a different character set.

TABLE A–5   Video BIOS (INT 10H) functions (pp. 834–837).

00H	SELECT VIDEO MODE
Entry	AH = 00H AL = mode number
Exit	Mode changed and screen cleared

01H	SELECT CURSOR TYPE
Entry	AH = 01H CH = starting line number CL = ending line number
Exit	Cursor size changed
02H	SELECT CURSOR POSITION
Entry	AH = 02H BH = page number (usually 0) DH = row number (beginning with 0) DL = column number (beginning with 0)
Exit	Changes cursor to new position
03H	READ CURSOR POSITION
Entry	AH = 03H BH = page number
Exit	CH = starting line (cursor size) CL = ending line (cursor size) DH = current row DL = current column
04H	READ LIGHT PEN
Entry	AH = 04H (not supported in VGA)
Exit	AH = 0, light pen triggered BX = pixel column CX = pixel row DH = character row DL = character column
05H	SELECT DISPLAY PAGE
Entry	AH = 05H AL = page number
Exit	Page number selected. Following are the valid page numbers: Mode 0 and 1 support pages 0–7 Mode 2 and 3 support pages 0–7 Mode 4, 5, and 6 support page 0 Mode 7 and D support pages 0–7 Mode E supports pages 0–3 Mode F and 10 support pages 0–1 Mode 11, 12, and 13 support page 0
Notes	Most modern displays use page 0 for most operations.

06H	SCROLL PAGE UP
Entry	AH = 06H AL = number of lines to scroll (0 clears window) BH = character attribute for new lines CH = top row of scroll window CL = left column of scroll window DH = bottom row of scroll window DL = right column of scroll window
Exit	Scrolls window from the bottom toward the top of the screen. Blank lines fill the bottom using the character attribute in BH.
07H	SCROLL PAGE DOWN
Entry	AH = 07H AL = number of lines to scroll (0 clears window) BH = character attribute for new lines CH = top row of scroll window CL = left column of scroll window DH = bottom row of scroll window DL = right column of scroll window
Exit	Scrolls window from the top toward the bottom of the screen. Blank lines fill from the top using the character attribute in BH.
08H	READ ATTRIBUTE/CHARACTER AT CURRENT CURSOR POSITION
Entry	AH = 08H BH = page number
Exit	AL = ASCII character code AH = character attribute
Notes	This function does not advance the cursor.
09H	WRITE ATTRIBUTE/CHARACTER AT CURRENT CURSOR POSITION
Entry	AH = 09H AL = ASCII character code BH = page number BL = character attribute CX = number of characters to write
Notes	This function does not advance the cursor.

0AH	WRITE CHARACTER AT CURRENT CURSOR POSITION
Entry	AH = 0AH AL = ASCII character code BH = page number CX = number of characters to write
Notes	This function does not advance the cursor.
0FH	READ VIDEO MODE
Entry	AH = 0FH
Exit	AL = current video mode AH = number of character columns BH = page number
10H	SET VGA PALETTE REGISTER
Entry	AH = 10H AL = 10H BX = color number (0–255) CH = green (0–63) CL = blue (0–63) DH = red (0–63)
Exit	Palette register color is changed. Note: The first 16 colors (0–15) are used in the 16-color, VGA text mode and other modes.
10H	READ VGA PALETTE REGISTER
Entry	AH = 10H AL = 15H BX = color number (0–255)
Exit	CH = green CL = blue DH = red
11H	GET ROM CHARACTER SET
Entry	AH = 11H AL = 30H BH = 2 = ROM 8 x 14 character set BH = 3 = ROM 8 x 8 character set BH = 4 = ROM 8 x 8 extended character set BH = 5 = ROM 9 x 14 character set BH = 6 = ROM 8 x 16 character set BH = 7 = ROM 9 x 16 character set
Exit	CX = bytes per character DL = rows per character ES:BP = address of character set

If an SVGA (super VGA), EVGA (extended VGA), or XVGA (also extended VGA) adapter is available, the super VGA mode is set by using INT 10H function call AX = 4F02H, with BX = to the VGA mode for these advanced display adapters. This conforms to the VESA standard for VGA adapters. Table A-6 shows the modes selected by register BX for this INT 10H function call. Most video cards are equipped with a driver called VVESA.COM or VVESA.SYS, which conforms the card to the VESA standard functions

INT 11H

This function is used to determine the type of equipment installed in the system. To use this call, the AX register is loaded with an FFFFH, and then the INT 11H instruction is executed. In return, an INT 11H provides information, as listed in Figure A–8.

INT 12H

The memory size is returned by the INT 12H instruction. After executing the INT 12H instruction, the AX register contains the number of 1K-byte blocks of memory (conventional memory in the first 1M bytes of address space) installed in the computer.

INT 13H

This call controls the diskettes ($5^1/4$" or $3^1/2$") and also fixed or hard disk drives attached to the system. Table A–7 lists the functions available to this interrupt via register AH. The direct control of a floppy disk or hard disk can lead to problems. Therefore we provide only a listing of the functions, without details of their usage. Before using these functions, refer to the BIOS literature available from the company that produced your version of the BIOS ROM. Never use these functions for normal disk operations.

TABLE A–6   Extended VGA functions.

BX	Extended Mode
100H	640×400 with 256 colors
101H	640×480 with 256 colors
102H	800×600 with 16 colors
103H	800×600 with 256 colors
104H	1024×768 with 16 colors
105H	1024×768 with 256 colors
106H	1280×1024 with 16 colors
107H	1280×1024 with 256 colors
108H	80×60 in text mode
109H	132×25 in text mode
10AH	132×43 in text mode
10BH	132×50 in text mode
10CH	132×60 in text mode

FIGURE A–8   The contents of AX as it indicates the equipment attached to the computer.

15	14	13	12	11	10	9	8	7	6	5	4	3	2	1	0
P1	P0		G	S2	S1	S0	D2	D1							

P1, P0 = number of parallel ports
G = 1 if game I/O attached
S2, S1, S0 = number of serial ports
D2, D1 = number of disk drives

TABLE A–7 Disk I/O function via INT 13H.

AH	Function
00H	Reset the system disk
01H	Read disk status to AL
02H	Read sector
03H	Write sector
04H	Verify sector
05H	Format track
06H	Format bad track
07H	Format drive
08H	Get drive parameters
09H	Initialize fixed disk characteristics
0AH	Read long sector
0BH	Write long sector
0CH	Seek
0DH	Reset fixed disk system
0EH	Read sector buffer
0FH	Write sector buffer
10H	Get drive status
11H	Re-calibrate drive
12H	Controller RAM diagnostics
13H	Controller drive diagnostics
14H	Controller internal diagnostics
15H	Get disk type
16H	Get disk changed status
17H	Set disk type
18H	Set media type
19H	Park heads
1AH	Format ESDI drive

INT 14H

Interrupt 14H controls the serial COM (communications) ports attached to the computer. The computer system contains two COM ports, COM1 and COM2, unless you have a newer AT-style machine, in which the number of communications ports are extended to COM3 and COM4. Communications ports are normally controlled with software packages that allow data transfer through a modem and the telephone lines. The INT 14H instruction controls these ports, as illustrated in Table A–8.

TABLE A–8 COM port interrupt INT 14H.

AH	Function
00H	Initialize communications port
01H	Send character
02H	Receive character
03H	Get COM port status
04H	Extended initialize communications port
05H	Extended communications port control

INT 15H

The INT 15H instruction controls many of the various I/O devices interfaced to the computer. It also allows access to protected mode operation and the extended memory system on an 80286–Pentium 4, but it is not recommended. Table A–9 lists the functions supported by INT 15H.

INT 16H

The INT 16H instruction is used as a keyboard interrupt. This interrupt is accessed by DOS interrupt INT 21H, but it can be accessed directly. Table A–10 shows the functions performed by INT 16H.

TABLE A–9   The I/O subsystem interrupt INT 15H.

AH	Function
00H	Cassette motor on
01H	Cassette motor off
02H	Read cassette
03H	Write cassette
0FH	Format ESDI periodic interrupt
21H	Keyboard intercept
80H	Device open
81H	Device closed
82H	Process termination
83H	Event wait
84H	Read joystick
85H	System request key
86H	Delay
87H	Move extended block of memory
88H	Get extended memory size
89H	Enter protected mode
90H	Device wait
91H	Device power on self test (POST)
C0H	Get system environment
C1H	Get address of extended BIOS data area
C2H	Mouse pointer
C3H	Set watchdog timer
C4H	Programmable option select

TABLE A–10   Keyboard interrupt INT 16H.

AH	Function
00H	Read keyboard character
01H	Get keyboard status
02H	Get keyboard flags
03H	Set repeat rate
04H	Set keyboard click
05H	Push character and scan code

INT 17H

The INT 17H instruction accesses the parallel printer port, usually labeled LPT1 in most systems. Table A–11 lists the three functions available for the INT 17H instruction.

DOS Low Memory Assignments

Table A–12 shows the low memory assignments (00000H–005FFH) for the DOS-based microprocessor system. This area of memory contains the interrupt vectors, BIOS data area, and the DOS/BIOS data.

TABLE A–11 Parallel
printer interrupt INT 17H.

AH	Function
00H	Print character
01H	Initialize printer
02H	Get printer status

TABLE A–12 DOS low memory assignments.

Location	Purpose
00000H–002FFH	System interrupt vectors
00300H–003FFH	System interrupt vectors, power on, and bootstrap area
00400H–00407H	COM1–COM4 I/O port base addresses
00408H–0040FH	LPT1–LPT4 I/O port base addresses
00410H–00411H	Equipment flag word, returned in AX by INT 11H (refer to Figure A–8)
00412H	Reserved
00413H–00414H	Memory size in K byte (0–640K)
00415H–00416H	Reserved
00417H	Keyboard control byte

Bit	Purpose
7	Insert locked
6	Caps locked
5	Numbers locked
4	Scroll locked
3	Alternate key pressed
2	Control key pressed
1	Left shift key pressed
0	Right shift key pressed

Location	Purpose
00418H	Keyboard control byte

Bit	Purpose
7	Insert key pressed
6	Caps lock key pressed
5	Numbers lock key pressed
4	Scroll lock key pressed
3	Pause key pressed
2	System request key pressed
1	Left alternate key pressed
0	Right control key pressed

(continued on the next page)

TABLE A–12   *(continued)*

Location	Purpose
00419H	Alternate keyboard entry
0041AH–0041BH	Keyboard buffer header pointer
0041CH–0041DH	Keyboard buffer tail pointer
0041EH–0043DH	32-byte keyboard buffer area
0043EH–00448H	Disk drive control area
00449H–00466H	Video control area
00467H–0046BH	Reserved
0046CH–0046FH	Timer counter
00470H	Timer overflow
00471H	Break key state
00472H–00473H	Reset flag
00474H–00477H	Hard disk drive data area
00478H–0047BH	LPT1–LPT4 time-out area
0047CH–0047FH	COM1–COM4 time-out area
00480H–00481H	Keyboard buffer start offset pointer
00482H–00483H	Keyboard buffer end offset pointer
00484H–0048AH	Video control data area
0048BH–00495H	Hard disk control area
00496H	Keyboard mode, state, and type flag
00497H	Keyboard LED flags
00498H–00499H	Offset address of user wait complete flag
0049AH–0049BH	Segment address of user wait complete flag
0049CH–0049FH	User wait count
004A0H	Wait active flag
004A1H–004A7H	Reserved
004A8H–004ABH	Pointer to video parameters
004ACH–004EFH	Reserved
004F0H–004FFH	Applications program communications area
00500H	Print screen status
00501H–00503H	Reserved
00504H	Single drive mode status
00505H–0050FH	Reserved
00510H–00521H	Used by ROM BASIC
00522H–0052FH	Used by DOS for disk initialization
00530H–00533H	Used by the MODE command
00534H–005FFH	Reserved

MOUSE FUNCTIONS

The mouse is controlled and adjusted with INT 33H function call instructions. These functions provide complete control over the mouse and information provided by the mouse driver program. Table A–13 lists the mouse INT 33H functions by number, and details the parameters required and any notes needed to use them. Refer to Chapter 7 for a discussion of the mouse driver and some example programs that access the mouse functions through INT 33H.

TABLE A–13   The mouse (INT 33H) functions (pp. 843–854).

00H	RESET MOUSE
Entry	AL = 00H
Exit	BX = number of mouse buttons Both software and hardware are reset to their default values
Notes	The default values are listed below: CRT Page = 0 Cursor = off Current cursor position = center of screen Minimum horizontal position = 0 Minimum vertical position = 0 Maximum horizontal position = maximum for display mode Maximum vertical position = maximum for display mode Horizontal mickey-to-pixel ratio = 1 to 1 Vertical mickey-to-pixel ration = 2 to 1 Double-speed threshold = 64 per second Graphics cursor = arrow Text cursor = reverse block Light-pen emulation = on Interrupt call mask = 0
01H	**SHOW MOUSE CURSOR**
Entry	AL = 01H
Exit	Displays the mouse cursor
02H	**HIDE MOUSE CURSOR**
Entry	AL = 02H
Exit	Hides the mouse cursor
Notes	When displaying data on the screen, it is important to hide the mouse cursor. If you don't, problems with the display will occur and the computer may reset and reboot.
03H	**READ MOUSE STATUS**
Entry	AL = 03H
Exit	BX = button status CX = horizontal cursor position DX = vertical cursor position
Notes	The right bit (bit 0) of BX contains the status of the left mouse button and bit 1 contains the status of the right mouse button. A logic 1 indicates the button is active.

04H	SET MOUSE CURSOR POSITION
Entry	AL = 04H CX = horizontal position DX = vertical position
05H	GET BUTTON PRESS INFORMATION
Entry	AL = 05H BX = desired button (0 for left and 1 for right)
Exit	AX = button status BX = number of presses CX = horizontal position of last press DX = vertical position of last press
06H	GET BUTTON RELEASE INFORMATION
Entry	AL = 06H BX = desired button
Exit	AX = button status BX = number of releases CX = horizontal position of last release DX = vertical position of last release
07H	SET HORIZONTAL BOUNDARY
Entry	AL = 07H CX = minimum horizontal position DX = maximum horizontal position
Exit	Horizontal boundary is changed.
08H	SET VERTICAL BOUNDARY
Entry	AL = 08H CX = minimum vertical position DX = maximum vertical position
Exit	Vertical boundary is changed.

09H	SET GRAPHICS CURSOR
Entry	AL = 09H BX = horizontal center DX = vertical center ES:DX = address of 16 x 16 bit map of cursor
Exit	New graphics cursor installed.
Notes	The center is where the mouse pointer position is set. For example, a center of 0,0 is the upper left corner and 15,15 is the lower right corner. The pixel bit map mask is stored at the address passed through ES:DX and is a 16×16 array. Following the pixel bit mask is the cursor mask, also 16×16. The contents of the bit map mask are ANDed with a 16×16 portion of the video display, after which the contents of the cursor mask are Exclusive-ORed with a 16×16 portion of the video display to produce a cursor.
0AH	SET TEXT CURSOR
Entry	AL = 0AH BX = cursor type (0 = software, 1 = hardware) CX = pixel bit mask or beginning scan line DX = cursor mask or ending scan line
Exit	Changes the text cursor.
0BH	READ MOTION COUNTERS
Entry	AL = 0BH
Exit	CX = horizontal distance DX = vertical distance
Notes	Returns the distance traveled by the mouse since the last call to this function.

0CH	SET INTERRUPT SUBROUTINE
Entry	AL = 0CH CX = interrupt mask ES:DX = address of interrupt service procedure
Exit	New interrupt handler is installed.
Notes	The interrupt mask defines the actions that request the installed interrupt handler. Following is a list of the actions that cause the interrupt when placed in the interrupt mask. Note that these actions can appear singly or in combination. For example, an interrupt mask of 8 plus 2 or 10 (0AH) causes an interrupt for either the left or right mouse buttons. 1 = any change in cursor position 2 = left mouse button pressed 4 = left mouse button released 8 = right mouse button pressed 16 = right mouse button released After the installation, interrupt is called by one of the selected changes. The following registers contain information about the mouse: AX = interrupt mask BX = button status CX = horizontal position DX = vertical position SI = horizontal change DI = vertical change

0DH	ENABLE LIGHT-PEN EMULATION
Entry	AL = 0DH
Exit	Light-pen emulation is enabled.
Notes	Used whenever the mouse must replace the action of a light-pen.

0EH	DISABLE LIGHT-PEN EMULATION
Entry	AL = 0EH
Exit	Light-pen emulation is disabled.
Notes	Used to disable light-pen emulation.

0FH	SET MICKEY-TO-PIXEL RATIO
Entry	AL = 0FH CX = horizontal ratio DX = vertical ratio
Exit	Mickey-to-pixel ratio changed.
Notes	This function allows the screen-tracking speed of the mouse cursor to be changed. The default value is 1. If it is changed to 2, the normal speed will be reduced by half.
10H	BLANK MOUSE CURSOR
Entry	AL = 10H CX = left corner DX = upper position SI = right corner DI = bottom position
Exit	Blanks the cursor in the window specified.
Notes	The mouse pointer is blanked in the area of the screen selected by registers CX, DX, SI, and DI.
13H	SET DOUBLE-SPEED
Entry	AL = 13H DX = threshold value
Exit	Changes the threshold for double-speed pointer movement.
Notes	The default threshold value for mouse pointer acceleration is 64. This default threshold is changed by this function to any other value.
14H	SWAP INTERRUPTS
Entry	AL = 14H CX = interrupt mask ES:DX = interrupt service procedure address
Exit	CX = old interrupt mask ES:DX = old interrupt service procedure address
Notes	As with mouse INT 33H function 0CH, function 14H also installs a new interrupt handler. The difference is that function 14H replaces the handler already installed with function 0CH.

15H	GET MOUSE STATE SIZE
Entry	AL = 15H
Exit	BX = size of buffer required to store mouse state
Notes	Indicates the amount of memory required to store the state of the mouse with mouse function INT 33H number 16H.

16H	SAVE MOUSE STATE
Entry	AL = 16H ES:DX = address where state of mouse is stored
Exit	Saves the current state of the mouse.

17H	RELOAD MOUSE STATE
Entry	AL = 17H ES:DX = address where state of mouse is stored
Exit	Reloads the saved state of the mouse.

18H	SET ALTERNATE INTERRUPT SUBROUTINE
Entry	AL = 18H CX = alternate interrupt mask ES:DX = address of alternate interrupt service procedure Carry = 0
Exit	Alternate interrupt handler is installed.
Notes	The alternate interrupt handler is accessed after the primary handler. The alternate interrupt mask defines the actions that request the alternate interrupt handler. Following is a list of the actions that cause an alternate interrupt when placed in the alternate interrupt mask: 1 = any change in cursor position 2 = left mouse button pressed 4 = left mouse button released 8 = right mouse button pressed 16 = right mouse button released 32 = shift key with mouse button 64 = control key with mouse button 128 = alternate key with mouse button After the installation, interrupt is called by one of the selected changes. The following registers contain information about the mouse: Carry = 1 AX = interrupt mask (see function 0CH) BX = button status CX = horizontal position DX = vertical position SI = horizontal change DI = vertical change

19H	GET ALTERNATE INTERRUPT ADDRESS
Entry	AL = 19H CX = alternate interrupt mask
Exit	AX = −1 if unsuccessful ES:DX = address of interrupt service procedure
Notes	Used to read the address of the alternate interrupt handler.

1AH	SET MOUSE SENSITIVITY
Entry	AL = 1AH BX = horizontal sensitivity CX = vertical sensitivity DX = double-speed threshold
Exit	Sensitivity is changed.
Notes	The default values for vertical and horizontal sensitivity are both 50. The default value for the double-speed threshold is 64. The values for sensitivity range between 1 and 100, which represent ratios of between 33 to 350 percent.

1BH	GET MOUSE SENSITIVITY
Entry	AL = 1BH
Exit	BX = horizontal sensitivity CX = vertical sensitivity DX = double-speed threshold

1CH	SET INTERRUPT RATE
Entry	AL = 1CH BX = number of interrupts per second
Exit	Changes the interrupt rate for the InPort mouse only.
Notes	The default value is 30 interrupts per second, but this can be changed to any value listed: 0 = interrupt off 1 = 30 interrupts per second 2 = 50 interrupts per second 3 = 100 interrupts per second 4 = 200 interrupts per second

1DH	SET CRT PAGE
Entry	AL = 1DH BX = CRT page number
Exit	Changes to a new CRT page for mouse cursor support.

1EH	GET CRT PAGE
Entry	AL = 1EH
Exit	BX = CRT page

1FH	DISABLE MOUSE DRIVER
Entry	AL = 1FH
Exit	AX = –1 if unsuccessful ES:BX = address of mouse driver

20H	ENABLE MOUSE DRIVER
Entry	AL = 20H
Exit	The mouse driver is enabled.

21H	SOFTWARE RESET
Entry	AL = 21H
Exit	AX = 0FFFFH if successful BX = 2 if successful

22H	SET LANGUAGE
Entry	AL = 22H BX = language number
Exit	Language is changed only with the International version.
Notes	The languages selected by BX are as follows: 0 = English 1 = French 2 = Dutch 3 = German 4 = Swedish 5 = Finnish 6 = Spanish 7 = Portuguese 8 = Italian

23H	GET LANGUAGE
Entry	AL = 23H
Exit	BX = language number

24H	GET DRIVER VERSION
Entry	AL = 24H
Exit	BH = major version number BL = minor version number CH = mouse type CL = interrupt request number

25H	GET DRIVER INFORMATION
Entry	AL = 25H
Exit	AX contains the following driver information in the bits indicated. Bits 13,12 00 = software text cursor active 01 = hardware text cursor active 1X = graphics cursor active Bit 14 0 = non-integrated mouse display driver 1 = integrated mouse display driver Bit 15 0 = driver is a .SYS file 1 = driver is a .COM file

26H	GET MAXIMUM VIRTUAL COORDINATES
Entry	AL = 26H
Exit	BX = mouse driver status CX = maximum horizontal coordinate DX = maximum vertical coordinate

27H	GET CURSOR MASKS AND COUNTS
Entry	AL = 27H
Exit	AX = screen mask or beginning scanning line BX = cursor mask or ending scan line CX = horizontal mickey count DX = vertical mickey count

28H	SET VIDEO MODE
Entry	AL = 28H CX = video mode DX = font size
Exit	CX = 0 if successful

29H	GET SUPPORTED VIDEO MODES
Entry	AL = 29H CX = search flag
Exit	BX:DX = address of ASCII description of video mode CX = mode number or 0 if unsuccessful
Notes	The search flag is 0 to search for the first video mode, and non-zero to search for the next video mode.

2AH	GET CURSOR HOT SPOT
Entry	AL = 2AH
Exit	AX = display flag BX = horizontal hot spot CX = vertical hot spot DX = mouse type
Notes	The hot spot is the position on the cursor that is returned when a mouse button is clicked. AX (display flag) indicates that the cursor is active (0) or inactive (1).

2BH	SET ACCELERATION CURVE
Entry	AL = 2BH BX = curve number ES:SI = address of the acceleration curve
Exit	AX = 0 if successful
Notes	Following are the contents of the acceleration curve table that contains curves 1 through 4: Offset Meaning 00H = curve 1 counts 01H = curve 2 counts 02H = curve 3 counts 03H = curve 4 counts 04H = curve 1 mouse count and threshold 24H = curve 2 mouse count and threshold 44H = curve 3 mouse count and threshold 64H = curve 4 mouse count and threshold 84H = curve 1 scale factor array A4H = curve 2 scale factor array C4H = curve 3 scale factor array E4H = curve 4 scale factor array 104H = curve 1 name 114H = curve 2 name 124H = curve 3 name 134H = curve 4 name

2CH	GET ACCELERATION CURVE
Entry	AL = 2CH BX = current curve ES:SI = address of current acceleration curves
Exit	AX = 0 if successful

2DH	GET ACTIVE ACCELERATION CURVE
Entry	AL = 2DH BX = curve number or −1 for current
Exit	AX = 0 if successful BX = curve number ES:SI = address of acceleration curves

2FH	MOUSE HARDWARE RESET
Entry	AL = 2FH
Exit	AX = 0 if unsuccessful

30H	SET/GET BALLPOINT INFORMATION
Entry	AL = 30H BX = rotation angle (+32K) CX = command (0 for read and 1 for write)
Exit	AX = FFFFH if unsuccessful or button state BX = rotation angle CX = button masks
31H	GET VIRTUAL COORDINATES
Entry	AL = 31H
Exit	AX = minimum horizontal BX = minimum vertical CX = maximum horizontal DX = maximum vertical
32H	GET ACTIVE ADVANCED FUNCTIONS
Entry	AL = 32H
Exit	AX = function flag
Notes	The function flags indicate which INT 33H advanced functions are available. The leftmost bit indicates function 25H through the rightmost bit that indicates function 34H.
33H	GET SWITCH SETTING
Entry	AL = 33H CX = buffer length ES:DI = address of buffer
Exit	AX = 0 CX = byte in buffer ES:DI = address of buffer
34H	GET MOUSE.INI
Entry	AL = 34H
Exit	AX = 0 ES:DX = buffer address

DPMI CONTROL FUNCTIONS

The DPMI (DOS protected mode interface) provides DOS and Windows applications with access to protected mode and also access to real and extended memory. Four functions are accessed through the DOS multiplex interrupt (INT 2FH) and the remaining functions are accessed via INT 31H. The functions accessed by the INT 2FH interrupt release a time slice (AX = 1680H), get the CPU mode (AX = 1686H), get the mode switch entry point (AX = 1687H), and get the API entry point (AX = 168AH). The INT 2FH functions are listed in Table A–14 and the INT 31H functions appear in Table A–15.

TABLE A–14   The INT 2FH, DPMI functions (pp. 855–856).

80H	RELEASE TIME SLICE
Entry	AX = 1680H
Exit	AX = 0000H if successful AX <> 1680H if unsuccessful
Notes	A time slice is a portion of time allowed to an application. If the application is idle, it should release the remaining portion of its time slice to other applications using this function.
86H	**GET CPU MODE**
Entry	AX = 1686H
Exit	AX = 0000H if CPU in protected mode AX <> 0000H if CPU in real or virtual mode
87H	**GET MODE ENTRY POINT**
Entry	AX = 1687H
Exit	AX = 0000H if successful AX <> 0000H if unsuccessful BX = support flag CL = processor type DX = DPMI version number SI = private paragraph count ES:DI = protected mode entry point
Notes	Used to locate the entry point required to switch the CPU into the protected mode using DPMI. BX is a logic 1 if 32-bit support is provided by DPMI. CL indicates the microprocessor (2, 3, 4, or 5 for the 80286, 80386, 80486, or Pentium/Pentium 4). DX = the DPMI version number returned binary value in DH (major) and DL (minor). SI = the number of paragraphs required by DPMI for proper operation. ES:DI = the address used to switch to protected mode.

8AH	GET API ENTRY POINT
Entry	AX = 168AH DS:DI = address of vendor name (ASCII-Z)
Exit	AX = 0000H if successful AX = 168AH if unsuccessful ES:DI = API extensions entry point
Notes	Used to reference a specific vendor's extensions to the DPMI interface.

TABLE A–15   The INT 31H, DPMI functions (pp. 856–875).

0000H	ALLOCATE LDT DESCRIPTOR
Entry	AX = 0000H CX = number of descriptors needed
Exit	AX = base selector, carry = 0 AX = error code, carry = 1
Notes	Allocates one or more local descriptors, with the base or starting descriptor number returned in AX.
0001H	**RELEASE LDT DESCRIPTOR**
Entry	AX = 0001H BX = selector
Exit	Carry = 0 if successful Carry = 1 if unsuccessful, AX = error code
Notes	Releases one descriptor selected by the selector placed in BX on the call to the INT 31H function.
0002H	**MAP REAL SEGMENT TO DESCRIPTOR**
Entry	AX = 0002H BX = segment address
Exit	Carry = 0 if successful, AX = selector Carry = 1 if unsuccessful, AX = error code
Notes	Maps a real mode segment to a protected mode descriptor. This cannot be released and should only be used to map the start of the descriptor table or other global protected mode segments.

0003H	GET SEGMENT INCREMENT
Entry	AX = 0003H
Exit	Carry = 0 if successful, AX = increment
Notes	Used with function AX = 0000H to determine the increment of the selector returned as the base selector.

0006H	GET SEGMENT BASE ADDRESS
Entry	AX = 0006H BX = selector
Exit	Carry = 0 if successful Carry = 1 if unsuccessful, AX = error code CX:DX = base address of segment
Notes	Returns the segment address as selected by BX in CX:DX where CX = the high-order word and DX = the low-order word expressed as a 32-bit linear address.

0007H	SET SEGMENT BASE ADDRESS
Entry	AX = 0007H BX = selector CX:DX = linear base address
Exit	Carry = 0 if successful Carry = 1 if unsuccessful, AX = error code
Notes	Sets the base address to the 32-bit linear address found in CX:DX where CX = high-order word and DX = low-order word.

0008H	SET SEGMENT LIMIT
Entry	AX = 0008H BX = selector CX:DX = segment limit in bytes
Exit	Carry = 0 if successful Carry = 1 if unsuccessful, AX = error code

0009H	SET DESCRIPTOR ACCESS RIGHTS
Entry	AX = 0009H BX = selector CX = access rights
Exit	Carry = 0 if successful Carry = 1 if unsuccessful, AX = error code
Notes	The access rights determines how a segment is accessed in the protected mode. The definition of each bit of CX follows: *Bit* *Rights* 0 Descriptor access (1) 1 Read/write (1) or Read-only (0) 2 Expand segment up (0) or expand segment down (1) 3 Code segment (1) or data segment (0) 4 Must be 1 5, 6 Descriptor desired privilege level 7 Present (1) or not present (0) 8–13 Must be 000000 14 32-bit instruction mode (1) or 16-bit mode (0) 15 Granularity bit is 0 for 1X multiplier and 1 for 4K multiplier for limit field Bits 15-8 must be 0000 0000 for the 80286

000AH	CREATE ALIAS DESCRIPTOR
Entry	AX = 000AH BX = selector
Exit	Carry = 0 if successful, AX = alias selector Carry = 1 if unsuccessful, AX = error code
Notes	Creates a new "carbon copy" as an alias local descriptor.

000BH	GET DESCRIPTOR
Entry	AX = 000BH BX = selector ES:DI = address of buffer
Exit	Carry = 0 if successful Carry = 1 if unsuccessful, AX = error code
Notes	Copies the 8-byte descriptor into the buffer addressed by ES:DI.

000CH	SET DESCRIPTOR
Entry	AX = 000CH BX = selector ES:DI = address of buffer
Exit	Carry = 0 if successful Carry = 1 if unsuccessful, AX = error code
Notes	Copies the 8-byte descriptor from the buffer to the descriptor table.

000DH	ALLOCATE SPECIFIC LDT DESCRIPTOR
Entry	AX = 000DH BX = selector
Exit	Carry = 0 if successful Carry = 1 if unsuccessful, AX = error code
Notes	Creates a descriptor based in the selector of your choice (4–7CH) with as many as 16 assigned by this function.

000EH	GET MULTIPLE DESCRIPTORS
Entry	AX = 000EH BX = number to get ES:DI = buffer address
Exit	Carry = 0 if successful, CX = number copied to buffer Carry = 1 if unsuccessful, AX = error code
Notes	Copies multiple descriptors to a buffer addressed by ES:DI.

000FH	SET MULTIPLE DESCRIPTORS
Entry	AX = 000FH BX = number to set ES:DI = buffer address
Exit	Carry = 0 if successful, number copied from buffer Carry = 1 if unsuccessful, AX = error code
Notes	Creates descriptors from the buffer at the location addressed by ES:DI. Each entry is 10 bytes in length with the first 2 bytes containing the selector number, followed by 8 bytes for descriptor data.

0100H	ALLOCATE DOS MEMORY BLOCK
Entry	AX = 0100H DX = paragraphs to allocate

Exit	Carry = 0 if successful
	    AX = real mode segment
	    DX = selector of descriptor for memory block
	Carry = 1 if unsuccessful
	    AX = error code
	    BX = size of available block

Notes	Allocates a local descriptor for the DOS memory block. Releases with function 0101H only.

0101H	RELEASE DOS MEMORY BLOCK
Entry	AX = 0101H DX = selector
Exit	Carry = 0 if successful Carry = 1 if unsuccessful, AX = error code

0200H	GET REAL MODE INTERRUPT
Entry	AX = 0200H BX = interrupt number
Exit	CX = vector segment DX = vector offset
Notes	Releases one descriptor selected by the selector placed in BX on the call to the INT 31H function.

0201H	SET REAL MODE INTERRUPT VECTOR
Entry	AX = 0201H BX = interrupt number CX = vector segment DX = vector offset

0202H	GET EXCEPTION HANDLER VECTOR
Entry	AX = 0202H BX = exception number 0–1FH

Exit	Carry = 0 if successful
	    CX = handler selector
	    DX = handler offset
	Carry = 1 if unsuccessful
	    AX = error code

0203H	SET EXCEPTION HANDLER VECTOR
Entry	AX = 0203H BX = exception number 0–1FH CX = handler selector DX = handler offset
Exit	Carry = 0 if successful Carry = 1 if unsuccessful, AX = error code

0204H	GET PROTECTED MODE INTERRUPT VECTOR
Entry	AX = 0204H BX = interrupt number
Exit	Carry = 0 if successful CX = handler selector DX = handler offset Carry = 1 if unsuccessful AX = error code

0205H	SET PROTECTED MODE INTERRUPT VECTOR
Entry	AX = 0205H BX = interrupt number CX = handler selector DX = handler offset
Exit	Carry = 0 if successful Carry = 1 if unsuccessful, AX = error code

0210H	GET EXTENDED REAL EXCEPTION HANDLER VECTOR
Entry	AX = 0210H BX = exception number 0–1FH
Exit	Carry = 0 if successful CX = handler selector DX = handler offset Carry = 1 if unsuccessful AX = error code

0211H	GET EXTENDED PROTECTED EXCEPTION HANDLER VECTOR
Entry	AX = 0211H BX = exception number 0–1FH
Exit	Carry = 0 if successful CX = handler selector DX = handler offset Carry = 1 if unsuccessful AX = error code

0212H	SET EXTENDED PROTECTED EXCEPTION HANDLER VECTOR
Entry	AX = 0212H BX = exception number 0–1FH CX = handler selector DX = handler offset
Exit	Carry = 0 if successful Carry = 1 if unsuccessful, AX = error code

0213H	SET EXTENDED REAL EXCEPTION HANDLER VECTOR
Entry	AX = 0213H BX = exception number 0–1FH CX = handler selector DX = handler offset
Exit	Carry = 0 if successful Carry = 1 if unsuccessful, AX = error code

0300H	EMULATE REAL MODE INTERRUPT
Entry	AX = 0300H BX = interrupt number CX = word copy count ES:DI = buffer address
Exit	Carry = 0 if successful, ES:DI = buffer address Carry = 1 if unsuccessful, AX = error code
Notes	Copies the number of words (CX) from the protected mode stack to the real mode stack. The ES:DI register combination addresses a memory buffer that specifies the contents of the register when the switch to the real mode interrupt occurs. The register set is stored as:

Offset	Size	Register
00H	doubleword	EDI
04H	doubleword	ESI
08H	doubleword	EBP
0CH	doubleword	must be 00000000H
10H	doubleword	EBX
14H	doubleword	EDX
18H	doubleword	ECX
1CH	doubleword	EAX
20H	word	flags
22H	word	ES
24H	word	DS
26H	word	FS
28H	word	GS
2AH	word	IP
2CH	word	CS
2EH	word	SP
30H	word	SS

0301H	CALL FAR REAL MODE PROCEDURE
Entry	AX = 0301H BH = 00H CX = word copy count ES:DI = buffer address
Exit	Carry = 0 if successful, ES:DI = buffer address Carry = 1 if unsuccessful, AX = error code
Notes	Calls a real mode procedure from protected mode. The buffer contains the register set as defined for function 0300H, which is loaded on the switch to real mode.

0302H	CALL REAL MODE INTERRUPT PROCEDURE
Entry	AX = 0302H BH = 00H CX = word copy count ES:DI = buffer address
Exit	Carry = 0 if successful, ES:DI = buffer address Carry = 1 if unsuccessful, AX = error code
Notes	Calls the real mode interrupt procedure from protected mode. The buffer contains the register set as defined for function 0300H, which is loaded on the switch to real mode.

0303H	ALLOCATE REAL MODE CALLBACK ADDRESS
Entry	AX = 0303H DS:SI = protected mode procedure address ES:DI = buffer address
Exit	Carry = 0 if successful, CX:DX = callback address Carry = 1 if unsuccessful, AX = error code
Notes	Calls a protected mode procedure from the real mode. The real mode address is found in CX (segment) and DX (offset).

0304H	RELEASE REAL MODE CALLBACK ADDRESS
Entry	AX = 0304H CX:DX = callback address
Exit	Carry = 0 if successful Carry = 1 if unsuccessful, AX = error code
Notes	Releases the callback address allocated by function 0303H.

0305H	GET STATE SAVE AND RESTORE ADDRESS
Entry	AX = 0305H

Exit	Carry = 0 if successful
	    AX = buffer size
	    BX:CX = real mode procedure address
	    SI:DI = protected mode procedure address
	Carry = 1 if unsuccessful, AX = error code

Notes	The procedure addresses returned by this function save the real or protected mode registers in memory before switching modes. If in the real mode, BX:CX contain the far address or a procedure that saves the protected mode registers. Likewise, if operating in the protected mode, save the real mode register with a call to the protected mode address in SI:DI. The value of AL dictates how the procedure called by the procedure address functions. If AL = 0, the registers are saved. If AL = 1, the registers are restored.
0306H	GET RAW CPU MODE SWITCH ADDRESS
Entry	AX = 0306H

Exit	Carry = 0 if successful
	    BX:CX = switch to protected mode address
	    SI:DI = switch to real mode address
	Carry = 1 if unsuccessful, AX = error code

Notes	To switch to protected mode from real mode, use a far JMP to the address found in BX:CX. To switch to real mode from protected mode, use a JMP to the address found in SI:DI.
	
	The register must be pre-loaded as follows before jumping to the switch address:
	
	    AX = new DS
	    BX = new SP
	    CX = new ES
	    DX = new SS
	    SI = new CS
	    DI = new IP

0400H	GET DPMI VERSION
Entry	AX = 0400H
Exit	Carry = 0 if successful AX = DPMI version number BX = implementation flag CL = processor type DH = master interrupt controller base address DL = slave interrupt controller base address Carry = 1 if unsuccessful, AX = error code
Notes	The interrupt controller vector numbers are usually returned as 08H for the master and 70H for the slave.

0401H	GET DPMI CAPABILITIES
Entry	AX = 0401H ES:DI = buffer address
Exit	Carry = 0 if successful, AX = capabilities Carry = 1 if unsuccessful, AX = error code
Notes	This function is supported under version 1.0 of DPMI and uses AX to return the following capabilities: Bit Purpose 6 write-protect host (DPMI) 5 write-protect client (DOS/WINDOWS) 4 demand zero fill 3 conventional memory mapping 2 device mapping 1 exceptions can be restarted 0 page dirty

0500H	GET FREE MEMORY
Entry	AX = 0500H ES:DI = buffer address
Exit	Carry = 0 if successful Carry = 1 if unsuccessful
Notes	Information about the system memory is returned by this function in the buffer addressed by the buffer address. The contents of the buffer are:

Offset	Size	Purpose
00H	doubleword	largest free block in bytes
04H	doubleword	maximum unlocked pages
08H	doubleword	maximum locked pages
0CH	doubleword	linear address in pages
10H	doubleword	total unlocked pages
14H	doubleword	total free pages
18H	doubleword	total physical pages
1CH	doubleword	free linear pages
20H	doubleword	size of paging file in pages
24H	12 bytes	reserved

0501H	ALLOCATE MEMORY BLOCK
Entry	AX = 0501H BX:CX = memory block size in bytes
Exit	Carry = 0 if successful BX:CX = linear base address of block SI:DI = memory block handle Carry = 1 if unsuccessful, AX = error code
Notes	Allocates a block of memory whose size in bytes is in BX:CX on the call to this function. On the return, BX:CX = the starting address of the memory block and SI:DI contains the block handle.

0502H	RELEASE MEMORY BLOCK
Entry	AX = 0502H SI:DI = memory block handle
Exit	Carry = 0 if successful Carry = 1 if unsuccessful, AX = error code

0503H	RESIZE MEMORY BLOCK
Entry	AX = 0503H BX:CX = new memory block size in bytes SI:DI = memory block handle
Exit	Carry = 0 if successful BX:CX = linear base address of block SI:DI = memory block handle Carry = 1 if unsuccessful, AX = error code

0504H	ALLOCATE LINEAR MEMORY BLOCK
Entry	AX = 0504H EBX = desired linear base address ECX = block size in bytes EDX = action code
Exit	Carry = 0 if successful EBX = linear base address of block ESI = memory block handle Carry = 1 if unsuccessful, AX = error code
Notes	Used with a 32-bit DPMI host to allocate a linear memory block. The action code is 0 to create an uncommitted block and 1 to create a committed block.

0505H	RESIZE LINEAR MEMORY BLOCK
Entry	AX = 0505H ES:EBX = buffer address ECX = block size in bytes EDX = action code (see function 0504H) ESI = memory block handle

Exit	Carry = 0 if successful
	    BX:CX = linear base address of block
	    SI:DI = memory block handle
	Carry = 1 if unsuccessful, AX = error code

0506H	GET PAGE ATTRIBUTES
Entry	AX = 0506H EBX = page offset within memory block ECX = page count ES:EDX = buffer address ESI = memory block handle
Exit	Carry = 0 if successful Carry = 1 if unsuccessful, AX = error code
Notes	The memory buffer contains a word of attribute information for each page in the memory block. The attribute word stored in the buffer indicates the following information: *Bit* *Function* 0 0 = uncommitted and 1 = committed 1 1 = mapped 3 0 = read-only and 1 = read/write 4 0 = dirty bit invalid and 1 = dirty bit valid 5 0 = page unaccessed and 1 = page accessed 6 0 = page unmodified and 1 = page modified

0507H	SET PAGE ATTRIBUTES
Entry	AX = 0507H EBX = page offset within memory block ECX = page count ES:EDX = buffer address ESI = memory block handle
Exit	Carry = 0 if successful Carry = 1 if unsuccessful, AX = error code
Notes	As with function 0506H, the memory buffer contains a word for each page that defines the attribute for each page. Following is the meaning of the bits in the page attribute word:

Bit	Function
0	0 = make uncommitted and 1 = make committed
1	1 = modify attributes, but not page type
3	0 = make read-only and 1 = make read/write
4	1 = modify dirty bit
5	1 = mark page accessed
6	1 = mark page modified

0508H	MAP DEVICE IN MEMORY BLOCK
Entry	AX = 0508H EBX = page offset within memory block ECX = page count EDX = device address ESI = memory block handle
Exit	Carry = 0 if successful Carry = 1 if unsuccessful, AX = error code
Notes	Assigns the physical address of a device (EDX) to a linear address in a memory block.

0509H	MAP CONVENTIONAL MEMORY
Entry	AX = 0509H EBX = page offset within memory block ECX = page count EDX = linear address of conventional memory ESI = memory block handle
Exit	Carry = 0 if successful Carry = 1 if unsuccessful, AX = error code
Notes	Allocates a linear (real) address to a memory block.

050AH	GET MEMORY BLOCK SIZE
Entry	AX = 050AH SI:DI = memory block handle
Exit	Carry = 0 if successful SI:DI = memory block size Carry = 1 if unsuccessful, AX = error code

050BH	GET MEMORY INFORMATION
Entry	AX = 050BH DI:SI = buffer address
Exit	Carry = 0 if successful Carry = 1 if unsuccessful, AX = error code
Notes	This function fills a 128-byte buffer addressed by DI:SI with information about the memory. Following is the contents of the buffer:

Offset	Size	Function
00H	doubleword	allocated physical memory
04H	doubleword	allocated virtual memory (host)
08H	doubleword	available virtual memory (host)
0CH	doubleword	allocated virtual memory (machine)
10H	doubleword	available virtual memory (machine)
14H	doubleword	allocated virtual memory (client)
18H	doubleword	available virtual memory (client)
1CH	doubleword	locked memory (client)
20H	doubleword	maximum locked memory (client)
24H	doubleword	maximum linear address (client)
28H	doubleword	maximum free memory block size
2CH	doubleword	minimum allocation unit
30H	doubleword	allocation alignment unit
34H	76 bytes	reserved

0600H	LOCK LINEAR REGION
Entry	AX = 0600H BX:CX = linear address of memory to lock SI:DI = number of bytes to lock
Exit	Carry = 0 if successful Carry = 1 if unsuccessful, AX = error code

0601H	UNLOCK LINEAR REGION
Entry	AX = 0601H BX:CX = linear address of memory to unlock SI:DI = number of bytes to unlock
Exit	Carry = 0 if successful Carry = 1 if unsuccessful, AX = error code

0602H	MARK REAL MODE REGION PAGABLE
Entry	AX = 0602H BX:CX = linear address of memory to mark SI:DI = number of bytes in region
Exit	Carry = 0 if successful Carry = 1 if unsuccessful, AX = error code

0603H	RELOCK REAL MODE REGION
Entry	AX = 0603H BX:CX = linear address to re-lock SI:DI = number of bytes to re-lock
Exit	Carry = 0 if successful Carry = 1 if unsuccessful, AX = error code

0604H	GET PAGE SIZE
Entry	AX = 0604H
Exit	Carry = 0 if successful BX:CX = page size in bytes Carry = 1 if unsuccessful, AX = error code

0702H	MARK PAGE AS DEMAND PAGING CANDIDATE
Entry	AX = 0702H BX:CX = linear address of memory region SI:DI = number of bytes in region
Exit	Carry = 0 if successful Carry = 1 if unsuccessful, AX = error code

0703H	DISCARD PAGE CONTENTS
Entry	AX = 0703H BX:CX = linear address of memory region SI:DI = number of bytes in region
Exit	Carry = 0 if successful Carry = 1 if unsuccessful, AX = error code
Notes	This releases the memory for other uses by DPMI. A discarded page still contains data, but it is undefined.

0800H	PHYSICAL ADDRESS MAPPING
Entry	AX = 0800H BX:CX = base physical address
Exit	Carry = 0 if successful BX:CX = base linear address Carry = 1 if unsuccessful, AX = error code
Notes	Converts a physical address to a linear address.

0801H	RELEASE PHYSICAL ADDRESS MAPPING
Entry	AX = 0801H BX:CX = linear address
Exit	Carry = 0 if successful Carry = 1 if unsuccessful, AX = error code

0900H	GET AND DISABLE VIRTUAL INTERRUPT STATE
Entry	AX = 0900H
Exit	Carry = 0 if successful Carry = 1 if unsuccessful, AX = error code

0901H	GET AND ENABLE VIRTUAL INTERRUPT STATE
Entry	AX = 0901H
Exit	Carry = 0 if successful Carry = 1 if unsuccessful, AX = error code

0902H	GET VIRTUAL INTERRUPT STATE
Entry	AX = 0902H
Exit	Carry = 0 if successful, AL = 0 if disabled Carry = 1 if unsuccessful, AX = error code

0A00H	GET API ENTRY POINT
Entry	AX = 0A00H DS:SI = vendor ASCII offset
Exit	Carry = 0 if successful, AL = 0 if disabled ES:DI = entry point address Carry = 1 if unsuccessful, AX = error code

0B00H	SET DEBUG WATCH-POINT
Entry	AX = 0B00H BX:CX = linear watch-point address DH = watch-point type DL = watch-point size
Exit	Carry = 0 if successful BX = watch-point handle Carry = 1 if unsuccessful, AX = error code
Notes	The BX:CX register provides the watch-point address to the function. The DH register provides the type of watch-point (0 = instruction executed at watch-point address, 1 = memory write to watch-point address, and 2 = a read or write to watch-point address). The DL register holds the size in byte of the watch-point address for types 1 and 2. When the watch-point is triggered, function AX = 0B02H is used to test for the trigger.
0B01H	CLEAR DEBUG WATCH-POINT
Entry	AX = 0B01H BX = watch-point handle
Exit	Carry = 0 if successful Carry = 1 if unsuccessful, AX = error code
Notes	Erase a watch-point assigned with function 0B00H.
0B02H	GET STATE OF DEBUG WATCH-POINT
Entry	AX = 0B02H BX = watch-point handle
Exit	Carry = 0 if successful AX = watch-point status Carry = 1 if unsuccessful, AX = error code
Notes	Used to detect a watch-point trigger. A status of 0 indicates that a watch-point has not been detected, and 1 indicates that it has.
0B03H	RESET DEBUG WATCH-POINT
Entry	AX = 0B03H BX = watch-point handle
Exit	Carry = 0 if successful Carry = 1 if unsuccessful, AX = error code
Notes	Clears the watch-point status, but does not erase the watch-point.

0C00H	INSTALL RESIDENT SERVICE PROVIDER CALLBACK
Entry	AX = 0C00H ES:DI = buffer address
Exit	Carry = 0 if successful Carry = 1 if unsuccessful, AX = error code
Notes	Used for a protected mode TSR to notify the DPMI to call the TSR whenever another machine is in the same virtual memory. The buffer contains the following information for DPMI:

Offset	Size	Function
00H	quadword	16-bit data segment descriptor
08H	quadword	16-bit code segment descriptor
10H	word	16-bit callback procedure offset
12H	word	reserved
14H	quadword	32-bit data segment descriptor
1CH	quadword	32-bit code segment descriptor
24H	doubleword	32-bit callback procedure offset

0C01H	TERMINATE AND STAY RESIDENT
Entry	AX = 0C01H BL = exit code DX = number of paragraphs to reserve
Exit	None

0D00H	ALLOCATE SHARED MEMORY
Entry	AX = 0D00H ES:DI = buffer address
Exit	Carry = 0 if successful Carry = 1 if unsuccessful, AX = error code
Notes	The buffer contains information for DPMI used to allocate shared memory. Its contents follow:

Offset	Size	Function
00H	doubleword	desired block size in bytes
04H	doubleword	actual memory block size in bytes
08H	doubleword	memory block handle
0CH	doubleword	linear address of memory block
10H	doubleword	offset of ASCII-Z string name
14H	word	selector of ASCII-Z string name
16H	word	reserved
18H	doubleword	must be set to 00000000H

0D01H	RELEASE SHARED MEMORY
Entry	AX = 0D01H SI:DI = memory block handle
Exit	Carry = 0 if successful Carry = 1 if unsuccessful, AX = error code

0D02H	SERIALIZE ON SHARED MEMORY
Entry	AX = 0D02H DX = option flag SI:DI = memory block handle
Exit	Carry = 0 if successful Carry = 1 if unsuccessful, AX = error code

0D03H	RELEASE SERIALIZATION ON SHARED MEMORY
Entry	AX = 0D03H DX = option flag SI:DI = memory block handle
Exit	Carry = 0 if successful Carry = 1 if unsuccessful, AX = error code
Notes	The DX register contains the option code as follows: *Bit* *Function* 0 0 = suspend until serialization and 1 = return error code 1 0 = exclusive serialization and 1 = shared serialization

0E00H	GET COPROCESSOR STATUS
Entry	AX = 0E00H
Exit	Carry = 0 if successful AX = status code Carry = 1 if unsuccessful, AX = error code

Notes	The following is the coprocessor status:
	
	Bit *Function*
	0 1 = coprocessor enabled
	1 1 = emulation of coprocessor enabled for client
	2 1 = coprocessor present
	3 1 = emulation of coprocessor enabled for host
	4–7 0000 = no coprocessor
	    0001 = 80287 present
	    0010 = 80387 present
	    0011 = 80487/Pentium present

0E01H	SET COPROCESSOR EMULATION
Entry	AX = 0E01H BX = action code
Exit	Carry = 0 if successful Carry = 1 if unsuccessful, AX = error code
Notes	The contents of register BX follows: *Bit Function* 0 1 = enable coprocessor for client 1 1 = client will supply emulation

APPENDIX B

Instruction Set Summary

The instruction set summary contains a complete alphabetical listing of the entire 8086–Pentium 4 instruction set. The coprocessor and MMX instructions are listed in Chapter 14 and are not repeated in this Appendix. The SIMD instructions are listed after the main instruction set summary.

Each instruction entry lists the mnemonic opcode plus a brief description of the purpose of the instruction. Also listed is the binary machine language coding of each instruction and any other data required to form the instruction, such as the displacement or immediate data. Listed to the right of each binary machine language version of the instruction are the flag bits and any change that might occur for the instruction. The flags are described in the following manner: a blank indicates no effect or change, a ? indicates a change with an unpredictable outcome, a * indicates a change with a predictable outcome, a 1 indicates the flag is set, and a 0 indicates that the flag is cleared. If the flag bits ODITSZAPC are not illustrated with an instruction, the instruction does not modify any of these flags.

Before the instruction listing begins, some information about the bit settings in binary machine language versions of the instructions is presented. Table B–1 lists the modifier bits, coded as oo in the instruction listings.

Table B–2 lists the memory-addressing modes available using a register field coding of mmm. This table applies to all versions of the microprocessor, as long as the operating mode is 16 bits.

Table B–3 lists the register selections provided by the rrr field in an instruction. This table includes the register selections for 8-, 16-, and 32-bit registers.

Table B–4 lists the segment register bit assignment (rrr) found with the MOV, PUSH, and POP instructions.

TABLE B–1 The modifier bits, coded as oo in the instruction listing.

oo	Function
00	If mmm = 110, then a displacement follows the opcode; otherwise, no displacement is used
01	An 8-bit signed displacement follows the opcode
10	A 16-bit signed displacement follows the opcode (unless it is a 32-bit displacement)
11	mmm specifies a register instead of an addressing mode

TABLE B–2 The 16-bit register/memory (mmm) field description.

mmm	Function
000	DS:[BX+SI]
001	DS:[BX+DI]
010	SS:[BP+SI]
011	SS:[BP+DI]
100	DS:[SI]
101	DS:[DI]
110	SS:[BP]
111	DS:[BX]

TABLE B–3 The register field (rrr) assignment.

rrr	W=0	W=1 (16-bit)	W=1 (32-bit)
000	AL	AX	EAX
001	CL	CX	ECX
010	DL	DX	EDX
011	BL	BX	EBX
100	AH	SP	ESP
101	CH	BP	EBP
110	DH	SI	ESI
111	BH	DI	EDI

TABLE B–4 Register field assignments (rrr) for the segment registers.

rrr	Segment Register
000	ES
001	CS
010	SS
011	DS
100	FS
101	GS

When the 80386–Pentium 4 are used, some of the definitions provided in Tables B–1 through B–3 change. See Tables B–5 and B–6 for these changes as they apply to the 80386–Pentium 4 microprocessors.

TABLE B–5 Index register specified with rrr for the advanced addressing mode found in the 80386–Pentium 4 microprocessors.

rrr	Index Register
000	DS:[EAX]
001	DS:[ECX]
010	DS:[EDX]
011	DS:[EBX]
100	No index (see Table B–6)
101	SS:[EBP]
110	DS:[ESI]
111	DS:[EDI]

TABLE B–6   Possible combinations of oo, mmm, and rrr for the 80386–Pentium 4 microprocessors using 32-bit addressing.

oo	mmm	rrr (base in scaled index byte)	Addressing Mode
00	000	—	DS:[EAX]
00	001	—	DS:[ECX]
00	010	—	DS:[EDX]
00	011	—	DS:[EBX]
00	100	000	DS:[EAX+scaled index]
00	100	001	DS:[ECX+scaled index]
00	100	010	DS:[EDX+scaled index]
00	100	011	DS:[EBX+scaled index]
00	100	100	SS:[ESP+scaled index]
00	100	101	DS:[disp32+scaled index]
00	100	110	DS:[ESI+scaled index]
00	100	111	DS:[EDI+scaled index]
00	101	—	DS:disp32
00	110	—	DS:[ESI]
00	111	—	DS:[EDI]
01	000	—	DS:[EAX+disp8]
01	001	—	DS:[ECX+disp8]
01	010	—	DS:[EDX+disp8]
01	011	—	DS:[EBX+disp8]
01	100	000	DS:[EAX+scaled index+disp8]
01	100	001	DS:[ECX+scaled index+disp8]
01	100	010	DS:[EDX+scaled index+disp8]
01	100	011	DS:[EBX+scaled index+disp8]
01	100	100	SS:[ESP+scaled index+disp8]
01	100	101	SS:[EBP+scaled index+disp8]
01	100	110	DS:[ESI+scaled index+disp8]
01	100	111	DS:[EDI+scaled index+disp8]
01	101	—	SS:[EBP+disp8]
01	110	—	DS:[ESI+disp8]
01	111	—	DS:[EDI+disp8]
10	000	—	DS:[EAX+disp32]
10	001	—	DS:[ECX+disp32]
10	010	—	DS:[EDX+disp32]
10	011	—	DS:[EBX+disp32]
10	100	000	DS:[EAX+scaled index+disp32]
10	100	001	DS:[ECX+scaled index+disp32]
10	100	010	DS:[EDX+scaled index+disp32]
10	100	011	DS:[EBX+scaled index+disp32]
10	100	100	SS:[ESP+scaled index+disp32]
10	100	101	SS:[EBP+scaled index+disp32]
10	100	110	DS:[ESI+scaled index+disp32]
10	100	111	DS:[EDI+scaled index+disp32]
10	101	—	SS:[EBP+disp32]
10	110	—	DS:[ESI+disp32]
10	111	—	DS:[EDI+disp32]

Notes: disp8 = 8-bit displacement and disp32 = 32-bit displacement.

In order to use the scaled index addressing modes listed in Table B–6, code oo and mmm in the second byte of the opcode. The scaled index byte is usually the third byte and contains three fields. The leftmost two bits determine the scaling factor (00 = X1, 01 = X2, 10 = X4, or 11 = X8). The next three bits toward the right contain the scaled index register number (this is obtained from Table B–5). The rightmost three bits are from the rrr field listed in Table B–6. For example, the MOV AL,[EBX+2*ECX] instruction has a scaled index byte of 01 001 011, where 01 = X2, the 001 = ECX, and 011 = EBX.

Some instructions are prefixed to change the default segment or to override the instruction mode. Table B–7 lists the segment and instruction mode override prefixes with append at the beginning of an instruction if they are used to form the instruction. For example, the MOV AL,ES:[BX] instruction used the extra segment because of the override prefix ES:.

In the 8086 and 8088 microprocessors, the effective address calculation required additional clocks that are added to the times in the instruction set summary. These additional times are listed in Table B–8. No such times are added to the 80286–Pentium 4. Note that the instruction set summary does not include clock times for the Pentium Pro through the Pentium 4. Intel has not released these times and has decided that the RDTSC instruction can be used to have the microprocessor count the number of clocks required for a given application. Even though the timings do not appear for these new microprocessors, they are very similar to the Pentium, which can be used as a guide.

TABLE B–7   Override prefixes.

Prefix Byte	Purpose
26H	ES: segment override prefix
2EH	CS: segment override prefix
36H	SS: segment override prefix
3EH	DS: segment override prefix
64H	FS: segment override prefix
65H	GS: segment override prefix
66H	Operand size instruction mode override
67H	Register size instruction mode override

TABLE B–8   Effective address calculations for the 8086 and 8088 microprocessors.

Type	Clocks	Example Instruction
Base or index	5	MOV CL,[DI]
Displacement	3	MOV AL,DATA1
Base plus index	7	MOV AL,[BP+SI]
Displacement plus base or index	9	MOV DH,[DI+20H]
Base plus index plus displacement	11	MOV CL,[BX+DI+2]
Segment override	ea + 2	MOV AL,ED:[DI]

INSTRUCTION SET SUMMARY   (pp. 880–963)

AAA	ASCII adjust AL after addition		

00110111		O D I T S Z A P C	
		? ? ? * ? *	
Example		Microprocessor	Clocks
AAA		8086	8
		8088	8
		80286	3
		80386	4
		80486	3
		Pentium	3

AAD	ASCII adjust AX before division		

11010101 00001010		O D I T S Z A P C	
		? * * ? * ?	
Example		Microprocessor	Clocks
AAD		8086	60
		8088	60
		80286	14
		80386	19
		80486	14
		Pentium	10

AAM	ASCII adjust AX after multiplication		

11010100 00001010		O D I T S Z A P C	
		? * * ? * ?	
Example		Microprocessor	Clocks
AAM		8086	83
		8088	83
		80286	16
		80386	17
		80486	15
		Pentium	18

AAS	ASCII adjust AL after subtraction		

00111111		O D I T	S Z A P C
		?	? ? * ? *
Example		Microprocessor	Clocks
AAS		8086	8
		8088	8
		80286	3
		80386	4
		80486	3
		Pentium	3

ADC	Addition with carry		

000100dw oorrrmmm disp		O D I T	S Z A P C
		*	* * * * *
Format	Examples	Microprocessor	Clocks
ADC reg,reg	ADC AX,BX	8086	3
	ADC AL,BL	8088	3
	ADC EAX,EBX	80286	3
	ADC CX,SI	80386	3
	ADC ESI,EDI	80486	1
		Pentium	1 or 3
ADC mem,reg	ADC DATAY,AL	8086	16 + ea
	ADC LIST,SI	8088	24 + ea
	ADC DATA2[DI],CL	80286	7
	ADC [EAX],BL	80386	7
	ADC [EBX+2*ECX],EDX	80486	3
		Pentium	1 or 3

ADC reg,mem	ADC BL,DATA1 ADC SI,LIST1 ADC CL,DATA2[SI] ADC CX,[ESI] ADC ESI,[2*ECX]	8086	9 + ea
		8088	13 + ea
		80286	7
		80386	6
		80486	2
		Pentium	1 or 2

| 100000sw oo010mmm disp data | | | |
Format	Examples	Microprocessor	Clocks
ADC reg,imm	ADC CX,3 ADC DI,1AH ADC DL,34H ADC EAX,12345 ADC CX,1234H	8086	4
		8088	4
		80286	3
		80386	2
		80486	1
		Pentium	1 or 3
ADC mem,imm	ADC DATA4,33 ADC LIST,'A' ADC DATA3[DI],2 ADC BYTE PTR[EBX],3 ADC WORD PTR[DI],669H	8086	17 + ea
		8088	23 + ea
		80286	7
		80386	7
		80486	3
		Pentium	1 or 3
ADC acc,imm	ADC AX,3 ADC AL,1AH ADC AH,34H ADC EAX,2 ADC AL,'Z'	8086	4
		8088	4
		80286	3
		80386	2
		80486	1
		Pentium	1

ADD	Addition		

000000dw oorrrmmm disp		O D I T S Z A P C	
		* * * * * *	
Format	**Examples**	**Microprocessor**	**Clocks**
ADD reg,reg	ADD AX,BX	8086	3
	ADD AL,BL		
	ADD EAX,EBX	8088	3
	ADD CX,SI	80286	2
	ADD ESI,EDI		
		80386	2
		80486	1
		Pentium	1 or 3
ADD mem,reg	ADD DATAY,AL	8086	16 + ea
	ADD LIST,SI		
	ADD DATA6[DI],CL	8088	24 + ea
	ADD [EAX],CL	80286	7
	ADD [EDX+4*ECX],EBX	80386	7
		80486	3
		Pentium	1 or 3
ADD reg,mem	ADD BL,DATA2	8086	9 + ea
	ADD SI,LIST3		
	ADD CL,DATA2[DI]	8088	13 + ea
	ADD CX,[EDI]	80286	7
	ADD ESI,[ECX+200H]	80386	6
		80486	2
		Pentium	1 or 2

100000sw oo000mmm disp data			
Format	**Examples**	**Microprocessor**	**Clocks**
ADD reg,imm	ADD CX,3	8086	4
	ADD DI,1AH		
	ADD DL,34H	8088	4
	ADD EDX,1345H	80286	3
	ADD CX,1834H	80386	2
		80486	1
		Pentium	1 or 3

ADD mem,imm	ADD DATA4,33
	ADD LIST,'A'
	ADD DATA3[DI],2
	ADD BYTE PTR[EBX],3
	ADD WORD PTR[DI],669H
		8086	17 + ea

		8088	23 + ea
		80286	7
		80386	7
		80486	3
		Pentium	1 or 3

ADD acc,imm	ADD AX,3
	ADD AL,1AH
	ADD AH,34H
	ADD EAX,2
	ADD AL,'Z'
		8086	4

		8088	4
		80286	3
		80386	2
		80486	1
		Pentium	1

AND   Logical AND

| 001000dw oorrrmmm disp | | O D I T   S Z A P C
0   * * ? * 0 | |
Format	Examples	Microprocessor	Clocks
AND reg,reg	AND CX,BX AND DL,BL AND ECX,EBX AND BP,SI AND EDX,EDI	8086	3
		8088	3
		80286	2
		80386	2
		80486	1
		Pentium	1 or 3
AND mem,reg	AND BIT,AL AND LIST,DI AND DATAZ[BX],CL AND [EAX],BL AND [ESI+4*ECX],EDX	8086	16 + ea
		8088	24 + ea
		80286	7
		80386	7
		80486	3
		Pentium	1 or 3

AND reg,mem	AND BL,DATAW	8086	9 + ea
	AND SI,LIST	8088	13 + ea
	AND CL,DATAQ[SI]		
	AND CX,[EAX]	80286	7
	AND ESI,[EOX+43H]	80386	6
		80486	2
		Pentium	1 or 2

100000sw oo100mmm disp data			
Format	Examples	Microprocessor	Clocks
AND reg,imm	AND BP,1	8086	4
	AND DI,10H	8088	4
	AND DL,34H		
	AND EBP,1345H	80286	3
	AND SP,1834H	80386	2
		80486	1
		Pentium	1 or 3
AND mem,imm	AND DATA4,33	8086	17 + ea
	AND LIST,'A'	8088	23 + ea
	AND DATA3[DI],2		
	AND BYTE PTR[EBX],3	80286	7
	AND DWORD PTR[DI],66H	80386	7
		80486	3
		Pentium	1 or 3
AND acc,imm	AND AX,3	8086	4
	AND AL,1AH	8088	4
	AND AH,34H		
	AND EAX,2	80286	3
	AND AL,'r'	80386	2
		80486	1
		Pentium	1

ARPL	Adjust requested privilege level			

| 01100011 oorrrmmm disp | | O D I T   S Z A P C | | |
| | | | | * |
Format	Examples	Microprocessor	Clocks	
ARPL reg,reg	ARPL AX,BX	8086	—	
	ARPL BX,SI			
	ARPL AX,DX	8088	—	
	ARPL BX,AX	80286	10	
	ARPL SI,DI			
		80386	20	
		80486	9	
		Pentium	7	
ARPL mem,reg	ARPL DATAY,AX	8086	—	
	ARPL LIST,DI			
	ARPL DATA3[DI],CX	8088	—	
	ARPL [EBX],AX	80286	11	
	ARPL [EDX+4*ECX],BP			
		80386	21	
		80486	9	
		Pentium	7	

BOUND	Check array against boundary		

| 01100010 oorrrmmm disp | | | |
Format	Examples	Microprocessor	Clocks
BOUND reg,mem	BOUND AX,BETS	8086	—
	BOUND BP,LISTG		
	BOUND CX,DATAX	8088	—
	BOUND BX,[DI]	80286	13
	BOUND SI,[BX+2]		
		80386	10
		80486	7
		Pentium	8

BSF	Bit scan forward		

00001111 10111100 oorrrmmm disp		O D I T S Z A P C ? ? * ? ? ?	
Format	Examples	Microprocessor	Clocks
BSF reg,reg	BSF AX,BX BSF BX,SI BSF EAX,EDX BSF EBX,EAX BSF SI,DI	8086	—
		8088	—
		80286	—
		80386	10 + 3n
		80486	6–42
		Pentium	6–42
BSF reg,mem	BSF AX,DATAY BSF SI,LIST BSF CX,DATA3[DI] BSF EAX,[EBX] BSF EBP,[EDX+4*ECX]	8086	—
		8088	—
		80286	—
		80386	10 + 3n
		80486	7–43
		Pentium	6–43

BSR	Bit scan reverse		

00001111 10111101 oorrrmmm disp		O D I T S Z A P C ? ? * ? ? ?	
Format	Examples	Microprocessor	Clocks
BSR reg,reg	BSR AX,BX BSR BX,SI BSR EAX,EDX BSR EBX,EAX BSR SI,DI	8086	—
		8088	—
		80286	—
		80386	10 + 3n
		80486	6–103
		Pentium	7–71

BSR reg,mem	BSR AX,DATAY	8086	—
	BSR SI,LIST		
	BSR CX,DATA3[DI]	8088	—
	BSR EAX,[EBX]	80286	—
	BSR EBP,[EDX+4*ECX]		
		80386	10 + 3n
		80486	7–104
		Pentium	7–72

BSWAP   Byte swap

00001111 11001rrr

Format	Examples	Microprocessor	Clocks
BSWAP reg32	BSWAP EAX	8086	—
	BSWAP EBX		
	BSWAP EDX	8088	—
	BSWAP ECX	80286	—
	BSWAP ESI	80386	—
		80486	1
		Pentium	1

BT   Bit test

00001111 10111010 oo100mmm disp data

O D I T   S Z A P C
                   *

Format	Examples	Microprocessor	Clocks
BT reg,imm8	BT AX,2	8086	—
	BT CX,4		
	BT BP,10H	8088	—
	BT CX,8	80286	—
	BT BX,2	80386	3
		80486	3
		Pentium	4

BT mem,imm8	BT DATA1,2 BT LIST,2 BT DATA2[DI],3 BT [EAX],1 BT FROG,6	8086	—
		8088	—
		80286	—
		80386	6
		80486	0
		Pentium	4

00001111 10100011 disp

Format	Examples	Microprocessor	Clocks
BT reg,reg	BT AX,CX BT CX,DX BT BP,AX BT SI,CX BT EAX,EBX	8086	—
		8088	—
		80286	—
		80386	3
		80486	3
		Pentium	4 or 9
BT mem,reg	BT DATA4,AX BT LIST,BX BT DATA3[DI],CX BT [EBX],DX BT [DI],DI	8086	—
		8088	—
		80286	—
		80386	12
		80486	8
		Pentium	4 or 9

BTC Bit test and complement

00001111 10111010 oo111mmm disp data		O D I T S Z A P C *	
Format	Examples	Microprocessor	Clocks
BTC reg,imm8	BTC AX,2 BTC CX,4 BTC BP,10H BTC CX,8 BTC BX,2	8086	—
		8088	—
		80286	—
		80386	6
		80486	6
		Pentium	7 or 8
BTC mem,imm8	BTC DATA1,2 BTC LIST,2 BTC DATA2[DI],3 BTC [EAX],1 BTC FROG,6	8086	—
		8088	—
		80286	—
		80386	7 or 8
		80486	8
		Pentium	8

00001111 10111011 disp			
Format	Examples	Microprocessor	Clocks
BTC reg,reg	BTC AX,CX BTC CX,DX BTC BP,AX BTC SI,CX BTC EAX,EBX	8086	—
		8088	—
		80286	—
		80386	6
		80486	6
		Pentium	7 or 13
BTC mem,reg	BTC DATA4,AX BTC LIST,BX BTC DATA3[DI],CX BTC [EBX],DX BTC [DI],DI	8086	—
		8088	—
		80286	—
		80386	13
		80486	13
		Pentium	7 or 13

BTR — Bit test and reset

00001111 10111010 oo110mmm disp data		O D I T S Z A P C *	
Format	**Examples**	**Microprocessor**	**Clocks**
BTR reg,imm8	BTR AX,2	8086	—
	BTR CX,4	8088	—
	BTR BP,10H	80286	—
	BTR CX,8	80386	6
	BTR BX,2	80486	6
		Pentium	7 or 8
BTR mem,imm8	BTR DATA1,2	8086	—
	BTR LIST,2	8088	—
	BTR DATA2[DI],3	80286	—
	BTR [EAX],1	80386	8
	BTR FROG,6	80486	8
		Pentium	7 or 8

00001111 10110011 disp			
Format	**Examples**	**Microprocessor**	**Clocks**
BTR reg,reg	BTR AX,CX	8086	—
	BTR CX,DX	8088	—
	BTR BP,AX	80286	—
	BTR SI,CX	80386	6
	BTR EAX,EBX	80486	6
		Pentium	7 or 13
BTR mem,reg	BTR DATA4,AX	8086	—
	BTR LIST,BX	8088	—
	BTR DATA3[DI],CX	80286	—
	BTR [EBX],DX	80386	13
	BTR [DI],DI	80486	13
	BTC [DI],DI	Pentium	7 or 13

BTS	Bit test and set		

00001111 10111010 oo101mmm disp data		O D I T S Z A P C *	
Format	Examples	Microprocessor	Clocks
BTS reg,imm8	BTS AX,2 BTS CX,4 BTS BP,10H BTS CX,8 BTS BX,2	8086	—
		8088	—
		80286	—
		80386	6
		80486	6
		Pentium	7 or 8
BTS mem,imm8	BTS DATA1,2 BTS LIST,2 BTS DATA2[DI],3 BTS [EAX],1 BTS FROG,6	8086	—
		8088	—
		80286	—
		80386	8
		80486	8
		Pentium	7 or 8

00001111 10101011 disp			
Format	Examples	Microprocessor	Clocks
BTS reg,reg	BTS AX,CX BTS CX,DX BTS BP,AX BTS SI,CX BTS EAX,EBX	8086	—
		8088	—
		80286	—
		80386	6
		80486	6
		Pentium	7 or 13
BTS mem,reg	BTS DATA4,AX BTS LIST,BX BTS DATA3[DI],CX BTS [EBX],DX BTS [DI],DI	8086	—
		8088	—
		80286	—
		80386	13
		80486	13
		Pentium	7 or 13

CALL — Call procedure (subroutine)

11101000 disp

Format	Examples	Microprocessor	Clocks
CALL label (near)	CALL FOR_FUN CALL HOME CALL ET CALL WAITING CALL SOMEONE	8086	19
		8088	23
		80286	7
		80386	3
		80486	3
		Pentium	1

10011010 disp

Format	Examples	Microprocessor	Clocks
CALL label (far)	CALL FAR PTR DATES CALL WHAT CALL WHERE CALL FARCE CALL WHOM	8086	28
		8088	36
		80286	13
		80386	17
		80486	18
		Pentium	4

11111111 oo010mmm

Format	Examples	Microprocessor	Clocks
CALL reg (near)	CALL AX CALL BX CALL CX CALL DI CALL SI	8086	16
		8088	20
		80286	7
		80386	7
		80486	5
		Pentium	2

CALL mem (near)	CALL ADDRESS CALL NEAR PTR [DI] CALL DATA1 CALL FROG CALL ME_NOW	8086	21 + ea
		8088	29 + ea
		80286	11
		80386	10
		80486	5
		Pentium	2

11111111 oo011mmm

Format	Examples	Microprocessor	Clocks
CALL mem (far)	CALL FAR_LIST[SI] CALL FROM_HERE CALL TO_THERE CALL SIXX CALL OCT	8086	16
		8088	20
		80286	7
		80386	7
		80486	5
		Pentium	2

CBW   Convert byte to word (AL $\Rightarrow$ AX)

10011000

Example		Microprocessor	Clocks
CBW		8086	2
		8088	2
		80286	2
		80386	3
		80486	3
		Pentium	3

CDQ	Convert doubleword to quadword (EAX $\Rightarrow$ EDX:EAX)		
11010100 00001010 Example		Microprocessor	Clocks
CDQ		8086	—
		8088	—
		80286	—
		80386	2
		80486	2
		Pentium	2

CLC	Clear carry flag		
11111000		O D I T S Z A P C 0	
Example		Microprocessor	Clocks
CLC		8086	2
		8088	2
		80286	2
		80386	2
		80486	2
		Pentium	2

CLD	Clear direction flag		
11111100		O D I T S Z A P C 0	
Example		Microprocessor	Clocks
CLD		8086	2
		8088	2
		80286	2
		80386	2
		80486	2
		Pentium	2

CLI	Clear interrupt flag		
11111010		O D I T S Z A P C	
		0	
Example		Microprocessor	Clocks
CLI		8086	2
		8088	2
		80286	3
		80386	3
		80486	5
		Pentium	7

CLTS	Clear task switched flag (CR0)		
00001111 00000110			
Example		Microprocessor	Clocks
CLTS		8086	—
		8088	—
		80286	2
		80386	5
		80486	7
		Pentium	10

CMC	Complement carry flag		
10011000		O D I T S Z A P C	
			*
Example		Microprocessor	Clocks
CMC		8086	2
		8088	2
		80286	2
		80386	2
		80486	2
		Pentium	2

CMOVcondition   Conditional move

00001111 0100cccc oorrrmmm

Format	Examples	Microprocessor	Clocks
CMOVcc reg,mem	CMOVNZ AX,FROG	8086	—
	CMOVC EAX,[EDI]	0000	—
	CMOVNC BX,DATA1		
	CMOVP EBX,WAITING	80286	—
	CMOVNE DI,[SI]	80386	—
		80486	—
		Pentium	—

Condition Codes	Mnemonic	Flag	Description
0000	CMOVO	O = 1	Move if overflow
0001	CMOVNO	O = 0	Move if no overflow
0010	CMOVB	C = 1	Move if below
0011	CMOVAE	C = 0	Move if above or equal
0100	CMOVE	Z = 1	Move if equal/zero
0101	CMOVNE	Z = 0	Move if not equal/zero
0110	CMOVBE	C = 1 + Z = 1	Move if below or equal
0111	CMOVA	C = 0 • Z = 0	Move if above
1000	CMOVS	S = 1	Move if sign
1001	CMOVNS	S = 0	Move if no sign
1010	CMOVP	P = 1	Move if parity
1011	CMOVNP	P = 0	Move if no parity
1100	CMOVL	S • O	Move if less than
1101	CMOVGE	S = 0	Move if greater than or equal
1110	CMOVLE	Z = 1 + S • O	Move if less than or equal
1111	CMOVG	Z = 0 + S = O	Move if greater than

CMP   Compare

001110dw oorrrmmm disp

	O D I T S Z A P C
	* * * * * *

Format	Examples	Microprocessor	Clocks
CMP reg,reg	CMP AX,BX	8086	3
	CMP AL,BL	8088	3
	CMP EAX,EBX	80286	2
	CMP CX,SI	80386	2
	CMP ESI,EDI	80486	1
		Pentium	1 or 2

CMP mem,reg	CMP DATAY,AL
	CMP LIST,SI
	CMP DATA6[DI],CL
	CMP [EAX],CL
	CMP [EDX+4*ECX],EBX
		8086	9 + ea

		8088	13 + ea
		80286	7
		80386	5
		80486	2
		Pentium	1 or 2

CMP reg,mem	CMP BL,DATA2
	CMP SI,LIST3
	CMP CL,DATA2[DI]
	CMP CX,[EDI]
	CMP ESI,[ECX+200H]
		8086	9 + ea

		8088	13 + ea
		80286	6
		80386	6
		80486	2
		Pentium	1 or 2

100000sw oo111mmm disp data

Format	Examples	Microprocessor	Clocks

CMP reg,imm	CMP CX,3
	CMP DI,1AH
	CMP DL,34H
	CMP EDX,1345H
	CMP CX,1834H
		8086	4

		8088	4
		80286	3
		80386	2
		80486	1
		Pentium	1 or 2

CMP mem,imm	CMP DATAS,3
	CMP BYTE PTR[EDI],1AH
	CMP DADDY,34H
	CMP LIST,'A'
	CMP TOAD,1834H
		8086	10 + ea

		8088	14 + ea
		80286	6
		80386	5
		80486	2
		Pentium	1 or 2

0001111w data Format	Examples	Microprocessor	Clocks
CMP acc,imm	CMP AX,3 CMP AL,1AH CMP AH,34H CMP EAX,1345H CMP AL,'Y'	8086	4
		8088	4
		80286	3
		80386	2
		80486	1
		Pentium	1

CMPS   Compare strings

1010011w		O D I T	S Z A P C
		*	* * * * *
Format	Examples	Microprocessor	Clocks
CMPSB CMPSW CMPSD	CMPSB CMPSW CMPSD CMPSB DATA1,DATA2 REPE CMPSB REPNE CMPSW	8086	32
		8088	30
		80286	8
		80386	10
		80486	8
		Pentium	5

CMPXCHG   Compare and exchange

00001111 1011000w 11rrrrrr		O D I T	S Z A P C
		*	* * * * *
Format	Examples	Microprocessor	Clocks
CMPXCHG reg,reg	CMPXCHG EAX,EBX CMPXCHG ECX,EDX	8086	—
		8088	—
		80286	—
		80386	—
		80486	6
		Pentium	6

0001111w data			
Format	Examples	Microprocessor	Clocks
CMPXCHG mem,reg	CMPXCHG DATAD,EAX CMPXCHG DATA2,EDI	8086	—
		8088	—
		80286	—
		80386	—
		80486	7
		Pentium	6

CMPXCHG8B   Compare and exchange 8 bytes

00001111 11000111 oorrrmmm		O D I T S Z A P C *	
Format	Examples	Microprocessor	Clocks
CMPXCHG8B mem64	CMPXCHG8B DATA3	8086	—
		8088	—
		80286	—
		80386	—
		80486	—
		Pentium	10

CPUID   CPU identification code

00001111 10100010			
Example		Microprocessor	Clocks
CPUID		8086	—
		8088	—
		80286	—
		80386	—
		80486	—
		Pentium	14

CWD — Convert word to doubleword (AX ⇒ DX:AX)

10011000
Example

CWD	Microprocessor	Clocks
	8086	5
	8088	5
	80286	2
	80386	2
	80486	3
	Pentium	2

CWDE — Convert word to extended doubleword (AX ⇒ EAX)

10011000
Example

CWDE	Microprocessor	Clocks
	8086	—
	8088	—
	80286	—
	80386	3
	80486	3
	Pentium	3

DAA — Decimal adjust AL after addition

00100111

O	D	I	T		S	Z	A	P	C
?					*	*	*	*	*

Example

DAA	Microprocessor	Clocks
	8086	4
	8088	4
	80286	3
	80386	4
	80486	2
	Pentium	3

DAS	Decimal adjust AL after subtraction		

| 00101111 | | O D I T S Z A P C | |
| | | ? * * * * * | |
Example		Microprocessor	Clocks
DAS		8086	4
		8088	4
		80286	3
		80386	4
		80486	2
		Pentium	3

DEC	Decrement		

| 1111111w oo001mmm disp | | O D I T S Z A P C | |
| | | * * * * * | |
Format	Examples	Microprocessor	Clocks
DEC reg8	DEC BL	8086	3
	DEC BH	8088	3
	DEC CL	80286	2
	DEC DH	80386	2
	DEC AH	80486	1
		Pentium	1 or 3
DEC mem	DEC DATAY	8086	15 + ea
	DEC LIST	8088	23 + ea
	DEC DATA6[DI]	80286	7
	DEC BYTE PTR [BX]	80386	6
	DEC WORD PTR [EBX]	80486	3
		Pentium	1 or 3

01001rrr Format	Examples	Microprocessor	Clocks
DEC reg16 DEC reg32	DEC CX DEC DI DEC EDX DEC ECX DEC BP	8086	3
		8088	3
		80286	2
		80386	2
		80486	1
		Pentium	1

DIV   Divide

1111011w oo110mmm disp		O D I T S Z A P C ? ? ? ? ? ?	
Format	Examples	Microprocessor	Clocks
DIV reg	DIV BL DIV BH DIV ECX DIV DH DIV CX	8086	162
		8088	162
		80286	22
		80386	38
		80486	40
		Pentium	17–41
DIV mem	DIV DATAY DIV LIST DIV DATA6[DI] DIV BYTE PTR [BX] DIV WORD PTR [EBX]	8086	168
		8088	176
		80286	25
		80386	41
		80486	40
		Pentium	17–41

ENTER   Create a stack frame

11001000 data

Format	Examples	Microprocessor	Clocks
ENTER imm,0	ENTER 4,0 ENTER 8,0 ENTER 100,0 ENTER 200,0 ENTER 1024,0	8086	—
		8088	—
		80286	11
		80386	10
		80486	14
		Pentium	11
ENTER imm,1	ENTER 4,1 ENTER 10,1	8086	—
		8088	—
		80286	12
		80386	15
		80486	17
		Pentium	15
ENTER imm,imm	ENTER 3,6 ENTER 100,3	8086	—
		8088	—
		80286	12
		80386	15
		80486	17
		Pentium	15 + 2n

ESC   Escape (obsolete–see coprocessor)

HLT	Halt		
11110100			
Example		Microprocessor	Clocks
HLT		8086	2
		0088	2
		80286	2
		80386	5
		80486	4
		Pentium	varies

IDIV	Integer (signed) division		

1111011w oo111mmm disp

		O D I T S Z A P C
		? ? ? ? ? ?

Format	Examples	Microprocessor	Clocks
IDIV reg	IDIV BL IDIV BH IDIV ECX IDIV DH IDIV CX	8086	184
		8088	184
		80286	25
		80386	43
		80486	43
		Pentium	22–46
IDIV mem	IDIV DATAY IDIV LIST IDIV DATA6[DI] IDIV BYTE PTR [BX] IDIV WORD PTR [EBX]	8086	190
		8088	194
		80286	28
		80386	46
		80486	44
		Pentium	22–46

IMUL	Integer (signed) multiplication		

| 1111011w oo101mmm disp | | O D I T   S Z A P C | |
| | | *   ? ? ? ? * | |
Format	Examples	Microprocessor	Clocks
IMUL reg	IMUL BL	8086	154
	IMUL CX	8088	154
	IMUL ECX		
	IMUL DH	80286	21
	IMUL AL	80386	38
		80486	42
		Pentium	10–11
IMUL mem	IMUL DATAY	8086	160
	IMUL LIST	8088	164
	IMUL DATA6[DI]		
	IMUL BYTE PTR [BX]	80286	24
	IMUL WORD PTR [EBX]	80386	41
		80486	42
		Pentium	10–11

| 011010s1 oorrmmm disp data | | | |
Format	Examples	Microprocessor	Clocks
IMUL reg,imm	IMUL CX,16	8086	—
	IMUL DI,100	8088	—
	IMUL EDX,20		
		80286	21
		80386	38
		80486	42
		Pentium	10
IMUL reg,reg,imm	IMUL DX,AX,2	8086	—
	IMUL CX,DX,3	8088	—
	IMUL BX,AX,33		
		80286	21
		80386	38
		80486	42
		Pentium	10

IMUL reg,mem,imm	IMUL CX,DATAY,99	8086	—
		8088	—
		80286	24
		80386	38
		80486	42
		Pentium	10

00001111 10101111 oorrmmm disp			
Format	Examples	Microprocessor	Clocks
IMUL reg,reg	IMUL CX,DX	8086	—
	IMUL DI,BX		
	IMUL EDX,EBX	8088	—
		80286	—
		80386	38
		80486	42
		Pentium	10
IMUL reg,mem	IMUL DX,DATAY	8086	—
	IMUL CX,LIST		
	IMUL ECX,DATA6[DI]	8088	—
		80286	—
		80386	41
		80486	42
		Pentium	10

IN Input data from port

1110010w port#			
Format	Examples	Microprocessor	Clocks
IN acc,pt	IN AL,12H	8086	10
	IN AX,12H		
	IN AL,0FFH	8088	14
	IN AX,0A0H	80286	5
	IN EAX,10H		
		80386	12
		80486	14
		Pentium	7

1110110w Format	Examples	Microprocessor	Clocks
IN acc,DX	IN AL,DX	8086	8
	IN AX,DX	8088	12
	IN EAX,DX	80286	5
		80386	13
		80486	14
		Pentium	7

INC   Increment

1111111w oo000mmm disp		O D I T S Z A P C * * * * *	
Format	Examples	Microprocessor	Clocks
INC reg8	INC BL	8086	3
	INC BH	8088	3
	INC AL	80286	2
	INC AH	80386	2
	INC DH	80486	1
		Pentium	1 or 3
INC mem	INC DATA3	8086	15 + ea
	INC LIST	8088	23 + ea
	INC COUNT	80286	7
	INC BYTE PTR [DI]	80386	6
	INC WORD PTR [ECX]	80486	3
		Pentium	1 or 3
INC reg16 INC reg32	INC CX	8086	3
	INC DX	8088	3
	INC BP	80286	2
	INC ECX	80386	2
	INC ESP	80486	1
		Pentium	1

INS — Input string from port

0110110w

Format	Examples	Microprocessor	Clocks
INSB	INSB	8086	—
INSW	INSW	8088	—
INSD	INSD	80286	5
	INS DATA2	80386	15
	REP INSB	80486	17
		Pentium	9

INT — Interrupt

11001101 type

Format	Examples	Microprocessor	Clocks
INT type	INT12H	8086	51
	INT15H	8088	71
	INT 21H	80286	23
	INT 2FH	80386	37
	INT 10H	80486	30
		Pentium	16–82

INT 3 — Interrupt 3

11001100

Example	Microprocessor	Clocks
INT 3	8086	52
	8088	72
	80286	23
	80386	33
	80486	26
	Pentium	13–56

INTO   Interrupt on overflow

11001110
Example

Microprocessor	Clocks
8086	53
8088	73
80286	24
80386	35
80486	28
Pentium	13–56

INTO

INVD   Invalidate data cache

00001111 00001000
Example

Microprocessor	Clocks
8086	—
8088	—
80286	—
80386	—
80486	4
Pentium	15

INTVD

IRET/IRETD   Return from interrupt

11001101 data

O	D	I	T		S	Z	A	P	C
*	*	*	*		*	*	*	*	*

Format	Examples	Microprocessor	Clocks
IRET	IRET	8086	32
IRETD	IRETD	8088	44
	IRET 100	80286	17
		80386	22
		80486	15
		Pentium	8–27

Jcondition   Conditional jump

0111cccc disp

Format	Examples	Microprocessor	Clocks
Jcnd label	JA ABOVE	8086	16/4
(8-bit disp)	JB BELOW	8088	16/4
	JG GREATER	80286	7/3
	JE EQUAL	80386	7/3
	JZ ZERO	80486	3/1
		Pentium	1

00001111 1000cccc disp

Format	Examples	Microprocessor	Clocks
Jcnd label	JNE NOT_MORE	8086	—
(16-bit disp)	JLE LESS_OR_SO	8088	—
		80286	—
		80386	7/3
		80486	3/1
		Pentium	1

Condition Codes	Mnemonic	Flag	Description
0000	JO	O = 1	Jump if overflow
0001	JNO	O = 0	Jump if no overflow
0010	JB/NAE	C = 1	Jump if below
0011	JAE/JNB	C = 0	Jump if above or equal
0100	JE/JZ	Z = 1	Jump if equal/zero
0101	JNE/JNZ	Z = 0	Jump if not equal/zero
0110	JBE/JNA	C = 1 + Z = 1	Jump if below or equal
0111	JA/JNBE	C = 0 • Z = 0	Jump if above
1000	JS	S = 1	Jump if sign
1001	JNS	S = 0	Jump if no sign
1010	JP/JPE	P = 1	Jump if parity
1011	JNP/JPO	P = 0	Jump if no parity
1100	JL/JNGE	S • O	Jump if less than
1101	JGE/JNL	S = 0	Jump if greater than or equal
1110	JLE/JNG	Z = 1 + S • O	Jump if less than or equal
1111	JG/JNLE	Z = 0 + S = O	Jump if greater than

JCXZ/JECXZ Jump if CX (ECX) equals zero

11100011

Format	Examples	Microprocessor	Clocks
JCXZ label JECXZ label	JCXZ ABOVE JCXZ BELOW JECXZ GREATER JECXZ EQUAL JCXZ NEXT	8086	18/6
		8088	18/6
		80286	8/4
		80386	9/5
		80486	8/5
		Pentium	6/5

JMP   Jump

11101011 disp

Format	Examples	Microprocessor	Clocks
JMP label (short)	JMP SHORT UP JMP SHORT DOWN JMP SHORT OVER JMP SHORT CIRCUIT JMP SHORT JOKE	8086	15
		8088	15
		80286	7
		80386	7
		80486	3
		Pentium	1

11101001 disp

Format	Examples	Microprocessor	Clocks
JMP label (near)	JMP VERS JMP FROG JMP UNDER JMP NEAR PTR OVER	8086	15
		8088	15
		80286	7
		80386	7
		80486	3
		Pentium	1

11101010 disp Format	Examples	Microprocessor	Clocks
JMP label (far)	JMP NOT_MORE JMP UNDER JMP AGAIN JMP FAR PTR THERE	8086	15
		8088	15
		80286	11
		80386	12
		80486	17
		Pentium	3

11111111 oo100mmm Format	Examples	Microprocessor	Clocks
JMP reg (near)	JMP AX JMP EAX JMP CX JMP DX	8086	11
		8088	11
		80286	7
		80386	7
		80486	3
		Pentium	2
JMP mem (near)	JMP VERS JMP FROG JMP CS:UNDER JMP DATA1[DI+2]	8086	18 + ea
		8088	18 + ea
		80286	11
		80386	10
		80486	5
		Pentium	4

11111111 oo101mmm Format	Examples	Microprocessor	Clocks
JMP mem (far)	JMP WAY_OFF JMP TABLE JMP UP JMP OUT_OF_HERE	8086	24 + ea
		8088	24 + ea
		80286	15
		80386	12
		80486	13
		Pentium	4

LAHF	Load AH from flags		

10011111 Example		Microprocessor	Clocks
LAHF		8086	4
		8088	4
		80286	2
		80386	2
		80486	3
		Pentium	2

LAR	Load access rights byte		

00001111 00000010 oorrmmmm disp		O D I T	S Z A P C *
Format	Examples	Microprocessor	Clocks
LAR reg,reg	LAR AX,BX LAR CX,DX LAR ECX,EDX	8086	—
		8088	—
		80286	14
		80386	15
		80486	11
		Pentium	8
LAR reg,mem	LAR CX,DATA1 LAR AX,LIST3 LAR ECX,TOAD	8086	—
		8088	—
		80286	16
		80386	16
		80486	11
		Pentium	8

LDS — Load far pointer to DS and register

11000101 oorrrmmm

Format	Examples	Microprocessor	Clocks
LDS reg,mem	LDS DI,DATA3	8086	16 + ea
	LDS SI,LIST2	8088	24 + ea
	LDS BX,ARRAY_PTR	80286	7
	LDS CX,PNTR	80386	7
		80486	6
		Pentium	4

LEA — Load effective address

10001101 oorrrmmm disp

Format	Examples	Microprocessor	Clocks
LEA reg,mem	LEA DI,DATA3	8086	2 + ea
	LEA SI,LIST2	8088	2 + ea
	LEA BX,ARRAY_PTR	80286	3
	LEA CX,PNTR	80386	2
		80486	2
		Pentium	1

LEAVE — Leave high-level procedure

11001001

Example	Microprocessor	Clocks
LEAVE	8086	—
	8088	—
	80286	5
	80386	4
	80486	5
	Pentium	3

LES   Load far pointer to ES and register

11000100 oorrrmmm

Format	Examples	Microprocessor	Clocks
LES reg,mem	LES DI,DATA3 LES SI,LIST2 LES BX,ARRAY_PTR LES CX,PNTR	8086	16 + ea
		8088	24 + ea
		80286	7
		80386	7
		80486	6
		Pentium	4

LFS   Load far pointer to FS and register

00001111 10110100 oorrrmmm disp

Format	Examples	Microprocessor	Clocks
LFS reg,mem	LFS DI,DATA3 LFS SI,LIST2 LFS BX,ARRAY_PTR LFS CX,PNTR	8086	—
		8088	—
		80286	—
		80386	7
		80486	6
		Pentium	4

LGDT   Load global descriptor table

00001111 00000001 oo010mmm disp

Format	Examples	Microprocessor	Clocks
LGDT mem64	LGDT DESCRIP LGDT TABLED	8086	—
		8088	—
		80286	11
		80386	11
		80486	11
		Pentium	6

LGS — Load far pointer to GS and register

00001111 10110101 oorrrmmm disp

Format	Examples	Microprocessor	Clocks
LGS reg,mem	LGS DI,DATA3 LGS SI,LIST2 LGS BX,ARRAY_PTR LGS CX,PNTR	8086	—
		8088	—
		80286	—
		80386	7
		80486	6
		Pentium	4

LIDT — Load interrupt descriptor table

00001111 00000001 oo011mmm disp

Format	Examples	Microprocessor	Clocks
LIDT mem64	LIDT DATA3 LIDT LIST2	8086	—
		8088	—
		80286	12
		80386	11
		80486	11
		Pentium	6

LLDT — Load local descriptor table

00001111 00000000 oo010mmm disp

Format	Examples	Microprocessor	Clocks
LLDT reg	LLDT BX LLDT DX LLDT CX	8086	—
		8088	—
		80286	17
		80386	20
		80486	11
		Pentium	9

LLDT mem	LLDT DATA1 LLDT LIST3 LLDT TOAD	8086	—
		8088	—
		80286	19
		80386	24
		80486	11
		Pentium	9

LMSW — Load machine status word (80286 only)

00001111 00000001 oo110mmm disp

Format	Examples	Microprocessor	Clocks
LMSW reg	LMSW BX LMSW DX LMSW CX	8086	—
		8088	—
		80286	3
		80386	10
		80486	2
		Pentium	8
LMSW mem	LMSW DATA1 LMSW LIST3 LMSW TOAD	8086	—
		8088	—
		80286	6
		80386	13
		80486	3
		Pentium	8

LOCK   Lock the bus

11110000

Format	Examples	Microprocessor	Clocks
LOCK:inst	LOCK:XCHG AX,BX LOCK:ADD AL,3	8086	2
		8088	3
		80286	0
		80386	0
		80486	1
		Pentium	1

LODS   Load string operand

1010110w

Format	Examples	Microprocessor	Clocks
LODSB LODSW LODSD	LODSB LODSW LODSD LODS DATA3	8086	12
		8088	15
		80286	5
		80386	5
		80486	5
		Pentium	2

LOOP/LOOPD   Loop until CX = 0 or ECX = 0

11100010 disp

Format	Examples	Microprocessor	Clocks
LOOP label LOOPD label	LOOP NEXT LOOP BACK LOOPD LOOPS	8086	17/5
		8088	17/5
		80286	8/4
		80386	11
		80486	7/6
		Pentium	5/6

LOOPE/LOOPED   Loop while equal

11100001 disp

Format	Examples	Microprocessor	Clocks
LOOPE label	LOOPE AGAIN	8086	18/6
LOOPED label	LOOPED UNTIL	8088	18/6
LOOPZ label	LOOPZ ZORRO	80286	8/4
LOOPZD label	LOOPZD WOW	80386	11
		80486	9/6
		Pentium	7/8

LOOPNE/LOOPNED Loop while not equal

11100000 disp

Format	Examples	Microprocessor	Clocks
LOOPNE label	LOOPNE FORWARD	8086	19/5
LOOPNED label	LOOPNED UPS	8088	19/5
LOOPNZ label	LOOPNZ TRY_AGAIN	80286	8/4
LOOPNZD label	LOOPNZD WOO	80386	11
		80486	9/6
		Pentium	7/8

LSL   Load segment limit

00001111 00000011 oorrrmmm disp

O D I T   S Z A P C
                *

Format	Examples	Microprocessor	Clocks
LSL reg,reg	LSL AX,BX	8086	—
	LSL CX,BX	8088	—
	LSL EDX,EAX	80286	14
		80386	25
		80486	10
		Pentium	8

LSL reg,mem	LSL AX,LIMIT LSL EAX,NUM	8086	—
		8088	—
		80286	16
		80386	26
		80486	10
		Pentium	8

LSS — Load far pointer to SS and register

00001111 10110010 oorrrmmm disp

Format	Examples	Microprocessor	Clocks
LSS reg,mem	LSS DI,DATA1 LSS SP,STACK_TOP LSS CX,ARRAY	8086	—
		8088	—
		80286	—
		80386	7
		80486	6
		Pentium	4

LTR — Load task register

00001111 00000000 oo001mmm disp

Format	Examples	Microprocessor	Clocks
LTR reg	LTR AX LTR CX LTR DX	8086	—
		8088	—
		80286	17
		80386	23
		80486	20
		Pentium	10

LTR mem16	LTR TASK LTR NUM	8086	—
		8088	—
		80286	19
		80386	27
		80486	20
		Pentium	10

MOV   Move data

100010dw oorrrmmm disp

Format	Examples	Microprocessor	Clocks

MOV reg,reg	MOV CL,CH
	MOV BH,CL
	MOV CX,DX
	MOV EAX,EBP
	MOV ESP,ESI
		8086	2

		8088	2
		80286	2
		80386	2
		80486	1
		Pentium	1

MOV mem,reg	MOV DATA7,DL
	MOV NUMB,CX
	MOV TEMP,EBX
	MOV [ECX],BL
	MOV [DI],DH
		8086	9 + ea

		8088	13 + ea
		80286	3
		80386	2
		80486	1
		Pentium	1

MOV reg,mem	MOV DL,DATA8
	MOV DX,NUMB
	MOV EBX,TEMP+3
	MOV CH,TEMP[EDI]
	MOV CL,DATA2
		8086	10 + ea

		8088	12 + ea
		80286	5
		80386	4
		80486	1
		Pentium	1

1100011w oo000mmm disp data			
Format	**Examples**	**Microprocessor**	**Clocks**
MOV mem,imm	MOV DATAF,23H	8086	10 + ea
	MOV LIST,12H	8088	14 + ea
	MOV BYTE PTR [DI],2	80286	3
	MOV NUMB,234H	80386	2
	MOV DWORD PTR[ECX],1	80486	1
		Pentium	1

1011wrrr data			
Format	**Examples**	**Microprocessor**	**Clocks**
MOV reg,imm	MOV BX,22H	8086	4
	MOV CX,12H	8088	4
	MOV CL,2	80286	3
	MOV ECX,123456H	80386	2
	MOV DI,100	80486	1
		Pentium	1

101000dw disp			
Format	**Examples**	**Microprocessor**	**Clocks**
MOV mem,acc	MOV DATAF,AL	8086	10
	MOV LIST,AX	8088	14
	MOV NUMB,EAX	80286	3
		80386	2
		80486	1
		Pentium	1
MOV acc,mem	MOV AL,DATAE	8086	10
	MOV AX,LIST	8088	14
	MOV EAX,LUTE	80286	5
		80386	4
		80486	1
		Pentium	1

100011d0 oosssmmm disp			
Format	**Examples**	**Microprocessor**	**Clocks**

MOV seg,reg	MOV SS,AX
	MOV DS,DX
	MOV ES,CX
	MOV FS,BX
	MOV GS,AX
		8086	2

		8088	2
		80286	2
		80386	2
		80486	1
		Pentium	1

MOV seg,mem	MOV SS,STACK_TOP
	MOV DS,DATAS
	MOV ES,TEMP1
		8086	8 + ea

		8088	12 + ea
		80286	2
		80386	2
		80486	1
		Pentium	2 or 3

MOV reg,seg	MOV BX,DS
	MOV CX,FS
	MOV CX,ES
		8086	2

		8088	2
		80286	2
		80386	2
		80486	1
		Pentium	1

MOV mem,seg	MOV DATA2,CS
	MOV TEMP,DS
	MOV NUMB1,SS
	MOV TEMP2,GS
		8086	9 + ea

		8088	13 + ea
		80286	3
		80386	2
		80486	1
		Pentium	1

00001111 001000d0 11rrrmmm			
Format	Examples	Microprocessor	Clocks

MOV reg,cr	MOV EBX,CR0
	MOV ECX,CR2
	MOV EBX,CR3
		8086	—

		8088	—
		80286	—
		80386	6
		80486	4
		Pentium	4

MOV cr,reg	MOV CR0,EAX
	MOV CR1,EBX
	MOV CR3,EDX
		8086	—

		8088	—
		80286	—
		80386	10
		80486	4
		Pentium	12–46

00001111 001000d1 11rrrmmm			
Format	Examples	Microprocessor	Clocks

MOV reg,dr	MOV EBX,DR6
	MOV ECX,DR7
	MOV EBX,DR1
		8086	—

		8088	—
		80286	—
		80386	22
		80486	10
		Pentium	11

MOV dr,reg	MOV DR0,EAX
	MOV DR1,EBX
	MOV DR3,EDX
		8086	—

		8088	—
		80286	—
		80386	22
		80486	11
		Pentium	11

00001111 001001d0 11rrrmmm			
Format	Examples	Microprocessor	Clocks
MOV reg,tr	MOV EBX,TR6 MOV ECX,TR7	8086	—
		8088	—
		80286	—
		80386	12
		80486	4
		Pentium	11
MOV tr,reg	MOV TR6,EAX MOV TR7,EBX	8086	—
		8088	—
		80286	—
		80386	12
		80486	6
		Pentium	11

MOVS   Move string data

1010010w			
Format	Examples	Microprocessor	Clocks
MOVSB MOVSW MOVSD	MOVSB MOVSW MOVSD MOVS DATA1,DATA2	8086	18
		8088	26
		80286	5
		80386	7
		80486	7
		Pentium	4

MOVSX   Move with sign extend

00001111 1011111w oorrrmmm disp

Format	Examples	Microprocessor	Clocks
MOVSX reg,reg	MOVSX BX,AL MOVSX EAX,DX	8086	—
		8088	—
		80286	—
		80386	3
		80486	3
		Pentium	3
MOVSX reg,mem	MOVSX AX,DATA34 MOVSX EAX,NUMB	8086	—
		8088	—
		80286	—
		80386	6
		80486	3
		Pentium	3

MOVZX   Move with zero extend

00001111 1011011w oorrrmmm disp

Format	Examples	Microprocessor	Clocks
MOVZX reg,reg	MOVZX BX,AL MOVZX EAX,DX	8086	—
		8088	—
		80286	—
		80386	3
		80486	3
		Pentium	3
MOVZX reg,mem	MOVZX AX,DATA34 MOVZX EAX,NUMB	8086	—
		8088	—
		80286	—
		80386	6
		80486	3
		Pentium	3

MUL   Multiply

1111011w oo100mmm disp		O D I T S Z A P C	
		* ? ? ? ? *	
Format	Examples	Microprocessor	Clocks
MUL reg	MUL BL	8086	118
	MUL CX	8088	143
	MUL EDX	80286	21
		80386	38
		80486	42
		Pentium	10 or 11
MUL mem	MUL DATA9	8086	139
	MUL WORD PTR [ESI]	8088	143
		80286	24
		80386	41
		80486	42
		Pentium	11

NEG   Negate

1111011w oo011mmm disp		O D I T S Z A P C	
		* * * * * *	
Format	Examples	Microprocessor	Clocks
NEG reg	NEG BL	8086	3
	NEG CX	8088	3
	NEG EDI	80286	2
		80386	2
		80486	1
		Pentium	1 or 3

NEG mem	NEG DATA9 NEG WORD PTR [ESI]	8086	16 + ea
		8088	24 + ea
		80286	7
		80386	6
		80486	3
		Pentium	1 or 3

NOP   No operation

10010000 Example	Microprocessor	Clocks
NOP	8086	3
	8088	3
	80286	3
	80386	3
	80486	3
	Pentium	1

NOT   One's complement

1111011w oo010mmm disp Format	Examples	Microprocessor	Clocks
NOT reg	NOT BL NOT CX NOT EDI	8086	3
		8088	3
		80286	2
		80386	2
		80486	1
		Pentium	1 or 3
NOT mem	NOT DATA9 NOT WORD PTR [ESI]	8086	16 + ea
		8088	24 + ea
		80286	7
		80386	6
		80486	3
		Pentium	1 or 3

OR	Inclusive-OR		

000010dw oorrrmmm disp		O D I T	S Z A P C
		0	* * ? * 0
Format	Examples	Microprocessor	Clocks
OR reg,reg	OR AX,BX	8086	3
	OR AL,BL	8088	3
	OR EAX,EBX	80286	2
	OR CX,SI	80386	2
	OR ESI,EDI	80486	1
		Pentium	1 or 2
OR mem,reg	OR DATAY,AL	8086	16 + ea
	OR LIST,SI	8088	24 + ea
	OR DATA2[DI],CL	80286	7
	OR [EAX],BL	80386	7
	OR [EBX+2*ECX],EDX	80486	3
		Pentium	1 or 3
OR reg,mem	OR BL,DATA1	8086	9 + ea
	OR SI,LIST1	8088	13 + ea
	OR CL,DATA2[SI]	80286	7
	OR CX,[ESI]	80386	6
	OR ESI,[2*ECX]	80486	2
		Pentium	1 or 3

100000sw oo001mmm disp data			
Format	Examples	Microprocessor	Clocks
OR reg,imm	OR CX,3	8086	4
	OR DI,1AH	8088	4
	OR DL,34H	80286	3
	OR EDX,1345H	80386	2
	OR CX,1834H	80486	1
		Pentium	1 or 3

OR mem,imm	OR DATAS,3 OR BYTE PTR[EDI],1AH OR DADDY,34H OR LIST,'A' OR TOAD,1834H	8086	17 + ea
		8088	25 + ea
		80286	7
		80386	7
		80486	0
		Pentium	1 or 3

0000110w data Format	Examples	Microprocessor	Clocks
OR acc,imm	OR AX,3 OR AL,1AH OR AH,34H OR EAX,1345H OR AL,'Y'	8086	4
		8088	4
		80286	3
		80386	2
		80486	1
		Pentium	1

OUT   Output data to port

1110011w port# Format	Examples	Microprocessor	Clocks
OUT pt,acc	OUT 12H,AL OUT 12H,AX OUT 0FFH,AL OUT 0A0H,AX OUT 10H,EAX	8086	10
		8088	14
		80286	3
		80386	10
		80486	10
		Pentium	12–26

1110111w Format	Examples	Microprocessor	Clocks
OUT DX,acc	OUT DX,AL OUT DX,AX OUT DX,EAX	8086	8
		8088	12
		80286	3
		80386	11
		80486	10
		Pentium	12–26

OUTS	Output string to port		

0110111w Format	Examples	Microprocessor	Clocks
OUTSB OUTSW OUTSD	OUTSB OUTSW OUTSD OUTS DATA2 REP OUTSB	8086	—
		8088	—
		80286	5
		80386	14
		80486	10
		Pentium	13–27

POP	Pop data from stack		

01011rrr Format	Examples	Microprocessor	Clocks
POP reg	POP CX POP AX POP EDI	8086	8
		8088	12
		80286	5
		80386	4
		80486	1
		Pentium	1

10001111 oo000mmm disp Format	Examples	Microprocessor	Clocks
POP mem	POP DATA1 POP LISTS POP NUMBS	8086	17 + ea
		8088	25 + ea
		80286	5
		80386	5
		80486	4
		Pentium	3

00sss111 Format	Examples	Microprocessor	Clocks
POP seg	POP DS POP ES POP SS	8086	8
		8088	12
		80286	5
		80386	7
		80486	3
		Pentium	3

00001111 10sss001 Format	Examples	Microprocessor	Clocks
POP seg	POP FS POP GS	8086	—
		8088	—
		80286	—
		80386	7
		80486	3
		Pentium	3

POPA/POPAD   Pop all registers from stack

01100001 Example	Microprocessor	Clocks
POPA POPAD	8086	—
	8088	—
	80286	19
	80386	24
	80486	9
	Pentium	5

POPF/POPFD	Pop flags from stack		

10010000		O D I T S Z A P C	
		* * * * * * * * *	
Example		Microprocessor	Clocks
POPF		8086	8
POPFD		8088	12
		80286	5
		80386	5
		80486	6
		Pentium	4 or 6

PUSH	Push data onto stack		

01010rrr			
Format	Examples	Microprocessor	Clocks
PUSH reg	PUSH CX	8086	11
	PUSH AX	8088	15
	PUSH EDI	80286	3
		80386	2
		80486	1
		Pentium	1

11111111 oo110mmm disp			
Format	Examples	Microprocessor	Clocks
PUSH mem	PUSH DATA1	8086	16 + ea
	PUSH LISTS	8088	24 + ea
	PUSH NUMBS	80286	5
		80386	5
		80486	4
		Pentium	1 or 2

00ss110 Format	Examples	Microprocessor	Clocks
PUSH seg	PUSH ES PUSH CS PUSH DS	8086	10
		8088	14
		80286	3
		80386	2
		80486	3
		Pentium	1

00001111 10sss000 Format	Examples	Microprocessor	Clocks
PUSH seg	PUSH FS PUSH GS	8086	—
		8088	—
		80286	—
		80386	2
		80486	3
		Pentium	1

011010s0 data Format	Examples	Microprocessor	Clocks
PUSH imm	PUSH 2000H PUSH 53220 PUSHW 10H PUSH ',' PUSHD 100000H	8086	—
		8088	—
		80286	3
		80386	2
		80486	1
		Pentium	1

PUSHA/PUSHAD   Push all registers onto stack

01100000
Example

	Microprocessor	Clocks
PUSHA	8086	—
PUSHAD	8088	—
	80286	17
	80386	18
	80486	11
	Pentium	5

PUSHF/PUSHFD   Push flags onto stack

10011100
Example

	Microprocessor	Clocks
PUSHF	8086	10
PUSHFD	8088	14
	80286	3
	80386	4
	80486	3
	Pentium	3 or 4

RCL/RCR/ROL/ROR   Rotate

1101000w ooTTTmmm disp

```
O   D  I  T      S  Z  A  P  C
*                            *
```

TTT = 000 = ROL, TTT = 001 = ROR, TTT = 010 = RCL, and TTT = 011 = RCR

Format	Examples	Microprocessor	Clocks
ROL reg,1	ROL CL,1	8086	2
ROR reg,1	ROL DX,1	8088	2
	ROR CH,1	80286	2
	ROR SI,1	80386	3
		80486	3
		Pentium	1 or 3

RCL reg,1 RCR reg,1	RCL CL,1 RCL SI,1 RCR AH,1 RCR EBX,1	8086	2
		8088	2
		80286	2
		80386	9
		80486	9
		Pentium	1 or 3
ROL mem,1 ROR mem,1	ROL DATAY,1 ROL LIST,1 ROR DATA2[DI],1 ROR BYTE PTR [EAX],1	8086	15 + ea
		8088	23 + ea
		80286	7
		80386	7
		80486	4
		Pentium	1 or 3
RCL mem,1 RCR mem,1	RCL DATA1,1 RCL LIST,1 RCR DATA2[SI],1 RCR WORD PTR [ESI],1	8086	15 + ea
		8088	23 + ea
		80286	7
		80386	10
		80486	4
		Pentium	1 or 3

1101001w ooTTTmmm disp			
Format	Examples	Microprocessor	Clocks
ROL reg,CL ROR reg,CL	ROL CH,CL ROL DX,CL ROR AL,CL ROR ESI,CL	8086	8 + 4n
		8088	8 + 4n
		80286	5 + n
		80386	3
		80486	3
		Pentium	4

RCL reg,CL RCR reg,CL	RCL CH,CL RCL SI,CL RCR AH,CL RCR EBX,CL	8086	8 + 4n
		8088	8 + 4n
		80286	5 + n
		80386	9
		80486	3
		Pentium	7–27
ROL mem,CL ROR mem,CL	ROL DATAY,CL ROL LIST,CL ROR DATA2[DI],CL ROR BYTE PTR [EAX],CL	8086	20 + 4n
		8088	28 + 4n
		80286	8 + n
		80386	7
		80486	4
		Pentium	4
RCL mem,CL RCR mem,CL	RCL DATA1,CL RCL LIST,CL RCR DATA2[SI],CL RCR WORD PTR [ESI],CL	8086	20 + 4n
		8088	28 + 4n
		80286	8 + n
		80386	10
		80486	9
		Pentium	9–26

1100000w ooTTTmmm disp data			
Format	Examples	Microprocessor	Clocks
ROL reg,imm ROR reg,imm	ROL CH,4 ROL DX,5 ROR AL,2 ROR ESI,14	8086	—
		8088	—
		80286	5 + n
		80386	3
		80486	2
		Pentium	1 or 3

RCL reg,imm	RCL CL,2	8086	—
RCR reg,imm	RCL SI,12	8088	—
	RCR AH,5	80286	5 + n
	RCR EBX,18	80386	9
		80486	8
		Pentium	8–27
ROL mem,imm	ROL DATAY,4	8086	—
ROR mem,imm	ROL LIST,3	8088	—
	ROR DATA2[DI],7	80286	8 + n
	ROR BYTE PTR [EAX],11	80386	7
		80486	4
		Pentium	1 or 3
RCL mem,imm	RCL DATA1,5	8086	—
RCR mem,imm	RCL LIST,3	8088	—
	RCR DATA2[SI],9	80286	8 + n
	RCR WORD PTR [ESI],8	80386	10
		80486	9
		Pentium	8–27

RDMSR   Read model specific register

00001111 00110010

Example	Microprocessor	Clocks
RDMSR	8086	—
	8088	—
	80286	—
	80386	—
	80486	—
	Pentium	20–24

REP   Repeat prefix

11110011 1010010w

Format	Examples	Microprocessor	Clocks
REP MOVS	REP MOVSB REP MOVSW REP MOVSD REP MOVS DATA1,DATA2	8086	9 + 17n
		8088	9 + 25n
		80286	5 + 4n
		80386	8 + 4n
		80486	12 + 3n
		Pentium	13 + n

11110011 1010101w

Format	Examples	Microprocessor	Clocks
REP STOS	REP STOSB REP STOSW REP STOSD REP STOS ARRAY	8086	9 + 10n
		8088	9 + 14n
		80286	4 + 3n
		80386	5 + 5n
		80486	7 + 4n
		Pentium	9 + n

11110011 0110110w

Format	Examples	Microprocessor	Clocks
REP INS	REP INSB REP INSW REP INSD REP INS ARRAY	8086	—
		8088	—
		80286	5 + 4n
		80386	12 + 5n
		80486	17 + 5n
		Pentium	25 + 3n

11110011 0110111w			
Format	Examples	Microprocessor	Clocks
REP OUTS	REP OUTSB	8086	—
	REP OUTSW		
	REP OUTSD	8088	—
	REP OUTS ARRAY	80286	5 + 4n
		80386	12 + 5n
		80486	17 + 5n
		Pentium	25 + 4n

REPE/REPNE   Repeat conditional

11110011 1010011w			
Format	Examples	Microprocessor	Clocks
REPE CMPS	REPE CMPSB	8086	9 + 22n
	REPE CMPSW		
	REPE CMPSD	8088	9 + 30n
	REPE CMPS DATA1,DATA2	80286	5 + 9n
		80386	5 + 9n
		80486	7 + 7n
		Pentium	9 + 4n

11110011 1010111w			
Format	Examples	Microprocessor	Clocks
REPE SCAS	REPE SCASB	8086	9 + 15n
	REPE SCASW		
	REPE SCASD	8088	9 + 19n
	REPE SCAS ARRAY	80286	5 + 8n
		80386	5 + 8n
		80486	7 + 5n
		Pentium	9 + 4n

| 11110010 1010011w | | | |
Format	Examples	Microprocessor	Clocks
REPNE CMPS	REPNE CMPSB	8086	$9 + 22n$
	REPNE CMPSW		
	REPNE CMPSD	8088	$9 + 30n$
	REPNE CMPS ARRAY,LIST	80286	$5 + 9n$
		80386	$5 + 9n$
		80486	$7 + 7n$
		Pentium	$8 + 4n$

| 11110010 101011w | | | |
Format	Examples	Microprocessor	Clocks
REPNE SCAS	REPNE SCASB	8086	$9 + 15n$
	REPNE SCASW		
	REPNE SCASD	8088	$9 + 19N$
	REPNE SCAS ARRAY	80286	$5 + 8n$
		80386	$5 + 8n$
		80486	$7 + 5n$
		Pentium	$9 + 4n$

RET	Return from procedure

| 11000011 | | |
Example	Microprocessor	Clocks
RET	8086	16
(near)	8088	20
	80286	11
	80386	10
	80486	5
	Pentium	2

11000010 data			
Format	Examples	Microprocessor	Clocks
RET imm	RET 4	8086	20
(near)	RET 100H	8088	24
		80286	11
		80386	10
		80486	5
		Pentium	3

11001011		
Example	Microprocessor	Clocks
RET	8086	26
(far)	8088	34
	80286	15
	80386	18
	80486	13
	Pentium	4–23

11001010 data			
Format	Examples	Microprocessor	Clocks
RET imm	RET 4	8086	25
(far)	RET 100H	8088	33
		80286	11
		80386	10
		80486	5
		Pentium	4–23

RSM	Resume from system management mode		

00001111 10101010	O D I T S Z A P C * * * * * * * * *		
Example	Microprocessor	Clocks	
RSM	8086	—	
	8088	—	
	80286	—	
	80386	—	
	80486	—	
	Pentium	83	

SAHF	Store AH into flags		

10011110	O D I T S Z A P C * * * * *		
Example	Microprocessor	Clocks	
SAHF	8086	4	
	8088	4	
	80286	2	
	80386	3	
	80486	2	
	Pentium	2	

SAL/SAR/SHL/SHR	Shift		

1101000w ooTTTmmm disp	O D I T S Z A P C * * * ? * *		
TTT = 100 = SHL/SAL , TTT = 101 = SHR, and TTT = 111 = SAR			
Format	Examples	Microprocessor	Clocks
SAL reg,1 SHL reg,1 SHR reg,1 SAR reg,1	SAL CL,1 SHL DX,1 SAR CH,1 SHR SI,1	8086	2
		8088	2
		80286	2
		80386	3
		80486	3
		Pentium	1 or 3

SAL mem,1	SAL DATA1,1	8086	15 + ea
SHL mem,1	SHL BYTE PTR [DI],1	8088	23 + ea
SHR mem,1	SAR NUMB,1	80286	7
SAR mem,1	SHR WORD PTR[EDI],1	80386	7
		80486	4
		Pentium	1 or 3

1101001w ooTTTmmm disp

Format	Examples	Microprocessor	Clocks
SAL reg,CL	SAL CH,CL	8086	8 + 4n
SHL reg,CL	SHL DX,CL	8088	8 + 4n
SAR reg,CL	SAR AL,CL	80286	5 + n
SHR reg,CL	SHR ESI,CL	80386	3
		80486	3
		Pentium	4
SAL mem,CL	SAL DATAU,CL	8086	20 + 4n
SHL mem,CL	SHL BYTE PTR [ESI],CL	8088	28 + 4n
SAR mem,CL	SAR NUMB,CL	80286	8 + n
SHR mem,CL	SHR TEMP,CL	80386	7
		80486	4
		Pentium	4

1100000w ooTTTmmm disp data

Format	Examples	Microprocessor	Clocks
SAL reg,imm	SAL CH,4	8086	—
SHL reg,imm	SHL DX,10	8088	—
SAR reg,imm	SAR AL,2	80286	5 + n
SHR reg,imm	SHR ESI,23	80386	3
		80486	2
		Pentium	1 or 3

SAL mem,imm SHL mem,imm SAR mem,imm SHR mem,imm	SAL DATAU,3 SHL BYTE PTR [ESI],15 SAR NUMB,3 SHR TEMP,5	8086	—
		8088	—
		80286	8 + n
		80386	7
		80486	4
		Pentium	1 or 3

SBB   Subtract with borrow

000110dw oorrrmmm disp

| | O D I T   S Z A P C |
| | *   * * * * * |

Format	Examples	Microprocessor	Clocks
SBB reg,reg	SBB CL,DL SBB AX,DX SBB CH,CL SBB EAX,EBX SBB ESI,EDI	8086	3
		8088	3
		80286	2
		80386	2
		80486	1
		Pentium	1 or 2
SBB mem,reg	SBB DATAJ,CL SBB BYTES,CX SBB NUMBS,ECX SBB [EAX],CX	8086	16 + ea
		8088	24 + ea
		80286	7
		80386	6
		80486	3
		Pentium	1 or 3
SBB reg,mem	SBB CL,DATAL SBB CX,BYTES SBB ECX,NUMBS SBB DX,[EBX+EDI]	8086	9 + ea
		8088	13 + ea
		80286	7
		80386	7
		80486	2
		Pentium	1 or 2

100000sw oo011mmm disp data

Format	Examples	Microprocessor	Clocks
SBB reg,imm	SBB CX,3 SBB DI,1AH SBB DL,34H SBB EDX,1345H SBB CX,1834H	8086	4
		8088	4
		80286	3
		80386	2
		80486	1
		Pentium	1 or 3
SBB mem,imm	SBB DATAS,3 SBB BYTE PTR[EDI],1AH SBB DADDY,34H SBB LIST,'A' SBB TOAD,1834H	8086	17 + ea
		8088	25 + ea
		80286	7
		80386	7
		80486	3
		Pentium	1 or 3

0001110w data

Format	Examples	Microprocessor	Clocks
SBB acc,imm	SBB AX,3 SBB AL,1AH SBB AH,34H SBB EAX,1345H SBB AL,'Y'	8086	4
		8088	4
		80286	3
		80386	2
		80486	1
		Pentium	1

SCAS   Scan string

| 1010111w | | O D I T | S Z A P C |
		*	* * * * *
Format	Examples	Microprocessor	Clocks
SCASB SCASW SCASD	SCASB SCASW SCASD SCAS DATAF REP SCASB	8086	15
		8088	19
		80286	7
		80386	7
		80486	6
		Pentium	4

SETcondition	Conditional set		

00001111 1001cccc oo000mmm

Format	Examples	Microprocessor	Clocks
SETcnd reg8	SETA BL	8086	—
	SETB CH	8088	—
	SETG DL	80286	—
	SETE BH	80386	4
	SETZ AL	80486	3
		Pentium	1 or 2
SETcnd mem8	SETE DATAK	8086	—
	SETAE LESS_OR_SO	8088	—
		80286	—
		80386	5
		80486	3
		Pentium	1 or 2

Condition Codes	Mnemonic	Flag	Description
0000	SETO	O = 1	Set if overflow
0001	SETNO	O = 0	Set if no overflow
0010	SETB/SETAE	C = 1	Set if below
0011	SETAE/SETNB	C = 0	Set if above or equal
0100	SETE/SETZ	Z = 1	Set if equal/zero
0101	SETNE/SETNZ	Z = 0	Set if not equal/zero
0110	SETBE/SETNA	C = 1 + Z = 1	Set if below or equal
0111	SETA/SETNBE	C = 0 • Z = 0	Set if above
1000	SETS	S = 1	Set if sign
1001	SETNS	S = 0	Set if no sign
1010	SETP/SETPE	P = 1	Set if parity
1011	SETNP/SETPO	P = 0	Set if no parity
1100	SETL/SETNGE	S • O	Set if less than
1101	SETGE/SETNL	S = 0	Set if greater than or equal
1110	SETLE/SETNG	Z = 1 + S • O	Set if less than or equal
1111	SETG/SETNLE	Z = 0 + S = O	Set if greater than

SGDT/SIDT/SLDT Store descriptor table registers

00001111 00000001 oo000mmm disp

Format	Examples	Microprocessor	Clocks
SGDT mem	SGDT MEMORY SGDT GLOBAL	8086	—
		8088	—
		80286	11
		80386	9
		80486	10
		Pentium	4

00001111 00000001 oo001mmm disp

Format	Examples	Microprocessor	Clocks
SIDT mem	SIDT DATAS SIDT INTERRUPT	8086	—
		8088	—
		80286	12
		80386	9
		80486	10
		Pentium	4

00001111 00000000 oo000mmm disp

Format	Examples	Microprocessor	Clocks
SLDT reg	SLDT CX SLDT DX	8086	—
		8088	—
		80286	2
		80386	2
		80486	2
		Pentium	2
SLDT mem	SLDT NUMBS SLDT LOCALS	8086	—
		8088	—
		80286	3
		80386	2
		80486	3
		Pentium	2

SHLD/SHRD Double precision shift

00001111 10100100 oorrrmmm disp data		O D I T	S Z A P C
		?	* * ? * *
Format	Examples	Microprocessor	Clocks
SHLD reg,reg,imm	SHLD AX,CX,10 SHLD DX,BX,8 SHLD CX,DX,2	8086	—
		8088	—
		80286	—
		80386	3
		80486	2
		Pentium	4
SHLD mem,reg,imm	SHLD DATAQ,CX,8	8086	—
		8088	—
		80286	—
		80386	7
		80486	3
		Pentium	4

00001111 10101100 oorrrmmm disp data			
Format	Examples	Microprocessor	Clocks
SHRD reg,reg,imm	SHRD CX,DX,2	8086	—
		8088	—
		80286	—
		80386	3
		80486	2
		Pentium	4
SHRD mem,reg,imm	SHRD DATAZ,DX,4	8086	—
		8088	—
		80286	—
		80386	7
		80486	2
		Pentium	4

00001111 10100101 oorrrmmm disp			
Format	Examples	Microprocessor	Clocks
SHLD reg,reg,CL	SHLD BX,DX,CL	8086	—
		8088	—
		80286	—
		80386	3
		80486	3
		Pentium	4 or 5
SHLD mem,reg,CL	SHLD DATAZ,DX,CL	8086	—
		8088	—
		80286	—
		80386	7
		80486	3
		Pentium	4 or 5

00001111 10101101 oorrrmmm disp			
Format	Examples	Microprocessor	Clocks
SHRD reg,reg,CL	SHRD AX,DX,CL	8086	—
		8088	—
		80286	—
		80386	3
		80486	3
		Pentium	4 or 5
SHRD mem,reg,CL	SHRD DATAZ,DX,CL	8086	—
		8088	—
		80286	—
		80386	7
		80486	3
		Pentium	4 or 5

SMSW	Store machine status word (80286)		

00001111 00000001 oo100mmm disp

Format	Examples	Microprocessor	Clocks
SMSW reg	SMSW AX SMSW DX SMSW BP	8086	—
		8088	—
		80286	2
		80386	10
		80486	2
		Pentium	4
SMSW mem	SMSW DATAQ	8086	—
		8088	—
		80286	3
		80386	3
		80486	3
		Pentium	4

STC	Set carry flag		

11111001

	O D I T S Z A P C
	1

Example	Microprocessor	Clocks
STC	8086	2
	8088	2
	80286	2
	80386	2
	80486	2
	Pentium	2

STD — Set direction flag

11111101		O D I T S Z A P C	
		1	
Example		Microprocessor	Clocks
STD		8086	2
		8088	2
		80286	2
		80386	2
		80486	2
		Pentium	2

STI — Set interrupt flag

11111011		O D I T S Z A P C	
		1	
Example		Microprocessor	Clocks
STI		8086	2
		8088	2
		80286	2
		80386	3
		80486	5
		Pentium	7

STOS — Store string data

1010101w			
Format	Examples	Microprocessor	Clocks
STOSB	STOSB	8086	11
STOSW	STOSW		
STOSD	STOSD	8088	15
	STOS DATA_LIST	80286	3
	REP STOSB	80386	40
		80486	5
		Pentium	3

STR	Store task register		

00001111 00000000 oo001mmm disp

Format	Examples	Microprocessor	Clocks
STR reg	STR AX STR DX STR BP	8086	—
		8088	—
		80286	2
		80386	2
		80486	2
		Pentium	2
STR mem	STR DATA3	8086	—
		8088	—
		80286	2
		80386	2
		80486	2
		Pentium	2

SUB	Subtract		

000101dw oorrrmmm disp

	O D I T S Z A P C * * * * * *

Format	Examples	Microprocessor	Clocks

SUB reg,reg	SUB CL,DL
	SUB AX,DX
	SUB CH,CL
	SUB EAX,EBX
	SUB ESI,EDI
		8086	3

		8088	3
		80286	2
		80386	2
		80486	1
		Pentium	1 or 2

SUB mem,reg	SUB DATAJ,CL
	SUB BYTES,CX
	SUB NUMBS,ECX
	SUB [EAX],CX
		8086	16 + ea

		8088	24 + ea
		80286	7
		80386	6
		80486	3
		Pentium	1 or 3

SUB reg,mem	SUB CL,DATAL
	SUB CX,BYTES
	SUB ECX,NUMBS
	SUB DX,[EBX+EDI]
		8086	9 + ea

		8088	13 + ea
		80286	7
		80386	7
		80486	2
		Pentium	1 or 2

100000sw oo101mmm disp data

Format	Examples	Microprocessor	Clocks

SUB reg,imm	SUB CX,3
	SUB DI,1AH
	SUB DL,34H
	SUB EDX,1345H
	SUB CX,1834H
		8086	4

		8088	4
		80286	3
		80386	2
		80486	1
		Pentium	1 or 3

SUB mem,imm	SUB DATAS,3
	SUB BYTE PTR[EDI],1AH
	SUB DADDY,34H
	SUB LIST,'A'
	SUB TOAD,1834H
		8086	17 + ea

		8088	25 + ea
		80286	7
		80386	7
		80486	3
		Pentium	1 or 3

0010110w data

Format	Examples	Microprocessor	Clocks

SUB acc,imm	SUB AL,3
	SUB AX,1AH
	SUB EAX,34H
		8086	4

		8088	4
		80286	3
		80386	2
		80486	1
		Pentium	1

TEST   Test operands (logical compare)

1000001w oorrrmmm disp

		O D I T S Z A P C
		0 * * ? * 0

Format	Examples	Microprocessor	Clocks
TEST reg,reg	TEST CL,DL TEST BX,DX TEST DH,CL TEST EBP,EBX TEST EAX,EDI	8086	5
		8088	5
		80286	2
		80386	2
		80486	1
		Pentium	1 or 2
TEST mem,reg reg,mem	TEST DATAJ,CL TEST BYTES,CX TEST NUMBS,ECX TEST [EAX],CX TEST CL,POPS	8086	9 + ea
		8088	13 + ea
		80286	6
		80386	5
		80486	2
		Pentium	1 or 2

1111011sw oo000mmm disp data			
Format	Examples	Microprocessor	Clocks
TEST reg,imm	TEST BX,3	8086	4
	TEST DI,1AH		
	TEST DH,44H	8088	4
	TEST EDX,1AB345H	80286	3
	TEST SI,1834H	80386	2
		80486	1
		Pentium	1 or 2
TEST mem,imm	TEST DATAS,3	8086	11 + ea
	TEST BYTE PTR[EDI],1AH		
	TEST DADDY,34H	8088	11 + ea
	TEST LIST,'A'	80286	6
	TEST TOAD,1834H	80386	5
		80486	2
		Pentium	1 or 2

1010100w data			
Format	Examples	Microprocessor	Clocks
TEST acc,imm	TEST AL,3	8086	4
	TEST AX,1AH	8088	4
	TEST EAX,34H	80286	3
		80386	2
		80486	1
		Pentium	1

VERR/VERW Verify read/write

00001111 00000000 oo100mmm disp		O D I T S Z A P C *	
Format	Examples	Microprocessor	Clocks
VERR reg	VERR CX VERR DX VERR DI	8086	—
		8088	—
		80286	14
		80386	10
		80486	11
		Pentium	7
VERR mem	VERR DATAJ VERR TESTB	8086	—
		8088	—
		80286	16
		80386	11
		80486	11
		Pentium	7

00001111 00000000 oo101mmm disp			
Format	Examples	Microprocessor	Clocks
VERW reg	VERW CX VERW DX VERW DI	8086	—
		8088	—
		80286	14
		80386	15
		80486	11
		Pentium	7
VERW mem	VERW DATAJ VERW TESTB	8086	—
		8088	—
		80286	16
		80386	16
		80486	11
		Pentium	7

WAIT	Wait for coprocessor		

10011011 Example		Microprocessor	Clocks
WAIT FWAIT		8086	4
		8088	4
		80286	3
		80386	6
		80486	6
		Pentium	1

WBINVD	Write-back cache invalidate data cache		

00001111 00001001 Example		Microprocessor	Clocks
WBINVD		8086	—
		8088	—
		80286	—
		80386	—
		80486	5
		Pentium	2000+

WRMSR	Write to model specific register		

00001111 00110000 Example		Microprocessor	Clocks
WRMSR		8086	—
		8088	—
		80286	—
		80386	—
		80486	—
		Pentium	30–45

XADD   Exchange and add

00001111 1100000w 11rrrrrr		O D I T S Z A P C * * * * * *	
Format	Examples	Microprocessor	Clocks
XADD reg,reg	XADD EBX,ECX XADD EDX,EAX XADD EDI,EBP	8086	—
		8088	—
		80286	—
		80386	—
		80486	3
		Pentium	3 or 4

00001111 1100000w oorrrmmm disp			
Format	Examples	Microprocessor	Clocks
XADD mem,reg	XADD DATA5,ECX XADD [EBX],FAX XADD [ECX+4],EBP	8086	—
		8088	—
		80286	—
		80386	—
		80486	4
		Pentium	3 or 4

XCHG   Exchange

1000011w oorrrmmm			
Format	Examples	Microprocessor	Clocks
XCHG reg,reg	XCHG CL,DL XCHG BX,DX XCHG DH,CL XCHG EBP,EBX XCHG EAX,EDI	8086	4
		8088	4
		80286	3
		80386	3
		80486	3
		Pentium	3

XCHG mem,reg reg,mem	XCHG DATAJ,CL XCHG BYTES,CX XCHG NUMBS,ECX XCHG [EAX],CX XCHG CL,POPS	8086	17 + ea
		8088	25 + ea
		80286	5
		80386	5
		80486	5
		Pentium	3

10010reg Format	Examples	Microprocessor	Clocks
XCHG acc,reg reg,acc	XCHG BX,AX XCHG AX,DI XCHG DH,AL XCHG EDX,EAX XCHG SI,AX	8086	3
		8088	3
		80286	3
		80386	3
		80486	3
		Pentium	2

XLAT   Translate

11010111 Example	Microprocessor	Clocks
XLAT	8086	11
	8088	11
	80286	5
	80386	3
	80486	4
	Pentium	4

XOR	Exclusive-OR		

000110dw oorrrmmm disp		O D I T S Z A P C	
		0 * * ? * 0	
Format	Examples	Microprocessor	Clocks
XOR reg,reg	XOR CL,DL	8086	3
	XOR AX,DX	8088	3
	XOR CH,CL		
	XOR EAX,EBX	80286	2
	XOR ESI,EDI	80386	2
		80486	1
		Pentium	1 or 2
XOR mem,reg	XOR DATAJ,CL	8086	16 + ea
	XOR BYTES,CX	8088	24 + ea
	XOR NUMBS,ECX		
	XOR [EAX],CX	80286	7
		80386	6
		80486	3
		Pentium	1 or 3
XOR reg,mem	XOR CL,DATAL	8086	9 + ea
	XOR CX,BYTES	8088	13 + ea
	XOR ECX,NUMBS		
	XOR DX,[EBX+EDI]	80286	7
		80386	7
		80486	2
		Pentium	1 or 2

100000sw oo110mmm disp data			
Format	Examples	Microprocessor	Clocks
XOR reg,imm	XOR CX,3	8086	4
	XOR DI,1AH	8088	4
	XOR DL,34H		
	XOR EDX,1345H	80286	3
	XOR CX,1834H	80386	2
		80486	1
		Pentium	1 or 3

XOR mem,imm	XOR DATAS,3 XOR BYTE PTR[EDI],1AH XOR DADDY,34H XOR LIST,'A' XOR TOAD,1834H	8086	17 + ea
		8088	25 + ea
		80286	7
		80386	7
		80186	8
		Pentium	1 or 3

0010101w data Format	Examples	Microprocessor	Clocks
XOR acc,imm	XOR AL,3 XOR AX,1AH XOR EAX,34H	8086	4
		8088	4
		80286	3
		80386	2
		80486	1
		Pentium	1

SIMD Instruction Set Summary (pp. 965–976)

The SIMD (single instruction multiple data) instructions add a new dimension to the use of the microprocessor for performing multimedia and other operations. The XXM registers are numbered from XMM0 to XMM7 and are each 128 bits in width. Data formats stored in the XXM registers and used by the SIMD instructions appear in Figure B–1.

Packed 64-bit double-precision floating-point data

Packed byte integer data

Packed word integer data

Packed 32-bit integer data

Packed 64-bit integer data

FIGURE B–1   Data formats for the 128-bit-wide XMM registers in the Pentium III and Pentium 4 microprocessors.

Data stored in the memory must be stored as 16-byte data in a series of memory locations accessed by using the OWORD PTR override when addressed by an instruction. The OWORD PTR override is used to address an octal word of data or 16 bytes. The SIMD instructions allow operations on packed and scalar double-precision floating-point numbers. The operation of both forms is illustrated in Figure B–2, which shows both packed and scalar multiplication. Notice that scalar only copies the left-most double-precision number into the destination register and does not use the left-most number in the source.

This section of the appendix details many of the SIMD instructions and provides examples of their usage.

FIGURE B–2   Packed and scalar double-precision floating-point operation.

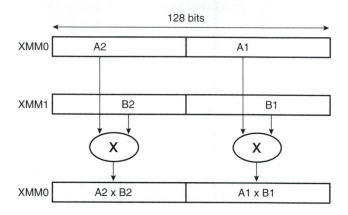

Packed double-precision multiplication MULPD XMM0, XMM1

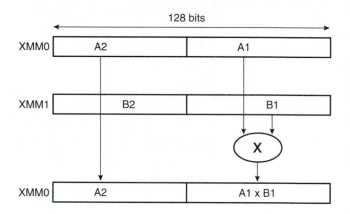

Packed double-precision multiplication MULSD XMM0, XMM1

DATA MOVEMENT INSTRUCTIONS

MOVAPD	Move aligned packed double-precision data, data must be aligned on 16 byte boundaries

Examples

    MOVAPD XMM0, OWORD DATA3   ;copies DATA3 to XMM0
    MOVAPD OWORD PTR DATA4, XMM2 ;copies XMM4 to DATA4

MOVUPD	Move unaligned packed double-precision data

Examples

    MOVUPD XMM0, OWORD DATA3   ;copies DATA3 to XMM0
    MOVUPD OWORD PTR DATA4, XMM2 ;copies XMM4 to DATA4

MOVSD	Move scalar packed double-precision data to low quadword

Examples

    MOVSD XMM0, DWORD DATA3   ;copies DATA3 to XMM0
    MOVSD DWORD PTR DATA4, XMM2 ;copies XMM4 to DATA4

MOVHPD	Move packed double-precision data to high quadword

Examples

    MOVHPD XMM0, DWORD DATA3   ;copies DATA3 to XMM0
    MOVHPD DWORD PTR DATA4, XMM2 ;copies XMM4 to DATA4

MOVLPD	Move packed double-precision data into low quadword

Examples

    MOVLPD XMM0, DWORD DATA3   ;copies DATA3 to XMM0
    MOVLPD DWORD PTR DATA4, XMM2 ;copies XMM4 to DATA4

MOVMSKPD	Move packed double-precision mask

Examples

    MOVMSKPD   EAX, XMM1   ;copies 2 sign bits to general purpose register

MOVAPS	Move 4 aligned packed single-precision data, data must be aligned on 16 byte boundaries

Examples

    MOVAPS XMM0, OWORD DATA3   ;copies DATA3 to XMM0

    MOVAPS OWORD PTR DATA4, XMM2 ;copies XMM4 to DATA4

MOVUPS	Move 4 unaligned packed single-precision data

Examples

    MOVUPS XMM0, OWORD DATA3   ;copies DATA3 to XMM0

    MOVUPS OWORD PTR DATA4, XMM2 ;copies XMM4 to DATA4

MOVLPS	Move 2 packed single-precision numbers to low-order quadword

Examples

    MOVLPS XMM0, OWORD DATA3   ;copies DATA3 to XMM0

    MOVLPS OWORD PTR DATA4, XMM2 ;copies XMM4 to DATA4

MOVHPS	Move packed single-precision numbers to high-order quadword

Examples

    MOVHPS XMM0, OWORD DATA3   ;copies DATA3 to XMM0

    MOVHPS OWORD PTR DATA4, XMM2 ;copies XMM4 to DATA4

MOVAPD	Move aligned packed double-precision data, data must be aligned on 16 byte boundaries

Examples

    MOVAPD XMM0, OWORD DATA3   ;copies DATA3 to XMM0

    MOVAPD OWORD PTR DATA4, XMM2 ;copies XMM4 to DATA4

MOVLHPS	Move 2 packed single-precision numbers from the low-order quadword to the high-order quadword

Examples

    MOVLHPS XMM0, XMM1   ;copies XMM1 low to XMM0 high

    MOVLHPS XMM3, XMM2   ;copies XMM2 low to XMM3 high

MOVHLPS	Move 2 packed single-precision numbers from high-order quadword to low-order quadword
Examples	
MOVHLPS XMM0, XMM2	;copies high XMM2 to low XMM0
MOVHLPS XMM4, XMM5	;copies high XMM5 to low XMM4

MOVMSKPS	Move 4-sign bits of 4 packed single-precision numbers to general purpose register
Examples	
MOVMSKPS EBX, XMM0	;copies sign bits of XMM0 to EBX
MOVMSKPS EDX, XMM2	;copies sign bits of XMM2 to EDX

ARITHMETIC INSTRUCTIONS

ADDPD	Adds packed double-precision data
Examples	
ADDPD XMM0, OWORD DATA3	;adds DATA3 to XMM0
ADDPD XMM2, XMM3	;adds XMM3 to XMM2

ADDSD	Adds scalar double-precision data
Examples	
ADDSD XMM0, OWORD DATA3	;adds DATA3 to XMM0
ADDSD XMM4, XMM2	;adds XMM2 to XMM4

ADDPS	Adds 2 packed single-precision numbers
Examples	
ADDPS XMM0, QWORD DATA3	;adds DATA3 to XMM0
ADDPS XMM3, XMM2	;adds XMM2 to XMM3

ADDLS	Adds scalar single-precision data

Examples

    ADDLS   XMM0, DWORD DATA3   ;adds DATA3 to XMM0
    ADDLS   XMM7, XMM2   ;adds XMM2 to XMM7

SUBPD	Subtracts packed double-precision data

Examples

    SUBPD   XMM0, OWORD DATA3   ;subtracts DATA3 from XMM0
    SUBPD   XMM2, XMM3   ;subtracts XMM3 from XMM2

SUBSD	Subtracts scalar double-precision data

Examples

    SUBSD   XMM0, OWORD DATA3   ;subtracts DATA3 from XMM0
    SUBSD   XMM4, XMM2   ;subtracts XMM2 from XMM4

SUBPS	Subtracts 2 packed single-precision numbers

Examples

    SUBPS   XMM0, QWORD DATA3   ;subtracts DATA3 from XMM0
    SUBPS   XMM3, XMM2   ;subtracts XMM2 from XMM3

SUBLS	Subtracts scalar single-precision data

Examples

    SUBLS   XMM0, DWORD DATA3   ;subtracts DATA3 from XMM0
    SUBLS   XMM7, XMM2   ;subtracts XMM2 from XMM7

MULPD	Multiplies packed double-precision data

Examples

    MULPD   XMM0, OWORD DATA3   ;multiplies DATA3 times XMM0
    MULPD   XMM3, XMM2   ;multiplies XMM2 times XMM3

MULSD	Multiplies scalar double-precision data

Examples

```
MULSD    XMM0, QWORD DATA3        ;multiplies DATA3 times XMM0
MULSD    XMM3, XMM6               ;multiplies XMM6 times XMM3
```

MULPS	Multiplies 2 packed single-precision numbers

Examples

```
MULPS    XMM0, QWORD DATA3        ;multiplies DATA3 times XMM0
MULPS    XMM0, XMM2               ;multiplies XMM2 times XMM0
```

MULSS	Multiplies a single-precision number

Examples

```
MULSS    XMM0, DWORD DATA3        ;multiplies DATA3 times XMM0
MULSS    XMM1, XMM2               ;multiplies XMM2 times XMM1
```

DIVPD	Divides packed double-precision data

Examples

```
DIVPD    XMM0, OWORD DATA3     ;divides XMM0 by DATA3
DIVPD    XMM3, XMM2            ;divides XMM3 by XMM2
```

DIVSD	Divides scalar double-precision data

Examples

```
DIVSD    XMM0, OWORD DATA3     ;divides XMM0 by DATA3
DIVSD    XMM3, XMM6            ;divides XMM3 by XMM6
```

DIVPS	Divides 2 packed single-precision numbers

Examples

```
DIVPS    XMM0, QWORD DATA3     ;divides XMM0 by DATA3
DIVPS    XMM0, XMM2            ;divides XMM0 by XMM2
```

DIVSS	Divides a single-precision number

Examples

| DIVSS | XMM0, DWORD DATA3 | ;divides XMM0 by DATA3 |
| DIVSS | XMM1, XMM2 | ;divides XMM1 by XMM2 |

SQRTPD	Finds the square root of packed double-precision data

Examples

| SQRTPD | XMM0, OWORD DATA3 | ;finds square root of DATA3, result to XMM0 |
| SQRTPD | XMM3, XMM2 | ;finds square root of XMM2, result to XMM3 |

SQRTSD	Finds the square root of scalar double-precision data

Examples

| SQRTSD | XMM0, OWORD DATA3 | ;finds square root of DATA3, result to XMM0 |
| SQRTSD | XMM3, XMM6 | ;finds square root of XMM6, result to XMM3 |

SQRTPS	Finds the square root of 2 packed single-precision numbers

Examples

| SQRTPS | XMM0, QWORD DATA3 | ;finds square root of DATA3, result to XMM0 |
| SQRTPS | XMM0, XMM2 | ;finds square root of XMM2, result to XMM0 |

SQRTSS	Finds the square root of a single-precision number

Examples

| SQRTSS | XMM0, DWORD DATA3 | ;finds the square root of DATA3, result to XMM0 |
| SQRTSS | XMM1, XMM2 | ;finds the square root of XMM2, result to XMM1 |

RCPPS	Finds the reciprocal of a packed single-precision number

Examples

| RCPPS | XMM0, OWORD DATA3 | ;finds the reciprocal of DATA3, result to XMM0 |
| RCPPS | XMM3, XMM2 | ;finds the reciprocal of XMM2, result to XMM3 |

RCPSS	Finds the reciprocal of a single-precision number

Examples

| RCPSS | XMM0, OWORD DATA3 | ;finds the reciprocal of DATA3, result to XMM0 |
| RCPSS | XMM3, XMM6 | ;finds the reciprocal of XMM6, result to XMM3 |

RSQRTPS	Finds reciprocals of packed single-precision data

Examples

| RSQRTPS | XMM0, OWORD DATA3 | ;finds reciprocal of square root of DATA3 |
| RSQRTPS | XMM3, XMM2 | ;finds reciprocal of square root of XMM2 |

RSQRTSS	Finds the reciprocal of square root of a scalar single-precision number

Examples

| RSQRTSS | XMM0, OWORD DATA3 | ;finds reciprocal of square root of DATA3 |
| RSQRTSS | XMM3, XMM6 | ;finds reciprocal of square root of XMM6 |

MAXPD	Compares and returns the maximum packed double-precision floating-point number

Examples

| MAXPD | XMM0, OWORD DATA3 | ;compares numbers in DATA3, largest to XMM0 |
| MAXPD | XMM3, XMM2 | ;compares numbers in XMM2, largest to XMM3 |

MAXSD	Compares scalar double-precision data and returns the largest

Examples

| MAXSD | XMM0, OWORD DATA3 | ;compares numbers in DATA3, largest to XMM0 |
| MAXSD | XMM3, XMM6 | ;compares numbers in XMM6, largest to XMM3 |

MAXPS	Compares and returns the largest packed single-precision number

Examples

| MAXPS | XMM0, QWORD DATA3 | ;compares numbers in DATA3, largest to XMM0 |
| MAXPS | XMM0, XMM2 | ;compares numbers in XMM2, largest to XMM0 |

MAXSS	Compares scalar single-precision numbers and returns the largest

Examples

```
    MAXSS    XMM0, DWORD DATA3  ;compares numbers in DATA3, largest to XMM0
    MAXSS    XMM1, XMM2         ;compares numbers in XMM2, largest to XMM1
```

MINPD	Compares and returns the minimum packed double-precision floating-point number

Examples

```
    MINPD    XMM0, OWORD DATA3  ;compares numbers in DATA3, least to XMM0
    MINPD    XMM3, XMM2         ;compares numbers in XMM2, least to XMM3
```

MINSD	Compares scalar double-precision data and returns the smallest

Examples

```
    MINSD    XMM0, OWORD DATA3  ;compares numbers in DATA3, least to XMM0
    MINSD    XMM3, XMM6         ;compares numbers in XMM6, least to XMM3
```

MINPS	Compares and returns the smallest packed single-precision number

Examples

```
    MINPS    XMM0, QWORD DATA3  ;compares numbers in DATA3, least to XMM0
    MINPS    XMM0, XMM2         ;compares numbers in XMM2, least to XMM0
```

MINSS	Compares scalar single-precision numbers and returns the smallest

Examples

```
    MINSS    XMM0, DWORD DATA3  ;compares numbers in DATA3, least to XMM0
    MINSS    XMM1, XMM2         ;compares numbers in XMM2, least to XMM1
```

LOGIC INSTRUCTIONS

ANDPD	Ands packed double-precision data

Examples		
ANDPD	XMM0, QWORD DATA3	;ands DATA3 to XMM0
ANDPD	XMM2, XMM3	;ands XMM3 to XMM2

ANDNPD	Nands packed double-precision data

Examples		
ANDNPD	XMM0, OWORD DATA3	;Nands DATA3 to XMM0
ANDNPD	XMM4, XMM2	;Nands XMM2 to XMM4

ANDPS	Ands 2 packed single-precision data

Examples		
ANDPS	XMM0, QWORD DATA3	;ands DATA3 to XMM0
ANDPS	XMM3, XMM2	;ands XMM2 to XMM3

ANDNPS	Nands 2 packed single-precision data

Examples		
ANDNPS	XMM0, DWORD DATA3	;Nands DATA3 to XMM0
ANDNPS	XMM7, XMM2	;Nands XMM2 to XMM7

ORPD	Ors packed double-precision data

Examples		
ORPD	XMM0, OWORD DATA3	;ors DATA3 to XMM0
ORPD	XMM2, XMM3	;ors XMM3 to XMM2

ORPS	Ors 2 packed single-precision numbers

Examples		
ORPS	XMM0, OWORD DATA3	;ors DATA3 to XMM0
ORPS	XMM3, XMM2	;ors XMM2 to XMM3

XORPD	Exclusive-ors packed double-precision data

Examples		
XORPD	XMM0, OWORD DATA3	;exclusive-ors DATA3 to XMM0
XORPD	XMM2, XMM3	;exclusive-ors XMM3 to XMM2

XORPS	Exclusive-ors packed double-precision data

Examples		
XORPS	XMM0, OWORD DATA3	;exclusive-ors DATA3 to XMM0
XORPS	XMM2, XMM3	;exclusive-ors XMM3 to XMM2

COMPARISON INSTRUCTIONS

CMPPD	Compares packed double-precision numbers

Examples		
CMPPD	XMM0, OWORD DATA3	;compares DATA3 with XMM0
CMPPD	XMM2, XMM3	;compares XMM3 with XMM2

CMPSD	Compares scalar double-precision data

Examples		
CMPSD	XMM0, QWORD DATA3	;compares DATA3 with XMM0
CMPSD	XMM3, XMM2	;compares XMM2 with XMM3

CMPISD	Compares scalar double-precision data and sets EFAGS

Examples		
CMPISD	XMM0, OWORD DATA3	;compares DATA3 with XMM0
CMPISD	XMM2, XMM3	;compares XMM3 with XMM2

UCOMISD	Compares scalar unordered double-precision numbers and changes EFLAGS

Examples

| UCOMISD | XMM0, QWORD DATA3 | ;compares DATA3 with XMM0 |
| UCOMISD | XMM3, XMM2 | ;compares XMM2 with XMM3 |

CMPPS	Compares packed single-precision data

Examples

| CMPPS | XMM0, OWORD DATA3 | ;compares DATA3 with XMM0 |
| CMPPS | XMM2, XMM3 | ;compares XMM3 with XMM2 |

CMPSS	Compares 2 packed single-precision numbers

Examples

| CMPSS | XMM0, QWORD DATA3 | ;compares DATA3 with XMM0 |
| CMPSS | XMM3, XMM2 | ;compares XMM2 with XMM3 |

COMISS	Compares scalar single-precision data and changes EFLAGS

Examples

| COMISS | XMM0, OWORD DATA3 | ;compares DATA3 with XMM0 |
| COMISS | XMM2, XMM3 | ;compares XMM3 with XMM2 |

UCOMISS	Compares unordered single-precision numbers and changes EFLAGS

Examples

| UCOMISS | XMM0, QWORD DATA3 | ;compares DATA3 with XMM0 |
| UCOMISS | XMM3, XMM2 | ;compares XMM2 with XMM3 |

DATA CONVERSION INSTRUCTIONS

SHUFPD	Shuffles packed double-precision numbers

Examples

SHUFPD	XMM0, OWORD DATA3	;shuffles DATA3 with XMM0
SHUFPD	XMM2, XMM2	;swaps upper and lower quadword in XMM2

UNPCKHPD	Unpacks the upper double-precision number

Examples

UNPCKHPD	XMM0, OWORD DATA3	;unpacks DATA3 into XMM0
UNPCKHPD	XXM3, XMM2	;unpacks XMM2 into XMM3

UNPCKLPD	Unpack the lower double-precision number

Examples

UNPCKLPD	XMM0, OWORD DATA3	;unpacks DATA3 into XMM0
UNPCKLPD	XMM3, XMM2	;unpacks XMM2 into XMM3

SHUFPS	Shuffles packed single-precision numbers

Examples

SHUFPS	XMM0, QWORD DATA3	;shuffles DATA3 with XMM0
SHUFPS	XMM2, XMM2	;swaps upper and lower quadword in XMM2

UNPCKHPS	Unpacks the lower double-precision number

Examples

UNPCKHPS	XMM0, QWORD DATA3	;unpacks DATA3 into XMM0
UNPCKHPS	XMM3, XMM2	;unpacks XMM2 into XMM3

UNPCKLPSD	Unpacks the lower double-precision number

Examples

UNPCKLPSD	XMM0, QWORD DATA3	;unpacks DATA3 into XMM0
UNPCKLPSD	XXM3, XMM2	;unpacks XMM2 into XMM3

APPENDIX C

Flag-Bit Changes

This appendix shows only the instructions that actually change the flag bits. Any instruction not listed does not affect any of the flag bits.

Instruction	O	D	I	T	S	Z	A	P	C
AAA	?				?	?	*	?	*
AAD	?				*	*	?	*	?
AAM	?				*	*	?	*	?
AAS	?				?	?	*	?	*
ADC	*				*	*	*	*	*
ADD	*				*	*	*	*	*
AND	0				*	*	?	*	0
ARPL						*			
BSF						*			
BSR						*			
BT									*
BTC									*
BTR									*
BTS									*
CLC									0
CLD		0							
CLI			0						
CMC									*
CMP	*				*	*	*	*	*
CMPS	*				*	*	*	*	*
CMPXCHG	*				*	*	*	*	*
CMPXCHG8B						*			
DAA	?				*	*	*	*	*
DAS	?				*	*	*	*	*
DEC	*				*	*	*	*	
DIV	?				?	?	?	?	?
IDIV	?				?	?	?	?	?
IMUL	*				?	?	?	?	*
INC	*				*	*	*	*	

Instruction	O	D	I	T	S	Z	A	P	C
IRET	*	*	*	*	*	*	*	*	*
LAR						*			
LSL						*			
MUL	*				?	?	?	?	*
NEG	*				*	*	*	*	*
OR	0				*	*	?	*	0
POPF	*	*	*	*	*	*	*	*	*
RCL/RCR	*								*
REPE/REPNE						*			
ROL/ROR	*								*
SAHF					*	*	*	*	*
SAL/SAR	*				*	*	?	*	*
SHL/SHR	*				*	*	?	*	*
SBB	*				*	*	*	*	*
SCAS	*				*	*	*	*	*
SHLD/SHRD	?				*	*	?	*	*
STC									1
STD		1							
STI			1						
SUB	*				*	*	*	*	*
TEST	0				*	*	?	*	0
VERR/VERW						*			
XADD	*				*	*	*	*	*
XOR	0				*	*	?	*	0

APPENDIX D

Answers to Selected Even-Numbered Questions and Problems

CHAPTER 1

2. Herman Hollerith
4. Konrad Zuse
6. ENIAC
8. Augusta Ada Bryon
10. A machine that stores its program in the memory system.
12. Over 100,000,000
14. 16M
16. 1993
18. 1999
20. Millions of instructions per second
22. 1 or a 0
24. 1024K
26. About 1,000,000
28. 640K
30. 1M
32. 80386, 80486, and Pentium (Note that the Pentium Pro through Pentium 4 address 64G)
34. The system BIOS (basic input/output system) contains programs to set up the computer and interface it with the hardware.
36. The AT model is designed to operate with the 80286 and the XT with the 8086/8088.
38. 8- or 16-bit interface cards (EISA holds up to 32-bit interface cards)
40. Advanced graphics port (data speeds up to 533M bytes per second)
42. Expanded memory system
44. Drivers may be stored in the TPA or upper memory.
46. The CONFIG.SYS file configures the personal computer by loading drivers and other devices into the memory on boot-up.
48. The COMMAND.COM program processes keyboard information at the DOS prompt.
50. High memory is located at memory locations 100000H–10FFEFH.
52. Upper memory blocks are placed in the systems area by DOS.
54. Refer to Figure 1–4 in Chapter 1.
56. The three buses are address, data, and control.
58. The memory read control signal ($\overline{\text{MRDC}}$) causes a memory read operation.
60. A memory read operation is performed.
62. (a) 13.25 (b) 57.1875 (c) 43.3125 (d) 7.0625
64. (a) 163.1875 (b) 297.75 (c) 172.859375 (d) 4,011.1875 (e) 3,000.0578125
66. (a) 0.101_2, 0.5_8, and $0.D_{16}$ (b) 0.00000001_2, 0.002_8, and 0.01_{16} (c) 0.10100001_2, 0.502_8, and $0.A1_{16}$ (d) 0.11_2, 0.6_8, and $0.C_{16}$ (e) 0.1111_2, 0.74_8, and $0.F_{16}$
68. (a) C2 (b) 10FD (c) B.C (d) 10 (e) 8BA
70. (a) 0111 1111 (b) 0101 0100 (c) 0101 0001 (d) 1000 0000
72. (a) 46 52 4F 47 (b) 41 72 63 (c) 57 61 74 65 72 (d) 57 65 6C 6C
74. MESS DB 'What time is it?'
76. (a) 0000 0011 1110 1000 (b) 1111 1111 1000 1000 (c) 0000 0011 0010 0000 (d) 1111 0011 0111 0100

78. (a) 34 12 (b) 22 A1 (c) 00 B1
80. DATA2   DW 123AH
82. (a) −128 (b) +51 (c) 110 (d) −118
84. (a) 0 01111111 10000000000000000000000
    (b) 1 10000010 01010100000000000000000
    (c) 0 10000101 10010001000000000000000
    (d) 1 10001001 00101100000000000000000

CHAPTER 2

2. 16 bits
4. EBX
6. The instruction pointer is used by the microprocessor to locate the next instruction in a program.
8. No
10. The interrupt flag (I)
12. In the real mode, a segment register locates the start of a 64K-byte memory segment.
14. (a) 12000H (b) 21000H (c) 24A00H (d) 25000H (e) 3F12DH
16. ES:DI
18. Stack segment plus the stack offset
20. (a) 12000H (b) 21002H (c) 26200H (d) A1000H (e) 2CA00H
22. Any location between and including 000000H–FFFFFFH
24. The segment register contains a selector that chooses a descriptor from either the local or global descriptor table. It also contains the requested privilege level.
26. A00000H–A01000H
28. Base address = 00280000H and end address = 00290FFFH
30. 3
32. 64K bytes
34.

03	10
90	00
00	00
2F	FF

36. The LDT is addressed through the local descriptor table register.
38. The program invisible register is the cache portion of the segment register, the task register, and also the descriptor table register.
40. 4K bytes
42. 1024
44. Page directory 000H and page table entry 200H
46. The TLB stores the last 22 linear-to-physical address translation from the paging unit.

CHAPTER 3

2. AH, AL, BH, BL, CH, CL, DH, and DL
4. EAX, EBX, ECX, EDX, ESP, EBP, EDI, and ESI
6. You may not mix register sizes.
8. (a) MOV EDX,EBX (b) MOV CL,BL (c) MOV BX,SI (d) MOV AX,DS (e) MOV AH,AL
10. #
12. .CODE
14. Opcode field
16. This is an assembly language directive that returns control to DOS.
18. The .STARTUP directive loads the DS register with the segment address of the data segment.
20. The [ ] symbols denote indirect addressing.
22. Memory-to-memory transfers are not allowed.
24. MOV WORD PTR [DI],3
26. The MOV BX,DATA instruction copies the contents of a data segment memory location DATA into BX, while the MOV BX,OFFSET DATA instruction loads BX with the offset address of DATA.
28. Nothing is wrong with this instruction; this is an alternate to MOV AL,[BX+SI].
30. (a) 11750H (b) 11950H (c) 11700H
32. BP/EBP

```
34. FIELDS  STRUC

    F1      DW  ?
    F2      DW  ?
    F3      DW  ?
    F4      DW  ?
    F5      DW  ?

    FIELDS  ENDS
```

36. Direct, indirect, and stack
38. The intrasegment jump is within a segment, while the intersegment jump is to any location in the memory system.
40. 32-bit
42. Short
44. JMP BX
46. Two bytes are stored for a 16-bit PUSH and 4 by a 32-bit PUSH.
48. AX, CX, DX, BX, SP, BP, DI and SI
50. PUSHFD

CHAPTER 4

2. The W bit selects either a byte (W = 0) or a word/doubleword (W = 1). The D bit selects the direction of flow between the register field and the register/memory field.
4. DL
6. DS:[BX+DI]
8. MOV AX,DI
10. 8B 77 02
12. You should never change CS without also changing IP. This instruction would most likely cause the system to crash because only the segment portion of the address of the next instruction is changed.
14. 16-bit
16. AX, CX, DX, BX, SP, BP, SI, and DI
18. (a) The PUSH AX instruction pushes the contents of AX onto the stack. (b) The POP ESI instruction removes a 32-bit number from the stack and places it into ESI. (c) The PUSH [BX] instruction pushes the 16-bit contents of the data segment memory location addressed by BX onto the stack. (d) The PUSHFD instruction pushes the EFLAG register onto the stack. (e) The POP DS instruction removes a 16-bit number from the stack and places it into the DS register. (f) The PUSHD 4 instruction places a 32-bit number 4 onto the stack.
20. The PUSH EAX instruction places bits 31–24 of EAX into memory location 20FFH, bits 23–16 into 20FEH, bits 15–8 into 20FDH, and bits 7–0 into 20FCH. After the data are stored, the contents of SP are decremented to four, which results in 20FCH.
22. One possibility is 200H in both registers.
24. The MOV using the OFFSET is more efficient than the LEA instruction for all microprocessors prior to the Pentium.
26. This instruction loads DS and BX with the 32-bit number stored at memory location NUMB.
```
28. MOV   BX,NUMB
    MOV   DX,BX
    MOV   SI,BX
```
30. The CLD instruction clears direction and the STD instruction sets it.
32. The LODSB instruction copies the contents of the data segment memory location addressed by SI into AL. Next, the contents of SI are either incremented or decremented by a 1, depending on the state of the direction flag.
34. The OUTSB instruction outputs the contents of the data segment memory location addressed by SI to the I/O port addressed by DX. Next, the contents of SI are either incremented or decremented by 1, depending on the state of the direction flag.
```
36. MOV   SI,OFFSET SOURCE
    MOV   DI,OFFSET DEST
    MOV   CX,12
    REP   MOVSB
```
38. XCHG EBX,ESI
40. The XLAT instruction adds the contents of AL to the contents of BX to form a data segment offset address that loads a byte of data from a table into AL.
42. The IN AL,12H instruction inputs a byte of data from I/O port 0012H into AL.
44. The segment override prefix allows the default segment to be changed to any other segment.
```
46. XCHG   AX,BX
    XCHG   ECX,EDX
    XCHG   SI,DI
```

48. This instruction must be encoded as a series of DB (define bytes) definitions.

50. The DB directive is used to define or store bytes, DW defines words, and DD defines doublewords.

52. The EQU directive allows one label to be equated to another or a constant.

54. The .MODEL directive identifies the type of memory model used to generate a program.

56. Full segment definitions

58. The PROC directive indicates the start of a procedure and the ENDP directive indicates the end of a procedure.

60. The directive USE16 is placed on the line following the SEGMENT directive for full segment definitions. If models are in effect, the .486 switch follows the model statement to select the 16-bit instruction mode.

62.
```
COPS   PROC   FAR
       MOV    AX,CS:DATA1
       MOV    BX,AX
       MOV    CX,AX
       MOV    DX,AX
       MOV    SI,AX
       RET
COPS   ENDP
```

CHAPTER 5

2. You may not mix register sizes.

4. Sum = 3100H, C = 0, A = 1, S = 0, Z = 0, and O = 0

6.
```
ADD    AX,BX
ADD    AX,CX
ADD    AX,DX
ADD    AX,SP
MOV    DI,AX
```

8. ADC DX,BX

10. The assembler cannot determine whether the memory location is a byte, word, or doubleword.

12. Difference = 81H, C = 0, A = 0, S = 1, Z = 0, and O = 0

14. DEC EBX

16. Both instructions are identical, except that the CMP instruction does not change the destination.

18. The product is found in DX:AX, where DX is the most significant part.

20. EDX:EAX

22.
```
MOV    DL,5
MOV    AL,DL
MUL    DL
MUL    DL
```

24. AX

26. If an overflow or divide by zero error occurs, the microprocessor executes a divide error interrupt.

28. AH

30. DAA (BCD addition) and DAS (BCD subtraction)

32. AAM converts AX to BCD by dividing it by a 10. The result (00–99) is found in AH and AL.

34.
```
PUSH   AX
MOV    AL,BL
ADD    AL,DL
DAA
MOV    DL,AL
MOV    AL,BH
ADC    AL,DH
DAA
MOV    DH,AL
POP    AX
ADC    AL,CL
DAA
MOV    CL,AL
MOV    AL,AH
ADC    AL,CH
DAA
MOV    CH,AL
```

36.
```
MOV    BH,DH
AND    BH,1FH
```

38.
```
MOV    SI,DI
OR     SI,1FH
```

40.
```
OR     AX,0FH
AND    AX,1FFFH
XOR    AX,0E0H
```

42. TEST CH, 4 or BT CH, 2

44. (a) SHR DI, 3 (b) SHL AL, 1 (c) ROL AL, 3 (d) RCR EDX, 1 (e) SAR DH, 1

46. Extra

48. The REPE prefix continues to compare while an equal outcome from the comparison occurs or while CX is not equal to a zero.

50. The CMPSB instruction compares the byte contents of two memory locations.

52. The letter C is displayed on the video screen.

CHAPTER 6

2. A near JMP

4. A far JMP

6. (a) near (b) short (c) far

8. EIP/IP

10. The JMP AX instruction is a near jump to the offset address loaded in AX.

12. The JMP [DI] instruction is a near jump that obtains the jump address from the data segment memory location addressed by DI. The other JMP is a far jump.

14. JA is the jump above instruction that jumps if the destination is above the source.

16. JE, JNE, JG, JGE, JL, and JLE

18. JBE and JA

20. SETZ

22. ECX

24.
```
          MOV     DI,OFFSET DATA
          CLD
          MOV     CX,150H
          MOV     AL,0
     AGAIN:
          STOSB
          LOOP    AGAIN
```

26.
```
          CMP     AL,3
          JNE     ?0000
          ADD     AL,2
     ?0000:
```

28.
```
     MOV    SI,OFFSET BLOCKA
     MOV    DI,OFFSET BLOCKB
     CLD
     .REPEAT
        LODSB
        STOSB
     .UNTIL AL==0
```

30.
```
     MOV    SI,OFFSET BLOCKA
     MOV    DI,OFFSET BLOCKB
     CLD
     MOV    AL,0
     .WHILE AL!=12H
         LODSB
         ADD    AL,ES:[DI]
         STOSB
     .ENDW
```

32. Both instructions store the return address on the stack, and then jump to the procedure. Note that the return address for a near CALL is EIP/IP, and the return address for the far CALL is EIP/IP and CS.

34. RET

36. By using NEAR or FAR to the right of the PROC directive.

38.
```
     CUBE   PROC    NEAR   USES AX DX
            MOV     AX,CX
            MUL     CX
            MUL     CX
            RET
     CUBE   ENDP
```

40.
```
     MOV    EDI,0
     ADD    EAX,EBX
     .IF CARRY?
         MOV     EDI,1
     .ENDIF
     ADD    EAX,ECX
     .IF CARRY?
         MOV     EDI,1
     .ENDIF
     ADD    EAX,EDX
     .IF CARRY?
         MOV     EDI,1
     .ENDIF
```

42. INT, INTO, and DIV (if an error occurs)

44. Interrupt type number 0 is used for a divide error.
46. The IRET pops the return address from the stack, as does a RET, but it also pops the flags, which RET does not.
48. INTO interrupts a program if the overflow flag is set.
50. The STI instruction enables the INTR pin and the CLI instruction disables INTR.
52. Interrupt vector number 9
54. The interrupt occurs when the boundary (in a 16- or 32-bit register) is outside of the boundary set by a pair of words or doublewords stored in the memory.
56. BP
58. A null string is a character string that ends with a 00H.
60. Change MES2 to the following:

```
MES2    DB      'Your Name',0
```

CHAPTER 7

2. The result depends on the options set for the assembler, but the TEST.OBJ file is always generated while the TEST.LST and TEST.CRF files can also be generated.
4. PUBLIC is used to declare that a label or even an entire segment is public to other modules.
6. NEAR PTR, FAR PTR, BYTE, WORD, or DWORD
8. The MACRO directive starts a macro and the ENDM ends it.
10. Parameters are indicated next to the MACRO statement. Parameters are placed in a program next to the name of the macro, as operands.
12. The LOCAL directive must immediately follow the macro statement and must contain all labels used within the macro sequence.

```
14. ADDM    MACRO   LIST,LENGTH
            MOV     CX,LENGTH
            MOV     SI,OFFSET LIST
            MOV     AX,0
            CLD
            .REPEAT
                ADD     AX,[SI]
                ADD     SI,2
            .UNTILCXZ
            ENDM
16. RANDOM  MACRO
            LOCAL   R1
    R1:
            INC     CL
            MOV     AH,6
            MOV     DL,-1
            INT     21H
            JZ      R1
            ENDM
18. DISP    MACRO   PARA
            IFB     <PARA>
                MOV     AH,6
                MOV     DL,13
                INT     21H
                MOV     DL,10
            ENDIF
            IFNB    <PARA>
                MOV     DL,PARA
            ENDIF
            MOV     AH,6
            INT     21H
            ENDM
20. DIPS    PROC    NEAR
            MOV     AH,6
            MOV     DL,-1
            INT     21H
            JZ      DISP
            .IF  AL==0
                PUSH    AX
                SHR     AL,4
                ADD     AL,30H
                .IF  AL>'9'
                    ADD     AL,7
                .ENDIF
                MOV     AH,6
                MOV     DL,AL
                INT     21H
```

```
            POP     AX
            AND     AL,0FH
            ADD     AL,30H
            .IF AL>'9'
                ADD     AL,7
            .ENDIF
            MOV   AH,6
            MOV   DL,AL
                INT   21H
            .ENDIF
            RET
    DISP    ENDP
```

22. INT 33H
24. If AH loaded with a 24H before INT 33H executes, the mouse type is returned in CH.
26. If not disabled, the mouse driver can, at worst, lock up the computer; at best, it can display repeated mouse arrows all over the screen.
28. A large number is converted by repeated divisions by 10.
30. 30H
32. A 30H is first subtracted from each digit, and then the most significant digit is multiplied by 100 and the middle digit is multiplied by 10. The three are then summed to generate a binary value.
34.
```
    HEXAS   PROC    NEAR        ;AL is converted
            AND     AL,0FH
            MOV     BX,OFFSET LOOK
            XLAT    CS:LOOK
            RET
    HEXAS   ENDP
    LOOK    DB      30H,31H,32H,33H
            DB      34H,35H,36H,37H
            DB      38H,39H,41H,42H
            DB      43H,44H,45H,46H
```

36. XLAT SS:LOOK
38. The boot sector contains a program that loads DOS or any operating system into the memory system. The FAT (file allocation table) holds information that indicates which clusters are in use and which are free. The root directory is the main directory on the disk drive.
40. The bootstrap loader is a program located in the boot sector.
42. The attribute byte indicates the type of directory entry, such as directory name, file name, volume name, and so forth.
44. 4G bytes
46. 256 bytes
48.
```
    RRNAM   PROC    NEAR    USES DS ES
            MOV     AX,CS
            MOV     DS,AX
            MOV     ES,AX
            MOV     DI,OFFSET NEW1
            MOV     SI,OFFSET OLD1
            MOV     AH,56H
            INT     21H
            RET
    RRNAM   ENDP
    NEW1    DB      'TEST.LIS',0
    OLD1    DB      'TEST.LST',0
```

50.
```
    .MODEL  TINY
    .CODE
    DISP    MACRO   PARA
            MOV     DL,PARA
            MOV     AH,6
            INT     21H
            ENDP
    .STARTUP
            MOV     CX,8
            .REPEAT
                DISP    13
                DISP    10
                DISP    ' '
                MOV     AL,8
                SUB     AL,CL
                ADD     AL,30H
                DISP    AL
                DISP    13
                DISP    10
                DISP    '2'
```

```
                    DISP     ' '
                    DISP     '='
                    MOV      AX,100H
                    SHR      AX,CL
                    .IF      AL>99
                         DISP     '1'
                         SUB      AL,100
                         AAM
                         ADD      AX,3030H
                         DISP     AH
                         DISP     AL
                    .ELSE
                         AAM
                         .IF      AH!=0
                              ADD      AH,30H
                              DISP     AH
                         .ENDIF
                         ADD   AL,30H
                         DISP  AL
                    .ENDIF


              .UNTILCXZ
         .EXIT
         END
52.  .MODEL TINY
     .386
     .CODE
     .STARTUP
              CALL     GETADR
              CALL     DISPA
     .EXIT
     DISP    MACRO    PARA
             MOV      DL,PARA
             MOV      AH,6
             INT      21H
             ENDM
     GETADR  PROC     NEAR
             MOV      EDI,0
             MOV      ESI,0
             .WHILE 1
                  CALL     GETN
                  .BREAK IF AL==13
                  SHL      EDI,4
                  MOV      AH,0
                  ADD      DI,AX
                  ADD      AL,30H
                  .IF AL> '9'
                       ADD      AL,7
                  .ENDIF
                  DISP     AL
             .ENDW
             MOV      SI,DI
             AND      SI,0FH
             SHR      EDI,4
             MOV      DS,DI
             SHL      EDI,4
             RET
     GETADR  ENDP
     DISPA   PROC     NEAR
             MOV      CX,256
             CALL     ADR
             .REPEAT
                  LODSB
                  ADD      AX,3030H
                  .IF AH> '9'
                       ADD      AH,7
                  .ENDIF
                  .IF AL> '9'
                       ADD      AL,7
                  .ENDIF
                  DISP     AH
                  DISP     AL
                  DISP     ' '
                  MOV      AX,SI
```

```
        AND     AX,15
        .IF ZERO?
            CALL    ADR
        .ENDIF
    .UNTILCXZ
    RET
DISPA   ENDP
ADR     PROC    NEAR    USES CX EDI
    DISP    13
    DISP    10
    MOV     CX,5
    ADD     EDI,ESI
    ROL     EDI,12
    .REPEAT
        ROL EDI,4
        MOV AX,DI
        AND AL,0FH
        ADD AL,30H
        .IF AL> '9'
            ADD     AL,7
        .ENDIF
        DISP AL
    .UNTILCXZ
    DISP        ' '
    RET
ADR     ENDP
END
```

CHAPTER 8

2. The in-line assembler does not accept data definition statements such as DB.
4. AX, BX, CX, DX, and ES for 16-bit applications; and EAX, EBX, ECX, EDX, and ES for 32-bit applications.
6. ENTER and LEAVE
8. This instruction uses SI as a pointer to a character in string1 that is copied into the DL register.
10. The compiled version of Example 8–6 is shorter; C/C++ libraries are large because they must handle many situations.
12. The in-line assemblers in both versions are essentially identical.
14. This allows the program to use console functions such as _putch and _getche.
16. An embedded application would use the IN and OUT assembly language commands for I/O.
18. A number is converted to any base by dividing by the base. For example, if you divide by an 8, the remainder can only contain a 0 through a 7. These are the only valid digits in a base 8 or octal number. The process of dividing by the number base will always convert a number to that number base.
20. The time for the CMOS memory is stored in BCD code.
22. The PUBLIC statement identifies a procedure or a label as being available to an outside program module.
24. It indicates that the GetIt function returns no value and it accepts one integer.
26. The edit screen can be used for both C/C++ modules and assembly language modules.
28. Scan uses byte-sized data.
30.
```
.386
.model flat, C
.code

public Myne

Myne proc , \
data1: byte

    mov     al,data1
    .if al >= 'a' && al <= 'z'
        sub     al,7
    .endif
    ret

Myne endp
end
```

CHAPTER 9

2. As long as you don't exceed a logic zero current of 2.0 mA, they are TTL-compatible.
4. Address bits A7–A0.
6. A logic 0 on $\overline{RD}$ indicates that the microprocessor is either reading data from memory or I/O.

8. The CLK input must be a TTL-compatible square-wave with a 33% duty cycle.
10. The $\overline{WR}$ signal indicates that the microprocessor has placed data on its data bus to be written to the memory or an I/O device.
12. If DT/$\overline{R}$ is a logic 1, it indicates that the microprocessor's data bus is transmitting data to the memory or I/O.
14. $\overline{S2}$–$\overline{S0}$
16. The queue tracking status bits indicate the condition of the queue within the microprocessor for the arithmetic coprocessor.
18. Three
20. 2.33 MHz
22. AD19–AD0
24. An eight-bit transparent latch (74LS373).
26. Buffers are required because of the drive current (2.0 mA) available at the output pins of the microprocessor.
28. Four
30. A read or a write
32. (a) State T1 is used by the microprocessor to provide the memory or I/O with the address. (b) State T2 provides access time to the memory and also is where the READY input is sampled. (c) State T3 is where the data are sampled or sent to the memory or I/O. (d) State T4 is used to deactivate the control signals.
34. 400 ns (2 clocks)
36. This input is used to request wait states.
38. Minimum mode is used unless the system must contain the arithmetic coprocessor; then we use the maximum mode.

CHAPTER 10

2. (a) 256 (b) 2048 (c) 4096 (d) 8192
4. The $\overline{CS}$ or $\overline{CE}$ pin on a memory device is used to select or enable the device so it can perform a read or a write operation.
6. The $\overline{WE}$ pin causes the memory to perform a write operation, provided that the $\overline{CS}$ or $\overline{CE}$ pin is also active.
8. The 5 MHz version of the 8088 allows 460 ns of time for the memory to access data. A 450 ns memory device will only function if the amount time required for the address decoder and buffers in a system is less than 10 ns, which is unlikely.
10. The SRAM (static RAM) is a device that retains data for as long as power is applied to the memory device. The SRAM can be read or written.
12. 250 ns
14. The address inputs to most DRAM devices are multiplexed, which allows a 16-bit address to be sent to the DRAM through address input pins.
16. 2–4 ms
18.

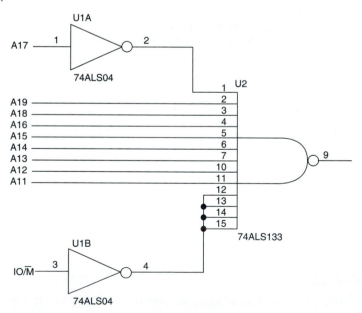

FIGURE D–1

20. One of the eight outputs becomes a logic 0, as dictated by the A, B, and C address inputs.

22.

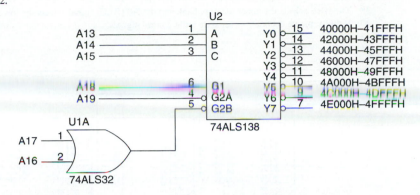

FIGURE D–2

24. The PROM address decoder is more suited to memory address decoding than the 74LS138 in many cases.

26.

OE	A8	A7	A6	A5	A4	A3	A2	A1	A0	O0	O1	O2	O3	O4	O5	O6	O7
0	0	0	0	0	1	1	0	0	0	0	1	1	1	1	1	1	1
0	0	0	0	0	1	1	0	0	1	1	0	1	1	1	1	1	1
0	0	0	0	0	1	1	0	1	0	1	1	0	1	1	1	1	1
0	0	0	0	0	1	1	0	1	1	1	1	1	0	1	1	1	1
0	0	0	0	0	1	1	1	0	0	1	1	1	1	0	1	1	1
0	0	0	0	0	1	1	1	0	1	1	1	1	1	1	0	1	1
0	0	0	0	0	1	1	1	1	0	1	1	1	1	1	1	0	1
0	0	0	0	0	1	1	1	1	1	1	1	1	1	1	1	1	0

28. EQUATIONS

```
/O1 = A19 * /A18 * A17 * A16 * /A15 * /A14 * /A13
/O2 = A19 * /A18 * A17 * A16 * /A15 * /A14 * A13
/O3 = A19 * /A18 * A17 * A16 * /A15 * A14 * /A13
/O4 = A19 * /A18 * A17 * A16 * /A15 * A14 * A13
/O5 = A19 * /A18 * A17 * A16 * A15 * /A14 * /A13
/O6 = A19 * /A18 * A17 * A16 * A15 * /A14 * A13
/O7 = A19 * /A18 * A17 * A16 * A15 * A14 * /A13
/O8 = A19 * /A18 * A17 * A16 * A15 * A14 * A13
```

30.

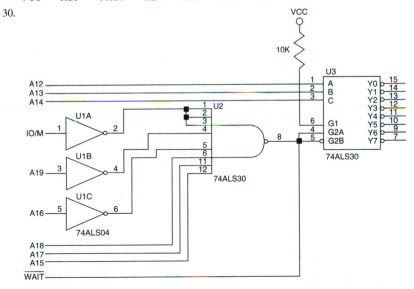

FIGURE D–3

34. 5
36. One
38. The $\overline{BHE}$ selects the upper memory bank and the A0 ($\overline{BLE}$) signal selects the lower memory bank.
40. Either separate decoders or separate write control signals.
42. Low
46.

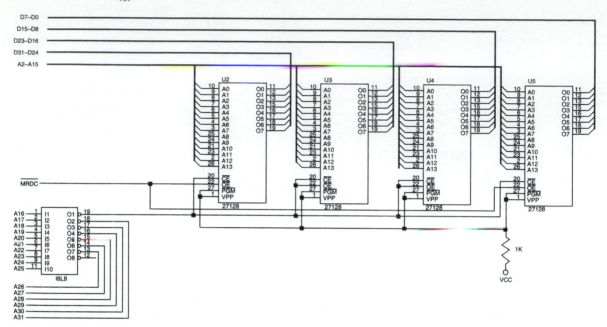

FIGURE D–4

48. The $\overline{RAS}$ cycle selects a new row address, but does not select the DRAM for a read or a write.
50. 15.625 μs
52. The BS pin selects a bank of memory.

CHAPTER 11

2. The fixed I/O port is stored in the memory immediately following the opcode.
4. Register DX
6. The OUTSB instruction copies the contents of the data segment memory location addressed by SI to the data bus, where it is written to the I/O device addressed by DX. After the transfer, the contents of SI are incremented or decremented, as dictated by the direction flag.
8. The difference between the memory-mapped I/O and the isolated I/O is that with isolated I/O, all memory locations are available to the system.
10. The basic output interface is a latch that holds data output from the microprocessor.
12. Low bank
14. The contact bounce eliminator removes mechanical bounces from a switch by using a latch or flip-flop.
16.

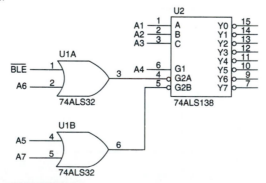

FIGURE D–5

18.

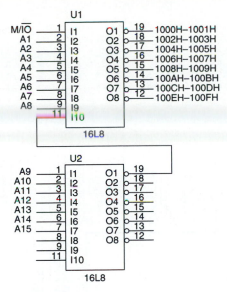

FIGURE D–6

;Equations for U1

EQUATIONS

```
/O1 = /U2 * /A4 * /A5 * /A6 * /A7 * A8 * /A3 * /A2 * /A1 * /MIO
/O2 = /U2 * /A4 * /A5 * /A6 * /A7 * A8 * /A3 * /A2 * A1 * /MIO
/O3 = /U2 * /A4 * /A5 * /A6 * /A7 * A8 * /A3 * A2 * /A1 * MIO
/O4 = /U2 * /A4 * /A5 * /A6 * /A7 * A8 * /A3 * A2 * A1 * /MIO
/O5 = /U2 * /A4 * /A5 * /A6 * /A7 * A8 * A3 * /A2 * /A1 * /MIO
/O6 = /U2 * /A4 * /A5 * /A6 * /A7 * A8 * A3 * /A2 * A1 * /MIO
/O7 = /U2 * /A4 * /A5 * /A6 * /A7 * A8 * A3 * A2 * /A1 * /MIO
/O8 = /U2 * /A4 * /A5 * /A6 * /A7 * A8 * A3 * A2 * A1 * /MIO
;Equation for U2

EQUATIONS

/U2 = /A9 * /A10 * /A11 * /A12 * /A13 * /A14 * /A15
```

20.

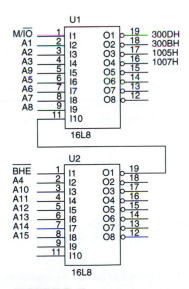

FIGURE D–7

```
;Equations for U1

EQUATIONS

/O1 = /U2 * /A9 * /A5 * /A6 * /A7 * A8 * A3 * A2 * /A1 * /MIO
/O2 = /U2 * /A9 * /A5 * /A6 * /A7 * A8 * A3 * /A2 * A1 * /MIO
/O3 = /U2 * A9 * /A5 * /A6 * /A7 * A8 * /A3 * A2 * /A1 * MIO
/O4 = /U2 * A9 * /A5 * /A6 * /A7 * A8 * /A3 * A2 * A1 * /MIO


;Equation for U2

EQUATIONS

/U2 = /BHE * /A4 * /A10 * /A11 * /A12 * /A13 * /A14 * /A15
```

22. D7–D0
24. 24
26. A1 and A0
28.

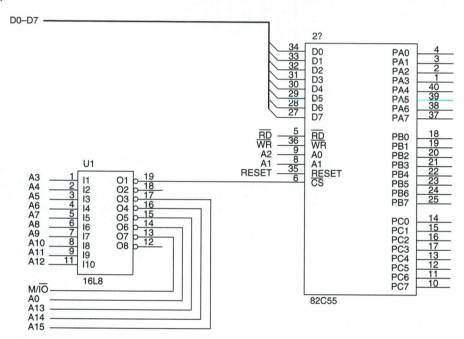

FIGURE D–8

```
EQUATION

/CS = /A15*/A14*/A13*/A12*/A11*/A10*A9*A8*A7*/A6*/A5*/A4*/A3*/MIO
```

30. Latched I/O (mode 0), strobed I/O (mode 1), and bi-directional I/O (mode 2)
32. Whenever a coil is energized, the current causes the permanent magnet armature to step (move) to the next position.
34. MOV   AL,0FH
    OUT   COMMAND,AL
36. The $\overline{\text{ACK}}$ signal is an output that signals that the data have been removed from the port.
38. IN   AL,PORTC
    TEST AL,16
    JNZ   FOR_A_ONE
40. PC0–PC2
42. A display position is selected by changing the display position address found in the set address command.
44. The busy line is tested by executing the busy flag command and then testing the most-significant bit.
46. 10 ms to 20 ms
48. Two wait states.
50. An overrun error occurs if the internal FIFO fills before the data are input to the microprocessor.

52.

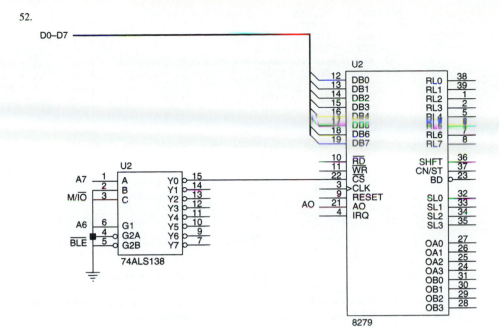

FIGURE D–9

54. 10 MHz
56.

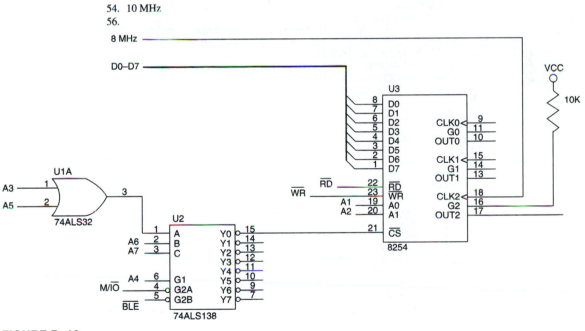

FIGURE D–10

```
MOV    AL,0B6H
OUT    16H,AL
MOV    AL,64H
OUT    14H,AL
MOV    AL,00H
OUT    14H,AL
```

58. Least-significant
60. ;using a 1 MHz clock
   ;
```
MOV    AL,74H
OUT    CONTROL,AL
```

```
        MOV     AL,65H          ;count of 101
        OUT     TIMER1,AL
        MOV     AL,0
        OUT     TIMER1,AL
```

62. Asynchronous serial data are data sent without a clock pulse.

64.
```
    LINE    EQU     023H
    LSB     EQU     020H
    MSB     EQU     021H
    FIFO    EQU     022H


            MOV     AL,10001010B    ;enable Baud divisor
            OUT     LINE,AL


            MOV     AL,60           ;program Baud rate
            OUT     LSB,AL
            MOV     AL,0
            OUT     MSB,AL


            MOV     AL,00011001B    ;program 7-data, odd
            OUT     LINE,AL         ;parity, one stop


            MOV     AL,00000111B    ;enable transmitter and
            OUT     FIFO,AL         ;and receiver
```

66. Simplex = sending or receiving, but not both. Half Duplex = sending and receiving, but only one direction at a time. Full Duplex = sending and receiving, both simultaneously.

68.
```
    SENDS   PROC NEAR

    MOV     CX,16
            .REPEAT
                .REPEAT
                    IN      AL,LSTAT    ;get line status register
                    TEST    AL,20H      ;test TH bit
                .UNTIL !ZERO?
                LODSB                   ;get data
                OUT  DATA,AL            ;transmit data
            .UNTILCXZ

        RET

    SENDS   ENDP
```

70. 0.01 V

72.
```
    .MODEL TINY
    .CODE
    .STARTUP
            MOV     DX,400H
            .WHILE 1
            MOV     CX,256
            MOV     AL,0
            .REPEAT
                OUT     DX,AL
                INC     AL
                CALL    DELAY
            .UNTILCXZ
            MOV  CX,256
            .REPEAT
                OUT     DX,AL
                DEC     AL
                CALL    DELAY
            .UNTILCXZ
            .ENDW
    DELAY   PROC    NEAR

    ;   39 microsecond time delay

    DELAY   ENDP
            END
```

74. The INTR pin indicates that the converter has completed a conversion.

76.

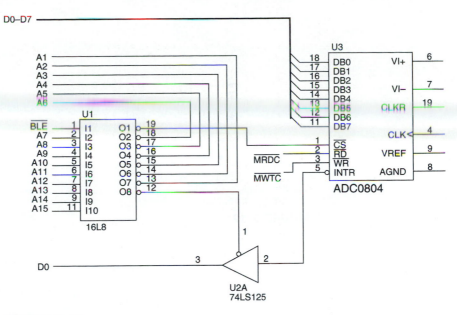

FIGURE D–11

```
;equations for U1
;
EQUATIONS

/O1 = /A15*/A14*/A13*/A12*/A11*/A10*A9*/A8*/A7*A6*A5*/A4*/A3*/A2*/A1*/BLE
/O8 = /A15*/A14*/A13*/A12*/A11*/A10*A9*/A8*/A7*A6*A5*A4*/A3*/A2*/A1*/BLE
```

CHAPTER 12

2. An interrupt is a hardware-initiated subroutine call.

4. Interrupts free processing time because the only time the processor is used is when the interrupt is active. This means that no time is wasted polling an I/O device.

6. INT, INTO, IRET/IRETD, CLI, STI

8. In the first 1K-byte at location 00000000H–000003FFH

10. Vectors 00H–1FH are reserved, even though some are used in the personal computer for other purposes.

12. The interrupt descriptor table is located anywhere in the memory system, as addressed by the interrupt descriptor-table register.

14. The main difference is the location of the interrupt vector. The real mode interrupt uses a vector from a table in the bottom 1K byte of memory, while the protected mode interrupt uses a descriptor from any location in the memory system. The other difference is that the protected mode interrupt service procedure can be placed anywhere in the memory system.

16. The INTO instruction interrupts a program only if the overflow flag bit is set.

18. The IRET instruction functions as a far RET, except that before the return occurs, data from the stack are popped into the flag register.

20. The flags are pushed onto the stack, the I and T flag bits are cleared to zero, and the interrupt service procedure is called using a vector from the interrupt vector table.

22. The trace or trap flag is set to enable tracing. Tracing is an interrupt that occurs after each instruction is executed to allow software to trace through a program.

24. The only way to clear or set the trace flag is to obtain an image of the flag register, and then change the trace bit to a zero or a one before returning the value back to the flag register. There is no instruction to set or clear the trace flag.

26. The $\overline{\text{INTA}}$ signal is only active in response to the INTR input.

28. The NMI input is both level and edge-sensitive.

30. A FIFO is a first-in, first-out memory system that stores data in the FIFO order. We sometimes also call a FIFO a *queue*.

32.

FIGURE D–12

34. A daisy chain is a method of connecting interrupts so that any active interrupt causes a logic 1 to be placed on the INTR input to the microprocessor. The daisy chain interrupt requires software to determine which interrupt is active.

36. The 8259A is a programmable interrupt controller that adds eight interrupt inputs to the microprocessor.

38. The IR inputs are the interrupt request input to the 8259A.

40. The slave INT pin connects to any IR pin on the master 8259A.

42. The OCW is an operational command word used to control the 8259A once it has been initialized by the ICW.

44. ICW2

46. ICW1, among other things, selects the level or edge triggering for the 8259A.

48. The priority rotation algorithm places the most recently serviced interrupt at the lowest priority level.

50. Ports 20H and 21H

CHAPTER 13

2. Whenever HOLD is placed at a logic 0 level, the microprocessor (a) stops executing a program within a few clocks, (b) places its address, data, and control buses at their high-impedance state, and (c) places a logic 1 on the HLDA to signal that the HOLD is in effect.

4. A DMA write transfers data from I/O to the memory.

6. DACK

8. If both HOLD and HLDA are a logic 1 level, the microprocessor is in its hold state.

10. 4

12. The command register

14.
```
    LATCHB   EQU    10H        ;latch B
    CLEAR_F  EQU    7CH        ;F/L flip flop
    CH0_A    EQU    70H        ;channel 0 address
    CH1_A    EQU    72H        ;channel 1 address
    CH1_C    EQU    73H        ;channel 1 count
    MODE     EQU    7BH        ;mode
    CMMD     EQU    78H        ;command
    MASKS    EQU    7FH        ;masks
    REQ      EQU    79H        ;request register
    STATUS   EQU    78H        ;status register

    TRANS    PROC   FAR USES AX

             MOV    AL,20H
             SHR    AL,4
             OUT    LATCHB,AL

             OUT    CLEAR_F,AL ;clear F/L flip-flop

             MOV    AX,1000H    ;program source address
             OUT    CH0_A,AL
             MOV    AL,AH
             OUT    CH0_A,AL
```

```
        MOV     AX,0000H     ;program destination address
        OUT     CH1_A,AL
        MOV     AL,AH
        OUT     CH1_A,AL

        MOV     AX,00FFH     ;program count
        OUT     CH1_C,AL
        MOV     AL,AH
        OUT     CH1_C,AL

        MOV     AL,88H       ;program mode
        OUT     MODE,AL
        MOV     AL,85H
        OUT     MODE,AL

        MOV     AL,1         ;enable block transfer
        OUT     CMMD,AL

        MOV     AL,0EH       ;unmask channel 0
        OUT     MASKS,AL

        MOV     AL,4         ;start DMA transfer
        OUT     REQ,AL

        .REPEAT              ;wait until DMA complete
           IN         AL,STATUS
        .UNTIL AL &1

TRANS   ENDP
```

16. Mini
18. Tracks
20. Cylinder
22.

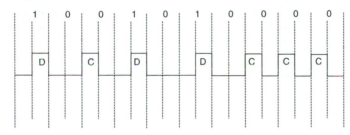

FIGURE D–13

24. The flying head is the name given to the head inside a hard disk drive because it floats on a cushion of air created by the spinning disk.
26. The stepper is not very accurate; the voice coil is because of feedback obtained from a timing track on the disk.
28. The CD-ROM is a device that stores data in pits on the surface of a plastic-coated metal disk. The pits are read by an optical system that uses a small LED laser diode.
30. Red, green, and blue
32. A pixel is the smallest picture element in a video display.
34. The TTL-level RGB monitor displays 16 levels because there is a high-intensity signal to the monitor to modify the eight colors sent through the R, G, and B video lines.
36. 128×128×128 levels or 2,097,152 colors
38. 560

CHAPTER 14

2. 16-bit (±32K), 32-bit (±2 G), and 64-bit (±9 × 10^{18})
4. Short (32-bits), long (64-bits), and extended (64-bits)
6. (a) −7.5 (b) +0.5625 (c) +320 (d) +2.0 (e) +10 (f) +0.0
8. The microprocessor is free to obtain and execute normal microprocessor instructions while the coprocessor executes a coprocessor instruction.
10. The FSTSW AX copies the status register into the AX register.
12. After executing the FCOMP ST(2) and FSTSW AX instructions, the SAHF instruction copies the AH register into the flag register (F7–F0). This allows the condition jump instruction JE to be used to test for equality.
14. FSTSW AX

16. Data are stored in eight 80-bit wide registers that are formed into a stack.
18. ST(0)
20. Affine allows positive or negative infinity, while projective does not.
22. Double-precision (64-bits)
24. The FST DATA instruction copies (does not pop) data from ST(0) into memory location data as a 64-bit floating-point number.
26. FADD ST,ST(3)
28. FSUB ST(2),ST
30. Forward division divides the operand into the top of the stack, while reverse division divides the top of the stack into the operand. If no operand appears, forward division divides ST(0) into ST(1), while reversion division divides ST(0) into ST(1).
32. The instruction does a move only if below.
34.
```
RECEP   PROC    NEAR

        MOV     TEMP,EAX
        FLD1
        FLD     TEMP
        FDIVR
        FSTP    TEMP
        MOV     EAX,TEMP
        RET
TEMP    DD   ?

RECEP   ENDP
```
36. The F2XMI instruction raises 2 to the integer power located at the stack top; then, a 1 is subtracted from the result before it replaces the power at the top of the stack.
38. FLDPI
40. It clears the contents of ST(2).
42. The FSAVE instruction saves all of the register in the coprocessor to memory.
44.
```
REAC    PROC    NEAR

        FLDPI
        FADD    ST,ST(0)
        FMUL    F
        FMUL    C1
        FLD1
        FDIVR
        FSTP    XC
        RET

REAC    ENDP
```
46. The FWAIT instruction is used to cause the microprocessor to wait for the coprocessor to complete its operation and is used before processing floating-point data.
48.
```
TOT     PROC    NEAR

        FLD1
        FDIV    R2
        FLD1
        FDIV    R3
        FLD1
        FDIV    R4
        FADD
        FADD
        FLD1
        FDIVR
        FADD    R1
        FSTP    RT
        RET

TOT     ENDP
```
50.
```
ARRAY1  DQ      100 DUP (?)
ARRAY2  DQ      100 DUP (?)
ARRAY3  DQ      100 DUP (?)

PROD    PROC    NEAR

        MOV     EBX,OFFSET ARRAY1-8
        MOV     EDX,OFFSET ARRAY2-8
        MOV     ESI,OFFSET ARRAY3-8
        MOV     ECX,100
AGAIN:
        FLD     QWORD PTR [EBX+8*ECX]
```

```
              FMUL     QWORD PTR [EDX+8*ECX]
              FSTP     QWORD PTR [ESI+8*ECX]
              LOOPD    AGAIN
              RET

     PROD     ENDP
52.  POW      PROC     NEAR

              MOV      TEMP,EBX
              FLD      TEMP
              F2XM1
              FLD1
              FADD
              MOV      TEMP,EAX
              FLD      TEMP
              FYL2X
              FSTP     TEMP
              MOV      ECX,TEMP
              RET

     POW      ENDP
54.  VOUT     DD       100 DUP (?)
     VIN      DD       100 DUP (?)
     DBG      DD       100 DUP (?)
     TWEN     DD       20.0

     GAIN     PROC     NEAR

              LEA      EBX,VOUT-4
              LEA      EDX,VIN-4
              LEA      ESI,DBG-4
              MOV      ECX,100
     AGAIN:
              FLD      DWORD PTR [EBX+4*ECX]
              FDIV     DWORD PTR [EDX+4*ECX]
              CALL     LOG10
              FMUL     TWEN
              FSTP     DWORD PTR [ESI+4*ECX]
              LOOPD    AGAIN
              RET

     GAIN     ENDP
```

56. The EMMS instruction clears the coprocessor stack to indicate that the MMX extension is finished using the co-processor registers.

58. Signed saturation occurs when numbers are added and have the values of 7FH for an overflow and a value of 80H for an underflow if the numbers are byte-sized.

60. The FSAVE instruction (coprocessor) could be used to save the state of the MMX registers in the memory system.

CHAPTER 15

2. The early ISA bus supports only 8-bit transfers, while the newest supports either 8- or 16-bit transfers.

4. (see Figure D–14 on page 1000)

6. (see Figure D–15 on page 1000)

8. (see Figure D–16 on page 1001)

10. See Instructor's Manual

12. 32-bits

14. 33 MHz

16. The main difference is that the PCI bus supports either a 32- or 64-bit data bus, which makes it suited to the Pentium or Pentium Pro microprocessor.

18. The configuration memory allows software to determine what board is plugged into a PCI slot, so the software can automatically set up the system for the PCI interface.

20. These pins are used to transfer the command to the PCI interface and also to select the memory or I/O bank.

22. See Instructor's Manual

24. 12M bits per second and 1.5M bits per second

26. 5 Meters

28. 127

30. A stuffed bit is a logic 0 placed in a data stream after a fixed number of ones are transmitted.

32. Up to 1024 bytes

34. The PCI bus transfers data at rates up to about 100M bytes per second. With the AGP 2X transfer rates are 528M bytes and with 4X transfers over 1G bytes.

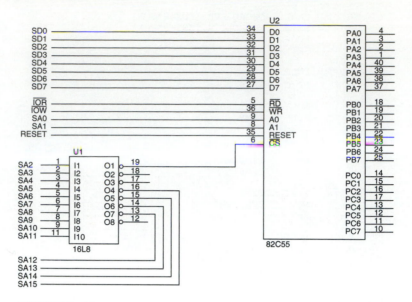

FIGURE D–14

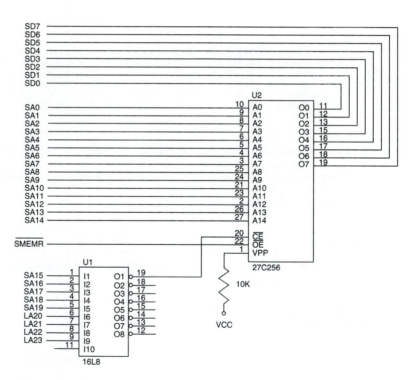

FIGURE D–15

CHAPTER 16

2. The 80186/80188 contain an internal clock generator, chip-selection logic, timers, programmable interrupt controller, DMA controller, power-down mode, serial interfaces, and parallel interfaces. Note that not all versions contain all features.

4. 10 MHz

6. 3.0 mA of current for a fan-out 7 TTLS parts or 10 CMOS components

8. The ALE signal appears $\frac{1}{2}$ clock earlier in the 80186/80188.

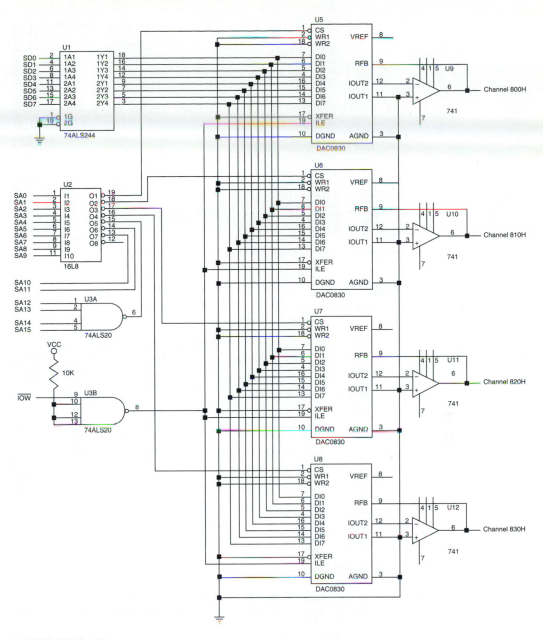

FIGURE D–16

10. 417 ns
12.
```
MOV    AX,1100H
MOV    DX,0FFFEH
OUT    DX,AL
```
14. Ten on most versions of the 80186/80188, including all the internal interrupts.
16. The interrupt-control registers each control a single interrupt input to the 80186/80188 interrupt controller.
18. The difference is that reading the interrupt poll register acknowledges the interrupt, while reading the interrupt poll-status register does not.
20. 3
22. Timer 2 always connects, and Timers 0 and 1 can be connected to the system clock.
24. The INH bit must be set to allow the EN bit to change.
26. Alternate operation using the two compare or maximum count registers.
28.
```
MOV    AX,    123
MOV    DX,    0FF5AH
```

```
OUT    DX,AL
MOV    AX,23
ADD    DX,2
OUT    DX,AL
MOV    AX,8007H
MOV    DX,0FF58H
OUT    DX,AL
```

30. 2
32. The channel is started by software control (control register) or by hardware control (timer 2 or the DRQ input).
34. 7
36. Base
38. 0 and 15
40. It selects the operation of the PCS5/A0 and PCS6/A1 pins.
42.
```
MOV    DX,0FF90H
MOV    AX,1001H
OUT    DX,AL
MOV    DX,0FF92H
MOV    AX,1008H
OUT    DX,AL
```
44. 1G
46. It verifies that a protected mode segment can be read.

CHAPTER 17

2. 64T
4.

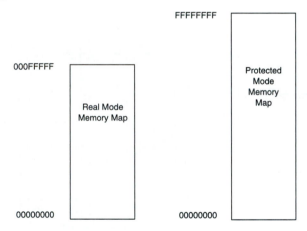

FIGURE D–17

6. The memory map for the 80386 contains 4G bytes of memory that is physically organized into a 32-bit wide memory system. Each of the four eight-bit wide memory banks is selected by using a bank enable signal labeled BE0–BE3.
8. The pipeline allows the microprocessor to send out an address while the data from a prior operation is being fetched. This allows the memory additional time for accessing the data.
10. 0000H–FFFFH
12. The only differences are a wider data bus (32 bits) and a wider address bus (32 bits).
14. The BS16 pin configures the microprocessor to operate with a 16-bit data bus.
16. EAX, EBX, ECX, EDX, WSP, EBP, ESI, EDI, EIP, and EFLAGS.
18. CR0 = selects paging, and enters or leaves the protected mode, CR1 = reserved for the future, CR2 = holds the linear address of any fault, and CR3 = holds the base address of the page directory.
20. Interrupt type 1
22. The BSR instruction scans a number from the right toward the left. If a 1 is found, the zero flag is set and the bit position of the logic one is placed into the destination register.
24. MOV FS:[DI],EAX
26. Yes
28. Interrupt type number 7 is used to emulate the arithmetic coprocessor.
30. A double fault interrupt occurs whenever more than one interrupt occurs within the microprocessor.
32. A descriptor is a sequence of eight bytes that describe the location, length, and attributes of a protected mode memory segment.

34. If the table indicator is a logic 1, the local descriptor table is chosen by the segment register.
36. 8192
38. The segment descriptor describes a data, code, or stack segment, while the system descriptor describes a CALL or interrupt gate or a task.
40. The TSS is addressed by a special descriptor that is accessed by the task register.
42. The 803786 is switched between real and protected mode by setting or clearing the rightmost bit of CR0.
44. CR3 holds the base address of the paging directory.
46. Linear address D0000000H addresses a physical page by accessing page directory entry 1101000000. In this entry, the address of the page table that describes 4M of memory is located. Page table entry 0000000000 holds memory address C0000000H to translate linear address D0000000H to C0000000H.
48. The FLUSH input erases the internal 80486 cache.
50. None, except for the 80486SX, which contains an alignment check flag used by the arithmetic coprocessor (80487).
52. Even parity
54. 16
56. A cache write-through occurs when the microprocessor writes data to the cache and to the memory system.
58. If paging is in effect, caching can be turned on and off for different page translations.
60. This instruction does nothing if AL = CL; if AL ≠ CL, then CL is copied into AL.
62. If PCD = 0, the cache is enabled for the current memory page.

CHAPTER 18

2. Up to 64G bytes.
4. These pins both generate parity for the ninth bit per byte and also check parity.
6. This pin signals the microprocessor that the bus is ready and is used to insert wait states into the timing.
8. 18.2 ns
10. T2
12. Two 8K-byte caches, one for data and the other for instructions.
14. Yes, as long as they are not dependent on each other.
16. The memory management mode is a special mode accessed through the memory management interrupt input to the Pentium and Pentium Pro.
18. 38000H
20. This instruction compares eight bytes of data stored in memory with EDX:EAX.
22. ID, VIF, and VIP
24. The Pentium and Pentium Pro access 4M-byte pages by using the page directory to store the base page address of a 4M-byte memory page instead of the address of a page table.
26. The Pentium and the Pentium Pro differ in address bus size (32 bits versus 36 bits on the Pentium Pro), the FCMOV and CMOV instructions are added to the Pentium Pro, and the Pentium Pro contains the level 2 cache with a size of either 256K or 512K bytes.
28. A35–A3
30. The access times are essentially the same on both microprocessors if operated with the same frequency bus clock.
32. 72-bit wide SDRAM called ECC SDRAM

CHAPTER 19

2. 512K, 1M, and 2M
4. The Pentium Pro level 2 cache is built into the integrated circuit, while the Pentium II level 2 cache is a separate integrated circuit mounted onto a printed circuit board called the Pentium II cartridge.
6. 64G bytes
8. 242
10. The read and write pins are replaced by request input used by the bus controller to request a read or a write operation.
12. 8 ns
14. SYSENTER_CS, SYSENTER_SS, and SYSENTER_ESP
16. The ECX register passes the address or register number to the RDMSR instruction. After executing the RDMSR instruction, EDX:EAX contains the contents of the register.
18.
```
TESTS   PROC NEAR
        CPUID
        TEST EDX,800H        ;test bit position 11
        .IF ZERO?
            CLC
        .ELSE
            STC
        .ENDIF
        RET
TESTS   ENDP
```
20. The return address is retrieved, placed in EDX, and then stored in EIP by the SYSEXIT instruction.
22. Level or ring 3

INDEX